NanoSafe2010: International Conference on Safe Production and Use of Nanomaterials

Journal of Physics: Conference Series Volume 304

Grenoble, France
16-18 November 2010

ISBN: 978-1-61839-048-6
ISSN: 1742-6588

Printed from e-media with permission by:

Curran Associates, Inc.
57 Morehouse Lane
Red Hook, NY 12571

Some format issues inherent in the e-media version may also appear in this print version.

Printed by Curran Associates, Inc. (2011)

Institute of Physics
Dirac House, Temple Back
Bristol BS1 6BE UK

Phone: 44 1 17 929 7481
Fax: 44 1 17 920 0979

techtracking@iop.org

Additional copies of this publication are available from:

Curran Associates, Inc.
57 Morehouse Lane
Red Hook, NY 12571 USA
Phone: 845-758-0400
Fax: 845-758-2634
Email: curran@proceedings.com
Web: www.proceedings.com

NanoSafe2010: International Conference on Safe Production and Use of Nanomaterials

Journal of Physics: Conference Series Volume 304

Grenoble, France
16-18 November 2010

ISBN: 978-1-61839-048-6
ISSN: 1742-6588

IOP Institute of Physics

TABLE OF CONTENTS

CONFERENCE OPENING

EXPOSURE ASSESSMENT

A. OCCUPATIONAL EXPOSURES

B. ENVIRONMENTAL EXPOSURES

CHARACTERIZATION, DETECTION AND MONITORING

A. DETECTION IN AIR

C. NEW THINKING APPROACH

D. ECOTOXICOLOGY

ENVIRONMENTAL IMPACT

NANOPARTICLE RELEASE FROM CONSUMER PRODUCTS

COMMERCIAL EQUIPMENT RELATED TO SAFE PRODUCTION AND USE OF NANOMATERIALS

PERSONAL PROTECTIVE EQUIPMENT

C. SAPHIR

D. TITNT

Journal of Physics: Conference Series **304** (2011) 011001 doi:10.1088/1742-6596/304/1/011001

16-18 NOVEMBER / MINATEC - FRANCE

Preface

The second edition of the international conference on the "Safe production and use of nanomaterials" – NanoSafe2010 – was held between 16–18 November at the Maison Minatec congress center located in Grenoble, France. As is recognised, the rapidly developing field of nanotechnologies presents many opportunities and benefits for new materials with significantly improved properties as well as revolutionary applications in the fields of energy, environment, medicine, etc. However, the potential impact of these new materials on human health and the environment is viewed with apprehension. All the security aspects have to be solved in order to reach an acceptable level of risk. The new nano industry can only develop dynamically if these legitimate concerns are satisfactory allayed.

Initiated within the framework of the Integrated European project NanoSafe2 and funded by the European Commission under FP6 programs, this biennial event was organized by the French Commissariat à l'Energie Atomique (CEA). NanoSafe2010 had broad international participation from 30 countries representing 4 continents, with registered delegates numbering over 300. The scientific program involved the presentation and discussion of 175 papers, classified as 13 plenary lectures, 117 oral presentations and 45 posters.

The conference attracted many of the best known leading scientists and specialists from almost all of the different fields of expertise covering nanosafety issues: Exposure, Characterisation, Detection, Monitoring, Life cycle, Toxicology, Environmental impact, Personal protection, Secure industrial production, Safety parameter evaluation, Standardization, Regulation, and Education.

An exhibition took place in parallel with the conference where 16 exhibitors presented the commercial equipment available today in relation to safety during the production or the use of nanomaterials.

We believe that this conference, with the above topics, has provided a forum where the many matters of interest to the NanoSafety community have been debated. The success of the conference confirms the need of such a global meeting with the aim of bridging the gap between the best scientists, engineers, exhibitors and participants from countries concerned with safe and responsible nanomaterial activities. Therefore, the third edition of the International NanoSafe Conference "**NanoSafe2012**" will be held in Grenoble at Minatec, **13–15 November 2012**.

The meeting was financially supported by generous contributions from the European Commission, La Région Rhône Alpes, la Ville de Grenoble, la Metro, and partners such as CEA, INERIS, Philips, Cordouan Technologies, Nanosight, TSI, Ecomesure, GRIMM Aerosol Technik, Particle Measuring Systems, Respirex, Lux Research, Dekati, Malvern Instruments, TSE Systems, INTERTEK ICS, NANO magazine and IOP Publishing.

We would like to express our thanks to all the authors for their time and genuine efforts and to the reviewers for their fruitful comments during the preparation of this volume.

The Conference Organizers
Carole Sentein, Frédéric Schuster and Francois Tardif

Nanosafe2010: International Conference on Safe Production and Use of Nanomaterials IOP Publishing
Journal of Physics: Conference Series **304** (2011) 011001 doi:10.1088/1742-6596/304/1/011001

Journal of Physics: Conference Series **304** (2011) 011001 doi:10.1088/1742-6596/304/1/011001

Local Organising Committee
Carole SENTEIN (CEA)
Dominique BAGUET (CEA)

Didier MOLKO (MINATEC)
Audrey SCARINGELLA (MINATEC)
Frédéric SCHUSTER (CEA)
François TARDIF (CEA)

International Advisory Committee
Chair: Frédéric SCHUSTER (CEA, FR), François TARDIF (CEA, FR)
Co-chair: Georgios KATALAGARIANAKIS (EC, BE)

Paul AMYOTTE (Dalhousie U., CA)
Mélanie AUFFAN (CEREGE, FR)
Daniel BERNARD (ARKEMA, FR)
Jorge BOCZKOWSKI (INSERM, FR)
Jean-Yves BOTTERO (CEREGE, FR)
Jacques BOUILLARD (INERIS, FR)
Jean-Philippe BOURGOIN (CEA, FR)
Marie CARRIERE (CEA, FR)
Krunal CHOPRA (LSFI, IN)
Rolf ECKHOFF (U. Bergen, NO)
Michael ELLENBECKER (U. Massachusetts, USA)
Claude EMOND (U. Montréal, CA)
François GENDARMES (IRSN, FR)
Peter HATTO (ISO, UK)
Peter HOET (KUL, BE)
Jean-Pierre KAISER (Empa, CH)
Olivier LE BIHAN (INERIS, FR)
Tinh NGUYEN (NIST, USA)
Bernd NOWACK (Empa, CH)
Günter OBERDÖRSTER (U. Rochester, USA)
David PUI (U. Minnesota, USA)
Eric QUEMENEUR (CEA, FR)
Robert SCHALLER (EPFL, CH)
Carole SENTEIN (CEA, FR)
Ken TAKEDA (U. Tokyo, JP)
Louis TREPIED (MINEFE, FR)
Jacques VENDEL (IRSN, FR)
Olivier WITSCHGER (INRS, FR)

Journal of Physics: Conference Series **304** (2011) 011001 doi:10.1088/1742-6596/304/1/011001

List of Participants

AGUY S	Cordouan, FR
ALLEMAN L	Ecole des Mines de Douai, FR
ALPER M	Univ. California Berkeley, US
ALVES O L	UNICAMP, BR
AMDAOUD M	Univ. Joseph Fourier, FR
AMODEO T	INERIS, FR
AMYOTTE P	Dalhousie Univ., CA
ARAKELIAN M C	Fundacion Argentina de Nanotecnologia, AR
ARZAMASTSEVA E	Tomsk Polytechnic Univ., RU
ASCHBERGER K	European Commission-JRC, IT
ASHTON A	JPCS, GB
ASIMAKOPOULOU A	CERTH/CPER, GR
AUBLANT J-M	LNE, FR
AUDE-GARCIA C	CEA, FR
AUFFAN M	CEREGE, FR
AUGER A	CEA, FR
BAEK M	Seoul Woman's Univ., KR
BAEZA A	Univ. Paris 7 Diderot, FR
BAGUET D	CEA, FR
BARONE F	Istituto Superiore di Sanità, IT
BAU S	INRS, FR
BELLEVILLE P	CEA, FR
BENAZET S	SNPE-Matériaux Energétiques, FR
BERGER L-M	Fraunhofer Institute, DE
BERGES M	DGUV-IFA, DE
BERNARD C	NIST, US
BERNARD D	ARKEMA, FR
BERNE P	CEA, FR
BERRET J-F	Univ. Paris 7 Diderot, FR
BIGORGNE E	CNRS, FR
BILLAT B	PMS, FR
BLOCH D	CEA, FR
BOGDANOVIC-GUILLON A	Freelance journalist, FR
BOIVIN D	RESPIREX France, , FR
BOLAND S	Univ. Paris 7 Diderot, FR
BOMBARDIER P	FAURE QEI, FR
BORASCHI D	Institute of Biomedical Technologies, IT
BOTTERO J-Y	CEREGE, FR
BRADLEY J	Lux Research, US
BREGOLI L	Veneto Nanotech, IT
BRIGNON J-M	INERIS, FR
BROCHOT C	IRSN, FR
BROUWER D	TNO Quality of Life, NL
BUDIMAN T	TSE-Systems, DE
CABOCHE J	Ecole des Mines de Douai, FR
CALLET G	CEA, FR
CALOGINE D	INERIS, FR
CAPONIGRO A	Loreal, FR
CARRIERE M	CEA, FR
CASSETTE S	Thales Research & Technology, FR
CHARPENTIER D	INERIS, FR
CHEN C-C	Taiwan Univ., TW
CHEN L	Health Canada, CA
CHEVALLET M	CEA, FR
CHOI S-J	Seoul Woman's Univ., KR

Journal of Physics: Conference Series **304** (2011) 011001 doi:10.1088/1742-6596/304/1/011001

CHOU C-C	Chung Cheng University, TW
CHUNG H-E	Seoul Woman's Univ., KR
CIMPAN M	Univ. Bergen, NO
CIOBANU N	TU Munich, DE
CLARK K	Institute for Work and Health, CH
CONJEAUD H	Univ. Paris 7 Diderot, FR
COSTA PEREIRA C	National Health Institute, PT
CROOKS M	TSI, GB
CUPERTINO D	C-Tech Innovation Ltd, GB
DASKALOS M	CERTH/CPERI, GR
DE BERARDIS B	Istituto Superiore di Sanità, IT
DE MIGUEL Y	TECNALIA, ES
DELAHAYE A	Ad Air Solutions, FR
DEMIDOVA T	Severtsov Inst. of Ecology and Evolution, RU
DENOO K	Solae, GR
DERROUGH S	CEA, FR
DOBRZYNSKA E B	Central Institute for Labour Protection, PL
DOLEZ P	École de technologie supérieure, CA
DOUKI T	CEA, FR
DRAIS E	INRS, FR
DUFOUR J-P	CILAS, FR
DURAN N	UNICAMP, BR
DURAND C	CEA, FR
DUTOUQUET C	INERIS, FR
DUVAL-ARNOULD G	Saint-Gobain, FR
ECKHOFF R K	Univ. Bergen, NO
ELLENBECKER M J	Univ. Massachusetts Lowell, US
EMOND C	Univ. Montreal, CA
ENGEL S	BASF, DE
ESTRELA-LOPIS I	Leipzig Univ., DE
FABBRI M	JRC, IT
FACCINI M	Leitat technological center, ES
FESSARD V	Anses, FR
FILIMUNDI E	TSI, DE
FIRSTOVA V	SRCAMB, RU
FLEURY D	INERIS, FR
FRABOULET D	CEA, FR
FRESNAY C	Thales Research & Technology, FR
GABORIEAU A	CEA, FR
GAFFET E	NanoMaterials Research Group, FR
GALLET S	Cefic, BE
GEIGER D	BASF, DE
GENSDARMES F	IRSN, FR
GERRITSEN-EBBEN R	TNO Quality of Life, NL
GKANIS V	Demokritos, GR
GLUSHKOVA A	RIHOPHE, RU
GONZALEZ-FERNANDEZ A	Univ. Vigo, ES
GOOSSENS H	Philips Research Aerasense, NL
GRAHNSTEDT S	Oslo Univ., NO
GREENHILL-HOOPER M	Rio Tinto Minerals, FR
GROSSEAU P	Ecole des Mines de Saint Etienne, FR
GUADAGNINI R	Univ. Paris 7 Diderot, FR
GUIOT A	CEA, FR
GUIZARD B	CEA, FR
HAASE A	Federal Institute for Risk Assessment, DE
HANINI A	Univ. Paris 7 Diderot, FR
HAYNES L	Univ. de los Andes, VE
HEJAZI M	Univ. Tehran , IR

HENRY F	INERIS, FR
HERRERA H	Institute for Work and Health, CH
HOET P	KU Leuven, BE
HOLE P	Nanosight, GB
HULME J	Univ. Cambridge, GB
JI X	INERIS, FR
JOUHANNAUD J	CEA, FR
JOUZEL J-N	Center for the Sociology of Organizations, FR
JURKSCHAT K	Oxford Univ., GB
KAISER J-P	Empa, CH
KANAYA F	Nat. Center for Global Health and Medicine, JP
KATALAGARIANAKIS G	European Commission, BE
KECK L	Grimm Aerosol Technik, DE
KELLER M	Fraunhofer Institute, DE
KHLEBNIKOVA N	RIHOPHE, RU
KHODABANDEH M	Univ.Tehran, IR
KHOLODENKO V	SRCAMB, RU
KOBAYASHI N	AIST, JP
KOPONEN I	NRCWE, DK
KOWAL S	INERIS, FR
KRYSANOV E	Severtsov Inst. of Ecology and Evolution, RU
KÜCK A	Univ. Bremen, DE
KUO Y-M	Chung Hwa University, TW
KVITEK L	Palacky Univ., CZ
LABILLE J	CEREGE, FR
LAMMINEN E	Dekati, FI
LARUE C	CEA, FR
LE BIHAN O	INERIS, FR
LE DUR D	Ecomesure, FR
LECERF P	CILAS, FR
LEGRAND M	Cordouan, FR
LELONG C	UJF CEA, FR
LIMOUSIN S	INERIS, FR
LINDELOEV J	GEA Process Engineering, DK
LIU P P	China Jiliang university, CN
LIU W	CEREGE, FR
MACHEREY A-C	CNRS, FR
MAGGA Y	CEA, FR
MAHLENDORF F	University Duisburg-Essen, DE
MANIER N	INERIS, FR
MANZO L	Univ. Pavia, IT
MARCHETTO A	CEA, FR
MARCONE G	UNICAMP, BR
MARI D	EPFL, CH
MARIE-DESVERGNE C	CEA, FR
MARIE-LOUISE A	PSA Peugeot-Citroen, FR
MARMUSE L	Nano-H S.A.S. , FR
MARRA J	Philips Research Aerasense, NL
MASION A	CEREGE, FR
MATEI E	Politehnica University Bucharest, RO
MATSUI Y	Kyoto Univ., JP
MATZKE M	Univ. Gothenburg, SE
MAYNE-L'HERMITE M	CEA, FR
MELINTE G A	Babes-Bolyai University, RO
MERINO C	Grupo Antolin Ingenieria, ES
MICHAUD-SORET I	CEA, FR
MICHELETTI C	JRC, IT
MONTIGEL E	Basler Versicherungen, CH

MONTOYA E	RAMEM, ES
MOSSUZ V	CEA, FR
MOTELLIER S	CEA, FR
MOTZKUS C	LNE, FR
MUIR B	Naneum, GB
NAKAMURA K	JAPAN NUS CO., JP
NEUBAUER N	Karlsruhe Institute of Technologie, DE
NEUMEISTER L	BG ETEM, DE
NGUYEN T	NIST, US
NIORT N	INTERTEK, FR
NOIRTIN A	INTERTEK, FR
NOWACK B	Empa, CH
NYEMBE D	Univ. Johannesburg, ZA
Ó CLAONADH N	Dublin Institute of Technology, IE
OBERDÖRSTER G	Univ. Rochester, US
OGURA I	AIST, JP
OSTIGUY C	IRSST, CA
OTSUKA K	JFE Techno-Research Corp., JP
OUF F-X	IRSN, FR
OUSACI S	ALMA, FR
PAGET V	CEA, FR
PAILLEUX M	Ecole des Mines de Saint Etienne, FR
PANDARD P	INERIS, FR
PANZER O	European Research Services , DE
PARISELLI F	CNRS, FR
PERLET J	NANO Magazine, GB
PETERS R	RIKILT, NL
PETIT A-N	CEA, FR
PETKOVIC J	National Institute of Biology, SI
PIMENOFF J	Beneq, FI
PINAULT M	CEA, FR
PIRET J-P	Univ. Namur, BE
PONTONE R	Tekna Plasma Systems, FR
POURCHEZ J	Ecole des Mines de Saint Etienne, FR
PRAETORIUS A	ETH Zurich, CH
PRAT O	CEA, FR
PREVENSLIK T	QED Radiations, CN
PREVOST C	IRSN, FR
PROY HUART D	France Nature Environnement, FR
PUI D Y H	Univ. Minnesota, US
QUARTARARO J	PSA Peugeot-Citroen, FR
RAULT S	Univ. Caen, FR
RAVANEL X	CEA, FR
RAVINDRAN GIRIJA A	Toyo Univ., JP
REBIB F	Pôle Technologique d'Auvergne, FR
RECHATIN J-L	IRELEC, FR
REYNIER M	INRS, FR
RICAUD M	INRS, FR
RIEDIKER M	Institut de Santé au Travail, CH
RMILI B	INERIS, FR
ROGERIEUX F	INERIS, FR
ROLANDO C	Univ. Lille 1, FR
ROMERO A	Freelance journalist, DE
ROUCHER B	CILAS, FR
SAEZ J-P	Nanosight, GB
SAFI M	Univ. Paris 7 Diderot, FR
SAKAI N	Kyoto Univ., JP
SALVI O	INERIS, FR

Journal of Physics: Conference Series **304** (2011) 011001 doi:10.1088/1742-6596/304/1/011001

SANFINS E	Univ. Paris 7 Diderot, FR
SAVOLAINEN K	FIOH, FI
SCHALLER R	EPFL, CH
SCHMID K	Institute for Work and Health, CH
SCHUBERT C	CEA, FR
SCHUSTER F	CEA, FR
SCHWEINBERGER F	TU Munich, DE
SCIFO L	TECNALIA, ES
SEABRA A	UNICAMP, BR
SEBASTIEN P	Saint-Gobain, FR
SENE C	STEPAN, FR
SENTEIN C	CEA, FR
SHEN J	Stockholm Univ., SE
SHINKAI Y	Tokyo Univ., JP
SICARD Y	UJF CEA, FR
SIKHWIVHILU L	Nat. Centre for Nanostructured Materials, ZA
SILLANPÄÄ M	Finnish Environment Institute, FI
SIMON J	CEA, FR
SINESI S	CRP, IT
SIRVEN J-B	CEA, FR
SIRVIÖ S	FIOH, FI
SOERENSEN S	Solae, DK
SON S W	Korea Univ., KR
SUBLEMONTIER O	CEA, FR
SYMON A	RUSNANO Metrology Center, RU
SYMONDS J	CAMBUSTION, GB
TAKEDA K	Tokyo Univ., JP
TARANTINI A	Anses, FR
TARDIF F	CEA, FR
TERRAY M	Malvern Instruments, FR
THABUIS C	Bureau Veritas, FR
THIERIET N	Anses, FR
TIKKANEN J	THL, FI
TROUILLER B	INERIS, FR
UKKONEN A	Dekati, FI
UMEMURA K	Tokyo University, JP
USTACHE A	INERIS, FR
VÄISÄNEN P	Rautaruukki Oyj, FI
VAN DEN BRINK W	Philips Research Aerasense, NL
VAN TONGEREN M	Institute of Occupational Medicine, GB
VANDELAC L	Univ. Québec, CA
VAQUERO C	TECNALIA, ES
VEERANARAYANAN S	Toyo Univ., JP
VENDEL J	IRSN, FR
VERBIST K	ArboUnie, NL
VIANA M	IDAEA-CSIC, ES
VIDAUD C	CEA, FR
VIGNES A	INERIS, FR
VOETZ M	Bayer Technology Services, DE
WAHLBERG M	GEA Process Engineering, DK
WALSER T D	ETH Zurich, CH
WARTENBERG N	CEA, FR
WEIGEL S	RIKILT, NL
WENZLER E	TSE-Systems, DE
WHITELEY C	Lancaster Univ., GB
WIJMA E	RIKILT, NL
WILDE C	AvantiCell Science Ltd, GB
WILLIAMS D	Manufacture Michelin, FR

WITSCHGER O	INRS, FR
WOHLLEBEN W	BASF, DE
WU Q	Det Norske Veritas, NO
YAGISHITA T	JFE Techno-Research Corporation, JP
YU J	Seoul Woman's Univ., KR
ZAPÓR L	CIOP-PIB, PL
ZIMMERMANN E	CEA, FR

All papers published in this volume of *Journal of Physics: Conference Series* have been peer reviewed through processes administered by the proceedings Editors. Reviews were conducted by expert referees to the professional and scientific standards expected of a proceedings journal published by IOP Publishing.

Journal of Physics: Conference Series **304** (2011) 012001 doi:10.1088/1742-6596/304/1/012001

Coordination and Collaboration in European Research towards Healthy and Safe Nanomaterials

Michael Riediker

Coordinator NanoImpactNet, Institute for Work and Health, Rue du Bugnon 21, CH-1011 Lausanne

E-mail: michael.riediker@alumni.ethz.ch

Abstract. Nanotechnology is becoming part of our daily life in a wide range of products such as computers, bicycles, sunscreens or nanomedicines. While these applications already become reality, considerable work awaits scientists, engineers, and policy makers, who want such nanotechnological products to yield a maximum of benefit at a minimum of social, environmental, economic and (occupational) health cost. Considerable efforts for coordination and collaboration in research are needed if one wants to reach these goals in a reasonable time frame and an affordable price tag. This is recognized in Europe by the European Commission which funds not only research projects but also supports the coordination of research efforts. One of these coordination efforts is NanoImpactNet, a researcher-operated network, which started in 2008 promote scientific cross-talk across all disciplines on the health and environmental impact of nanomaterials. Stakeholders contribute to these activities, notably the definition of research and knowledge needs. Initial discussions in this domain focused on finding an agreement on common metrics, and which elements are needed for standardized approaches for hazard and exposure identification. There are many nanomaterial properties that may play a role. Hence, to gain the time needed to study this complex matter full of uncertainties, researchers and stakeholders unanimously called for simple, easy and fast risk assessment tools that can support decision making in this rapidly moving and growing domain. Today, several projects are starting or already running that will develop such assessment tools. At the same time, other projects investigate in depth which factors and material properties can lead to unwanted toxicity or exposure, what mechanisms are involved and how such responses can be predicted and modelled. A vision for the future is that once these factors, properties and mechanisms are understood, they can and will be accounted for in the development of new products and production processes following the idea of "Safety by Design". The promise of all these efforts is a future with nanomaterials where most of their risks are recognized and addressed before they even reach the market.

1. Introduction: Nanoparticles are reality in today's industry

Nanomaterials have structures in the nanoscale, i.e. below 100 nm [1]. The physico-chemical properties of such materials often are different from the same, but larger structured materials. Prominent examples are altered interactions with light such as invisibility or photocatalytic effects, different conductivity or increased chemical reactivity due to increased surface and altered material structure and configuration [2].

These changing properties of materials in the nanoscale make them interesting for numerous kinds of applications and products. Research in the field of nanotechnology is developing fast and has generated already a considerable number of novel manufactured nanomaterials (MNM) and related products, as evidenced by the large number of research publications and patents filed. Industries already started with the manufacturing of products and goods that are based on these nanotechnological accomplishments. For example, a recent representative survey [3] in Switzerland showed that about 0.6% of all companies already handle, use or produce nanoparticles (MNM that are in three dimensions smaller than 100 nm). However, many of these companies still seemed to be in an early phase of production, which is characterized by manual handling steps and heavy reliance on personal protective equipment, as also evidenced in a more detailed, targeted study among the same type of companies [4].

The presence of novel physico-chemical properties makes it plausible that these materials will show interactions with biological systems and behavior in the environment that are different from larger-structured materials composed of the same substances. To what degree this will manifest in real effects is something that needs to be evaluated and that is subject to research programmes worldwide. Such a risk assessment needs information about material characteristics, potential exposures and hazards. This information can then be used to develop strategies for the management and control of the associated risks. Seen the multitude of aspects that need to be studied, it makes sense to address these issues in a multidisciplinary and well coordinated approach. Otherwise, valuable time towards safe nanoproducts would be lost to prevent human and environmental harm and loss of investments by governments, industry and the financial sector.

2. The need for exposure assessment

Information on potential human exposure and liberation into the environment is currently being generated and many aspects need to and will be further studied. To evaluate the human health risks, one needs to know the number of exposed people, the type and quantity of exposure and the associated health effects. For environmental risk assessment the quantities of permanently or accidentally released nanoparticles need to be known.

An ongoing study sponsored by the European Commission's 7th Framework Programme (NanEx - Development of Exposure Scenarios for Manufactured Nanomaterials, project number 247794) collects and reviews available exposure information, evaluates the applicability of existing models for occupational and consumer exposure assessment and for environmental release from these scenarios, and carries out a small number of specific case illustrations and gap analyses of the available knowledge and data.

The project already identified challenges and roadblocks towards the creation of exposure scenarios. Notably, exposure information available in scientific literature seems insufficient to create specific exposure scenarios, and the currently published literature focuses mostly on small, laboratory scale manufacturing processes, even though the majority of applications are in the use of these materials [3] with only few companies producing relatively large quantities of MNM. The case illustrations of the NanEx study revealed that, if companies have direct access to workplaces and consumer product information, there are relatively few gaps in the data needed to generate exposure scenarios (even though finding a coherent and consistent approach to build them is a challenge even for experts in exposure assessment).

Exposure to nanoparticles is special compared to "normal" chemical exposure in that it cannot be described by one single number such as an airborne concentration. Instead, in addition to the information about the material itself, size distribution, mass and probably also surface area, particle number, morphology and in some cases even more metrics are needed to sufficiently describe the exposure (Figure 1).

This information alone, however, is not sufficiently informative, because the approach how this information was generated also needs to be known. Notably the involved type of activities, the processes, the context, the sampling strategy and probably more information is needed. This is information that can easily be collected during a measurement campaign. However, to compare exposures and to allow subsequent pooling of data for meta-analyses, common metrics and a standardized approach to describe this secondary information are needed.

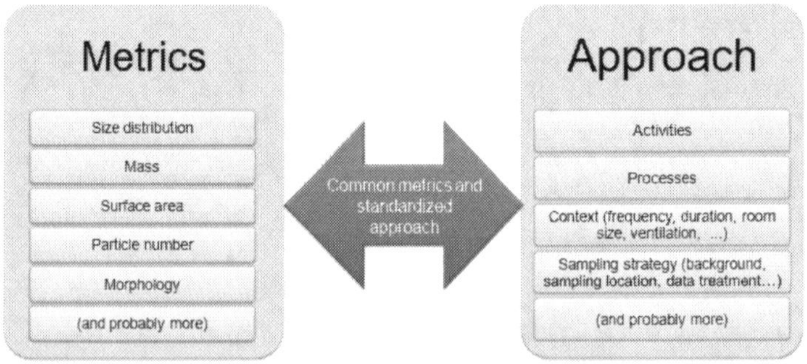

Figure 1. Key metrics and information about the approach to obtain the data that are needed for a good comparison of exposure data and for pooling data for meta-studies (conclusions of the FP7 NanEx-study).

Reports are expected to be published in 2011 that will provide a detailed description of the findings of the NanEx study, notably the case studies, consumer and occupational exposure scenarios, the inclusion of environmental aspects and the integration and gap analysis.

3. Hazards of nanomaterials and what to learn from ambient air pollution

Nanomaterials have new properties, which are expected to result in different biological responses as compared to existing materials with larger structure. However, the basic principles established for particles seem to apply also to particulate nanomaterials whether nanoparticles or larger nanostructured entities (Figure 2). In the most simple case, (nano)materials overload the target organ: even completely inert materials can disturb the function of an organ if there are excessive quantities deposited. This overload effect can also be expected for nanomaterials. Second, particle can be the carrier of toxic substances. This applies again to MNM with the addition that nanoparticles were reported to easily transfer biological barriers implying that they can carry the toxic substances through membranes [5] into cells and, by translocation [6] to other organs even though only very small quantities do translocate. Thus, this opens the way to novel effects in cells or organs.

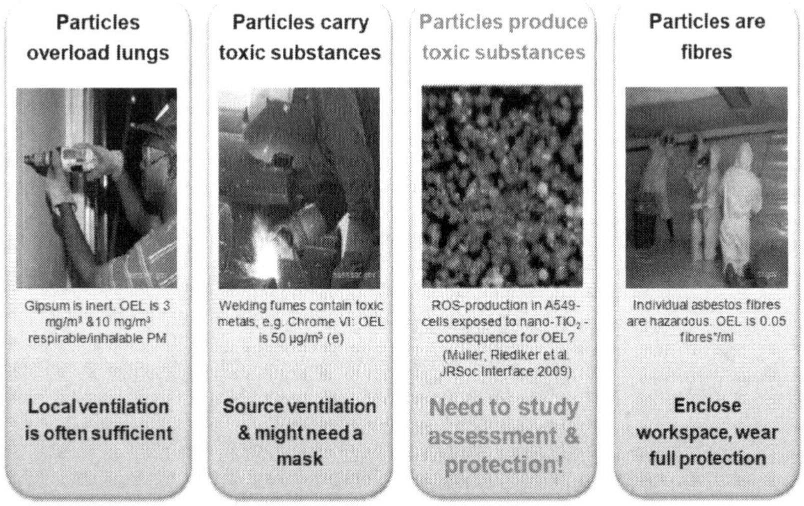

Figure 2. Established paradigms in particle toxicology. The paradigm that particles can produce toxic substances is relatively new. This is a lesson learnt mostly from ambient air pollution research.

The morphology of material can play a critical role in the hazardousness of biopersistent materials, which is well known for asbestos fibres. Initial research suggests that multi-walled carbon nanotubes provoke a similar response as asbestos fibres provided they are sufficiently long [7].

The concept of toxic substances being produced by the particles (as opposed to a mere activation by metabolic processes) comes from epidemiological studies of airborne particulate matter. Ambient air pollution is associated to adverse health effects such as cancer, cardiovascular and pulmonary diseases [8]. The production of Reactive Oxygen Species (ROS) and the generation of oxidative stress is believed to be one of the key mechanisms. Oxidative stress is an imbalance between oxidants and antioxidants. In the presence of particulate matter, the natural antioxidant defenses may be overwhelmed (Halliwell and Gutteridge, 2007). A model [9] to explain how this is possible is shown in Figure 3.

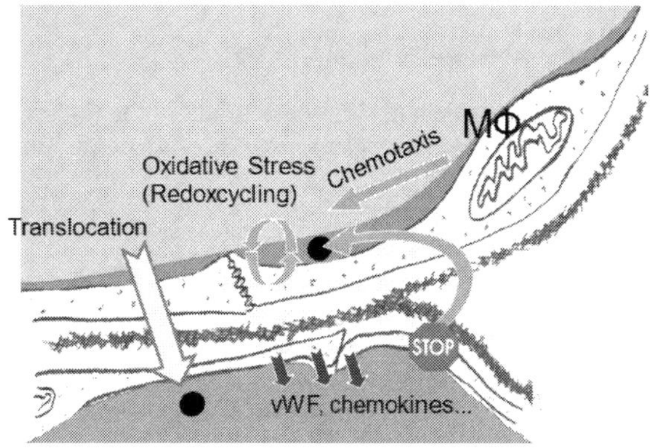

Figure 3. Proposed model how particles deposited on the lung surface generate oxidative stress that causes direct damage and initiates an endothelial cell response that is pro-thrombotic, pro-inflammatory and can release substances that can interfere with the cardiac rhythm.

It proposes that oxidative stress can build up quickly in the thin liquid layer of the alveolar region surrounding the deposition site of the particle by the production of ROS on the particle surface but also the adsorption of antioxidants on particles [10]. This is a site with a high oxygen concentration and antioxidants can quickly be overwhelmed. The presence of the ROS leads to epithelial cell damage and to the release of pro-thrombotic and pro-inflammatory mediators from the adjacent endothelial cells [9]. Some of these factors, such as endothelin can also influence the heart rate [11]. This stress also can lead to DNA-damage in the surrounding cells and subsequent DNA-repair [12]. Ambient air pollution particles and also nanoparticles are small and can easily reach all parts of the lungs. Consequently, the above-described cellular responses to oxidative stress can occur over large parts of the surface of the lungs, resulting in a response that is visible and relevant on a systemic level.

In conclusion, the existing paradigms of particle toxicology can be used to explain many of the toxic effects of nanomaterials. However, they may need to be adapted to account for the special characteristics of nanomaterials. It is also unclear what consequences can result from translocation of particles into cells and to novel target organs, to what extent one has to consider morphological effects of nanomaterials on large biological molecules such as DNA and proteins, and what potential interactions can occur with the immunological system. The answer to these questions will most likely require close multidisciplinary collaborations.

4. Risk management tools and safety by design

Risk management starts with the assessment of the hazards and the extent of contact to the hazard - the exposure level and frequency and duration of exposure. Once the risk is know, appropriate management and control strategies can be developed. This concept also can be applied to nanomaterials. In an ideal world, the potential risks and strategies to manage them are known already in the early phase of product development, when concepts and product design are developed. In the idea of Safety by design, product and material developers will define the material characteristics and the design of a product to be inherently safe, or at least they will give priority to approaches that avoid characteristics where health and environmental problems are likely to occur. Figure 4 illustrates the benefits of this concept. In the beginning of the development, the design options are very large and the developers have considerable flexibility in their choice. At the same time, the costs for safety approaches are quite reasonable. However, further down the development and production chain, the costs for correcting an earlier decision towards "unsafe design" increase more and more, which is paralleled by a decrease in design flexibility.

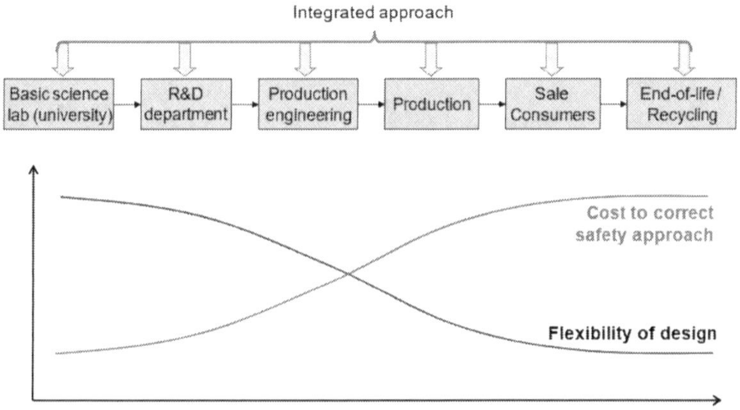

Figure 4. Schematic representation of the value chain including the flexibility of the design to address health and safety concerns and the increase of cost to correct safety approaches down the value chain.

This is a classical problem of step-wise optimization along the value chain, where one step can generate problems for later steps. In theory, an overall system optimisation will yield a higher benefit. There are several challenges with this approach, for example that there are different actors in along the value chain with different interests, risks and benefits. The ones carrying the risk might be not the ones having the benefits and finally, monetary benefit maximization might yield different outcomes than health or environmental benefit optimization. There is again nothing special "nano" in this concept per se. The challenge lays in the many knowledge gaps and the multidisciplinarity that is required to correctly design the products to be safe. There is not only uncertainty (and disagreement) about the most relevant characteristics to explain the hazards and exposure potential of nanomaterials; there is also considerable complexity in the system where the links between the elements must be well understood to make clear statements about which parameters a product developer should optimize.

While a refined risk assessment and safe design approach does not yet seem possible, there are already simplified approaches (toolkits) to evaluate and manage the risks at the different levels. The recommended health and safety approach to new, unknown substances is to treat them as if they were very hazardous. These approaches are then refined as novel knowledge is generated about the hazards, exposure potential and appropriate strategies to protect workers, consumers and the environment. This tiered approach is already in use for "normal" chemicals and is also reflected and supported by the REACH-regulation, which requires an extensive dossier for substances exceeding certain production quantities. For nanomaterials, initial tools exist. For example the precautionary matrix [13] allows testing the potential for nano-specific risks. Also, ISO started a working group to develop control banding strategies that are specifically adapted to nanomaterials.

The challenge is now to develop sufficiently accurate toolkits in a situation where the number of applications and the extent of industrial production are rapidly growing, and therefore where the toolkits need to anticipate the next technological development steps. This requires the joint efforts of researchers, engineers, producers and distributors who must ensure that products are inherently safe, that the potential for acute but also long-term effects is appropriately understood, and that production and use are accompanied by adequate and appropriate good working practices.

5. Coordination and collaboration efforts in Europe

There are clear knowledge-gaps that need to be addressed. They are of multidisciplinary nature and should be addressed on a European and global level. On this background, scientists started to collaborate, first independently, now also with support by the European Commission. One of these efforts is NanoImpactNet, the European Network on the Health and Environmental Impact of Nanomaterials (FP7 project number CSA-CA 218539). The objective of the NanoImpactNet is to create a widely supported scientific basis to ensure the safe and responsible development of engineered nanomaterials and products containing them. This includes a strong two-way communication with the various stakeholder groups, notably policy makers, industry and SMEs, and the public interest groups. This communication provides a way to disseminate scientific information while at the same time obtaining input from these groups about their knowledge needs and questions. Another special feature is that NanoImpactNet is open for participation to all scientists and stakeholders with an interest in the topic, thereby guaranteeing open and transparent discussions. Funding for NanoImpactNet will be provided until early 2012. Already now, new structures are being built that will allow continue NanoImpactNet.

There is still considerable work to be done on the road towards healthy, safe and environmental nanomaterials. Coordination and collaboration is especially needed for the development of best practice to ensure that studies are comparable in terms of basic parameters such as particle type, cell types, dispersion and characterisation protocols and appropriate testing methodologies. Also, close and continued collaboration and discussions are needed if one wants to draw policy-relevant recommendations from the activities mentioned, especially in terms of the occupational and public health and environmental impact potentially associated with the findings of the studies. The European Commission already funds over 20 projects in this field and encourages the researchers to collaborate through networks such as NanoImpactNet or the NanosafetyCluster [14].

NanoImpactNet was able to bring together the leading European groups in the fields of nanosafety, nanorisk assessment and nanotoxicology. This unique opportunity already allowed to reach consensus on some of the issues [15, and www.nanoimpactnet.eu]. It also allowed giving recommendations to the research agenda and proposing priorities in this arena for the coming years. Finally, the development of standardized guidelines and protocols will ensure that the research performed in the individual groups (whether funded by other EU programs or nationally) is of the highest quality. Overall, the exchange of ideas and information between researchers and stakeholders leads to very constructive discussions and it makes way to the development of strategies that can ensure the safe and responsible development of nanomaterials.

References

[1] ISO/TC 229 - Nanotechnologies, Nanotechnologies - Terminology and definitions for nanoparticles, ISO/TS 27687:2008(E), 2008.
[2] HM government, Response to the Royal Society and Royal Academy of Engineering Report: Nanosciences and Nanotechnologies: Opportunities and Uncertainties, Department of Trade and Industry: 2005.
[3] Schmid K, Danuser B, and Riediker M: Nanoparticle Usage and Protection Measures in the Manufacturing Industry--a Representative Survey. J Occup Environ Hyg 2010, 7 (4): 224-232.
[4] Schmid K and Riediker M: Use of Nanoparticles in Swiss Industry: A Targeted Survey. Environ Sci Technol 2008, 42 (7): 2253-2260.
[5] Geiser M, Rothen-Rutishauser B, Kapp N, Schurch S, Kreyling W, Schulz H, Semmler M, Im

H, V, Heyder J, Gehr P: Ultrafine particles cross cellular membranes by nonphagocytic mechanisms in lungs and in cultured cells. Environ Health Perspect 2005, 113:1555-1560.

[6] Kreyling WG, Semmler M, Erbe F, Mayer P, Takenaka S, Schulz H, Oberdorster G, Ziesenis A: Translocation of ultrafine insoluble iridium particles from lung epithelium to extrapulmonary organs is size dependent but very low. J Toxicol Environ Health A 2002, 65:1513-1530.

[7] Duffin R, Tran L, Brown D, Stone V, Donaldson K: Proinflammogenic effects of low-toxicity and metal nanoparticles in vivo and in vitro: highlighting the role of particle surface area and surface reactivity. Inhal Toxicol 2007, 19: 849-856.

[8] Brook RD, Rajagopalan S, Pope CA, III, Brook JR, Bhatnagar A, Diez-Roux AV, Holguin F, Hong Y, Luepker RV, Mittleman MA, Peters A, Siscovick D, Smith SC, Jr., Whitsel L, Kaufman JD: Particulate matter air pollution and cardiovascular disease: An update to the scientific statement from the American Heart Association. Circulation 2010, 121:2331-2378.

[9] Riediker M, Devlin RB, Griggs TR, Herbst MC, Bromberg PA, Williams RW, Cascio WE: Cardiovascular effects in patrol officers are associated with fine particulate matter from brake wear and engine emissions. Part Fibre Toxicol 2004, 1:2.

[10] Sauvain JJ, Sánchez Sandoval Hohl M, Riediker M. Differentiated chemical reactivity of nanoparticles toward DTT. American Journal of Respiratory and Critical Care Medicine, 2009:179, A5258

[11] Isaka M, Kudo A, Imamura M, Kawakami H, Yasuda K: Endothelin receptors, localized in sympathetic nerve terminals of the heart, modulate norepinephrine release and reperfusion arrhythmias. Basic Res Cardiol 2007, 102:154-162.

[12] Setyan A, Sauvain JJ, Guillemin M, Riediker M, Demirdjian B, Rossi MJ: Probing Functional Groups at the Gas-Aerosol Interface Using Heterogeneous Titration Reactions: A Tool for Predicting Aerosol Health Effects? Chemphyschem 2010.

[13] Höck J, Hofmann H, Krug H; Lorenz C, Limbach L, Nowack B, Riediker M, Schirmer K, Som C, Stark W, Studer C, von Götz N, Wengert S, Wick P. Grille de précaution pour les nanomatériaux synthétiques Berne. : Office fédéral de la santé publique ; Office fédéral de l'environnement, 2008.

[14] Riediker M, Katalagarianakis G (Eds.). Compendium of Projects in the European NanoSafety Cluster. Lausanne, Switzerland: Institute for Work and Health, 2010 - 196 pp

[15] Bouwmeester H, Lynch I, Marvin HJP, Dawson KA, Berges M, Braguer D, Byrne HJ, Casey A, Chambers G, Clift MJD, Elia G, Fernandes TF, Fjellsbo LB, Hatto P, Juillerat L, Klein C, Kreyling WG, Nickel C, Riediker M, Stone V. Minimal analytical characterization of engineered nanomaterials needed for hazard assessment in biological matrices. Nanotoxicology 2010. [EPub ahead of print doi: 10.3109/17435391003775266]

Visualisation of exposure to nanoparticles using PIMEX

Koen Verbist and Petra Beurskens

ArboUnie, PO Box 6990, 6503 GL Nijmegen, The Netherlands

E-mail: koen.verbist@arbounie.nl

Abstract. In 2010 the Video Exposure Monitoring method PIMEX was applied to visualize exposure to nanoparticles for several tasks. Next to personal observations the video captured the working conditions and registered possible changes in operational conditions. Simultaneously and coupled with the video both the personal exposure (in the breathing zone) and the background exposure were monitored. In this way the personal exposure to synthetic particles can be corrected for the background exposure to natural occurring nanoparticles or confounding factors. The resulting videos can be edited and used for risk communication purposes. Overall, PIMEX can be of added value for the risk assessment of engineered nanoparticles.

1. Introduction

Nanotechnology is a rapidly growing business, whether it is for sunscreen applications, on car windows, in socks or as nano-chemotherapy. The possibilities seem almost unlimited for these new techniques whereby materials are being designed or manipulated on the nanoscale. The term 'nano' is usually used for materials with one or more dimensions between 1-100 nanometers (a nanometer is a millionth of a millimeter). This can be a single particle, but also a fiber or a plate.

The applications of nanotechnology are many because with these techniques it is possible to create materials with new properties. These properties may, however, differ substantially from the properties that these same materials have at a larger scale [1]. Due to these other properties, it is possible that the hazardous properties, compared to the parent material, are also different. For instance, decreasing particle size might lead to more reactive particles per unit mass and result in a larger surface specific area. As a result other health effects than the known human health effects of the parent material might occur. At the moment the possible hazardous properties of nanoparticles are not yet fully characterized.

Besides the uncertainty concerning the hazardous properties of the nanoparticles there is also discussion on the exposure potential of the so-called synthetic nanoparticles. How and when does exposure happen, via which exposure routes and how should exposure be measured and characterized? Typically, exposure to particles is expressed in terms of mass concentration (e.g. mg/m^3). However, there is increasing evidence that exposure to (water insoluble) nanoparticles can be better expressed in terms of particle number (e.g. p/cm^3), or surface area concentration [2]. At the moment there is no standardized measurement strategy for exposure assessment of nanoparticles in the workplace. In addition there is no (portable / personal air sample) measurement device available yet that is able to simultaneously measure all these different exposure parameters.

Because of the uncertainty related to the hazardous properties of nanoparticles and the possible health effects following exposure, in 2009 the Dutch government decided to apply the precautionary principle [3]. This means that policies and implementation measures should focus on preventing or minimizing the exposure of workers to nanoparticles. It is up to the employers to apply the precautionary principle and to implement it in the risk inventory and the corresponding action plans. To help implement the precautionary principle good practices have been developed. In 2008 an investigation was carried out in the Netherlands to obtain an overview of the good uses of engineered nanoparticles in Dutch companies and research facilities [4]. Part of the deliverables was a best-practice guidance sheet (good-use-decision-tree) with a step-wise approach of risk characterization (figure 1).

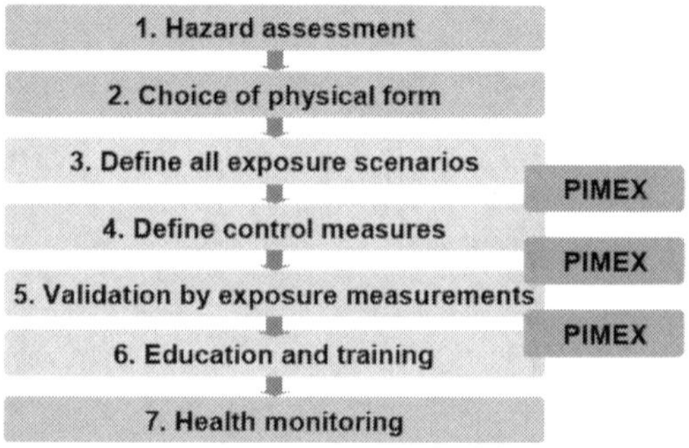

Figure 1. Steps of the good-use-decision-tree.

2. PIMEX

PIMEX (Picture Mix Exposure) is a visualization technique that can support employers in carrying out the exposure assessment and in complying with the precautionary principle. It is a video exposure monitoring method that had been broadly applied within occupational hygiene practice for over 20 years. PIMEX was originally developed in the eighties in Sweden by the National Institute for Working Life [5]. Similar developments took place around the same time in the USA (NIOSH) under the name Video Exposure Monitoring (VEM) [6]. In 2004 PIMEX was introduced in the Netherlands and applied for different purposes, with currently >100 movies freely available in Dutch. Movies cover broad topics such as chemicals, physical strain, vibration, noise, heat stress and ionizing radiation. These movies have been developed for various industries and private companies.

The PIMEX set is coupled with measurement equipment, a computer and a video camera. The work situation or action is filmed and simultaneously real-time measurements are performed (e.g. with a DustTrak or PID). The devices are connected to a laptop with PIMEX software. The measurement data are being sent to the laptop via an analogue signal and integrated with the video so that both the work situation and measurement data are presented on the screen. In this way, activities where for instance exposure to hazardous substances are possible, and specific moments where exposure occurs, are visualized. With the latest version (PIMEX-2008) up to four different devices can be connected to simultaneously monitor and visualize exposure.

3. PIMEX and Nano

In 2009 a survey was carried out by Arbo Unie and funded by the Ministry of Social Affairs and Employment to identify different types of nano measuring equipment presently available and to examine if these devices can be coupled with PIMEX. In order to be useful for workplace (personal) sampling and for coupling with the PIMEX set several technical requirements were composed. The technical requirements are a) analogue output, b) direct readable, c) portable and d) useful for personal air sampling. Two measurement devices were selected that were available and met two or more of the technical requirements. These were connected with the PIMEX set and test-movies were made for situations were exposure to nanoparticles occurred.

The practical tests were successful and have shown that it is possible to connect nano-measuring equipment with PIMEX and thus visualize exposure to nanoparticles. This offers a number of practical features and benefits that can assist companies with the implementation of the precautionary principle and with some steps of the good-use-decision tree (figure 1). These steps are mentioned below.

3.1. Identify all tasks and activities with potential exposure

PIMEX can assist in detecting and visualizing possible sources of exposure to nanoparticles. In monitoring exposure to nanoparticles the observation and recording of workplace conditions is a very important aspect of the risk assessment, which increasingly becomes a standard procedure. With video recording a thorough observation and interpretation of the intended processes and the surrounding work conditions are possible and simplified. One of the advantages of video monitoring is that all visible details of the work situation are captured and stored (more than with only written registration), and can be viewed and analyzed afterwards. In this way these situations or specific processes are identified were exposure to nanoparticles occurs. This information can be further used for the implementation of risk management measures.

3.2. Personal air sampling to determine exposure

Besides the detection of sources related to process emissions, it is possible to assess exposure to nanoparticles in the breathing zone of the employee. This can be accomplished with portable measurement equipment. At the moment there are a few devices on the market that are portable and useful for personal air sampling. In this way it is clear which tasks result in exposure of the employee to nanoparticles. Based on this information risk management measures can be implemented if necessary.

3.3. Determination of and correction for the background concentration

There are numerous sources of nanoparticles. In the natural environment and in the ambient air outside the workplace nanoparticles occur manifold, ranging from traffic emissions to emission from the climate system or devices used for housekeeping. The contribution of a task or process compared with this background concentration may be relatively small, thus making it more difficult to correctly interpret the measurement data. The PIMEX set can be coupled to multiple devices, enabling the simultaneously assessment and recording of the process emission as well as the background concentration. In this way it is possible to correct exposure to manufactured nanoparticles for the background concentration. When doing this it is important to select good reference locations for background measurements (e.g. measuring incoming air in the breathing zone, in the same industrial setting, but located further away or a measurement in another room, e.g. an office).

3.4. Identification of confounding factors

The background concentration of nanoparticles can fluctuate considerably, e.g. due to opening / closing of doors or switching on of other equipment (e.g. a forklift). These additional sources of nanoparticles can strongly affect the exposure measurements of synthetic particles, especially when the expected exposure concentrations are low. When the contextual information during the measurements is not recorded it is impossible to determine whether increased concentrations are due to release of synthetic particles or due to a change in background levels. With PIMEX the work situation is recorded and possible confounding factors that might influence the exposure measurements are identified.

3.5. Validation of risk management measures

PIMEX can be used to determine the effect of risk management measures in reducing exposure to nanoparticles. Situations with and without control measures can be assessed and recorded (e.g. with/without local exhaust ventilation). During the recording the video and graphic images can be viewed on the screen. Immediately after the recording the images can be displayed again, so that those involved (operator, manager or technical assistant and occupational hygienist) can see and discuss the effect of the control measure. If the effect of the measure is insufficient the situation can directly be adjusted and again recorded with PIMEX. In this way PIMEX can be used as an intervention method.

3.6. Education and training of employees
The recordings can be viewed on the spot. In addition they can also be edited for education and training purposes and viewed with a simple movie player. PIMEX has proven itself as an ideal tool for education and training of employees, because workers can see for themselves the effect of their actions on exposure. Seeing is believing!

4. Conclusion
In the Netherlands employers are required to implement the precautionary principle when synthetic nanoparticles are used or manufactured. The good use decision tree gives occupational hygiene professionals directions to get started with the risk assessment. An occupational hygiene background with knowledge of characterization of exposure specific to nanoparticles is essential. PIMEX is a valuable contribution in carrying out a number of steps of the good use decision tree. The PIMEX recordings can be used for different purposes or/and be edited into films for education or training purposes.

References
[1] Oberdörster, G. Oberdörster E, Oberdörster J. Nanotoxicology: an emerging discipline evolving from studies of ultrafine particles. Env. Health Persp, 113(7), 823-839, 2005.
[2] Workplace exposure to nanoparticles - literature review. European Risk Observatory Report, European Agency for Safety and Health at work, 2009.
[3] Dutch Ministry of Social Affairs and Employment: reaction on the Social Economic Council report on handling nanoparticles on the workfloor. http://www.rijksoverheid.nl/documenten-en-publicaties/kamerstukken/2009/06/08/reactie-van-minister-donner-op-het-advies-van-de-ser-over-nanodeeltjes-op-de-werkplek.html
[4] Borm P, Houba R, Linker F. Omgaan met nanodeeltjes op de werkplek. Survey naar goede praktijken in omgaan met nanomaterialen in de Nederlandse industrie en kennisinstellingen. 2009. (http://www.arboportaal.nl/downloads/Omgaan%20met%20nanomaterialen-final%20version11%2007%2008.pdf/view)
[5] Rosén G, Andersson I-M, Walsh PT, Clark RDR, Säämänen A, Heinonen K, Riipinen H, Pääkkönen R. A review of video exposure monitoring as an occupational hygiene tool. Ann. Occup. Hyg, 49(3), pp 201-217, 2005 (http://annhyg.oxfordjournals.org/content/49/3/201.full.pdf+html)
[6] CDC. Analyzing workplace exposure using direct reading instruments and video exposure monitoring techniques. Cincinnati, OH: Centers for Disease Control, Department of Health and Human Services, 1992.

Journal of Physics: Conference Series **304** (2011) 012003 doi:10.1088/1742-6596/304/1/012003

Characterization of manufactured nanoparticles at workplace

C. Durand, X. Ravanel, S. Derrough, E. Zimmermann, JL. Fromaget, J. Garrione, D. Locatelli, E. Lefranc, T. Farouz

CEA-Grenoble, DSP/SMR, Nanoparticles Expertise Center, 17 avenue des Martyrs, 38054, Cedex 9, Grenoble, France

E-mail: catherine.durand@cea.fr

Abstract. French government has launched a dedicated program on nanotechnologies in May 2009. One of the topics was to develop measurements and characterisations at workplace, with the aim of getting the experience over 100 of them, either in research centres, universities and industries. At this time, about 80 have been realized. The methodology is based on the study of nanoparticles emission, during the whole duration of the process and / or operation, in real time, by comparison between background level and process ones. Parameters such as number of particles, size and developed surface are obtained. Shape, agglomerate state and chemical composition are obtained by further microscopy analysis. Currently, dustiness and surface state are not characterized and future developments are needed. It has to be pointed out that, at this time, and thanks to very various scenarios met, data collected from this program lead in a field database already used in different European research programs.

1. Introduction

Nanoparticles are or are about to be present in various and numerous fields and activities (medicine, cosmetology, building materials…). As a result, the number of workers in contact with nanoparticles is increasing and it becomes necessary to be able to measure and prevent potential exposure during processes. Different initiatives' and work groups are already in process, either at international, European or national scale.

In May 2009, French government has announced an important and specific program dedicated to Nanotechnologies: NanoINNOV. The aim of this plan is to give to industry all the abilities in order to catch all the benefits of "nanorevolution", while research would continue to explore intimate properties of matter, and find technical applications.

Over 3 integration centres, based in Paris, Toulouse and Grenoble, NanoINNOV topics are covering 4 axis:

- build a national device joining international standards,
- focus means on technology, heart of innovation process,
- bring this French ambition to European and international level,
- inscribe this process in a societal and timeless dynamism.

Nanoparticles's safety is one of the topics involved in this last item, such as the building of a dedicated nanocenter or training as well.

Since the beginning of 2010, CEA, INERIS and CNRS have joined their efforts and have worked together, in order to be able to propose an innovatory offer to industrials, research centres and universities: measure and characterize manufactured nanoparticles in workplace air, and improve safety on work stations.

2. Experimental Setup

2.1. Campaign sequence

Each campaign sequence obeys different compulsory steps. The first one is a pre-assessment on site in which the adequation between operating conditions and our methodology and equipments range is established. For example, the workstation layout, aerodynamics, the timing of each operation, the safety settings are checked. Moreover, during this meeting, data concerning the manufactured nanoparticles are also collected: scale, shape, chemical composition. The second step, which is the measurement on site, is completely described in point 2.2 and 2.3. The final stage consists in a restitution which is done through a technical report including all the results obtained on site completed by a list of recommendations in order to improve the safety and environment working conditions.

2.2. Equipments

To study the nanoparticles liable to be released in air at workplace, we used several techniques to characterize particles aerosol in terms of volume concentration, size distribution, specific surface area shape and chemical properties.

Particles counting is performed using water-based Condensation Particle Sizers (CPC 3785) manufactured by TSI Inc. [1] operating on the principle of enlarging small particles using a condensation technique to a size allowing optical detection. The CPC detects particles in the size range from 5 nm to 3 µm and provides measurements over a particle concentration range from 0 to 10^7 p.cm^{-3} with time resolutions in the order of one second.

This tool is continuously used during the measurement at workplace because of its high sensitivity and its quick response time. As it will be detailed in the methodology section, we made the choice of using two CPC in parallel to evaluate the influence of surrounding environment on each operation at the workplace.

To go further and get information on size distribution of particles, different granulometric techniques were used.

We used a GRIMM 5.416 Scanning Mobility Particle Sizer (SMPS) [2] consisting of a particle neutralizer (Am241), a short-column Differential Mobility Analyzer (DMA) and a CPC. In such a configuration, the SMPS provides the size distribution of nanoparticles in the range 5-350 nm up to a concentration of 10^7 p.cm^{-3}.

We performed real-time measurements by using a Fast Mobility Particle Sizer (FMPS). The FMPS 3091 from TSI measures the particle mobility diameter in the range 5.6-560 nm with a time resolution of 1 s using a flow rate of 10 l.min^{-1}. A detailed description of the FMPS can be found in reference [3].

An Electrical Low Pressure Impactor (ELPI) [4,5] manufactured by DEKATI Ltd was used. It combines both size distribution measurements with time resolutions in the order of seconds and the possibility of size resolved particle collection on impaction substrates for subsequent chemical analysis. The 12 stages cover a size range from 30 nm to 10 µm and the lower size limit can be extended to a cut-off diameter of 7 nm by using an electrical filter stage.

In order to get an information of the specific surface area of particles, we sometimes used a Nanoparticle Surface Area Monitor (NSAM) that indicates the human lung-deposited surface area of particles in units of square micrometers per cubic centimetre (µm^2.cm^{-3}), corresponding both to tracheobronchial and alveolar regions of the lung. The instrument used is the NSAM 3550 manufactured by TSI

The NSAM 3550 matches the corresponding lung deposition criteria of particles for a reference worker as predicted by human lung deposition models published by the International Commission on Radiological Protection (ICRP, 1995). The lung deposition is calculated for a reference worker as defined in a publication by the American Conference of Governmental Industrial Hygienists (ACGIH, *ed.* Vincent J.H., 1999).

For the qualitative side of the characterization, we use a Scanning Electronic Microscopy (SEM) 5500 from HITACHI. The high resolution of this tool allows the visualization of the nanoparticles for size measurement and shape determination. Associated to the SEM, we used an EDX (Energy Dispersive X-Ray Spectroscopy) tool from THERMO SCIENTIFIC Inc. to get information on the chemical composition of the nanoparticles collected during the previous step.

2.3. Methodology

When emission of nanoparticles is suspected during an operation at a workplace, three characterization steps are conducted. The first one is a real time quantitative characterization, as close as possible to the operation taking place, to have access to the number of particles emitted per unit of volume (CPC), the size distribution (SMPS, FMPS, ELPI) and the associated specific surface area (NSAM). This initial step also includes sampling on dedicated membranes (within the ELPI). The second step of the methodology is the chemical characterization of the nanoparticles emitted, which is done offline from the samples collected in step one and using dedicated analysis instruments (SEM, EDX). This second set of characterization gives access to new and very important data. Are we in front of isolated species or agglomerates? What is the chemical composition? Those fundamental questions are answered through this qualification step.

It is important to point out that our methodology was built from the beginning with the rule of conduct that each operation at the workplace should not be disturbed from the possible variations of the surrounding ambiance. For this purpose, we established a protocol where each operation measurement should be completed by a real time comparison with the air of the local.

One CPC is set to a fixed position to measure continually the ambient air in the workplace. Another one is linked to the sampling stick to follow the operator activity and to measure the potential sources of nanoparticles release in air. The measurements performed by the CPC tools in parallel (so called CPC_{source} and $CPC_{ambient\ air}$) bring information about the global concentration variations, over the whole duration of the process, between the ambient air of the workplace and the different steps of the process that could be sources of nanoparticles release in air

This is different from the NEAT method (Nanoparticle Emission Assessment Technique) developed by the National Institute for Occupational Safety and Health (NIOSH) [6,7]. This last method includes a first run to draw a baseline or a background picture of the nanoparticles already in the environment before any operation. The second step is then to run the same measurements but this time with the operation taking place. If measurements during the process indicate particles concentration higher than the background level, then further analytical information are conducted to identify of the nanoparticles. The complete comparison of the two methodologies will be the topic of a coming publication in the following months.

3. Results - Example

A workstation is defined as a process or a process step, involving nanomaterials and liable to release nanoparticles in air due to operator activity.

The case we consider here is the preparation of a solution for CMP (Chemical and Mechanical Polishing) (Figure **1**). This process consists of several steps among them we were particularly interested in the dry ones: the weighing phase of a nanometric AlO_2 powder and the decanting of the powder in a bowl. These operations were performed in a class 10000 - ISO 4 clean room environment under a laboratory fume hoods.

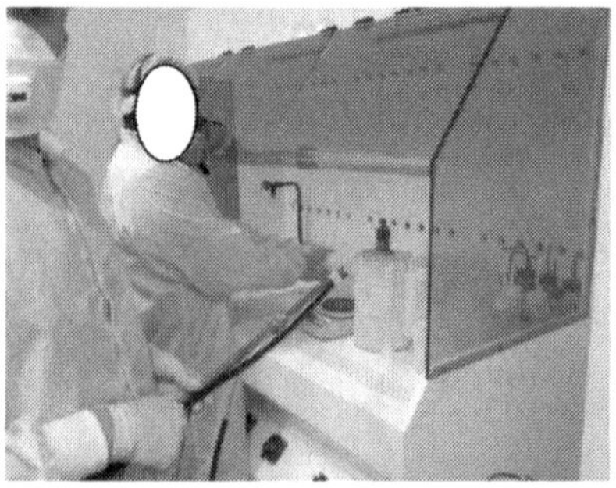

Figure 1. Preparation of a CMP solution involving a nano-powder.

First, we sampled in the lab near the polishing zone and both the CPC measured a concentration of 0.1 p.cm^{-3} (Figure **2**). Then, a measurement with the so called CPC_{source} was performed in the introduction airlock (number 1 on figure 2) while the other CPC was still in the lab but next to half-opened airlock. We observe a concentration of 2800 p.cm^{-3} in the airlock while the second CPC indicates a concentration increase from to 0.1 to 1 p.cm^{-3}. As a comparison, the ambient concentration level generally measured in an office with normal human activity is around 10^4 p.cm^{-3}.

Respectively at 11:28 and 11:32 two peaks were detected (79 p.cm^{-3} and 27 p.cm^{-3}) and seem to be related to equipments movements. At 11:33 the sampling under the fume hoods was started. During this phase, the fume hoods aspiration was on (light blue on figure 1). The particulate concentration before the beginning of the operator activity is constant between 1 and 10 p.cm^{-3}. At 11:41 the decanting of the AlO_2 powder in a bowl was started by the operator (number 2 on figure 2). During this short operation, a concentration peak of 179 p.cm^{-3} was measured while the ambient air particles concentration keeps constant around 1 p.cm^{-3}. Then, a sampling was immediately performed very close to the operator respiratory tracts. A level equal to the ambient air one was detected. This may reveal that the particles release in air keeps very close to the source emission point and that the fume hoods provides a well-adapted air recirculation.

At 11:47, the aspiration of the fume hoods was turned off. During the first five minutes of this phase (light orange on figure 2), no operations from the operator were done and we measured under the fume hoods with the CPC_{source} while the other CPC was still measuring the ambient air nearby the workplace of interest. We observe an increase of the particles concentration under the fume hoods, from 1 to several 10 p.cm^{-3}. The ambient air particles concentration is constant and lower than 1 p.cm^{-3}.

From 12 a.m., the same operations than with the fume hoods turned on were performed. A general increase (factor 10) of the concentration of particles present in the ambient air was observed. However, no significant increase of the concentration due to the different steps of the process was emphasized, as shown by the CPC_{source} signal. It seems that under no air aspiration, the fact that the air is more charged in particles entails a lack of sensitivity whereas a cleaner environment allows the identification of more variations phases.

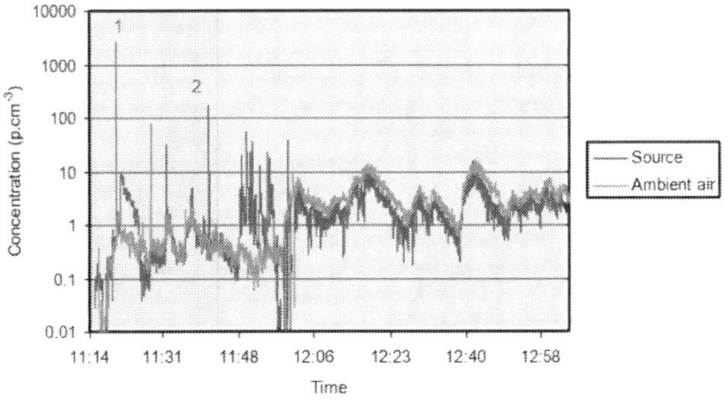

Figure 2. Concentration (p.cm^{-3}) of particles within the size range 5-3000 nm as measured in parallel by two CPC. The blue line corresponds to the potential source of particles release, the pink one corresponds to the ambient air of the workplace. The first part of the graph (light blue) corresponds to measurements with the fume hoods on, the second part (light orange) with fume hoods off

The decanting phase of the powder in a bowl was identified by the CPC measurements as a potential source of nanoparticles release in air. The distribution size of those particles was studied by means of FMPS (Figure **3**). Considering the quickness of the decanting of the powder, the SMPS which needs 4 minutes to record a spectrum, does not provide a granulometry representative of the phase of interest. The FMPS measurement shows a monodispersed aerosol with a size distribution centred on 250 nm.

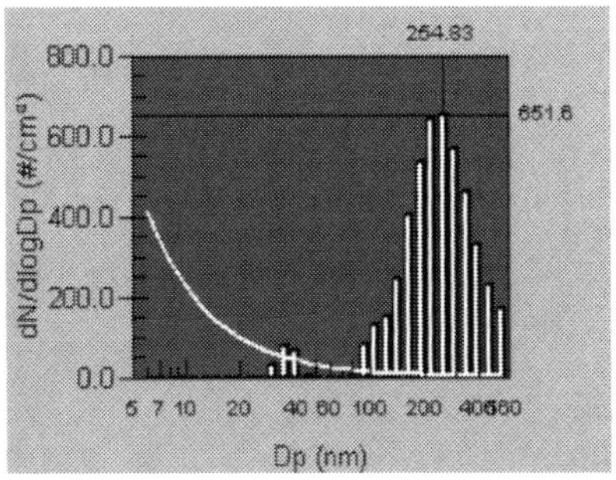

Figure 3. Size distribution of the aerosol as measured by FMPS during the decanting of the powder in the bowl

To get information about the specific surface area of particles, we performed a real-time NSAM measurement indicating the human lung-deposited surface area of particles, corresponding in that case to alveolar regions (Figure **4**). At the day of redaction of this paper, NSAM data are still not totally understood in some cases. Nevertheless, presently, we identified three principal peaks that all correspond to well-known events. Indeed, the two first significant peaks at 11:50 and 11:52 (28 and 51 $\mu m^2.cm^{-3}$) correspond to the moment when the fume hoods aspiration was turned off, as shown by the CPC measurements. The highest peak at 12:02 (77 $\mu m^2.cm^{-3}$) corresponds to the moment when the operator started to sand the powder under the fume hoods off. Here we point out the fact that surface deposition of particles in the alveolar region of lung seems correlated to the presence or not of air aspiration and so collective protection.

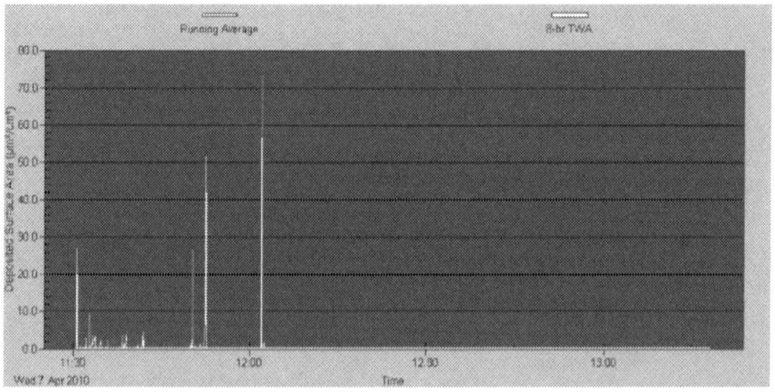

Figure 4. Deposited Surface Area ($\mu m^2.cm^{-3}$) of particles as measured by NSAM in alveolar mode

A batch of Aluminium impaction substrates were placed on the ELPI stages for subsequent chemical analysis. We specially concerned our interest on samples to stage 4 and 5, corresponding respectively to theoretical aerodynamic diameters 155 nm and 260 nm. These samples were then analyzed by means of SEM-EDX giving access to both structural and chemical information on the particles. The SEM measurements revealed that particles impacted on stage 4 had aerodynamic diameters between 170 and 210 nm and on stage 5 between 210 and 280 nm. A SEM observation of a particle impacted on stage 5 of the ELPI is shown on Figure **5**. This particle has a critical size of 226 nm and EDX spectra were recorded inside and outside of it to get information about its chemical composition (Figure **6**).

The EDX analysis revealed the presence of oxygen only inside the particle. Moreover, the same Al peak was detected both on the Al substrate and in the particle. The EDX analysis confirmed that the particle consists of alumina and oxygen.

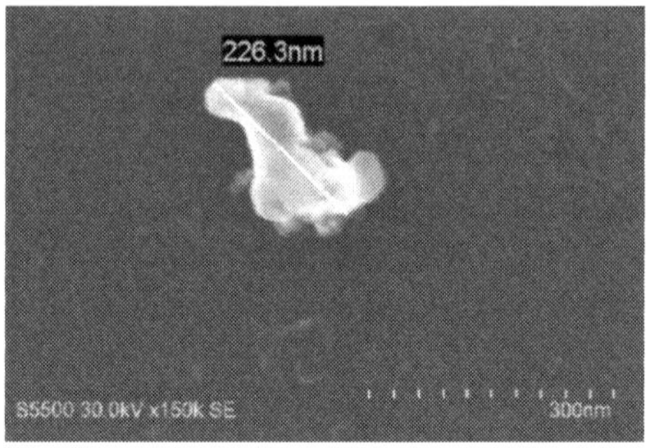

Figure 5. SEM observation of a particle measured on stage 5 of the ELPI.

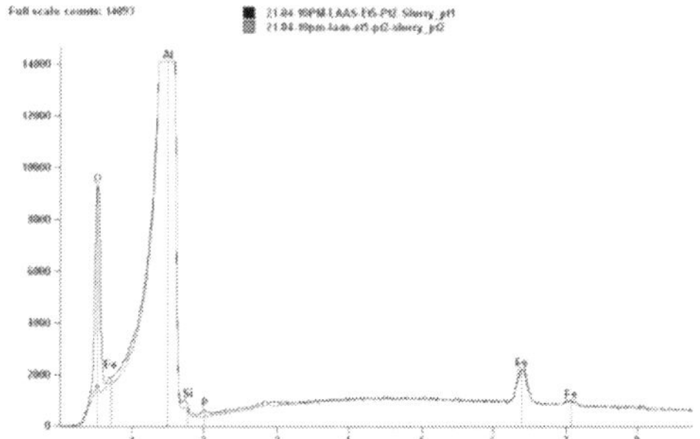

Figure 6. EDX spectrum inside the AlO_2 particle (black) and outside, on the Al substrate (red)

Our methodology allowed us to provide a complete characterization of a workplace in terms of nanoparticles released in air in the form of aerosol and liable to be inhaled by the operator.

4. Conclusion

Over the last 10 months, 76 workstations and / or operations have been studied and characterized following our strategy. The objective of NanoINNOV is to join the aim of 100 till the end of 2010 and this target is about to be reached.

Different scenarios and type of nanoparticles have been characterized that way:

- As type of conditions:
  - Processes such as weighing, grinding, pelleting, repackaging, scraping, evaporation, cleaving, ...
  - Operations such as ink preparation, maintenance, cleaning, furnace opening, ...
  - Use of equipments such as sandblaster...
- As type of nano-objects : nanopowders, nanotubes, nanowires,
- As type of substances: titanium dioxide, carbon, carbon black, silicon, silicon dioxide, titanium, alumina, zinc oxide, copper, iron phosphates, arsenic, ...
- As amounts used : from few micrograms to some kilograms

For the next months, one of our objectives is to continue to collect on-site data on a national scale while continuing to improve the methodology in order to be able to measure complementary parameters such as surface properties, dustiness. In the same way, measurement on colloïds could be developed.

Acknowledgments

We want to thanks all the persons that have helped us in this work, and particularly those from INERIS and CEA / LITEN / LCSN labs.

References

[1] Hermanna, M., Wehner, B., Bischofb, O., Hanc, H.-S., Krinkeb, T. Liuc, W. Zerrathc, A. Wiedensohlera, A., Journal of Aerosol Science, vol. 38, Issue 6, 2007, pages 674-682

[2] Wang, S. C., Flagan, R. C., Aerosol Science and Technology, vol. 13, 1990, pages 230-240

[3] Johanson, T., Caldow, R., Pöcher, A., Mirme, A., Kittelson, D., Society of Automotive Engineers Technical Paper Series 2004-01-1341

[4] Keskinen, J., Pietarinen, K., Lehtim ki M.,. J. Aerosol Sci., vol. 23, 1992, pages 353–360

[5] Moisio, M. (1999) Real time size distribution measurement of combustion aerosols. Ph.D. Thesis Tampere University of Technology publications 279, Tampere Finland.

[6] Methner M., Hodson L., Geraci C., Journal of Occupational and environmental Hygiene, vol. 7, 2010, pages 127-132

[7] Methner M., Hodson L., Dames A., Journal of Occupational and environmental Hygiene, vol. 7, 2010, Pages 163-176

Onsite aerosol measurements for various engineered nanomaterials at industrial manufacturing plants

I Ogura, H Sakurai and M Gamo

National Institute of Advanced Industrial Science and Technology (AIST), 16-1 Onogawa, Tsukuba, Ibaraki 305-8569, Japan

E-mail: i-ogura@aist.go.jp

Abstract. Evaluation of the health impact of and control over exposure to airborne engineered nanomaterials (ENMs) requires information, inter alia, the magnitude of environmental release during various industrial processes, as well as the size distribution and morphology of the airborne ENM particles. In this study, we performed onsite aerosol measurements for various ENMs at industrial manufacturing plants. The industrial processes investigated were the collection of SiC from synthesis reactors, synthesis and bagging of $LiFePO_4$, and bagging of ZnO. Real-time aerosol monitoring using condensation particle counters, optical particle counters, and an electrical low-pressure impactor revealed frequent increases in the number concentrations of submicron- and micron-sized aerosol particles, but few increases in the number concentrations of nanoparticles. In the SEM observations, a large number of submicron- and micron-sized agglomerated ENM particles were observed.

1. Introduction

Although engineered nanomaterials (ENMs) have many potential benefits for society, there is growing concern about their potential impact on health [1]. A significant exposure route is inhalation of airborne ENMs during manufacture and handling. Information on airborne ENMs, such as the magnitude of release and the industrial processes involved, as well as particle size distribution and morphology, is required in order to evaluate and control ENM exposure [2].

In this study, we performed onsite aerosol measurements for various ENMs at industrial manufacturing plants. The processes investigated were the collection of silicon carbide (SiC) from synthesis reactors, synthesis and bagging of lithium iron phosphate ($LiFePO_4$), and bagging of zinc oxide (ZnO). The investigations were conducted in collaboration with the National Institute of Occupational Safety and Health, Japan (JNIOSH). However, this paper presents only the results obtained by the National Institute of Advanced Industrial Science and Technology (AIST).

2. Methods

2.1. Real-time aerosol monitoring

The number concentrations and size distributions of aerosol particles were measured using real-time aerosol monitoring instruments, viz., condensation particle counters (CPCs; model 3007; TSI Inc., USA; 10 to >1000 nm), optical particle counters (OPCs; Handheld 3016; Lighthouse Worldwide Solutions Inc., USA; optical diameters of >300, >500, >1000, >3000, >5000, and >10000 nm), and an electrical low-pressure impactor (ELPI; Dekati Ltd., Finland; aerodynamic diameters of 7–10000 nm, 12 channels). A stainless pipe and conductive silicone tubing were used to transmit the particles from the sampling point around the work area to the measuring instruments. The sampling flow rates of the CPC, OPC, and ELPI were 0.7, 2.8, and 10 L/min, respectively, and the average time was 10 s per sample.

2.2. Filter sampling and electron microscope observation

The aerosol particles were collected on a Nuclepore membrane filter (diameter of 25 mm; pore size of 0.080 μm; Nomura Micro Science Co. Ltd., Japan) with a stainless steel filter holder (available filtration area of 380 mm^2; KS-25F; Advantec Toyo Kaisha Ltd., Japan) for electron microscopy analysis. The sampling flow rate was set at approximately 1–2 L/min. A stainless pipe and conductive silicone tubing were used to transmit the particles from the sampling point around the work area to the sampling filter.

The aerosol particles collected on the Nuclepore filters were observed using a field-emission scanning electron microscope (SEM; S-4300; Hitachi High-Technologies Corporation, Japan) under an accelerating voltage of 1 kV. Before the SEM observation, the samples mounted on SEM grids were coated with platinum–palladium (approximately 1 nm) to avoid image charging.

2.3. Processes investigated

We performed onsite aerosol measurement for ENMs at industrial plants manufacturing ENMs during the following processes.

2.3.1. Collection of SiC from synthesis reactors. During the collection of SiC from two reactors after synthesis by thermal plasma chemical vapor deposition, the aerosol particles at a site near to the work area (approximately 50–100 cm) and at a site far from the work area (approximately 5 m) were measured simultaneously. Because the room had a ventilation system with a high efficiency particulate air (HEPA) filter, the aerosol particle concentrations in the room during nonworking periods were generally low. The representative characteristics of SiC provided by the manufacturer are as follows: the average primary particle diameter is 30 nm and the specific surface area is 50 m^2/g.

2.3.2. Synthesis and bagging of $LiFePO_4$. $LiFePO_4$, which is used as the cathode materials in lithium-ion batteries, is synthesized by sintering a precursor material; the nature of this material is confidential. Aerosol particles were measured at a distance of approximately 60 cm from the work area where the synthesized $LiFePO_4$ was being bagged (11:00–13:20) and approximately 100 cm from the work area where the precursor material was being filled into containers and weighed (13:20–15:20). In addition, the aerosol particles in the air entering through the outside air intake were simultaneously measured. Both $LiFePO_4$ and precursor material mainly consist of the primary particles with a diameter of approximately a hundred to a few hundred nanometers.

2.3.3. Bagging of ZnO. When the synthesized ZnO was being bagged, aerosol particles at a distance of approximately 50–100 cm from the work area and outdoors were measured simultaneously. Although the hoppers of ZnO were automatically emptied into drums, which were lined with plastic bags, the processes of disconnecting the bags containing the synthesized ZnO from the hopper outlets, sealing the bags, and connecting the empty bags in the drums to the hopper outlets were performed manually. The bagging began at 10:00 and ZnO was filled into ten drums during a period of three hours, *i.e.*, till 13:00. The primary particle diameter was a few tens of nanometers as observed through the SEM (see section 3.2).

3. Results

3.1. Particle number concentrations and size distributions of released airborne ENMs

3.1.1. *Collection of SiC from synthesis reactors.* Figure 1 shows the changes in the particle number concentrations measured by the CPCs and OPCs. When collecting SiC from the two synthesis reactors (first: 10:54–11:05 and second: 11:10–11:17), the number concentrations of micron-sized particles increased, whereas those of nano- and submicron-sized particles did not.

Figure 2 shows the particle size distributions during the SiC collection (highest 10-s-average concentration at 11:10) and before it (i.e., the average concentration in the period 10:42–10:54) at the site near to the work area measured by the ELPI and OPC. Figure 2 indicates that the number concentrations of particles with a diameter greater than a few hundred nanometers were higher during the collection process than before.

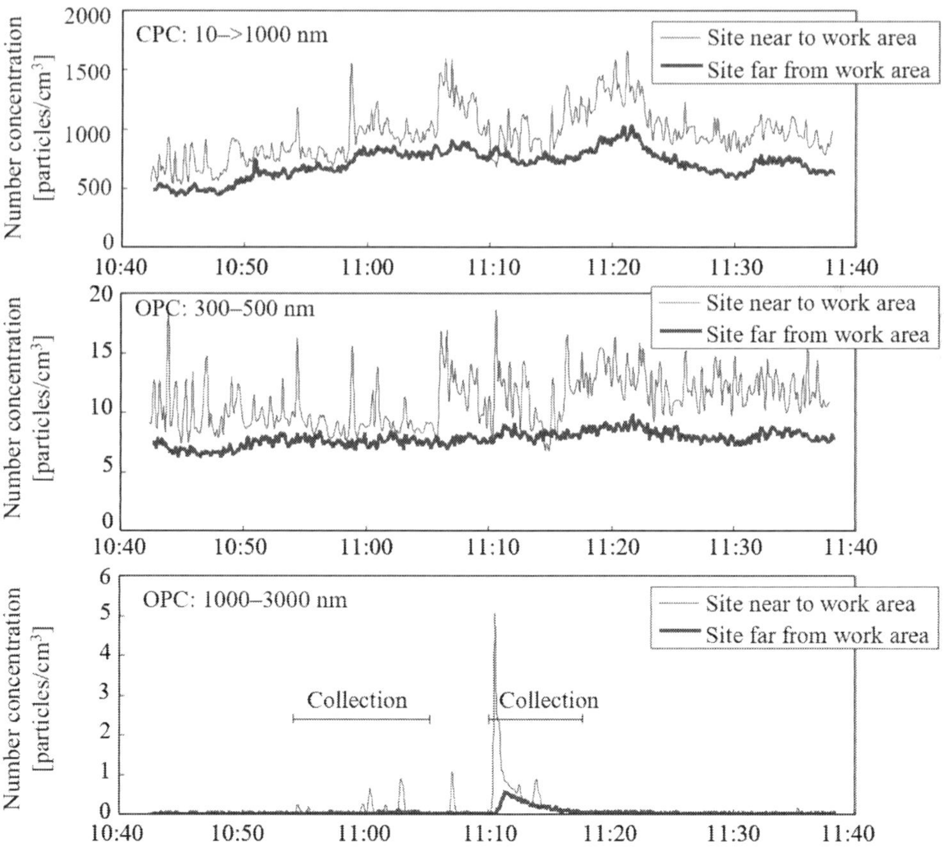

Figure 1. Changes in the particle number concentrations at a site near to and at a site far from the work area where SiC was collected from the synthesis reactors.

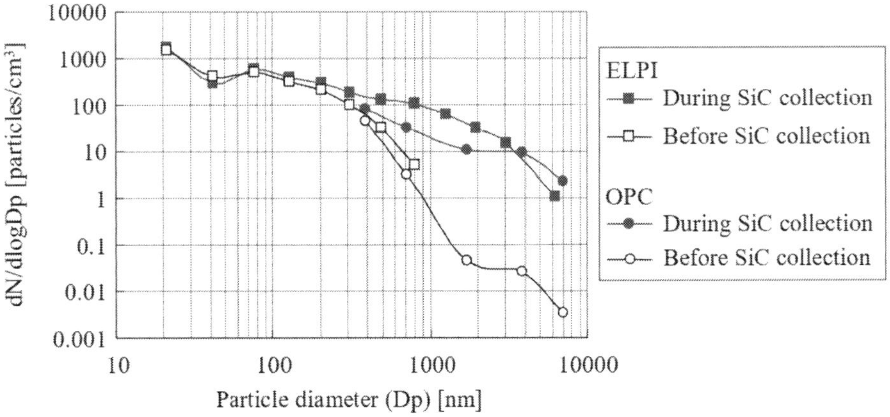

Figure 2. Particle size distributions during and before the collection of SiC. Particle sizes are expressed in terms of aerodynamic diameter for ELPI and optical diameter for OPC.

3.1.2. *Synthesis and bagging of $LiFePO_4$.* Figure 3 shows the changes in the particle number concentrations measured by the CPCs and OPCs near the work areas where the synthesized $LiFePO_4$ was bagged (11:00–13:20) and where the precursor material was filled into containers and weighed (13:20–15:20) and outdoors (11:00–15:20). The number concentrations of submicron- and micron-sized particles increased when a bag containing $LiFePO_4$ was being disconnected from the hopper outlet and sealed (11:35 and 12:50) and when filling and weighing the precursor material (11:40–11:52, 14:00–14:24).

Figure 4 shows the particle size distributions during the bagging of $LiFePO_4$ (highest 10-s-average concentration at 11:35) and before it (i.e., the average concentration in the period 11:20–11:34) near the work area measured by ELPI and OPC. Figure 5 shows the particle size distributions during the filling and weighing of the precursor material (*i.e.*, the average concentration in the period 14:00–14:24) and before it (i.e., the average concentration in the period 13:50–14:00) near the work area measured by the ELPI and OPC. Figures 4 and 5 indicate that the number concentrations of particles with a diameter greater than several hundred nanometers increased when $LiFePO_4$ was being bagged and when filling and weighing the precursor material.

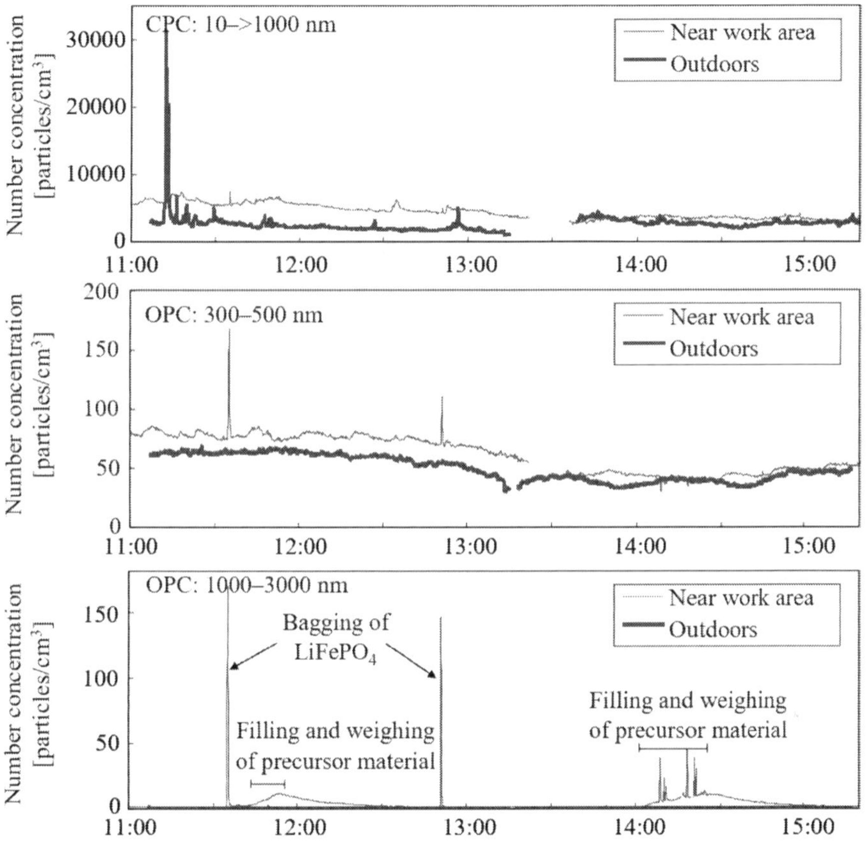

Figure 3. Changes in the particle number concentrations near the work areas where synthesized $LiFePO_4$ was bagged (11:00–13:20) and where the precursor material was filled into containers and weighed (13:20–15:20) and outdoors (11:00–15:20)

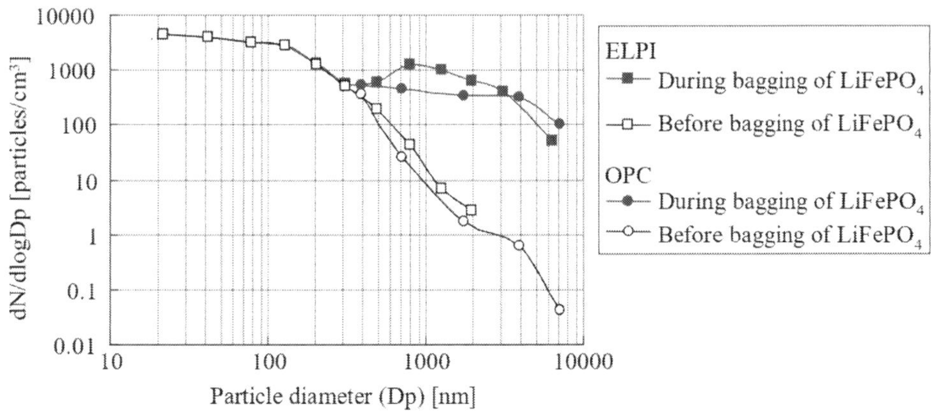

Figure 4. Particle size distributions during and before the bagging of $LiFePO_4$. Particle sizes are expressed in terms of aerodynamic diameter for ELPI and optical diameter for OPC.

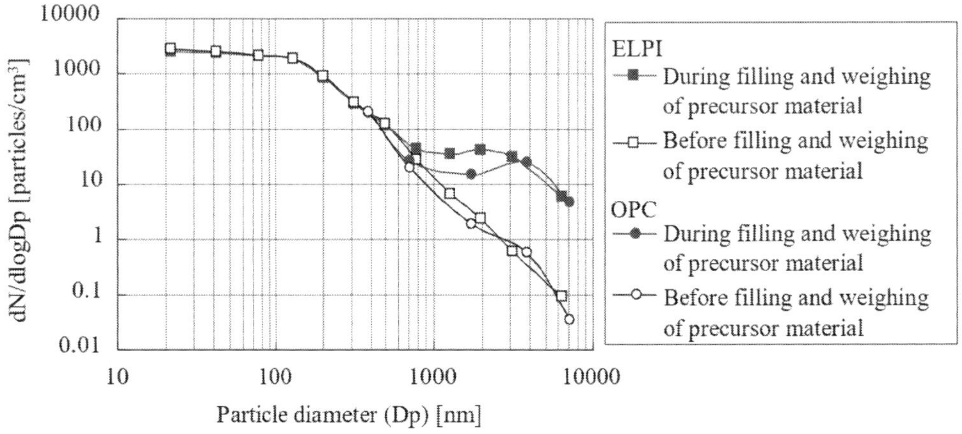

Figure 5. Particle size distributions during and before the filling and weighing of the precursor material. Particle sizes are expressed in terms of aerodynamic diameter for ELPI and optical diameter for OPC.

3.1.3. *Bagging of ZnO.* Figure 6 shows the changes in the particle number concentrations measured by the CPCs and OPCs near the work area and outdoors during the bagging of ZnO (10:00–13:00). The number concentrations of micron-sized particles were slightly higher near the work area than outdoors.

Figure 7 shows the particle size distributions during the bagging of ZnO (highest 10-s-average concentration at 11:15) and before it (*i.e.*, the average concentration in the period 9:55–10:00) near the work area measured by the ELPI and OPC. The number concentrations of micron-sized particles during the bagging of ZnO were slightly higher than those before.

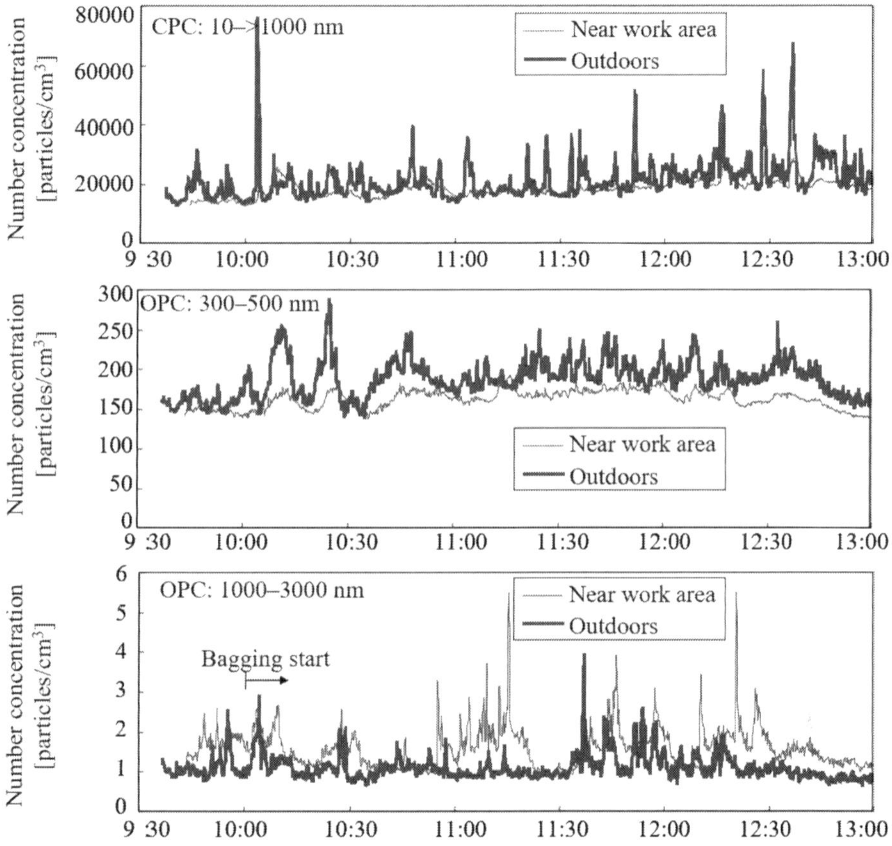

Figure 6. Changes in the particle number concentrations near the work area and outdoors during the bagging of ZnO.

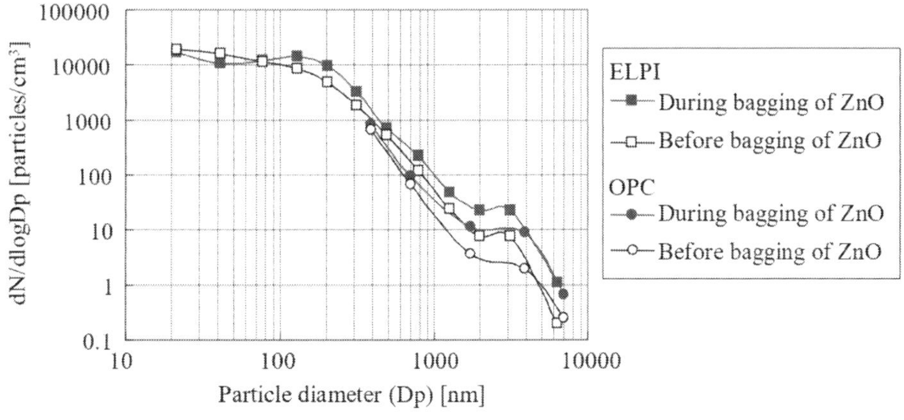

Figure 7. Particle size distributions during and before the bagging of ZnO. Particle sizes are expressed in terms of aerodynamic diameter for ELPI and optical diameter for OPC.

3.2. Morphology of released airborne ENMs
Figure 8 shows the SEM images of the aerosol particles collected during each process. A large number of submicron- and micron-sized ENM aggregates/agglomerates were observed.

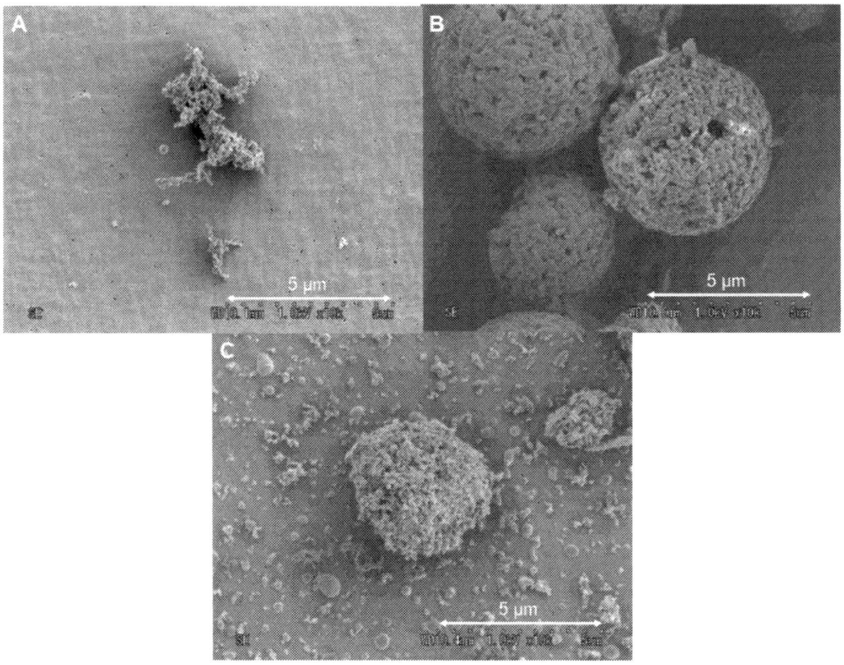

Figure 8. SEM images of the aerosol particles collected during each process. A: Collection of SiC from synthesis reactors, B: Synthesis and bagging of $LiFePO_4$, C: Bagging of ZnO

4. Conclusions
In onsite aerosol measurements for ENMs at industrial manufacturing plants, increases in the number concentrations of submicron- and micron-sized aerosol particles were generally observed, whereas few increases in the number concentrations of nanosized particles were observed. Although the relatively high background concentration of nanosized aerosol particles may have prevented the detection of a small amount of released nanosized aerosol particles, we can at least be certain that the number concentration of nanosized particles released during these processes was substantially lower than the background concentration. In the SEM observations of the aerosol particles collected during each process, a large number of submicron- and micron-sized ENM aggregates/agglomerates were observed.

Acknowledgment
This work was funded by the New Energy and Industrial Technology Development Organization of Japan (NEDO) under a grant for "Evaluating risks associated with manufactured nanomaterials" (no. P06041). We would like to express our gratitude to Sumitomo Osaka Cement Co. Ltd, for their cooperation during the onsite measurements. We wish to thank the researchers participating in the JNIOSH project "Occupational Health Issue of Nanotechnology Industry," for their cooperation.

References
[1] Oberdörster G, Oberdörster E and Oberdörster J 2005 Environ. Health Perspect. 113 823
[2] Maynard A D 2006 Nano Today 1 22

Characterization of a multiculture in-vitro cell exposure chamber for assessing the biological impact of diesel engine exhaust

Akrivi Asimakopoulou[1], Manos Daskalos[1], Leonidas Chasapidis[1], Theofilaktos Akritidis[1], Nickolaos D. Vlachos[1], Eleni Papaioannou[1] and Athanasios G. Konstandopoulos[1,2]

[1]Aerosol and Particle Technology Laboratory, CPERI/CERTH, P.O. Box 60361, 57001, Thessaloniki, Greece

[2]Department of Chemical Engineering, Aristotle University, P.O. Box 1517, 54006, Thessaloniki, Greece

E-mail: asimak@cperi.certh.gr

Abstract. In order to study the various health influencing parameters related to particulate as well as to gas-phase pollutants emitted by Diesel engine exhaust, there is an urgent need for appropriate sampling devices and methods for cell exposure studies and associated biological and toxicological tests. In a previous paper [1], a specific concept for a cell culture exposure chamber was introduced to allow the uniform exposure of cell cultures to diesel aerosols. In the present work, this cell culture exposure chamber is evaluated and characterized with state-of-the-art nanoparticles measurement instrumentation to assess the local deposition of soot aggregates on the cell cultures and any losses due to particle deposition on the cell culture exposure chamber walls, and in addition an upgraded Multiculture Exposure Chamber (MEC) for in vitro continuous flow cell exposure tests is introduced with improved, compared to the previous version, features. Analysis and design of the MEC employs CFD and true to geometry representations of soot particle aggregates.

1. Introduction

Primary health concerns from airborne pollutants include lung carcinogenicity and non-malignant respiratory effects such as irritation, inflammation, and exacerbation or initiation of allergic hypersensitivity. The latter especially is an emerging area of concern [2]. As the prevalence of asthma and other allergic diseases has increased throughout the industrialized world in recent decades, air pollution, including exhaust emissions, especially in urban areas has been suggested as one possible cause. Therefore, the environmental and health related impacts of Diesel exhaust emissions continue to attract the attention of researchers, industry and legislative bodies. The Diesel engine effluent is a complex mixture of particles and gases with hundreds of chemicals, including many organics, present both in the gaseous and condensed phase. Even though advanced technology Diesel engines and emission control devices have been developed to improve their environmental performance and therefore to reduce the expected emissions, the currently prevailing approach does not guarantee that a decrease in regulated emissions will not generate compounds that might have more deleterious health effects, since the Diesel exhaust is experiencing a highly active catalytic environment as it passes through the different emission control devices such as Diesel Oxidation Catalysts (DOC), Diesel Particulate Filters (DPF), Selective Catalytic Reduction (SCR) systems, etc.

Specifically the particle size distribution is considered as a very pertinent factor influencing the toxic effect of exhaust emissions. The influence of particle size on the deposition rate in each region of the respiratory tract is well known. Especially particles with diameters less than 300 nm deposit with significant rates in the alveolar region. In addition, it is self-evident that the composition of particles can be also distributed with respect to size. Aerosol particle size distribution can be dramatically altered by biased sampling and/or storage/resuspension of particles. This observation justifies the strategy that toxicological assays must be carried out with "real" exhaust, including particulates sampled directly from the exhaust flow pipe of a running engine and maintained in suspension in the diluted gases, in order to minimize changes of physicochemical properties of Particulate Matter (PM) and pollutant bio-availability [3]. Measuring exposure to Diesel exhaust aerosol is challenging due to the physical characteristics and chemical complexity of particulate matter: with a mean diameter of ~100 nm, Diesel PM is composed primarily of organic elemental carbon, adsorbed and condensed hydrocarbon, sulfate and metals [4]. The ratio of organic to inorganic carbon depends upon a number of factors that include fuel, engine type, duty cycle, engine maintenance, operator habits, use of emission control devices, and lubricant oil consumption [5]. Research is highly required to assess the actual fate and bioreactivity of exhaust components with focus on particles and also on particle-associated compounds in conditions which do not alter their physicochemical properties and bio-availability.

2. Objectives

In order to study the various health influencing parameters related to particulate as well as to gas-phase pollutants created in the Diesel engine exhaust, there is an urgent need for appropriate sampling devices and methods for cell exposure studies and associated biological and toxicological tests.

In a previous paper [1], a specific concept for a cell culture exposure chamber had been introduced to allow the uniform exposure of cell cultures to diesel aerosols. In the present work, the characterization and evaluation of this cell culture exposure chamber is performed with state-of-the-art nanoparticle measurement instrumentation to assess the local deposition of soot aggregates on the cell cultures and any losses due to particle deposition on the exposure chamber walls, in order to address the establishment of the requisite cell exposure conditions and operational parameters, i.e. dilution and pretreatment of the Diesel engine exhaust. In addition, an upgraded Multiculture Exposure Chamber (MEC) for in-vitro continuous flow cell exposure tests is introduced with the following improved, compared to the previous version, features:

- An improved geometrical arrangement of exposed cell cultures to accommodate testing of up to thirty six cell cultures per MEC.
- On-line local measurement of particle number concentration (hence direct dose).

In the next sessions, a description of the methodology employed to evaluate the cell culture exposure chamber is given and the results obtained regarding the assessment of the impact of the electrostatic phenomena on the soot particles' deposition on the walls of the exposure chamber and the overall average efficiency in capturing soot particles in the exposure chamber are presented. Finally, the analysis and design of the new generation MEC is described based on CFD and true to geometry representations of soot particle aggregates.

3. Methodology

In order to characterize and evaluate the cell culture exposure chamber, experiments with state-of-the-art nanoparticles' measurement instrumentation, which provide the essential data to assess the local deposition of soot aggregates on the cell cultures and any losses due to particle deposition on the exposure chamber walls, are needed and therefore, the following experimental set up is considered **(Figure 1)**.

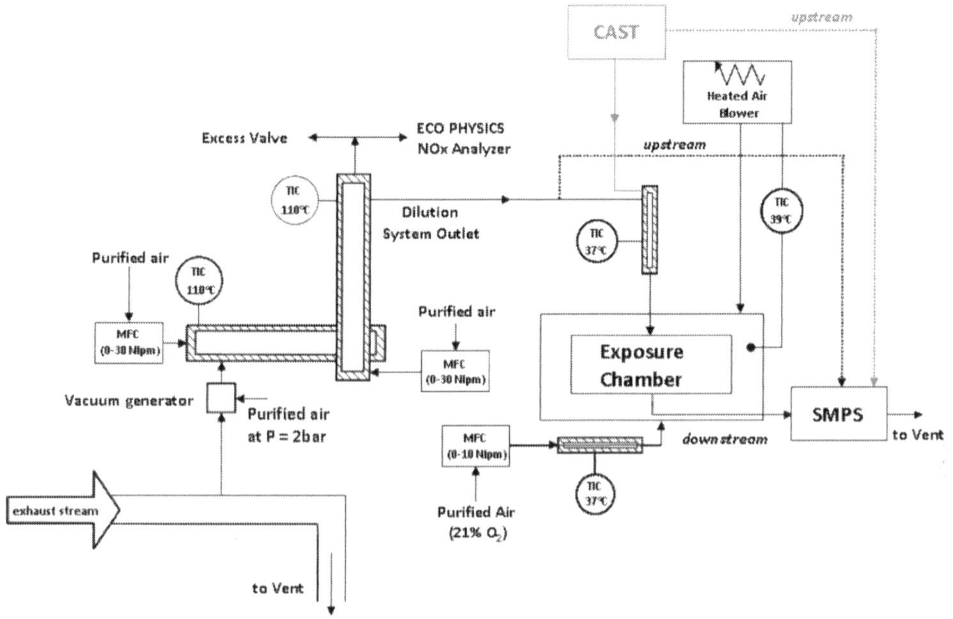

Figure 1. Flow diagram of the experimental setup

In this experimental setup, a sample either from the Diesel engine exhaust (at a steady engine operation point: speed and torque, 2250 rpm and 101 Nm, respectively), or from the exit aerosol of the soot generator Combustion Aerosol Standard – CAST (at several operation points delivering thus aerosol streams with different average soot particle diameter at each operating point) is being diluted and thermally pretreated before it enters the exposure chamber. The dilution of the sample is considered as a crucial experimental factor and the final dilution ratio is selected in order to adequately simulate the "real" exposure of human lung cells to the environmental particulate pollution [1]. The temperature of the exposure chamber is constantly kept at 37±1°C with the aid of a specially constructed oven. Soot particles concentration and size distribution in the aerosol phase, from both Diesel engine exhaust and the CAST soot generator, are being recorded simultaneously "upstream" and "downstream" of the exposure chamber with a Scanning Mobility Particle Sizer (SMPS, TSI model 3936). The experimental conditions under which the measurements were conducted are given in Table **1**.

Table 1. Experimental conditions for each case of measurement

Experimental conditions	**CELL**	**CAST**
Exhaust flowrate	2.2 lpm	2.2 lpm
Upstream average particle concentration	$3*10^7$#/cm^2	$2*10^5$#/cm^2
Temperature	37°C	38 °C
Duration	3 hr	3 hr

3.1. Assessment of particle losses due to particle deposition on the walls of the exposure chamber

Diesel nanoparticle formation may be attributed to nucleation, absorption or condensation mechanisms as well as to ion-induced nucleation mechanism. Therefore, charges on Diesel soot particles are expected especially if ion-induced nucleation is the dominant mechanism in the formation of nuclei mode particles during exhaust dilution. Metal additives in the fuel contribute also in the increase of the charged fraction of soot particles especially in the accumulation mode [6]. The existence of charge on soot particles may vary with the particle diameter. However, it is expected that the electric forces would affect the deposition mechanism of soot particles in the exposure chamber. Therefore, to assess the impact of electrostatic phenomena on soot particles deposition on the walls of the cell culture exposure chamber and eventually to estimate the degree of particle losses, the whole inner surface of the chamber is covered with a grounded conductive film (aluminum foil). **Figure 2** depicts the cell culture exposure chamber in a laminar flow hood that was employed for "real" in-vitro cell exposure tests in [1] is depicted, while **Figure 3** illustrates the same chamber when its whole inner surface is covered with a grounded conductive aluminum foil. Under these conditions, an aerosol stream of soot particles from CAST, properly diluted and pretreated, is delivered to the exposure chamber, while soot particles concentration and size distribution are being recorded simultaneously with an SMPS "upstream" and "downstream" of the exposure chamber with:

- a non grounded inner surface
- a grounded aluminum foil covered inner surface

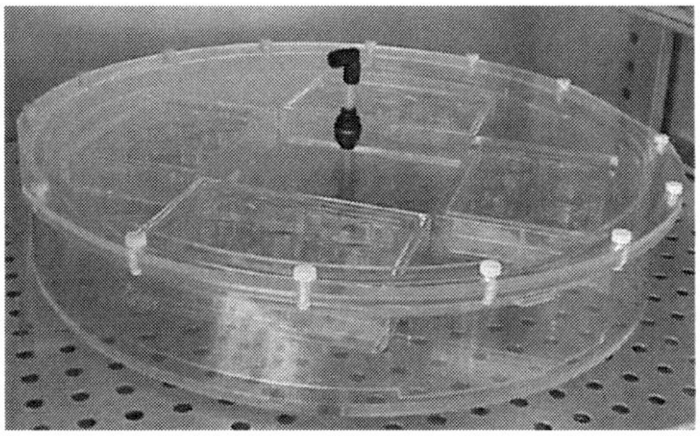

Figure 2. The cell culture exposure chamber in the laminar flow hood [1]

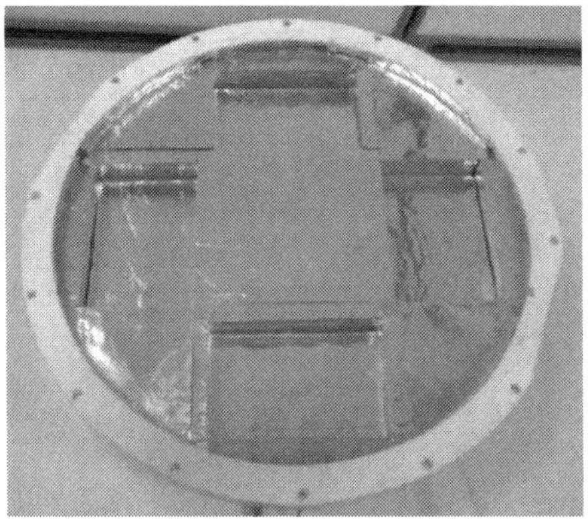

Figure 3. Exposure chamber inner surface covered with a grounded conductive film

3.2. Dose of soot particles delivered to cell cultures

In order to estimate the dose of soot particles to which cell cultures are being exposed in the normal operation of the in-vitro cell culture exposure chamber, measurements of the size specific soot particle collection efficiency of a free of cells device are performed. In this situation of exposure tests, particles from two different sources of soot (Diesel Engine exhaust and CAST) are delivered to the exposure chamber, and soot particles concentration and size distribution are being recorded simultaneously with an SMPS, as described previously.

4. Results

In **Figure 4**, the size specific collection efficiency in the case of the exposure chamber with grounded and not-grounded inner surface as a function of the average soot particle size d_p, is illustrated. The existence of electrostatic field results in the deposition of fewer soot particles in the multiculture exposure chamber, especially in the case of smaller particle diameters (d_p< 60 nm). The average collection efficiency of the not-grounded exposure chamber is reduced more than 26% for particles with diameters lower than 60 nm, due to the effect of electrostatic forces, in addition to diffusion mechanisms.

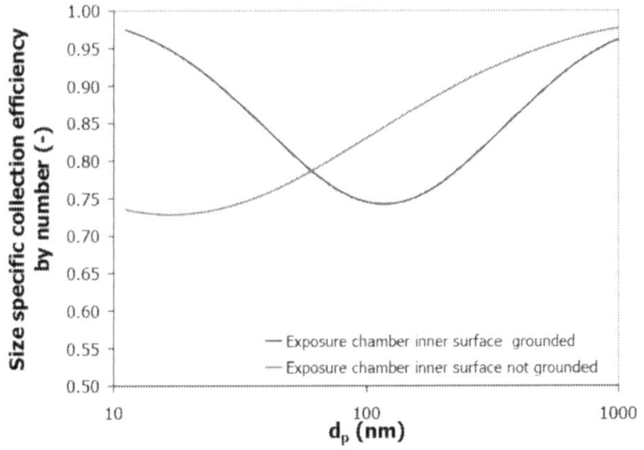

Figure 4. Assessment of the impact of electrostatic phenomena on soot particles deposition in the multiculture exposure chamber

From the collection efficiency measurements of soot particles generated by CAST at various modes of operation (variation of the size of the soot particles delivered in the exposure chamber), it is obvious that the overall collection efficiency of the exposure chamber is approximately 76% (**Figure 5**). These results were being confirmed by respective measurement with a sample from the real Diesel engine exhaust (**Figure 6**).

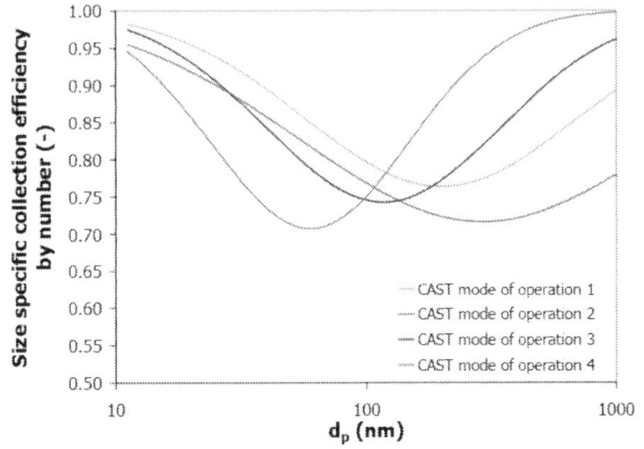

Figure 5. Collection efficiency of the exposure chamber with respect to soot particles generated by CAST at various modes of operation

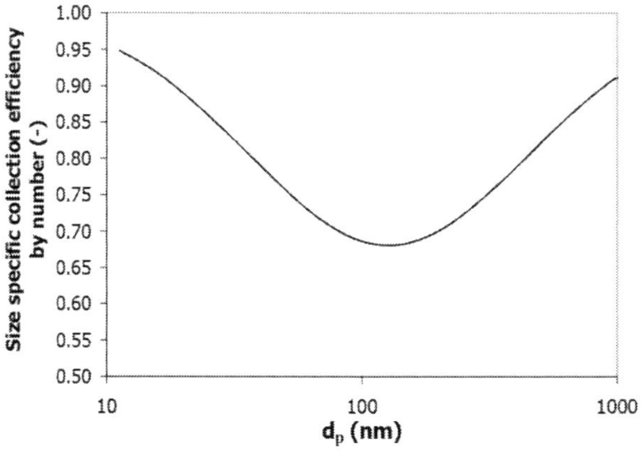

Figure 6. Collection efficiency of the exposure chamber with respect to soot particles from Diesel engine exhaust

5. Future work

In order to optimize the operation of the cell exposure chamber, an upgraded design for multiculture exposure tests is proposed with the following improved features:

- an improved geometrical arrangement of exposed cell cultures to accommodate testing of up to thirty six cell cultures
- on-line local measurement of particle number concentration (hence direct dose estimation)
- on-line particle sampling for TEM analysis of particles depositing on the cell cultures.

CFD analysis

The performance of the proposed cell exposure chamber geometry is being studied using Computational Fluid Dynamics (CFD) and particle transport simulation within the chamber volume. The CFD analysis is developed with the aim to assess the uniformity of soot deposition among the culture medium wells, particularly the difference between upstream and downstream samples, and to study the effect of geometry modifications of the flow inlet and outlets, especially with respect to soot deposition uniformity. A number of results already obtained with this analysis indicate the potential of the updated exposure chamber configuration proposed.

The CFD analysis of the updated exposure chamber needs to consider only a sub-unit of the total geometry corresponding to two half-inserts (see **Figure 7**). This is possible due to symmetry of the culture wells and periodicity of the overall geometry, given a small / negligible gap between each set of (six) inserts. The CFD geometry was constructed in two versions: a baseline geometry with approx. 15 mm from inflow slit to the first culture wells and an elongated geometry with an increase of 20 mm in the distance between wells and inflow slit. The flow boundary conditions consisted of prescribed inflow velocity along the inflow slit and prescribed (atmospheric) pressure at the outlet. The inflow velocity corresponded to a flow of 0.2 lpm per culture well. Analysis of the soot particle exposure dose of each culture well was performed by using an Eulerian representation of the soot particle concentration field, i.e. the particle transport mechanisms of convection and diffusion were modeled. A boundary condition of perfect sticking, i.e. zero particle concentration, was applied on all solid and liquid surfaces within the exposure chamber. An overall grid resolution of 0.2 mm was employed throughout the model, with refinement to a 0.1 mm resolution in proximity to the culture well liquid surface. Example results for the flow and particle concentration fields can be seen in **Figure 8** and **Figure 9**.

Despite the fine spatial resolution used for the discretisation of the flow and soot particle concentration fields, the direct calculation of deposition rate on the surfaces can be ill conditioned because it involves the multiplication of diffusivities of the order of 10^{-9} kg/m-s with a concentration gradient of the order 10^5 i.e. the very thin particle concentration boundary layer (of thickness ≈ 0.1 mm) can only be marginally resolved in a model encompassing the exposure chamber geometry (size of order 100 mm). Therefore, soot deposition is best obtained by indirect interpretation of the concentration field and so particle concentration profiles normal to the culture medium surface are extracted at the center of each culture dish for all geometric configurations considered, with the resulting profiles shown in **Figure 10**. Interpretation of the concentration gradients so obtained indicates only a minor / indiscernible variation in soot deposition rate between the upstream – downstream culture wells. This result can be explained to a large extent by the small (only ≈ 5%) reduction in soot concentration "outside" the concentration boundary layer over the culture medium surfaces, between upstream and downstream wells, irrespective of inlet geometry. It is believed that, in future work, this difference in particle concentration can be reduced further by manipulation of the flow field through geometric structures on the exposure chamber cover, i.e. the surface facing the culture medium wells.

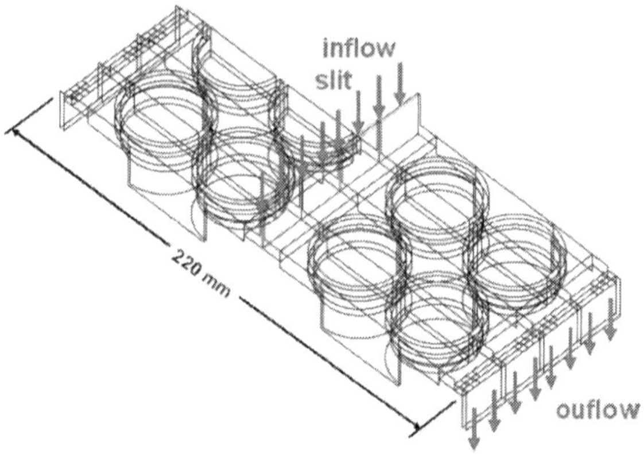

Figure 7. Overview of the exposure chamber geometry showing symmetric and periodic repetitions of the geometry sub-unit considered in the CFD analysis

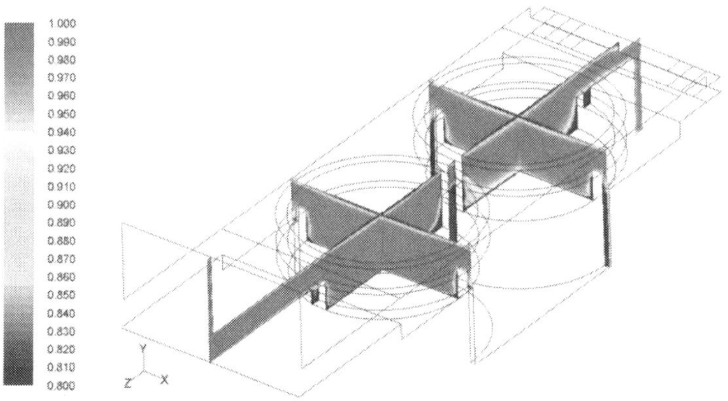

Figure 8. Visualisation of the flow velocity field on symmetry and periodic sections through the exposure chamber (baseline) geometry. Velocity is in m/s on a logarithmic scale.

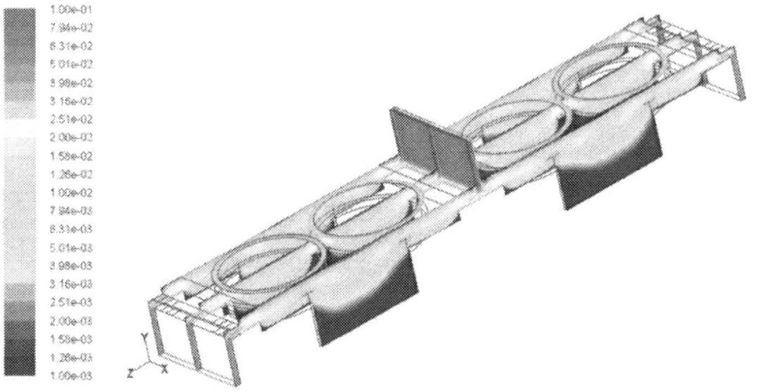

Figure 9. Visualisation of the soot concentration (normalised against inflow concentration) on longitudinal and lateral sections through the cell culture wells. The result is shown for the elongated geometry.

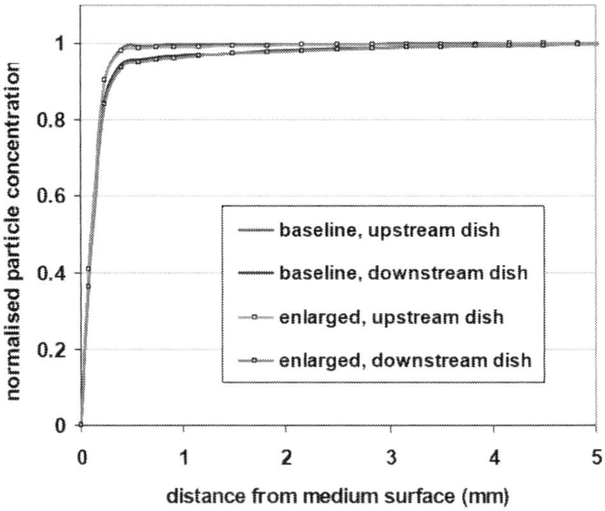

Figure 10. Particle concentration profiles at the center of the culture medium wells for all geometry cases and well positions considered.

6. Conclusions

A cell culture exposure chamber, designed and built in a previous work [1], was characterized and evaluated with respect to the local deposition of soot aggregates on the cell cultures and possible losses due to particle deposition on the cell culture exposure chamber walls. According to the results obtained when measuring the size specific collection efficiency of the exposure chamber with its inner surface grounded and not grounded, it was noticed that a greater selectivity in collecting smaller soot nanoparticles can be achieved, if the existence of electrostatic field in the exposure chamber is eliminated. Based on simulating exposure tests with two sources of soot particles (real Diesel engine exhaust and synthetic soot generator CAST), the average collection efficiency of the proposed exposure chamber is estimated to be approximately 76%, independently of the specific ranges of the particle size distribution delivered to the exposure chamber.

An upgraded Multiculture Exposure Chamber (MEC) for in vitro continuous flow cell exposure tests is proposed with improved, with respect to the previous version, features. Preliminary CFD analysis of the updated/proposed exposure chamber configuration already indicates satisfactory soot deposition uniformity and the scope for further improvement by flow field manipulation.

The evaluation of the upgraded multiculture exposure chamber will address the establishment of the requisite cell exposure conditions and operational parameters, i.e. dilution and pretreatment of the diesel engine exhaust.

References

[1] Papaioannou E., Konstandopoulos A.G., Morin J.P., Preterre D., (2006) "A Selective Particle Size Sampler Suitable for Biological Exposure Studies of Diesel Particulate", *SAE Tech. Paper* No. 2006-01-1075 (SP-2024), 389-399.

[2] Oberdörster, G., Ferin J., Lehnert B.E., (1994) "Correlation between Particle Size, in-Vivo Particle Persistence, and Lung Injury", *Environ. Health Perspect.*, **102** (S5), 173-179.

[3] Le Prieur, E., Morin J.P., Bion A., Dionnet F, (2000) "Toxicity of Diesel Engine Exhaust: Induction of Pro-inflammatory Response and Apoptosis in an in-Vitro Model of Lung Slices in bi-phasic Organotypic Culture", Archives . *Toxicology*, **74**, 460-466.

[4] Graskow B.R., Kittelson D.B., Abdul-Khalek I.S., Ahmadi M.R., Morris J.E., (1998) "Characterization of Exhaust Particulate Emissions from Spark Ignition Engine", *SAE Tech. Paper* No. 980528 (SP-1326), 1-10.

[5] Perez J.M., Williams R.L., (1989) "A Study of Particulate Extracts from 1980s Heavy Duty Diesel Engines Run on Steady State and Transient Cycles", *SAE Tech. Paper* No. 892491.

[6] Jung H., Kittelson D.B., (2005) " Measurement of Electrical Charge on Diesel Particles", *Aerosol Sci and Technol*, **39**, 1129–1135.

Nanosafe2010: International Conference on Safe Production and Use of Nanomaterials IOP Publishing
Journal of Physics: Conference Series **304** (2011) 012006 doi:10.1088/1742-6596/304/1/012006

Workplace exposure to traffic-derived nanoscaled particulates

M Viana, S Díez, A Alastuey, X Querol and C Reche
Institute for Environmental Assessment and Water Research (IDAEA-CSIC),
C/ Jordi Girona 18, 08034 Barcelona, Spain.

E-mail: mar.viana@idaea.csic.es

Abstract. Workplace exposure to traffic-derived nanoscaled particulates was determined at a chemical research facility. Sub-micron particles were monitored by means of a multi-angle absorption photometer (MAAP) and a laser spectrometer (GRIMM 1107), providing 10-minute black carbon (BC) concentrations and 15-minute PM_1 concentrations, respectively, over a 4-month period (22/03/2010 - 28/07/2010). BC levels were simultaneously monitored during 1-day periods using a handheld aethalometer (Magee AE51), with excellent agreement between both techniques (MAAP and AE51, r^2=0.96, y=0.84x).The studied laboratory is located on the 5th floor of an 8-storey building in an urban background environment in Barcelona, Spain. The laboratory was not in use during the study period, and both of its doors were kept open at all times in order to ensure air circulation between the study laboratory and the remaining offices and laboratories on the same floor (where workers were exposed). Windows were kept closed at all times. Indoor BC and PM_1 concentrations were compared with ambient BC and PM_1 levels from an outdoor monitoring station located at <150 m away from the research facility. Results evidenced the major impact of outdoor vehicular traffic emissions on the levels of nanoscale particulates monitored in the workplace, with clear daily cycles coinciding with traffic rush hours, especially during week days. Penetration ratios were calculated for BC which showed that, even ensuring that all windows were closed, at least 82% of indoor BC concentrations originate from outdoor emissions. Outdoor/indoor penetration ratios were stable for BC (ranging between 1.20 and 1.35) but not for PM_1 (1.76 to 1.02), suggesting that it is necessary to monitor the variability of penetration factors as a function of time. BC emission sources in the workplace still need to be determined, but could be related to printer/photocopier toner emissions and laboratory work. Potential contamination due to the monitoring instruments (pumps) was discarded through the analysis of daily indoor BC cycles.

1. Introduction

Numerous research works available in the literature have reported significant health risks associated with exposure to particulate matter (PM) [1]. In recent years research interest has been directed towards fine and ultrafine particles (particulates with aerodynamic diameter <1 μm). Epidemiological studies [2] have indicated that fine and ultrafine may have a greater potency to cause adverse health effects than larger particles, given that they have shown increased deposition rates in the lower respiratory tract [3].

Population exposure to both supra-micron and nanoscaled particulate concentrations is usually considered to take place in outdoor environments. However, it is estimated that on average adults spend 60-90% of their time indoors [4]. Whereas data on outdoor PM levels and chemical composition are widely available in Europe [5], relatively few data are available in the literature regarding the levels and characteristics of atmospheric particulates in indoor air. This is generally attributed to the complexity in the determination of PM levels indoors, given the broad variability of direct indoor sources (cleaning, cooking, smoking, etc.) as well as the multiplicity of microenvironments (households, work environments, commerce, etc.). Studies [6] have shown that measured particle levels in indoor environments are greatly influenced by indoor activities, leading to high-concentration but short-term events. The examination of short-term variations in particulate concentrations is, therefore, essential in exposure studies.

Assessment of the risk to the community resulting from exposure to airborne pollutants should ideally include measurements of concentration levels of the pollutants in all microenvironments where people spend their time. However, due to the vast range of different microenvironments it is not possible to conduct measurements in all of them. Over the past several decades, exposure to indoor air pollutants in work environments is believed to have increased due to a variety of factors [7], including the construction of more tightly sealed buildings, reduced ventilation rates to save energy, the use of synthetic building materials and furnishings, and the use of chemically formulated cleaning products. In work environments, indoor air quality problems can be traced to sources including office equipment, office supplies, construction materials (renovations and new furnishings), maintenance work (cleaning products, etc.) and outdoor sources.

In this respect, the influence of outdoor PM sources on indoor environments is a key issue to be taken into consideration. In recent years, indoor/outdoor (I/O) relationships have been studied in the US and Australia with a special focus on the penetration rates of traffic-related particulates [8]. Conversely, fewer studies are available in Europe and they mostly focus on modelling of ultrafine particulate concentrations in indoor environments [9]. Studies are especially rare in Spain [10].

While the focus of most of the available studies reported has been on particle mass, there is still very little information on the I/O relationship as a function of the geochemistry, size or sources of particulate matter. Theoretically, the indoor particle concentration is a function of a number of factors, the most important of which are the generation rate of particles indoor, the outdoor particle concentration, air exchange rate, particle penetration efficiency from the outdoor to the indoor environment, and the particle deposition rate on indoor surfaces [11]. However, in practice, it is usually very difficult to assess the exposure due to the lack of data and information on the correlation between indoor and outdoor particles, which are building and environment specific. Understanding the relationship between indoor and outdoor aerosol particles under different environmental conditions is of importance for improving exposure estimates and in turn for developing efficient control strategies to reduce human exposure and thus health risk.

Based on the above considerations, the main objective of the present work is to characterise exposure to indoor particulate matter of outdoor origin in the work environment. Specifically, our work focuses on traffic-derived aerosols. Taking one step further from the studies currently available in the literature, this study characterises workplace exposure to aerosols in terms of nano-scaled particulate mass (<1 micron, PM_1) and black carbon concentrations.

2. Methodology

Monitoring of indoor PM_1 and black carbon (BC) concentrations was carried out at a chemical research facility belonging to the Spanish National Research Council (Institute for Environmental Assessment and Water Research, IDAEA-CSIC), in the urban area of Barcelona (NE Spain). The building is >30 years old. The study location was a vacant laboratory located on the 5th floor of the building, adjacent to other laboratories and offices which were in use on a daily basis. The study laboratory had the same dimensions, geometry, windows and doors as the laboratories in use, and was connected to them by means of a corridor. The doors of the vacant laboratory were kept open at all times in order to ensure air circulation with the remaining offices and laboratories on the same floor (where workers were exposed). Windows were kept closed at all times. Indoor BC and PM_1 concentrations were compared with outdoor BC and PM_1 levels from an urban background monitoring station located at <150m from the research facility.

The same instrumentation was deployed at the indoor and outdoor monitoring stations. Black carbon concentrations were monitored using two multi-angle absorption photometers (MAAP) with a 10-minute (indoor) and 1-hour (outdoor) time resolution. PM_1 concentrations were monitored by means of two laser spectrometers (GRIMM 1107) with a 15-minute (indoor) and 30-minute (outdoor) time resolution. In addition, in the indoor environment BC levels were also monitored using a portable aethalometer (Magee AE51), simultaneously with the MAAP instrument. Monitoring of BC and PM_1 was carried out during a 4-month period between 22/03/2010 and 28/07/2010, with breaks due to technical issues.

In addition to these instruments, one low volume $PM_{2.5}$ sampler was located in the indoor environment, which collected 12-hr samples on 47-mm quartz fibre filter substrates. In the outdoor site, $PM_{2.5}$ samples were collected over 24-hr periods by means of a high-volume sampler (MCV SA, 30 m^3/h). The chemical composition of the particulates collected on the filters (indoor and outdoor) was characterised by means of ICP-MS, ICO-OES, IC and organic and elemental carbon analysis. The results of the chemical characterisation are not included in the present work.

3. Results and Discussion

Mean, minimum and maximum levels of BC and PM_1 at the indoor and outdoor locations are summarised in Table 1. PM_1 levels, determined by laser spectrometry, were underestimated when compared with results obtained by means of the reference gravimetric method (outdoor PM_1=18.6 $\mu g/m^3$; unpublished data). Automatic data should be corrected by comparison with the reference gravimetric data for air quality monitoring purposes (which are not the focus of the present study).

On average, BC and PM_1 levels were higher in the outdoor environment than indoors (19% and 37% higher, respectively), with mean levels for the entire monitoring period of 1572 $ngBC/m^3$ and 7.5 $\mu gPM_1/m^3$ indoors, and 1871 $ngBC/m^3$ and 10.3 $\mu gPM_1/m^3$ outdoors. Maximum BC concentrations (15- or 30-minute means) were, however, higher in the indoor environment (7% higher), as a result of the high-concentration but short-duration events described above. Indoor/outdoor (I/O) ratios were calculated on a sample-by-sample basis and then averaged for the entire period, and thus were 0.97 and 0.92 for BC and PM_1, respectively, which are much higher than values found in the literature for urban environments (0.60-0.65 for BC and 0.38-0.46 for PM_1, [12]). These ratios evidence a major influence from outdoor particles in the indoor environment, probably as a result of inadequate insulation of the building due to its old age (>30 years).

Table 1. Mean, minimum and maximum levels of black carbon (BC) and PM_1 at the indoor and outdoor locations, for the entire monitoring period.

	Mean	Min.	Max.
BC Indoor (ng/m^3)	1572	178	10740
BC Outdoor (ng/m^3)	1871	115	10060
BC Indoor/Outdoor	0.97	0.14	3.42
PM_1 Indoor ($\mu g/m^3$)	7.5	1.9	41.5
PM_1 Outdoor ($\mu g/m^3$)	10.3	0.7	65.0
PM_1 Indoor/Outdoor	0.92	0.21	6.0

The time series of simultaneous indoor and outdoor PM_1 and BC concentrations were analysed in search for emission patterns. As shown in Figure 1, the correlation between indoor and outdoor measurements was significantly high, with good correspondence in absolute values (y=1.07x) but a relatively high dispersion (r^2=0.35) in the case of BC. The results evidenced a clear daily pattern coinciding with vehicular traffic emissions, with maximum levels during traffic rush hours (06:00-08:00 and 18:00-20:00 UTC), especially visible for BC. The strong correlation between the indoor and outdoor levels of both parameters, as well as the similarity in the absolute concentrations, suggests vehicular traffic as the major and common emission source in both environments.

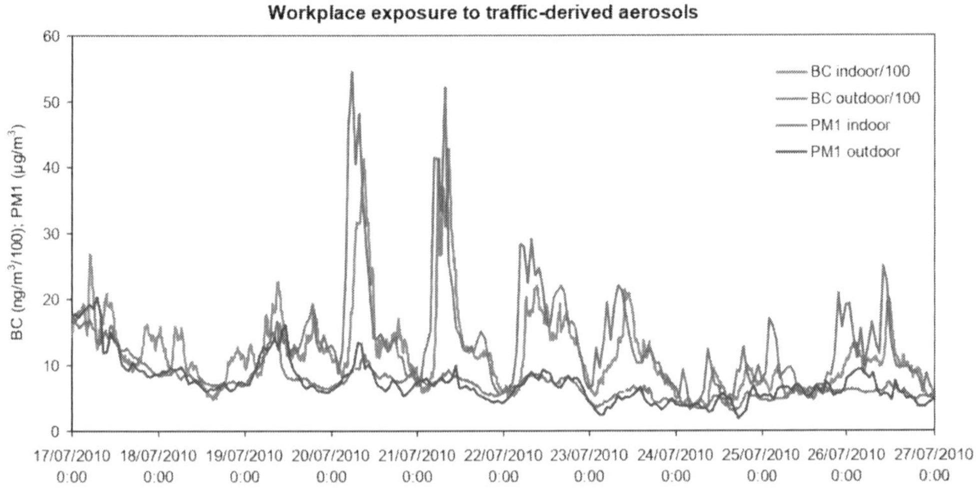

Figure 1. Indoor BC and PM_1, and outdoor BC levels, monitored in a chemical research facility.

In order to quantify the influence of outdoor vehicular emissions of PM_1 and BC, penetration factors were calculated based on the methodology described in [13]. This methodology aims to determine the penetration ratio of outdoor particles into the indoor environment, in order to be able to determine the indoor particle load which is originated by indoor emissions, as opposed to the particles generated outdoors but infiltrated and detected indoors. To this end, outdoor-to-indoor ratios (O/I) are calculated for time periods during which no indoor PM emissions are generated, typically weekends or Sundays in industrial facilities and workplaces. In the chemical research facility under study, weekends were representative of periods with zero indoor-generated emissions. Penetration ratios for BC are shown in Table 2, calculated for the 3 weekends for which simultaneous indoor and outdoor BC data were available. According to our results, the O/I ratio remained relatively constant between May and July, with a mean value of 1.29 (ranging between 1.20 and 1.35, standarc deviation = 0.08), suggesting that the infiltration of BC particles was relatively constant and independent of the season (spring vs. summer). Conversely, the O/I ratios calculated for PM_1 showed a decrease from April to July (Table 2, 7 weekends available for calculation of the ratios for PM_1). Whereas the mean O/I ratio for PM_1 was 1.76 in April, it was close to 1 (0.95-1.02) in July, indicating a much higher degree of particle infiltration during the summer period than in spring. Taking into account that windows remained closed at all times, this increase in particle infiltration can only be interpreted as a change in particle characteristics, either particle composition and/or size. Particulate sulphate formation is enhanced in summer due to faster oxidation rates of SO_2, which results in higher concentrations of fine sulphate particles in summer [14]. One possible interpretation for the different O/I ratios obtained for BC and PM_1 could be the increase in sulphate particles in ambient air, which would result in a finer grain size distribution of the PM_1 fraction and thus an enhanced infiltration from outdoors to the indoor environment. In addition, this would not influence the penetration ratio of BC particles, with a grain size distribution which is unaffected by seasonal trends. The chemical speciation data of $PM_{2.5}$ particles collected on filters (not included in the present work) will provide further insights into this issue.

Consequently, our results evidence that penetration factors do not depend only on the physical barriers (e.g., windows) determining the outdoor-indoor exchange of particulates, but also on the physico-chemical properties of those particles. Therefore, in order to accurately discriminate the contribution of indoor particles originating from indoor and outdoor emissions, it is necessary to monitor the variability of the calculated penetration factors as a function of time.

Table 2. BC and PM_1 penetration ratios.

	Outdoor/Indoor		
Dates (weekends)	**Mean**	**Min**	**Max**
BC			
1-2/05/2010	1.30	0.45	2.61
10-11/07/2010	1.20	0.38	2.30
24-25/07/2010	1.35	0.58	3.23
Mean BC O/I	**1.29 (±0.08)**		
PM_1			
3-4/04/2010	1.76	0.81	2.95
15-16/05/2010	1.63	0.66	2.44
29-30/05/2010	1.44	0.83	2.68
12-13/06/2010	1.20	0.62	2.37
19-20/06/2010	1.20	0.32	2.66
10-11/07/2010	0.95	0.60	1.29
24-25/07/2010	1.02	0.57	1.39
Mean PM_1 O/I	**1.61 (±0.30)**		

Indoor BC emissions were then estimated based on the penetration factors described above and the outdoor and indoor BC concentrations [13]. For the sake of brevity, results are only shown for BC. Figure 2 shows an example of the indoor BC concentrations monitored (total), as well as the calculated indoor BC of outdoor origin and indoor BC of indoor origin, for the month of July (when simultaneous indoor and outdoor BC data were available). Results showed that, on average, 82% of the BC concentrations measured in the indoor environment were originated outdoors, and only 18% were linked to indoor BC emissions. Even after the subtraction of indoor BC concentrations of outdoor origin, indoor BC emissions still showed peaks coinciding with vehicular traffic rush hours (Figure 2). This is probably a consequence of the application of mean weekend O/I ratios to hourly data. The daily cycle of total indoor BC concentrations (Figure 3) showed a clear traffic-derived trend for total indoor BC, whereas indoor BC emissions of indoor origin showed one major increase between 9:00-12:00 UTC, probably coinciding with characteristic workplace activities linked to indoor BC sources (printing, photocopying, etc). The potential contamination due to the monitoring instruments (pumps) was discarded based on the clear daily cycles observed. In addition, the influence of outdoor traffic-derived aerosols was also detected as a secondary BC peak at 7:00 UTC. Once again, this indicates that particles infiltrated from the outdoor environment cannot be fully segregated by applying O/I ratios. When the daily cycle of indoor BC emissions was analysed separately for working days and weekends (Figure 4), the morning traffic peak was detected throughout the entire week (with different intensities) given that traffic emissions only decrease on weekends. Conversely, the 9:00-12:00 UTC peak (due to workplace activities) was only present during weekdays, and it was completely absent on weekends when the building was empty. In terms of mass, the BC increase due to workplace activities represents a 71% increase (616 ng/m^3) on weekdays with respect to weekends (on average 254 ng/m^3 on weekends and 870 ng/m^3 on weekdays, for the period 09:00-13:00 h).

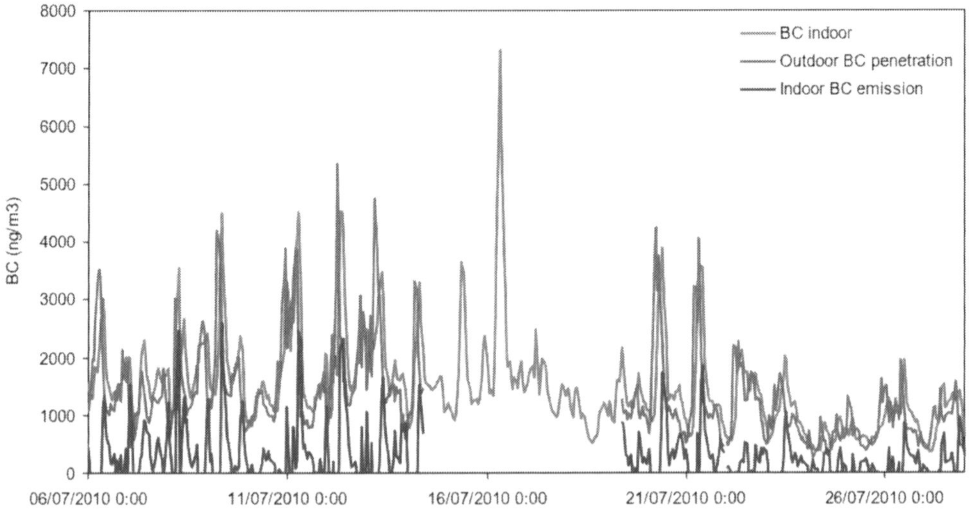

Figure 2. Indoor BC concentrations measured, and indoor BC emissions inferred from calculated BC penetration ratios.

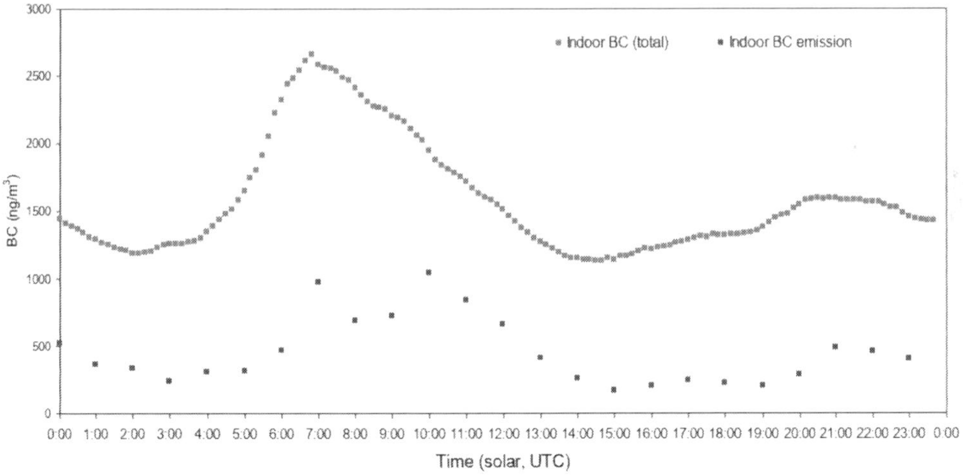

Figure 3. Daily cycle of total indoor BC concentrations and indoor BC concentrations of indoor origin.

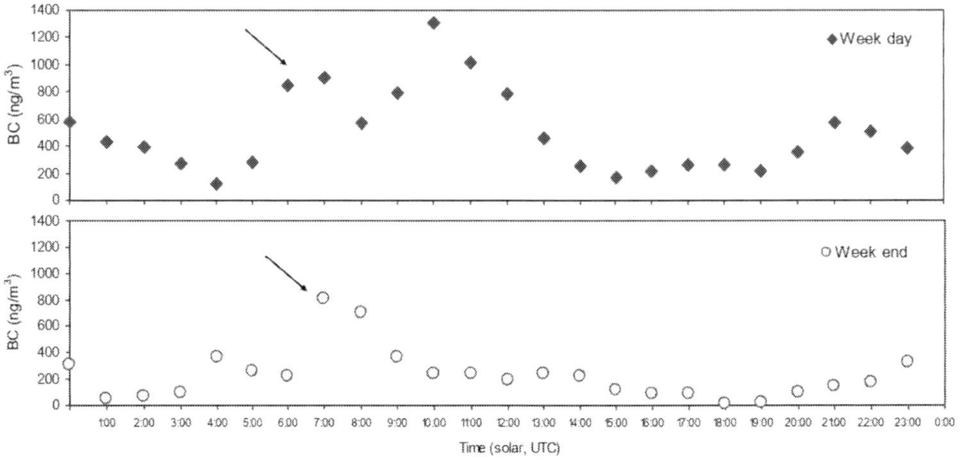

Figure 4. Indoor BC emission during working days (Monday to Friday) and on weekends (Saturday-Sunday). The arrows highlight the traffic-derived indoor BC peak.

Finally, semi-hourly BC data obtained with the MAAP instrument were compared with the data obtained with the mini aethalometer AE51 (Figure 5), for the purpose of validating the results provided by the handheld instrument. Results showed that 24 hours is the maximum monitoring period in order to obtain an acceptable comparison between the handheld and the online instruments, even if the battery life of the AE51 instrument is longer. Monitoring times >24h with the handheld instrument resulted in filter saturation which decreased the correlation dramatically (r^2=0.96 for <24h periods vs. r^2=0.61 for >24h).

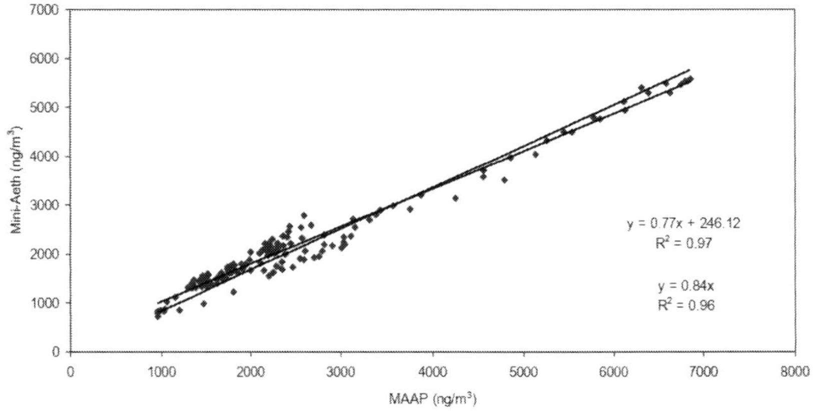

Figure 5. Comparison between 10-minute BC data obtained by means of an online absorption photometer (MAAP) and a handheld aethalometer (AE51).

4. Conclusions

Workplace exposure to nanoscaled particulates of indoor and outdoor origin was determined at a chemical research facility by means of a multi-angle absorption photometer (MAAP) and a laser spectrometer (GRIMM 1107), providing black carbon (BC) and PM_1 concentrations. Indoor BC and PM_1 concentrations were compared with ambient BC and PM_1 levels from an outdoor monitoring station located at <150m away from the research facility. Results evidenced the major impact of outdoor vehicular traffic emissions on the levels of nanoscale particulates monitored in the workplace, with clear daily cycles coinciding with traffic rush hours, especially during week days. Penetration ratios of 1.29 (with 0.08 standard deviation) and 1.61 (with 0.30 standard deviation) were obtained for BC and PM_1, respectively. While no seasonal trend was observed for these ratios for BC, they decreased strongly for PM_1 from March to July (1.76 to 1.02). This indicates a higher degree of particle infiltration during the summer period than in spring for PM_1 (not for BC), which can only be interpreted as a change in particle characteristics, either particle composition and/or size. One possible interpretation for the different O/I ratios obtained for BC and PM_1 could be the increase in sulphate particles in ambient air, which would result in a finer grain size distribution of the PM_1 fraction and thus an enhanced infiltration from outdoors to the indoor environment. Even ensuring that windows were closed at all times, at least 82% of indoor BC concentrations originated from outdoor emissions due to inadequate insulation of the building materials. BC emission sources in the workplace still need to be determined, but should be related to printer/photocopier toner emissions and laboratory work.

Acknowledgements
This work was funded by the Spanish Ministry of the Environment and Rural and Marine Affairs (010/PC08/3-04.1) and the Ramón y Cajal Programme. The authors gratefully acknowledge J. Pey and J. Gil from IDAEA-CSIC for their technical support.

References

[1] Brunekreef B., Janssen N.A.H., Harssema H., Knape M., Vliet P.v., 1997. Air Pollution from truck traffic and lung function in children living near motorways. *Epidemiology* **8**, 298-303.
Pekkanen J., Timonen K., Ruuskanen J., Reponen A., Mirme A., 1997. Effects of ultrafine and fine particles in urban air on peak expiratory flow among children with asthmatic symptoms. *Environ. Res.* **74**, 24-33.
Lippmann M., Schlesinger R.B., 2000. Annual Review of Public Health - Toxicological bases for the setting of health-related air pollution standards. Nelson Institute of Environmental Medicine, New York University School of Medicine, Tuxedo, New York 10987, Vol. **21**: 309-333 pp.
Künzli N., Kaier R., Medina S., Studnicka M., Chanel O., Filliger P., Herry M., Jr, F.H., Puybonnieux-Texier V., Quénel P., Schneider J., Seethaler R., Vergnaud J.C., Sommer H., 2000. Public health impact of outdoor and traffic related air pollution: a European assessment. *The Lancet* **356**, 795-801.
Pope C.A., Dockery D.W., 2006. Health effects of fine particulate air pollution: Lines that connect. *J. Air & Waste Manag. Assoc.* **56**, 709-742.
[2] Donaldson K., MacNee W., 2001. Potential mechanisms of adverse pulmonary and cardiovascular effects of particulate air pollution (PM10). *Intl. J. Hyg.Environ. Health* **203**, 411-415.
Oberdorster G., 2001. Pulmonary effects of inhaled ultrafine particles. *Int Arch Occup Environ Health*, 1–8.
[3] Jaques P.A., Kim C.S., 2000. Measurement of total lung deposition of inhaled ultrafine particles in healthy men and women. *Inhal.Toxicology*, 715–731.
[4] Jenkins P.L., Phillips T.J., Mulberg E.J., Hui S.P., 1992. Activity patterns of Californians: Use of and proximity to indoor pollutant sources. *Atmos. Environ.* Part A. General Topics: Fifth International Conference on Indoor Air Quality and Climate Indoor Air '90: Characterization of Indoor Air **26**, 2141-2148.
Klepeis N.E., Nelson W.C., Ott W.R., Robinson J.P., Tsang A.M., Switzer P., Behar J.V., Hern S.C., Engelmann W.H., 2001. The national human activity pattern survey (NHAPS): a resource for assessing exposure to environmental pollutants. *J. Exposure Anal. & Environ. Epidem.*, 231–252.
[5] Putaud J.P., Baltensperger U., Brüggemann E., Facchini M., Fuzzi S., Gehrig R., Hansson H.C., Harrison R.M., Jones A., Laj P., Maenhaut W., Mihalopoulos N., Müller K., Palmgren F., Querol X., Rodriguez S., Spindler G., Brink H., Tunved P., Dingenen R., Wehner B., Weingartner E.,

Wiedensohler A., Wåhlin P., Raes F., 2004. A European aerosol phenomenology II: physical and chemical characteristics of particulate matter at kerbside, urban, rural and background sites in Europe. *Atmos. Environ.* **38**, 2579-2595.

Querol X., Alastuey A., Ruiz C.R., Artíñano B., Hansson H.C., Harrison R.M., Buringh E., Brink H.M.t., Lutz M., Bruckmann P., Straehl P., Schneider J., 2004. Speciation and origin of PM10 and PM2.5 in selected European cities. *Atmos. Environ.* **38**, 6547-6555.

Andersen Z.J., Wahlin P., Raaschou-Nielsen O., Scheike T., Loft S., 2007. Ambient particle source apportionment and daily hospital admissions among children and elderly in Copenhagen. *J. Exposure Sci. & Environ. Epidem* 1-12.

Lonati G., Ozgen S., Giugliano M., 2007. Primary and secondary carbonaceous species in PM2.5 samples in Milan (Italy). *Atmos. Environ.* **41**, 4599–4610.

Viana M., Maenhaut W., Chi X., Querol X., Alastuey A., 2007. Comparative chemical mass closure of fine and coarse aerosols at two sites in South and West Europe: implications for EU air pollution policies. *Atmos. Environ.***41**, 315-326.

[6] Long C.M., Suh H., Koutrakis P., 2000. Characterization of indoor particle sources using continuous mass and size monitors. *J. Air & Waste Manag. Assoc.*, 1236–1250.

[7] Heinrich J., 2010. Influence of indoor factors in dwellings on the development of childhood asthma. *Intl. J. Hyg.& Environ. Health*, in press.

[8] Lunden M.M., Kirchstetter T.W., Thatcher T.L., Hering S.V., Brown N.J., 2008. Factors affecting the indoor concentrations of carbonaceous aerosols of outdoor origin. *Atmos. Environ.*, 5660–5671.

Martuzevicius D., Grinshpun S.A., Lee T., Hub S., Biswas P., Reponen T., LeMasters G., 2008. Traffic-related PM2.5 aerosol in residential houses located near major highways: Indoor versus outdoor concentrations. *Atmos. Environ.* 6575–6585.

Morawska L., Jamriska M., Guo H., Jayaratne E.R., Cao M., Summerville S., 2009. Variation in indoor particle number and PM2.5 concentrations in a radio station surrounded by busy roads before and after an upgrade of the HVAC system. *Building & Environ.* **44**, 76–84.

[9] Matson U., 2005. Comparison of the modelling and the experimental results on concentrations of ultra-fine particles indoors. *Building & Environ.*, 996–1002.

Diapouli E., Chaloulakou A., Spyrellis N., 2007. Levels of ultrafine particles in different microenvironments — Implications to children exposure. *Sci. Tot. Environ.*, 128–136.

Bohlin P., Jones K.C., Tovalin H., Strandberg B., 2008. Observations on persistent organic pollutants in indoor and outdoor air using passive polyurethane foam samplers. *Atmos. Environ.* **42**, 7234–7241.

[10] Orosa J.A., Baaliña A., 2008. Passive climate control in Spanish office buildings for long periods of time. *Building & Environ.* **43**, 2005–2012.

[11] Kamens R., Lee C.T., Weiner R., Leith D., 1991. A study to characterize indoor particles in three non-smoking homes. *Atmos. Res.*, 939-948.

Thatcher T.L., Layton D.W., 1995. Deposition, resuspension, and penetration of particles within a residence. *Atmos. Environ.*, 1487-1497.

[12] Lunden M.M., Kirchstetter T.W., Thatcher T.L., Hering S.V., Brown N.J., 2008. Factors affecting the indoor concentrations of carbonaceous aerosols of outdoor origin. *Atmos. Environ.*, 5660–5671.

[13] Kuhlbusch T.A.J., Fissan H., Asbach C., 2009. Nanotechnologies and Environmental Risks: Measurement Technologies and Strategies, in: Verlag, S. (Ed.), Nanomaterials: Environmental Risks and Benefits and Emerging Consumer Products, pp. 233-243.

[14] Querol X., Alastuey A., Moreno T., Viana M., Castillo S., Pey J., Rodríguez S., Artíñano B., Salvador P., Sánchez M., Garcia Dos Santos S., Herce Garraleta M.D., Fernandez-Patier R., Moreno-Grau S., Minguillón M.C., Monfort E., Sanz M.J., Palomo-Marín R., Pinilla-Gil E., Cuevas E., 2008. Spatial and temporal variations in airborne particulate matter (PM10 and PM2.5) across Spain 1999-2005. *Atmos. Environ.* **42**, 3964–3979.

Effects of physicochemical properties of zinc oxide nanoparticles on cellular uptake

J Yu, M Baek, HE Chung and SJ Choi

Department of Food Science and Technology, Seoul Women's University, 126 Gongneung 2-Dong, Nowon-Gu, Seoul 139-774, South Korea

E-mail: sjchoi@swu.ac.kr

Abstract. Zinc oxide (ZnO) nanoparticles have been used as a source of zinc, an essential trace element in food industry and also widely applied to various cosmetic products. However, there are few researches demonstrating that the cellular uptake behaviours of ZnO with respect to the physicochemical characteristics such as particle size and surface charge in human cells. In this study, we evaluated the cellular uptake of ZnO with two different sizes (20 and 70 nm) and different charges (positive and negative). Human lung epithelial cells were exposed to ZnO for a given time, and then the uptake amount of ZnO was measured with inductively coupled plasma-atomic emission spectroscopy (ICP-AES). The results showed that the smaller sized ZnO could more easily enter the cells than the larger sized ZnO. In terms of surface charge, positively charged ZnO showed high cellular uptake compared to ZnO with negative charge. The internalization pathway of positively charged ZnO nanoparticles was determined to be primarily related to the energy-dependent endocytosis. It is, therefore, concluded that the particle size and surface charge of ZnO nanoparticles are critical factors influencing on their cellular uptake. Understanding the cellular uptake behaviours of nanoparticles with respect to physicochemical properties may be important to predict their toxicity potential on human.

1. Introduction

ZnO has been used as a source of zinc, an essential trace element in food industry. ZnO has been also widely applied to cosmetics such as personal care and sunscreen products, since it acts as an invisible barrier that scatters UV radiation away from the skin rather than allowing its harmful energy to be absorbed [1-4]. But, absorption rate of zinc is low via oral ingestion and bulk-sized ZnO has poor dispersion property, resulting in low UV blocking capacity, low transparency on skin, and hard agglomeration compared to small-sized ZnO. Thus, nano-sized ZnO has recently attracted much attention in order to enhance the uptake of zinc and increase UV filtering efficiency with cosmetic clarity. It is now generally accepted that nanoparticles efficiently penetrate the cell membrane compared to micro-sized particles. However, few researches were performed to demonstrate the effects of physicochemical parameters of ZnO nanoparticles on cellular uptake, which may provide critical information on their toxicity potential [5-7]. In this study, we evaluated cellular uptake behaviours of ZnO nanoparticles with different sizes (20 and 70 nm) and different charge (positive: +, and negative: -) in human lung epithelial cells. Moreover, the uptake mechanism of each nanoparticle was further investigated to understand its uptake behaviours.

2. Experimental methods

2.1. Materials
ZnO nanoparticles of two different sizes (20 and 70 nm) were purchased from Sumitomo (Japan) and American Elements (U.S.A.), respectively. For surface modification of ZnO nanoparticles with negative charge, ZnO (10 g) was suspended in 20 mM HEPES buffer (pH 7.0) containing 1% sodium citrate [8-10]. The particle size and surface charge (zeta potential) of ZnO nanoparticles were determined by transmission electron microscopy (TEM: JEM-1010, JEOL) and a zeta potentiometer (Zetasizer Nano ZS system, Malvern Instruments), respectively.

2.2. Cell culture
Human lung epithelial cells (A549) were purchased from the Korean Cell Line Bank and cultured in RPMI1640 medium supplemented with 10% heat inactivated fetal bovine serum (Welgene, Ltd., South Korea), 100 units/ml penicillin, and 100 μg/ml streptomycin, under a humidified atmosphere (5% CO_2 plus 95% air).

2.3. Cellular uptake
The cells (5×10^5/ml) were incubated overnight at 37°C under a 5% CO_2 atmosphere and the medium in the wells was then replaced with fresh medium containing nanoparticles. After incubation for a given time, the cells were washed twice with phosphate buffered saline (PBS), harvested by scraping, and then re-suspended in deionized and distilled water. The cell pellets were then digested in 3 ml of ultrapure nitric acid. After adding 0.5 ml of H_2O_2, each mixture was heated at about 160°C until the cell pellets were completely digested. Then, the remaining nitric acid was removed by heating until the solutions were colorless and clear. The solutions were finally diluted to 3 ml with ultrapure water and ICP-AES (Optima 3300DV) was used to analyze Zn concentration in the samples. The cells incubated in the absence of particles were used as control.

2.4. Uptake mechanism
Energy-dependent endocytic mechanism of ZnO nanoparticles was evaluated by incubating the cells (5×10^5/ml) overnight at 37°C under a 5% CO_2 atmosphere, and replacing the medium in the wells with fresh medium containing nanoparticles. The cells were incubated with 125 μg/ml at 37°C and 4°C, respectively, for 1 h. And the cells were washed twice with PBS, harvested by scraping, and then the pellets were digested for ICP-AES analysis as described above.

2.5. Statistical analysis
Statistical analyses were performed using Student's t test for unpaired data and p values of less than 0.05 were considered significant. All data are presented as mean ± standard error of the mean (S.E.M.).

3. Results and discussion

3.1. Characterization of ZnO nanoparticles
Particle size and surface charge of ZnO nanoparticles were measured by TEM and zeta potentiometer, respectively. As shown in Table 1, net charge of ZnO nanoparticles in aqueous solution at pH 7.0 was determined to be positive, about 25 mV, thus surface modification to obtain negatively charged ZnO nanoparticles was performed with citrates. Citrates are widely used as capping agents for inorganic nanoparticles, giving rise to negative surface coating. This is based on the fact that citrates are important biological ligands for metal ions to form strong metal complexes. The measured particle size of 20 nm was well distinguished from that of 70 nm (Figure 1) and the surface charge for

positively or negatively charged ZnO was well prepared for comparative cytotoxicity study in the next step.

Table 1. Particle size and surface charge of ZnO nanoparticles as measured by TEM and zeta potentiometer, respectively.

Sample		Particle size [nm][a]	Zeta potential [mV]
20 nm	Positive charge	24 ± 5.57	25.0 ± 0.5[b]
	Negative charge		-44.6 ± 0.7[c]
70 nm	Positive charge	67 ± 12.41	29.1 ± 0.6[b]
	Negative charge		-45.2 ± 0.8[c]

[a] Average particle size was measured by randomly selecting 500 particles from TEM images.
[b] Zeta potential was measured in distilled water (pH 7.0).
[c] Zeta potential was measured in 20 mM HEPES buffer (pH 7.0) containing 1% sodium citrate.

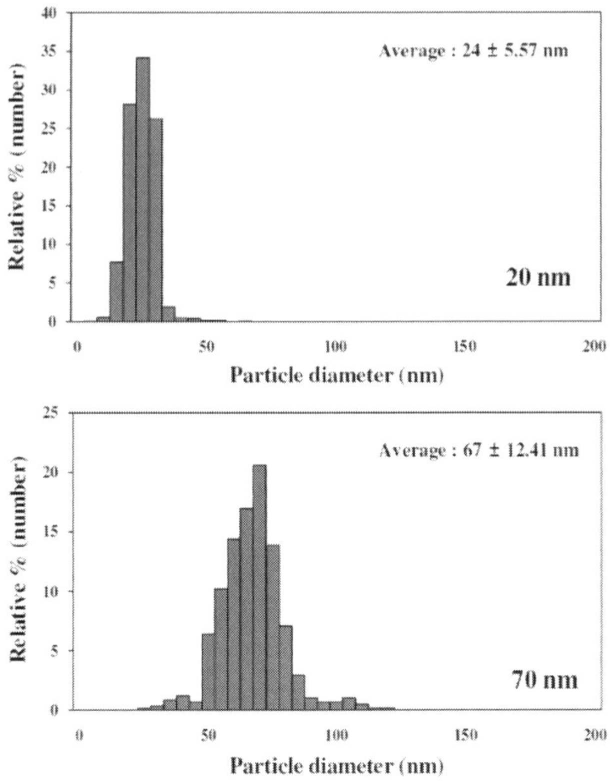

Figure 1. Histogram of particle size distribution for ZnO nanoparticles.

3.2. Uptake behaviors of ZnO nanoparticles: effect of incubation time

A549 cells were exposed to different ZnO nanoparticles with a concentration of 125 μg/ml for a given time and their intracellular uptake amount was evaluated by measuring Zn concentration with ICP-AES. Figure 2 showed a clear difference between nanoparticles tested; higher uptake was found in the cells exposed to positively charged nanoparticles than negatively charged ones. Moreover, size-dependent uptake behaviours of ZnO nanoparticles were remarkable; 20 nm was determined to largely enter the cells compared to 70 nm. Thus, ZnO with 20 nm (+) was the most efficient one in terms of high cellular uptake. The cellular uptake of different ZnO nanoparticles reached a maximum level at 4 h post-incubation and decreased as incubation time increased, suggesting the possible uptake saturation of ZnO nanoparticles in the cells.

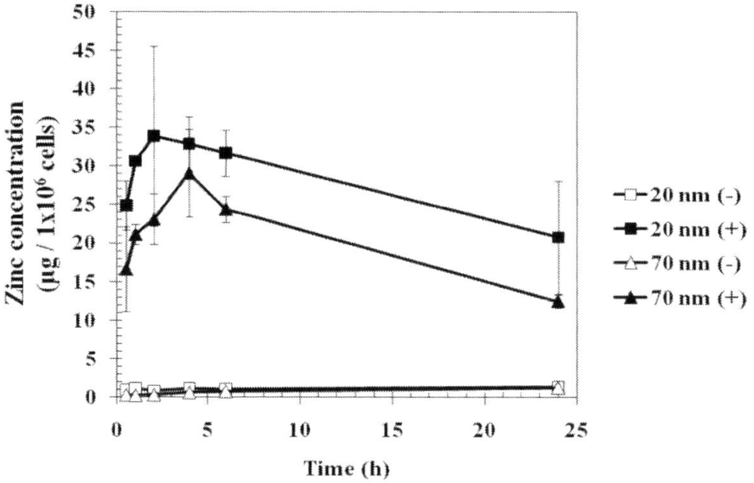

Figure 2. Cellular uptake of ZnO nanoparticles with respect to incubation time.

3.3. Uptake behaviors of ZnO nanoparticles: effect of ZnO concentration

When the cellular uptake of each ZnO nanoparticle was evaluated with respect to nanoparticle concentration after 4 h incubation, charge- and size-dependent uptake behaviours were also remarkable (Figure 3). Smaller size, 20 nm, and positively charged ZnO nanoparticles showed higher cellular uptake than larger size, 70 nm, and negatively charged ones, which was well consistent with the result presented in Figure 2. The cellular uptake of ZnO nanoparticles reached a plateau at 500 μg/ml, indicating that their cellular uptake was probably saturated. According the results as shown in Figure 2 and Figure 3, it seems that surface charge of ZnO nanoparticles is more important factor affecting their cellular uptake rather than their particle size. It can be explained by high interaction of positively charged ZnO nanoparticles with negatively charge plasma membrane, resulting in high and easy cellular uptake.

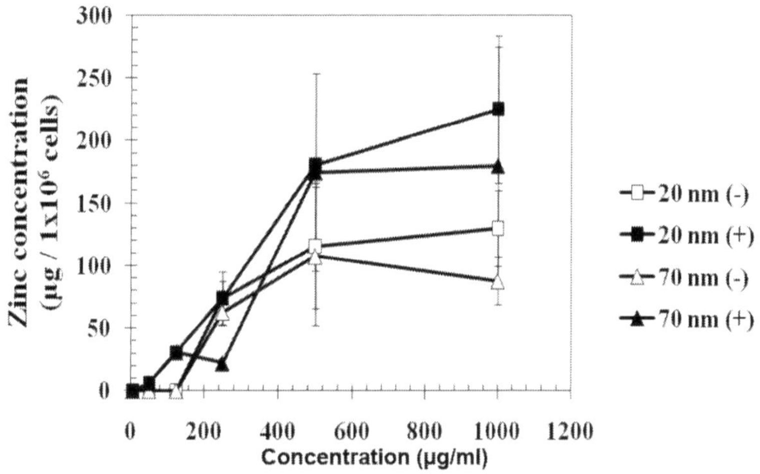

Figure 3. Cellular uptake of ZnO nanoparticles with respect to nanoparticle concentration.

3.4. Uptake mechanism of ZnO nanoparticles

Energy-dependent endocytic mechanism of different ZnO nanoparticles was evaluated by comparing cellular uptake amount of ZnO after incubation at 37°C with that at 4°C. This is based on the fact that endocytic uptake mechanism is a typical energy-driven, in other words, ATP-dependent process, thus can be blocked when ATP synthesis is inhibited [11]. Therefore, the effect of energy depletion on cellular uptake process of ZnO nanoparticles was examined by incubating the cells with nanoparticles at 37°C and at 4°C, respectively, followed by comparison of each uptake level. Figure 4 demonstrated that the uptake amount of positively charged ZnO nanoparticles significantly decreased after incubation at 4°C compared to 37°C, while the uptake of negatively charged ZnO nanoparticles was not considerably affected by energy depletion, suggesting that energy-dependent endocytosis is the primary uptake mechanism for the internalization of positively charged ZnO nanoparticles. It seems that uptake amount of negatively charged ZnO nanoparticles at 37°C is too low to be distinguished from that at 4°C. It is worth noting that particle size did not influence on the cellular uptake mechanism of ZnO nanoparticles. It is, therefore, concluded that high cellular uptake of positively charge ZnO nanoparticles is strongly associated with their energy-dependent endocytic pathway.

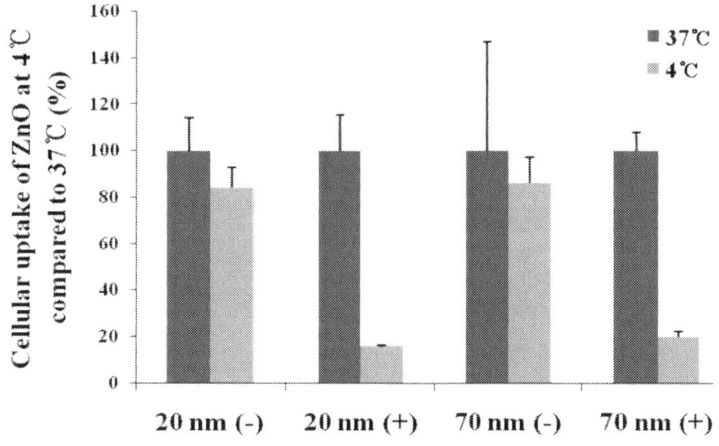

Figure 4. Effect of energy depletion on cellular uptake of ZnO nanoparticles in A549 cells after 1 h. Uptake amount of each ZnO at 37°C was calculated as 100%.

4. Conclusion

The effects of physicochemical properties of ZnO nanoparticles such as particle size and surface charge were evaluated in human lung cells. As a result, small-sized ZnO nanoparticles, 20 nm, with positive charge showed the highest cellular uptake with a following order of 70 (+), 20 (-), and 70 (-). It seems that surface charge of ZnO nanoparticles plays a critical role in their internalization, probably due to strong attractive force between positive surface charge of ZnO nanoparticles and the plasma membrane with negative charge. Moreover, the cellular uptake of positively charged ZnO nanoparticles decreased remarkably by inhibition of ATP synthesis, suggesting the main energy-dependent uptake mechanism for their internalization. It is, therefore, concluded that the particle size and surface charge of ZnO nanoparticles are critical factors affecting their cellular uptake. Understanding the cellular uptake behaviour of nanoparticles with respect to physicochemical properties may be important to predict their toxicity potential on human.

Acknowledgement

This research was supported by a grant (10182KFDA991) from Korea Food & Drug Administration in 2010.

References

[1] Fan Z and Lu J G 2005 *J. Nanosci. Nanotechnol.* **5** 1561
[2] Djurišić A B and Leung Y H 2006 *small* **2** 944
[3] Chen W and Zhang J 2006 *J. Nanosci. Nanotechnol.* **6** 1159
[4] Serpone N, Dondi D and Albini A 2006 *Inorg. Chim. Acta* **360** 794
[5] Huang C C, Aronstam R S, Chen D R and Huang Y W 2010 *Toxicol. In Vitro* **24** 45
[6] Barardis B D, Civitelli G, Condello M, Lista P and Pozzi R 2010 *Toxicol. Appl. Pharm.* **246** 116
[7] Heng B C, Zhao X, Xiong S, Ng K W, Boey F Y-C and Loo J S-C 2010 *Food Chem. Toxicol.* **48** 1762
[8] Cho S, Jung S H and Lee K H 2008 *J. Phys. Chem.* **112** 12769
[9] Kuo C L, Kuo T J and Huang M H 2005 *J. Phys. Chem.* **109** 20115
[10] Tian Z R, Voigt J A, Liu J, Mckenzie B, Mcdermott M J, Rodriguez M A, Konishi H and Xu H F 2003 *Nat. Mater.* **2** 821
[11] Letoha T, Goal S, Somlai C, Venkei Z, Glavinas H, Kusz E, Duda E, Czajlik A, Petak F and Penke B 2005 *J. Pept. Sci.* **12** 805

Characterization, Exposure Measurement and Control for Nanoscale Particles in Workplaces and on the Road

Jing Wang[1,2] and David Y.H. Pui[3]

[1]Institute of Environmental Engineering, ETH Zürich, 8093, Switzerland,

[2]Empa, Analytical Chemistry, 8600 Dübendorf, Switzerland

[3]Particle Technology Laboratory, University of Minnesota, 55414, USA

E-mail: jing.wang@ifu.baug.ethz.ch

Abstract. The amount of engineered nanoparticles is increasing at a rapid rate and more concerns are being raised about the occupational health and safety of nanoparticles in the workplace, and implications of nanotechnology on the environment and living systems. At the same time, diesel engine emissions are one of the serious air pollution sources in urban area. Ultrafine particles on the road can result in harmful effects on the health of drivers and passengers. Research on characterization, exposure measurement and control is needed to address the environmental, health and safety issues of nanoscale particles. We present results of our studies on airborne particles in workplaces and on the road.

1. Introduction

The quantity of engineered nanomaterials is expected to grow significantly in the next several years [1]. Analysts have estimated that the worldwide market for nanomaterials will be 700-1000 billion Euros in 2011 (www.nanosafe.org). Nanomaterial manufacturers are in need of fast and online instruments to characterize and measure the produced nanomaterials, and ensure that the concentration, size and morphology are in the expected ranges. Instruments are also needed to monitor emission and release of nanoparticles in workplaces to facilitate occupational safety and health. A number of recent toxicological studies suggest the total surface area of airborne nanoscale particles as a health-relevant exposure metric [2,3]. Researchers argue that the particles react with the biological matrix via the particle surface. The instrument Nanoparticle Surface Area Monitor (NSAM) was designed to measure the lung-deposited surface area of nanoparticles [4,5], and to provide real-time results. Many engineered nanomaterials are in the form of nanoparticle agglomerates [6]. Fast and online measurement for agglomerates is required in many scenarios including measuring fast changing agglomerates, quality control for material manufacturing, monitoring potentially toxic air-borne agglomerates, etc. We have developed the instrument Universal Nanoparticle Analyzer (UNPA), for online measurement of gas-borne nanoparticle agglomerates [6,7]. The principle and applications of UNPA are presented here.

Emission and exposure play important roles for societal impact of nanoparticles. Data are needed about the release of nanoparticles into the environmental and exposure to human. The US National Institute of Occupational Safety and Health (NIOSH) plans to develop recommended exposure limits

(RELs) for ultrafine titanium dioxide and carbon nanotubes (CNTs) for the period of 2009 – 2012 (NIOSH 2009) [8]. We present emission measurement in an industrial production facility for CNT-imbedded nanocomposites. Exposure to ultrafine particles from engine emissions has drawn increasing concern because of its association with adverse health effects and its potential impact on daily commuters [9,10]. Particle concentrations on the road could exceed 10^6 particles/cm^3, much higher than the ambient background concentration [11-13].

Filtration is one of the most effective ways for particle emission control. To reduce commuters' exposure to airborne particles, an increasing number of automobiles have been equipped with a cabin air filter. The filters are characterized using particles larger than 300nm following the current International Standards Organization (ISO) standard (ISO 11155-1). However, particles smaller than 300nm comprise the majority of freeway particulate matter by number and surface area [11,14]. Therefore, Qi et al. [15] suggested that the ISO 11155-1 standard [16] to be extended to include particles smaller than 300 nm. We present the results of on-road evaluations of cabin air filters here.

2. Measurement and Characterization of Nanoscale particles

2.1. Nanoparticle Surface Area Monitor (NSAM)

Relatively few techniques are available to monitor exposures with respect to aerosol surface area. The BET measurement based on a gas adsorption method is not suited for a rapid evaluation of aerosol surface area at lower concentration [17]. It can be used only for powders, not for particles in the gasborne state. It does not have on-line capabilities. The epiphaniometer [18] was designed to measure aerosol surface area. This device measures the Fuchs or active surface-area of the aerosols by measuring the attachment rate of radioactive ions. The epiphaniometer is not well suited to widespread use in the workplace because of the inclusion of a radioactive source and toxic lead, and the lack of effective temporal resolution. One of other possible methods is diffusion charging (DC) which measures the surface area indirectly (e.g. [19]). When nanoparticles are exposed to a unipolar ion environment, they acquire a charge level which at least in a certain particle size range is theoretically proportional to the particle surface area [20]. If the particles are deposited on a filter, the electric current, induced by the rate of deposited particles, can be directly proportional to the total particle surface area [21]. The current can be measured by connecting the filter to an electrometer. Instruments using DC include LQ1-DC diffusion charger (Matter Engineering, Switzerland) and TSI model 3070a Electrical Aerosol Detector (EAD). While the LQ1-DC aims at measuring the Fuchs surface area, the EAD is designed for the particle length. Several studies using atmospheric field data, however, have shown that EAD can be used as a useful indicator for the quantity of particle surface area deposited in the lung [22-24].

Fissan et al. [4] observed that the response of EAD as a function of particle diameter is deviating from the charging theory and is closer to the lung deposited surface area, which is the geometric surface area weighted with the lung deposition curves. Based on these results, TSI Nanoparticle Surface Area Monitor (NSAM) model 3550 has been developed. Figure 1 shows a schematic of NSAM that uses a corona discharge to produce positively charged ions and mixes these ions with particles in an opposed flow mixing chamber. An ion trap is located downstream of the mixing chamber. The particles are deposited on a HEPA filter inside a Faraday cup and the current, induced by the deposited particles, is measured with an electrometer.

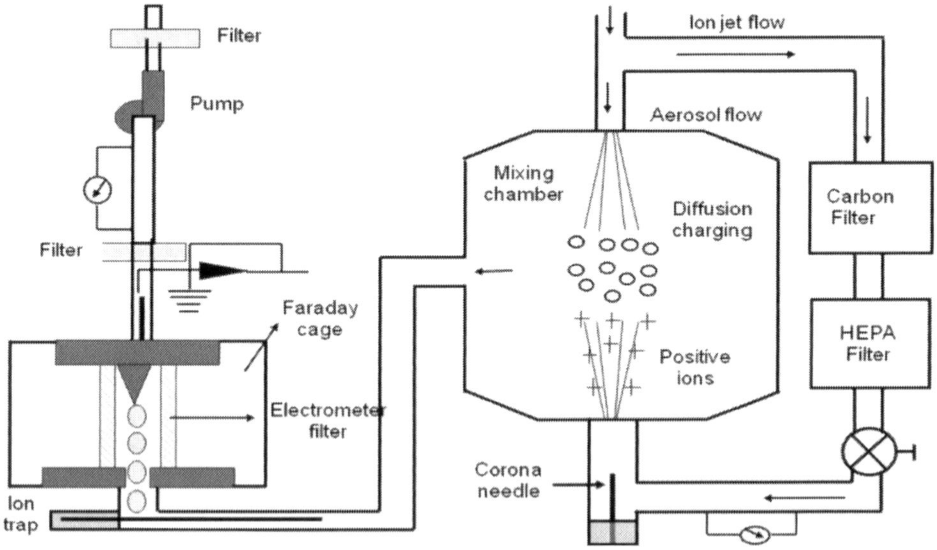

Figure 1. Schematic of the Nanoparticle Surface Area Monitor (NSAM, TSI Model 3550).

Fissan et al. [4] found the NSAM response can be adjusted. Between the mixing chamber and particle filter, all excess ions are removed in the ion trap by means of an electric field. Due to the high electric mobility of ions, the voltage can be relatively low. Even though the electric field strength within the ion trap is very low, some charged particles near the electrode of opposite polarity are also removed. Since in the human respiratory tract some particles also get lost before they reach the lung, the ion trap voltage can be adjusted such that the particle losses in the ion trap match those in certain areas of the human inhalation system. It is found that the response function of NSAM matches the surface area deposited in the tracheobronchial region with an ion trap voltage of 100 V, whereas the response function simulates the deposition in the alveolar region with an ion trap voltage of 200 V. The NSAM was calibrated [7] to measure the nanoparticle surface area deposited in two regions, trancheobronchial and alveolar of the human lung.

2.2. Measurement of Agglomerates using the Universal NanoParticle Analyzer (UNPA)

We have developed the instrument UNPA, for online measurement of gas-borne nanoparticle agglomerates [6,7]. UNPA utilizes Differential Mobility Analyzer (DMA), Condensation Particle Counter (CPC) and NSAM to characterize airborne nanoparticle morphology and measure the number, surface area and volume distributions of loose nanoparticle agglomerates. The key parameter measured is the UNPA sensitivity, which is defined as the current I (fA) measured by the NSAM divided by the number concentration N (#/cm^3) measured by the CPC.

$$S = I/N \text{ (fA cm}^3\text{)}. \tag{1}$$

Charging theories of Chang [25] for aerosol particles of arbitrary shape indicate that the geometric surface area and electrical capacitance of the particles are two important parameters to determine the mean charge of non-spherical particles. The electrical capacitance of agglomerates may be computed using a variational method proposed by Brown and Hemingway [26]. The surface area of loose agglomerates may be calculated using a mobility analysis developed by Lall and Friedlander [27]. Shin et al. [7] combined the above analyses to show that the electrical capacitance of loose agglomerates is larger than that of spherical particles with the same mobility, and loose agglomerates can gain more charges from unipolar charging.

The primary particle size plays an important role in determination of the surface area and electrical capacitance, thus the charges on agglomerates. The UNPA sensitivity is related to the primary particle diameter d_p. We found that the UNPA sensitivity can be correlated to the primary particle size through a power law relation [6]

$$S = c_2 \left(\frac{12\pi\lambda}{c^* d_p^2} \frac{d_m}{C_c} \right)^k c_1 (d_p)^h \tag{2}$$

where c^* is a constant regarding particle orientation, λ is the mean gas free path, c_1, c_2, k and h are constants which can be determined from the experimental data. Then the sensitivity data from the experiments can be fitted into (2) to determine the primary particle diameter d_p. Once the primary particle size is determined, surface area and volume of the agglomerates can be calculated.

We used the UNPA to measure silver spheres, and agglomerates of silver, SiO_2 and TiO_2 [6]. SiO_2 and TiO_2 agglomerates were generated by a diffusion burner. The sensitivities for silver spheres and loose agglomerates are the lowest and highest, respectively, among our sensitivity data (Fig. 2). The sensitivities for TiO_2 and SiO_2 agglomerates are between the upper and lower bounds. We also used UNPA to measure gold and nickel nanoparticle agglomerates from a spark discharge generator [28]. We determined the primary particle sizes for agglomerates using the sensitivity data and equation (2). We also analyzed electron micrographs and obtained the primary particle sizes. A comparison of the primary particle sizes from the two methods is shown in Table 1. It can be seen that the agreement for the primary particle size is good and UNPA can measure primary particles in a rather wide size range.

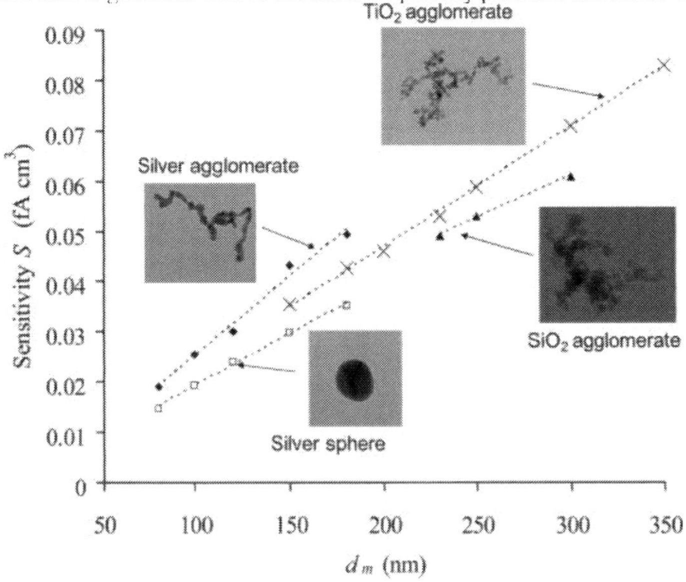

Figure 2. Measured UNPA sensitivity for silver agglomerates and spheres, SiO_2 and TiO_2 agglomerates.

Table 1. A comparison for the primary particle sizes obtained from the UNPA sensitivity analysis and electron microscopy (EM).

	d_p from UNPA sensitivity (nm)	d_p from EM (nm)	standard deviation of d_p from EM (nm)
TiO_2 agglomerates	22.8	19.9	7.5
SiO_2 agglomerates	32.5	27.3	9.3
Au agglomerates	5.97	7.90	1.47
Ni agglomerates	4.09	6.56	0.98

Diesel engine exhaust is a major source for on-road ultrafine particles and has profound impact on environment and ecosystem. Diesel aerosols are mainly agglomerates with open structures and composed of primary particles (Fig. 3). We measured diesel aerosols generated from a 4045 John Deere Diesel Engine using UNPA [28]. A schematic of the experiment is shown in Fig. 4. Diesel aerosols were sampled form the exhaust pipe and mixed with dilution air in the residence chamber. Then the diesel aerosols were sent to Nanometer Aerosol Sampler (NAS) for collection of electron micrograph samples, and to UNPA for agglomerate measurement.

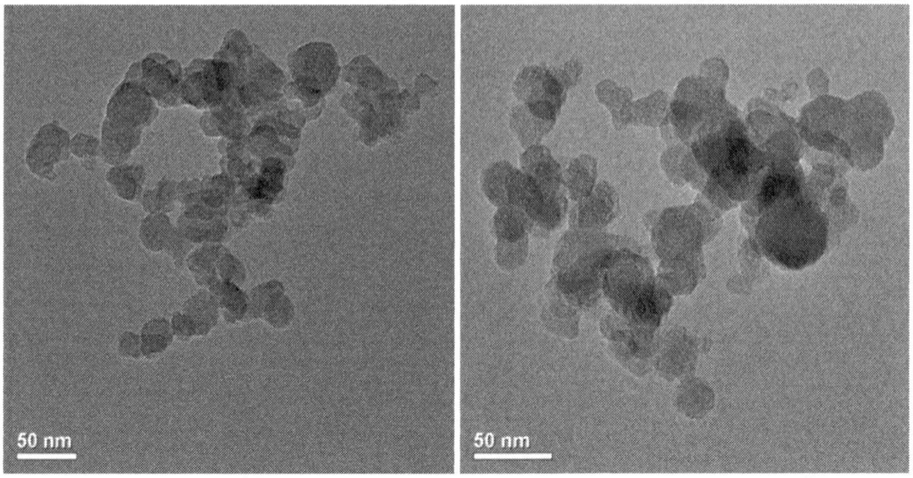

(a) (b)

Figure 3. TEM images of diesel particles under the following conditions (a) light load, (b) heavy load.

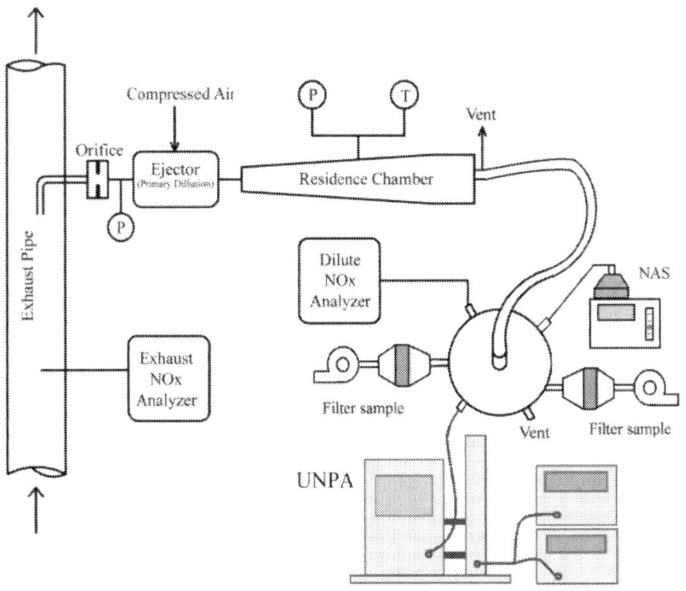

Figure 4. A schematic of the setup for diesel aerosol sampling and measurement.

We sampled the diesel aerosols under two engine conditions: light load and heavy load. The engine speed, torque and exhaust temperature increased with the load. A catalytic stripper was used to remove sulfur compounds and the soluble organic fraction by passing the diluted diesel exhaust over two heated catalysts. The primary particle sizes of diesel particles measured by UNPA are listed in Table 2. The primary sizes were also measured using a large number of TEM images. A total of 212 primary particles were analyzed in the case of heavy load, and 189 primary particles in the case of light load. The primary particle size from TEM agreed well with UNPA measurement for the heavy load case. In the case of light load, there was about 6 nm difference between UNPA and TEM results, which was still within the standard deviation of the TEM result.

Table 2. Comparison of the primary particle size in diesel agglomerates measured by UNPA and TEM.

Engine load	Primary particle sizes from UNPA (nm)	Primary particle size from TEM (nm)
Light	29.5	23.2 ± 6.6
Heavy	24.2	23.9 ± 6.8

For loose agglomerates, UNPA can measure the number, surface area and volume distributions. We show the results for the volume of silver agglomerates in Fig. 5. The distributions based on spheres are also plotted for comparison. The difference of the volume distributions is substantial. The results show that the UNPA measurement can potentially improve distribution measured based on spheres significantly.

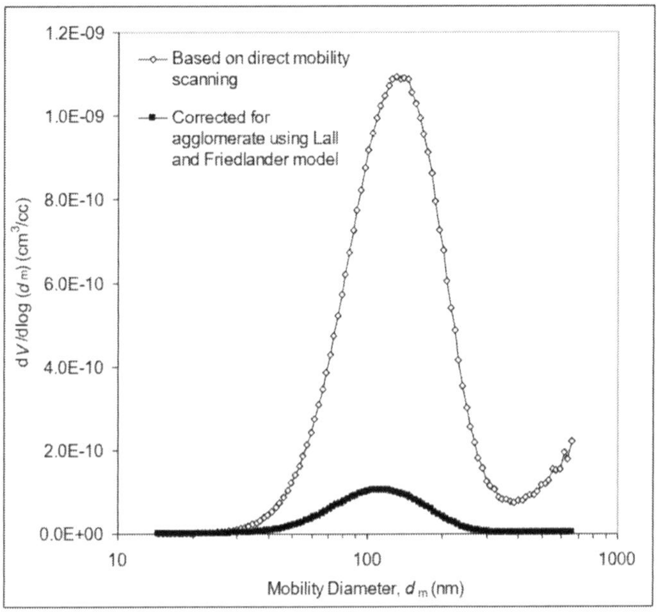

Figure 5. Agglomerate volume distribution from direct mobility scanning and after correction by Lall and Friedlander model (silver agglomerates, primary particle diameter measured by UNPA: 8.4 nm).

3. Exposure Measurement for Production of CNT-imbedded Nanocomposites

The unique properties of CNTs have led to their increased usage in advanced materials. CNTs are used in structural composites for sporting equipment, conductive plastics, electron field emitters, semiconductor devices, etc. As more industries incorporate CNTs into consumer products the opportunities for worker exposure will rise. NIOSH specified identification of the long-term health effects of CNTs in its strategic plan for 2009 – 2012 (NIOSH 2009, [8]). We review our exposure measurement of CNTs in the production of nanocomposites.

We measured exposure in an industrial production facility for CNT-imbedded nanocomposites [28]. In the extrusion operation, CNTs were mixed with melted polymer and other compounds to produce nanocomposite pellets. First polymer was melted and mixed with CNTs. The mixture was extruded from a die into a water bath. Cooled strands passed through a dryer before entering a cutter. The pieces were fed into a second extruder with CNT dispersing agent. At the exit of the second extruder the strands were pelletized. The pellets passed through a shaker and a cyclone for size selection.

A suite of instruments was used to obtain air-borne particle number concentration, surface area concentration, and size distribution data. The instruments were placed in the closed room near the extrusion system. The measurement results are shown in Fig. 6. The measurement duration can be divided into five periods. Period I (from 12:50 to 13:40) accounted for when there was no activity in the room. The total number concentration was low, about 6,000 #/cm^3. In Period II (from 13:40 to 13:55), the system was warming up. The particle concentration increased, possibly due to evaporation of polymeric residuals in the extruder. In Period III (from 13:55 to 14:15), the warm up continued and a spike of 10 – 40 nm particles appeared. These particles did not cause significant increase in the lung-deposited surface area because of their small sizes, and were likely polymer residuals from the warm up. In Period IV (from 14:15 to 14:50), the first extruder was running. The particle number concentration increased to about 40,000 #/cm^3. In Period V (from 14:50 to 15:00), both extruders were running in the full production mode. The particle number concentration increased to about 90,000 –

100,000 #/cm^3 and the particles were mainly in the range of 30 – 90 nm. We believe that volatile polymer fumes were a major particle source. This is corroborated by the observation that the particle concentrations were highest near the extrusion barrel and vapor was visible at the extrusion barrels. Further research is needed to find out the fraction of CNTs in the total particle concentration.

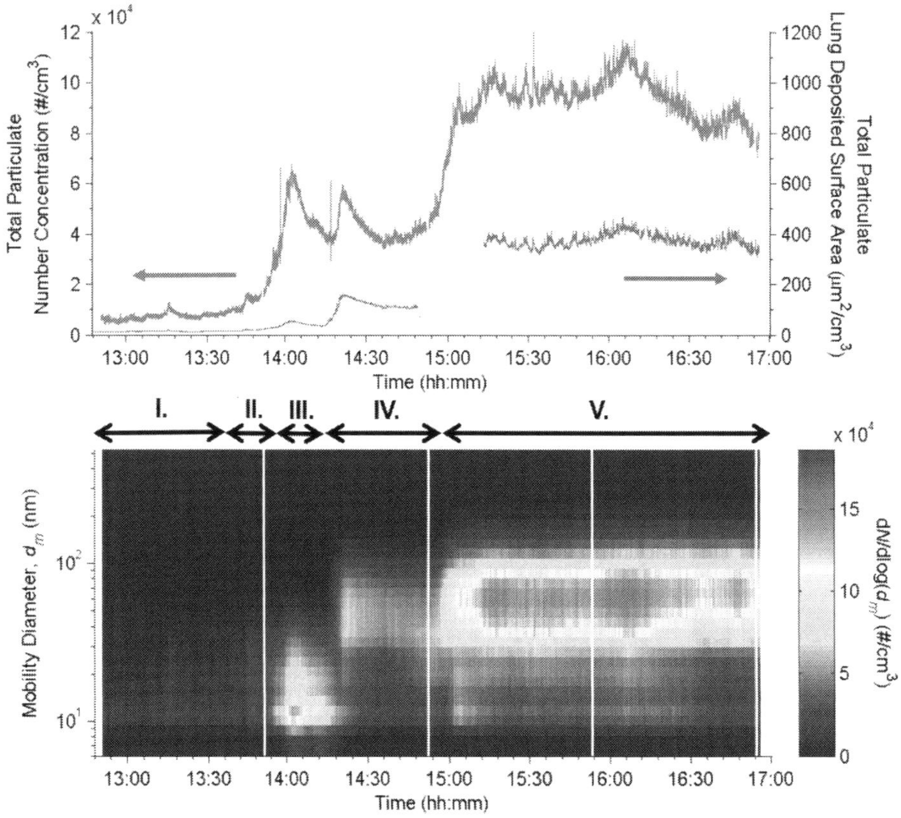

Figure 6. Measurement for air-borne particles during the extrusion operation for CNT-imbedded nanocomposites. Upper panel shows the total number concentration and lung-deposited surface area concentration; lower panel shows the evolution of the particle number-size distribution.

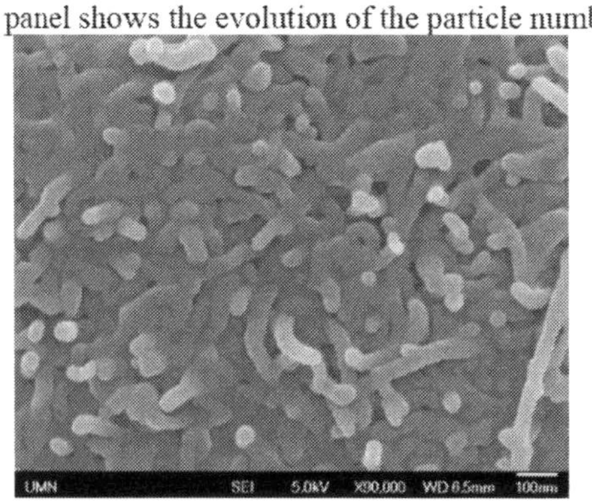

Figure 7. Examples of SEM images of the CNTs agglomerates collected on filters during the extrusion process. The picture on the right is a magnified local view of the one on the left.

We also used a filter to collect samples for electron microscopic analysis. Examples of the SEM pictures are shown in Fig. 7. We found a number of CNT agglomerates on the filter. The sizes of the CNT agglomerates are in the range of 1 μm to 20 μm. The individual CNTs are much smaller, with the diameter of about 40 nm. The CNTs were already in the agglomerated form when provided by the manufacturer. Such large agglomerates have limited mobility, and are easier and safer to handle and transport. Question remains whether the CNTs agglomerates can be disintegrated in the extrusion process and whether single CNTs can be released. Thus far we have not found single CNTs on the filter samples. Further research will be performed to identify possible released single CNTs.

4. On-road exposure and recirculation control in car cabins

With the filter installed in the test vehicle and driving in freeway traffic, we measured particle number and surface area concentrations of exterior freeway air and cabin air simultaneously and determined the exposure reduction based on the two concentration metrics [15,30]. The detailed testing protocol is as follows. New cabin air filters were installed in the test vehicle before each on-road test to make sure the test data were comparable. Two water-based condensation particle counters (WCPC, TSI 3785) were used to measure the particle number concentration. The WCPC 3785 is able to measure the particle number concentration up to 10^7 particles/cm^3 and has $d_{50\%}$ (particle size with 50% counting efficiency) at 5 nm, making it very suitable for such on-road ultrafine particle measurements. A pair of AeroTrak 9000 (TSI Inc.) instruments were used to measure the human lung deposited surface area concentrations. The AeroTrak 9000 is the portable version of the NSAM. NSAM/AeroTrak 9000 was calibrated against neutralized particles [5], thus a neutralizer (Po^{210} source) was used upstream of each AeroTrak 9000 to make sure the sampled particles are neutralized. The AeroTrak 9000 was set to measure the lung deposited surface area concentration in the alveolar region only. The two WCPCs and AeroTrak 9000 units were tested before being deployed in the on-road tests and showed a high internal precision (R^2 value of 0.99 for both particle number and surface area concentrations). Data collected from both WCPCs and AeroTrak 9000 were in 1 s intervals to provide a high temporal resolution.

All the windows were closed during the on-road tests. For the test in a Saab 93 (2003 model, cabin volume of 2.55 m^3), when the fan is on and the recirculation air is off, outside air is brought in after passing through the cabin air filter and then released to the outside through gaps around the doors and windows due to the positive pressure in the cabin. In this study, fan speed levels 2 (medium) and 5 (high) were tested, with flow rates at 3600 and 7200 liters per minute (LPM), respectively. A WCPC and an AeroTrak 9000 were used to measure particle concentrations of exterior freeway air sampled through a probe mounted on the car window. All on-road tests were conducted when driving on interstates I-35 and I-94 during rush hour (3-6 pm), providing representative data for typical commuters on Twin Cities, USA freeways.

Figure 8 shows the particle concentrations measured outside and inside the vehicle when driving on the freeway with the fan speed at level 5 (high) and the recirculation air off. In the same time span, the data from WCPC (Fig. 8a) and AeroTrak 9000 (Fig. 8b) followed very similar trends for both outside and inside concentrations. The outside concentration varied rapidly due to the change of the vehicle's position in the traffic and emission status of the surrounding vehicles, while the inside concentration fluctuated much less. The more stable inside concentration was explained by the in-cabin space serving as a chamber that damped down the outside variations.

The effect of the cabin air filter can be quantified by the exposure reduction (ER), which can be defined as

$$\mathrm{ER} = 1 - C_i / C_o \tag{3}$$

where C_i and C_o are time averaged particle concentrations inside and outside the car, respectively (15 min duration in this study). As shown in Fig. 8, data from both WCPC and AeroTrak 9000 indicate an obvious average exposure reduction in the cabin. At high fan speed, the average ER based on number concentrations is 55.0 ± 3.3% (from data of more than 10 replications, the same for all subsequent data), while the ER based on lung deposited surface area concentrations is 48.9 ± 8.8%. The number- and surface area-based ER are 65.6 ± 6.0% and 60.6 ± 9.4% at medium fan speed, higher than the corresponding values at high fan speed due to a higher filtration efficiency at lower face velocity. The lower ER value for surface area concentration is reasonable because for freeway particles, the maximum particle concentration for the surface area-based size distribution is closer to the most penetrating particle size of the cabin air filter (about 350 nm) than the maximum particle concentration for the number based size distribution.

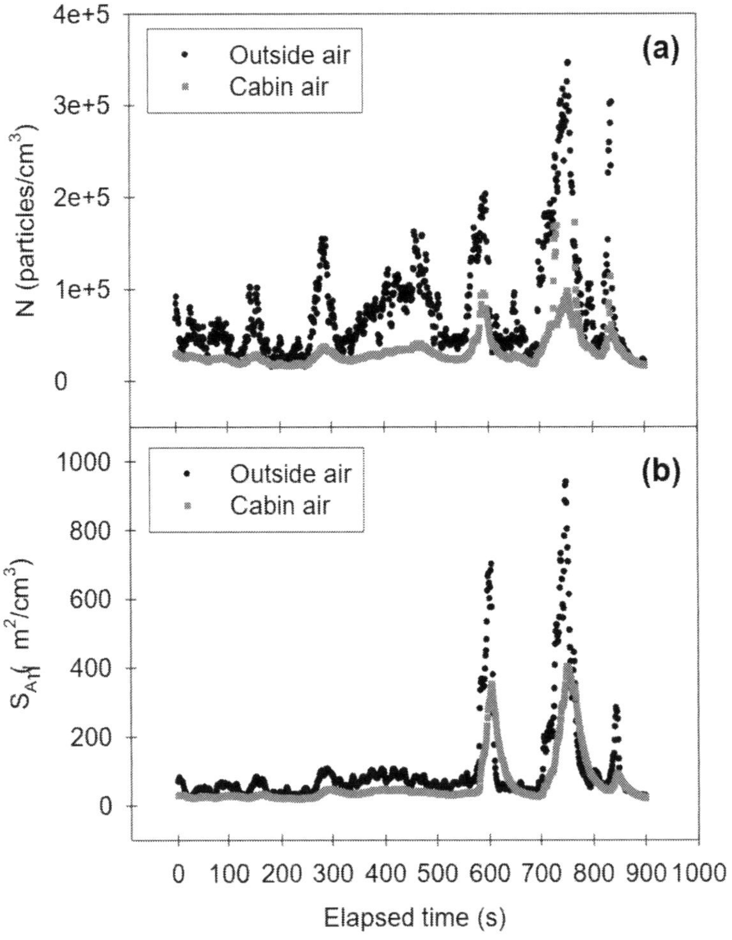

Figure 8. Particle (a) number and (b) lung deposited surface area concentrations of freeway air and cabin air observed on December 6, 2007 while driving the Saab 93 on interstate I-35 with the recirculation air off and fan speed at level 5.

The effect for exposure reduction is more pronounced when the recirculation mode is used in the car cabin. The tests were performed in a Toyota Camry (2007 model) with the cabin air filter in the recirculation loop. In-cabin aerosol concentration in the Camry was reduced to below typical office air concentrations (~ 4,000 particles/cm^3) in approximately 3 min. Figure 9 shows the time-resolved in-cabin aerosol size distribution during recirculation. It can be seen that particle concentrations of all the sizes were reduced continuously during the recirculation. Aerosol particles < 100 nm were rapidly removed from the air by the recirculation ventilation system. As would be expected from filtration theory, particles close to the filter's most penetrating particle size were removed with the least efficiency, but still experienced significant reductions. This result demonstrates that the recirculating air filtration is especially effective in reducing exposure to ultrafine particles.

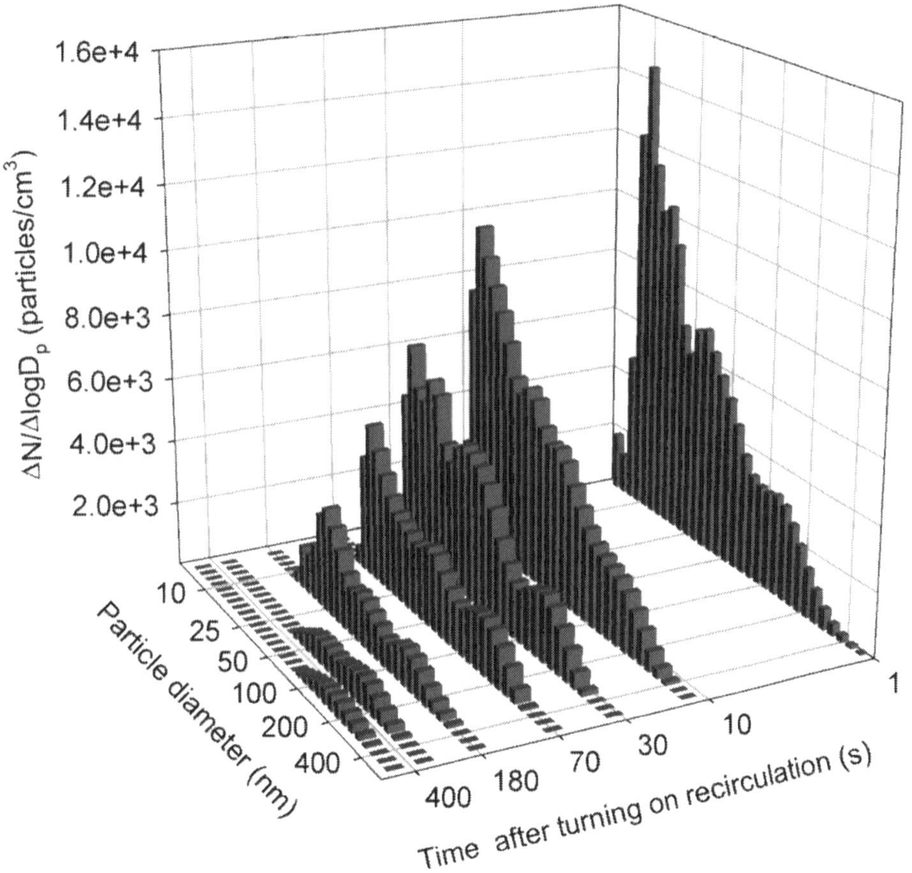

Figure 9. Measured variations in particle size distribution inside a Toyota Camry, with the air ventilation system in recirculation mode and the air filter in place. D_p, particle diameter; N, particle number concentration in size bin $\Delta \log D_p$.

5. Summary

Nanoscale particles exist in workplaces and on the road and their effects on environment and health of workers and passengers are of concern. We performed research covering particle characterization, exposure measurement and control. NSAM provides real-time measurement for the lung deposited surface area of airborne nanoparticles, which is a health-related metric. UNPA can differentiate compact particles from particles with open structures and can measure the number, surface area and volume distributions of loose nanoparticle agglomerates. We have applied UNPA to measure engineered nanoparticles and diesel exhaust particles and obtained satisfactory results. Measurement at a CNT composite production facility showed that CNT agglomerates were released and there was possible exposure. On-road test results demonstrated that using air recirculation can substantially and rapidly reduce exposure to airborne ultrafine particles.

Acknowledgement
The work was partially supported by the National Institute of Environmental Health Sciences grant # 1RC2ES018741-01 (sub-grant 100029-D) on "Hazard Assessment and Risk Estimation of Inhaled Nanomaterials Exposure" and by the NSF grant (Award ID: 1056479) on "Real Time Measurement of Agglomerated or Aggregated Airborne Nanoparticles Released From a Manufacturing Process and Their Transport Characteristics".

References

[1] Roco, M. (2006) Nanotechnology's Future, *Scientific American*, 24 July 2006, 7 Oct. 2006.
[2] Oberdörster, G.; Stone, V.; Donaldson, K. Toxicology of nanoparticles: A historical perspective. Nanotoxicology 2007, 1, 2–25.
[3] Donaldson, K., X.Y. Li & W. MacNee (1998) Ultrafine (nanometer) particle mediated lung injury, *Journal of Aerosol Science* 29, 553-560.
[4] Fissan, H.; Neumann, S.; Trampe, A.; Pui, D. Y. H.; Shin, W. G. Rationale and principle of an instrument measuring lung deposited nanoparticle surface area. J. Nanopart. Res. 2007, 9, 53–59.
[5] Shin, W. G.; Pui, D. Y. H.; Fissan, H.; Neumann, S.; Trampe, A. Calibration and numerical simulation of nanoparticle surface area monitor (TSI Model 3550 NSAM). J. Nanopart. Res. 2007, 9, 61–59.
[6] Wang, J., Shin, W-G, Mertler, M., Sachweh, B., Fissan, H., Pui, D.Y.H. (2010) Measurement of Nanoparticle Agglomerates by Combined Measurement of Electrical Mobility and Unipolar Charging Properties, Aerosol Sci. & Technology, 44, 97–108.
[7] Shin, W.G., Wang, J., Mertler, M., Sachweh, B., Fissan, H., and Pui, D.Y.H. (2010) The Effect of Particle Morphology on Unipolar Diffusion Charging of Nanoparticle Agglomerates in the Transition Regime, J. Aerosol Science, 41, 975–986.
[8] NIOSH (2009) Strategic Plan for NIOSH Nanotechnology Research and Guidance - Filling the Knowledge Gaps, DHHS (NIOSH) Publication No. 2010–105.
[9] Mills, N. L.; Törngvist, H.; Gonzalez, M. G.; Vink, E.; Robinson, S. D.; Söderberg, S.; Boon, N. A.; Donaldson, K.; Sandström, T.; Blomberg, A.; Newby, D. E. Ischemic and thrombotic effects of dilute diesel-exhaust inhalation in men with coronary heart disease. N. Engl. J. Med. 2007, 357, 1075–1082.
[10] Oberdörster, G., (2000) Toxicology of ultrafine particles: in vivo studies, *Philosophical Transactions of the Royal Society, London* 358:2719-2740.
[11] Kittelson, D. B.; Watts, W. F.; Johnson, J. P. Nanoparticle emissions on Minnesota highways. Atmos. Environ. 2004, 38, 9–19.
[12] Fruin, S.; Westerdahl, D.; Sax, T.; Sioutas, C.; Fine, P. M. Measurements and predictors of on-road ultrafine particle concentrations and associated pollutants in Los Angeles. Atmos. Environ. 2008, 42, 207–219.
[13] Zhu, Y. F.; Eiguren-Fernandez, A.; Hinds, W. C.; Miguel, A. H. In-cabin commuter exposure to ultrafine particles on Los Angeles freeways. Environ. Sci. Technol. 2007, 41, 2138–2145.
[14] Yao, X. H.; Lau, N. T.; Fang, M.; Chan, C. K.Onthe time-averaging of ultrafine particle number size spectra in vehicular plume. Atmos. Chem. Phys. 2006, 6, 4801–4807.
[15] Qi, C., Stanley, N., Pui, D.Y.H. & Kuehn, T. (2008) Laboratory and On-Road Evaluations of Cabin Air Filters Using Number and Surface Area Concentration Monitors, Environ. Sci. Technol. 2008, 42, 4128–4132.
[16] ISO. Road Vehicles-Air Filters for Passenger Compartments. Part 1: Test for Particulate Filtration; International Standards Organization: Geneva, 2001.
[17] Brunauer, S. PH Emmett, E Teller (1938) Adsorption of Gases in Multimolecular Layers. *Journal of the American Chemical Society*, 60, 309 – 319.
[18] Baltensperger, U; Gaeggeler, H W; Jost, D T (1988) The epiphaniometer, a new device for

continuous aerosol monitoring. *J. Aerosol Science*. 19(7), 931-934.
[19] Hinds W.C. (1999) Aerosol technology: properties, behavior, and measurement of airborne particles, 2nd ed. New York: Wiley-Interscience.
[20] Rogak S.N., R.C. Flagan & H.V. Nguyen (1993) The mobility and structure of aerosol agglomerates, *Aerosol Science & Technology*, 18:25-47.
[21] Kaufman S.L., A. Medved, A. Pöcher, N. Hill, R. Caldow & F.R. Quant (2002) An electrical aerosol detector based on the corona-jet charger, AAAR conference (poster).
[22] Woo, K., D. Chen, D.Y.H. Pui and W.E. Wilson (2001) Use of continuous measurements of integral aerosol parameters to estimate particle surface area, *Aerosol Science and Technology* 34, 57–65.
[23] Wilson W.E., H.-S. Han, J. Stanek, J. Turner & D.Y.H. Pui, (2003) The Fuchs surface area measured by charge acceptance of atmospheric particles may be a useful indicator of the quantity of particle surface area deposited in the lung. Abstracts of the European aerosol conference. S421-S422. Madrid, Spain.
[24] Wilson W.E., H.-S. Han, J. Stanek, J. Turner, D.-R. Chen & D.Y.H. Pui, (2004) Use of electrical aerosol detector as an indicator for the total particle surface area deposited in the lung. Paper #37, Symp. On air quality measurement methods and technology sponsored by air and waste management association. Research triangle park, NC.
[25] Chang, J.-S. (1981) Theory of diffusion charging of arbitrarily shaped conductive aerosol particles by unipolar ions. Journal of aerosol science 12: 19-26.
[26] Brown, R. C. and Hemingway, M. A. (1995) Electric charge distribution and capacitance of agglomerates of spherical particles: theory and experimental simulation. Journal of aerosol science 26: 1197-1206.
[27] Lall AA, Friedlander SK (2006) On-line measurement of ultrafine aggregate surface area and volume distributions by electrical mobility analysis: I. Theoretical analysis. J. Aerosol Sci 37: 260-271.
[28] Wang, J., Asbach, C., Fissan, H., Hülser, T., Kuhlbusch, T.A.J., Thompson, D. and Pui, D.Y.H. (2010) How can nanobiotechnology oversight advance science and industry? Accepted, J. Nanoparticle Research.
[29] Wang, J. and Pui, D.Y.H. (2010) Dispersion, Measurement and Filtration of Carbon Nanotubes (CNTs) and Nanoparticle Agglomerates. Submitted, *Chemical Engineering Sciences*.
[30] Pui, D.Y.H., Qi, C., Stanley, N. Oberdörster, G. & Maynard, A. (2008) Recirculating Air Filtration Significantly Reduces Exposure to Airborne Nanoparticles, Environ Health Perspect 116:863–866, 2008.

Direct quantification of airborne nanoparticles composition by TXRF after collection on filters

S Motellier[1], K Lhaute[1], A Guiot[1], L Golanski[1], C Geoffroy[2] and F Tardif[1]

[1]CEA Grenoble, DRT, LITEN, DTNM, Laboratory of Nanochemistry and Nanosafety, 17 Avenue des Martyrs, Cedex 9, F-38054 Grenoble, France.

[2]Elexience, 9 rue des petits ruisseaux, BP 61, 91371 Verrières-le-Buisson Cedex, France.

E-mail : sylvie.motellier@cea.fr

Abstract. Direct TXRF analysis of nanoparticles deposited on filters was evaluated. Standard filters spiked with known amounts of NP were produced using an atomizer which generates an aerosol from a NP containing-liquid suspension. Polycarbonate filters provided the highest fluorescence signals and black polycarbonate filters containing chromium were further selected, Cr being used as internal standard for elemental quantification of the filter contaminants.
Calibration curves were established for various NP (TiO_2, ZnO, CeO_2, Al_2O_3). Good linearity was observed. Low limits of detection were in the tens to the hundreds of ngs per filter, the method being less adapted to Al_2O_3 due to the poor TXRF sensitivity for light elements. The analysis of MW-CNTs was attempted by quantification of their metal (Fe) catalyst impurities. Problems like CNT dispersion in liquids, quantification of the deposited quantity and high Fe-background contamination.

1. Introduction

Occupational exposure limits to aerosols do not presently take into account the size of airborne particles [1], although their potential toxicity may well depend on this characteristic, particularly when talking of nanoparticles (NP). Numbers of techniques (Scanning Mobility Particle Sizer (SMPS), Electrical Low Pressure Impactor (ELPI),...) based on dynamic light scattering and/or electric mobility measurements are now commercially available to characterise NP aerosols [2]. Although very useful when the size distribution of the NP population is to be instantaneously estimated, they are not suited for continuous follow-up of the exposure of workers handling specific engineered NPs; information on the chemical composition of the particles is required to discriminate them from the ambient background.

For such purpose, a two-step method can be proposed, which consists in first collecting the particles on filters, and then analysing the deposited particles by an elemental analytical technique. Several kinds of sampler devices are commercially available and have been tested for their efficiency [3]. They all showed satisfactory performances, although some loss of particles could be observed on the inner wall of non-treated polystyrene cassettes.

Various techniques can be proposed to achieve the second step-analytical challenge, among which Atomic Absorption Spectrometry (AAS) or Inductively Coupled Plasma-Mass Spectrometry (ICP-MS). These elemental techniques need time-consuming sample digestion steps prior to analysis that may contaminate the sample, particularly with omnipresent elements like aluminium and zinc [4]. Furthermore NP containing elements like refractory metals and metalloids may not be fully dissolved leading to poor recoveries.

More recently, Total reflection X-Ray Fluorescence (TXRF) has been applied to environmental analyses [5-13]. TXRF is a surface and near-surface multi-elemental spectro-chemical analytical technique [14] which makes it an interesting alternative particularly well-suited here, provided that the NPs deposited on the filters can be analysed without any extra sample-handling. However, most applications include pre-treatment of the filter, whether by a separate acid digestion [5,6,8,11] or by cold plasma ashing directly on the sample carrier [7,10]; as with other techniques that require sample preparation, contamination problems were reported [11] as well as loss of material [12]. An interesting alternative was proposed by some authors [9,13]: it consists in collecting the aerosol particles by impaction on a polished Si or quartz carrier that could subsequently be analysed by TXRF with no further treatment. However, impaction methods require that the carrier be coated with grease to avoid bounce-off effects and use of Si-based carriers may be too expensive for a wide use of this technique.

Recently, Bontempi et al. [12] have investigated the possibility of analysing the chemical composition of atmospheric aerosol particles by TXRF directly on the collection filter. This is an interesting alternative of characterization since it is not destructive and allows subsequent analysis of the filter by other techniques, typically scanning electron microscopy (SEM).They have proven the effectiveness of such simplified procedure that gives higher signal intensities compared with the usual procedure (liquid deposition of a few μl of the digested filter on a quartz glass sample carrier). Recommendations concerning the positioning of the filter in the TXRF chamber must be followed (it must be inserted in a very flat manner) and limitations concerning quantification were set due to the lack of internal standard.

In the present study, we propose a method of characterization and quantification of the chemical composition of particles sampled on a filter by direct TXRF measurement.

2. Experimental

The filters on which NP were deposited were from Millipore or Whatman. All metal oxide NPs were from commercial aqueous suspensions: TiO_2 (Mixture of rutile and anatase) was purchased from Aldrich; ZnO was from Evonik, Al_2O_3 (BA15, α-form) was from Baikowski and CeO_2 was from Rhodia. MW-CNT (Graphistrength C100 from Arkema) were crushed by the supplier and used as received. Their average diameter was 10-15 nm; the length was less than 10 μm. They contain impurities from catalyst residues: 2.3 % Al and 1.9 % Fe on a weight basis. Ultrapure (UP) water was used throughout of the study; it was obtained from a MilliQ system (Millipore).

The NP suspensions were prepared by weighing a known amount of commercial NP powder or concentrated suspension in a known volume of UP water. Dispersion was assured by ultrasound treatment (2 – 5 min) prior to each atomisation procedure. Concentrations of 1 mg/l or 10mg/l of metals were used for metal oxide NPs. Because of the low metal content of MW-CNT, suspensions with higher concentrations were prepared with these NPs (280 mg/l MW-CNT).

The system set up to produce standard filters containing known amounts of NP is given in figure 1. A constant output atomizer (TSI, model 3076) equipped with a diffusion dryer (TSI, model 3062) was used to produce aerosols from NP dispersed suspensions. The nano-aerosol was generated in a flow stream of purified air and then passed through a filter membrane embedded in a cassette sampler (25 mm polypropylene conductive black from SKC Omega, including a support pad). The air flow rate was measured by a thermal mass flowmeter (TSI, model 4000).

The generated aerosol was characterized with the help of a Scanning Mobility Particle Sizer (SMPS Grimm, series 5400): their particle size distribution was measured, based on electrical-mobility particle size classification. The total mass deposited on the filter was calculated as the mass-based size distribution given by the SMPS (knowing the density of the tested particles) times the flow rate times the duration of exposure.

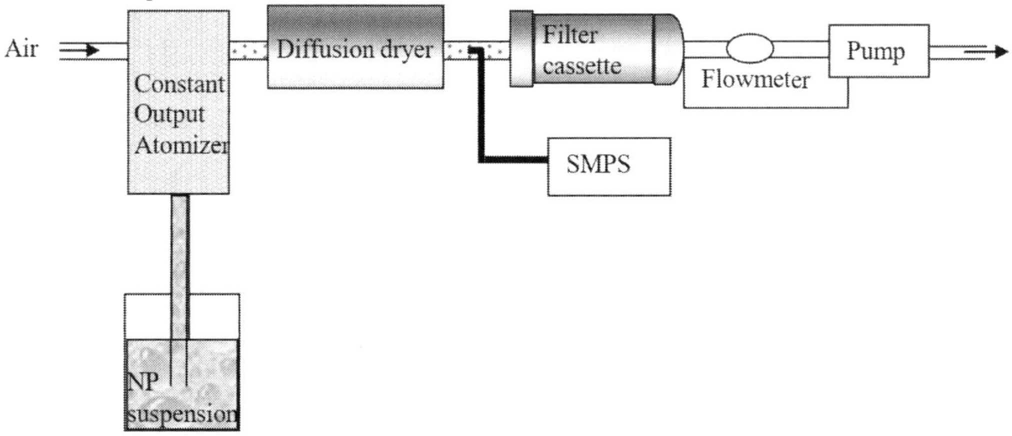

Figure 1. Experimental set-up for the preparation of standard filters.

The spiked filters were then withdrawn from their cassettes and analysed with a benchtop Total reflection X-Ray Fluorescence spectrometer (TXRF NanoHunter from Rigaku) featuring two different X-ray tubes (Cu and Mo). The sample spectra were collected for 200 s with an angle of the incident beam set at 0.75 °. A special carrier was designed in order to set the filter in a flat position beneath the detector so that no hindrance occurs and its surface be hit by the low glancing angle incident X-rays. No other specific handling was needed.

For the in-situ sampling procedure in an alumina manufactory, a portable sampling device was used to collect the ambient airborne particles generated during grinding and sieving operations. The air flow rate was 0.9 l/min.

3. Results and discussion

3.1. Homogeneity of the deposition

Personal devices used for exposure monitoring are traditionally based on a portable vacuum pump that draws air up through a filter embedded inside a special leak-free holder. Such design ensures that collection of the particles be homogeneous throughout the surface of the filter. Consequently, depositions made on purpose for our application should be as homogeneous as possible to mimic the usual procedure.

Two ways have been investigated. First, a liquid deposition of an aqueous NP suspension of known concentration was carried out on the filter. After drying, clogging of the particles would always occur, as illustrated for CNTs in figure 2.a. Second, an aerosol was generated from an aqueous NP suspension (with the help of a constant output atomizer) and sprayed through the filter.

a) 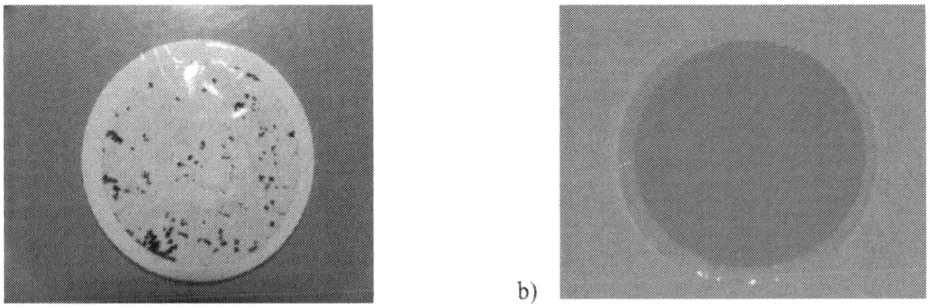 b)

Figure 2. Two MW-CNT deposition processes on filters. a: liquid deposition. b: aerosol deposition

With this configuration, homogeneous deposition was observed (figure 2.b), confirming that only the atomization process would be useful when dealing with direct measurement of the filter surface by TXRF.

3.2. Choice of the filter

The first condition the filter should meet is efficiency. Indeed, nearly all fibrous and sponge-like structure filters would meet this requirement as Brownian motion favours interception and diffusive deposition of the particles on the fibers. Polycarbonate (PC) membranes do not show the same structure (they are made of plain material with calibrated poreholes) and their efficiency was to be confirmed.

Penetration tests were conducted on various filters with different pore sizes. Figure 3 shows that, for all kinds of filters that were tested, the efficiency is higher than 99.5 %, whatever the size of the particles, limited to 300 nm in diameter.

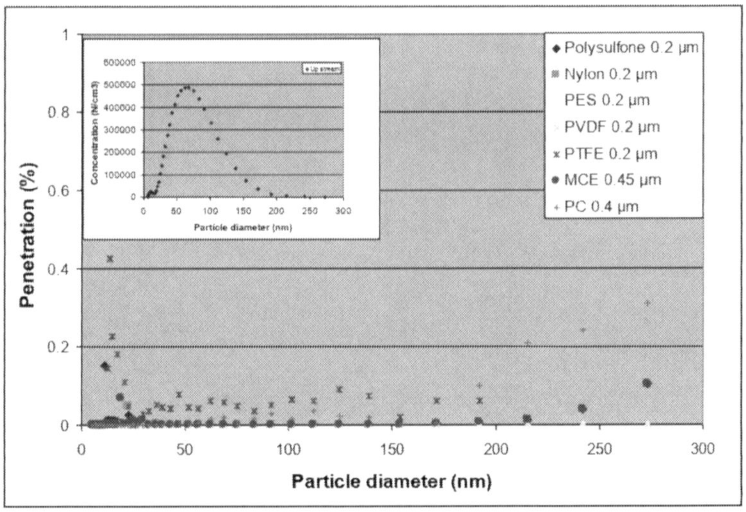

Figure 3. Efficiency of various filters towards NP retention. Porosity is indicated in the legend.

The second condition the filter should meet is reflectivity. Indeed, the smoother the surface of the substrate, the higher the sensitivity of the fluorescence signal. Various filters have been tested to check this property. Each of them was spiked with TiO_2 NP (14 µg/cm^2 of Ti) generated by the aerosol technique. The TXRF signal for Ti is reported in figure 4.

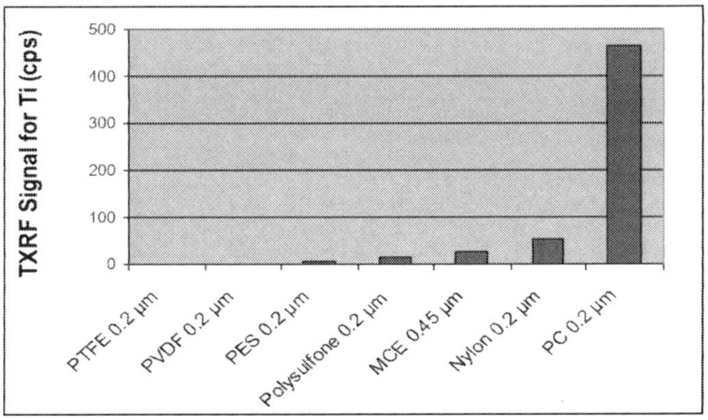

Figure 4. Ability of the filter material to be a sensitive TXRF substrate

The best sensitivity was obtained with polycarbonate filters (by at least one order of magnitude compared with the second best candidate). Fibrous or sponge-like materials like mixed cellulose ester, PVDF, PTFE or Nylon show a rough surface structure that allows the NP penetration deep into the filter thickness. On the contrary, polycarbonate filters show a smooth and flat surface with calibrated poreholes. Nanoparticles are deposited mainly on the reflective surface; they are hardly incorporated within the filter material, except for those that pass through the pore holes and are deposited on the pore surface. Both aspects (smooth surface and limited incorporation inside of the substrate) account for the high level of signal obtained with polycarbonate.

3.3. Calibration

Quantification of the deposition was first attempted with white PC filters, by plotting the TXRF signal of the element of interest (deposited as NPs) as a function of the quantity deposited on the filter. The signal was dependent on the deposited quantity of NP but the data were somewhat scattered. Black polycarbonate filters were also tested. Because of the occurrence of chromium in their PC matrix (figure 5), an attempt to use this constitutive element as internal standard was proposed.

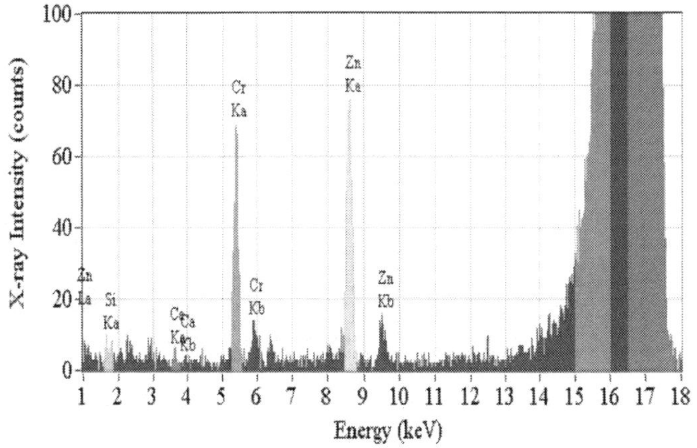

Figure 5. TXRF spectrum of a black PC filter spiked with ZnO NP (0.3 μg/cm² Zn).
Filter: black PC; porosity: 0.4 μm; diameter 25 mm.

Figure 6 gives an example of the calibration curves that were obtained either considering the gross signal for Zn (Figure 6.a) or including the Cr correction and plotting the signal ratio Zn/Cr (figure 6.b). Linearity is better in the second case, showing that Cr contained inside of the filter matrix can account for the differences in the positioning of the filter surface under the detector.

a) 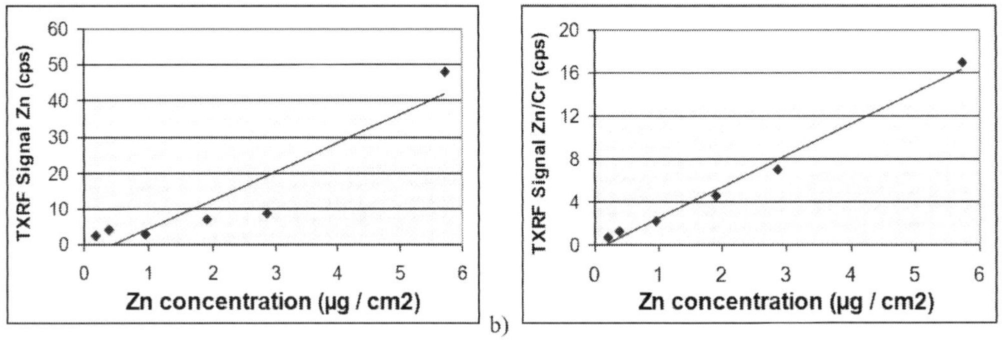

Figure 6. Calibration curves for Zn by direct TXRF measurements on PC filters spiked with ZnO NP.
a: without correction for Cr. b: with correction for Cr.
Filter: black PC; porosity: 0.4 μm; diameter 25 mm.

3.4. Quantification of NP / limit of detection

Standard filters were prepared by aerosol deposition of increasing amounts of various NPs. ZnO, TiO_2, CeO_2, Al_2O_3 and MW-CNT were tested.

Figure 7 gives the calibration curves for the metal oxide NPs. Ti, Ce, Si and Al were analysed with the Cu-X-ray tube while Zn was analysed with the Mo-X-Ray tube. In all cases, Cr was analysed with the same X-ray tube as that used for the element of interest.

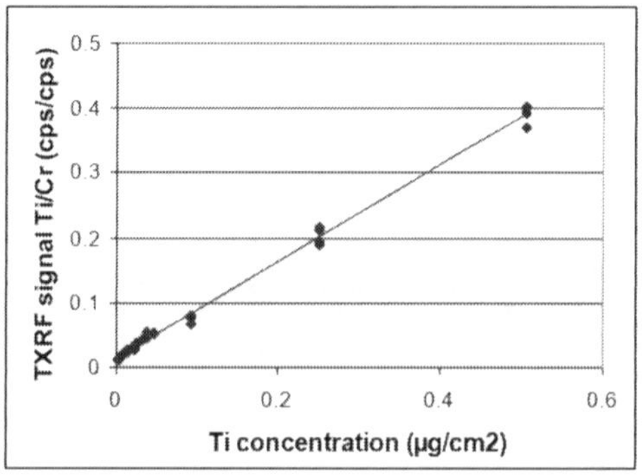

a

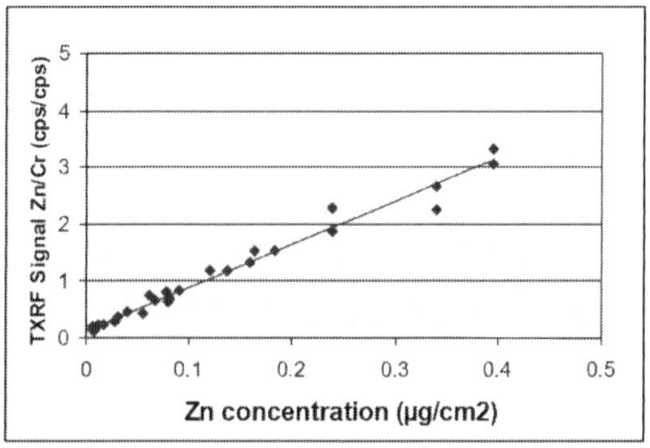

b

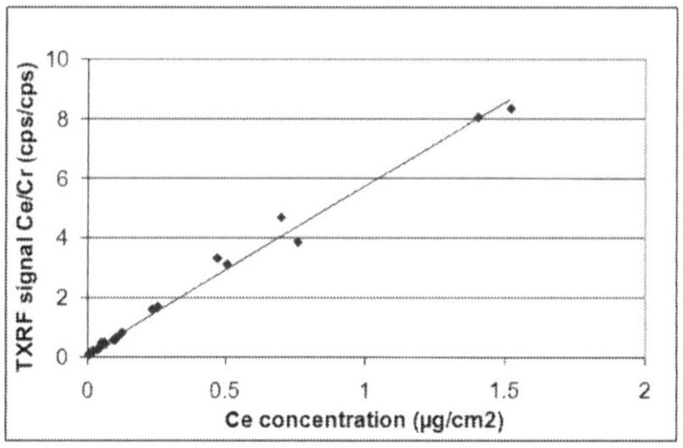

c

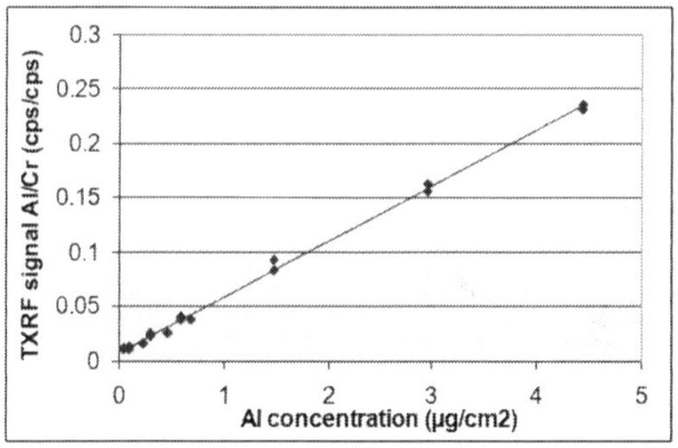

d

Figure 7. Calibration curves for a: TiO_2, b: ZnO, c: CeO_2 and d: Al_2O_3.
Filter: black PC; porosity: 0.4 µm; diameter 25 mm.

Good sensitivity was obtained for Ti, Zn, and Ce. Low limits of detection (LLD) for these elements were calculated as three times the standard deviation of the background signal of 6 different bare filters from the same batch. They are gathered in table 1.

Table 1. Low limits of detection of the method.

Particle	ng of element / cm^2	ng of particles / filter
TiO_2	5	38
ZnO	10	58
CeO_2	26	159
Al_2O_3	80	737

TXRF is not a sensitive technique for light elements; it is therefore not appropriate for trace analyses of Al, as revealed by the calibration curve in figure 7d. The LLD for this element is significantly higher than for the other metal oxides.

Atomisation of a CNT suspension was undertaken using the same procedure as that used with metal oxide NPs. The visual aspect of the suspension included agglomerates; the suspension was not steady and clogging of the atomizer head would occur after a few hours of continuous aerosol generation. Such instability has already been observed for CNTs [15] with this type of aerosol generation [16] and is essentially due to the high hydrophobic properties of these particles that favours their clustering.

A first series of CNT-spiked PC filters was analysed by TXRF. Carbon is too light to be analysed by TXRF. Hence, CNTs are spotted looking at Fe, the metallic catalyst. CNT concentration-dependent Fe signal was obtained (figure 8). However, due to the omnipresence of this element in the environment, the data are very scattered and the background level is high.

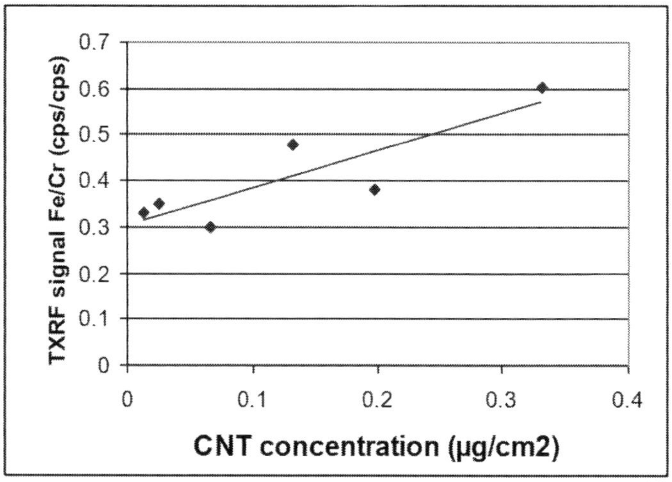

Figure 8. Calibration curve for Fe-containing CNTs

Furthermore, a second series was prepared and did not show concentration dependence calibration curve. It was suspected that segregation might occur by preferential sedimentation of the Fe-containing CNTs (of higher density) over those without catalyst. Further investigations need to be undertaken in order to assess the relevance of this elemental analytical method applied to such

4. Limitations

One of the trickiest points here is the preparation of standard filters. Particularly, evaluation of the deposited quantity is not obvious. At present, it is ensured by SMPS measurements. This technique is based on electrical mobility properties of the particles and gives access only to their equivalent diameter, considering that particles be spherical. However, for a number of NPs, this equivalent diameter is not representative of their real size. This is particularly true for specific morphologies like needles or tubes and models have been proposed to try to convert the electrical mobility size given by SMPS measurements to that of cylindrical particles [17]. It was found that the cylindrical particles have a significantly smaller volume than the spherical particles with similar mobility. Further investigations must thus be undertaken to evaluate the concentration of particles deposited on standard filters more precisely.

5. Conclusions

The method proposed here is an interesting alternative to conventional analytical procedures that require time-consuming and potentially contaminating sample treatments. The feasibility of the chemical detection of metallic particles directly on filters was demonstrated using TXRF and PC filters for good reflectivity. Semi-quantification was established for a number of nanoparticles (TiO_2,

ZnO, CeO_2) with low limits of detections in the tens to hundreds of particle ngs per filter, except for light elements like Al that shows poor sensitivity inherent to the method.
Evaluation of the deposited quantity, as measured by SMPS, still needs to be clarified, and particularly size response of such aerosol characterisation device need to be checked, with ICP-MS controls, for instance.

Acknowledgements
This work is part of the GENESIS project (Global Evaluation of Nanocomposites for Emerging Safe Industrial Solutions) financed by OSEO.

References

[1] Schulte P A, Murashov V, Zumvalde R, Kuempel E D and Geraci C L 2010 *J. Nanopart. Res.* **12** 1971-87
[2] Hagendorfer H, Lorenz C, Kaegi R, Sinnet B, Gehrig R V, Goetz N, Scheringer M, Ludwig C and Ulrich A 2010 *J. Nanopart. Res.* **12** 2481-94
[3] Görner P, Simon X, Wrobel R, Kauffer E and Witschger O 2010 *Ann. Occup. Hyg.* **54** 165-87
[4] Rasmussen P E, Wheeler A J, Hassan N M Filiatreault A and Lanouette M 2007 *Atmosph. Environ.* **41** 5897-5907
[5] Leland D J, Bilbrey D B, Leyden D E, Wobrauschek P, Aiginger H and Puxbaum H 1987 *Anal. Chem.* **59** 1911-14
[6] Klockenkämper R, Bayer H, von Bohlen A, Schmeling M and Klockow D 1995 *Anal. Sci.* **11** 495-98
[7] Theisen M and Niessner R 1999 *Fresenius J. Anal. Chem.* **365** 332-37
[8] Schmeling M 2001 *Spectrochim. Acta B* **56** 2127-36
[9] Esaka F, Watanabe K, Onodera T, Taguchi T, Magara M and Usuda S 2003 *Spectrochim. Acta Part B* **58** 2145-55
[10] Wagner A and Mages M 2010 *Spectrochim. Acta B* **65** 471-77
[11] Boman J, Wagner A and Gatari M J 2010 *Spectrochim. Acta B* **65** 478-82
[12] Bontempi E, Zacco A, Benedetti D, Borgese L, Colombi P, Stosnach H, Finzi G, Apostoli P, Buttini P and Depero L E 2010 *Environ. Technol.* **31** 467-77
[13] Wagner A, Boman J and Gatari M J 2008 *Spectrochim. Acta B* **63** 1426-31
[14] Klockenkämper R and von Bohlen A 2001 *Spectrochim. Acta B* **56** 2005-18
[15] Lee J, Kim M, Hong C K and Shim S E 2007 *Meas. Sci. Technol.* **18** 3707-12
[16] Lee S-B, Lee J-H and Bae G-N 2010 *J. Nanopart. Res.* **12** 501-12
[17] Moisala A, Nasibulin A G, Shandakov S D, Jiang H and Kauppinen E I 2005 *Carbon* **43** 2066-74

Journal of Physics: Conference Series **304** (2011) 012010 doi:10.1088/1742-6596/304/1/012010

Using the Aerasense NanoTracer for simultaneously obtaining several ultrafine particle exposure metrics

J Marra

Philips Research Laboratories, HTC 4, 5656 AE Eindhoven, The Netherlands

E-mail: johan.marra@philips.com

Abstract. The expanding production and use of nanomaterials increases the chance of human exposure to engineered nanoparticles (NP), also referred to as ultrafine particles (UFP; $\leq$ 100 – 300 nm). This is particularly true in workplaces where they can become airborne and thereafter inhaled by workers during nanopowder processing. Considering the suspected hazard of many engineered UFPs, the general recommendation is to take measures for minimizing personal exposure while monitoring the UFP pollution for assessment and control purposes. The portable Aerasense NanoTracer accomplishes this UFP monitoring, either intermittently or in real time. This paper reviews its design and operational characteristics and elaborates on a number of application extensions and constraints. The NanoTracer's output signals enable several UFP exposure metrics to be simultaneously inferred. These include the airborne UFP number concentration and the number-averaged particle size, serving as characteristics of the pertaining UFP pollution. When non-hygroscopic particles are involved, the NanoTracer's output signals also allow an estimation of the lung-deposited UFP surface area concentration and the lung-deposited UFP mass concentration. It is thereby possible to distinguish between UFP depositions in the alveolar region, the trachea-bronchial region and the head airway region, respectively, by making use of the ICRP particle deposition model.

1. Introduction

Ultrafine particles (UFPs) are widely used for manufacturing nanostructured materials and for improving certain characteristics of more conventional materials. In recent years, stakeholders in the nanotech industry have expressed concern about the safety aspects associated with the use of engineered UFPs [1]. They can become airborne during their manufacture or handling and may subsequently find their way into the body through inhalation or skin contact. Evidence exists that after UFP inhalation, they may enter from the lung into the blood stream, thereby allowing them to reach, penetrate and subsequently damage the cells of vital organs [2-5]. Toxicological studies indicated that, per unit particle mass, exposure to particles becomes relatively more hazardous at decreasing particle size [6]. This applies in particular to water-insoluble solid particles and for particles comprising hazardous or reactive adsorbates on their surface. It has therefore been proposed to involve a particle surface area concentration metric in case the severity of the UFP air pollution is to be assessed, thereby accounting for all particles up to a size limit which is at least 100 nm but possibly as high as 300 nm [6-8]. This may be done in addition to the more commonly used PM_{10} particle mass concentration metric for assessing ambient air pollution with respirable fine particles ($d_p \leq 10$ μm).

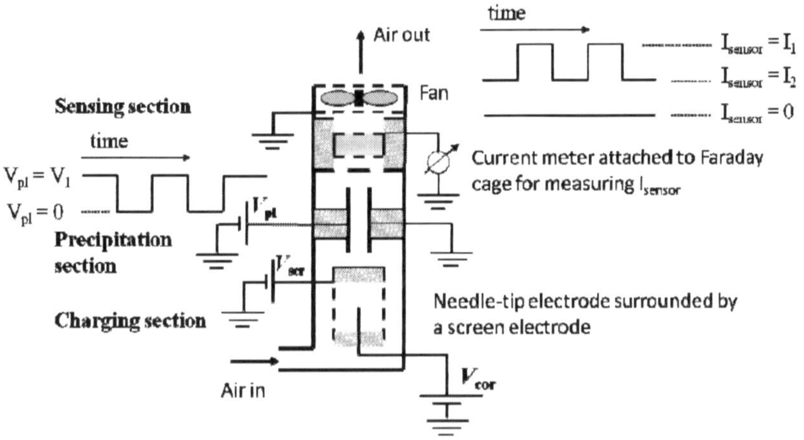

Figure 1. The Aerasense NanoTracer.

The intrinsic toxicity of engineered UFPs for humans is often difficult to assess and this complicates the setting of specific safety standards and regulations that involve one or more UFP concentration metrics [9-10]. The currently recommended safety policy is therefore a precautionary one aiming at exposure minimization in general and advocating in-situ control by monitoring both the workplace pollution level and the personal exposure to UFPs. This implies a need for robust and reliable UFP monitors that lend themselves for routine use by workers in the nano-industry. Preferably, they are portable and battery-powered while being able to relay relevant metrics in real time about the pertaining UFP pollution level. The Aerasense NanoTracer meets these demands and was described and benchmarked before in [11]. Ref. [12] discusses several of its applications. The present paper extends on [11] and presents additional features and conceptual constraints of the NanoTracer. To put them into perspective, a brief introduction about the NanoTracer is given first.

2. The NanoTracer's design and operation characteristics

Figure 1 shows the schematic design of the NanoTracer. It comprises a ventilator drawing a controlled airflow ϕ through the monitor. Particles in this airflow first enter the charging section, which comprises a porous screen electrode, set at V_{scr}, surrounding a high-voltage needle electrode that is set at a sufficiently high voltage V_{cor} to emit a constant corona current I_{cor} over time. The transmitted fraction of I_{cor} through the screen electrode induces diffusion charging of airborne particles through ion attachment. Charged particles subsequently enter the precipitation section wherein they are subjected to a block-shaped electric field pulse between two electrodes varying between $E_{pl} = V_1/d_{pl}$ and $E_{pl} = 0$. The electrodes have a travelling length L_{pl} and spacing d_{pl}. V_1 is chosen such that only a partial precipitation of charged particles with $d_p = 10$ - 15 nm is obtained. This implies that all larger particles will also only partially precipitate. The airflow carries all non-precipitated particles from the precipitation section into the sensing section where a fibrous filter captures them with their charge inside a Faraday cage. A sensitive current meter connects the Faraday cage to a reference potential and records the currents I_1 and I_2 that correspond with the application of $E_{pl} = 0$ and $E_{pl} = V_1/d_{pl}$, respectively, in the precipitation section. I_1 and I_2 are the sensor signals and equal the captured particle charge per unit time.

The difference $I_1 - I_2$ accounts for the precipitated amount of particle charge per unit time inside the precipitation section and is represented by

$$I_1 - I_2 = \int_{dp=0}^{\infty} \sum_{q=0}^{\infty} q e \phi f_q(N_{ion} t_r, d_p) \xi_q(d_p, E_{pl}) d(N(d_p)) \quad (1)$$

$\xi_q(d_p,E_{pl})$ is the precipitated fraction of particles with diameter d_p carrying "q" elementary charges according to

$$\xi_q(d_p,E_{pl}) = 1 \quad \text{when} \quad \frac{qeC_c(d_p)E_{pl}L_{pl}}{3\pi\eta_{air}d_p v_{air}d_{pl}} \geq 1 \qquad \text{(full precipitation)}$$

$$\xi_q(d_p,E_{pl}) = \frac{qeC_c(d_p)E_{pl}L_{pl}}{3\pi\eta_{air}d_p v_{air}d_{pl}} \qquad \text{otherwise} \qquad (2)$$

v_{air} is the average airspeed between the electrodes in the precipitation section, $N(d_p)$ the number concentration of particles with diameter d_p, η_{air} the viscosity of air, and "e" the elementary charge. $N_{ion}t_r$ denotes the product of the average ion concentration in the charging section and the particle residence time therein. $f_q(N_{ion}t_r,d_p)$ is the fraction of all particles of diameter d_p charged with "q" elementary charges and relates to the average particle charge $Q(d_p)$ according to

$$Q(d_p) = \sum_{q=0}^{\infty} qf_q(N_{ion}t_r,d_p) \qquad (3)$$

Numerical values for $f_q(N_{ion}t_r,d_p)$ are predicted by the Fuchs theory for particle diffusion charging [13].

When $E_{pl} = 0$, the signal I_1 is obtained which represents the total airborne particle charge as

$$I_1 = \int_{dp=0}^{\infty} \sum_{q=0}^{\infty} qe\phi f_q(N_{ion}t_r,d_p)d\big(N(d_p)\big) \qquad (4)$$

The NanoTracer's design and process parameters in combination with the measured ion current transmitted through the screen electrode were found to be consistent with a numerical value for the charging parameter $N_{ion}t_r = 0.5 - 0.7\times10^{13}$ ions.m^{-3}.s. According to the Fuchs theory, the latter value implies a substantially linear dependence of $Q(d_p)$ on d_p. This was experimentally confirmed [11].

3. UFP pollution and exposure assessment with the NanoTracer

For characterizing the UFP pollution level, it is desirable to have knowledge about the total UFP number concentration N and the size distribution $N(d_p)$. For exposure assessment, both the deposited UFP surface area concentration and the deposited UFP mass concentration can be relevant metrics.

3.1. Obtaining the UFP number concentration N and average particle size $d_{p,av}$

The proportionality $Q(d_p) \propto d_p$ leads to a direct proportionality between I_1 and the particle length concentration $L = \int_{dp=0}^{\infty} d_p d\big(N(d_p)\big) = Nd_{p,av}$, with $d_{p,av}$ the number-averaged particle size, because

$$I_1 = \int_{dp=0}^{\infty} \sum_{q=0}^{\infty} qe\phi f_q(N_{ion}t_r,d_p)dN(d_p) \propto \int_{dp=0}^{\infty} \sum_{q=0}^{\infty} qf_q(N_{ion}t_r,d_p)dN(d_p)$$

$$\propto \int_{dp=0}^{\infty} Q(d_p)dN(d_p) \propto \int_{dp=0}^{\infty} d_p dN(d_p) = Nd_{p,av} \qquad (5)$$

Another consequence (see ref. [11]) is that both N and $d_{p,av}$ can be inferred from I_1 and I_2 according to

$$N = S_N\big(I_1 - I_2\big) \qquad (6)$$

$$d_{p,av} = S_{dp}\frac{I_1}{I_1 - I_2} \qquad (7)$$

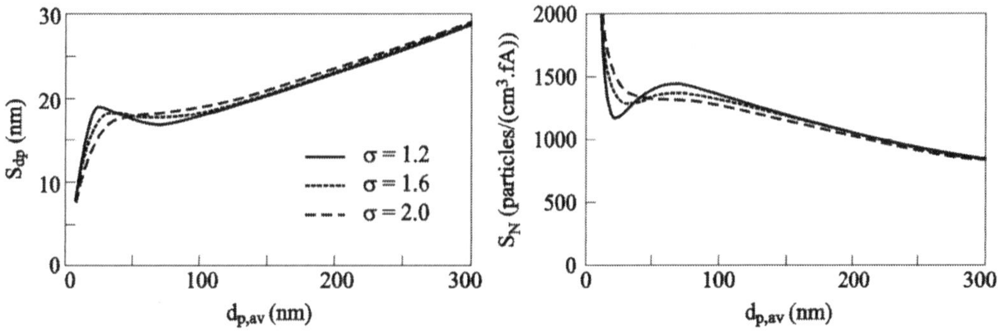

Figure 2. S_{dp} and S_N values according to the ICRP model for a mono-modal log-normal particle size distribution at $N_{ion}t_r = 0.6\times10^{13}$ ions.m^{-3}.s and $\phi = 0.45$ liter/min, as existing in the NanoTracer.

within the average particle size domain wherein the proportionality constants S_N and S_{dp} are substantially independent of $d_{p,av}$ and the particle size distribution characteristics. It was shown in [11] that this size domain exists in the 20 nm $\leq d_{p,av} \leq$ 100 nm region which includes particles with $d_p > 100$ nm. Knowledge of $d_{p,av}$ gives some insight in the UFP size distribution and, as will be shown below, is often sufficient for deriving appropriate UFP exposure metrics.

It is of interest to investigate the validity of equations (6) and (7) when $d_{p,av} > 100$ nm. This was done by substituting equations (1) and (4) into (6) and (7) at a given N as a function of $d_{p,av}$ and σ. σ is the standard deviation of the used log-normal particle size distribution $N(d_p)$ according to

$$dN(d_p) = \frac{N}{\sqrt{2\pi}\ln\sigma}\exp\left[-\frac{\left(\ln d_p - \ln d_{p,av} + 0.5\ln^2\sigma\right)^2}{2(\ln\sigma)^2}\right]d\ln(d_p) \quad (8)$$

Results are shown in Figure 2 for $d_{p,av} \leq 300$ nm at the existing ratio $\frac{E_{pl}L_{pl}}{v_{air}d_{pl}} = 1.927\times10^6$ V.s/m^2 inside the NanoTracer. They reveal that σ only marginally affects S_N and S_{dp}. At a commonly encountered value $\sigma = 1.6$, the outcome for S_N and S_{dp} in equations (6) and (7) is that

$$S_{dp} = S^*_{dp} \qquad S_N = S^*_N \qquad 20\text{ nm} \leq d_{p,av} \leq 100\text{ nm} \quad (9)$$

$$S_{dp} \approx S^*_{dp}\left(0.714 + 2.86\times10^{-3}d_{p,av}\right) \qquad S_N \approx \frac{S^*_N}{0.714 + 2.86\times10^{-3}d_{p,av}} \qquad d_{p,av} \geq 100\text{ nm} \quad (10)$$

At $\frac{E_{pl}L_{pl}}{v_{air}d_{pl}} = 1.927\times10^6$ V.s/m^2, one has $S^*_{dp} = 18$ nm and $S^*_N = 1340$ particles.cm^{-3}.fA^{-1}, almost independent of $d_{p,av}$. For $d_{p,av} > 100$ nm, the combination of equations (6), (7), (9) and (10) results in

$$d_{p,av} = \frac{0.714 S^*_{dp} I_1}{(I_1 - I_2) - 2.86\times10^{-3}S^*_{dp}I_1} \qquad N = \frac{S^*_N(I_1 - I_2)}{0.714 + 2.86\times10^{-3}d_{p,av}} \quad (11)$$

I_1 and I_2 in (11) and (12) must be expressed in "fA', yielding N in "particles/cm^3" and $d_{p,av}$ in "nm".

3.2. Obtaining the lung-deposited UFP surface area and mass concentrations

As mentioned above, evidence exists that when solid water-insoluble particles are involved, much of the health risk associated with UFP inhalation is probably associated with the actually deposited UFP surface area concentration in the lung. UFP deposition in the deep alveolar (AL) region of the lung

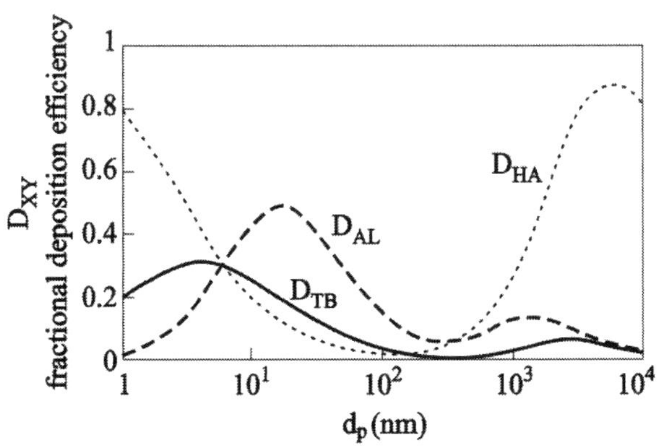

Figure 3. Fractional particle depositions D_{AL}, D_{TB} and D_{HA} of inhaled (spherical) particles in the AL, TB and HA regions of the respiratory tract, respectively, according to the ICRP model for a reference person.

where gas exchange with the blood occurs is particularly relevant, because they can persist there for extended periods and even find their way into the blood stream. UFP deposition in the head airways (HA) is not *a priori* harmless either, because of possible UFP transport from the nose via the olfactory nerve into the brain [5].

On the other hand, the deposited mass concentration remains a relevant exposure metric for water-soluble or liquid-phase UFPs. Many toxicologists still prefer the use of the mass concentration metric for any type of particle in combination with the average particle size. Hence, it would be convenient to be able to infer both the UFP surface area concentration and the UFP mass concentration that deposit in the lung from inhaled air. This might sometimes deal with the problem that the NanoTracer itself cannot recognise the chemical composition, shape and phase state of airborne particles.

The ICRP model (International Commission on Radiological Protection, Publication 66, 1994) for a reference person describes the fractional deposition of inhaled non-hygroscopic spherical particles in the head airway (HA) region, the trachea-bronchial (TB) region and the alveolar (AL) region of the respiratory tract. Analytical equations for the corresponding deposited fractions $D_{HA}(d_p)$, $D_{TB}(d_p)$ and $D_{AL}(d_p)$ are available from the literature [14] and depicted in Figure 3. They can be used to derive the deposited surface area concentrations S_{XY} (XY = HA, TB or AL) as

$$S_{XY}(d_{p,\max}) = \int_{dp=0}^{dp,\max} \pi d_p^2 D_{XY}(d_p) dN(d_p) \qquad (13)$$

for particles in the $0 \le d_p \le d_{p,max}$ size region. The deposited mass concentration M_{XY} is given by

$$M_{XY}(d_{p,\max}) = \int_{dp=0}^{dp,\max} \rho_p \frac{\pi d_p^3}{6} D_{XY}(d_p) dN(d_p) \qquad (14)$$

ρ_p is the particle density. Figure 4 present calculated results for $S_{XY}(d_{p,max})$ as a function of $d_{p,av}$ in the AL and TB lung compartments, respectively, for $d_{p,max} \rightarrow \infty$ and several mono-modal log-normal particle size distributions according to equation (13). Within the size interval 20 nm $\le d_{p,av} \le$ 100 nm, they can be estimated with the linear equations (see Figure 4)

$$S_{AL} = 4.7 \times 10^{-5} N d_{p,av} = 4.7 \times 10^{-5} S_N^* S_{dp}^* I_1 \qquad (15)$$

$$S_{TB} = 0.95 \times 10^{-5} N d_{p,av} = 0.95 \times 10^{-5} S_N^* S_{dp}^* I_1 \qquad (16)$$

Results for S_{HA} are given in Figure 5 for $d_{p,max} \rightarrow \infty$ and $d_{p,max}$ = 200 nm. It is evident that when only the deposition of UFPs is concerned up to d_p = 200 nm, the deposition results in the HA region can also be estimated with a linear equation according to

$$S_{HA} = 0.6 \times 10^{-5} N d_{p,av} = 0.6 \times 10^{-5} S_N^* S_{dp}^* I_1 \qquad (17)$$

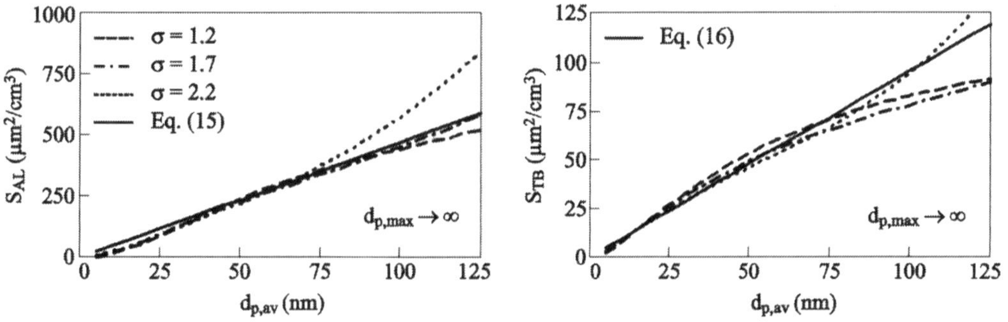

Figure 4. Surface area concentrations S_{XY} that deposit in the AL and TB lung regions, respectively, at $N = 10^5$ particles/cm^3 and $d_{p,max} \to \infty$ according to the ICRP model.

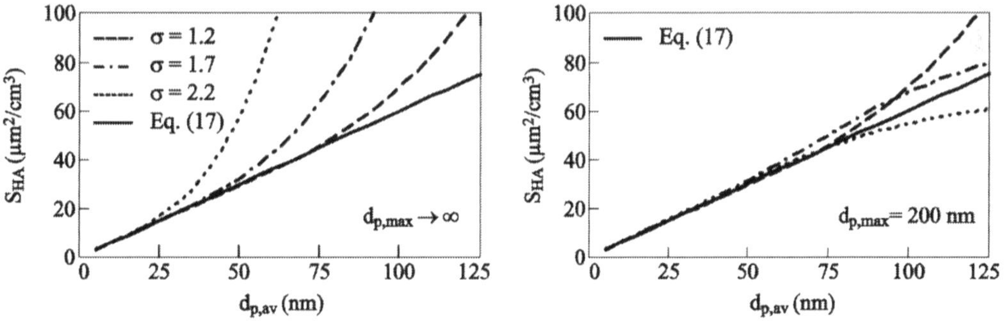

Figure 5. S_{HA} at $N = 10^5$ particles/cm^3 for $d_{p,max} \to \infty$ and $d_{p,max} = 200$ nm, respectively.

In equations (15) – (17), S_{XY} is obtained in "μm^2/cm^3" when I_1 is substituted in "fA", S^*_{dp} in "nm", S^*_N in "particles/(cm^3.fA)", N in "particles/cm^3" and $d_{p,av}$ in "nm". Deposited surface area concentrations are therefore in all cases approximately proportional to the output signal I_1. The numerical values of and ratios between the various S_{XY} values are comparable to those obtained in [15] with the Nanoparticle Surface Area Monitor (NSAM, TSI model 3550). One difference is that the S_{XY} results obtained for the NanoTracer are explicitly obtained in terms of N and $d_{p,av}$ whereas separate values for N and $d_{p,av}$ cannot be obtained with the NSAM. The NSAM operates according to a different particle charging and precipitation strategy.

Examples of calculated values $M_{XY}(d_{p,max})$ are presented in Figures 6 - 8. Here, σ plays a prominent role when σ > 1.7 at $d_{p,max} \to \infty$, particularly concerning the results for $M_{HA}(d_{p,max})$. This is due to the sharp upswing in $D_{HA}(d_p)$ at d_p > 300 nm (see Figure 3). Less influence from σ exists when $d_{p,max}$ is limited to 250 nm for $d_{p,av} \le 75$ nm in the AL and TB regions. The solid lines in Figures 6 - 8 obey the relationships:

for $d_{p,max} \to \infty$:

$$M_{AL} = N\rho_p \times 10^{-5}\left[-3.28\times10^{-7}d^3_{p,av} + 8.79\times10^{-4}d^2_{p,av} + 1.08\times10^{-3}d_{p,av}\right] \quad (18)$$

at $\sigma \le 1.7$; $d_{p,av} \le 100$ nm

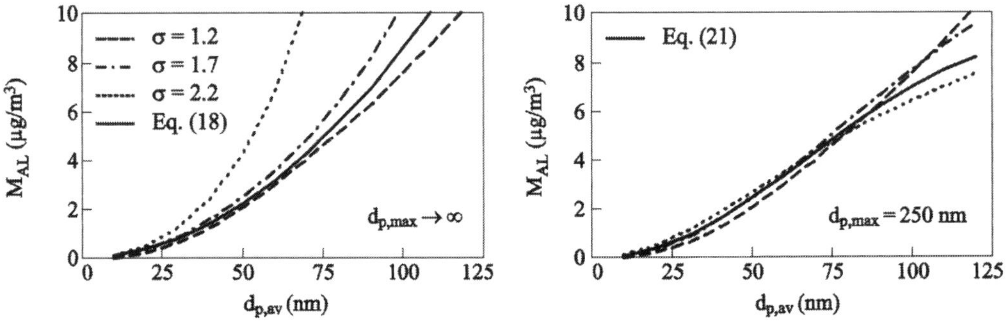

Figure 6. Deposited mass concentrations M_{AL} at N = 10^5 particles/cm^3 and ρ_p = 1 gram/cm^3 for $d_{p,max} \rightarrow \infty$ and $d_{p,max}$ = 250 nm, respectively, according to the ICRP model.

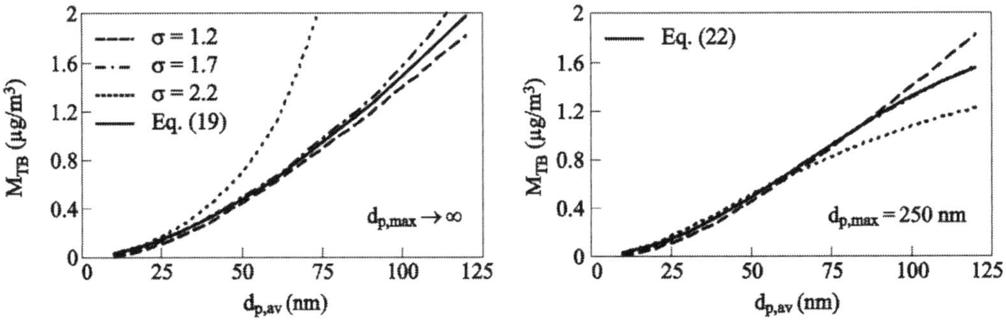

Figure 7. Deposited mass concentrations M_{TB} at N = 10^5 particles/cm^3 and ρ_p = 1 gram/cm^3 for $d_{p,max} \rightarrow \infty$ and $d_{p,max}$ = 250 nm, respectively, according to the ICRP model.

$$M_{TB} = N\rho_p \times 10^{-5}\left[-4.415\times10^{-7} d_{p,av}^3 + 1.74\times10^{-4} d_{p,av}^2 + 1.916\times10^{-3} d_{p,av}\right] \quad (19)$$

at $\sigma \leq 1.7$; $d_{p,av} \leq 120$ nm

$$M_{HA} = N\rho_p \times 10^{-5}\left[9.05\times10^{-7} d_{p,av}^3 - 2.63\times10^{-6} d_{p,av}^2 + 2.538\times10^{-3} d_{p,av}\right] \quad (20)$$

at $\sigma \leq 1.2$; $d_{p,av} \leq 120$ nm

for $d_{p,max}$ = 250 nm:

$$M_{AL} = N\rho_p \times 10^{-5}\left[-7.516\times10^{-6} d_{p,av}^3 + 1.547\times10^{-3} d_{p,av}^2 - 9.42\times10^{-3} d_{p,av}\right] \quad (21)$$

at $\sigma \leq 2.2$; $d_{p,av} \leq 100$ nm

$$M_{TB} = N\rho_p \times 10^{-5}\left[-1.144\times10^{-6} d_{p,av}^3 + 2.39\times10^{-4} d_{p,av}^2 + 7.107\times10^{-4} d_{p,av}\right] \quad (22)$$

at $\sigma \leq 2.2$; $d_{p,av} \leq 80$ nm

$$M_{HA} = N\rho_p \times 10^{-5}\left[9.05\times10^{-7} d_{p,av}^3 - 2.63\times10^{-6} d_{p,av}^2 + 2.538\times10^{-3} d_{p,av}\right] \quad (23)$$

at $\sigma \leq 1.2$; $d_{p,av} \leq 120$ nm

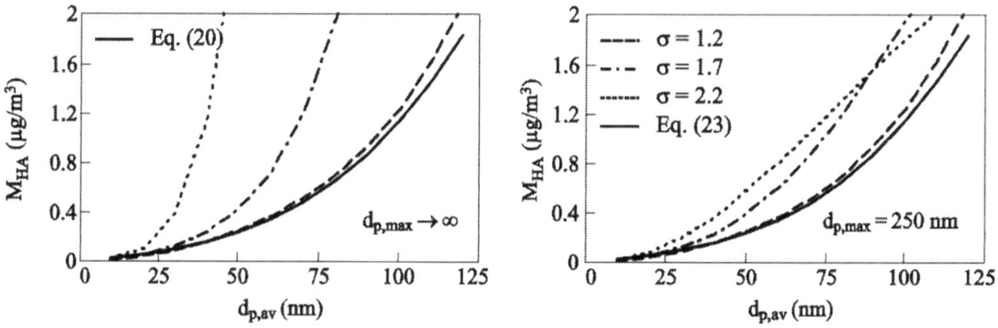

Figure 8. Deposited mass concentrations M_{HA} at $N = 10^5$ particles/cm^3 and $\rho_p = 1$ gram/cm^3 for $d_{p,max} \to \infty$ and $d_{p,max} = 250$ nm, respectively, according to the ICRP model.

In equations (18) – (23), $d_{p,av}$ must be expressed in "nm", N in "particles/cm^3" and ρ_p in "gram/cm^3", yielding M_{XY} in "µg/m^3". Equations (18) – (23) show that when certain conditions are met w.r.t. σ and $d_{p,av}$, it is possible to deduct estimates for M_{XY} from the inferred values for N and $d_{p,av}$ obtained from I_1 and I_2 according to equations (6) – (12). More precise predictions are obtained when the particle size distribution is known upfront. It must be noted though that the accuracy of any calculated exposure metric is inherently limited due to person-to-person variability in the functions $D_{XY}(d_p)$ shown in Figure 3.

4. The NanoTracer's sensitivity towards UFP and FP mass concentration changes

Other than for UFP monitoring in workplaces, the NanoTracer can also be used for more general environmental air pollution monitoring. Particle pollution in outdoor air is traditionally quantified in terms of the particle mass concentration PM_{10} of all airborne fine particles (FP) with $d_p \leq 10$ µm, which can potentially reach and deposit in the alveolar region (see Figure 3). These include UFPs with 10 nm $\leq d_p \leq$ 300 nm, but their mass is normally small in comparison with the mass of FPs with 300 nm $\leq d_p \leq$ 10 µm. It was argued before that the UFP mass concentration metric for inhaled solid UFPs is not likely to be the most suitable exposure metric for quantifying their exposure risk. It would therefore make sense to distinguish UFPs from FPs as two distinct classes of respirable particles, each being assessed with a different metric.

It is instructive to investigate the relative sensitivity of the NanoTracer signals to changes in the airborne particle mass concentration M at various values of $d_{p,av}$ in order to assess the NanoTracer's application constraints. M is obtained from

$$M = \int_{dp=0}^{\infty} \rho_p \frac{\pi d_p^3}{6} dN(d_p) \tag{24}$$

A sensitivity number $S_1(d_{p,av},\sigma)$ can be defined according to

$$S_1(d_{p,av},\sigma) = \left(\frac{d(I_1)}{dM}\right)_{d_{p,av},\sigma} \times 10^6 = \left(\frac{d(I_1)}{dN}\cdot\frac{dN}{dM}\right)_{d_{p,av},\sigma} \times 10^6 = \frac{I_1}{M}\times 10^6 \tag{25}$$

wherein a constant log-normal particle size distribution is assumed to exist (i.e. constant values for $d_{p,av}$ and σ) and wherein changes in M are therefore only due to changes in N. For monodisperse particles (σ = 1), the explicit result is

$$S_1(d_{p,av},\sigma) = S_1(d_p,1) = \frac{6Q(d_p)e\phi}{\rho_p \pi d_p^3}\times 10^6 \tag{26}$$

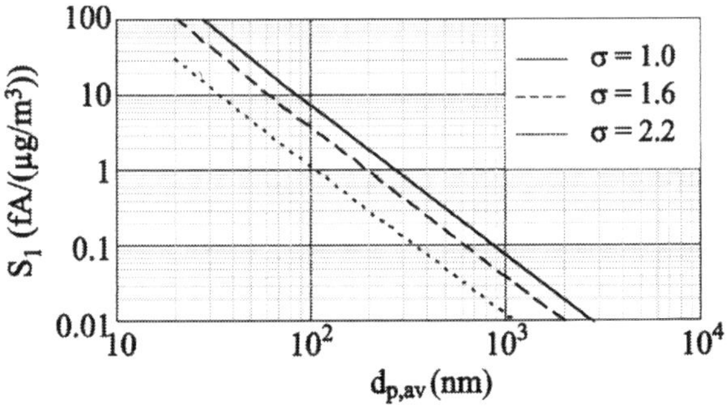

Figure 9. The NanoTracer sensitivity number S_1 as a function of $d_{p,av}$ and σ.

wherein all parameters must be substituted in SI units, yielding $S_1(d_{p,av},\sigma)$ in "fA/(µg.m^{-3})". The NanoTracer's specifications include $N_{ion}t_r \approx 0.6\times10^{13}$ ions.s.m^{-3} and $\phi = 0.45$ liter/min, leading to

$$Q(d_p) = 0.0326\times10^9 d_p \tag{27}$$

(d_p in "m").

The fine particle PM_{10} concentration in moderately polluted air is about 10 µg/m^3 and may increase to over 100 µg/m^3 in seriously polluted air. It appears reasonable therefore to evaluate $S_1(d_{p,av},\sigma)$ as the measured change ΔI_1 when $\Delta M = 1$ µg/m^3. Results for monodisperse and two heterodisperse particle distributions are shown in Figure 9. Figure 9 shows a markedly decreasing $S_1(d_{p,av},\sigma)$ at increasing $d_{p,av}$ and σ. A satisfactory measuring sensitivity towards FPs ($d_p > 300$ nm) exists when $S_1(d_p,1) \geq 1$ fA/(µg.m^{-3}), also in view of the NanoTracer's noise level $\Delta I_1 = \pm 1$ fA. Figure 9 indicates that the NanoTracer falls short of this demand at $d_{p,av} > 300$ nm, thereby highlighting the NanoTracer's primary function as a UFP monitor for particles with $d_{p,av} \leq 300$ nm. Note that $S_1(d_{p,av},\sigma) \propto \phi$. $S_1(d_{p,av},\sigma)$ can therefore in theory be increased up to any desired level by increasing ϕ albeit at the expense of a proportional increase in the equipment size. But I_1 will always remain more sensitive to UFP mass concentration changes than to FP mass concentration changes.

5. Application constraints of the NanoTracer

The presented equations for the deduction of N, $d_{p,av}$, S_{XY} and M_{XY} hold for single particles with a more-or-less spherical shape whose charging behavior follows the predictions of the Fuchs theory [13]. Several application examples are described in [12] concerning (N, $d_{p,av}$) measurements in the HVAC unit of an office building, in a workplace, a home, a driving automobile and outdoors. However, shape and composition of the measured particles remain unknown and the existence of a spherical particle shape is not guaranteed. Part of the UFPs and FPs in workplaces may be encountered as loosely structured aggregates composed of many primary particles [16]. The size distribution of aggregate mobility diameters d_m can be readily inferred with a SMPS but d_m cannot simply replace the geometric diameter d_p used by the Fuchs theory. The aggregate's charging behavior also depends on its fractal dimension and the primary particle size d_{prim} [16-17]. These are usually not known upfront. Interpreting sensor signals in the presence of mixtures of single particles and particle aggregates is therefore also problematic, although this depends on the desired accuracy. Within a restricted size range for $d_{p,av}$, the relationship between I_1 and N is foreseen [16] to have the general form

$$N = \frac{C}{d^a_{p,av}} I_1 \tag{28}$$

"a" and "C" are parameters that depend on d_{prim} and the fractal particle dimension. For single spherical particles, a = 1.0 and $C = S^*_N S^*_{dp}$. The proportionality $N \propto I_1$ holds at constant $d_{p,av}$, σ, and d_{prim}.

An informative example in this regard is given in Figure 10 where preliminary NanoTracer measurements on airborne fractal carbon aggregates are shown as a function of N and $d_{p,av}$. The carbon aggregates were produced by a Combustion Aerosol STandards generator (CAST) manufactured by

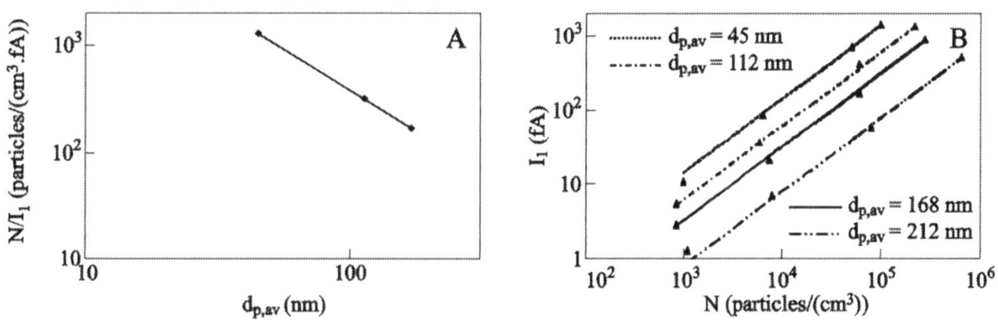

Figure 10. Measured ratio N/I_1 and signal I_1 for carbon aggregates produced by the CAST generator.

Jing Ltd., Switzerland. Data for N, σ and $d_{p,av}$ were obtained with a SMPS (Model 3936, TSI Inc.) and compared with the NanoTracer's I_1 data. The aggregates had a log-normal size distribution with $\sigma \approx 1.6$. At each average aggregate mobility diameter $d_{p,av}$, Figure 10B shows that the relationship $N \propto I_1$ indeed holds. The slope of the curve in Figure 10A is consistent with a = 1.5, thus equations (6) and (7) no longer apply. It is unknown to which extent possible changes in d_{prim} may have affected the slope of the curve in Figure 10A when changing $d_{p,av}$. The CAST product literature shows that the produced carbon aggregates have a pronounced fractal character, more so than that of typical soot aggregates encountered in diesel exhaust. The deviation of the parameter values "C" and "a" in equation (28) from the corresponding values predicted by the Fuchs theory for spherical particles are therefore likely to be larger for the CAST aggregates than for diesel soot aggregates, though this awaits experimental confirmation. Apart from that, the deduction of reasonably accurate (N, $d_{p,av}$) data with equations (6) and (7) is expected to remain possible when dealing with a defined source of particles (e.g. diesel exhaust) that emit aerosols of a more-or-less reproducible composition, size distribution and aggregate characteristics. The source-specific parameters S_N^*, S_{dp}^* can then be ad-hoc determined through calibration with other aerosol instruments, notably the CPC and/or the SMPS.

Another issue concerns the particle hygroscopicity. Inferred numbers for S_{XY} and M_{XY} in the presence of spherical particles can be relied upon only when non-hygroscopic particles are involved. Hygroscopic particles substantially increase in size when entering the high-humidity environment of the respiratory tract, which drastically changes their deposition behavior [18 – 19]. When inhaled particles do not comprise water-soluble (ionic) species, substantial hygroscopic growth inside the lung will not occur. This is true for many engineered UFPs.

The particle size d_p in Figure 3 refers to the aerodynamic particle diameter. For single particles, the aerodynamic diameter equals the electrical mobility diameter and, for particles with $d_p < 200$ nm, also equals the geometric particle diameter irrespective of ρ_p [15]. It is therefore usually justified to neglect the influence of ρ_p on S_{xy}. On the other hand, M_{xy} always remains proportional to ρ_p.

6. Conclusion

The Aerasense NanoTracer is a portable electrical UFP monitor device designed to serially generate two independent electrical signals I_1 and I_2 from diffusion-charged airborne particles. Under the assumption that the extent of particle charging reasonably agrees with the predictions of the Fuchs theory, these signals can be used to determine the N and $d_{p,av}$ of airborne particles in the 10 nm < d_p < 500 nm size region, thereby partly characterizing the existing airborne UFP pollution. Larger FPs are also measured but their contribution to the generated signals is usually only small. The interpretation of I_1 and I_2 in terms of N and $d_{p,av}$ becomes less accurate when the UFP pollution is substantially comprised of particles with a pronounced fractal aggregate structure. No evidence exists that this occurs in typical ambient or workplace environments but is can be relevant when the particle emission from specific sources must be characterized. When such sources emit aggregates with a reasonably constant size distribution and composition, reliable characterization with a NanoTracer remains possible after ad-hoc calibration with a CPC and/or SMPS. Use of a NanoTracer has specific advantages because of its easy use, portability, small size and weight, its capability of almost real-time measuring, and the absence of a radioactive source and consumable liquids.

When solid non-hygroscopic water-insoluble particles are present, as is the case with most engineered UFPs, the measurement of only I_1 is sufficient for estimating the deposited UFP surface area concentrations S_{XY} from inhaled air in the various lung compartments. This follows from the ICRP deposition model. S_{XY} is a relevant exposure metric considering that the toxicity of deposited UFPs per unit particle mass increases when their surface area concentration increases.

The deposited UFP mass concentrations M_{XY} in the various lung compartments can simultaneously be estimated from the separately determined values for N and $d_{p,av}$. Reasonable estimates for M_{HA} are only possible when the UFP size distribution parameter σ is known upfront. The mass concentration metric is relevant when the deposited material is soluble and/or in the liquid state.

Complete risk assessment not only involves exposure assessment but also knowledge about the toxicity of the inhaled UFPs and the total inhaled amount of material over a certain period. Furthermore, when the (suspected) toxicity of the UFPs released in air at the workplace differs from that of the ambient UFPs produced by other sources (e.g. from outdoor air pollution), it is necessary to somehow distinguish between these two kinds of UFPs before an assessment of exposure and risk concerning the workplace-generated NPs can be carried out. An off-line chemical speciation of sampled UFPs remains necessary to assess the chemical characteristics of the polluting particles and can help to distinguish workplace-released engineered particles from ambient particles.

References

[1] Balbus J M, Florini K, Denison R A and Walsh S A 2007 *J. Nanoparticle Res.* **9** 11
[2] Kreyling W G, Semmler M, Erbe F, Mayer P, Takenaka S, Schultz H, Oberdörster G and Ziesenis A 2002 *J. Toxicol. Environ. Health* **A65** 1513
[3] Oberdörster G 2000 *Philos. Trans. R. Soc. Lond.* **A358** 2719
[4] Oberdörster G, Gelein R M, Ferin J and Weis B 1995 *Inhal. Toxicol.* **7** 111
[5] Oberdörster E 2004 *Environ. Health Perspect.* **112** 1058
[6] Oberdörster G, Oberdörster E and Oberdörster J 2005 *Environ. Health Perspect.* **113** 823
[7] Donaldson K, Li X Y and MacNee W 1998 *J. Aerosol Sci.* **29** 553
[8] [Oberdörster G 1996 *Part. Sci. Technol.* **14** 135
[9] Howard J and Murashov V 2009 *J. Nanoparticle Res.* **11** 1673
[10] Murashov V, Engel S, Savolainen K, Fullam B, Lee M and Kearns P 2009 *J. Nanoparticle Res.* **11** 1587
[11] Marra J, Voetz M and Kiesling H J 2010 *J. Nanoparticle Res.* **12** 21
[12] Marra J, van den Brink W, Goossens H and Kessels S 2009 *IEEE Nanotech. Magazine* **3 (2)** 6
[13] Adachi M, Kousaka Y and Okuyama K 1985 *J. Aerosol Sci.* **16** 109
[14] Hinds W C 1999 *Aerosol technology: properties, behaviour and measurement of airborne particles* Chapter 11, 2nd Ed. (New York: Wiley-Interscience)
[15] Asbach C, Fissan H, Stahlmecke B, Kuhlbusch T A J and Pui D Y H 2009 *J. Nanoparticle Res.* **11** 101
[16] Wang J, Shin W G, Mertler M, Sachweh B, Fissan H and Pui D Y H 2010 *Aerosol Science and Technology* **44** 97
[17] Shin W G, Wang J, Mertler M, Sachweh B, Fissan H and Pui D Y H 2010 *J. Aerosol Sci.* **41** 975
[18] Löndahl J, Massling A, Pagels J, Swietlicki E, Vaclavik E and Loft S 2007 *Inhal. Toxicol.* **19** 109
[19] Asgharian B 2004 *Aerosol Science and Technology* **38** 938

A highly sensitive technique for detecting catalytically active nanoparticles against a background of general workplace aerosols

N Neubauer, F Weis, A Binder[1], M Seipenbusch and G Kasper

Institut für Mechanische Verfahrenstechnik und Mechanik, Karlsruhe Institute of Technology, Straße am Forum 8, 76131 Karlsruhe, Germany

[1] current address: BASF SE, GCP/TR-L549, 67056 Ludwigshafen, Germany

E-mail: nicole.neubauer@kit.edu

Abstract. A new measurement technique was studied using catalysis to specifically detect airborne nanoparticles in presence of background particles in the workplace air. Catalytically active nanoparticles produced by spark discharge were used as aerosol catalysts. According to these particles suitable catalytic test reactions were chosen and investigated by two different approaches: catalysis on airborne nanoparticles and catalysis on deposited nanoparticles. The results indicate that catalysis is applicable for the specific measurement of nanoparticles in the workplace air. Catalysis on airborne particles is suitable for the specific detection of very active nanoparticles, e.g. platinum or nickel, at high concentrations of about 10^7 #/cm^3. The approach of catalysis on deposited particles is better suited for nanoparticle aerosols at low concentrations, for slow catalytic reactions or less active nanoparticles like iron oxide (Fe_2O_3). On the basis of the experimental results detection limits in the range of μg or even ng were calculated which assure the good potential of catalysis for the specific detection of nanoparticles in the workplace air based on their catalytic activity.

1. Introduction

The inhalation of nanoparticles is considered to be the primary route of exposure to engineered nanoparticles (ENP) in the workplace [1]. However, a quantitative exposure assessment depends strongly on the exposure scenario. Airborne ENP in the environment as well as in workplace air are normally mixed with background particles. The concentration and the composition of these background particles depend on the location and the operations which are carried out in the workplace [2,3] and may be higher than the ENP concentration [4]. Aside from difficulties in isolating the ENP contribution to the total concentration, another uncertainty arises from the possible adhesion of ENP to the background aerosol particles due to coagulation [5], and the resulting disappearance of these particles from the size distribution.

A variety of devices is available to measure the particle size distribution, the number or mass concentration and the specific surface area of nanoparticles in workplace air [6,7]. However, these techniques are not generally capable of distinguishing between a target ENP and the background particles of the workplace [8-10]. Furthermore, the determination of particle properties such as the morphology or the chemical composition requires off-line techniques for physical or chemical analysis. New, material specific on-line analytical methods are therefore needed to identify ENP in background aerosols.

Catalysis offers significant potential for the substance-specific detection of catalytically active particles as well as for the discrimination of trace amounts of such airborne ENP against an inactive background aerosol. In [11] and [12] it is shown that metal nanoparticles like palladium and nickel supported on SiO_2 or TiO_2 are still catalytically active. In the case of palladium the supports have nearly no influence on the catalytic behaviour of the particles. So we assume that ENP can be detected via their catalytic activity even if they are coagulated with the background particles of the workplace. We therefore report preliminary results of an investigation into the suitability of such a technique for measuring the catalytic activity of engineered catalyst nanoparticles, both in their airborne state and after deposition from the aerosol onto a substrate. The experiments reported below were conducted with known amounts of freshly generated aerosol with a precisely known size distribution, which were exposed for defined amounts of time to gaseous educts; the gaseous reaction products were detected by infrared spectroscopy. On the basis of these measurements we give first estimates of the sensitivity of the technique for several materials and demonstrate that it is indeed possible to very rapidly detect very small amounts of nanoaerosols on the basis of their activity. In Section 2 we describe our choices of ENP material and the corresponding chemical model reactions. Section 3 contains the methods of catalyst particle preparation and the experimental setup for measuring their activity. Section 4 presents experimental results and calculated detection limits. Finally Section 5 concludes and points out future work.

2. Model Reactions

A well understood chemical reaction was selected for each of several industrially relevant catalyst substances according to the following criteria: Both educts and reaction products should be gaseous; the sensitivity and rate of the reaction should be as high as possible in order to achieve low detection limits.

2.1. Oxidation of H_2 over platinum: $2\ H_2 + O_2 \rightarrow 2\ H_2O$

The catalytic oxidation of hydrogen over platinum nanoparticles is very fast [13]. The reaction already takes place at ambient air temperature and is therefore at best suitable for the detection of platinum at workplaces. The catalytic activity of the platinum during the oxidation of H_2 is known to depend on the primary particle size of the platinum agglomerates [14].

2.2. Methanation of CO over nickel: $CO + 3\ H_2 \rightarrow CH_4 + H_2O$

The formation of methane from carbon monoxide and hydrogen over nickel nanoparticles represents a rather slow catalytic reaction [15] however reaching sufficient velocity at elevated temperature. The maximum catalytic activity was observed at 450°C [12]. Nickel catalysts have a high selectivity to the formation of methane so this model reaction shows good potential for the specific detection of nickel nanoparticles in the workplace air. However, the catalytic behaviour of nickel is also dependent on the primary particle size of the nickel agglomerates.

2.3. Oxidation of CO over iron oxide (Fe_2O_3): $2\ CO + O_2 \rightarrow 2\ CO_2$

The CO oxidation represents a catalytic reaction which is catalyzed by oxidic nanoparticles. It is a well researched reaction due to its importance for fuel cells, with an activity maximum at 300°C and a primary particle size of the iron oxide agglomerates of 3 nm [16].

2.4. Hydrogenation of ethene over palladium: $C_2H_4 + H_2 \rightarrow C_2H_6$

Palladium nanoparticles are frequently used in hydrogenation or dehydrogenation reactions of hydrocarbons. Especially the hydrogenation of ethene is an extensively studied catalytic reaction [11] which already runs at ambient air temperature. It was therefore chosen as a model reaction specific to the detection of palladium. As with the other materials, the catalytic activity of palladium is dependent on the primary particle size of the metal [17].

3. Experimental Procedures

Nano-aerosols of the above mentioned materials were produced by a spark discharge technique described frequently in the literature [18,19]. These aerosols had very well defined concentrations and particle morphologies. The catalytic activity of the nanoparticles was then measured directly in the aerosol stream by adding educt gases or – in the case of less active materials and slower reactions – by first depositing the particles onto filters. Details of these techniques are described below.

3.1. Catalyst preparation and characterization

Catalyst nanoparticles of nickel, platinum, palladium, and iron were produced in a spark discharge generator using high purity electrodes of the desired material. The carrier gas was nitrogen with a purity of 99.99% or a 4:1 mixture of Ar and O_2 with a purity of 99.5%, each at a flow rate of 1.0 liter/min. The addition of O_2 oxidizes the iron nanoparticles to Fe_2O_3. All flow rates were controlled by mass flow controllers (MKS, Brooks).

In the spark discharge, relatively uniform primary particles, usually in the size range of a few nanometers, are formed by homogeneous nucleation of small amounts of metal vapour ejected from the electrodes during the discharge. These primary particles rapidly undergo collisions to form agglomerates which can additionally be sintered in the aerosol state in a second process step, if so desired, to modify their structure in defined ways.

The catalyst aerosols were characterized on-line using an electrical mobility spectrometer (SMPS, TSI 3071A and 3775) to provide information about the mobility size distribution and concentration of the agglomerates. The diameters of the primary particles as well as the shape of the agglomerates were determined by transmission electron microscopy (TEM, Philips CM12, 120 kV). The surface area was measured by nitrogen BET (Quantachrome Nova 1200). (Note that the morphological characterization of the agglomerates does not take any size changes occurring at elevated temperatures into account. This issue is well understood, but not addressed in this paper.)

3.2. Catalysis on airborne nanoparticles

Catalysis on airborne nanoparticles - also known as "aerosol catalysis" [14,15,20] - offers the possibility to determine the catalytic activity of nanoparticles on very short time scales in the range of seconds to minutes. It is therefore a true on-line technique for measurements of the catalytic activity of ENP. In addition the method requires only small catalyst mass concentrations in the order of ng/cm^3. Another advantage is the avoidance of artefacts due to contamination occurring more commonly in off-line methods.

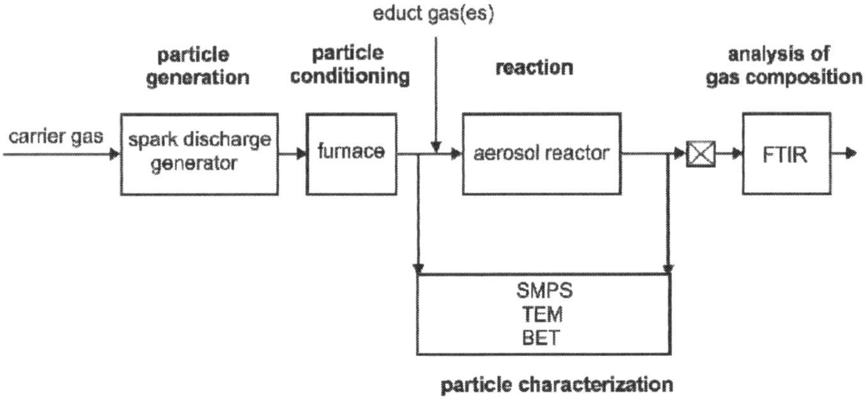

Figure 1. Schematic diagram of the experimental setup for measuring the catalytic activity and structure of airborne nanoparticles.

Figure 1 shows a schematic diagram of the experimental setup used for catalysis on airborne nanoparticles and for structural analysis. Following its generation in the spark discharge generator and any structural modifications, the educt gases are added to the aerosol before entering a heated glass reactor where the catalytic reaction is initiated. The concentration of gas phase products generated from the reaction is monitored online with a FTIR Spectrometer (Bruker Vector 22) equipped with a cell of suitable path length depending on their concentration. A particle filter just upstream of the cells protects the mirrors from particle deposition. It was determined that this filter and any particles deposited on it over time have no influence on the gas composition.

3.3. Catalysis on deposited nanoparticles

By sampling the nanoaerosol on a filter prior to running the catalytic reaction, it is possible to accumulate small amounts of catalyst material (in the order of micrograms). This approach is better suited for ENP aerosols at low concentrations, for slow catalytic reactions and/or less active ENP. The catalytic activity is nevertheless determined on-line, in the sense that educt gases are added directly to the filter, thereby avoiding filter handling and potential contaminations of the sample by exposure to ambient air.

A schematic diagram of the experimental setup for catalysis on deposited nanoparticles is shown in Figure 2. Catalyst aerosols are produced by spark discharge as before and then sampled onto a 37 mm diameter glass fiber filter (Machery-Nagel) placed downstream of the spark discharge generator. After sampling a defined amount of aerosol, the carrier gas is switched off and pre-mixed educt gases are added. If necessary the filter housing is heated to initiate the catalytic reaction. The deposited mass of ENP material is determined gravimetrically. The gas phase products of the catalytic reaction are also detected by FTIR. As before, the nanoaerosols are characterized on-line by SMPS, and off-line by TEM and BET for concentration, size and particle morphology.

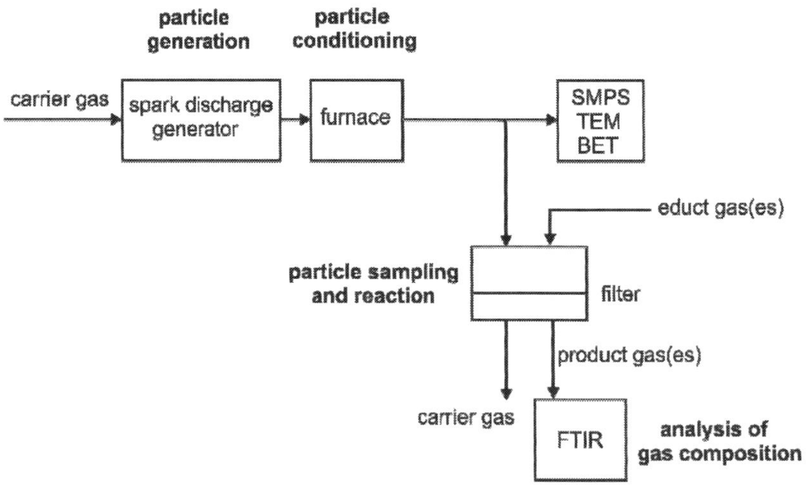

Figure 2. Schematic diagram of the experimental setup for measuring the catalytic activity and structure of deposited nanoparticles.

4. Results and Discussion

The results of the catalyst preparation and the catalytic experiments on airborne as well as on deposited nanoparticles are presented in the following. The turn-over frequency (TOF) was used as a measure for the catalytic activity of particles. Where possible, it was calculated as a function of the number of product molecules produced per time and per number of active sites. (Note that the active sites per unit surface area were not experimentally investigated values but literature data). At the end of the chapter a summary of calculated detection limits for each catalyst is presented.

4.1. Catalysis on airborne nanoparticles

4.1.1. Oxidation of H_2 over platinum. The platinum agglomerates (Figure 3) produced by spark discharge had primary particle diameters in the range of 5 nm and a BET surface area of 58 m^2/g. The particle number concentration in the aerosol immediately downstream of the spark discharge generator was very stable in the range of 10^7 agglomerates per cm^3; the corresponding mass concentration $c_{m,Pt}$ was determined gravimetrically to 56 ng/cm^3.

To initiate the oxidation of H_2 over the platinum nanoparticles, 5 vol-% H_2 (99.9%) and 2 vol-% O_2 (99.5%) were added to the platinum aerosol at 25°C. The reaction product H_2O was detected by FTIR in a concentration of 665 ppm, giving a calculated TOF of 5.36 1/s. Assuming a lower detection limit of 10 ppm H_2O for the FTIR and average particle parameters in the same range as the one used here, one can derive a lower detectable mass concentration for platinum of 1 ng/cm^3.

4.1.2. Methanation of CO over nickel. After the production of the nickel agglomerates (Figure 3) in the spark discharge generator, the primary particle size was determined to 5 nm and the BET surface of the nickel agglomerates was 132 m^2/g. The mass concentration $c_{m,Ni}$ was identified gravimetrically to 30 ng/cm^3 at a stable number concentration in the range of 10^7 nickel agglomerates per cm^3.

To initiate the reaction, 5 vol-% CO (99.5%) and 25 vol-% H_2 (99.9%) were added to the nickel aerosol. The glass reactor was heated to a reaction temperature of 450°C. The concentration of the reaction product CH_4 measured by FTIR was 0.25 ppm. The TOF under these conditions was calculated to be 1.34 1/s. Thus, the CO methanation over nickel is a slower catalytic reaction than the oxidation of H_2 over platinum. Experiments with other primary particle sizes resulted in a similar production rate of CH_4 in the lower ppm range. These results show that nickel nanoparticles can be detected in the air by direct aerosol catalysis, however only at relatively high concentrations.

4.1.3. Oxidation of CO over iron oxide (Fe_2O_3). The iron oxide aerosol produced by spark discharge consisted of agglomerates (Figure 3) with a primary particle diameter of 3 nm and a BET surface area of 160 m^2/g. The aerosol mass concentration $c_{m,Fe2O3}$ immediately after generation was determined to 64 ng/cm^3; the number concentration of the Fe_2O_3 agglomerates was approximately 10^7 #/cm^3.

To initiate catalysis, 2.8 vol-% CO (99.5%) were added to the carrier gas consisting of a 4:1 mixture of Ar and O_2 (99.5%, 1.0 liter/min). The reaction temperature was varied between 100°C and 400°C. These reaction conditions resulted in no reliable detection of the anticipated reaction product CO_2. Iron oxide does not seem to have sufficient catalytic activity for its detection by aerosol catalysis.

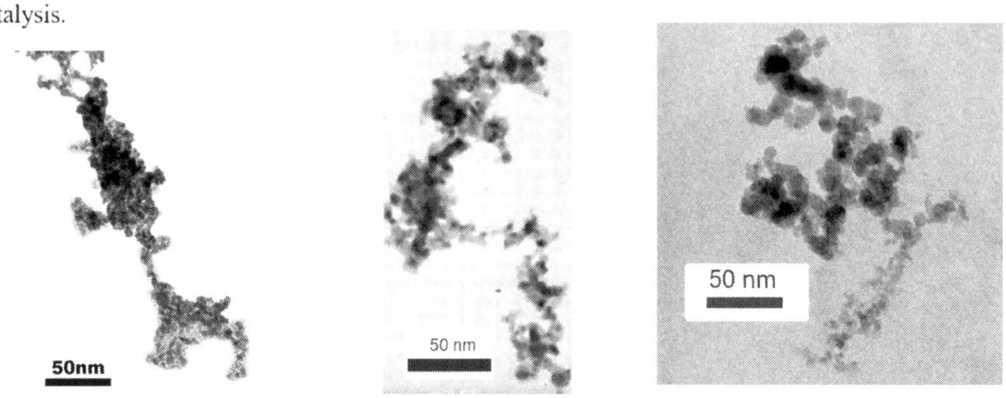

Figure 3. TEM images of different ENP produced by spark discharge, from left to right: platinum, nickel, iron oxide (Fe_2O_3).

4.2. Catalysis on deposited nanoparticles

Experiments were performed with those nanoparticles where the direct catalysis on gasborne particles gave insufficient signals, using the same aerosols and reaction conditions as described before in Chapter 4.1.

4.2.1. *Methanation of CO over nickel.* Nickel nanoparticles were sampled on a filter for a fixed time interval of 1 to 30 minutes at a mass concentration of 98 ng/cm^3. The filter afterwards was heated rapidly to 450°C; then 5 vol-% CO (99.5%) and 25 vol-% H_2 (99.9%) were added to initiate the catalytic reaction. The reaction product CH_4 was detectable by FTIR. Its concentration was related to the mass of the accumulated nickel nanoparticles which was determined by gravimetry. Figure 4 shows the resulting calibration curve for the concentration of CH_4 vs. the particle mass. At this early stage, each data point represents a single measurement. Nevertheless, it is evident that the concentration of the reaction product CH_4 increases proportionally to the mass of the nickel nanoparticles. The lower detection limit still remains to be investigated, but can be calculated on the basis of the curve linearity. Currently the lowest studied nickel mass of 220 μg results in 1450 ppm CH_4. Assuming a lower detection limit of 10 ppm CH_4 by FTIR a detection limit of 2 μg nickel nanoparticles can be calculated. This indicates that deposited nickel nanoparticles in the range of a few μg can be detected by this technique, making it a very fast and near-real time method.

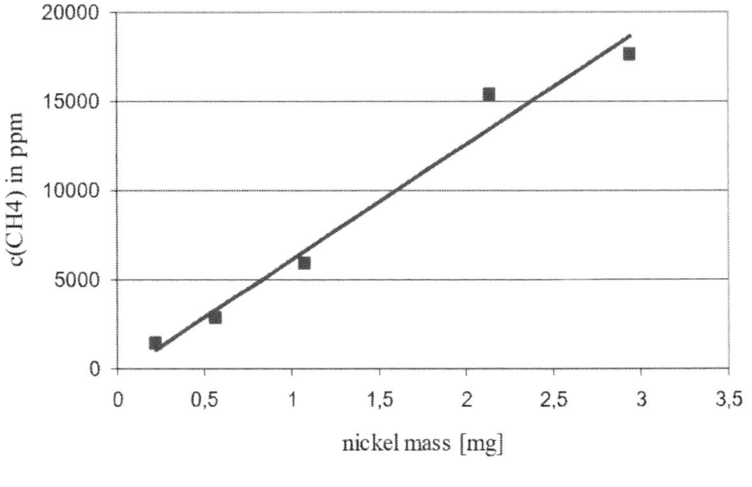

Figure 4. Calibration curve for the concentration of CH_4 against the mass of the nickel nanoparticles during the CO-methanation at 450°C.

4.2.2. *Oxidation of CO over iron oxide (Fe_2O_3).* A sampling time of 10 hours at ambient air temperature and a mass concentration of about 60 ng/cm^3 led to 38.8 mg of iron oxide nanoparticles on the filter. The addition of a 4:1 mixture of Ar and O_2 (99.5%, 1.0 liter/min) and 2.8 vol-% CO (99.5%) gases at a temperature of 250°C resulted in 14000 ppm of CO_2. A calibration curve similar to Figure 4 as well as the lower detection limit still need to be established. Nevertheless, iron oxide can be detected by the method of catalysis on deposited nanoparticles. Assuming a lower detection limit of 10 ppm CO_2 by FTIR and curve linearity, this hints that even iron oxide nanoparticles with a mass of 28 μg can be detected via this technique.

4.2.3. *Hydrogenation of ethene over palladium.* The palladium agglomerates (Figure 5) produced by spark discharge and sintered at 200°C had primary particle diameters in the range of 3 nm. The particle number concentration in the aerosol immediately after the sintering process was in the range of 10^7 agglomerates per cm^3; the corresponding mass concentration $c_{m,Pd}$ was determined gravimetrically to 1.1 ng/cm^3. Palladium nanoparticles were sampled on a filter for 10 minutes resulting in a mass of 11 µg.

To initiate the hydrogenation of ethene, 2 vol-% C_2H_4 (99.9%), 2 vol-% H_2 (99.9%) and 0.05 liter/min N_2 (99.99%) for dilution were added at 20°C. The reaction product ethane was detected at a concentration of 4020 ppm. Thus, palladium can be detected easily by this method. The detection limit was not experimentally investigated so far. Calculations result in a preliminary lower detection limit of 0.03 µg palladium nanoparticles.

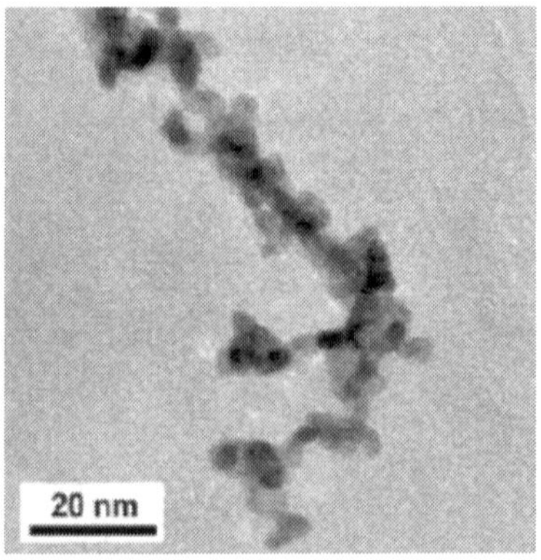

Figure 5. TEM image of a platinum agglomerate after sintering at 200°C.

4.3. Summary of preliminary detection limits

The calculated detection limits for palladium, nickel and iron oxide on the basis of the experimental results of deposited particles are summarized in Table 1. A lower detection limit of 10 ppm for the reaction products by FTIR was assumed. The limit of detection is of course dependent on the device used. For platinum the results of the direct measurement on airborne particles were transferred into a corresponding mass of deposited platinum nanoparticles of 157 µg for 665 ppm of the reaction product H_2O. Assuming a detection limit of 10 ppm by FTIR and curve linearity, this results in a lower detection limit of 2 µg deposited platinum particles.

Table 1. Calculated detection limits for the selected model ENP platinum, nickel, palladium and iron oxide (Fe_2O_3).

Model reaction	Model catalyst	Detection limit	Sampling time at 1 mg/m^3
Hydrogenation of C_2H_4	Pd	0.03 µg	< 3 s
Oxidation of H_2	Pt	2 µg	< 3 min
Methanation of CO	Ni	2 µg	< 2 min
Oxidation of CO	Fe_2O_3	28 µg	< 30 min

The detection limit for palladium by our method is 2 orders of magnitude lower than that for platinum or nickel which illustrates the known extreme catalytic activity of palladium. The sampling time therefore is in the order of seconds at a concentration of 1 mg/m^3. Thus, the described method can be considered a true real-time method for palladium, and a near-real-time method for catalysts with similar activities as nickel or platinum. Iron oxide reacts somewhat more slowly, but then one has to keep in mind that nano-sized iron oxide aerosol particles are not a primary target of concern for ENP, given that it is a very common pollutant in welding and metal treatment fumes, with often unregulated concentrations well above the mg range.

5. Summary and Conclusions

A new, substance specific technique for the detection of airborne nanoparticles via their catalytic activity is presented. The method consists of two different experimental approaches. In one case the catalytic activity is determined directly on the airborne particles; in the other, the aerosol is first deposited onto a filter.

In each case, the detection of a specific particle material is based on a specific catalyzed chemical reaction. Four such reactions are investigated here for the materials platinum, nickel, iron oxide (Fe_2O_3) and palladium. Direct measurements on airborne particles have shown so far that this on-line approach is suitable for very active ENP aerosols in high concentrations, such as platinum or nickel. The detection of platinum and palladium is even possible with a room temperature reaction, whereas the other catalysts require temperatures between about 250°C and 450°C. The filter based technique shows a good potential for the detection of less concentrated or less active ENP aerosols based on their catalytic activity. In addition, the filter method can be considered a true real-time method for the detection of palladium and a near-real-time technique for catalysts with similar activities as platinum or nickel.

Tentative detection limits are given on the basis of our preliminary experimental results and assuming a certain sensitivity of the gas analyzers used to measure the concentration of reaction products. It appears possible to detect concentrations in the range of 1 ng/cm^3 on a time scale of a few seconds for palladium, a time scale of a few minutes for platinum or nickel, and a time scale of about 30 min for iron oxide. An inherent advantage of the method is that its sensitivity increases in proportion to level concern for a given catalyst material.

Future investigations will be directed toward a more accurate definition of detection limits as a function of primary particle size and support material including typical background aerosol particles, as well as a refinement and down-sizing of the test equipment toward a mobile or portable device.

Acknowledgement

The research leading to these results has received funding from the European Commission's Seventh Framework Programme (FP7/2007-2013) under grant agreement n° 211464-2 (Nanodevice).

References

[1] Evans DE, Ku BK, Birch ME and Dunn KH "Aerosol characterization during carbon nanofiber production: Mobile direct-reading sampling" *Ann. Occup. Hyg.* **54**(5), 514-531, 2010

[2] Wake D, Mark D and Northage C "Ultrafine aerosols in the workplace". *Ann. Occup. Hyg.* **46** (1), 235-238, 2002

[3] Berges M, Möhlmann C, Swennen B, Van Rompaey Y and Berghmans P (2007). "Workplace exposure characterisation at TiO2 nanoparticle production" *3rd International Symposium on Nanotechnology, Occupational and Environmental Health, Taipei*, 183-184, 2007

[4] Muir R, Gorbunov B, Bello D, Gnewuch H, LeCloux A, Luizi F, Del Tedesco M and Green D "Chemically selective technique for health risk evaluation of exposure to airborne engineered nanoparticles (ENP)" *Conference on Workplace Aerosols, Karlsruhe*, 2010

[5] Seipenbusch M, Binder A and Kasper G "Temporal Evolution of Nanoparticle Aerosols in Workplace Exposure" *Ann. Occup. Hyg.* **52**(8), 707-716, 2008

[6] Brouwer DH, Gijsbers JHJ and Lurvink MC "Personal Exposure to Ultrafine Particles in the Workplace: Exploring Sampling Techniques and Strategies" *Ann. Occup. Hyg.* **48**(5), 439-453, 2004
[7] Maynard A "Monitoring of nanoparticles in the workplace" *Proceedings First International Symposium on Occupational Health Imlications of Nanomaterials*, 233-237, 2004
[8] Kuhlbusch TAJ, Fissan H and Asbach C "*Nanotechnologies and Environmental Risks – Measurement Technologies and Strategies*" Nanomaterials: Risks and Benefits, Springer, 233-243, 2009
[9] Murashov V, Engel S, Savolainen K, Fullam B, Lee M and Kearns P "Occupational safety and health in nanotechnology and Organisation for Economic Cooperation and Development" *J. Nanopart. Res.* **11**, 1587-1591, 2009
[10] Kuhlbusch TAJ, Neumann S and Fissan H "Number size distribution, mass concentration, and particle composition of PM1, PM2.5 and PM10 in bag filling areas of carbon black production" *J. Occup. Environ. Hyg.* **10**(1), 660-671, 2004
[11] Binder A, Seipenbusch M, Muhler M and Kasper G „Kinetics and particle size effects in ethene hydrogenation over supported palladium catalysts at atmospheric pressure“ *Journal of Catalysis* **268**, 150-155, 2009
[12] Weber AP, Davoodi P, Seipenbusch M and Kasper G „Catalytic behaviour of nickel nanoparticles: gasborne vs. supported state" *J. Nanopart. Res.* **8**, 445-453, 2006
[13] Weber AP, Seipenbusch M and Kasper G "Size effects in the catalytic activity of unsupported metallic nanoparticles" *J. Nanopart. Res.* **5**, 293-298, 2003
[14] Seipenbusch M, Binnig J, Heim M, Weber AP and Kasper G „Aerosol Catalysis: the influence of particle structure on the catalytic activity of platinum-nanoparticle agglomerates" *Helv. Chim. Acta*, **84**, 3686-3701,2001
[15] Weber AP, Seipenbusch M, Thanner C and Kasper G "Aerosol catalysis on nickel nanoparticles" *Journal of Nanoparticle Research* **1**, 253-265, 1999
[16] Ping L, Miser DE, Rabiei A, Yadav RT and Hajaligol MR "The removal of carbon monoxide by iron oxide nanoparticles" *Applied Science B: Environmental* **43**, 151-162, 2003
[17] Bond GC "The origins of particle-size effects in heterogeneous catalysis" *Surf. Sci.* **156** (JUN), 966-981, 1985
[18] Schwyn S, Garwin E and Schmidt-Ott A "Aerosol Generation by Spark Discharge" *J. Aerosol Sci.* **19**(5), 639-642, 1988
[19] Kruis FE, Fissan H and Peled A "Synthesis of nanoparticles in the gas phase for electronic, optical and magnetic applications – a review." *J. Aerosol Sci* **29**(5/6), 511-535, 1999
[20] Weber AP, Seipenbusch M and Kasper G „Application of Aerosol Techniques to Study the Catalytic Formation of Methane on Gasborne Nickel Nanoparticles“ *J. Phys. Chem. A* **105**, 8958-8963, 2001

Journal of Physics: Conference Series **304** (2011) 012012 doi:10.1088/1742-6596/304/1/012012

Airborne Nanoparticle Detection By Sampling On Filters And Laser-Induced Breakdown Spectroscopy Analysis

Pascale Dewalle[1], Jean-Baptiste Sirven[1], Audrey Roynette[2], François Gensdarmes[2], Luana Golanski[3] and Sylvie Motellier[3]

[1]CEA Saclay, DEN, Department of Physical Chemistry, F-91191 Gif-sur-Yvette, France.

[2]IRSN, DSU, Aerosol Physics and Metrology Laboratory, F-91192 Gif-sur-Yvette, France.

[3]CEA Grenoble, DRT, LITEN, Laboratory of Nanomaterial Chemistry and Security, F-38054 Grenoble, France.

E-mail: jean-baptiste.sirven@cea.fr

Abstract. Nowadays, due to their unique physical and chemical properties, engineered nanoparticles are increasingly used in a variety of industrial sectors. However, questions are raised about the safety of workers who produce and handle these particles. Therefore it is necessary to assess the potential exposure by inhalation of these workers. There is thereby a need to develop a suitable instrumentation which can detect selectively the presence of engineered nanoparticles in the ambient atmosphere. In this paper Laser-Induced Breakdown Spectroscopy (LIBS) is used to meet this target. LIBS can be implemented on site since it is a fast and direct technique which requires no sample preparation. The approach consisted in sampling Fe_2O_3 and TiO_2 nanoparticles on a filter, respectively a mixed cellulose ester membrane and a polycarbonate membrane, and to measure the surface concentration of Fe and Ti by LIBS. Then taking into account the sampling parameters (flow, duration, filter surface) we could calculate a detection limit in volume concentration in the atmosphere. With a sampling at 10 L/min on a 10 cm^2 filter during 1 min, we obtained detection limits of 56 $\mu g/m^3$ for Fe and 22 $\mu g/m^3$ for Ti. These figures, obtained in real time, are significantly below existing workplace exposure recommendations of the EU-OSHA and of the NIOSH. These results are very encouraging and will be completed in a future work on airborne carbon nanotube detection.

1. Introduction

Nowadays, due to their unique physical and chemical properties [1-4], engineered nanoparticles are increasingly used in a variety of industrial sectors, with more than 1000 products commercialized in the world. However, questions are now raised about the safety of workers who produce and handle these particles [5-7]. Indeed, the effect of nanoparticles on human health is not fully understood yet, and their toxicity is the subject of many researches [8-9]. In addition there is presently no regulation at the world level related to the synthesis and the safe handling of nanomaterials, nor any recommendation on the admissible exposure levels for workers [10]. Besides, to properly assess the risk of nanoparticles, it is necessary to measure the exposure of workers. The main route of exposure is the inhalation of these particles [11], which are then distributed in the whole respiratory tract [12-13]. There is thereby a need to develop a suitable instrumentation to detect the presence of engineered nanoparticles in the ambient atmosphere.

Usual instruments for aerosol measurements were tested for that purpose [14-15]. Results show that aerosol particle sizers (mostly ELPI, SMPS, APS, FMPS) allow detecting in real time an increase of the particles concentration. However as the implemented techniques give no information on the particles shape or composition, it is sometimes difficult to ensure that the detected particles are those of interest. Therefore sampling and off-line analysis are often required. Furthermore, the natural background can be high (10^4-10^6 part/cm^3) compared to the concentration of engineered nanoparticles, and it can fluctuate a lot in time and in space, which makes their detection still more difficult.

In this context, we propose to use the Laser-Induced Breakdown Spectroscopy (LIBS) technique to meet this target [16]. The principle is to focus a high energy pulsed laser beam on a sample, causing its vaporization and a plasma formation. The plasma light is collected and spectrally analyzed. It is then possible to identify the chemical elements of the sample and their concentrations after calibration. The LIBS technique is fast and can be used directly for on-site measurements without sample preparation. It is very versatile and can be implemented on solids, liquids, gases and aerosols.

In this work we chose to sample airborne nanoparticles on a filter and to analyze the filter by LIBS to measure the particle mass surface concentration, and then their volume concentration in the ambient atmosphere. This approach has been already used by a few authors for particle measurements. Its first application was the detection of beryllium particles [17]. It was also implemented for environmental purposes to monitor heavy metals aerosols, with detection limits of the order of several tens to several hundreds of ng/cm² depending on the element [18-19]. In two interesting papers LIBS was used to analyze the particles collected on the stages of an ELPI to have a size-resolved quantitative chemical information [20-21]. And recently authors coupled LIBS to an aerodynamic lens depositing the particles on a rotating substrate, and they obtained a volume detection limit of 80 ng/cm^3 of copper for a 5 min sampling time [22].

This paper presents our first results obtained on Fe_2O_3 and TiO_2 nanoparticles. A known mass of these particles was deposited, respectively, on mixed cellulose ester membranes and on polycarbonate membranes, in order to make calibration samples. Then we traced calibration curves for both elements, Fe and Ti, and we determined the limits of detection.

2. Experimental

2.1. Filters enrichment

The experimental setup used to enrich filters with a controlled amount of Fe_2O_3 nanoparticles is shown on figure 1. A spark discharge aerosol generator was used (PALAS GFG-1000), producing nanoparticles with a mass median diameter around 100 nm. A flow splitter was placed after the generator to divide the aerosol in four lines. Two of them were used for sampling on filters, a third one enabled to do real-time mass deposition measurements with a tapered-element oscillating microbalance (TEOM 1105) and the excess flow was evacuated through the last one. The flow rate in the sampling lines and in the TEOM line was the same, equal to 3 L/min, so that the TEOM mass measurement was directly representative of the particles mass deposited on the filters. We made replicates of a series of filters and we checked that the TEOM data was consistent with ICP-OES measurements performed after complete dissolution of the filter. The particle surface concentration on the filters was changed by varying the exposure time. The filters used for Fe_2O_3 nanoparticles were mixed cellulose ester membranes (HAWP Millipore) with a diameter of 47 mm and a pore diameter of 0.45 µm.

The enrichment of filters with TiO_2 nanoparticles was performed differently. We used an atomizer which generated aerosols by atomization of a liquid dispersion of nanoparticles. The particle size distribution was measured with a Scanning Mobility Particle Sizer and the mass median diameter was of the order of 100 nm. The particle mass concentration on the filters was controlled by subsequent TXRF measurements. The filters used were polycarbonate membranes (PC Osmonics) with a diameter of 37 mm and a pore diameter of 0.20 μm.

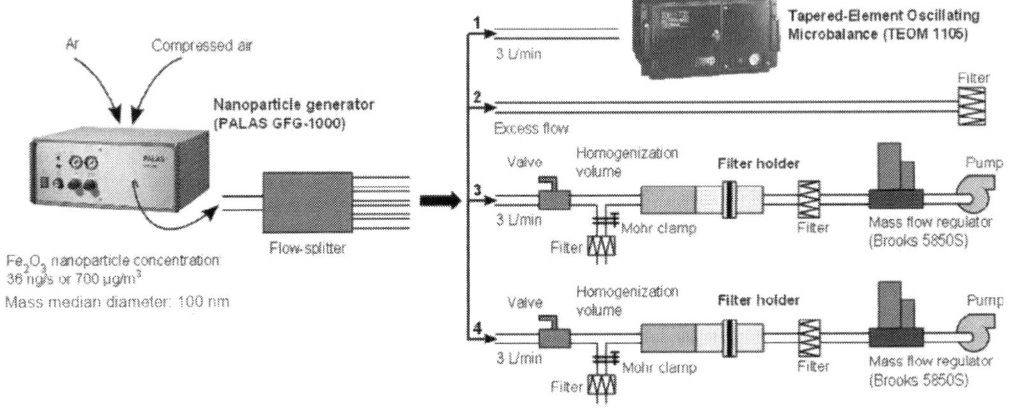

Figure 1. Experimental setup for the enrichment of filters with Fe_2O_3 nanoparticles.

2.2. LIBS setup

The LIBS setup is represented on figure 2. A Nd:YAG laser was used (Quantel Brio), delivering pulses of 5 ns duration and 25 mJ maximum energy at 355 nm. A shutter enabled to control the number of laser shots on the sample. The beam was then shaped by means of a home made optical system so that the energy profile at the sample surface was homogeneous ("top-hat" profile), for this configuration is well suited for surface, single-shot analysis. The spot diameter on the filter was 220 μm, the corresponding irradiance was 13 GW/cm^2. The working distance of the focusing lens was about 230 mm. The light emitted by the plasma was collected by a telescope and injected into a 3 m optical fiber transmitting the UV radiation, with a core diameter of 910 μm. This fiber was directly connected at the entrance of the spectrometer, a monochromator equipped with gratings of 2400 grooves/mm (for Fe_2O_3 nanoparticles) and 1200 grooves/mm (for TiO_2 nanoparticles). The detector used was an intensified CCD camera (Andor iStar). The gate delay of the detection was 1 μs for both types of particles, the integration time was 1.5 μs for Fe_2O_3 nanoparticles and 1 μs for TiO_2 ones.

The acquisition consisted in scanning the filter surface by performing a laser shot on one spot, without overlapping between two successive shots. The step between two shots was adjusted to 1 mm for that purpose. A spectrum was recorded for each shot. For Fe_2O_3 nanoparticles we recorded 529 spectra per filter, for TiO_2 ones, for which the filters used were smaller, we recorded 220 spectra.

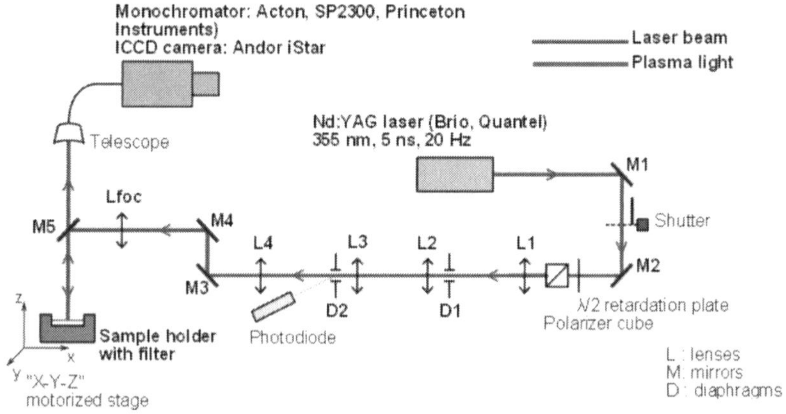

Figure 2. Experimental setup for the LIBS analysis of filters.

3. Results and discussion

For Fe_2O_3 nanoparticles, figure 3 shows the spectra averaged over all laser shots for each filter, and the calibration line traced between 0 and 600 ng/cm^2 (Fe concentration). Figure 4 shows the same for TiO_2 nanoparticles, with a calibration line between 0 and 200 ng/cm^2 (Ti concentration). In both cases a good linearity was obtained. For the Fe and Ti elements, a limit of detection (LOD) of, respectively, 56 and 22 ng/cm^2 was determined.

To convert these figures into volume detection limits (in µg/m^3), let us consider for example a particle sampling at 10 L/min during 1 min on a 10 cm^2 surface filter. Then the volume LOD is 56 µg/m^3 for Fe and 22 µg/m^3 for Ti. These values, obtained in real time, are significantly below existing recommendations on workplace exposure: 330 µg/m^3 for Fe nanoparticles (EU-OSHA [23]) and 60 µg/m^3 for Ti nanoparticles (NIOSH [24]). These results are very encouraging and can be still improved by changing the sampling parameters. The LOD will be inversely proportional to the sampling duration for example. We deliberately took the case of real time monitoring here, but this can be discussed. In particular the technical implementation of a filter enrichment over such a short time (1 min) and at a high frequency has to be investigated. On the other hand if we accept a sampling over a day for instance, we can expect an improvement of more than two orders of magnitude in the sensitivity of the detection. LIBS can be implemented in various configurations to answer different types of problems. The results reported here show the possibility to perform real time quantitative measurements of workplace exposure to nanoparticles with a sufficient limit of detection, at least for Fe_2O_3 and TiO_2 nanoparticles, but this approach is not exclusive and can also be extended to other types of nanoparticles.

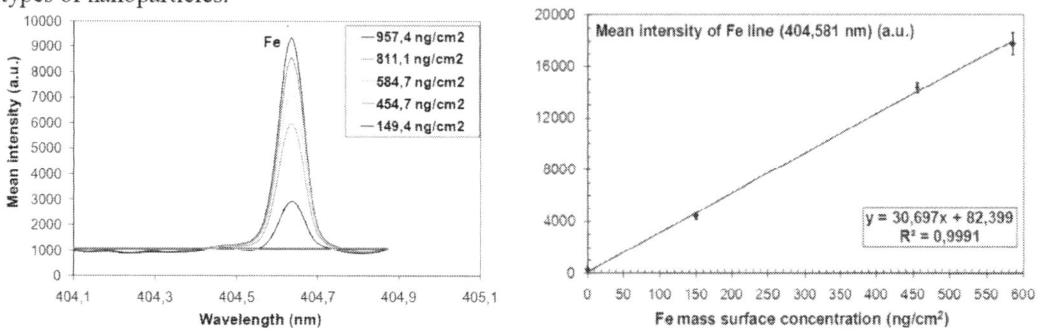

Figure 3. Average spectra for filters enriched with Fe_2O_3 nanoparticles (left) and calibration line of the Fe surface concentration (right). The error bars represent one standard deviation calculated for 23 spectra accumulated over 23 laser shots.

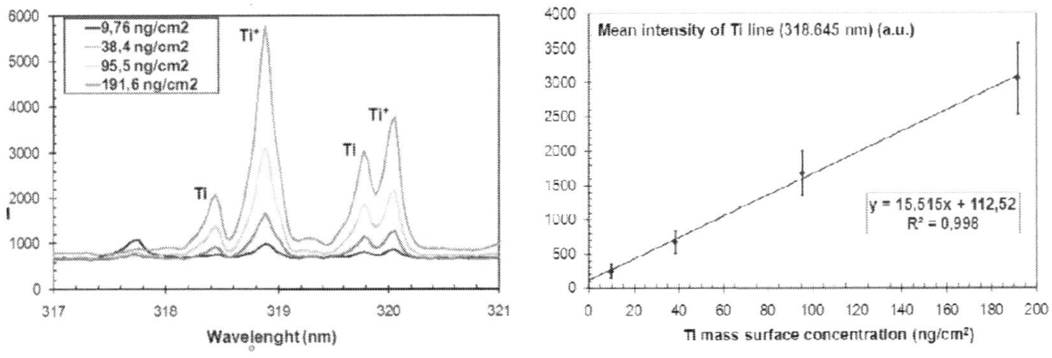

Figure 4. Average spectra for filters enriched with TiO_2 nanoparticles (left) and calibration line of the Ti surface concentration (right). The error bars represent one standard deviation calculated for 11 spectra accumulated over 20 laser shots.

4. Summary

In this paper Laser-Induced Breakdown Spectroscopy (LIBS) is used to make quantitative measurements of the concentration of Fe_2O_3 and TiO_2 airborne nanoparticles. This technique was chosen since it is fast, direct and requires no sample preparation. Nanoparticles were generated by a spark discharge generator or by nebulisation of a liquid dispersion. The approach consisted in sampling the particles on a filter, respectively a mixed cellulose ester membrane and a polycarbonate membrane, and to analyze the filter by LIBS. Then taking into account the sampling parameters (flow, duration, filter surface) we could calculate a detection limit in volume concentration in the atmosphere. With a sampling at 10 L/min on a 10 cm² filter during 1 min, we obtained detection limits of 56 µg/m³ for Fe and 22 µg/m³ for Ti. These figures, obtained in real time, are significantly below existing workplace exposure recommendations of the EU-OSHA and of the NIOSH. In addition the sampling parameters can be adjusted, particularly the sampling duration, to potentially decrease the detection limit of more than two orders of magnitude. This opens the way to the monitoring of workplace exposure over a longer time scale with a high sensitivity. These results are very encouraging and will be completed in a future work on airborne carbon nanotube detection.

Acknowledgement

This work is part of the GENESIS project (Global Evaluation of Nanocomposites for Emerging Safe Industrial Solutions) financed by OSEO (France).

References

[1] Lu L, Sui M L and Lu K 2000 Superplastic extensibility of nanocrystalline copper at room temperature *Science* **287** 1463-66
[2] Kamat P V 2002 Photophysical, photochemical and photocatalytic aspects of metal nanoparticles *J. Phys. Chem. B* **106** 7729-44
[3] Qi H J, Teo K B K, Lau K K S, Boyce M C, Milne W I, Robertson J and Gleason K K 2003 Determination of mechanical properties of carbon nanotubes and vertically aligned carbon nanotube forests using nanoindentation *Journal of the Mechanics and Physics of Solids* **51** 2213-22
[4] Zhan G D, Kuntz J D, Garay J E and Mukherjee A K 2003 Electrical properties of nanoceramics reinforced with ropes of single-walled carbon nanotubes *Appl. Phys. Lett.,* **83** 1228-30
[5] Oberdörster G, Maynard A, Donaldson K, Castranova V, Fitzpatrick J, Ausman K, Carter J, Karn B, Kreyling W, Lai D, Olin S, Monteiro-Riviere N, Waeheit D and Yang H 2005 Principles for characterizing the potential human health effects from exposure to nanomaterials: elements of a screening strategy *Particle and Fibre Toxicology* **2** 1-35

[6] Maynard A D 2006 Nanotechnology: assessing the risks *Nano Today* **1** 22-33
[7] Kandlikar M, Ramachandran G, Maynard A, Murdock B and Toscanao W A 2007 Health risk assessment for nanoparticles: A case for using expert judgment *Journal of nanoparticles* **22** 969-76
[8] Simon-Deckers A, Gouget B, Mayne-L'Hermite M, Herlin-Boime N, Reynaud C and Carrière M 2008 In vitro investigation of oxide nanoparticle and carbon nanotube toxicity and intracellular accumulation in A549 human pneumocytes *Toxicology* **253** 137-46
[9] Barillet S, Simon-Deckers A, Herlin-Boime N, Mayne-L'Hermite M, Reynaud C, Cassio D, Gouget B and Carrière M 2010 Toxicological consequences of TiO_2, SiC nanoparticles and multi-walled carbon nanotubes exposure in several mammalian cell types: an in vitro study *Journal of Nanoparticle Research* **12** 61-73
[10] Schulte P A, Murashov V, Zumwalde R, Kuempel E D and Geraci C L 2010 Occupational exposure limits for nanomaterials: state of the art *Journal of Nanoparticle Research* **12** 1971-87
[11] Wang S M, Inthavong K, Wen J, Tu J Y and Xue C L 2009 Comparison of micron- and nanoparticles deposition patterns in a realistic human nasal cavity *Respiratory Physiology & Neurobiology* **166** 142-51
[12] Borm P J, Robbins D, Haubold S, Kuhlbusch T, Fissan H, Donaldson K, Schins R, Stone V, Kreyling W, Lademann J, Krutmann J, Warheit D and Oberdörster E 2006 The potential risks of nanomaterials: a review carried out for ECETOC *Particle and Fibre Toxicology* **3** 1-35
[13] Nel A, Xia T, Madler N and Li N 2006 Toxic potential of materials at the nanolevel *Science* **311** 622-27
[14] Brouwer D, van Duuren-Stuurman B, Berges M, Jankowska E, Bard D and Mark D 2009 From workplace air measurement results toward estimates of exposure? Development of a strategy to assess exposure to manufactured nano-objects *Journal of Nanoparticle Research* **11** 1867-81
[15] Brouwer D 2010 Exposure to manufactured nanoparticles in different workplaces *Toxicology* **269** 120-7
[16] Miziolek A W, Palleschi V and Schechter I 2006 *Laser-Induced Breakdown Spectroscopy (LIBS) Fundamentals and applications* (Cambridge: Cambridge University Press)
[17] Cremers D A and Radziemski L J 1985 Direct Detection of Beryllium on Filters Using the Laser Spark *Appl. Spectrosc.* **39** 57– 63
[18] Neuhauser R E, Panne U and Niessner R 1999 Laser-induced plasma spectroscopy (LIPS): a versatile tool for monitoring heavy metal aerosols *Analytica Chimica Acta* **392** 47-54
[19] Panne U, Neuhauser R E, Theisen M, Fink H and Niessner R 2001 Analysis of heavy metal aerosols on filters by laser-induced plasma spectroscopy *Spectrochim. Acta Part B* **56** 839-50
[20] Kuhlen T, Fricke-Begemann C, Strauss N and Noll R 2008 Analysis of size-classified fine and ultrafine particulate matter on substrates with laser-induced breakdown spectroscopy *Spectrochim. Acta Part B* **63** 1171–76
[21] Yoshiie R, Yamamoto Y, Uemiya S, Kambara S and Moritomi H 2008 Simple and rapid analysis of heavy metals in sub-micron particulates in flue gas *Powder Technology* **180** 135–39
[22] Park K, Cho G and Kwak J H 2009 Development of an aerosol focusing-laser induced breakdown spectroscopy (aerosol focusing-LIBS) for determination of fine and ultrafine metal aerosols *Aerosol Sci. Technol.* **43** 375–86
[23] EU-OSHA 2009 Literature review – Workplace exposure to nanoparticles
[24] NIOSH 2005 Current Intelligence Bulletin: Evaluation of Health Hazard and Recommendations for Occupational Exposure to Titanium Dioxide (Draft for public comment)

An electrical sensor for long-term monitoring of ultrafine particles in workplaces

Timo Lanki[1], Juha Tikkanen[2], Kauko Janka[2], Pekka Taimisto[1] and Matti Lehtimäki[3]

[1]National Institute for Health and Welfare (THL), Department of Environmental Health, P.O.Box 95, FIN-70701 Kuopio, Finland.

[2]Navaro Oy, Rautapellonkatu 37, FIN-33700 Tampere, Finland

[3]Technical Research Centre of Finland (VTT), P.O. Box 1300, FIN-33101 Tampere, Finland

E-mail: timo.lanki@thl.fi

Abstract. Pegasor Oy Ltd. (Finland) has developed a diffusion charging measurement device that enables continuous monitoring of fine particle concentration at a low initial and lifecycle cost. The innovation, for which an international process and apparatus patent has been applied for, opens doors for monitoring nanoparticle concentrations in workplaces.
The Pegasor Particle Sensor (PPS) operates by electrostatically charging particles passing through the sensor and then measuring the current caused by the charged particles as they leave the sensor. The particles never touch the sensor and so never accumulate on its surfaces or need to be cleaned off. The sensor uses an ejector pump to draw a constant sample flow into the sensing area where it is mixed with the clean, charged pump flow air (provided by an external source). The sample flow containing charged particles passes through the sensor. The current generated by the charge leaving the detection volume is measured and related to the particle surface area. This system is extremely simple and reliable – no contact, no moving parts, and all critical parts of the sensor are constantly cleaned by a stream of fresh, filtered air. Due to the ejector pump, the sample flow, and respectively the sensor response is independent of the flow and pressure conditions around the sampling inlet.
Tests with the Pegasor Particle Sensor have been conducted in a laboratory, and at a workplace producing nanoparticles for glass coatings. A new measurement protocol has been designed to ensure that process workers are not exposed to unusually high nanoparticle concentrations at any time during their working day. One sensor is placed inside the process line, and a light alarm system indicates the worker not to open any protective shielding or ventilation systems before concentration inside has reached background levels.
The benefits of PPS in industrial hygiene are that the same monitoring technology can be used at the source as well as at the worker breathing zone. Up to eight sensors can be installed in series for centralized monitoring of the whole process in real time.

1. Introduction

Health risks caused by particle exposures at workplaces are typically underestimated. Epidemiological studies conducted during the last 15 years have yet established the role of particulate air pollution as a risk factor for not only respiratory but also cardiovascular diseases [1]. Recently, researcher have linked particle exposure also e.g. with diabetes and rheumatoid arthritis [2, 3]. Together with improved exposure assessment methods, risk estimates have gradually increased. Highest risks are associated with long-term (years) exposures, but even very short-term exposures may affect health [4].

There is evidence that all major size fractions of particles – ultrafine, fine and coarse – have adverse effects on health. However, ultrafine particles have some unique properties compared to larger particles: very high number challenging macrophage based immune system, large surface area enhancing chemical reactions, and extremely small size enabling efficient penetration into body [5]. Thus, there are concerns that the expansion of nanotechnology and rapidly increasing nanoparticle production will increase the health risks at workplaces [6].

Particle exposures encountered at work are often several magnitudes higher than elsewhere [7]. In contrast to urban environments, particles are yet rarely monitored at workplaces. This is mainly due to the high costs of monitoring. High spatial variability of particles, especially coarse and ultrafine, further complicates exposure assessment, because ideally particles should be monitored simultaneously in multiple locations. The electrical sensor presented here enables particle monitoring at low initial and maintenance costs using several sensors.

2. Pegasor Particle Sensor

The patented [8] Pegasor Particle Sensor (PPS) operates by electrostatically charging particles passing through its body and then measuring the current produced by the charged particles as they leave the sensor. The basic principle of directly measuring the current escaping with charged particles was first described by Lehtimäki [9].

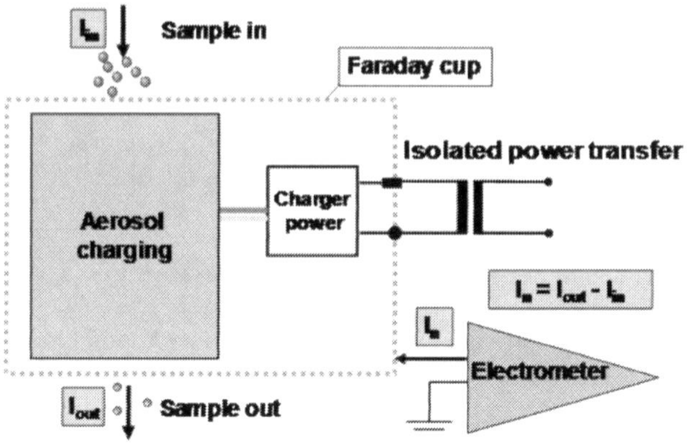

Figure 1. The basic principle of the electrical aerosol sensor based on the measurement of the escaping current.

The particles are charged by ions generated by a corona discharge, which is generated around a sharp tip at high voltage (Figure 1). After charging, the particles are neither collected on any filter nor accumulate on any part of the sensor. However, as charged particles leave the sensor, they produce an electrical current escaping the Faraday cup. Measurement of this current is proportional to the particle concentration.

Even minimal soot deposition on critical surfaces, as electrical insulators, may cause leakage currents altering the measurement result. Further, the contamination of high-voltage insulators reduces the discharge voltage, preventing ion production. The discharge may also be fused by contamination of the corona tip. In order to get around these issues, the sensor protects all sensitive surfaces from soot deposition. Figure 2 shows how this is achieved. Clean filtered air is fed through an air nozzle, where the corona tip is located. In the area of high flow velocity, the corona discharge effectively generates ions at sufficiently low voltage. The nozzle discharges sonic flow functioning as a "pump flow" of an ejector. The resulting ejector pump sucks the sample flow from the inlet. The charged pump and sample flows are mixed prior to entering the ejector throat, which allows efficient particle charging. The ion trap followed by the ejector throat removes the free ions, so that only charged particles leave the sensor with the sample flow. The functional part of the sensor taking care of the charging, ion trapping and sample flow guiding are electrically insulated. The electric current from the insulated part to ground is measured by an electrometer. This measured current represents the charging rate of the particles in the sample flow.

The sensor electronics contain a wide-dynamic range electrometer measuring the current carried by charged particles and a high-voltage unit powering the corona discharge with constant current. The system control and data processing and transfer are performed by a digital control unit. In addition to the pure processing of the measurement signal, the control unit has also functions to diagnose the operating condition and cleaning needs of the sensor. The sensor data and control signals (RS485 bus) are transferred through an interface unit to and from a laptop computer (USB bus). Optionally the sensor can be used as a stand alone instrument without PC using a 4-channel, 20-bit digital-to-analog conversion unit. The time response of the current sensor version is 0.3 s, and data sampling rate is 10 Hz. Specialized sensor software enables the sensor control and data logging on a PC. It contains basic data processing functions (user defined time constants and data averaging), as well as the graphical data presentation and data storage.

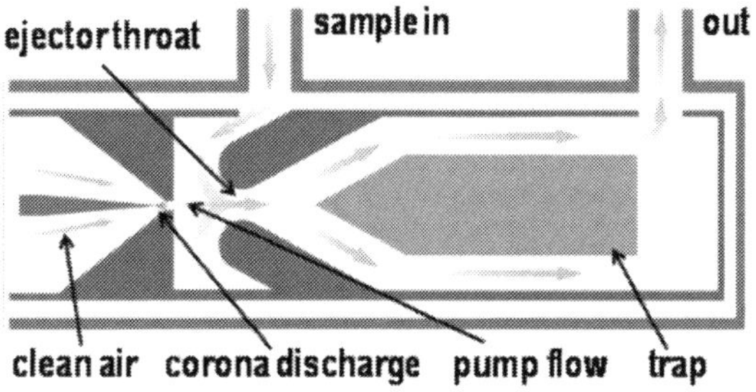

Figure 2. Sensor schematic. The air jet ionized in the corona discharge functions as a pump flow of an ejector. The ejector draws the sample flow, and the particles are charged prior to the ion trap. The inner part of sensor is electrically isolated, and the current measured from it equals the charging rate of particles exiting with the outlet flow.

3. Experimental setup

3.1. Comparison to other methods

The performance of the Pegasor sensor was studied as a part of the measurements which were focused on the behaviour of nanosized particles generated through laser ablation. These measurements were made at VTT Lappeenranta (Factory of the future) by using a femtosecond (fs) laser workstation. During the experiments nickel disc target was continuously scanned with a constant power laser beam and the released particles were captured with an exhaust hood. The principle of the set-up used in the study is illustrated in Figure 3.

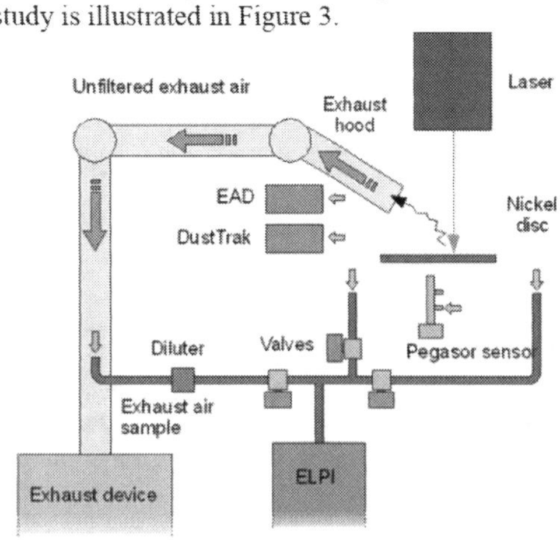

Figure 3. Principle of the measurement system used when studying nanoparticles generated by the fs laser work station.

The purpose of these measurements was to provide information about the behaviour of nanoparticles generated by fs laser work station, and clarify how the local exhaust system can be utilized to prevent nanoparticles from dispersing into the workplace air. The particle measurements were made with a system which included ELPI (Electrical low-pressure impactor) equipped with a sampling system, which was used to continuously monitor particles from three sampling points. Two of the sampling points were located in the vicinity of the nickel disc processed by the laser beam and the third sampling line was used to sample air from the exhaust duct. Besides ELPI, the quality of air was monitored with Pegasor sensor, DustTrak aerosol photometer and electrical aerosol detector (EAD).

The major aim of experiments with different aerosol sensors was to demonstrate how they can be used to detect possible emissions of engineering nanoparticles from a work process. In many practical applications, the existing background concentration of ultrafine particles is so high that nanoparticle emissions may contribute only slightly on the total concentration. The sudden nanoparticle release was simulated here by switching the local exhaust system off for a short period of time.

3.2. Trap voltage scanning

An interesting feature in the Pegasor sensor is the possibility to modify its characteristics by varying the trap voltage, i.e. the voltage which is primarily for preventing ions from escaping the sensor. This means that by increasing the trap voltage to a certain level, particles below the corresponding size limit can be collected before they leave the sensor. Thus, these particles do not contribute to the output of the sensor. It is even possible to scan the trap voltage in a wide range in such a way that the relationship between sensor output and trap voltage can be determined. In principle, this relationship includes information about the particle size distribution.

The trap voltage scanning was tested by measuring the sensor output vs. trap voltage for three polydisperse aerosols. The test aerosols were generated with an aerosol atomizer from very diluted solutions of DEHS in isopropanol. The count median particle sizes of these aerosols were 8 - 15 nm.

The relationship between sensor output and trap voltage was also studied at the laser work station. In the first phase, trap voltage scanning was made when measuring the background aerosol. The corresponding experiment with trap voltage scanning was also made from the exhaust air. In this case the measured aerosol represents the nanoparticles generated by the laser ablation.

3.3. Feasibility study

The feasibility of the method for monitoring simultaneously particle concentrations at a source and in the breathing air was evaluated at an industrial facility producing nanoparticles for glass coating. One sensor was placed inside the nanoparticle production chamber, and another one in the working space.

4. Results

4.1. Comparison to other methods

Figure 4 shows the responses of the ELPI and Pegasor particle sensor. In this figure the total current collected by the ELPI impaction stages are shown together with the sensor output. This result illustrates that responses of ELPI total current and Pegasor sensor are identical. This is an expected result because the particle charging in both devices is accomplished with the unipolar diffusion charging process. This result also indicates that the sensor responded quickly to the nanoparticle release caused by the switch off of the local exhaust.

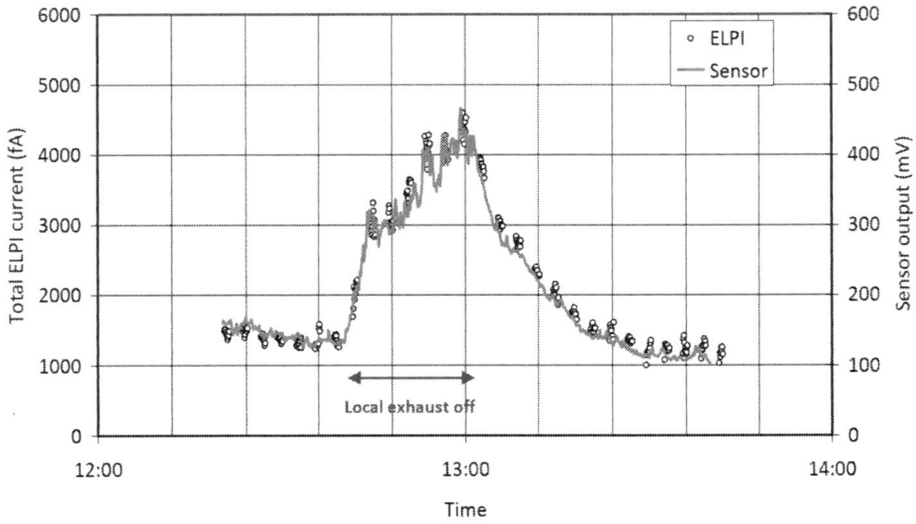

Figure 4. ELPI total current and the Pegasor output voltage. Effect of switching off the local exhaust.

The corresponding comparison between the electrical aerosol detector (EAD) and Pegasor sensor is shown in Figure 5. Also in this case the devices, the principle of which is based on the same physical process, provide almost identical responses. However, the output of the EAD also contains additional concentration peaks. These peaks could be due to different locations of the devices, i.e. EAD may have been impacted more by the sporadic nanoparticle puffs from the laser processing.

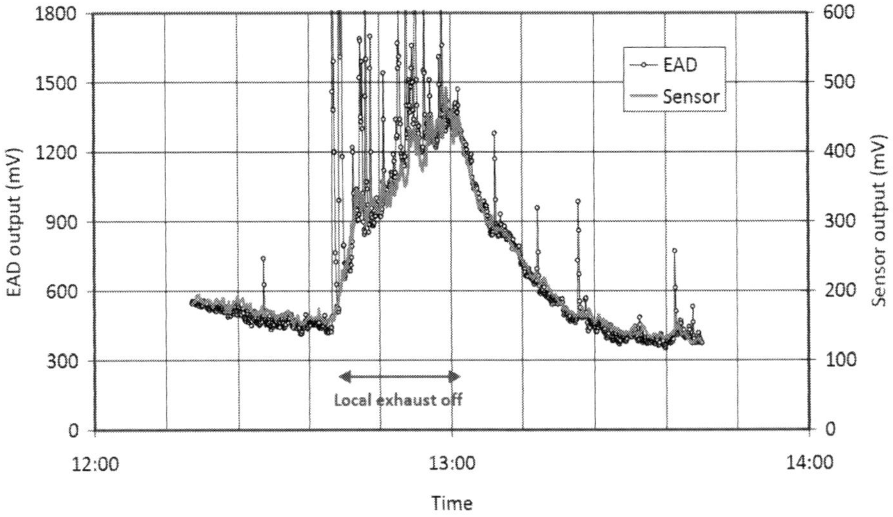

Figure 5. Responses from the EAD and Pegasor sensor. Effect of switching off the local exhaust.

The response of the Pegasor sensor was also compared with the aerosol photometer (DustTrak), which is a widely used tool in industrial hygiene. In this case the two devices provide quite different responses (Figure 6). The concentration indicated by DustTrak increases after switching off the local exhaust. However, the decrease of the concentration (after the local exhaust was switched on) is quite modest indicating that the output of the DustTrak may be affected by some other particles from the other processes in the building.

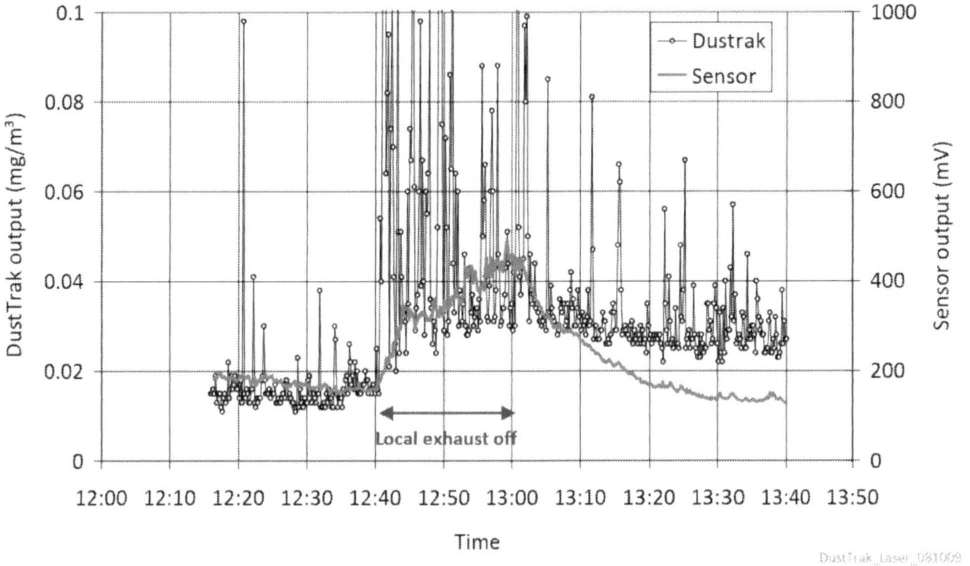

Figure 6. Responses from the DustTrak aerosol photometer and Pegasor sensor. Effect of switching off the local exhaust.

4.2. Trap voltage scanning

The results shown in Figure 7 illustrate that the aerosols with different size distributions provide totally different characteristics. The lower the count median particle size is the steeper is the effect of trap voltage.

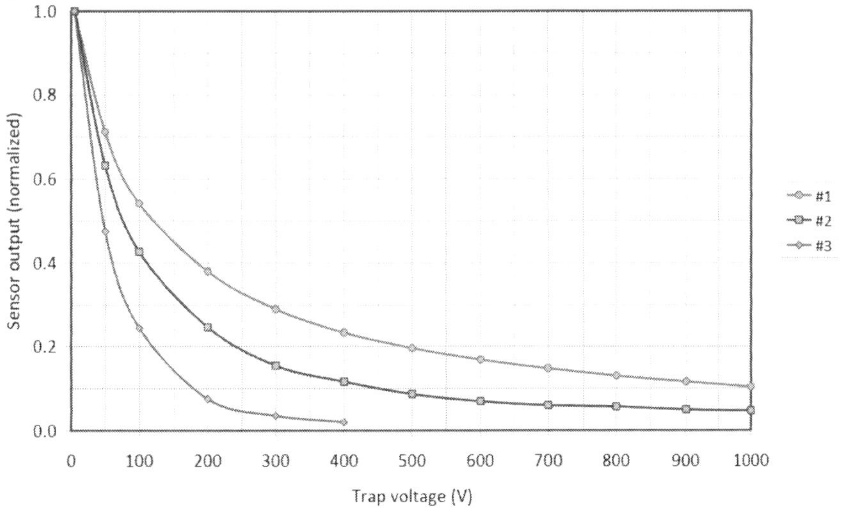

Figure 7. The normalized sensor output vs. trap voltage characteristics measured with three nanosized test aerosols (#1: cmd=15 nm, #2: cmd=11 nm, #3: cmd=8 nm).

Measurements of background aerosol at the laser work station (Figure 8) show that the sensor voltage decreases less steeply with trap voltage than in the case of very fine laboratory aerosols (Figure 7).

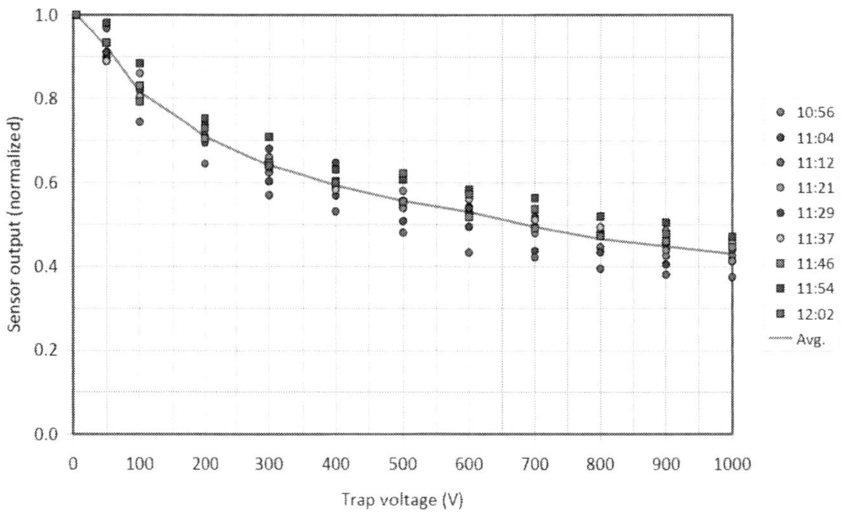

Figure 8. The normalized sensor output vs. trap voltage characteristics measured from the background aerosol.

The corresponding experiment with trap voltage scanning was also made from the exhaust air. In this case the measured aerosol represents the nanoparticles generated by the laser ablation. The results (Figure 9) show that the generated aerosol is somewhat finer than the aerosol found from the surrounding air.

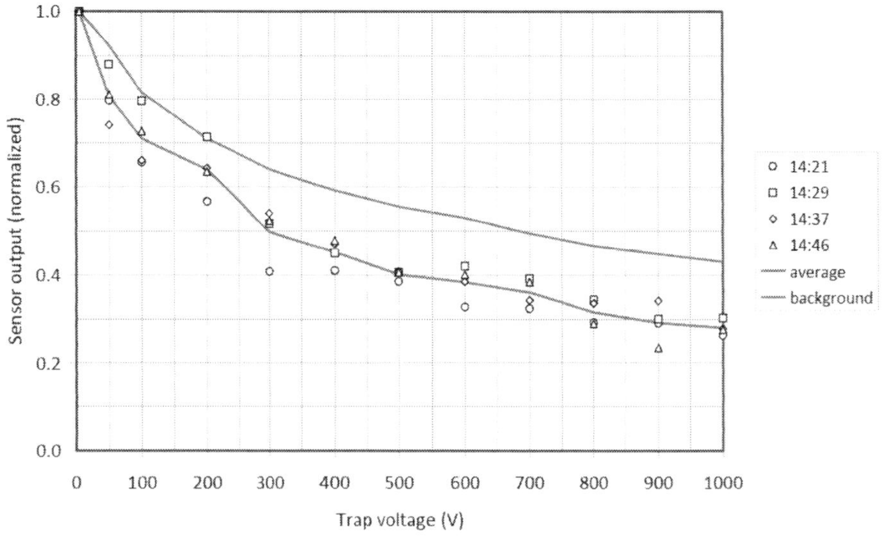

Figure 9. The normalized sensor output vs. trap voltage characteristics measured from the exhaust air. The average from the results with background aerosol included.

4.3. Feasibility study

Median (max) concentration inside the nanoparticle production chamber was 0.90 (1000) mV and outside 0.42 (2.1) mV. Production cycles were reflected as very high but short-term concentration peaks within the chamber (Figure 10).

New measurement protocol was developed where background level is measured at the workers' breathing zone for one week continuously. The average of this measurement period is used as the zero level – green light indication. Light indicating alarm is used to inform the worker, if the average background level is exceeded five times (yellow light) or ten times (red light). One sensor is placed inside the process line to indicate the worker not to open any protective shielding or ventilation systems before concentration inside has reached background levels.

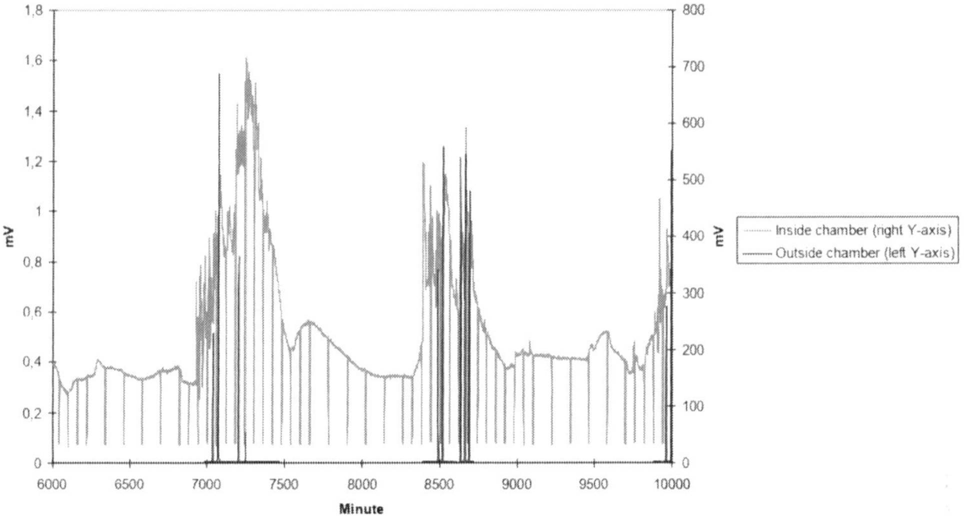

Figure 10. Output voltages of Pegasor sensors placed within and outside a nanoparticle production line (note the very different scales of the axes).

5. Conclusions

Diffusion chargers have been investigated for many years [10], but it is only in the last few years that the available technology and the technical requirements of various applications have come together to offer a practical solution. The electrical sensor developed by Pegasor Oy Ltd. (Finland) enables particle monitoring at low initial and maintenance costs using several sensors. Clean air is utilized both for the charged pump flow and for cooling and shielding the critical insulators. As a result, all contamination-sensitive parts are shielded by clean air flow. The fact that the ion-containing pump flow is mixed extremely rapidly prior to the ejector throat enables efficient particle charging in minimal volume. This leads to low internal gas volume in the sensor and rapid response time. Due to the ejector pump, the sample-flow rate and sensor response is independent of the exhaust flow conditions around the sampling inlet.

Concentrations measured with the Pegasor sensor closely followed those measured with an electrical low-pressure impactor and an electrical aerosol detector. Concentrations differed more from the results of a photometer, most likely because Pegasor sensor is more sensitive to smallest particles.

Our results within a nanoparticle production facility demonstrated the usefulness of the sensor, together with a carefully designed monitoring program, in ensuring that process workers are not exposed to unusually high nanoparticle concentrations at any time during their working day. High exposures to nano-sized particles are common in a wide variety of occupations. Particle monitoring in nanoparticle industry may be considered especially important because exposure is not necessarily associated with odorous volatile components which enable identification of exposure to many other types of ultrafine particles such as welding fumes.

Acknowledgements
This project was funded by the Finnish Work Environment Fund (109121).

References

[1] Brook RD, Rajagopalan S, Pope CA 3rd, Brook JR, Bhatnagar A, Diez-Roux AV et al. 2010 Particulate matter air pollution and cardiovascular disease: An update to the scientific statement from the American Heart Association *Circulation* **121** 2331-78
[2] Pearson JF, Bachireddy C, Shyamprasad S, Goldfine AB, Brownstein JS 2010 Association between fine particulate matter and diabetes prevalence in the U.S. *Diabetes Care* **33**:2196-201
[3] Hart JE, Laden F, Puett RC, Costenbader KH, Karlson EW 2009 Exposure to traffic pollution and increased risk of rheumatoid arthritis *Environ Health Perspect* **117**:1065-9
[4] Lanki T, Hoek G, Timonen KL, Tiittanen P, Vanninen E, Pekkanen J 2008 Hourly variation in fine particle exposure is associated with transiently increased risk of ST segment depression *Occup Environ Med* **65**:782-6
[5] Delfino RJ, Sioutas C, Malik S 2005 Potential role of ultrafine particles in associations between airborne particle mass and cardiovascular health *Environ Health Persect* **113**:934-46
[6] Schulte PA, Salamanca-Buentello F 2007 Ethical and scientific issues of nanotechnology in the workplace *Environ Health Perspect* **115**:5-12
[7] Miller A, Drake PL, Hintz P, Habjan M 2010 Characterizing exposures to airborne metals and nanoparticle emissions in a refinery *Ann Occup Hyg* **54**:504-13
[8] Pegasor Oy, Tikkanen J 2009 Particle Measurement Process and Apparatus, Patent publication no WO 2009/109688.
[9] Lehtimäki M 1983 Modified Aerosol Detector *Aerosols in Mining and Industrial Work Environment 3* Ed V.A. Marple and B.Y.H. Liu (Michigan U.S.: Ann Arbor Science Publishers) pp 1135–43.
[10] Ntziachristo L, Giechaskiel B, Ristimäki J, Keskinen J. Use of a corona charger for the characterisation of automotive exhaust aerosol 2004 *J Aerosol Sci* **35**:943-63

CAIMAN: a versatile facility to produce aerosols of nanoparticles

Jonathan Jacoby, Sébastien Bau and Olivier Witschger

Institut National de Recherche et de Sécurité (INRS), Laboratoire de Métrologie des Aérosols, Rue du Morvan, CS 60027, 54519 Vandoeuvre Cedex, France

E-mail: sebastien.bau@inrs.fr

Abstract. This work aims at presenting a nanoparticle generation non-transportable facility in aerosol phase called "CAIMAN" (acronym for Characterization of Instruments Measuring Aerosols of Nanoparticles) and its performances. This facility delivers primary nanoaerosols from electrodes made of C, Al, Cu (and mixtures containing Be), Ag, Constantane (a mixture of Cu-55wt% and Ni-45wt%) particles at known concentrations, sizes, shapes and mean charge levels. It is also capable to deliver well-known particle mixture containing combinations of the "primary" nanoaerosols and particles representatives of background aerosols (in the present work NaCl). The output of the CAIMAN facility is very consistent over long time intervals when operating under similar conditions. It indicates that repeatability is also one of the important assets of the facility.

1. Introduction

Rationally designed, manufactured nano-objects (nanoparticles and nanostructured particles – NP) attract a great deal of interest, due to their many technologically interesting properties. The properties of NP and their applications have given birth to technological and economic expectations for industries using NP or materials/end products containing NP (nanomaterials). However, some of these properties have given rise to concern that they may be harmful to humans.

In the workplaces where NP are manufactured or used, exposure to airborne NP is plausible at all phases of the nanomaterial life cycle, although the nature, level of exposure and the number of person involved could differ greatly. One major uncertainty in the health risk assessment of NP arises from the lack of knowledge of their physicochemical properties and behavior in the airborne state (aerosol phase).

Among the technical challenges ahead for assessing workers exposures to NP are to:

- redesign and evaluate "NP-capable" instruments already in laboratory use into portable and affordable devices, and
- expand the sensing technology available for NP detection by adopting new options with realistic potential for real-time measurement and compact design.

To investigate the performances of these instruments and technologies, the aim of this work was to design an experimental facility able to produce aerosols of NP with controlled properties in various situations. This paper describes the CAIMAN setup and its main characteristics.

2. Description of the experimental setup (CAIMAN)

CAIMAN (for 'Characterization of Instruments Measuring Aerosols of Nanoparticles') is a versatile experimental facility that was designed and built to generate stable and reproducible aerosols of nanoparticles (nanoaerosols) with controlled properties (concentration, size, shape, mean charge level). It was initially developed in a collaborative research work within the Aerosol Physics and Metrology Laboratory from the Institut de Radioprotection et de Sûreté Nucléaire (IRSN) and the Aerosol Metrology Laboratory from INRS, which holds the non-transportable facility in its research centre near Nancy, France.

Figure 1 shows a schematic of the CAIMAN experimental facility in its standard configuration. A different arrangement has been used for producing mixture of NP and particles representing background aerosols (see paragraph 3.7).

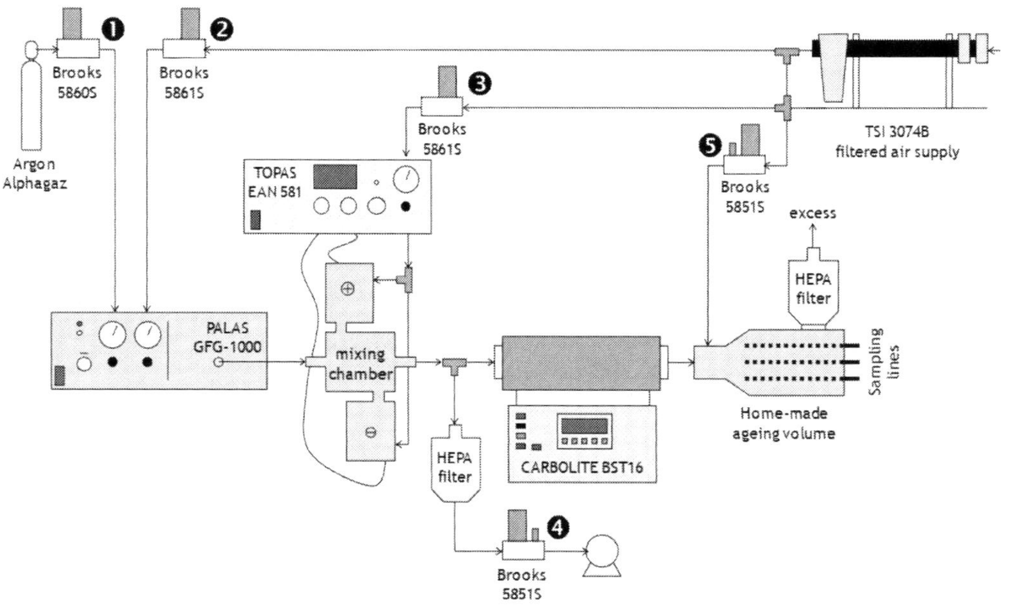

Figure 1. Schematic of the CAIMAN facility in its standard configuration

Within CAIMAN, the particle-free air introduced into different points of the set-up comes from a purification unit (TSI model 3074B), and the excess aerosol is filtered by means of HEPA filters (CAMFIL, model 'filtre BAG').

Following sections briefly describe each of the elements constituting the CAIMAN facility, *i.e.* an airborne NP generator, a bipolar ion generator, a high-temperature furnace and a home-made ageing volume.

2.1. Airborne NP generation

Within CAIMAN, airborne NP are produced by means of a spark-discharge generator (GFG-1000, PALAS). The production of airborne NP by spark discharge has been studied extensively by several authors in different situations [1-15].

Aerosol production rate varies by adjusting the spark-discharge frequency ω, which is directly controlled by the potentiometer setting of the GFG-1000 PALAS, and the flow rates of argon shielding and primary dilution by air. Consequently, this NP generation system allows a variation of both the particle concentration (via the argon and air flow rates, spark-discharge frequency) and their chemical nature by changing the material of the electrodes. Presently, seven different electrode materials have been studied, which chemical composition was determined by ICP-MS and gathered in table 1.

Table 1. Characteristics of the electrodes used within the NP generator

Electrode	Composition
Carbon[1]	Graphite (pure)
Aluminum (type 2071A)	Al(93.1%) – Cu(4.8%) – Fe(0.7%) – Mn(0.6%) – Si(0.5%) – Mg(0.3%)
Copper	Cu(99.45%) - Al(0.5%) – Si(0.04%)
Silver[2]	Ag > 99.99%
Constantane[2]	Cu(55%) – Ni(45%)
Cu/Be[2]	Cu(98%) – Be(1.9%) – Co(0.2%) – Pb(0.3%)
Cu/Co/Be[2]	Cu(95.6%) – Be(0.5%) – Co(2.4%)

[1] Delivered by Palas
[2] Delivered by Goodfellow

2.2. Airborne NP state-of-charge
To cover a wide range of electrical mean charge level, a bipolar ion generator (EAN 581, TOPAS) is included in CAIMAN at the outlet of the NP generator. Its principle is based on air ionisation by corona charging in two independent chambers. Positive and negative ions can be produced separately in each chamber, and then mixed consecutively with the particles within a mixing chamber which geometry was optimised to increase particle charging efficiency. Both the air flow rate and the corona charger parameters (voltage and current) can be adjusted, allowing a wide range of NP electrical state-of-charge. Further elements can be found in the work of Marquard *et al.* [16, 17] in unipolar mode.

2.3. Airborne NP morphology
Morphology being another key-parameter of the NP produced within CAIMAN, a high-temperature furnace (CARBOLITE BST16 – maximum operating temperature 1500°C) was inserted in the facility. Indeed, several studies [18-28] have shown particle restructuration by sintering under high temperatures. The high-temperature furnace can be operated at different temperatures to provide either partial sintering of agglomerates or coalescence, allowing some degree of control over particle morphology of particles with constant chemistry.

2.4. Airborne NP sampling and measurement
The aerosol leaving the furnace is allowed to age in a 2-liter volume located at the end of the facility (see figure 1) from which airborne NP can be sampled and diluted by means of an additional air flow injected at the outlet of the device. It should be noted that this ageing volume is equipped with four sampling lines in parallel, which allows simultaneous measurements. Obviously, more instruments can be connected if flow splitters are put into line. For testing with monodisperse particles, a DMA can be connected at one of the four outputs. Connections are in metric but any other connections can easily be used.
It should be noted that CAIMAN can only be used with aspirating aerosol instruments.

2.5. Safety issues
Due to strong nuclear safety regulation in France, no radioactive source can be brought to the Institute. The only radioactive source hold by the Aerosol Metrology Laboratory is an Am^{241} sealed source devoted to the Grimm DMA.
The entire CAIMAN facility is confined as shown in figure 2.

The first part of the facility is contained within a box containment system (4m x 0.8m x 0.8m) that limits any worker exposure from accidental release from the generator to the entrance of the ageing chamber. The second part of the facility is located within a constant velocity 2-m large fume hood.

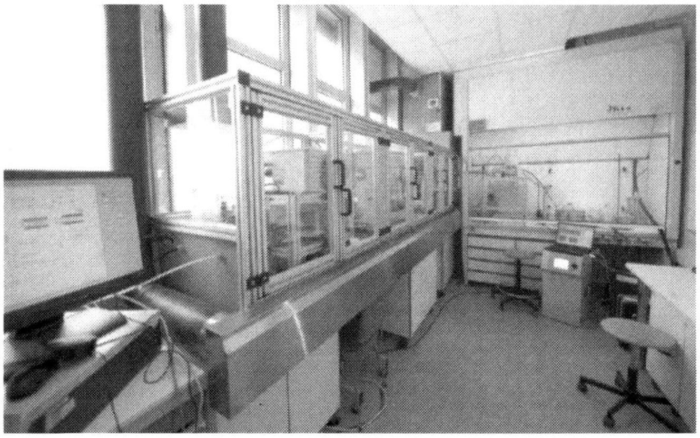

Figure 2. Picture of the CAIMAN facility

3. Features of CAIMAN

In this section, key properties of the aerosols of NP produced within CAIMAN are exposed. For all experiments, argon flow Q_1 was fixed to 2.5 L/min and ageing volume flow to Q_5 = 10 L/min; configuration 1 refers to air flow Q_2 = 0 while configuration 2 refers to Q_2 = 20 L/min. According to PALAS, the frequency indication (ω) is related to spark discharge frequency f by $f = 0.3\ \omega$.

3.1. Generation stability

Before comprehensive characterization of the airborne NP produced within CAIMAN, the generation stability was tested over 170 minutes for carbon electrodes in configuration 2 and ω = 1000 a.u. The 46 scans measured by the SMPS showed a variation of ± 4 % in terms of count median mobility diameter (CMMD) relative to its mean value over all scans as calculated by:

$$\text{ratio} = \frac{\text{CMMD (scan i)}}{\text{CMMD (all scans)}}$$

Figure 3 shows the results obtained.

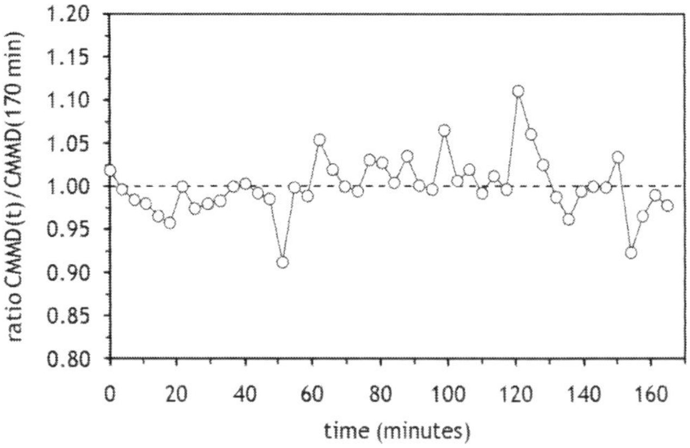

Figure 3. CMMD for Carbon electrodes in configuration 2 and ω = 300 a.u. over 170 min

The mass concentration of the aerosol produced was also monitored by means of a TEOM microbalance, leading to a variability of ± 7 % during a measurement period of 1 hour, as shown in figure 4.

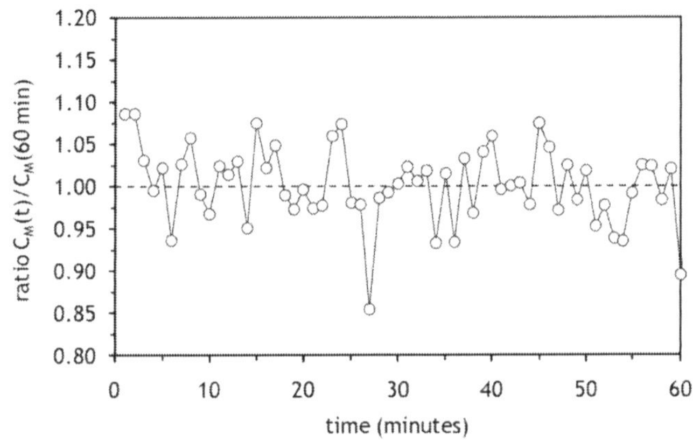

Figure 4. Mass concentration for Carbon electrodes in configuration 2 and ω = 300 a.u. over 60 min

In terms of total number concentration of airborne NP, variations below 4 % were observed over 4 hours of generation. Furthermore, reproducibility of the generation was tested (same operating parameters set with one week delay) and showed a close agreement between NP size distributions (less than 10 % variation in terms of total concentration and CMMD).

3.2. Number size distributions

The size distribution of the generated airborne NP was measured by a SMPS system (GRIMM SMPS+C, composed of a DMA 'Vienna Type' and a CNC model 5.403) operating at a flow rate of 0.3 L/min. This system allows the measurement of the number of particles according to their electrical mobility diameter (from 5.5 to 350 nm with the middle size DMA, and from 11 to 1080 nm with the long DMA). The choice of the DMA was adapted depending on the median size of the aerosols measured to provide most of the distribution.

Figure 5 presents the count median mobility diameters (CMMD) for all types of electrode as a function of the spark-discharge frequency for configurations 1 and 2 respectively.

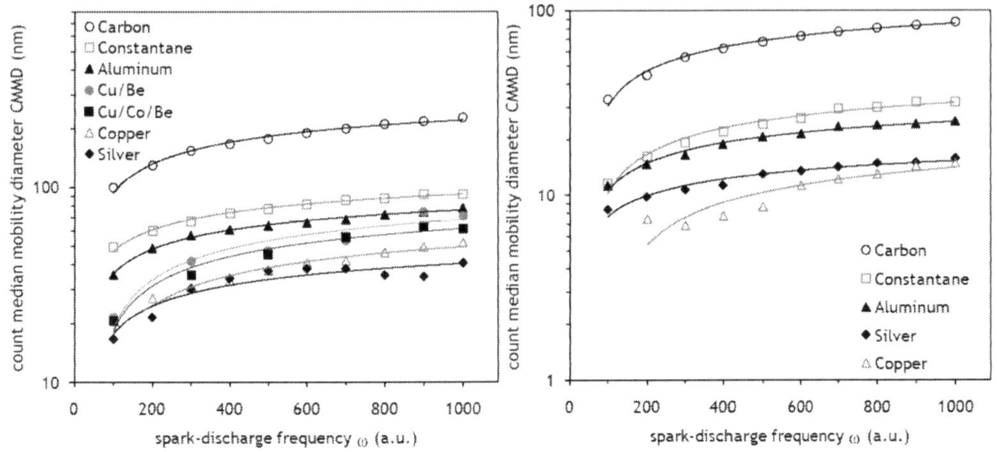

Figure 5. Airborne NP count median mobility diameter for the different materials (derived from SMPS data)

It can be observed a significant evolution of the CMMD as a function of the spark frequency and additional air flow conditions imposed. In configuration 1, CMMD were found between 17 nm and 228 nm. In configuration 2, the range of CMMD extends from 7 to 87 nm. This decrease in particle size can be linked to lower coagulation effects by (1) dilution due to additional 20 L/min air flow rate, and (2) decrease of residence time (by a factor of 9).
From TEM micrographs, it can be seen that all collected NP are aggregates, except those stemming from Silver which are spherical. This might be due to the non oxidation of Silver-based NP, which enables primary particles to coalesce in absence of an oxidation layer, leading to metal-metal contacts, as stated by Tabrizi *et al.* [11].

3.3. Mass concentrations

A previous study carried out on another CAIMAN configuration allowed this determination by means of a TEOM system. If the data cannot be directly used because of configuration modification of CAIMAN, table 2 presents the range of mass concentrations of airborne NP produced and maximum mass production rate.

Table 2. Range of mass concentrations and mass rates of NP

Electrode	Range of mass concentration (μg/m³)	Maximum mass rate (μg/h)
C	800 – 6400	4800
Al	< 20 – 450	340
Cu	< 20 – 240	180
Ag	< 20 - 680	510
Cu/Be	110 – 1800	1350
Cu/Co/Be	140 – 740	555

3.4. Electrical state-of-charge

The mean charge per airborne NP generated within our facility was determined by the simultaneous measurement of the number concentration (GRIMM CNC model 5.403) and current (electrometer TSI model 3068B). The mean charge per particle ($\bar{p}$) is given by:

$$\bar{p} = \frac{I}{C_N \cdot e \cdot Q}$$

where C_N corresponds to the mean number concentration of airborne NP (measured by the CNC), I is the mean current (measured by the electrometer), e represents the elementary charge ($e = 1.6 \cdot 10^{-19}\,C$) and Q is the flow rate of the electrometer (adjusted from 0.3 to 5 L/min to measure a current well above the limit of quantification of the electrometer).
The same behavior can be noted for each electrode material. For sparking frequencies below 200-300 a.u., the raw mean charge per particle is negative, and becomes positive for sparking frequencies above this limit. The addition of ions produced within the ion generator lead to a wide range of possibilities in terms of electrical state-of-charge of airborne NP, summarized in table 3.

Table 3. Range of variation of the mean charge per particle of airborne NP

Electrode material	Range of variation of raw $\bar{p}$		$\bar{p}$ with ⊕ ions		$\bar{p}$ with ⊖ ions		Configuration
C	-1.61	0.94	1.11	3.09	-1.59	-2.46	1
	-0.25	0.18	0.40	0.58	-0.33	-0.88	2
Al	-0.53	0.48	0.43	1.11	-0.32	-1.60	1
	-0.11	0.05	0.09	0.21	-0.13	-0.24	2
Cu	-0.08	0.05	0.07	0.39	-0.19	-0.46	1
	0.00	0.01	0.06	0.13	-0.08	-0.21	2
Ag	-0.02	0.02	0.19	0.33	-0.16	-0.49	1
	-0.01	0.01	0.03	0.05	-0.04	-0.07	2
Constantane	-0.44	0.31	0.89	2.04	-0.78	-1.99	1
	-0.03	0.04	0.13	0.26	-0.13	-0.29	2

3.5. Particle shape

Partial sintering of primary particles constituting aggregates of Aluminum-based NP was observed by use of the high-temperature furnace. As a consequence, airborne NP morphology was varied from chain-like aggregates to "compact" aggregates when increasing the temperature within the furnace. In parallel of NP sampling, SMPS number size distributions were measured, as presented in figure 6.

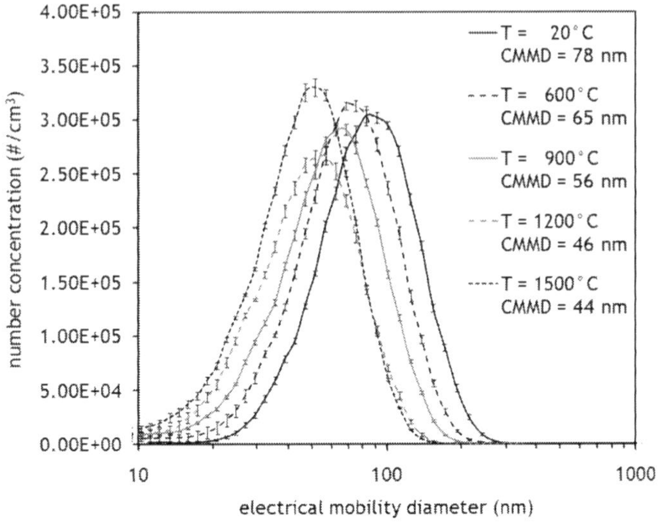

Figure 6. Effect of the furnace temperature on the number size distributions measured for Aluminum electrodes in configuration 1, ω = 1000 a.u.

Figure 7 presents TEM pictures of sampled Aluminum-based NP for the different temperatures. Particle restructuring can be highlighted in these pictures when increasing temperature.

From figure 7, it can be observed that particles tend to be more compact when temperature increases. This feature may be useful for studying the effect of particle morphology with constant chemistry on instruments behavior.

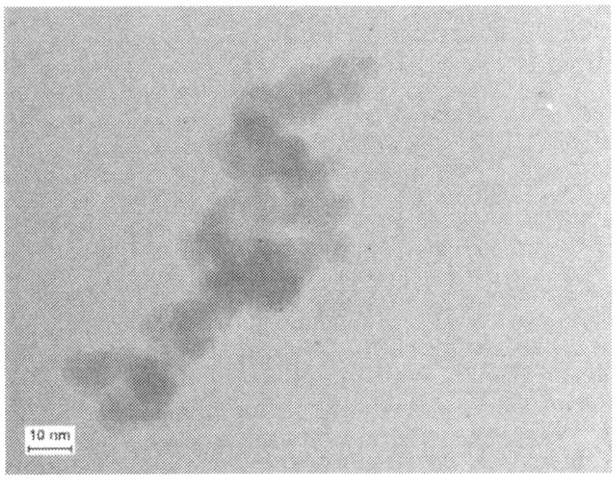

T = 20°C

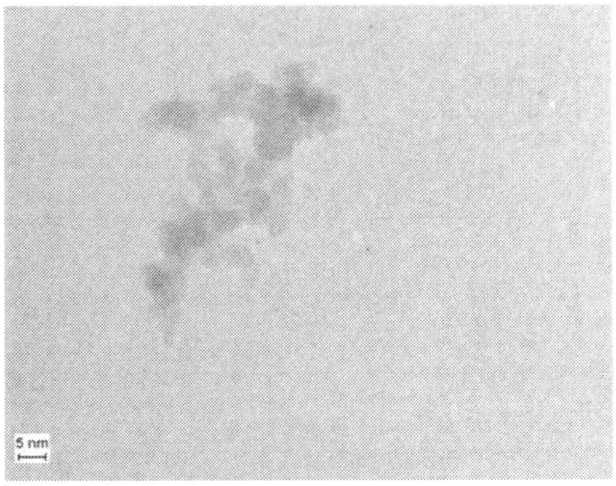

T = 900°C

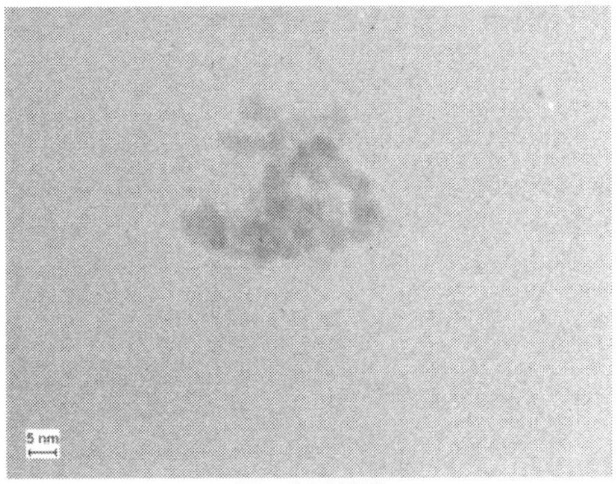

T = 1200°C

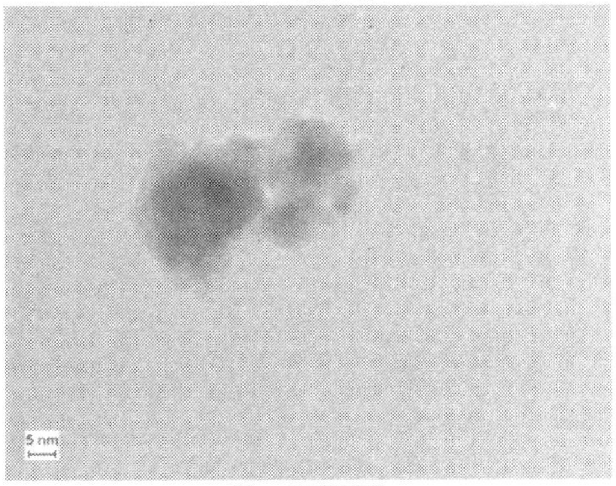

T = 1500°C

Figure 7. Change of morphologies for airborne NP for Aluminum electrodes at various sintering temperatures: ambient temperature (20°C), sintering at 900°C; sintering at 1200°C; sintering at 1500°C.

3.6. Aerosol dilution

To avoid saturating the instruments in study, airborne NP were produced from Carbon electrodes with ω = 1000 a.u. The additional air flow rate injected within the ageing volume (notated Q_5) was successively set to 10 (normal configuration), 20, 30, 40 and 50 L/min. Table 4 presents, for both configurations, the evolution of aerosol CMMD and the number concentrations ratio calculated by:

$$\text{ratio} = \frac{C_N(Q_5)}{C_N(Q_5 = 10\ \text{L/min})}$$

Table 4. Effect of the dilution flow rate (Q_5) on the relative number size distributions measured for Carbon electrodes with ω = 1000 a.u. and varying additional air flow rates Q_5

	Configuration 1		Configuration 2	
Q_5 (L/min)	CMMD (nm)	Ratio	CMMD (nm)	Ratio
10	248	-	93	-
20	234	0.69	90	0.88
30	233	0.50	86	0.69
40	232	0.39	83	0.58
50	231	0.32	81	0.51

It can be observed from this table that the aerosol characteristics are almost independent from the additional air flow, dilution factors up to 3.1 can be obtained without significant change of the airborne NP number size distribution.

3.7. Particle mixture (NP together with background aerosols)

One requirement of test facilities is the possibility to deliver well-known particle mixture containing combinations of the "primary" nanoaerosols together with particles representatives of background aerosols. The initial CAIMAN configuration has been redesigned to produce not only the "primary" nanoaerosol (standard configuration) but also a known particle mixture containing combinations of the "primary" nanoaerosol and particles representatives of background aerosols. In the present work, preliminary tests have been carried out with NaCl particles as being representatives of background aerosols. The Laskin-type nebulizer was chosen as generation system. The aerosol of NP produced within CAIMAN was mixed to an aerosol of NaCl crystals at the inlet of the ageing volume.
The preliminary experiments were performed with an aqueous solution of NaCl at a concentration of 0.5 g/L and an air flow rate of 20 L/min within the Laskin-type nebulizer. SMPS measurements are shown in figure 9, where the finer mode (23 nm) corresponds to the NaCl particles and the larger one (350 nm) to the carbon particles.

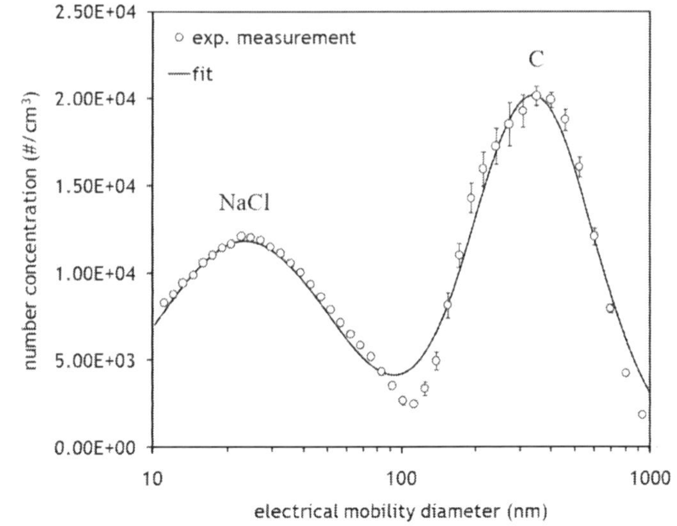

Figure 8. Number size distribution of the particle mixture obtained from Carbon electrodes and NaCl particles. The finer mode (23 nm) corresponds to NaCl particles and the larger one (350 nm) to C particles

4. Conclusion

In this work, the performances of the CAIMAN non-transportable facility have been evaluated. The CAIMAN facility provides now well-characterized "primary" nanoaerosols at known concentrations, sizes, shapes and mean charge levels. Within CAIMAN, nanoaerosols are produced by means of a spark-discharge generator (GFG-1000, PALAS). Until now, seven material electrodes have been used within CAIMAN: Carbon (delivered by PALAS), Aluminum (type 2071A), Copper, Silver, Constantane, Cu/Be and Cu/Co/Be.
The airborne NP characterized in this work present count median mobility diameters from 7 to 228 nm, with number concentrations varying from 1.7 10^6 to 1.8 10^7 #/cm^3.

The standard configuration of CAIMAN has been redesigned to produce not only the "primary" nanoaerosol but also a known particle mixture containing combinations of the "primary" nanoaerosol and particles representatives of background aerosol. In the present work, tests have been made with NaCl.
The output of the CAIMAN facility is now very consistent over long time intervals when operating under similar conditions. It indicates that repeatability is one of the important assets of the CAIMAN facility.

Acknowledgments
The research leading to these results has received funding from the European Community's Seventh Framework Programme (FP7/2007-2013) under grant agreement n°211464-2.
Authors would like to thank S. Veissiere from INRS for the TEM pictures.

References

[1] Schwyn S, Garwin E and Schmidt-Ott A 1988. *J. Aerosol Sci.* **19**(5): 639.
[2] Helsper C, Molter W, Loffler F, Wadenpohl C, Kaufmann S and Wenninger G 1993. *Atmos. Environ.* **27**: 1271.
[3] Roth C, Karg E and Heyder J 1998. *J. Aerosol Sci.* **29**(S1): S679.
[4] Brown J S, Kim C S, Reist P C, Zeman K L and Bennett W D 2000. *Aerosol Sci. & Technol.* **32**: 325.
[5] Evans D E, Harrison R M and Ayres J G 2003. *J. Aerosol Sci.* **34**: 1023.
[6] Evans D E , Harrison R M and Ayres J G 2003. *Aerosol Sci. & Technol.* **37**: 975.
[7] Horwath H and Gangl M 2003. *J. Aerosol Sci.* **34**: 1581.
[8] Roth C, Ferron G A, Karg E, Lentner B, Schumann G, Takenaka S and Heyder J 2004. *Aerosol Sci. & Technol.***38**: 228.
[9] Borra J P 2006. *J. Phys. D*, **39**: R19.
[10] Byeon J H, Park J H and Hwang J 2008. *J. Aerosol Sci.* **39**: 888.
[11] Tabrizi N S, Ullmann M, Vons V A, Lafont U and Schmidt-Ott A 2009. *J. Nanopart. Res.* **11**(2): 315.
[12] Tabrizi N S, Xu Q, van der Pers N M, Lafont U and Schmidt-Ott A 2009. *J. Nanopart. Res.* **11**(5): 1209.
[13] Tabrizi N S, Xu Q, van der Pers N M and Schmidt-Ott A 2010. *J. Nanopart. Res.* **12**(1): 247.
[14] Bau S, Witschger O, Gensdarmes F and Thomas D 2009. *J. Phys., Conf. Series* **170**: 012006. DOI 10.1088/1742-6596/170/1/012006.
[15] Bau S, Witschger O, Gensdarmes F, Thomas D and Borra J P 2010. *J. Nanopart. Res.* **12**: 1989.
[16] Marquard A, Meyer J and Kasper G 2006. *J. Aerosol Sci.* **37**: 1052.
[17] Marquard A, Meyer J and Kasper G 2006. *J. Aerosol Sci.* **37**: 1069.
[18] Schmidt-Ott A 1988. *J. Aerosol Sci.* **19**: 553.
[19] Schmidt-Ott A, Baltensperger U, Gäggeler H W and Jost D T 1990. *J. Aerosol Sci.* **21**(6): 711.
[20] Shimada M, Seto T and Okuyama K 1994. *J. Chem. Eng. Japan* **27**: 795.
[21] Weber A P, Baltensperger U, Gäggeler H W and Schmidt-Ott A 1996. *J. Aerosol Sci.* **27**: 915.
[22] Weber A P and Friedlander S K 1997. *J. Aerosol Sci.* **28**: 179.
[23] Jang H D and Friedlander S K 1998. *Aerosol Sci. & Technol.***29**: 81.
[24] Nakaso K, Shimada M, Okuyama K and Deppert K 2002. *J. Aerosol Sci.* **33**(7): 1061.
[25] Ku B K and Maynard A D 2005. *J. Aerosol Sci.* **36**(9): 1108.
[26] Ku B K and Maynard A D 2006. *J. Aerosol Sci.* **37**: 452.
[27] Cho K, Hogan C J and Biswas P 2007. *J. Nanopart. Res.* **9**: 1003.
[28] Wang Y Q, Liang W S and Geng C Y 2010. *J. Nanopart. Res.* **12**: 655.

Response of three instruments devoted to surface-area for monodisperse and polydisperse aerosols in molecular and transition regimes

Sébastien Bau[1], Olivier Witschger[1], François Gensdarmes[2] and Dominique Thomas[3,4]

[1]Institut National de Recherche et de Sécurité (INRS), Laboratoire de Métrologie des Aérosols, Rue du Morvan, CS 60027, 54519 Vandoeuvre Cedex, France

[2]Institut de Radioprotection et de Sûreté Nucléaire (IRSN), Laboratoire de Physique et de Métrologie des Aérosols, BP 68, 91192 Gif-sur-Yvette, France

[3]Laboratoire Réactions et Génie des Procédés (LRGP), groupe SAFE, 1 rue Grandville, BP 20041, 54001 Nancy Cedex, France

[4]Nancy Université, Université Henri Poincaré, F-54500 Vandoeuvre, France

E-mail: sebastien.bau@inrs.fr

Abstract. An increasing number of experimental and theoretical studies focus on airborne nanoparticles (NP) in relation with many aspects of risk assessment. Indeed, our understanding of the hazards, the actual exposures in the workplace and the limits of engineering controls and personal protective equipment with regard to NP are still under development.
Several studies have already identified surface-area as an important determinant of low solubility nanoparticles toxicity. As a consequence, the concept that surface-area could be a relevant metric for characterizing exposure to low solubility airborne NP has been proposed [1]. To provide NP surface-area concentration, some direct-reading instruments have been designed, based on diffusion charging. The actual available instruments providing airborne NP surface-area concentration are studied in this work: LQ1-DC (Matter Engineering), AeroTrak™ 9000 (TSI) and NSAM (TSI model 3550). Their performances regarding monodisperse carbon NP have been investigated by Bau *et al.* [2].
This work aims at completing the instruments characterization regarding monodisperse NP of other chemical composition (aluminium, copper, silver) and studying their performances against polydisperse aerosols of NP.

1. Introduction

The novel properties of engineered nanostructured particles (NP) and their applications have given birth to technological and economic expectations for industries using NP or products containing NP. However, their unique properties may also make them harmful to humans. Consequently, assessing inhalation exposure to airborne NP constitutes an important challenge.

Several studies have already identified surface area as an important determinant of low solubility nanoparticles toxicity [3-5]. Although there is still considerable uncertainty over how exposure to airborne NP should be should be most appropriately evaluated, it has been proposed that surface-area measurements should be relevant to a broad range of NP class/attribute combinations [1].
Direct-reading instruments based on diffusion charging have been designed to measure airborne NP active surface-area concentration (LQ1-DC, Matter Engineering) or lung-deposited surface-area concentrations (NSAM model 3550 and AeroTrak™ 9000, TSI Inc.). Their performances regarding monodisperse carbon NP have been investigated by Bau *et al.* [2].
The aims of this work were (1) to complete previous data on the instrument response functions for monodisperse aerosols with three other chemical substances (aluminum, copper, silver) and (2) to characterize the instruments responses regarding polydisperse aerosols of NP.

2. Materials and methods

2.1. Aerosol production

CAIMAN (for 'Characterization of Instruments Measuring Aerosols of Nanoparticles') is a versatile experimental facility that was designed and built to generate stable and reproducible aerosols of nanoparticles (nanoaerosols) with controlled properties (concentration, size, shape, mean charge level). It was initially developed in a collaborative research work within the Aerosol Physics and Metrology Laboratory from the Institut de Radioprotection et de Sûreté Nucléaire (IRSN) and the Aerosol Metrology Laboratory from INRS, which holds the non-transportable facility in its research centre near Nancy, France.
In this facility, airborne NP are produced by means of a spark-discharge generator (GFG-1000, PALAS). In this work, four different electrode materials have been studied: carbon, aluminum, copper and silver.
After being generated, further conditioning of the aerosols of NP produced within CAIMAN is achieved by means of:

- a bipolar ion generator (EAN 581, TOPAS) to cover a wide range of electrical mean charge level,
- a high-temperature furnace (CARBOLITE BST16 – maximum operating temperature 1500°C) to provide either partial sintering of agglomerates or coalescence, allowing some degree of control over particle morphology of particles with constant chemistry,
- a 2-liter volume located at the end of the facility from which airborne NP can be sampled and diluted by means of an additional air flow injected at the outlet of the device. It should be noted that this ageing volume is equipped with four sampling lines in parallel, which allows simultaneous measurements.

In this study, neither the ion generator nor the high-temperature furnace were used to modify the properties of the aerosols produced. By varying the operating parameters of the generation, aerosols of NP with differing properties have been produced within CAIMAN. The main characteristics of these aerosols are gathered in Table 1: count median mobility diameter – CMMD – (SMPS), mean agglomerate and primary particle size (TEM-images analysis) and 2D-fractal dimension (TEM-images coupled to an optimized Box Counting procedure described in [2]). The ranges presented in Table 1 were obtained by varying the sparking frequency from 30 to 300 Hz and the air flow introduced into the NP generator from 0 to 20 L/min. All runs were performed with an argon flow rate of 2.5 L/min and a dilution flow rate of 10 L/min introduced into the ageing volume.

Table 1. Characteristic properties of the aerosols of nanoparticles produced within CAIMAN.

Electrode material	C	Al	Cu	Ag
Range of CMMD (nm)	27 – 163	8 – 50	6 – 18	6 – 18
Mean size of agglomerates (nm)	205 ± 95	550 ± 46	93 ± 26	12 ± 3
Primary particle diameter (nm)	16 ± 2	16 ± 3	9 ± 2	N.A.
2D-fractal dimension (-)	1.67 ± 0.03	1.59 ± 0.02	1.55 ± 0.06	N.A.

N.A.: not applicable

Table 1 suggests that the CAIMAN facility provides now well-characterized nanoaerosols at known concentrations, sizes and shapes. In this work, four material electrodes have been used within CAIMAN, presenting count median mobility diameters (CMMD) from 6 to 163 nm. It should be noted that silver electrodes lead to spherical particles; consequently the primary particle size and agglomerate fractal dimensions were not determined for this electrode material.
The output of the CAIMAN facility is very consistent over long time intervals when operating under similar conditions. It indicates that repeatability is one of the important assets of the CAIMAN facility. A more complete description of the features of CAIMAN is available in this issue [6].

2.2. Study of the instruments in monodisperse mode: determination of the response functions

To investigate the response of the instruments in study regarding monodisperse aerosols, a Differential Mobility Analyzer (DMA Grimm Vienna Type, operating flow rate of 0.3 L/min) was placed at the outlet of the ageing volume. Figure 1 is a schematic of the experimental setup.

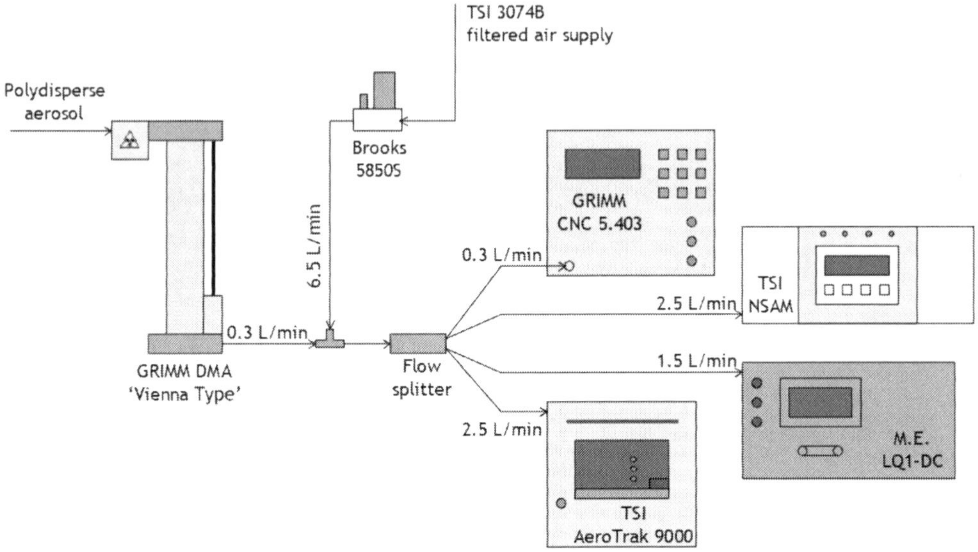

Figure 1. Experimental setup for the measurement of the response functions of the instruments on monodisperse aerosols at the outlet of CAIMAN.

For a selected particle size d_m, the experiments consisted in measuring the lung-deposited (NSAM and AeroTrak™ 9000) and active (LQ1-DC) surface-area concentrations, notated $C_S(d_m,t)$; in parallel, the number concentration was measured by means of a condensation nuclei counter (CNC Grimm model 5.403), notated $C_N(d_m,t)$.

The instrument response, for the selected size d_m, is then related to the mean concentrations measured over 10 minutes:

$$\mathcal{R}(d_m) = \frac{\overline{C_S(d_m)}}{\overline{C_N(d_m)}}.$$

It should be mentioned that the majority of particles exiting the DMA carried one negative charge and a small fraction carried two negative charges, they have not been neutralized before measurement with surface area instruments. Particles carrying multiple charges exiting the DMA and measured simultaneously were taken into account according to a classical procedure described in Bau *et al.* [2].
The slight polydispersion (geometric standard deviations from 1.4 to 1.8) of the aerosols produced allowed the selection and measurement of monodisperse NP depending on their nature: from 20 to 520 nm for carbon, from 15 to 120 nm for aluminum electrodes and from 15 to 100 nm for copper and silver electrodes.

2.3. Study of the instruments in polydisperse mode
The second objective of that study was to investigate the instruments response regarding polydisperse aerosols of NP. This step is particularly interesting because it represents the situations encountered by the instruments in the case of workplace exposure assessment or even environmental monitoring.
The setup used in this configuration is schematized in figure 2.

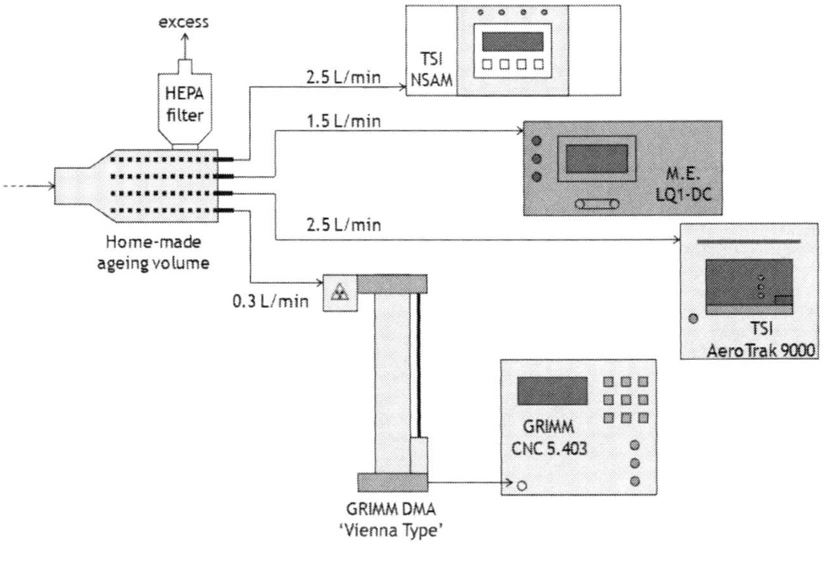

Figure 2. Experimental setup for the measurement of the response functions of the instruments on polydisperse aerosols at the outlet of CAIMAN.

In polydisperse mode, the experiments consisted in measuring in parallel the surface-area concentrations (lung-deposited, active) and the number size distribution of the aerosol (SMPS Grimm composed of Vienna Type DMA and a CNC 5.403). Data from experiments carried out in polydisperse mode were then used to compare surface-area concentrations measured C_S to those calculated (C_S^*) from the response function and the number size distribution $C_N(d_m)$ according to:

$$C_S^* = \sum_{d_m} \mathcal{R}(d_m) C_N(d_m).$$

Different operating parameters have been used within CAIMAN to produce aerosols of NP of four substances with count median mobility diameters from 64 to 177 nm (geometric standard deviations from 1.45 to 1.54) and number concentrations between 10^4 and 15.10^4 #/cm^3. It should be noticed that in this case, the majority of particles carry zero charge as they are used directly from the outlet of the spark discharge generator, as stated in Bau *et al.* [7].

2.4. Description of the instruments under study

Direct-reading instruments based on diffusion charging have been designed to measure airborne nanostructured particles active surface-area concentration (LQ1-DC, Matter Engineering) or lung-deposited surface-area concentrations (NSAM model 3550 and AeroTrak™ 9000, TSI Inc.).

The principle of these instruments is identical: airborne NP entering the device are first electrically charged within a diffusion charger where positive ions attach the particles. An ion trap, positioned at the exit of the charger, removes the ions produced in excess, while the charged particles are collected on a filter. The filter current measured in real-time is then directly correlated to the aerosol flow rate and the number of charges carried by the particles, which is related to their surface-area.

The next subsections briefly describe the instruments examined in this work.

2.4.1. NSAM model 3550 and AeroTrak™ 9000

NSAM and AeroTrak™ 9000 were designed to measure the lung-deposited surface-area concentration of airborne NP either in the alveolar (alv) or in the trachea-bronchiolar (tb) region. This is enabled by adjusting the voltage in the ion trap to manipulate the particle size distribution and therefore the response function (200 V alv, 100 V tb).

Lung-deposited surface-area concentrations correspond to the product of the geometric surface-area and the deposition probability in the considered region:

$$C_{S_{dep,i}} = C_{S_{geo}} \cdot \eta_{dep,i} .$$

The deposition probabilities $\eta_{dep,i}$ are defined by the lung deposition model developed by ICRP [8] using physiological, worker activity related and aerosol parameters considered by the manufacturer of the NSAM and Aerotrak™ 9000 [9].

2.4.2. LQ1-DC

The LQ1-DC is an instrument designed to measure the active surface-area concentration of airborne NP in real-time.

First introduced by Fuchs [10], the concept of active surface-area represents the fraction of the geometric surface-area of a particle that interacts with surrounding species. In the case of diffusion chargers, the latter correspond to the positive ions produced within the charger and that attach to the surface of the particles.

Jung & Kittelson [11] proposed a simplified approach to calculate the active surface-area of a spherical particle according to:

$$S_{act} = S_{geo} \cdot \frac{Cu^*(d_p, \lambda_{ion})}{Cu(d_p, \lambda_{ion})},$$

where:

$$Cu^*(d_p, \lambda_{ion}) = 2\frac{\lambda_{ion}}{d_p}(\alpha + \beta) \text{ and } Cu(d_p, \lambda_{ion}) = 2\frac{\lambda_{ion}}{d_p}\left[\alpha + \beta \cdot \exp\left(-\frac{d_p \cdot \gamma}{2 \cdot \lambda_{ion}}\right)\right].$$

In these equations, α, β and γ are the slip correction coefficients, taken for 1.165, 0.483 and 0.997 respectively according to Kim *et al.* [12]. The ion mean free path λ_{ion} was 15.2 nm.

3. Experimental results

3.1. Monodisperse mode

The response functions measured with monodisperse aerosols were found in good agreement with theoretical curves in the range from 15 to 520 nm for the three devices in the case for carbon [2].

The same behavior was observed for the three other substances (copper, aluminum, silver); the biases between experimental data and theory are in the range -45 % to +50 %.

More precisely, the biases observed for LQ1-DC are in the range –40 % to +30 %, for AeroTrak™ 9000 between –45 % and +30 %, and for NSAM between –10 % and +30 %. It should be noted that the three-quarters of the experimental data are in a range of biases to theory of ± 25 %.

3.2. Polydisperse mode

As stated in paragraph 2.3, the surface-area concentrations were measured and calculated from the SMPS number size distribution coupled to the experimental response functions measured for monodisperse NP. Figure 3 presents the comparison between measured and calculated surface-area concentrations (active and lung-deposited) for all conditions.

Whatever the instrument considered, it can be observed from figure 3 that there are deviations between measured and calculated surface-area concentrations. Over all experiments, the relative discrepancies are found between -35% and +150%. They are calculated by:

$$\Delta(\%) = \frac{C_S^* - C_S}{C_S} \cdot 100 \ .$$

Table 2 presents the range of biases for each instrument in study.

Table 2. Range of relative biases observed between measured and calculated surface-area concentrations regarding polydisperse aerosols.

Instrument	Range of bias
NSAM – alv. deposition	-11% - +103%
AeroTrak™ 9000 – alv. deposition	-6% - +43%
NSAM – tb. deposition	0% - +107%
AeroTrak™ 9000 – tb. deposition	+11% - +150%
LQ1-DC	-35% - +117%

The discrepancies observed are quite important, which led us to formulate hypothesis to explain these deviations.

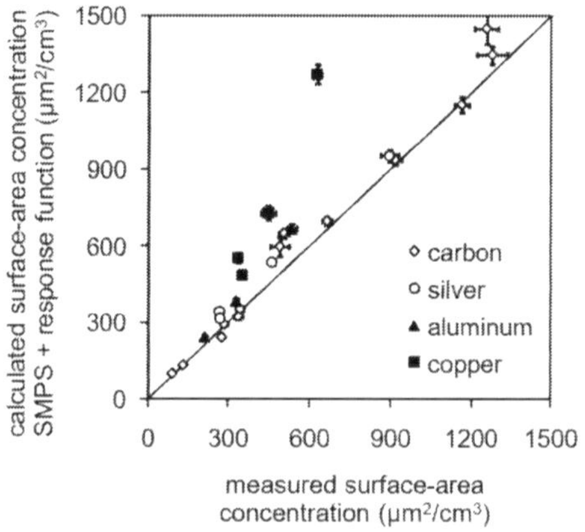

(a) NSAM alv

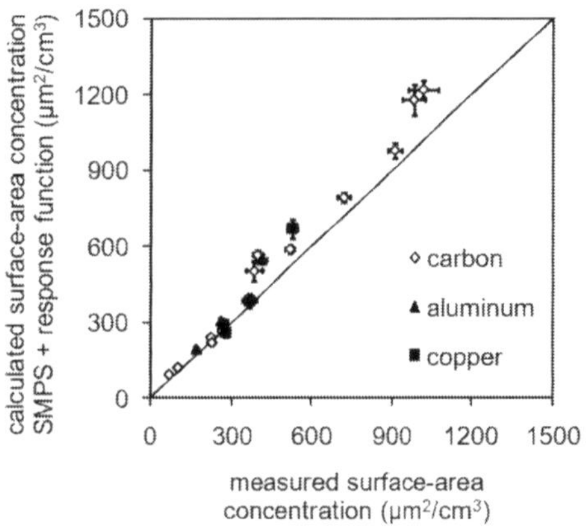

(b) AeroTrak 9000 alv

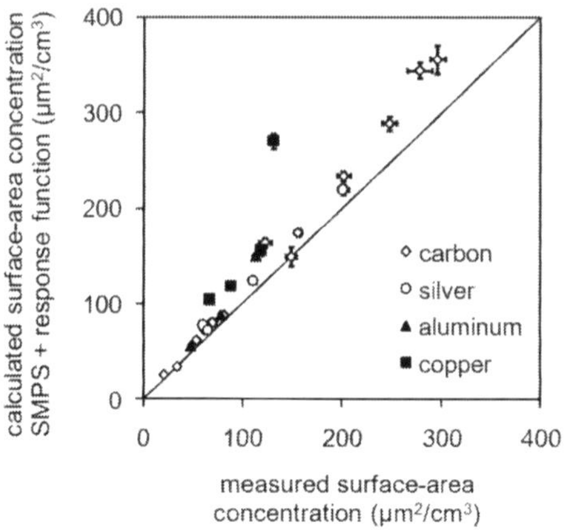

(c) NSAM tb

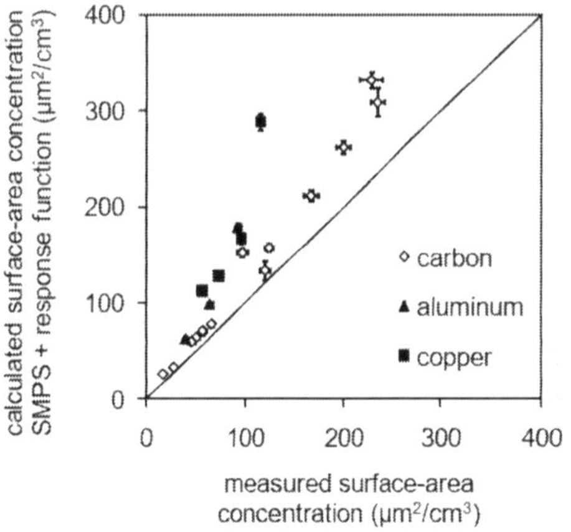

(d) AeroTrak 9000 tb

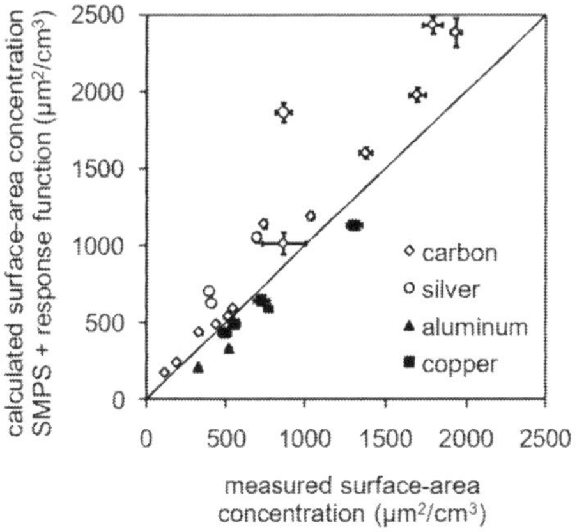

(e) LQ1-DC

Figure 3. Experimental results for polydisperse aerosols.

3.2.1. Influence of airborne NP number concentration

The first parameter that might influence the response of the instruments is the total concentration of the aerosol. Indeed, the charging efficiency of the diffusion chargers can be varied, leading to different number of elementary charges carried by the particles. To examine this effect, two conditions (named "A" and "B") with identical relative number size distribution have been compared. Table 3 presents the main characteristics of these polydisperse aerosols.

Table 3. Properties of the polydisperse aerosols of NP in conditions A & B

Configuration	A	B
CMMD (nm)	64.5	65.7
GSD (-)	1.53	1.53
Total number concentration (#/cm^3)	1.493.10^5	4.905.10^4
Electrode material	Ag	Ag

For these configurations, the ratio (r) of the number concentrations per channel is constant over the whole range of particle size:

$$r(m) = \frac{C_N(m)_A}{C_N(m)_B} = 3.03 \pm 0.15 \approx K .$$

As a consequence, the ratio of the surface-area concentrations should be found near the value of 3. However, our results indicate that the experimental ratios range from 3.33 for NSAM in "alv" configuration to 1.89 for AeroTrak™ 9000 in "tb" configuration. LQ1-DC is placed in between with a ratio of 2.13. This indicates that measured surface-area concentration is not proportional to the number concentration of the particles, whatever the instrument used.

This observation then led us to investigate the influence of the charge distribution of the particles on the signal measured by the diffusion chargers.

3.2.2. Influence of airborne NP electrical charge distribution

Within a diffusion charger, particles carry p elementary charges, depending on (1) the ion concentration N_i, (2) their residence time within the charger t, and (3) their size d_p. The current I measured on the filter of the instruments corresponds to the total amount of electrical charges deposited, *i.e.*:

$$I = e \cdot Q \cdot \sum_{d_p} p(d_p) C_N(d_p) .$$

From this current, surface-area concentration is obtained by means of a linear relationship established by the manufacturer:

$$C_S = \alpha \cdot I ,$$

where α is a constant determined after calibration. Physically, this parameter is a complex function depending mainly on attachment coefficients of ions to particles surface, $N_i t$ product of the charger, number concentration, shape, and size of the particles.

Because particles of size d_p do not all carry the same number of elementary charges, there are configurations where the total amount of electrical charges is the same while the surface-area is different. The numeric values taken as a reference were published by Biskos *et al.* [13] and were obtained with a unipolar diffusion charger with $N_i t = 1.97\ 10^{13}$ ions.s/m^3.

For example, a concentration of 10 #/cm^3 of 40-nm particles will lead to the measurement of the same current than a concentration of 4 #/cm^3 of 120-nm particles. According to the last equation, the instrument will indicate the same surface-area concentration. However, the 120-nm particles, even less in terms of number concentration, represent 3.6 times more in terms of geometric surface-area. This effect can result in differences in the measurement of polydisperse aerosols; nevertheless it is difficult to conclude whether this effect will over or underestimate the true surface-area concentration.

4. Conclusion

This experimental study provides new results leading to a better understanding and characterization of three devices devoted to the real-time measurement of airborne NP surface-area concentration. It should be reminded that this parameter has been highlighted by several toxicological studies for the case of insoluble NP but is not associated to standardized measurement method yet.

The first part of this work was devoted to the measurement of the response function of the instruments regarding monodisperse NP. For the instruments NSAM and LQ1-DC, our experimental data enrich and enlarge the size range of the previously published data [14-16]. Concerning the third instrument AeroTrak™ 9000, these data constitute the first ever published.

In a second step, instruments were used to measure the surface-area concentration of polydisperse airborne NP. Calculations from SMPS size distributions and response functions for monodisperse NP were compared to responses of the instruments obtained for polydisperse NP, showing discrepancies from -36% to +150%. Hypotheses were formulated to interpret the deviations observed; in particular, the influence of the total number concentration of NP and their electrical charge distribution were identified as sources of bias.

References

[1] Maynard A D and Aitken R J 2007. *Nanotoxicol.* **1**(1):26.

[2] Bau S, Witschger O, Gensdarmes F and Thomas D 2009. *J. Phys., Conf. Series* **170**: 012006. DOI 10.1088/1742-6596/170/1/012006.

[3] Tran C L, Buchanan D, Cullen R T, Seal A, Jones A D and Donaldson K 2000. *Inhal. Toxicol.* **12**:1113.

[4] Stoeger T, Reinhard C, Takenata S, Schroeppel A, Karg E, Ritter B, Heyder J and Schulz H 2006. *Environ Health Perspectives* **114**(2):328.

[5] Rabolli V, Thomassen LCJ, Princen C, Napierska D, Gonzales L, Kirsch-Volders M, Hoet P H, Huaux F, Kirschhok C E A, Martens J A and Lison D 2010. *Nanotoxicol.* **4**(3):307.

[6] Jacoby J, Bau S and Witschger O. *J. Phys., Conf. Series*. This issue.

[7] Bau S, Witschger O, Gensdarmes F, Thomas D and Borra J-P 2010. *J. Nanopart. Res.* **12**:1989.

[8] ICRP 1994. *Publication 66: Human respiratory tract model for radiological protection*. Oxford: Pergamon.

[9] TSI Application Note 2008. *Measuring Nanoparticle Exposure*. Application Note NSAM-001. 4/11/2008: 8p.

[10] Fuchs N A 1964. *The mechanics of aerosols* (1st Edition). New York, Dover Publications Inc.

[11] Jung H and Kittelson D B 2005. Characterization of aerosol surface instruments in transition regime. *Aerosol Sci. and Technol.* **39**: 902.

[12] Kim J H, Mulholland G W, Kukuck S R and Pui D Y H 2005. *J. Res. Nat. Inst. of Standards and Technol.* **110**(1): 31.

[13] Biskos G, Reavell K and Collings N 2005. *J. Aerosol Sci.* **36**:247.

[14] Shin W G, Pui D Y H, Fissan H, Neumann S and Trampe A 2007. *J. Nanopart. Res.* **9**:61.

[15] Asbach C, Fissan H, Stahlmecke B, Kuhlbusch T A J and Pui D Y H 2008. *J. Nanopart. Res.* **11**:101.

[16] Ku B K and Maynard A D 2005. *J. Aerosol Sci.* **36**(9):1108.

Journal of Physics: Conference Series **304** (2011) 012016 doi:10.1088/1742-6596/304/1/012016

Development of a novel method to detect a potential inhalation of nanoaerosols

C Marie-Desvergne[1], F Tardif[2], S Motellier[2] and V Mossuz[1]

[1]LABM, Medical Biology Analyses Laboratory, CEA, 17 rue des Martyrs, 38054 Grenoble Cedex 9, France

[2]NanoChemistry and NanoSafety Laboratory, LITEN, CEA, 17 rue des Martyrs, 38054 Grenoble Cedex 9, France

E-mail: veronique.mossuz@cea.fr

Abstract. In the context of the precautionary principle for engineered nanoparticles, a test able to verify the non exposure of individuals to nanoaerosols seems to be of great interest, especially for workers involved with nanomaterials. A novel method is proposed to detect the presence of nanoparticles in the nose. Different materials, including paper and polycarbonate, have been tested for their sampling efficiency towards ZnO nanoparticles. The detection of nanoparticles is performed by elemental analysis using Total Reflection X-Ray Fluorescence (TXRF). For an optimal sensitivity of the TXRF signal, samples have to be deposited on a smooth surface. Therefore, different materials have been compared as sample carrier. The results indicate that pure paper is the most efficient material for sampling nanoparticles (ZnO), while the use of polycarbonate, and even more flexible polyester, as sample carrier allows a better sensitivity for TXRF analysis. Therefore, the sampling device made of paper is digested under slightly acidic conditions in a microwave oven after collection, and the resulting solution is deposited on polyester for TXRF analysis. Internal calibration is performed thanks to an addition of Rb to the samples. The proposed method allows the chemical identification of nanoparticles deposited in the nose: a qualitative signature of a potential nanoaerosol inhalation.

1. Introduction

The market of engineered nanoparticles is expanding very fast worldwide, and should reach a thousand billion dollars in 2011-2012, according to the estimates of the National Science Foundation in the United Stated [1]. It has been estimated in 2009 that more than a thousand products containing nanomaterials were already internationally commercialized [2].

Along with the increasing use of nanomaterials, there is a growing concern for potential health impact following inhalation due to the known adverse health effects of atmospheric particles and more precisely of ultrafine particles [3-5]. Toxicological studies have increased over the past few years, but there is still a lack of clear answer on the potential toxicity of nanomaterials.

The precautionary principle can still be applied by lowering the exposure to zero, and currently, its application is recommended in France [6-7]. In this approach, exposure monitoring of individuals occupationally exposed to nanomaterials is of major importance. The assessment of external exposure thanks to air sampling is essential, but this type of measurement does not take into account the use of individual protective devices, and individual parameters such as absorption, distribution and elimination. This is why, it is admitted that biomonitoring is the best approach to assess individual exposure to occupational and environmental toxicants [8]. Concerning nanoparticles, as shown by the congress on Nanomaterials and Worker Health held in Keystone (Colorado, USA) in July 2010, there is now a general agreement for the necessity to develop research studies on the establishment of biomonitoring [6, 9-10]. Pending the results of toxicological studies, a first step in this research area is the determination of biomarkers of exposure in order to identify the situations of exposure.

In general, biomonitoring is based on blood and urine analyses, but in the case of insoluble nanoparticles, toxicokinetics data would tend to prove that very few nanoparticles would be found in those matrixes following inhalation [6, 11-14]. This is why other methods have to be explored.

In the first place, a test able to verify the non exposure to nanoaerosols seems to be of great interest, especially at work places where workers might be exposed to nanomaterials. Therefore, nasal sampling could be proposed for the evaluation of potential inhalation to nanoparticles.

The objective of the present work is therefore to develop a method that would allow the sampling and the analysis of nanoparticles deposited in the nose. Ideally, this method should be fast and convenient in order to give a quick indication on the existence, or not, of an inhalation situation.

Microscopic analyses allow nanoparticles identification according to their shape, size, aggregation state, and even their composition using EDX additional technique, but such analyses are time-consuming and hardly adequate for routine measurements. Chemical analyses, such as atomic adsorption spectrometry or inductively coupled plasma mass spectrometry, allow the identification of particle composition but they often require fastidious pretreatment protocols due to the necessity to mineralize particles before analysis. Interestingly, total reflection X-ray fluorescence (TXRF) allows elemental analysis of surfaces without fastidious pretreatment procedures. This method is already used for diverse biological applications as referred in the review of Szoboszlai et al. [15]. Therefore, TXRF was evaluated for its ability to analyze nanoparticles after nasal sampling, according to their chemical composition. In the present work, this evaluation was performed on ZnO nanoparticles.

2. Materials and methods

2.1. Nanoparticles, chemicals and materials

Nanoparticles used in the study were ZnO nanoparticles in a commercial aqueous suspension from Evonik (Antwerp, Belgium). Scanning electron microscopy indicated that these nanoparticles are spherical with a diameter around 40 nm, and nanoparticles are essentially aggregated in water (data not shown).

Ultrapure water was used for solution preparation (MilliQ, Millipore). Suprapure nitric acid (65%) from VWR (Fontenay-sous-Bois, France) and H_2O_2 ($\geq$ 30%) from Sigma-Aldrich (Lyon, France) were used for microwave digestion. Rb ICP standard (1000 ppm) from Merck (Martillac, France) was used as an internal standard. Polycarbonate membranes were from Ecomesure (Janvry, France). Rigid polyester was cut in laser transparency film from 3M (Cergy-Pontoise, France), and flexible polyester was from Avery (Ecully, France). A new paper without impurities and a recycled paper used as sampling materials were from Fiorini (Ingre, France). They were pasted on wood sticks forming sampling devices in the shape of paper flags.

2.2. Determination of nanoparticle sampling efficiency

Ten µL of ZnO nanoparticles in water (1000 ppm) were deposited on glass slides and air-dried at room temperature. Glass slides were then rubbed 10 times with a polycarbonate membrane, a new paper flag, or a recycled paper flag, in triplicate for each material.

In a second experiment, 10 µL of ZnO nanoparticles in water at different concentrations (500, 1000, 2000, 4000, and 6000 ppm) were deposited on glass slides in duplicate, and air-dried at room temperature. Each series was then rubbed with either new paper flags or recycled paper flags.

Afterward, the polycarbonate membranes and the two types of paper were washed in 1 mL of HNO_3 0.65% in a three dimensional agitator (Turbula) during 2 hours in order to dissolve ZnO nanoparticles. The washing solutions were then quantitatively analyzed by inductively coupled plasma mass spectrometry (ICP-MS) so as to determine the percentages of nanoparticles recovered by each material.

2.3. TXRF analysis

TXRF analyses were performed on a Rigaku NanoHunter bench top apparatus. Prior to TXRF analysis, the sampling devices were digested in Teflon vessel containing 1 mL of HNO_3 1M + 1 mL of H_2O_2, in a microwave oven during 1.5 min. Afterward, 10 µL of the samples were deposited at the centre of the sample carrier and dried at 50°C on a hot plate. The materials used as sample carrier (paper, polycarbonate and polyester) were pasted on glass slides for rigidity. Zn and Rb analyses were performed on a Mo X-ray tube with an angle of the incident beam at 0.75° during 200 sec, and both elements were analyzed on their Kα lines.

2.4. Comparison of sample carriers for TXRF analysis

Prior to ZnO analysis all materials were tested for their absence of background fluorescence on Zn Kα line. Ten µL of nanoparticles of ZnO in water (1000 ppm) were deposited on different materials in triplicate, including polycarbonate membrane, pure and recycled paper, rigid and flexible polyester. All the samples were dried on a hotplate at 50°C. The intensities of the obtained TXRF responses were compared in order to determine which material shows the best sensitivity.

2.5. Calibration and limit of detection of the method

Calibration was performed using Rb as an internal standard. Ten µL of Rb (1000 ppm) were added to the sample prior to the microwave digestion step. A calibration curve was obtained by the TXRF analysis of solutions containing ZnO and Rb at different ratios in water.

The detection limit of the method (signal to noise = 3) was determined following the analysis of six blank paper flags. The limit of detection was calculated as three times the standard deviation of the six Zn/Rb blank values.

3. Results

3.1. Selection of the material for the nanoparticle sampling device

In order to choose the material for the sampling device, we compared the sampling efficiency of a polycarbonate membrane, known to be an adequate material as a substrate for TXRF analysis, and of two types of paper flags, a pure and a recycled one.

The polycarbonate membrane allowed a recovery of only 5 ± 2% of 10 µg of ZnO deposited on glass slides while the pure paper and the recycled paper allowed recoveries of 46 ± 4% and 55 ± 6%, respectively, for the same quantity of ZnO. The polycarbonate membrane, as a sampling material, was therefore eliminated for the rest of the study.

The pure paper and the recycled one were further studied for their sampling efficiency using increasing quantities of ZnO deposited on glass slides. The results shown in Figure 1 indicate that the sampling recovery was almost constant, around 50%, whatever the quantity of ZnO, for both paper flags, the recycled paper being a little more efficient than the pure one.

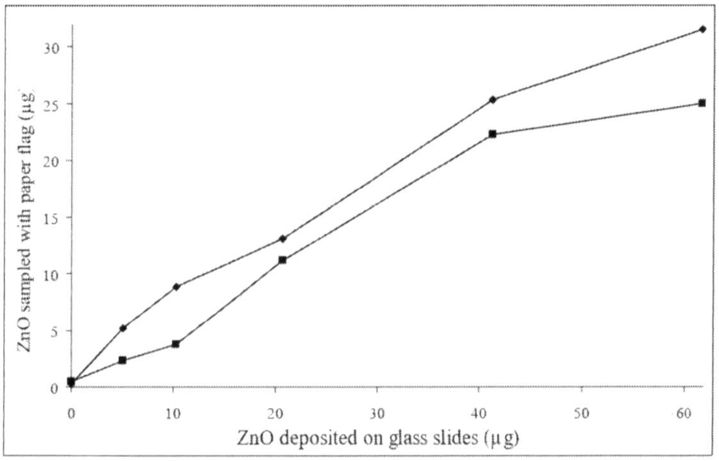

Figure 1. Sampling efficiency of recycled (♦) and pure (■) paper flags for increasing quantities of ZnO deposited on glass slides.

3.2. Selection of a sample carrier for best TXRF sensitivity

Different materials were compared as sample carrier for TXRF analysis of ZnO nanoparticles. The results of the comparison are shown in Figure 2 and indicate that the flexible polyester showed the most intense signal for ZnO detection with a two orders of magnitude higher signal than the signals obtained with the two types of paper. Therefore, the flexible polyester was chosen as the sample carrier for TXRF analysis in the final method.

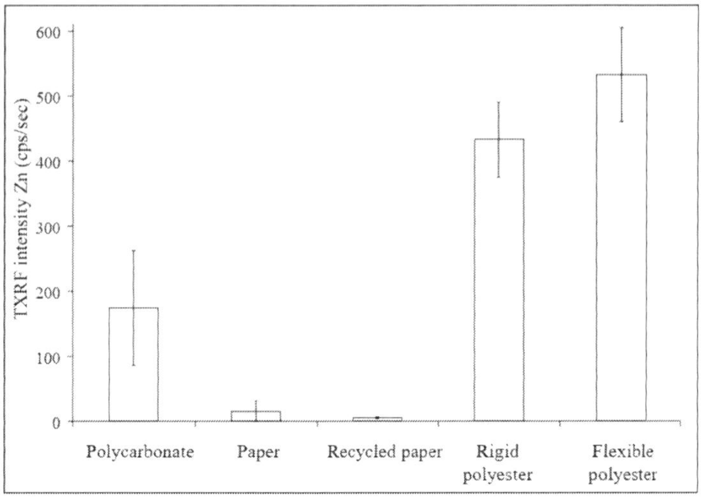

Figure 2. Comparison of different materials as sample supports for TXRF analysis after deposition of 10 µL of ZnO nanoparticles in water (1000 ppm) and drying at 50°C.

3.3. Proposition of an entire procedure

The entire protocol consisted in collecting nanoparticles with paper flags, digesting the paper flags and depositing the resulting liquid on flexible polyester for TXRF analysis. For that purpose, the digestibility of the two types of paper was tested (data not shown) and it appeared that the pure paper was more easily digested than the recycled one in acidic conditions described in part 2.3. Therefore, the pure paper flag was selected as the sampling device. Following paper digestion, the resulting solution was deposited on flexible polyester for TXRF analysis.

3.4. Calibration of TXRF measurement

The calibration of ZnO TXRF measurements was obtained using internal calibration with Rb. In Figure 3, a calibration curve of ZnO in water is shown, in addition to a calibration curve obtained following the analysis of paper flags spiked with 4 to 80 μg of ZnO. It appears that the two calibration curves are perfectly superimposed, the slope of the two curves being very similar. This means that no matrix effect due to the digestion step of the paper flags is observed, and therefore, the calibration in water can be used for the analysis of paper flags.

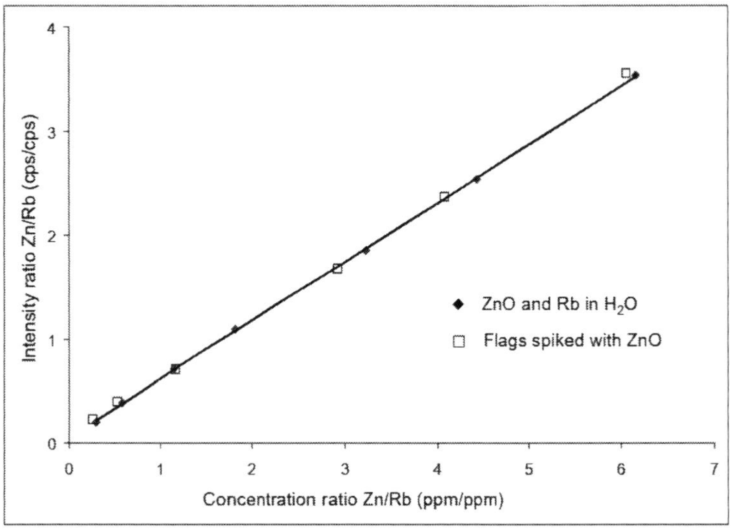

Figure 3. TXRF analysis of ZnO nanoparticles in water (◆, $y = 0.5642x + 0.0502$, $R^2 = 0.9998$) and of paper flags spiked with ZnO nanoparticles and treated according to the entire protocol (□, $y = 0.5702x + 0.0639$, $R^2 = 0.9991$), using Rb as an internal standard.

3.5. Sensitivity of the method

The limit of detection of the method was calculated to be a little less than 1 μg of ZnO on paper flags. The TXRF spectrum obtained for 2 μg of ZnO is shown in Figure 4 in comparison with a blank spectrum.

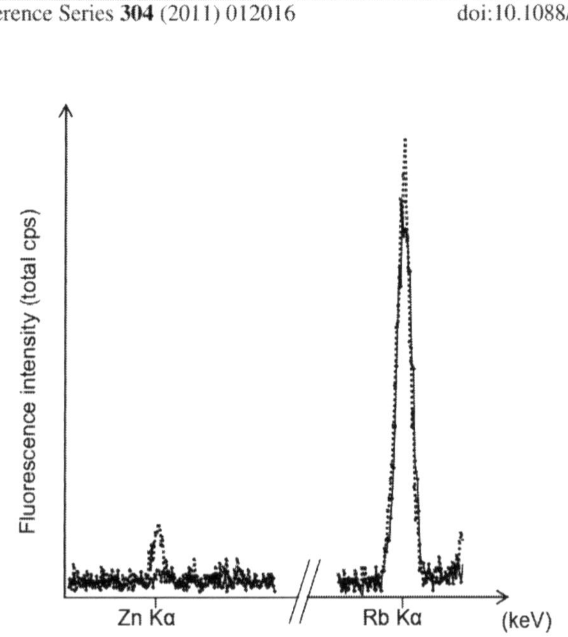

Figure 4. Superposition of fluorescence spectra obtained for the TXRF analysis of a blank paper flag (—) and of a paper flag spiked with 2 μg of ZnO (.....).

4. Discussion

The objective of the present work was to develop a method including the sampling and the analysis of nanoparticles potentially deposited in the nose after inhalation. In order to develop a fast and convenient method, the analytical technique of TXRF was chosen. This technique is an elemental surface analysis that requires no pre-treatment of the samples. The main requirement of the technique is the use of a very smooth and flat sample carrier to allow a sensitive detection.

The first step of the study was to identify an adequate material for nanoparticles collection. The choice was guided by the fact that paper flags are already used and found to be efficient for radioactive particle collection [16]. Polycarbonate membranes were also tested because of their suitability for TXRF analysis as sample substrate. It was shown that the paper flags were more efficient for nanoparticles collection than the polycarbonate membranes. Moreover, pieces of paper were easily pasted on little wood sticks to form sampling devices, whereas polycarbonate membranes were very fragile and would easily tear.

The second step was to identify the best sample carrier for TXRF analysis. Indeed, the developed method had to be adequate for the analysis of traces, and therefore, had to be as sensitive as possible. It was shown that flexible polyester pasted on a glass slide allowed the best sensitivity of detection for ZnO nanoparticles. Glass slides themselves were also tested (data not shown) but the detection was less sensitive than with flexible polyester, probably due to the fact that glass surface is less hydrophobic than polyester.

Ideally, the easiest protocol would have been to directly analyze the sampling device after collection. However, we have shown that the sampling devices made of paper were efficient for nanoparticles sampling but they induced a poor sensitivity of TXRF signal when used as sample carriers. On the opposite, the polycarbonate membranes had a very poor sampling efficiency whereas they were better sample substrates than paper for TXRF analysis. This is directly due to the fact that sampling requires a fibrous material whereas TXRF analysis requires a very smooth material. As a consequence, the proposed protocol combines the use of paper as a sampling device and the use of flexible polyester as sample carrier for TXRF analysis. Polyester was shown to be even more interesting for TXRF sensitivity than polycarbonate.

As a consequence, the sampling device was digested in order to release the nanoparticles in liquid after collection. Then, the liquid was deposited on a flexible polyester piece pasted on a glass slide to make a rigid sample carrier for TXRF. Interestingly, it was not an obligation to dissolve nanoparticles since it was shown that TXRF response is the same whether an element is dissolved in solution or constitutive of a particle, provided that the size of the particle is very small, and that nanoparticle concentration is not too high.

Another important requirement of TXRF is the necessity to use an internal calibration for quantification purpose. Indeed, without internal calibration, no linear calibration curve could be obtained, even when homogeneous conditions of sample deposition were used prior to TXRF analysis. This is why internal calibration was performed using Rb. The choice of Rb resulted from the necessity to use a non-toxic element which is not ubiquitous in the environment, not used as nanoparticles, and with fluorescence lines distinct from those of Zn. With internal calibration, a linear calibration curve was obtained ant it was confirmed that this calibration was similar whether Zn was analysed as ZnO in water or dissolved in acidic conditions (see Figure 3). The only restraint for quantification was the necessity to use low concentration samples because the linearity is deteriorated when using concentrated samples. This is in line with general recommendation for TXRF which is a technique that was first developed for ultra-trace analyses in the field of micro-electronics [17].

Concerning the sensitivity of the method (1 µg of ZnO collected on the paper flag), a rapid calculation was done to assess the magnitude of this value. Of course, the result has to be considered with precaution due to all the approximations it includes. First, we considered the occupational exposure limit of 0.1 mg/m^3 proposed by NIOSH for TiO_2 on a working week of 40 hours (8 hours per day). Then, we applied this value for ZnO, and we considered the respiratory flow rate of an adult (11 L/min) on an 8 hour post shift. According to the CIPR model, the extra-thoracic fraction of nanoparticles deposited after inhalation is between 5 and 80%, depending on the size of the nanoparticles, which finally leads to 26 to 422 µg of ZnO deposited in the extra-thoracic part of the respiratory apparatus. We could therefore assume that at least 2 µg would be deposited in the nostrils (two times the detection limit, the sampling efficiency being around 50%). Of course, this assumption will have to be verified in the future with adequate studies.

As a conclusion, the developed method has the main advantage to be suitable for several kinds of nanoparticles, even those which are difficult to dissolve, like TiO_2 for instance. Nanoparticles are specifically detected according to their chemical composition. One resulting drawback can be that no information on the morphological characteristics of the nanoparticles is obtained, but in the case of positive detection with TXRF analysis, the second flag obtained in the second nostril could be used for scanning electron microscopy. Otherwise, the second flag could be used to perform a more sensitive TXRF analysis.

Also, it has to be kept in mind that this type of sampling can not be used for quantitative dose assessment. The sampling efficiency determined on glass slides should be different in the nose, and the sampling by itself is non-quantitative. However, the purpose of the method is to propose a qualitative indicator of potential inhalation which is a first step in the identification of exposure scenarios.

One limitation of the TXRF method is that some elements can not be analyzed by TXRF such as carbon, and some elements present a poor sensitivity, including alumina and silica. Other methodologies should be developed in the future for these types of nanoparticles. Improvements of the current method could also be envisioned such as the lowering of the limit of detection, or the improvement of the sampling device in terms of sampling efficiency.

The next step will consist in applying the method to other engineered nanoparticles, and then, in the validation of the method on non-exposed volunteers in order to determine the background levels in the population at large. This method provides a qualitative indicator of inhalation that could be useful in the case of accidental scenarios to verify the absence, or not, of any exposure.

References

[1] Thomas J 2006 An Introduction to Nanotechnology: The next small big thing.*Development* **49** 39-46

[2] PEN 2009 The Project on Emerging Nanotechnologies.*The Woodrow Wilson International Center for Scholars and the Pew Charitable Trusts*

[3] Moller P, Folkmann JK, Forchhammer L, Brauner EV, Danielsen PH, Risom L and Loft S 2008 Air pollution, oxidative damage to DNA, and carcinogenesis.*Cancer Lett* **266** 84-97

[4] Stone V, Johnston H and Clift MJ 2007 Air pollution, ultrafine and nanoparticle toxicology: cellular and molecular interactions.*IEEE Trans Nanobioscience* **6** 331-40

[5] Valavanidis A, Fiotakis K and Vlachogianni T 2008 Airborne particulate matter and human health: toxicological assessment and importance of size and composition of particles for oxidative damage and carcinogenic mechanisms.*J Environ Sci Health C Environ Carcinog Ecotoxicol Rev* **26** 339-62

[6] AFSSET 2008 Les nanomatériaux. Sécurité au travail.*Rapport d'expertise collective. Agence française de sécurité santiaire de l'environnement et du travail* **Saisine Afsset n° 2006/006**

[7] HCSP 2009 Avis relatif à la sécurité des travailleurs lors de l'exposition aux nanotubes de carbone.*AVIS du Haut Conseil de Santé Publique*

[8] Angerer J, Ewers U and Wilhelm M 2007 Human biomonitoring: state of the art.*Int J Hyg Environ Health* **210** 201-28

[9] NIOSH 2009 Interim Guidance for Medical Screening and Hazard Surveillance for Workers Potentially Exposed to Engineered Nanoparticles.*Current Intelligence Bulletin 60*

[10] Schulte PA, Trout D, Zumwalde RD, Kuempel E, Geraci CL, Castranova V, Mundt DJ, Mundt KA and Halperin WE 2008 Options for occupational health surveillance of workers potentially exposed to engineered nanoparticles: state of the science.*J Occup Environ Med* **50** 517-26

[11] Kreyling WG, Semmler-Behnke M, Seitz J, Scymczak W, Wenk A, Mayer P, Takenaka S and Oberdorster G 2009 Size dependence of the translocation of inhaled iridium and carbon nanoparticle aggregates from the lung of rats to the blood and secondary target organs.*Inhal Toxicol* **21 Suppl 1** 55-60

[12] Mills NL, Amin N, Robinson SD, Anand A, Davies J, Patel D, de la Fuente JM, Cassee FR, Boon NA, Macnee W, Millar AM, Donaldson K and Newby DE 2006 Do inhaled carbon nanoparticles translocate directly into the circulation in humans?*Am J Respir Crit Care Med* **173** 426-31

[13] Moller W, Felten K, Sommerer K, Scheuch G, Meyer G, Meyer P, Haussinger K and Kreyling WG 2008 Deposition, retention, and translocation of ultrafine particles from the central airways and lung periphery.*Am J Respir Crit Care Med* **177** 426-32

[14] Sarlo K, Blackburn KL, Clark ED, Grothaus J, Chaney J, Neu S, Flood J, Abbott D, Bohne C, Casey K, Fryer C and Kuhn M 2009 Tissue distribution of 20 nm, 100 nm and 1000 nm fluorescent polystyrene latex nanospheres following acute systemic or acute and repeat airway exposure in the rat.*Toxicology* **263** 117-26

[15] Szoboszlai N, Polgari Z, Mihucz VG and Zaray G 2009 Recent trends in total reflection X-ray fluorescence spectrometry for biological applications.*Anal Chim Acta* **633** 1-18

[16] Castagnet X, Cazoulat A, Briot F, Bohand S, Bohand X and Laroche P 2008 [Alpha emitters inhalation: detection by a process with flags].*Ann Pharm Fr* **66** 140-5

[17] Klockenkämper R, ed. *Total-Reflection X-Ray Fluorescence Analysis*. Chemical Analysis: a series of monographs on analytical chemistry and its applications, ed. Winefordner JD. Vol. 140. 1997, Wiley-Interscience.

Journal of Physics: Conference Series **304** (2011) 012017 doi:10.1088/1742-6596/304/1/012017

Nanoparticle uptake and their co-localization with cell compartments - a confocal Raman microscopy study at single cell level

I Estrela-Lopis[1], G Romero[2], E Rojas[2], S E Moya[2] and E Donath[1]

[1]Institute of Medical Physics and Biophysics, Leipzig University, Härtelstraße 16, 04107 Leipzig, Germany

[2]CIC biomaGUNE, Paseo Miramón 182 Edificio Empresarial C, 20009 San Sebastián, Gipuzkoa, Spain.

E-mail: Irina.Estrela-Lopis@medizin.uni-leipzig.de

Abstract. Confocal Raman Microscopy, a non-invasive, non-destructive and label-free technique, was employed to study the uptake and localization of nanoparticles (NPs) in the Hepatocarcinoma human cell line HepG2 at the level of single cells. Cells were exposed to carbon nanotubes (CNTs) the surface of which was engineered with polyelectrolytes and lipid layers, aluminium oxide and cerium dioxide nanoparticles. Raman spectra deconvolution was applied to obtain the spatial distributions of NPs together with lipids/proteins in cells. The colocalization of the NPs with different intracellular environments, lipid bodies, protein and DNA, was inferred. Lipid coated CNTs associated preferentially with lipid rich regions, whereas polyelectrolyte coated CNTs were excluded from lipid rich regions. Al_2O_3 NPs were found in the cytoplasm. CeO_2 NPs were readily taken up and have been observed all over the cell. Raman z-scans proved the intracellular distribution of the respective NPs.

1. Introduction

Nanotechnology has seen a rapid development in recent years. Novel products appeared at the market ahead of that nanoparticle and nanomaterials related ecological and health risks have been addressed by the respective regulations. A proper risk management of nanomaterials requires more information on the human and ecological effects of exposure to various nanomaterials [1]. Understanding the effects of nanoparticles (NPs) and nanostructured materials on human health (and ecology) is thus an issue of major importance [2]. Studies with cell culture model systems may contribute to the understanding of the effect of nanoparticles on human health [3].

In contrast to toxicity at the molecular level, where toxic compounds directly interfere with the metabolism, the toxicity of NPs may be caused by the interaction of the NPs with various cellular components. NPs may interfere with the complex intracellular machinery by disturbing the conformation of cellular constituents, the intracellular hierarchy and transport routes. For a deeper understanding of the mechanism of the interaction of NPs with biological matter it is therefore desirable to study the localization of NPs within cells and organs. However, the possibilities to monitor NPs at the cellular level are limited. Fluorescent techniques as a rule require labeling, which, given the small size of the nanoparticles, may seriously interfere with their properties. Electron microscopy has an excellent resolution but it is time consuming and restricted to the analysis of a limited number of images.

An alternative technique capable of measuring the presence and distribution of NPs and nanostructured materials in cells is Raman Microscopy, since a wide variety of NPs and nanostructures shows Raman active vibration modes. A powerful nonlinear Raman spectroscopy technique is the Coherent Anti-Stokes Raman Scattering (CARS) technique. The anti-Stokes signal is significantly enhanced by means of focusing the excitation energy on a single Raman band. This permits, combined with confocal microscopy, the imaging of cells within a very short time [4]. The drawbacks of CARS are the lack of spectral information and the presence of nonresonant contributions to the CARS signal. Spontaneous confocal Raman microscopy, on the other hand, records the full vibrational spectrum in each spot. It thus allows for subsequent background-free spectral analysis at every scanned spot. This is an important advantage, because, for example, spectral decomposition allows for separating and extracting various spectral components related to the presences of different materials in the confocal spot, which would be very difficult with CARS. Spontaneous confocal Raman microscopy, however, requires long measuring times, orders of magnitude larger than required for CARS.

We show here, employing spectral decomposition in the various regions of the vibrational spectrum that spatially resolved Raman microscopy at the single cell level can be used to simultaneously study nanoparticle uptake and their colocalization with intracellular compartments. Being strictly a label free technique, Confocal Raman Microscopy (CRM) [5, 6] thus is a valuable tool to study the interaction of NPs with biological matter at the cellular level. In this work we studied the uptake of CeO_2 and Al_2O_3 NPs, as well as of CNTs into Hepatocarcinoma cells by means of Confocal Raman Spectroscopy. The possibility of simultaneously detecting NPs together with the identification of their respective intracellular environment provided information about the three dimensional NP distribution in and their co-localization with different cell compartments, such as lipid bodies (LB), cytoplasm and nucleus.

2. Materials and Methods

2.1. Nanoparticles

2.1.1. *Metal oxide nanoparticles.* Al_2O_3 NPs were donated from PlasmaChem GmbH, Berlin, Germany. An average diameter between 2 and 5 nm was measured by TEM. CeO2 NPs were purchased from Evonik-Degussa GmbH. TEM studies revealed an average diameter of about 15 nm.

2.1.2. *Carbon Nanotubes (CNTs).* Multiwalled CNTs were purchased from Proforma (USA). To improve their stability in aqueous solutions CNTs were oxidised as described by Zhang et al. [7]. CNTs were sonicated with an ultrasonicator (SONICS® VCX 500) at 20% amplitude applying pulses of 30 seconds for 2 minutes to break the nanotubes into smaller pieces. After sonication TEM images showed an average length of 500 nm for the CNTs. Oxidized CNTs were silanized and endowed with initiator functionality for subsequent Atom Transfer Radical Polymerization (ATRP) following the method described by Llarena et al. [8]. Poly-sulfo propyl methacrylate (PSPM) was synthesized as a brush on the surface of CNTs (CNT/PSPM). The polyelectrolyte corona around the CNTs ensured stability in aqueous solution and served also as a support for the subsequent adsorption of a lipid layer. To this aim a layer of poly (allyl amine hydrochloride) (PAH) was assembled on top of the PSPM brush followed by adsorption of lipid vesicles composed of (1,2-Dioleoyl-sn-glycero-3-choline/1,2-Dioleoyl-sn-glycero-3-phospho-L-serine sodium salt) (DOPC/DOPS) in a molar ratio 3:1 [9].

2.2. Cell culture and NP exposure

The Hepatocarcinoma human cell line HepG2 (from American Type Culture Collection) was cultured in Dulbecco's Modified Eagle's Medium (DMEM) with 10% Fetal Bovine Serum (FBS), 1000 U penicillin and 10 mg/ml streptomycin (all of them purchased from Sigma) at 37°C and 5 % CO_2. When cell confluence reached around 70%, the culture was trypsinized and 100000 cells were seeded in glass bottom of Petri dishes. After 24 hours UV sterilized NPs were added to the Petri dishes. Details of the incubation protocol are the following: 50 μg/ml CeO_2 incubated for 48 h, 1 mg/ml Al_2O_3 NPs, 150 μg/ml CNTs/lipid and 50 μg/ml CNT/PSPM, all of them incubated for 24 h. Afterwards, the plate was rinsed with PBS several times and the cells were fixed with 3.7% formaldehyde in PBS during 30 minutes and rinsed again.

2.3. Confocal Raman Microscopy

Micro-Raman analyses were performed using a Renishaw inVia Raman Microscope. An excitation wavelength of 532 nm with a grating of 1800 mm^{-1} was used. Microspectroscopy measurements were conducted using the ×50 objective. The size of the focal spot was approximately 1 μm. Raman spectra were recorded from adherent cells placed into Petri dishes with a glass bottom filled with PBS. Spectra were measured in the region 300–3600 cm^{-1} with a resolution of approximately 7 cm^{-1}. The system was calibrated to the spectral line of crystalline silicon at 520.7 cm^{-1}. At least 8-20 accumulation scans in confocal mode at different spots in the various cell compartments were used to reduce the spectral noise. Control spectra of the different cell compartment were taken in different cells and averaged. All spectra were corrected for the PBS solution and Petri dish glass bottom baseline. After NPs exposure and repeated washings with PBS the Raman spectra were taken only from cells, where no visible NP aggregates could be observed.

3. Results and Discussion

Raman microspectroscopy allows for identification and localization of lipid bodies, cytoplasm and the nucleus in Hepatocarcinoma cells [10]. In brief, lipid bodies are identified by their strong CH_2 signal as compared with the CH_3 related bands. The nucleus shows a minimum intensity of CH_2 related vibrational bands, but comparatively intensive amide bands. In addition a specific nucleotide vibrational band can be seen, which is not present in the cytoplasm. Further details are given below.

3.1. Characteristic Raman spectra of NPs

Raman spectra of the two dried NPs and of the CNTs were recorded in non-confocal mode. Figure 1 represents the characteristic Raman spectra of CeO_2, Al_2O_3 NPs and CNTs.

CeO_2 NPs observed a single strong band located around 465 cm^{-1}. Cerium dioxide has a fluorite-type structure lattice and belongs to the space group O_h^5. Six optical-phonon branches of space group O_h^5 yield three zone center frequencies. The related triply degenerate Raman-active F_{2g} mode is centred at about 465 cm^{-1}. This mode represents a symmetric breathing vibrational mode of O ions around each cation [11]. It is rather sensitive to the oxygen sublattice disorder. One of the reasons of lattice disorder is a decrease of crystal dimensions. The Raman band shape of F_{2g} mode and their position thus change with decreasing size of NPs [12]. The corresponding wavenumber shift and asymmetrical broadening as a function of the crystal size can be theoretically described using the phonon confinement model [13]. The measured width of the F_{2g} mode of the studied CeO_2 NPs was about 8 cm^{-1} (Figure 1). Therefore the average size of the CeO_2 NPs can be estimated as about 18 nm. This value correlates very well with TEM studies of the used CeO_2 NPs.

Al_2O_3 belongs to the D_{3d} space group. From the 27 possible optical modes in the crystal structure of Al_2O_3 7 modes are Raman active and 6 modes are infrared (IR) active [14]. The seven arrows in Figure 1 denoted the seven Raman active modes in dried NPs. One of the IR active modes at 600 cm^{-1} can be also detected. This observation suggests a distorted microcrystalline structure of the NPs. Other bands observed in Al_2O_3 NPs can be assigned to the capping agent, which has been used by the supplier to avoid NP aggregation.

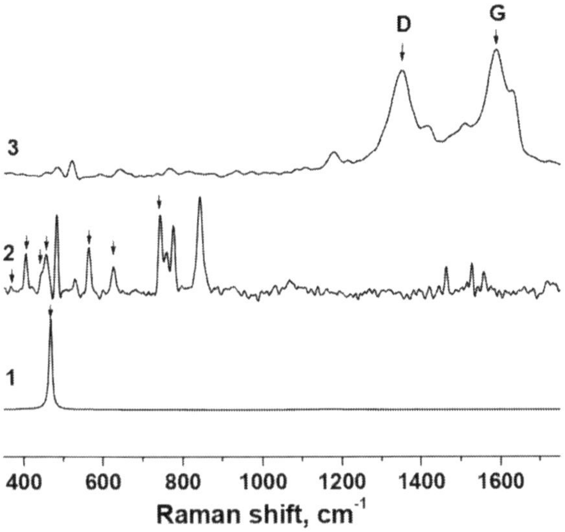

Figure 1. Raman spectra of CeO_2 (1), Al_2O_3 (2) and CNT (3) NPs.

The Raman spectrum of CNTs observes two prominent bands located at 1350 cm^{-1}, known as D-band, and at 1585 cm^{-1}, known as the G-band, see Figure 1. The D-band is an indicator for disorder in the graphene sheet and is also called "disorder-included" band. The G-band is a tangential mode originating from tangential oscillations of the carbon atoms in CNTs [15].

3.2. NPs localization in Hepatocarcinoma cells

The Raman spectra recorded from Hepatocarcinoma cells after their exposure to CeO_2, Al_2O_3 and CNT NPs all taken in confocal mode are shown in Figures 2, 3 and 4, respectively.

After exposure of the cells to CeO_2 Raman spectra were taken in various regions of the cells. Representative Raman spectra are displayed in Figure 2. They correspond to three different regions of the cell: lipid bodies, cytoplasm and nucleus. The assignments [5,16] of typical intracellular vibration bands are denoted in Figure 2. The three spectra reveal the different chemical nature of each of the chosen spots. The Raman spectrum in the lipid rich region was recorded from LBs, which can be seen in the microscopic transmission image as small spherical objects. The comparison of the spectra taken at different regions of the cell reveals that the intensity ratios of the CH_2 (2850 cm^{-1}) to the CH_3 (2935 cm^{-1}) symmetric stretching bands is much higher in LBs than in the cytoplasm or the nucleus, respectively. This follows from the considerably higher CH_2 group density in lipids than in proteins. The cytoplasmic compartment is characterized by a large amount of CH_3 groups and a comparatively small amount of lipids representing intracellular membranes. A comparison of the spectra in the cytoplasm and the nucleus shows some common bands, i.e. amid I (1655 cm^{-1}), and amid III (1250 cm^{-1}) bands, as well as different protein CH deformation modes. The band at 1338 cm^{-1} assigned to vibrations of adenine (A) and guanine (G) is quite specific for the nucleus and can be used to discriminate between cytoplasm and nucleus.

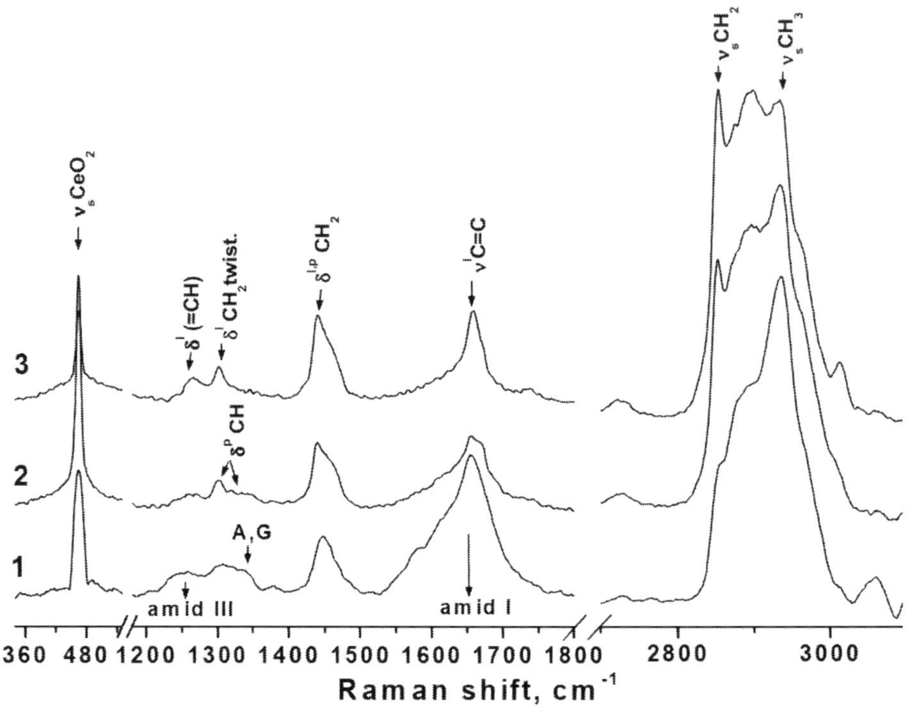

Figure 2. Raman spectra of cells exposed to CeO_2 NPs. The curves 1, 2 and 3 represent spectra recorded from different cell compartments: nucleus, cytoplasm and lipid bodies, respectively. The symbols ν and δ indicate stretching and deformation vibration modes, respectively. The indexes: l and p denote the vibrations of lipids and protein, respectively.

Regardless of the specific region of the cells in all spectra the CeO_2 phonon mode (465 cm^{-1}) was seen in parallel with the typical intracellular bands. This result demonstrates that CeO_2 NPs are distributed all over the cytoplasm. This does not necessarily mean that CeO_2 NPs enter the nucleus because NPs located just beneath or above the nucleus would contribute to the Raman spectra recorded from the nucleus. This is in agreement with the high but nevertheless limited spatial resolution of the confocal mode. A closer inspection reveals that the phonon band F_{2g} increases in its width to about 12 cm^{-1} after NPs uptake compared to pure NPs. From this follows an estimated NPs size of about 7 nm. This value is two times smaller compared to that of bare NPs. One of possible explanations may be that NPs with smaller sizes were preferentially taken by the cells. The other explanation is a relative dissolution of CeO_2 due to a relative reduction of Ce-IV atoms in Ce-III ones [17].

Figure 3 shows a Raman spectrum taken in a spot close to the nucleus of cell, which had been exposed to Al_2O_3 NPs. The typical seven Raman active bands resulting from Al_2O_3 in the low frequency region, denoted by arrows in Figure 3, were observed simultaneously with the signal from the cellular environment seen in the high frequency region. The pronounced CH_3 symmetric stretching band (2935 cm^{-1}) originated mainly from proteins. A weak lipid related CH_2 symmetric band indicates that the Al_2O_3 NPs were present in the cytoplasm being co-localized with proteins.

As a further example Figure 4 shows a spectrum recorded from cells treated with CNTs, which had been coated with lipids (CNT/lipid). Two characteristic vibrational bands of CNTs at 1350 cm^{-1} and 1585 cm^{-1}, D-band and G-band, can be clearly seen superimposed on the cellular spectrum. The spectrum can be decomposed into its two components, one from the CNTs and the other from the cell, taking advantage of the spectrum of pure CNTs. After decomposition it was found that extracted cellular spectrum was very similar to spectra taken in the lipid-rich region of control cells. Clearly pronounced symmetric bands can be observed in the CH_2 stretching region. From the comparison of these spectra follows that lipid bodies and CNTs are both present within the 1 μm spot of the laser beam. From this observation one may conclude that lipid coated CNTs are likely colocalized with the lipid bodies in the cytoplasm. This may indicate a lipid facilitated way of uptake or an association with lipid bodies post uptake.

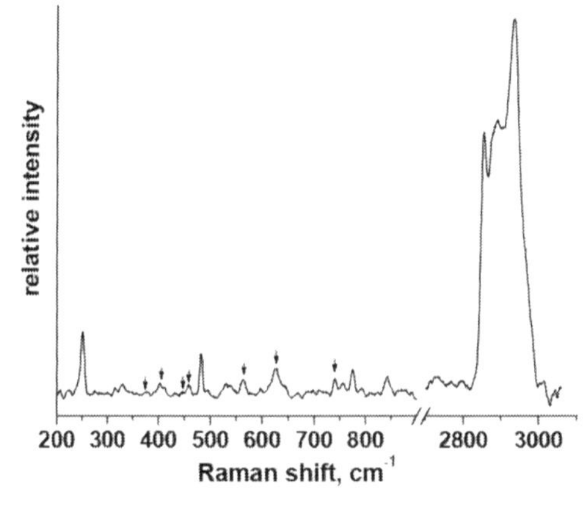

Figure 3. Spot Raman spectra taken close to the nucleus in cells exposed to Al_2O_3 NPs.

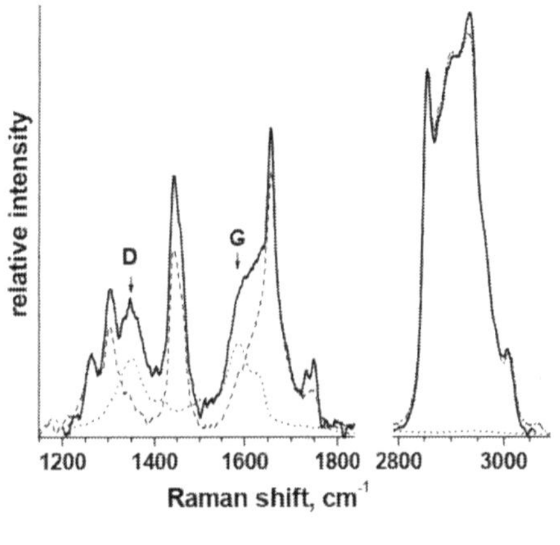

Figure 4. Spot Raman spectrum in cells exposed to CNT/lipid NPs (solid line). Dash and dot lines denote the contribution in recorded spectrum from NPs and LBs in cells, respectively.

In summary, it can be concluded that Raman spectra of NPs in cells do not only provide information about their presence but also on the intracellular environment in the vicinity of the nanoparticles.

3.3. NPs internalization and distributions across treated cells

In principle, confocal microscopy allows for the recording of Raman spectra in all three dimensions with an accuracy limited by the resolution of the device. However, since the signal intensity is small, a multidimensional spectral mapping would require a rather long time. Therefore, the spatial mapping was restricted to line scans in z-direction and in the xy-plane. The confocal resolution in z direction is not as good as that in the xy-plane. The accuracy in the z-direction is nevertheless sufficient to prove NPs internalization, but for colocalization studies it will be more appropriate to perform line scans in the xy-plane

Figures 5 and 6 show two examples of z-scans from cells exposed to CeO_2 NPs and CNTs covered with PSPM (CNT/PSPM). Z-scans were recorded in confocal mode along the z-axes with an increment of 2-5 μm. The z-plane denoted by zero is the reference plane assumed to be close to the equatorial plane of the cell. This plane was manually adjusted guided by imaging in transmission mode.

The signals from both CeO_2 NPs in low frequency region and from cells in the methyl/methylene stretching region can be seen in the z-scan spectra (Figure 5). The ratio between the phonon band intensity to the methyl/methylene band intensity depends on the z-plane. The maximum is found at z=5 μm. The pronounced lipid band at 2852 cm^{-1} evidences furthermore the colocalization of CeO_2 NPs with lipid-rich regions in the cell. From these z-scans clearly follows the internalization of the

CeO_2 NPs. The difference in intensity of the phonon band indicates an inhomogeneous distribution of the CeO_2 NPs within the cytoplasm.

In the case of cells exposed to CNT/PSPM both signals from NPs and cell were simultaneously recorded in the Raman spectra as observed in Figure 6. The typical D and G bands of the CNT were detected at different planes within the cells. The high intensity of the protein CH_3 stretching band and the apparent absence of lipid related Raman bands prove *i)* the cytoplasmic distribution of CNT/PSPM that the polyelectrolyte coated CNTs, together with *ii)* the absence of a specific affinity to the lipid body rich parts of the cytoplasm. As it can be seen the ratio between the signal intensity of the CNTs and that from the protein ($\nu_s(CH_3)$) is different for different planes. This implies that the CNTs are not homogeneously distributed in the cytoplasm.

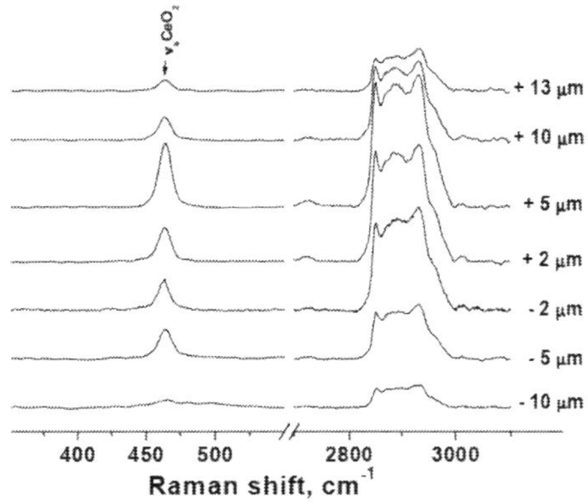

Figure 5. Raman spectra taken at different z-planes of the cell exposed to CeO_2 NPs

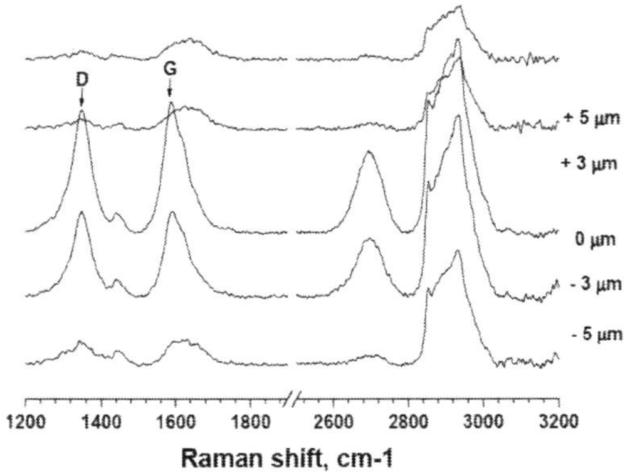

Figure 6. Raman spectra taken at different z-planes of the cell exposed to CNT/PSPM NPs

Line scans in the xy-plane of cells were performed to get more detailed information about the NPs distribution and their colocalization. Raman spectra were recorded along lines crossing through the exposed cells with a distance increment of 0.5 μm.

The recorded spectrum was decomposed into two components, one from the NPs and the other from the cell as outlined above (Figure 4). The relative integral intensities of the NP contribution in the Raman spectrum were calculated. The lipid CH_2 symmetrical stretching band at 2852 cm^{-1} was chosen as the reference band to obtain the lipid distribution along the scanned line. The integral intensities of $\nu_s(CH_2)$ was plotted as a function of the distance along the line. The lipid distribution across the cell reflects the presence of mainly lipid bodies. Other lipid fractions of the cytoplasm contribute to a lesser extent to the spectrum because their concentration is small compared with that of lipids in lipid bodies. A well pronounced maximum in the lipid distribution would thus indicate the presence of lipid bodies at this particular location.

Figure 7 displays the distributions of lipid bodies and NPs in cells exposed to CNTs, decorated with PSPM and to CNTs coated with lipids. For the purpose of displaying both distributions at the same graph in a comparable scale the component intensities of CNTs and cellular lipids were scaled to the intensity of a well pronounced maximum. The analysis revealed two regions where CNT/PSPM NPs concentration was large (Figure 7(a)). In the case of cells exposed to CNT/lipid NPs three maxima along the scan line in the xy-plane were found as shown in Figure 7(b). Of course, these maxima of the CNT distribution do not bear a specific relevance on its own. Only when compared to the lipid distribution conclusions may be drawn. Indeed, in Figure 7b all three maxima of the lipid distribution coincide with or are at least close to the maxima of the CNT distribution. This behaviour was regularly observed for cells exposed to lipid coated CNTs. In the case of CNT/PSPM NPs a spatial correlation between the localization of lipid bodies and the NPs was, however, not observed (Figure 7(a)). The lipid body intensity distribution had its maximum where the CNT signal intensity lowest. That fact can be interpreted as a tendency of the CNT/PSPM NPs to avoid the site where the lipid bodies are preferentially located. These findings corroborate the conclusion drawn from Figure 4 that lipid coated CNTs within cells are found in close proximity of lipid bodies. This finding may point to either a lipid associated mechanism of uptake or to a retrospective arrangement of the CNTs with the lipid rich compartments in hepatocytes. In any case our finding suggests strongly that the nature of the surface of CNTs is important concerning their possible effects on cells.

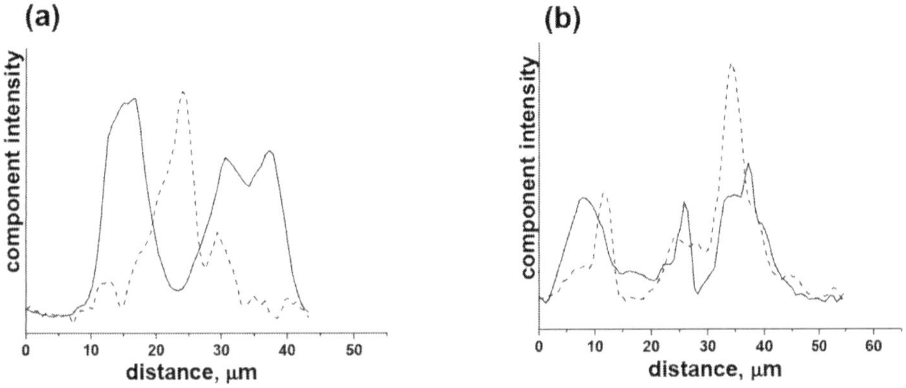

Figure 7. (a) CNT/PSPM, (b) CNT/lipids NPs distributions (solid lines) and lipids (dashed lines) distribution across CNT treated cell.

The analysis of a typical Raman microspectroscopy line scan after Al_2O_3 NPs exposure is shown in Figure 8. In this particular case the Al_2O_3 NPs were found close to the nucleus. The corresponding distributions of the components are presented as integral intensity distributions as well as in rainbow presentations overlaid on the transmission image. The protein distribution was calculated from the integral intensity of protein CH_3 stretching region as a function of distance along the line.

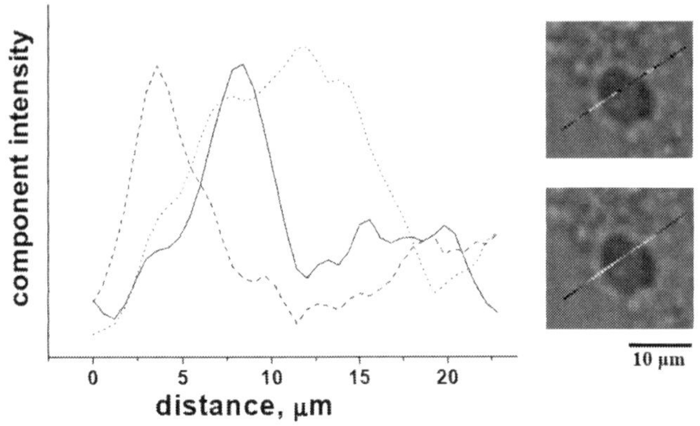

Figure 8. Al_2O_3 NPs (solid line), proteins (dot line) and lipids (dash line) distribution across the selected cell. The figure inserts represent NPs (top) and protein (bottom) line mapping overlaid on cell transmission images in rainbow presentation. The red colour corresponds to the maximum intensity.

The maximum of the protein distribution is correlated with the minimum of the lipid distribution. The maximum of the protein signal intensity or the minimum of lipid related signal intensity both could serve as a spectroscopic criterion of the nucleus position. The lower insert of Figure 8 proves that, for example, the protein signal intensity distribution is a good match of the optically observed nucleus as can be inferred from the overlap of the red dots with the optically dense nucleus region.

The analysis of lipid, protein and NPs distribution behaviour revealed the preferential association of the Al_2O_3 NPs with a protein rich region. The upper insert of Figure 8 displays an Al_2O_3 accumulation in the lower left sector close to the nucleus. On the other hand, maximum of NPs distributions differs by about 4 μm from the maximum of the protein related Raman signal, which is expected within the nucleus. From this analysis it most likely follows that in this particular case the Al_2O_3 NPs found along the line scan resided in the cytoplasm yet in a region close to nucleus. This conclusion is in agreement with the spectrum displayed in Figure 3, which in the same spot shows the signal from the Al_2O_3 NPs and the cytoplasm after Al_2O_3 NPs uptake.

4. Conclusions

It has been shown that Confocal Raman Microscopy is capable of providing detailed data on uptake and distribution of NPs in culture cells at single cell level. A specific advantage of the approach is that it does not require any labeling allowing in this way the investigation of the interaction of authentic nanoparticles with cells. Since most nanoparticles and nanomaterials observe Raman active vibration modes Confocal Raman Microscopy is a rather universal technique. The colocalization of nanoparticles with intracellular compartments can be elaborated applying spectral deconvolution. Another interesting aspect in the context of the interaction of nanoparticles with biological matter is that the size and surface composition of NPs has an influence on the phonon modes. This allows for monitoring changes in size and composition of nanoparticles while they are residing inside cells.

It was found that lipid coated CNTs associated preferentially with lipid bodies, while CNT/PSPM NPs avoided lipid-rich cell regions. This finding indicates that the surface properties of the nanoparticles may play a significant role concerning the uptake and intracellular distribution. Al_2O_3 and CeO_2 NPs were found in cytoplasm. The CeO_2 NPs signal was strong and found almost everywhere in the cytoplasm. This indicates a high rate of uptake of the latter and a non-specific distribution inside Hepatocarcinoma cells.

Acknowledgement

This work was supported by the European Commission in the framework of FP7 Theme 4 – NMP - Nanosciences, Nanotechnologies, Materials and New Production Technologies, Proposal No: CP-FP 28825-2 HINAMOX.

References

[1] Wardak A, Gorman ME, Swami N, Deshpande S 2008 *J Ind Ecol* **12** 435-48; Gwinn MR, Tran L 2010 *Wiley Interdiscip Rev Nanomed Nanobiotechnol* **2** 130-7; Tervonen T, Linkov I, Figueira JR, Steevens J, Chappell M, Merad M 2009 *J Nanopart Res* **11** 757-66; Schulte PA, Salamanca-Buentello F 2007 *Environ Health Perspect* **115** 5-12

[2] Maynard AD, Aitken RJ, Butz T, Colvin V, Donaldson K, Oberdorster G, Philbert MA, Ryan J, Seaton A, Stone V, Tinkle SS, Tran L, Walker NJ, Warheit DB 2006 *Nature* **444** 267-9; Gao XH, Cui YY, Levenson RM, Chung LWK, Nie SM 2004 *Nat Biotechnol* **22** 969-76; Oberdörster G, Maynard A, Donaldson K, Castranova V, Fitzpatrick J, Ausman K, Carter J, Karn B, Kreyling W, Lai D, Olin S, Monteiro-Riviere N, Warheit D, Yang H 2005 *Part Fibre Toxicol* **2** 8

[3] Hillegass JM, Shukla A, Lathrop SA, MacPherson MB, Fukagawa NK, Mossman BT 2010 *Wiley Interdiscip Rev Nanomed Nanobiotechnol* **2** 219-31; Lee J, Lilly GD, Doty RC, Podsiadlo P, Kotov NA 2009 *Small* **5** 1213-21

[4] Chan J, Fore S, Wachsman-Hogiu S, Huser T 2008 *Laser Photonics Rev* **2** 325-49; Moger J, Johnston BD, Tyler CR 2008 *Opt Express* **16** 3408-19; Tong L, Lu Y, Lee RJ, Cheng JX 2007 *J Phys Chem B* **111** 9980-5

[5] Puppels GJ, Demul FFM, Otto C, Greve J, Robertnicoud M, Arndtjovin DJ, Jovin TM 1990 *Nature* **347** 301-3; Deng H, Bloomfield VA, Benevides JM, Thomas GJ 1999 *Biopolymers* **50** 656-66; Chan JW, Taylor DS, Zwerdling T, Lane SM, Ihara K, Huser T 2006 *Biophys J*

90 648-56

[6] Chan JW, Lieu DK, Huser T, Li RA 2009 *Anal Chem* **81** 1324-31; Uzunbajakava N, Lenferink A, Kraan Y, Volokhina E, Vrensen G, Greve J, Otto C 2003 *Biophys J* **84** 3968-81; Takai Y, Masuko T, Takeuchi H 1997 *BBA-Gen Subjects* **1335** 199-208; Notingher I, Bisson I, Bishop AE, Randle WL, Polak JMP, Hench LL 2004 *Anal Chem* **76** 3185-93; Chernenko T, Matthaus C, Milane L, Quintero L, Amiji M, Diem M 2009 *ACS Nano* **3** 3552-9

[7] Zhang NY, Xie JN, Guers M, Varadan VK 2003 *Smart Mater Struct* **12** 260-3

[8] Llarena I, Romero G, Ziolo RF, Moya SE 2010 *Nanotechnology* **21** # 055605

[9] Fischlechner M, Zaulig M, Meyer S, Estrela-Lopis I, Cuellar L, Irigoyen J, Pescador P, Brumen M, Messner P, Moya S, Donath E 2008 *Soft Matter* **4** 2245-58; Moya S, Donath E, Sukhorukov GB, Auch M, Baumler H, Lichtenfeld H, Mohwald H 2000 *Macromolecules* **33** 4538-44

[10] Romero G, Estrela-Lopis I, Zhou J, Rojas E, Franco A, Espinel CS, Fernandez AG, Gao CY, Donath E, Moya SE 2010 *Biomacromolecules* **11** 2993-9

[11] Lacina WB, Pershan PS 1970 *Phys Rev B* **1** 1765-85

[12] Popovic ZV, Dohcevic-Mitrovic Z, Scepanovic M, Grujic-Brojcin M, Askrabic S 2011 *Annalen Der Physik* **523** 62-74; Weber WH, Hass KC, Mcbride JR 1993 *Phys Rev B* **48** 178-85; Kostic R, Askrabic S, Dohcevic-Mitrovic Z, Popovic ZV 2008 *Appl Phys A - Mater* **90** 679-83

[13] Richter H, Wang ZP, Ley L 1981 *Solid State Commun* **39** 625-9; Campbell IH, Fauchet PM 1986 *Solid State Commun* **58** 739-41

[14] Misra A, Bist HD, Navati MS, Thareja RK, Narayan J 2001 *Mat Sci Eng B-Solid* **79** 49-54; Porto SPS, Krishnan RS 1967 *J Chem Phys* **47** 1009-12

[15] Bose SM, Gayen S, Behera SN 2005 *Phys Rev B* **72** #153402; Keszler AM, Nemes L, Ahmad SR, Fang X 2004 *J Optoelectron Adv M* **6** 1269-74; Kuzmany H, Burger B, Hulman M, Kurti J, Rinzler AG, Smalley RE 1998 *Europhys Lett* **44** 518-24

[16] Peticolas WL. Raman Spectroscopy of DNA and Proteins. In: Biochemical Spectroscopy; 1995:389-416; Naumann D. Infrared and Raman spectroscopy of biological materials. New York: Marcel Dekker; 2001; Prescott B, Steinmetz W, Thomas GJ 1984 *Biopolymers* **23** 235-56

[17] Auffan M, Rose J, Orsiere T, De Meo M, Thill A, Zeyons O, Proux O, Masion A, Chaurand P, Spalla O, Botta A, Wiesner MR, Bottero JY 2009 *Nanotoxicology* **3** 161-U15

Journal of Physics: Conference Series **304** (2011) 012018 doi:10.1088/1742-6596/304/1/012018

Aggregation and deposition of engineered TiO_2 nanoparticles in natural fresh and brackish waters

Markus Sillanpää, Tiina-Mari Paunu and Pirjo Sainio

Finnish Environment Institute, Research and innovation laboratory, P.O. Box 140, FIN-00251 Helsinki, Finland

E-mail: markus.sillanpaa@ymparisto.fi

Abstract. The use and thus environmental release potential of metal-based nanoparticles have rapidly increased. Due to their size-dependent new properties, the fate and effect of nanomaterial may differ from those of the conventional form of corresponding material. The agglomeration and sedimentation were studied by spiking the TiO_2-P25 particles in natural fresh and brackish water samples. The natural waters were determined for conductivity, pH, salinity, total organic carbon, turbidity, common nutrients and trace elements. The hydrodynamic diameter and concentration of TiO_2-P25 particle dispersions were monitored by using a dynamic light scattering and a spectrophotometer, respectively. The experiments were performed at two particle concentrations 100 mg/l and 1 mg/l (10 mg/l for deposition studies). The aggregation rates in brackish waters were clearly higher in higher initial concentration and the sedimentation of aggregates decreased the TiO_2 concentration down to 20% and 80% of initial higher and lower concentrations, respectively. One fresh water sample favoured the destabilisation of TiO_2-P25 particles whereas another fresh water sample stabilised the TiO_2 particle dispersion. The aggregation had a strong dependence on the particle concentration. High ionic content of brackish water probably explains the formation of aggregates, whereas organic substances and pH may account for the different agglomeration behaviour in fresh waters.

1. Introduction

Nanosized materials have attracted the science community due to their unique physicochemical properties. These materials have been used in various scientific fields including engineering, biotechnology, biomedical, environmental and materials sciences. This has resulted in a number of new technical innovations and applications. Currently, there are hundreds of consumer products containing nanomaterials in global markets [1-2] and the number is increasing annually. Due to new properties associated with conventional or surface modified materials at nanoscale, some of nanomaterials may have exceptional environmental behavior and effects. The products containing nanomaterials can be divided in three groups on a basis of the location of nanomaterials: nanomaterials in bulk (e.g. composite materials), nanomaterials on the surface (e.g. textiles, catalysts, sun energy cells) and nanomaterials in dispersion (e.g. paints, sprays, cosmetics) [3]. The last two of the groups are expected to have potential to release nanomaterials to environment during the use of products.

Due to the commercial relevance, high production and potential adverse effects, engineered TiO_2 nanomaterials have been chosen as one of fifteen nanomaterials for extensive testing program coordinated by the OECD Working Group of Manufactured Nanomaterials [4]. TiO_2 nanoparticles have already been utilized in numerous applications. They have been widely used in sun creams to improve the UV blocking. In addition, TiO_2 nanoparticles have mixed as a dye into paints and attached as photocatalysts onto the different surfaces. Kaegi et al. [5] reported the release TiO_2 particles from the facade paints by natural weather conditions. They measured the concentration of TiO_2 in runoff water to be 13.4 µg/l and a part of this appeared at a nanoscale. Based on the modeling study of ENP emissions in Switzerland, the predicted environmental concentration (PEC) of engineered TiO_2 nanomaterials is 16 µg/l in surface waters [6].

There are a number of studies dealing with the aggregation and deposition of engineered nanoparticles [e.g. 7-12]. These studies have been conducted nearly without exception under the controlled test condition by adjusting the pH, ionic strength, composition of mono- and divalent ions as well as the concentration of dissolved organic matter. However, while providing valuable knowledge, particle behavior in model laboratory systems might not be representative of that observed in far more complex natural environments. This paper focuses on the aggregation and transport potential of engineered TiO_2 particles in four different natural waters. The differences in particle behavior observed during the experiments are compared with the physicochemical water quality characteristics of tested natural waters. All the experiments have been conducted at two initial concentrations of nanoparticles. To our best knowledge, this is the first paper that focuses on the behavior and fate of engineered TiO_2 particles in natural waters typical to Nordic conditions.

2. Materials and Methods

2.1. Nanomaterial and preparation of dispersion

Titanium dioxide nanoparticles (TiO_2-P25; Evonik Degussa GmbH, Germany) were purchased as a white powder with a nominal primary diameter of 21 nm. Figure 1 shows the image of the TiO_2-P25 powder by using a Field Emission Scanning Electron Microscopy (see section 2.3.4.). Stock dispersions were prepared by adding 100 mg of TiO_2-P25 powder into 100 ml of deionised Milli-Q water. Test samples were prepared by spiking a 30-min sonicated stock dispersion in test matrix at a ratio of 1:9. The particle size was measured prior to each experiments.

2.2. Natural water samples

The agglomeration and deposition of engineered TiO_2 nanoparticles were studied in two different fresh and brackish waters. The natural waters were sampled at a depth of 1 meter between May 18 and 20, 2010. The samples were stored at 4°C under dark conditions until needed. The sampling sites are described briefly below.

Simijärvi (60°10 N, 23°33 E). Simijärvi is located at a village of Antskog about 75 km west of Helsinki. It represents a unique lake of esker area that is characterized by clear water and the scarcity of aquatic vegetation. Simijärvi is mainly surrounded by rocky coniferous forest. The water sample was collected on May 19, 2010.

*Iso lehmälampi (*60°21 N, 24°36 E*).* Iso Lehmälampi (60°21 N, 24°36 E) is a small lake (0.03 km^2) that is situated on an upland area about 25 km from the southern coast of Finland. The lake itself consists of two parts; the shallower north-western part and the deeper south-eastern part. Due to the hilly topography of the area, which is characterized by bedrock outcrops, the drainage area is very small. The modern forest around the lake is composed of Scots pines, Norway spruces, birches and Alder trees. Iso Lehmälampi water was sampled on May 20, 2010. The surroundings of the sampling site have been described in detail elsewhere [13].

*Tvärminne (59°*50 *N, 23°*15 *E).* The Tvärminne Zoological Station (University of Helsinki) is located on a shore of western Gulf of Finland at Hanko, the southmost town in Finland. The Tvärminne area is not directly affected by large sewage outlets, but nutrients are unintentionally, but frequently introduced to the productive surface layer. The Tvärminne sample was collected at

Zoological Station about 10 meters from the coastline. Coastal seawater was sampled on May 18, 2010.

Lånskär (59°82 N, 23°28 E). Långskär is the southernmost island of outer archipelago of Hanko area. The sample was collected on open sea zone 2 km south of Långskär. The Gulf of Finland is the easternmost part of the Baltic Sea. Its salinity at the surface is typically clearly below 10‰. Open and coastal seawaters were sampled at the same day (May 18, 2010). The surroundings of the sampling site have been described in detail Elsewhere [14].

The water samples were filtered (cellulose acetate membrane filter, pore size 0.45 μm, Whatman, Germany) to remove the large colloidal particles that may interrupt the measurements. The agglomeration of engineered TiO_2 nanoparticles were studied by spiking nanoparticle suspension in natural water samples at concentration of 1 mg/l and 100 mg/l. Due to the detection limit of spectrophotometer, the deposition studies were done at the concentrations of 10 mg/l and 100 mg/l. The measurements were started immediately after each spiking and all the experiments were done in duplicate samples.

2.3. Analytical instruments

2.3.1. Characterisation of natural water samples. All analyses were carried out at the SYKE Environment Measurement and Testing Laboratory, which is accredited as a testing laboratory (T003) by the Finnish Accreditation Services, FINAS (SFS-EN ISO/IEC 17025). Natural waters were analysed by using the analytical instruments as follows: a 815 Robotic USB Sample Processor XL (Metrohm Ion analysis, Switzerland) for conductivity, a 814 USB Sample Processor (Metrohm Ion analysis, Switzerland) for pH, a Salinometer Autosal 8400B (Guildline Instruments Ltd, UK) for salinity, a titrator (Titrator Mettler DL70Es, Mettler Toledo, USA) for alkalinity, a Total Organic Carbon Analyzer (TOC-V CPH/CPN; Shimadzu, Japan) for total inorganic carbon (TIC) and total organic carbon (TOC), a turbidimeter (HACH 2100 AN IS, Hach company, USA) for turbidity, a continuous flow analyser (BRAN+LUEBBE AutoAnalyzer3, BRAN+LUEBBE, Germany) for silicon dioxide and nitrogen, a comparator (Hellige Neo comparator, Nessler, Germany) for water colour, an ion chromatography (Dionex DX500, Dionex corporation, USA) for chloride, fluoride and sulphate and a spectrophotometer (Hitachi U-2000, Japan) for phosphorus. The chemical oxygen demand (COD_{Mn}) and dissolved oxygen were determined with titrimetric method by following the Finnish standards SFS3036 and SFS3040, respectively. The trace elements were analysed by using either an inductively coupled plasma optical emission spectrometry (ICP-OES; Varian Vista PRO Radial, Varian Incorporation, USA) or an inductively coupled plasma mass spectrometry (ICP-MS; Perkin Elmer ELAN 6000, Perkin Elmer corporation, USA).

2.3.2. Dynamic Light Scattering. A dynamic light scattering (DLS; Malvern Zetasizer Nano S, model ZEN 1600, Malvern Instruments, United Kingdom) was used to monitor the hydrodynamic diameter of TiO_2-P25 particles spiked in different natural waters. The increase of particle size indicates the formation of aggregates or agglomerates (hereafter only the term aggregates is used in the text). The DLS was operated with a He-Ne laser at a wavelength of 633 nm and light scattering was detected at an angle of 173°. Prior to analysis, 1 ml of sample added in polystyrene cuvette was stabilised at 25°C. The data presented in this paper are based on the CUMULANTS analysis that converts the intensity autocorrelation functions to intensity weighted particle mean Brownian diffusion motion (and polydispersity index). The hydrodynamic diameter is determined with Stokes-Einstein equation by assuming that the particles are spherical. Three measurements of each sample were run and all the experiments were done in duplicates.

2.3.3. Spectroscopy. A UV-Vis spectroscopy (Shimadzu UV-1601 PC, Shimadzu Corporation, Japan) was applied to monitor the absorbance of TiO_2-P25 particles. At the studied concentrations, the relative absorbance is highly consistent with the relative particle concentration. The test samples were

Journal of Physics: Conference Series **304** (2011) 012018 doi:10.1088/1742-6596/304/1/012018

prepared in 20 mm quartz cuvettes. Absorbance of dispersed TiO_2 nanoparticles were measured at a wavelength of 378 nm every 5 min over the 300-min period. The detection limit of the method was about 1 mg/l. All the experiments were done as duplicates.

2.3.4. Electron microscopy. The TiO_2-P25 powder was imaged with a Field emission scanning electron microscopy (FESEM; JEOL JSM-6335F) operated at secondary mode. The optimal resolution of FESEM is 1.5 nm and its measurement uncertainty is approximately 1%. The elemental composition of the investigated powder was confirmed with an Energy-dispersive X-ray analysis (EDS). The detection limit of the EDS varied in the range of 0.1-0.5 % according to an analyte.

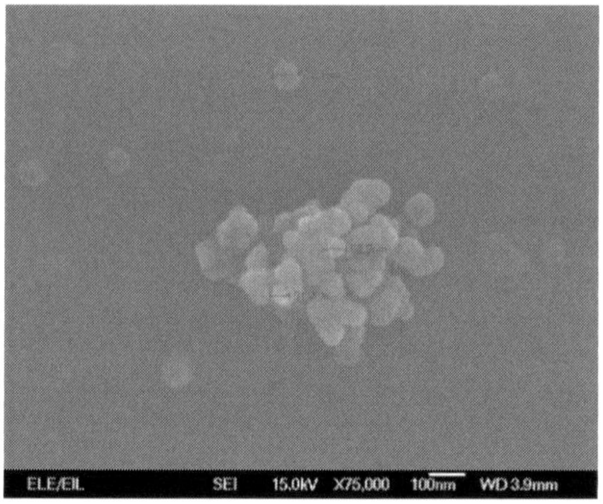

Figure 1. FESEM image of titanium dioxide (P25-TiO_2) powder.

3. Results and discussion

3.1. Water quality parameters of natural water samples

The selected physicochemical water quality parameters are presented for two brackish water and two fresh water samples in Table 1. The brackish water samples was basic (pH ~ 8), whereas Iso Lehmälampi water sample was more acidic (pH ~ 5.0) than Simijärvi water sample (pH 6.1). The conductivity describing the ion strength was predictably much higher in brackish waters than in fresh waters. The concentration of total organic carbon (TOC) varied in a relatively narrow range. The highest TOC concentration was measured in Iso Lehmälampi sample (6.1 mg/l) and the lowest in Simijärvi sample (4.1 mg/l). The water colour of the analysed water samples was consistent with the TOC concentration which reflects the content of natural organic matter. In addition, the fresh waters were analysed for several elements and common nutrients. The concentrations of most abundant divalent ions were above 2-fold higher in Simijärvi than in Iso Lehmälampi.

3.2. Aggregation of TiO_2 nanoparticles

The stability of TiO_2-P25 particles were tested at two initial concentrations of 100 mg/l and 1 mg/l (Figure 2). As expected, the agglomeration rate was very high in high ion strength conditions, i.e. brackish water samples. In three minutes, that is the time between the spiking and the first measurement, the intensity weighted mean hydrodynamic diameter had already increased from 200 nm to 1500 nm at the initial concentration of 100 mg/l. During the 60-min experiment, the mean hydrodynamic diameter increased close to 3000 nm in brackish waters. Contrary to brackish waters, there was difference in the agglomeration rates between the two fresh water samples. After the immediate increase of particle size (from 200 nm to 500 nm), the intensity weighted mean hydrodynamic diameter grew only slightly in Iso Lehmälampi water, whereas the mean particle size in Simijärvi sample was approximately 1600 nm at the end of 60-min experiment. The higher conductivity and concentration of divalent ions as well as lower TOC in Simijärvi sample may result in the faster agglomeration than in Iso Lehmälampi. In addition, the pH value of Simijärvi sample may be close to the isoelectric point of TiO_2-P25 particle dispersion, which would also favour the agglomeration.

Table 1. Physicochemical water quality characteristics of the tested natural waters.

	Långskär	Tvärminne	Simijärvi	Iso Lehmälampi
Al (µg/l)	–	–	73	160
Alcalinity (mmol/l)	–	–	0.011	0.012
Ba (µg/l)	–	–	6.3	4.7
Ca (mg/l)	–	–	1.7	0.7
Cl (mg/l)	–	–	1.8	1.5
Water colour (mg Pt/l)	15	–	15	50
COD_{Mn} (mg/l)	–	–	2.7	5.1
Conductivity (mS/m)	978	704	2.7	1.9
F (µg/l)	–	–	41	47
Fe (µg/l)	–	–	24	400
K (mg/l)	–	–	0.3	0.4
Mg (mg/l)	–	–	0.7	0.3
Mn (µg/l)	–	–	7.5	15
Na (mg/l)	–	–	1.5	1.1
N-NH_4 (µg/l)	<2	–	<2	48
N-NO_2+NO_3 (µg/l)	<2	–	154	41
N-total (µg/l)	350	–	300	310
Dissolved O_2 (mg/l)	13.4	–	11.7	4.9
pH	8.1	8.0	6.1	5.0
P-PO_4 (µg/l)	3	–	<2	<2
P-total (µg/l)	25	–	3	8
Salinity (‰)	5.55	3.89	–	–
SiO_2 (mg/l)	0.49	0.52	2.5	1.6
SO_4 (mg/l)	–	–	5.4	2.7
Sr (µg/l)	–	–	15	5.3
TIC (mg/l)	–	–	<0.5	3.0
TOC (mg/l)	4.2	5.6	4.1	6.1
Turbidity (FNU)	1.9	4.7	0.40	1.1

At concentration of 1 mg/l, the initial particle size was close to that of stock dispersion and it remained stable over 60-min period in fresh water samples. The aggregates formed in brackish waters but the intensity weighted mean particle size was only about 400 nm at the end of 60-min experiment. A comparison between the agglomerate size in the two concentrations shows that the concentration plays an important role in the particle aggregation. This is due to the decreasing distance between the suspended particles at the increasing concentrations, which results in the increased probability of collisions between particles. Therefore, to get more realistic understanding on the environmental fate of engineered nanoparticles in different environmental samples, some of the experiments should be conducted close to the predicted environmental levels in the future studies. However, this requires the development of more sensitive analytical methods.

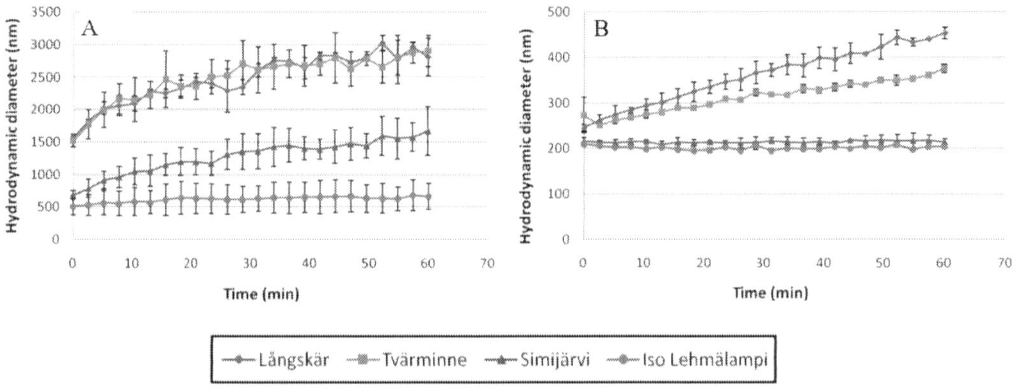

Figure 2. The mean±SD hydrodynamic diameter of TiO_2-P25 spiked in four different natural waters at the initial concentrations of 100 mg/l (A) and 1 mg/l (B).
Brackish water = Långskär and Tvärminne; Fresh water = Simijärvi and Iso Lehmälampi.

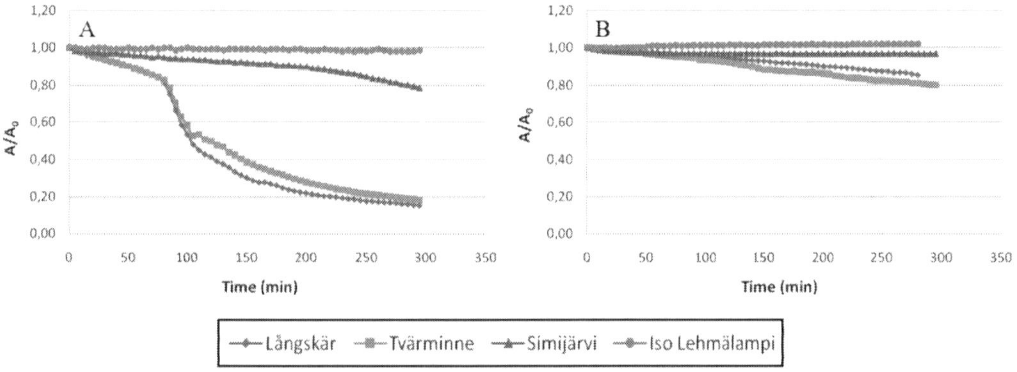

Figure 3. The relative absorbance of TiO_2-P25 spiked in four different natural waters at the initial concentrations of 100 mg/l (A) and 10 mg/l (B).
Brackish water = Långskär and Tvärminne; Fresh water = Simijärvi and Iso Lehmälampi.

3.3. Deposition of TiO_2 nanoparticles

The TiO_2 concentration was monitored over a 5-h period by using the UV-Vis spectrophotometer. As the formation of aggregates indicated, the dispersed particle concentration decreased strongly in brackish waters. At the concentration of 100 mg/l in brackish waters, the relative absorbance dropped down to 20% of initial absorbance in 5 hours. The relative absorbance showed a similar trend in both brackish waters: first it decreased slowly, followed by a steep drop of absorbance between 80 and 100 min and then the decrease slowed down. The similar deposition pattern was shown earlier for reactive nanosized iron particles [12]. The steep drop in the relative absorbance is likely due to the gravitational settling of large aggregates. The sedimentation lowers the concentration of dispersed particles and thus the settling slows down. At the higher concentration, the relative absorbance decreased in Simijärvi water, but it remained stable in Iso Lehmälampi water over the whole 5-hour experiment. The stable particle size and absorbance in Iso Lehmälampi water indicate that the deposition observed in other water samples was due to gravitational settling and the deposition on the walls due to the particle diffusion was minor. The aggregation and deposition observations support strongly each other.

The deposition was strongly dependent on the particle concentration. Since the detection limit of UV-Vis spectroscopy for the TiO_2 particles was 1 mg/l, the lower concentration for deposition experiments had to be chosen as 10 mg/l. At the lower concentration, the deposition rate was clearly lower in brackish waters. During the 5-hour experiments, the relative absorbance decreased about 20% in the brackish waters and about 5% in Siimjärvi water, whereas the relative absorbance remained stable in Iso Lehmälampi water.

4. Summary and conclusions

Nanoparticle transport, persistence and bioavailability in the environment are essential aspects to consider in risk assessment and management. The stabilization of nanoparticles in aquatic environment is the sum of the intrinsic properties of material itself (e.g. size, surface chemistry and charge and hydrophilicity) and the physicochemical characteristics of receiving environment (e.g. pH, ionic strength, ratio of divalent to monovalent ions and organic matter). As shown in this study, the particle concentration also plays an important role. Nanoparticles that undergo aggregation will sediment and thus they become less mobile. Consequently, the released engineered nanoparticles occur close to their sources and benthic organisms have higher potential to expose these aggregated nanoparticles. On the other hand, the stabilized nanoparticles are dispersed over larger area and may have impact on aquatic organisms. The brackish water samples tested in this study destabilized the TiO_2 particle dispersion which resulted in the remarkable sedimentation within a few hours. The fresh water led either the destabilisation or stabilisation of TiO_2 particles depending on the water quality variables. While the environmental concentrations of TiO_2 nanoparticles will probably be lower than those tested in this study, the initial source may discharge at these levels. The further studies should focus on the nanoparticle concentrations close to the PEC-values, which challenges the current analytical techniques.

Acknowledgements

The authors wish to thank the Finnish Standards Association SFS and the association of Maa- ja Vesitekniikan Tuki ry for financial support. In addition, Ms. Heljä Haapala is acknowledged for her technical help during the experiments and the laboratory staff of SYKE Environment measurement and testing group for the analysis of water quality.

References

[1] Hansen S F, Michelson E S, Kamper A, Borling P, Stuer-Lauridsen F and Baun A 2008 Categorization framework to aid exposure assessment of nanomaterials in consumer products *Ecotocixol.* **17** 438–47

[2] ANEC (European consumer voice in standardisation) and BEUN (European Consumers' Organisation) 2010 ANEC/BEUC inventory of products claiming to contain nanoparticles available on the EU market Link to the inventory on http://www.anec.eu or http://www.beuc.eu/

[3] Sellers K., Mackay C., Bergeson L.L., Clough S.R., Hoyt M., Chen J., Henry K. & Hamblen J. (2009) *Nanotechnology and the Environment.* CRC Press, Boca Raton.

[4] OECD-WPMN (OECD Working Party on Manufactured Nanomaterials) (2008) *List of manufactured nanomaterials and list of endpoints for phase one of the OECD testing programme.* Series on the safety of manufactured nanomaterials, No. 6. http://www.olis.oecd.org/olis/2008doc.nsf/LinkTo/NT000034C6/$FILE/JT03248749.PDF

[5] Kaegi R., Ulrich A., Sinnet B., Vonbank R., Wichser A., Zuleeg S., Simmler H., Brunner S., Vonmont H, Burkhardt M ja Boller M 2008 Synthetic TiO_2 nanoparticle emission from exterior facades into the aquatic environment *Environ. pollution* **156** 233–9

[6] Mueller N C ja Nowack B 2008 Exposure modeling of engineered nanoparticles in the environment *Environ. Sci. Technol.* **42** 4447–53

[7] Baalousha M 2009 Aggregation and disaggreagtion of iron oxide nanoparticles: influence of

particle concentration, pH and natural organic matter *Sci. Tot. Environ.* **407** 2093–101
[8] Domingos R F, Tufenkji N and Wilkinson K J 2009 Aggregation of titanium dioxide nanoparticles: role of a fulvic acid *Environ. Sci. Technol.* **43** 1282–6
[9] Dunphy Guzman K A, Finnegan M P and Banfield J F 2006 Influence of surface potential on aggregation and transport of titania nanoparticles *Environ. Sci. Technol.* **40** 7688–93
[10] French R A, Jacobson A R, Kim B, Isley S L, Penn R L and Baveye P C 2009 Influence of ionic strength, pH and cation valence on aggregation kinetics of titanium dioxide nanoparticles *Ennviron. Sci. Technol.* **43** 1354–9
[11] Keller A A, Wang H, Zhou D, Lenihan H S, Cherr G, Cardinale B J, Miller R and Ji Z 2010 Stability and aggregation of metal oxide nanoparticles in natural aqueous matrices *Environ. Sci. Technol.* **44** 1962–7
[12] Phenrat T, Saleh N, Sirk K, Tilton R D and Lowry G V Aggregation and sedimentation of aqueous nanoscale zerovalent iron dispersions *Environ. Sci. Technol.* **41** 284–90
[13] Sarmaja-Korjonen K and Alhonen P 1999 Cladoceran and diatom evidence of lake-level fluctuations from a Finnish lake and the effect of a aquatic-moss layers on microfossil assemblages. *J Paleolimnol.* **22** 277–90
[14] Lignell R, Hoikkala L and Lahtinen T 2008 Effects of inorganic nutrients, glucose and solar radiation on bacterial growth and exploitation of dissolved organic carbon and nitrogen in the northern Baltic Sea *Aquat. Microb. Ecol.* **51** 209–21

Gaseous effluents from the combustion of nanocomposites in controlled-ventilation conditions

D Calogine[1], G Marlair[1], J-P Bertrand[1], S Duplantier[1], J-M Lopez-Cuesta[2], R Sonnier[2], C Longuet[2], B Minisini[3], C Chivas-Joly[4], E Guillaume[4] and D. Parisse[5]

[1]INERIS – Parc Technologique Alata, BP2, 60550 Verneuil-en-Halatte / France

[2]Ecole des Mines d'Alès - CMGD, 6 Avenue de Clavières, 30319 Alès / France

[3]ISMANS, 44 Avenue F. A. Bartholdi ,72000 Le Mans / France

[4]LNE – 29 Avenue Roger Hennequin, 78197 Trappes Cedex / France

[5]PlasticsEurope – 14 Rue de la République, 92800 Puteaux / France

E-mail: didier.calogine@ineris.fr

Abstract. Composite materials are more and more used every day. In order to further enhance their attractive mechanical and physico chemical performances, the last generation of these materials largely makes use of nanomaterials.

Various nanofillers are eligible for such a purpose, the best ones depending on the associated matrices. One favorite field of application of these nanomaterials is fire retardancy and fire behavior of nanocomposites. In the context of the ANR research project NanoFeu, various technical analyses have been performed [1]. One focuses on the characterization of the dispersion of nanofillers in the matrix; another deals with the characterization of the fire behavior of samples including the study of the composition of the gaseous effluents, the characterization of the emitted soot [2]. A third part of the work focused on molecular modeling of observed phenomena within the matrices. This paper focuses mainly on the combustion of nanocomposite samples under various ventilation conditions.

Tests have been performed with the Fire Propagation Apparatus (FPA). Samples are based on poly(methyl methacrylate); various nanofillers were used: carbon nanotubes, alumina and silica. Efficiency of fillers is compared to the classical ammonium polyphosphate in equal proportions. During testing, the ventilation-controlled conditions were obtained by adjusting the combustion air flow rate entering the apparatus. Gaseous effluents were analyzed by Fourier Transform Infra-Red spectrometer.

Fire behavior is characterized in terms of fire parameters and chemical composition of gaseous effluents. The influence of ventilation conditions is especially significant in terms of amount of gases released: much more important production of specific gases is generally observed in case of under ventilation regime as compared to the well ventilated case.

1. Introduction

Nanotechnologies and the relating use of nanomaterials represent one of the major economic break-through for the developed societies today, with headways which have already allowed important evolutions in various domains such as health, energy, information, or transport... In the field of polymeric materials various types of additives are used to improve their behavior. Indeed, to respond to more stringent demands in terms of overall performances of plastic materials, including reaction to fire, new fire retardancy systems based on introduction of nanoparticles may become a suitable alternative to traditional approaches. The alternative brought by nanoparticles must also be assessed and confronted to the traditional systems, in terms of environmental impact. It is in this context that the NanoFeu project has been initiated. The objective of these works is to study the fire behaviour of materials containing nanoparticles, to define their role and their fate during the combustion. To perform this study, nanocomposites with a polymer matrix were elaborated; experiments were performed to assess the reaction to fire of the test samples and the composition of the emitted gaseous effluents. Section 2 presents the NanoFeu project; section 3 is devoted to the description of experimental devices. Main results are given in section 4; conclusion is presented in section 5.

2. The *NanoFeu* project

The elaboration of nanocomposites using a polymer matrix is one of the most active domains in nanomaterials development. The properties assigned to nanoparticles are highly varying and concern in particular the mechanical intensification, electrical and thermal conduction characteristics, the barrier properties to gases and liquids, as well as the fire behavior that is thought to result from barrier properties efficiency [3].

The ANR project called *NanoFeu*, started in 2008 for three years. It focused on the impact of fillers on the fire behavior of nanocomposites. Coordinated by INERIS (Institut national de l'environnement industriel et des risques), four other partners have taken part to the research consortium, namely: Laboratoire National de métrologie et d'Essais (LNE), Ecole des Mines d'Alès (EMA), Institut supérieur des matériaux et mécaniques avancés (ISMANS) and PlasticsEurope. PlasticsEurope supports this academic group as an association of nanocomposite and polymer producers.

The *NanoFeu* project focuses on nanofillers impact on the smoke composition (gases and aerosols), by comparing polymers own responses to the ones of polymers containing nanofillers and polymers containing both nanofillers and conventional flame retardant systems. Another aspect is to evaluate the morphological modifications of nanoparticles produced during the combustion of such materials and to assess relating potential health hazards. The consortium gathers experts in several fields like fire physics, chemistry and numerical modeling.

3. Fire behaviour tests

3.1. Samples composite setup

In order to evaluate the fire behaviour of nanocomposites, a series of formulations have been elaborated at Ecole des Mines d'Alès for this project. Dimensions of samples are 100 mm x 100 mm x 4 mm. Blends of polymers with nanofillers were compounded using a twin screw extruder (Clextral BC 21, L ¼ 1200 mm, L/D ¼ 48, 2.7 kg/h, 200 rpm). PMMA and commercial PMMA/MWNT 5%wt. masterbatches were mechanically mixed prior to feeding the extruder.

Main characteristics of samples tested in this paper are given in the following table:

Table 1. Table of samples

Sample	Formulation	Designation
1	Poly(methyl methacrylate);	PMMA
2	PMMA + 0.2%wt Carbon Nanotubes	PMMA - 0.2 CNT
3	PMMA + 1%wt Carbon Nanotubes	PMMA - 1 CNT
4	PMMA + 15%wt Silica	PMMA - 15 SiO_2
5	PMMA + 15%wt Alumina	PMMA - 15Al_2O_3
6	PMMA + 15%wt Ammonium Polyphosphate	PMMA - 15 APP
7	PMMA + 5% wt Alumina +10%wt Ammonium Polyphosphate	PMMA - 5Al_2O_3 – 10 APP

3.2. The Fire Propagation Apparatus

Fire behavior of nanocomposites is evaluated using a fire propagation apparatus (ASTM E2058:09, ISO DIS 12136 [4]). Under a thermal irradiance of 35kW/m^2, various samples are tested for ventilation controlled condition of air flows of 50L/min and 350 L/min.

Figure 1. The Fire Propagation Apparatus

3.3. The infrared spectrometer instrument

In order to analyze the composition of gases emitted during combustion, infrared spectroscopy is used with a Thermofisher Nicolet 6700. This spectrometer is equipped with a detector of type MCT-A cooled with liquid nitrogen, scanning a spectral range from 650 to 4200 cm^{-1}. Its spectral resolution is 0,5 cm, the cells of optical path of 2 meters and 10 meters are heated at 180°C. Data acquisition is performed every 4s. Gases at present measured for quantification are: H_2O, CO_2, CO, NO, NO_2, SO_2, N_2O, CH_4, C_3H_6, HBr, HCl, HF, HCN, NH_3. A semi-quantitative measure of following species is performed too: formaldehyde, acetaldehyde, ethylene, acetylene.

4. Experimental results

4.1. Fire behaviour of nanocomposites

For the various regimes of ventillation, combustion parameters are given in the following table. Variations for instance in terms of ignition time, effective heat of combustion and mass loss rates are noticed.

Table 2. Combustion parameters of sample

Formulations	Ignition Time		Effective heat of combustion				mass loss rates	
	s		kJ/kg		MJ/m^2		$g/m^2/s$	
Test condition	Well-ventilated	Under-ventilated	Well-ventilated	Under-ventilated	Well-ventilated	Under-ventilated	Well-ventilated	Under-ventilated
PMMA	52	45	24880	16120	121	78	34	29
PMMA-0.2 CNT	42	45	25100	15950	121	77	35	31
PMMA-1 CNT	39	42	24130	17220	116	83	28	25
PMMA - 15 SiO_2	47	43	24205	23964	109	108	18	16
PMMA-15 Al_2O_3	60	50	23878	23280	110	107	20	15
PMMA-15 APP	72	55	23840	18510	108	84	25	23
PMMA - 5Al_2O_3 – 10 APP	50	39	23745	20830	109	96	25	19

In the well-ventilated case, the latest ignition times are obtained with the formulations PMMA-15 Al_2O_3 and PMMA-15 APP. The heat of combustion obtained for the PMMA is in accordance with the 24±2 MJ/kg measured more recently with the fire propagation apparatus [5]. As compared to PMMA without any additive (first sample in table 2), all tested formulations including fillers and or fire retardants (except PMMA with 0.2%CNT) show in our tests lower values for the effective heat of combustion and for the average mass loss rate. Nevertheless, the most significant decrease of heat of combustion is about 6%. This value is lower than corresponding weight fraction of incorporated additives and is obtained for the formulation containing PMMA – 5% Al_2O_3 – 10% APP.

In the under-ventilated case, the variations are less significant as regard times of ignition. The most 'retarded' times of ignition are observed for the formulations PMMA-15 Al_2O_3 and PMMA-15 APP. The shortest time is obtained with the formulation combining alumina and ammonium polyphosphate. The highest effective heats of combustion are obtained with the formulations PMMA - 15 SiO_2, PMMA-15 Al_2O_3 and PMMA - 5Al_2O_3 – 10 APP. The most largest mass loss rates are obtained with formulations containing CNT.

The kinetics of heat release versus time is represented in figures 2 and 3. The peaks of HRR vary according to formulations, as well as the general shape of HRR overall responses.

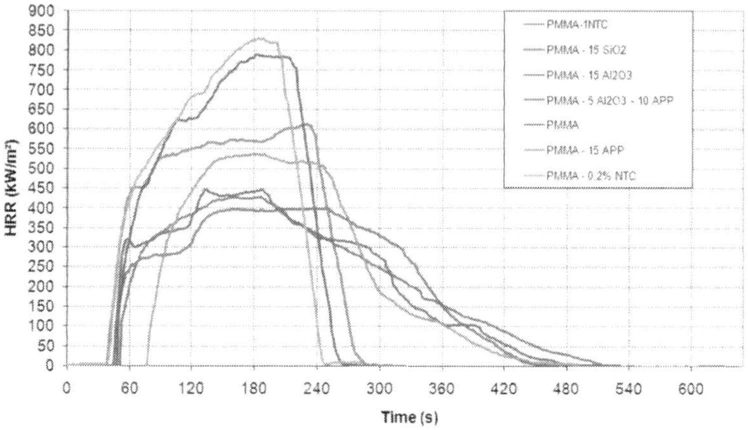

Figure 2. Heat Release Rate measuring during tests in well-ventilated conditions

The impact of nanoparticles, as well as APP on the decrease of peak of heat released (pHRR) during combustion is observed. Compared to PMMA with a pHRR include between 750 and 800 kW/m^2, the presence of CNT in concentration of 1%wt CNT decreases this peak of 25%. Our results are in agreement with Costache et al. who obtained a decreasing of 29% in the pHRR with matrix containing 3%wt multiwall carbon nanotubes (MWCNT) [6]. On the other hand Kashiwagi et al. [7] have obtained a decrease superior to 50% with 1%wt singlewall CNT. They also showed that this decrease was depending on the quality of nanotubes dispersion into the matrix. Indeed, with 0,5% of nanotubes badly scattered, the relating decrease in the pHRR is only about 10%, and this value can reach up to more than 50% in the case of a good dispersion.

Identical tests are performed in the well-ventilated case.

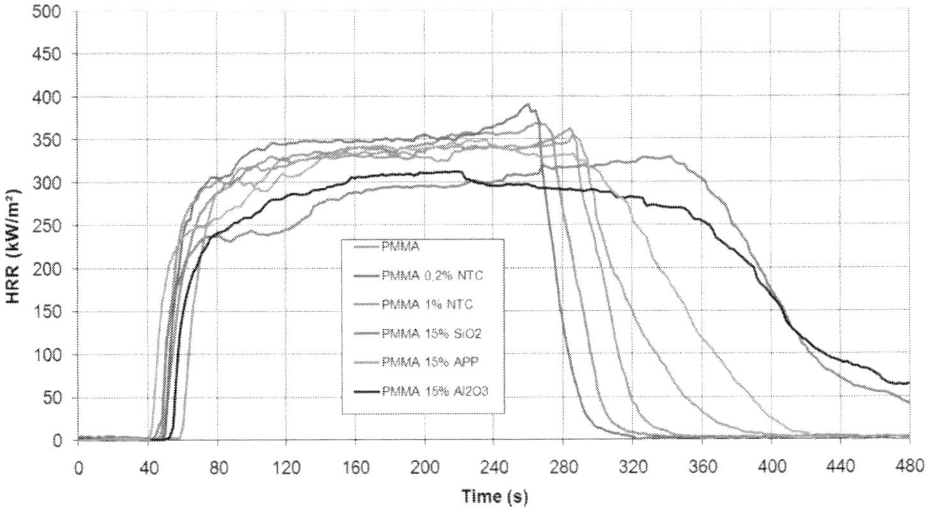

Figure 3. Heat Release Rate measuring during tests with under-ventilated conditions.

In this case, a significant decrease of HRR for some formulations is noticed, as well as a variation of the shapes of HRR curves.

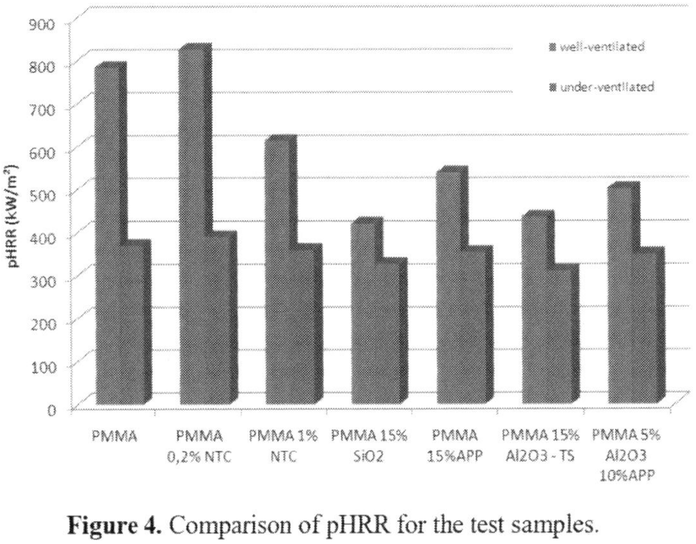

Figure 4. Comparison of pHRR for the test samples.

Figure 4. illustrates more particularly that for various formulations, a decrease more or less significant of pHRR is observed, according to the composition of nanofillers for combustion tests in the well-ventilated case. In the underventilated case, the peak of HRR vary in a very limited manner (between 300 and 400 kW/m²).

4.2. Composition of gaseous effluents

In addition to the observation of fire behavior of samples, the composition of gaseous effluents from combustion of nanocomposites is analyzed. The obtained results are given in the following figures for the various identified gases for the two regimes of ventilation. In fig. 5 to 10, THCs refer to overall hydrocarbons CxHy, including methane, as measured by a dedicated analysor (flame ionization detector).

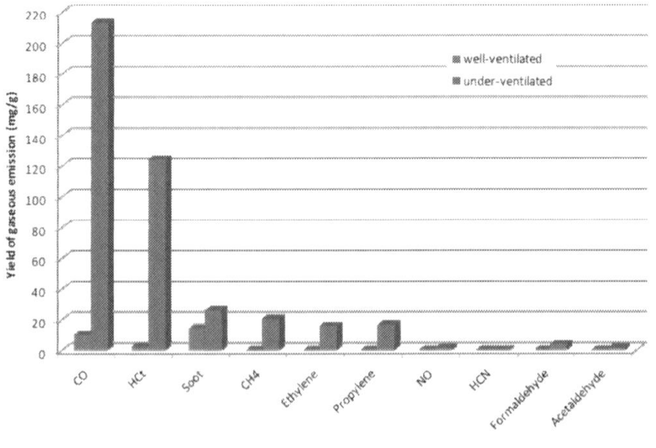

Figure 5. Emission yields of various gases from combustion of PMMA in controlled-ventilation conditions.

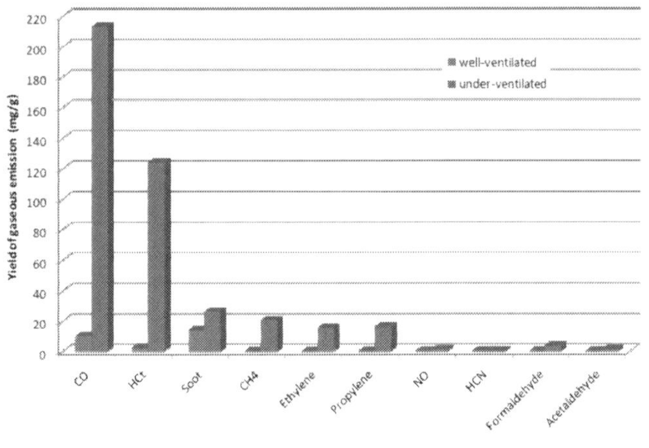

Figure 6. Emission yields of various gases from combustion of PMMA-0.2 CNT in controlled-ventilation conditions

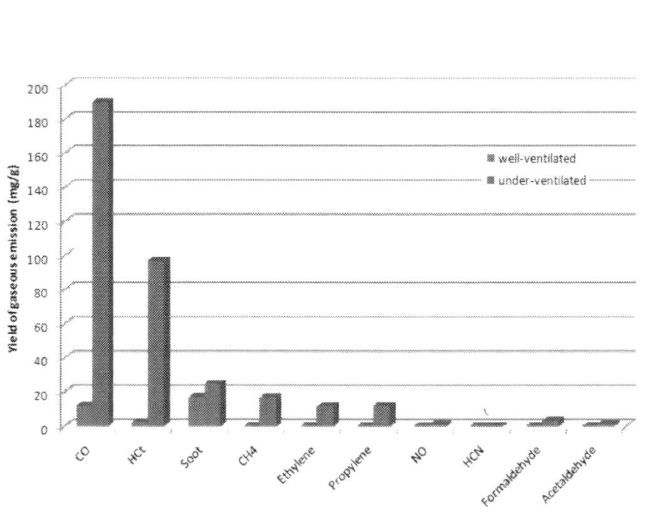

Figure 7. Emission yields of various gases from combustion of PMMA-1 CNT in controlled-ventilation conditions

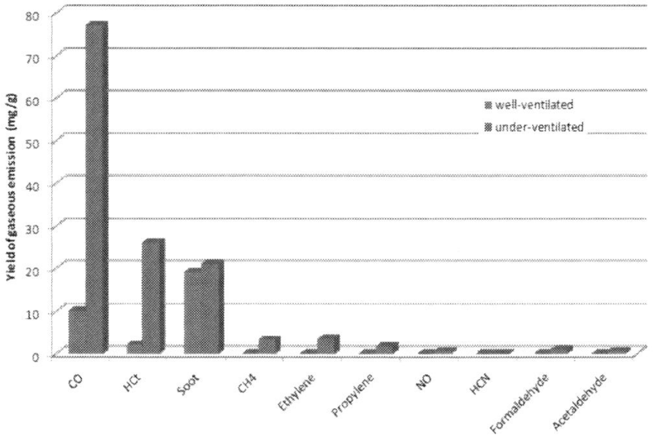

Figure 8. Emission yields of various gases from combustion of PMMA - 15 SiO2 in controlled-ventilation conditions

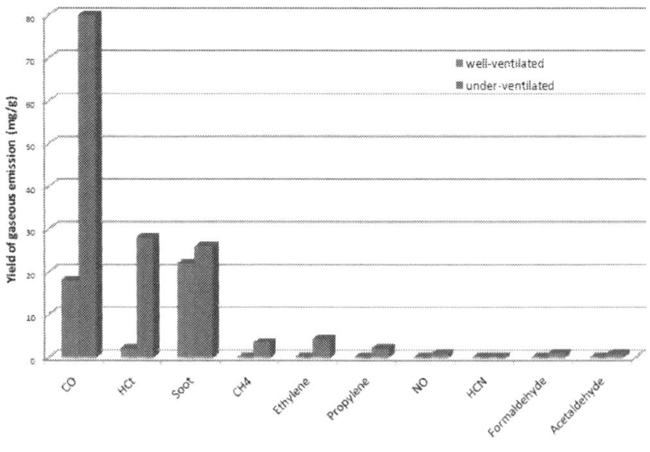

Figure 9. Emission yields of various gases from combustion of PMMA-15 Al2O3 in controlled-ventilation conditions

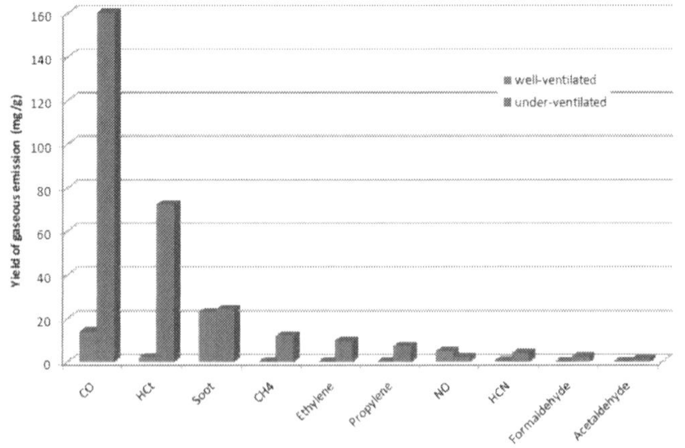

Figure 10. Emission yields of various gases from combustion of PMMA-15 APP in controlled-ventilation conditions

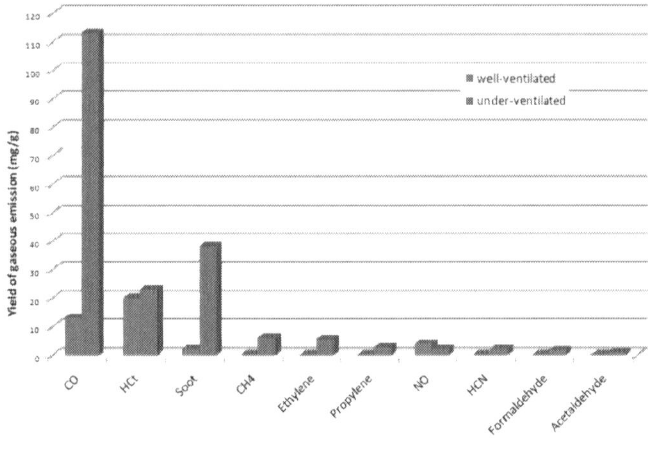

Figure 11. Emission yields of various gases from combustion of PMMA - 5Al2O3 – 10 APP in controlled-ventilation conditions

Significant concentrations in NO are obtained in presence of samples containing APP. A measurable value of HCN is also observed for samples containing APP. This statement highlights reactive processes that take place in gaseous phase involving the ammonia resulting of decomposition of APP and carbon monoxide produced by the combustion of PMMA. This production of HCN is lower for PMMA - $5Al_2O_3$ – 10 APP than for PMMA-15 APP.

In the under-ventilated case, the introduction of nanofillers in the PMMA matrix can significantively reduce the yield of CO emission. The more the samples contain nanofillers, the more significant the decrease is. This decrease is also observed for the yield of THCs (total hydrocarbons) emission. The yields of soots production are more significant for the formualtions PMMA-15 Al_2O_3, PMMA - 15 SiO_2 as well as for the PMMA - $5Al_2O_3$ – 10 APP.

Globally, from the overall gas emission sprectrum, 3 classes of combustion related emission yields are identified. The first one deals with the species CO, THCs and Soot: these species are emitted in relative large quantities whatever the samples and the combustion conditions are (say yields may reach more than 10 mg/g. The second class concerns the species CH_4, ethylene, propylene and NO that are globally produced in moderate yields for both well and under-ventilated regimes. The last class containing HCN, Formaldehyde and Acetaldehyde refers to toxic products emitted in a still lower yield than the other species during combustion. A global view of compared overall toxicity would require however require a more detailed analysis integrating individual contribution of toxic species as well as cumulative effects that is out of scope of this paper.

5. Conclusion

Experimental tests for the combustion of nanocomposites samples have been performed in the context of *NanoFeu* project. Various regimes of ventilation have been tested in order to observe the influence of nanofillers on the composition of gaseous effluents.

Gaseous effluents spectrums have been established for the combustion of PMMA and PMMA containing nanofillers. Combustion parameters seem to be coherent with others references on the subject, the impact of the filler is observed in terms of HRR curves and peak of values. Nanofillers modified heat release rate in the two regimes of ventilation. pHRR values are less significant in under-ventilated case than in well-ventilated case. Concerning gaseous effluents, differences are observed between the two regimes. This difference is more or less significant for 3 classes of species identified during the combustion of samples, and that are ranked according to levels of mass yields observed. High emissions are expected for the first class (including CO_2, which yield is not given here), moderate ones for the second (light hydrocarbons + NO) and still significantly less for the last class (specific species like HCN). Nevertheless this observation does not completely inform about the global toxicity of gaseous effluents. Others complementary studies are in progress to further address the question.

Acknowledgments

The authors gratefully acknowledge ANR for their financial support of the NanoFeu project.

References

[1] Calogine D., Duplantier S., Guillaume E., Chivas C., Lopez-Cuesta J.-M., Longuet C. Minisini B., Gensous F., Chesné E., "The Nanofeu project : objectives and tools", Journal of Physics: conference series, vol. 170, n°1, (2009).
[2] Chivas-Joly C., Guillaume E., Ducourtieux S., Saragoza L., Lesenechal D., Lopez-Cuesta J-M., Longuet C., Sonnier R., Duplantier S., Bertrand J.P., Calogine D., Minisini B., Parisse D., "Influence of carbon nanotubes on fire behaviour and on decomposition products of thermoplastic polymers", Interflam 12th international conference on fire science and engineering (2010).
[3] Alexandre M., Dubois P., Polymer-layered silicate nanocomposite: preparation, properties and uses of a new class of materials, Mat. Sci. Eng. 28, 1-63 (2000).
[4] "ISO/DIS 12136: Reaction to fire tests – Measurement of fundamental material properties using a Fire Propagation Apparatus" (2010).
[5] Kashiwagi T., "Flame retardant mechanism of the Nanotube-based nanocomposite- final report" NIST GCR 07-912 (2007).
[6] Costache M.C., Wang D., Heidecker M.J., Manias E., Wilkie C.A., "The thermal degradation of PMMA nanocomposites with montmorillonite, layered double hydroxides and carbon nanotubes », Polym Adv. Tech. 17, 272-280 (2006).
[7] Kashiwagi T., Du F., Winey K. I., Groth K. M., Shields J. R., Bellayer S. P., Kim H., Douglas J. F., Polymer, "Flammability properties of polymer nanocomposites with single-walled carbon nanotubes: effects of nanotube dispersion and concentration" 46, 471–481 (2005).

Characterization of aerosol emitted by the combustion of nanocomposites

C Motzkus, C Chivas-Joly, E Guillaume, S Ducourtieux, L Saragoza, D Lesenechal and T Macé

LNE, 29 Avenue Roger Hennequin, 78197 Trappes Cedex, France,

E-mail: charles.motzkus@lne.fr

Abstract. Day after day, new applications using nanoparticles appear in industry, increasing the probability to find these particles in the workplace as well as in ambient air. As epidemiological studies have shown an association between increased particulate air pollution and adverse health effects in susceptible members of the population, it is particularly important to characterize aerosols emitted by different sources of emission, during the combustion of composites charged with nanoparticles for example. The present study is led in the framework of the NANOFEU project, supported by the French Research Agency (ANR), in order to characterize the fire behaviour of polymers charged with suitable nanoparticles and make an alternative to retardant systems usually employed. To determine the impact of these composites on the emission of airborne particles produced during their combustions, an experimental setup has been developed to measure the mass distribution in the range of 30 nm – 10 µm and the number concentration of submicrometric particles of the produced aerosol. A comparison is performed on the aerosol emitted during the combustion of several polymers alone (PMMA, PA-6), polymers containing nanofillers (silica, alumina, and carbon nanotubes) and polymers containing both nanofillers and a conventional flame retardant system (ammonium polyphosphate). The results on the morphology of particles were also investigated using AFM.

1. Introduction

There is a growing interest of natural or synthesized nanoparticles due to the fact that nanoscale materials are known to exhibit significantly different physical, chemical, electrical, and optical properties, compared to their properties at the macroscale. From an industrial point of view, it is an opportunity to develop new products in various fields such as health, energy, information, and transport. However, there is a concern that nanoscale materials and nanomanufactured products may endanger human beings, as different public debates on the safety of nanosciences and nanotechnologies take place. Both assessments of these risks, preventive actions, and remediation, when needed, will rely on the measurements and characterization of nanoparticles. Consequently, it is necessary to improve our knowledge in this field to assess the potential toxicology risks of nanotechnology [1]. As epidemiological studies have shown an association between increased particulate air pollution and adverse health effects in susceptible members of the population, it is particularly important to characterize aerosols emitted by different sources of emission; for example during the combustion of composites filled with nanoparticles The material in accidental fires could then become a new source of nanoparticles. In this context, we present this study, which is led in the framework of the NANOFEU [2] project, started in January 2008 for 3 years, supported by the French

Research Agency (ANR). In order to know the impact of these composites on the emission of airborne particles produced during their combustion, an experimental set-up has been developed to measure the mass distribution and the number concentration of the produced aerosol.

2. Experimental methods

The presented experimental methods deal with the determination of the nanofillers impact on the smoke composition of aerosols, by comparing polymers alone, polymers containing nanofillers and polymers containing both nanofillers and a flame retardant (ammonium polyphosphate). Three different nanofillers (silica, alumina, and carbon nanotubes) have been selected, with or without surface treatment based on silane. These nanofillers are incorporated in Polymethyl Methacrylate (PMMA) and Polyamide-6 (PA-6) matrices [3]. This work also focuses on the study of the influence of the matrix on such modifications.

An experimental setup has been developed to measure the mass distribution and the number concentration of the produced aerosol with a cone calorimeter according to standard ISO 5660-1 [4]. This setup is composed of a cascade impactor [5] and a Condensation Nuclei Counter (CNC) coupled with a cone calorimeter (see figure 1). The concentration of submicrometer particles was measured by a CNC [6]. The CNC used (TSI 3022) detects particles down to 7 nm (minimum particle Size corresponding to an efficiency of 50%) at concentrations up to 10^7 particles/cm^3 [7]. In our experimental conditions, the flow rate of the CNC is equal to 0.3 L/ min. A Low Pressure cascade Impactor (DLPI: DEKATI, with the electrical configuration disabled) has been used to measure the mass distribution of the airborne particles in the range of 30 nm – 10 µm by post-gravimetric analysis.

The cone calorimeter main duct is modified in order to avoid the trapping of soot on the 90° angle of the exhaust duct, and also to allow sampling at a point where the flow of effluents can be considered as homogeneous and laminar. The heat flux is set at 50 kW/m^2 with a standardized ventilation rate. Two ventilation rates have been used depending on the associated metrology, equal to 38 L/s with the CNC alone (A, see figure 1) or equal to 24 L/s for the measurement with the Dilutor FPS (DEKATI) coupled at CNC (B, figure 1) and with the DLPI alone (C, figure 1).

An Atomic Force Microscope (AFM) is used to investigate the morphological differences of the particles produced during the combustion of the various formulations. To prepare appropriate samples for AFM measurements, particles were collected using the DLPI where polycarbonate circular filter membranes have been replaced by optical microscope glass cover slips. Those cover slips have been chosen for their particular dimensions: a glass square of 18 mm x 18 mm that perfectly fits into the impactor and the smallest thickness of 150 µm. The glass slip roughness is of a few nanometers. The cover slips were imaged using an AFM with a scan range of 90 µm x 90 µm x 8 µm and an accurate three-axis closed-loop (NanoMan VS from Digital Instruments equipped with the hybrid XYZ scanner). The AFM is protected with an acoustic/vibration isolation enclosure and equipped with a vibration isolation table. The system is placed on a massive concrete block dissociated from the building in a laboratory controlled in humidity and temperature (50 % Hr and 20°C). The measurements were conducted in air using tapping mode and OTESPA probes (Veeco probes). A low scan speed (2 µm/s) was applied to the scanner to get high quality and high resolution images.Topographic and phase images were recorded without real-time filter.

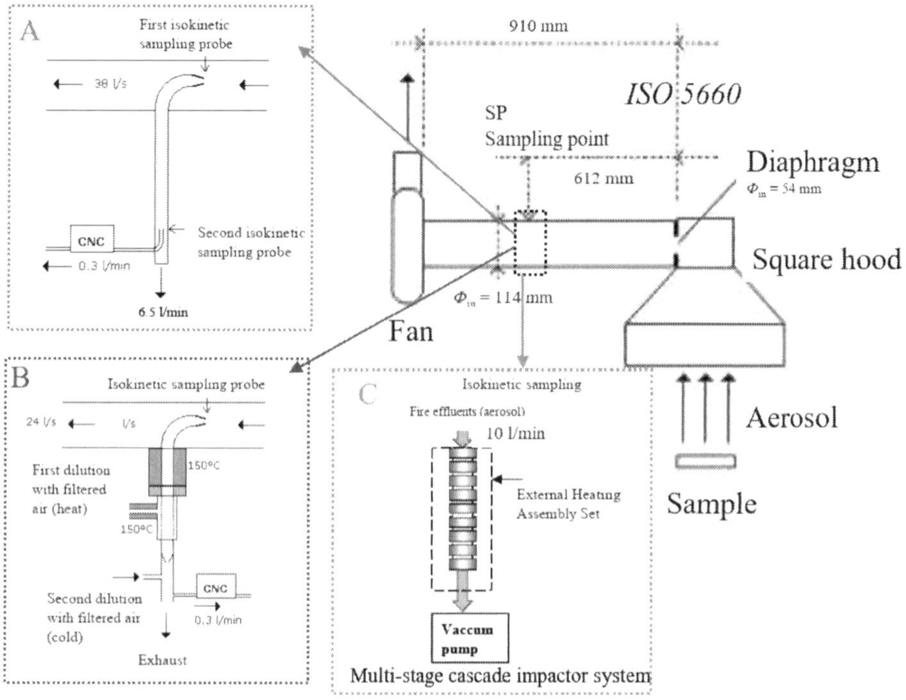

Figure 1. Experimental setup : Cascade impactor and a Condensation Nuclei Counter coupled with a cone calorimeter according to standard ISO 5660-1.

In the configuration where the CNC is directly linked to the cone calorimeter (A, figure 1), two constraints must be taken into account: not operating the CNC in ambient temperature above to 40°C and measuring an aerosol with a concentration below the limit concentration of 10^7 particles/cm^3. To lower the temperature of the effluent, the length of the primary probe is increased of 80 cm. An experiment with a sample of PMMA with a dimension of 100 x 100 mm was used to check that the temperature doesn't exceed 40 ° C at the secondary probe sampling of the CNC. The ventilation rate is increased to 38 L/s in order to dilute the combustion aerosol and thus to not saturate the detector of the CNC. The sampling probe (diameter) had been selected to be in isokinetic sampling conditions. A small quantity of the sample was chosen to reduce the amount of aerosol produced : an average mass of 1.4 g for samples of all formulations (with a dimension of 17x17x4 mm^3).

In the case of certain experiments, we observe that the concentration is above the limit concentration of the CNC. Thus, it was necessary to use a diluter in order to measure the emitted aerosol and compare the results obtained by different formulations of nanocomposite: a configuration with the Diluter FPS-4000 (DEKATI) coupled at CNC (B, figure 1). This diluter is adapted for combustion dilution with two-stages (double diluter setup), where the first diluter is heated and the second operates at ambient temperature. This system allows to dilute the aerosol avoiding the condensation and the particle perturbation with regard to a simple cold dilution. The nanocomposite sample size is fixed at 25x25x4 mm^3 (corresponding to 3 g of materials).

For the measurement with the DLPI linked to the cone calorimeter (C, figure 1), an external heating assembly has been used to prevent the effect of humidity on the aerosol size distribution. The experimental conditions (C, figure 1) are the following: the nanocomposite sample size is fixed at 50x50x4 mm^3 (12 g) and the sample flow rate is equal to 10 l/ min for the DLPI, during 5 minutes. This multi-stage aerosol-sampling device separates particles by size, according to their inertial properties in a moving air stream. Each stage of the impactor is characterized by a "cut-off size" (D50%). DLPI has a 13-stages particles impactor (table 1).

Table 1. Cut-off diameter D50% of each stage of Low Pressure cascade Impactor (DLPI: DEKATI) for a sample flow rate equal to 10 l/ min.

N°of stage	1	2	3	4	5	6	7	8	9	10	11	12	13
Cut-off size $D_{50\%}$. (µm)	0.03	0.06	0.09	0.16	0.26	0.38	0.62	0.95	1.60	2.39	4.00	6.69	9.93

The "cut-off size" corresponds to the aerodynamic diameter of trapped particles with an efficiency of 50% on a given stage. The aerodynamic particle diameter (d_{ae}) is defined as the diameter of a spherical particle with a reference density equal to 1000 kg.m^{-3} which has the same gravitational settling velocity as the particle under consideration [8].

The experiments were performed with the nanocomposite described in the table 2.

Table 2. Used Samples: polymers alone, polymers containing nanofillers and polymers containing both nanofillers and a conventional flame retardant system.

Sample	**Composition of the sample**	**Information on the nanofillers**
A	PMMA	
B	PMMA containing nanofillers of 15 % wt.SiO_2	Average primary particle size of 12 nm
C	PMMA containing nanofillers of 15 % wt.Al_2O_3	Average primary particle size of 13 nm
D	PMMA containing nanofillers of 15 % wt.Al_2O_3 with a surface treatment (silane)	Average primary particle size of 13 nm
E	PMMA containing nanofillers of 5 % wt.Al_2O_3 and 10 % of Ammonium Polyphosphate	Average primary particle size of 12 nm
F	Polyamide-6 (PA-6) containing 1% wt.CNT	Multi – wall carbon nanotubes

3. Result on the aerosol measurements

3.1. Results with the Low Pressure cascade Impactor

The aerosol size distribution is obtained by measuring the mass of the particles collected at each stage of the impactor with the balance Metler (with a 10 µg precision). The doubled weight gravimetric method is used: the gravimetric measurement of each filter before and after aerosol sampling is performed two times.

3.1.1. Discussion on the uncertainty of the mass distribution in the case of PA-6

Figure 2 presents the mass distribution obtained with the DLPI in the case of three combustions of a Polyamide-6 (PA-6) containing multi – wall carbon nanotubes (MWNT with 1%). This figure shows the distribution of the mass plotted on the logarithmic diameter scale. The Y – axis (dM/dlog (D_{ae})) represents the mass of the particles collected divided by the width in the logarithmic scale of various channels of measurement of the diameter, with the channel determined by the cut-off sizes ($D_{50\%}$).

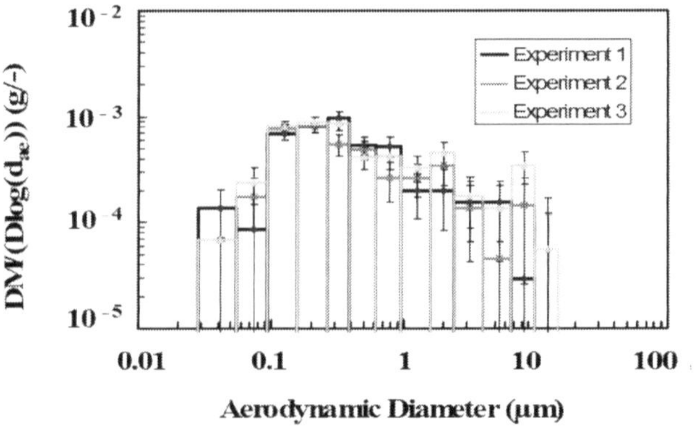

Figure 2. Mass distribution obtained with the DLPI in the case of three combustions of a PA-6 filled with 1% carbon nanotubes.

The results presented in figure 2 illustrate that the measurements are repeatable: we observe that the deviation obtained between different experiments could be explained by the standard deviation calculated only from the gravimetric measurement method. All the measured masses are significant for stages with a cut-off diameter between 0.1 µm and 4 µm. These results reveal an highlight presence in particle mass in these ranges. For other measurement channels above 4 µm, and below 100 nm, the calculated standard deviation of all the measurements couldn't always determine the significant presence in mass. However, the measurements always give a positive difference between the masses of blank and loaded filters (filters before and after the aerosol sampling). These results seem to suggest the presence of particles in all the measurement range of 30 nm-10 microns. The aerodynamic diameter of each size class midpoint is determined by the square root of the multiplication of the two successive cut-off diameters.

Figure 3 presents the average mass distribution on the results of three replicates of combustion of a Polyamide-6 filled with 1% carbon nanotubes (MWNT).

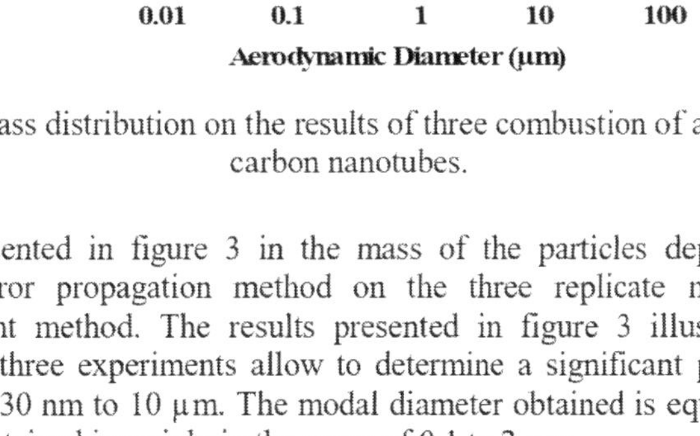

Figure 3. Average mass distribution on the results of three combustion of a PA-6 filled with 1% carbon nanotubes.

The uncertainty presented in figure 3 in the mass of the particles deposed on each stage is calculated using an error propagation method on the three replicate measurements from the gravimetric measurement method. The results presented in figure 3 illustrate that the standard deviation calculated on three experiments allow to determine a significant presence in mass of the particles in the range of 30 nm to 10 µm. The modal diameter obtained is equal to 0.21 µm. Figure 3 shows that the aerosol obtained is mainly in the range of 0.1 to 2 µm.

3.1.2. Results on the mass distributions in the case of several formulations with PMMA

Results are given below for PMMA alone, PMMA without flame retardants, but containing nanofillers of SiO_2 or Al_2O_3. Figure 4 presents the mass distribution obtained with the DLPI in the case of the combustion of (A) PMMA alone, (B) PMMA containing nanofillers of 15 % SiO_2 (with an average primary particle size of 12 nm), (C) PMMA containing nanofillers of 15 % Al_2O_3 (with an average primary particle size of 13 nm) and (D) PMMA containing nanofillers of 15 % Al_2O_3 with a silane surface treatment. Each mass distribution presented is the result of the average of three replicate of combustion.

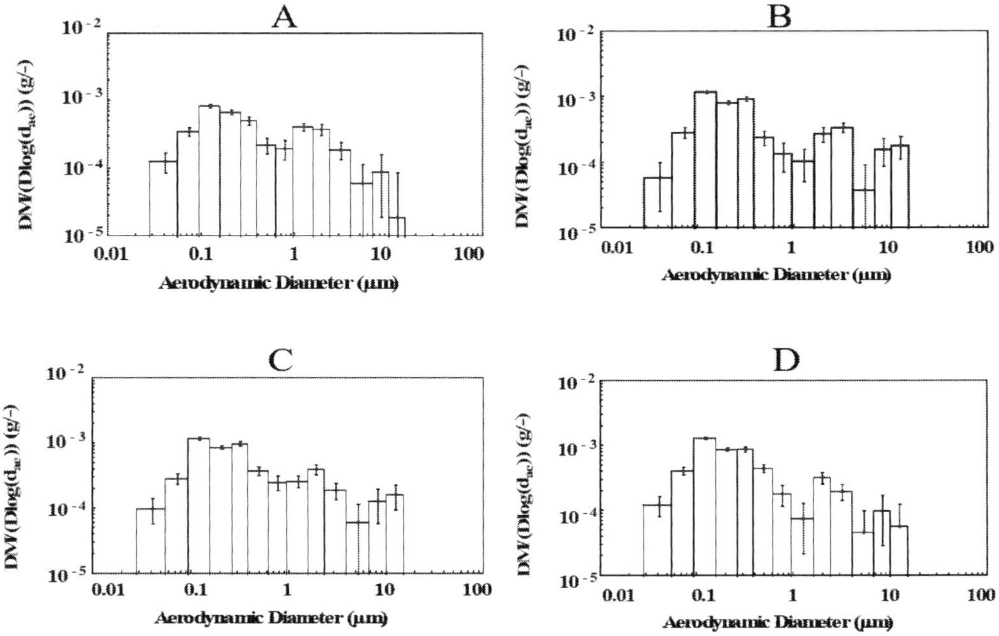

Figure 4. Mass distributions obtained with the DLPI in the case of the combustion of (A) PMMA alone, (B) PMMA containing nanofillers of 15 % SiO_2, (C) PMMA containing nanofillers of 15 % Al_2O_3 and (D) PMMA containing nanofillers of 15 % Al_2O_3 with a silane surface treatment.

Figure 4 shows that the aerosol obtained are not monodispersed. The modal diameter obtained, equal to 0.12 µm, is the same for the 4 types of PMMA,. A comparison of the graphs obtained for all types of PMMA shows a more significant presence of larger particles, greater than 4 µm (three last size classes), in all formulations containing nanofillers of SiO_2, Al_2O_3 with and without silane surface treatment compared to the PMMA matrix alone. Our experiments show that the nanofillers with or without silane surface treatment could have an influence on the mass distribution in the range of 1 nm to 4 µm. A slight decrease is observed for class 1.13 µm in the case of PMMA containing nanofillers of 15 % SiO_2 in regard to the results for PMMA alone. In the range of 30 nm - 1 µm, the form of the size distribution is the same in the four cases, whatever the formulation is. However, the value of the mode for PMMA with nanofillers is greater than the value of the mode for PMMA alone: the quantities in mass near the modal diameter (0.12 µm) are greater for PMMA with nanofillers compared to the PMMA matrix.

Figure 5 shows the cumulative distribution of mass for the 4 types of PMMA described above (A, B, C and D).

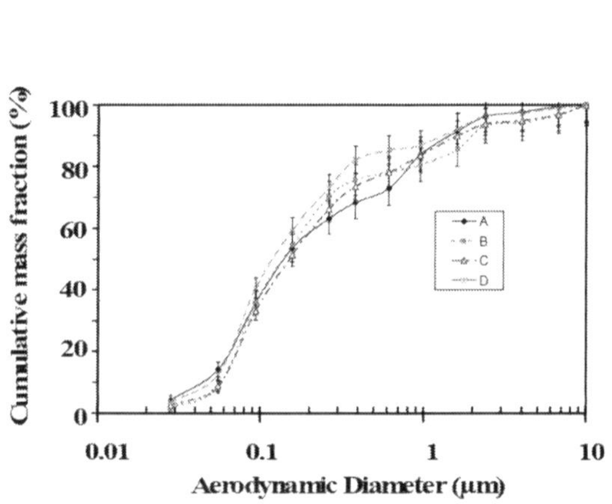

Figure 5. Cumulative distributions of mass for the 4 types of PMMA with nanofillers and PMMA alone (A, B, C and D).

The mass median diameter (50 %) is between 100 and 200 nm for the 4 types of PMMA (figure 5). The results show that respectively, the mass fractions of submicrometric particles (< 1 µm) and ultrafine particles (< 0.1 µm) are in the range of 75 – 92 % and 30 - 45 %. These results illustrate that the large mass quantities emitted have been obtained with submicrometric particles (close to 80 %) and ultrafine particles (close to 40 %).

3.2. Results with the CNC alone

Figure 6 presents the number concentration measured by the CNC as function of the time of nanocomposites combustion. This figure shows the kinetics of the number concentration of submicrometric particles emitted for the 4 formulations of PMMA described above (A, B, C and D).

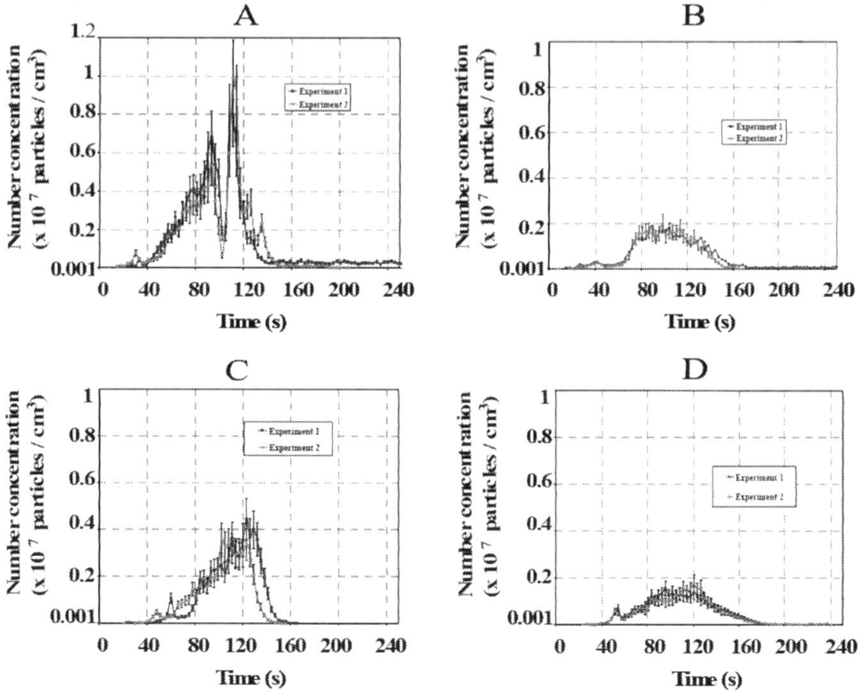

Figure 6. Number concentration measured by the CNC as function of the time of nanocomposites combustion (A, B, C and D).

The emission rate obtained for the PMMA matrix is very different from those obtained for other formulations. The presence of nanofillers has a significant effect on the rate of particle emission with a decrease of the number concentration of the submicrometric particles emitted. For all the formulations, a peak representing the emitted particles preceding or at the moment of the inflammation of the sample is observed around 40 s. At this peak, the number concentration is similar for all the formulations of PMMA. The comparison of graphics C and D illustrate the influence of the surface treatment on the same matrix composed of PMMA containing nanofillers of 15 % Al_2O_3. The surface treatment involves a decrease of the number concentration.

Another formulation with Ammonium Polyphosphate (APP) has been tested. However, a rapid saturation of the number concentration is observed. Thus, it was necessary to change the setup in order to characterize the airborne particles (setup B, figure 1).

3.3. Results with a Diluter coupled at CNC

Figure 7 illustrates the kinetics of the number concentration measured by the CNC coupled with the diluter as function of the time for the combustion of 3 g of PMMA without fillers and of PMMA containing nanofillers of 5 % Al_2O_3 and 10 % APP. The dilution ratio used, equal to 54, is taken into account in the results presented in this figure.

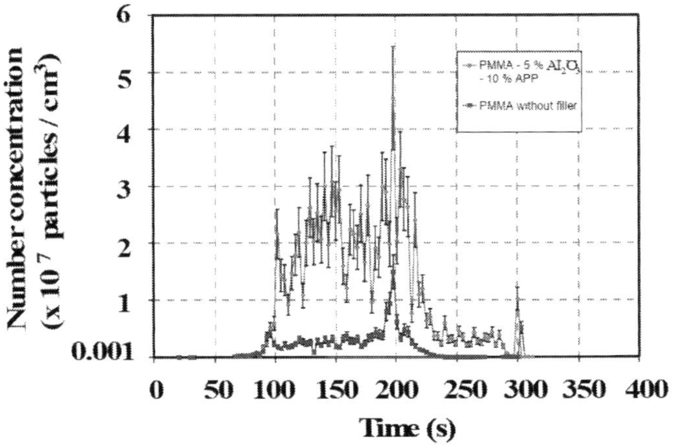

Figure 7. Number concentration measured by the CNC coupled with the diluter as function of the time for the combustions of PMMA alone and of PMMA containing nanofillers of 5 % Al2O3 and 10 % APP.

The comparison of figures 6 and 7 allows to illustrate the influence of the Ammonium Polyphosphate added in the matrix composed of PMMA containing nanofillers of Al_2O_3. The Ammonium Polyphosphate involves a strong increase of the number concentration of the particles emitted.

3.4. Results on the morphology of particles

Figure 8 presents the AFM images (2 μm x 2 μm) of the particles collected on the first stage of the impactor (cut off aerodynamic diameter: 30 nm) for two formulations: PMMA (1st column) and PMMA + Al_2O_3 + APP (2nd column) with the topographic image on top and the phase image on the bottom.

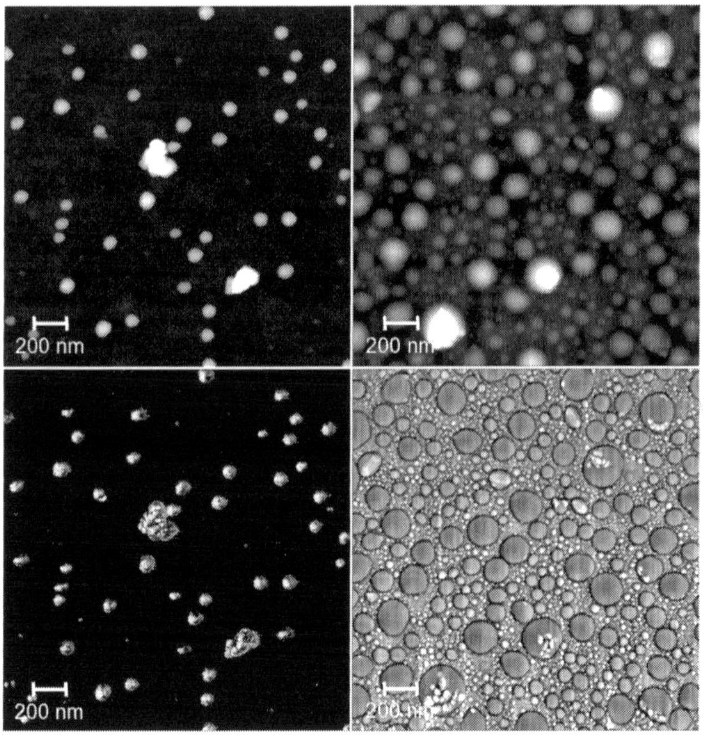

Figure 8. 2 µm x 2 µm AFM images of the particles collected on the first stage of the impactor (cut off aerodynamic diameter: 30 nm) for two formulations: PMMA (1st column) and PMMA + Al_2O_3 + APP (2nd column) with the topographic and phase images.

For the PMMA formulation, the surface is mainly populated with nanoparticles. The few present aggregates do not exceed a diameter of 200 nm. In the case of PMMA + Al_2O_3 + APP, the images seem to show that the population of nanoparticles has increased compared to PMMA alone. This highest density of particles observed on the impactor stage with a cut off aerodynamic diameter of 30 nm could be expected because the measurements obtained by the CNC, including all sub-micron particles (<1 um), showed a greater concentration in the case of PMMA + Al_2O_3 + APP (see figure 7). We also observe the physico-chemical modifications of the particles suspected as tar or liquid phase coalesce on particles.

4. Conclusion

As epidemiological studies have shown an association between increased particulate air pollution and adverse health effects in susceptible members of the population, it is particularly important to characterize aerosols emitted by different sources of emission; for example, during the combustion of composites charged with nanoparticles. The present study is led in the framework of the NANOFEU project, supported by the French Research Agency (ANR), in order to characterize the fire behaviour of polymers charged with suitable nanoparticles and make an alternative to usually employed retardant systems. To evaluate the impact of these composites on the emission of airborne particles produced during their combustion, an experimental setup has been developed to measure the mass distribution and the number concentration of the produced aerosol. In this paper, a comparison of the mass distributions of the aerosol emitted during the combustion for several polymer alone (PMMA, PA-6), polymers containing nanofillers (silica, alumina, and carbon nanotubes) and polymers containing -both nanofillers and APP as a flame retardant was presented. In the case of PMMA with and without nanofillers of SiO_2 or Al_2O_3, the results illustrate that the large mass quantities emitted have been obtained with submicrometric particles (close of 80 %) and ultra fine particles (close of 40 %). The presence of nanofillers in PMMA shows a significant effect on the rate of particle emission with a decrease of the number concentration of the emitted submicrometric particles. The results show that a silane surface treatment of the nanofillers contained in the PMMA matrix could involve a decrease of the number concentration of submicrometric particles. Conversely, the Ammonium Polyphosphate involves an high increase of this number concentration. The morphology of these particles collected by the impactor has been studied using Atomic Force Microscopy.

Acknowledgments

The authors are grateful to the French Research Agency (ANR) for the financial support provided for the NANOFEU project. Thanks to Dr. François Xavier OUF and Dr. François GENSDARMES of the Aerosol Physics and Metrology Laboratory of IRSN (The French Institute for Radiological Protection and Nuclear Safety) for their help in the implementation of the dilution system.
All the NanoFeu partners (EMA, ISMANS, INERIS and PlasticsEurope) are also gratefully acknowledged for their collaboration in this project.
The authors would like to thank Digital Surf SARL for making Digital Surf's MountainsMap licenses available in the framework of the Digital Surf's Mount Shasta technical cooperation programme.

References

[1] Maynard A D 2006 Nanotoday *2* 22-32
[2] NANOFEU Web Site - http://www.ineris.fr/nanofeu/
[3] Chivas-Joly C, Guillaume E, Ducourtieux S, Saragoza L, Lopez-Cuesta J M, Longuet C, Duplantier S, Bertrand J, Calogine D, Minisini B 2010 Influence Of Nanoparticles On Fire Behavior And Composition Of Decomposition Products Of Thermoplastic Polymers, *INTERFLAM 2010*, Nottingham.
[4] ISO 5660-1 2002 Reaction to fire tests - Heat release, smoke production and mass loss rate - Part 1: Heat release rate (cone calorimeter method)
[5] ISO/CD 29904 Chimie de la combustion - Génération et mesure des aérosols du feu
[6] Hinds W 1999 *Aerosol Technology- Properties, behavior, and measurement of airborne particles* (Wiley-interscience, second edition) chapter 13 pp 292-294
[7] Sem GJ 2002 Design and performance characteristics of three continuous-flow condensation particle counters: a summary. *Atmos. Res.* **62** 267–294
[8] Baron P and Willeke K 2001 *Aerosol measurement - principles, techniques, and applications* (New York: Wiley inter-science) chapter 3 pp 50-51

On Line Characterization of SiC Nanoparticles Produced by Laser Pyrolysis

G Wattieaux[1,2], P Lecerf[1], L Meyer[2], L Boufendi[2], Y Leconte[3], O Sublemontier[3], N Herlin[3], A Asimakopoulou[4], A Tsakis[4], M Daskalos[4], A G Konstandopoulos[4,5] and F X Ouf[6]

[1]CILAS, 8 Avenue Buffon 45063, Orléans cedex, France.

[2]GREMI, 14 rue d'Issoudun BP 6744, 45067 Orleans cedex 2, France.

[3]IRAMIS-SPAM/LFP CEA CNRS URA 2453 CEA Saclay, 91192 Gif-sur-Yvette cedex, France.

[4]APTL, Aerosol and Particle Technology Laboratory, CPERI/CERTH, P.O. Box 60361, 57001, Thessaloniki, Greece.

[5]AUTH, Department of Chemical Engineering Thessaloniki, Aristotle University of Thessaloniki, Univ. Box 116, 54124, Greece.

[6]IRSN, Aerosol Physics and Metrology Laboratory, BP 68, 91192 Gif-sur-Yvette Cedex, France.

E-mail: lecerf@cilas.com

Abstract. On Line Measurements campaigns have been carried out for the first time in IRAMIS/SPAM CEA's laser gas-phase pyrolysis nanoparticles production facilities. The produced aerosol is composed of Argon and/or Helium laden with SiC nanoparticles concentrated up to ~14 mg/l.
Different commercial apparatus were used for sampling, and characterisation of size, and morphology of the particles. A series of experiment is performed by IRSN with a DMS500 from Cambustion Ltd™, which gives electrical mobility equivalent diameter distribution in real time. Particles are sampled with an ejector-diluter VKL-10 from Palas™. On the same sampling line, APTL performed aerodynamic equivalent diameter mass distribution using a Nanomoudi from MSP Corp.™ and TEM post analysis, as well as post elemental analysis (EDS) after collection on TEM grid with a Thermophoretic Precipitator (TP). A previously developed model is used to obtain the aggregate morphology and primary particles sizes.
In addition to these commercial or well-known techniques, a new patent-pending technique called RFPM (Radio Frequency Plasma Metrology developed by CILAS + GREMI in the frame of NANOCARA program) is tested for the first time off the lab. The principle is based on levitation of particles in plasma RF. The entire set up is a prototype, including the sampling line. Experiments showed that this technique is very promising for on line gas phase monitoring, even though future improvements are needed, especially on the direct injection of the sample in the measuring chamber.

First results of size distribution on pyrolysis line were obtained: three modes of different geometric mean diameter D50 were measured. Further analysis of TEM micrographs gave insight on these three modes, they can be interpreted as primary particles, aggregated particles, and a third one which could be composed of silica oxide nanoparticles issued from the combustion of the remaining silane initiated by an air leak in the exhaust of the reactor.
Furthermore, a parametric study was undertaken. Helium addition in reactor and laser power were authorized to vary within a significant range. There was an evidence of influence of these parameters on the size distributions.
Thus, we have demonstrated the high interest of implementing a size monitoring set up on gas phase nanoparticles production line for safety, yield improvement, and cost reduction purposes. This set up can run in a safe and non invasive way, and require low maintenance.

1. Introduction

The worldwide consumption of nanomaterials was estimated at more than 10 millions of tons and its annual growth is evaluated at 2.7% in volume in 2006 [1]. Their domain of interest coincides with the emergence of research activities like biomedical, food packaging, coatings or new surface properties, energy, and catalysis applications. The manufacturing of nanoparticles opens new economical developments as well as highlights a social concern about their control.

One may distinguish two kind of technique for nanoparticles elaboration: Liquid-phase and Gas-phase. We focus on the second one. Gas-phase nanoparticles manufacturing is nowadays widely employed. A topic of annual Aerosol Conference (European and International) is dedicated to this [2]. Major benefits of this technique can be summarized as:

- High chemical purity (no surfactant, purity of the gas)
- Continuous operation, high throughput (in the order of kg/h).
- Low waste of chemical, solvent, or other products = low environmental impact, cost effective, large scale production.

The most widely used gas-phase manufacturing techniques for mass production are based on heating (vaporization, cracking) and cooling (condensation, coagulation) process steps: Hot wall furnace, flame, spark-discharge, plasma, or laser pyrolysis can be cited as the most representative techniques [3].

The size and aggregation state of the nanoparticles is a crucial parameter regarding their properties and process control (for development, as well as for routine production). Monitoring this parameter On Line is of prime importance to achieve a high-yield, safe, and reproducible mass production.

Up to now, the best techniques for size distribution and counting nanoparticles in gas phase in real or quasi real time are:

- Electrical mobility spectrometers: Scanning Mobility Particle Sizer (SMPS) [4], and Differential Mobility Spectrometer (DMS) [5].
- Aerodynamic equivalent diameter: Electrical Low Pressure Impactor (ELPI) [6].

The major drawback of these techniques is that they show limitation in continuous operation on a production line at high concentration values. Indeed, whatever the technique, clogging, deposits on sensors or other parts cause a drift or malfunction. Maintenance is then necessary at interval period of the order of hours, which is not compatible with continuous process operation.

Thus, reliable, low maintenance, secure, and non invasive technique for continuous real time size measurements has to be developed.

Development of innovative techniques for Gas-phase On Line characterization tools is the main objective of NANOCARA program [7]-[8], which is currently in progress. In the frame of this program, a new patent-pending technique for particle sizing is being developed. We will not give details the principle of operation, for confidentiality reasons.

This innovative technique has been successfully tested in the lab with micron and submicron particles [9], but never on a gas-phase nanoparticles production line. The complementarities between NANOCARA and SAPHIR goals were an opportunity to assess this technique in parallel with other established ones. Therefore, the work presented in this paper is carried out in the frame of SAPHIR program, which aims at implementing direct production of nanoparticles through the development of a global integrated concept, with safe oriented processes, and global monitoring strategy.

The objective of this paper is to report the two measurements campaigns performed at IRAMIS/SPAM CEA Saclay at different time:

- CILAS and GREMI: integration of RFPM prototype device on the pyrolysis line. Sampling and size distributions measurements for nominal synthesis conditions. A few series of measurements with laser power variation and helium addition in the reactor were done.
- IRSN and APTL: counter measurements with DMS500 Cambustion Ltd™, Thermophoretic Precipitator for TEM analysis, and Impactor NanoMoudi MSP Corp. ™. These measurements were carried out with the same synthesis conditions, and parametric variations.

2. Experimental set up

2.1. Overview of experimental set up

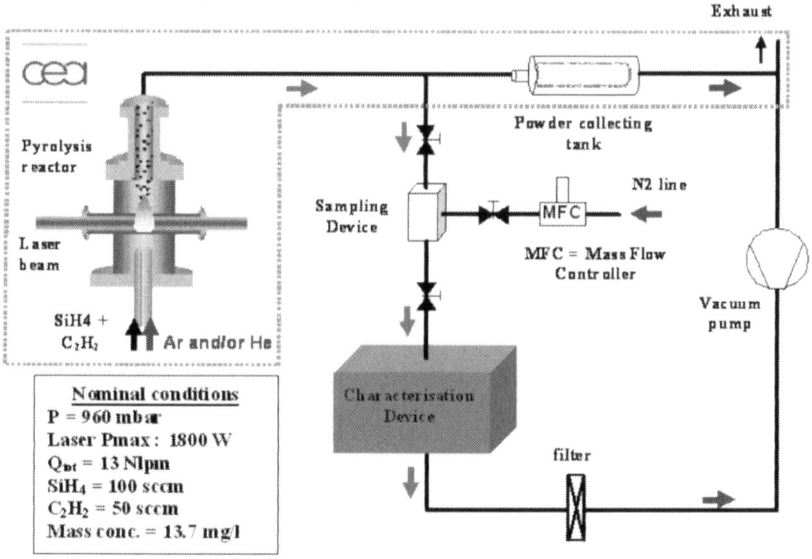

Figure 1. Overview of the experimental set up installed on line.

Table 1. Sampling parameters. DMS500, Nanomoudi, and TP were installed on the same sampling line simultaneously, with a flow splitter (section 0). RFPM was operated at a different period (section 2.4.).

Sampling	DMS500	Nanomoudi	TP	RFPM
Flow Nlpm	8	8.2-9.2	3.5-4.1	0.15-0.7
Duration	Contin.	15-20 min	7-9 min	60 - 90 s
Dilution ratio	62:1	62:1	62:1	2:1 - 10:1

Table 2. Pyrolysis parameters. During experiments campaign, a parametric study of fundamental parameters has been carried out.

Laser power	Helium dilution
10 - 30 % of F.S.(*) 180-540 W	0 - 40% of F.S. MFC (F.S = 3.5 Nlpm)

(*): F.S. = Full Scale

Figure 1 is a diagram representing the main parts of the experimental set up. The pyrolysis set up and principle of operation are described in section 2.2. Sampling and characterization devices are depicted in sections 2.3 and 2.4. An additional gas line fed with nitrogen is used for dilution, cooling, and carrying sample to the instrument. A mass flow controller (0 to 2 Nlpm) sets the flow of nitrogen. Each set up has its own vacuum pump and filter to bring the sampled aerosol into the instrument and toward the pyrolysis exhaust.

2.2. Laser Pyrolysis technique

Laser pyrolysis is a gas phase process currently used for the synthesis of various kinds of nanoparticles [10]. It is based on the resonance between the emission of a laser, usually a CO_2 laser at 10.6 μm, and the absorption of a gaseous (or liquid) precursor molecule. In this thermal process, the CO_2 laser is used as an easily adjustable heat source for the precursor molecules collision-induced decomposition. The precursor's flow, guided by a coaxial inert gas flow, intersects orthogonally the laser beam in a reaction chamber kept under controlled pressure. The temperature increases rapidly in the laser-crossing region. If the particle formation temperature is reached, nucleation takes place from dissociated species and particles are growing rapidly. The reaction zone shows a luminescent flame originated by the thermal emission of the particles and possible emission contributions from excited radicals. The confinement by the inert gas flow avoids any reaction with the chamber walls and thus prevents particles from pollution.

Nucleation and growth processes in laser pyrolysis show some similarities with those observed in combustion flames. Both methods essentially differ in the way precursor molecules are dissociated. This may leads to the formation of different chemical intermediate species and/or different electronic energy states of these species. Chemical reactions that are involved at least in the beginning of the process may then have different pathways. Particles grow either by atom addition to their surface induced by vapor flow condensation or by chemical heterogeneous reactions on their surface. If the temperature is sufficiently high, coagulation can occur simultaneously. In that case, particles are growing by melted particle-particle collision or by solid-state diffusion between two hot solid particles. Primary particles may agglomerate or aggregate in larger fractal-like objects. The growth process, the primary particle size, structure and the agglomeration degree are strongly influenced by the synthesis parameters such as precursor molecules flow, pressure in reaction chamber and laser intensity. In addition to these usual parameters, it is possible to dilute the reacting precursor flow.

2.3. IRSN + APTL sampling and characterization set up

Nanoparticles were sampled in the exhaust duct of the pyrolysis reactor with an ejector type dilution device (PALAS™ VKL 10) fed with nitrogen for inerting the remaining silane. Prior to sampling, a dilution ratio of 62:1 was measured using two thermal mass flowmeters (4043 TSI™). The size distribution was determined continuously in terms of electrical mobility diameter using a differential mobility spectrometer (DMS 500 Cambustion Ltd.™) and in terms of mass aerodynamic diameter according to a substrates sampling (aluminum foil) of 15-20 min with a rotating micro-orifice impactor NanoMoudi (MSP Co.™). The size range covered by these two spectrometers was respectively from 5 to 1000 nm for the DMS 500 and 10 nm to 10 μm for the NanoMoudi. Nanoparticles have been also deposited on electronic microscopy grids (Formvar and Holey carbon) using a thermophoretic precipitator developed by APTL [11]. In parallel to the size distribution, the morphology and chemical composition have been investigated on grids by transmission electronic microscopy [11], energy dispersive spectroscopy and on NanoMoudi samples by x-ray diffraction.

The morphology of collected nanoparticles samples is studied via TEM image analysis in order to extract important structural information such as: (i) the primary particle distribution (Dpr), (ii) the equivalent projected area diameter (Dea) distribution of the aggregates, which relates straightforward to the corresponding mobility diameter (dm) distribution, (iii) the fractal dimension (Df), (iv) the fractal prefactor (kg), (v) the coefficient of overlap (Cov) in-between primary particles, that leads to an estimation of (vi) the depth of penetration (dij) and, finally, vii) the fill factor (f(Cov)).

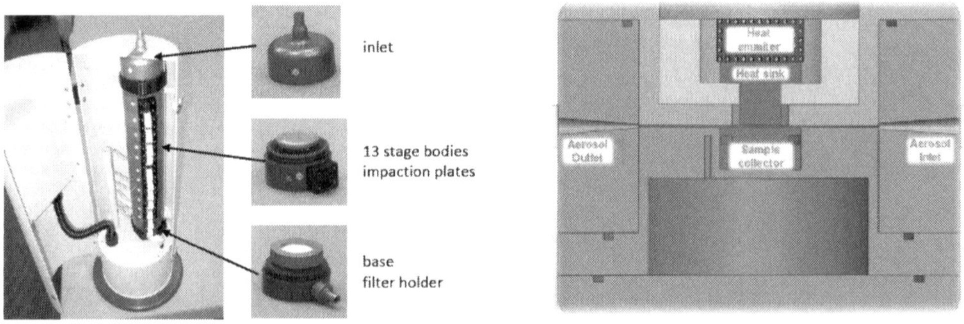

(a) (b)

Figure 2. (a) Nano-range cascade impactor NanoMoudi, MSP: internal view and partially disassembled impactor (b) A cross section of the thermoprecipitator with its individual parts

2.4. RFPM set up

2.4.1. *RFPM brief description.* This new device is now patent pending [12], thus the operating principle cannot be described here in details.

This technique takes advantage of the low pressure and power plasma RF as follows:

- Levitation of particles in plasma allows trapping of particles.
- De-agglomeration of particles weakly bonded (Van der Waals) in dry phase.
- Low pressure mobility of particles is high even for nanoparticles.

Figure 3 shows a CAD view of the main parts of the prototype in-house device that was used for lab and On Line experiments. Other parts are described in section 2.4.2.

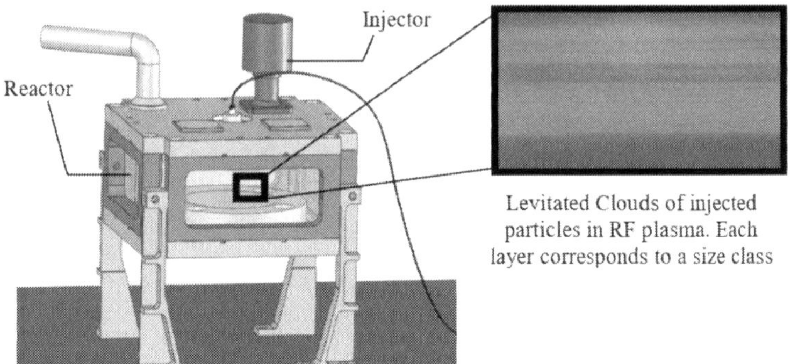

Figure 3. CAD view of RFPM plasma reactor and injection device

Particles to be measured are charged in the plasma by free electrons, the arising electrostatic force counterbalances the gravitational one, and thus lead to levitation of particles arranged in layers for diameters under approximately 10 μm depending on the material density.

The measured diameter could be related to a kind of Low Pressure Mobility equivalent diameter. At this range of pressure settling velocity of nanoparticles are high enough to be exploitable with reasonable time response (few seconds).

2.4.2. *Sampling On Line.* The sampling system used for experiments is a prototype made in the lab. The nozzle is Swagelok™ stainless steel tubing 4 mm internal diameter, and adjustable in position. It is adapted to an ISO KF cross, installed on the exhaust duct of the pyrolysis. The flow in the instrument is controlled by a downstream critical orifice (500 µm diameter) set at 2 Nlpm of nitrogen at atmospheric pressure. Mixing ratio and sampled aerosol flow was adjusted by changing the setpoint value of the nitrogen mass Flow Controller. The flow of sample extracted from the line was controlled between 0.17 and 0.7 Nlpm, corresponding to mixing ratios between 2:1 to 10:1.

Temperature measurements in situ at the inlet of the instrument have been performed in these conditions. Temperature of the diluted aerosol never exceeded 25°C, which is compliant with RFPM device (no thermophoresis deposits in the chamber).

The sample to be measured is not directly injected in the plasma chamber, but thanks to a sequential injection technique. Particles (only a few milligrams or less) are collected in a tank at sampling pressure, then, the pressure is lowered by pumping with a dry primary pump to the working value (a few tenth of millibar). By means of actuation, the sampled particles are released in the low power Radio Frequency Plasma.

The main issue is the pressure adaptation: in fact, between two measurements, device must be blanked out to minimize memory effect due to remaining particles in the collecting tank after injection. Thus, the entire 4-steps sequence Sampling/Collection - Injection - Measurement – Blank leads to delay of approximately 15 minutes between two samples.

3. Results and discussion

3.1. Size and mode distribution

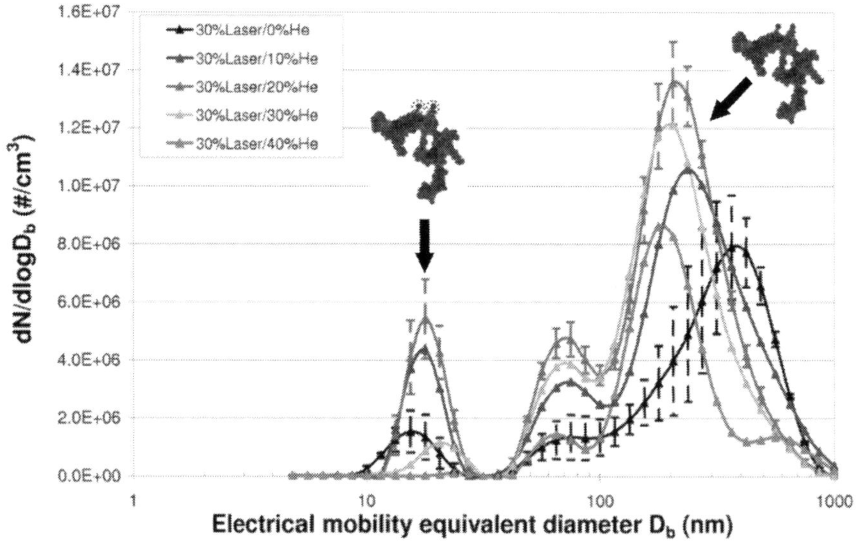

Figure 4. DMS 500 size distributions for 30 % of laser power and various helium flow rates

The size distributions measured in real time by the DMS 500 are presented on Figure 4 for 30 % of laser power and various helium dilution flow-rates. Three well defined modes can be identified on these size spectrums. The smallest one, close to 17 nm, is associated to the primary nanoparticles whose the diameter measured by TEM is in good agreement (cf. part 3.2). The largest mode (ranging from 194 to 340 nm) can be attributed to the nanostructured agglomerates observed by TEM (cf. part 3.2). Furthermore, a third intermediate mode (close to 69 nm) can be clearly identified but has not yet been fully described. This mode may be due to an air leak observed detected after the experiments. This leak may cause the combustion of the remaining silane and, as a consequence, a significant production of silica oxide nanoparticles. One must notice that a large amount of oxygen has been clearly identified on the EDS spectrum of nanoparticles (cf. part 3.2).

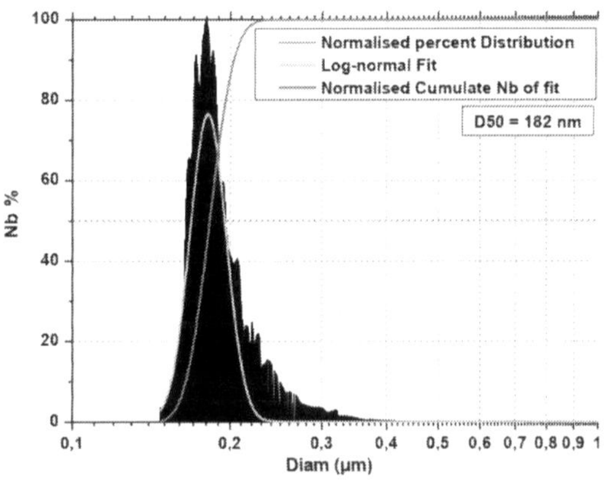

Figure 5. Example of RFPM calculated Size Distribution in normalized number. To statically improve the D50 measurements, and mitigate noise influence, distributions are fitted with log-normal function. Thus, results are more reliable.

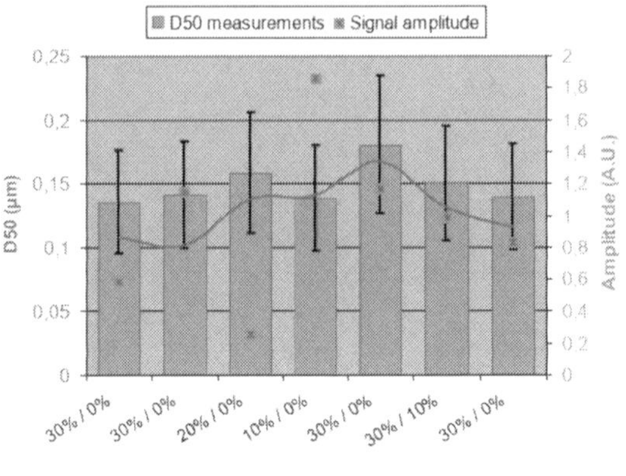

Figure 6. Significant RFPM D50 (Blank / Signal ratio lower than 5%). X axis scale represents the experimental conditions. 30%/10% means CO_2 laser is operated at 30% of maximum power, and 10% = Helium flow-rate is set at 10% of full scale (cf. Table **2**).

RFPM raw data consists of two signals coming from two sensors. Further calculations allow constructing the size distribution. This methodology is patent pending, and we will not explain it in details here. It is worth to note that hardware encountered a malfunction in the pressure recording during all the experiments. It is an important parameter, as it is used in formula for size calculation. Therefore, **size distribution and median diameter D50 (count median diameter mode) are calculated with an error of 20 to 30%**, because pressure is set at 0.25 mbar for all data (this is the nominal value for each measurement, but manually adjusted and varying within ~20%).

Fluctuations of the signals are not artefacts (**Figure 5**). The resolution is very high. This is a strong advantage of the technique when multimodal and polydisperse populations are to be measured. Due to the hardware design and signal processing electronics (lab prototype version), S/N ratios are not optimized, especially in the range 10-100 nm.

It is worth to note that **it is not a physical limitation of the technology RFPM**. Hardware upgrade and sensor gains adjustments in the low diameter range can drastically improve the S/N ratio. In fact, the larger the diameter is, the larger is the measured signal, with a predictable function. Thus, this can be overcome by adjusting the gain of sensor response with respect to the diameter.

The RFPM size distribution corresponds to aggregate state of nanoparticles. Calculations give a D50 between 135 nm and 185 nm for all conditions (**Figure 6**). D50 values are not dependent on the injected and measured quantity.

The results obtained with the impactor NanoMoudi are expressing the mass of each size range as a function of the aerodynamic diameter of the particles. In the following figure (Figure 7), the results obtained from NanoMoudi at three typical points of laser pyrolysis operation (30%Laser - 0%He, 30%Laser - 20%He and 30%Laser - 40%He) are presented, expressed as the calculated number concentration dN versus the average aerodynamic particle diameter d_p. SiC particle density employed for raw data assessment is 3.02 g/cm^3.

3.2. TEM and EDS analysis

The samples collected on the TEM grids by the thermoprecipitator representing the whole size distribution of each experiment, as well as selected samples collected by NanoMoudi representing fractionated parts of the size distribution, were studied at the TEM microscope and several pictures are taken at various magnifications. Most of the samples that were studied at TEM were also analyzed by EDS (Energy Dispersive Spectroscopy) elemental analysis. In the following representative TEM pictures, important characteristics of the nanoparticles produced by laser pyrolysis are depicted, such as representative values of the primary particle size, the aggregate size and the number of the aggregates as well as representative chemical analyses in selected spots of the samples.

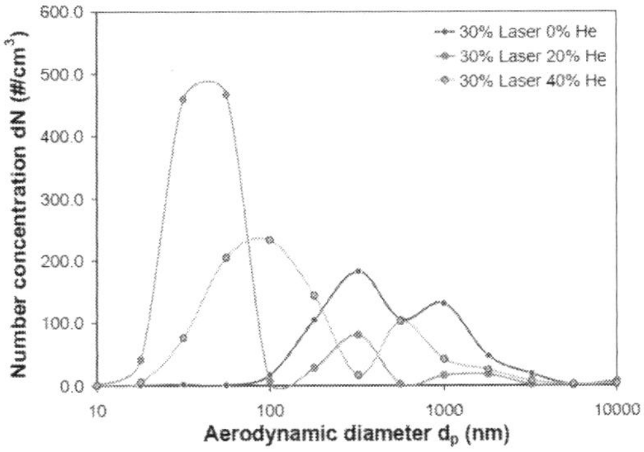

Figure 7. NanoMoudi results at three typical points of laser pyrolysis operation (30%Laser - 0%He, 30%Laser - 20%He and 30%Laser - 40%He

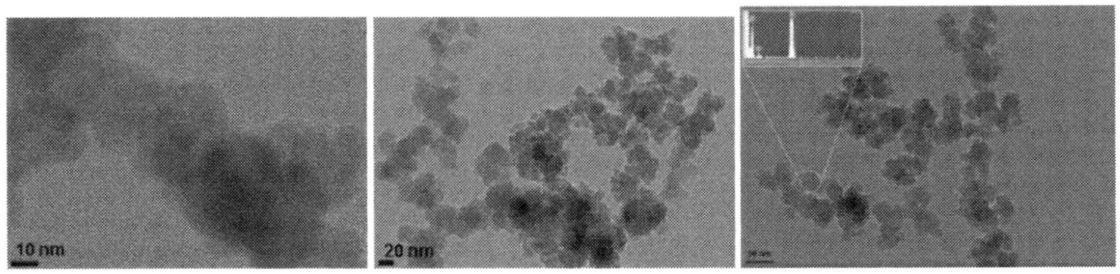

Figure 8. Representative TEM pictures of typical point of operation 30% Laser-0% He

Figure 9. Representative TEM pictures of typical point of operation 30% Laser-20% He

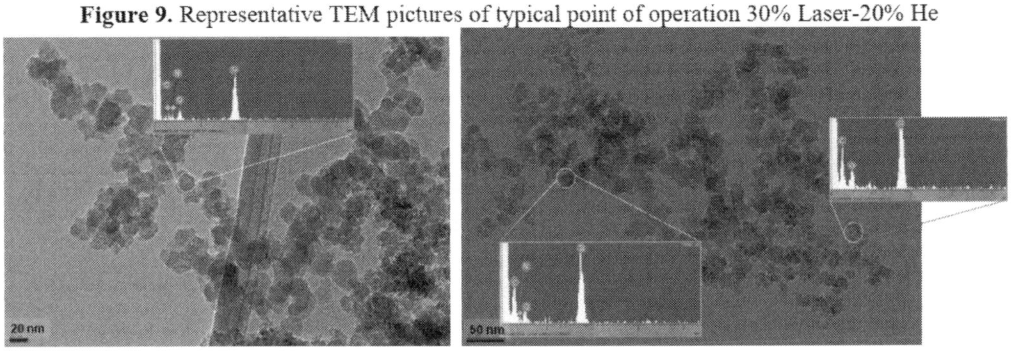

Figure 10. Representative TEM pictures of typical point of operation 30% Laser-40% He

It is worth noticing that mainly Si, C and O were identified by EDS and in limited cases also N was identified. However, the quantities measured by EDS are really small and that is the reason why the EDS diagrams are produced after zooming a lot in the measuring range and therefore they are not of the highest quality.

3.3. Fractal analysis, primary particles and aggregate size distribution analysis

Image analysis was performed on samples collected from the impaction stage of NanoMoudi that corresponds to the aerodynamic range of 1000nm to 1800nm, while for each sample, the diameters of a satisfactory number (above 100) of primary particles were measured using high magnification images. The results are exhibited on the following table. For two of the samples it was found that the primary particles are following a bimodal normal distribution. For the rest of the analysis, the set of measuring quantities were the overall projected area (Aagg) and the length of the bounding box that could fully enclose the aggregate. From the equivalent projected area diameter Aagg, Dea was obtained for each aggregate. Those values were used to reconstruct the size distribution of the aggregates assuming that they follow a lognormal distribution. The overall results are summarized in the graph of Figure 11 and Table 3.

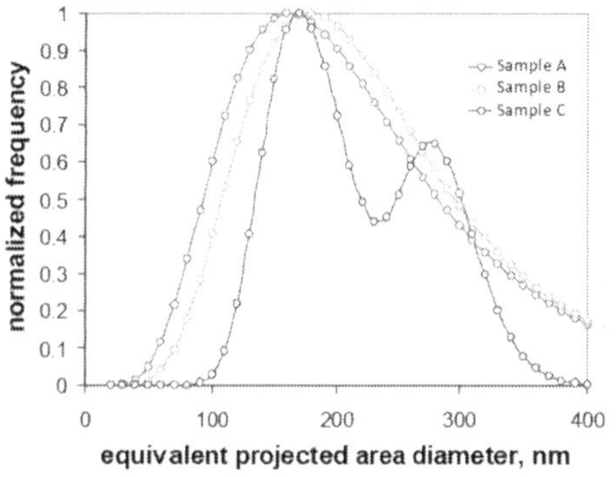

Figure 11. The lognormal distributions that were reconstructed via TEM image analysis for each sample

Table 3. Structural parameters of the samples obtained via adequate fractal analysis

Sample	Dpr (nm)	Dea (nm)	Df	kg	Cov	dij (nm)	f(Cov)	N/aggregate
0%He - 30%Laser	9.9	161	1.77	4.1	0.88	1.17	0.03	1041
20%He - 30%Laser	6.8 & 11.3	127	1.71	4.83	0.90	0.67	0.03	748
40%He - 30%Laser	5.6 & 15.2	99 & 159	1.79	3.86	0.88	0.7	0.03	864

3.4. Pyrolysis parameters study

The influence of two main pyrolysis parameters on the size distribution and physico-chemical characteristics of nanoparticles has been investigated, and results are exposed in Figure 12.

While the size of the primary nanoparticles (from 5 to 30 nm) and the 69 nm particles (from 30 to 100 nm) do not depend significantly of the experimental conditions, the median diameter of the aggregates is significantly smaller for 10 % of laser power than 30 %. Moreover, for a constant laser power, as the helium flow rate increases, the median diameter of aggregates strongly decreases. This significant evolution is mainly due to the helium dilution which mitigates the agglomeration of nanoparticles in the reaction zone.

Simple dilution in an inert gas induces several changes in the reaction. First, if the amount of reacting precursor molecules and total pressure are kept constant, the total gas flow is increased, decreasing the residence time in the laser zone and so decreasing the reaction duration and temperature. Second, if the global density is kept constant in the reaction zone, the dilution gas acts as a thermal bath. The collisions frequency between precursor-based dissociated species is reduced, as the reactant molecules partial pressure is decreased, which increases the cooling rate of the reaction by gas diffusion and quenches more rapidly the particles growth. With a high diffusion coefficient, helium appears to be an efficient dilution gas to quench the particles growth. The growth process is essentially driven by diffusion.

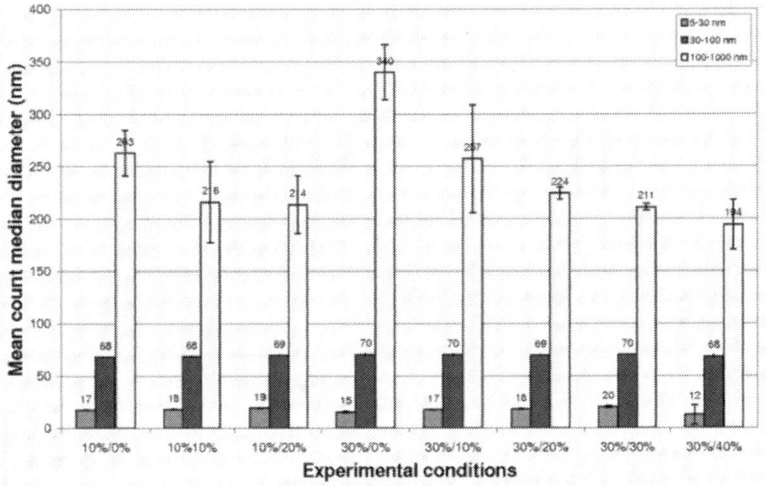

Figure 12. Evolution of the count median diameter in terms of electrical mobility diameter (DMS 500), as a function of the helium flow-rates and laser power

4. Conclusion

On-line measurements of particle size distribution at CEA/Saclay laser pyrolysis line have been achieved by employing various state-of-the-art as well as innovative measurement instrumentation and techniques. The correlation between various equivalent diameters is under development.

Three median diameter modes were identified. The first around 15-20 nm is attributed to primary particles, the second ranging from 150 to 350 nm is the aggregation/agglomeration mode, and the last one around 70 nm is still to be investigated but may be composed of silica oxide nanoparticles produced by the combustion of remaining silane. It is almost interesting to notice that nanoparticles are not strongly linked between each other and denote less than 14 % of overlapping while nanoparticles produced by combustion are strongly agglomerated [13]. Finally, the fractal dimensions observed for all the different operational point are closed to 1.80 and confirm that the aggregation/agglomeration processes are mainly governed by the diffusion.

Since these experimental campaigns are the first ones carried out on the IRAMIS/SPAM laser pyrolysis reactor, present results are promising and several issues need to be deeply investigated. The first one deals with the pyrolysis parameters study which may be further analyzed to understand the helium dilution and laser power effects on the overall production and aggregation/agglomeration of nanoparticles. The second key issue will be to verify the possible production of silica oxide by silane combustion in case of leak in the reactor. Finally, further investigations are needed on the monitoring of nanoparticles by such laser pyrolysis and on the way to retrieve the physic-chemical parameters of these particles in the most representative and continuous way.

Monitoring size of nanoparticles on production line in real time gave valuable information that could be used in a full automated synthesis set up, as a feedback for process parameters adjustment.

Acknowledgements:
SAPHIR Program (http://www.saphir-project.eu/) supported by the European Commission through the Sixth Framework Programme (FP6) for Research and Development.
NANOCARA Program (framework PNANO), supported by French National Research Agency ANR, and European Funds for Regional Development FEDER of Region Centre France.

References

[1] Les nanomatériaux, effet sur la santé de l'homme et sur l'environnement, Afsset report, 2006
[2] Aerosol Based Nanotechnology Sessions in *Proceedings of the International Aerosol Conference* (IAC 2010), Helsinki, Finland (Aug. 29-Sep. 3).
[3] G. Biskos, V. Vons, C. U. Yurteri and A. Schmidt-Ott, *KONA Powder and Particle Journal,* No.26 (2008). pp. 13-33
[4] Wang, S. C., and Flagan, R. C. (1990): Scanning electrical mobility spectrometer, *Aerosol Science and Technology*, 13, pp.230-240.
[5] Biskos, G., Reavell, K., and Collings, N. (2005): Description and theoretical analysis of a differential mobility spectrometer, *Aerosol Sci. Technol.*, 39, pp.527-541.
[6] Keskinen, J., Pietarinen, K. and Lehtimäki, M. (1992) Electrical Low Pressure Impactor, *J. Aerosol Sci.* 23, 353-360.
[7] http://europe-centre.eu/fr/32/0/125/Projets_finances.html
[8] http://www.agence-nationale-recherche.fr/documents/aap/2008/finances/pnano-resumes-2008.pdf
[9] Journées Nationales en Nanosciences et Nanotechnologies J3N Toulouse, France, 2009. Poster Session http://www.rmnt.org/com/J3N2009/posters/08-NANO-044_NANOCARA.pdf.
[10] N. Herlin-Boime, M. Mayne-LHermite and C. Reynaud, *Encyclopedia of Nanoscience and Nanotechnology*, Naiwa Ed. (2004) Vol. 10, pp.1-26
[11] E. Daskalos and A.G. Konstandopoulos, Evaluation of in-house constructed thermophoretic sampler for soot particle analysis, *Proceedings of the International Aerosol Conference* (EAC 2008), Thessaloniki, Greece (Aug 24 - 29).
[12] Patent pending / submission number F644/284 FR
[13] F.X. Ouf, J. Yon, P. Ausset, A. Coppalle and M. Maillé. Influence of sampling and storage protocol on fractal morphology of soot studied by Transmission Electron Microscopy. Aerosol Science Technology, 44: 11, 1005-1017, 2010

Properties of magnetic iron oxides used as materials for wastewater treatment

E Matei[1], A Predescu[1], E Vasile[2] and A Predescu[1]

[1]Politehnica University of Bucharest, Faculty of Science and Engineering Materials, 313 Splaiul Independentei, 060042, Bucharest, Romania

[2]METAV CD, 31 C.A. Rosetti Street, 020011, Bucharest, Romania

E-mail: ecaterina.matei@ecomet.pub.ro

Abstract. The paper describes the properties of some nanopowders obtained by coprecipitation and used as adsorbent for wastewater treatment. The Fe_3O_4 and γ-Fe_2O_3 nanopowders were obtained using iron salts and NaOH as precipitation agents. D-sorbitol was used to prevent the agglomeration between the nanoparticles. The particle size and distribution were detected using a transmission electron microscopy (TEM) and a scanning electron microscope (SEM) equipped with dispersive analyze system in X radiation energy (EDS). The structure of the iron oxide nanoparticles was characterized by X-ray powder diffraction. Thus, the nanoparticles were characterized and compare in terms of particle size and chemical composition and used for adsorption studies in order to removal hexavalent chromium from waste waters.

1. Introduction

The recent advanced researches regarding the use of nanoscale materials in different applications have made possible the separation and detection of various analytes from complex environmental and biological matrices [1]. For environmental application, the attention has been focused on the synthesis and functionalization of magnetic nanoparticles with diameter < 50 nm. These nanomaterials have the potential to remove heavy metals and organic compounds. Magnetic nanomaterials are adsorbents with high chemical specificity, being capable to retain the pollutant from the contaminated water system.

It is important to mention that the variations regarding shape, core size and surface functionality can decide the magnetic properties of the nanomaterials. In these conditions, the synthesis method plays a critical role for the final product used as adsorbent. The industrial waste waters consists of different heavy metals, with a great environmental and public health impact, because their toxicity and persistence in the environment. Chromium and its compounds are presented in wastewaters from iron and steel, metal finishing, leather industries or nuclear power plants [2].

There are many types of nanomaterials but magnetic iron oxide nanoparticles are an efficient adsorbent, which couples magnetic separation with ionic exchange capacity for removal of heavy metals pollutants [3].

In this paper the synthesis and characterization of two magnetic iron oxide nanoparticles (Fe_3O_4 and γ-Fe_2O_3) and batch adsorption studies are described. The batch experiments were made for adsorption of hexavalent chromium from aqueous solutions. The chromium is a common contaminant, being detected in soil and waters. Also, chromium species with trivalent and hexavalent oxidation state are usually presented in industrial processes such as electroplating, leather tanning or metallurgical applications. The hexavalent form is highly mobile and soluble in environments and has been widely known to be toxic, carcinogenic, and mutagenic. Therefore, remediation technologies have been developed to reduce Cr(VI) in subsurface to Cr(III) which is immobile, insoluble and less toxic. [4]

2. Materials and methods

The iron oxide nanoparticles were synthesized, characterized and evaluated in order to be used as adsorbents for hexavalent chromium. The magnetite (Fe_3O_4) nanoparticles were synthesized by coprecipitation method and transformed by air oxidation at 200°C into maghemite (γ-Fe_2O_3).

2.1. Synthesis of iron oxide nanoparticles

In the literature, various methods are known for preparation of iron oxide nanoparticles, the first coprecipitation method being described by Massart [5] achieved by the addition of NaOH or NH_4OH to an aqueous mixture of $FeCl_2$ and $FeCl_3$ as it is indicated by reaction:

$$FeCl_2 + 2FeCl_3 + 8NaOH \rightarrow Fe_3O_4 + 4H2O + 8NaCl$$

Some of the most used methods applied for synthesis of iron oxide nanoparticles are presented in the table 1.

Table 1. Methods for iron oxide nanoparticles synthesis.

Synthesis method	Reagents	Characteristics of the synthesis	Reference
Coprecipitation	$FeCl_2$, $FeCl_3$, NaOH or NH_4OH	fast method; without unprotected shell is possible the agglomeration; low costs.	Massart (1981), Qu et al. (1999)
Hydrothermal / solvothermal method	Iron salts, surfactants	narrow size distribution and a good shape control; long time of the reaction.	Wang et al. (2006), Liang et al. (2009)
Microemulsion	Oil/water/surfactant	control of particle size; it is necessary a large quantity of solvent.	Hayashi et al. (2002)

The synthesis method, for this paper, was coprecipitation and all chemicals were analytical grade purchased from Merck. The used reagents were chloride of ferrous ion (Fe^{2+}) and ferric ion (Fe^{3+}) with NaOH solution. D-sorbitol was used to prevent the agglomeration between the nanoparticles. The molar ratio was 1:2:8 for $FeCl_2$: $FeCl_3$: NaOH [7]. The experiments were carried out in a laboratory installation consists of: a four necked rounded flask equipped with a mechanical stirrer, a dropping funnel, a thermometer and a refrigerator. The iron solutions were strongly stirred in water, after adding NaOH solution. After mixing of the solutions, the color of the solution changes into black indicating the nucleation of Fe_3O_4 nanoparticles [8]. The precipitate was separated by magnetic decantation and washed several times with distilled water and ethanol. The magnetite nanoparticles were dried into oven at 60°C. In order to obtain maghemite (γ-Fe_2O_3), a part from obtained magnetite nanoparticles was heated at 200 °C, for 3 hours and finally, red-brown maghemite nanoparticles were collected.

2.2. Characterization of the iron oxide nanoparticles

The particle size and distribution were detected by high resolution transmission electron microscopy (HRTEM). The samples were prepared in ethanol and ultrasonicated and placed onto carbon grids in order to be analyzed at a TECNAI F30 G^2 transmission electron microscope. The crystalline structure was verified by X-rays diffraction (XRD) at room temperature. The Panalytical X' PERT MPD X-ray diffractometer was equipped with a copper anode generating high-intensity Cu Kα radiation (λ = 1.54065 Å) with the 2θ range from 10° to 90°. Also, the scanning electron microscopy (SEM) measurements were performed, after batch adsorption studies in order to detect the chromium presence onto iron oxides surface. A QUANTA INSPECT F microscope was used equipped with field emission gun (FEG) with 1.2 nm resolution and spectrometer with energy dispersive spectra (EDS) with 133 ev resolution at MnK.

2.3. Adsorption studies of the hexavalent chromium onto iron oxide nanoparticles surface

The adsorption studies were made by adding of 0.1 g γ-Fe_2O_3 into about 100 mL solution of hexavalent chromium at different concentrations: 10, 20, 40, 100 mg/L Cr^{6+}. In this paper, the results for 40 and 100 mg/L are presented, the others results being already presented in another paper [6].
The optimum pH was 2.5. The maghemite (γ-Fe_2O_3) was analyzed after adsorption of the hexavalent chromium onto the surface in order to identify the presence of the chromium by SEM and EDS techniques.

3. Results

3.1. Structural and morphological investigations

The XRD investigations showed the presence of magnetite (Fe_3O_4) and maghemite (γ-Fe_2O_3). Unfortunately, the data diffraction patterns indicate that the two phases are very similar as it can be seen in Figure 1. The difference between magnetite and maghemite appears with position and intensities of peaks. Also, characteristic peaks of impurities were not detected in XRD investigations. The final samples were not a mixture of the two nano-phases.

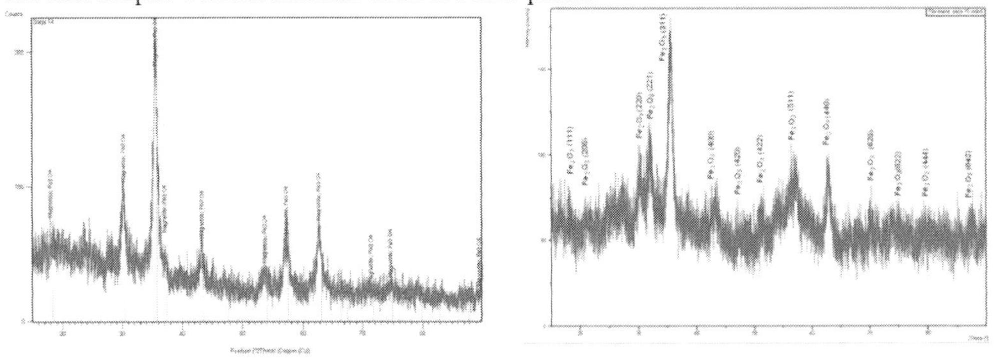

(a) (b)

Figure 1. XRD patterns obtained for magnetite nanoparticles (a) and maghemite nanoparticles (b).

All the detected diffraction peaks were indexed. The analyzed samples showed very broad diffraction lines, in accordance with their small particle size and high specific surface area. The figure 2(a) indicates the small dimension for synthesized magnetite, with diameters smaller than 20 nm. Combined with XRD patterns, the data indicates the inverse cubic spinel structure of Fe_3O_4.

The morphology of the Fe_3O_4 samples was investigated by transmission electron microscopy through bright field (TEMBF). The image is presented in Figure 2(a). Spherical particles with a reduced size and a good dispersion can be observed in this figure. Selected area electron diffraction (SAED) image from Figure 2(b) confirms the presence of Fe_3O_4 in analyzed sample.

The image is associated with the surface area presented in Figure 2(a).

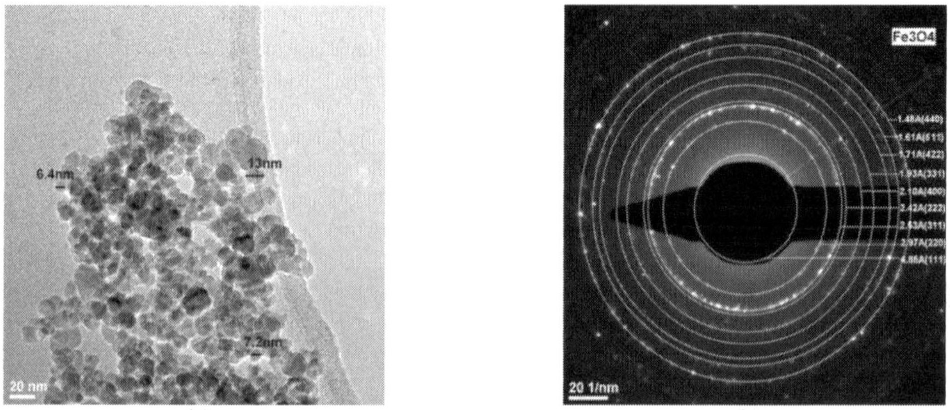

(a) (b)

Figure 2. (a) TEMBF image for Fe_3O_4; (b) SAED image associated to the sample from figure (a).

The magnetite (Fe_3O_4) was oxidized in air at 200°C for 3 hours and the obtained product was γ-Fe_2O_3. The product was a red-brown precipitate and TEMBF and SAED images are shown in the Figure 3 (a) and (b) respectively.

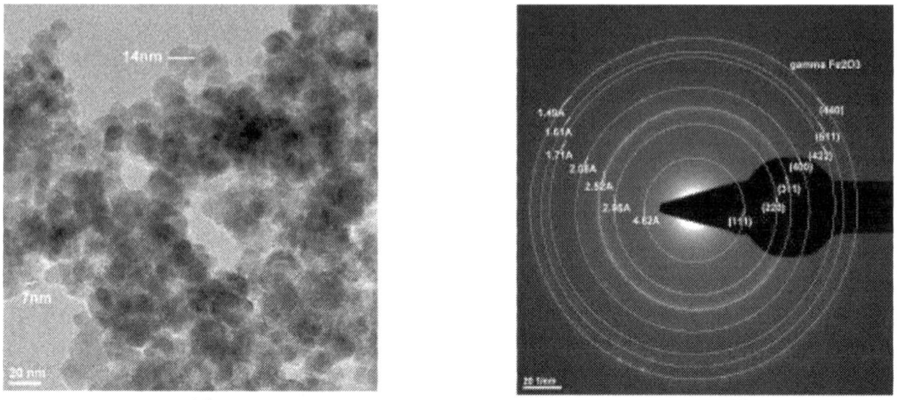

(a) (b)

Figure 3. (a) TEMBF image for γ-Fe_2O_3; (b) SAED image associated to the sample from figure (a).

Table 2 shows the most intense powder diffraction peaks of the Fe3O4 and γ-Fe_2O_3 samples. The d-spacing values (nm) calculated from the electron diffraction patterns inserted in Figure 2 and Figure 3 and the standard atomic spacing for Fe_3O_4 and γ-Fe_2O_3 along with respective *hkl* indexes from the JCPDS card (19-629) and (39-1346) respectively.

Table 2 The d-spacing values (nm) calculated from the electron diffraction patterns inserted in Figures 2 and 3 and the standard atomic spacing for Fe_3O_4 and γ-Fe_2O_3 along with respective *hkl* indexes.

Ring	Calculated d spacing	JPCDS data for Fe_3O_4	*hkl*	Calculated d spacing	JPCDS data for γ-Fe_2O_3	*hkl*
1	0.486	0.483	111	0.482	0.482	111
2	0.297	0.296	220	0.295	0.295	220
3	0.253	0.252	311	0.252	0.252	311
4	0.210	0.209	400	0.208	0.208	400
5	0.193	0.192	331			
6	0.171	0.170	422	0.171	0.170	422
7	0.162	0.162	333			
8	0.161	0.161	511	0.161	0.161	511
9	0.147	0.148	440	0.149	0.148	440

3.2. Adsorption studies for hexavalent chromium

0.1 g/L γ-Fe_2O_3 nanoparticles were mixed with 50 mL of Cr (VI) solution, the solution concentrations being 40 and 100 mg/L. The pHs of solutions were 2.54 and 6.46 for each concentration. The rate of removal for hexavalent chromium decreases gradually with an increase of pH from 2.54 at 6.46. The contact time was between 0 and 120 minutes. First samples were analyzed after 10 minutes and the last after 120 minutes. About 90% from hexavalent chromium was removed, after first minute of the reaction.

The effect of pH on adsorption of 40 mg/L Cr (VI) can be seen in the Figure 4.

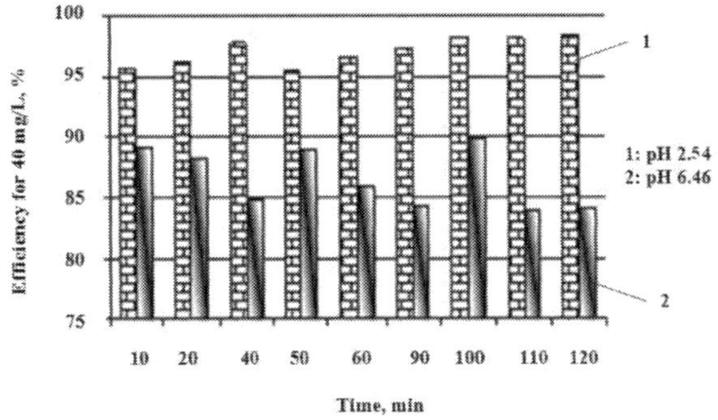

Figure 4. The time effect of the adsorption of Cr (VI) at different pH values.

It is known from the literature [9] that the surface charge is neutral at the zero point of charge (pHpzc). The pHpzc value is 6.3 for maghemite. Below this value, the adsorbent surface is positively charged and the anions are adsorbed by electrostatic attraction. Above this value of pHpzc, the adsorbent surface is negative charged and the cations are adsorbed on the maghemite. From the

Figure 4 it is observed that the optimum pH value is 2.5 this value being an expected results, in these conditions. With the increase in pH, the uptake of Cr (VI) ions decreased, because in the aqueous phase, the surface of the metal oxides is covered with hydroxyl groups that vary at different pH values.

Hexavalent chromium in wastewaters is presented in forms as chromates (CrO_4^{2-}), dichromate ($Cr_2O_7^{2-}$) and bichromate ($HCrO_4^-$), in these conditions, the OH^- groups from mixture will compete with chromium species for adsorptions sites.

Also, if the adsorption surface is negatively charged, the electrostatic repulsions increase between negatively charged Cr (VI) species and negatively charged nanoparticles, leading to the removal of some adsorbed species such as bichromate ($HCrO_4^-$) or chromate (CrO_4^{2-}) [2].

The results regarding adsorption efficiency, at pH 2.5, for initial concentrations of 40 and 100 mg/L Cr(VI) are presented in Figure 5.

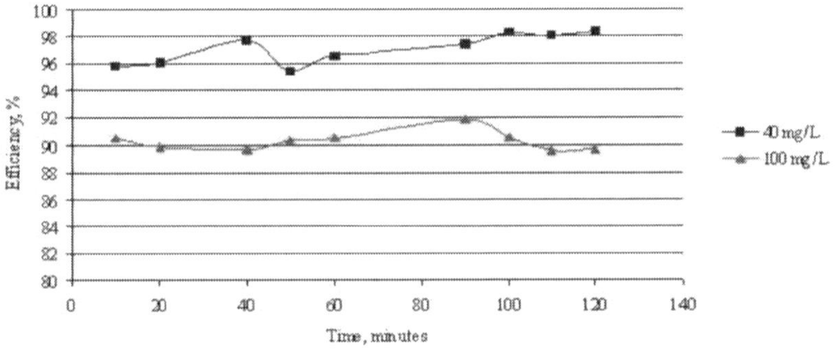

Figure 5. The adsorption efficiency for different concentrations of Cr (VI) onto maghemite nanoparticles.

It can be seen the rapid adsorption of chromium on maghemite nanoparticles, initially quite high, followed by a much slower removal rate this leading to an equilibrium state. Also, it is obviously that the removal of 40 mg/L Cr (VI) solution is more efficiently than the 100 mg/L Cr (VI) solution, but for the two concentrations the removal is very efficiently.

The Langmuir isotherm was chosen for estimating the maximum adsorption capacity corresponding to adsorbent surface saturation and the linear equation is expressed as (1):

$$\frac{C_e}{q_e} = \frac{1}{\theta} \cdot C_e + \frac{1}{\theta \cdot b} \tag{1}$$

where:
C_e – equilibrium concentration, mg/L;
q_e – adsorbed amount at equilibrium, mg/g;
θ *and* b – constants related with adsorption capacity and apparent heat change.

The fundamental characteristics of Langmuir equation can be interpreted in terms of a dimensionless constant separation factor (R_L), defined as (2):

$$R_L = \frac{1}{1 + bC_0} \tag{2}$$

where: C_0 is initial concentration, mg/L and b was defined above.

The value R_L indicates the types of Langmuir isotherm of irreversible ($R_L = 0$), favorable ($0 < R_L < 1$), linear ($R_L = 1$) or unfavorable ($R_L > 1$).

The adsorbed amount at equilibrium was 19.16 mg Cr (VI) adsorbed /g adsorbent after 10 minutes and it remained almost the same after 120 minutes. Calculation of the data was presented in another work [6]. This value indicates a good adsorption capacity in comparison with others common adsorbent such as activated carbon or diatomite, according to the results from literature, namely: 11.5 mg/g for diatomite and 15.47 mg/g for activated carbon, respectively [9]. The adsorption data were fitted well with Langmuir monolayer adsorption isotherm, considering a homogenous surface of adsorbent, without interaction between adsorbate molecules. The linearized Langmuir isotherm for hexavalent chromium removal is shown in Figure 6.

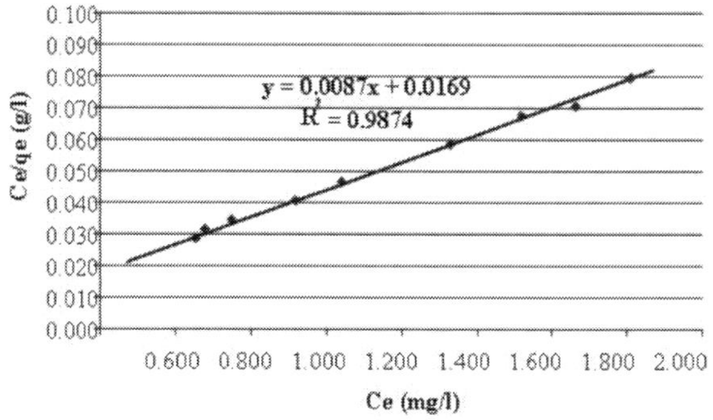

Figure 6. Linearized Langmuir isotherm for Cr (VI) adsorption on the maghemite nanoparticles.

The R_L factor can be calculated, from obtained data, according to the equation (2), thus R_L is 0.0462 for an initial concentration of hexavalent chromium of 40 mg/L. This result indicates that the uptake of Cr (VI) by maghemite is favorable.

The values of Langmuir constants and correlation factor are shown in Table 3.

Table 3. Langmuir constants for Cr (VI) adsorption on maghemite.

pH values	**Langmuir constants**		**R^2**	**R_L**
	q_e (mg/L)	**b (L/mg)**		
2.54	19.16	0.51	0.9874	0.0462
6.46	17.85	0.12	0.9352	0.1701

These values indicate a favorable Langmuir isotherm according to R_L factor (0.0462) and a good adsorption capacity at 19.16 mg/g for pH 2.54. These values are available for an initial concentration at 40 mg/L Cr (VI). Also, for a better correlation factor (R^2), more data are necessary regarding the time contact and the rotation speed between chromium solution and maghemite. In this experiment, the solution was manually mixed with maghemite nanoparticles. The magnetic separation can be seen in the Figure 7.

Figure 7. Magnetic separation of the maghemite from Cr (VI) solutions.

The hexavalent chromium was analyzed by molecular absorption spectrometry (diphenylcarbazide method) and the total chromium was analyzed with flame atomic absorption spectrometry. The results are shown in table 4.

Table 4. Chromium concentration analyzed before and after adsorption process at pH 2.54.

Before adsorption	***After adsorption***	
***Co* Cr (VI), mg/L**	***Ce* Cr (VI), mg/L**	***Ce* Cr tot, mg/L**
40	1.675	1.671
100	4.300	4.310

Dissolved hexavalent chromium form a colored compound with diphenylcarbazide detectable at wavelength of 530 nm. The detection method was molecular absorption spectrometry. Total chromium (trivalent and hexavalent) was detected by atomic absorption spectrometry. Small differences indicate that there is no chemical redox reaction during the adsorption process. Also, the adsorption of the hexavalent chromium onto maghemite surface can be seen by EDAX analysis from Figure 8.

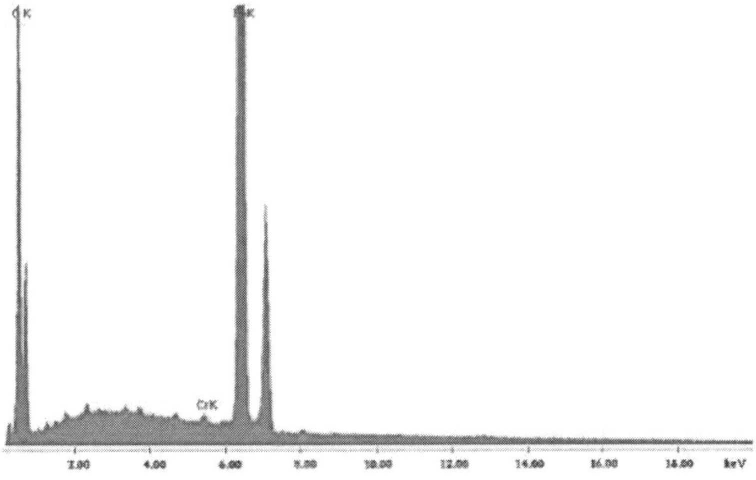

Figure 8. EDAX spectra for Cr (VI) adsorbed onto maghemite.

The hexavalent chromium adsorbed onto maghemite was analyzed by atomic absorption spectrometry, after mineralization. The results obtained as average were about 0.098 mg Cr(VI) for 1 g γ-Fe_2O_3.

4. Conclusions

The nanoparticles used as adsorbent for Cr (VI) were synthesized by coprecipitation method, having 10 nm diameters. These nanoparticles were characterized by XRD, TEM and SEM methods. The adsorption studies indicated a good correlation between chromium concentration and maghemite quantity at pH 2.54 and adsorption data were fitted well with Langmuir isotherm. The hexavalent chromium was analyzed before and after adsorption onto maghemite by atomic absorption spectrometry and molecular absorption spectrometry and also an EDAX analysis was provided for identification of chromium onto maghemite surface during adsorption process. The obtained data showed that the electrostatic attraction was responsible for the hexavalent chromium removal.

Acknowledgements

Authors recognize financial support from the European Social Fund through POSDRU/89/1.5/S/54785 project: "Postdoctoral Program for Advanced Research in the field of nanomaterials".

References

[1] Kumar C 2009 *Magnetic Nanomaterials* (Weinheim: Wiley-VCH Verlag GmbH & Co.) 648
[2] Hu J, Lo I and Chen G 2007 Comparative study of various magnetic nanoparticles for Cr (VI) removal *Separation and Purification Technology* **56** 249
[3] Wang X Z, C Dou P, Ding Y and Xu P 2009 Gellan gel beads containing magnetic nanoparticles: An effective biosorbent for the removal of heavy metals from aqueous system *Bioresource Technology* **100** 2301
[4] Covaliu C I, Georgescu G, Jitaru I, Neamtu J, Malaeru T, Oprea O and Patroi E 2009 Synthesis and characterization of a hydroxyapatite coated magnetite for potential cancer tratment *Revista Chimie* **60 (12)** 1254
[5] Massart R 1981 Preparation of aqueous magnetic liquids in alkaline and acidic media *IEEE Transactions on Magnetics* **17** 1247
[6] Matei E, Predescu A, Predescu A, Vasile E 2010 New alternatives for water pollution treatment using nanomaterials, *Proc. 4th International Conference "Biomaterials, Tissue Engineering & Medical Devices BIOMMEDD 2010"* 23 – 25th September 2010, Sinaia, p.189
[7] Predescu A 2010 Researches regarding the dimension and distribution of magnetic iron oxide nanopowders and their characterization *Bulletin of the Polytechnic Institute of Jassy* **LVI (LX)** 95
[8] Legodi M A and de Waal D 2007 The preparation of magnetite, goethite, hematite and maghemite of pigment quality from mill scale iron waste *Dyes and Pigments* **74** 161
[9] Hu J, Chen G and Lo I 2005 Removal and recovery of Cr (VI) from wastewater by maghemite nanoparticles *Water Research* **39** 4528

Bacterial remediation from effluent containing multi-walled carbon nanotubes

A P Lemes[1,2], L Cordi[1], A Santos[3] and N Durán[1,4]

[1]Instituto de Química, Universidade Estadual de Campinas (UNICAMP), Brazil

[2] Universidade Federal de São Paulo Campus São José dos Campos (UNIFESP), Brazil

[3]Hemocenter-FCM-Universidade Estadual de Campinas (UNICAMP), Brazil

[4] Universidade Federal do ABC, Santo André (UFABC), Brazil.

E-mail: aplemess@gmail.com

Abstract. Multi-wall carbon nanotubes (MWCNT) were functionalized with functional groups containing oxygen, mainly carboxylic groups (-COOH), through reaction with a mixture of H_2SO_4/HNO_3 (3:1 v/v). The oxidized multi-wall carbon nanotubes (MWCNTOOH) were used to prepare an effluent, 2 mg L^{-1} in a saline solution of NaCl (0.9%), to study of remediation of MWCNTOOH in aqueous suspension by utilization of *Escherichia coli*. The suspensions of *E. coli* (4.5 x 10^5 CFU mL^{-1} and 4.5 x 10^8 CFU mL^{-1}) in test tubes with MWCNTOOH effluent caused the precipitation of a large amount of MWCNTOOH and supernatant clearing. The scanning electron microscopy (SEM) analysis of the precipitate and supernatant showed the adhesion and interlace of MWCNTOOH in bacteria surface. Although the precipitate consist of a large quantity of MWCNTOOH and bacteria, it was verified their presence in the supernatant. The spread plate technique showed that MWCNTOOH caused no cellular death of *E. coli* in the supernatant.

1. Introduction

The carbon nanotubes (CNT) have been one of the most interesting nanomaterials due their unique optical, electrical, mechanical and thermal properties, which lead to numerous applications [1]. However, many researchers have verified a high toxicity of CNT dispersed in solution against microorganisms. The CNT can absorb significant amounts of bacteria with serious impact on the cellular membrane, metabolic activity, and their morphology [2,3].

The results from toxicity studies with CNT and other nanomaterials are often contradictory, because of the use of CNTs of different purities and functionalizations. Therefore, detailed purification and physicochemical characterization of CNT are extremely important before testing their toxicity [3,4].

Many data in the literature have been using carbon nanotubes to remove pathogenic organisms from aqueous solutions [2,5-8]. The bacteria are adhered in the CNT surfaces and they are removed of solution by CNT precipitation.

Most of these studies using carbon nanotubes without any modification and then the precipitation of CNT occur spontaneously in aqueous solution due to their high hydrophobicity. The high van der Waals attractive force of CNT provide the production of wide aggregates that precipitate in aqueous solution after periods that depend on CNT dimensions and the dispersion method used [9].

However, the use of surfactants and the functionalization of carbon nanotubes can produce suspensions that are stables for a very long period, which may consist in environmental risks due the CNT toxicity. In these cases, efficient and low cost methods for CNT removal of aqueous suspensions are desirable.

Thus, based on results of researches which no modified CNT are used to eliminate pathogenic organisms by precipitation, would the opposite way be viable? Would it be possible to use bacteria to remove carbon nanotubes of stable aqueous suspension?

This work evaluated the utilization of *Escherichia coli* to remove of oxidized multi-wall carbon nanotubes (MWCNTOOH) in stable aqueous suspensions, as a model of bioremediation for this nanomaterial.

2. Experimental

2.1. Functionalization of Carbon Nanotubes

The carbon nanotubes used in this work were multi-wall (MWCNT) with 10-40 nm in outer diameter, 5-20 μm length and 93% purity. The MWCNT were produced by thermal chemical (CVD) process and supplied by CNT Co. Ltd. Korea. In order to produce a stable suspension the MWCNT were treated in a mixture of H_2SO_4 /HNO_3 (3:1 v/v) and sonicated in an ultrasonic bath for 1 h. This suspension was filtered in a PVDF membrane and washed with distilled water until neutral pH. The chemical oxidation with H_2SO_4/HNO_3 mixture promote the functionalization of MWCNT with oxygen-containing groups, mainly carboxylic groups (-COOH). These functional groups present in the oxidized carbon nanotubes (MWCNTOOH) allow that their suspension remain stable by long periods.

2.2. Remediation tests with Escherichia coli

The MWCNTOOH were used to prepare an effluent with concentration of 2 mg L^{-1} in a solution of NaCl (0.9%). A suspension of *E. coli*, previously grown in a culture medium (LB broth) was inoculated into 10 mL of MWCNTOOH effluent. The test tubes with MWCNTOOH effluent and *E. coli* 4.5 x 10^5 and 4.5 x 10^8 CFU mL^{-1} were incubated at 37°C for 24 h at 120 rpm.

2.3. Characterization

The precipitates of test tubes were characterized by Scanning Electron Microscopy (SEM). A low mass of precipitate was deposited in a PVDF membrane (0.22 μm) and immersed for 1 hr in 1.5% glutaraldehyde solution at 5-8°C and the samples were dehydrated by different ethanol volumes starting; 30%, 50%, 70%, 80%, 90% and 100%, the period in each ethanol solution was 10 minutes. After, the samples were sputtered with Pd/Au alloy in a Balzer MED 020 machine and analyzed in a Jeol JSM-T300.

For supernatant analysis, 100 μL of supernatant were deposited in the PVDF membrane and after the same process with glutaraldehyde solution (1.5%) and ethanol gradient was realized before SEM analysis.

The supernatant was removed and centrifuged at 2000 rpm by 20 min. Thus, 100 μL of this 2nd supernatant (after centrifugation) was characterized by SEM, as well as, the 1st supernatant (before centrifugation).

Aliquots of 100 μL were used to evaluate the *E. coli* growth in the test tubes by spread plate method in LB agar. The incubation of plates occurred at 37°C for 16h.

3. Results and Discussion

The Figure 1 shows the test tubes with MWCNTOOH effluent (A) and the mixture of MWCNTOOH and bacteria (B) after the period of incubation.

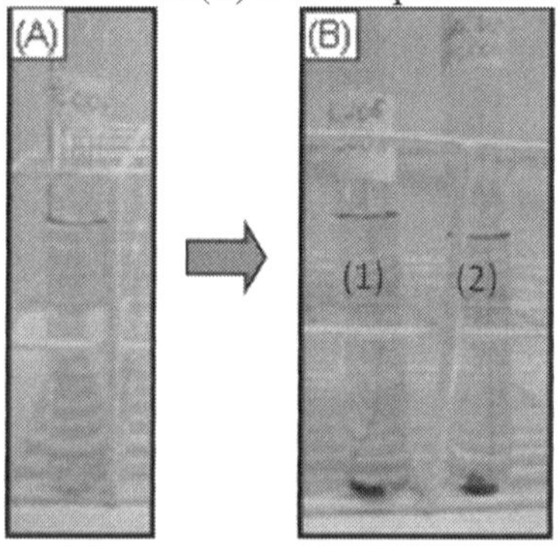

Figure 1. (A) MWCNTOOH effluent; (B) MWCNTOOH effluent with (1) 4.5 x 10^5 *E. coli* CFU mL^{-1} and (2) 4.5 x 10^8 *E. coli* CFU mL^{-1}, after incubation at 37°C, 120 rpm for 24h.

The test tube with MWCNTOOH effluent (without bacteria) did not show any significant change after incubation. However, the test tubes with MWCNTOOH effluent and bacteria showed the formation of precipitate and the supernatant clearing in both concentration of bacteria. This indicates that the presence of bacteria promoted the precipitation of a high amount of MWCNTOOH, and consequently, the MWCNTOOH removal from aqueous suspension.

The adhesion of bacteria in the MWCNTOOH surface provides the formation of large agglomerates that decant in solution due to their high weight.

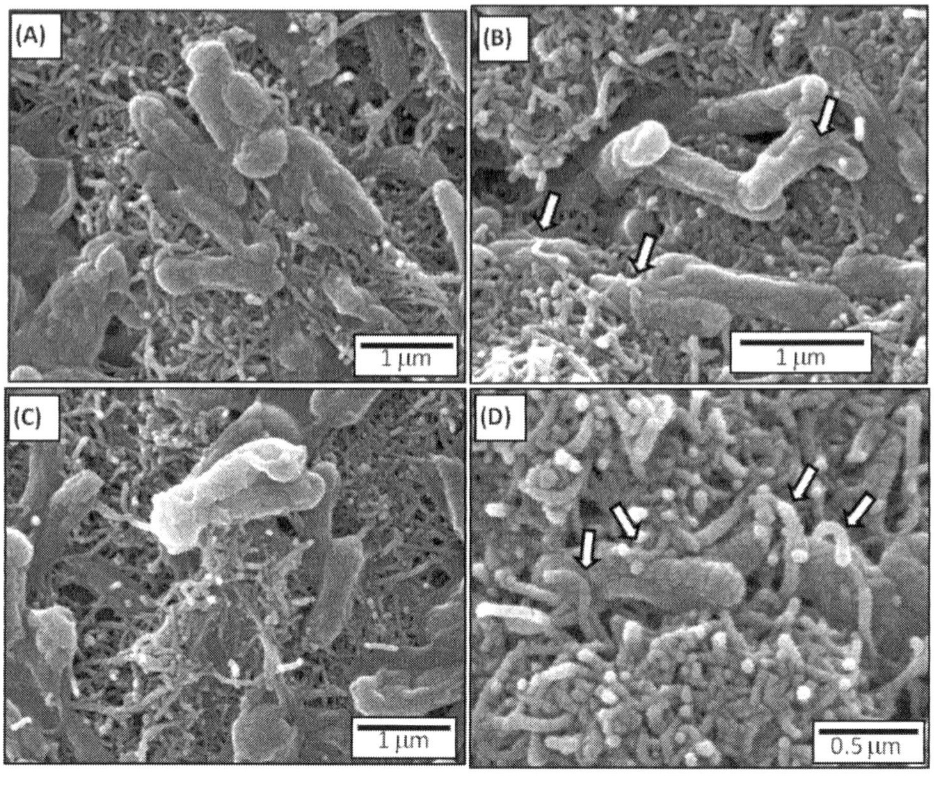

Figure 2. SEM images of precipitate present in the tubes with MWCNTOOH effluent and *E.* coli: 4.5 x 10^5 CFU mL^{-1}(A e B) and 4.5 x 10^8 CFU mL^{-1} (C e D).

The characterization of precipitate by SEM can be observed in the Figure 2. Both concentrations of *E. coli* resulted in the formation of agglomerates composed by MWCNTOOH and bacteria (Figure 2). In these agglomerates we can observe possible points of adhesion and interlace between MWCNTOOH and bacteria, as indicated by the arrows in Figure 2-B and D.

The LB agar plates incubated with 100 μL of 1st and 2st supernatants came from test tube containing only *E. coli* 4.5 x 10^5 CFU mL^{-1} (Figure 3 A and B) show a significant reduction of CFU mL^{-1} after the centrifugation at 2000 rpm for 20 min. Therefore, the centrifugation condition was efficient to promote the removal of a high concentration of *E. coli* from aqueous solution.

In the LB agar plates of 1st and 2nd supernatants of samples with MWCNTOOH effluent and *E. coli* 4.5 x 10^5 CFU mL^{-1} (Figure 3 C and D) we can verify again the reduction of bacteria in the supernatant after centrifugation and the high concentration of bacteria in the 1st supernatant. This result indicates that the MWCNTOOH did not cause the cellular death of bacteria present in the supernatants.

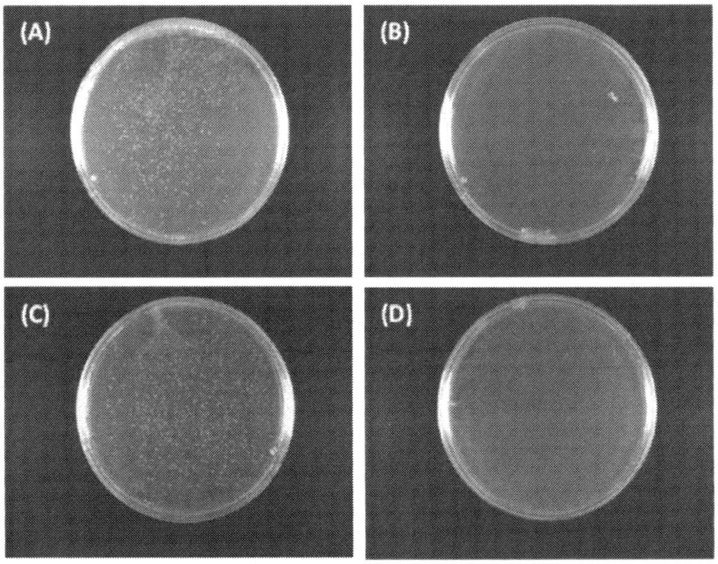

Figure 3. Photographs of LB agar plate of test tubes with *E. coli* 4.5 x 10^5 CFU mL^{-1} (A) 1st supernatant and (B) 2nd supernatant (after centrifugation), and test tubes with MWCNTOOH effluent and *E. coli* 4.5 x 10^5 CFU mL^{-1} (C) 1st supernatant and (D) 2nd supernatant (after centrifugation).

In the case of test tubes with higher *E. coli* CFU mL^{-1} (4.5 x 10^8 CFU mL^{-1}), although it was not possible to clearly visualize the CFU due to high concentration of colonies in Figure 4 (A) and (C), it was possible to verify again that MWCNTOOH did not promote the bacterial death.

This agrees with the results observed by Kang *et al.* [10]. The authors observed that single- wall carbon nanotube (SWCNT) promoted the bacterial death in aqueous suspension, while the bacteria in the presence of MWCNT remained alive.

The centrifugation of supernatant provided a decrease of *E. coli* in the aqueous medium but it remained still high, comparing the 1st supernatant to 2nd supernatant of each sample with *E. coli* 4.5 x 10^8 CFU mL^{-1}. Therefore, it is necessary more severe conditions of centrifugation to remove efficiently the bacteria from supernatant at that *E. coli* culture.

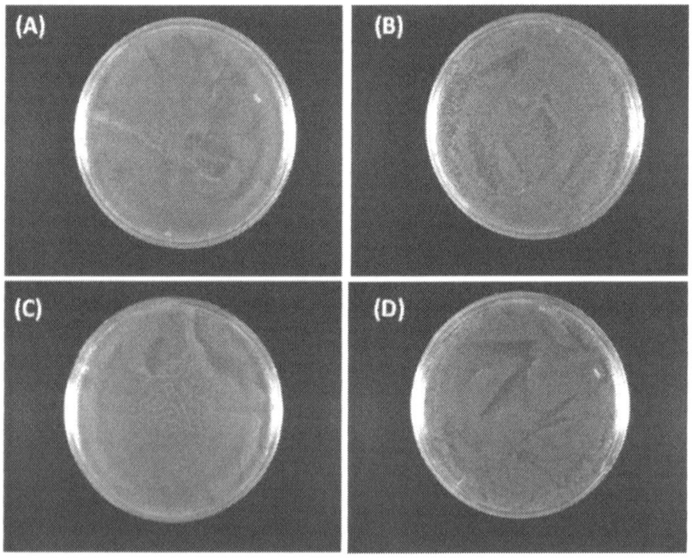

Figure 4. Photographs of LB agar plates of test tubes with *E.* coli 4.5 x 10^8 CFU mL^{-1} (A) 1st supernatant and (B) 2nd supernatant (after centrifugation), and test tubes with MWCNTOOH effluent and *E. coli* 4.5 x 10^5 CFU mL^{-1} (C) 1st supernatant and (D) 2nd supernatant (after centrifugation).

MWCNTOOH and bacteria were verified only in the SEM images of 1st and 2nd supernatants of sample with MWCNTOOH effluent and *E. coli* 4.5 x 10^8 CFU mL^{-1} (Figure 5 A and B). In the SEM images of all supernatants of other samples nothing was observed.

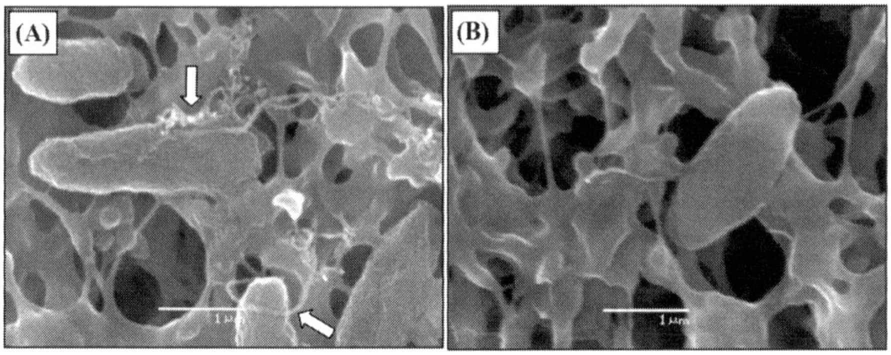

Figure 5. SEM images of 1st supernatant (A) and 2nd supernatant (B) of MWCNTOOH effluent with *E. coli* 4.5 x 10^8 CFU mL^{-1}.

A low number of bacteria were evidenced on the 100 μL of 1st supernatant (Figure 5 A) and only one bacterium was found in the case of 2nd supernatant (Figure 5 B). In both SEM images we can verify the presence of MWCNTOOH, however in Figure 5 (A) the arrows show adhesion points between MWCNTOOH and isolated bacteria. In this case the adhesion was better visualized than in the agglomerates that compound the precipitate, as shown in Figure 2. The SEM images in the Figure 5 show that a determined amount of MWCNTOOH and bacteria remain suspended in the supernatants, even after of centrifugation. In this case, higher centrifugation conditions can lead to better removal of MWCNTOOH.

4. Conclusion

The addition of *E. coli* in the MWCNTOOH effluent promoted the precipitation of MWCNTOOH. The *E coli* showed a normal growth in the MWCNTOOH effluent, indicating that the MWCNTOOH did not cause their cellular death.

It was possible to observe bacteria and MWCNTOOH in the supernatants of test tubes with MWCNTOOH and *E. coli* 4.5 x 10^8. Thus, although the removal of MWCNTOOH was not total in this case, the presence of bacteria *E.* coli resulted in a high decrease of MWCNTOOH concentration in aqueous suspension, indicating that the growth of *E. coli* in MWCNTOOH effluent can be a promissory method for MWCNTOOH bioremediation.

Acknowledgements
FAPESP, CNPq and Brazilian Network of Carbon Nanotubes (MCT/CNPq).

References

[1] Baughman R H, Zakhidov A A and de Heer W A 2002 Carbon nanotubes the route toward applications *Science* **297** 787
[2] Kang S, Pinault M, Pfefferle L D and Elimelech M 2007 Single-Walled Carbon Nanotubes Exhibit Strong Antimicrobial Activity *Langmuir* **23** 8670
[3] Lam C W, James J T, McCluskey R, Arepalli S and Hunter R L 2006 A review of carbon nanotube toxicity and assessment of potential occupational and environmental health risks *Crit. Rev. Toxicol.* **36** 189
[4] Jia G, Wang H, Yan L, Wang X, Pei R, Yan T, Zhao Y and Guo X 2005 Cytotoxicity of Carbon Nanomaterials: Single-Wall Nanotube, Multi-Wall Nanotube, and Fullerene *Environ. Sci. Technol.* **39** 1378
[5] Moon H M and Kim J W 2010 Carbon Nanotube Clusters as Universal Bacterial Adsorbents and Magnetic Separation Agents *Biotechnol. Prog.r* **26** 179
[6] Gu L, Elkin T, Jiang X, Li H, Lin Y, Qu L, Tzeng T R J, Josepha R and Sun Y P 2005 Single-walled carbon nanotubes displaying multivalent ligands for capturing pathogens *Chem. Commun.* **7** 874
[7] Arias L R and Yang L 2009 Inactivation of Bacterial Pathogens by Carbon Nanotubes in Suspensions *Langmuir* **25** 3003–3012.
[8] Akasaka T and Watari F 2009 Capture of bacteria by flexible carbon nanotubes *Acta Biomateralia* **5** 607
[9] Tan H, Jiang L Y, Huang Y, Liu B and Hwang K C 2007 The effect of van der Waals-based interface cohesive law on carbon nanotube-reinforced composite materials *Compos. Sci. Tech.* **67** 2941
[10] Kang S, Herzberg M, Rodrigues D F and Elimelech M 2008 Antibacterial Effects of Carbon Nanotubes: Size Does Matter! *Langmuir* **24** 6409

Hydrotalcites: a highly efficient ecomaterial for effluent treatment originated from carbon nanotubes chemical processing

O L Alves[1], D Stéfani[1], N V Parizotto[1] and A G Souza Filho[1,2]

[1]Solid State Chemistry Laboratory, Institute of Chemistry, University of Campinas – UNICAMP, P. O. Box 6154, 13083-970, Campinas-SP, Brazil.

[2]Departament of Physics, Universidade Federal do Ceará – UFC, P. O. Box 6030, 60455-900, Fortaleza-CE, Brazil.

E-mail: oalves@iqm.unicamp.br

Abstract. It has been reported that a mixture of carboxylated carbonaceous fragments (CCFs), so called *oxidation debris*, are generated during carbon nanotubes chemical processing using oxidant agents such as HNO_3. The elimination of these fragments from carbon nanotubes surface has been point out to be a crucial step for an effective functionalization of the nanotubes as well as for improving the material. However, this process can introduce a potential environmental problem related water contamination because these CCFs can be viewed as a mixture of carbonaceous polyaromatic systems similar to humic substances and dissolved organic matter (DOM). The negative aspects of humic substances and DOM to water quality and wastewater treatment are well known. Since carbon nanotubes industry expands at high rates it is expected that effluent containing oxidation debris will increase since HNO_3 chemical processing is the most applied method for purification and functionalization of carbon nanotubes. In this work, we have demonstrated that Hydrotalcites (HT) are highly efficient to remove oxidation debris from effluent solution originated from HNO_3-treated multiwalled carbon nanotubes. The strategy presented here is a contribution towards green chemistry practices and life cycle studies in carbon nanotubes field.

1. Introduction

Safety issues are important in nanotechnology and they need to be addressed before the products originating from this emerging technology platform come to the market on a large scale. Many aspects including ethics and new ways of preventing waste generation either during fabrication or manipulation of nanomaterials are key points. The starting step for any nanotechnology outcome begins with nanomaterials fabrication and/or processing (purification, functionalization, etc). Safe disposal and recycling strategies should be present in order to achieve sustainable development of this emerging technology. In this regard, one should think about the full life cycle of the nanomaterials and this assessment demands green strategies that would prevent society from facing serious health and environmental problems in the future [1]. Therefore, the sustainable development of nanotechnologies will depend on developing the ability to control both health and environmental risks, thus not creating or at least minimizing future problems similar to those that people have faced from many of the technologies developed in the last century.

Carbon nanotubes are among the most promising materials for nanotechnologies and their applications in electronics, materials, and biotechnology demands chemical treatments (purification and oxidation) of the as-prepared samples and nitric acid treatment is one of the most applied methods for achieving this goal [2]. Recent work has pointed out that the treatment of carbon nanotubes with concentrated nitric acid generates a mixture of oxygenated polyaromatic-like fragments. These polyaromatic-like compounds have been called "oxidation debris" or carboxylated carbonaceous fragments (CCFs) and they play an important role in the functionalization of carbon nanotubes because these fragments adsorb on the nanotubes surface [3-5]. The removal of these fragments from the carbon nanotubes surface is crucial for developing nanotubes surface chemistry and studies about their removal from the nanotubes have been reported [6-8].

Although the removal of oxidation debris from the carbon nanotubes surface is not a currently standard procedure for applications of oxidized carbon nanotubes, this procedure can be largely applied in carbon nanotubes field to improvement of the material quality. However, the industrial implementation of this procedure can introduces a potential environmental problem because these CCFs can be viewed as a mixture of carbonaceous polyaromatic systems similar to dissolved organic matter (DOM) and humic substances which play an important role in water quality and wastewater treatment [9, 10] as well as on the interaction with many environmental pollutants [11-13]. Since the carbon nanotubes mass production and application is growing very fast it is very important to find a safe way of handling this emerging residue that would be generated in the chemical processing of these oxidized carbon nanotubes. Therefore, the investigation of removal process of oxidation debris compounds is important in view of contributing to the development of a proactive, safe and environmental friendly carbon nanotubes-based technology.

In this paper we address this problem using a layered double hydroxide material, the so called Hydrotalcites (HT), for removing the oxidation debris from sodium hydroxide solution (effluent originated from carbon nanotubes chemical processing). These layered materials are important ecomaterials [14] and opened up many possibilities of removing pollutant species from water not only through adsorption on the external surface of the layers but also through an intercalation process [15-17]. The use of this ecofriendly material is a strategy that contributes to address the issue of a safe cycle-end nanomaterials disposal [18, 19].

2. Experimental

All chemicals (reagent grade, Synth or Vetec) were used as received, without further purification. Commercially available chemical vapour deposition (CVD) grown multiwall carbon nanotubes (MWCNT) [*Ctube 100*, CNT Co. Ltd., Incheon – South Korea] were used.

2.1. Nitric acid treatment of Multiwalled carbon nanotubes.

The MWCNT (1.0 g) were refluxed in HNO_3 7.0 mol/L (200 mL) for 24 h at 140 °C. After cooling to room temperature, they were filtered through a 0.2 μm PTFE membrane and washed with deionized water until neutral pH of the filtrate was reached. The oxidized MWCNT were dried in vaccum system for 24 h.

2.2. Preparation of oxidation debris effluent stock-solution.

The oxidation debris stock-solution was prepared according to the method described by Verdejo et al [5]. Briefly, the nitric acid oxidized MWCNT (0.25 g) were homogenized in 0.1 mol/L of sodium hydroxide solution (200 mL). Then the mixture was sonicated in an ultrasound bath [Cole-palmer 8891] for 15 min. After it was submitted to strong magnetic stirring for 1 h. The colored filtrate (orange-brown) was obtained using standard filtration system through a 0.2 μm PVDF membrane. This filtrate has considered to be oxidation debris or CCFs compounds (effluent stock solution) and was used in further adsorption studies.

2.3. Hydrotalcites synthesis.
Mg–Al-based hydrotalcite with a Mg^{2+}/Al^{3+} molar ratio equal to 2 was prepared by the co-precipitation method at variable pH. An aqueous solution (100 ml) containing 0.1 mol of $Mg(NO_3)_2$•$6H_2O$ and 0.05 mol of $Al(NO_3)_3$•$9H_2O$ was added dropwise under vigorous mechanical stirring into a 2.0 mol/L NaOH solution (150 ml). After addition of the nitrates the mixture was kept at room temperature and maintained for 2 h under stirring for ageing. The solid product was separated by centrifugation and washed several times with deionized water until pH 7–8. Finally, the solids were dried at 40 °C for 48 h and crushed above 150 meshes. The same procedure was adopted for the cloride and sulfate samples, using magnesium and aluminium chlorides and sulphates solutions respectively. The Mg^{2+}/Al^{3+} molar ratio was kept equal to 2 and so were the synthesis, ageing and drying conditions.

2.4. Adsorption experiments.
Adsorption experiments of oxidation debris onto hydrotalcites were carried out in centrifuge tubes containing 5.0 mL of effluent stock-solution at room temperature under magnetic stirring. Then, the effect of three parameters were evaluated: (i) interlayer anion SO_4^{2-}, NO_3^- and Cl^-, (ii) contact time: 30 min, 60 min and 90 min and (iii) HT/ effluent stock-solution volume ratio = 1.0, 5.0 and 10 g/L. These tubes were subsequently centrifuged at 3500 rpm for 10 min. The adsorbed oxidation debris or CCFs concentration was evaluated using UV-vis spectrometer at 500 nm wavelength and expressed as percentage of the removal.

2.5. Characterization methods.
Absorption spectra of solutions were obtained using a UV-visible spectrophotometer [Shimadzu model UV-1650 PC]. X-ray powder diffraction (XRD) patterns were obtained with a Shimadzu XRD6000 diffractometer, using Cu Kr (1.5406 Å) radiation operating with 30 mA and 40 kV. A scan rate of 1°/min was employed. The Fourier transform infrared (FT-IR) spectra were recorded using a Bomem MB spectrometer in the 4000-400 cm^{-1} frequency range. A total of 34 scans and a resolution of 4 cm^{-1} were employed in obtaining the spectra. The KBr solid was used to prepare the samples for FT-IR analysis.

3. Results and Discussions
Chemical oxidation of carbon nanotubes using HNO_3 has been used for purifying, cutting and functionalizing carbon nanotubes aiming to improve their dispersion in water and other solvents. Dillon et al. [20] observed that treatment of carbon nanotubes with nitric acid produces a uniform functionalized coating of disordered carbon on the nanotubes sidewalls. Recently it has been shown that this functionalized coating (oxidation debris), is indeed the major carrier of the COOH groups present in the sample, and that it can be easily removed from tube surface by base aqueous wash such as NaOH [21, 22]. The colored (orange-brown) sodium hydroxide filtrate shown in the insert to Figure 1 contains oxidation debris compounds resulting from the oxidative fragmentation of MWCNT using 7.0 mol/L HNO_3 for 24 h at 140 °C. UV-visible absorption spectrum (Figure 1) of the alkaline filtrate (effluent originated from carbon nanotubes chemical processing) exhibits a broad adsorption profile, with a shoulder at about 280 nm. The spectrum indicates the presence of a molecular chromophore containing a system of conjugated double or triple bonds or polycyclic aromatic substances. This spectral characteristic supports the identification that CCFs or oxidation debris were generated during acid oxidation [23].

The colored oxidation debris dispersion is a potential water contaminant and it must be remediated before being returned to the aquatic environment. The implementation of green chemistry principles seems to be an appropriated strategy towards sustainability, which includes the use of ecofriendly materials for treating wastewater or effluents such as the oxidation debris. Hydrotalcites have these characteristic and they appear as natural candidates to be tested as sorbent material for removal of oxidation debris from wastewater. HT compounds have positively charged layers and the interlayer space is negatively charged thus leading to a system with a rich adsorption and intercalation chemistry.

Many studies have been devoted to investigating the ability of HT to remove harmful oxyanions such as, humic substances, phenols, herbicides, pesticides, boron, textile dyes, among other anionic pollutants from water [24, 25]. In this sense, the HT materials were synthesized by co-precipitation method of nitrate, sulfate and chloride solutions of magnesium and aluminium. The obtained solids were named HT-NO_3, HT-SO_4 and HT-Cl, respectively. XRD and FT-IR analysis have confirmed that Hydrotalcites structures were obtained (Figure 2).

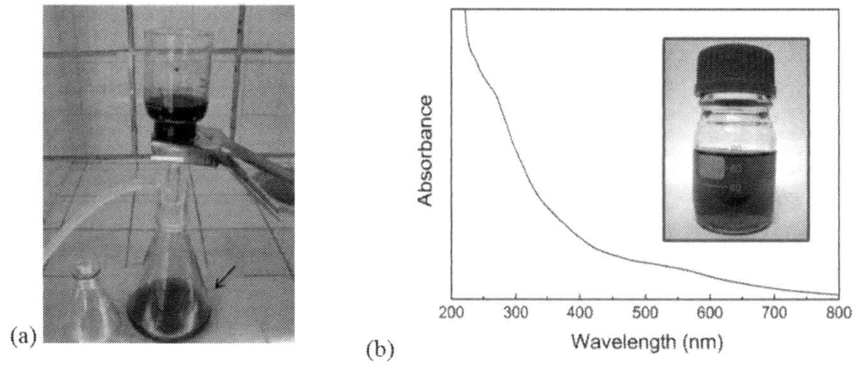

Figure 1. (a) Filtration system used to separation of oxidation debris from oxidized multiwalled carbon nanotubes. The arrow indicates the oxidation debris effluent; (b) UV-vis absorption spectrum of oxidation debris effluent stock-solution filtrate wash originated from oxidized multiwall carbon nanotubes (*see text*). [Inset: Oxidation debris effluent stock-solution].

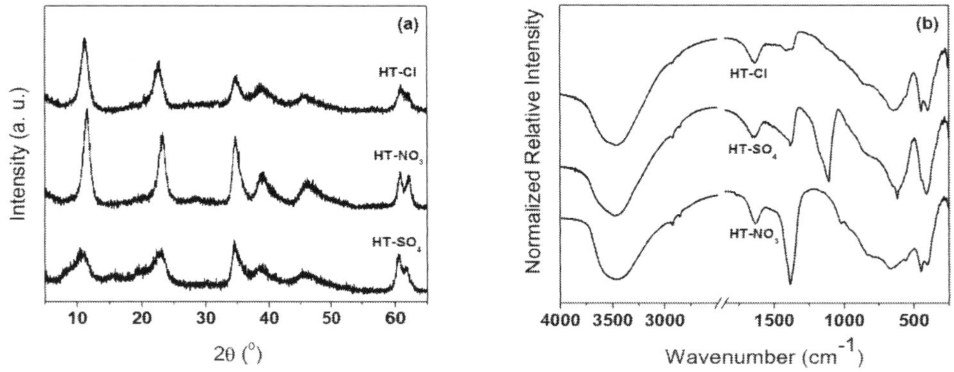

Figure 2. (A) X-ray diffraction patterns and (B) FT-IR spectroscopy of the as-prepared Hydrotalcites.

Table 1. Percentage of oxidation debris removal as a function of interlayer anion, contact time and HT/effluent stock-solution volume ratio.

HT/ Volume Ratio	**HT-NO_3 (%)**			**HT-SO_4 (%)**			**HT-Cl (%)**		
	30 min	60 min	90 min	30 min	60 min	90 min	30 min	60 min	90 min
1.0 g/L	63.8	90.4	94.4	82.5	88.1	94.9	70.1	67.8	85.9
5.0 g/L	86.5	100	97.2	94.4	100	100	88.1	100	92.7
10 g/L	95.0	100	99.4	100	100	100	100	100	100

Since oxidation debris is a mixture of carboxylated carbonaceous fragments (negatively charged) we have evaluated the potential of our synthetic Hydrotalcites (positively charged) to remove these fragments dispersed in NaOH solution (effluent stock-solution). The amount of oxidation debris removed by the hydrotalcites from an effluent stock-solution increase by increasing the HT quantity and contact time as shown in Table 1. All HT samples prepared were found to be highly efficient for removing oxidation debris compounds reaching the complete removal after 60 min. For 30 min test it was observed the highest removal efficient levels for HT-SO4 (superior to 80%). Quantitative (Table 1) and qualitative data inspection (Figure 3), indicate that Hydrotalcites are very efficient as sorbents for these oxidation debris compounds from carbon nanotubes.

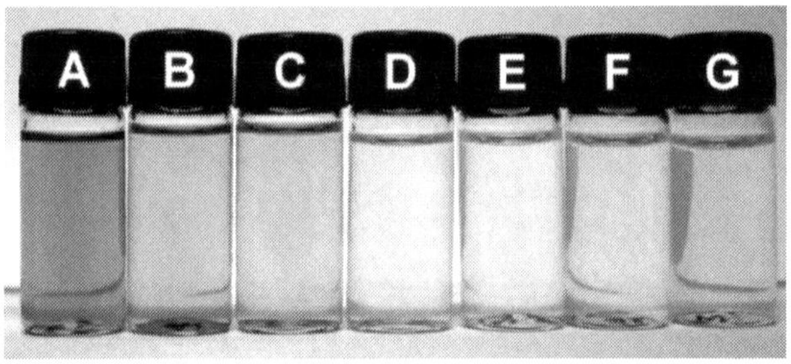

Figure 3. Qualitative assessment of oxidation debris removal as a function of Hydrotalcite ($HT-NO_3$) content: (A) 0.1 g/L; (B) 1.0 g/L; (C) 5.0 g/L; (D) 7.0 g/L; (E) 10 g/L; (F) 15 g/L; (G) 20 g/L. All systems were submitted to 30 minutes of magnetic stirring starting with a 5 mL of oxidation debris effluent stock-solution.

In order to get further insights into the adsorption mechanism we have performed X-ray diffraction (for probing interlayer distance) and FT-IR spectroscopy (for probing interlamelar species) for as-prepared and debris-adsorbed $HT-NO_3$ (Figure 4). We can observe that after the interaction with oxidation debris residue the X-ray pattern does not exhibit significant change. Both angular position and relative intensity of the various diffraction peaks are the same thus indicating that the oxidation debris residues are likely to be adsorbed on the $HT-NO_3$ surface rather than intercalated in the interlayer space. The infrared spectra for both samples are quite similar thus suggesting that the species in the interlamelar space are not affected by the presence of oxidation debris.

By comparing the FT-IR spectra for as-prepared samples and debris-adsorbed $HT-NO_3$ we can observe a change in the band about 1640 cm^{-1} that is related to H-O-H deformation mode from water. For the debris-adsorbed sample we observe a new peak located at 1590 cm^{-1} (marked with a down arrow) which can be attributed to C=C vibration from the fragments (oxidation debris) adsorbed on the Mg-Al $HT-NO_3$. However, a peak related to C=O from the carboxylic groups, which are expected to saturate the edges of carbon fragments, was not observed. By combining X-ray and FT-IR data we came to the conclusion that the basic mechanism of removing oxidation debris from the base aqueous solution is by adsorption rather than by intercalation.

The results showed that Mg-Al-based HT can be efficiently used for removing oxidation debris dispersed in base aqueous solution and this protocol offers to both the scientific and technological community a safe and clean strategy for not releasing in the environment these dissolved organic matter (DOM) species look like humic substances and polyaromatic compounds. These HTs materials are already available in the market with competitive costs so the remediation of this effluent can be applied in both laboratory and industrial scales. Hydrotalcites also exhibit a striking property which is called *memory effect* in which some the products of their thermal decomposition at a given temperature experience a spontaneous structural reconstruction when placed in aqueous medium and this phenomenon is well-known for Mg-Al hydrotalcites [26, 27]. We have performed recycling tests using $HT-NO_3$ (10 g/L for 90 min), with thermal treatment for 2h at 500 °C after each cycle. The solid was reused for three cycles with oxidation debris effluent stock-solution and the removal efficiency was superior to 97% (data not showed). These results open up the possibility of removing the oxidation debris from the HTs (through thermal decomposition) because the HT has the ability to reconstruct back to their parent structure makes the HT suitable materials for producing recyclable sorbents. In addition, it was found that both pH and molar concentration of the NaOH solution after removing the oxidation debris with HT remains practically unchanged. These results suggest that the reuse of this solution (process water) is possible for subsequent treatments of oxidized carbon nanotubes. The reuse of this process water can leads to: saving water and diminishing effluent waste.

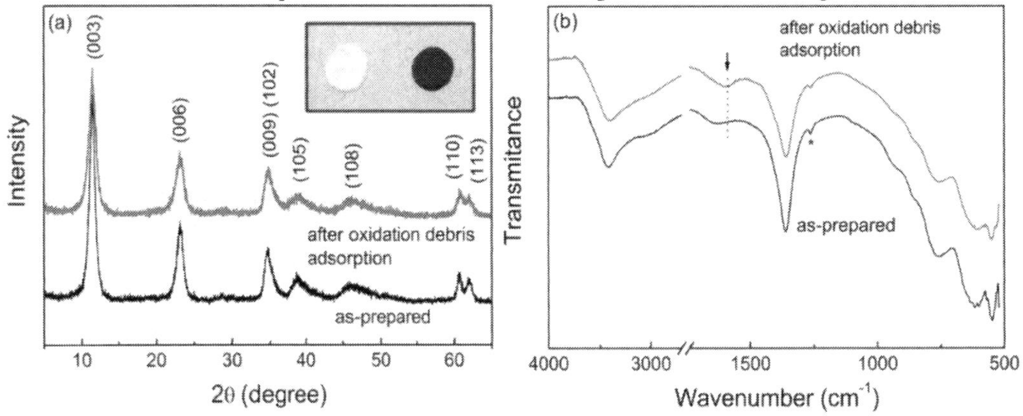

Figure 4. (a) X-ray diffraction patterns and (b) FT-IR spectroscopy data of as-prepared (lower trace) and debris-adsorbed (upper trace) HT-NO_3.
[Inset: samples of powdered on HT-NO_3 as-prepared (left) and with debris-absorbed (right)]

4. Conclusions

In summary, we have proposed a very efficient, simple and ecofriendly strategy using Hydrotalcites for effluent treatment originated from carbon nanotubes chemical processing. Since the industrial production of carbon nanotubes has already reached tons per year this work point out a way of removing the dissolved carbonaceous residues that nanotubes-based technology will generate. Finally, the results described here can be straightforwardly transferred to graphene oxide technology which is pointed out to be an important industry in the near future.

Acknowledgments

The authors are grateful to CNPq and FAPESP for financial support. This is a contribution of the National Institutes of Science, Technology and Innovation on Complex Functional Materials (Inomat) and NanoBioSimes.

References

[1] Albrecht M A, Evans C W, Raston C L 2006 Green chemistry and the health implications of nanoparticles *Green Chemistry* **8** 417-432
[2] Datsyuk V, Kalyva M, Papagelis K, Parthenios J, Tasis D, Siokou A, Kallitsis I, Galiotis C 2008 Chemical oxidation of multiwalled carbon nanotubes *Carbon* **46** 833-840
[3] Salzmann C G, Llewellyn S A, Tobias G, Ward M A, Huh Y, Green M L H 2007 The role of carboxylated carbonaceous fragments in the functionalization and spectroscopy of a single-walled carbon-nanotube material *Advanced Materials* **19** 883-887
[4] Fogden S, Verdejo R, Cottam B, Shaffer M 2008 Purification of single walled carbon nanotubes: The problem with oxidation debris *Chemical Physics Letters* **460** 162-167
[5] Verdejo R, Lamoriniere S, Cottam B, Bismarck A, Shaffer M 2007 Removal of oxidation debris from multi-walled carbon nanotubes *Chemical Communications* 513-515

[6] Yu H, Jin Y G, Peng F, Wang H J, Yang J 2008 Kinetically controlled side-wall functionalization of carbon nanotubes by nitric acid oxidation *Journal of Physical Chemistry C* **112** 6758-6763
[7] Worsley K A, Kalinina I, Bekyarova E, Haddon R C 2009 Functionalization and Dissolution of Nitric Acid Treated Single-Walled Carbon Nanotubes *Journal of the American Chemical Society* **131** 18153-18158
[8] Wang Z, Korobeinyk A, Whitby R L D, Meikle S T, Mikhalovsky S V, Acquah S F A, Kroto H W 2010 Direct confirmation that carbon nanotubes still react covalently after removal of acid-oxidative lattice fragments *Carbon* **48** 916-918
[9] Liang L, Singer P C 2003 Factors influencing the formation and relative distribution of haloacetic acids and trihalomethanes in drinking water *Environmental Science & Technology* **37** 2920-2928
[10] Nikolaou A D, Golfinopoulos S K, Lekkas T D, Kostopoulou M N 2004 DBP levels in chlorinated drinking water: Effect of humic substances *Environmental Monitoring and Assessment* **93** 301-319
[11] Chien S W C, Wang M C, Huang C C 2006 Reactions of compost-derived humic substances with lead, copper, cadmium, and zinc *Chemosphere* **64** 1353-1361
[12] Garbin J R, Milori D M B P, Simoes M L, da Silva W T L, Neto L M 2007 Influence of humic substances on the photolysis of aqueous pesticide residues *Chemosphere* **66** 1692-1698
[13] Durjava M K, ter Laak T L, Hermens J L M, Struijs J 2007 Distribution of PAHs and PCBs to dissolved organic matter: High distribution coefficients with consequences for environmental fate modeling *Chemosphere* **67** 990-997
[14] Oh J M, Biswick T. T, Choy J. H 2009 Layered nanomaterials for green materials. *Journal of Materials Chemistry* **19** 2553-2563
[15] Ferreira O P, Alves O L, Macedo J D, Gimenez I D, Barreto L S 2007 Ecomaterials: Development and application of functional porous materials for environmental protection *Quimica Nova* **30** 464-467
[16] Ferreira O P, de Moraes S G, Duran N, Cornejo L, Alves O L 2006 Evaluation of boron removal from water by hydrotalcite-like compounds *Chemosphere* **62** 80-88
[17] Goh K H, Lim T T, Dong Z L 2009 Enhanced Arsenic Removal by Hydrothermally Treated Nanocrystalline Mg/Al Layered Double Hydroxide with Nitrate Intercalation. *Environmental Science & Technology* **43** 2537-2543
[18] Helland A, Scheringer M, Siegris M, Kastenholz H G, Wiek A, Scholz R W 2008 Risk assessment of engineered nanomaterials: A survey of industrial approaches *Environmental Science & Technology* **42** 640-646
[19] Meyer D E, Curran M A, Gonzalez M A 2009 An Examination of Existing Data for the Industrial Manufacture and Use of Nanocomponents and Their Role in the Life Cycle Impact of Nanoproducts *Environmental Science & Technology* **43** 1256-1263
[20] Dillon A C, Gennett T, Jones K M, Alleman J L, Parilla P A, Heben M J 1999 A simple and complete purification of single-walled carbon nanotube materials *Advanced Materials* **11** 1354-1358
[21] Shao L, Tobias G, Salzmann C G, Ballesteros B, Hong S Y, Crossley A, Davis B G, Green M L H 2007 Removal of amorphous carbon for the efficient sidewall functionalisation of single-walled carbon nanotubes *Chemical Communications* 5090-5092
[22] Tobias G, Shao L D, Ballesteros B, Green M L H 2009 Enhanced Sidewall Functionalization of Single-Wall Carbon Nanotubes Using Nitric Acid *Journal of Nanoscience and Nanotechnology* **9** 6072-6077
[23] Wang Z W, Shirley M D, Meikle S T, Whitby R L D, Mikhalovsky S V 2009 The surface acidity of acid oxidised multi-walled carbon nanotubes and the influence of in-situ generated fulvic acids on their stability in aqueous dispersions *Carbon* **47** 73-79
[24] Seida Y, Nakano Y 2000 Removal of humic substances by layered double hydroxide containing

iron *Water Research* **34** 1487-1494
[25] Goh K H, Lim T T, Dong Z 2008 Application of layered double hydroxides for removal of oxyanions: A review *Water Research* **42** 1343-1368
[26] Rocha J, del Arco M, Rives V, Ulibarri M A 1999 Reconstruction of layered double hydroxides from calcined precursors: a powder XRD and Al-27 MAS NMR study *Journal of Materials Chemistry* **9** 2499-2503
[27] Rajamathi M, Nataraja G D, Ananthamurthy S, Kamath P V 2000 Reversible thermal behavior of the layered double hydroxide of Mg with Al: mechanistic studies *Journal of Materials Chemistry* **10** 2754-2757

Gene expression analysis in rat lungs after intratracheal exposure to nanoparticles doped with cadmium

Teresa Coccini[1], Marco Fabbri[2],Elisa Roda[3], Maria Grazia Sacco[2], Luigi Manzo[1,3] and Laura Gribaldo[2]

[1]Toxicology Division, Salvatore Maugeri Foundation IRCCS, Pavia, Italy.

[2]European Commission, Joint Research Centre, IHCP- 21027 Ispra, Italy.

[3]European Centre for Nanomedicine, University of Pavia, 27100 Pavia, Italy.

Abstract. Silica nanoparticles (NPs) incorporating cadmium (Cd) have been developed for a range of potential application including drug delivery devices. Occupational Cd inhalation has been associated with emphysema, pulmonary fibrosis and lung tumours. Mechanistically, Cd can induce oxidative stress and mediate cell-signalling pathways that are involved in inflammation.This in vivo study aimed at investigating pulmonary molecular effects of NPs doped with Cd (NP-Cd, 1 mg/animal) compared to soluble $CdCl_2$ (400 μg/animal), in Sprague Dawley rats treated intra-tracheally, 7 and 30 days after administration. NPs of silica containing Cd salt were prepared starting from commercial nano-size silica powder (HiSil™ T700 Degussa) with average pore size of 20 nm and surface area of 240 m^2/g. Toxicogenomic analysis was performed by the DNA microarray technology (using Agilent Whole Rat Genome Microarray 4x44K) to evaluate changes in gene expression of the entire genome. These findings indicate that the whole genome analysis may represent a valuable approach to assess the whole spectrum of biological responses to cadmium containing nanomaterials.

1. Introduction

Among nanomaterials, silica nanoparticles (NPs) have found extensive applications in several commercial fields. Recently, their use in biomedical area has been extended to DNA delivery [1-4] and drug delivery [5][5].

Silica exists in several crystalline and amorphous forms all characterised by different physicochemical properties as particle shape and dimension, hydrophilicity, presence of surface radicals. The surface properties also depend on the mechanical, thermal and chemical history of a given dust and on the presence of metallic impurities. The NP fate in body and related adverse effects depend on these properties. Crystalline silica dust inhalation may cause silicosis and lung cancer. Amorphous silicas are not generally associated with progressive lung fibrosis, but animal inhalation studies showed reversible lung inflammation, granuloma formation and emphysema.

Silica NPs may contain metals as contaminants, but still unknown is the effect of metals on NP toxicity [6] and if these NPs act as carriers of toxic metals.

Certain inhaled NPs cause higher lung toxicity (e.g., inflammation, cancer) than larger particles of similar chemical composition at equivalent mass concentration. Besides the NP size, additional factors, such as shape, surface charge, coatings and surface reactivity, and particle aggregation/

disaggregation potential, may considerably contribute to NP toxicity [7-9]. Furthermore, the surface can carry large amounts of adsorbed or condensed toxic air pollutants, including the pneumotoxic metals such as cadmium (Cd).

Silica NPs incorporating cadmium [6] have been developed for potential application in informatics and as drug delivery devices. Fluorescent, radio-opaque, and paramagnetic CdS-containing "quantum dots" have been produced as multifunctional probe for bioimaging.

Cd is toxic to the respiratory tract and other organ systems when inhaled. Inhalation of Cd can result in acute injury such as oedema, or in chronic injuries like emphysema, pulmonary fibrosis or adenocarcinomas Although, the mechanisms of Cd toxicity are not yet fully understood, several publications reported on interference with essential metals, induction of oxidative stress, disruption of cadherins, inhibition of DNA repair and interference with apoptosis [10]. Also some of the molecular sensors of Cd intoxication are known such as metallothionein (MT), heme oxygenase-1 (HMOX-1), and heat shock protein 70 (HSP70) [11]. Moreover inhaled $CdCl_2$ resulted in pulmonary inflammatory and cell proliferative responses in vitro and in laboratory animals [12,13].

Nanomaterials fall between molecular species and physical entities and their mode of action may vary greatly based on their physicochemical parameters. The scientific basis for performing a toxicological evaluation of these materials can be different from the corresponding nanoscale chemicals.

Conventional procedures for assessing chemical toxicity are not completely suitable for application to nanomaterials [14].

The present study was focalized on the in vivo effects, in laboratory animals, of the test silica nanoparticles incorporating cadmium compared with the metal cadmium chloride.

Nanomaterial testing was focused on specific endpoints and pathological outcomes of lung tissues. Genomic analyses have been performed to provide information about specific mechanisms at the molecular level.

2. Materials and Methods

2.1. Preparation of NP of silica containing Cd salt.

Impregnation of SiO_2 (silica nanosize HiSilTM T700 commercial Degussa, average pore size 20 nm, surface area 240 m^2/g) with cadmium nitrate dehydrate was obtained from a preparation of aqueous solution of $CdNO_3$ 3.56 x 10^{-2} M, in which the silica was dispersed in concentration ratios which produces a sample containing 40% Cd by weight. Dispersion was brought to dryness and treated at 600 °C for 2 hr in air. This heat treatment leads to the decomposition of nitrate Cd to Cd oxide, dispersed in silica. Powder was later subjected to grinding mills with high energy (200 rpm for 1.5 hr, 400 rpm for 1.5 hr, 600 rpm for 2 hr) to get the most equal distribution of particle size and/or aggregates of particles. The thermal treatment followed by heating the particles at higher temperatures (ca. 500-600 °C) was performed in order to remove the residual organic groups. Attention was devoted to tune the thermal treatments in order to retain the grains dimensions, pores size, and distributions, and particles surface areas.

Cd concentration released from NP of silica in the physiological solution was determined by Flame-atomic absorption analysis for trace metals. Quantitative analysis was also performed on the NPCd sample for the presence of trace metals.

2.2. In vivo study.

Sprague Dawley male adult rats were treated with a single intratracheal instillation (i.t.) as following: (i) Control group (100 µL physiological solution-ps); (ii) Cadmim Chloride ($CdCl_2$: 400 µg/rat/100 µL ps), (iii) NP doped with Cadmium (NP-Cd: 1 mg/rat/100 µL ps). Animal lung evaluations were performed 7 and 30 days after i.t. administration.

All experimental procedures involving animals were performed in compliance with the European Council Directive 86/609/EEC on the care and use of laboratory animals (European Council directive of 24 November 1986 on the approximation of low, regulations, and administrative provisions of the Member States regarding the protection of animals used for experimental and scientific purposes. Off. J. Eur. Communities L358, 1-28)

2.3. RNA Preparation

Rat lung samples were collected and stored at -80 °C in RNA Later (SIGMA Aldrich) according to manufacturer instructions.

Samples of about 30 mg were homogenized by adding a stainless steel bead followed by 600 ml of Buffer RLT (Quiagen, Germantown, MD, USA)using the TissueLyser (Quiagen, Germantown, MD, USA) at 30HZ for 5 min. The lysate was centrifuged at maximum speed for 5 min and the supernatant was transferred to a G-Column (Quiagen, Germantown, MD, USA) to remove DNA and RNA was then purified using the RNeasy Plus kit (Qiagen, Germantown, MD, USA). RNA was quantified using a ND-1000 UV–Vis Spectrophotometer (NanoDrop Technologies), and the integrity of the RNA was assessed with the Agilent 2100 Bioanalyzer (Agilent), according to the manufacturer's instructions. RNA samples used in this study all had a 260/280 ratio $\geq$1.9 and a RNA Integrity Number (RIN) $\geq$8.

2.4. Microarray expression profiling.

The microarray experiment was designed to perform three replicates for each treatment. All the sample-labeling, hybridization, washing, and scanning steps were conducted following the manufacturer's specifications. In brief, Cy3-labeled cRNA was generated from 500 ng input total RNA using Quick Amp Labeling Kit, One-color (Agilent).

For every sample 1.65 μg cRNA from each labeling reaction (with a specific activity above 9.0) was hybridized using the Gene expression Hybridization Kit (Agilent) to the Agilent Whole Rat Genome Oligo Microarray (Agilent), which is a 4 x 44k 60mer slide format where each of the 4 arrays represents about 44,000 unique rat genes and transcripts. After hybridization, the slides were washed and then scanned with the Agilent G2565BA Microarray Scanner (Agilent). The fluorescence intensities on scanned images were extracted and preprocessed by Agilent Feature Extraction Software (v10.5.1.1).

Quality control and array normalization was performed in the R statistical environment (http://www.r-oject.org) using the Agi4x44PreProcess package downloaded from the Bioconductor web site (http://bioconductor.org/). The normalization and filtering steps were based on those described in the Agi4x44PreProcess reference manual.

In order to detect expression differences among different treatments a moderated t test was applied. Moderated t statistics were generated by limma bioconductor package [15]. Modulated genes were chosen as those with a fold of induction greater than 1.5 and a p-value smaller than 0.01 (NP doped with Cadmium versus control and Cadmium versus control). All of the above computations were conducted using the R statistics programming environment available at www.r-project.org.

After removal of redundant genes, figures-of-merit analysis was applied to define the optimal number of clusters [16]. Gene trends were organized by K-means clustering using squared Pearson correlation by mean of TIGR MultiExperiment Viewer (Mev) [17][17]. Each clusters generated by K-mean analysis was analyzed in the Metcore package to identify genes with similar functions. The MetaCore software (GeneGo, St. Joseph, MI) is a computational resource that uses logic operations for identifying biological processes that are altered because of changes in gene expression [18].

3. Results and Discussion

3.1. Evaluation of the release time of Cd from silica nanoparticles (NP-Cd).

Cd concentration released from NP of silica in the physiological solution was linear during a period of 16-232 hr (evaluated by linear of regression: y= 473.24+0.7235x, R2=0.9975). The maximum release (about 15%) was observed after 16 hr in physiological solution.

3.2. Purity of NP doped with cadmium.
Quantitative analysis was performed on the NPCd, to verify the amount of doping effectively dispersed in the silica and the presence of any metal pollutants from possible impurities in the reagents or the grinding process in jars of tungsten carbide alloy with cobalt. The presence of Cd and Si has been highlighted (32.5% and 24.1% respectively), the main pollutants were Ca (0.3%), Na (0.2%), K (0.2%), Fe (0.04%) and Mn (0.001%). Other metals were present in quantities less than 1 ‰.

3.3. Preliminary characterization of NPCd.
The crystalline structure of the product was investigated by X-ray powder diffraction; an amorphous compound was obtained. The diffractogram shows only a broad band in the range of 10-35° angle/2θ, typical of components of low crystallinity and low particle size (Fig. 1). NPCd showed stability in suspension > 30 days; amorphous structure; spherical form; average size of 20 nm; size distribution mono-modal, asymmetric, with modal value of 10 nm; 200 m^2/g surface area; tendency to form aggregates and agglomerates of about 150 nm; and zeta potential of -30 mV (in dehyonized water).

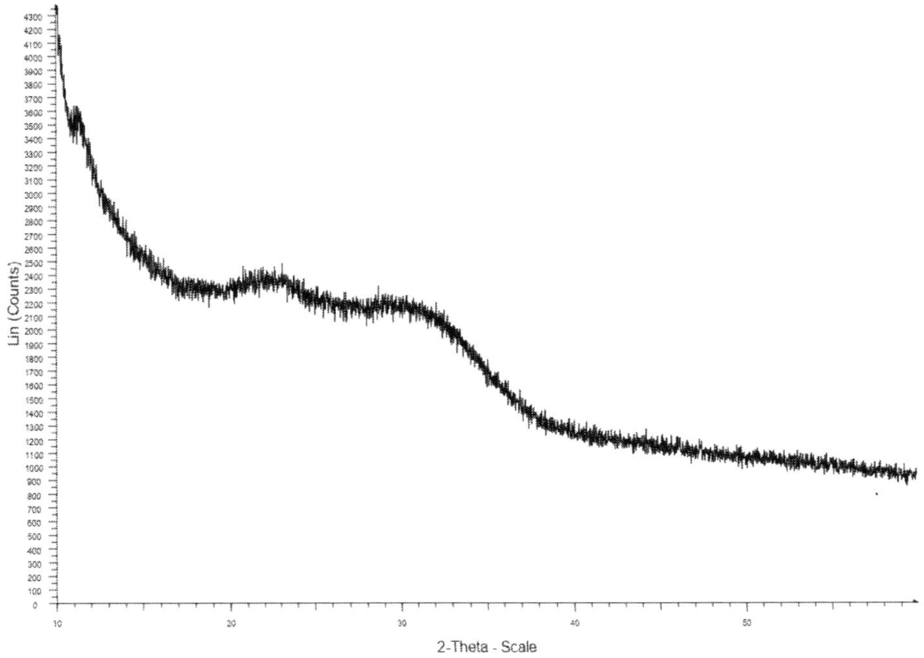

Figure 1. X-ray diffractogram of a NPCd sample.

3.4. Pulmonary Toxicogenomic data.
The study approach was not to investigated the well know effects of Cd toxicity for which the bulk of experimental and epidemiological data have abundantly addressed them till to define the limit value in the human exposure scenario[19], but that to evaluate the effects of engineered NPs (ENPs) namely cadmium-doped NPs using a new model of screening toxicity study, such as whole genome analysis, by means of the Cd as a tool of lung toxicity model. Specifically, the aim of the present work was to use Cd, a well know pneumotoxicant, as a tool to investigate its counterpart-doped nanoparticles. Consequently, we employed a Cd dosage that is known to induce lung toxicity in rodents. The dose selected was based on previous experiments showing lung dysfunction, cytotoxicity, edema and fibrosis after application of cadmium chloride [20-22]. In particular, lung injury induced by $CdCl_2$ given i.t. at the dose of 400 μg /rat (≡ 245 μg Cd) was shown to represent a good model of human interstitial lung disease [22]. In other studies, the dose of 0.3 mg Cd corresponding to 490 μg $CdCl_2$,

caused leukocytic infiltration and edema in lung 24 hr after i.t. dosing in guinea pig [21], and 400 μg of i.t. administered $CdCl_2$ caused various forms of intraluminal fibrosis 4 and 7 days after treatment in rat [22]. Persistent lung damage has been reported in an animal model following a single i.t. exposure to 25, 100, or 400 μg $CdCl_2$/kg bw [23] with histopathological alterations including inflammation and fibrosis still present 90 days post-exposure and the incidence and severity of the lesions greater at 90 days compared to 28 days.

Differences exist in the dose rate and distribution of material delivered by instillation versus inhalation, however, the data base on pulmonary toxicity of various particulate materials indicates that the two modes of exposure yield qualitatively similar results for a variety of biological endpoints, including pulmonary inflammation, fibrosis , susceptibility to infection, allergic sensitization and lung cancer in rats after exposure to poorly soluble particle [24]. In particular for Cd, as discussed in Oberdörster [25-27], the pulmonary retention half-time of both inhaled and intratracheal instilled $CdCl_2$ resulted in similar Cd retention and similar clearance kinetics (i.e. 60-85 days) in rats.

For the present study the i.t. was applied as a useful method to compare the lung effects in rats of a new material (i.e., counterpart-doped nanoparticles) against similar compounds (i.e. $CdCl_2$) for which an extensive inhalation database is available [19]. Moreover, i.t. model may serve to a screening tool to determine the approximate dose range that may be appropriate for later inhalation studies as well as to address specific endpoints regarding the respiratory toxicity of nanomaterials [24].

Based on the physiology of the human respiratory tree, a comprehensive model to predict the kinetics of inhaled cadmium in humans [28] has suggested that only about 5% of particles >10 μm in diameter will be deposited, up to 50% of particles <0.1 μm will be deposited, and between 50 and 100% of cadmium deposited in the alveoli will ultimately be absorbed.

Since, nanoparticles have unique physicochemical properties and functionalities that are different from their bulk counterparts, their mode of action may vary greatly based on their physicochemical parameters. This is therefore crucial to determine the appropriate way to compare physicochemical properties with fate and toxicity. Moreover, once they have entered the body (e.g. through inhalation), engeneered NPs have the ability to translocate and to be distributed to distant target organs. The harmful effects associated with this process have not yet sufficiently investigated. Translocation is also an important component in the toxicity of Cd which has been shown to accumulate in kidney and affect renal function after chronic inhalation exposure other than produce toxic effects on the primary target organ, i.e. lung, being the site of entry.

The detrimental effects of silica NPs containing metals as contaminants are still unknown [29] and still largely undefined is whether these NPs act as carriers of toxic metals. However, engineered cadmium-containing silica NPs show great promise for targeted drug delivery [30].

Transcriptomics was applied to study whole genome expression in the lung tissues collected from particle-exposed animals in order to identify early adverse events at transcriptional level. Such early changes may be predictive of delayed irreversible responses and may be useful as early biomarkers of organ toxicity.

After 7 days we identified 22 probes regulated by cadmium and 71 by NP doped with cadmium, whereas 138 probes are regulated by cadmium and 156 by NP doped with cadmium after 30 days. The few genes regulated at 7 days were still modulated at 30 days, therefore, we decided to further analyse only the set of genes regulated at 30 days.

After removal of duplicated genes and probes without Entrez ID, 129 genes were identified as the union set of genes regulated. This set was subjected to figures-of-merit analysis, where K-means clustering performed optimally for 4 clusters, with no increase in the predictive value of the algorithm for additional behavioral categories (data not shown). Pearson correlation was used as similarity measurement, and generating a total of 4 gene clusters (Figure 2). To gain insight into the biological processes involved, each cluster was then subjected to GO (gene ontology) analysis as included in the Metacore package (www.genego.com).

Cluster I is identified by genes that are up-regulated by both stimuli, and as a result of meta-analysis the most overrepresented category was circadian rhythm, that is represented by Per1, Per2,

Per3 and Nr1d1. Circadian variation has been associated with inflammatory response as it has been described in patients affected by asthma [31][15], therefore such an unexpected result could actually fit with the induction of inflammation in our model. The second category represented in terms of relevance is unfolding protein response, confirming previous published data on cadmium as an inducer of endoplasmic stress and of unfolding protein response [32].

Cluster II contains genes that are upregulated only by cadmium, the main categories enriched in this cluster are inflammation, blood coagulation and phagocytosis. This is probably due to the cellular damage induced by Cadmium, that ultimately causes inflammation and macrophage recruitments.

Cluster III represents genes that are upregulated only by NP doped with cadmium. This group is enriched in genes involved in the inflammatory process. Neutrophils and macrophages are probably activated by damaged cells through the expression of chemokines: interestingly in this cluster CXCL9 and CXC3 are represented. It has been described that CXCL9 interaction with cells expressing CXCR3 has an important role in the recruitment of mononuclear cells [33].

Cluster IV is populated by genes that are downregulated by both stimuli. Again, we can observe here an enrichment of genes involved in circadian rhythm Arntl and Npas2. This category is also enriched in cluster I where Per1, Per2, Per3 and Nr1d1 are upregulated. This pathway will be further investigated since the data suggest that cadmium and NP doped with Cd have an assorted effect on this pathway. Moreover, genes involved in cell cycle regulation are downregulated and this can probably induce inhibition in the cell cycle processes.

The overall data show a complex regulation induced by nanoparticles doped with cadmium, only partially overlapping with cadmium treated, but modulating some relevant specific pathways.

In summary, the present preliminary results indicate that whole genome analysis of pulmonary tissue may represent a valuable approach to assess early molecular events in response to model NPs in a biological system.

In this context, one strategy in the application of microarray technology is the correlation to toxic end results. Specifically, combining the present gene expression results with biochemical and/or histopathological evaluation at the target tissues could be profitable in providing more direct linkages between a change in gene expression and the type and outcome of the toxic response. In this manner, patterns of gene expression can be used as biomarkers of exposure and as a method of identifying molecular mechanisms of toxicity to predict detrimental effects.

This technology should be further applied to verify its suitability to detect subtle effects for example in low exposure situation mimicking the real world scenario in a context of environmental persistent nanomaterials.

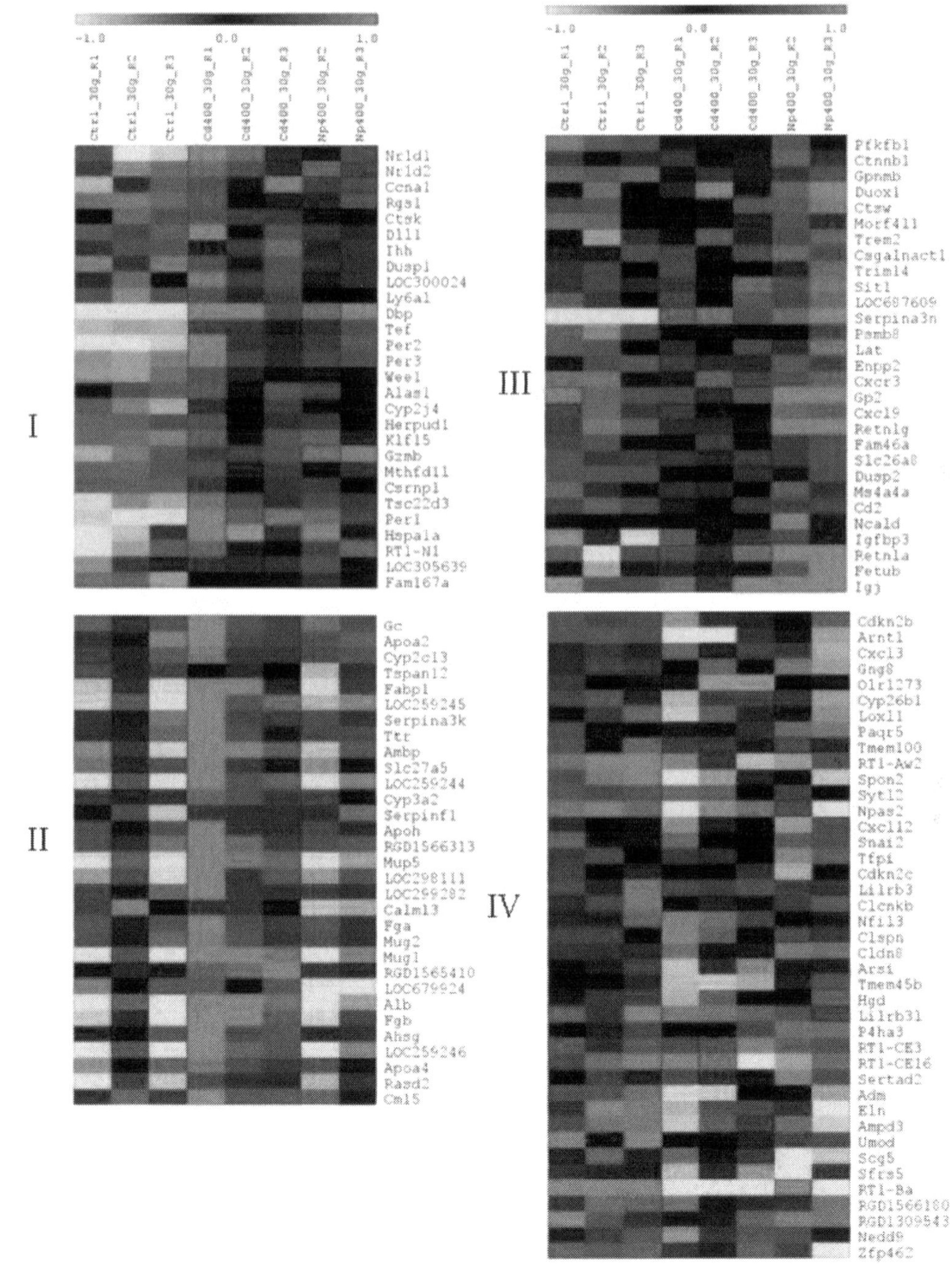

Figure 2. K-mean clustering of genes differentially expressed. Modulated genes were organized in 4 clusters by K-means clustering. Each column corresponds to a single biological replicate. Each row represents a gene, with red and green for high and low expression levels, respectively.

Table 1. Top ranked pathways from GeneGo MetaCore pathway analysis.

Cluster I

name	pValue
Neurophysiological process_Circadian rhythm	1.88324E-05
Protein folding_Response to unfolded proteins	5.36253E-05
Reproduction_Male sex differentiation	5.80875E-05
Development_Ossification and bone remodeling	8.89511E-05
Protein folding_Folding in normal condition	0.005492635
DNA damage_Checkpoint	0.00652288
Protein folding_Protein folding nucleus	0.013729699
Cell adhesion_Amyloid proteins	0.018599377
Cell cycle_G2-M	0.025183908
Cell cycle_Meiosis	0.038738286

Cluster II

name	pValue
Inflammation_IL-6 signaling	0.000712405
Transport_Bile acids transport and its regulation	0.001804934
Blood coagulation	0.009247253
Inflammation_Kallikrein-kinin system	0.023610214
Immune response_Phagocytosis	0.038935501
Regulation of metabolism_Bile acid regulation of lipid metabolism and negative FXR-dependent regulation of bile acids concentration	0.091342713
Apoptosis_Anti-Apoptosis mediated by external signals via MAPK and JAK/STAT	0.227018485
Development_Ossification and bone remodeling	0.228417535
Cell adhesion_Platelet-endothelium-leucocyte interactions	0.250494565
Apoptosis_Anti-Apoptosis mediated by external signals via PI3K/AKT	0.272001439

Cluster III

name	pValue
Cell adhesion_Leucocyte chemotaxis	3.78124E-05
Chemotaxis	0.000147037
Proliferation_Negative regulation of cell proliferation	0.007075679
Inflammation_NK cell cytotoxicity	0.036375402
Reproduction_FSH-beta signaling pathway	0.046560893
Immune response_TCR signaling	0.057686767
Reproduction_Feeding and Neurohormone signaling	0.08168305
Inflammation_IL-13 signaling pathway	0.144983618
Immune response_Th17-derived cytokines	0.17443034
Proteolysis_ECM remodeling	0.186759724

Cluster IV

name	pValue
Neurophysiological process_Circadian rhythm	0.013089908
Cell cycle_G0-G1	0.017940113
Immune response_Th17-derived cytokines	0.029541451
Cell cycle_Core	0.051104212
Chemotaxis	0.063016076
Cell adhesion_Glycoconjugates	0.06979147
Cell cycle_G1-S	0.092543148
Immune response_Antigen presentation	0.106901061
Proliferation_Negative regulation of cell proliferation	0.114877047
Cytoskeleton_Actin filaments	0.11602948

Acknowledgements
This work was supported by the Italian Ministries of Health, Research and Education - PRIN 2007. Preparation and preliminary characterization of NP with Cd were gently provided by Prof Piercarlo Mustarelli, University of Pavia, Department of Physical Chemistry.
The authors wish to acknowledge Mr. Davide Acerbi for his excellent technical assistance.

References

[1] Bharali D J, Klejbor I, Stachowiak E K, Dutta P, Roy I, Kaur N, Bergey E J, Prasad P N, and Stachowiak M K 2005 Organically modified silica nanoparticles: a nonviral vector for in vivo gene delivery and expression in the brain. *Proc. Natl. Acad. Sci. U.S.A.* **102**, 11539-11544.
[2] Gemeinhart R A , Luo, D and Saltzman, W M 2005 Cellular fate of a modular DNA delivery system mediated by silica nanoparticles. *Biotechnol. Progr.* **21**,532–537
[3] Oberdörster G, Oberdörster E and Oberdörster J 2005 Nanotoxicology: an emerging discipline evolving from studies of ultrafine particles. *Environ. Health Perspect.* **113**, 823-839.
[4] De Jong WH and Borm PJ 2008 Drug delivery and nanoparticles:applications and hazards. Int. J. *Nanomedicine.* **3(2)**,133-149.
[5] Slowing I I, Vivero-Escoto J L, Wu C W and Lin V S 2008 Mesoporous silica nanoparticles as controlled release drug delivery and gene transfection carriers. *Adv. Drug Deliv. Rev.* **60**,1278-1288.
[6] Lai C Y, Trewyn B G, Jeftinija D M, Jeftinija K, Xu S, Jeftinija S and Lin VS 2003 A mesoporous silica nanosphere-based carrier system with chemically removable CdS nanoparticle caps for stimuli-responsive controlled release of neurotransmitters and drug molecules. *J. Am. Chem. Soc.* **125**,4451-4459.
[7] Tsuji J S, Maynard A D, Howard P C, James J T, Lam C, Warheit D B and Santamaria A B 2006 Risk assessment of nanoparticles. *Toxicol. Sci.* **89**,42-50.
[8] Sayes C M , Wahi R, Kurian P A, Liu Y, West J L , Ausman K D , Warheit D B and Colvin V L 2006 Correlating nanoscale titania structure with toxicity: a cytotoxicity and inflammatory response study with human dermal fibroblasts and human lung epithelial cells. *Toxicol. Sci.* **92**,174-185.
[9] Schins R P, Duffin R, Hohr D, Knaapen A M, Shi T, Weishaupt C, Stone V, Donaldson K and Borm P J 2002 Surface modification of quartz inhibits toxicity, particle uptake, and oxidative DNA damage in human lung epithelial cells. *Chem Res Toxicol.* **15**,1166-73.
[10] Bertin G and Averbeck D 2006 Cadmium: cellular effects, modifications of biomolecules, modulation of DNA repair and genotoxic consequences (a review). *Biochimie* **88**,1549-1559.
[11] Forti E, Bulgheroni A, Cetin Y, Hartung T, Jennings P, Pfaller W and Prieto P 2010 Characterisation of Cadmium Chloride Induced Molecular and Functional Alterations in Airway Epithelial Cells. *Cell. Physiol. Biochem.* **25**,159-168.
[12] Låg M, Rodiono D, Øvrevik J, Bakke O, Schwarze P E and Refsnes M 2010 Cadmium-induced inflammatory responses in cells relevant for lung toxicity: Expression and release of cytokines in fibroblasts, epithelial cells and macrophages. *Toxicol. Lett.* **193**,252-260.
[13] Oberdörster G, Cherian M G and Baggs R B 1994 Correlation between cadmium-induced pulmonary carcinogenicity, metallothionein expression, and inflammatory processes: a species comparison. *Environ. Health Perspect.* **102 Suppl 3**,257-263.
[14] Thomas K and Sayre P 2005 Research strategies for safety evaluation of nanomaterials. Part I: Evaluating the human health implications of exposure to nanoscale materials. *Toxicol. Sci.* **87**,316-321.
[15] Smyth G K 2004 Linear models and empirical bayes methods for assessing differential expression in microarray experiments. *Stat. Appl. Genet. Mol. Biol.* **3(1)**,Article3.
[16] Yeung K Y, Haynor D R and Ruzzo W L 2001 Validating clustering for gene expression data. *Bioinformatics* **17**,309-318.

[17] Saeed A I, Bhagabati N K, Braisted J C, Liang W, Sharov V, Howe E A, Li J, Thiagarajan M, White J A and Quackenbush J 2006 TM4 microarray software suite. *Methods Enzymol.* **411**,134-193.

[18] Ekins S, Bugrim A, Brovold L, Kirillov E, Nikolsky Y, Rakhmatulin E, Sorokina S, Ryabov A, Serebryiskaya T, Melnikov A, et al. 2006 Algorithms for network analysis in systems-ADME/Tox using the MetaCore and MetaDrug platforms. *Xenobiotica* **36**,877-901.

[19] ATSDR (Agency for Toxic Substances and Disease Registry) 2008 Toxicological profile for Cadmium *(Draft for Public Comment)*. Atlanta, GA: U.S. Department of Health and Human Services, Public Health Service.

[20] Bell R R, Nonavinakere V K, and Soliman M R 2000 Intratracheal exposure of the guinea pig lung to cadmium and/or selenium: a histological evaluation. *Toxicol Lett.* **114(1-3)**,101-109.

[21] Bell R R, Soliman M M, Nonavinakere V K, Hammerbeck D M, and Early J L 2nd 1997 Selenium and cadmium induced pulmonary functional impairment and cytotoxicity. *Toxicol Lett.* **90(2-3)**,107-114.

[22] Damiano VV, Cherian P V, Frankel F R, Steeger J R, Sohn M, Oppenheim D, Weinbaum G 1990 Intraluminal fibrosis induced unilaterally by lobar instillation of CdCl2 into the rat lung. *Am. J. Pathol.* **137(4)**,883-894.

[23] Driscoll K E, Maurer J K, Poynter J, et al. 1992 Stimulation of rat alveolar macrophage fibronectin release in a cadmium chloride model of lung injury and fibrosis. *Toxicol. Appl. Pharmacol.* **116**,30-37.

[24] Driscoll K E, Costa D L, Hatch G, Henderson R, Oberdörster G, Salem H, and Schlesinger R B 2000 Intratracheal instillation as an exposure technique for the evaluation of respiratory tract toxicity: uses and limitations. *Toxicol. Sci.* **55**,24-25.

[25] Oberdörster G, Baumert H-P, and Hochrainer D 1979 The clearance of cadmium aerosols after inhalation exposure. *Am. Ind. Hyg. Assoc. J.* **40(6)**,443-450.

[26] Oberdörster G 1992 Pulmonary deposition, clearance and effects of inhaled soluble and insoluble cadmium compounds. In: Nordberg GF, Herber RFM, Alessio L, eds. Cadmium in the human environment: Toxicity and carcinogenicity. Lyon: International Agency for Research on Cancer, 189-204.

[27] Oberdörster G, Cherian M G, and Baggs R B 1994 Importance of species differences in experimental pulmonary carcinogenicity of inhaled cadmium for extrapolation to humans. *Toxicol. Lett.* **72**,339-343.

[28] Nordberg G F, Kjellström T, and Nordberg M 1985 Kinetics and metabolism. In: Friberg L, Elinder CG, Kjellström T, et al. eds. Cadmium and health: A toxicological and epidemiological appraisal. Vol. I. Exposure, dose, and metabolism. Boca Raton, FL: CRC Press, 103-178.

[29] Maynard A D, Aitken R J, Butz T, Colvin V, Donaldson K, Oberdorster G, Philbert M A, Ryan J, Seaton A, Stone V, Tinkle S S, Tran L, Walker N J, and Warheit D B 2006 Safe handling of nanotechnology. *Nature* **444**,267-269.

[30] Barik T K, Sahu B, and Swain V 2008 Nanosilica-from medicine to pest control. *Parasitol. Res.* **103**,253-258.

[31] Kelly E A, Houtman J J and Jarjour N N 2004 Inflammatory changes associated with circadian variation in pulmonary function in subjects with mild asthma. *Clin. Exp. Allergy.* **34(2)**,227-233.

[32] Masanori K 2008 Endoplasmic reticulum stress and unfolded protein response in renal pathophysiology: Janus faces. *Am. J. Physiol. Renal. Physiol.* **295**,F323-F334.

[33] Belperio J A, Keane M P, Burdick M D, Lynch J P 3rd, Zisman D A, Xue Y Y, Li K, Ardehali A, Ross D J and Strieter R M 2003 Role of CXCL9/CXCR3 chemokine biology during pathogenesis of acute lung allograft rejection. *J. Immunol.* **171(9)**,4844-4852.

Aquatic toxicity assessment of single-walled carbon nanotubes using zebrafish embryos

Huichin Pan[1], Yu-Jun Lin[1], Meng-Wei Li[1], Han-Ni Chuang[2] and Cheng-Chung Chou[2]

[1]Department of Biomedical Sciences, Chung Shan Medical University, Taichung 40201, Taiwan

[2]Department of Life Science, National Chung Cheng University, Min-Hsiung, 62102 Taiwan

Email: bioccc@ccu.edu.tw and hp29@csmu.edu.tw

Abstract. Zebrafish embryos selected at the 64-cell stage were exposed to various concentrations of amide functionalized single-walled carbon nanotubes (SWCNTs) ranging from 1 to 10 μg/ml dissolved in 1% Pluronic F-68 (a cell culture grade surfactant), and the development of embryos was examined from 24 to 120 hours post fertilization (hpf). Incubation of embryos in 1% F-68 did not induce overt abnormal phenotype as compared to the wild-type; neither did it cause significant mortality during the exposure period. Generally, there was a slight developmental delay in larvae treated with SWCNTs of 5 μg/ml or above. Only larvae exposed to ≥ 5 μg/ml SWCNTs showed significantly reduced survival rates. About 50% of the embryos exposed to 5 μg/ml showed abnormal phenotypes at 24 hpf as compared to the control group. As development proceeds to 120 hpf, more embryos displayed defective morphology. A slight hatching delay was observed in embryos exposed to concentrations above 5 μg/ml. There was a general reduction of body axes, including narrowed somite and shortened yolk stalk. In addition, pigmentation in the ventral trunk area was less than that observed in control group. The body lengths of the exposed embryos were decreased significantly at 48 hpf (3.11 mm in control vs. 3.00 mm in SWCNTs-exposed embryos). However, exposure to SWCNTs did not affect the number of somites. Other features that were noticed in the SWCNTs-exposed embryos included edema and shrinkage and blebbling of the epidermal lining. Most of these observed phenotypes persisted from 48 hpf through 120 hpf. Overall, the aforementioned results indicate that soluble amide-functionalized SWCNTs are toxic to zebrafish embryos at a minimum concentration of 5 μg/ml.

1. Introduction

Carbon nanotubes (CNTs) are one of the most widely explored nanomaterials with great potential applications in electronics, chemistry and biomedicine [1]. With the increasing demand of the nanomaterial in industry, a growing public concern about their potential health risks is emerging dramatically [2,3]. CNTs seem to have unusual toxicity properties due to their structural resemblance to asbestos. Many studies have indicated that, *in-vitro*, CNTs exhibit substantial cytotoxicity, including the induction of oxidative stress, the inhibition of cellular proliferation and the induction of apoptosis/necrosis [5,6]. The toxic effects are also observed in exposed mice and rats [7,8]. Raw CNTs are extremely hydrophobic and can form large agglomerates in solutions. Functionalization of CNTs with different chemical molecules could enhance their solubility, but may thus result in differential cytotoxicity and biodistribution in the body than pristine ones. For example, intravenous administration of diethylentriaminepentaacetic (DTPA) functionalized single-walled CNTs (SWCNTs) into mice followed by indium radioactivity tracing indicated that they are not retained in liver or spleen, but are rapidly cleared from systemic blood circulation through the renal excretion route [9]. Taken together, CNTs may produce toxicity and the toxic effects could be influenced by surface modifications.

CNTs may be possibly released into the environment during the processes of manufacturing, modification or use, which could lead to noxious interactions with biological systems to generate toxicity [10]. However, studies investigating the impact of CNTs on the environment are still limited. A recent study showed that raw SWCNTs induced slight hatching delay in zebrafish embryos [11], indicating that SWCNTs have potential toxic effects on aquatic environment. Zebrafish has been widely used in developmental and genetic research; their value in toxicology as well as drug discovery has also been recognized [12]. With the aid of large-scale and high-throughput screening method, zebrafish can provide a suitable vertebrate model for straightforward *in vivo* toxicity assessment and serve as an intermediate step between cell-based and mammalian testing. It has been reported that CNTs could be readily dispersed as an aqueous suspension by natural organic matter in the environment [12]. The dissolved CNTs may display an unexpected toxicity than pristine ones. Therefore, the aim of this study is to evaluate the potential toxic effect of soluble amide-functionalized SWCNTs on the development of zebrafish embryos.

2. Materials and methods

2.1. SWCNTs

Amide functionalized SWCNTs (Sigma) was used in this study. The SWCNTs are labelled with amide groups per 4 atoms. The average size of amide-SWCNTs is about 4-6 nm in diameter and 0.7-1 μm in length. Amide-SWCNTs powder are completely dissolved in 1% Pluronic F68 at a concentration of 1 mg/ml using sonicator for overnight and then ultrasonic cell disruptor for 5 min.

2.2. Zebrafish

Wild-type zebrafish (AB strain) embryos were raised and maintained at 28°C in a 14h:10h light:dark cycle according to standard laboratory conditions [14] and staged at 28°C according to Kimmel et al. [15]. At 2 hours post fertilization (hpf), the embryos were examined under a dissecting microscope, and those embryos that developed normally and reached the 64-cell stage were selected for subsequent experiments.

2.3. Embryo toxicity test

A stock solution of 2 mg/ml SWCNTs was prepared by dissolving SWCNTs in 2% Pluronic F-68 (sigma). The selected embryos were finally exposed to 0, 1, 3, 5, 7.5 and 10 μg/ml of SWCNTs in 1% F-68. At least 10 embryos were randomly distributed into each well of the 24-well plate, 4 replicates were set up for each concentration. Exposure time was from 2 hpf to 120 hpf. The embryos were incubated at 28°C and the solutions were replaced every 48 hours. The embryos/larva were examined under a dissecting stereomicroscope at 12, 24, 48, 72, 96, 120 hpf for morphological abnormalities, and defective rates and survival rates were recoded for each replicate at all time points.

2.4. Whole-mount in situ hybridization (WISH)

myoD gene products were amplified from recombinant plasmids (pGEM-T easy) using PCR with SP6 and T7 primers. They were then used as template to synthesize digoxigenin (DIG)-labeled riboprobes using T7 RNA polymerases (DIG-labeling kit, Roche). The specificity of antisense probes was verified by running control experiments with sense probes. WISH experiments were performed as described [16] on embryos fixed at 19 hpf. Hybridization was performed in 50% formamide and DIG-labeled probes were detected with anti-DIG alkaline phosphatase-coupled Fab fragments (1:5000) and NBT/BCIP (Roche). Embryos were cleared in graduate concentration of glycerol up to 50% before observations.

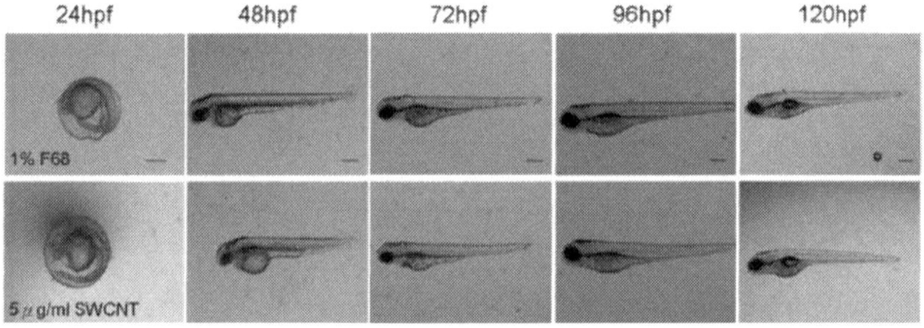

Figure 1. Morphological comparison of zebrafish embryos and larvae exposed to 1% F68 and 5 μg/ml of SWCNTs. A slight developmental delay was observed in SWCNTs treated larvae.

3. Results and discussion

3.1. Dose-dependent toxicity of SWCNTs on zebrafish embryos

Raw SWCNTs are highly insoluble. Functionalized SWCNTs, such as amide-CNTs, are more hydrophilic and dissolvable in 1% F68. Zebrafish embryos selected at the 64-cell stage were exposed to five concentrations of amide-SWCNTs (1, 3, 5, 7.5, 10 μg/ml), and the development of embryos was examined from 24 to120 hours post fertilization (hpf) (Figure 1). Generally, there was a slight developmental delay in larvae treated with 5 μg/ml of SWCNTs or above. Incubation of embryos in 1% F68 did not cause significant mortality during the exposure period (Figure 2). As the concentration of SWCNTs increased, the survival rates of exposed embryos were significantly decreased. At 24 hpf, 95% of the embryos exposed to 5 μg/ml of SWCNTs were viable whereas embryos exposed to 7.5 and 10 μg/ml SWCNTs had survival rates of 83% and 62%, respectively (Figure 2). After five days of exposure, the survival rates of zebrafish larvae were significantly decreased to 75%, 20% and 42% in 5, 7.5 and 10 μg/ml of SWCNTs (*p* values were less than 0.05 at least vs. F68 control), respectively.

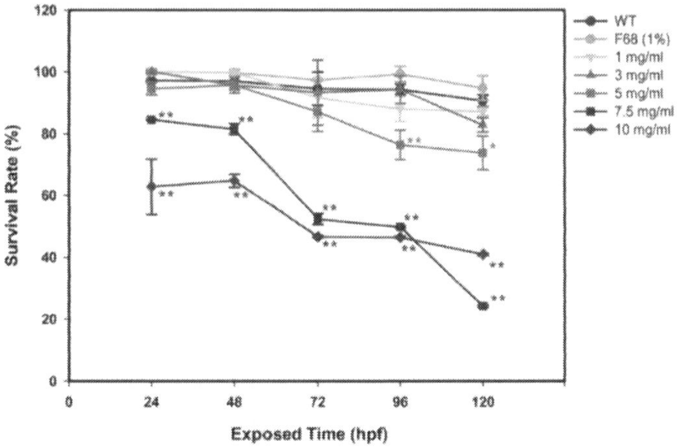

Figure 2. Survival rates of zebrafish larvae exposed to different concentrations of SWCNTs at different time points. Four replicate experiments were performed for each concentration. The data are presented as mean ± standard deviation. "*" and "**" represent $p < 0.05$ and $p < 0.01$, respectively, using the Student's *t* test compared to the control group (F68).

Among the survived embryos, embryos exposed in 1% F68 control did not induce overt abnormal phenotype as compared to the wild-type. However, the percentage of embryos with normal phenotype was significantly lower in embryos exposed to ≥ 3 μg/ml of SWCNTs at 24 hpf ($p < 0.01$), and the normal rates continued to decrease at later post fertilization (Figure 3). Notably, the lowest normal phenotype rate occurred for embryos exposed to 5 μg/ml of SWCNTs for 120 hours. For instance, about 50% of the embryos exposed to 5 μg/ml showed abnormal phenotype at 24 hpf as compared to the control group. As development proceeds to 120 hpf, more embryos displayed defective morphology. This result could be explained by the higher mortality rates in embryos exposed to higher concentrations of SWCNTs (e.g., 7.5 and 10 μg/ml).

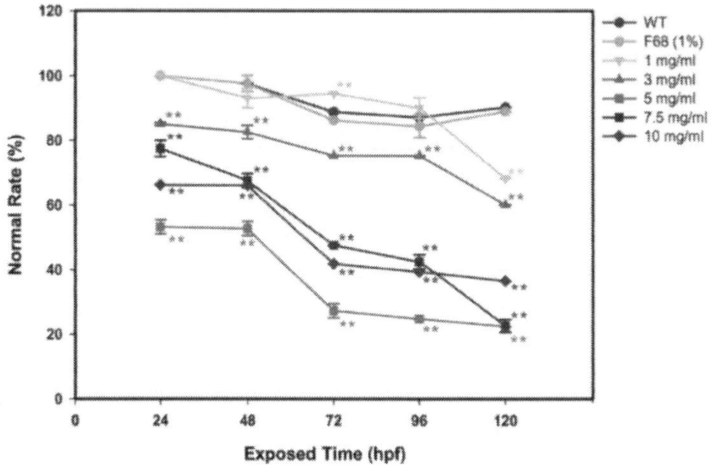

Figure 3. Normal rates of zebrafish larvae exposed to different concentrations of SWCNTs at different time points. "**" represents $p < 0.01$ compared to the control group (F68).

Previous studies have reported that exposure of zebrafish embryos to dispersed SWCNTs agglomerates in the aquatic environment delayed hatching but did not influence embryonic development and survival of exposed embryos [11]. Because the size of the pores on the embryo chorion was nanoscaled and the size of the SWCNTs agglomerates was microscaled, the authors suggested that the chorion of zebrafish embryos was an effective barrier for protection from SWCNTs agglomerates. In our study, the surfaces of SWCNTs were functionalized with amide groups to increase their hydrophilicity. We found that there was an increased dissolution for amide-SWCNTs in aquatic environment, which also correlates with an increased toxicity. Although a dose-dependent toxicity was observed on the survival of embryos exposed to 1-10 μg/ml of SWCNTs, the maximal defective rate was observed when embryos were exposed to 5 μg/ml of SWCNTs for 120 hours. This is likely due to a reduced dissolution of SWCNTs at higher concentrations because agglomerates of larger sizes were found in solutions of concentration 7.5 μg/ml or higher. Results from our study showed a much higher toxicity of SWCNTs than the previous one (toxic effects ranging from 5 to 10 μg/ml in this study vs. minimal toxic effects ranging from 20 to 360 μg/ml for the previous study), indicating that the toxic response of zebrafish embryos with chorion is dependent on the degree of dissolution of SWCNTs.

3.2. Phenotypes of zebrafish embryos exposed to SWCNTs

Under normal conditions (28.5°C in egg water), zebrafish embryos complete the rapid morphogenesis of primary organ systems and start to hatch out of the chorion from 48 hpf. We observed several developmental features, including somite formation, eye development, heart beat, blood circulation, pigmentation in head-body and in tail, pectoral fin formation, hatching and protruding mouth, of the treated-embryos at different time points in comparison to the control group. Some of them are shown in Table 1. The abnormal phenotypes in embryos exposed to 5 μg/ml of SWCNTs were most obvious at 48 hpf.

Table 1. Observed phenotype changes in the developmental features of zebrafishs treated with 5 μg/ml of SWCNTs at different exposure times. The abnormal phenotypes in embryos were most obvious at 48 hpf. "*" indicates $p < 0.05$ from the control group (1% F68).

Phenotype	Exposuure Time	WT	F68 (1%)	SWCNTs (5 μg/ml)
Hatching	48 hpf	2/39 (5.1%)	8/40 (20%)	8/37 (21.6%)
	72 hpf	36/38 (94.7%)	39/40 (97.5%)	32/36 (88.9%)*
Pigmentation	48 hpf	31/39 (79.5%)	20/40 (50%)	10/37 (27.0%)*
	72 hpf	35/38 (92.1%)	38/40 (95%)	34/36 (94.4%)
Edema	72 hpf	2/38 (5.3%)	1/40 (2.5%)	6/36 (16.7%)*
Yolk stalk shrinkage	48 hpf	4/38 (10.5%)	6/40 (15%)	11/36 (30.6%)*

There was a general reduction of body axes, including narrowed somite and shortened yolk stalk (Figure 4). In addition, pigmentation in the ventral trunk area was less than that observed in control group (Figure 4). The body lengths of the exposed embryos were decreased significantly at 48 hpf (3.11 mm in control v.s. 3.00 mm in SWCNTs-exposed embryos) (Figure 5). However, exposure to SWCNTs did not affect the number of somites. Other features that were noticed in the SWCNTs-exposed embryos included shrinkage and blebbling of the epidermal lining, hatching delay and edema (Table 1). Most of these observed phenotypes persisted from 48 hpf through 120 hpf.

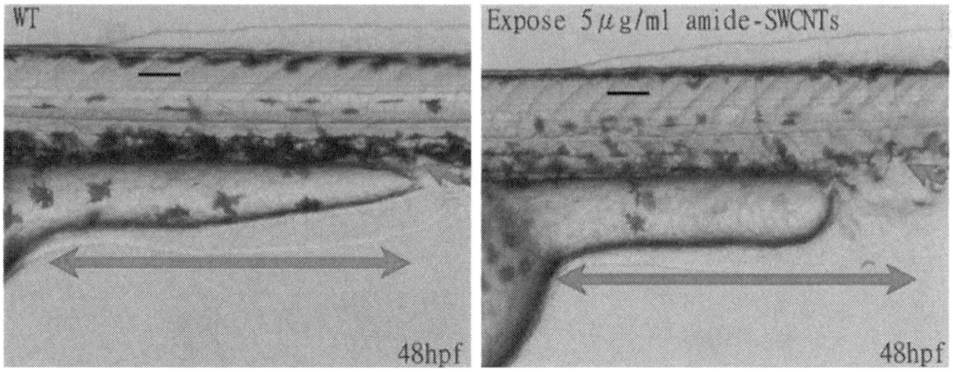

Figure 4. Comparison of morphology between the wild-type and SWCNTs-exposed embryos at 48 hpf. Note that the lengths of individual somites (indicated by the black bars) and of the yolk stalks (indicated by red lines with double arrows) were reduced in embryos exposed to 5 μg/ml of SWCNTs. Also there was significantly less pigmentation along the ventral trunk area (red arrows).

Although the body lengths and individual somite widths were reduced at 48 hpf, the analysis of the muscle differentiation markers myoD and myogenin by RT-PCR revealed that both genes were expressed in SWCNTs-exposed and control embryos at comparable levels (Figure 6). The whole mount in situ hybridization of embryos exposed to 5μg/ml of SWCNTs did not also indicate a clear difference in the organization and morphology of somites from controls, suggesting that SWCNTs did not affect the formation of somites.

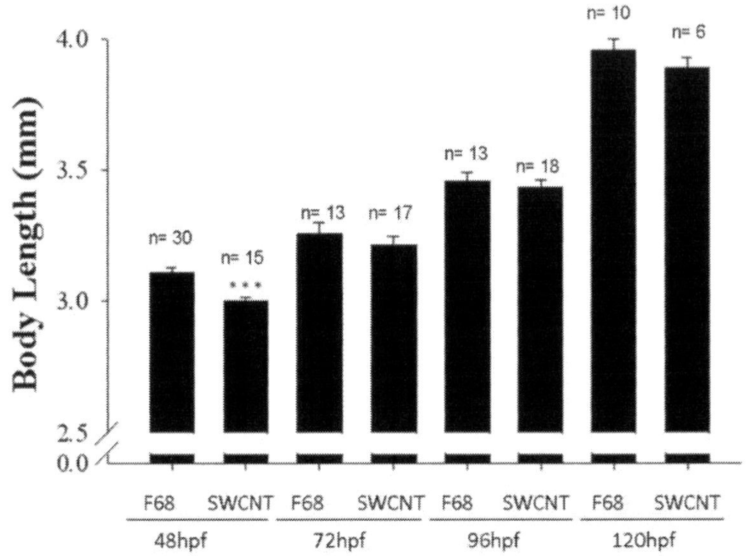

Figure 5. The body lengths of zebrafish embryos exposed to 5 μg/ml of SWCNTs dissolved in 1% F68 at different time points. The body lengths were significantly reduced at 48 hpf in SWCNTs-exposed embryos as compared to the control group.

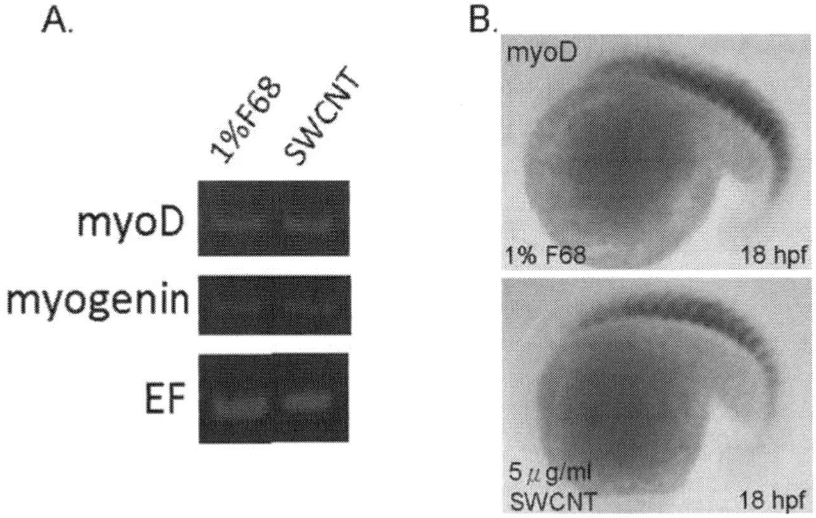

Figure 6. Expression of myoD and myogenin in control and SWCNTs-exposed embryos. (A) RT-PCR analysis of RNA isolated from embryos exposed to 1% F68 control or 5 μg/ml of SWCNTs at 48 hpf. EF, elongation factor, was amplified as a loading control. (B) Whole-mount in situ hybridization of embryos exposed to 1% F68 or 5 μg/ml of SWCNTs at 18 hpf.

4. Conclusion

In summary, we examined the effect of different concentrations of amide-SWCNTs on zebrafish embryos and found that solubilized SWCNTs could induce toxicity in zebrafish embryos by reducing their survival and normal morphology rates. The present study may help define a no observable effect concentration for SWCNTs in the developing zebrafish embryos and thus provide a better understanding of their use in potential impacts on aquatic environment.

References

[1] Klumpp C, Kostarelos K, Prato M and Bianco A 2006 *Biochimica et biophysica acta* **1758** 404
[2] Cheng J, Chan C M, Veca L M, Poon W L, Chan P K, Qu L, Sun Y P and Cheng S H 2009 *Toxicol Appl Pharmacol* **235** 216
[3] Colvin V L 2003 *Nat Biotechnol* **21** 1166
[4] Shi Kam N W, Jessop T C, Wender P A and Dai H 2004 *J Am Chem Soc* **126** 6850
[5] Ding L, Stilwell J, Zhang T, Elboudwarej O, Jiang H, Selegue J P, Cooke P A, Gray J W and Chen F F 2005 *Nano Lett* **5** 2448
[6] Jia G, Wang H, Yan L, Wang X, Pei R, Yan T, Zhao Y and Guo X 2005 *EnViron. Sci. Technol* **39** 1378
[7] Lam C W, James J T, McCluskey R and Hunter R L 2004 *Toxicol. Sci* **77** 126
[8] Elgrabli D, Floriani M, Abella-Gallart S, Meunier L, Gamez C, Delalain P, Rogerieux F, Boczkowski J and Lacroix G 2008 *Part Fibre Toxicol* **5** 20
[9] Singh R, Pantarotto D, Lacerda L, Pastorin G, Klumpp C, Prato M, Bianco A and Kostarelos K 2006 *Proc Natl Acad Sci U S A* **103** 3357
[10] Helland A, Wick P, Koehler A, Schmid K and Som C 2007 *Environ Health Perspect* **115** 1125
[11] Cheng J, Flahaut E and Cheng S H 2007 *Environ Toxicol Chem* **26** 708
[12] Hyung H, Fortner J D, Hughes J B and Lim J-H 2007 *Environ Sci Technol* **41** 179
[13] Heiden T C, Dengler E, Kao W J, Heideman W and Peterson R E 2007 *Toxicology and applied pharmacology* **225** 70
[14] Westerfield M 2000 The zebrafish book. A guide for the laboratory use of zebrafish (Danio rerio) 4th edn Eugene, University of Oregon Press
[15] Kimmel C B, Ballard W W, Kimmel S R, Ullmann B and Schilling T F 1995 *Dev Dyn* **203** 253
[16] Thisse C and Thisse B 2008 *Nature protocols* **3** 59

Nanoparticles in treatment of thermal injured rats: Is it safe?

P S Melo[1,2], P D Marcato[3], S C Huber[2,3], I R Ferreira[1,2], L B de Paula[3], A B A Almeida[4], N Durán[3,5], S Torsoni[6], A B Seabra[7] and O L Alves[3]

[1]Biochemistry Department, Universidade Estadual de Campinas (UNICAMP), SP, Brazil

[2]VERIS, Campinas, SP, Brazil

[3] Chemistry Institute, UNICAMP, SP, Brazil

[4] SOBRAPAR, Campinas, SP, Brazil

[5] Universidade Federal do ABC (UFABC), Santo André, SP, Brazil,

[6] FCA, UNICAMP, SP, Brazil

[7] Universidade Federal de São Paulo, SP (UNIFESP), Diadema, SP, Brazil

E-mail: pmelo@unicamp.br

Abstract. The aim of this study was to assess whether thermal trauma induced oxidative stress altered the balance between oxidant and antioxidant systems in the blood of burn wound rats in the absence and presence of silver nanoparticles and S-nitrosoglutathione, GSNO. Free silver nanoparticles, free GSNO and silver nanoparticles + GSNO had no cytotoxic effects. Under anesthesia, the shaved dorsum of the rats was exposed to 90°C (burn group) water bath. Studied compounds were administered topically immediately and at 28 days after the burn injury, four times a day. Silver nanoparticles and silver nanoparticles + GSNO were no toxic *in vitro* and *in vivo*. There were no significant differences in the levels of urea, creatinine, aminotransferases and hematological parameters, in control-burn groups (free silver nanoparticles) and treated-burn groups (free GSNO or silver nanoparticles + GSNO). There were no differences in lipid peroxidation and in the levels of protein carbonyls and glutathione, used as oxidative stress markers. A little inflammatory cell response, papillary dermis vascularization, fibroblasts differentiated into contractile myofibroblasts and the presence of a large amount of extracellular matrix were evidenced in treated groups following skin injury. These results indicate that silver nanoparticles and GSNO may provide an effective action on wound healing.

1. Introduction

The aim of wound healing is a speedy recovery with minimal scarring and maximal function. Wound healing proceeds through an overlapping pattern of events including coagulation, inflammation, proliferation, matrix and tissue remodelling. For this efficient and highly controlled repair process to take place, numerous cell-signalling events are required [1]. Silver has been used for centuries to prevent and treat a variety of diseases including pleurodesis, cauterisation, and healing of skin wound [1]. The use of silver in the past has been restrained by the need to produce silver as a compound, thereby increasing the potential side effects. It has been reported [1] that the overall decrease in the inflammatory response at local wound sites in the silver nanoparticles group mimic the events observed in the further confirming the detrimental effects of inflammation on healing. They conclude that silver nanoparticles can modulate local and systemic inflammatory response following burn injury by cytokine modulation.

Nanotechnology includes the production of nanoparticles, defined as particles with three dimensions of less than 100 nm. Due to the their small size, nanoparticles exhibit greater specific surface areas and surface energies, and generally increased surface reactivity than those of the conventional (larger) forms, leading to vastly different properties. For these reasons nanoparticles are being increasingly employed in a variety of consumer products, including paints, cosmetics, medicines, food and sun block lotions [2].

Silver ions have been used for centuries to prevent and treat a variety of diseases including pleurodesis, cauterization, and healing of skin wounds. The antibacterial property is important to prevent bacteria growth in the wound. Furthermore, some authors have reported possible pro-healing properties of silver [3]. For example, Tian and collaborators [1] verified that silver nanoparticles can promote wound healing and reduce scar appearance in a dose-dependent manner. Furthermore, the authors showed that silver nanoparticles act by decreasing inflammation through cytokine modulation. GSNO is an endogenous found nitric oxide (NO) donor that exhibit NO-like activities such as the inhibition of platelet adhesion, vasodilation, microbicidal actions, and wound healing [4-8]. Thus, combination of GSNO and silver nanoparticles has a great potential to promote wound healing.

The central role of liver and kidney in drug metabolism predisposes them to toxic injury. Hepatic metabolism is a mechanism that converts drugs and other compounds into products that are more easily excreted. A metabolite may have higher activity and/or greater toxicity than the original drug. Metabolites of the drugs that are excreted form kidneys may also cause cellular damage leading to kidney dysfunction.

The aim of this study was to assess whether thermal trauma induced oxidative stress altered the balance between oxidant and antioxidant systems in the blood of burn wound rats in the absence and presence of compounds (silver nanoparticles and GSNO). Previously, it was evaluated the cytotoxic effects of these substances in fibroblast V79 cell culture. Moreover, it was evaluated if these substances induce toxic effects in kidneys (urea and creatinine measurements), in liver (alanine and aspartate aminotransferases) and in hematological parameters.

2. Methods

2.1. Silver nanoparticles production

The fungal inoculates were prepared in a malt extract 2% and yeast extract 0.5% at 28°C in Petri plates. Silver reduction was carried out as described: approximately 10 g of F. oxysporum biomass was taken in a conical flask containing 100 mL of distilled water, kept for 72 h at 28°C and then the aqueous solution components were separated by filtration. To this solution, AgNO3 (10-3M) was added and kept for several hours at 28°C [7,8].

2.2. Transmission Electron Microscopy (TEM) analysis

The silver nanoparticles were characterized by Transmission Electron Microscopy (TEM) (Carl Zeiss CEM-902 transmission electron microscope, 80 KeV). For the examination of the silver particle, one drop of the particle dispersion was deposited on carbon-coated parlodion films supported in 300 mesh copper grids (Ted Pella).

2.3. GSNO synthesis

GSNO was synthesised as previously reported [9, 10]. In brief, glutathione (GSH) was nitrosated by adding equimolar amount of sodium nitrite ($NaNO_3$) in acidified aqueous solution.

2.4. Gel preparation

Carbopol 940® gel (0.1%) with propyleneglycol (5%) was prepared. Silver nanoparticles or GSNO (10-50% w/w) was added to the Carbopol gel.

2.5. Cytotoxicity of silver nanoparticles and silver nanoparticles + GSNO

The cytotoxicity of silver nanoparticles and silver nanoparticles + GSNO was assessed in a permanent lung fibroblast cell line (V79) culture derived from Chinese hamster [11]. These cells are commonly used for cytotoxicity studies [10]. V79 fibroblasts were grown as monolayers in Dulbecco's modified Eagle's medium (DMEM) supplemented with 10% fetal bovine serum, 100 IU of penicillin/ml and 100 Ag of streptomycin/ml in a humidified incubator with 5% CO_2 in air at 37 °C. The cells were plated at a density of 3 x10^4 cells/ml in 96-well plates. Forty-eight hours after cell seeding, semiconfluent cultures were exposed to silver nanoparticles and silver nanoparticles + GSNO, at different concentrations (0 - 22 μM of silver nanoparticles). The maximum concentration of silver nanoparticles + GSNO was 22 μM of silver nanoparticles + 8 μM of GSNO. The cells were exposed for 24h to the test medium with or without the compounds studied (control). Each concentration was tested in six replicates in each of three separate experiments. At the end of the incubation, two independent endpoints for cytotoxicity (MTT reduction and neutral red uptake) were evaluated.

2.6. Endpoint tests for cytotoxicity

2.6.1. *Neutral red uptake (NRU):* The NRU was measured by the method of Borefreund and Puerner [11]. Briefly, cells were washed once with PBS after removal of the culture medium. After 4 h of incubation with serum-free medium containing neutral red (50 μmol/L), the cells were washed in PBS, followed by the addition of 0.1 ml of a solution of 1% acetic acid and ethanol (50%) to each well to fix the cells and to remove the neutral red from the solution. The plates were shaken gently for 20 min on a plate shaker, and the absorbance of the solution was read at 540 nm (VersaMaxk, Tunable Microplate Reader).

2.6.2. *Methylthiazoletetrazolium (MTT) reduction:* The MTT reduction assay was performed as described by Denizot and Lang [14]. Briefly, cells were washed once with PBS before adding 0.1 ml of serum-free medium containing 0.05% of MTT salt to each well. After incubation for 5h, the culture medium was removed and 0.1 ml of ethanol was added to each well to solubilize the formazan formed. The plates were shaken gently for 10 min and the absorbance was measured at 570 nm (VersaMaxk, Tunable Microplate Reader, Molecular Devices, Co., Sunnyvale, CA, USA).

2.7. Thermal injury animal and treatment

Male Wistar rats (250-300g; CEMIB/UNICAMP, Brazil) were used. Throughout the experiment, animals were housed in the same care facility with food and water available during a 12 hour light/dark cycle. Animal care was carried out in accordance with guidelines approved by the Ethical Animal Committee (Unicamp). Major thermal injury model has been previously described as a suitable model demonstrating the response associated with thermal injury [1]. Briefly, rats were anesthetized with ketamine/xylazine intraperitoneal 2% 20 mg/Kg; their dorsal body surface hair shaved off, and placed in an appropriately sized template device such that the shaved area of the skin on the animals' back was exposed. The template device was then lowered into a hot water bath (90°C) to immerse the exposed skin area in hot water for 10 seconds. With this technique, full thickness third degree thermal injury comprising 30% of the total body surface were obtained. Sham rats were subjected to identical anesthesia and other treatments, except that they were immersed in 37°C water. After the thermal injury, five treatments were used on the dorsal skin: free Carbopol gel (positive control group, n=7, and negative control group, n=7); silver nanoparticles (1% m/v) (Ag group, n=7); free GSNO (GSNO group, n=7) (2 mM); silver nanoparticles + GSNO (1% m/v) (Ag-GSNO group, n=7) and silver sulfadiazine (1% m/v) (sulfadiazine group, n=7), four time a day. After 28 days, the animals were euthanized and its blood was collected for the biochemistry analyses and hematological parameters (toxic effects *in vivo*) and oxidative stress analysis. The burned dorsal skin was carefully removed for the histological analysis.

2.8. Toxic effects *in vivo*

2.8.1. *Biochemistry analyses:* Urea, creatinine, aspartate aminotransferase (AST) and alanine aminotransferase (ALT) values were analyzed in blood serum, using a Bioplus biochemical analyzer and LabTest reagents.

2.8.2. *Hematological parameters:* Leukocyte, erythrocyte and platelet counts and hemoglobin concentration were analyzed using KX21 system.

2.9. Oxidative stress analysis

Thiobarbituric acid reactive substances (TBARS), protein carbonyls and reduced glutathione (GSH) were measured as previously described [15].

2.10. Histological analysis

After 28 days of thermal injury and treatment the damaged skins were removed, fixed in 4% phosphate-buffered formalin, and embedded in paraffin. From the paraffin embedded tissue-blocks, 5 μm sections were serially cut and stained with hematoxylin-eosin (H&E). The sections were examined by light microscopy using a Leica FW 4500 B microscope.

3. Results and Discussion

3.1. Silver nanoparticles production

Spherical nanoparticles with size between 10-30 nm were observed by TEM analysis as shown the Figure 1.

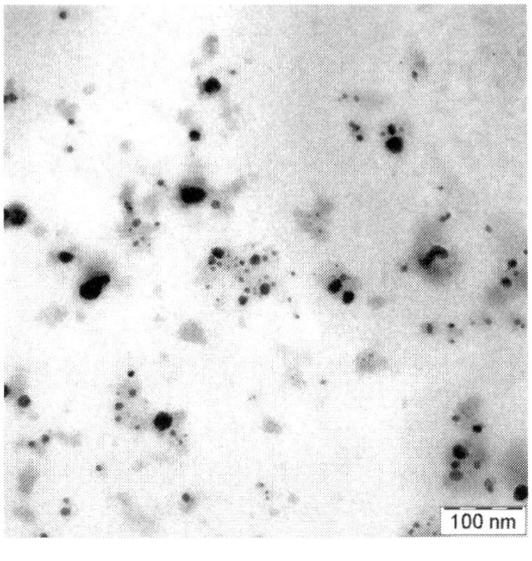

Figure 1. TEM micrograph of silver nanoparticles

3.2. Cytotoxicity of silver nanoparticles and silver nanoparticles + GSNO

Free glutathione (GSH) has shown to be no cytotoxic to fibroblast V79 cells in the lysosomal (NRU) assay and slightly cytotoxic to mitochondrial (MTT) assay (15-20% around to 10 μM – data not shown). Silver nanoparticles have shown to be no cytotoxic until 16 μM, evaluated by NRU and MTT assays (Figure 2A). Silver nanoparticles + GSNO also have shown to be no cytotoxic in the same assays (100% of viability until 22 μM concentration of silver nanoparticles and 8 μM of GSNO) (Figure 2B).

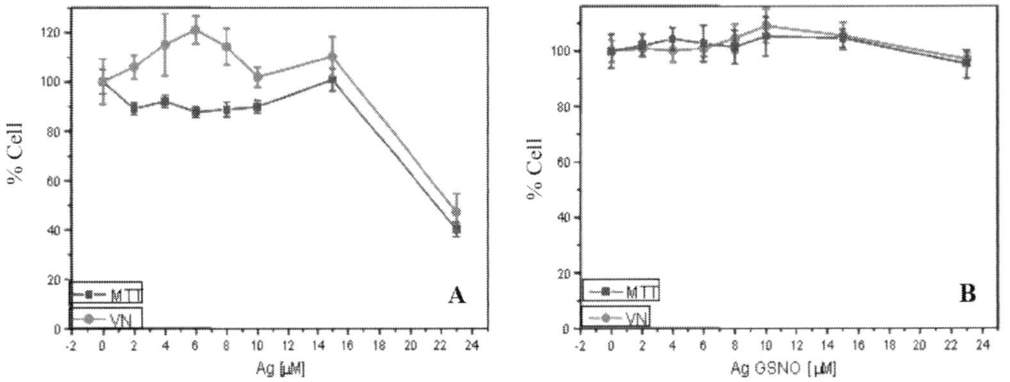

Figure 2. Cytotoxicity in V79 fibroblast cell culture in the presence of silver nanoparticles (Figure 2A) and silver nanoparticles + GSNO (Figure 2B). Concentration of silver nanoparticles 0-22 μM.

3.3. Toxic effects *in vivo*

The effects of silver nanoparticles and silver nanoparticles + GSNO on selected renal and liver function biochemistry tests (urea and creatinine concentrations, AST and ALT activities) in the animals submitted to thermal injury were presented in Table 1.

Table 1. Effects of thermal injury and/or treatment with silver nanoparticles and silver nanoparticles + GSNO *in vivo*

	Sham	Negative Control	Positive Control	Ag	GSNO	Ag-GSNO	Sulfadiazine	p*
Urea (mg/dL)	3.85 ± 0.50	4.14 ± 0.67	4.00 ± 0.37	3.83 ± 0.30	3.17 ± 0.48	5.28 ± 0.68	4.14 ± 0.45	p>0.05
Creatinine (mg/dL)	0.56 ± 0.03	0.55 ± 0.01	0.54 ± 0.02	0.45 ± 0.02	0.63 ± 0.02	0.54 ± 0.04	0.49 ± 0.01	p>0.05
AST (U/L)	60.43 ± 16.99	62.43 ± 15.82	75.71 ± 27.15	81.67 ± 25.93	96.83 ± 15.82	89.57 ± 8.24	81.57 ± 5.54	p>0.05
ALT (U/L)	38.29 ± 15.72	32.57 ± 9.40	30.86 ± 6.44	34.00 ± 16.02	42,50 ± 14.52	35.00 ± 6.51	44.71 ± 1.12	p>0.05
Leukocyte (cells)	6128.5 ± 344.8	5971.4 ± 353.7	7442.8 ± 481.0	7457.1 ± 642.4	7414.2 ± 953.3	6771.4 ± 437.9	7357.1 ± 427.5	p>0.05
Erythrocyte (cells)	7.75 ± 0.22	7.94 ± 0.12	6.98 ± 0.17	7.06 ± 0.21	7.12 ± 0.16	7.26 ± 0.25	7.70 ± 0.13	p>0.05
Platelet (cells)	779.28 ± 70.67	783.00 ± 50.12	894.14 ± 29.17	689.14 ± 92.17	744.28 ± 88.14	934.00 ± 58.04	870.71 ± 42.37	p>0.05
Hb (g/dL)	13.72 ± 0.32	14.07 ± 0.21	12.37 ± 0.22	12.55 ± 0.22	12.61 ± 0.27	12.90 ± 0.28	13.82 ± 0.17	p>0.05

*Compared to the Positive Control Group. One-Way ANOVA + Tukey Test at OriginPro 8.

Kidneys and liver are considered to be the most susceptible organs in the case of exposure and absorption of silver. The toxicity in kidneys can be evaluated by urea and creatinine measurements. The damaging effect of hepatotoxic substances on the liver is manifested by an increase of aspartate aminotranferase (AST) and alanine aminotransferase (ALT). There were no differences between the groups in relation to these tests ($p > 0.05$), meaning that the thermal injury and/or the treatment have not shown toxicity in these biomarkers. The effects of silver nanoparticles and silver nanoparticles + GSNO on the hematological parameters were also presented in Table 1. There were no differences between the groups in relation to these parameters.

3.4. Oxidative stress

The thiobarbituric acid reactive substances (TBARS), protein carbonyls and reduced glutathione (GSH) measurements have not shown any significant changes. These results indicate that the studied substances (silver nanoparticles and GSNO) have not induced an unbalance between oxidant and antioxidant systems.

3.5. Histological analysis

Animals exposed to thermal injury, treated only with Carbopol gel, presented inflammatory cells in higher amounts, few neovascularization and the extracellular matrix was poorly deposited. In Ag, GSNO, Ag-GSNO and Sulfadiazine groups, the histological analysis showed an extended area of connective tissue, where collagen and other extracellular matrix molecules were laid down. Moreover it was possible to observe a presence of myofibroblasts, cells involved with the generation of contractile force into the wound, in pappilay dermis, but only a few leukocytes infiltration. A neovascularization in papillary dermis was evident. The negative control and sham groups have shown preserved epithelium and dermis, without morphological changes (Figure 3).

The ultimate goal for wound healing is a speedy recovery with minimal scarring and maximal function. Wound healing is a complex biological process that integrates the function of a variety of cell types and repair phases, including inflammation, proliferation, reepithelialization, remodeling, and maturation [8]. Rapid wound closure involves the migration of fibroblasts/keratinocytes to the wound site, where extracellular matrix (ECM) molecules are deposited. Histological analysis of tissue in MRL/MpJ mice lesions showed a delay in critical wound-healing events that culminated in a impairment of wound contraction [16]. Silver nanoparticles can modulate local and systemic inflammatory response following burn injury by cytokine modulation, that may be important in providing a permissive environment for scarless wound repair to proceed [1]. For many years, silver sulfadiazine has been the standard treatment for burns, but some of the benefits of pure silver appear to be lost. In the present study, we observed that silver nanoparticles could improve the healing of burn wounds by accelerating the process. Histologically, a delay in main events of healing was observed when wound was not treated. However a similar healing wound response in groups treated with Ag, Ag-GSNO, GSNO and those treated with sulfadiazine could be observed.

4. Conclusions

Our results demonstrated that silver nanoparticles and silver nanoparticles + GSNO were no toxic *in vitro* and *in vivo*. There were no significant differences in urea, creatinine and aminotransferases levels and in the hematological parameters between the control-burn groups (silver nanoparticles) and treated-burn groups (free GSNO or silver nanoparticles + GSNO forms). Moreover, there were no significant differences in lipid peroxidation (TBARS), protein carbonyls and reduced glutathione (GSH) between the groups. A little inflammatory cell response, papillary dermis vascularization, fibroblasts differentiated into contractile myofibroblasts and the presence of a large amount of extracellular matrix could be evidenced in treated groups following skin injury.

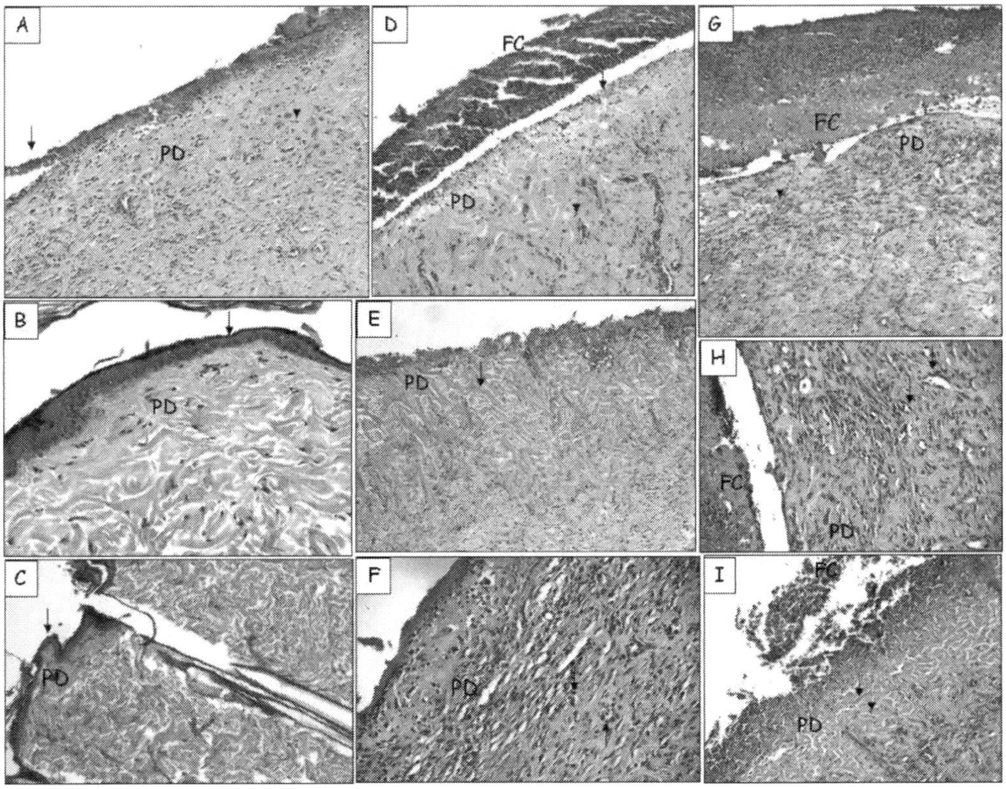

Figure 3. Photomicrography of dorsal skin from rats 28 days after thermal injury. **A**: Burned region was treated only with Carbopol gel after thermal injury (positive control). Arrowhead indicates presence of fibroblasts in hyperactive state in the papillary dermis (PD). Arrow indicates necrosis area (fibrin cap). H&E 100X; **B**: Healthy skin was treated only with Carbopol gel (negative control). It is possible to observe stratified epithelium preserved (arrow) and unchanged papillary dermis. H&E 100X; **C**: Sham. Epithelium and dermis similar to negative control. H&E 100X; **D**: Burned region was treated with silver nanoparticles after thermal injury. Arrowhead indicates presence of extracellular matrix with few fibroblasts in hyperactive state in the papillary dermis (PD) and basal cells (arrow). Tick area of fibrin cap (FC). H&E 100X; **E**: Burned region was treated with GSNO after thermal injury. Arrows indicates a large amount of extracellular matrix in papillary dermis and some myofibroblasts. H&E 100X; **F**: Burned region was treated with GSNO after thermal injury. It is possible to observe increases in dermis vascularity and some leukocyte migration (arrow). H&E 200X; **G**: Burned region was treated with silver nanoparticles + GSNO after thermal injury. Arrowhead indicates the presence of miofibroblasts in papillary dermis. Fibrin cap can be observed (FC). H&E 100X; **H**: Burned region was treated with silver nanoparticles + GSNO after thermal injury. Arrow shows increases in dermis vascularity and some leukocyte migration. Fibrin cap (FC) evident. H&E 200X; **I**: Burned region was treated with silver sulfadiazine after thermal injury. Presence of extracellular matrix (arrow) and miofibroblasts (arrowhead). H&E 100X.

Acknowledgements
Support from Brazilian Network of Nanocosmetics (MCT/CNPq), FAPESP (process number 2010/14949-4) and CNPq are acknowledged.

References

[1] Tian J, Wong K K Y, Ho C M, Lok C N, Yu W Y, Che C M, Chiu J F and Tam P K H 2007 *J. Chem. Med. Chem.* **2**, 129
[2] Teli M K, Mutalik S and Rajanikant G K 2010 *Cur. Pharm. Design.* **16**, 1882
[3] Schaller M, Laude J, Bodewaldt H, Hamm G and Korting H C 2004 *Skin Pharmacol. Physiol.* **17**, 31
[4] Seabra A B, da Silva R, de Souza G F P and de Oliveira M G 2008 *Artif. Organs* **32,** 262
[5] Seabra A B, Pankotai E, Fehér M, Somlai A, Kiss L, Bíró L, Szabó C, Kollai M, de Oliveira M G and Lacza Z 2007 *Br. J. Dermatol.* **156**, 814
[6] Seabra A B, Martins D, Simões M M S G, da Silva R, Brocchi M and de Oliveira M G 2010 *Artif. Organs* **34,** E204
[7] Durán N, Marcato P D, Alves O L, De Souza G I H, Esposito E 2005. *J. Nanobiotechnol.* **3**, 1.
[8] Durán N, Marcato P D, De Souza G I H, Alves OL, Esposito E 2007. *J. Biomed. Nanotechnol.* **3**, 203.
[9] Amadeu T P, Seabra A B, de Oliveira M G and Costa A M A 2007 *J. Eur. Acad. Dermatol. Venereol.* **21**, 629
[10] Amadeu T P, Seabra A B, de Oliveira M G and Costa A M A 2008 *J. Surg Res.* **149**, 84
[11] [Correa D H A, Melo P S, de Carvalho C A A, de Azevedo M B M, Durán N and Haun M 2005 *Eur. J. Pharmacol.* **510**, 17
[12] de Conti R, Oliveira D A, Fernandes A M A P, Melo P S, Rodriguez J A, Haun M, Castro S L, Souza-Brito A R M and Durán N 1998 *In vitro Mol. Toxicol* **11**, 153
[13] Borefreund E and Puerner J A 1984 *J. Tissue Cult. Methods* **9**, 7
[14] Denizot F and Lang R 1986 *J. Immunol. Methods,* **89**, 271
[15] Michailidis Y, Jamurta A Z, Nikolaidis M G, Fatouros I G, Koutedakis Y and Papassotiriou I 2007 *Med. Sci. Sport Exerc.* **39**, 1107
[16] Davis T A, Amare M, Naik S, Kovalchuk A L and Tadaki D 2007 *Wound Repair Regen.* **15**, 577

Assessment of dermal exposure and histopathologic changes of different sized nano-silver in healthy adult rabbits

Mohammad kazem Koohi[1], Marzie Hejazy[1], Farzad Asadi[2] and Peyman Asadian[3]

[1]Toxicology division, Basic science department, Faculty of Veterinary Medicine, University of Tehran ,Qareeb Street, Azadi Av. P.O.Box: 14155-6453, Tehran – Iran

[2]Biochemistry division, Basic science department, Faculty of Veterinary Medicine, University of Tehran, Qareeb Street, Azadi Av. P.O.Box: 14155-6453, Tehran – Iran

[3]Clinical pathology department, Faculty of Veterinary Medicine, University of Khorramabad, Iran

E-mail: mkkoohi@ut.ac.ir

Abstract. The purpose of this study is to evaluate the dermal toxicity (Irritation/Corrosion) of three sizes of nanosilver particles (10, 20 and 30 nm) during 3 min, 1 and 4 hours according to the OECD/OCDE guideline Histopathological effects in secondary organs from liver, kidney, heart, spleen and brain 14 day post dermal administration are also reported. 10 and 20 nm Ag nanoparticles treated group showed well defined dermal erythema and oedema. Histopathological findings of 10 and 20 nm (4 hours exposure) on 14-day post dermal administration showed hyperkeratosis, acanthosis, hair-filled follicles and papillomatosis in an irregular epidermis, fibrosis, hyperemia, erythema, intracellular oedema and hyalinisation of collagen in dermis of skin. Liver revealed midzonal and periacinar necrosis, portal mononuclear infiltration, liver fatty change, liver congestion and hyperemic central vein. Splenic red pulp congestion and white pulp hyperreactivity, splenic trabeculae and sinusoidal congestion and hyaline change were found in spleen. Fatty degeneration in some cardiovascular cells and subendocardial hemorrhage without inflammation was perceived. Picnotic appearance of pyramidal neurons in the brain cortex, gliosis and mild perineuronal oedema ischemic cell change and hyperemic meninges was observed in brain. Our research concluded that dermal exposure to lesser sizes of silver nanoparticles is more disastrous than greater ones.

1. Introduction

Nanotoxicology is a new branch in toxicological research which has emerged with the aim to assess risk of novel nanotechnology products. Nanotechnology seems to have a lucrative prospect for the future. However, as the outcome of unique properties of nanoparticles, and because of increasing potential for exposure to manufactured nanoparticles, there is an increasing public concern about possible side effects of exposure to nanoparticles on human, environment and ecosystem [1-4].Silver is a white and brilliant metallic element which has been used as a precious metal to make jewelry, dental alloy, silver kitchenware, conductors, in mirrors and in catalysts. Some silver compounds and silver ions are used as antiseptics, disinfectants and microbiocides, algaecide and fungicide [5], thus silver was used in cloths, silver alloy catheters, topical gels and impregnated into bandages because of its wide-spectrum antimicrobial activity [6]. However, irreversible pigmentation due to silver deposition in the skin and in the eyes, argyria or argyrosis, may develop after chronic exposure to silver. Here, a question remained unanswered whether toxic effects of nano sized silver increase as a consequence of reaching nanoscale. At the recent years, nanosilver production have had remarkably growing trend due to its potent anti-microbial activity. Nanosilver coated or embedded instruments and devices such as contraceptive devices, surgical instruments, bone prostheses, wound dressings and laundry detergents are some examples of nanosilver application [7]. Dermal exposure to nanosilver has increased through nanosilver coated / impregnated textile fibers of nano-silver manufactured clothing, nanosilver-based dressings and surgical sutures [7-9]. Skin is permeable to some specific nanoparticles and very small fraction of what is applied to the skin may penetrate. It may act as a portal of entry for localized, and possibly systemic, exposure of humans to specific nanoparticles [10]. Intradermal nanoparticles could access to systemic circulation through subcutaneous lymphatics [11] and may cause argyria like symptoms following exposure to injured skin [12]. Since rising application of nano-silver, possible toxic effects of nanosilver on human, animal and environment need to be considered.

The aim of this study is to evaluate the dermal toxicity (Irritation / Corrosion) of nano- silver particles according to the guidelines of OECD (OECD, 1992). Additionally, histopathological examination on skin, liver, kidney, heart, spleen and brain were also investigated after dermal administration of silver nanoparticles.

2. Materials and methods

Nanosilver particles (10, 20 and 30 nm) were provided from Iran Nanotechnology Initiative Council. The size of the nanosilver particles was determined by transmission electron microscopy (TEM). The purity was analyzed by X-ray fluorescence technique. The analytical results show that the purity of the nano silver particles is more than 99%.

Trace metal grade Ag powder (CAS: 7440-22-4) and $AgNO_3$ (Merck, Germany) was used in trace size.

2.1. Preparation of particle suspension

Ethanol (100%) was used as solvent for 10 nm nano-silver and mono ethylene glycol (100%) was used as solvent for 20 and 30 nm nano-silver. The concentration of nano silver in the suspension was 8000 mg/L. Solutions containing nano silver particles were stirred on vortex agitator for 15 min and before each dermal exposure with single dose of 0.5 mL/animal.

Trace metal grade Ag powder in ethanol/ mono ethylene glycol and $AgNO_3$ in ethanol was used (see Table 1).

2.2. Animals and treatment

72 healthy adult male and female albino rabbits individually housed in steel cages in 20±2°C, 50–70% relative humidity room with a 12-h light/dark cycle. For feeding, conventional laboratory diets including 2390 kcal/kg^{-1} metabolic energy, 10320 kcal/kg^{-1} digestible energy, crude protein 19.5%, crude fiber 10%, phosphor 0.69% and calcium 0.76% with an unrestricted supply of drinking water provided.

After 2 weeks acclimation, animal were divided into 6 groups (see Table 1) for treatment with 10 nm, 20 nm, 30 nm nano particles and $AgNO_3$, pure trace metal grade Ag powder Ag powder (CAS: 7440-22-4). The potential influence of the carrier/ vehicle on irritation/ corrosion of the skin was investigated in parallel with the nano particle samples. Each group was divided into 3 subgroups (n: 4) for 1 minute, 1 hour and 4 hours exposure to nano particles and carrier/vehicle. Approximately 24 hours before the test substances administration, fur was removed by closely clipping the dorsal area of the trunk of the animals according OECD 404 guideline.

The test substance was first applied in a single dose according OECD 404 guideline.

The test substance was first applied on a 6 cm^2 gauze patch and then was applied on a small area (approximately 6 cm^2) of skin. Gauze patch was held in place with non-irritating tape. The patch was loosely held in contact with the skin by means of a suitable semi-occlusive dressing for the duration of the exposure period. Gauze patch was attached to the skin in such a manner that there should be a good contact and uniform distribution of the substance on the skin. Access by the animal to the patch and ingestion or inhalation of the test substance was prevented by means of Elizabeth collar.

Untreated skin areas of the test animal served as the control. At the end of the exposure period residual test substance were removed using water.

Table 1. Details of the suspensions prepared for the different treatments applied.

Particle	Size	Solvant	Time of exposure
Nano silver	10 nm	ethanol	1 minute, 1 hour, 4 hours
Nano silver	20 nm	mono ethylene glycol	1 minute, 1 hour, 4 hours
Nano silver	30 nm	mono ethylene glycol	1 minute, 1 hour, 4 hours
$AgNO_3$	Trace	ethanol	1 minute, 1 hour, 4 hours
Pure Ag powder	Trace	ethanol	1 minute, 1 hour, 4 hours
Pure Ag powder	Trace	mono ethylene glycol	1 minute, 1 hour, 4 hours

2.3. Clinical observation and histopathological examination

The test site was examined immediately after skin treatment period (1 minute, 1 hour and 4 hours) and then after 60 minutes, 24, 48 and 72 hours and until day 14 of the last administration for signs of erythema and oedema, local toxic effects and clinical signs of toxicity. Dermal reactions were graded according to the OECD instructions. On 14 day post dermal administration, all animals were sacrificed after being anaesthetized and skin, heart, liver, spleen, brain and striated muscles under the administration site were excised and histopathological tests were performed using standard laboratory procedures. The tissues were embedded in paraffin blocks, then sliced into 5 μm in thickness and placed onto glass slides. After hematoxylin–eosin (HE) staining, the slides were observed.

3. Results

3.1. Clinical manifestation, dermal symptoms observation and histopathological findings of nano-silver particles

No mortality, abnormal behavior and obvious body weight differences were found after 2 weeks of nanoparticles dermal exposure of all groups. Dermal reactions was graded and recorded according to OECD/ OCDE 404 guideline.

3.1.1. *Dermal toxicity symptoms of 10 nm silver particles.* All 4 rabbits which were treated with 10 nm nano-silver during 3 minutes did not show any symptoms of oedema, erythema and eschar formation. However, rabbits which were treated during 1 hour showed well defined erythema during days 2-8 which decreased toward very slight erythema till to day 14 of after exposure. Slight oedema also was observed during days 2-8 after exposure. Rabbits of the subgroup which were treated during 4 hours showed definite erythema rising from moderate to severe and an increasing slight oedema with defined edges.

3.1.2. *Histopathological findings of 10 nm silver particles.* On 14-d post dermal administration, the pathological observation was performed on skin, liver, spleen, muscle, heart and brain of exposed rabbits to 10 nm Ag nano particles for 4 hours. Treated skin showed hyperkeratosis, hair-filled follicles and papilomatosis in an irregular epidermis and fibrosis, hyperemia, erythema, intracellular edema and hyalinisation of collagen in dermis (Figure 1). The histological examination of the liver revealed midzonal heptic necrosis, mononuclear infiltration in the portal area and liver fatty change (Figure 2). As shown in Figure 3, splenic red pulp congestion, splenic white pulp hyperreactivity, splenic trabecule and sinusoid congestion was found in 10 nm Ag nano particle on 14-day post dermal administration. Fatty degeneration in some cardiovascular cells and subendocardial hemorrhage was induced by 10 nm Ag treatment (Figure 4). Picnotic appearance of pyramidal neurons, ischemic cell change and hyperemic meninges was observed in histopathological examination of brain (Figure 5). No sign of necrosis, inflammation, fatty degeneration or specific degenerative changes was seen in striated muscles located under the administration site.

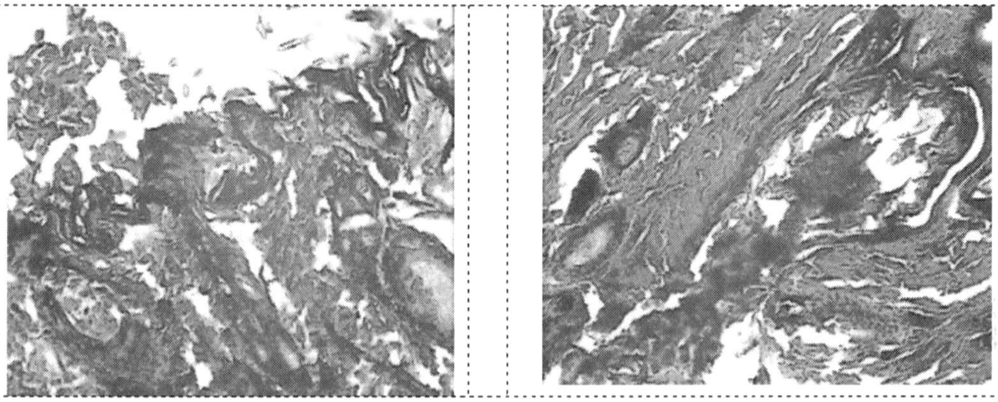

Figure 1. Hyperkeratosis, hair-filled follicles, papillomatosis in the epidermis and fibrosis, hyperemia, erythema, intracellular edema and hyalinisation of collagen in dermis of rabbits exposed to 10 nm Ag nano particle on 14-d post dermal administration (magnification = 100).

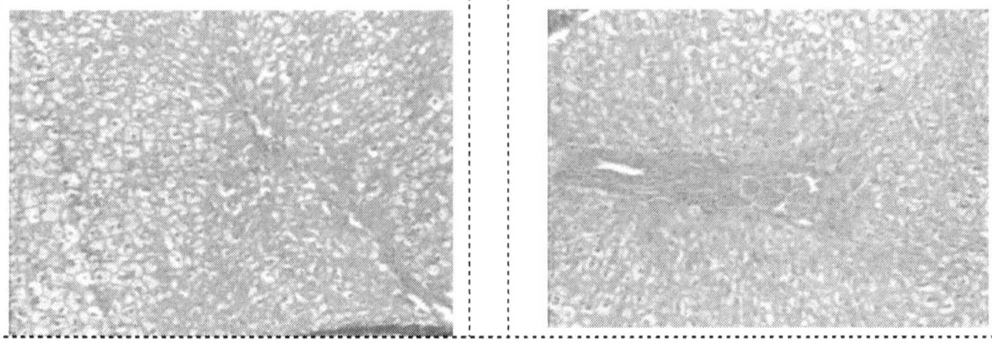

Figure 2. Oedema and degeneration, fatty change in hepatocytes of the rabbit e exposed to 10 nm Ag nano particle on 14-d post dermal administration (magnification = 40).

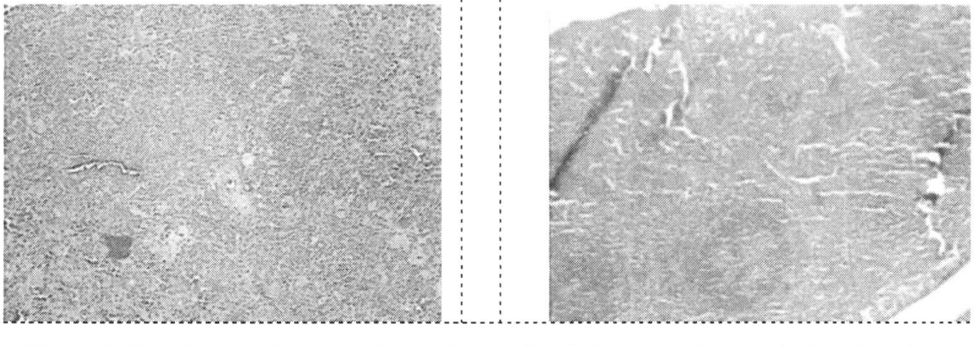

Figure 3. Splenic red pulp congestion, splenic red pulp hyperreactivity, splenic trabeculae and sinusoid congestion was found in 10 nm Ag nano particle on 14-day post dermal administration.

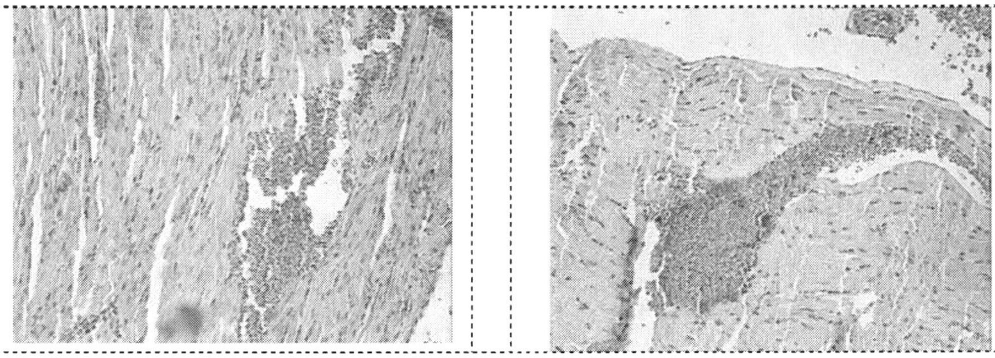

Figure 4. Fatty degeneration in cardiovascular cells, subendocardial hemorrhage found in 10 nm Ag nano particle on 14-day post dermal administration.

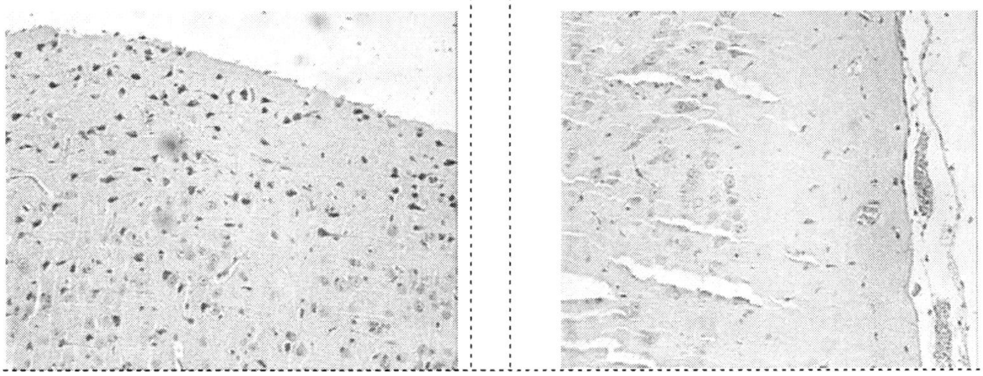

Figure 5. Pyramidal neurons necrosis, ischemic cell change and hyperemic meninges found in 10 nm Ag nano particle on 14-day post dermal administration.

3.1.3. *Dermal toxicity symptoms of 20 nm silver particles.* All treated animals with 20 nm nano-silver during 3 minutes did not show any symptoms of oedema, erythema and eschar formation. However, treated animals during 1 hour showed barely perceptible erythema during days 3- 4 and well defined erythema till end of experiment.

4 hours treated animals showed definite progressive symptoms rising from erythema to eschar formation preventing grading in appearance and beef redness in color. Very slight oedema also was seen during 2-14 days.

3.1.4. *Histopathological findings of 20 nm silver particles.* On day 14 post 4 hours dermal administration, treated skin showed hyperkeratosis, acanthosis hair-filled follicles and papillomatosis in an irregular epidermis and fibrosis, hyperemia, erythema, intracellular edema and hyalinisation of collagen in dermis (Figure 6). The histological examination of the liver revealed midzonal and periacinar hepatic necrosis. Liver fatty change, liver hyperemia/congestion and hyperemic central vein were detected with no sign of inflammation (Figure 7). As shown in Figure 8, splenic red pulp congestion, splenic white pulp hyperreactivity, splenic trabeculae, sinusoidal congestion and hyaline change was found in spleen 14-day post 4 hours dermal exposure. Fatty degeneration in some cardiovascular cells and subendocardial hemorrhage without inflammation was perceived. Picnotic (necrotic) appearance of pyramidal neurons in the brain cortex, gliosis and mild perineuronal oedema was observed in histopathological examination of brain. No encephalitis or inflammation was seen (Figure 9). No sign of necrosis, inflammation, fatty degeneration or specific degenerative changes was seen in striated muscles located under the administration site.

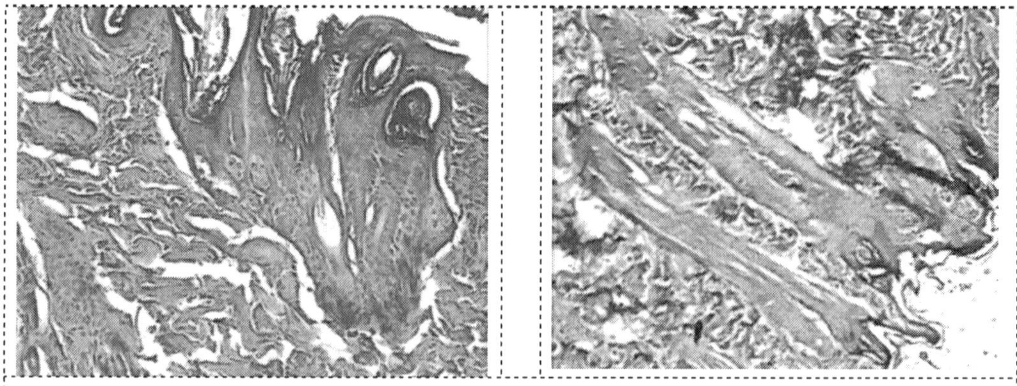

Figure 6. Histopathological photograph of dermis exposed to 20 nm Ag nano particle on 14-d post dermal administration (magnification = 40).

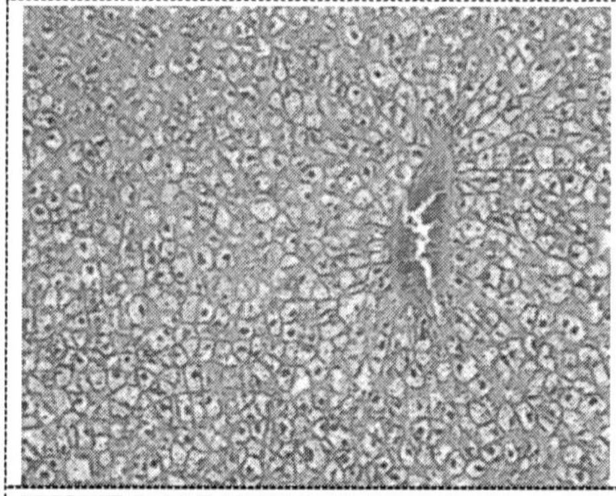

Figure 7. Histopathological photograph of liver exposed to 20 nm Ag nano particle on 14-d post dermal administration (magnification = 100).

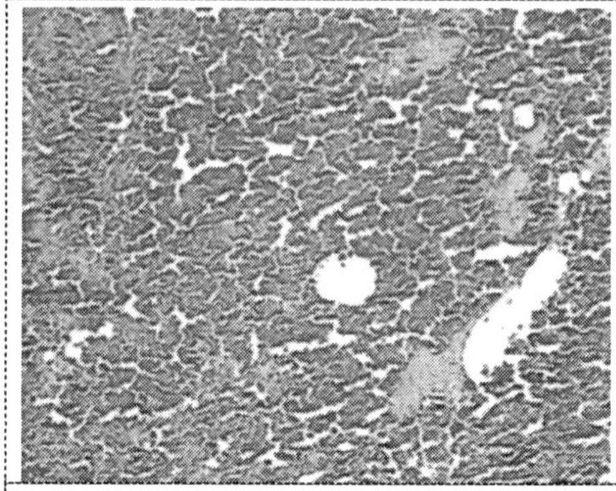

Figure 8. Histopathological photograph of spleen exposed to 20 nm Ag nano particle on 14-d post dermal administration.

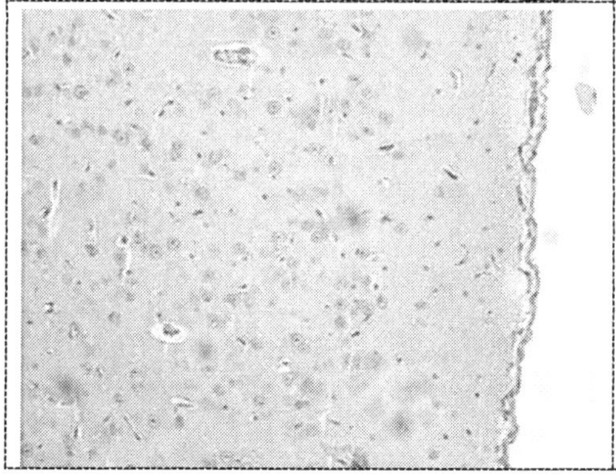

Figure 9. Histopathological photograph of brain exposed to 20 nm Ag nano particle on 14-d post dermal administration.

3.1.5. *Dermal toxicity symptoms of 30 nm silver particles.* No sign of oedema, erythema and eschar was observed in the treated rabbits with 30 nm silver particles during 3 minutes, 1 hour and 4 hours, even 14 days after dermal treatment.

3.1.6. *Histopathological findings of 30 nm silver particles.* Dermal epidermis and dermis did not show any sign of pathological lesions, inflammation (dermatitis), necrosis or hyperplasia.

3.2. Dermal toxicity symptoms and histopathological findings of Ag powder, Ag NO_3
4 hours treated rabbits with Ag NO_3 demonstrated mild erythema on exposure area. They did not show any sign of pathological lesions.
However, rabbits treated with pure trace grade Ag powder dispersed in ethanol and mono ethylene glycol did not show any sign of oedema, erythema and eschar formation. There were no abnormal pathology lesions in the skin tissue.

4. Conclusion

Silver is a relatively rare and poorly soluble element that has been used as an antibacterial agent. Silver compounds are absorbed via the oral, inhalation and both intact and damaged skin (1% of topically applied) routes of exposure [13-14] After absorption, excess silver is stored in the reticuloendothelial cells of the skin, liver, spleen, bone marrow, renal glomerulus, lymph nodes, mucous membranes, basement membranes [15-18], central nervous system (CNS) [19], blood vessels, choroid plexus, mesenteric glands and thyroid, adrenals, lungs, dura mater, bones, cartilage, muscles and gastrointestinal tract [19,20]. Argyria, gray or blue-gray discoloration of the skin and mucous membranes, has been occurred in chronic case of silver ingestion. Also, silver-containing granules, particularly in basement membranes and elastic fibers have been observed during histopathologic examination of the skin [13]. The blue color of argyria is not only due to the deposition of metallic silver, but also due to an increased deposition of melanin because of silver melanocyte-stimulating property [21].

Silver is a natural biocide, but compared with many nonessential heavy metals, silver nanoparticles show the highest antimicrobial efficacy against bacteria, viruses, and other eukaryotic microorganisms [22].Widespread usage of silver nanoparticles because of their antibacterial properties in varied products including textiles, wound dressings, hospital and laboratory gowns and other products which come in direct contact with the skin, dermal exposure must be carefully evaluated. Nanosilver particles may cause inflammatory, oxidative, genotoxic, and cytotoxic effects [21]. In our experiment, the acute dermal toxicity (acute dermal irritation/ corrosion) of 10, 20 and 30 nm silver particles was investigated according to the standard procedure (OECD Guidelines, No. 404) for the testing of chemicals. No obvious dermal toxicity was observed after exposure to $AgNO_3$, trace grade Ag, ethanol and mono ethylene glycol. However, our research concluded that dermal exposure to lesser sizes of silver nanoparticles is more disastrous than greater ones. However, we did not analyze the skin or secondary organs for determination of silver/silver compounds to support that the observed effects are caused by translocated Ag nanoparticles.

Silver nanoparticles with average particle size of 25 nm were shown to penetrate into the upper layers of the epidermis in excised human skin in static diffusion cells [23]. Gopee *et al* showed that intradermal nanoparticles could enter subcutaneous lymphatics and Tinkle *et al* [24] reported phagocytosis of the nanoparticles in the skin by macrophages and Langerhans cells and immune system disorders by these particles. Also silver nanoparticles could activate mast cells [25,26]. Monteiro-Riviere *et al* reported nanoparticles phagocytosis ability of epidermal keratinocytes and provoking inflammatory responses by nanoparticles [27]. Several studies on some types of nanosilver coated dressings showed their cytotoxicity on keratinocytes and fibroblasts which affected on their viability, proliferation and their morphology [28-29]. Nanosilver treated cells exhibited chromosome aberrations, mitotic arrest, cytoskeleton deformations, cell surface ruffling, significant cell morphology alterations, up regulation of metallothionein, heme oxygenase-1 gene [30] decreased mitochondrial function [31]. Some clinical observations also showed delayed wound healing and inhibition of wound reepithelialization after the use of certain topical silver dressings [32].

However, some *in vivo* and *in vitro* studies confirmed biocompatibility of nanosilver-coated dressings [7]. Samberg *et al* evaluated toxicity of silver nanoparticle in porcine skin and keratinocytes cell culture. They did not observe gross erythema or oedema in macroscopic observations during the entire 14-day *in vivo* study. However, microscopic observations showed focal inflammation and localization of silver nanoparticle on the surface and in the upper stratum corneum layers of the skin. A significant increase in IL-1β, IL-6, IL-8, and TNF-α concentration was observed. Skin treated with the dosing concentration of 0.34, 3.4 and 34 μg/mL of 20 nm silver nanoparticles showed slight, moderate and severe intracellular and intercellular epidermal edema and focal epidermal and dermal inflammation (spongiosis), epidermal hyperplasia, and parakeratosis. Also, the extension of the rete pegs increased into the superficial papillary layer of the dermis. They also dose-dependent decrease in viability of nano silver treated HEK cells. Silver nanoparticles may interact with proteins and enzymes with thiol groups like glutathione, thioredoxin, SOD and thioredoxin peroxidase within cells which are substantial components of the cell's antioxidant defense mechanism. In the absence of these substantial components ROS can initiate an inflammatory response and perturbation and destruction of the mitochondria take place. Then apoptogenic factors are released and apoptosis is occurred. Damage to cell membranes appears to be another part of nanosilver's mechanism of cytotoxicity [7]. Evaluation of acute dermal Irritation / Corrosion of silver nano particles will provide further context of nano particle studies and will provide method for a better production and use of nano- particles. Since staying the nanoparticles onto the skin will change their shape and activity, it is necessary to check their effect in the earliest time. Since dermal contact time should be less than the proposed time in OECD 404 or other guidelines (e.g. EPA) different criteria should be considered for the checking nanoparticles acute dermal Irritation/Corrosion effects. Monitoring time either will be depend on vehicle characterization and type of nanoparticles. However, 4 hours checking time cannot be acceptable because of nanoparticles aggregation. We propose further single oral / parenteral administration of silver nanoparticles to compare dermal administration in rabbits and verification of Ag nanoparticles across skin and into secondary organs.

References

[1] Warheit DB, Borm PJ, Hennes C, Lademann 2007 *Inhal. Toxicol.* **19** 631–643
[2] Colvin VL 2003 *Nat. Biotechnol.* **21** (10) 1166–1170.
[3] Dagani R 2003 Chem. *Eng. News* **81** 30–33.
[4] Hoet PHM, Bruske-Hohlfeld I, Salata OV 2004 *J. Nanobiotechnol.* **2** 12
[5] Chopra I 2007 *J. antimicrob. Chemotherap.* **59** 587–90
[6] Slawson RM, Van Dyke MI, Lee H, Trevors JT 1992 *Plasmid* **27** 72–9
[7] Chen X, Schluesener HJ 2008 *Toxicol. Lett.* **176** 1–12
[8] Lee KJ, Nallathamby PD, Browning LM 2007 *ASC. Nano.* **1** 133-43
[9] Vigneshwaran N, Kathe AA, Varadarajan PV, Nachane RP, Balasubramanya RHJ 2007 *Nanosci. Nanotechnol.* **7** 1893–1897
[10] Ryman-Rasmussen JP, Riviere JE, Monteiro-Riviere NA 2006.*Tox. Sci.* **91** 159–165.
[11] Gopee NV, Roberts DW, Webb P, Cozart CR, Siitonen PH, Warbritton AR, Yu WW, Colvin VL, Walker NJ, Howard, PC 2007 *Toxicol. Sci.* **98** 249–257
[12] Trop M, Novak M, Rodl S, Hellbom B, Kroell W, Goessler W 2006 *J. Trauma* **60** 648–652.
[13] ATSDR 1990 Toxicological Profile for Silver ATSDR/TP-90-24
[14] Snyder W S 1975. Report of the Task Group on Reference Man. Pergamon Press, Oxford, England,407-708. (Cited in ATSDR, 1990)
[15] Ham K N and Tange J D 1972 *Aust. J. Biol. Med. Sci.* **50** 423-434
[16] Bader K F 1966 *Food Cosmetics Toxicol.* **5** 435
[17] Anderson W A D 1966 Pathology, C. V. Mosby, Saint Louis **1** 73
[18] Olcott C T 1948 *Am. J.Path.* **24** 813-833

[19] Rungby J, Danscher G 1984 *Acta Pharmacol. Toxicol.* **55** 398-401
[20] Fung MC, Bowen DL 1997 J. Toxicol. Clin. Toxicol. 34 119-126
[21] Faust RA 1992 Oak ridge reservation environmental restoration program
[22] Samberg ME, Oldenburg SJ, Monteiro-Riviere NA 2009 *Environ. Health Perspect* **118** 407–413
[23] Larese Filon F, D'Agostin F, Bovenzi M, Crosera M, Adami G, Romano C, Maina G 2009 *Toxicology* **255** 33–37
[24] Tinkle SS, Antonini JM, Rich BA, Roberts JR, Salmen R, DePree K, Adkins EJ 2003 *Environ. Health Perspect.* **111** 1202–1208
[25] Kakurai M., Demitsu T, Umemoto N, Ohtsuki M, Nakagawa H 2003 *Br.J. Dermatol.* **148** 822
[26] Suzuki Y, Yoshimaru T, Yamashita K, Matsui T, Yamaki M, Shimizu K 2001 Biochem. *Biophys. Res. Commun.* **283** 707–714
[27] Monteiro-Riviere NA, Nemanich RJ, Inman AO, Wang YY, Riviere JE 2005 *Toxicol. Lett.* **155** 377–384
[28] Paddle-Ledinek JE, Nasa Z, Cleland HJ 2006 *Plast. Reconstr. Surg.* **117** 110S–118S
[29] Poon V K, Burd A 2004 *Burns* **30** 140–147
[30] AshaRani PV, Hande MP, Valiyaveettil S 2009 *BMC Cell Biology* **10:**65
[31] Arora S, Jain J, Rajwade JM, Paknikar KM 2009 Toxicolo. App. Pharmacol. **236** 310–318
[32] Burd A, Kwok CH, Hung SC, Chan HS, Gu H, Lam WK, Huang L 2007 Wound Repair and Regeneration **15** 94 – 104

Antibacterial activity and toxicity of silver - nanosilver versus ionic silver

L Kvitek[1], A Panacek[1], R Prucek[1], J Soukupova[1], M Vanickova[1], M Kolar[2] and R Zboril[1]

[1]Regional Centre of Advanced Technologies and Materials, Department of Physical Chemistry, Faculty of Science, Palacky University, 17. Listopadu 12, 77146 Olomouc, Czech Republic

[2]Department of Microbiology, Faculty of Medicine and Dentistry, Palacky University, Hnevotinska 3, 77520 Olomouc, Czech Republic

E-mail: ales.panacek@upol.cz

Abstract. The *in vitro* study of antibacterial activity of silver nanoparticles (NPs), prepared via modified Tollens process, revealed high antibacterial activity even at very low concentrations around several units of mg/L. These concentrations are comparable with concentrations of ionic silver revealing same antibacterial effect. However, such low concentrations of silver NPs did not show acute cytotoxicity to mammalian cells - this occurs at concentrations higher than 60 mg/L of silver, while the cytotoxic level of ionic silver is much more lower (approx. 1 mg/L). Moreover, the silver NPs exhibit lower acute ecotoxicity against the eukaryotic organisms such as *Paramecium caudatum*, *Monoraphidium* sp. and *D. melanogaster*. The silver NPs are toxic to these organisms at the concentrations higher than 30 mg/L of silver. On contrary, ionic silver retains its cytoxicity and ecotoxicity even at the concentration equal to 1 mg/L. The performed experiments demonstrate significantly lower toxicity of silver NPs against the eukaryotic organisms than against the prokaryotic organisms.

1. Introduction

Fundamental and applied physico-chemical research in the field of nanomaterials has witnessed rather great boom in the last few years. Nanomaterials attract attention due to their unique physico-chemical properties that are rooted in their diameter, eventually in their large surface area. These unique properties cannot be additionally found for the chemically identical material in its bulk form. The nanoparticles (NPs) are not nowadays only the target of scientific research but they can be continuously more and more frequently found not only in scientific laboratories, industrial applications, and chemical technologies but also as a part of common life due to their usage in commercially available products.

Silver NPs represent material, which can be used as a potential antibacterial agent in medical applications and different commercial products due to its biological activity [1,2]. Recently, silver NPs as well as various silver-based compounds containing ionic silver (Ag^+), exhibiting high antimicrobial activity, have been synthesized [3-11]. Especially silver NPs, with magnificent bactericidal activity against both gram-positive and gram-negative bacteria including multi-resistant strains, can be considered as potential disinfection agents usable in variety practical applications [3,5,10-12]. Silver NPs can effectively eliminate bacteria and yeasts even at rather low concentrations in units of mg/L [5,7]. These low concentrations are not additionally toxic against higher organisms [7,13]. Furthermore, there has not been proved any bacterial resistance against silver NPs, which however represents rather crucial problem in the case of conventionally prescribed antibiotics [14,15]. The silver NPs can be, due to its high antibacterial activity, low toxicity against higher organisms and unproved bacterial resistance, considered one of the greatest antibacterial agents for the treatment of burns [12] or for the prevention of bacterial colonization on catheters, prosthetics and dental materials [3,16-21]. Since the beginning of the 21st century, silver NPs have been widely used as antibacterial agents in many devices of everyday usage, e.g. textile coating or various cosmetic or disinfectant products. However, toxicity of silver NPs against living organisms and/or environment still remains an unsolved question.

2. Experimental Part

2.1. Synthesis, Stabilization, and Characterization of Silver Nanoparticles

The modified Tollens process was used for the preparation of silver nanoparticles (NPs) [5,22]. This process is based on the reduction of the complex cation $[Ag(NH_3)_2]^+$ by a suitable raducing substance. Two disaccharides, D-(+)-maltose monohydrate (p.a., Riedel-de Haën), D(+)-lactose (p.a., Fluka), and two monosaccharides D-glucose (p.a., Fluka) and D(+)-galactose (p.a., Fluka), were used as the reducing agents. The concentrations of the reaction components were as follows: 10^{-3} mol/L for silver nitrate (99.9 %, Tamda), $5 \cdot 10^{-3}$ mol/L for ammonia (p.a., 25% [w/w] aqueous solution, Lachema), 10^{-2} mol/L for sodium hydroxide (p.a., Lachema) and 10^{-2} mol/L for the reducing agents. Sodium hydroxide was added to the reaction mixture for the initiation of the reduction and for the purpose of the reduction completed within few minutes. pH of the reaction system was established at the value 11.5. The reduction process was continually monitored by the measurement of turbidity of the colloid system as the formation proceeded. All the measurements were performed at laboratory temperature (23 °C). Subsequent stabilization of the as-prepared silver NPs was performed by an addition of the chosen stabilizer in the final concentration of 1% (w/w). The tested stabilizers involved sodium dodecyl sulfate (SDS; > 98%, Aldrich), polyoxyethylenesorbitan monooleate (Tween 80; > 98, Aldrich), and polyvinylpyrrolidone (PVP) of the average molecular weight 360,000 (PVP 360, Aldrich).

The prepared silver NPs were characterized by means of dynamic light scattering (DLS) at Zeta Plus analyzer (Brookhaven, USA). The average values of the particle diameters were determined from the DLS measurements. Microscopic TEM images of the silver particles were obtained at JEM 2010 (Jeol, Japan) transmission electron microscope. UV/VIS absorption spectra of the silver NP dispersions were obtained at Specord S600 (Analytic Jena AG, Germany) spectrophotometer.

2.2. Antibacterial and Antifungal Assay

The antibacterial and antifungal activity of the synthesized silver NPs were tested using a standard dilution micromethod, determining the minimum inhibitory concentration (MIC) leading to inhibition of bacterial and yeast growth. The prepared silver NP dispersions were diluted 2-2048 times with 100 μL of Mueller-Hinton broth Difco (Bedston Dickinson) inoculated with the tested bacteria or yeast at a concentration of 10^5 to 10^6 CFU/mL. The obtained concentrations of silver in the dispersion ranged from 54 mg/L to 0.05 mg/L. The minimum inhibitory concentration (MIC) was read after 24 h (antibacterial assay) and after 36 h (antifungal assay) of incubation at 37 °C. The silver NP dispersions were used in the form, in which they had been prepared and modified. In addition, reference antibacterial tests with the solution of ionic silver and the used modifiers were also performed. As the standard reference strains (labeling according to the Czech Collection of Microorganisms, Czech Republic) there were used the following ones: *Staphylococcus aureus* CCM 3953, *Enterococcus faecalis* CCM 4224. Further, the following bacterial strains isolated from human clinical material at the University Hospital in Olomouc were used: *Pseudomonas aeruginosa*, methicillin-resistant *Staphylococcus aureus* (MRSA), and vancomycin-resistant *Enterococcus faecium* (VRE). Antimycotic activity was tested using *Candida albicans* I, *Candida tropicalis* and *Candida parapsilosis* strains isolated from the blood of those patients at the University Hospital Olomouc who had confirmed candida sepsis.

2.3. Cytotoxicity and Ecotoxicity Assay

In vitro cytotoxicity of the silver NPs and ionic silver was determined using the BJ cell lines (human fibroblasts). The cells, cultured in Dulbecco's Modified Eagle's Medium (supplemented with 10% fetal calf serum, 4 mM glutamine, 100 IU/mL penicillin, 100 µg/mL streptomycin) in a humidified CO_2 incubator at 37°C, were redistributed into 96-well microtiter plates at the appropriate densities for their respective cell sizes and growth rates. After 12-hour preincubation, the solution of ionic silver and dispersion of the silver NPs were added in order to get the final concentrations ranging from 0.3 mg/L to 60 mg/L. Following 72 hours of incubation, the cells were treated for 1 hour with Calcein AM, and live-cell fluorescence was measured at 485 nm/538 nm (ex/em) with the Fluoroskan Ascent microplate fluorometer (Labsystems). LC100 values, the concentrations of silver lethal to 100% of the cells, were determined from the dose-response curves. The cytotoxic effects of surfactant-stabilized silver NPs were not determined because the pure solutions of the SDS itself proved to be highly cytotoxic. As the used cell lines had no cell walls they could not adequately resist changes in the surface activity caused by the presence of the used surfactants.

Environmental toxicity of the silver NPs and ionic silver was tested also on the chosen aquatic organisms such as *Paramecium caudatum* (supplied by Faculty of Medicine, Masaryk University in Brno, Czech Republic), algae *Monoraphidium* sp. (obtained from Faculty of Science, Palacky University kindly provided by dr. P. Hasler) and also soil organism such as *Drosophila melanogaster* (wild type laboratory strain of *Drosophila melanogaster* was obtained from the Drosophila strains' collection of Masaryk University kindly provided by prof. J. Relichova). The silver NPs in the form of aqueous dispersion were added in to the cultivation medium required by the particular organism. The cultivation medium involved 5g of yeast per 1L of distilled water was used for *Paramecium caudatum* cultivation, standard cornmeal medium was used for *Drosophila melanogaster* cultivation, and *Monoraphidium* sp. was cultivated in BBM medium (Bold-Basal/Bristol Medium). The silver concentrations, which impact was tested, were in its final value in the interval from 0.3 mg/L up to 60 mg/L. The values of LC100, i.e. the values of lethal concentration leading to death of 100% of the tested organisms, were determined from the dependency of the toxic effect on the dose (dose-dependent study). The toxic effect against *Monoraphidium* sp. was determined on the decrease of the amount of chlorophyll a in the dependency of silver concentration. Chlorophyll a was extracted into acetone and its concentration was determined using spectrophotometry. The toxic effect of silver NPs against *D. melanogaster* was determined on the bases of the number of hatched adults in the dependency on the silver concentration. In the case of ecotoxicity against *P. caudatum*, the toxic index LT50 (i.e. the time in which 50% of the tested organisms died) was determined.

3. Results and discussion

3.1. Synthesis of Silver NPs Using the Modified Tollens Process

The silver NPs were prepared using the modified Tollens process, which is based on the reduction of the $[Ag(NH_3)_2]^+$ by a suitable reducing agent. Using D-maltose the silver NPs of spherical morphology and with the diameter of 25 nm were prepared. The silver NPs with the diameter of 35 nm, 44 nm, and 50 nm were also prepared using the modified Tollens process. However in this case different reducing sugars with different redox potential were used – lactose, glucose, and galactose [23]. The decreasing reducing power of the saccharides was reflected in larger particle diameter. The as-prepared silver NPs were characterized by the following methods: dynamic light scattering (DLS) and transmission electron microscopy (TEM). The images of the as-prepared silver NPs are listed in figure 1. The images only confirmed that the tested particles were of the following diameters: 25nm (D-maltose) and 50 nm (D-galactose). The increase in the average diameter of the silver NPs was indirectly confirmed by the UV/VIS absorption spectra. The absorption peak, characteristic for metal silver, is called surface Plasmon and is shifter towards higher wavelength and broadens with the growing diameter of the silver NPs. Such changes in the absorption spectrum are characteristic for the silver NPs with increasing particle diameter (figure 1).

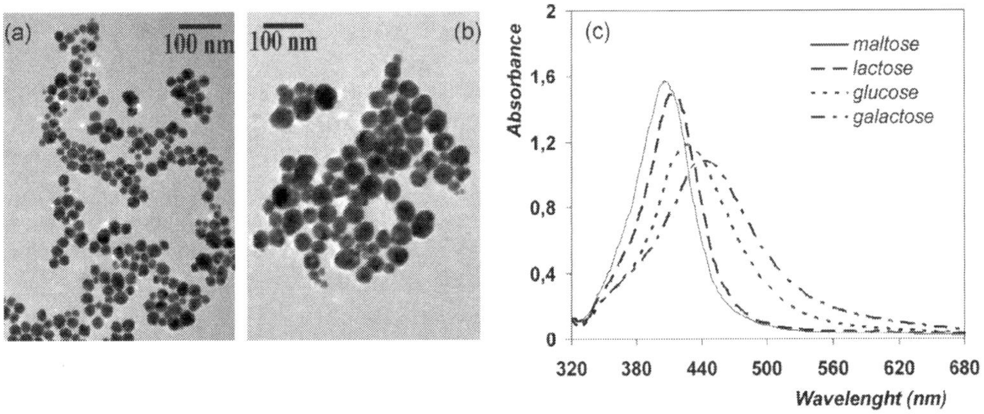

Figure 1. TEM images of silver NPs prepared by reduction via (a) maltose and (b) galactose and (c) absorption spectra of silver NPs prepared by various reducing substances.

3.2. Antibacterial and Antifungal Activity of Silver NPs

The silver NPs, with different diameters, prepared via the modified Tollens process were used for the determination of the dependency of the antibacterial effect on their diameter. The obtained values of MICs are listed in table 1. Based on these results it can be concluded that the antibacterial effect decreases with the growing particle diameter. The greatest antibacterial effect was gained with the smallest tested particles, having the diameter of 25 nm, which were prepared using D-maltose as the reducing agent. On contrary the lowest antibacterial effect was achieved with the particles prepared using D-galactose as the reducing agent. These large particles did not even reveal any antibacterial effect against *Enterococcus faecalis* CCM 4224 and *Enterococcus faecium* (VRE). The decrease in the antibacterial effect is evidently connected with the decreasing surface area, which is also reflected in reduced effective surface, which can come into contact with the present bacteria. Additionally, the 50 nm silver NPs cannot pass through the cell wall as easily as the particles having the diameter of 25 nm. Even though the antibacterial activity of the silver NPs, with the diameter equal to 25 nm, was rather high and their MICs were in the interval from 3.38 mg/L to 6.75 mg/L, in comparison with the antibacterial activity of ionic silver was slightly lower.

Table 1. Minimum inhibitory concentrations of silver (MICs), in the form of NPs with different particle diameter, against the tested bacteria.

Tested strain	Minimum inhibitory concentration (mg/L Ag)				
	Ionic silver	Size of Ag NPs/reducing agent			
		25 nm (Maltose)	35 nm (Lactose)	44 nm (Glucose)	50 nm (Galactose)
Enterococcus faecalis CCM 4224	1.69	6.75	54	-	-
Staphylococcus aureus CCM 3953	1.69	3.38	6.75	6.75	54
Pseudomonas aeruginosa	0.84	3.38	13.5	13.5	27
Staphylococcus aureus MRSA	0.84	3.38	27	27	54
Enterococcus faecium VRE	1.69	6.75	54	-	-

The silver NPs, with 25 nm diameter, were considered as the most effective and therefore all of the consequential experiments considering antifungal activity were performed just with these particles. Comparably with the antibacterial effect, also the antifungal effect of these silver NPs was reasonably high. It was proved by the obtained MICs values tested against the selected yeast. The MICs were equal to approx. 1 mg/L (table 2). The MICs values against the yeasts are lower than against the tested bacterial strains, which only confirmed that the tested yeasts are more sensitive on the presence of silver NPs. Interestingly, the antifungal activity of silver NPs and ionic silver equals. Yeasts revealed similar sensitivity against silver without any dependency on its form - nanoparticulate or ionic.

Although the antibacterial and antifungal activity of the silver NPs, with the average particle diameter equal to 25 nm, is reasonably high, it can be even enhanced. The enhancement of the aggregation stability of the silver NPs represents one of the ways. The silver NPs are highly sensitive on the presence of electrolytes, which can under the "suitable concentration" reflected in partial of complete aggregation of the originally well-dispersed particles that would be followed by sedimentation of the solid part of the dispersion. Partial aggregation proceeds also in the case when the dispersion of silver NPs is added to the bacterial cultivation medium in the course of the dilution test. The used cultivation medium involved a variety of electrolytes that are necessary for the optimal growth of the bacterial strains but on the other hand cause the unwanted aggregation of the tested particles. Partial aggregation of the silver NPs is then reflected in the existence of larger aggregates of the primarily prepared silver NPs and subsequently also in the decrease in the antibacterial effect. The problem of partial aggregation can be prevented by a suitable surface modification of the silver NPs, which can be performed by surfactants or polymers. In this study, the modification was performed by anionic (SDS) and non-ionic (Tween 80) surfactants and by polymer PVP 360. Majority of the differently modified silver NPs revealed lower MICs against strains than the unmodified silver NPs (table 2). Only in four cases the MICs of the modified silver NPs corresponded with the values determined for the unmodified ones. The antibacterial activity of the stabilized NPs was not negatively influenced by their partial aggregation, which was then reflected also in enhanced antibacterial activity of these particles. The unmodified silver NPs revealed the MICs in the interval from 3.38 to 6.75 mg/L; in the case of the modified particles the MICs values decrease a ranged from 1.69 mg/L to 3.38 mg/L. Comparable effect was achieved also in the tests evaluating the toxic effect of silver NPs against yeasts. The stabilized silver NPs showed antifungal effect in the concentration range from 0.052 mg/L to 0.84 mg/L, with respect to the tested sort of yeast and the type of the used stabilizer. On contrary, the unmodified silver NPs suppress the yeast growth in the concentrations interval from 0.42 mg/L to 1.69 mg/L.

The solutions of pure stabilizers (SDS, Tween 80, and PVP 360) did not reveal any antibacterial effect in the tested concentration which was equal to 1% (w/w). In the case of yeasts, the solution of SDS revealed partial suppressive effect. The SDS modified silver NPs can therefore exhibit better antifungal effect due to the synergic effect of the silver NPs and SDS.

Table 2. Minimum inhibitory concentrations (MICs) of the non-stabilized and stabilized silver NPs against the tested bacteria and yeasts.

Tested strain	Minimum inhibitory concentrations (mg/L Ag)				
	Ionic silver	Non-stabilized silver NPs	Stabilized silver NPs		
			SDS	Tween 80	PVP 360
Staphylococcus aureus CCM 3953	1.69	3.38	1.69	3.38	3.38
Pseudomonas aeruginosa	0.84	3.38	3.38	1.69	1.69
Staphylococcus aureus MRSA	0.84	3.38	1.69	3.38	1.69
Enterococcus faecium VRE	1.69	6.75	3.38	3.38	3.38
C. albicans I	0.42	0.42	0.052	0.1	0.1
C. parapsilosis	1.69	1.69	0.84	0.84	0.84
C. tropicalis	0.84	0.84	0.42	0.42	0.42

3.3. Cytotoxicity and Ecotoxicity of Silver NPs

Cytotoxcity of the silver NPs, with the diameter of 25 nm, in different concentrations was performed *in vitro* against human fibroblasts. The cytotoxicity was evaluated as the number of living cells using the fluorescent spectroscopy. The obtained dependency of the toxic effect of silver NPs on their concentrations can be found in figure 2a. It can be seen that the silver NPs cause death of the human fibroblasts at the concentrations higher than 60 mg/L. On contrary ionic silver cause death of all the tested fibroblasts already at the concentration equal to 2 mg/L.

Comparable results were gained also for the case of ecotoxicity tests of silver NPs against the organisms of *Monoraphidium* sp. and *D. melanogaster*. The silver NPs cause death of all the tested organisms at the silver concentration higher than 60 mg/L (figure 2b and 2c). On contrary ionic silver revealed to be more toxic against *Monoraphidium* sp., when all of the tested cells were death at the concentration equal to 3 mg/L of silver. Ecotoxicity of ionic silver against *D. melanogaster* was not monitored. The organism of *P.caudatum* was more sensitive to silver NPs. For this organism, the silver NPs were toxic at the concentrations equal to 30 mg/L and higher. However, in comparison with ionic silver, the silver NPs were further less toxic. Ionic silver cause death of the organisms of *P. caudatum* nearly immediately even at such low concentrations as 0.4 mg/L (figure 2d).

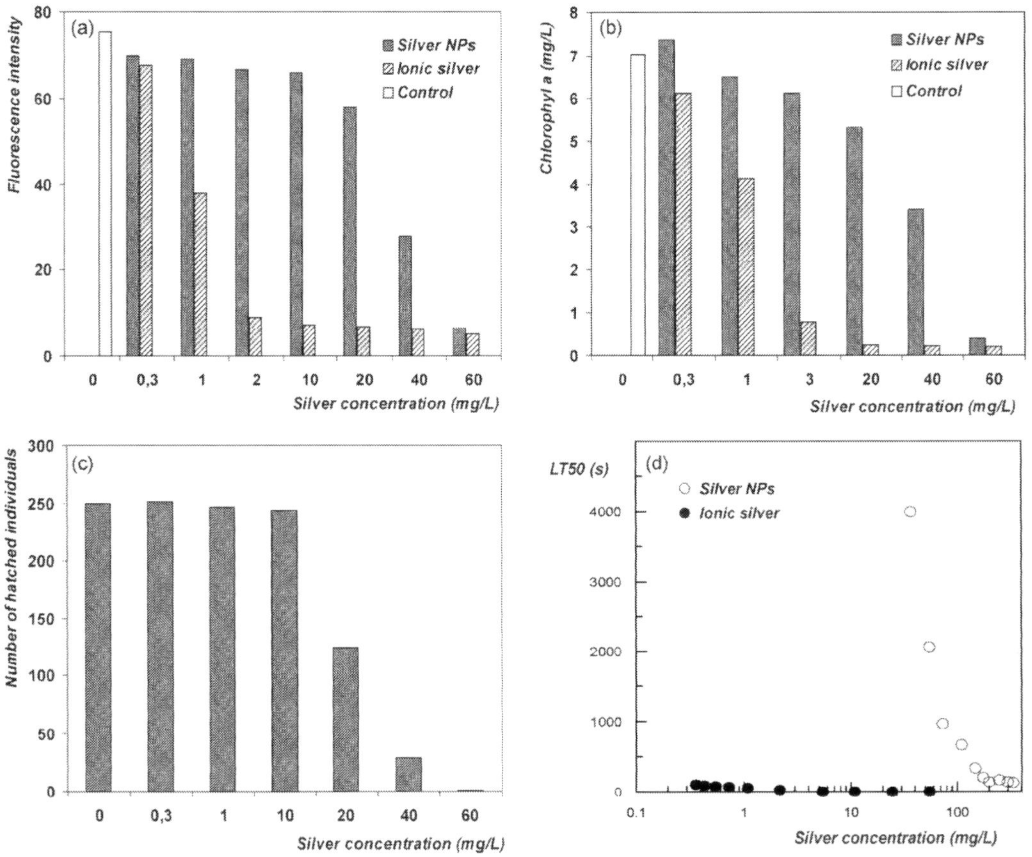

Figure 2. Toxic effect of silver NPs and ionic silver against (a) human fibroblasts, (b) *Monoraphidium* sp., (c) *D. melanogaster*, and (d) *P. caudatum*.

3.4. Comparison of Antimicrobial Activity and Toxicity of Silver NPs and Ionic Silver

The results of the antibacterial activity and toxicity of both of the silver forms, i.e. ionic and nanoparticulate, intermediate comparison of concentrations of silver leading to growth inhibition of pathogenic organisms with the concentrations evoking toxic effect against higher organisms (table 3). Due to this comparison it is possible to pronounce, which of both of the silver forms is more effective against the pathogenic organisms with respect to the low cytotoxicity and ecotoxicity. The silver NPs suppress the bacterial and yeast growth at the concentration interval from 1 mg/L to 3 mg/L with respect to the tested microorganism and to the applied modifier. Ionic silver inhibits the bacterial and yeast growth at the concentration approx. equal to 1 mg/L, which is the concentration comparable with the inhibition concentration of silver NPs. However, the cytotoxic (against human fibroblasts) and ecotoxic (against *Monoraphidium* sp., *D. melanogaster*, and *P. caudatum*) concentrations of ionic silver leading to the death of all of the tested organisms are significantly lower than the toxic concentrations of silver in the form of NPs. Therefore it can be concluded that silver NPs represent more significant antimicrobial agent than silver in its ionic form because silver NPs suppress the growth of microorganisms at such low concentrations that are not toxic against human cells and that do not also represent any risk for the environment.

Table 3. Comparison of antimicrobial activity and toxicity of silver NPs and ionic silver.

	Antimicrobial effect (MICs; mg/L)		Cytotoxicity and ecotoxicity (LC100; mg/L)			
	bacteria	yeasts	Human fibroblasts	*Monoraphidium sp.*	*Drosophila melanogaster*	*Paramecium caudatum*
Silver NPs	1-3*	1	60	60	60	30
Ionic silver	1	1	2	3	-	0.4

*MICs of modified silver NPs

4. Conclusion

Silver NPs has proved to be further more suitable antibacterial agent than the substances on the bases of ionic silver especially because of its significantly lower toxicity. The silver NPs, comparably with ionic silver, suppress bacterial and yeast growth at comparable concentrations equal to approx. 1 mg/L - 3 mg/L. Nevertheless, in the case of ionic silver these concentrations are also toxic against eukaryotic organisms including human cells. On contrary silver NPs effectively suppress bacterial and yeast growth at these concentrations that are not toxic to the tested human fibroblasts and also against *P. caudatum*, *D. melanogaster* or *Monoraphidium* sp. Based on these findings, the silver NPs do not represent any risk for human beings, when used in medical applications and commercially available products, but only under the condition that the silver concentration is retained at units of mg/L, which is sufficient for the suppression of bacterial and yeast growth.

Acknowledgment

The authors gratefully acknowledge the supports by the Operational Program Research and Development for Innovations – European Social Fund (project CZ.1.05/2.1.00/03.0058). The authors also gratefully acknowledge the supports by the Ministry of Education, Youth and Sports of the Czech Republic (MSM6198959218, MSM6198959201, MSM6198959223) and by the Czech Science Foundation (GAP304/10/1316).

References

[1] Chen X and Schluesener H J 2008 *Toxicol. Lett.* **176** 1-12
[2] Rai M, Yadav A and Gade A 2009 *Biotechnol. Adv.* **27** 76-83
[3] Alt V, Bechert T, Steinrucke P, Wagener M, Seidel P, Dingeldein E, Domann E and

Schnettler R 2004 *Biomaterials* **25** 4383-4391
[4] Lok C N, Ho C M, Chen R, He Q Y, Yu W Y, Sun H Z, Tam P K H, Chiu J F and Che C M 2006 *J. Proteome Res.* **5** 916-924
[5] Panacek A, Kvitek L, Prucek R, Kolar M, Vecerova R, Pizurova N, Sharma V K, Nevecna T and Zboril R 2006 *J. Phys. Chem. B* **110** 16248-16253
[6] Pal S, Tak Y K and Song J M 2007 *Appl. Environ. Microb.* **73** 1712-1720
[7] Panacek A, Kolar M, Vecerova R, Prucek R, Soukupova J, Krystof V, Hamal P, Zboril R and Kvitek L 2009 *Biomaterials* **30** 6333-6340
[8] Sharma V K, Yngard R A and Lin Y 2009 *Adv. Colloid Interface Sci.* **145** 83-96
[9] Dallas P, Tucek J, Jancik D, Kolar M, Panacek A and Zboril R 2010 *Adv. Funct. Mater.* **20** 2347-2354
[10] Baker C, Pradhan A, Pakstis L, Pochan D J and Shah S I 2005 *J. Nanosci. Nanotechnol.* **5** 244-249
[11] Morones J R, Elechiguerra J L, Camacho A, Holt K, Kouri J B, Ramirez J T and Yacaman M J 2005 *Nanotechnology* **16** 2346-2353
[12] Ip M, Lui S L, Poon V K M, Lung I and Burd A 2006 *J. Med. Microbiol.* **55** 59-63
[13] Kvitek L, Vanickova M, Panacek A, Soukupova J, Dittrich M, Valentova E, Prucek R, Bancirova M, Milde D and Zboril R 2009 *J. Phys. Chem.C* **113** 4296-4300
[14] Neu H C 1992 *Science* **257** 1064-1073
[15] Stewart P S and Costerton J W 2001 *Lancet* **358** 135-138
[16] Gosheger G, Hardes J, Ahrens H, Streitburger A, Buerger H, Erren M, Gunsel A, Kemper F H, Winkelmann W and von Eiff C 2004 *Biomaterials* **25** 5547-5556
[17] Johnson J R, Kuskowski M A and Wilt T J 2006 Ann. Intern. Med. **144** 116-126
[18] Li Y, Leung P, Yao L, Song Q W and Newton E 2006 *J. Hosp. Infect.* **62** 58-63
[19] Shi Z L, Neoh K G, Kang E T and Wang W 2006 *Biomaterials* **27** 2440-2449
[20] Kollef M H, Afessa B, Anzueto A, Veremakis C, Kerr K M, Margolis B D, Craven D E, Roberts P R, Arroliga A C, Hubmayr R D, Restrepo M I, Auger W R and Schinner R 2008 *Jama-J.Am. Med. Assoc.* **300** 805-813
[21] Roe D, Karandikar B, Bonn-Savage N, Gibbins B and Roullet J B 2008 *J.Antimicrob. Chemoth.* **61** 869-876
[22] Kvitek L, Prucek R, Panacek A, Novotny R, Hrbac J and Zboril R 2005 *J. Mater. Chem.* **15** 1099-1105
[23] Blackburn R S and Harvey A 2004 Environ. Sci. Technol. 38 4034-4039

Toxicity of silver nanoparticles in human macrophages: uptake, intracellular distribution and cellular responses

A Haase[1], J Tentschert[1], H Jungnickel[1], P Graf[2], A Mantion[3], F Draude[4], J Plendl[5], M E Goetz[1], S Galla[4], A Mašić[6,7], A F Thuenemann[3], A Taubert[6,7], H F Arlinghaus[4] and A Luch[1]

[1]BfR — Federal Institute for Risk Assessment, Department of Product Safety, Thielallee 88-92, 14195 Berlin, Germany

[2]University of Basel, Department of Chemistry, Klingelbergstrasse 80, 4056 Basel, Switzerland

[3]BAM — Federal Institute for Materials Research and Testing, Richard-Willstätter-Strasse 11, 12489 Berlin, Germany

[4]University of Münster, Institute of Physics, Wilhelm Klemm Strasse 10, 48149 Münster, Germany

[5]Free University of Berlin, Department of Veterinary Medicine, Institute of Veterinary Anatomy, Koserstrasse 20, 14195 Berlin, Germany

[6]University of Potsdam, Institute of Chemistry, Karl- Liebknecht- Strasse 24-25, 14476 Potsdam-Golm, Germany

[7]Max-Planck-Institute of Colloids and Interfaces, Am Mühlenberg 2, 14476 Potsdam-Golm, Germany

E-mail: andrea.haase@bfr.bund.de; alexandre.mantion@bam.de

Abstract. Silver nanoparticles (SNP) are among the most commercialized nanoparticles worldwide. They can be found in many diverse products, mostly because of their antibacterial properties. Despite its widespread use only little data on possible adverse health effects exist. It is difficult to compare biological data from different studies due to the great variety in sizes, coatings or shapes of the particles. Here, we applied a novel synthesis approach to obtain SNP, which are covalently stabilized by a small peptide. This enables a tight control of both size and shape. We applied these SNP in two different sizes of 20 or 40 nm (Ag20Pep and Ag40Pep) and analyzed responses of THP-1-derived human macrophages. Similar gold nanoparticles with the same coating (Au20Pep) were used for comparison and found to be non-toxic. We assessed the cytotoxicity of particles and confirmed their cellular uptake via transmission electron microscopy and confocal Raman microscopy. Importantly a majority of the SNP could be detected as individual particles spread throughout the cells. Furthermore we studied several types of oxidative stress related responses such as induction of heme oxygenase I or formation of protein carbonyls. In summary, our data demonstrate that even low doses of SNP exerted adverse effects in human macrophages.

1. Introduction

Due to their unique optical, catalytical and disinfectant properties silver nanoparticles (SNP) gain high commercial and scientific interest. They are used for many different applications and in a wide range of different products [1,2] ranging from wound dressing, coatings of surgical instruments and prostheses [3,4] to the use in food container systems or as coating material for certain household devices such as washing machines. They are incorporated into textiles [5,6] and also added to cosmetics [7]. Besides that they are highly attractive for creation of novel and advanced functional materials.

Serious concerns about putative toxicological and environmental effects have been raised [8-10]. Recently different reports demonstrate that incorporation of SNP into organisms is possible via different routes and systemic dissemination can occur. Thus SNP are likely to reach secondary target organs. In an inhalation study in rats SNP could be detected in lung tissue, but also in liver and brain [11]. After subcutaneous injection SNP were found in kidney, liver, spleen, lung and brain [12]. In contrast to nanoscaled titanium dioxide dermal uptake has been proven for SNP [13]. Although acute toxic effects could not be detected *in vivo*, systemic distribution and accumulation in certain tissues could lead to chronic toxicity. Different *in vitro* studies could clearly demonstrate adverse effects of SNP in various types of cells including human mesenchymal stem cells [14], THP-1-derived macrophages [15], human lung fibroblasts and glioblastoma cells [16] as well as alveolar macrophages [17]. The development of oxidative stress and in particular the generation of reactive oxygen species (ROS) represents the model mostly used to explain the *in vitro* toxicity of nanoparticles[18]. Generation of ROS is also assumed to be underlying the toxicity of SNP [17,19]. This could occur via different mechanisms. First, nanoparticles could catalyze redox reactions directly on their surface leading to ROS formation. Another mechanism that might be of particular relevance for silver is based on direct interactions with mitochondrial membrane proteins, which often contain sulfur-containing amino acids. Disturbance of mitochondrial functions can lead to increased production of ROS.

SNP belong to the group of nanoparticles with moderate solubility. As consequence, silver ions are being released and thus could also contribute to toxic reactions in cells [20]. It is still an ongoing discussion whether toxicity is caused by particles as such, or by silver ion release from silver nanoparticles, or both [21]. Usually nanotoxicity studies remain to suffer from several discrepancies and irreproducibilities. SNP can be produced via different synthesis routes in a great variety of forms with different coatings, shapes and sizes. Often characterization data of those particular nanoparticles applied in a specific study are incomplete. However it is well known that each of these particle properties strongly affects the biological responses, as has been already shown for size-dependent reactivity [17,22], coating- or surface-dependent reactivity [23], or shape-dependent properties [24] of SNP. With classical synthesis approaches, such as citrate reduction, the properties of the resultant particles cannot be sufficiently controlled or modified. This usually results in preparations with broad variance in nanoparticle sizes and large polydispersity indices. Significant batch-to-batch variations are often detected as well. Thus the relationship between particle properties and toxicological responses usually remains ambiguous and structure-reactivity correlations cannot be trustfully demonstrated. In nanotoxicology it thus becomes crucial to tightly control and direct syntheses routes of particles toward desired well-defined properties and to fully characterize the resulting particles regarding their physicochemical properties prior to its application in biological systems. Recently the application of peptides as structure-directing agents gained highest interest. Certain peptides can be used to control the growth of inorganic materials [25,26] or of complex silver-peptide hybrids [27]. Here we used a novel type of SNP, which are made of a silver nanoparticle core surrounded by a small peptidic shell [28]. These particles display a narrow size distribution and possess a defined shape. The particles can be easily obtained in a reproducible way and in large quantities. In aqueous preparations they constitute stable and monodispersed suspensions, thus making them highly suitable for toxicological testing in cell culture systems. In the present study we use peptide-coated SNP in toxicity assays *in vitro* and compare the outcome with effects of corresponding gold nanoparticles of same size and coating.

2. Materials and methods

2.1. Nanoparticle preparation

Chemicals were purchased from Fluka (Buchs, Switzerland) or Bachem (Bubendorf, Switzerland) and used without further purification. All amino acids were L-isomers. The peptide sequence is shown in Figure 1. Nanoparticles were prepared according to an in-house protocol, and characterized as published [28]. The nanoparticles became available in two sizes, *i.e.*, 20 nm (Ag20Pep) and 40 nm (Ag40Pep), respectively. The peptide-coated gold nanoparticles (Au20Pep) were prepared by ligand exchange of 20 nm citrate-coated gold nanoparticles with the peptide CKK.

2.2. Cell culture

The THP-1 cell line was obtained from the German Collection of Microorganisms and Cell Cultures GmbH (DSMZ, Braunschweig, Germany). Cells were grown at 37°C with 5% CO_2 in RPMI medium supplemented with 10% fetal calf serum, 2 mM L-glutamine, 10 mM HEPES, 1 mM pyruvate, 100 U/ml penicillin and 0.1 mg/ml streptomycin. Differentiation into macrophage-like cells was performed by adding 100 ng/ml phorbol-12-myristate-13 acetate as described in the literature [35,36].

2.3. Cytotoxicity tests

Cell vitality after nanoparticle treatment was determined using WST-1 assay (Roche Applied Biosystems) according to the manufacturer's instructions with modifications to adapt for nanoparticle-treated cells. Briefly, cells were seeded in a 96-well plate with a density of 1 x 10^4 cells per well, differentiated and incubated with nanoparticles (4 replicates per concentration). After 24 or 48 h, WST-1 reagent was added to the cells, and after color reaction the resulting solution was centrifuged to remove the physically interfering nanoparticles. Finally spectrophotometric evaluation was performed. The relative viability (% viability compared to untreated cells) was calculated as mean value ± standard error of the mean (SEM) as a result of at least 3 independent experiments. The LDH assay was performed using a standard LDH assay from Promega (Mannheim, Germany) according to the manufacturer's instructions. The results shown represent mean values ± SEM of at least 3 independent experiments.

2.4. Transmission electron microscopy

TEM analysis was performed as previously described [37]. Cells in the culture dish were washed with phosphate buffered saline (PBS) and fixed by immersion with Karnovsky's fixative at 4°C. After three washing steps in 0.1 M cacodylate buffer, postfixation was performed with 2% osmium tetroxide in 0.1 M cacodylate buffer for 1 h at 4°C. After another three washing cycles in 0.1 M cacodylate buffer, cells were removed from the culture dish and centrifuged at 2000 x g for 5 min. The pellet was then coated with 1.5% agar (Merck Eurolab, Darmstadt, Germany) for 30 min at 4°C. Subsequently the agar with the attached cell layer was removed from the wells. The samples were dehydrated in an ascending ethanol series (30-100% alcohol v/v) and embedded in Epon using beem capsules (Plano, Marburg, Germany). Polymerization was carried out at 60°C for 24 h. Semithin sections (1 µm) were cut on an Ultracut E ultramicrotome (Reichert-Jung, Vienna, Austria) with a diamond knife, stained as published elsewhere [34] and analyzed by light microscopy. Ultrathin sections (60 nm) were cut with a diamond knife, mounted on copper grids (Plano, Marburg, Germany) and examined with a Zeiss 10CR electron microscope (Jena, Germany).

2.5. Confocal Raman microscopy

After fixation with 4% paraformaldehyde cells were investigated with a confocal Raman microscope (CRM300, WITec, Ulm, Germany) equipped with a piezo-scanner (P-500, Physik Instrumente,

Karlsruhe, Germany), a 60x objective, and a 532 nm Nd:YAG laser. Spectra were acquired with an air-cooled CCD detector (DU401-BV, Andor, UK) with 600 gratings/mm (UHTS 300, WITec, Germany). ScanCtrlSpectroscopyPlus (version 2.04, WITec) was used for data acquisition and processing. Power was adjusted to provide a good signal-to-noise ratio and to avoid sample destruction. Typically, less than 1 mW full beam power was applied at the sample.

2.6. Cell lysates, SDS-PAGE, immunoblot and detection of protein carbonyls

Cells were washed with PBS three times and lysed by adding a modified RIPA buffer (50 mM Tris/HCl pH 7.4; 150 mM NaCl, 1 mM EDTA, 1% Igepal, 0.25% Na-deoxycholate). Lysates were centrifuged and stored at –80°C. Protein concentrations were determined through Bradford assay (BioRad, München, Germany) and SDS-PAGE was performed according to standard protocols. SDS-PAGE gels were transferred onto nitrocellulose membranes with a semidry blotting system. For 2D gel electrophoresis we used the system consisting of an IPGPHOR 3 unit and an EttanDALTtwelf of GE Healthcare (Freiburg, Germany). Proteins were precipitated and recovered in 2D lysis buffer containing 7 M urea, 2 M thiourea and 4% chaps. For the first dimension we used 24 cm IPG strips (pH 3-10, NL) and further steps were according to manufacturer's instructions. The gels were blotted with a semi-dry system as well. For detection of protein carbonyls OxyBlot Kit (Millipore, Schwalbach, Germany) was used. Antibodies against HO-1, tubulin or actin were obtained from Abcam (Abcam, Cambridge, UK). Images were obtained with a GelDoc system (BioRad, München, Germany) and analyzed with QuantityOne software.

2.7. Analysis of the protein corona of nanoparticles

Ag20Pep nanoparticles were incubated with complete cell culture medium containing 10% serum in a final concentration of 125 µg/ml for 2h. The particles were pelleted with table top centrifugation at 13.000 rpm for 15 min, washed once with PBS and eluted with Laemmli Buffer at 95°C for 5 min. The eluted proteins were separated via SDS-PAGE and stained with Coomassie dye. Images were captured with a BioRad GelDoc system and further analyzed with Quantity One Software (BioRad).

3. Results

3.1. Characterization of nanoparticles

Here we used the structure-directing activity of the central cystine residue as part of a dimer of the tripeptide CKK, along with ascorbate as reducing agent, to form SNP of well-defined sizes, shapes and coatings. The resultant nanoparticles and the structure of the peptide are explained schematically in Figure 1. Representative TEM pictures are shown in Figure 2. A single nanoparticle is composed of a silver metal core and a covalently attached peptidic layer made of the peptide CKK. X-ray photoelectron spectroscopy (XPS) revealed that this peptide becomes covalently linked to the metal surface through the *in situ* cleaved cystine bridge. Moreover, surface-enhanced Raman spectroscopy showed that the peptide is attached in a brush-like orientation, rather than enwrapping the metal core in a concentric manner. Thus, the silver nanoparticle is coated with a dense vertical layer of peptide molecules and only marginally covered at the surface by ascorbate. This indicates that the surface of the particles is tightly covered by a dense and covalently bound peptide, thus making a simple disintegration into silver ions unlikely.

Small-angle X-ray scattering (SAXS) and dynamic light scattering (DLS) were applied to characterize the colloidal properties of the nanoparticles in aqueous suspensions (Figure 3). SAXS demonstrated that the particle suspension is mainly monodisperse. The polydispersity index is only about 18% as it can be deduced from the minimum in the scattering curve. All nanoparticles synthesized have a narrow size range. Interestingly, the reason for this moderate polydispersity can be directly connected to the structural information obtained from Rietveld refinement of the powder diffraction pattern. The single crystallites are of an average size of 5 nm. This relates to both a controlled size (dependent on the reaction conditions) and a controlled polydispersity, as the crystallite size is "pre-programmed" by the structure-directing properties of the peptide.

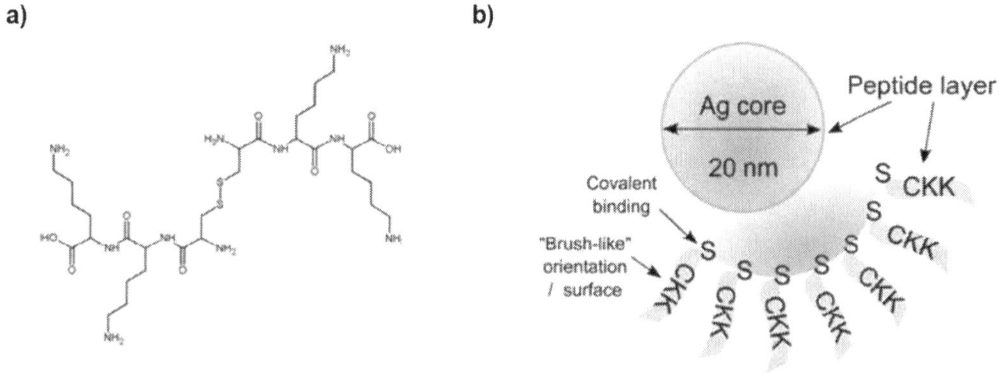

Figure 1. Structure of the peptide used for coating (a) and a scheme of the resultant nanoparticle (b).

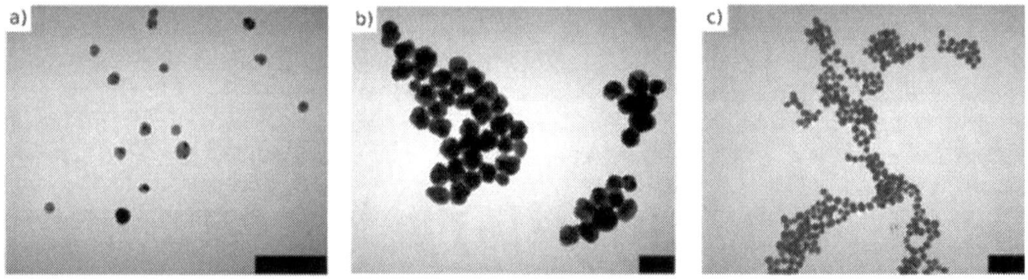

Figure 2. Representative TEM images of the particles produced: (a) Ag20Pep, (b) Ag40Pep, and (c) Au20Pep. The scale bar equals 100 nm in all images shown.

Both SAXS and DLS indicate a very similar size distribution of these particles (Figure 3, inset). DLS data in particular reveal that there are no larger aggregates present in solution. Interestingly, DLS also points to the presence of loose but concentration-dependent interactions between the nanoparticles, as could be inferred from the low q region in the SAXS curve. SAXS provides information about the electron-dense core as the organic shell is transparent to X-rays, while DLS provides information about the hydrodynamic radius of the nanoparticles. For round-shaped nanoparticles, the difference between both radii is directly related to the thickness of the shell. However, DLS results are influenced by (repulsive or attractive) interactions between the particles. Both techniques clearly show that the nanoparticles can be dispersed as single entities and that they do not coalesce to larger objects once being in solution.

Furthermore the covalently linked peptide helps to ensure several properties such as stability in the dry state, and the prevention of Ostwald ripening and ligand exchange at the surface. Importantly the interface is not passivated for further interactions with living organisms, as would be a polyethylene glycol layer repellant to proteins [29]. Furthermore the peptide could be used as a bio-interface to specifically target these nanoparticles or it can be useful for detection and purification of the particles.

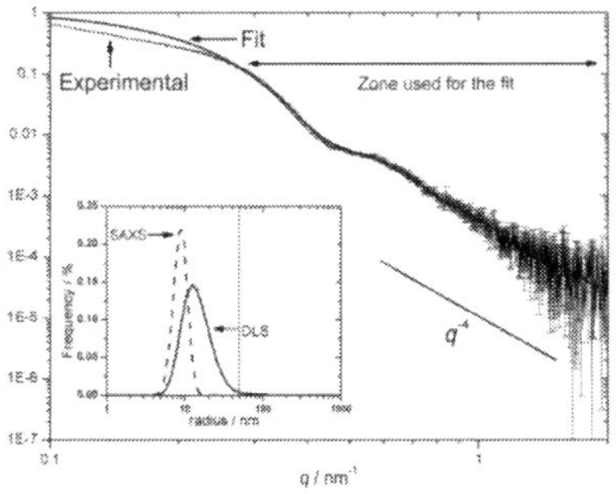

Figure 3. Small-angle X-ray scattering (SAXS) curve of Ag20Pep. Results of the fit are: 9.4 ± 0.2 nm and a polydispersity of $\sigma = 18\%$ using a Schultz sphere population model. Inset: size distribution as inferred from the SAXS curve and hydrodynamic radii as inferred from the DLS data, both volume-averaged. The straight line shows the limit upon which an object cannot be seen anymore at our SAXS settings.

3.2. Cytotoxicity of silver nanoparticles

We used macrophages derived from the human monocytic leukemia cell line THP-1 as a cellular model for toxicity testing. Since macrophages usually contribute to a first-line defence *in vivo* operating against foreign intruders such as systemically distributed nanoparticles they are of high relevance as model system. In addition they trigger subsequent immunological processes via secretion of cytokines. First we analyzed the nanoparticle-mediated cytotoxicity with a WST-1 viability assay. Cells were exposed for 24 or 48 h with increasing concentrations of silver or gold nanoparticles (Figure 4). Both types of SNP displayed strong cytotoxicity, which was dependent on dose and time of treatment. The smaller SNP (Ag20Pep) were slightly more toxic when compared to the larger counterparts (Ag40Pep) with respect to mass doses. The resultant IC(50) values after 24 h of exposure were 110 µg/ml (Ag20Pep) and 140 µg/ml (Ag40Pep), respectively. After 48 h of exposure the IC(50) values strongly decreased and were determined at 18 µg/ml (Ag20Pep) and 30 µg/ml (Ag40Pep), respectively. However, if the doses were calculated based on particle surface the viability curves and IC(50) values were in a similar range for both sizes of SNP (Figure 5). By contrast, gold nanoparticles proved mainly inert. Only at the highest concentrations tested little cytotoxicity could be observed, but viability was always higher than 80% (Figure 4).

In parallel we performed another, independent assay to measure cytotoxicity via quantifying the release of lactate dehydrogenase (LDH) into the supernatant (data not shown). Both assays provided comparable information. In addition we were interested in the mechanism of cell death. Therefore we measured the percentage of apoptotic cells after treatment with the different nanoparticles through propidium iodide/annexin V staining. Since apoptotic cells were virtually absent irrespective of the treatment conditions applied, we conclude that SNP-mediated cell death occurs via non-apoptotic pathways (data not shown).

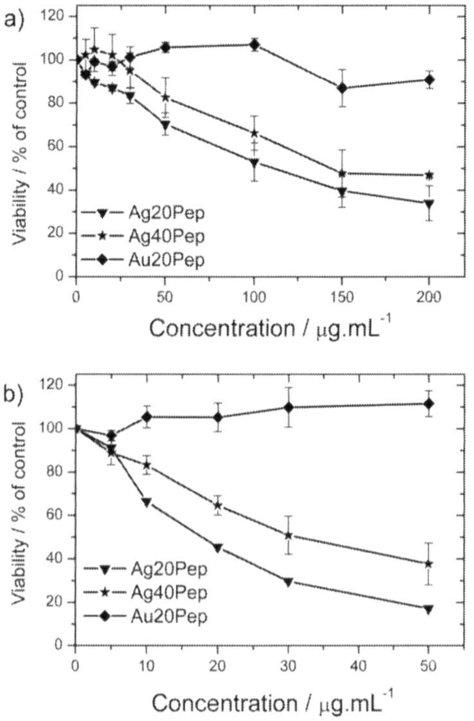

Figure 4. Measurement of viability after treatment of THP-1-derived macrophages with Ag20Pep, Ag40Pep or Au20Pep particles for 24 h (a) or 48 h (b).

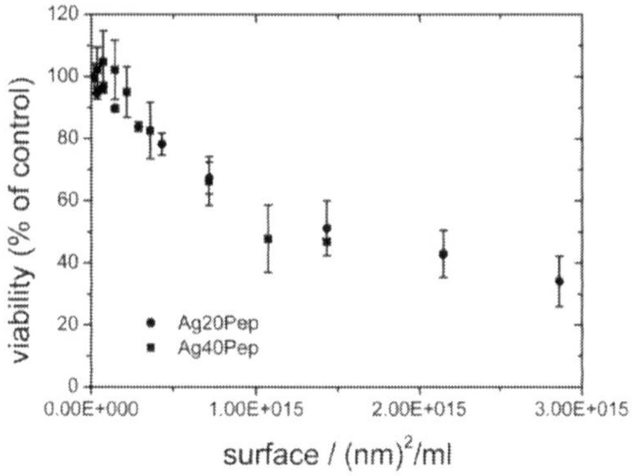

Figure 5. Calculation of cell viability after treatment with Ag20Pep or Ag40Pep particles. Data are derived from Figure 4 (WST-1 assay, 24 h exposure) and doses were calculated based on nanoparticle's surface and the assumption that particles are spherically shaped.

3.3. Visualization of particle uptake

We used transmission electron microscopy (TEM) to study nanoparticle-treated macrophages on an ultrastructural level (Figure 6). With TEM we could demonstrate that both sizes of SNP (20 nm and 40 nm) are efficiently taken up into macrophages. Virtually each cell analyzed was packed with SNP. We could detect some larger aggregates of SNP in the cytoplasm. However, the majority of SNP appeared as small and individual particles freely distributed throughout the cytoplasm, and without being surrounded by membrane envelopes. Furthermore SNP were detected in nuclei and even deep in the nucleoli, and inside lysosomes. Other organelles such as mitochondria or the endoplasmic reticulum were free of nanoparticles. Directly underneath the cellular membrane SNP were found highly concentrated as individual particles, indicating that these particles have been freshly taken-up through a non-phagocytic mechanism.

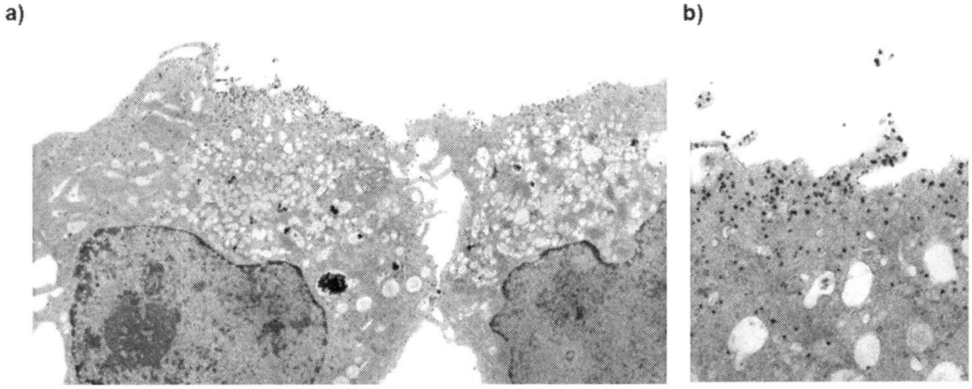

Figure 6. Transmission electron microscopy (TEM) images of macrophages upon exposure to Ag40Pep particles. (a) Overview of a treated cell (2.500x). (b) Detailed picture of particles directly underneath the cell membrane (20.000x). In each experiment cells were treated with 20 μg/ml SNP for 20 h.

Based on these findings we suggest that macrophages may ingest SNP through at least two different pathways. Aggregated SNP are possibly incorporated via phagocytosis while single nanoparticles might enter cells rather via non-phagocytic routes. Membrane flip-flop mechanisms or direct penetration via ion-channels are possible routes for passive and non-phagocytic uptake of SNP, but active transport routes might exist as well. Further experiments are needed to analyze these different possibilities.

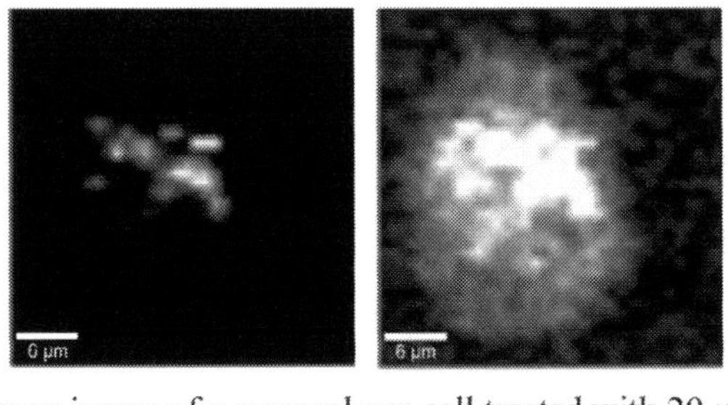

Figure 7. Confocal Raman image of a macrophage cell treated with 20 μg/ml Ag20Pep for 20 h. On the left side only signals from the nanoparticles are shown. The right side shows an overlay of signals originating from both nanoparticles and organic material inside the cell.

In addition, nanoparticles inside macrophages were visualized with confocal Raman microscopy (Figure 7). Due to the covalently linked peptide to the metal core the SNP prepared offered the advantage of being specifically detectable via induced metal-sulphur stretching vibrations monitored through confocal Raman microscopy.

Confocal Raman microscopy can be applied very fast without tedious sample preparation. Again, we detected a few number of large aggregates together with many smaller aggregates and countless individual SNP, the latter being spread throughout the entire cell.

3.4. Analysis of oxidative stress responses

Since exposure of cells to metal nanoparticles most likely results in the generation of oxidative stress we used different biochemical approaches to analyze this in more detail (Figure 8). Oxidative stress results from of an imbalance between the generation and elimination of reactive oxygen species (ROS) within cells. Once generated ROS may then react fastly with cellular components and proteins [30]. Protein carbonyls are among the most common non-enzymatic reaction products and very sensitive methods for their detection exist. The detection of protein carbonyls can thus serve as very sensitive indicator for ROS production. Here we show that protein carbonyls can be detected in cells after treatment with Ag20Pep but only marginally after treatment with Au20Pep (Figure 8A). In parallel we also followed the induction of the cytoprotective enzyme heme oxygenase 1 (HO-1), which is another established oxidative stress marker (Figure 8B) [31]. The carbonylated proteins could be enriched via immunoprecipitation (Figure 8C) and subsequent quantification demonstrated that treatment with Ag20Pep particles led to approximately 2.7-fold higher levels of protein carbonyls compared to control cells (Figure 8D).

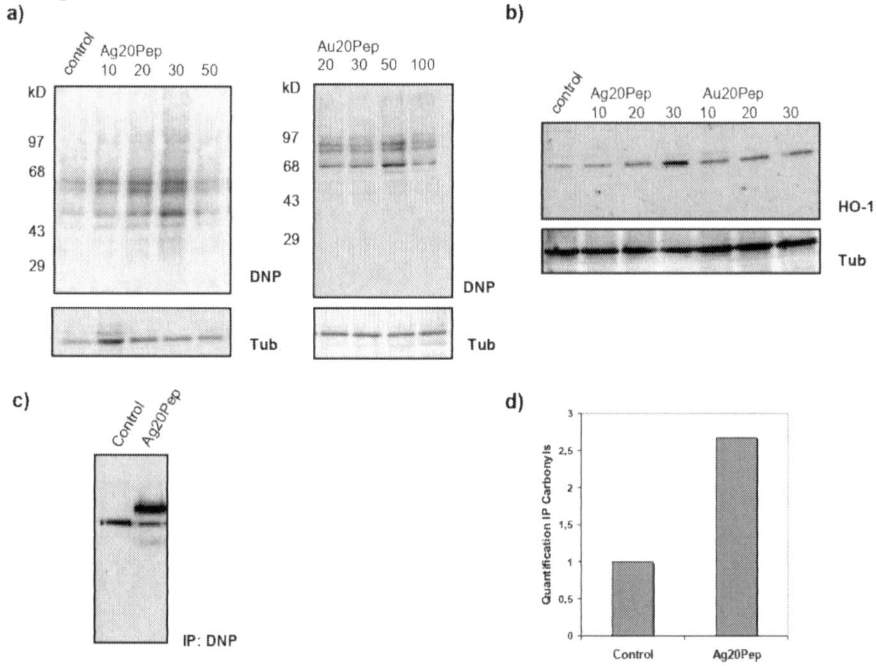

Figure 8. Oxidative stress responses in THP-1-derived macrophages after treatment with SNP. (a) Formation of protein carbonyls in cells after treatment with Ag20Pep (left panel) or Au20Pep particles (right panel). Protein carbonyls can be detected via reaction with 2,4-dinitrophenylhydrazine (DNP) to form hydrazone adducts. The doses are given in μg/ml and the treatment period was 20 h. (b) Significant induction of heme oxygenase I (HO-1) was detected in cells upon treatment with Ag20Pep particles for 20 h, but not with Au20Pep particles. All doses are given in μg/ml. (c) After formation of DNP-hydrazone adducts the carbonylated proteins were immunoprecipitated (IP) using an anbibody directed against DNP. Treatment was with 20 μg/ml Ag20Pep for 20 h. (d) Quantification of protein carbonyl levels after immunoprecipitation as shown in (c).

Protein carbonyls can also be analyzed on a 2D gel, which enables higher resolution. The differentiation between cellular reaction patterns on different types of nanoparticles can be performed with higher accuracy. Typically in a 2D gel up to 3500 different protein spots can be separated by an IPG strip approach. A typical example for a particle-induced 2D protein carbonyl pattern is shown in Figure 9.

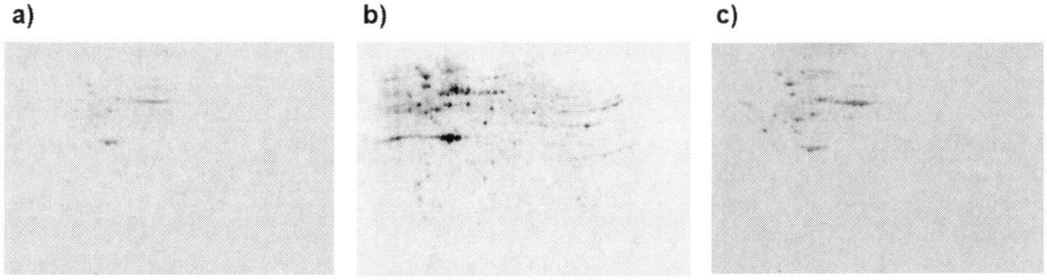

Figure 9. Detection of protein carbonyls in a 2D blot. Protein carbonyls are detectable as specific 2,4-dinitrophenylhydrazone (DNP) adducts. (a) Control cells without nanoparticle treatment. (b) 2D pattern obtained after treatment of cells with 20 μg/ml Ag20Pep particles for 20 h. (c) 2D pattern obtained after treatment of cells with 20 μg/ml Au20Pep particles for 20 h. In all gels the same amount of protein has been loaded and the resulting immunoblots were exposed for the same time period.

Finally we were interested whether or not the peptide-coated SNP do form a protein corona in cell culture media. Protein corona formation has been recently shown to occur upon contact of nanoparticles with biological fluids such as cell culture medium and there is evidence that the components of this corona affect the uptake, fate and toxicity of the particles in living systems [32,33]. In Figure 10 we show that our Ag20Pep particles indeed engage in complex protein corona formation and that some of the serum proteins become specifically enriched on the surface of the nanoparticles.

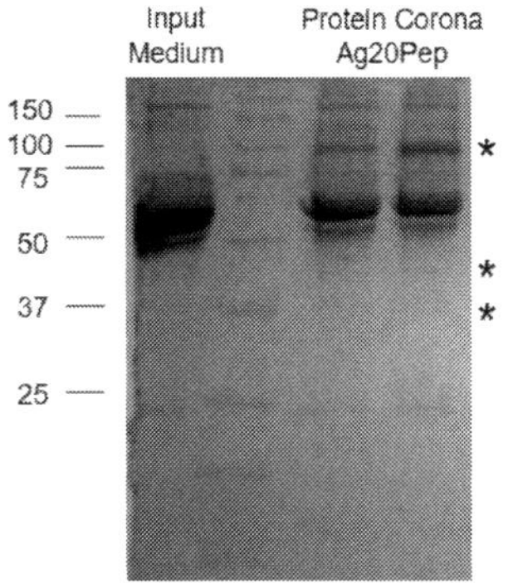

Figure 10. Protein corona formation of the Ag20Pep particles in cell culture medium. Several serum proteins are specifically enriched on the surface of the particles (marked with asterisk).

4. Discussion

Here we introduce a novel type of SNP synthesized to achieve well defined sizes and shapes and excellent colloidal properties. These SNP are nicely homogenous and display a narrow size distribution as indicated by their low polydispersity index (0.18). Classical citrate-coated particles in comparison are typically synthesized with polydispersity indices of about 0.38 and higher and thus are not as accurately defined in terms of sizes and shapes. Furthermore the new SNP form stable dispersions in aqueous environments, thereby allowing for application in toxicological studies. Indeed, the low polydispersity along with high stability against coagulation and irreversible aggregation are advantages when compared to loosely bound ligand shells. We could prove that the majority of the SNP remain individual particles inside macrophages and spread throughout cellular compartments including nuclei and nucleoli. We suggest that at least two different mechanisms of uptake exist. The larger aggregates could represent material that has been taken up via phagocytosis. By contrast, the individual nanoparticles, which are found enriched directly underneath the cell membrane without membrane envelopes, might have been taken up by another non-phagocytic mechanism.

The peptide applied in our synthesis approach does not only play the role of a ligand, but also is crucial during first stages of nanoparticle growth, thereby allowing for both production of large scale batches and high batch-to-batch reproducibility. Especially the latter point is a very important aspect for their application in toxicity testing. Furthermore the synthesis is performed under mild and biomimetic aqueous conditions, which eliminates the possibility that organic solvents or other chemicals are subsequently transferred into cell culture systems. The surface of our nanoparticles is completely covered by the peptide, which is covalently bound. Since dissolution of the particles under these conditions seems unlikely we suggest that the majority of toxicity observed in cells is related to the nanoparticles as such and not due to the release of silver ions. This aspect will need further attention in future studies.

In addition we analyzed basic toxicity of the peptide-coated SNP. The cytotoxicity detected revealed dose- and time-dependent. The smaller SNP (*i.e.*, diameter of 20 nm) were more cytotoxic compared to the larger particles. However, if doses were calculated based on particle surfaces both types of SNP (*e.g.*, 20 nm and 40 nm diameter) appeared similar cytotoxic. We applied three independent cytotoxicity tests which delivered comparable albeit not identical results. In recent days, nanosilver becomes available in a variety of different shapes, coatings and sizes. Since we are particularly interested to compare the toxicity of the peptide-coated nanoparticles with other types of SNP that can be obtained through rather conventional syntheses, we started to look into the cytotoxicity of peptide-coated SNP *versus* citrate-coated SNP of the same size (20 nm). Preliminary results show that citrate-coated SNP exert only little cytotoxicity in THP-1 cell cultures up to concentrations of 200 μg/ml (data not shown). It will be interesting to include other types of SNP including pristine nanosilver to better characterize the influence of the different coating on the cellular responses observed.

We also demonstrate that exposure to 20 nm SNP triggers significant oxidative stress insides macrophages, thereby clearly contrasting the absence of such effects after treatment with nanogold. At all endpoints addressed the peptide-coated nanogold only induced very faint responses, if any, and thus may be considered biologically inert. Oxidative stress was demonstrated by analyzing two different responses. Besides particle-mediated induction of HO-1, we could indirectly show the formation of ROS via detection of protein carbonyls. The protein carbonyl species generated within cells could be further separated by means of 2D gel chromatography. With this approach we show that nanosilver treatment leads to significant increases in the intracellular levels of protein carbonyls when compared to nanogold or cells without treatment. The patterns of protein carbonyls obtained with nanogold-treated cells were similar to untreated cells.

In nanotoxicology there is some debate on the issue of whether or not it may be justifiable to compare results obtained *in vitro* with those expected to occur *in vivo*. This is due to the fact that in *in vitro* studies one bolus dose is delivered, which usually is rather high. By contrast, during *in vivo* studies typically lower doses are delivered over a longer period of time. In our study we observed strongest effects in terms of oxidative stress in a concentration range of 20-30 μg/ml SNP. Taking the density of silver (0.01049 ng/cm^3) under the assumption that particles are ideally spheric, this concentration range equals a particle number range of 5.7 – 8.5 x 10^{10}/ml. A regular cell density of 1.5 x 10^5 cells/ml then results in the highest observed effect at the external dose of 0.38 to 0.57 x 10^6 particles per cell. To become clear about the concentration range, we compare this to a regular 28-days rat inhalation study published in 2007 [11]. In this study rats were exposed for 6 h per day and 5 days a week to SNP with sizes ranging between 1.9 and 65 nm diameter. The doses applied were low (1.74 x 10^4 particles/cm^3), medium (1.27 x 10^5 particles/cm^3) or high (1.32 x 10^6 particles/cm^3), with the highest dose equaling 61 μg/m^3. Considering the average weight and the breath volume (0.29 m^3/kg) of the animals results in an exposure level of approximately 1.4 x 10^9 (low dose) to 1.1 x 10^{11} particles (high dose) per day. For the complete 4-week period this would equal 2.83 x 10^{10} (low dose) to 2.16 x 10^{12} particles (high dose). Taking these numbers into account we conclude that the concentrations applied in our study *in vitro* to obtain maximal effects (5.7 – 8.5 x 10^{10} particles/ml) were in a comparable range to what will get inhaled within one day (6 hours) by the lungs of animals exposed to a dose of 1.32 x 10^6 particles/cm^3. Certainly, for the *in vivo* situation this is only a rough estimation. Typically in the bronchoalveolar lavage of one rat a total of 5 x 10^6 cells can be found [34]. In our study we used 1.5 x 10^5 cells for each treatment. In comparison to other *in vitro* studies that regularly use treatment doses of up to 100-200 μg/ml, we already observed strongest effects in the comparably low range of 20-30 μg/ml (10-15 μg/cm^2) that was proven well below the threshold level of cytotoxicity.

Acknowledgments
The authors acknowledge excellent technical assistance of Annika-Mareike Gramatke, Antje Bergert, Doreen Wittke and Monika Sachtleben. A. Mantion thanks BAM for an Adolf-Martens Fellowship. The authors are grateful to the following institutions for financial support: Swiss National Science Foundation, Federal Institute for Risk Assessment (BfR), Federal Institute for Material Research and Testing (BAM), The Free University of Berlin, University of Potsdam, Fonds der Chemischen Industrie and MPI of Colloids and Interfaces (Colloid Chemistry Department).

References

[1] Edward-Jones V 2009 The benefits of silver in hygiene, personal care and healthcare *Lett. Appl. Microbiol.* **49** 147-52
[2] Chen X and Schluesener H J 2008 Nanosilver: a nanoproduct in medical application *Toxicol. Lett.* **176** 1-12
[3] Jain J, Arora S, Rajwade J M , Omray P, Khandelwal S and Paknikar K M 2009 Silver nanoparticles in therapeutics: development of an antimicrobial gel formulation for topical use *Mol. Pharm.* **6** 1388-1401
[4] Eby D M, Luckarift H R and Johnson G R 2009 Hybrid antimicrobial enzyme and silver nanoparticle coatings for medical instruments *ACS Appl. Mater. Interfaces* **1** 1553-60
[5] Lee H Y, Park H K, Lee Y M, Kim K and Park S B 2007 A practical procedure for producing silver nanocoated fabric and its antibacterial evaluation for biomedical applications *Chem. Commun.* **28** 2959-61
[6] Vigneshwaran N, Kathe A A, Varadaraian P V, Nachane R P and Balsubramanva, R H J 2007 Functional finishing of cotton fabrics using silver nanoparticles *Nanosci. Nanotechnol.* **7** 1893-97
[7] Kokura S, Handa O, Takagi T, Ishikawa T, Naito Y and Yoshikawa T 2010 Silver nanoparticles as a safe preservative for use in cosmetics. *Nanomedicine* **6** 570-74
[8] Hoet P, Brueske-Hohlfeld I and Salata O V 2004 Nanoparticles— known and unknown health risks *J. Nanobiotechnol.* **2** 12-27
[9] Seaton A, Tran L, Aitken R and Donaldson K 2010 Nanoparticles, human health hazard and regulation. *J. R. Soc. Interface* **7** Suppl. 1 S119-29
[10] Xia T, Li N and Nel A E 2009 Potential health impact of nanoparticles *Annu. Rev. Public Health* **30** 137-50
[11] Ji J H, Jung J H, Kim S S, Yoon J U, Park J D, Choi B S, Chung Y H, Kwon I H, Jeong J, Han B S, Shin J H, Sung J H, Song K S and Yu I J 2007 Twenty-eight-day inhalation toxicity study of silver nanoparticles in Sprague-Dawley rats *Inhal. Toxicol.* **19** 857-71
[12] Tang J, Xiong L, Wang S, Wang J, Liu L, Li J, Yuan F and Xi T 2009 Distribution, translocation and accumulation of silver nanoparticles in rats *J. Nanosci Nanotechnol.* **9** 4924-32
[13] Larese F F, D'Agostin F, Grosera M, Adami G, Renzi N, Bovenzi M and Maina G 2009 Human skin penetration of silver nanoparticles through intact and damaged skin *Toxicology* **255** 33-7
[14] Greulich C, Kittler S, Epple M, Muhr G and Koller M 2009 Studies on the biocompatibility and

the interaction of silver nanoparticles with human mesenchymal stem cells (hMSCs) *Langenbecks Arch. Surg.* **394** 495-502

[15] Foldbjerg R, Olesen P, Hougaard M, Dang D A, Hoffmann H J and Autrup H 2009 PVP-coated silver nanoparticles and silver ions induce reactive oxygen species, apoptosis and necrosis in THP-1 monocytes *Toxicol. Lett.* **190** 156-62

[16] AshaRani P V, Low Kah Mun G, Hande M P and Valiyaveettil S 2009 Cytotoxicity and genotoxicity of silver nanoparticles in human cells *ACS Nano* **3** 279-90

[17] Carlson C, Hussain S M, Schrand A M, Braydich-Stolle L K, Hess K L, Jones R L and Schlager JJ 2008 Unique cellular interaction of silver nanoparticles: size-dependent generation of reactive oxygen species *J. Phys. Chem. B.* **112** 13608-19

[18] Li N, Xia T and Nel A E 2008 The role of oxidative stress in ambient particulate matter-induced lung diseases and its implications in the toxicity of engineered nanoparticles *Free Radic. Biol. Med.* **44** 1689-1699

[19] Arora S, Jain J, Rajwade J M and Paknikar K M 2008 Cellular responses induced by silver nanoparticles: *in vitro* studies *Toxicol. Lett.* **179** 93-100

[20] Ivask A, Bondarenko O, Jepihhina N and Kahru A 2010 Profiling of the reactive oxygen species-related ecotoxicity of CuO, ZnO, TiO2, silver and fullerene nanoparticles using a set of recombinant luminescent *Escherichia coli* strains: differentiating the impact of particles and solubilized metals *Anal. Bioanal. Chem.* **398** 701-16

[21] Lubick N 2008 Nanosilver toxicity: ions, nanoparticles— or both? *Environ. Sci. Technol.* **42** 8617

[22] Lankveld D P, Oomen A G, Krystek P, Neigh A, Troost-de Jong A, Noorlander C W, Van Eijkeren J C, Geertsma RE and De Jong W H 2010 The kinetics of the tissue distribution of silver nanoparticles of different sizes *Biomaterials* **31** 8350-61

[23] Sur I, Cam D, Kahraman M and Baysal A 2010 Interaction of multi-functional silver nanoparticles with living cells *Nanotechnology* **21** 175104

[24] Pal S, Tak Y K and Song J M 2007 Does the antibacterial activity of silver nanoparticles depend on the shape of the nanoparticle? A study of the gram-negative bacterium *Escherichia coli. Appl. Environ. Microbiol.* **73** 1712-20

[25] Mantion, A, Massuger L, Rabu P, Palivan C, McCusker L B and Taubert A 2008 Metal-peptide frameworks (MPFs) : « Bioinspired » metal organic frameworks *J. Am. Chem. Soc.* **130** 2517-26

[26] Dickerson M B, Sandhage K H and Naik R R 2008 Protein- and peptide-directed syntheses of inorganic materials. *Chem. Rev.* **108** 4935-78

[27] Mantion A, Guex A G, Foelske A, Mirolo L, Fromm K M, Painsi M and Taubert A 2008 Silver nanoparticle engineering via oligovaline organogels *Soft Matter* **4** 606-617

[28] Graf P, Mantion A, Foelske A, Shkilnyy A, Masic A, Thuenemann A F and Taubert A 2009 Peptide-coated silver nanoparticles: synthesis, surface chemistry, and pH-triggered, reversible assembly into particles assemblies *Chem. Eur. J.* **15** 5831-44

[29] Nehring R, Palivan C G, Moreno-Flores S, Mantion A, Tanner P, Toca-Herrera J L, Thuenemann A, and Meier W 2010 Protein decorated membranes by specific molecular interactions *Soft Matter* **6** 2815-24

[30] Dalle-Donne I, Carini M, Vistoli G, Ragazzoni L, Colombo G, Rossi R, Milzani A and Aldini G. 2009 Protein carbonylation: 2,4-dinitrophenylhydrazine reacts with both aldehydes/ketones and sulfenic acids *Free Radic. Biol. Med.* **46** 1411-19

[31] Xia T, Kovochnich M, Brant J, Hotze M, Sempf J, Oberley T, Sioutas C, Yeh J, Wiesner M R and Nel A E 2006 Comparison of the abilities of ambient and manufactured nanoparticles to induce cellular toxicity according to an oxidative stress paradigm *Nano Lett.* **6** 1794-1807

[32] Lundqvist M, Stigler J, Elia G, Lynch I, Cedervall T, Dawson K A 2008 Nanoparticle size and surface properties determine the protein corona with possible implications for biological impacts *Proc. Natl. Acad. Sci. USA* **105** 14265-14270

[33] Lynch I, Dawson K A 2008 Protein-nanoparticle interactions *Nano Today* **3** 40-48
[34] Pauluhn J 2007 Pulmonary toxicity and fate of agglomerated 10 and 40 nm aluminium oxyhydroxides following 4-week inhalation exposure of rats: toxic effects are determined by agglomerated, not primary particle size *Toxicol. Sci.* **109** 152-167
[35] Auwerx J 1991 The human leukaemia cell line, THP-1: a multifaceted model for the study of monocyte-macrophage differentiation *Experientia* **47** 22-31
[36] Park E K, Jung H S, Yang H I, Yoo M C, Kim C and Kim K S 2007 Optimized THP-1 differentiation is required for the detection of responses to weak stimuli *Inflamm. Res.* **56** 45-50
[37] Richardson K C, Jarett L and Finke E H 1960 Embedding in epoxy resins for ultrathin sectioning in electron microscopy *Stain Technol.* **35** 313-23

Nanoparticles used in medical applications for the lung: hopes for nanomedicine and fears for nanotoxicity

S Boland, R Guadagnini, A Baeza-Squiban, S Hussain and F Marano

Laboratory of Functional and Adaptative Biology, unit of Réponses Moléculaires et Cellulaires aux Xénobiotiques (RMCX), CNRS EAC 4413, University Paris Diderot-Paris 7, CC 7073, 75205 Paris Cedex 13, France.

E-mail: boland@univ-paris-diderot.fr

Abstract. Nanotechnology is a promising tool for the development of innovative treatment strategies allowing to overcome obstacles encountered by classical drug delivery. This has led to the development of nanomedicine. Indeed, nano-delivery systems (NDS) may allow the controlled release of therapeutics, protection of drugs against degradation, targeted drug delivery and facilitated transport across barriers. All these advantages of NDS are particularly interesting for treatments of the lung which is a challenging organ in respect to drug delivery. However, for the development of nanomaterials aimed to transport therapeutics, there is also a need to assess the potential health hazards of these new materials, especially as a variety of nanoparticles have been shown to induce toxicity related to their nanometer size leading to the new field of nanotoxicology. We will address both aspects of NDS, specifically in respect to lung treatments: their potential benefits and the possible adverse health effects of these materials.

1. Introduction

Nanoscience and Nanotechnology are developing very rapidly. The European Commission's strategy for nanotechnology has highlighted the need to invest in knowledge, technology and experience in this field in Europe (COM (2004) 338). An emerging component in this field is nanomedicine which is rapidly developing. However, a comparison of the number of publications on nanomedicine with nanotechnology indicates that nanomedicine still accounts for a little amount of nanotechnology research worldwide [1]. Actually there are two areas in which the impacts of nanomedicine are likely to be most significant: first the diagnostic and medical records sector and second new treatment options [2]. Indeed, various nanoparticles (NPs) are in clinical use or under development for therapeutic purposes with the aim to develop systems for imaging and drug delivery. Nanovectors may in fact allow to overcome several challenges in drug delivery. Embedding of pharmaceuticals into nano-delivery sytems (NDS) could protect them from degradation. Furthermore a controlled release of therapeutics would enable persistent drug delivery and treatments at diminished doses. A great advantage of nanomedicine will be targeted drug delivery which allows reducing administration doses and thus lowering side-effects, toxicity and exposure of non-target organs. Certain drugs are difficult or impossible to be efficiently transported through physiological barriers and are administered systemically, which results in possible metabolism and clearance in the liver. Thus higher dosing is necessary to achieve therapeutic effects with possible target and non-target organ toxicity. An important field of nanomedicine research is consequently the development of new treatment strategies for a controlled, selective and efficient transport of drugs through biological barriers such as the lung. However, the safety of NDS is a principal issue of concern for the development of nanomedicine as nanomaterials will interact with complex biological systems. Several *in vivo* and *in vitro* studies have indeed demonstrated that nanomaterials could induce toxicity in multiple organ systems and NDS may thus have adverse health effects which have been poorly studied thus far.

2. Promising benefits of nanovectors for improved drug delivery to the lungs

Several recent reviews have highlighted the promising role of nanomedicine for the treatment of respiratory diseases [3-5]. Indeed, the treatment of the airways could be limited by the anatomical and physiological barriers of the respiratory tract, which may be bypassed by the use of NDS. In the conducting airways, the mucociliary clearance allows the transport of foreign substances such as bacteria and particles to the pharynx, for their elimination by expectoration or ingestion. In the distal lung, the alveolar clearance is achieved by alveolar macrophages that are present in the lumen of alveoli and which are professional phagocytes. An important challenge of nanomedicine will be to circumvent these elimination roads. Furthermore, the use of NDS will allow to protect protein drugs from proteases present in the mucus and surfactant, especially in inflamed lungs. Another promising feature of NDS, beside their potential to cross the lung lining fluids, is their capacity to be taken up by respiratory epithelial cells [6] as modern treatment strategies aim to introduce expression vectors, genes, antibodies or siRNA into cells to induce or inhibit local production of proteins involved in pulmonary diseases [7-10]. As the airway epithelium is highly resistant to transfection, due to the extracellular lining fluids [11], the use of NDS is a promising tool to achieve penetration. Attempts using NDS to increase transfection efficiencies have recently been conducted for pulmonary treatments with promising results. In this regard, it is also interesting to note that endocytosis of NPs by epithelial cells has been shown to be size dependent and smaller NPs are more efficiently taken up [6,12]. The epithelium of the conducting airways also presents tight junctions, preventing paracellular crossing of foreign substances, but NPs have been shown to translocate through the epithelium in *in vitro* and animal studies, with smaller NPs being more efficient [13,14] thus allowing to foresee treatment strategies of the submucosa. Indeed, pathophysiological modifications, such as fibrosis and infiltration of inflammatory cells, are important features of several respiratory diseases. However, body distribution studies of inhaled NPs show low levels of translocation [15]. It is important to note that translocation studies have only been performed after a sole exposure to NPs and a more significant translocation is thus likely to occur in case of chronic exposure and coating of NPs may also allow to increase their translocation. Interestingly, long-term studies over 6 months have shown initial interstitial uptake and a return of NPs to the alveolar lumen, where they were cleared by macrophages [16]. In pathological conditions, the administration may even be more difficult. For instance, the epithelium in chronic bronchitis is characterized by mucus hypersecretion and squamous cell differentiation at late stages. Due to the hypersecretion, the delivery of drugs targeting the alveolar region may also be more difficult in COPD or asthma. Pathological situations could also increase the translocation, as shown in animal models of inflammation [17,18] which may be due to increased paracellular translocation. The differences in route of administration (inhalation versus instillation) may explain the divergence in the extent of translocation observed between studies. Instillations are generally performed at high doses as a bolus whereas inhalation studies are usually carried out at much lower doses. *In vitro* studies have furthermore highlighted the role of NP physicochemical characteristics in their ability to cross the epithelial barrier [19]. However, interpretation of *in vitro* studies has to be done with great caution since they are generally carried out at high doses. In addition, to direct translocation through the epithelial barrier, an indirect transfer via macrophages is plausible as migration of NP loaded-alveolar macrophages has been observed towards the blood circulation and extrapulmonary organs [20]. Due to the high surface area and small air-blood distance of the alveolar epithelium, the lower respiratory tract may be treated by intravenous administration of pulmonary drugs. This treatment route is already studied for antitubercular drugs and NDS have been shown to increase their pulmonary delivery [21]. The translocation through the alveolar epithelium of drugs, by the use of NDS, is also considered for systemic treatment strategies for non pulmonary diseases. Insulin is already administered by inhalation (Exubera®) and NDS have been shown to increase plasma levels [22,23]. Indeed the alveolar clearance by macrophages may be bypassed by NPs, as it has been shown that these are less efficiently taken up by macrophages than micrometer sized particles [13,24], whereas epithelial and interstitial uptake is increased for smaller particles [13]. Furthermore this administration route allows circumventing first-pass metabolism and side effects seen in systemic oral or intravenous delivery. Indeed the lung presents lower levels of drug efflux transporters and metabolizing enzymes than the gastrointestinal tract.

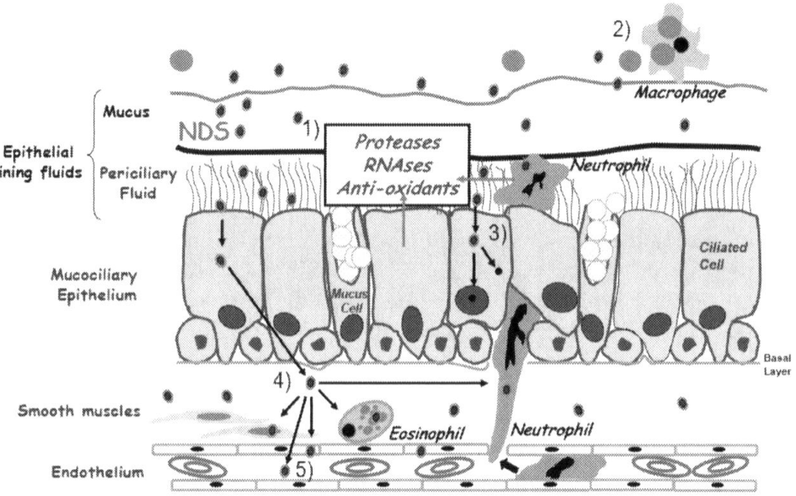

Figure 1. Nanovectors for improved drug delivery to the lungs.

Nanodelivery systems (NDS) present several possibilities to improve treatments of the upper airways:

1) NDS could protect the drugs from degradation by enzymes present in the epithelial lining fluids.
2) The nanometre size of NDS may enable to circumvent the clearance by macrophages.
3) Facilitated crossing of the lining fluids and endocytosis of NDS will improve transfection efficiencies. Particularly important target cells are epithelial cells, neutrophils (eg in COPD) and eosinophils (eg in asthma).
4) Transcytosis of NDS will allow the treatment of the submucosa.
5) Crossing of the endothelium will allow systemic treatment by NDS as well as intravenous administration of pulmonary drugs. The crossing of the air/blood barrier is however even more prominent at the alveolar epithelium.

2.1. Benefits of NDS for innovative treatment strategies of the lung: monoclonal antibody therapies for COPD

Chronic obstructive pulmonary disease (COPD) is currently the fourth leading cause of death in the world, and if current trends continue, it is likely to rank as the third most important cause of death by the year 2030 (WHO, World Health Statistic 2008). COPD has a well recognised inflammatory component that is central to the development and progression of the illness and contributes also to many of the extra-pulmonary features. As cytokines and chemokines play a critical role in the orchestration of COPD, the use of monoclonal antibodies are valuable treatment options for more specific therapies [8,9]. Special emphasis has been put on the inhibition of neutrophil infiltration, which plays an important role in the pathogenesis of COPD. A multitude of antibodies targeting cytokines, chemokines, growth factors and their receptors are considered for the treatment of COPD [8,9,25,26] and some have already been studied in clinical trials with COPD patients. The anti IL-8 antibodies Abgenix [27] improved dyspnea, but not lung function and this limited clinical benefit has been attributed to the suboptimal dosing in the airways, due to intravenous administration. Thus, the use of NDS may allow administration by inhalation, increasing local concentrations and avoiding possible side effects. Moreover, several clinical trials, with the anti TNFalpha antibody infliximab have been conducted but have shown no effects in COPD patients so far [28-30]. A clinical trial using the EGFR inhibitor BIBW 2948 in COPD patients has been conducted but was poorly tolerated. Therefore a higher local dosage is needed to allow a better inhibition of this receptor [31] and NDS should enable a better dosage at the site of action and a prolonged release of drugs. The use of monoclonal antibodies could have higher selectivity and reduced side effects compared to the use of inhibitors [9]. Thus, delivery of monoclonal antibodies such as Cetuximab (already in clinical trials for cancer therapy) could be a valuable option for the treatment of COPD.

3. Toxicity of nanomaterials

While a lot of attention has been paid to the development of new engineered nanomaterials and to new applications of nanotechnologies, comparatively less research has been performed to assess the potential hazard of these new materials. Concerns about the potential impacts of NPs on health effects have been expressed since NPs of same chemical composition possess different physicochemical properties than particles of larger size [6,13]. Indeed, when the particle size decreases, the proportion of atoms on the surface is increased to manifolds. Thus, NPs could have a greater reactivity and increased catalytic potential. This increase in the surface of engineered particles provides thus benefits compared to larger particles but could also result in much higher biological activities than bigger particles [34]. Answers to these questions are essential to identify the hazards related to nanomaterials and to accept or to compel knowingly the expected explosion of nanotechnology in general, and nanomedicine in particular. NP toxicity studies are mostly conducted with materials that are not designed for medical use (CB, CNT, metal oxides...) and which are more relevant for environmental and occupational exposures. However, in regard to development of NDS for clinical applications, the present knowledge of NP toxicity from studies on occupationally relevant NPs could be of interest. However, most of the toxicity studies have been carried out at very high doses which may not represent real-world conditions. Nevertheless these studies have allowed to identify several potential health hazards of NPs. Especially these studies have put forward the need of nanomaterial specific toxicity testing strategies and the role of physico-chemical characteristics (surface area, surface reactivity, charge as well as size and shape) in the toxicity of NPs (for review see [32]).

3.1. Pulmonary toxicity of nanomaterial

Respiratory effects of NPs are mainly inflammation, oxidative stress and functional disturbances. A variety of *in vivo* studies demonstrates inflammatory effects of NPs as transitory responses and assess whether the degree of the inflammatory response is related to the exposure dose of NPs. In fact, inhalation [33-35] or instillation exposures [36-38] of various types of nanomaterials have shown to induce inflammation in lungs. These studies revealed local invasion of leukocytes, increased numbers of inflammatory cells in broncho-alveolar lavage fluid (BALF), release of LDH and increased cytokine production. In addition, typical granulomatous reactions are observed after pulmonary exposures to CNTs [39-41]. Furthermore functional disturbances could be observed after NP exposure including airway hyper-reactivity (AHR) to non-specific stimuli, tissue injury leading to disturbance in respiratory functions [42-45] and effects on existing pulmonary inflammation [46]. These disturbances are generally studied in pathologic animal models, like asthma etc. Translocation could be increased by lung injury and thus the presence of cadmium NPs in the liver, after inhalation of high doses, has been attributed to their pulmonary toxicity [47]. In the context of pulmonary toxicity, it is important to note that computational models predict increased deposition of inhaled NPs in diseased or constricted airways [48] and obstructive lung diseases have indeed been shown to increase pulmonary retention [49].

The carcinogenic potential of low toxicity low solubility dusts (LTSD), like CB and TiO_2, has been established in experimental animals [50] but these were generally performed at high concentrations and in sensitive species. There are very few epidemiological studies on human exposures and thus lung cancer risks could not directly be associated with engineered nanomaterials. The mechanisms leading to NP carcinogenicity are unknown and may involve primary genotoxic insult or secondary genotoxic response due to particle induced inflammation [51].

3.2. Cellular mechanisms

The cellular mechanisms, responsible for these effects, have been postulated to involve an oxidative stress induction by nanomaterials. Nel *et al* and Xia *et al* proposed a hierarchical oxidative stress model to explain the biological effect of NPs, stating that "minor levels of oxidative stress induce protective effects that may yield to more damaging effects at higher levels of oxidative stress" [52,53]. The induction of a cascade of cellular events, due to this oxidative stress, leads to the induction of redox sensitive pathways in the cells. Modifications of cell cycle, proliferation, pro-inflammatory response and cell death by apoptosis or necrosis could be the consequences. Many recent reviews on the mechanisms of action of NPs sustain this oxidative stress paradigm [52-54] while few have questioned the central role of oxidative stress in these responses to NP [55,56]. Several studies pointed out the roles of size and surface reactivity in the ability to induce an oxidative stress [33,53]. The adverse effects on the target tissues could either be due to abiotic or biotic reactive oxygen species (ROS) induced by NP or to ROS produced indirectly from the inflammatory process induced by NP exposures. The activation of redox sensitive transcription factors, such as NF-kappaB, AP1 or Nrf2, by this oxidative stress could lead to the transcription of a number of pro-inflammatory genes and the increased release of numerous pro-inflammatory factors. However, adaptive responses may diminish these effects which make the extrapolation on health effects in humans very difficult.

A modulation of intracellular calcium by carbon black NPs has also been demonstrated [57]. This transient increase of intracellular calcium by NPs has been shown to trigger impaired phagosome transport and cytoskeletal dysfunctions. The underlying mechanism may involve direct effects of NPs on ion channels which control the cellular calcium homeostasis. Higher NP concentrations have also been shown to induce apoptosis in a variety of cells [58-62]. Differential signalling events are responsible for these apoptotic effects, involving death receptor, mitochondria or lysosomes.

4. Conclusions

Nanomedicine presents a great opportunity for the radical improvement of current therapies and development of new treatment options for diseases previously thought difficult or impossible to treat. In particular the treatment of lung disorders could be achieved by the use of specially designed NDS. In this paper we reviewed the multiple advantages of NDS to improve lung treatments, particularly the special feature of NPs to be taken up by cells and to translocate through the epithelial barrier. However, the need exists of a balance between therapeutic efficacy and safety of the NDS with the aim to increase the benefit to risk ratio. Studies of pulmonary toxicity of environmental and occupational NPs allow to foresee possible adverse health effects of NDS. There is now a great need to perform studies with chronic exposures at low doses which are more realistic to be encountered in real live conditions. The understanding of the mechanisms behind these NP specific effects is essential for the development of safe NDS allowing the rapid expansion of new applications of nanomedicine for innovative treatment strategies.

Acknowledgement
Supported by EC FP7 (Health-2007-1.3-4) Contract no: 201335

References

[1] Wagner V, Dullaart A, Bock AK, and Zweck A. 2006 The emerging nanomedicine landscape. *Nature Biotechnology* **24** Number 10 October 1211-17.

[2] Allhoff F. The Coming Era of Nanomedicine *The American Journal of Bioethics*, **9**(10) 3–11.

[3] Card JW, Zeldin DC, Bonner JC, Nestmann ER. 2008 Pulmonary applications and toxicity of engineered nanoparticles. *Am J Physiol Lung Cell Mol Physiol.* **295**(3) L400-11.

[4] Smola M, Vandamme T, Sokolowski A. 2008 Nanocarriers as pulmonary drug delivery systems to treat and to diagnose respiratory and non respiratory diseases. *Int J Nanomedicine* **3**(1) 1-19.

[5] Mansour HM, Rhee YS, Wu X. 2009 Nanomedicine in pulmonary delivery. *Int J Nanomedicine* **4** 299-319

[6] Hussain S, Boland S, Baeza-Squiban A, Hamel R, Thomassen LC, Martens JA, Billon-Galland MA, Fleury-Feith J, Moisan F, Pairon JC, Marano F. 2009 Oxidative stress and proinflammatory effects of carbon black and titanium dioxide nanoparticles: role of particle surface area and internalized amount. *Toxicology* 16;260(1-3):142-9.

[7] Tanaka M, Nyce JW. 2001 Respirable antisense oligonucleotides: a new drug class for respiratory disease. *Respir Res.* **2**(1) 5-9.

[8] Yamagata T, Ichinose M. Agents against cytokine synthesis or receptors. 2006 *Eur J Pharmacol* 8;533(1-3):289-301.

[9] Li L, Das AM, Torphy TJ, Griswold DE. 2002 What's in the pipeline? Prospects for monoclonal antibodies (mAbs) as therapies for lung diseases. *Pulm Pharmacol Ther* **15**(5) 409-16.

[10] Gill DR, Bazzani RP, Hyde SC. 2010 Strategies for long-term expression of transgenes in the respiratory epithelium. *Curr Opin Mol Ther* **12**(4) 386-93

[11] Sanders N, Rudolph C, Braeckmans K, De Smedt SC, Demeester J. 2009 Extracellular barriers in respiratory gene therapy. *Adv Drug Deliv Rev.* 27;61(2):115-27

[12] Limbach LK, Wick P, Manser P, Grass RN, Bruinink A and Stark WJ. 2007 Exposure of engineered nanoparticles to human lung epithelial cells: influence of chemical composition and catalytic activity on oxidative stress. *Environ. Sci. Technol.* **41** 4158–63.

[13] Oberdörster G, Oberdörster E, Oberdörster J. 2005 Nanotoxicology: an emerging discipline evolving from studies of ultrafine particles. *Environ Health Perspect* **113**(7) 823-39.

[14] Semmler-Behnke M, Kreyling W, Lipka J, Fertsch S, Wenk A, Takenaka S, Schmid G, Brandau W. 2008 Biodistribution of 1.4- and 18-nm gold particles in rats. *Small* **4**(12) 2108-11.

[15] Kreyling W, Semmler M, Erbe F, Mayer P, Takenaka S, Schulz H, Oberdörster G, Ziesenis A. 2002 Translocation of ultrafine insoluble iridium particles from lung epithelium to extrapulmonary organs is size dependent but very low. *J Toxicol Environ Health* A **65**(20) 1513-30.

[16] Semmler-Behnke M, Takenaka S, Fertsch S, Wenk A, Seitz J, Mayer P, Oberdörster G, Kreyling W. 2007 Efficient elimination of inhaled nanoparticles from the alveolar region: evidence for interstitial uptake and subsequent reentrainment onto airways epithelium. *Environ Health Perspect* **115**(5) 728-33.

[17] Chen J, Tan M, Nemmar A, Song W, Dong M, Zhang G, Li Y. 2006 Quantification of extrapulmonary translocation of intratracheal-instilled particles in vivo in rats: effect of lipopolysaccharide. *Toxicology* **222**(3) 195-201.

[18] Meiring J, Borm P, Bagate K, Semmler M, Seitz J, Takenaka S, Kreyling W. 2005 The influence of hydrogen peroxide and histamine on lung permeability and translocation of iridium nanoparticles in the isolated perfused rat lung. *Part Fibre Toxicol* **2** 3.

[19] Yacobi N, Demaio L, Xie J, Hamm-Alvarez S, Borok Z, Kim K, Crandall E. 2008 Polystyrene nanoparticle trafficking across alveolar epithelium. *Nanomedicine* **4**(2) 139-45.

[20] Furuyama A, Kanno S, Kobayashi T, Hirano S. 2009. Extrapulmonary translocation of intratracheally instilled fine and ultrafine particles via direct and alveolar macrophage-

associated routes. *Arch Toxicol* **83**(5) 429-37

[21] Pandey R, Zahoor A, Sharma S, Khuller GK. 2003 Nanoparticle encapsulated antitubercular drugs as a potential oral drug delivery system against murine tuberculosis. *Tuberculosis (Edinb)* **83**(6) 373-8

[22] Kawashima Y, Yamamoto H, Takeuchi H, Fujioka S, Hino T. 1999 Pulmonary delivery of insulin with nebulized DL-lactide/glycolide copolymer (PLGA) nanospheres to prolong hypoglycemic effect. *J Control Release* 1;62(1-2):279-87

[23] Zhang Q, Shen Z, Nagai T. 2001 Prolonged hypoglycemic effect of insulin-loaded polybutylcyanoacrylate nanoparticles after pulmonary administration to normal rats. *Int J Pharm* 7;218(1-2):75-80

[24] Geiser M, Casaulta M, Kupferschmid B, Schulz H, Semmler-Behnke M, Kreyling W. 2008 The role of macrophages in the clearance of inhaled ultrafine titanium dioxide particles. *Am J Respir Cell Mol Biol* **38**(3) 371-6.

[25] Church LD, McDermott MF. 2009 Canakinumab, a fully-human mAb against IL-1beta for the potential treatment of inflammatory disorders. *Curr Opin Mol Ther* **11**(1) 81-9

[26] Barnes PJ. 2003 Cytokine-directed therapies for the treatment of chronic airway diseases. *Cytokine Growth Factor Rev* **14**(6) 511-22

[27] Mahler DA, Huang S, Tabrizi M, Bell GM. 2004 Efficacy and safety of a monoclonal antibody recognizing interleukin-8 in COPD: a pilot study. *Chest* **126**(3) 926-34.

[28] Rennard SI, Fogarty C, Kelsen S, Long W, Ramsdell J, Allison J, Mahler D, Saadeh C, Siler T, Snell P, Korenblat P, Smith W, Kaye M, Mandel M, Andrews C, Prabhu R, Donohue JF, Watt R, Lo KH, Schlenker-Herceg R, Barnathan ES, Murray J. 2007 The safety and efficacy of infliximab in moderate to severe chronic obstructive pulmonary disease. *Am J Respir Crit Care Med* 1;175(9):926-34

[29] Dentener MA, Creutzberg EC, Pennings HJ, Rijkers GT, Mercken E, Wouters EF. 2008 Effect of infliximab on local and systemic inflammation in chronic obstructive pulmonary disease: a pilot study. *Respiration* **76**(3) 275-82.

[30] Van der Vaart H, Koëter GH, Postma DS, Kauffman HF, ten Hacken NH. 2005 First study of infliximab treatment in patients with chronic obstructive pulmonary disease. *Am J Respir Crit Care Med* 15;172(4):465-9.

[31] Woodruff PG, Wolff M, Hohlfeld JM, Krug N, Dransfield MT, Sutherland ER, Criner GJ, Kim V, Prasse A, Nivens MC, Tetzlaff K, Heilker R, Fahy JV. 2010 Safety and efficacy of an inhaled epidermal growth factor receptor inhibitor (BIBW 2948 BS) in chronic obstructive pulmonary disease. *Am J Respir Crit Care Med* 1;181(5):438-45.

[32] Baeza A, Boland S, Hussain S, Marano F. 2011 Health effects of nanoparticles. *Systems Toxicology : from omics technology to nanotechnology »* John Wiley and Sons Ltd.

[33] Bermudez E, Mangum JB, Wong BA, Asgharian B, Hext PM, Warheit DB, Everitt JI. 2004. Pulmonary responses of mice, rats, and hamsters to subchronic inhalation of ultrafine titanium dioxide particles. *Toxicol.Sci.* **77**(2) 347-357.

[34] Elder A, Gelein R, Finkelstein JN, Driscoll KE, Harkema J, Oberdorster G. 2005 Effects of subchronically inhaled carbon black in three species. I. Retention kinetics, lung inflammation, and histopathology. *Toxicol.Sci.* **88**(2) 614-629.

[35] Grassian VH, O'Shaughnessy PT, Mcakova-Dodd A, Pettibone JM, Thorne PS. 2007. Inhalation exposure study of titanium dioxide nanoparticles with a primary particle size of 2 to 5 nm. *Environ.Health Perspect.* **115**(3) 397-402.

[36] Warheit DB, Webb TR, Reed KL. 2006. Pulmonary toxicity screening studies in male rats with TiO2 particulates substantially encapsulated with pyrogenically deposited, amorphous silica. *Part Fibre.Toxicol.* **3** 3.

[37] Warheit DB, Webb TR, Reed KL, Frerichs S, Sayes CM. 2007. Pulmonary toxicity study in rats with three forms of ultrafine-TiO2 particles: differential responses related to surface properties. *Toxicology* **230**(1) 90-104.

[38] Warheit DB, Webb TR, Sayes CM, Colvin VL, Reed KL. 2006b. Pulmonary instillation studies with nanoscale TiO2 rods and dots in rats: toxicity is not dependent upon particle size and surface area. *Toxicol.Sci.* **91**(1) 227-236.
[39] Lam CW, James JT, McCluskey R, Hunter RL. 2004. Pulmonary toxicity of single-wall carbon nanotubes in mice 7 and 90 days after intratracheal instillation. *Toxicol.Sci.* **77**(1) 126-134.
[40] Mercer RR, Scabilloni J, Wang L, Kisin E, Murray AR, Schwegler-Berry D, Shvedova AA, Castranova V. 2008. Alteration of deposition pattern and pulmonary response as a result of improved dispersion of aspirated single-walled carbon nanotubes in a mouse model. *Am.J.Physiol Lung Cell Mol.Physiol* **294**(1) L87-L97.
[41] Warheit DB, Laurence BR, Reed KL, Roach DH, Reynolds GA, Webb TR. 2004. Comparative pulmonary toxicity assessment of single-wall carbon nanotubes in rats. *Toxicol.Sci.* **77**(1) 117-125.
[42] Alessandrini F, Schulz H, Takenaka S, Lentner B, Karg E, Behrendt H, Jakob T. 2006. Effects of ultrafine carbon particle inhalation on allergic inflammation of the lung. *J.Allergy Clin.Immunol.* **117**(4) 824-830.
[43] Inoue K, Takano H, Yanagisawa R, Koike E, Shimada A. 2009. Size effects of latex nanomaterials on lung inflammation in mice. *Toxicol.Appl.Pharmacol.* **234**(1) 68-76.
[44] Inoue K, Takano H, Yanagisawa R, Sakurai M, Abe S, Yoshino S, Yamaki K, Yoshikawa T. 2007. Effects of nanoparticles on lung physiology in the presence or absence of antigen. *Int.J.Immunopathol.Pharmacol.* **20**(4) 737-744.
[45] Hussain S, Vanoirbeek JA, Luyts K, De Vooght V, Verbeken E, Thomassen LC, Martens JA, Dinsdale D, Boland S, Marano F, Nemery B, Hoet PH. 2010. Lung exposure to nanoparticles modulates an asthmatic response in a mouse model of asthma. *Eur Respir J.* **6** 11.
[46] Nygaard UC, Samuelsen M, Aase A, Lovik M. 2004. The capacity of particles to increase allergic sensitization is predicted by particle number and surface area, not by particle mass. *Toxicol.Sci.* **82**(2) 515-524.
[47] Takenaka S, Karg E, Kreyling WG, Lentner B, Schulz H, Ziesenis A, Schramel P, Heyder J. 2004. Fate and toxic effects of inhaled ultrafine cadmium oxide particles in the rat lung. *Inhal Toxicol.* **16** Suppl 1:83-92.
[48] Farkas A, Balásházy I, Szocs K. 2006. Characterization of regional and local deposition of inhaled aerosol drugs in the respiratory system by computational fluid and particle dynamics methods. *J Aerosol Med.* **19**(3) 329-43.
[49] Anderson PJ, Wilson JD, Hiller FC. 1990. Respiratory tract deposition of ultrafine particles in subjects with obstructive or restrictive lung disease. *Chest.* **97**(5) 1115-20.
[50] Roller M. 2009. Carcinogenicity of inhaled nanoparticles. *Inhal.Toxicol.* **21**(S1) 144-157.
[51] Schins RP, Knaapen AM. 2007. Genotoxicity of poorly soluble particles. *Inhal.Toxicol.* **19** Suppl 1:189-198.
[52] Nel A, Xia T, Mädler L, Li N. 2006. Toxic potential of materials at the nanolevel. *Science* **311**(5761) 622-7.
[53] Xia T, Kovochich M, Brant J, Hotze M, Sempf J, Oberley T, Sioutas C, Yeh J, Wiesner M, Nel A. 2006. Comparison of the abilities of ambient and manufactured nanoparticles to induce cellular toxicity according to an oxidative stress paradigm. *Nano Lett* **6**(8) 1794-807.
[54] Hoet P, Nemmar A, Nemery B. 2004. Health impact of nanomaterials? *Nat Biotechnol* 22(1):19.
[55] Donaldson K, Borm PJ, Castranova V, Gulumian M. 2009. The limits of testing particle-mediated oxidative stress in vitro in predicting diverse pathologies; relevance for testing of nanoparticles. *Part Fibre Toxicol.* **27**;6:13.
[56] Marano F, Hussain S, Rodrigues-Lima F, Baeza-Squiban A, Boland S. 2010. Nanoparticles: molecular targets and cell signalling. *Arch Toxicol.* **5** 26.
[57] Möller W, Brown D, Kreyling W, Stone V. 2005. Ultrafine particles cause cytoskeletal dysfunctions in macrophages: role of intracellular calcium. *Part Fibre Toxicol* **2**:7.
[58] Pan Y, Leifert A, Ruau D, Neuss S, Bornemann J, Schmid G, Brandau W, Simon U, Jahnen-

Dechent W. 2009. Gold nanoparticles of diameter 1.4 nm trigger necrosis by oxidative stress and mitochondrial damage. *Small* **5**(18) 2067-76.

[59] Pan Y, Neuss S, Leifert A, Fischler M, Wen F, Simon U, Schmid G, Brandau W, Jahnen-Dechent W. 2007. Size-dependent cytotoxicity of gold nanoparticles. *Small* **3**(11) 1941-9.

[60] Sydlik U, Bierhals K, Soufi M, Abel J, Schins R, Unfried K. 2006. Ultrafine carbon particles induce apoptosis and proliferation in rat lung epithelial cells via specific signaling pathways both using EGF-R. *Am J Physiol Lung Cell Mol Physiol* **291**(4) L725-33.

[61] Vamanu C, Cimpan M, Høl P, Sørnes S, Lie S, Gjerdet N. 2008. Induction of cell death by TiO2 nanoparticles: studies on a human monoblastoid cell line. *Toxicol In Vitro* **22**(7) 1689-96.

[62] Hussain S, Thomassen LC, Ferecatu I, Borot MC, Andreau K, Martens JA, Fleury J, Baeza-Squiban A, Marano F, Boland S. 2010. Carbon black and titanium dioxide nanoparticles elicit distinct apoptotic pathways in bronchial epithelial cells. *Part Fibre Toxicol.* 16;7:10.

In vitro cytotoxicity assays of solid lipid nanoparticles in epithelial and dermal cells

D M Ridolfi[1], P D Marcato[1], D Machado[2], R A Silva[2], G Z Justo[3] and N Durán[1,4]

[1]Instituto de Química, Universidade Estadual de Campinas, Brazil

[2]Instituto de Biologia, Universidade Estadual de Campinas, Brazil

[3]Departamento de Bioquímica, Universidade Federal de São Paulo, Brazil

[4]Centro de Ciências naturais e Humanas, Universidade Federal do ABC, Brazil

E-mail: daniela_ridolfi@hotmail.com

Abstract. In recent years, the interest in nanostructured systems to drug delivery has increased because they offer several advantages over conventional dosage forms. Solid Lipid Nanoparticles (SLN) have been highlighted among these systems because they have advantages such as high physical stability, protection against drug degradation and ease of scale-up and manufacturing, without using organic solvent. The aim of this work was to evaluate the potential of SLN, by in vitro cytotoxicity assays, for dermal drug delivery. SLN of three different lipids were prepared by hot high pressure homogenization and the cytotoxicity was assessed by 3-(4,5-dimethylthiazol- 2-yl)-2,5-diphenyl tetrazolium bromide (MTT) test in mouse 3T3 fibroblasts and human HaCaT keratinocytes. SLN showed no cytotoxic potential suggesting a great potential for dermal application.

1. Introduction

In recent years, the interest in nanostructured systems to drug delivery has increased because they offer several advantages over conventional dosage forms such as increased therapeutic efficacy and reduction of collateral effects and toxicity [1]. Solid Lipid Nanoparticles (SLN) were developed at the beginning of the 1990s as an alternative carrier system to emulsions, liposomes and polymeric nanoparticles. They offer several advantages compared to other colloidal systems such as high physical stability, protection against drug degradation due to the solid state of lipid matrix and ease of scale-up and manufacturing, without using organic solvent. SLN are produced by replacing the liquid lipid (oil) of an o/w emulsion by a solid lipid (at both room and body temperature) or a blend of solid lipids [2].

Nanoparticles can cause cytotoxicity by adherence of the particle to the cell membrane, degradation and subsequent release of cytotoxic degradation products. Another mechanism is the internalization of nanoparticles by cells, intracellular degradation and subsequent toxic effects inside the cell [3]. The first step towards assessing the safety of a new formulation often involves *in vitro* cytotoxicity test methods. The majority of cytotoxicity assays used throughout published nanoparticles studies measure cell death via colorimetric methods that assess membrane integrity or mitochondrial activity.

Mitochondrial activity can be tested using tetrazolium salts as mitochondrial dehydrogenase enzymes cleave the tetrazolium ring. Only active mitochondria contain these enzymes; therefore, the reaction only occurs in living cells. The most widely used test is the MTT viability assay. MTT is pale yellow in solution but produces a dark-blue formazan product within live cells [4].

Schöler et al. found that the cytotoxicity of SLN assessed by MTT test on murine peritoneal macrophages was concentration dependent and influenced by the lipid matrix [5]. The cytotoxicity of SLN was also found to be dependent on the type and concentration of surfactant used [6-9]. Therefore, the choice of the lipid matrix and the surfactant is essential in order to formulate a safe preparation.

The aim of this work was to assess the cytotoxic potential of SLN, by in *vitro* cytotoxicity assays, using fibroblasts and keratinocytes that are the predominant resident cells in dermis and epidermis, respectively, in order to evaluate the potential of SLN for dermal drug delivery.

2. Methods

2.1. Solid Lipid Nanoparticles (SLN) preparation

The solid lipids used, cetyl palmitate (CP), myristyl myristate (MM) and cetyl esters (SS), were kindly donated by Croda (Brazil). SLN were produced by hot high pressure homogenization. The solid lipid heated to around 10°C above its melting point. Afterwards, the mixture was added to a hot aqueous solution of Pluronic F68 under high agitation in an Ultra-turrax® T18 to form a pre-emulsion. The pre-emulsion was homogenized using a Panda 2k (Niro Soavi, Italy), applying three homogenization cycles at 600 bar and cooled to form the SLNs.

2.2. Particle Size and Zeta Potential

The mean diameter (z-average diameter) and size distribution were measured by photon correlation spectroscopy (PCS) (Nano ZS Zetasizer, Malvern Instruments Corp, UK) at 25°C in polystyrene cuvettes with path length of 10 mm. The zeta potential was measured in capillary cells with path lengths of 10 mm, using the Nano ZS Zetasizer. Measurements were performed in deionized water obtained by a MilliQ system. All the measurements were performed in triplicate and data were expressed as the means ± standard deviation (SD).

2.3. Physical stability of SLN

The physical stability of the SLN was evaluated by examining changes of mean particle size and zeta potential during storage at 4°C.

2.4. Cytotoxicity assays

2.4.1. *Cell lines and culture.* BALB/c 3T3 cell line was purchased from National Institute of Health-Baltimore, USA (NIH) and, HaCaT cell line was kindly gifted from Dr. Liudmila L. Kodach (Academic Medical Center, Amsterdam University). Both cell lines were routinely grown in DMEM supplemented with fetal bovine serum, 15% for BALB/c 3T3 and 10% for HaCaT and, antibiotics (100 U/mL penicillin, 10 µg/mL streptomycin) in a humidified incubator with 5% carbon dioxide, at 37oC.

2.4.2. *Treatment of cells with CP, MM and SS.* Cells, BALB/c 3T3 10^4 cells/well and HaCaT $7x10^3$ cells/well, were incubated in 96-well plates until reach the semiconfluence and then treated with different concentrations of SLN (10–500 µg/mL final concentrations) for 24 h. Cell viability was assessed by the MTT reduction assay.

2.4.3. *MTT reduction assay.* The medium containing riboflavin was removed and 100 µL of 3-(4,5-dimethylthiazol-2-yl)-2,5-diphenyl tetrazolium bromide (MTT) solution (0.5 mg/mL of culture medium) was added to each well. After incubation for 4 h at 37°C, the medium was removed and the formazan crystals solubilized in 100 μL of ethanol. The plate was shaken for 5 min on a plate shaker and the absorbance was measured at 570 nm in a microplate reader (ELx800, BioTek Instruments, Inc., Winooski, VT) [10]

2.4.4. *Statistical analysis.* Cell viability data were expressed as the means ± standard deviation (SD) of experiments. A percentage of viability compared with control wells (the means optical density of untreated cells was set to 100% viability) was calculated from the concentration-response curves by linear regression analysis. Data from each assay were analyzed statistically by ANOVA. Multiple comparisons among group mean differences were checked with the Tukey test. Differences were considered significant when the $p < 0.05$.

3. Results and discussion

3.1. SLN characterization

SLN were prepared with three different lipids, cetyl palmitate (CP), myristyl myristate (MM) and cetyl ester (SS). The mean diameter, polydispersity index (PI) and zeta potential of SLN are listed in Table 1. The SLN dispersions were obtained with high homogeneity (low PI) and both mean diameter and zeta potential of the SLN dispersion were close. As shown in Figure 1, SLN dispersions did not show any significant changes of particle size and zeta potential when maintained at 4°C in test period (six months), indicating a high physical stability. This can be explained by high zeta potential of the SLN dispersions and by the steric stabilization provided by surfactant used (Pluronic F68).

Table 1. z-average diameter, polydispersity index (PI) and zeta potential and of SLN

SLN	Diameter (nm)	PI	Zeta Potential (mV)
CP	189.0 ± 1.8	0.149 ± 0.037	-34.7 ± 4.1
MM	185.4 ± 6.3	0.188 ± 0.012	-31.9 ± 2.1
SS	197.5 ± 3.5	0.234 ± 0.023	-30.5 ± 1.5

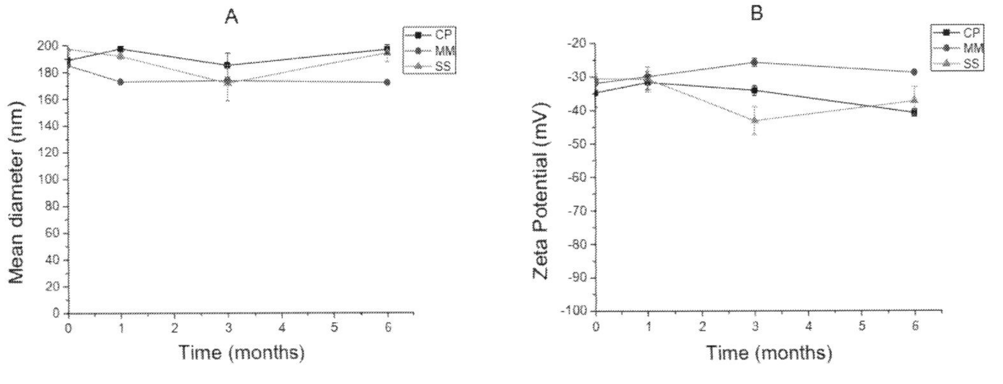

Figure 1. Physical stability of SLN during storage at 4 °C. (A) Mean diameter; (B) Zeta potential.

3.2. Cytotoxicity of SLN

The results obtained in cytotoxicity assays are presented in Figure 2. The cytotoxic effect was influenced by the lipid matrix, as verified by statistical analysis. The lipid that showed smaller reduction in cell viability in the tested concentration range was the SS. Although in some concentrations of SLN cell viability was reduced, until the highest concentration tested (500 μg/mL) the IC_{50} (half maximal inhibitory concentration) was not reached for the three lipids in both cell showing no cytotoxic potential of the tested SLN dispersions. When compared to results obtained by

Marcato et al. [11] in MTT test with polymeric nanoparticles, using the same cell lines, the SLN were less cytotoxic.

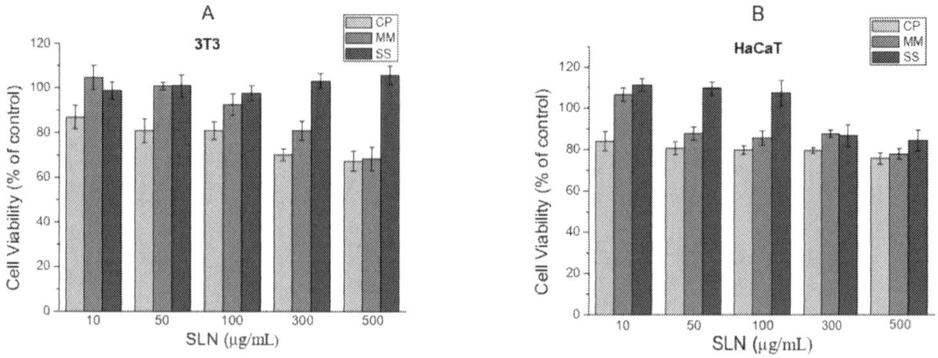

Figure 2. Cytotoxicity of SLN in (A) 3T3 fibroblasts and (B) HaCaT keratinocytes.

4. Conclusions

The results obtained in this work suggest that SLN have a great potential for dermal drug delivery because they showed no cytotoxic potential in fibroblasts and keratinocytes cells.

Acknowledgements

FAPESP, CNPq and Brazilian Network of Nanocosmetics (MCT/CNPq).

References

[1] Mishra B, Patel B B and Tiwari S 2010 Nanomed.-Nanotechnol. Biol. Med. 6 9
[2] Mehnert W and Mäder K 2001 Adv. Drug Deliv. Rev. 47 165
[3] Pardeike J, Hommoss A and Müller R H 2009 Int. J. Pharm. 366 170
[4] Lewinski N, Colvin V and Drezek R 2008 Small 4 26
[5] Schöler N, Hahn H, Müller R H and Liesenfeld O 2002 Int. J. Pharm. 231 167
[6] Müller R H, Maassen S, Schwarz C and Mehnert W 1997 J. Control. Release 47 261
[7] Shöler N, Hahn H, Müller R H and Liesenfeld O 2002 Int. J. Pharm. 231 167
[8] Weyenberg W, Filev P, Plas D V, Vandervoort J, Smet K D, Sollie P and Ludwig A 2007 Int. J. Pharm. 337 291
[9] Kristl J, Teskac K, Milek M and Rascan I M 2008 Toxicol. Appl.Pharmacol. 232 218
[10] Mosmann T. 1983 J. Immunol. Methods 65 55
[11] Marcato P D , Caverzan J, Rossi-Bergmann B, Pinto E F, Machado D, Silva R A, Justos G Z, Ferreira C V and Durán N 2011. J. Nanosci. Nanotechnol., in press.

In vitro toxicity and uptake of magnetic nanorods

M Safi, S Clowez, A Galimard and J-F Berret

Matière et Systèmes Complexes, UMR 7057 CNRS Université Denis Diderot Paris-VII, Bâtiment Condorcet, 10 rue Alice Domon et Léonie Duquet, 75205 Paris (France)

E-mail: jean-francois.berret@univ-paris-diderot.fr

Abstract. In this paper we investigate the internalization and cytotoxicity of nanostructured materials having the form of elongated rods, with diameter of 200 nm and lengths 1 – 10 µm. The rods were made from the controlled aggregation of sub-10 nm iron oxide nanoparticles. Recently, we have shown that the nanorods inherited the superparamagnetic property of the particles. These rods can actually be moved by the application of an external magnetic field. Here we evaluate the in vitro toxicity of the magnetic nanorods by using MTT assays on NIH/3T3 mouse fibroblasts. The toxicity assays revealed that the nanorods are biocompatible, as exposed cells remained 100% viable relative to controls over a period of a few days. Optical microscopy allows to visualize the rods inside the cells and to determine their number per cell. Roughly 1/3 of the total incubated rods were uptaken by the fibroblasts.

1. Introduction

Inorganic nanomaterials and particles with enhanced optical, mechanical or magnetic attributes are currently being developed for a wide scope of applications, including catalysis, photovoltaics, coating and nanomedicine [1]. More generally, nanomaterials of different shapes and sizes are regarded as promising tools at the scale of the cell for manipulation, diagnostic and therapy. Interactions between nanomaterials and living organisms were also investigated extensively because of the concerns raised by their potential toxicity. As shown in several reviews [2, 3], the risk assessment of nanomaterials towards living cells and tissues have not been fully evaluated. Among the wide variety of nanomaterials available, magnetic nanowires have received considerable attention [4-8]. Nanowires are anisotropic colloidal objects with submicronic diameters and lengths in the range 1 – 100 µm. In specific applications such as cell separation, ferromagnetic nickel nanowires made by electrodeposition were shown to outperform magnetic beads of comparable volume [5]. However, one major drawback encountered with nickel or iron nanowires is that these objects carry a permanent magnetic moment, and are thus susceptible to aggregate in solution because of magnetic dipolar attraction [4].

In the present paper, we exploit a simple and versatile waterborne synthesis process to generate magnetic nanorods [9, 10]. Highly persistent rods of diameters around 200 nm and lengths comprised between 1 µm and 30 µm were fabricated by controlling the assembly of sub-10 nm iron oxide nanoparticles. These magnetic nanorods are different from the ferrimagnetic electrodeposited wires mentioned previously. The rods are superparamagnetic *i.e.* they do not carry a permanent magnetic moment, which prevents their spontaneous aggregation in a dispersion [9, 11]. Since these rods are designed to be used as micromechanical tools at the cellular level, their interactions with living cells need to be investigated. Here we explore their interactions with NIH/3T3 mouse fibroblasts and show that the rods were internalized in the cellular matrix. Toxicity assays show that the rods do not display acute toxicity towards the cells on a short-term basis. In this study, the effects of the rods on the fibroblasts and those of the nanoparticles from which they are made were also compared.

2. Materials and Methods

2.1. Chemicals, synthesis and characterization

The synthesis of superparamagnetic nanoparticles (NP) used the technique of 'soft chemistry' based on the polycondensation of metallic salts in alkaline aqueous media elaborated by R. Massart [12]. This technique resulted in the formation of magnetite (Fe_3O_4) NPs of sizes comprised between 4 and 15 nm. Magnetite nanocrystals were further oxidized into maghemite (γ-Fe_2O_3) and sorted according to their size. Adding large amount of nitric acid to the dispersions down to pH 0.5 initiated a liquid-gas type phase separation, resulting in coexisting dilute and concentrated phases of γ-Fe_2O_3 particles. The concentrated phase which contained the largest particles was separated by magnetic sedimentation. The process was repeated 5 times, reducing the initial polydispersity from 0.4 to 0.2. In the synthesis conditions, the magnetic dispersions were stabilized by electrostatic interactions arising from the native cationic charges at the surface of the particles. Vibrating sample magnetometry was performed to characterize the size distribution of the maghemite cores [13, 14]. Fig. 1 shows the evolution of the macroscopic magnetization M(H) normalized by its saturation value M_S for the present γ-Fe_2O_3 batch. Here, $M_S = \phi m_S$, where m_S is the volumetric magnetization of maghemite ($m_S = 2.9\times10^5$ A m^{-1}).

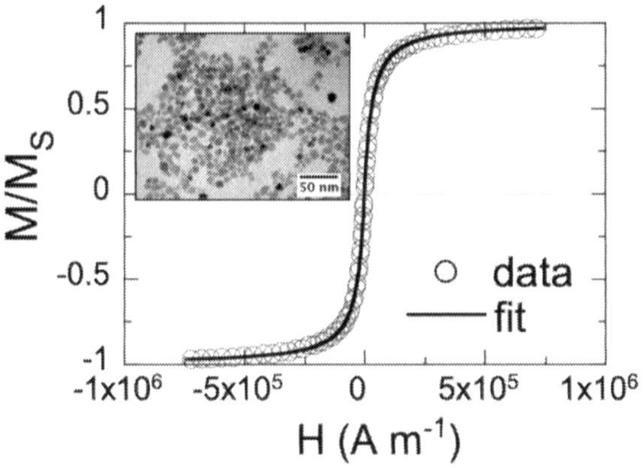

Figure 1. Magnetic field dependence of the macroscopic magnetization M(H) of a γ-Fe_2O_3 dispersion normalized by its saturation value M_S. The solid curve was obtained using the Langevin function convoluted with a log-normal distribution with median diameter D_0 = 6.7 nm and polydispersity s_{NP} = 0.21.
Inset: Transmission electron microscopy of iron oxide at the magnification of ×120000.

The solid curve in Fig. 1 was obtained using the Langevin function convoluted with a log-normal distribution function of the particle size. The parameters of the distribution are the median diameter D_0 = 6.7 nm and the polydispersity s_{NP} = 0.21. The polydispersity was defined as the ratio between the standard deviation and average diameter. A transmission electron microscopy image obtained from the same γ-Fe_2O_3 dispersion is illustrated in the inset of Fig.1. The iron oxide NPs exhibited compact and spherical structures, with size and polydispersity in excellent agreement with the VSM data [14].

In order to improve their colloidal stability, the cationic particles were coated by poly(acrylic acid) oligomers. Poly(sodium acrylate), the salt form of poly(acrylic acid) with a molecular weight M_W = 2000 g mol^{-1} and a polydispersity of 1.7 was purchased from Sigma Aldrich and used without further purification. In order to adsorb polyelectrolytes on the surface of the nanoparticles, we followed the *Precipitation–Redispersion* protocol, as described elsewhere [14, 15]. The precipitation of the iron oxide dispersions by PAA_{2K} was performed in acidic conditions (pH 2). The precipitate was separated from the solution by centrifugation and its pH was increased by the addition of ammonium hydroxide. The precipitate redispersed spontaneously at pH 7–8, yielding a clear solution that contained the polymer coated particles. The hydrodynamic sizes found in coated NPs were 6 nm above that of the bare particles, indicating the presence of a 3 nm PAA_{2K} brush surrounding the particles. This simple technique allowed us to produce large quantities of coated particles (> 1 g of oxides) within a relatively short time (< 1 h) [16]. The fabrication of the rods was performed using these coated particles, noted PAA_{2K}–γ-Fe_2O_3 in the following.

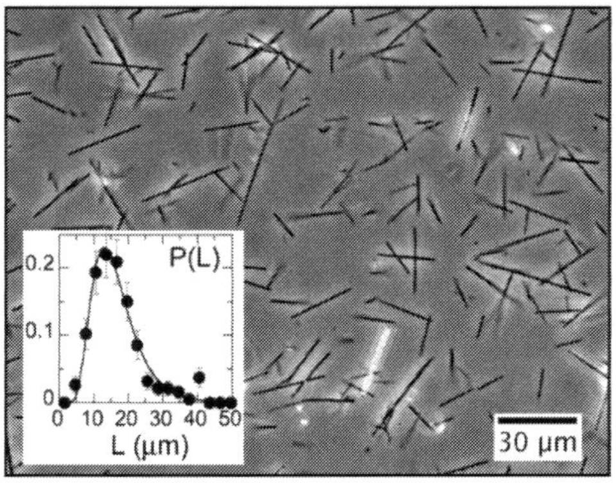

Figure 2. Phase-contrast optical microscopy images (40×) of a dispersion of nanostructured rods made from 6.7 nm γ-Fe_2O_3 particles in absence of magnetic field.

Inset: Length distribution of the rods. The distribution was found to be log-normal with median length and polydispersity noted $L_0 = 15$ µm and $s_R = 0.40$ respectively.

The fabrication of the nanorods is based on the NPs co-assembly in the presence of a magnetic field. The adopted strategy involved in a first step the preparation of two separate 1 M NH_4Cl solutions containing respectively *i)* the anionic iron oxide NPs (PAA_{2K}–γ-Fe_2O_3) and *ii)* the $PTEA_{11K}$-*b*-PAM_{30K} diblock copolymers. These copolymers (PTEA-*b*-PAM stands for poly(trimethylammonium ethylacrylate)-*b*-poly(acrylamide)) were synthesized by MADIX® controlled radical polymerization which is a Rhodia patented process [17]. The initial concentrations were adjusted to c = 0.1 wt. % for both NPs and copolymers. In a second step, the two solutions were mixed with each other and it was checked by dynamic light scattering that the two components remained dispersed [9, 11]. In a third step, the ionic strength of the mixture was progressively diminished by dialysis in the presence of an external magnetic field (0.1 T). Dialysis was performed against de-ionized water using a Slide-a-Lyzer® cassette with MWCO of 10 kD (Thermo Scientific). The volume of the dialysis bath was 300 larger than that of the samples. The electrical conductivity (Cyberscan PC6000) of the dialysis bath was measured during the ion exchange and served to monitor the desalting kinetics. In the conditions described here, the whole process reached a stationary and final state within one hour. Once the ionic strength of the bath reached its stationary value (3×10^{-3} M), the external magnetic field was removed and the dispersions were studied by optical microscopy (Fig. 2). In order to determine the length distribution of the rods, pictures were digitized and treated by the ImageJ software (http://rsbweb.nih.gov/ij/). The inset of Fig. 2 displays the length distribution function that was determined from a panel of 250 objects. It was found to be log-normal with a median length of $L_0 = 15$ µm and a polydispersity $s_R = 0.4$ (continuous line). Rods were further sonicated to reduce their average length down to 3 µm.

2.2. Experimental Methods

2.2.1. *Cell culture.* NIH/3T3 fibroblast cells from mice were grown as a monolayer in Dulbecco's modified Eagle's medium (DMEM) with high glucose (4.5 g L^{-1}) and stable glutamine (PAA Laboratories GmbH, Austria). This medium was supplemented with 10% fetal bovine serum (FBS) and 1% penicillin/streptomycin (PAA Laboratories GmbH, Austria), referred to as cell culture medium. Exponentially growing cultures were maintained in a humidified atmosphere of 5% CO_2 and 95% air at 37°C, and under these conditions the plating efficiency was 70 – 90% and the cell duplication time was 12 – 14 h. Cell cultures were passaged twice weekly using trypsin–EDTA (PAA Laboratories GmbH, Austria) to detach the cells from their culture flasks and wells. The cells were pelleted by centrifugation at 1200 rpm for 5 min.

2.2.2. *Transmission optical Microscopy.* For optical microscopy observations, phase-contrast images of the cells containing rods were acquired on an IX71 inverted microscope (Olympus) equipped with 40× and 60× objectives. 2×10^4 NIH/3T3 fibroblasts cells were first seeded onto a 96-well plate for 24 h prior being incubated with nanorods. µl-aliquots containing nanorods were added to the supernatant. The incubation of the rods lasted 24 hours. The third day, the excess medium was removed and the cells were washed with PBS solution (with calcium and magnesium, Dulbecco's, PAA Laboratories), trypsinized and centrifuged. The cell pellets were resuspended in DMEM. For optical microscopy, 20 µl of the previous cell suspension were deposited on a glass plate and sealed into to a Gene Frame® (Abgene/Advanced Biotech) dual adhesive system. The sample was then left for 4 h in the incubator to let the cells adhere onto the glass plate. Images were observed using a Photometrics Cascade camera (Roper Scientific) and Metaview software (Universal Imaging Inc.) as acquisition system.

2.2.3. *MILC protocol.* UV-visible spectrometry was performed in the MILC protocol (**M**ass of metal **I**nternalized/Adsorbed by **L**iving **C**ells) which consist in the measurement of the mass of nanoparticles incorporated into living cells. The quantity to be determined is the mass of iron expressed in the unit of picogram per cell. Cells were seeded onto 3.6 cm Petri dishes, and incubated until reaching 60% confluence and then incubated with nanomaterials, including the nanorods at different concentrations for 24 h. After the incubation period, the supernatant was removed and the layer of cells washed thoroughly with PBS. The cells were then trypsinised and mixed again with white DMEM without serum. 20 µl-aliquots were taken up for counting using a Malassez plate. The cells were finally centrifuged and pellets were dissolved in 35% HCl. When incubated with γ-Fe_2O_3, the HCl dissolved NIH/3T3 exhibited a yellow color. The absorbance of the liquid was studied using a Variant spectrophotometer (Cary 50 Scan), and it was compared to a calibrated reference. The reference was obtained by dissolving directly γ-Fe_2O_3 dispersions of known concentrations into 35% HCl.

2.2.4. *MTT toxicity assays.* MTT assays were performed with PAA_{2K}-coated iron oxide nanoparticles and nanorods for metal molar concentrations [Fe] between 10 µM and 10 mM. To avoid contamination from synthesis residues and free ions, nanoparticle and nanorod dispersions were dialyzed against DI-water which pH was adjusted to 8 by addition of sodium hydroxide (Spectra Por 2 dialysis membrane with MWCO 12 kD). Cells were seeded into 96-well microplates, and the plates were placed in an incubator overnight to allow for attachment and recovery. Cell densities were adjusted to 2×10^4 cells per well (200 µl). After 24 h, the nanoparticles and nanorods were applied directly to each well using a multichannel pipette. Cultures were incubated in triplicate for 24 h at 37°C. The MTT assay depends on the cellular reduction of MTT (3- (4,5-dimethylthiazol-2-yl)-2,5-diphenyl tetrazolium bromide, Sigma Aldrich Chemical) by the mitochondrial dehydrogenase of viable cells forming a blue formazan product which can be measured spectrophotometrically [18]. MTT was prepared at 5 mg ml^{-1} in PBS (with calcium and magnesium, Dulbecco's, PAA Laboratories) and then diluted 1–5 in medium without serum and without Phenol Red. After 24 h of incubation with nanoparticles, the medium was removed and 200 µl of the MTT solution was added to the microculture wells. After 4 h incubation at 37°C, the MTT solution was removed and 100 µl of 100% DMSO was added to each well to solubilize the MTT–formazan product. The absorbance at 562 nm was then measured with a microplate reader (Perkin Elmer). Prior to the microplate UV–vis spectrometry, MTT assays without particles were carried out with cell populations ranging from 5×10^3 to 5×10^5 cells.

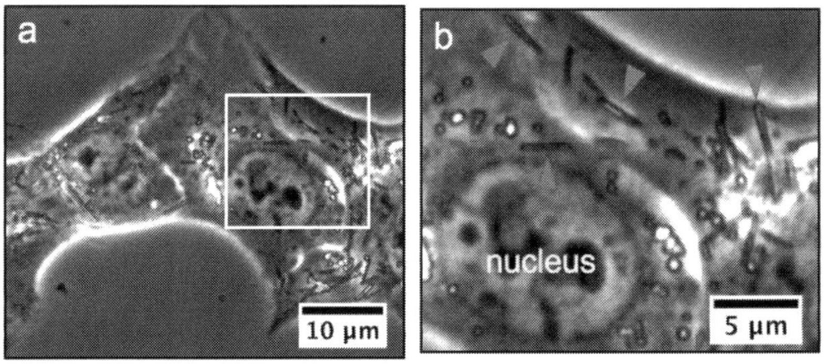

Figure 3. a) Phase contrast microscopy images of NIH/3T3 fibroblasts cells treated with nanorods for 24 h at a concentration of 30 rods/cell. b) Zoom of the previous image to emphasize the perinuclear region of a unique cell. Rods of various lengths and orientations are indicated by arrows.

3. Results and discussion

3.1. Nanorods are internalized by the cells

Fig. 3a shows a cluster of cells that were incubated with magnetic nanorods of average length 10 μm and at 30 rods per cell. In this work, the nanorod concentration was defined by the ratio of the number of rods incubated per cell. To allow comparison with data on γ-Fe_2O_3 NPs or with data from the literature, it was also expressed in terms of iron molar concentration [Fe] (30 rods per cell corresponds to [Fe] = 0.15 mM). The cells were found to maintain their morphology and adherence properties after a 24 h exposure. Fig. 3b displays a zoom of the previous image and emphasizes the perinuclear region of a single cell. There, elongated threads with lengths comprised between 1 to 7 μm and with random orientations are observed (arrows). The location of the rods into the cytoplasm and not outside in the supernatant or attached at the cellular membrane was confirmed by following their Brownian motions. The orientational fluctuations of the rods were recorded as a function of the time and compared with those of rods dispersed in water. It was found that the rotational diffusion constant was much slower for the internalized objects, in good agreement with recent estimations of the viscosity of the intracellular matrix [19]. These results also suggest that the internalization of nanorods was optimized when their length corresponded to the average size of the adherent cells [5].

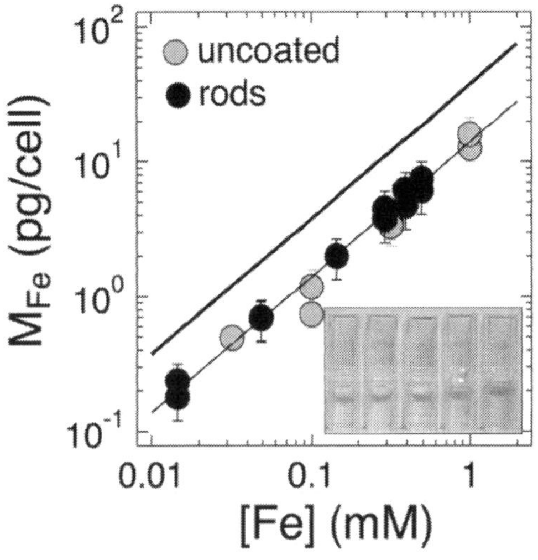

Figure 4. Amount of internalized or adsorbed iron oxide M_{Fe} (pg/cell) determined following the MILC protocol (see text for details). The upper thick straight line depicts the maximum amount of iron that can be uptaken by the cells for a molar concentration [Fe] in the supernatant. Grey bullets: uncoated γ-Fe_2O_3; closed bullets: nanorods (L_0 = 3 μm, s_R = 0.5). Inset: UV-cells containing increasing concentrations of iron oxide dissolved in HCl.

3.2. Quantitative determination of internalized iron oxide

The amount of internalized or adsorbed iron oxide was determined following the protocol MILC (**M**ass of metal **I**nternalized/Adsorbed by **L**iving **C**ells) which makes use of UV spectrophotometry to calculate the iron concentration from pelleted cells dissolved in concentrated HCl. The results (M_{Fe} *versus* [Fe]) were expressed in picograms of iron per cell. Here, [Fe] denotes the iron oxide molar concentration in the supernatant during the 24 h of incubation. Fig. 4 compares the [Fe]-dependences of internalized/adsorbed amounts for uncoated NPs and for the nanorods (L_0 = 3 µm, s_R = 0.5). In the inset, UV-cells containing increasing concentrations of iron oxide dissolved in HCl are shown for illustration. The thick straight line in Fig. 4 depicts the maximum amount of iron that could be uptaken by the cells. For instance, 3×10^6 of fibroblasts exposed to a [Fe] = 1 mM solution, the maximum amount lies at 37 pg/cell. For the uncoated particles and for the nanorods, M_{Fe} increases linearly and represents about 35% of the maximum value. Optical microscopy carried out 24 h after incubation and thorough washing shown that none of the rods were adsorbed at the cell membranes, leading to the conclusion that the M_{Fe}-data in Fig. 4 represent the amount of internalized iron under the form of nanorod. For uncoated γ-Fe_2O_3, the large quantities detected up to [Fe] = 10 mM were explained by the precipitation of the particles in the culture medium. This precipitation produced large and compact clusters in the micrometer range (1 – 20 µm) [16] that for the most part were adsorbed on the plasma membrane. Thorough washing with PBS buffer did not desorb these aggregates, probably because of their positive surface charges.

3.3. Toxicity assays

Cytoxicity studies determine the cellular alterations or damages of vital functions induced by xenobiotics. MTT viability assays were conducted on NIH/3T3 cells for the uncoated γ-Fe_2O_3 particles ([Fe] = 10 µM - 10 mM) and for the nanorods (L_0 = 3 µm, [Fe] = 0.1 µM - 1 mM). The comparison between this two types of nanomaterials aimed to identify the role of size and morphology on the cell survival. A rod concentration of 1 mM corresponded to 100 rods per cell. Exposure times were set at 24 h. As shown in Figs. 5 for both studies, the viability remained at a 100% level within experimental accuracy. These findings indicate a normal mitochondrial activity for the cultures tested. The results on the precipitated γ-Fe_2O_3 are in good agreement with earlier reports [20, 21]. As mentioned before, for this system most of the NPs were precipitated into large clusters that were not internalized because of their size (> 10 µm), but rather adsorbed on the plasma membranes. For the nanorods, the data also confirm those obtained recently on parent micron-size materials, such as silica [22] or carbon nanotubes [23]. In conclusion, the MTT assays show conclusive evidences and ensure the suitability of the rods for biomedical and biophysical applications.

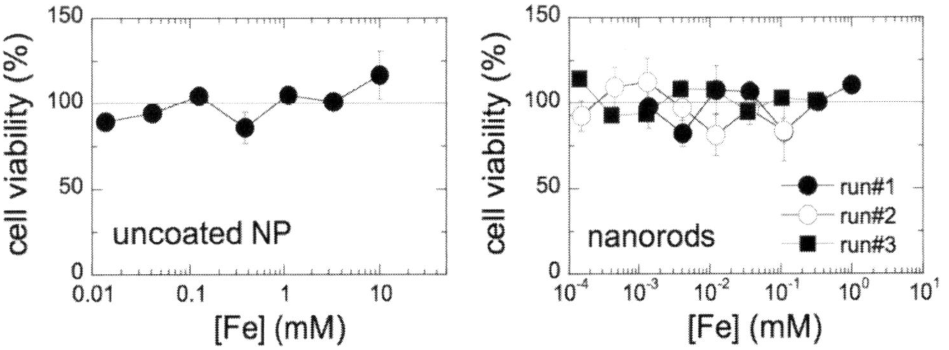

Figure 5. MTT (3-(4,5-dimethylthiazol-2-yl)-2,5-diphenyl tetrazolium bromide) viability assays conducted on NIH/3T3 cells incubated during 24 h with a) uncoated γ-Fe_2O_3 NPs and b) 3 µm-nanorods. In this work, the nanorod concentration was defined by the ratio of the number of rods incubated per cell. To allow comparison with data from the literature, it was also expressed in terms of iron molar concentration [Fe].

4. Conclusion

In this paper, we have evaluated the interactions and toxicity of magnetic nanorods with respect to NIH/3T3 mouse fibroblasts. Magnetic nanorods are a particularly promising class of nanomaterials since they can be used in confined geometries for cell manipulation, microrheology and microfluidics. The magnetic properties of the rods were inherited from the iron oxide particles, and allow to rotate the rods in a propeller-like motion by the application of an external field [9, 10]. Our approach with cell culture consisted to show that the rods were internalized and biocompatible with living fibroblasts. Concerning the first point, quantitative measurements of internalized amounts revealed that 3 µm-nanorods were uptaken in large proportion, typically 35% of the initial seeded quantity. Direct visualization of the rods inside the cytoplasm by optical microscopy also confirmed these conclusions: a large number of incubated rods were able to cross the cell plasma membranes. In the future, we plan to use transmission electron microscopy to identify if the rods are located in intracellular compartments or directly dispersed in the cytosol, and to have a better insight about the portals of entry into the cells. From these initial studies, it is concluded that the iron oxide based nanorods can be used safely with living cells, e.g. as microtools for *in vitro* and *in vivo* applications.

Acknowledgments

This research was supported in part by Rhodia (France), by the Agence Nationale de la Recherche under the contracts BLAN07-3_206866 and ANR-09-NANO-P200-36, by the European Community through the project : "NANO3T—Biofunctionalized Metal and Magnetic Nanoparticles for Targeted Tumor Therapy", project number 214137 (FP7-NMP-2007-SMALL-1) and by the Région Ile-de-France in the DIM framework related to Health, Environnement and Toxicology (SEnT).

References

[1] Kim J, Piao Y, Hyeon T. Chem Soc Rev. 2009;38(2):372-90.
[2] Lewinski N, Colvin V, Drezek R. Small. 2008;4(1):26-49.
[3] Singh N, Manshian B, Jenkins GJS, Griffiths SM, Williams PM, Maffeis TGG, et al. Biomaterials. 2009;30(23-24):3891-914.
[4] Johansson F, Jonsson M, Alm K, Kanje M. Experimental Cell Research. 2010;316(5):688-94.
[5] Hultgren A, Tanase M, Felton EJ, Bhadriraju K, Salem AK, Chen CS, et al. Biotechnology Progress. 2005;21(2):509 - 15.

[6] Goubault C, Jop P, Fermigier M, Baudry J, Bertrand E, Bibette J. Physical Review Letters. 2003;91(26):260802.
[7] Cappallo N, Lapointe C, Reich DH, Leheny RL. Physical Review E. 2007;76(3):6.
[8] Singh H, Laibinis PE, Hatton TA. Langmuir. 2005;21(24):11500 - 9.
[9] Fresnais J, Berret J-F, Frka-Petesic B, Sandre O, Perzynski R. Adv Mater. 2008;20(20):3877-81.
[10] Fresnais J, Ishow E, Sandre O, Berret J-F. Small. 2009;5(22):2533-6.
[11] Fresnais J, Berret J-F, Frka-Petesic B, Sandre O, Perzynski R. Journal of Physics: Condensed Matter. 2008;(49):494216.
[12] Massart R, Dubois E, Cabuil V, Hasmonay E. J Magn Magn Mat. 1995;149(1-2):1 - 5.
[13] Berret J-F, Sehgal A, Morvan M, Sandre O, Vacher A, Airiau M. J Colloid Interface Sci. 2006;303(1):315-8.
[14] Berret J-F, Sandre O, Mauger A. Langmuir. 2007;23(6):2993-9.
[15] Sehgal A, Lalatonne Y, Berret J-F, Morvan M. Langmuir. 2005;21(20):9359-64.
[16] Chanteau B, Fresnais J, Berret J-F. Langmuir. 2009;25(16):9064-70.
[17] Jacquin M, Muller P, Talingting-Pabalan R, Cottet H, Berret J-F, Futterer T, et al. J Colloid Interface Sci. 2007;316(2):897-911.
[18] Denizot F, Lang R. Journal of Immunological Methods. 1986;89(2):271-7.
[19] Kuimova MK, Botchway SW, Parker AW, Balaz M, Collins HA, Anderson HL, et al. Nat Chem. 2009;1(1):69-73.
[20] Safi M, Sarrouj H, Sandre O, Mignet N, Berret J-F. Nanotechnology. 2010;21(14):10.
[21] Petri-Fink A, Steitz B, Finka A, Salaklang J, Hofmann H. European Journal of Pharmaceutics and Biopharmaceutics. 2008;68(1):129-37.
[22] Nan AJ, Bai X, Son SJ, Lee SB, Ghandehari H. Nano Lett. 2008;8(8):2150-4.
[23] Dumortier H, Lacotte S, Pastorin G, Marega R, Wu W, Bonifazi D, et al. Nano Lett. 2006;6(7):1522-8.

Nanosafe2010: International Conference on Safe Production and Use of Nanomaterials IOP Publishing
Journal of Physics: Conference Series **304** (2011) 012034 doi:10.1088/1742-6596/304/1/012034

Effects of nanoparticles on murine macrophages

M Chevallet[1*], C Aude-Garcia[1*], C Lelong[1], S Candéias[1], S Luche[1], V Collin-Faure[1], S Triboulet[1], D Diallo[1], H Diemer[2], A van Dorsselaer[2] and T Rabilloud[1]

[1]CEA/DSV/IRTSV, laboratoire de Biochimie et Biophysique des Systèmes Intégrés, Unité Mixte CNRS UMR5092, Université Joseph Fourier - Grenoble, CEA Grenoble, 17 rue des martyrs, 38054 Grenoble, France.

[2]IPHC, Institut Pluridisciplinaire Hubert Curien, CNRS UMR7178, Université Louis Pasteur – Strasbourg I, France.

E-mail: thierry.rabilloud@cea.fr

Abstract. Metallic nanoparticles are more and more widely used in an increasing number of applications. Consequently, they are more and more present in the environment, and the risk that they may represent for human health must be evaluated. This requires to increase our knowledge of the cellular responses to nanoparticles. In this context, macrophages appear as an attractive system. They play a major role in eliminating foreign matter, e.g. pathogens or infectious agents, by phagocytosis and inflammatory responses, and are thus highly likely to react to nanoparticles. We have decided to study their responses to nanoparticles by a combination of classical and wide-scope approaches such as proteomics. The long term goal of this study is the better understanding of the responses of macrophages to nanoparticles, and thus to help to assess their possible impact on human health.

We chose as a model system bone marrow-derived macrophages and studied the effect of commonly used nanoparticles such as TiO_2 and Cu. Classical responses of macrophage were characterized and proteomic approaches based on 2D gels of whole cell extracts were used. Preliminary proteomic data resulting from whole cell extracts showed different effects for TiO_2-NPs and Cu-NPs. Modifications of the expression of several proteins involved in different pathways such as, for example, signal transduction, endosome-lysosome pathway, Krebs cycle, oxidative stress response have been underscored. These first results validate our proteomics approach and open a new wide field of investigation for NPs impact on macrophages.

1. Introduction

Nanoparticles (NPs) are increasingly used for developing new tools for biotechnology, as well as manufacturing widely-used products. They offer interesting properties, regarding for example electromagnetic characteristics, mechanical properties, chemical and/or biological reactivity. Nanoparticles can be used for medical imaging and drug delivery, as well as for electronic devices [1-5]. Moreover, many household products contain NPs of metal oxides, such as sunscreens, cosmetic products, or clothes [6,7]. New tools have also been developed for biotechnology and life sciences,

[*] These authors contributed equally to this work.

using metallic NPs as well as nanotubes, nanowires or quantum dots [8]. This massive development of nanotechnologies requires studies to assess potential associated biological risks.

Thus, there is an important corpus of literature assessing the effects of nanoparticles on living organisms. This literature can be split in two types of studies: on the one hand, there are *in vivo* studies on animal models where macroscopic results can be obtained [9-12]. However, because of the short lifespan of laboratory animals, long-term effects are difficult to derive. Moreover, the *in vivo* systems are so complex that it is almost impossible to get mechanistic data that would help to predict long term effects. On the other hand, there are in vitro studies on cells [6,8,13-17]. This simplified system can be used to derive mechanistic data. However, most of this literature focuses on a few cellular parameters (survival, oxidative stress, cytokine productions) so that very little is known on the possible responses elicited by nanoparticles on cells.

In order to bring new, mechanistic knowledge, we have decided to use a wide-scope, proteomics approach. As a cell type, we have chosen the macrophage, i.e. the specialized cell type that will scavenge bacteria or viruses [18] as well as any type of particulate material, nanoparticles [19]. Indeed, macrophages have been shown to be key players in the toxicity of asbestos fibers and silica nanoparticles, and thus in the etiology of asbestosis and silicosis, by the virtue of their stress response to the phagocytosis of these particles [20].

Thus, we have decided to apply proteomics to investigate the response of macrophages to TiO_2 considering that this NP is widely produced and therefore, potential widespread exposure may occur during both manufacturing and use. We also compared the effects of TiO_2 with Cu, which at the difference of TiO_2 is a redox metal and can release cupric ions which can be toxic for cells. Moreover, Cu-NPs are also widely-used NPs because of their antibacterial properties. This work has shown changes in the central cellular metabolism, as well as a peculiar oxidative stress response.

2. 2. Materials and methods

2.1. Mice
C57BL/6 mice were bred in the animal facility of the Commissariat à l'Energie Atomique, Grenoble, France. For hind legs recovery, 10-20 week-old mice were sacrificed by cervical dislocation in accordance with Atomic Energy Commission Care and Use Committee.

2.2. Primary macrophages
Unless otherwise stated, all chemical reagents and cell culture products were purchased from Sigma-Aldrich (Saint Quentin Fallavier, France). Murine primary macrophages were generated from bone marrow as described [21]. Briefly, the bone marrow cavities of femurs and tibiae were flushed several times with BM20 medium (DMEM supplemented with 10% FCS, 20% L929 cell-conditioned medium as a source of M-CSF, 5% horse serum, 1% glutamine and 1% sodium pyruvate). 40-60 x 10^6 cells were recovered per mouse and cultured in 50 ml of the same medium. At day 7, the supernatant was removed. Adherent cells were incubated for 10 min at 4°C in cold 0.02%EDTA in PBS. Cells were then recovered and transferred in 50 ml BM10 medium (same as BM20 but with 10% L929 cell-conditioned medium) for 3 additional days. 20-30 x 10^6 cells were generated at the end of the differentiation process and most of the resulting cells (> 90%) were CD11b and CD14 positive, as assessed by flow cytometry. These mature macrophages, derived from bone marrow exactly as occurs *in vivo*, are therefore a good model of all the macrophages that exist *in vivo*, alveolar as well as peritoneal, dermal or hepatic.

2.3. Flow cytometry analysis
For the analysis of cell surface protein expression, bone marrow-derived macrophages were stained for 30 min at 4°C with conjugated antibodies in PBS containing 3% fetal calf serum and 0.16% sodium azide. Anti-CD11b and PE-conjugated anti-CD14 were from CliniScience (Montrouge, France), FITC-conjugated anti-CD86 (B7-2) and PE-conjugated anti IA^b were from BD Pharmingen (le Pont de

Claix, France). Data acquisition and analysis were performed with a FacsCalibur flow cytometer equipped with CellQuest software (Becton Dickinson).

2.4. Preparation of nanoparticle suspensions

Titanium oxide (TiO_2, anatase, < 25nm, 99.7% purity) and Copper (Cu, < 50nm, 99.5% purity) were purchased from Sigma-Aldrich (France). For their sterilization, the NPs were suspended in 95% ethanol, at a concentration of 10 mg/ml. After 1h incubation at room temperature, the NPs were diluted to a final concentration of 1 mg/ml in polyvinylpyrrolidone 40 (PVP40) 0.2%, sonicated in a bath sonicator for 1h and stored at 4°C. PVP40 increases the dispersions of the NPs but there is still some aggregation in the culture media. Characterization of the particle size distribution was performed by Dynamic Light Scattering measurements (DLS) which indicated a Z-average of 318.9 nm for TiO_2 NPs and of 427.8 nm for Cu NPs after 24h incubation in culture media. We chose to perform this polymer coating to mimic the opposite phenomena occurring when nanoparticles are in contact with a biological fluid, whether it is pulmonary surfactant, saliva, lymph or blood. On the one hand the salt concentration present in all these fluids promote aggregation, so that nanoparticles are never as dispersed in biological media as they can be in air or in pure water. On the other hand, the polymeric materials present in biological fluids (polysaccharides, proteins) will act as coating agents and limit particle aggregation. We have used PVP 40 for its lack of interference with our analyses, while being perfectly biocompatible and able to limit particles aggregation.

2.5. Exposure of macrophages to nanoparticles

Cells ($5x10^6$) were cultured in Petri dishes and treated for 24h with either 100 μg/ml TiO_2 or 10 μg/ml Cu-NPs. Two types of negative controls were done with and without PVP40.

2.6. Two-dimensional electrophoresis

Primary macrophage cells were harvested by scraping, rinsed 2 times in phosphate-buffered saline and pellets were suspended in homogeneization buffer (0.25 M sucrose, 10 mM Tris-HCl, pH 7.5, 1 mM EDTA). A buffer volume approximately equal to the packed cell volume was used. The suspension was transferred to polyallomer ultracentrifuge tubes, and cells were lysed by the addition of 4 volumes (respective to the suspension volume) of 8.75 M urea, 2.5 M thiourea, 25 mM spermine and 13 mM Tris carboxyethylphosphine. After 1 hour at room temperature, the extracts were centrifuged (30 min at 200,000 x g). The supernatant was collected and the protein content was determined by the Bradford protein assay using bovine serum albumin as a standard. The protein extracts were stored at –20 °C. The first dimension of electrophoresis was performed with immobilized pH gradients for isoelectric focusing. Linear 4–8 pH gradients were used. Home-made pH gradient plates were cast and cut into 4-mm-wide strips [22, 23]. A total of 150 μg of proteins were diluted in 560 μl of rehydratation buffer (7 M urea, 2 M thiourea, 4% CHAPS, 0.4% ampholytes, 100 mM dithiodiethanol). The samples were applied onto the strips by in-gel rehydration overnight using a thiourea-urea mixture as denaturing agent [24]. IEF was carried out for 60,000 Vh at a maximum of 3000 V using the Multiphor II system (Amersham-Pharmacia, Sweden). Strips were then equilibrated for 20 min in 0.15 M BisTris/0.1 M HCl, 6 M urea, 2.5% SDS, 30% glycerol. Strips were placed on top of a SDS-polyacrylamide gel. After migration, the gels were stained with silver [25]. Expression ratios were estimated by image analysis of pairs of gels with the Delta2D software (DECODON GmbH, Greifswald, Germany). Spots of interest (ratio >1.5) were cut and analysed by MS/MS.

2.7. MS/MS analysis

In gel digestion was performed with an automated protein digestion system, MassPrep Station (Waters Corp., Milford, USA). The gel plugs were washed twice with 50 μL of 25 mM ammonium hydrogen carbonate (NH_4HCO_3) and 50 μL of acetonitrile. The cysteine residues were reduced by 50 μL of 10 mM dithiothreitol at 57°C and alkylated by 50 μL of 55 mM iodoacetamide. After dehydration with acetonitrile, the proteins were cleaved in gel with 10 μL of 12.5 ng/μL of modified porcine trypsin

(Promega, Madison, WI, USA) in 25 mM NH_4HCO_3. The digestion was performed overnight at room temperature. The generated peptides were extracted with 60% acetonitrile in 5% acid formic. NanoLC-MS/MS analysis was performed using an Agilent 1100 series nanoLC-Chip system (Agilent Technologies, Palo Alto, USA) coupled to an HCT Plus ion trap (Bruker Daltonics, Bremen, Germany). The chip was composed of a Zorbax 300SB-C18 (43 mm × 75 μm, with a 5 μm particle size) analytical column and a Zorbax 300SB-C18 (40 nL, 5 μm) enrichment column. Elution of the peptides was performed at a flow rate of 300 nL/min with a 8-40% linear gradient (solvent B, 98%ACN/0.1%FA) over 7 first minutes. The voltage applied to the capillary cap was optimized to -1750V and the mass range was 250-2500 *m/z*. For tandem MS experiments, the system was operated Data-Dependent-Acquisition (DDA) mode with automatic switching between MS and MS/MS. The three most abundant precursor ions were selected to be further isolated and fragmented. The MS/MS scanning was performed in the ultrascan resolution mode at a scan rate of 26.000 *m/z* per second. A total of six scans were averaged to obtain an MS/MS spectrum. The system was fully controlled by the ChemStation (v 5.3) and EsquireControl (Rev B.01.03) software (Agilent Technologies and Bruker Daltonics, respectively). For protein identification, the MS/MS data were interpreted using the MASCOT 2.2.0. algorithm (Matrix Science, London, UK). Spectra were searched with a mass tolerance of 0.2 Da in MS and MS/MS modes, allowing a maximum of one trypsin missed cleavage. Carbamidomethylation of cysteine residues and oxidation of methionine residues were specified as variable modifications. Following the guidelines for proteomic data publication [26, 27], and to avoid consideration of poor quality data, filtering criteria based on probability-based scoring of the identified peptides have been taken into account for high confidence identification and a decoy database was generated and searched to determine false positive rate in protein identification [28]. The Scaffold software (v.3, Proteome software Inc., Portland, OR) was used in order to validate identification using the following criteria: spectra were searched against a target-decoy version of the UniProtKB/SwissProt database restricted to Mammalia downloaded in January 2010 (1028424 target and decoy entries). Protein identifications were validated when at least two peptides with Mascot identity scores of greater than 10 were detected.

3. Results and discussion

Before adding the NPs onto the bone marrow-derived macrophages, we checked by flow cytometry that these cells were mature macrophages, expressing both CD11b and CD14. Moreover, the cells obtained could be activated by LPS, as shown by release of TNFα and NO (data not shown). Using fluorescent beads, we also verified by flow cytometry that they were phagocytic. The concentration of NPs used was determined beforehand by viability tests (LD20). These doses are not intended to represent doses for a given body weight and a typical exposure. However, they are legitimate for a mechanistic study to exemplify what deleterious mechanisms can be at play during pathogenesis. Moreover, it should be kept in mind that doses per body weight are also an approximation, and that "overloads" can exist locally *in vivo*, e.g. in the dermis just below a wound in the case of cutaneous exposure, and that a single dose *in vitro* can also mimic the accumulation of low doses occurring *in vivo* during the lifespan of a macrophage, as macrophages are known to accumulate nanoparticles [10]. Furthermore, in the case of cancers where a single event can lead to the pathology, this "hotspot" model is relevant.

We performed two comparative proteomics analysis: control versus TiO_2 treated cells and control versus Cu treated cells.

3.1. Analysis of total proteome changes after NPs-TiO_2 exposure

A total of 9 protein spots with an increased intensity as well as 1 spot with a decreased intensity after TiO_2 treatment could be detected (Figure 1, Table 1). They were excised from gels and analyzed by MS. All these spots were identified. TiO_2 treatment altered metabolism as showed by increase of spots corresponding to malate deshydrogenase, aldehyde deshydrogenase and glucosidase. We also observed an increase in VDAC (mitochondrial porin), peroxiredoxin 4 (Prx 4), normal form and proteasome alpha6 subunit, which are usually seen in various stress conditions. All these proteins belong to the "déjà vu in proteomics" [29, 30].

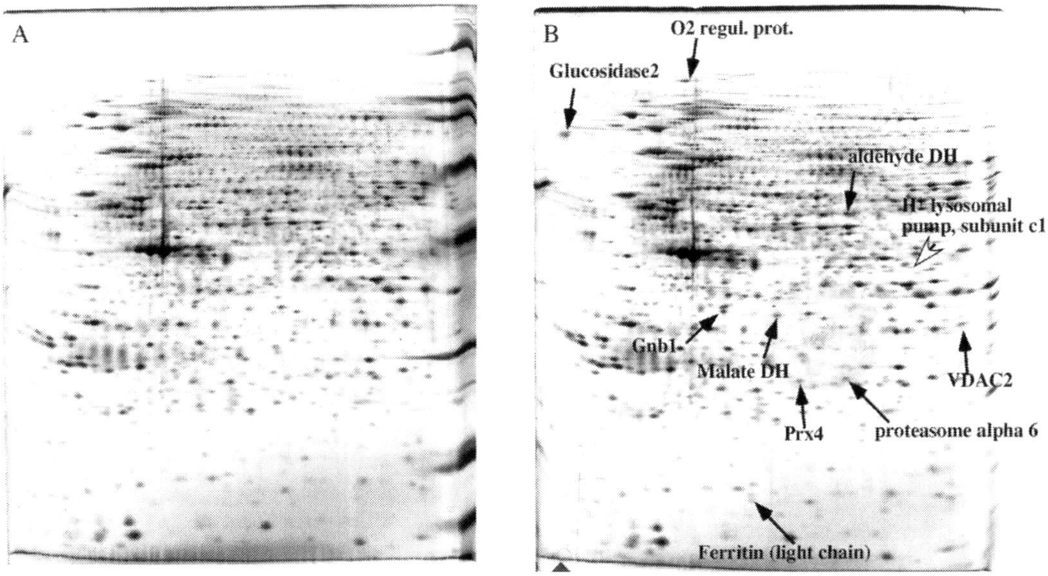

Figure 1. Comparison of the total extract patterns of control primary macrophages and TiO_2-NPs treated cells. Primary macrophages were treated with 100 µg/ml TiO_2-NPs for 24h. 150 µg of total proteins were separated by 2-D and detected with silver staining as described in Section 2. Gel images were analyzed by Delta-2D. (A) control cells, (B) TiO_2-NPs treated cells. Solid and empty arrows point at increased ($\geq$1.50-fold) and decreased ($\leq$0.67) proteins, respectively.

Cells responded to such a "light stress" by increasing general metabolism and some classical proteins. We identified some more original proteins: Gnb1, a G-protein implicated in signal transduction, ferritin (light chain) implicated in iron homeostasis and O_2-regulated protein, which is activated by hypoxia. This suggests that TiO_2 treatment may induce hypoxia-like conditions. At last, we noted that the subunit c1 of H^+ lysosomal pump was decreased. It is not surprising that the lysosomal pathway is altered by NPs phagocytosis as it is known that NPs can use the endosome-lysosome pathway [19].

3.2. Analysis of total proteome changes after NPs-Cu exposure

The same process was repeated for the 6 protein spots increased after Cu treatment and for the one with decreased intensity (Figure 2, Table 1). Again, Cu treatment increased some proteins spots belonging to metabolism and "déjà vu" in proteomics, such as ATPase subunit d, prohibitin and HSP70 (chaperone). Cells responded to stress by increasing metabolism, energy production and by increasing chaperones for protein refolding. More interestingly, an increase in heme oxygenase and peroxiredoxin1 (Prx 1), oxidized form was a signature of an oxidative stress response. In basal condition, this spot is very weak but increased almost 4 times upon Cu treatment. The oxidized spot level of Prx is a well-known indicator of the oxidative injury to the cell [31]. Heme oxygenase is generally induced when the level of free heme increases and this is associated to an oxidative stress. Moreover, heme oxygenase appears to be a potential defense against oxidative stress [32]. This result is consistent with a metal redox stress. In addition, two proteins implicated in endosome-lysosome pathway were also modified. MVB prot 4b was decreased and the d1 subunit of H^+ lysosomal pump was increased.

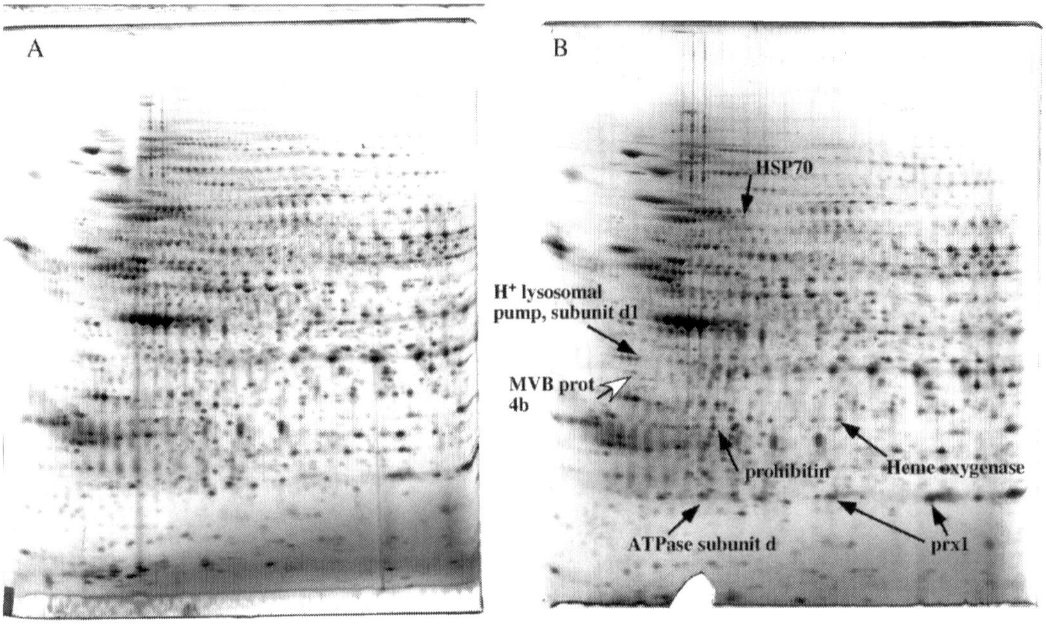

Figure 2. Comparison of the total extract patterns of control primary macrophages and Cu-NPs treated cells. Primary macrophages were treated with 10 μg/ml Cu-NPs for 24h. 150 μg of total proteins were separated by 2-DE and detected with silver staining as described in Section 2. Gel images were analyzed by Delta-2D. (A) control cells, (B) Cu-NPs treated cells. Solid and empty arrows point at increased (≥1.50-fold) and decreased (≤0.67) proteins, respectively.

3.3. Comparison of total proteome changes after TiO_2-NPs and Cu-NPs exposure

When we compared the two lists of proteins, we observed that there was no protein in common (Table 1). In both cases, we found metabolic proteins and classical stress proteins. We found more of these groups of proteins after TiO_2 treatment than after Cu treatment. However, Cu treatment appeared to be more toxic as it clearly induced oxidative stress. With TiO_2 treatment, we just noted an increase in the normal form of Prx 4, while Cu induced an important increase in oxidized forms of Prx 1 and of heme oxygenase. Peroxiredoxins are enzymes detoxifying reactive oxygen species, but they can get inactivated by oxidation in case of severe oxidative stress [31]. Presence of oxidized form is thus a signature of a deleterious oxidative stress. It is not surprising that Cu, which is a redox metal, generated such an oxidative stress. After TiO_2 treatment, ferritin was increased, possibly to protect the cell against free iron and oxidative stress. In addition, the two NPs affected the endosome-lysosome pathway even if we did not find the same proteins and found that two subunits of the same complex (H^+ lysosomal pump) varied in an opposite way.

The results obtained were different in terms of proteins but consistent in terms of function (metabolism, core stress response and endosome-lysosome pathway). Cu-NPs appeared in our conditions to be more toxic than TiO_2-NPs, and responsible of an oxidative stress. We observed with proteomics specific effects of Cu or TiO_2-NPs treatment. The fact that we did not observe the same list of proteins with these two different NPs showed that NPs toxicity is depending primarily of the nature of the NPs.

Table 1. Ratio NPs/control of the spots intensities measured by Delta 2D software on silver stained 2D gels.

TiO_2		Cu	
VDAC 2	2.420	HSP 70	2.800
Aldehyde deshydrogenase	2.240	Prohibitin	1.733
Prx 4, normal form	1.793	ATPase subunit d	1.700
Proteasome α 6	1.763		
Glucosidase	1.737		
Malate deshydrogenase	1.498		
Ferritin (light chain)	5.630	Prx 1, oxidized form	3.920
O_2 regulated protein	1.892	Heme oxygenase	3.250
Gnb1	1.468	H+ lysosomal pump d1	1.670
H+ lysosomal pump c1	0.580	MVB Prot 4b	0.652

4. Perspectives

From the above, proteomics appears to be a relevant approach for mechanistic studies of cellular responses to nanoparticles. However, it must be kept in mind that such mechanistic studies are hypothetical and must be ultimately verified on relevant *in vivo* models. Nevertheless, this preliminary study showed specific effects of Cu-NPs and TiO_2-NPson murine bone-marrow macrophages. In this study, we looked at total cell extracts and thus only at more abundant and soluble proteins. To get further insights into the cell response to NPs, it will be of great interest to investigate different compartments and/or sub fractions of the cells. Such studies will bring more information on NPs toxicity pathways. We also consider to investigate toxicity of other NPs of different size and nature, to determine if a general pattern emerges or if each NP elicits a unique response.

Acknowledgments

This project belongs to the NanoBioMet consortium and is partially financed by the Toxicology transversal program.

References

[1] Govender T, Ojewole E, Naidoo P and Mackra, I 2008 Polymeric nanoparticles for enhancing antiretroviral drug therapy *Drug Deliv* **15** 493-501
[2] Karlsson H L, Cronholm P, Gustafsson J and Moller L 2008 Copper oxide nanoparticles are highly toxic: a comparison between metal oxide nanoparticles and carbon nanotubes *Chem Res Toxicol* **21** 1726-32
[3] Maesaki S 2002 Drug delivery system of anti-fungal and parasitic agents *Curr Pharm Des.* **8** 433-40
[4] Ravi Kumar M, Hellermann G, Lockey R F and Mohapatra S S 2004 Nanoparticle-mediated gene delivery: state of the art *Expert Opin Biol Ther* **4** 1213-24
[5] Wang L, O'Donoghue M B and Tan W 2006 Nanoparticles for multiplex diagnostics and imaging *Nanomedicine* **1** 413-26
[6] Fahmy B and Cormier S A 2009 Copper oxide nanoparticles induce oxidative stress and cytotoxicity in airway epithelial cells *Toxicol In Vitro* **23** 1365-71
[7] Kasemet K, Ivask A, Dubourguier H C and Kahru A 2009 Toxicity of nanoparticles of ZnO, CuO and TiO2 to yeast Saccharomyces cerevisiae *Toxicol In Vitro* **23** 1116-22
[8] Hussain S M, Hess K L, Gearhart J M, Geiss K T and Schlager J J 2005 In vitro toxicity of nanoparticles in BRL 3A rat liver cells *Toxicol In Vitro* **19** 975-83
[9] Briley-Saebo K, Bjørnerud A, Grant D, Ahlstrom H, Berg T and Kindberg GM 2004 Hepatic cellular distribution and degradation of iron oxide nanoparticles following single intravenous

injection in rats: implications for magnetic resonance imaging *Cell Tissue Res* **316** 315-23

[10] Sadauskas E, Danscher G, Stoltenberg M, Vogel U, Larsen A and Wallin H 2009 Protracted elimination of gold nanoparticles from mouse liver *Nanomedicine* **5** 162-9

[11] Lam CW, James J T, McCluskey R and Hunter R L 2004 Pulmonary toxicity of single-wall carbon nanotubes in mice 7 and 90 days after intratracheal instillation *Toxicol Sci.* **77** 126-34

[12] Oberdorster, E 2004 Manufactured nanomaterials (fullerenes, C60) induce oxidative stress in the brain of juvenile largemouth bass *Environ Health Perspect* **112** 1058-62

[13] Jin CY, Zhu BS, Wang XF and Lu QH 2008 Cytotoxicity of titanium dioxide nanoparticles in mouse fibroblast cells *Chem Res Toxicol* **21** 1871-7

[14] Kwon Y M, Xia Z, Glyn-Jones S, Beard D, Gill H S and Murray D W 2009 Dose-dependent cytotoxicity of clinically relevant cobalt nanoparticles and ions on macrophages in vitro *Biomed Mater.* **4** 025018

[15] Lanone S, Rogerieux F, Geys J, Dupont A *et al.* 2009 Comparative toxicity of 24 manufactured nanoparticles in human alveolar epithelial and macrophage cell lines *Part Fibre Toxicol* **6** 14

[16] Wagner A J, Bleckmann C A, Murdock R C, Schrand A M *et al.* 2007 Cellular interaction of different forms of aluminum nanoparticles in rat alveolar macrophages *J Phys Chem B* **111** 7353-59.

[17] Waters K M, Masiello L M, Zangar R C, Tarasevich B J *et al.* 2009 Macrophage responses to silica nanoparticles are highly conserved across particle sizes *Toxicol Sci* **107** 553-69

[18] Mosser D M, Edwards J P 2008 Exploring the full spectrum of macrophage activation *Nat Rev Immunol* **8** 958-69

[19] Shukla R, Bansal V, Chaudhary M, Basu A, Bhonde R R and Sastry M 2005 Biocompatibility of gold nanoparticles and their endocytotic fate inside the cellular compartment: a microscopic overview *Langmuir* **21** 10644-54

[20] Dostert C, Petrilli V, Van Bruggen R, Steele C, Mossman BT and Tchopp J 2008 Innate immune activation through Nalp3 inflammasome sensing of asbestos and silica *Science* **320** 674-7.

[21] Schleicher U and Bogdan C 2009 Generation, culture and flow-cytometry characterization of primary mouse macrophages *Methods Mol Biol.* **531** 203-24

[22] Gorg A, Boguth G, Obermaier C and Weiss W 1998 Two-dimensional electrophoresis of proteins in an immobilized pH 4-12 gradient *Electrophoresis* **19** 1516-19

[23] Rabilloud T, Valette C and Lawrence J J 1994 Sample application by in-gel rehydration improves the resolution of two-dimensional electrophoresis with immobilized pH gradients in the first dimension *Electrophoresis* **15** 1552-58

[24] Rabilloud T, Adessi C, Giraudel A and Lunardi J J 1997 Improvement of the solubilization of proteins in two-dimensional electrophoresis with immobilized pH gradients *Electrophoresis* **18** 307-16

[25] Sinha P, Poland J, Schnolzer M and Rabilloud T 2001 A new silver staining apparatus and procedure for matrix-assisted laser desorption/ionization-time of flight analysis of proteins after two-dimensional electrophoresis *Proteomics* **1** 835-40

[26] Carr S, Aebersold R, Baldwin M, Burlingame A, Clauser K and Nesvizhskii A 2004 The need for guidelines in publication of peptide and protein identification data: working group on publication guidelines for peptide and protein identification data *Mol Cell Proteomics* **3** 531-33

[27] Wilkins M R, Appel R D, Van Eyk J E *et al.* 2006 Guidelines for the next 10 years of proteomics *Proteomics* **6** 4-8

[28] Elias J E and Gygi S P 2007 Target-decoy search strategy for increased confidence in large-scale protein identifications by mass spectrometry *Nat Methods* **4** 207-14

[29] Petrak J, Ivanek R, Toman O, Cmejla R, Cmejlova J, Vyoral D, Zivny J and Vulpe C D 2008 Déjà vu in proteomics. A hit parade of repeatedly identified differentially expressed proteins *Proteomics* **8** 1744-49

[30] Wang P, Bouwman F G and Mariman E C M 2009 Generally detected proteins in comparative proteomics – A matter of cellular stress response? *Proteomics* **9** 2955-66

[31] Rabilloud T, Heller M, Gasnier F, Luche S, Rey C, Aebersold R, Benahmed M, Louisot P and Lunardi J 2002 Proteomics analysis of cellular response to oxidative stress. Evidence for in vivo overoxidation of peroxiredoxins at their active site *J Biol Chem* **277** 19396-401

[32] Stocker R 1990 Induction of haem oxygenase as a defense against oxidative stress *Free Radic Res Commun* **9** 101-12

Nanosafe2010: International Conference on Safe Production and Use of Nanomaterials IOP Publishing
Journal of Physics: Conference Series **304** (2011) 012035 doi:10.1088/1742-6596/304/1/012035

Interference between nanoparticles and metal homeostasis

AN Petit[1], C Aude Garcia[2], S Candéias[2], A Casanova[3], P Catty[1], P Charbonnier[1], M Chevallet[2], V Collin-Faure[2], M Cuillel[1], T Douki[4], N Herlin-Boime[3], C Lelong[2], S Luche[2], E Mintz[1], JM Moulis[1], V Nivière[1], S Ollagnier de Choudens[1], T Rabilloud[2], JL Ravanat[4], S Sauvaigo[4], M Carrière[5] and I Michaud-Soret[1]

[1]Laboratoire de Chimie et Biologie des Métaux UMR 5249 CEA-CNRS-UJF, 17 rue des Martyrs, 38054 Grenoble Cedex 09, France

[2]Laboratoire de Biochimie et Biophysique des Systèmes Intégrés UMR 5092 CNRS-CEA-UJF, 17 rue des martyrs, 38054 Grenoble Cedex 09, France

[3]Laboratoire Edifices Nanométriques URA 2453 CEA-CNRS-IRAMIS, 91191 Gif-sur-Yvette, France

[4]Laboratoire Lésions des Acides Nucléiques UMR E3 CEA-UJF, 17 rue des Martyrs, 38054 Grenoble Cedex 09, France

[5]Laboratoire de Structure et Dynamique par Résonance Magnétique UMR 9956 CEA-CNRS-IRAMIS, 91191 Gif-sur-Yvette, France

E-mail: isabelle.michaud-soret@cea.fr

Abstract. The TiO_2 nanoparticles (NPs) are now produced abundantly and widely used in a variety of consumer products. Due to the important increase in the production of TiO_2-NPs, potential widespread exposure of humans and environment may occur during both the manufacturing process and final use. Therefore, the potential toxicity of TiO_2-NPs on human health and environment has attracted particular attention. Unfortunately, the results of the large number of studies on the toxicity of TiO_2-NPs differ significantly, mainly due to an incomplete characterization of the used nanomaterials in terms of size, shape and crystalline structure and to their unknown state of agglomeration/aggregation. The purpose of our project entitled **NanoBioMet** is to investigate if interferences between nanoparticles and metal homeostasis could be observed and to study the toxicity mechanisms of TiO_2-NPs with well-characterized physicochemical parameters, using proteomic and molecular approaches. A perturbation of metal homeostasis will be evaluated upon TiO_2-NPs exposure which could generate reactive oxygen species (ROS) production. Moreover, oxidative stress consequences such as DNA damage and lipid peroxidation will be studied. The toxicity of TiO_2-NPs of different sizes and crystalline structures will be evaluated both in prokaryotic (*E. coli*) and eukaryotic cells (A549 human pneumocytes, macrophages, and hepatocytes). First results of the project will be presented concerning the dispersion of TiO_2-NPs in bacterial medium, proteomic studies on total extracts of macrophages and genotoxicity on pneumocytes.

1. Introduction

Titanium dioxide (TiO_2) nanoparticles (NPs) are manufactured nanosized materials that are already used for a wide range of applications [1]. These applications are principally related to their strong ability to absorb UV-light and their photocatalytic activity. In addition, TiO_2 is chemically stable and resistant to photocorrosion and has a high ability to break molecular bonds leading to degradation of many organic compounds. Compared to bulk TiO_2, nanosized particles are significantly more reactive due to their larger specific surface area. These properties, together with its high abundancy and low cost, have generated widespread use of TiO_2-NPs in popular products such as sunscreens, toothpastes and cosmetics [2-4]. Moreover, TiO_2 is used at the nanoscale level to decontaminate air, soil, and water by destruction of microorganisms and pesticide [5-8]. The proliferation of applications and products containing TiO_2-NPs will inevitably result in release into the environment and may cause health problems to human beings.

Therefore, the potential toxicity of TiO_2-NPs on human health and environment has attracted particular attention. Despite the large number of studies on the toxicity of TiO_2-NPs, results are often contradictory, certainly because the tested nanoparticles differ in their physico-chemical characteristics or in their agglomeration status. Since these characteristics are often not related in the articles, reported results are difficult to compare. First studies have revealed induction of inflammatory responses, cytotoxicity, and reactive oxygen species (ROS) formation in a variety of cell types and tissues [9-15]. These studies have identified either size or crystalline phase as the key determinant for TiO_2-NPs toxicity and suggest a larger toxicity of the smaller NPs in anatase cristalline phase. By contrast, other studies have shown no induction of toxic event following TiO_2-NPs exposure [16,17]. However, accurate characterization of NPs (e.g. chemical composition, size, morphology, surface area and crystallinity) has not always been described and it seems still essential to establish the relationship between properties of NPs and their toxic potential to biological systems and to ensure that results are reproducible and relevant [18-22]. Many of these characteristics can be determined using dry, powdered NPs but aggregation of TiO_2-NPs in culture media and salted buffers should also be studied. Indeed, NPs have been shown to agglomerate immediately upon addition to cell culture media that would result in inaccurate results of toxicity [23-25]. Hydrodynamic size of NPs dispersed in liquids is thus often larger than the nominal primary size of the particles. Based on these considerations, increasing attention is being paid to optimize the dispersion and the stability of suspensions of TiO_2-NPs. Toward this objective, several studies have investigated the influence of different types of dispersing agents to favour the slurrying of TiO_2-NPs in solution and to produce stable suspensions [23,26-28].

Several studies have suggested that toxic effects of TiO_2-NPs may be related to the generation of ROS, leading to damage on cellular and subcellular structures [10,11,29-32]. Elevated amounts of free radicals in cells exposed to TiO_2-NPs have also been shown to induce oxidative DNA damage [33,34]. Despite these clear effects identified in cells treated with TiO_2-NPs, the precise mechanisms of toxicity and their consequences on the metabolism and adaptation are largely unknown. The interrelationship between disruptions in metal homeostasis (primarily iron and copper) and increased oxidative damage have been demonstrated in both prokaryotic [35] and eukaryotic cells [36].

For example, iron is a trace element of crucial importance to living cells that is redox active and may exist into mainly two oxidative states, ferrous and ferric ions. Iron is essential to many metalloenzymes and metalloproteins involved in vital biological processes (oxygen transport, energy production, respiratory process, DNA biosynthesis, and xenobiotic metabolism....), that require electron transfers. However, excess of available iron can result in a number of cellular disturbances such as oxidative stress because of its propensity to react with oxygen, generating harmful free radicals *via* the Fenton reaction. Because of this dual nature, control of iron homeostasis is critical for cell survival and should be strictly regulated in relation with oxidative stress responses.

Because of the above drawbacks, transition metals levels are tightly regulated by metal homeostasis to maintain the appropriate amount of metals in the cell. Oxidative stress represents an unbalance between the production of ROS and the potency of antioxidant defenses. The enzymes that detoxify ROS often are metalloenzymes. In addition oxidative stress conditions and metal homeostasis dysfunction may cause metalloenzymes inactivation or breakdown in endogenous antioxidant defences with potential relevance for the toxicity of some NPs.

Therefore, exposure to TiO_2-NPs may lead to disruption of metal homeostasis and oxidative stress that are closely related.

2. Description of the project

The **NanoBioMet** project will focus on the interference between oxidative stress and changes of metal homeostasis induced by TiO_2-NPs exposure in both prokaryotic and eukaryotic cells (figure 1). The toxicity of several TiO_2-NPs either synthesized by laser pyrolysis (N. Herlin-Boime and coworkers) or purchased from manufacturers such as Degussa or Sigma-Aldrich will be studied. Laser pyrolysis technique produces particles of high chemical purity, low mean sizes and remarkable narrow size distributions [37].

Different groups of the Commissariat à l'Energie Atomique et aux Energies Alternatives (CEA) with complementary competences are involved in this project. The key factors taken into account for this project will be:

- synthesis by laser pyrolysis of TiO_2-NPs then physico-chemical characterization and state of agglomeration/dispersion status of the NPs in cell culture media;
- effects of TiO_2-NPs on *Escherichia coli* prokaryotic cells: iron homeostasis, siderophore (ferric ion specific chelating agents) synthesis, enzymatic activities of iron-sulfur proteins and oxidative stress enzymes in *E. coli*;
- effects of TiO_2-NPs on several types of eukaryotic cells (macrophages, pneumocytes and hepatocytes): global stress response by proteomic approaches, iron and copper homeostasis, activation of the Unfolded Protein Response (UPR), oxidative stress enzymes, genotoxicity and DNA damages.

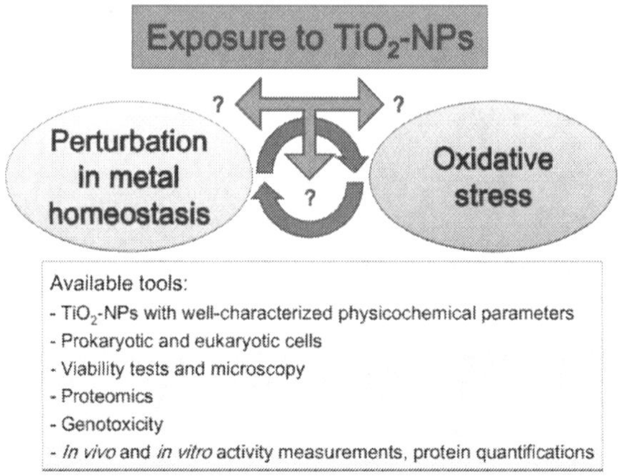

Figure 1. Schematic representation of the objectives of the NanoBioMet project.

3. Preliminary results

3.1. Characterization of TiO_2-NPs and dispersion in aqueous solution

The morphology, the primary size, the specific surface area and the crystal form of two nanoparticle powders have been characterized (table 1). The commercial DegussaP25 is a widely used industrial photocatalyst while CEA12A has been synthesized at the CEA by laser pyrolysis. Both NPs are spherical. The mean primary diameter of CEA12A-NPs is 12±3 nm while DegussaP25 are twice larger. On the contrary, specific surface area is higher for the CEA12A-NPs. Finally, crystalline structure of CEA12A-NPs is 95% anatase and this percentage is lower for DegussaP25 with 75%.

Table 1. Nanoparticles physico-chemical characteristics.

	Supplier	Morphology	Size (TEM)	SSA (m^2/g)	Crystalline structure
P25	Degussa	spherical	25±7 nm	46±1	75% anatase
12A	CEA	spherical	12±3 nm	82±4	95% anatase

Size and morphology were determined by transmission electron microscopy (TEM, Philips EM208, accelerating voltage 80 kV). Specific surface areas (SSA, m^2/g) were measured using Brunauer, Emmet and Teller (BET) method consisting in isothermal gas-adsorption [38] using a Micromeritics Flowsorb 2300 (Norcross, USA). Crystalline structure was determined by X-ray diffraction (XRD).

The dispersion of NPs in aqueous suspension is also a prerequisite for the adequate assessment of their potential biological effects. Ultrasonication is a commonly used technique to prepare highly dispersed suspensions [27,39]. Suspensions of DegussaP25-NPs (10 g/L) were prepared in ultrapure sterile water. Then the influence of different ultrasonication methodologies on the resultant hydrodynamic size of particles was investigated. Two types of probes were compared on the same sonicator (Autotune 750 W, Bioblock Scientific, Illkirch, France): a titanium tapered microtip (3 mm Ø) and a Cup Horn device (64 mm Ø). The microtip, which produces ultrawaves, is in direct contact with the solution. With the Cup Horn sonicator, the suspension is instead kept in tubes that are put into a water bath through which the waves are transmitted. Therefore, the Cup Horn delivers high energy indirect sonication: it avoids exposure of the solution to a titanium probe and provides a level of containment of NPs. All sonications were carried out at 4°C, using a pulsed mode and a circulating water bath. Two sonication times have been tested: 15 and 30 min.

Characterization of the particle size distribution was performed by Dynamic Light Scattering measurements (DLS, Dynapro™ NanoStar, Wyatt Technology). This technique analyzes, through photon correlation spectroscopy, fluctuation with time of the particles light scattering intensity. Dynamics software (Wyatt Technologies Corp.) was used to calculate and fit an autocorrelation function plotted from the random fluctuations in scattering intensity. The intensity autocorrelation functions before and after sonication either with tapered microtip or a Cup Horn device, are presented (figure 2). The scattered light can be related to the particle translational diffusion coefficient (Dt). Finally, the hydrodynamic diameter (d_H) of the NPs, which corresponds to the diameter of a sphere that has the same diffusion coefficient as the particles, can be estimated using the Stokes-Einstein equation $d_H = kT/3\pi\eta Dt$; where k is the Boltzmann's constant, T is the absolute temperature, and η is the viscosity of the solvent. Comparison of results was made with another DLS instrument: the ZetaSizer 3000HS (Malvern, Worcestershire, UK). The correlation functions from DLS were analyzed by the constrained regularized CONTIN method. For both apparatus, the mean size for the sample and the size of major population were determined.

Table 2. NPs sizes obtained after 30 min of ultrasonication with either a microtip or a Cup Horn device. Measurements of NPs in ultrapure sterile water (10 g/L) were carried out using either the Dynapro (Wyatt) or the Zetasizer (Malvern).

	mean size (nm)		size of major population (nm)	
	Microtip	**Cup Horn**	**Microtip**	**Cup Horn**
Dynapro (Wyatt)	159	163	183	162
ZetaSizer (Malvern)	128	129	46	23

Table 2 presents the summary of the results obtained for DegussaP25. After 30 min of sonication, the particle size measured with the Wyatt instrument was about 160 nm using either the microtip or the Cup Horn device. Previous articles [27,40] reported similar sizes after ultrasonication of DegussaP25 NPs in aqueous suspension. To obtain these results with the Cup Horn, optimization has been done: a water bath circulating system was used to avoid overheating and a 12 mL tube was put in the centre of the Cup Horn to increase the sonication energy. It can be concluded that the use of a Cup Horn sonicator seems a useful way of dispersing nanoparticles in aqueous suspension.

Measurements were also carried out with the Zetasizer from Malvern on the same NPs solutions. However, the sizes measured with this second apparatus were different. This may be explained by the fact that the same raw data may be treated in different ways by the software packages provided with the two DLS instruments. Therefore, comparison of size data obtained with the two DLS instruments is difficult.

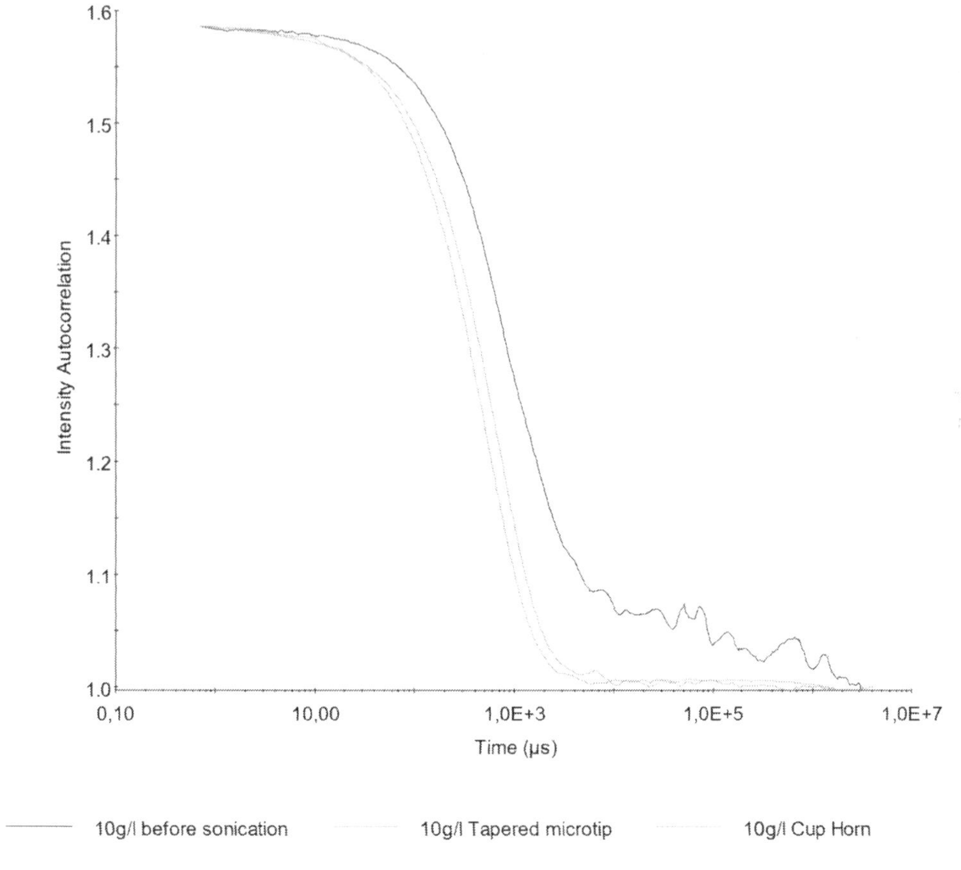

Figure 2. Intensity autocorrelation functions of TiO_2-NPs suspensions in water before sonication or after 30 min of sonication either with tapered microtip or a Cup Horn device (Autotune 750 W, Bioblock Scientific). The concentration of suspensions was 10 g/L.

3.2. Effects of TiO_2-NPs on E. coli growth

Before studying the effects of NPs on bacterial growth, NPs (DegussaP25 and CEA12A) suspension stability was evaluated in bacterial media. Indeed, the objective is to select a medium allowing bacterial growth but avoiding NPs aggregation. In commonly used bacterial media, the Luria-Bertani

Broth (LB) rich medium and the M9 minimal medium, TiO_2-NPs showed a strong aggregation. Therefore, a minimal medium containing low concentration of salts and allowing bacterial growth was optimized. It consisted in: 0.4% Glucose, 50 μM $MgSO_4$, 0.1% casamino acids. Nevertheless, NPs aggregation was still observed in this medium. Therefore, a dispersing agent sodium pyrophosphate ($Na_4P_2O_7$), pH 7.0, was added. It increased NPs surface charge and the electrostatic repulsive force between particles (Jiang et al, 2009). Although this agent was not able to prevent aggregation in LB and M9 media, it efficiently prevented NPs aggregation in this optimized minimal medium. No increase in the NPs size in this culture medium was observed compared to stock solution in water and this solution was stable for several days.

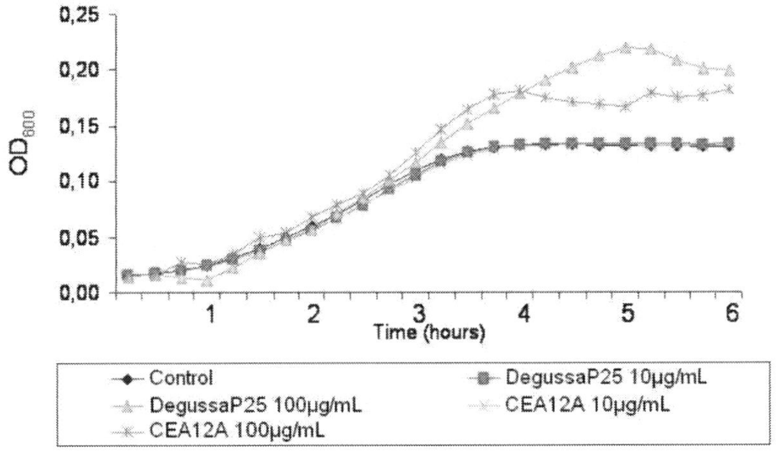

Figure 3. TiO_2-NPs effects on growth of *E. coli* MG1655. Cultures were grown in medium consisting in 0.4% Glucose, 50μM $MgSO_4$, 0.1% casamino acids, 5mM $Na_4P_2O_7$ pH 7. NPs were added at t=0 at final concentrations of 10 or 100 μg/mL. OD_{600} were recorded.

The effects of TiO_2-NPs (DegussaP25 and CEA12A) at concentrations of 10 and 100 μg/mL on the growth of *E. coli* were then evaluated (figure 3). The growth rates were measured at 37°C by monitoring the optical density at 600 nm in a microplate reader (Tecan Infinite F200, Männedorf, Switzerland). Surprisingly, no decrease in bacterial growth was noticed following exposure to 10 or 100 μg/mL of both NPs. An increase was even observed after exposure to 100μg/mL. On the contrary, a previous study showed that survival of *E. coli* was affected when exposed to these TiO_2-NPs [41]. However, these experiments have been done in water which could explain the differences with our study.

3.3. Effects of TiO_2-NPs on murine macrophages

Macrophages play a major role in eliminating foreign matter, e.g. pathogens or infectious agents, by phagocytosis and inflammatory responses, and are thus highly likely to react to NPs. As a model system, we have utilized macrophages derived from murine bone marrow [42]. Mice were killed by cervical dislocation and bone marrow was obtained by flushing femura and tibiae with BM20 medium (DMEM supplemented with 10% FCS, 20% L929 cell-conditioned medium as a source of M-CSF, 5% horse serum, 1% glutamine and 1% sodium pyruvate). Before adding TiO_2-NPs on the macrophages, flow cytometry was used to check if cells were mature macrophages, expressing both CD11b and CD14. Cells ($5x10^6$) were cultured in Petri dishes and treated for 24h with 100 μg/ml of TiO_2 (anatase,<25 nm from Sigma-Aldrich) diluted in polyvinylpyrrolidone 40 (PVP40) 0.2% that increased the dispersion of NPs but did not prevent their aggregation.

Proteomic analysis based on 2D gels was used and performed on whole cell extracts. This study revealed that TiO_2-NPs induced an increase in the expression of proteins involved in different metabolic pathways, some of them related to the iron homeostasis control and oxidative stress response (See communication by M Chevallet *et al.* for more details about the all set of proteins increasing in expression upon NPs treatment). Indeed, an induction of peroxiredoxin 4 and ferritin (light chain) was observed. Peroxiredoxins (Prxs), a family of thioredoxin-dependent peroxidases, are known to protect aerobic organisms against oxidative stress by degradation of peroxides and other substrates [43]. Mammalians ferritins are composed of 24 subunits that form an almost spherical shell, delimiting a large cavity that can accommodate up to 4000 Fe atoms. They are known to play a role in the regulation of available cellular iron and storage in the response to oxidative stress. These preliminary results suggest that the effects of TiO_2-NPs may be mediated through oxidative stress and a putative modification of iron homeostasis. Further experiments have to be done with characterized and dispersed solutions of DegussaP25 and CEA12A NPs in order to see if they trigger the same changes in ferritin and peroxiredoxin contents.

3.4. Cytotoxicity and genotoxicity of TiO_2-NPs to A549 human pneumocytes

Toxicity of TiO_2-NPs was evaluated in pneumocytes since NPs might be inhaled by populations and reach pulmonary alveoli. To model this situation, we studied the response of A549 human pneumocytes after exposure to DegussaP25-NPs diluted in DMEM medium supplemented with 50 UI/mL penicillin and 50 µg/mL streptomycin. Cell metabolic activity, reflecting toxicity of NPs, was evaluated by using 3-(4,5-dimethylthiazol-2-yl)-2,5-dipheny-tetrazolium bromide (MTT) reduction method [44]. Mitochondrial dehydrogenases of viable cells reduce MTT to water insoluble blue formazan crystals which are then solubilised by dimethyl sulfoxide (DMSO). Therefore, this assay indicates cell mitochondrial activity impairment. After exposure to 100 µg/mL of TiO_2-NPs, the cell death rate measured with the MTT assay was 17% and 48% after 48h and 72h, respectively.

To determine whether these observation could be explained by induction of oxidative stress, the formation of ROS was measured *via* monitoring the increased fluorescence of 2',7'-dichlorodihydrofluorescein diacetate acetyl ester (H_2DCF-DA, Invitrogen [45] (figure 4). This hydrophobic non-fluorescent chemical can get across the cell membrane easily, in order to be converted into hydrophilic non-fluorescent form by esterases which remove acetate groups. This compound is oxidized by intracellular ROS and turned into hydrophilic green fluorescent form, DCF. Fluorescence count was significantly higher than in the negative control after exposure to 100 µg/ml of TiO_2-NPs. It was about 6 times higher than control after both 4h and 24h of exposure.

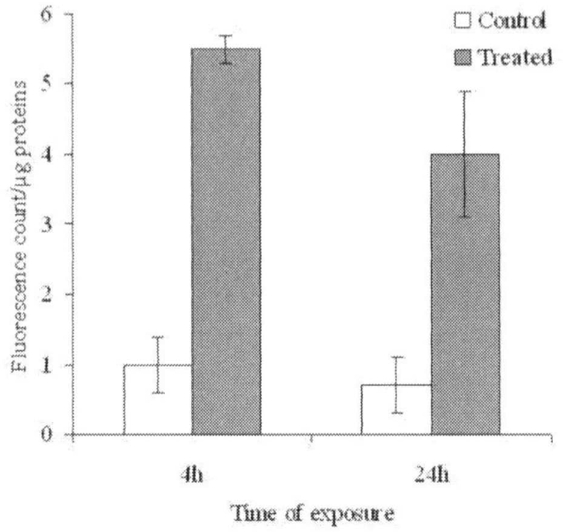

Figure 4. Fluorescence count per µg of proteins quantified in A549 human pneumocytes exposed to TiO_2-NPs at 100 µg/mL after 4h or 24h of exposure.

After showing the occurrence of oxidative stress upon incubation with NPs, we studied its potential genotoxic properties. We thus first quantified the level of 8-oxo-dG formation as a biomarker of oxidative damage of DNA after ROS production. The level of 8-Oxo-7,8-dihydro-2'-deoxyguanosine (8-oxo-dG) was measured in cells exposed to TiO_2-NPs using high-performance liquid chromatography coupled to tandem mass spectrometry (figure 5). Production of 8-oxo-dG residues in DNA leads to a GC→TA translocation unless repaired prior to DNA replication [46]. Therefore, the presence of 8-oxo-dG may lead to mutagenesis. Furthermore, many observations indicated a direct correlation between *in vivo* 8-oxo-dG formation and carcinogenesis [47,48]. The percentage of 8-oxo-dG was statistically higher for cells exposed to TiO_2-NPs, which is a sensitive marker of ROS-induced DNA damage.

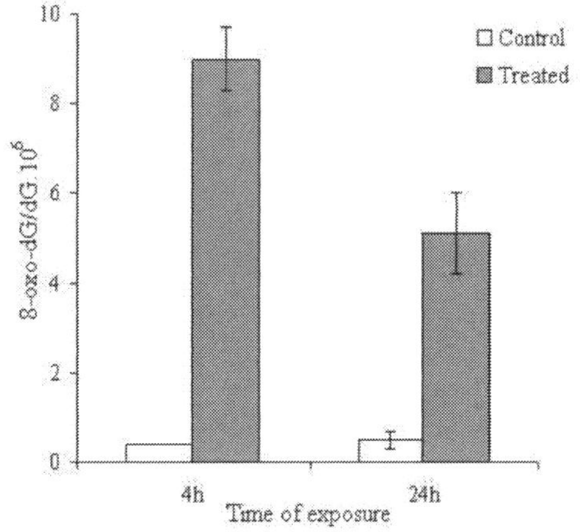

Figure 5. 8-oxo-dG formation in A549 human pneumocytes treated with DegussaP25 TiO_2-NPs at 100 μg/mL after 4h or 24h of exposure.

Finally, the comet assay was performed under alkaline conditions to assess NPs-induced DNA damage [49]. The comet assay (single cell gel electrophoresis) detects DNA strand breaks and alkaline labile sites in individual cells. These classes of DNA lesions mostly result from oxidative processes. At least 50 comets per sample were analyzed under a fluorescence microscope. The comet tail length was increased by a factor 4.3 and 11.6 as compared to unexposed cells following 4h and 24h of exposure, respectively. These results indicated that TiO_2-NPs caused DNA breakage. The genotoxicity may come from direct interaction between NPs and DNA or through the attack of DNA by over-accumulated ROS, which may gain access to cell nucleus by diffusion through nuclear pores and then attack DNA.

4. Conclusion

These first results confirmed the importance of the NPs characterization (size, stability...) after dispersion in culture media in order to obtain reproducible and meaningful results. We obtained TiO_2-NPs dispersion in a minimal bacterial growth medium that we optimized. Therefore, we will be able to follow modification of protein expression and enzymatic activity of metalloenzymes caused by both NP-TiO_2 (CEA12A and DegussaP25) studied in this article, even if no effect on the bacterial growth was observed in the tested conditions. Another experiment done for the NanoBioMet project revealed that aggregated TiO_2-NPs (anatase, <25 nm, Sigma-Aldrich) induced an increase in the expression of proteins including a ferritin subunit and a peroxiredoxin in macrophages. Finally, toxicity of DegussaP25-NPs was evaluated in A549 pneumocytes. Decrease in viability, formation of 8-oxo-dG, DNA strand breaks and ROS, were observed suggesting oxidative stress conditions upon TiO_2-NPs treatment. Further work is currently in progress in order to identify and quantify putative metallomic perturbations following treatments of various cellular models by well-characterized NPs.

Acknowledgements

We thank Roger Miras for his technical help in bacterial growth experiments. This project and the postdoctoral fellowship of Anne-Noëlle Petit are funded by the Transversal Program Nanosciences and the Transversal Program of Toxicology of the CEA.

References

[1] Usepa U, Nanotechnology White Paper. Prepared for the U.S. Environmental Protection Agency by Members of the Nanotechnology Workgroup, a Group of Epa's Science Policy Council Science Policy Council, U.S. Environmental Protection Agency, Washington, Dc (2007) Http://Www.Epa.Gov/Osa/Pdfs/Nanotech/Epa-Nanotechnology-Whitepaper-0207.Pdf Pp. 20460(April 25, 2010).
[2] Ellsworth D K, Verhurst D, Spitler T M and Sabacky B.J.2000 *Chem. Innov.* **30** 30
[3] Fisher J, Egerton T and Kirk-Othmer 2001 *Titanium Compounds,* Inorganic (New York: John Wiley & Sons)
[4] Kaida T, Kobayashi K, Adachi M and Suzuki F 2004 *J. Cosmet. Sci.* **55** 219
[5] Esterkin C R, Negro A C, Alfano O M and Cassano A E 2005 *Aiche J.* **51** 2298
[6] Choi H, Stathatos E and Dionysiou D D 2006 *Appl. Catal. B-Environ.* **63** 60
[7] Higarashi M M and Jardim W E 2002 *Catal. Today* **76** 201
[8] Konstantinou I K and Albanis T A 2004 *Appl. Catal. B Environ.* **49** 1
[9] Grassian V H, O'shaughnessy P T, Adamcakova-Dodd A, Pettibone J M and Thorne P S 2007 *Environ. Health Perspect.* **115** 397
[10] Gurr J R, Wang A S, Chen C H and Jan K Y 2005 *Toxicology* **213** 66
[11] Long T C, Saleh N, Tilton R D, Lowry G V and Veronesi B 2006 *Environ. Sci. Technol.* **40** 4346
[12] Renwick L C, Brown D, Clouter A and Donaldson K 2004 *Occup. Environ. Med.* **61** 442
[13] Sayes C M, Wahi R, Kurian P A, Liu Y, West J L, Ausman K D, Warheit D B and Colvin V L 2006 *Toxicol. Sci.* **92** 174
[14] Warheit D B, Webb T R, Sayes C M, Colvin V L and Reed K L 2006 *Toxicol. Sci.* **91** 227
[15] Simon-Deckers A, Gouget B, Mayne-L'hermite M, Herlin-Boime N, Reynaud C and Carrière M 2008 *Toxicology* **253** 137
[16] Park S, Lee Y K, Jung M, Kim K H, Chung N, Ahn E K, Lim Y and Lee K H 2007 *Inhal. Toxicol.* **19 Suppl 1** 59
[17] Singh S, Shi T, Duffin R, Albrecht C, Van Berlo D, Hohr D, Fubini B, Martra G, Fenoglio I, Borm P J and Schins R P 2007 *Toxicol. Appl. Pharmacol.* **222** 141
[18] Jiang J, Oberdorster G, Elder A, Gelein R, Mercer P and Biswas P 2008 *Nanotoxicology* **2** 33
[19] Nel A, Xia T, Madler L and Li N 2006 *Science* **311** 622
[20] Oberdorster G, Maynard A, Donaldson K, Castranova V, Fitzpatrick J, Ausman K, Carter J, Karn B, Kreyling W, Lai D, Olin S, Monteiro-Riviere N, Warheit D and Yang H 2005 *Part Fibre Toxicol.* **2** 8
[21] Oberdorster G, Oberdorster E and Oberdorster J 2005 *Environ. Health Perspect.* **113** 823
[22] Powers K W, Brown S C, Krishna V B, Wasdo S C, Moudgil B M and Roberts S M 2006 *Toxicol. Sci.* **90** 296
[23] Sager T M, Porter D W, Robinson V A, Lindsley W G, Schwegler-Berry D E and Castranova V 2007 *Nanotoxicology* **1** 118
[24] Bihari P, Vippola M, Schultes S, Praetner M, Khandoga a G, Reichel C A, Coester C, Tuomi T, Rehberg M and Krombach F 2008 *Part. Fibre Toxicol.* **5** 14
[25] Porter D, Sriram K, Wolfarth M, Jefferson A, Schwegler- Berry D, Andrew M and Castranova V 2008 *Nanotoxicology* **2** 144
[26] Buford M C, Hamilton R F Jr and Holian A 2007 *Part. Fibre Toxicol.* **4** 6
[27] Jiang J, Oberdorster G and Biswas P 2009 *J. Nanopart. Res.* **11** 77
[28] Ji Z, Jin X, George S, Xia T, Meng H, Wang X, Suarez E, Zhang H, Hoek E M V, Godwin H, Nel E and Zink J I 2010 *Environ. Sci. Technol.* **44** 7309
[29] Dick C A, Brown D M, Donaldson K and Stone V 2003 *Inhal. Toxicol.* **15** 39
[30] Vileno B, Lekka M, Sienkiewicz A, Jeney S, Stoessel G, Lekki J, Forro L and Stachura Z 2007 *Environ. Sci. Technol.* **41** 5149
[31] Minhua X, Jianmin M, Jianhua G and Zuhong L 1998 *Supramol. Sci.* **5** 511

[32] Yeung K L, Leung W K, Yao N and Cao S 2009 *Catal. Today* **143** 218
[33] Bhattacharya K, Davoren M, Boertz J, Schins R P, Hoffmann E and Dopp E 2009 *Part. Fibre Toxicol.* **6** 17
[34] Ghosh M, Bandyopadhyay M and Mukherjee A 2010 *Chemosphere* **81** 1253
[35] Touati D 2000 *Arch. Biochem. Biophys.* **373** 1
[36] Halliwell B and Gutteridge J M C 1989 *Free Radicals in Biology and Medecine* pp 131
[37] Pignon B, Maskrot H, Guyot Ferreol V, Leconte Y, Coste S, Gervais M, Pouget T, Reynaud C, Tranchant J F and Herlin-Boime N 2008 *Eur. J. Inorg. Chem.* **6** 833
[38] Brunauer S, Emmett P H and Teller E 1938 *J. Am. Chem. Soc.* **60** 309
[39] Sato K, Li J G, Kamiya H and Ishigaki T 2008 *J. Am. Ceram. Soc* **91** 2481
[40] Mandzy N, Grulke E and Druffel T 2005 *Powder Technol.* **160** 121
[41] Simon-Deckers A, Loo S, Mayne-L'hermite M, Herlin-Boime N, Menguy N, Reynaud C, Gouget B and Carriere M 2009 *Environ. Sci. Technol.* **43** 8423
[42] Schleicher U and Bogdan C 2009 *Methods Mol. Biol.* **531** 203
[43] Hall A, Karplus P A and Poole L B 2009 *FEBS J.* **276** 2469
[44] Mosmann T 1983 *J. Immunol. Methods* **65** 55
[45] Oyama Y, Hayashi A, Ueha T and Maekawa K 1994 *Brain Res.* **635** 113
[46] Cheng K C, Cahill D S, Kasai H, Nishimura S and Loeb L A 1992 *J. Biol. Chem.* **267** 166
[47] Loft S and Poulsen H E 1996 *J. Mol. Med.* **74** 297
[48] Olinski R, Jaruga P and Zastawny T H 1998 *Acta Biochim. Pol.* **45** 561
[49] Singh N P, Mccoy M T, Tice R R and Schneider E L 1988 *Exp. Cell. Res.* **175** 184

Journal of Physics: Conference Series **304** (2011) 012036 doi:10.1088/1742-6596/304/1/012036

Toxicity of nano- and micro-sized silver particles in human hepatocyte cell line L02

Pengpeng Liu[1], Rongfa Guan[1], Xingqian Ye[2], Jiaxin Jiang[1], Mingqi Liu[1], Guangrong Huang[1] and Xiaoting Chen[1]

[1]Zhejiang Provincial Key Laboratory of Biometrology and Inspection & Quarantine, College of Life Sciences, China Jiliang University, Hangzhou 310018, China.

[2]Department of Food Science and Nutrition, School of Biosystems Engineering and Food Science, Zhejiang University, Hangzhou 310029, China.

E-mail: rfguan@163.com

Abstract. Silver nanoparticles (Ag NPs) previously classified as antimicrobial agents have been widely used in consumers and industrial products, especially food storage material. Ag NPs used as antimicrobial agents may be found in liver. Thus, examination of the ability of Ag NPs to penetrate the liver is warranted. The aim of the study was to determine the optimal viability assay for using with Ag NPs in order to assess their toxicity to liver cells. For toxicity evaluations, cellular morphology, mitochondrial function (3-(4, 5-dimethylazol-2-yl)-2, 5-diphenyl-tetrazolium bromide, MTT assay), membrane leakage of lactate dehydrogenase (lactate dehydrogenase, LDH release assay), Oxidative stress markers (malonaldehyde (MDA), glutathione (GSH) and superoxide dismutase (SOD)), DNA damage (single cell gel eletrophoresis, SCGE assay), and protein damage were assessed under control and exposed conditions (24 h of exposure).The results showed that mitochondrial function decreased significantly in cells exposed to Ag NPs at 25 $\mu g{\cdot}mL^{-1}$. LDH leakage significantly increased in cells exposed to Ag NPs (≥ 25 μg mL^{-1}) while micro-sized silver particles tested displayed LDH leakage only at higher doses (100 $\mu g{\cdot}mL^{-1}$). The microscopic studies demonstrated that nanoparticle-exposed cells at higher doses became abnormal in size, displaying cellular shrinkage, and an acquisition of an irregular shape. Due to toxicity of silver, further study conducted with reference to its oxidative stress. The results exhibited significant depletion of GSH level, increase in SOD levels and lead to lipid peroxidation, which suggested that cytotoxicity of Ag NPs in liver cells might be mediated through oxidative stress.

The results demonstrates that Ag NPs lead to cellular morphological modifications, LDH leakage, mitochondrial dysfunction, and cause increased generation of ROS, depletion of GSH, lipid peroxidation, oxidative DNA damage and protein damage. Though the exact mechanism behind Ag NPs toxicity is suggested oxidative stress and lipid peroxidation playing an important role in Ag NPs elicited cell membrane disruption, DNA damage, protein damage and subsequent cell death. Our preliminary data suggest that oxidative stress might contribute to Ag NPs cytotoxicity. To reveal whether apoptosis involved in Ag NPs toxicity, further studies are underway.

1. Introduction

Ag nanoparticles (Ag NPs) were previously classified as antimicrobial agents, and have been widely used in consumers and industrial products, especially food storage material [1,2]. However, the cytotoxicity of Ag NPs has caused wide concerns by scientists and engineers in the last decades.

In vitro cell line studies have shown decreased mitochondrial function after exposure to Ag NPs in murine neuroblastoma cells [3], rat liver cells [4], germline stem cells [5], human skin carcinoma cells [6], human hepatoma cells [7], human alveolar epithelial and macrophage cell lines [8], human epidermal keratinocytes (HEKs) and fibroblasts [9], human lung fibroblast cells (IMR-90) and human glioblastoma cells (U251) [10]. Ingestion of Ag can cause agyria, the benign condition characterized by the bluish-graying of the skin that occurs through the preferential deposition of Ag in the basal lamina of soft tissues such as the liver, and spleen [11]and blood vessels, gastrointestinal tract, liver, and kidney [12]. Thus, examination of the ability of Ag NPs to penetrate the liver is warranted.

Our objectives in this study were to determine the optimal viability assay for using with Ag NPs in order to assess their toxicity to liver cells. In this paper, we evaluated the toxicity of Ag NPs and analyzed cellular viability, oxidative stress, and DNA damage in AgNPs-treated cells.

2. Materials and methods

2.1. Materials

The nano- and micro-sized silver particles used in this study were purchased from Qinhuangdao Taiji Ring Nano-Products Co., Ltd, Republic of China. The nano- and micro-sized silver particles were dissolved in Milli-Q water at a final concentration of 10 $mg \cdot mL^{-1}$.

2.2. Synthesis of silver particles

The nano- and micro-sized silver particles were characterized by Scanning Electron Microscope (SEM) imaging (JEOL JSM-5610LV).

2.3. Cell culture and treatment

L02 cells were cultured in RPMI 1640 medium with fetal calf serum (10 %), L-glutamine (2.9 $mg \cdot mL^{-1}$), streptomycin (1 $mg \cdot mL^{-1}$), and penicillin (100 $units \cdot mL^{-1}$). Cells were cultured at 37°C in water-saturated air supplemented with 5% CO_2. Culture media were changed every 3 days. Cells were passaged once a week.

After the monolayer of cells became conXuent in 6, 24 or 96-well plate, L02 cells were treated with a range of concentrations of nano- and micro-sized silver particles suspended in RPMI 1640 medium without serum for 24 h. After the 24 h treatment, the various toxicity end points were evaluated in control and silver particles exposed cells.

2.4. Cell morphology

L02 cells were exposed as mentioned above at various concentrations of nano- and micro-sized silver particles for 24 h. After completion of the exposure period, cells (control, nano-Ag and micro-Ag exposed) were washed with PBS and observed by phase contrast inverted microscopy at 200 × magnification.

2.5. Cell activity

LDH leakage due to membrane damage was assessed by measuring the activity of LDH in the cells and medias described elsewhere with some modifications [13]. Mitochondrial function was evaluated spectrophotometrically by 3-(4, 5-dimethylazol-2-yl)-2, 5-diphenyl-tetrazolium bromide (MTT) assay. The MTT assay helps in cell-viability assessment by measuring the enzymatic reduction of yellow tetrazolium MTT to a purple formazan, as measured at 570 nm using Enzyme-labeled instrument (Tecan Co.).

2.6. DNA damage

DNA damage by Ag NPs was further studied using comet assay. After treatment with nano- and micro-sized silver particles for 24 h, the cells were rinsed with ice-cold 1×PBS and trypsinized. Then the cells were washed once in ice-cold 1×PBS and resuspended at 1×10^5 cells mL^{-1} in ice-cold 1×PBS. An aliquot of 10 μL cell suspension was mixed with 100 μL molten agarose (37°C), and 75μL of this mixture was immediately applied to a glass slide. The slide was held horizontal at 4 °C for 30 min to improve adherence. Then the slide was immersed in cold lysis solution to lyse the cells. After 50 min at 4 °C in the dark, the slide was immersed in an alkaline solution (300 mM NaOH, 1 mM EDTA, pH>13) at room temperature in the dark to denature the DNA. After 30 min, the slide was placed on a horizontal electrophoresis unit and the unit was filled with fresh buffer (300 mM NaOH, 1 mM EDTA, pH>13) to cover the slide. Electrophoresis was conducted at 27 V (300 mA) for 40 min at 4°C in the dark. The slide was then washed gently with distilled water and immersed in 70% ethanol for 5 min. After the slide was air dried, 50μL of Ethidium bromide (EB) working solution was applied to each circle of dried agarose. All steps described above were conducted under yellow light to prevent additional DNA damage.

Slides were viewed using an epifluorescence Leica DMI 4000B microscope equipped with a fluorescein filter. Observations were made at a final magnification of 400×. Fifty randomly selected cells per experimental point were imaged and analyzed using CASP software (download from http://www.casp.of.pl/). Results were reported as tail moment, a parameter describing the number of migrated fragments, represented by the fluorescence intensity in the tail, expressed as the mean of the 50 cells.

2.7. Oxidative stress

Cells were cultured in 75-cm^2 culture flask and exposed to Ag NPs (5-100μg·mL^{-1}) for 24 h. After exposure, the cells were harvested in chilled PBS by scraping and washed twice with 1× PBS at 4°C for 6 min at 1500 rpm. The cell pellet was then sonicated at 15 W for 10 s (3 cycles) to obtain the cell lysate.

Oxidative stress markers (MDA, GSH and SOD) were estimated by Nanjingjiancheng Kit (China) according to manufacturer's protocol. Protein content was measured by the method of Lowry [14] using BSA as the standard.

2.8. Statistical analysis

The data were expressed as mean ± standard deviation of three independent experiments. Wherever appropriate, the data were subjected to statistical analysis by one-way analysis of variance (ANOVA) followed by Dunnett's method for multiple comparisons. A value of $p<0.05$ was considered significant. SPSS 16.0 software was used for the statistical analysis.

3. Results

3.1. Characterization studies

The average size of nano-Ag and micro-Ag reported by SEM was 37.8±6.7 nm and 985.4±283.5 nm. Figure **1** shows a representative SEM image recorded nano-Ag and micro-Ag.

3.2. Cell morphology

Figure **2** shows the comparative morphologies of unexposed, and nano- and micro-sized silver particles exposed cells. At the low dose of nano-Ag (10 μg·mL^{-1}), the cells appear similar to control cells with brownish particles most likely associated with the cell membranes. With increasing doses of nano-Ag, the cells started to shrink and became irregular in shape (Figure **2**.*e*). The microscopic studies demonstrated that nano-Ag exposed cells at higher doses became abnormal in size, displaying cellular shrinkage, and an acquisition of an irregular shape.

Figure 1. SEM images of nano-Ag and micro-Ag
a: nano-Ag, 37.8±6.7 nm; b: micro-Ag, 985.4±283.5 nm.

3.3. Cell activity

L02 cells were exposed to nano- and micro-sized silver particles (5-100 μg·mL^{-1}) for 24h and cytotoxicity was determined with MTT assay and LDH release assay.

The MTT assay results demonstrated a concentration dependent cytotoxicity after exposure to Ag NPs (Figure **3**). The percentage (%) MTT reduction (relative to control) observed after nano-Ag exposure at concentrations 25, 50 and 100 μg·mL^{-1} was 49.35%, 26.61% and 14.44% respectively with reduction to 83.61%, 68.64% and 58.82% after micro-Ag exposure. It was noted that there is a statistically significant difference between different sizes of silver particle, where the nano-Ag showed higher toxicity at 25-100 μg·mL^{-1} than the micro-Ag($p<0.05$).

The LDH release assay results demonstrated that exposure to Ag nanoparticles for 24h resulted in concentration-dependent increase in LDH leakage and exhibited a significant ($p<0.05$) cytotoxicity at 25-100 μg·mL^{-1} (Figure 4). This is only ture for nano-Ag.

3.4. DNA damage

DNA damage by Ag NPs was further studied using comet assay. Chromosome abnormalities are a direct consequence of DNA damage such as double-strand breaks and misrepair of strand breaks in DNA, resulting in chromosome rearrangement.

Extensive and dose-dependent damage to DNA was observed after treatment of the cells with Ag NPs. Comet assay of Ag-NPs treated cells showed a concentration-dependent increase in tail moments (Figure **5**.*c*) as compared to control cells (Figure **5**.*a*), which gave the extent of DNA damage (Figure **5**.*c*). A comet-like tail implies presence of a damaged DNA strand that short DNA fragments are migrating faster than the intact nucleus. The length of the tail increases with the extent of DNA damage. Tail moments of control DNA was compared with nano-Ag treated cells, and extent of damage was assessed. An increase in DNA damage with increase in nano-Ag concentration was observed, whereas the cell showed no further increase in DNA damage beyond a micro-Ag concentration of 100 μg·mL^{-1} (Figure **5**.*d*).

3.5. Oxidative stress markers

3.5.1. *Effect of Ag NPs on MDA.* Lipid peroxidation was examined by measuring Malondialdehyde (MDA) concentration. A significant increase ($p<0.05$) in MDA formation was observed at all concentrations above 10 μg· mL^{-1} of Ag NPs as evident from Table 1.

3.5.2. *Effect of Ag NPs on GSH level.* Cells exposed to Ag NPs showed depletion of GSH level in a dose dependent manner, exposure concentrations exhibiting statistically significant ($p<0.05$) depletion of 57.51% and 26.53% at 50 and 100 μg·mL^{-1} respectively after 24 h (Table 1).

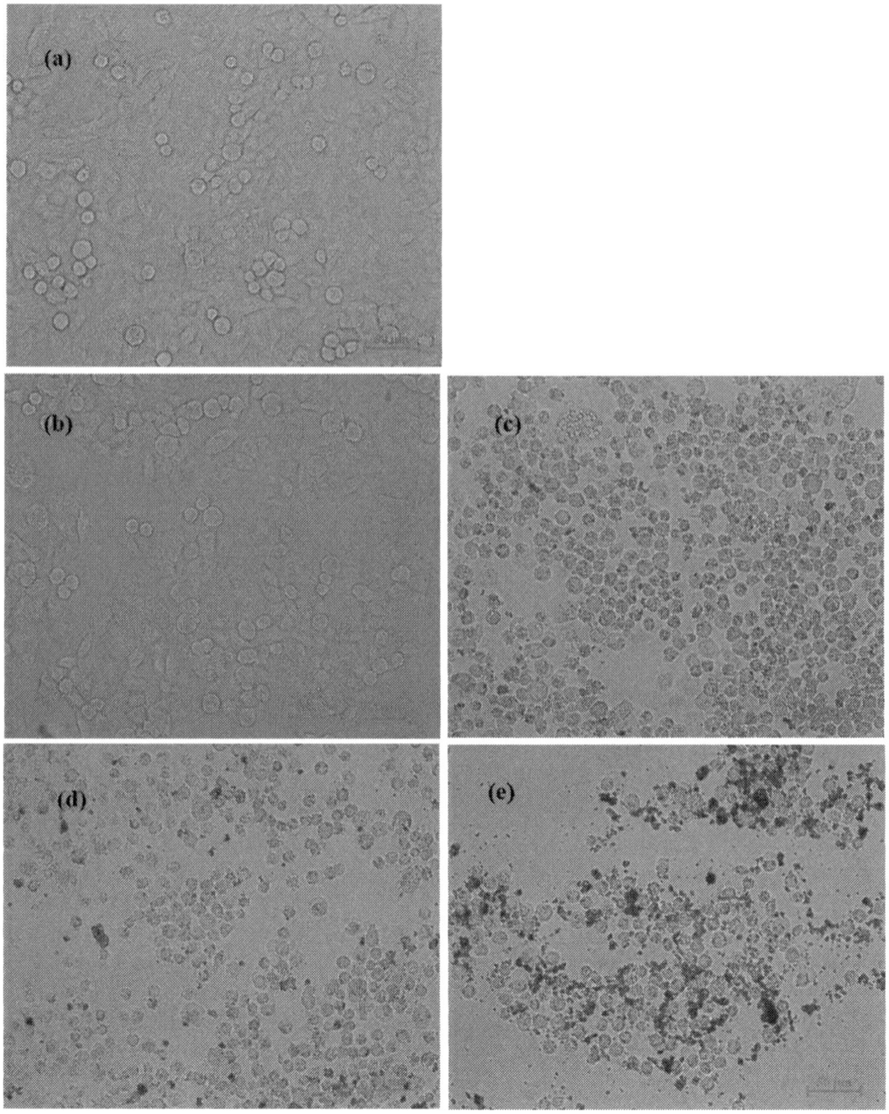

Figure 2. Morphology of human liver cells (L02)
(a) normal, and micro-Ag treated (b) 10 μg·mL^{-1} for 24 h, (c) 100 μg·mL^{-1} for 24 h, and nano-Ag treated (d) 10 μg·mL^{-1} for 24 h, (e) 100 μg·mL^{-1} for 24 h.

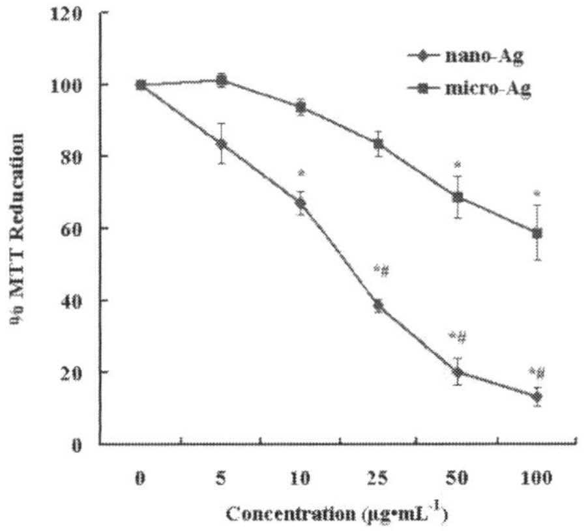

Figure 3. Cytotoxicity of Ag NPs on human liver cells (L02). MTT assay

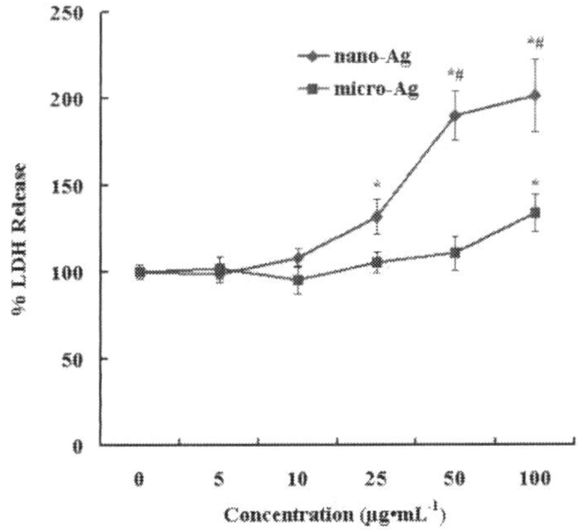

Figure 4. Cytotoxicity of Ag NPs on human liver cells (L02).LDH release assay

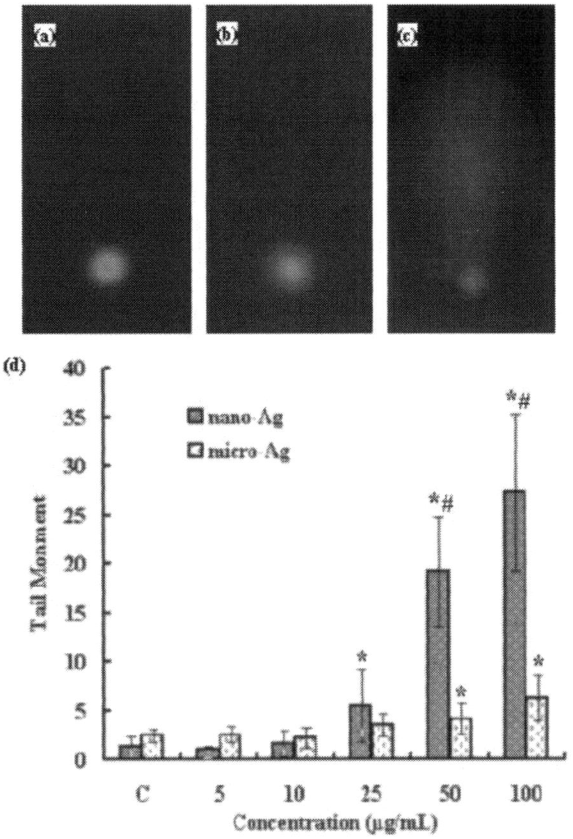

Figure 5. Comet analysis:
(a) control
(b) micro-Ag treated
(c) nano-Ag treated
(d) the Tail Moments of DNA

* $p<0.05$, compared with the control group.
# $p<0.05$, compared with micro-Ag group.

3.5.3. *Effect of Ag NPs on SOD activity.* SOD activity in cells treated with 25-100μg·mL^{-1} Ag NPs was significantly (p<0.05) reduced after 24 h of exposure when compared to unexposed cells as evident from Table 1.

Table 1. Effect of Ag NPs on Oxidative stress.

Conc.(μg·mL^{-1})	MDA(nmol·mgprot^{-1})	GSH(mg·gprot^{-1})	SOD(U·mgprot^{-1})
Control	0.188±0.017	4.26±0.62	31.56±2.56
5	0.195±0.037	4.31±0.26	30.62±1.26
10	0.248±0.022*	3.49±0.36	28.53±2.45
25	0.267±0.029**	3.00±0.23*	25.12±1.02*
50	0.312±0.036**	2.45±0.34**	22.83±0.95**
100	0.404±0.043**	1.13±0.25**	18.23±1.34**

* p<0.05, compared with control group.
** p<0.01, compared with control group.

4. Discussion

The field of nanotechnology has rapidly been growing worldwide for the past few years showing tremendous applicability in a variety of different sectors for the betterment of human life. However, a major and simultaneous outcome of these increasing nano-based applications is that humans now are at a greater risk of exposure to nanomaterials which can enter the biological systems through different routes [15,16].

Ag nanoparticles (Ag NPs) were previously classified as antimicrobial agents, and have been widely used in consumer and industrial products, especially food storage material. However, the cytotoxicity of Ag NPs has caused wide concerns by scientists and engineers in the last decades. Therefore in this study, cellular responses on exposure of Ag NPs to human liver cells (L02) were investigated. Our data demonstrate that exposure of Ag NPs to human liver cells causes cytotoxicity and oxidative stress. We have also observed DNA damaging effects of Ag NPs on L02 cells for which lipid peroxidation and oxidative stress may be attributed as one of the probable cause.

The cytotoxicity of Ag NPs was evident by morphological changes that appeared in L02 cells. Loss of normal morphology started appearing even at 10 μg·mL^{-1} after 24 h of exposure. With a consequent increase in exposure time, cells retracted into spherical shape and formed clusters in media after detachment from surface. A high tendency of Ag NPs adhering to cell membrane was observed at higher magnification. There is an earlier report that Ag NPs exposed human hepatoma cells reflect abnormal morphology, cellular shrinkage, detachment from the surface of the flask as well as decreased mitochondrial function and significantly increased LDH release at concentrations 0.7-2μg·mL^{-1} after 24 h exposure [7].

The production of free radicals has been found in a diverse range of nanomaterials which is one of the primary mechanisms of NPs toxicity [15]. It may result in oxidative stress, inflammation and consequent damage to proteins, membranes and DNA [17]. Thus in our study we investigated the levels of GSH and other antioxidant marker enzymes in the cells exposed to Ag NPs. A depletion in the GSH and SOD level were found at 25 μg·mL^{-1} on 24 h exposure. This indicates a condition of oxidative stress in cells which may arise due to imbalance in the ROS formation and antioxidant defence system of cells. As formation of ROS by Ag NPs is still unclear, the mechanism of ROS formation by Ag NPs needs further investigations. We also studied lipid peroxidation which represents another marker of oxidative stress. An increase in the formation of hydroperoxide compared to control was observed indicating lipid peroxidation on exposure to Ag NPs. This increase in lipid peroxidation may lead to membrane damage indicated by enhanced LDH released on Ag NPs exposure.

Our results also demonstrate that Ag NPs cause statistically significant DNA damage at concentrations of 25 $\mu g{\cdot}mL^{-1}$ after an exposure period of 24 h. This is supported by a previous study conducted by Ahamed *et al* [6] who have shown that Ag NPs caused different levels of DNA damage in mouse embryonic stem (mES) and mouse embryo fibroblast (MEF) cells, induced p53 protein expression, DNA double strand breakage and apoptosis.

As there is a well documented link between NPs and oxidative stress, one of the possible modes that can be suggested for Ag NPs induced DNA damage may be lipid peroxidation and oxidative stress [18]. ROS are known to react with DNA molecule causing damage to both purine and pyrimidine bases as well as DNA backbone [19]. Another important outcome of ROS production is lipid peroxidation which generates a variety of products reactive towards cellular macro molecules including DNA. One of the major products of lipid peroxidation malondialdehyde, is a proven mutagen and carcinogenic compound which reacts with DNA to form adducts to deoxyguanosine, deoxyadenosine and deoxycytidine [20].

This Ag NPs induced genotoxicity, if in actuality mediated by oxidative stress can be further linked to different pathways [21,22]. Principle modes that can be considered for ROS generation by Ag NPs are—intrinsic ROS generation from particles or their adsorbed species and ROS generation in target cells on exposure to Ag NPs. DNA damage resulting from any of these probable mechanisms may trigger signal transduction pathways leading to apoptosis or cause interferences with normal cellular processes thereby causing cell death [23]. A schematic diagram has been depicted in Figure 6 for the plausible mechanisms of Ag NPs induced cellular toxicity. However, further studies are needed to unravel the exact mechanism behind Ag NPs induced genotoxicity.

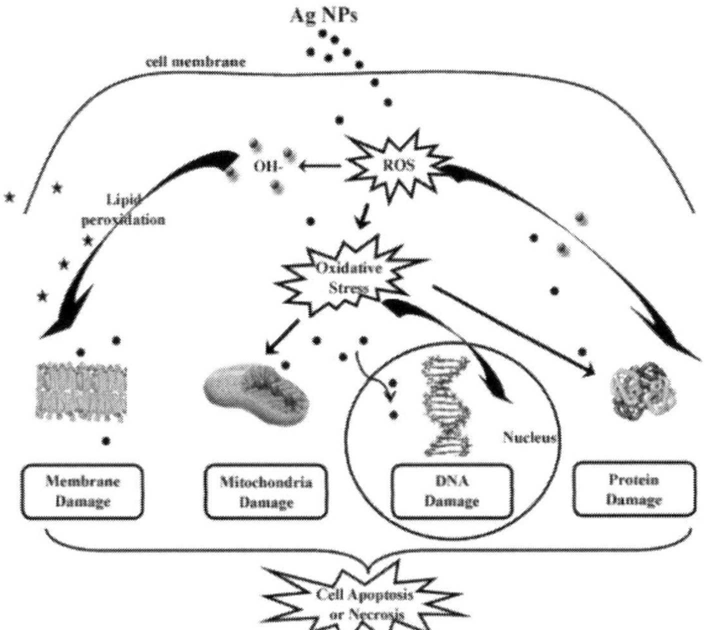

Figure 6. A schematic representation of results from this study representing the Ag NPs induced cytotoxicity.

Acknowledgement

The project supported by National High Technology Research and Development Program of China (2007AA100403), Food safety Supervision Program of General Administration of Quality Supervision,

Inspection and Quarantine of the People's Republic of China (AQSIQ2009SP26), Science and Technology Development Program of Hangzhou (20091832B50) and Science and Technology Development Program of Hangzhou (20101032B18).

References

[1] Siegrist M, Stampfli N, Kastenholz H and Keller C 2008 *Appetite* ***51*** 283-90
[2] Morones J R, Elechiguerra J L, CamaCho A, Holt K, Kouri J B, Ramirez J T, Yacaman M J, Jose Luis E, Alejandra C, Katherine H, Juan B K, Jose Tapia R and Miguel Jose Y 2005 *Nanotechnology* ***16*** 2346-53
[3] Schrand A M, Braydich-Stolle L K, Schlager J J, Dai L M and Hussain S M 2008 *Nanotechnology* ***19*** 13-19
[4] Hussain S M, Hess K L, Gearhart J M, Geiss K T and Schlager J J 2005 *Toxicol. in Vitro* ***19*** 975-83
[5] Braydich-Stolle L, Hussain S, Schlager J J and Hofmann M-C 2005 *Toxicol. Sci.* ***88*** 412-19
[6] Ahamed M, Karns M, Goodson M, Rowe J, Hussain S M, Schlager J J and Hong Y 2008 *Toxicol. Appl. Pharmacol* ***233*** 404-10
[7] Kim S, Choi J E, Choi J, Chung K-H, Park K, Yi J and Ryu D-Y 2009 *Toxicol. in Vitro* ***23*** 1076-84
[8] Lanone S, Rogerieux F, Geys J, Dupont A, Maillot-Marechal E, Boczkowski J, Lacroix G and Hoet P 2009 *Part. Fibre. Toxicol* ***6*** 14-25
[9] Andrew B, Chi H K, Siu C H, Hui S C, Hua G, Wai K L and Lin H 2007 *Wound. Repair. Regen.* ***15*** 94-104
[10] AshaRani P V, Mun G L K, Hande M P and Valiyaveettil S 2009 *Acs Nano* ***3*** 279-90
[11] Fung M C and Bowen D L 1996 *Clin. Toxicol* ***34*** 119-26
[12] Danscher G 1981 *Histochem. Cell. Biol* ***71*** 177-86
[13] Hussain S M and Frazier J M 2002 *Toxicol. Sci.* ***69*** 424-32
[14] Lowry O, Rosebrough N, Farr A and Randall R 1951 *J. Biol. Chem* ***193*** 265-75
[15] Nel A, Xia T, Madler L and Li N 2006 *Science* ***311*** 622-27
[16] Oberdörster G, Maynard A, Donaldson K, Castranova V, Fitzpatrick J, Ausman K, Carter J, Karn B, Kreyling W, Lai D, Olin S, Monteiro-Riviere N, Warheit D and Yang H 2005 *Part. Fibre. Toxicol* ***2*** 8-42
[17] Donaldson K and Stone V 2003 *Ann Ist Super Sanita* ***39*** 405-10
[18] Xia T,Kovochich M, Brant J, Hotze M, Sempf J, Oberley T, Sioutas C, Yeh J I, Wiesner M R and Nel A E 2006 *Nano. Lett* ***6*** 1794-807
[19] Yang W J, Shen C C, Ji Q L, An H J, Wang J J, Liu Q D and Zhang Z Z 2009 *Nanotechnology* ***20*** 1-7
[20] Marnett L J 2002 *Toxicology* ***181-182*** 219-22
[21] Kumari M, Mukherjee A and Chandrasekaran N 2009 *Sci.Total. Environ* ***407*** 5243-46
[22] Singh N, Manshian B, Jenkins G J S, Griffiths S M, Williams P M, Maffeis T G G, Wright C J and Doak S H 2009 *Biomaterials* ***30*** 3891-914
[23] Hsin Y-H, Chen C-F, Huang S, Shih T-S, Lai P-S and Chueh P J 2008 *Toxicol.Lett* ***179*** 130-39

Influence of TiO_2 nanoparticles on cellular antioxidant defense and its involvement in genotoxicity in HepG2 cells

Jana Petković, Bojana Žegura and Metka Filipič

Department of Genetic Toxicology and Cancer Biology, National Institute of Biology, SI-1000 Ljubljana, Slovenia

E-mail: metka.filipic@nib.si

Abstract. We investigated the effects of two types of TiO_2 nanoparticles (<25 nm anatase, TiO_2-An; <100 nm rutile, TiO_2-Ru) on cellular antioxidant defense in HepG2 cells. We previously showed that in HepG2 cells, TiO_2 nanoparticles are not toxic, although they induce oxidative DNA damage, production of intracellular reactive oxygen species, and up-regulation of mRNA expression of DNA-damage-responsive genes (*p53*, *p21*, *gadd45α* and *mdm2*). In the present study, we measured changes in mRNA expression of several antioxidant enzymes: catalase, superoxide dismutase, glutathione peroxidase, nitric oxide synthase, glutathione reductase and glutamate-cysteine ligase. As reduced glutathione has a central role in cellular antioxidant defense, we determined the effects of TiO_2 nanoparticles on changes in the intracellular glutathione content. To confirm a role for glutathione in protection against TiO_2-nanoparticle-induced DNA damage, we compared the extent of TiO_2-nanoparticle-induced DNA damage in HepG2 cells that were glutathione depleted with buthionine-(*S,R*)-sulfoximine pretreatment and in nonglutathione-depleted cells. Our data show that both types of TiO_2 nanoparticles up-regulate mRNA expression of oxidative-stress-related genes, with TiO_2-Ru being a stronger inducer than TiO_2-An. Both types of TiO_2 nanoparticles also induce dose-dependent increases in intracellular glutathione levels, and in glutathione-depleted cells, TiO_2-nanoparticle-induced DNA damage was significantly greater than in nonglutathione-depleted cells. Interestingly, the glutathione content and the extent of DNA damage were significantly higher in TiO_2-An- than TiO_2-Ru-exposed cells. Thus, we show that TiO_2 nanoparticles cause activation of cellular antioxidant processes, and that intracellular glutathione has a critical role in defense against this TiO_2-nanoparticle-induced DNA damage.

1. Introduction

The genotoxicity of nanomaterials is believed to be due to oxidative stress as the underlying key mechanism [1,2]. This involves alterations in redox homeostasis, which includes both increased production of reactive oxygen species (ROS) and depletion of antioxidant defenses. ROS are highly cytotoxic and can cause DNA damage, impair protein function, and result in peroxidation of lipids [3] and in apoptosis [4]. Generation of ROS can also be promoted by large surface area, which is one of the main characteristics of nanoparticles (NPs). Consequently, the smaller the NPs, the greater the oxidative stress they can induce. In addition, with their high surface area, transition metal ions that can be released from certain NPs, such as titanium, cadmium, chromium, cobalt, copper, iron, nickel, and zinc, have the potential to cause the conversion of cellular oxygen metabolic products, such as H_2O_2 and superoxide anions, to hydroxyl radicals ($^{\bullet}OH$), which is one of the primary DNA damaging species in cells [5]. Numerous studies have demonstrated that TiO_2 NPs can induce ROS production and promote oxidative DNA damage [6,7]. In our previous study, we showed that in a test system with HepG2 human hepatoma cells, TiO_2 NPs are not toxic, although they induce oxidative DNA damage, which is associated with intracellular ROS production. Exposure of these cells to TiO_2 NPs also induced changes in mRNA expression of DNA-damage-responsive genes (*p53*, *p21*, *gadd45*α and *mdm2*), which provides additional evidence that TiO_2 NPs are genotoxic [8]. However, although considerable attention has been given to the involvement of ROS in the genotoxicity of TiO_2 NPs, their influence on the antioxidant defense systems has been largely overlooked.

Antioxidant enzymes, such as superoxide dismutase (SOD), catalase (CAT) and nitric oxide synthase (NOS), convert ROS into less damaging compounds, thus providing the first line of antioxidant defense [9]. The second tier of defense is provided by enzymes such as glutathione peroxidase (GPx), glutathione S-transferase, aldo-keto reductase and aldehyde dehydrogenase, which detoxify the reactive intermediates. Glutathione (GSH) has a central role in these intracellular antioxidant defense processes as it is a free-radical scavenger and can conjugate reactive intermediates via glutathione S-transferase and GPx [3]. However, because of its importance, GSH is also a vulnerable target in cell defense, as different agents (environmental and other) can directly deplete or inactivate GSH. For example, depletion of intracellular GSH has been shown to occur by direct conjugation to or reaction with GSH by environmental agents, or indirectly by the activity of GSH-transferases [10]. The consequence is the facilitated accumulation of ROS and their by-products [4]. Despite the importance of GSH in defense against oxidative stress, its role in TiO_2-NP-induced genotoxicity has not been investigated yet.

The aim of this study was to determine the effects of two types of TiO_2 NPs on cellular antioxidant defense in HepG2 cells: <25 nm anatase (TiO_2-An) and <100 nm rutile (TiO_2-Ru). We measured changes in mRNA expression of the key antioxidant enzymes: CAT, SOD, GPx and NOS. CAT catalyzes the production of water and molecular oxygen from H_2O_2 and has a very important role in oxidative-stress tolerance [11]. SOD has a critical role in efficient dismutation of superoxide anions [12], and GPx detoxifies H_2O_2 or organic hydroperoxides using GSH as the reducing equivalent. NOS catalyzes the production of nitric oxide, which leads to the formation of peroxynitrite and reactive nitrogen species [13]. In addition, we measured expression of glutathione reductase (GSR), which reduces oxidized GSH (glutathione disulfide) to the reduced, sulfhydryl form, and of glutamate-cysteine ligase (GCL), which catalyzes the final step in GSH synthesis [14]. As reduced GSH has a central role in cellular antioxidant defense, we determined the effects of TiO_2 NPs on changes in the intracellular content of GSH. The role of GSH in protection against TiO_2-NP-induced DNA damage was confirmed by comparing the extent of TiO_2-NP-induced DNA damage in HepG2 cells that were GSH depleted with buthionine-(*S*,*R*)-sulfoximine (BSO) pretreatment to that of the same cells without GSH depletion.

2. Materials and methods

2.1. Chemicals

Eagle's minimal essential medium, penicillin/ streptomycin, L-glutamine, phosphate-buffered saline (PBS), trypsin, fetal bovine serum, non-essential amino-acid solution (100×), D,L-buthionine-[S,R]-sulfoximine (BSO), *tert*-butyl hydroperoxide (*t*-BOOH), ethidium-bromide solution, dimethyl sulfoxide were obtained from Sigma-Aldrich (St. Louis, USA). Monochlorobimane (mBCl) was from Fluka (Steinheim, Germany). TRIzol was from Invitrogen (Carlsbad, USA), and cDNA High Capacity Archive kits, TaqMan Universal PCR Master Mix, and TaqMan Gene Expression assays were from Applied Biosystems (Foster City, USA). Normal melting-point agarose and low melting-point agarose were from Gibco BRL (Paisley, Scotland).

2.2. Cell culture
HepG2 human hepatoma cells were obtained from the European Collection of Cell Cultures. They were grown in Eagle's minimal essential medium containing 10% fetal bovine serum, 1% non-essential amino acid solution, 2 mM L-glutamine, and 100 U/ml penicillin plus 100 μg/ml streptomycin, at 37°C in a humidified atmosphere with 5% CO_2.

2.3. Characteristics of TiO_2 nanoparticles
Two grades of TiO_2 powders with different average particle sizes (d <25 nm; d <100 nm) and crystalline structures (anatase; rutile) were used in this study, which were obtained from Sigma-Aldrich (St. Louis, USA). These are abbreviated as TiO_2-An (Cat. No. 637254: anatase; particle size <25 nm; 99.7% trace metals basis) and TiO_2-Ru (Cat. No. 637262: rutile; particle size ~10 × 40 nm; 99.5% trace metals basis). Their size, crystalline structure, specific surface area, and agglomeration/ aggregation were confirmed experimentally in our previous study [8].

2.4. Preparation of TiO_2 nanoparticles
Powdered TiO_2 NPs were suspended in PBS at a concentration 10 mg/ml and sonicated for 30 min at a frequency of 30 kHz in an ultrasonic bath (Sonorex, Bandelin Electronic, Berlin, Germany), to ensure uniform suspensions. These stock solutions were subsequently diluted in complete cell-growth medium to yield final concentrations ranging from 1 μg/ml to 250 μg/ml. These samples were then sonicated for 30 min to produce stable, less-aggregated suspensions before exposure of the cells in culture.

2.5. mRNA expression analysis
Cells were seeded at a density of 1×10^6 cells in T-25 flasks (Corning Costar Corporation, Corning, New York, USA) and incubated for 20 h at 37°C in 5% CO_2 for their attachment. The growth medium was then replaced with fresh medium containing 0, 1, 10, or 100 μg/ml TiO_2 NPs, and the cells were incubated for a further 4 h or 24 h. Each experiment included a negative control. Total RNA from these cells was isolated using TRIzol reagent, and the cDNA was synthesized using 2 μg total RNA with cDNA High Capacity Archive kits (Applied Biosystems), according to the manufacturer protocol. Gene expression levels of *CAT*, *SOD*, *GPx*, *NOS*, *GSR* and *GCL* were quantified using quantitative real-time PCR (ABI 7900 HT Sequence Detection System, Applied Biosystems). TaqMan Universal PCR Master Mix and the following TaqMan Gene Expression Assays were used (all from Applied Biosystems): *CAT* (catalase), Hs00937387_m1; *SOD* (superoxide dismutase 1, soluble), Hs00166575_m1; *GPx* (glutathione peroxidase 1), Hs01028922_g1; *NOS* (nitric oxide synthase 2, inducible), Hs01075529_m1; *GSR* (glutathione reductase), Hs00167317_m1; and *GCL* (glutamate-cysteine ligase, catalytic subunit), Hs00155249_m. Amplification of a glyceraldehyde 3-phosphate dehydrogenase probe was performed as an internal control. The conditions for the quantitative real-time PCR were 50°C for 2 min, 95°C for 10 min, 40 cycles of 95°C for 15 s, and 60°C for 1 min. The data obtained from TaqMan Gene Expression assays were analyzed using the ΔΔCt algorithm. Statistical significance between controls and treated groups was determined using two tailed Student's t-tests, and $P < 0.05$ was considered significant. Independent experiments were performed in duplicate, and each was repeated at least three times.

2.6. Determination of intracellular reduced glutathione
The level of GSH in HepG2 cells was quantified using a fluorogenic bimane probe, mBCl, as described by [15], with minor modifications [10]. The nonfluorescent probe mBCl forms a stable, fluorescent adduct with GSH in a reaction catalyzed by GSH S-transferases [15].

The cells were seeded into 96-well, black, tissue-culture-treated microtiter plates (Nunc, Naperville, USA) at a density of 1.25×10^5 cells/ml. After a 20 h incubation at 37°C in 5% CO_2, the cells were treated with 0, 1, 10, 100 or 250 μg/ml TiO_2 NPs in PBS. Negative (nonNP-treated) and positive (0.5 mM *t*-BOOH treated) cell-treatment controls were included in each experiment. After 2 h, 4 h and 6 h, 100 μM mBCl in PBS was added directly to the cells in the microtiter plates. After 15 min, the fluorescence of the wells was measured at 485 nm, with excitation at 390 nm, using a microplate-reading spectrofluorimeter (Tecan, Genios, Austria). The data are expressed as percentage fluorescence compared with the relevant negative controls. Statistical significance between control and treated groups was determined by two tailed Student's t-tests, and $P < 0.05$ was considered as statistically significant. Three independent experiments were performed, each with five replicates.

2.7. Comet assay

The HepG2 cells were seeded at a density of $\approx 6 \times 10^4$ cells/ml in 12-well microtiter plates (Corning Costar Corporation, Corning, New York, USA). They were then pre-cultured in the absence and presence of 150 μM BSO in complete cell-growth medium for 20 h. The GSH levels in these HepG2 cells pre-treated with BSO were determined using the fluorogenic bimane probe mBCl, as described in the previous subsection. After pretreatment with BSO, the growth medium containing BSO was replaced with fresh medium containing 0, 1, 10, 100 or 250 μg/ml TiO_2 NPs and incubated for 4 h. Each experiment included negative and positive controls (0.25 mM *t*-BOOH). After 4 h, the comet assays were performed as described previously [8]. Fifty cells were analyzed per experimental point, in each of three independent experiments. The percentages of tail DNA were used to measure the levels of DNA damage. One-way analysis of variance (ANOVA, Kruskal–Wallis) was used to analyze the differences between the treatments within each experiment. Dunnet's tests were used for comparing median values of percentage tail DNA; $P < 0.05$ and $P < 0.001$was considered as statistically significant.

3. Results

3.1. Effects of TiO_2 NPs on expression of oxidative-stress-responsive genes

The mRNA expression of selected oxidative-stress-responsive genes (*CAT*, *SOD*, *GPx*, *NOS*, *GSR*, *GCL*) was determined after 4 h and 24 h exposure of HepG2 cells to 0, 1, 10 or 100 μg/ml TiO_2 NPs, using quantitative real-time PCR.

In cells exposed to TiO_2-An for 4 h (Figure 1A) there were significant increases in expression of *CAT* and *GSR* at the highest TiO_2-An concentration (100 μg/ml), which after 24 h returned to control levels. After 24 h of exposure to TiO_2-An, the expression of *NOS* was significantly decreased at the highest TiO_2-An concentration (100 μg/ml) (Figure 1B).

In cells exposed to TiO_2-Ru, the expression of all six measured genes (*CAT*, *SOD*, *GPx*, *NOS*, *GSR*, *GCL*) were significantly increased at the highest TiO_2-Ru concentration (100 μg/ml) (Figure 1C). After 24 h of exposure, only expression of *GPx* and *GCL* remained significantly elevated at 100 μg/ml TiO_2-Ru, while that of *NOS* significantly decreased at 10 μg/ml TiO_2-Ru. Under the same conditions, mRNA expression of the other genes (*CAT*, *SOD*, *GSR*) remained unchanged (Figure 1D).

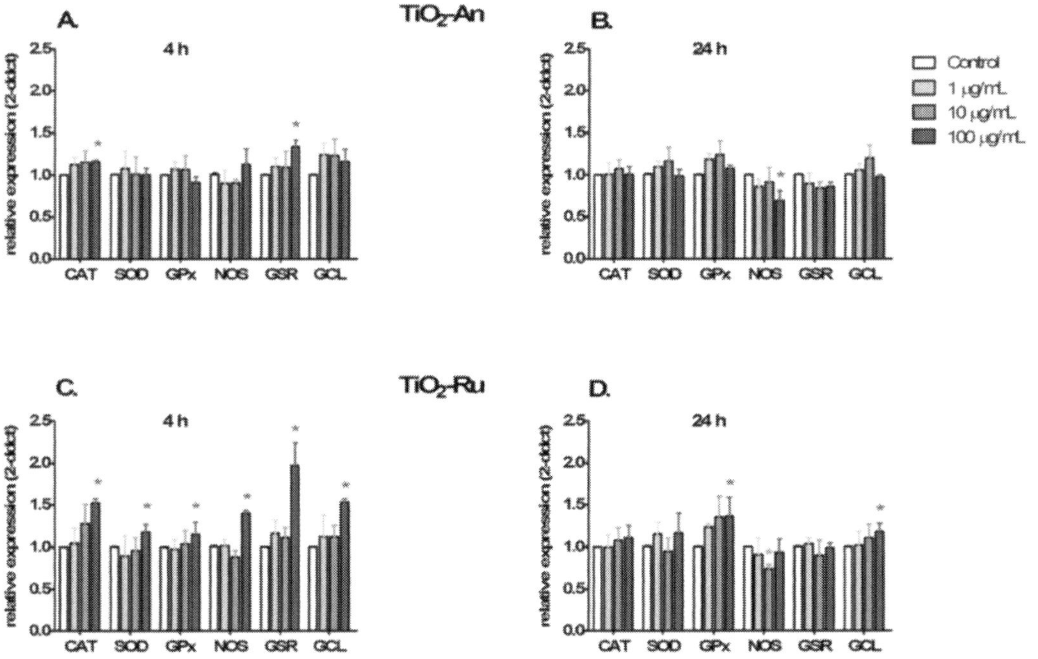

Figure 1. Quantitative real-time PCR analysis of changes in *CAT, SOD, GPx, NOS, GSR,* and *GCL* gene expression after exposure of HepG2 cells to TiO_2-An (A, B) and TiO_2-Ru (C, D) for 4 h (A, C) and 24 h (B, D). The quantitative real-time PCR was performed as described in Materials and methods. The data are expressed as relative mRNA expression normalized to the untreated controls. Duplicate experiments were repeated at least three times.
*, significant difference of TiO_2-NP-treated groups over controls (Student's t-tests; $P < 0.05$).

3.2. Changes in intracellular GSH

The intracellular levels of reduced GSH in HepG2 cells were quantified using the fluorogenic bimane probe mBCl, after 2 h, 4 h and 6 h exposure to 0, 1, 10, 100 and 250 µg/ml TiO_2 NPs.

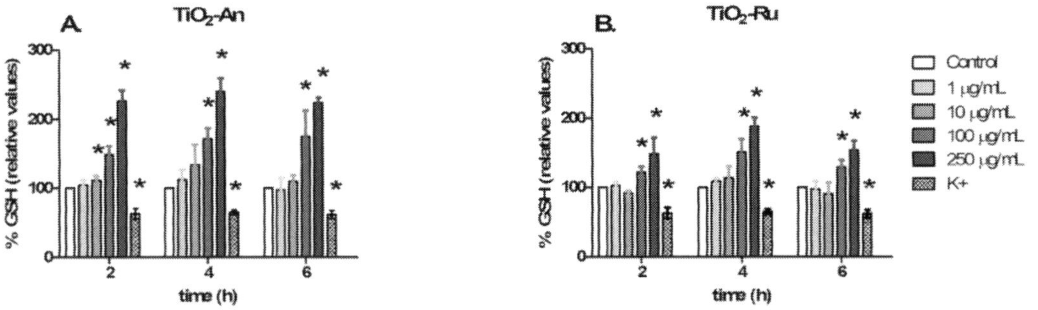

Figure 2. Intracellular reduced GSH levels in HepG2 cells after treatment with TiO_2-An (A) and TiO_2-Ru (B) at 0, 1, 10, 100 and 250 µg/ml TiO_2 NPs for 2 h, 4 h and 6 h. The positive control was 0.5 mM *t*-BOOH (K+). GSH levels were quantified using the fluorogenic bimane probe mBCl. Each bar represents the mean (±SD) of three independent experiments.
*, significant difference of TiO_2-NP-treated groups over controls (Student's t-tests; $P < 0.05$).

At 10 μg/ml, 100 μg/ml, and 250 μg/ml, exposure of HepG2 cells to TiO_2-An for 2 h, 4 h, and 6 h resulted in significant, dose-dependent increases in intracellular GSH levels. The maximal increase in the GSH levels was observed at the highest TiO_2-An concentration (250 μg/ml) after 4 h of exposure (to 2.4-fold the control) (Figure 2A).

In the cells exposed to TiO_2-Ru, there were significant increases in GSH levels at 100 μg/ml and 250 μg/ml TiO_2-Ru at all times of exposure (2 h, 4 h, 6 h). The maximal GSH level was observed after 4 h at 250 μg/ml TiO_2-Ru (to around 1.9-fold the control), which declined slightly after 6 h exposure, but remained significantly greater than the control (at 1.5-fold the control) (Figure 2B).

As the positive control for this test system, 0.5 mM *t*-BOOH led to a 60% decrease in intracellular GSH levels, compared to control cells.

3.3. The role of GSH in TiO_2-NP-induced DNA damage

To explore the role of GSH in TiO_2-NP-induced DNA damage, the cells were cultured in the absence and presence of BSO in complete cell-growth medium for 20 h. The intracellular levels of reduced GSH were then quantified using the fluorogenic bimane probe mBCl, after 4 h exposure to 0, 1, 10, 100 and 250 μg/ml TiO_2 NPs. The positive control was 0.25 mM *t*-BOOH. In these BSO-pretreated HepG2 cells, the GSH levels were reduced to 60% compared to the control cells (data not shown).

In the nonBSO-pretreated cells exposed to the increasing concentrations of TiO_2-An, there was a small but dose-dependent increase in DNA strand breaks in comparison to the nonTiO_2-An-treated control, which was significant only at the highest concentration of TiO_2-An (250 μg/ml; $P < 0.05$). On the other hand, in the BSO-pretreated cells, TiO_2-An induced a further significant dose-dependent increase in the percentage tail DNA, from TiO_2-An concentrations of 10 μg/ml and above ($P < 0.001$; Table 1).

The percentage DNA damage in nonBSO-pretreated cells exposed to the increasing TiO_2-Ru concentrations did not significantly differ from the control cells. In the BSO-pretreated cells, there was a small, but not statistically significant, increase in the DNA strand breaks over the control with the 100 μg/ml, and 250 μg/ml exposure to TiO_2-Ru (Table 2).

Table 1. Levels of TiO_2-An-induced DNA strand breaks in control (nonBSO-pretreated) and BSO-pretreated HepG2 cells.

TiO_2-An	**%Tail DNA (mean ±SD)**	
(μg/ml)	**NonBSO-pretreated**	**BSO-pretreated**
0	3.67 ±0.76	5.15 ±0.52
1	4.40 ±1.73	5.29 ±0.86
10	5.24 ±1.19	8.90 ±1.51**
100	5.65 ±0.56	12.63 ±2.13**
250	6.17 ±1.85*	13.75 ±2.32**
K+	44.56 ±9.62**	39.87 ±5.67**

*, $P < 0.05$; **, $P < 0.001$: significant difference between TiO_2-NP-treated and control groups using ANOVA, and Kruskal-Wallis with Dunnet's post test.

Table 2. Levels of TiO_2-Ru-induced DNA strand breaks in control (nonBSO-pretreated) and BSO-pretreated HepG2 cells.

TiO_2-Ru	**%Tail DNA (mean ±SD)**	
(μg/ml)	**NonBSO-pretreated**	**BSO-pretreated**
0	3.58 ±0.76	3.67 ±0.80
1	3.96 ±0.49	3.78 ±0.92
10	3.74 ±0.74	3.89 ±1.15
100	4.14 ±1.16	4.49 ±0.37
250	3.99 ±0.82	5.27 ±1.62
K+	41.06 ±7.40**	34.32 ±2.50**

**, $P < 0.001$: significant difference between TiO_2-NP-treated and control groups using ANOVA, and Kruskal-Wallis with Dunnet's post test.

4. Discussion and conclusions

In this study, we have shown for the first time that different crystalline forms of TiO_2 NPs can differently activate cellular antioxidant processes at the gene-expression level, and that intracellular GSH has a critical role in defense against TiO_2-NP-induced DNA damage. From our previous study [8], it appeared that ROS formation and the induction of oxidative DNA damage by TiO_2 NPs is a consequence of deregulation of the endogenous antioxidant enzymes, including CAT, SOD and others, that synergize with the GSH redox system.

After short-term exposure (4 h) to TiO_2-Ru, we observed up-regulation of all of the genes monitored, while TiO_2-An up-regulated only the expression of *CAT* and *GSR*. These observations indicate that the cells responded to the exposure to these TiO_2 NPs by activation of their antioxidant defense system, which provides indirect evidence that they induce oxidative stress. This finding would also explain our previously reported higher ROS production and oxidative DNA damage induced by TiO_2-An compared to TiO_2-Ru NPs [8]. After longer exposure to TiO_2 NPs (24 h), expression of all of these genes monitored returned to their control levels, and some were even slightly decreased. This pattern of elevated expression after short-term exposure followed by normal levels of expression after longer exposure is probably the normal cellular response to exogenous stress caused by agents like these TiO_2 NPs. As during the prolonged exposure the mRNA levels in these cells returned to normal levels even though the TiO_2 NPs were not removed, this indicates that the cells adapted to such stressors, and in this case TiO_2 NPs.

Another marker of response to oxidative stress that we monitored in the present study was the change in the intracellular GSH levels, which has been often shown to be involved in cellular responses to toxins and xenobiotics [3,10]. Exposure of HepG2 cells to TiO_2 NPs resulted in dose-dependent and time-dependent increases in the intracellular GSH levels. Gerloff et al. [16] recently reported that in Caco-2 cells exposure to TiO_2 NPs (80% anatase, 20% rutile) for 4 h did not change the cellular GSH content. On the contrary, in human keratinocyte HaCaT cells, Horie et al. [17] reported a significant decrease in intracellular GSH levels after 24 h of exposure to nanosized TiO_2, but not to fine-sized TiO_2 particles. The reasons for these contradictory results might be because Horie et al. [17] measured the GSH levels after 24 h of exposure, while in the present study, we measured GSH levels only up to 6 h of exposure; at the same time, these different cell types might respond differently to this exposure to TiO_2 NPs.

Our demonstration that HepG2 cells responded to exposure to TiO_2 NPs by increases in their intracellular GSH levels, and taking into account the evidence that TiO_2-NP-induced genotoxicity is ROS mediated, led us to the assumption that GSH is involved in cellular protection against

TiO_2-induced DNA damage. To confirm this hypothesis, we used an approach with BSO, as a GSH-depleting agent that inhibits GCL [18]. The data here clearly showed that exposure of the GSH-depleted cells to TiO_2-An resulted in significantly higher levels of DNA damage, as compared to nondepleted cells. TiO_2-Ru however, did not induce DNA damage, neither in GSH-depleted nor in nondepleted cells.

This study has demonstrated that the cell redox status is disturbed after exposure to TiO_2 NPs. Both crystalline types of TiO_2-NP induced changes in gene expression of the enzymes involved in the cellular defense against oxidative stress, with TiO_2-Ru NPs being significantly more efficient. However, TiO_2-An was a more efficient inducer of increases in cellular GSH levels, and only TiO_2-An NPs induced DNA damage, with intracellular GSH shown to have a role in protection against TiO_2-An-induced genotoxicity. Similar to this study, our previous study showed that TiO_2-Ru NPs induce the expression of DNA-damage-responsive genes more efficiently, while TiO_2-An induced more intracellular ROS and oxidative DNA damage [8]. The lack of genotoxic effects of TiO_2-Ru can therefore be explained, such that in cells exposed to TiO_2-Ru, an early defense response triggers the antioxidant enzymes that reduce the formation of ROS, and repair processes that eliminate DNA damage before it can be detected. Our studies thus demonstrate that these anatase and rutile crystalline types of TiO_2 NPs have different toxic potentials, while also showing differences in their toxicity mechanisms.

References

[1] Kim S, Choi JE, Choi J, Chung K-H, Park K, Yi J and Ryu D-Y 2009 *Toxicol. in Vitro* **23** 1076-84
[2] Schins RP and Knaapen AM 2007 *Inhal. Toxicol.* **19** 189-98
[3] Hayes JD and McLellan LI 1999 *Free Radical Res.* **31** 273-300
[4] Franco R, Sánchez-Olea R, Reyes-Reyes EM and Panayiotidis M 2009 *Mutat. Res.-Gen. Tox. En.* **674** 3-22
[5] Singh N, Manshian B, Jenkins GJS, Griffiths SM, Williams PM, Maffeis TGG, Wright CJ and Doak SH 2009 *Biomaterials* **30** 3891-914
[6] Karlsson HL, Cronholm P, Gustafsson J and Moller L 2008 *Chem. Res. Toxicol.* **21** 1726-32
[7] Gurr JR, Wang AS, Chen CH and Jan KY 2005 *Toxicology* **213** 66-73
[8] Petković J, Žegura B, Stevanović M, Drnovšek N, Uskoković D, Novak S and Filipič M 2010 *Nanotoxicology*
[9] McCord JM and Fridovich I 1988 *Free. Radic. Biol. Med.* **5** 363-9
[10] Zegura B, Lah TT and Filipic M 2006 *Mutat. Res.* **611** 25-33
[11] Brown MR, Miller FJ Jr, Li W-G, Ellingson AN, Mozena JD, Chatterjee P, Engelhardt JF, Zwacka RM, Oberley LW, Fang X, Spector AA and Weintraub NL 1999 *Circ. Res.* **85** 524-33
[12] Zelko IN, Mariani TJ and Folz RJ 2002 *Free. Radic. Biol. Med.* **33** 337-49
[13] Halliwell B and Gutteridge JMC 1999 *Free Radicals in Biology and Medicine* (Oxford: Clarendon Press)
[14] Krohne-Ehrich G, Schirmer RH and Untucht-Grau R 1977 *Eur. J. Biochem.* **80** 65-71
[15] Fernández-Checa J and Kaplowitz N 1990 *Anal. Biochem.* **190** 212-9
[16] Gerloff K, Albrecht C, Boots AW, Förster I and Schins RPF 2009 *Nanotoxicology* **3** 355-64
[17] Horie M, Nishio K, Fujita K, Kato H, Endoh S, Suzuki M, Nakamura A, Miyauchi A, Kinugasa S, Yamamoto K, Iwahashi H, Murayama H, Niki E and Yoshida Y 2010 *Toxicol. in Vitro* **24** 1629-38
[18] Griffith OW and Meister A 1979 *J. Biol. Chem.* **254** 7558-60

Nano-enhanced food contact materials and the in vitro toxicity to human intestinal cells of nano-ZnO at low dose

Niall Ó Claonadh[1], Alan Casey[1], Sean Lyons[2], Clement Higginbotham[2], Sanchali Gupta Mukherjee[1], Gordon Chambers[1]

[1] Nanolab Research Centre, Focas Institute, Dublin Institute of Technology, Dublin, Ireland

[2]Materials Research Institute, Athlone Institute of Technology, Westmeath, Ireland

E-mail: Niall.OClaonadh@DIT.ie; Alan.Casey@DIT.ie

Abstract: Nano Zinc Oxide (nZnO) has been shown to display antimicrobial effects which have lead to its application in a number of areas such as antimicrobial surface coatings, anti bacterial wound dressings and more recently in polymer composite systems for use in food contact materials. Concerns have been raised due to the incorporation of nanoparticles in food packaging stemming from the possibility of repeated low dose direct exposure, through ingestion, primarily due to degradation and nanoparticle leaching from the polymer composite. To address these concerns, composites consisting of nZnO and polyethylene were formed using twin screw extrusion to mimic commercial methods of food contact material production. A leaching study was performed using Atomic Absorption Spectroscopy in order to determine the concentration of nZnO leached from the composite. Composite stability studies were performed and a leached nZnO concentration was evaluated. This concentration range was then utilised in a series of tests aimed at determining the toxicity response associated with nZnO when exposed to an intestinal model. In this study two human colorectal carcinoma cell lines, HT29 (ATCC No: HTB-38) and SW480 (ATTC No: CCL-228), were employed as a model to represent areas exposed by ingestion. These lines were exposed to a concentration range of nZnO which incorporated the concentration leached from the composites. The cytotoxic effects of nZnO were evaluated using four cytotoxic endpoints namely the Neutral Red, Alamar Blue, Coomassie Blue and MTT assays. The results of these studies are presented and their implications for the use on nano ZnO in direct food contact surfaces will be discussed.

1. Introduction

The rapid expansion of nano based technology into consumer goods represents a new dynamic between research and manufacturing with a move away from the traditionally slow incorporation of frontline research into commercial markets. According to the project on emerging nanotechnologies there are currently 806 identified 'nano' products, produced by 484 companies, located in 24 countries (www.nanotechproject.org). This 'nano' advance gives rise to impressive projections and statistics highlighting the importance of nanotechnology to national and international economies. However hidden behind the commercial and popular statistics is a significant lack of data on the toxicology of many of the commercially relevant nano-particles. Indeed the determination of the toxicity of a nanoparticle raises many questions ranging from physicochemical characterization protocols and standards, through to the relevance of certain toxicological methods to specific particle[1]. Early nanotoxicology studies suffered difficulties in dose determination and in considering the effect of physical size and surface area [2,3], a significant flaw since these nano-properties can be directly related to many other properties such as reactivity, mobility and absorption which can influence the mechanism and the degree of toxicity [4]. To date, studies in the area have moved quickly to address these flaws however much of the reported literature has focused upon occupational exposure rather than risks facing consumers with the respiratory tract receiving considerable attention [5, 6]. While these studies are of great importance in promoting safe handling of nanomaterials, knowledge gaps still remain in the area of exposure to engineered nanomaterials by ingestion. To date only few studies explore the potential exposure to engineered nanoparticles by ingestion[7] and many of the current knowledge surrounding this type of exposure focus on very high dose exposures. Bing Wang et al [6] reported in 2006 an in vivo study of nano zinc oxide which was orally administered at high doses (5 g/kg body weight) to healthy adult mice. The results showed that nano zinc oxide gave rise to vomiting and lethargy in the mice with death occurring in two within one week which was deciphered to be due to intestinal obstruction. They also reported mice exposed to nano ZnO were found to have lesions in the heart and liver indicating the possible translocation in the body of the nanoparticles. It should be noted that these effects were the result of excessively high doses which are arguably not indicative of realistic exposure scenarios [8]. However the use of nanoparticles in consumer food packaging [9] and in some instances as an ingredient in food products has given rise to concerns about the possibility of repeated low dose direct exposure, through ingestion.

This study concentrates on nano Zinc Oxide, which is currently being utilised in novel food packaging materials as both a UV blocker and an anti-microbial agent. The anti-microbial activity of nZnO within the food contact material is said to hinder food spoilage by reducing the number of microbes in the food environment [9-12]. The ability of the nano material to leach from the food contact material and subsequently be ingested has yet to be addressed. In this study a composite of nZnO and Polyethylene was formed using twin screw extrusion in order to mimic one of the standard commercial methods of polymer based composite formation. Polyethylene was utilised due to its ubiquitous use in the production of consumer goods. From this a leaching study was performed on the material and the determined concentrations due to leaching were then used in a cytotoxic assessment. The low dose exposures directly mimic those that may be encountered during use of a nZnO activated food contact material.

The effects of low dose nZnO exposure to two colorectal cell lines were examined taking care to include the determined concentrations which occur by leaching. The cell lines used were SW480 and HT29 both of which originate from the lower intestine. These two lines were chosen due to their ease of culture and their reported use in other studies [13]. A number of assays namely Alamar Blue, Neutral Red, Coomassie Blue and MTT, were utilised to monitor any observed cellular response. Each assay measures a unique cellular function such as mitochondrial action, lysosomal action and general respiration therefore the use of a battery of assays allows for a greater understanding of the cytotoxic impact of nZnO on the two cell lines.

2. Methods

2.1. Characterisation of pristine materials

Nano particulate zinc oxide (ZnO) was purchased from Sigma Aldrich Ltd (Dublin, Ireland) and the particle size was characterised using three different methods namely Dynamic Light Scattering (Malvern nanoZS Zetasizer), Atomic Force Microscopy (Asylum Research Bio-AFM) and Transmission Electron Microscopy (Joel 100x). For DLS analysis nZnO was suspended in de-ionised water and measurements recorded at 10oC. For TEM analysis the nZnO was suspended in ethanol and drop cast on to Formvar copper grids (AGAR Scientific). Finally to perform AFM studies the nZnO was again dispersed in alcohol and for analysis samples were drop cast onto silicon wafers.

2.2. Composite formation

Composites were formed by hot melt extrusion. A Micro 27 Labscale Twin Screw Extruder (Leistritz Ltd) was used. A temperature of 150°C was set at the load-in point, the barrel containing the screws was set to 175°C and the temperature at the die was set to 180°C. The physical action of the screws along with the temperature in the barrel ensured a uniform dispersion of nZnO in polyethylene. The extrudate was cooled in water and then pelleted. Composites were produced using a range of nZnO loading fractions to produce materials of increasing percentage weight/weight ratios. The ratios were 0.25/0.5/1/2.5/5/10 % w/w.

2.3. Characterisation of Composites

The aforementioned composites were characterised using Atomic Force Microscopy and Fourier Transform Infrared Spectroscopy. For the AFM analysis the composites were heat pressed into thin disks which would provide a smooth surface for analysis. The FTIR analysis was also performed on the heat pressed composites.

2.4. Leaching Study

Leaching studies were performed by submersing 10g of the composite material in pellet form, in 20ml of 0.005M acetic acid which had a pH value of 4. This was done for each of the composites of 0.25-2.5% w/w nZnO. Acetic acid was used as it is a common additive in foods as well as an ingredient in some acidic foods (food additive code E260). These were stored in clean glass vials which were kept in normal light conditions at room temperature for 6 days. The concentration of the metal leached into the acid solution was measured using a Varian SpectrAA-200 Atomic Absorption Spectrometer. The samples were filtered through Whatmann number 5 filter paper in order to remove any large composite particles from solution which may interfere with the instrument. The instrument nebulises the analyte solution into an acetylene/air flame (with a flow rate of 2 dm^3/minute and 13.5dm^3/minute respectively) and any of the metal present is burned off and the absorption measured. The response from the instrument is measured against a standard curve attained by analysing solutions of known concentration.

2.5. Cytotoxicity assays.

For cytotoxicity evaluation cells were seeded in 96-well micro plates (Nunc, Denmark) in triplicate for each of the four time points studied 24, 48, 72, 96 hr. The plates were seeded at a density of 1.5×10^5 cells/ml for 24hr, 5×10^4 cells/ml for 48hr, 3×10^4 cells/ml for 72hr and 2×10^4 cells/ml for 96hr exposure. These densities were found to be optimal to achieve the desired confluence at the end of the exposure period. After an initial 24 hr of cell attachment, the media was removed and the plates were washed with 100 μl/well phosphate buffered saline (PBS). The cells were then treated with increasing concentrations of each nanomaterial and with a positive control of a 10% DMSO 90% media solution. These test solutions were prepared as follows: dispersions of the nano particles were created in fresh RPMI-F12 media supplemented with 10% foetal bovine serum (FBS) and 45 IU/ml penicillin and 45 g/ml streptomycin. An initial stock concentration of 1000 ppm was prepared by dispersing the nano ZnO using a sonic tip operating at 30 % amplitude in ten second burst followed by a ten second rest time to a total sonication time of 1 minute. Solutions of 500, 250, 100, 80, 40, 10, 0.1 ppm were then prepared from the 1000 ppm stock by serial dilution.

The cells were then incubated for the desired time period and the cytotoxic effects evaluated. For each independent experiment six replicate wells were used for control, six replicate wells were employed for the positive control and six replicate wells were used for each test concentration per micro plate. For cytotoxicity evaluation, fluorescence and absorbance were all quantified using a microplate reader (TECAN GENios, Grödig, Austria). Cytotoxicity was assessed using a number of assays as outlined below.

2.5.1. Alamar blue, neutral red, coomassie blue and MTT assays. The Alamar Blue, Neutral Red and Coomassie Blue assays were conducted subsequently on the same set of plates in order to measure metabolic activity, lysosomal activity and protein concentration according to the methods laid out by Casey et al [14, 15] A second series of plates were set up for the MTT assay. These plates were seeded and exposed identically to the first series of plates prepared for the Alamar Blue, Neutral Red, and Coomassie Blue assays and again the methods described by Casey et al used [14, 15].

2.6. Statistics

At least three independent experiments were conducted in triplicate for each cell line and toxicity endpoint. Test results for each assay were expressed as percentage of the unexposed control ± standard deviation (SD). Control values were set as 100%. Differences between samples and the control were evaluated using the statistical analysis package SPSS 14.0. Statistically significant differences were set at $p \leq 0.05$. Normality of data was confirmed with Q-Q percentile plots and Kolmogorov-Smirnov tests. One-way analysis of variances [16] followed by Dunnett's multiple comparison tests were carried out for normally distributed samples with homogeneous variances.

3. Results

3.1. Characterisation

3.1.1. Pristine Materials. The nano Zinc Oxide particles were size characterised by three separate techniques DLS/AFM/TEM. Each one of these analyses showed that the average size of the particles fell well below the 100nm threshold to be considered on the nanoscale.

Table 1. Results of sizing of nano zinc oxide particles and the associated technique.

Method	Size (nm)
Dynamic Light Scattering	22 +/- 1.1
Transmission Electron Microscopy	42.7 +/- 9.9
Atomic Force Microscopy	51 +/- 9.1

3.1.2. Composite Characterisation. Atomic Force Microscopy was performed on the heat pressed composites. It was noted that the composites with a higher concentration of nano ZnO maintained a smoother topography. This could be due to the ZnO present altering the properties of the polymer matrix. A further investigation into this phenomenon is underway. IR spectroscopic analysis was performed on all the produced nano composites, the presence of the nZnO was seen to alter the spectroscopic band associated with a carbon carbon double bond in polymer (1740 Cm^{-1}), this perhaps giving an indication that the properties of the polymer are indeed altered by the presence of the nZnO. Further spectral and thermal studies will be perform to monitor any alterations in the polymer due to the presence of nZnO.

3.2. Leaching study

Composites were formed as described in section 2.2, the composite extrudate was noted to be an opaque white solid taking on a more brilliant white colour as the concentration of nZnO was increased. The leaching study consisted of analysing a series of standards on known concentration in order to produce a standard curve. From this the concentration of a number of unknowns could be calculated using the equation of the standard curve line. Composites of various mass loading fractions were analysed and a leached concentration of nZnO in the rage of approximately 50 to 135ppm was determined, the amount of leached nZnO was also noted to be dependent on the loading fraction. In general a higher loading fraction resulted in a lower leached nZnO concentration. This perhaps giving an indication that the nZnO improved the chemical robustness of the composite but further studies are required to verify this theory.

3.3. Cytotoxicity Evaluation.

The following section will present the cytotoxicity data obtained from the colorimetric assays employed, namely Alamar Blue, Neutral Red, Coomassie Blue, MTT, to evaluate the effects of nano zinc oxide exposure to the two tested cell lines SW480 and HT29. Figure two gives an example of the type of cytotoxic response curves evaluated for nZnO as determined by the Alamar blue assay for the SW480 cell line.

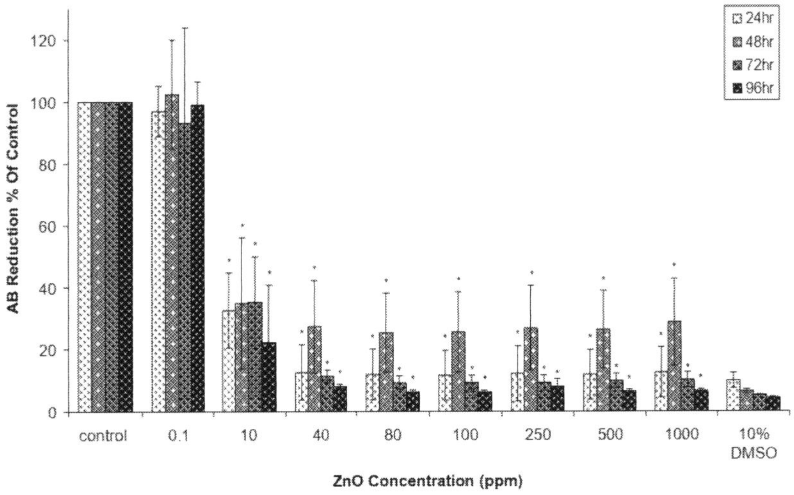

Figure 1. Cytotoxicity of nano ZnO to SW480 cells after 24, 48, 72 and 96 hour exposures determined by the Alamar Blue assay. Data are expressed as percent of control mean ± SD of three independent experiments * denotes a statistically significant effect ($p \leq 0.05$).

Alamar Blue measures the innate metabolic activity of cultured cells and is routinely used in cytotoxicity screening, as can be seen in figure 1 there is an immediate cytotoxicological response in SW480 cells after exposure to nZnO. Cellular survival drops to approximately 30% after a 24 hour exposure to concentrations as low as 10 ppm, this survival rate drops further to approximately 10% at higher concentrations and longer exposure periods. Following the AB the neutral red assay was perform this assay monitors the Lysosomal activity of cultured cells. Again in this assay we see an immediate cytotoxicological effect at concentrations of 10 ppm and above (data not shown) resulting from a 24 hour exposure to nZnO with a maximum cellular survival rate of 20% being observed. Comassie Blue is a broad spectrum protein stain which measures protein levels in cultured cells; it is detected by monitoring the absorbance of the Comassie Blue after an exposure. In this assay remnant test particles increased the background absorbance so for this reason the highest test concentration cells were exposed to was that of 100 ppm nZnO. However in agreement with the previous AB and NR assays a cytotoxicity was noted at concentrations of 10 ppm and above after 24 hour exposure. The final assay performed was that of the MTT assay again this assays detection method is absorbance based so for this reason the highest tested nZnO concentration was 100 ppm. This assay measures the mitochondrial activity of cultured cells and is highly sensitive

The second cell line tested was that of the HT29 cell line overall the associated cytotoxicity resulting from nZnO exposure in this line was substantially lower than that of the SW480 cell line. Again the first assay performed was that of the AB assay (data not shown) in contrast to the results seen from the SW480 cells here cellular survival only reduced to 60% at the top tested concentration of 1000 ppm nZnO

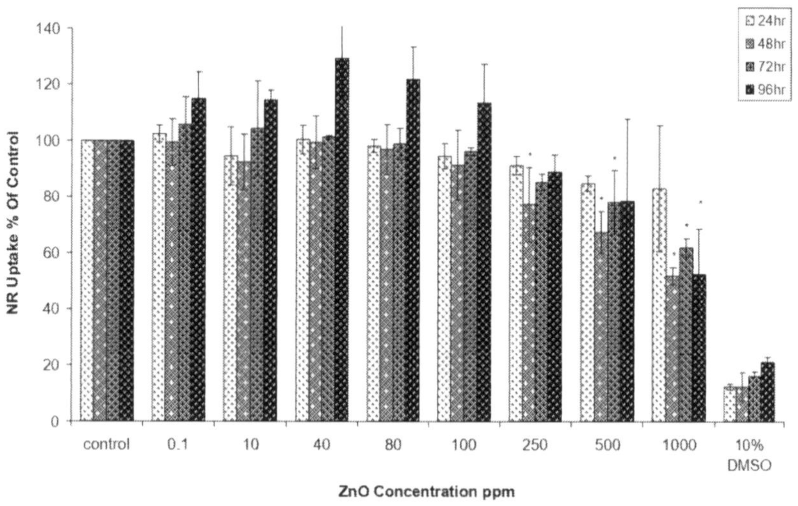

Figure 2. Cytotoxicity of nano ZnO to HT29 cells after 24, 48, 72 and 96 hour exposures determined by the NR assay. Data are expressed as percent of control mean ± SD of three independent experiments, * denotes a statistically significant effect ($p \leq 0.05$).

Figure 2 displays the cytotoxic response resulting from nZnO exposure in the HT29 cell line as determined by the NR assay, again with this assay the HT29 appears to be more resilient to nZnO exposure. However cytotoxicity was observed at a lower concentration than that of the AB assay for this cell line, with statistically significant cytotoxicity being observed at concentrations' of 250 ppm and above after 48 hour exposures. No data could be evaluated with this assay for the HT29 inconsistent results were continually acquired. The final assay performed for the HT29 cell line was that of the MTT (data not shown) as with the SW480 the highest test concentration studied with this assay was 100 ppm and no cytotoxicity was observed with this assay, which is in agreement with the AB and NR assays which indicated that a concentration above 250 ppm was required in the HT29 cell line to induce any cytotoxicity.

3.4. Light Microscopy.

A number of images were taken of both the SW480 and HT29 cells lines using a standard light microscope equipped with Nikon digital camera in order to assess any physical changes that may have occurred in the cells when exposed to nano ZnO. Furthermore these images were recorded to verify the experimental results of the cytotoxicity screening. For imaging a selection of the test concentrations (10 and 100 ppm) were prepared by sonication in cell culture media and exposed to each cell line for 24 hr. The cells were then washed with PBS and imaged. Figure 3 displays the light micrographs recorded figure 3 A and C display the unexposed control cells (HT29 and SW480). Upon exposure to the test particle clear differences became apparent; in both cases at the test concentration of 100 ppm (Figure 3 C and D) there were considerable morphological changes in both the tested cell lines with an almost complete loss of the cell monolayer. At the lower test concentration of 10 ppm differences between the responses of the two tested cell lines were noted with the test particles have a lesser effect on the HT29 cells in compassion to that of the SW480 which were significantly altered, this being in broad agreement with the results obtained from the cytotoxicity screening. Interestingly in the HT29 cell line higher amounts of cellular debris was noted in the exposures, this debris would be proteinaceous in natures and its presence may account for the inconsistent data acquired for the Comassie Blue assay in the HT29 cell line tests.

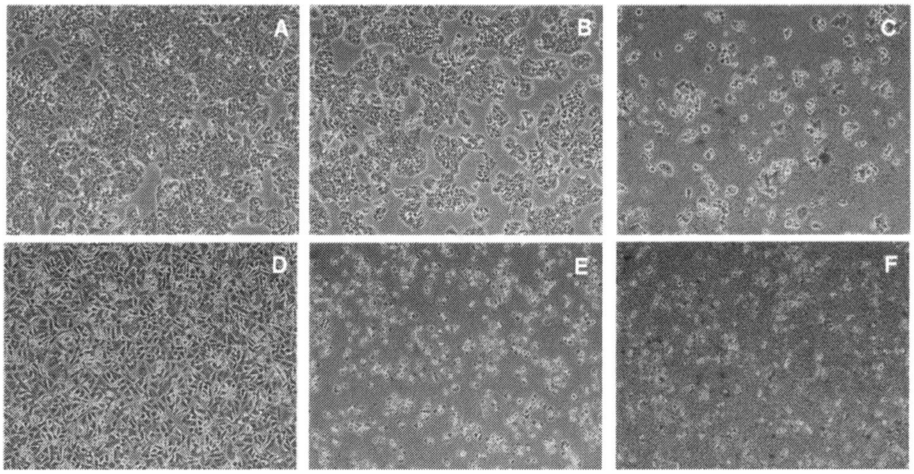

Figure 3. Light micrographs of (A) Unexposed HT29 cells (B) HT29 cells exposed to 10ppm ZnO for 24 hr (C) HT29 cells exposed to 100 ppm ZnO for 24 hr (D) Unexposed SW480 cells (E) SSW480 cells exposed to 10 ppm ZnO for 24 hr (F) SW480 cells exposed to 100 ppm ZnO for 24 hr.

4. Discussion and conclusion

A series of nZnO/Polyethylene composites were formed. Polyethylene was chosen as due to its ubiquitous use in consumer goods. A leaching study was then performed by submersing a sample of the composite in a weak acid solution and analysing the leachate using AAS. Factoring in dilution the concentration of Zinc leached from the composite was seen to be between 50-135ppm for the low loading fraction composites. The higher loading fraction of 2.5% was seen to leach a concentration of approximately 50ppm. This lower leaching concentration may be due to the nZnO present altering the properties of the polymer matrix making it less susceptible to chemical stress however it should be noted that this leaching study was an initial investigation and supplementary experimentation is required in order to assess the leaching potential further. Cytotoxicity testing indicated a toxic response to the presence of nZnO in both HT29 and SW480 cell lines. The SW480 cell line was seen to be more susceptible to toxic effects. A significant response was noted in concentrations of 10ppm and above at exposure times as little as 24hrs. Both the MTT and Neutral Red assays confirm the low dose toxicity of the material to the cell line. It can be seen from the collected toxicological data the nZnO displays a cytotoxic effect to both the HT29 and SW480 cell lines at low concentrations, lower than those seen to leach from the composite materials. The concentration range of the leached Zinc is seen to be between 50-135 ppm which falls in the middle of the tested toxicity range.

Over the course of this study it has been clearly demonstrated that that there is significant leaching of zinc from composite materials that mimic commercially available food contact materials and at the concentration ranges determined that nano particulate ZnO elicits a strong cytotoxic response. The Alamar blue and neutral red assays were noted to be the most sensitive of the colorimetric endpoints. These assays indicated that nano ZnO elicited a cytotoxic effect at concentrations below 10ppm in the SW480 line and above 40 ppm (Alamar Blue assay) in the HT29 cell line. These findings suggest further studies are needed to understand other underlying mechanisms by which nano ZnO elicits this cytotoxicity in intestinal cell lines. A recent study [17] has indicated that nano ZnO can cause significant genotoxic damage to human epithelial cells and work is ongoing to determine if a similar effect is occurring in the tested cell lines of this study. There is a possibility of the stability of the nano enhanced food contact material being compromised resulting in nanoparticle leaching which in turn may facilitate direct exposure to the intestinal tract by ingestion. The results of this study would lend themselves to the recommendation of a cautionary approach regarding the usage of nano ZnO in direct food contact materials.

Acknowledgements
N. Ó Claonadh was funded under the Technological Sector Research Strand I, Post Graduate R&D Skills Programme. This work was partially enabled by the framework of the INSPIRE programme, funded by the Irish Government's Programme for Research in Third Level Institutions, Cycle 4, National Development Plan 2007-2013, supported by the European Union Structural Fund.

References
[1] Ireland, F.S.A.o., The Relevance for Food Safety of Applications of Nanotechnology in the Food and Feed Industries. FSAI, 2008.
[2] Oberdorster, G., V. Stone, and K. Donaldson, Toxicology of nanoparticles: A historical perspective. Nanotoxicology, 2007. **1**(1): p. 2-25.
[3] Lewinski, N., V. Colvin, and R. Drezek, Cytotoxicity of nanoparticles. Small, 2008. **4**(1): p. 26-49.
[4] Meng, H., et al., Ultrahigh reactivity provokes nanotoxicity: Explanation of oral toxicity of nano-copper particles. Toxicology Letters, 2007. **175**(1-3): p. 102-110.
[5] Maynard, A.D. and R.J. Aitken, Assessing exposure to airborne nanomaterials: Current abilities and future requirements. Nanotoxicology, 2007. **1**(1): p. 26-41.
[6] Wang, B., et al., Acute toxicity of nano- and micro-scale zinc powder in healthy adult mice. Toxicology Letters, 2006. **161**(2): p. 115-123.
[7] Koeneman, B.A., et al., Toxicity and cellular responses of intestinal cells exposed to titanium dioxide. Cell Biology and Toxicology, 2010. **26**(3): p. 225-238.
[8] Oberdorster, G., Safety assessment for nanotechnology and nanomedicine: concepts of nanotoxicology. Journal of Internal Medicine, 2010. **267**(1): p. 89-105.
[9] Chaudhry, Q., et al., Applications and implications of nanotechnologies for the food sector. Food Additives and Contaminants, 2008. **25**(3): p. 241-258.
[10] Nanotechnologies, T.P.o.E. Nano Plastic Wrap - SongSing Nano Technology Co. Ltd. 2009 [cited December 1st 2009]; Available from: http://www.nanotechproject.org/inventories/consumer/browse/products/nano_plastic_wrap/.
[11] Jin, T., et al., Antimicrobial Efficacy of Zinc Oxide Quantum Dots against Listeria monocytogenes, Salmonella Enteritidis, and Escherichia coli O157:H7. Journal of Food Science, 2009. **74**(1): p. M46-M52.
[12] Xu, L.G., et al., Applications and toxicological issues surrounding nanotechnology in the food industry. Pure and Applied Chemistry, 2010. **82**(2): p. 349-372.
[13] Duranton, B., et al., Polyamine metabolism in primary human colon adenocarcinoma cells (SW480) and their lymph node metastatic derivatives (SW620). Amino Acids, 2003. **24**(1-2): p. 63-72.
[14] Casey, A., et al., Spectroscopic analysis confirms the interactions between single walled carbon nanotubes and various dyes commonly used to assess cytotoxicity. Carbon, 2007. **45**(7): p. 1425-1432.
[15] Casey, A., et al., Single walled carbon nanotubes induce indirect cytotoxicity by medium depletion in A549 lung cells. Toxicology Letters, 2008. **179**(2): p. 78-84.
[16] Ayres, J.G., et al., Evaluating the toxicity of airborne particulate matter and nanoparticles by measuring oxidative stress potential - A workshop report and consensus statement. Inhalation Toxicology, 2008. **20**(1): p. 75-99.
[17] Sharma, V., et al., DNA damaging potential of zinc oxide nanoparticles in human epidermal cells. Toxicology Letters, 2009. **185**(3): p. 211-218.

Nanoparticle-protein interactions: from crucial plasma proteins to key enzymes

Elodie Sanfins, Julien Dairou, Fernando Rodrigues-Lima and Jean-Marie Dupret

Université Paris Diderot-Paris 7, Unité de Biologie Fonctionnelle et Adaptative, CNRS EAC 4413, 75013 Paris, France

Email: fernando.rodrigues-lima@univ-paris-diderot.fr ; jean-marie.dupret@univ-paris-diderot.fr

Abstract. Studying the effects of NPs on proteins may help understanding potential biological injuries such as changes in protein fibrillation, exposure of new antigenic epitopes, and loss of function such as enzymatic activity impairment. In this mini-review we present recent data which help understand the basis of NP-protein interactions and their subsequent potential effects on key mediators of biological functions such as enzymes.

1. Introduction

In the recent past nanomaterials emerged and took their place in lots of fields, such as electronics, optics, cosmetics, food... Nanomaterials were also used in medicine and diagnostics. This dramatic increase of industrial production and use of nanomaterials has lead to investigate their potential health effects.

Indeed, the nanotoxicology is a field well established. Studies of the effects of nanoparticles (NPs) from different industry branches on cells and pathways are emerging, and most of the biological effects of NPs seem due to their interactions with proteins. Here, we review some articles showing the interaction between NPs and proteins, and the effects of these nanomaterials upon biological actors such as plasma proteins and enzymes.

2. Nanoparticle-protein interactions, corona and plasma proteins

When entering the body NPs are rapidly in contact with biological fluids such as plasma. In these physiological environments, NPs selectively bind proteins to form a coat known as the 'protein corona'. Since this corona is likely to determine the fate of the nanomaterials in vivo [1], understanding how and why proteins are adsorbed to NPs is of prime importance.

The affinity of a protein towards NPs is modulated by surface properties of NP. In a given medium the surface properties are determined by NPs characteristics such as chemical composition, surface functionalization, the shape and angle of curvature, the porosity and surface crystallinity, heterogeneity, roughness and hydrophilicity or hydrophobicity. An important determinant of the nature of proteins adsorbed onto the surface of NPs is chemical composition of the nanomaterials which further modifies the biological fate and interactions inside living organisms. The study of the plasma proteins bound to single walled carbon nanotubes (SWCNT) and nano-sized silica indicated different patterns of adsorption. Serum albumin was found to be the most abundant protein coated on SWNCT but not on silica NPs, suggesting the importance of NPs composition in the formation of the corona [2]. Moreover, it has been shown that TiO_2, SiO_2, ZnO NPs of similar surface charge, bind to different plasma proteins. For example transferin was found to bind only on the ZnO NPs and neither on TiO_2 nor SiO_2. Other important plasma proteins such as immunoglobulins, lipoproteins, coagulation and complement proteins were also found to bind differentially to these NPs. Furthermore the size and shape of TiO_2 NPs seemed to be important for the determination of the proteins bound to the NPs suggesting that the agglomeration state largely affected the protein binding pattern [3].

The forces that seem to be involved in the protein-NPs interactions are Van der Waals forces (related to charges that can be measured by the zeta potential of the nanoparticles), London dispersion forces (hydrophobic interactions), hydrogen bond acidity and basicity, polarization and lone-pair electrons. Recently it has been shown that the NPs-protein interaction mechanism can be predicted with the help of a mathematical model that is based on the assumption of an equal contribution of all inter-moleculars interactions [4]. Nanodescriptors can be created using the adsorption coefficients of probe compounds. These nanodescriptors represent the contribution and relative strengths of each molecular interaction. As an example, this model helped to predict for example the adsorption of various small molecules onto carbon nanotubes [4]. In summary, the affinity of nanoparticles to biomolecules is determined by the contribution of multiple site of adsorption on the NPs surfaces (that are close to the amino-acids residues of the proteins), and the contribution of various forces that depend on the protein and the type of nanoparticles. Predicting this kind of interaction could be very helpful in nanomedicine to eliminate certain type of nanoparticles based on their relative nanodescriptors [4]. Another example of a mathematical model has been described by Dell'Orco et al. This model describes the kinetics formation of the protein corona around copolymer NPs. Three plasma proteins were used: albumin, fibrinogen and high density lipoprotein. This model tends to be used to predict protein corona formation on NPs with other biological fluids [5].

Furthermore, the nanoparticles bound to protein can undergo dynamic changes as the particles enter onto or into cells. The NP-protein complex can persist in body fluids and cells from microseconds to days. As the environment of the NPs change while moving to another compartment or fluid, the composition of the protein corona is likely to also vary. It is interesting to note that high affinity proteins can displace the binding of low affinity proteins. Göppert et al. studied the NP-plasma protein interaction using 2-D gels approaches and showed that the protein adsorption pattern on solid lipid NPs was dependent on contact time [6]. This study confirmed the Vroman effect [7] that explains the time evolution of the corona: proteins with high affinities but in low concentration in the plasma can displace highly concentrated proteins that have low affinity for the NPs. When using N-isopropylacrilamide/N-ter-butylacrylamide NPs as a model, human serum albumin (HAS) was rapidly adsorbed onto NPs but soon replaced by apolipoproteins such as AI AII AIV and AV. In addition, given the high number of protein on the plasma, the equilibrium of the corona formation will not be reached immediately. [8]

Using Isothermal calorimetry (ITC), it seems that the HSA-NPs association is an exothermic reaction. Interestingly the highest surface coverage of HSA is achieved for the more hydrophobic particles. This shows that the surface coverage is strongly dependent on the particle hydrophobicity. It has been shown also that surface curvature of N-isopropylacrilamide/N-ter-butylacrylamide interferes with HSA binding. Indeed the protein adsorbed on flat surfaces tends to accumulate in multi-layers and can formed two dimensional structures. However for high curvatures, proteins are far from each other and tend to form one-layer around the NPs. [9]

Studying the effects of NPs on proteins give a plausible explanation and help understanding potential biological injuries, such as changes in protein fibrillation, exposure of new antigenic epitopes, and loss of function such as enzymatic activity impairment [10].

As pointed above, the biological fate and biodistribution of NPs are strongly dependent on the proteins that NPs meet in the competitive world of body fluid proteins and in particular in plasma. Interestingly, the proteins coated on the NPs depend on the composition, the surface and the size of the NPs. Therefore, it is important to study, which specific proteins bound to a certain kind of NPs, and mathematical models could help to predict it. It is then the NPs and the proteins coated to them that we have to consider for biological effects and not the NPs alone. Maybe this "corona" could helping classify the NPs and even engineering NPs which give better responses when used as nanomedical applications.

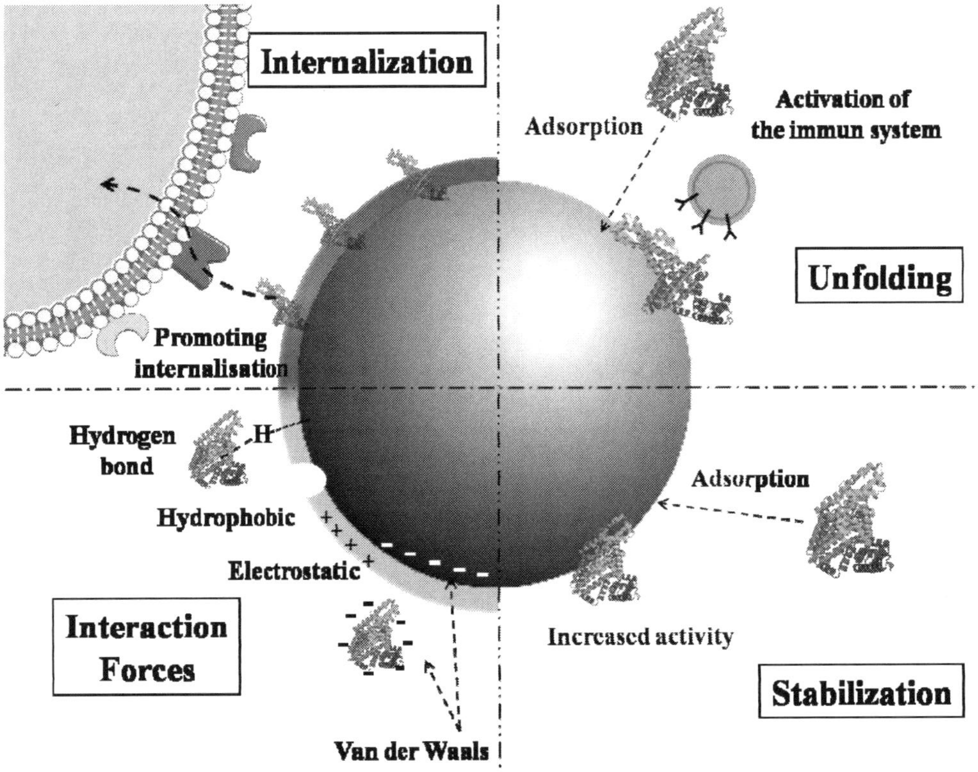

Figure 1. Effects of NPs on proteins, and the forces of interactions. (Adapted from [10])

3. Nanoparticle-enzyme interactions

Like serum albumin, lysozyme is often used as a model in protein structural studies. In addition, the activity of lysozyme can be easily monitored. Studies of the interaction NPs-lysozyme can demonstrate that either the structure or /and the activity changes. This fact gives a better understanding and leads us to measure the interaction of NPs with the biological actors: enzymes.

Lysozyme is a small monomeric globular protein, formed of 129 amino acids. This protein has the ability of disrupting bacterial functions, including their membrane structure. The protein has α helix and β sheet domains with 4 disulfide bound. It contains 6 tryptophan (trp) residues and three of them in its active site. Two of them are located in a hydrophobic environment and one is separated from the others.

Studies show an evident conformational change when lysozyme interacts with NPs. For example, the lysozyme adsorbed on silver colloids show a loss in conformation, and more precisely the Ag NPs seems to interact with a trp and phenylalanine residues, situated in a loop in close proximity. Moreover it seems that Ag NPs interact rapidly with lysozyme, and as the time goes by, the molecule stabilizes itself on the NP. It seems also that the active site of the lysozyme is not affected by the adsorption, because the area adsorbed on the NPs is the N-terminal end of the protein. [11] ZnO NPs have been reported to modify the secondary structure of lysozyme. Indeed, with circular dichroism the content of α helix and β sheet has been determined in presence of ZnO NPs. As a matter of fact, α helix content increases at the expense of random coil. It also seems that lysozyme keeps its catalytic activity when adsorbed to the ZnO NPs. Moreover, the protein resist to 8 M urea denaturation in presence of these NPs. NPs can stabilize the structure of the protein, increasing the α helix content, in the case of ZnO and lysozyme, preventing the urea to denature it and foreclosing the oligomerization of the protein. [12]

On the other hand, the interaction of lysozyme with NPs has also been described for TiO_2 NPs. Lysozyme seems to form bridges between the NPs and enhance the formation of aggregates. A change in the conformation of lysozyme was observed in the presence of TiO_2. Indeed the content of α helix decreased while the content of β sheet increased, resulting in a loss of activity. [13]

With a different kind of NPs, CdTe coated with thioglycolic acid, it has also been seen that the secondary structure of lysozyme was changed during the binding process.[14]

Other proteins with catalytic activities were also studied in the presence of NPs. While NPs are often the cause of toxicity or loss of secondary structure, it seems that promising issues can be found in the NP-enzymes interactions. It has been shown that Horsereadish peroxidase and egg white lysozyme retained a high fraction of their activity upon binding to SWNTs. Furthermore the SWNT-enzymes conjugates seem to be more stable than the non conjugated enzyme in guanidine hydrochloride and at elevated temperatures [15] In addition a model enzyme, soybean peroxidase, adsorbed onto highly curved surface of C60 fullerenes had an half time 13-fold higher than the native enzyme. Furthermore, this phenomenon is not unique to fullerenes, but can also be extended to other nanoscale supports including silica and gold NPs [16].

This stabilization versus inhibition of enzymes by nanomaterials has been studied with chymotrypsin [17]. Functionalized gold NPs exhibited different effects on chymotrypsin. There was no inhibition from the cationic NPs while anionic NPs were effective inhibitors, because of the electrostatic complementarities between the NPs and the protein [18]. Moreover, chymotrypsin has been found inactivated by carbon nanotubes. Not only those functionalized multiwalled carbon nanotubes bind to the chymotrypsin active site but they also competitively inactivate this protein enzymatic function [19]. In another study, Shang et al show that silica NPs induce the unfolding of ribonuclease A (Rnase A). Moreover, this team showed that the thermodynamic stability of the Rnase A decreased upon adsorption onto the silica NPs, and that the size and curvature of this kind of NPs play a role in the stability of adsorbed proteins.

Interestingly, NPs have been shown to target the central nervous system [20] and can inactivate enzymes which have crucial catalytic activities. Acetylcholinesterase is a hydrolase of the neurotransmitter, acetylcholine, in cholinergic synapses and is one of the most important enzymes in the nervous systems. Wang et al. demonstrated the inactivation/inhibition of both acetylcholinesterase and butyrylcholinesterase by NPs in a dose dependant manner. Moreover this inhibition is due to the adsorption of the enzyme on the NPs [21].

More recently, the effects of NPs on major xenobiotic metabolizing enzyme have gained interest in the scientific community since these enzymes catalyze reactions of pharmacological and toxicological importance. The effects of polystyrene and silver NPs on certain cytochrome P450, (phase I XME), have been studied. Fröhlich et al. show that there is a size dependent effect of polystyrene NPs on biotransformation activity of various cytochrome P450 enzymes. They have also shown the interference of NPs on the cleavage of different synthetic substrates in the presence or in absence of cytochrome P450 inhibitors. Smallest polystyrene NPs (20 nm) have the highest effect on these xenobiotic metabolizing enzymes, due to the hydrophobicity of this kind of NPs and the adsorption of the enzymes on them [22]. In another study it has been shown that silver NPs of 15 nm diameter show an inhibition of 95% of activities of human CYP1A2, CYP2C19 and CYP3A4 [23].

Current studies in our laboratory confirm that certain NPs may have an impact on the biotransformation of pollutants by others XMEs.

4. Conclusion

The NP-induced protein modifications are promising fields for future research. Proper understanding of such phenomenon is further emphasized by the fact that these materials are utilized for diagnostic and therapeutic purposes. In this article we reviewed the scientific data on the deleterious effects of NPs on key mediators of biological functions such as enzymes. Only a few studies are conducted on the crucial metabolism enzymes and enzyme dysfunctions, which are related to various pathologies [23-27]. It is crucial to understand how NPs are able to bind proteins, and how this corona can help NP internalization and interaction of NPs with intracellular compounds.

Acknowledgments
We thank Dr Salik Hussain and Georges-Philippe Moreira for reading the manuscript, and this work was supported by a grant from Ministère de l'Enseignement Supérieur et de la Recherche.

References

[1] Cedervall T. et al., "Understanding the nanoparticle–protein corona using methods to quantify exchange rates and affinities of proteins for nanoparticles," *Proceedings of the National Academy of Sciences of the United States of America*, vol. 104, n°. 7, p. 2050-2055, Fév. 2007.
[2] Dutta D. et al., "Adsorbed Proteins Influence the Biological Activity and Molecular Targeting of Nanomaterials," *Toxicological Sciences*, vol. 100, n°. 1, p. 303 -315, Nov. 2007.
[3] Deng Z. J., Mortimer G., Schiller T., Musumeci A., Martin D. and Minchin R. F., "Differential plasma protein binding to metal oxide nanoparticles," *Nanotechnology*, vol. 20, n°. 45, p. 455101, 2009.
[4] Xia X., Monteiro-Riviere N. A. and Riviere J. E., "An index for characterization of nanomaterials in biological systems," *Nature Nanotechnology*, vol. 5, n°. 9, p. 671-675, Sep. 2010.
[5] Dell'Orco D., Lundqvist M., Oslakovic C., Cedervall T. and Linse S., "Modeling the Time Evolution of the Nanoparticle-Protein Corona in a Body Fluid," *PLoS ONE*, vol. 5, n°. 6, p. e10949, Juin. 2010.
[6] Göppert T. M. and Müller R. H., "Adsorption kinetics of plasma proteins on solid lipid nanoparticles for drug targeting," *International Journal of Pharmaceutics*, vol. 302, n°. 1, p. 172-186, Sep. 2005.
[7] "The Vroman Effect - Proteins at Interfaces II - ACS Symposium Series (ACS Publications)." [Online]. Available: http://pubs.acs.org.gate1.inist.fr/doi/abs/10.1021/bk-1995-0602.ch008. [Accessed: 25-Oct-2010].
[8] Cedervall T. et al., "Detailed Identification of Plasma Proteins Adsorbed on Copolymer Nanoparticles," *Angewandte Chemie International Edition*, vol. 46, n°. 30, p. 5754-5756, 2007.
[9] Lindman S., Lynch I., Thulin E., Nilsson H., Dawson K. A. and Linse S., "Systematic Investigation of the Thermodynamics of HSA Adsorption to N-iso-Propylacrylamide/N-tert-Butylacrylamide Copolymer Nanoparticles. Effects of Particle Size and Hydrophobicity," *Nano Letters*, vol. 7, n°. 4, p. 914-920, Avr. 2007.
[10] Nel A. E. et al., "Understanding biophysicochemical interactions at the nano-bio interface," *Nat Mater*, vol. 8, n°. 7, p. 543-557, Juil. 2009.
[11] Chandra G., Ghosh K. S., Dasgupta S. and Roy A., "Evidence of conformational changes in adsorbed lysozyme molecule on silver colloids," *International Journal of Biological Macromolecules*, vol. 47, n°. 3, p. 361-365, Oct. 2010.
[12] Chakraborti S. et al., "Structure and activity of lysozyme on binding to ZnO nanoparticles," *Langmuir: The ACS Journal of Surfaces and Colloids*, vol. 26, n°. 5, p. 3506-3513, Mar. 2010.
[13] Xu Z., Liu X., Ma Y. and Gao H., "Interaction of nano-TiO2 with lysozyme: insights into the enzyme toxicity of nanosized particles," *Environmental Science and Pollution Research*, vol. 17, n°. 3, p. 798-806, 2009.
[14] Wu Y., He F., He X., Li W. and Zhang Y., "Spectroscopic studies on the interaction between

CdTe nanoparticles and lysozyme," *Spectrochimica Acta Part A: Molecular and Biomolecular Spectroscopy*, vol. 71, n°. 4, p. 1199-1203, Déc. 2008.

[15] Asuri P., Bale S. S., Pangule R. C., Shah D. A., Kane R. S. and Dordick J. S., "Structure, Function, and Stability of Enzymes Covalently Attached to Single-Walled Carbon Nanotubes," *Langmuir*, vol. 23, n°. 24, p. 12318-12321, Nov. 2007.

[16] Asuri P., Bale S. S., Pangule R. C., Shah D. A., Kane R. S. and Dordick J. S., "Structure, Function, and Stability of Enzymes Covalently Attached to Single-Walled Carbon Nanotubes," *Langmuir*, vol. 23, n°. 24, p. 12318-12321, Nov. 2007.

[17] You C., Arvizo R. R. and Rotello V. M., "Regulation of alpha-chymotrypsin activity on the surface of substrate-functionalized gold nanoparticles," *Chemical Communications (Cambridge, England)*, n°. 27, p. 2905-2907, Juil. 2006.

[18] Fischer N. O., McIntosh C. M., Simard J. M. and Rotello V. M., "Inhibition of chymotrypsin through surface binding using nanoparticle-based receptors," *Proceedings of the National Academy of Sciences of the United States of America*, vol. 99, n°. 8, p. 5018 -5023, Avr. 2002.

[19] Zhang B., Xing Y., Li Z., Zhou H., Mu Q. and Yan B., "Functionalized Carbon Nanotubes Specifically Bind to α-Chymotrypsin's Catalytic Site and Regulate Its Enzymatic Function," *Nano letters*, vol. 9, n°. 6, p. 2280, Juin. 2009.

[20] Medina C., Santos-Martinez M. J., Radomski A., Corrigan O. I. and Radomski M. W., "Nanoparticles: pharmacological and toxicological significance," *British Journal of Pharmacology*, vol. 150, n°. 5, p. 552-558, Mar. 2007.

[21] Wang Z., Zhang K., Zhao J., Liu X. and Xing B., "Adsorption and inhibition of butyrylcholinesterase by different engineered nanoparticles," *Chemosphere*, vol. 79, n°. 1, p. 86-92, Mar. 2010.

[22] Fröhlich E., Kueznik T., Samberger C., Roblegg E., Wrighton C. and Pieber T. R., "Size-dependent effects of nanoparticles on the activity of cytochrome P450 isoenzymes," *Toxicology and Applied Pharmacology*, vol. 242, n°. 3, p. 326-332, Fév. 2010.

[23] Lamb J. G., Hathaway L. B., Munger M. A., Raucy J. L. and Franklin M. R., "Nano-silver Particle Effects on Drug Metabolism in Vitro," *Drug Metabolism and Disposition: The Biological Fate of Chemicals*, Sep. 2010.

[24] D'Armiento J., "Matrix metalloproteinase disruption of the extracellular matrix and cardiac dysfunction," *Trends in Cardiovascular Medicine*, vol. 12, n°. 3, p. 97-101, Avr. 2002.

[25] Longthorne V. L. and Williams G. T., "Caspase activity is required for commitment to Fas-mediated apoptosis," *The EMBO Journal*, vol. 16, n°. 13, p. 3805-3812, Juil. 1997.

[26] Ozbek Y. K. et al., "Combined effect of CYP1B1 codon 432 polymorphism and N-acetyltransferase 2 slow acetylator phenotypes in relation to breast cancer in the Turkish population," *Anticancer Research*, vol. 30, n°. 7, p. 2885-2889, Juil. 2010.

[27] Shultz M. D., Ham Y., Lee S., Davis D. A., Brown C. and Chmielewski J., "Small-molecule dimerization inhibitors of wild-type and mutant HIV protease: a focused library approach," *Journal of the American Chemical Society*, vol. 126, n°. 32, p. 9886-9887, Aoû. 2004.

[28] Singh I., "Biochemistry of peroxisomes in health and disease," *Molecular and Cellular Biochemistry*, vol. 167, n°. 1, p. 1-29, Fév. 1997.

Inflammation response at the transcriptional level of HepG2 cells induced by multi-walled carbon nanotubes

Jean-Pascal Piret[1], Sébastien Vankoningsloo[1], Florence Noël[1], Jorge Mejia Mendoza[2], Stéphane Lucas[2], Christelle Saout[1] and Olivier Toussaint[1]

[1]Research Unit in Cellular Biology (URBC), Narilis, University of Namur, 5000 Namur, Belgium.

[2]Research Center for the Physics of Matter and Radiation (PMR), Narilis, University of Namur, 5000 Namur, Belgium.

E-mail: olivier.toussaint@fundp.ac.be

Abstract. Poor information are currently available about the biological effects of multi-walled carbon nanotubes (MWCNT) on the liver. In this study, we evaluated the effects of MWCNT at the transcriptional level on the classical *in vitro* model of HepG2 hepatocarcinoma cells. The expression levels of 96 transcript species implicated in the inflammatory and immune responses was studied after a 24h incubation of HepG2 cells in presence of raw MWCNT dispersed in water by stirring. Among the 46 transcript species detected, only a few transcripts including mRNA coding for interleukine-7, chemokines receptor of the C-C families CCR7, as well as Endothelin-1, were statistically more abundant after treatment with MWCNT. Altogether, these data indicate that MWCNT can only induce a weak inflammatory response in HepG2 cells.

1. Introduction

Depending on the publications, parameters influencing the carbon nanotubes (CNT) toxicity include surface structure [1-3]; length [4]; agglomeration state [5,6], presence of a dispersing agent [7-10] ; and contamination with metal catalyst and amorphous carbon [11]. Being single-walled or multi-walled, is also crucial to determine the biological effects of CNT. Consideration of the cell line tested also seems important when dealing with the toxic potential of CNT. We have shown previously that multi-walled carbon nanotubes (MWCNT) can be toxic and pro-inflammatory for telomerase-immortalized human keratinocytes [7,8]. However, MWCNT exerted no cytotoxicity on SV-40 immortalized sebocytes and *in vitro* reconstituted epidermises [7]. Moreover, MWCNT did not modify Caco-2 cells or HepG2 cells viability and did not induce release of pro-inflammatory interleukin-8 (IL-8) [12] (unpublished data). MWCNT were nevertheless able to abnormal induce production of reactive oxygen species (ROS) in HepG2 cells (unpublished data). Overproduction of ROS seems to be an important biological effect of CNT [9] even in absence of cell death [11]. Still the pro-oxidant effect has to be confirmed and deeper investigated to determine the possible mechanisms leading to ROS overproduction. Indeed, other studies showed that MWCNT did not induce oxidative stress [13] or were a remarkable radical scavengers when in contact with an external source of hydroxyl or superoxide radicals [14]. Oxidative stress can contribute to the initiation and progression of chronic inflammation by promoting cell proliferation, expression of adhesion molecules and matrix metalloprotenase and production of cytokines/chemokines. Oxidative signals are able to modulate the expression of a variety of cytokines and chemokines, such as Tumor Necrosis Factor-α (TNF-α), IL-1β, IL-6, IL-8, and monocyte chemotactic protein-1 (MCP-1), and adhesion molecules, such as vascular cell adhesion molecule 1 (VCAM-1) and intercellular adhesion molecule-1 (ICAM-1) [15]. In this study, we investigated the effect of MWCNT on the level of 96 transcript species implicated in the inflammatory and immune responses after a 24h incubation of HepG2 cells in presence of raw MWCNT dispersed in water by stirring.

2. Materials and Methods

2.1. MWCNT characterisation and dispersion

Physico-chemical characterisation of MWCNT (Nanocyl™ NC7000 multi-walled carbon nanotubes, from Nanocyl s.a.) was described previously [7,16]. Briefly, MWCNT have an average diameter of 12 nm and variable lengths in the micron range. Their specific surface area is very high (323.7 m^2/g). They are mainly composed of carbon but presence of catalytic metal residues was detected (Al 2.1%, Fe 0.1%, Co 0.1%) [7, 16]. Raw MWCNT were dispersed into pure water at 1 mg/ml by stirring with a magnetic bar for 90 min. MWCNT suspension were then diluted at 100 μg/ml into cell culture medium before incubation of the cells for 24h. The size distribution (calculated as hydrodynamic diameters) of MWCNT suspended in culture medium was analysed by centrifugal sedimentation using a Disc Centrifuge DC24000 (CPS) at 24,000 rpm. MWCNT formed big agglomerates with an average size of 3.8 μm.

2.2. Cell culture

Human hepatocarcinoma cells (HepG2) were maintained in culture (20 passages) in 75-cm^2 polystyrene flasks (Costar) with 15 ml of Dulbecco's modified Eagle's medium (DMEM) containing 5 ml/500 ml penicillin-streptomycin (BioWhittaker) and 10% fetal calf serum and incubated under an atmosphere of 5% CO_2.

2.3. Taqman low density array

Total RNA from biological triplicate was extracted using the "TRI Reagent Soln" (Ambion). mRNA contained in 1 μg of total RNA was reverse transcribed using the "High-Capacity cDNA Reverse Transcription" kit (Applied Biosystems). Then multiplex qPCR amplifications were performed with an ABI PRISM 7000 SDS thermal cycler, using 384-well "microfluidic card" plates (Human Immune Array, Applied Biosystems) containing various primer pairs for parallel analyses of 96 transcripts on 4 samples. GAPDH mRNA was used as the reference for normalization and relative mRNA steady-state level quantification based on cycle thresholds (Ct). Melting curves were generated after amplification, and data were analyzed using the thermal cycler SDS 2.2.2 software by the $2^{-\Delta\Delta Ct}$ method [17]. CTL mean values were calculated from triplicate ct values for each mRNA. Each MWCNT-treated triplicate value was then calculated by the $2^{-\Delta\Delta Ct}$ method in comparison with the CTL mean value and S.D. were determined. When calculable, statistical significance was considered for $P < 0.05$ using Student's *t*-tests.

3. Results

The possibility of a pro-inflammatory response of HepG2 cells incubated for 24h with MWCNT was investigated in triplicate by reverse transcription (RT) real time TaqMan PCR low density arrays (TLDA). These arrays allow the quantitative TaqMan PCR amplification of 96 transcripts involved in human immunity and inflammation. Among the 46 transcript species detected (listed in Table 1), only a few transcripts were statistically more abundant after treatment with MWCNT (Table 2). In this case, 3 independent biological replicates could be considered with a mean ratio calculated with standard deviations (s.d.). High variability is frequently obtained when profiling gene expression in triplicates with arrays [18,19]. Other changes of abundance were not statistically significant when large or lacking (mean ratios calculated from only two values) s.d. were obtained despite all efforts. A third type of data are qualitative data, representing situations where a transcript is not abundant enough and below the detection threshold in the basal state, but activated upon treatment with the MWCNT [19]. In the case of LRP2 the ratio obtained from two quantitative values is indicated, the third value being qualitative. Consideration of these three types of data is equally important since they may represent a tip for further functional studies at protein level. We observed an increased abundance of transcripts of pro-inflammatory cytokines or proteins able to recruit and activate immune cells, such as interleukine-7 (IL-7), chemokines receptors of the C-C families (CCR7, CCR4), as well as colony stimulated factor-3 (CSF3) or lymphotoxin alpha (LTA). IL-8 was among all the transcript whose expression level is not modified following MWCNT exposure (data not shown), confirming previous other studies [11].

4. Discussion

Poor data are currently available on the impacts of CNT at the hepatic level. Only a few studies described the effects of MWCNT on HepG2 cells demonstrating poor or no cytotoxic effects [12].

In this study, we investigated the effects of MWCNT on the abundance of transcripts coding for proteins implicated in the inflammatory and immune responses. Several transcripts of pro-inflammatory cytokines or proteins able to recruit and activate immune cells were overexpressed after an incubation of 24h in presence of MWCNT. For example, the abundance of IL-7, chemokines CCR7 and CCR4 and CSF3 transcripts was increased after incubation with MWCNT. CCR4 and CCR7 belong to the superfamily of G-protein-coupled receptors. These chemokine receptors are membrane-bound proteins composed of 7-transmembrane domains and are coupled to G-proteins (for a review [20]). After interaction with their specific chemokine ligands, chemokine receptors trigger a release of calcium ions (Ca^{2+}) from intracellular stores leading to the activation of protein kinase C (PKC) and other calcium-sensitive protein kinases. This causes cell responses, including the onset of a process known as chemotaxis that traffics the cell to a desired location within the organism. Recently, CCR7 was shown to be involved in the survival of metastatic squamous cell carcinoma of the neck and head [21] and to promote migration and invasion of some tumours [22]. By inducing its expression, MWCNT could favor the survival of the hepatocarcinoma HepG2 cell line.

IL-7 is a cytokine important for B and T cell development [23] which is described to be produced by hepatocytes in normal adult human liver forming a suitable microenvironment for T cell differentiation [24]. This cytokine and the hepatocyte growth factor form a heterodimer that functions as a pre-pro-B cell growth-stimulating factor.

Endothelin-1 (EDN1) is a key mediator of vascular tone and renal homeostasis through antagonistic vasoactive effects. EDN1 induce vasoconstriction by binding to its receptor ET_A associated with vascular smooth muscle in the arteries walls. It has also a vasodilation property on endothelial cells by inducing the production of the potent vasodilator, nitric oxide (NO) following its binding with the ET_B receptor (for a review [25]). EDN1 plays also an important role in wound healing in a variety of parenchymal organs including liver [26].

Low density lipoprotein-related protein 2 (LRP2 or megalin) is a multiligand binding receptor found in the plasma membrane of many absorptive epithelial cells. LRP2 plays important role in the regulation of levels of diverse circulating compounds like vitamins A and D and cholesterol homeostasis by receptor-mediated endocytosis of lipoprotein particles.

LTA is a member of the tumor necrosis factor family (also called TNF-β) produced mainly by lymphocytes mediating a large variety of inflammatory, immunostimulatory, and antiviral responses. Interestingly, LTA was shown to mediate both hepatocyte and progenitor oval cell-associated liver regeneration after acute or chronic liver injuries respectively [27].

Table 1. Alphabetical list of the 46 transcripts detected in HepG2 cells by RT real time TaqMan PCR array analysis, including one transcript used as housekeeping gene for data normalisation (GAPDH). Gene symbols are given with full gene names and general functions of the encoded proteins.

Symbol	Name	Function
ACE	angiotensin I converting enzyme (peptidyl-dipeptidase A) 1	Renin-angiotensin system regulation
ACTB	actin, beta	Cell cytoskeleton and mobility
AGTR1	angiotensin II receptor, type 1	Vasoconstriction, aldosterone synthesis
BAX	BCL2-associated X protein	Apoptosis
BCL2	B-cell CLL/lymphoma 2	Apoptosis
BCL2L1	BCL2-like 1	Apoptosis
C3	complement component 3	Complement system activation
CCL5	chemokine (C-C motif) ligand 5	Basophil and eosinophil activation - Chemoattraction
CCR4	chemokine (C-C motif) receptor 4	C-C chemokines receptor - Immune signal transduction
CCR7	chemokine (C-C motif) receptor 7	C-C chemokines receptor - Immune signal transduction
CD3E	CD3E antigen, epsilon polypeptide (TiT3 complex)	T cell receptor- CD3 complex - Immune transduction
CD4	CD4 antigen (p55)	Immune response - Antigen presentation
CD8A	CD8 antigen, alpha polypeptide (p32)	Immune response - Antigen recognition
CD40	CD40 antigen (TNF receptor superfamily member 5)	TNF receptor - Immune transduction
CD68	CD68 antigen	Macrophage recruitment and activation
COL4A5	collagen, type IV, alpha 5	Basement membrane structure
CSF1	colony stimulating factor 1 (macrophage)	Macrophage differentiation
CSF3	colony stimulating factor 3 (granulocyte)	Granulocyte differentiation
CYP7A1	cytochrome P450, family 7, subfamily A, polypeptide 1	drug metabolism and cholesterol catabolism
ECE1	endothelin converting enzyme 1	Endothelin processing
EDN1	endothelin 1	Vasoconstriction and vasodilatation
FAS	Fas (TNF receptor superfamily, member 6)	Apoptosis
FASLG	Fas ligand (TNF superfamily, member 6)	Apoptosis
FN1	fibronectin 1	Cell adherence - Connective tissue structure
GAPDH	glyceraldehyde-3-phosphate dehydrogenase	Glycolysis
GUSB	glucuronidase, beta	Glycosaminoglycan metabolism
HMOX1	heme oxygenase (decycling) 1	Heme catabolism - Anti-oxidant defenses
ICAM1	intercellular adhesion molecule 1 (CD54)	Cell adherence
IKBKB	inhibitor of NFKB, kinase beta	NF-kappaB regulation
IL7	interleukin 7	B and T cell development
IL8	interleukin 8	Inflammation - Chemoattraction - Angiogenesis
IL12A	interleukin 12A	T cell differentiation, T cell stimulating factor, anti-angiogenic activity
IL18	interleukin 18	T cell and B cell activation
LRP2	low density lipoprotein-related protein 2	Uptake of lipoproteins, sterols, vitamin-binding proteins and hormones
LTA	lymphotoxin alpha (TNF superfamily, member 1)	Inflammation, apoptosis, immune and antiviral responses
NFKB2	nuclear factor of kappa light polypeptide gene enhancer in B-cells 2	Inflammation - Anti-oxidant defenses
NOS2A	nitric oxide synthase 2A (inducible, hepatocytes)	Cell signal transduction
PGK1	phosphoglycerate kinase 1	Glycolysis
PRF1	perforin 1 (pore forming protein)	Cell death, apoptosis, cell membrane pore-forming protein
RPL3L	ribosomal protein L3-like	Ribosomal protein
SKI	v-ski sarcoma viral oncogene homolog (avian)	TGF-beta signaling repression
SMAD3	SMAD, mothers against DPP homolog 3 (Drosophila)	Cytokine signaling
SMAD7	SMAD, mothers against DPP homolog 7 (Drosophila)	Cytokine signaling
STAT3	signal transducer and activator of transcription 3	Apoptosis - Cell cycle
TFRC	transferrin receptor (p90, CD71)	Iron metabolism
TGFB1	transforming growth factor, beta 1	Cell cycle - Differentiation - Cell adherence
TNF	tumor necrosis factor (TNF superfamily, member 2)	Cell cycle - Differentiation - Apoptosis
VEGF	vascular endothelial growth factor	Angiogenesis - Vascular permeability
18S	18S ribosomal RNA	Ribosomal RNA

Table 2. Results of the transcriptomic profiling (RT real time TaqMan PCR arrays) of HepG2 cells considering 96 transcripts involved in immunity or pro-inflammatory response, after treatment of 24 h with 100 µg/ml of raw H_2O-stirred MWCNT. Results were expressed as the ratios of the levels of transcripts in treated cells versus untreated cells (mean ratios ± s.d. for n = 3). Ratios between 0.7 and 1.3 were cut off. The means lacking s.d. values were obtained from 2 quantitative ratios. When a given mRNA was detected in MWCNT-treated cells and was below the detection threshold in the untreated cells, these ratios were qualified as qualitative. Grey boxes indicate statistically significant effects with $P < 0.05$.

	Mean	**S.D.**
FASLG	16.62	
LTA	15.08	
CCR4	4.00	
LRP2	3.96 (qualitative)	/
CSF3	3.64	
ACE	3.18	
CCR7	2.92	1.06
EDN1	2.91	0.47
IL-7	1.45	0.17
CCL5	0.59	
CD8A	0.21	

Currently, almost no data are available concerning the putative role of these different proteins in MWCNT-treated HepG2 cells. Moreover, these data indicate altogether that MWCNT can only induce a weak inflammatory response in HepG2 cells. Indeed, among the 46 detected mRNA, only these few transcripts were shown to be more abundant in comparison with the control cells. By comparison, water-stirred MWCNT induced overexpression of 26 transcripts coding for immune and inflammatory proteins in telomerase-immortalised human keratinocytes [8]. Furthermore, IL-8, which is among the main cytokines produced during a hepatic inflammation [28] did not show a significant increase of the abundance of their mRNA, confirming previous observation [11]. Recently, subcutaneously injected oxidated water-soluble MWCNT were shown to act as an immunostimulatory substance in the tumor-bearing mice. MWCNT induced significant activation of the complement system, promoted inflammatory cytokines production and activated macrophage cells leading to the inhibion of the tumor growth. A similar inflammatory response was not observed in our study. The origin of MWCNT (pristine MWCNT in our study and oxidized MWCNT in Meng *et al.* [29]) is likely to explain these differences. Oxidation is a process which modifies MWCNT surface and decreases their length [30]. These alterations can in turn change the biological effects of MWCNT. In addition, the experimental context (*in vitro* or *in vivo*) is also an important parameter to explain these differences in inflammatory responses induced by MWCNT.

In conclusion, this preliminary study described the impact of MWCNT at the transcriptional level on hepatocarcinoma HepG2 cell line showing very few effects of MWCNT on expression of mRNA coding for immune and inflammation proteins.

References

[1] Bottini M, Bruckner S, Nika K, Bottini N, Bellucci S, Magrini A, Bergamaschi A and Mustelin T 2006 *Toxicol Lett* **160** 121-126.

[2] Magrez A, Kasas S, Salicio V, Pasquier N, Seo J W, Celio M, Catsicas S, Schwaller B and Forro L 2006 *Nano Lett* **6** 1121-1125.

[3] Tian F, Cui D, Schwarz H, Estrada G G and Kobayashi H 2006 *Toxicol In Vitro* **20** 1202-1212.
[4] Sato Y, Yokoyama A, Shibata K, Akimoto Y, Ogino S, Nodasaka Y, Kohgo T, Tamura K, Akasaka T, Uo M, Motomiya K, Jeyadevan B, Ishiguro M, Hatakeyama R, Watari F and Tohji K 2005 *Mol Biosyst* **1** 176-182.
[5] Wick P, Manser P, Limbach L K, Dettlaff-Weglikowska U, Krumeich F, Roth S, Stark W J and Bruinink A 2007 *Toxicol Lett* **168** 121-131.
[6] Smart S K, Cassady A I, Lu G Q and Martin D J 2006 *Carbon* **44** 1034-1047.
[7] Vankoningsloo S, Piret J P, Saout C, Noel F, Mejia J, Zouboulis C C, Delhalle J, Lucas S and Toussaint O 2010 *Nanotoxicology* **4** 84-97.
[8] Vankoningsloo S, Piret J P, Saout C, Noel F, Mejia J, Coquette A, Zouboulis C C, Delhalle J, Lucas S and Toussaint O 2011 *Nanotoxicology* 10.3109/17435390.17432011.17558642.
[9] Herzog E, Byrne H J, Davoren M, Casey A, Duschl A and Oostingh G J 2009 *Toxicol Appl Pharmacol* **236** 276-281.
[10] Monteiro-Riviere N A, Inman A O, Wang Y Y and Nemanich R J 2005 *Nanomedicine* **1** 293-299.
[11] Pulskamp K, Diabate S and Krug H F 2007 *Toxicol Lett* **168** 58-74.
[12] Ponti J, Colognato R, Rauscher H, Gioria S, Broggi F, Franchini F, Pascual C, Giudetti G and Rossi F 2010 *Toxicol Lett* **197** 29-37.
[13] Tabet L, Bussy C, Amara N, Setyan A, Grodet A, Rossi M J, Pairon J C, Boczkowski J and Lanone S 2009 *J Toxicol Environ Health A* **72** 60-73.
[14] Fenoglio I, Tomatis M, Lison D, Muller J, Fonseca A, Nagy J B and Fubini B 2006 *Free Radic Biol Med* **40** 1227-1233.
[15] Luchoomun J, Sinibaldi D and Chen X-L 2006 *Antioxidant therapy for chronic inflammatory diseases*, *Vol.* (Ed. H. V. Panglossi), Nova Publishers, New York, p. 145–177.
[16] Piret J-P, Detriche S, Vigneron R, Vankoningsloo S, Rolin S, Mejia Mendoza J H, Masereel B, Lucas S, Delhalle J, Luizi F, Saout C and Toussaint O 2010 *J Nanopart Res* **12** 75-82.
[17] Livak K J and Schmittgen T D 2001 *Methods* **25** 402-408.
[18] Satih S, Savinel H, Rabiau N, Fontana L, Bignon Y J and Bernard-Gallon D J 2009 *J Mol Signal* **4** 3.
[19] Vankoningsloo S, De Pauw A, Houbion A, Tejerina S, Demazy C, de Longueville F, Bertholet V, Renard P, Remacle J, Holvoet P, Raes M and Arnould T 2006 *J Cell Sci* **119** 1266-1282.
[20] Murdoch C and Finn A 2000 *Blood* **95** 3032-3043.
[21] Liu F Y, Zhao Z J, Li P, Ding X, Zong Z H and Sun C F 2009 *Br J Oral Maxillofac Surg* **48** 291-296.
[22] Ding Y, Shimada Y, Maeda M, Kawabe A, Kaganoi J, Komoto I, Hashimoto Y, Miyake M, Hashida H and Imamura M 2003 *Clin Cancer Res* **9** 3406-3412.
[23] Fry T J and Mackall C L 2002 *Blood* **99** 3892-3904.
[24] Golden-Mason L, Kelly A M, Traynor O, McEntee G, Kelly J, Hegarty J E and O'Farrelly C 2001 *Cytokine* **14** 143-151.
[25] Khimji A K and Rockey D C 2010 *Cell Signal* **22** 1615-1625.
[26] Rockey D C, Fouassier L, Chung J J, Carayon A, Vallee P, Rey C and Housset C 1998 *Hepatology* **27** 472-480.
[27] Knight B and Yeoh G C 2005 *Cell Tissue Res* **319** 61-70.
[28] Gomez-Quiroz L, Bucio L, Souza V, Escobar C, Farfan B, Hernandez E, Konigsberg M, Vargas-Vorackova F, Kershenobich D and Gutierrez-Ruiz M C 2003 *Hepatol Res* **26** 134-141.
[29] Meng J, Yang M, Jia F, Kong H, Zhang W, Wang C, Xing J, Xie S and Xu H 2010 *Nanotechnology* **21** 145104.
[30] Hong C E, Lee J H, Kalappa P and Advani S G 2007 *Composites science and technology* **67** 1027-1034.

Study on the toxicity of inhaled alumina nanoparticles: impact of physicochemical properties and adsorption artifacts on the measurement of biological responses.

M. Pailleux[1,2], J. Pourchez[1,3], D. Boudard[1,4,5], P. Grosseau[2] and M. Cottier[1,4,5]

[1]LINA Laboratoire Interdisciplinaire d'étude des Nanoparticules Aérosolisées, F-42023, Saint-Etienne, France

[2]Ecole Nationale Supérieure des Mines de Saint-Etienne, Centre SPIN, LPMG, UMR CNRS 5148, F-42023, Saint-Etienne, France

[3]Ecole Nationale Supérieure des Mines de Saint-Etienne, Centre Ingénierie et Santé, Saint-Etienne, France

[4]PRES Lyon - Université Jean Monnet, Faculté de Médecine, IFR Inserm 143, F-42023, Saint-Etienne, France

[5]Laboratoire d'Histologie CHU de Saint-Etienne, F-42055, Saint-Etienne, France

E-mail :pailleux@emse.fr

Abstract. This work aims at developping a multidisciplinary approach to highlight the correlation between the toxicity of alumina engineered nanoparticles (NP) and their physicochemical characteristics. Accuracy of measurements depends on cell production after contact with particles, but also depends on the ability of biomolecules to get adsorbed on the NP [1]. That's why, mechanisms of biomolecules adsorption on NP must be fully understood to avoid misinterpretation of data.

1. Introduction

Aluminum hydroxide is a material used in various applications such as abrasive materials, refractories, substrates for electronic circuits...

At each step of the manufacturing process (synthesis, collection and purification of the powders), workers could be exposed to fine dust suspended in the atmosphere. Inhalation of some types of particles may cause effects on the health such as reduced lung function, respiratory symptoms and increased risk of tumors. Therefore, it is important to evaluate intrinsic toxicity of fine and ultrafine particles of aluminum hydroxide particles.

2. Materials and methods

2.1. Physicochemical characterization

We work on an aluminium hydroxide powder. The physicochemical characterizations were done by laser granulometry (Malvern Mastersizer 2000), XRD (D5000, Siemens) and TEM (JEOL 2010). The specific surface area of powders was determined by the BET method using a nitrogen sorption apparatus (MicroMetrics) and the Photon Correlation Spectroscopy (Zetasizer Nano S, Malvern) was performed in solutions of different pH (pH3.5, pH7 and pH9.4).

2.2. Toxicity assessment

To assess the biological activity of aluminum hydroxide powder, cells from the RAW 264.7 macrophage cell line were incubated with aluminum hydroxyde nanoparticles.

Three parameters were examined:

- The lactate deshydrogenase (LDH) release: LDH is an enzyme found in the cytoplasm of living cells. In case of alteration of the plasma membrane, LDH is released into the environment.
- The Tumor Necrosis Factor alpha (TNF-α) production: TNF-α is a cytokine secreted by macrophages. It reflects the activation of macrophage and acute phase of the inflammatory response.
- The Reactive Oxygen Species (ROS) production, reflecting the oxidative stress [2-3].

The interactions can be established between these cells and nanoparticles in specific experimental conditions. This study will highlight the correlation between the level of toxicity induced by nanoparticles and their physicochemical properties.

Moreover, biomolecules-nanoparticles interactions were evaluated on a cellular medium measuring both affinity and amount of adsorbed TNF-α on aluminum hydroxyde nanoparticles, at equilibrium state (ELISA kit, R&D Systems®).

3. Results and conclusion

The different physicochemical characterizations show coherent results, the crystallites size is 10nm. Photon correlation spectroscopy shows that the aggregation of particles in water increases with pH.

Biological activity of aluminum hydroxide was less important than that of the positive control of toxicity (quartz) but also, and unexpectedly, less important than that of cells alone. To determine if this phenomenon was linked with a cytokine adsorption process, we added 75 pg of TNF-α to aluminum nanoparticles, and after a 24h incubation we found 65 pg of TNF-α in cell culture medium, suggesting a significant adsorption of TNF-α on aluminum hydroxide nanoparticles (13%).

In conclusion, cytokine adsorption on nanoparticles could represent an important artifact to take into account for an accurate evaluation of biological responses, especially in order to reliably compare nanoparticles with different physicochemical characteristics.

References

[1] Val, S., 2009. Carbon black and titanium dioxide nanoparticles induce pro-inflammatory responses in bronchial epithelial cells: Need for multiparametric evaluation due to adsorption artifacts. Inhalation Toxicology, 21, 115-122

[2] Bruch, J. et al., 2004. Variation of biological responses to different respirable quartz flours determined by a vector model. Int J Hyg Environ Health, 207(3), 203-16

[3] De la Harpe J, Nathan CF. A semi-automated micro-assay for H2O2 release by human blood monocytes and mouse peritoneal macrophages. J Immunol Methods 1985;78(2):323-36

Toxicity test: Fluorescent silicon nanoparticles

K Fujoka[1], S Hanada[2], F Kanaya[2], A Hoshino[1,2], K Sato[3,4], S Yokosuka[3], Y Takigami[3], K Hirakuri[3], A Shiohara[5], R D Tilley[5], N Manabe[6], K Yamamoto[2] and Y Manome[1,7]

[1]Department of Molecular Cell Biology, Jikei University School of Medicine, Minatoku, Tokyo 105-8461, Japan

[2]Research Institute, National Centre for Global Health and Medicine, Shinjuku-ku, Tokyo 162-8655, Japan

[3]Department of Electrical and Electronic Engineering, Tokyo Denki University, Chiyoda-ku, Tokyo 101-8457, Japan

[4]Department of Physics, University of Bologna, Viale Berti Pichat 6/2, 40127 Bologna, Italy

[5]MacDiarmid Institute of Advanced Materials and Nanotechnology, Victoria University of Wellington, Wellington, New Zealand

[6]Institute of Multidisciplinary Research for Advance Materials, Tohoku University, Sendai, Miyagi 980-8577, Japan

[7]Core Research Facilities, Jikei University School of Medicine, Minato-ku, Tokyo 105-8461, Japan

E-mail: kfujioka@jikei.ac.jp

Abstract. Semiconductor nanoparticles ('quantum dots', QDs) are useful fluorescent materials because of their high fluorescent stability compared with existing organic fluorescent dyes. QDs were tested in many biochemical experiments, and the reported results suggested their advantages. However, when we consider their application at the clinical level, their large-scale use may be problematic because of their influence on the environment and the living body as a result of cadmium contained in existing mainstream QDs. Here we report on the characteristics of silicon particles (synthesised using the gas phase method and liquid phase method, currently in the development stage) as a substitute material, focusing on cell-level safety and the potential mechanisms of toxicity.

1. Introduction

Semiconductor quantum dots (QDs) exhibit brighter and longer fluorescence than organic dyes. Taking advantage of these characteristics, QDs are used as fluorescent labels in many biomedical studies [1-3]. Currently, a CdSe core with a ZnS shell is one of the most popular QDs used in

4. Potential toxicity mechanisms of silicon particles

We report the toxic mechanism of silicon particles that we have clarified. As we mentioned earlier, the toxicity of silicon particles synthesized in gas phase method is lower compared with that of mercaptoacetic acid coated QDs in MTT assay method [5]. However, toxicity of silicon particles can be observed at higher concentrations. We confirmed that from the silicon particles using the gas phase method, radicals were produced in small quantities and the cell membrane was oxidised [5]. We suggest that the cell membrane is damaged by the radical at higher concentrations, leading to cell death (Figure 3(A)). In addition, by using superoxide dismutase and *N*-acetylcysteine, we succeeded in suppressing radicals that would have been generated from silicon particles synthesised using the gas phase method [10]. Thus, we may reduce toxicity by adding materials with radical inhibitory activity in the surface modification step or in use of silicon particles.

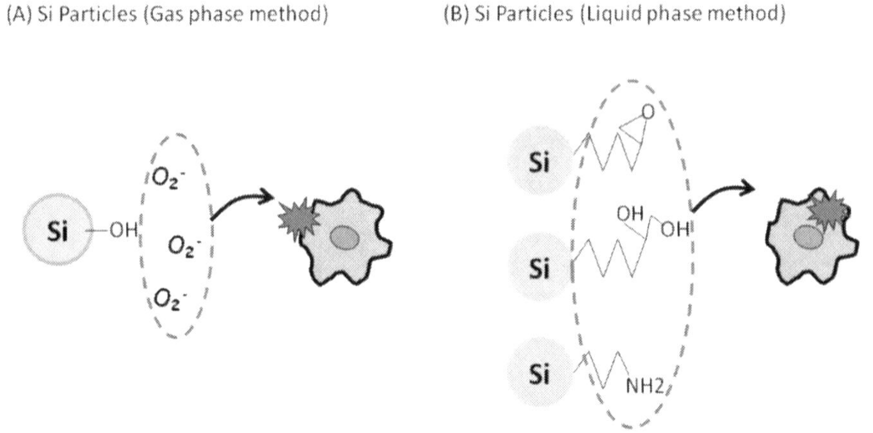

Figure 3. Scheme of potential toxicity mechanisms by silicon particles synthesised using gas phase method (A) or liquid phase method (B).

On the other hand, we suggest that particles synthesised using the liquid phase method have a different toxic mechanism (Figure 3(B)). We did not observe any radical generation from the liquid-phase synthesized particles (data not shown). Limited objective data show that toxicity is higher in epoxy in which chemical activity is high than in diol, suggesting that surface molecules affect toxicity [7].

5. Conclusion

In this report, we compared the characteristics of silicon particles synthesised using the gas phase method and the liquid phase method, focusing on the toxicity test using cells and the toxic mechanism. Radicals and surface molecules were identified as potential causes of toxicity. At present, despite the limitation of lower luminance relative to existing QDs, silicon particles have good potential as fluorescent materials considering the relatively low toxicity and the abundance of silicon in the earth. Given their small influence on the environment and the living body, they may have a bright future if we develop particles with increased fluorescence.

6. Materials and Methods

6.1. Assays for Cellular toxicity

We have reported the detail method in previous papers [5,7]. Briefly, the cell lines were cultured in each well of 96-well plates for 48 hr at 37 °C, and then, the silicon nanoparticles were co-cultured in indicated concentration for 48 hr. Finally, for the measurement of mitochondrial activity, Cell

Dispersion of ZrO_2 and Y_2O_3 nanopowders in physiological suspensions

A Godymchuk[1], E Arzamastseva[1], D Kuznetsov[2] and S Milyaeva[2]

[1]Tomsk Polytechnic University, Division of nanomaterials and nanotechnologies, 30, Lenina ave., Tomsk, Russia, tel./fax: +7(3822)427242

[2]National Research Technological University "MISIS", Inter-chair of teaching and research laboratory "Nanomaterials", 2, Leninsky ave., Moscow, Russia, tel. / fax: +7 (495) 2372226

E-mail: Godymchuk@tpu.ru

Abstract. The dispersion of suspensions based on nanostructured powders (ZrO2 and Y2O3) and the simplest physiological fluids (phosphate buffering saline, glucose solution, water) in terms of their use in ecotoxicological tests was studied. Using the scanning electron microscopy and laser diffraction, an increase in the size of particles and agglomerates in the suspensions in 2-11 times was shown. The particle size distribution in the suspensions was characterized by a single peak, but its magnitude and width varied ambiguously upon standing suspensions in time. It was shown experimentally that the average size of the dispersed phase could rapidly grow in the DW-suspension, could be stabilized in PBS-suspension and might decrease with time in Gl-suspension. By electro-acoustic method it was shown that the Zeta-potential on the solid/liquid boundary in the study varied over a wide range of values from – 200 to + 200, and the nature of change - a jump. This demonstrated the instability of the suspensions based on NP-ZrO2 and NP-Y2O3.

1. Introduction

Nanosized powders of metal oxides are widely used in electronics and optics (SiO_2, Nd_2O_3, Eu_2O_3, Dy_2O_3, Au, Pt) [1], chemical and cosmetic industries (Al_2O_3, ZnO, TiO_2, Ag, Pt) [2], water treatment technologies (Al_2O_3, Fe_2O_3), energetic (Dy_2O_3, Au), in agriculture and food industry (Fe_2O_3) [3].

The life cycle of nanopowders includes the manufacture, use and disposal of powders or products containing nanopowders. In this case, employees of nanotechnology industries and laboratories are primarily at risk, as they are exposed to nanoparticles [4]. Intensively developing nanotech production of powdered materials requires consideration of safety and risk assessment of the impact of nanomaterials on humans and other living organisms.

Nanopowders possess properties that often radically differ from their counterparts in the form of macroscopic dispersions or solid phase [2]. Small size and high physico-chemical activity of nanoparticles allow them to penetrate through biological membranes and bind to cells, tissues and organs easier than microsized particles. In contact with living organisms, nanoparticles can penetrate through cellular barriers, as well as to overcome the blood-brain barrier in the central nervous system, to circulate and accumulate in tissues and organs [5,6], thus causing toxic effects on organs and tissues

[6-8]. According to several studies, even non-toxic by nature TiO_2, ZnO, Ag in nanoscale condition can cause a cytotoxic effect [9-12].

This phenomenon demonstrates that the development of methods assessing the risks associated with the production of nanopowders and the use of products containing nanoparticles is one of the leading trends of nanotechnology development [13].

2. Application of dispersion in the toxicological study

Identification of nanoparticles in the environment, determination of their composition, concentration, structure are influenced by a number of physico-chemical factors: the number of particles, their charge, size, size distribution, surface area, structure and shape, degree of agglomeration and element composition [7]. Therefore, toxicological study of nanomaterials in various environmental media has many analytical subtleties. One of the biggest problems in the study of nanomaterials is their tendency to aggregation [14]. In practice, to prevent aggregation or deposition of particles, dispersion is used.

Dispersion of solid particles is often achieved by the addition of solvents and surfactants (stabilizers), ultrasonic action and prolonged agitation. The main advantages and disadvantages of currently used methods for dispersing the nanoparticles are shown in table 1 [15]. Here it should be given more precise definition to the terms used in the paper – agglomerates and aggregates. So, the particles aggregates are the particular case of particles agglomerates but differ from the last with less force between the particles.

Table 1. Advantages and disadvantages of nanomaterials dispersion

<table>
<tr><th>№</th><th>Method Name</th><th>Advantages</th><th>Disadvantages</th></tr>
<tr><td>1.</td><td>Without additional dispersion</td><td>No chemical additives
No additional toxicants
Rapid sample preparation
Close to the real conditions</td><td>Uncontrolled dispersion
Study of toxicity, behavior, activity of the agglomerates/aggregates</td></tr>
<tr><td>2.</td><td>Adding solvents and stabilizers</td><td>Rapid stabilization of suspensions
Stability of suspensions</td><td>Possible toxicity of additives
Adding of toxic contaminants from additives
Surplus supplements can alter the structure and influence the toxicity
Limitations in obtaining the appropriate information about the real toxicity potential of an unknown nanomaterial
Need to control the concentration</td></tr>
<tr><td>3.</td><td>Sonication*</td><td rowspan="2">No chemical additives
No additional toxicants
Not necessary to control additives</td><td>Need a time to process
Processing time affects the concentration of aggregates
Processing can affect the shape and, consequently, the toxicity of the samples
Sound for natural samples in the presence of electron-donor promotes oxidation of the samples in solution
Unstable systems</td></tr>
<tr><td>4.</td><td>Stirring and shaking</td><td>Prolonged mixing
Stirring can affect the shape and, consequently, the toxicity of the samples
Unstable systems</td></tr>
</table>

** In this way "sonication" means particularly the water bath sonication.*

Despite the shortcomings of the second method "Adding solvents and stabilizers", it is the only one method that allows maintaining the dispersion of suspensions during the whole experiment. However, the nature of the solvent is one of the most important criteria for assessing its suitability. For example, tetrahydrofuran (THF), one of the most effective stabilizer of the fullerene, promotes solvation of the fullerene molecules in the water, as THF makes the system «solvent – water» less polar. This radically changes the behavior of C_{60}-nanoparticles in water. For example, when water fleas were exposed to THF-stabilized by and stabilized pre-mixing (2 months) fullerene suspensions, their lethal dose LD_{50} for 48 hours reached 0.8 mg/l and 35 mg/l, respectively [16]. This suggests that the toxicity of the fullerene particles could increase in the presence of THF. So many researchers have questioned the use of THF as a non-toxic preservative. Chemical additives (amines and furans) are also invited to use as stabilizers, but the lack of experimental data leaves the question open [7,15].

Also one of the most common media for toxicological testing is phosphate buffering saline (PBS). PBS's composition is very simple; its pH-value is very close to the blood serum pH. Moreover, PBS is not toxic to living organisms. However, experiments show that this medium is not always appropriate, because oxide and metal particles can be strongly agglomerated and can interact with the formation of oxide-hydroxide forms on the boundary of the particle/solution [17]. As a consequence, the use of stabilizers can uniquely affect the interpretation of physical and chemical reactivity of nanomaterials during the tests. What is more, the nanoparticles are not delivered in its original form to into tested organisms.

Initial experiments have shown that nanoparticles of metals and metal oxides demonstrate chemical stability and their dispersion can be well preserved in physiological solutions of organic nature. Therefore, **the purpose of this study** was to estimate the dispersity/stability of suspensions based on the metal oxide nanopowders and simple saline solution for their use in ecotoxicological tests in the further study.

3. Objects of research

Nanopowders of zirconium oxide (NP-ZrO_2) and yttrium oxide (NP-Y_2O_3), obtained by plasma-chemical method, were taken for the study. These nanopowders are the promising materials for the manufacturing of functional and structural ceramic with the help of a dry compaction technique under the influence of an ultrasonic action [18]. In the Nano-Centre of Tomsk Polytechnic University (Russia, Tomsk) [http://portal.tpu.ru/departments/centre/nano/eng] a lab-scale production line of nanostructured ceramic has been working for more than 5 years. Presently, the Nano-Centre is introducing a line of nanopowder production (as a raw material for industry) using spray-drying methods of production. In the production process (obtaining, drying, packing, transport and use) nanopowders form an aerosol in the occupational area, which, somehow, makes contact with the staff.

4. Determination of the composition and morphology of the powder particles

The phase composition of powders was determined by X-ray diffraction (XRD) method on the powder diffractometer XRD-7000 (Shimadzu). Recording X-ray was performed using CuK_α radiation in the range of angles 2θ from 10 to 120 °. To identify X-ray data it was used ICPDS (International Center for Diffraction Date) data.

Determining the characteristics of nanopowders microstructure surface was carried out with a scanning electron microscope (SEM) JSM-7500FA (JEOL), equipped with a microanalyzer EDS (JEOL), which allows to determine the elemental composition of the surface depth of 50 Å from boron to uranium.

5. Preparation of physiological solutions

To study the effect of the stabilizer nature, phosphate buffering saline (PBS) and glucose solution (Gl) were chosen. Distilled water was used for comparison. Chemical composition and pH values of the solutions are presented in table 2. Chemical vessels before preparing the solution were kept at least for 24 h in 10%-solution of HNO_3, the vessels were cleaned twice in distilled water. On the base of prepared solutions and chosen nanopowders the suspensions were prepared, the concentration of particles was 0.10-0.15 wt. %.

Table 2. The composition and pH of the solutions

№	Name of the solution (their name in the paper)	Contents of chemicals, g/l				pH
		NaCl	Na_2HPO_4	KH_2PO_4	$C_6H_{12}O_6$	
1.	Distilled water (DW)	---	---	---	---	7
2.	Phosphate buffering saline (PBS)	8.77	1.28	1.36	---	6.4 – 7.2
3.	Glucose solution (Gl)	---	---	---	44.39	5.5 – 6.5

PBS is the base of almost all biological fluids (serum, lung fluid, sweat, saliva, etc.); it is traditionally used for toxicological study; and it is easy to maintain pH in PBS. Gl-solution is a valuable source of easily digestible nourishing material by a human body. Furthermore, glucose is a component of blood and interstitial fluid. The selected media are not only nontoxic to living organisms, but also used for the therapeutic treatment (update and replenish fluids in diseases).

6. Determination of powder dispersion and dispersed phase in suspensions

The specific surface of dry powders (S) was determined using the low temperature adsorption of nitrogen (BET theory) with the instrument «SorbiPrep» (META, Russia). The accuracy of measurements was ± 0.20 m^2/g. Assuming that the particles are spherical, average diameter of the particles was calculated with the experimental formula:

$$d = \frac{6}{\rho \cdot S},$$

where d - average particle's diameter, m; ρ - material density, kg/m^3; S - specific surface area of powder (BET data), m^2/kg.

Particle size distribution in the suspensions was obtained using the particle analyzer SALD-7101 (Shimadzu) basing on the method of laser diffraction. The method of laser diffraction is based on the particle scattering light, since the scattering angle of light is universally proportional to the size of the particles. The measurement was carried out for suspensions of nanopowders volume of 220 ml for 60-90 minutes with an interval of 2-10 minutes. The experiment was carried out while shaking without ultrasound exposure, since ultrasound promotes fragmentation of agglomerates [15]. All vessels and tools used for the experiments were acid-cleaned (10%-HNO_3) and twice cleaned in distilled water. The experimental results are curves surround the particle size distribution in the range of 0.01-300 µm. The results were used to calculate the average diameter (D_{av}) as follows:

$$D_{av} = \sum d \times \frac{q(\%)}{100\%},$$

where d - particle size in dispersions, µm; q - number of the particles with diameter d, %.

7. Determination of electrokinetic properties of the suspension

Zeta-potential at the solid/liquid boundary in suspensions were determined with the electro-acoustic spectrometer Zeta-ASP (Matec Applied Sciences, USA). The method is based on electro-acoustic effects arising as a result of transmission of ultrasound through a Newtonian liquid, dispersion medium or a porous body. According to the method, the sensor analyzer sends short pulses of high frequency (pulse duration 30 μs), to the sample located in the space around the electrode sensor, with the pulse frequency of 0.5-3.5 MHz, the amplitude of 100-600 V.

The measurements were carried out for 20-90 minutes at intervals of 1 minute in 200 ml of suspension. Each experiment was repeated twice, vessels were acid-cleaned at least for 24 h (10%-HNO_3).

8. Results and discussion

X-ray data of the NP-ZrO_2 and NP-Y_2O_3 nanopowders are demonstrated in figure 1. It shows that the nanopowder of NP-ZrO_2 was stabilized with small amounts of yttrium oxide, which is a common practice in the use of this nanopowder for the manufacture of ceramics.

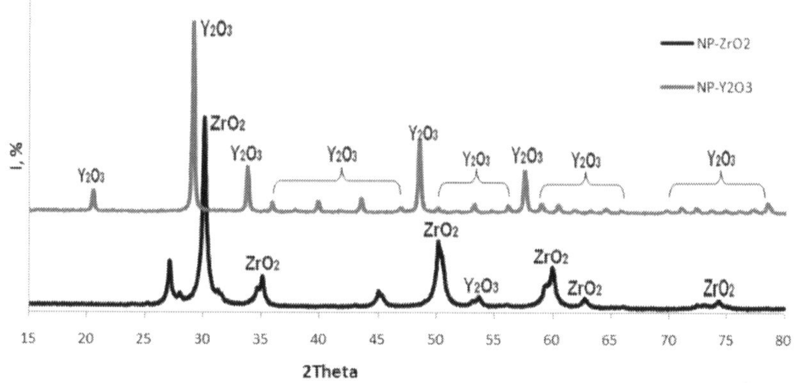

Figure 1. X-ray data of the nanopowders of ZrO_2 and Y_2O_3

According to SEM the particles of dry NP-ZrO_2 nanopowder form the agglomerates with the size of 3-6 μm (figure 2a). Individual particles (size 100-500 nm) are spherical in shape, but inside they are hollow (figure 2b) with a wall thickness of 20-30 nm (figure 2b). According to the forecasts particles of such a form can be carriers of substances due to the presence of the internal cavity. Agglomerates of dry nanopowder NP-Y_2O_3 are blocks with dimensions of 0.5...0.8 × 0.8...2.0 μm (figure 3a). Particles constituting the agglomerates are in the form of grains with a size of 20-250 nm (figure 3b).

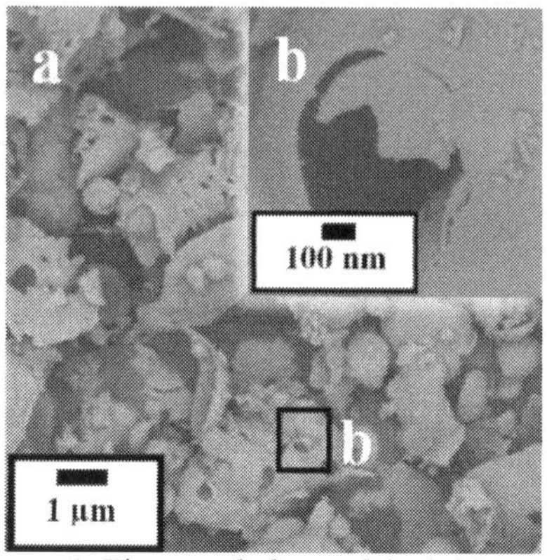

Figure 2. The morphology of the initial ZrO_2 powder: a) agglomerates of the powder, and b) the individual particles. According to the BET specific surface S = 8.10 m^2/g

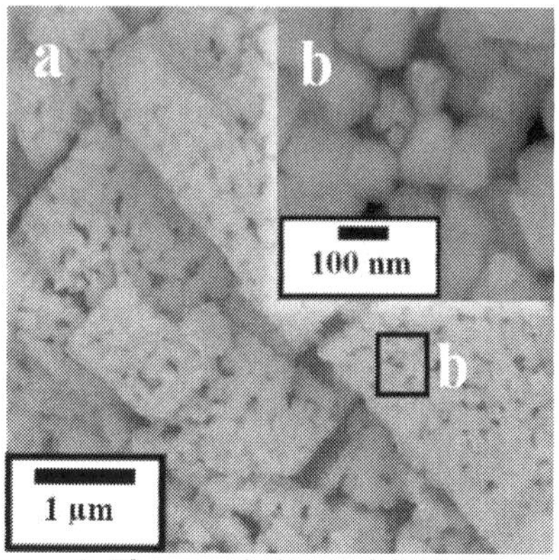

Figure 3. The morphology of the initial Y_2O_3 powder: a) agglomerates of the powder, and b) the individual particles. According to the BET specific surface S = 10.45 m^2/g

Data of the dispersion obtained with SEM are confirmed by the BET-data: surface average particle diameter as a rough approximation for NP-ZrO_2 (density 6.02 g/cm^3) d_{av} = 123 nm, and for NP- Y_2O_3 (density 4.84 g/cm^3) - d_{av} = 118 nm.

Stability of studied suspensions may be evaluated with the dispersion changing (the size distribution and/or average size) and the Zeta-potential at the boundary of the dispersed phase and the dispersion medium. To analyze the dispersion of nanopowders in the suspension based on the simplest physiological solutions, the experimental data on the amount of dry agglomerates (SEM) and the data on the agglomerates size of the dispersed phase (the method of laser diffraction) were compared.

According to the method of laser diffraction, the particles distribution is characterized with a single peak in all the suspensions. It is important to note that, despite the fact that the number of peaks does not change over time, the peak value can vary ambiguously upon testing the suspensions. Thus, for the both powders the value of the peak is gradually changing in the Gl-suspensions: for NP-ZrO_2 the peak value (the value of q, the wt.%) increases in time from 11 to 19 wt.%, while the maximum particle size is significantly reduced from 45 to 31 μm (Figure 4a). While for the sample NP-Y_2O_3 magnitude of the peak decreases from 13 to 10 wt.%, the peak itself is expanding, and the maximum particle size increases from 47 to 70 μm (Figure 4b). Apparently, the opposite trend to a dispersion change can be related to particle shapes and, consequently, to agglomerate forms.

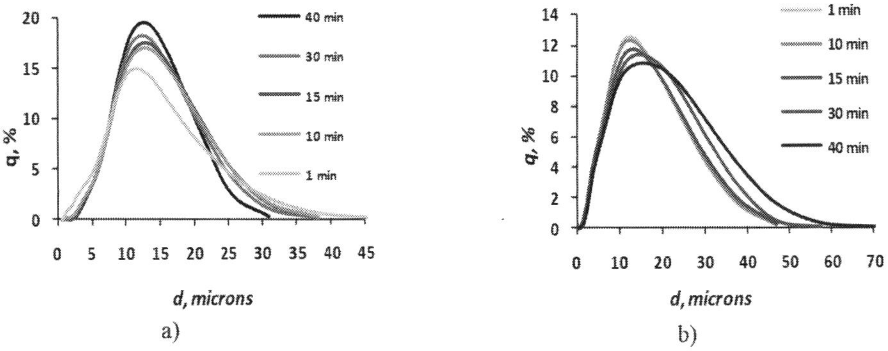

Figure 4. The particles size distribution in the Gl-suspension: a) the suspension of the ZrO_2 nanopowder, b) the suspension of the Y_2O_3 nanopowder (method of laser diffraction)

The dispersion of powders in PBS- and DW-suspensions varies ambiguously, with the clear trend to a change observed neither for the magnitude of the peak, nor for the maximum aggregates.

Data on changes in the calculated average particle size (D_{av}) in the suspensions are shown in figure 5. Comparison with the initial size of the powders agglomerates, determined by SEM, reveals that the average size of dry aggregates (3-6 μm) for NP-ZrO_2 under suspension stirring slightly increased in the PBS- and DW-suspension - up to 7.5-8.0 μm, and when released into the Gl-solution D_{av} it increased almost in 2 times - up to 12-13 μm (figure 5a). However, after a 30-minute exposure in the DW-suspension D_{av} is intensively growing; in PBS-suspension D_{av} is stabilized, and in Gl-suspension it is dropping slightly. The opposite is typical for NP-Y_2O_3: the size of dry aggregates increases significantly in all the solutions: from 2 μm to 13 μm (Gl-suspension), 17 μm (DW-suspension) and up to 18-19 μm (for PBS-suspension) (figure 5b). Moreover, within an hour aggregates size of this sample tends to increase, and in the Gl-solution to increase significantly.

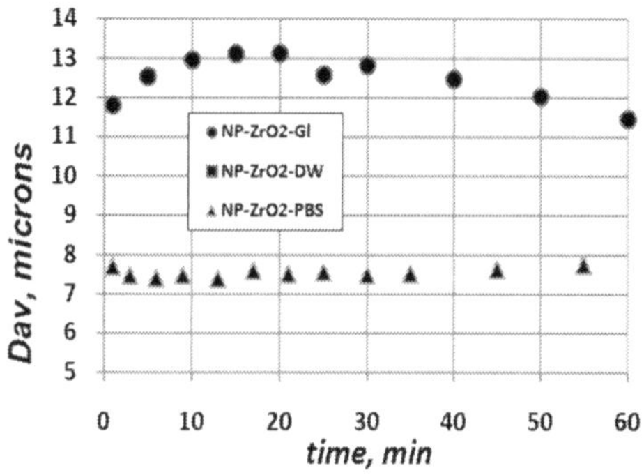

a)

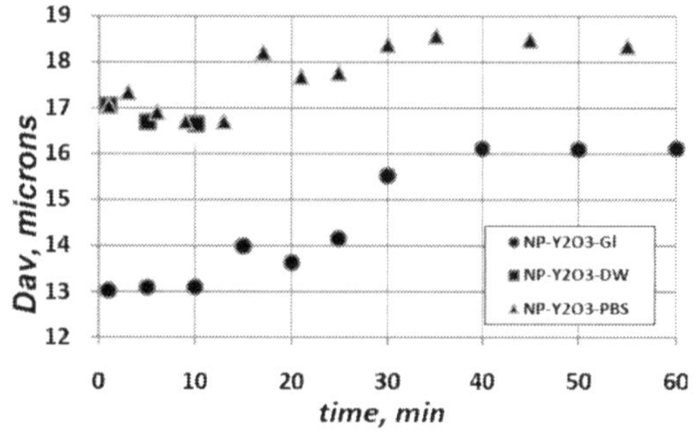

b)

Figure 5. Change in the average size of agglomerates D_{av} in the suspensions on the basis of: a) ZrO_2 nanopowder, b) Y_2O_3nanopowder.

Surface properties of nanoparticles determine their behavior in suspensions. Thus, in the formation of suspensions based on nanopowders the particles will be characterized by coagulation or aggregative stability, as evidenced by the experimental results using the method of laser diffraction. However, traditionally the stability of colloids is characterized by the electrokinetic potential or Zeta-potential that arises in a diffuse layer of counterions on the particle/fluid boundary. Since Zeta-potential is proportional to the charge of colloidal particles, the aggregative stability of the soles increases as its size grows [7].

Electro-acoustic method, used for the determining Zeta-potential of colloidal solutions and suspensions, has revealed that the Zeta-potential of suspensions changes in a large range of values from -200 to +200, with jump pattern of change (figure 6). On the one hand, these data may indicate that in the suspensions on the basis of the studied nanopowders it is difficult to reach equilibrium, so the measurement of Zeta-potential in these conditions may be not reasonable. On the other hand, the measured potential can carry the information about the processes of adsorption and desorption, that can explain the abrupt change in Zeta-potential difference that occurs between the dispersed particles and the disperse medium with their mutual displacement. However, changing of the Zeta-potential in such a large interval testifies the instability of the suspensions based on NP-ZrO_2 and NP-Y_2O_3 and chosen physiological solutions.

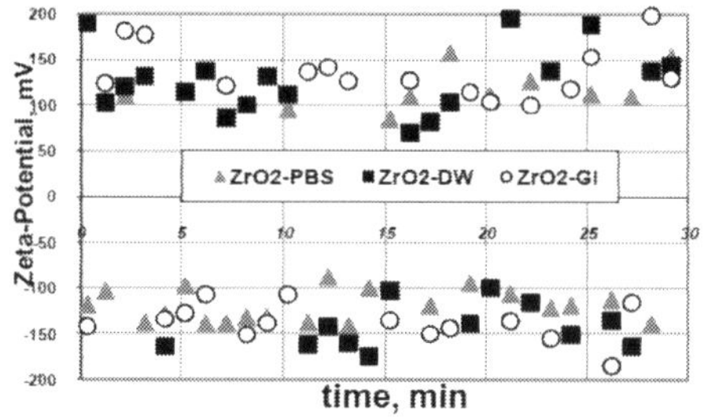

a)

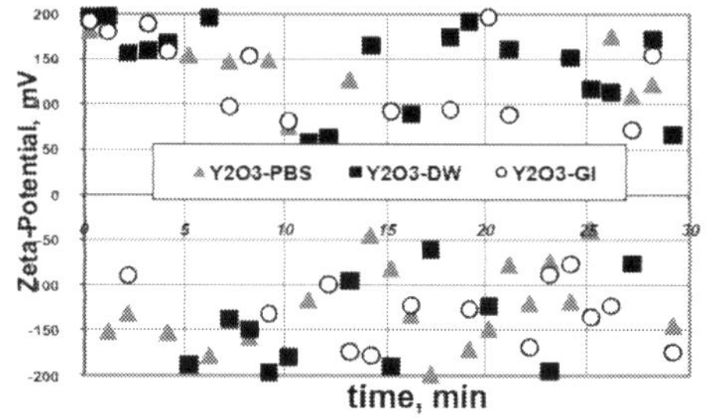

b)

Figure 6. Zeta-potential of suspensions at the solid/liquid boundary in the suspensions of: a) ZrO_2 nanopowder, b) Y_2O_3nanopowder.

9. Conclusion

The experimental data about the state of the dispersed phase in suspensions formed by nanopowders of zirconia dioxide and yttrium oxide in the simplest physiological fluids are ambiguous. Obtained by the scanning electron microscopy and laser diffraction data indicate that Gl-suspension of the dispersed phase can be characterized by a desire to monodispersity. Therefore, it is possible to predict the most probable size of the particles in the solutions for its application in ecotoxicological tests. While for water inorganic solutions, probability of obtaining stable suspensions is very small, that according to a restriction of their use for research in aqueous objects.

Electro-acoustic method confirmed that the magnitude of Zeta-potential in the suspensions based on NP-ZrO_2 and NP-Y_2O_3 varied over a wide range (-200 / +200 mV) and had a jump character. It is likely that in the ecotoxicological tests, which evaluated the effect of dose and particle size on the toxic effect, electrokinetic values will not affect because they change too quickly. But to interpret the physical, chemical and biochemical processes, the magnitude of Zeta-potential is an important magnitude, so the interpretation of phenomena occurring on these particles may be limited by the impossibility of achieving a balance in the forming systems.

Acknowledgments

This work was supported with *Russian Fund for Basic Research* (project 09-03-99043). Language processing work was performed under the supervision of *Sergey Fomin, PhD*.

References

[1] Borisov G and Kumakhov M 2004 *Nucl.Instr& Meth.* **A529(1-3)** 98-101
[2] Andrievsky R and Ragulya A 2005 *Nanostructed materials* Moscow: Academiya 192
[3] Kovalenko L 2006 *Biologically active iron nanopowders* Moscow: Nauka 124
[4] Mark D 2004 *Proc.Int.Symp. Occupational Health Implications of Nanomaterials* 77-83
[5] Oberdörster G, Oberdörster E, Oberdörster J. 2005 *Env. Health Persp.* **113**(7) 823–839
[6] Rasmussen R, Jessica P., et al. 2006, *Tox Sciences* **91**(1) 159–165
[7] Handy R, von der Kammer F, et al. 2008 *Ecotoxicology* **17** 287-314.
[8] Hoet P, Bruske-Hohlfeld I, Salata O 2004 *Jour of Nanobiotechn* **2** 12
[9] Sahoo S, Parveen S, Panda J. 2007 *Nanomedicine: Nanotech, Bio and Med* **3** 20-31
[10] Liu H, Ma L, Zhao J, et al. 2009 *Biol Trace Elem Res* **129**(1-3) 170-80
[11] Hund-Rinke K, Simon M 2006 *Env Science and Pollution Res* **13** 225-232
[12] Lin W, Xu Y, Huang C-C, et al 2008 *Jour of Nanopart Res* **11** 25-39
[13] EPA Nanotechnology White Paper 2007 *EPA/100/B-07/001* 120
[14] Buffle J. 2006 *Env and Chem* **3** 155-158
[15] Farre M, Gajda-Schrantz K, Kantiani L, Barcelo D. 2009 Anal and Bioanal Chem **393** 81-95
[16] Zhu S, Oberdorster E, Haasch M 2006 *Marine Env Res* **62** S5-S9
[17] Godymchuk A, Savel'ev G, Gorbatenko D 2010 *Russian Jour of General Chem* **80**(5) 881-888.
[18] Khasanov O. 2006 *Innovations (in Russia)* **7** 35-36

Factors influencing the cytotoxicity of zinc oxide nanoparticles: particle size and surface charge

M Baek, MK Kim, HJ Cho, JA Lee, J Yu, HE Chung and SJ Choi

Department of Food Science and Technology, Seoul Women's University, 126 Gongneung 2-dong, Nowon-gu, Seoul 139-774, South Korea

E-mail: sjchoi@swu.ac.kr

Abstract. Zinc oxide (ZnO) nanoparticle is one of the most important materials in diverse applications, since it has UV light absorption, antimicrobial, catalytic, semi-conducting, and magnetic properties. However, there is little information about the toxicological effects of ZnO nanoparticles with respect to physicochemical properties. The aim of this study was, therefore, to evaluate the relationships between cytotoxicity and physicochemical properties of ZnO nanoparticle such as particle size and surface charge in human lung cells. Two different sizes of ZnO nanoparticles (20 and 70 nm) were prepared with positive (+) or negative (-) charge, and then, cytotoxicity of different ZnO nanoparticles was evaluated by measuring cell proliferation in short-term and long-term, membrane integrity, and generation of reactive oxygen species (ROS). The results demonstrated that smaller particles exhibited high cytotoxic effects compared to larger particles in terms of inhibition of cell proliferation, membrane damage, and ROS generation. In addition, positively charged ZnO showed greater ROS production than ZnO with negative charge. These findings suggest that the cytoxicity of ZnO nanoparticles are strongly affected by their particle size and surface charge, highlighting the role of the physicochemical properties of nanoparticles to understand and predict their potential adverse effects on human.

1. Introduction

ZnO nanoparticle is one of the most widely used engineered nanomaterials in commercial products due to its UV light absorption, antimicrobial, catalytic, semi-conducting, and magnetic properties [1]. ZnO nanoparticle is, therefore, widely applied to personal care products, sunscreen, paints, electronic materials, rubber manufacture, food additives, and medicine [2-4]. In particular, nano-sized ZnO exhibits unique features that may completely differ from bulk-sized ZnO. As the particle size of ZnO decreases, its transparency to visible light and chemical reactivity increase, contrary to micro-sized ZnO having opacity and low reactivity. These unique characteristics are related to high proportion of atoms on the surface of nano-sized materials compared to bulk-sized ones. Thus, ZnO nanoparticles have been extensively applied to diverse products where transparency or great reactivity is required. On the other hand, the high reactivity of ZnO nanoparticles gives rise to increase biological responses such as cellular uptake and delivery efficiency, and thereby raising concern about their toxicity potential on biological systems. Therefore, the safety aspect of ZnO nanoparticles should be assumed to further expand their industrial applications with safe levels. Many studies on the toxicity of ZnO nanoparticles were performed in cell lines [5-6] as well as in animal models [7-8], but some contradictory results were reported, in particular, in terms of physicochemical parameters affecting their toxicity. Thus, toxicity of ZnO nanoparticles still remains to be elucidated, needing more vigorous toxicological evaluation by applying several methods. In this study, we evaluated the cytotoxicity of ZnO nanoparticles of two different sizes (20 and 70 nm) and different charges (positive and negative), respectively, in human lung A549 cells to determine the correlation between physicochemical properties of nanoparticles and their cytotoxicity.

2. Experimental methods

2.1. Materials

ZnO nanoparticles of two different sizes (20 and 70 nm) were purchased from Sumitomo (Japan) and American Elements (U.S.A.), respectively. For surface modification of ZnO nanoparticles with negative charge, ZnO (10 g) was suspended in 20 mM HEPES buffer (pH 7.0) containing 1% sodium citrate. The particle size and surface charge (zeta potential) of ZnO nanoparticles were determined by transmission electron microscopy (TEM: JEM-1010, JEOL) and a zeta potentiometer (Zetasizer Nano ZS system, Malvern Instruments), respectively.

2.2. Cell culture

Human lung epithelial cells (A549) were purchased from the Korean Cell Line Bank and cultured in RPMI1640 medium supplemented with 10% heat inactivated fetal bovine serum (Welgene, Ltd., South Korea), 100 units/ml penicillin, and 100 μg/ml streptomycin, under a humidified atmosphere (5% CO_2 plus 95% air).

2.3. Cell proliferation and viability

Cells (2×10^3 cells/100 μl) were seeded onto 96-well plates and incubated overnight at 37°C under a 5% CO_2 atmosphere. The medium in the wells was then replaced with fresh medium containing nanoparticles (0.5~1000 μg/ml) and incubation continued for 48 h. The effect of the nanoparticles on cell proliferation and viability was determined by WST-1 assay (Roche). Briefly, 10 μl of WST-1 solution (Roche) was added to each well and the plates were further incubated. After 4 h, the absorbance was measured with a plate reader at 440 nm. Cells incubated without nanoparticles were used as a control.

2.4. Lactate dehydrogenase (LDH) leakage assay

The release of LDH was monitored with the CytoTox 96 Non-Radioactive Cytotoxicity Assay (Promega). Cells (2×10^4 cells/ml) grown on 24-well plates were incubated with nanoparticles (0.5~1000 μg/ml) for 48 h. The plates were centrifuged, and aliquots (50 μl) of cell culture medium were collected from each well and placed in new microplates. Finally, 50 μl of substrate solution was added to each well and the plates were further incubated for 30 min at room temperature. The absorbance at 490 nm was measured with a microplate reader. Cytotoxicity is expressed relative to the basal LDH release by untreated control cells.

2.5. ROS production

The generation of intracellular ROS was measured with a cell-permeant fluorescent probe, carboxy-2',7'-dichlorofluorescein diacetate (carboxy-H_2DCFDA) (Molecular Probes) according to the manufacturer's guideline. Cells (2×10^3 cells/ 100 μl) were exposed to nanoparticles (0.5~1000 μg/ml) for 48 h, washed several times with phosphate-buffered saline (PBS) and incubated with 20 μM carboxy-H_2DCFDA for 60 min at 37°C. After washing, DCF florescence was immediately measured with a fluorescence microplate reader (SpectraMas M3, Molecular Devices). Basal ROS generation in cells treated without nanoparticles was used as a control.

2.6. Clonogenic assay
Cells were seeded in 35 mm dish at a density of 5×10^2 and incubated overnight under a standard condition. The medium in the dish (2 ml) was then replaced with fresh medium containing nanoparticles (5, 50, and 125 μg/ml) and incubation continued for 9 days. For colonies counting, cells were washed with PBS, fixed with 90% methanol for 30 min at 4°C and stained with 0.5% crystal violet solution (in 20% methanol, Sigma) for 1 h. After cells were washed with deionized water and air-dried, colonies consisted of more than 50 cells were counted. Each experiment was done in triplicate and colony number in the absence of nanoparticles was used as a control.

2.7. Statistical analysis
Statistical analyses were performed using Student's t test for unpaired data and p values of less than 0.05 were considered significant. All data are presented as mean ± standard error of the mean (S.E.M.).

3. Results and discussion

3.1. Characterization of ZnO nanoparticles
Particle size and surface charge of ZnO nanoparticles were measured by TEM and zeta potentiometer, respectively. As shown in Table 1, net charge of ZnO nanoparticles in aqueous solution was determined to be positive, about 43 mV, thus surface modification to obtain negatively charged ZnO nanoparticles was performed with citrates. Citric acid or citrates are widely used capping agents for inorganic nanoparticles, giving rise to negative surface coating. This is based on the fact that citrates are important biological ligands for metal ions to form strong metal complexes [9]. The measured particle size of 20 nm was well distinguished from that of 70 nm and the surface charge for positively or negatively charged ZnO was well prepared for comparative cytotoxicity study in the next step.

Table 1. Particle size and surface charge of ZnO nanoparticles as measured by TEM and zeta potentiometer, respectively.

Sample		Particle size [nm]	Zeta potential [mV]
20 nm	Positive charge	24 ± 5.57	43.8 ± 0.5
	Negative charge		-44.6 ± 0.7
70 nm	Positive charge	67 ± 12.41	44.1 ± 0.6
	Negative charge		-45.2 ± 0.8

3.2. Effect of particle size and surface charge on cell proliferation and viability
Cytotoxicity of different-sized ZnO nanoparticles with different charges were tested in human lung cancer A549 cells by measuring inhibition of cell proliferation and viability with the WST-1 assay (Figure 1). This colorimetric assay is based on the conversion of tetrazolium salt WST-1 into soluble colored formazan by cellular enzymes like mitochondrial dehydrogenase. Thus, an increase in the amount of formazan products is directly related to the number of metabolically active cells in cell culture system. Figure 1 showed that all the ZnO nanoparticles greatly inhibited cell proliferation and viability after 48 h incubation in a concentration-dependent manner. But, slight difference in cytotoxicity between nanoparticles with different sizes and charges was found; 20 nm (+) exhibited the highest toxicity compared to 20 nm (-) and larger particle 70 nm. IC_{50} (inhibition concentration 50%) values for 20 nm (+), 20 nm (-), 70 nm (-), and 70 nm (+) were about 68.47, 77.48, 94.67, 150.78 μg/ml, respectively.

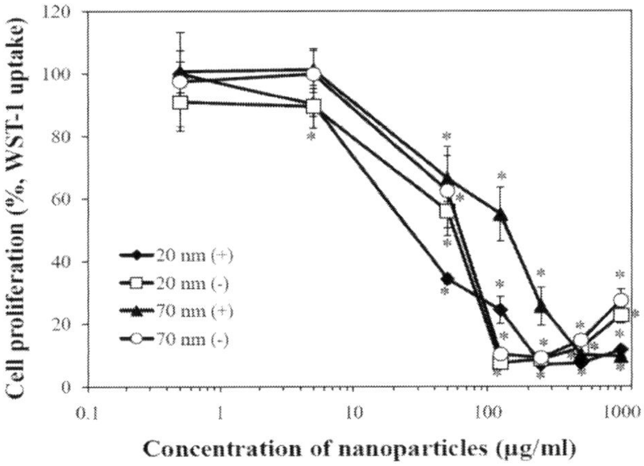

Figure 1. Effect of ZnO nanoparticles on cell proliferation of A549 cells after 48 h as measured by WST-1 assay. * denotes a significant difference from the control ($p < 0.05$).

3.3. Effect of particle size and surface charge on LDH leakage

LDH is an intracellular enzyme normally presented in the cytoplasm and released into the culture medium after cell membrane damage, thus increased LDH levels in the extracellular medium reflect reduced integrity of the plasma membrane. As shown in Figure 2, all the ZnO nanoparticles remarkably induced LDH leakage from A549 cells after 48 h incubation, in particular, at concentration above 50 μg/ml. Particle size and surface charge of ZnO nanoparticles highly affected cytotoxicity; the highest LDH was released from the cells incubated with 20 nm (+), which is consistent with the WST-1 result (Figure 1). In terms of particle size, 20 nm was determined to be more toxic than 70 nm, suggesting size-dependent toxicity of ZnO nanoparticles.

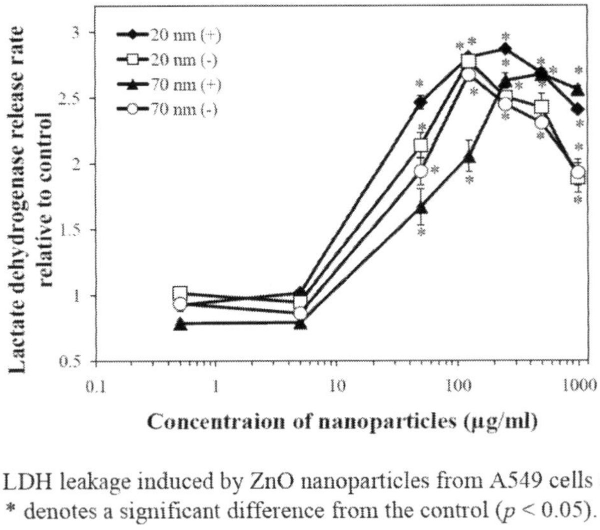

Figure 2. LDH leakage induced by ZnO nanoparticles from A549 cells after 48 h.
* denotes a significant difference from the control ($p < 0.05$).

3.4. Effect of particle size and surface charge on colony formation

According to the WST-1 assay, ZnO nanoparticles greatly affect cell proliferation and viability of A549 cells at the concentration above 5 μg/ml. However, it is possible that an exposure to even low concentration of ZnO nanoparticles could cause cytotoxicity after incubation for long time. Thus, the long-term cytotoxicity of different-sized or charge ZnO nanoparticles was evaluated by employing clonogenic assay, a method used to evaluate the ability of a single cell to grow into a colony. Figure 3 showed that all the ZnO nanoparticles tested completely inhibited cell proliferation of A549 cells at concentration above 50 μg/ml. But, interestingly, low concentration of 5 μg/ml of ZnO nanoparticles did not affect the colony formation of the cells even after long-term exposure for 9 days. This result is highly consistent the WST-1 assay as shown in Figure 1. It is worthy to note here that size or charge-dependent cytotoxicity of ZnO nanoparticles was not found, indicating exposure concentration and time to ZnO are more important factor affecting the long-term cytotoxicity rather than their physicochemical properties.

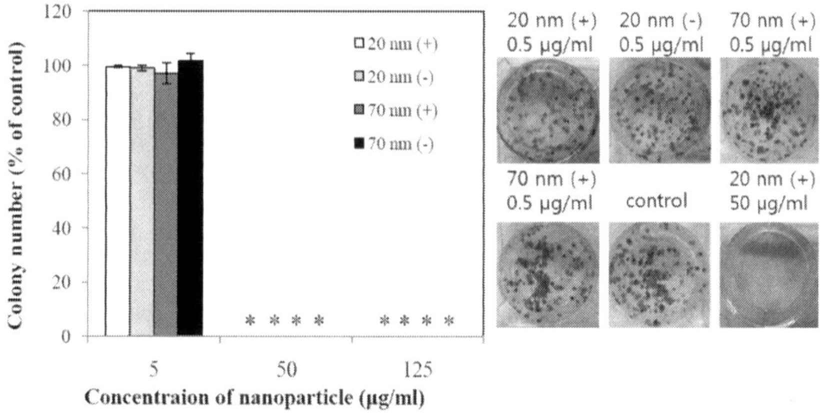

Figure 3. Clonogenic assay of A549 cells exposed to ZnO nanoparticles for 48 h exposure. * represents a significant difference from the control ($p < 0.05$).

3.5. Effect of particle size and surface charge on ROS generation

Several researches reported that oxidative stress is one of the main toxicological effects caused by ZnO nanoparticles in cell culture systems [10-14]. Thus, intracellular production of ROS was evaluated with a non-fluorescent dye, carboxy-H_2DCFDA until it is converted to green fluorescent dichlorofluorescein (DCF) upon oxidation within cells. The result represented that all different ZnO nanoparticles could generate ROS in a concentration dependent-manner (Figure 4). Interestingly, particle size and surface charge of ZnO nanoparticle were strongly associated with ROS production; 20 nm (+) generated the highest ROS at concentration range of 250 – 1000 μg/ml with a following order of 70 nm (+) > 20 nm (-) > 70 nm (-). In terms of ROS generation, it seems that surface charge of ZnO nanoparticles is more critical parameter than its particle size. In addition, size-dependent cytotoxicity was also measured, indicating high toxicity of 20 nm compared to 70 nm.

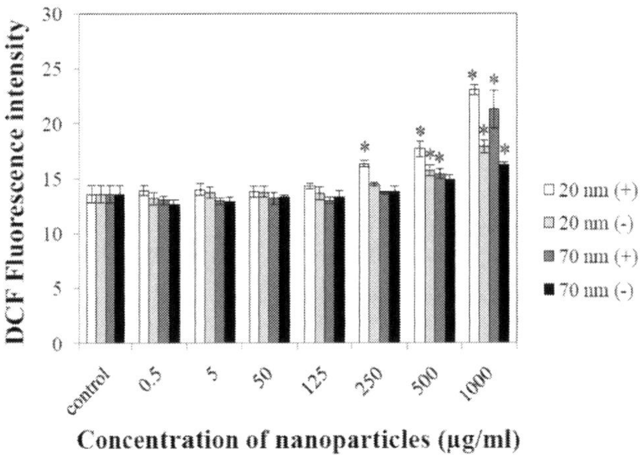

Figure 4. ROS induced by ZnO nanoparticles in A549 cells after 48 h exposure. * denotes a significant difference from the control ($p < 0.05$).

4. Conclusion

In this work, toxicological effects of ZnO nanoparticles with respect to physicochemical properties such as particle size and surface charge were evaluated in cell culture systems by measuring inhibition of cell proliferation and viability in short-term as well as in long-term, membrane damage, and ROS generation. Among ZnO nanoparticles tested with different sizes and charges, 20 nm (+) exhibited the highest cytotoxicity in terms of inhibition of short-term cell proliferation/viability, membrane damage, and ROS production. Small-sized ZnO nanoparticles, 20 nm, were more toxic than larger sized ones, 70 nm, and moreover, positively charged ZnO nanoparticles caused more marked ROS production compared to negatively charged ones. It seems that positively charged ZnO nanoparticles of small size strongly interact with the negatively charged plasma membrane, and thereby induce more toxicological effects. However, it should be noted here that size- or charge-dependent cytotoxicity was not always found and differed with the assessing methods used, indicating essential necessity of several methodological evaluations to conclude the toxicity of nanoparticles. The result demonstrated that the cytotoxic effects of ZnO nanoparticles are highly related to particle size and surface charge, suggesting an important role of the physicochemical parameters of nanoparticles to understand and possibly minimize their toxicity potential on human.

Acknowledgement

This research was supported by a grant (10182KFDA991) from Korea Food & Drug Administration in 2010.

References

[1] Djurišić A B and Leung Y H 2006 *Small* **2** 944-961
[2] Fan Z and Lu J G 2005 *J. Nanosci. Nanotechnol.* **5** 1561
[3] Serpone N, Dondi D and Albini A 2006 *Inorg. Chim. Acta* **360** 794-802
[4] Chen W and Zhang J 2006 *J. Nanosci. Nanotechnol.* **6** 1159-66
[5] Yang H, Liu C, Yang D, Zhang H and Xi Z 2009 *J. Appl. Toxicol.* **29** 69-78
[6] Xia T, Kovechich M, Liong M, Madler L, Gilbert B, Shi H, Yeh J, Zink J and Nel A 2008 *ACS Nano* **2** 2121-34
[7] Wang B, Feng W, Wang M, Wang T, Gu Y, Zhu M, Oryang H, Shi J, Zhang F and Zhao Y 2008 *J. Nanopart. Res.* **10** 263-276
[8] Sayes C M, Reed K L and Warheit D B 2007 *Toxicol. Sci.* **97** 163-180

[9] Cho S, Jang J W, Jung S H, Lee B R, Oh E and Lee K H 2009 *Langmuir* **25** 3825-31
[10] Muller J, Huaux F, Fonseca A, Nagy J B, Moreau N, Delos M, Raymundo-Piñero E, Beguin F, Kirsch-Volders M, Fenoglio L, Fubini B and Lison D 2008 *Chem. Res. Toxicol.* **21** 1698-1705
[11] Karlsson H L, Cronholm P, Gustafsson J and Moller L 2008 *Chem. Res. Toxicol.* **21** 1726-32
[12] Huang C C, Aronstam R S, Chen D R and Huang Y W 2010 *Toxicol. In Vitro* **24** 45-55
[13] Berardis B D, Civitelli G, Condello M, Lista P, Pozzi R, Arancia G and Meschini S 2010 *Toxicol. Appl. Pharm.* **246** 116-127
[14] Heng B C, Zhao X, Xiong S, Ng K W, Boey F Y-C and Loo J S-C 2010 *Food Chem. Toxicol.* **48** 1762-66

Glutathione and S-nitrosoglutathione in alginate/chitosan nanoparticles: Cytotoxicity

P D Marcato[1], L F Adami[1], P S Melo[2,3], L B de Paula[3], N Durán[1,3] and A B Seabra[4]

[1]Instituto de Química, Universidade Estadual de Campinas, Campinas, SP, Brazil

[2]Metrocamp, Campinas, SP, Brazil

[3]Departamento de Bioquímica, Universidade Estadual de Campinas, SP, Brazil

[3]Centro de Ciências Naturais e Humanas, Universidade Federal do ABC, Santo André, Brazil

[4]Departamento de Ciências Exatas e da Terra, Universidade Federal de São Paulo, Diadema, SP, Brazil

E-mail: amedea.seabra@gmail.com

Abstract. Nitric oxide (NO) is involved in several physiological processes, such as the control of vascular tone, the immune response and the wound healing process. Thus, there is a great interest in the development of NO-releasing drugs and in matrices which are able to stabilize and release NO locally in different tissues. Thiols, such as glutathione (GSH), are ready nitrosated to form the NO donors S-nitrosothiols (RSNOs). In this work, GSH, a precursor of the NO donor S-nitrosoglutathione (GSNO), was encapsulated into a mucoadhesive combination of alginate/chitosan nanoparticles. The encapsulated GSH was nitrosated in the alginate/chitosan nanoparticles by adding sodium nitrite, leading to the formation of encapsulated GSNO. The cytotoxicity characterization of the nanoparticles containing either GSH or GSNO showed that these materials were completely non cytotoxic to cellular viability. These results show that this novel nanostructure biomaterial has a great potential to be use in biomedical applications where NO has a therapeutical effect.

1. Introduction

Nitric oxide (NO) is an endogenous free radical that plays a key role in the regulation of several physiological processes, such as the control of vasodilation, platelet interaction with the vessel wall, immune responses, wound healing and cellular communication [1-5]. The endogenous synthesis of NO is mediated through nitric oxide synthase (NOS) which catalyzes the conversion of *L*-arginine to NO and citrulline. At low concentrations, NO is quite stable and minimally reactive, however, at higher concentrations NO can interact with transition metals, heme-containing proteins, thiol groups leading to oxidation of functional groups on RNA, DNA and proteins [6]. The biological actions of NO have been shown to be dependent on its site and source of production, as well as its concentration.

Due to the multifaceted role of NO *in vivo*, there is a great interest on the developing of NO releasing molecular systems for biomedical applications, especially to treat important diseases [7]. In this context, many different classes of NO donors have been prepared and applied in biological systems [8]. Some studies suggested that cysteine thiols (RSHs) of bioactive peptides such as glutathione (GSH), the most abundant non-protein thiol found *in vivo*, act as NO carriers and donors in tissues [9]. Thiols are directly involved in the metabolism and mobilization of NO, and GSH forms S-nitrosoglutathione (GSNO) *in vivo* [10], which belongs to a class of compounds named S-nitrosothiols (RSNOs). GSNO and others RSNOs exhibit NO-like activities such as the inhibition of platelet adhesion, vasodilation, microbicidal actions, and wound healing [1,2,4,5,11,12]. However, the potential pharmacological applications of RSNOs, and in particular GSNO, require a vehicle since RSNOs are considered unstable in aqueous solution [9].

A promising approach that has been considered to promote the uses of RSNOs as NO donors in biomedical applications is either the direct incorporation of NO donors, such as RSNOs, into biopolymers or nanoparticles [7,13-15]. In both cases, NO can be released directed to the target site for prolonged periods of time and exerts its therapeutic effects, such as the promotion of cutaneous wound healing and dermal vasodilation, upon application *in vivo* [1,2,4,5].

However, a crucial issue with regards to propose the clinical uses of NO-releasing materials is the evaluation of the cytotoxicity of the NO donor molecules (RSNOs) and RSNOs-loaded to the polymers. It was reported that GSNO decreased neutrophil viability in a time and concentration-dependent manner, while GSH exposure alone did not enhance cell death or DNA fragmentation at any concentration studied, as assessed by a fluorescent viability/cytotoxicity assay [16].

In this work the tri-peptide GSH, precursor of the NO donor GSNO, was encapsulated into a mucoadhesive combination of alginate/chitosan. GSH-encapsulated alginate/chitosan nanoparticles were characterized, and the thiol was nitrosated to form encapsulated GSNO. The kinetics of NO release from GSNO-containing alginate/chitosan nanoparticles were monitored at physiological temperature, and compared to free GSNO. Finally, the cytotoxicity of either GSH- or GSNO-encapsulated were evaluated. The results showed that encapsulated GSH can be nitrosated to form GSNO-containing nanoparticles. The results suggest that this formulation is a potent candidate for pharmacological applications that require the release of controlled amounts of NO for prolonged periods of time, where NO can have its physiological actions, without severe side effects.

2. Methods

2.1. Preparation of polymeric nanoparticles with GSH

The alginate/chitosan nanoparticles containing GSH were prepared thorough ionic gelation method [17]. Briefly, it was prepared an aqueous chitosan solution (1.33%) with GSH. A ratio of alginate/chitosan solutions of 0.75 was obtained by dropping wise previous prepared chitosan aqueous solution into the alginate aqueous solution (0.005 g/100 mL) (pH 4.0), under magnetic stirring.

2.2. Characterization, encapsulation efficiency and physical stability of the alginate/chitosan nanoparticles

The average particle size (number average size) and size distribution of prepared nanoparticles as the zeta potential of alginate/chitosan ratio (0.75) were measured by photon correlation spectroscopy (PCS) (Nano ZS Zetasizer, Malvern Instruments Corp). The encapsulation efficiency of GSH in nanoparticles was determined by DTNB titration after ultrafiltration of the nanoparticles solution [12]
The physical stability alginate/chitosan nanoparticles at 0.75 ratio was monitored over 37 days at 4°C, by measuring the average particles sizes and surface charges, as described previously.

2.3. Nitrosation of GSH encapsulated in alginate/chitosan nanoparticles

GSH (400 μM) encapsulated in alginate/chitosan nanoparticles at ratio 0.75 and free GSH (400 μM) were nitrosated by adding equimolar amounts of sodium nitrite directly into polymeric aqueous solution (pH 4.0) at 336 nm (ε = 940 mol/L/cm), using an UV-Vis spectrophotometer [12].

2.4. Kinetic measurements of NO release from encapsulated GSNO and free GSNO

The kinetic curves of GSNO (400 μM) decomposition with NO release were monitored for free GSNO (GSNO in aqueous solution) and encapsulated GSNO in alginate/chitosan nanoparticles from the absorption changes at 336 nm.

2.5. Cytotoxicity of GSH and GSNO free and encapsulated in nanoparticles

The cytotoxicity of GSH and GSNO, free and encapsulated in alginate/chitosan nanoparticles, was assessed in a permanent lung fibroblast cell line (V79) culture derived from Chinese hamster [18]. These cells are commonly used for cytotoxicity studies [19]. The cells were plated at a density of 3×10^4 cells/ml in 96-well plates. Forty-eight hours after cell seeding, semiconfluent cultures were exposed to free GSH, free GSNO, encapsulated GSH and encapsulated GSNO, at different concentrations (0 – 18 μM). The cells were exposed for 24 h to the test medium with or without the compounds studied (control). Each concentration was tested in six replicates in each of three separate experiments. At the end of the incubation, two independent endpoints tests for cytotoxicity (Methylthiazoletetrazolium reduction- MTT and neutral red uptake were evaluated [20,21].

3. Results and Discussion

3.1. Preparation and characterization of GSH-containing alginate/chitosan nanoparticles

Preparation of alginate/chitosan nanoparticles at ratio of 0.75, with and without GSH, showed average particles sizes of 361 and 301 nm, respectively, and positive surface charges with average values of +27 (with GSH) and + 22 mV (without GSH) (Table 1). Alginate/chitosan nanoparticles prepared showed the encapsulation efficiency of 27% (Table 1). These positively charged particles exhibited high stability after 37 days.

Table 1. Particles sizes, surface charge (Zeta potential), and encapsulation efficiency of alginate/chitosan nanoparticles.

Alginate/Chitosan ratios	GSH (mg)	Zeta Potential (mV)	Particle sizes (nm)	PdI	Encapsulation efficiency (%)
0.75	0	+22	301	0.316	---
0.75	5	+27	361	0.330	27

3.2. Nitrosation of GSH-encapsulated in alginate/chitosan nanoparticles

GSH encapsulated in nanoparticles of alginate/chitosan was directly nitrosated by reacting with equimolar amount of sodium nitrite at pH 4.0. The formation of GSNO was confirmed by the detection of the characteristic GSNO absorption band at 336 nm ($\pi \rightarrow \pi^*$ transition) (data not shown). Thus, a NO-donor molecule (GSNO) could be prepared directly in alginate/chitosan suspension by directly nitrosation of encapsulated GSH in the nanoparticles.

3.3. Kinetics of GSNO decomposition with NO release from free GSNO and encapsulated GSNO in polymeric nanoparticles

The kinetic data of GSNO decomposition with NO release at 37°C for free GSNO, which corresponds to aqueous solution of GSNO (400 μM) at pH 4.0 and GSNO (400 μM) encapsulated in alginate/chitosan nanoparticles (alginate/chitosan ratio 0.75) at pH 4.0 were analyzed. The decrease of the intensity of the absorption band at 336 nm is assigned to the homolytic cleavage of the S-N bond, with GSNO decomposition leading to the release of NO and the formation of oxidized glutathione [9]. Either free GSNO or encapsulated were able to spontaneously decomposes, releasing NO, under physiological temperature, for hours. However, for the first 3 h of the kinetic, the GSNO encapsulate in the nanoparticles can be considered stable (no change in the absorption at 336 nm). Only after 3 h of the kinetic, it was observed an increase in the rate of decomposition of encapsulated GSNO, with NO release. On the other side, the rate of free GSNO decomposition was observed to be continuously and linear from time zero to 7 h. Indeed, the rate of encapsulated GSNO decomposition was found to be $1.2 \times 10^{-2} \pm 0.2 \times 10^{-2}\ h^{-1}$ for 3 to 7 h. On the other hand, the rate of free GSNO decomposition was found to be $1.5 \times 10^{-2} \pm 0.2 \times 10^{-2}\ h^{-1}$ for 0 to 7 h. Therefore, encapsulation of GSNO in alginate/chitosan nanoparticles was responsible to decrease the rate of NO release, especially during the initial hours. This delay on the NO release may be due to the low diffusion rates of GSNO from the nanoparticles to the solution.

3.4. Cytotoxicity of free GSH, free GSNO and GSNO and GSH encapsulated in alginate/chitosan nanoparticles

Free glutathione showed to be non cytotoxic to fibroblast V79 cells in the lysosomal (Neutral red) and slightly cytotoxic to mitochondrial (MTT) assays (15-20% around to 10 μM concentration), as is observed in Figure 1A. In the Figure 1B is shown that cytotoxicity in the case of GSNO was also not cytotoxic to fibroblast V79 cells in the lysosomal (Neutral red) and slightly cytotoxic to mitochondrial (MTT) assays (20% around to 16 μM concentration), as is observed in Figure 1B.

Alginate/chitosan nanoparticles with or without GSH or GSNO exhibited non cytotoxicity (till 18 μM concentration).

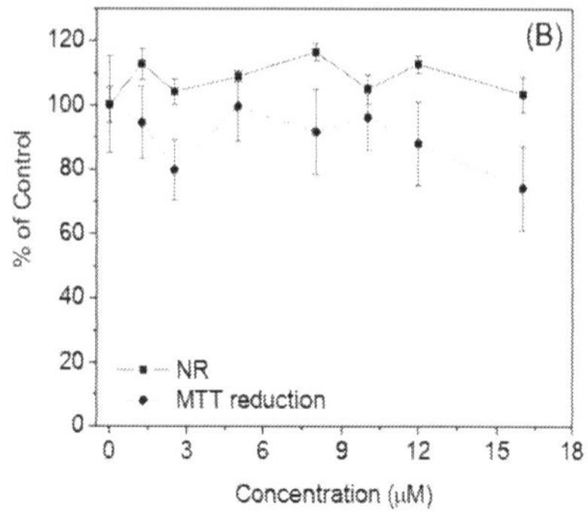

Figure 1. Viability graph in V79 cells with the cellular targets as MTT and NR of the free GSH (A) and free GSNO (B).

4. Conclusions

The kinetic of NO release from free GSNO compared with GSNO-encapsulate in alginate/chitosan reveled that the encapsulation of the NO donor resulted in a thermal and hydrolytic protection of GSNO leading to an enhancement of NO activity by a controlled and sustained NO release from this new biomaterial. The higher stability of encapsulated GSNO, compared to free GSNO enable the use of GSNO in pharmaceutical applications, such as wound healing, or others uses where the biological effects of NO are required.

Acknowledgements

Support from Brazilian Network of Nanobiotechnology, and Nanocosmetics (MCT/CNPq) and FAPESP (process number 2010/14949-4) are acknowledged.

References

[1] Seabra A B, Fitzpatrick A, Paul J, De Oliveira M G and Weller R 2004 *Br. J. Dermatol.* **151**, 977

[2] Seabra A B, Pankotai E, Fehér M, Somlai A, Kiss L, Bíró L, Szabó C, Kollai M, de Oliveira M G and Lacza Z 2007 *Br. J. Dermatol.* **156**, 814

[3] Moore C, Tymvios C and Emerson M 2010 *Thromb. Haemost.* **104**, 342

[4] Amadeu T P, Seabra A B, de Oliveira M G and Costa A M A 2007 *J. Eur. Acad. Dermatol. Venereol.* **21**, 629

[5] Amadeu T P, Seabra A B, de Oliveira M G and Costa A M A 2008 *J. Surg Res.* **149**, 84

[6] Nahrevanian H and Amini M 2008 *Iranian J. Basic Med. Sci.* **11**, 197

[7] Seabra A B and Durán N 2010 *J. Mater. Chem.* **20**, 1624

[8] Kapadia M R, Chow L W, Tsihlis N D, Ahanchi S S, Eng J W, Murar J, Martinez J, Popowich D A, Jiang Q, Hrabie J A, Saavedra J E, Keefer L K, Hulvat J F, Stupp S I and Kibbe M R 2008 *J. Vasc. Surg.* **47**, 173.

[9] de Oliveira M G, Shishido S M, Seabra A B and Morgon M H 2002 *J. Phys. Chem.* **106**, 8963.

[10] Hogg N, Singh R J and Kalyanaraman B 1996 *FEBS Lett.* **382**, 223

[11] Seabra A B, da Silva R, de Souza G F P and de Oliveira M G 2008 *Artif. Organs* **32,** 262

[12] Seabra A B, Martins D, Simões M M S G, da Silva R, Brocchi M and de Oliveira M G 2010 *Artif. Organs* **34,** E204

[13] Seabra A B, de Souza G F P, da Rocha L L, Eberlin M N, de Oliveira M G 2004 *Nitric Oxide* **11**, 263.

[14] Seabra A B, da Silva R and de Oliveira M G 2005 *Biomacromolecules* **6,** 2512

[15] Frost M C, Reynolds M and Meyerhoff M E 2005 *Biomaterials* **26**, 1685

[16] Fortenberry J D, Owens M L and Brown L A S 1999 *Am. J. Physiol. Lung Cell Mol. Physiol.* **276**, 435

[17] Douglas K L, Piccirillo C A and Tabrizian M 2006 *J. Control. Release* **115**, 354

[18] Correa D H A, Melo P S, de Carvalho C A A, de Azevedo M B M, Durán N and Haun M 2005 *Eur. J. Pharmacol.* **510**, 17

[19] de Conti R, Oliveira D A, Fernandes A M A P, Melo P S, Rodriguez J A, Haun M, Castro S L, Souza-Brito A R M and Durán N 1998 *In vitro Mol. Toxicol* **11**, 153

[20] Borefreund E and Puerner J A 1984 *J. Tissue Cult. Methods* **9**, 7

[21] Denizot F and Lang R 1986 *J. Immunol. Methods,* **89**, 271

Journal of Physics: Conference Series **304** (2011) 012046 doi:10.1088/1742-6596/304/1/012046

Cytotoxicity effects of metal oxide nanoparticles in human tumor cell lines

T Lozano[1], M Rey[1], E Rojas[2], S Moya[2], J Fleddermann[3], I Estrela-Lopis[3], E Donath[3], B Wang[4], Z Mao[4], C Gao[4] and África González-Fernández[1]

[1] Immunology Area, Biomedical Research Center, University of Vigo, Campus Lagoas Marcosende, 36310 Vigo, Spain

[2] CIC biomaGUNE, Paseo Miramón, 182 Edificio Empresarial C, E-20009 San Sebastián, Gipuzkoa, Spain

[3] Institute of Medical Physics and Biophysics, University of Leipzig, Härtelstrasse 16-18, D-04107 Leipzig, Germany

[4] Department of Polymer Science and Engineering, Zhejiang University, Hangzhou 310027, China

E-mail: africa@uvigo.es

Abstract: Metallic and metal oxide nanoparticles (Nps) have a wide range of applications in various settings including household, cosmetics and chemical industries, as well as for coatings. Nevertheless, an in-depth study of the potential toxic effects of these Nps is still needed, in order to fulfill the mandatory requirement of ensuring the safety of workers, patients and the general public. In this study, Quick Cell colorimetric assays were used to evaluate the *in vitro* toxicity of different metal oxide Nps [Fe(II,III)O_x, TiO_x, ZnO and CeO_2] in several cell lines. The ZnO Nps were found to be highly toxic, with a lethal dose ≥100 μg/ml for all the cell lines studied. Western blot was also used to test the ability of the different Nps to activate the complement pathway. However, no activation of this cascade was observed when the Nps were added. In addition, the aggregation state and charge of the Nps in culture media was studied by dynamic light scattering (DLS) and measurement of zeta potential. Transmission Electron Microscopy was used to analyze Np uptake and localization at the cellular level.

1. Introduction

The term nanoparticle (Np) applies to particles between 1 and 100 nm in at least two dimensions [1]. These particles have specific physicochemical properties that are not shown in the bulk form [2], most of the unique properties of Nps being defined by their high ratio of surface to volume, which implies that almost all the material is at the surface, and for the presence of quantum confining phenomena at the nanoscale. Nps offer, compared to bulk materials, a larger surface for adsorption, and sometimes a higher reactivity, interfering as catalizators in many processes [3]. The huge potential of Nps for various applications makes the in-depth analysis of their potential toxicity in humans essential. In this context, a new discipline called Nanotoxicology has arised, a branch of the toxicology and of the nanotechnology dealing with the interaction of nanomaterials, nanostructures and devices with biological molecules and organisms, aiming to understand possible toxicological side-effects of the

Nps [2]. One of the scopes of nanotoxicology is the evaluation of the safety of Nps for industrial applications, for providing information about the undesirable effects of the Nps, and for developing the tools to prevent such effects.

Two main mechanisms might be responsible for an eventual toxicity of metallic and metal oxide Nps. First, the intrinsic catalytic activity of such Nps can disturb several processes and intracellular signaling pathways. Second, ions can also be released from the Np, affecting the finely regulated concentration of metallic ions inside the cell. The Nps could induce toxicity in several organs, even leading to systemic toxicity [1]. They could also affect the immune system, activating or inhibiting it, by inducing allergic responses or hypersensitivity, the generation of antibodies against the Nps or their coatings, or by activating cells from the endothelial reticulum system [4].

In this study, using different Np concentrations in several cell lines, *in vitro* cytotoxicity analyses and endocytosis assays were undertaken to evaluate the potential toxicity of metal oxide Nps.

2. Materials and methods

2.1. Nps
The Fe(II,III)O_x and TiO_x Nps were supplied by PlasmaChem (Berlin, Germany), and the ZnO and CeO_2 Nps by Evonik Degussa (GmbH, Germany) (Fig.1). The concentration of Fe(II,III)O_x Nps was 72 mg/ml, and for the remaining Nps, stock solutions of 72 mg/ml in PBS were prepared. The sterility of the Nps was preserved in all cases.

2.2. Cells
The toxicity assays were performed in the human tumoral cell lines A549, NCI-H460, SK-MES-1 and HeLa, all purchased from ATCC (American Type Culture Collection). All cell types were cultured in RPMI medium (GIBCO Invitrogen Corp., Grand Island, NY), supplemented with 10% FBS, at 37°C, 5% CO_2.

2.3. Aggregation studies
For evaluating the behavior of the Nps in the physiological medium RPMI with 10% FBS, a 24-well plate was used, in which aggregation was assessed for the sonicated or unsonicated Nps. The 20 min sonication was carried out in a Branson ultrasound bath (Branson 1510, Danbury, CT), at low frequency (47 kHz), preventing Np exposure to potential contaminating agents. For these assays, the final concentration of the Nps was 8 μg/ml, in 1 ml final volume. The Nps were incubated overnight at 37°C and the formation of aggregates was evaluated by optical microscopy in an inverted microscopy model IX50 from Olympus with 20× and 40× objectives (Olympus Optical Co, GMBH, Germany).

2.4. Dynamic light scattering (DLS)
A Malvern NanoZS device(UK) was used for DLS measurements.

2.5. Cellular proliferation colorimetric assay
To measure the effect of the Nps on cellular viability, the colorimetric kit Quick Cell Proliferation Testing Solution (GenScript Corporation, Piscataway, NJ, USA) was used following the instructions of the manufacturer. Once the optimal cell number for each cell line was determined, the viability assay was performed including the Nps at three different concentrations, 0.5 μg/ml, 50 μg/ml and 1 mg/ml, respectively. Briefly, cells were incubated with 200μl RPMI 10% FBS, both in the absence and in the presence of Nps, for 24 and 48 h. The plates were then centrifuged at 1000 g for 1 min, 100 μl of the supernatant were discarded and 50 μl of Quick Cell reagent added. Subsequently, the plates were incubated for 4h, centrifuged again and the supernatants were transferred to clean plates, to avoid possible interference due to the Nps. Finally, the absorbance was measured at 450 nm in an Envision multidetector (Perkin Elmer Inc., Norwalk, Connecticut, USA). As a positive toxicity control, cells were incubated with 5% Triton X-100 (Sigma-Aldrich, Steinheim, Germany), which was

removed 2 hours before 24/48 h; in wells used as a control of cellular death, 100 μl of medium was added. As a negative control, RPMI and Nps alone were used. The results were then analyzed by the following equation:

% cellular viability= Abs Cells+Nps-Abs Nps/Abs Cells-Abs RPMI×100

For measuring the lethal dose of the ZnO Nps, the same assay was performed, incubating the cells with intermediate concentrations of these Nps, ranging from 0.5 up to 500 μg/ml.

2.6. Analysis of apoptosis and necrosis

The FITC Annexin V Apoptosis Detection Kit I from BD Biosciences (San Diego, USA) was used according to the manufacturer's instructions. Briefly, cells cultivated in exposure media were stained with Annexin V-FITC, which binds to phosphatidylserine at the outer cell membrane and can, therefore, be used as a marker to detect apoptosis. Propidiumiodide (PI) was used to quantify necrotic cells. The dye stains DNA after the membrane becomes permeable for PI. Stained cells were then analyzed by flow cytometry (FACSCalibur, Becton Dickinson, Mountain View, USA) at an excitation wavelength of 488 nm. FITC and PI fluorescence were detected in the green and red fluorescence channels, respectively.

2.7. Complement activation

Western blot with an anti-C3 antibody was carried out to analyze the degree of degradation of this factor upon Np addition. A pool of human sera from healthy donors was incubated with two different concentrations of Fe(II,III)O_x, TiO_x, CeO_2 and ZnO Nps (0.5 and 50μg/ml). Cobra venom (CVF, Quidel Corporation, San Diego, CA, USA), and PBS were used as positive and negative controls, respectively. The membrane was revealed with an antibody specific for C3b from Abcam (Cambridge, UK).

2.8. Scanning electron microscopy

Cells were prefixed in 2.5% glutaraldehyde/0.1 mol/L sodium cacodylate (pH 7.4). After postfixation in 1% osmiun tetraoxide (in 0.2 mol/l cacodylate buffer), cells were dehydrated in a series of increasing ethanol concentrations and at the critical point dried using carbon dioxide. After coating with gold, cells were examined with a JEOL JSM-6700F scanning electron microscope.

3. Results and discussion

Metal oxide Nps are widely employed in creams, implants, drug carriers or as contrast agents [5], but for the potential use of these Nps in biomedicine strict toxicity rules should be followed. The Nps should also be stable in physiological conditions, not forming aggregates that could obtrude the capillaries. To test aggregation, we immersed the Nps in physiological medium and found that all Nps were aggregated in these conditions (not shown), with the smallest aggregates corresponding to the ZnO Nps. To disperse the NPs these were sonicated, but this proved to be only partially successful (Figure 1a). To further address the aggregation status of the Nps, dynamic light scattering (DLS) experiments were carried out, dispersing the particles in different media, with some Nps being sonicated, while others were left unsonicated. Concurring with the results of previous studies, a strong aggregation effect was observed in all the Nps analyzed (Figure 1b and not shown). This clearly represents an important limitation for the *in vivo* use of these Nps.

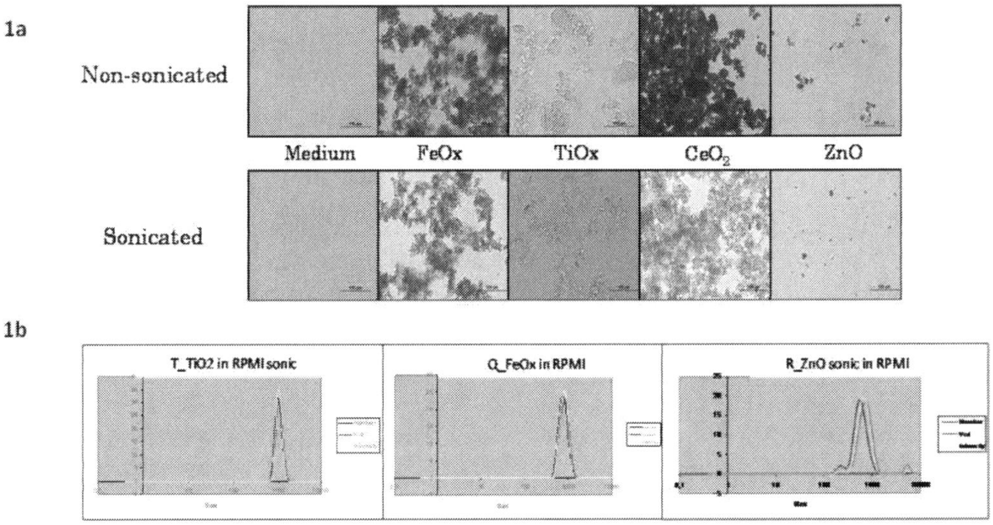

Figure 1. Aggregation of the Nps. (1a) Optical microscopy images depicting the aggregation of the Nps before and after sonication. The Nps were used at 8-μg/ml concentration and incubated overnight. Bar: 100 μm. (1b) Aggregation status of TiO_x, FeO_x and ZnO Nps measured by DLS. All the Nps show a tendency to form aggregates in RPMI medium.

The use of *in vitro* assays of cellular viability in different cell lines is essential for evaluating the potential toxicity of the Nps. In this study, we employed toxicity studies based on colorimetric methods. For an Np to be considered non-toxic, we followed the criteria of the Nanotechnology Characterization Laboratory (NCL) in Frederick (Maryland, USA), especially that cell viability at 48 h should be higher than 75%. Using the Quick Cell assay, it was seen that ZnO Nps induced a massive toxicity in all the cell lines studied at 24 and 48 hours (Figures 2 a-d), in agreement with previous studies by other research groups [6-8]. This could be due to either the presence of small aggregates formed by ZnO Nps, which could have greater interaction with the cell causing its death, or to the release of Zn^{+2} ions from the Nps, which are also toxic.

These assays were optimized, as cellular metabolism varies significantly among cell lines and viability can be affected by factors, such as cell density, percentage of living cells with respect to dead ones, and different proliferation rates. Once the optimal amount of cells to be analyzed for all the cell lines studied was determined (data not shown), we were able to carry out dose-response viability assays with these Nps, and found that the lethal dose 50 (LD 50), able to cause the mortality of at least 50% of the cells, was ≥ 100 mg/ml for all the cell types studied (Figures 3a-d).

In addition, the population of necrotic and apoptotic A549 cells exposed to ZnO Nps, and stained with annexin V-FITC/PI was analyzed by flow cytometry. The percentage of viable and dead cells (apoptotic, necrotic and late apoptotic cells) as a function of exposure time and concentration are shown in Figure 3e. With higher Np concentrations and longer exposure times the number of dead cells increased, mainly late apoptotic and apoptotic ones, with the latter only seen after 24-h exposure time. Although 3 μg/ml ZnO Nps only had a small impact on A549 cells, 30 and 100 μg/ml ZnO Nps considerably reduced cell viability. Even after 12-h exposure at Np concentrations of 30 and 100 μg/ml, treated cells did not match the "Acceptance Criteria". Yet, ZnO concentrations of 3 μg/ml are considered non-toxic for A549 cells, even after 72-h treatment with the Nps.

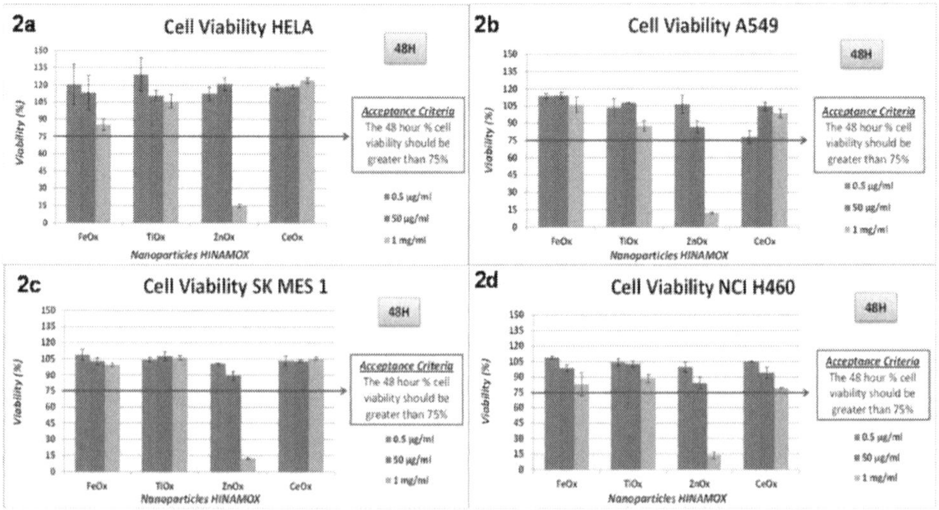

Figure 2. Effect of the different Nps on cell viability. Graphs showing the effect on cell viability of the different Nps, by Quick cell colorimetric assay, at 48 h, in different cell types: HeLa (2a), A549 (2b), SK-MES-1 (2c) and NCI-H460 (2d). Nps were considered non-toxic if at 48h cell viability reached at least 75%. All the assays were performed twice for each cell line, at 24 and 48 h, and in triplicate for each Np concentration.

Another important limitation to be taken into account regarding the potential *in vivo* use of Nps is that they are quickly removed from the circulation by the endothelial reticulum system, which would make the use of these compounds as drug carriers or contrast agents more difficult. Nevertheless, covering the Nps with agents such as polyethilenglycol can improve their permanence in the circulation, thereby avoiding recognition by the cells of the endothelial reticulum system [9]. In our study, transmission electronic microscopy was used to evaluate the intracellular localization of TiO_x Nps (Figure 4). After 6-h incubation, a large amount of particles adhered to the cell membrane, and the cells appeared to have ingested some of these particles (Figure 4c). Most of the particles seemed to be confined inside vesicles distributed across the cytoplasm, not crossing into the nucleus. At 24-h incubation time with the Nps, particles were also seen to be attached to the cell membrane, suggesting that the cells had ingested a relatively large amount of these compounds. The vesicles moved towards the nuclear membrane, but few had penetrated inside the nucleus (Figure 4d).

Possible activation of the complement, a key component of the innate immune system consisting of a cascade of more than 30 factors, should also be taken into consideration before *in vivo* use of Nps. It is well established that, *in vitro*, many types of particle are able to activate the alternative pathway of the complement [10]. In our study, western blot was used to evaluate the ability of the Fe(II,III)O_x, ZnO, CeO_2 and TiO_x Nps to activate the complement. The use of a specific antibody against the C3 factor makes it possible to evaluate the degradation of this protein, which corresponds to activation of the complement. It was seen that none of the Nps assessed was able to induce complement activation (Figure 5), as the C3 factor remained largely intact, with basal degradation similar to that of the negative control.

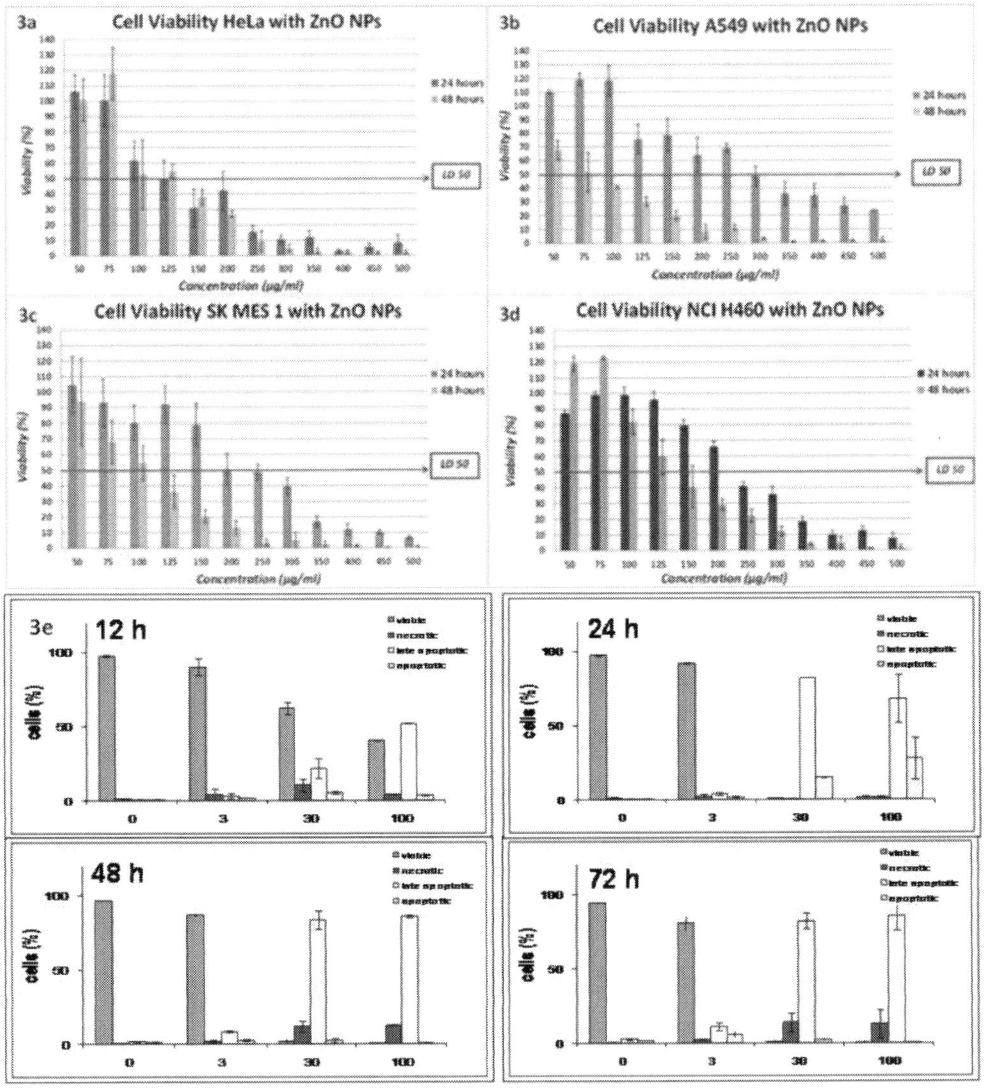

Figure 3. Determination of the lethal dose 50 (LD 50) for ZnO Nps in the different cell lines. Effect on cell apoptosis and necrosis of ZnO Nps: HeLa (3a), A549 (3b), SK-MES-1 (3c) and NCI- H460 (3d) by Quick cell colorimetric assay, at 24 and 48 h, using increasing concentrations of these Nps, as indicated. The LD 50 was reached at ZnO Np concentrations ≥ 100 μg/ml for all the cell types. (3e) Impact of ZnO nanoparticles on the viability of A549 cells stained with annexin V-FITC/PI and measured by flow cytometry. Cells were incubated with the indicated Np concentrations, ranging from 3 to 100 μg/ml, at different time points (12, 24, 48 and 72 h). Fractions of viable, apoptotic, late apoptotic and necrotic cells (%) were determined.

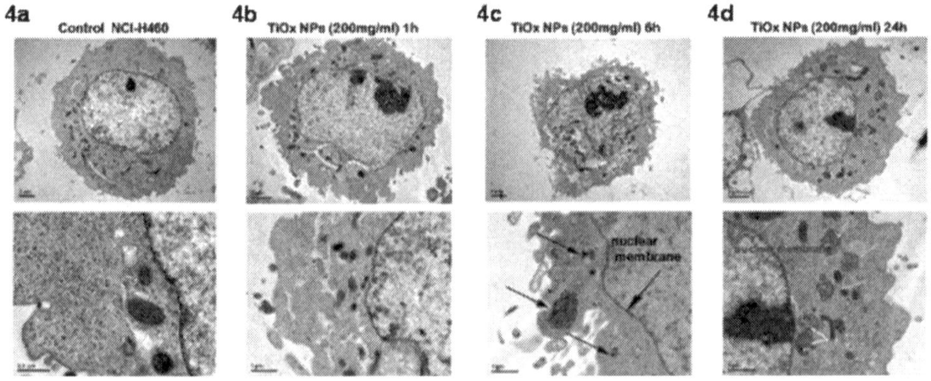

Figure 4. Subcellular distribution of TiO_x Nps in NCI-H460 cells. Transmission electron microscopy (TEM) images of microtomed sections of HCI-H460 cells untreated (4a) or incubated with 200 μg/ml TiO_x nanoparticles at different time points, from 1 up to 24 hours (4b-d). The small black dot and the large black regions correspond respectively to glycogen and mitochondria. Almost no particles were found around or inside the cells after 1-h Np incubation (4b). After 6-h incubation, a large amount of particles adhered to the cell membrane (left), and the cells seem to have ingested some particles (4c). At higher magnification, many particles can be seen attached to the membrane, observing clusters with the indented cell membrane in the process of endocytosis (top), and endosomes filled with particles. Most of the Nps seem to be confined inside the vesicles distributed across the cytoplasm, not crossing into the nucleus (4c). At 24-h incubation time, particles are also observed attached to the cell membrane, suggesting that the cells have ingested a relatively large amount of particles, and two of the vesicles close to the nucleus seem to be in the process of fusing together. The vesicles moved towards the nuclear membrane, but few have penetrated inside the nucleus (4d).
Bars: 2, 1 and 0.5 μm.

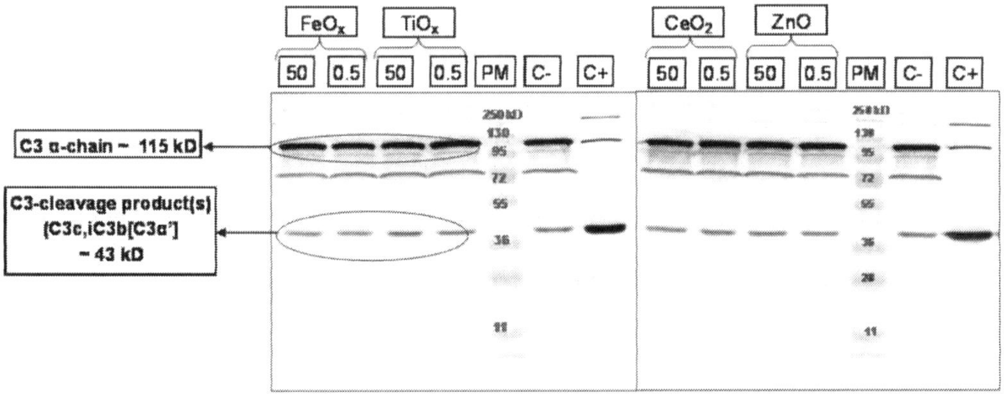

Figure 5. The complement system is not activated by the Nps. The degradation of the complement factor C3 was analyzed by Western blot and revealed with an anti-C3 antibody. A pool of sera from different healthy donors was incubated with the Nps indicated, at 0.5 and 50μg/ml, with cobra venom as a positive control of C3 degradation or with PBS as negative control. The bands of 115 kD and 43 kD correspond to the intact C3 factor and main degradation products of this protein, respectively.

4. Conclusion

In summary, taking into account the information available and the data obtained in this study, the Nps analyzed show a tendency to form aggregates in media containing serum. In addition, the ZnO Nps were found to be toxic, with a lethal dose ≥100 μg/ml for all the cell lines studied. The amount of dead cells increased with rising Np concentration and exposure time, but an increase of apoptotic cells was seen only after 24h. Moreover, most of the Nps seemed to be confined inside the vesicles distributed across the cytoplasm, not crossing into the nucleus and none of the Nps assessed were able to induce complement activation.

Further in-depth research of these aspects is, therefore, still required to determine the potential toxic effects on human health.

Acknowledgements

Authors would like to acknowledge PlasmaChem GmbH for the samples of iron oxide and titanium oxide nanopowders. This work was performed under financial support of EU Project HINAMOX NMP4-SL-2009-228825.

References

[1] Hussain, S.M., Toxicity Evaluation for Safe Use of Nanomaterials:Recent Achievements and Technical Challenges. *Advanced Materials*, 2009. **21**: p. 1549-1559.

[2] Arruebo, M., M. Valladares, and Á. González-Fernández, Antibody-Conjugated Nanoparticles for Biomedical Applications. *Journal of Nanomaterials*, 2009.

[3] Borm, P.J. and W. Kreyling, Toxicological hazards of inhaled nanoparticles--potential implications for drug delivery. *J Nanosci Nanotechnol*, 2004. **4**: p. 521-31.

[4] Hu, X., et al., In vitro evaluation of cytotoxicity of engineered metal oxide nanoparticles. *Sci Total Environ*, 2009. **407**(8): p. 3070-2.

[5] Moghimi, S.M., A.C. Hunter, and J.C. Murray, Nanomedicine: current status and future prospects. *FASEB J*, 2005. **19**(3): p. 311-30.

[6] Heng, B.C., et al., Toxicity of zinc oxide (ZnO) nanoparticles on human bronchial epithelial cells (BEAS-2B) is accentuated by oxidative stress. *Food Chem Toxicol*, 2010. **48**(6): p. 1762-6.

[7] Moos, P.J., et al., ZnO particulate matter requires cell contact for toxicity in human colon cancer cells. *Chem Res Toxicol*, 2010. **23**(4): p. 733-9.

[8] De Berardis, B., et al., Exposure to ZnO nanoparticles induces oxidative stress and cytotoxicity in human colon carcinoma cells. *Toxicol Appl Pharmacol.*, 2010. In press.

[9] Zahr, A.S., C.A. Davis, and M.V. Pishko, Macrophage uptake of core-shell nanoparticles surface modified with poly(ethylene glycol). *Langmuir*, 2006. **22**(19): p. 8178-85.

[10] Wilson, M.R., Gaumer, H.R., Salvaggio, J.E., Activation of the alternative complement pathway and generation of chemotactic factors by asbestos. Journal of Allergy and Clinical Immunology, 1977. 60(4): p. 218-222.

Physico-chemical characteristics and cyto-genotoxic potential of ZnO and TiO_2 nanoparticles on human colon carcinoma cells

F Barone[1], B De Berardis[2], L Bizzarri[1], P Degan[3], C Andreoli[1], A Zijno[1] and I De Angelis[1]

[1]Department of Environment and Primary Prevention, Istituto Superiore di Sanità, Viale Regina Elena 299, 00161 Rome, Italy

[2]Department of Technology and Health, Istituto Superiore di Sanità, Viale Regina Elena 299, 00161 Rome, Italy

[3]Molecular Mutagenesis and DNA Repair, Istituto Nazionale per la Ricerca sul Cancro, L.go R. Benzi 10, 16132 Genova, Italy

E-mail: barone@iss.it

Abstract The aim of the present study is to investigate the role of the physico-chemical properties of ZnO and TiO_2 NPs in the potential cytotoxicity, genotoxicity and oxidative DNA damage induction on Caco-2 cell line. As negative control, fine TiO_2 particles were used. The characterization of particles was carried out by electron microscopy (SEM, TEM) using a Soft Imaging System. To evaluate the effects of ZnO and TiO_2 NPs induced on Caco-2 viability, Neutral Red assay was performed after treatment with different particle concentrations. Our results showed a significant dose and time dependent effect after treatment with ZnO NPs. On the contrary, no effect was observed on Caco-2 cells exposed to TiO_2 particles either in micro- and in nano-size. The role of surface in the cytotoxicity induced on Caco-2 was also considered. The levels of DNA 8-oxodG, as the main marker of oxidative DNA damage, were measured by high-performance liquid chromatography with electrochemical detection (HPLC/EC). A significant increase in the 8-oxodG levels was observed after 6 h exposure for both NPs. The estimation of the potential genotoxicity of the two NPs is ongoing by the cytokinesis-block micronucleus assay. Our preliminary results showed that a slight micronucleus increase in binucleated cells was detected in the dose range applied only for ZnO.

1. Introduction

The unique nature of nanoparticles (NPs) expressed in their size, shape and surface characteristics is expected to play an important role in their possible toxicological effects [1,2].

Moreover, in the last years several studies indicated that NPs may interfere with cellular system by the interaction with proteins, DNA, lipids, membranes, organelles and biological fluids. Although the mechanisms underlying the NPs toxicity are not yet to be elucidated, it has been suggested that oxidative stress and lipid peroxidation play an important role in DNA damage, cell membrane disruption and cell death [3-5].

Zinc oxide (ZnO) and titanium dioxide (TiO_2) nanoparticles are widely used in sunscreens because of their less scattering visible light ability. Many studies have been performed on cellular lines representative of dermal and pulmonary exposure [6,7], while few data are available on gastrointestinal exposure [8].

The aim of the present study is to investigate the ZnO and TiO_2 NP potential cytotoxicity, genotoxicity and oxidative DNA damage induction on Caco-2 cell line, a well known *in vitro* intestinal model. As negative control, fine TiO_2 particles were used. Moreover, the role of the physico-chemical characteristics of these particles on the observed toxic effects has been considered.

2. Materials and methods

2.1. Particles

ZnO NPs, TiO_2 anatase NPs e TiO_2 fine were obtained from Sigma Aldrich Company LTd. The nominal sizes of particles as purchased are 50-70 nm for ZnO, < 25 nm for TiO_2 NPs and about 1 μm for TiO_2 fine. The manufacturer makes no statement on purity of ZnO NPs and TiO_2 fine, while a purity of 99.7% for TiO_2 anatase NPs was reported.

2.2. Sample preparation for electron microscopy characterization

Few milligrams of particles were weighted with a Mettler H54 AR electric balance (precision 0.01 mg). The particles were suspended in cell culture medium in order to evaluate the state of agglomeration of particles. Stock suspension of particles was probe sonicated (Vibracell, Sonics & Materials Inc., USA) and immediately after sonication few millilitres of suspensions were filtered through 0.05 μm pore polycarbonate membranes.

2.3. Electron microscopy for single-particle characterization

Scanning electron microscopy and Transmission electron microscopy were used to characterize the single NPs. Morphological analysis of single particles was performed by a SEM FEI XL30 (FEI Company, The Netherlands) equipped with Soft Imaging System [9]. More than 1000 particles were analyzed. Known the volume of suspensions deposited on the filters, the area of exposed filters and the portion of them analysed by SEM, the number of particles in the suspensions was determined. From size distribution obtained by morphological analysis, the percentage of particles in granulometric size range of 100 nm and the average diameter of particles belonging to these intervals were determined. Therefore the surface area of particles was estimated, assuming them spherical. Morphology and primary size of the particles were also determined by a TEM (FEI Company, The Netherlands).

2.4. Cell culture

Caco-2 cells (ATCC collection) were routinely cultured in an atmosphere of 5% carbon dioxide in air at 37°C in Dulbecco's modified Eagle's medium with high glucose (4.5 g/L), supplemented with 100 IU/mL penicillin and 100 μg/mL streptomycin, 4 mmol/L glutamine, 1% non-essential amino acids, 10 mM HEPES and 10% heat inactivated foetal calf serum (FCS) (GIBCO BRL, Gaithersburg, MD).

2.5. Exposure conditions

TiO_2 fine, TiO_2 NPs and ZnO NPs were suspended in culture medium (with or without FCS) and sonicated. For cytotoxicity assay, Caco-2 cells were plated on 24-multiwell plates at a density of 8×10^4 cells/well and treated after 4 days culture with 1, 2.5, 5, 10 and 20 μg/cm^2 particles for 6 and 24 h.

2.6. Viability assay

Viability of Caco-2 cells was measured as neutral red uptake (NRU). Treated cells were processed for NRU determination according to Borenfreund and Puerner [10].

2.7. 8-oxoguanine determination

To evaluate the oxidative DNA damage, 1 x 10^6 Caco-2 cells were exposed to 2.5 μg/cm^2 particles for 2- 4- 6 and 24 h. 8-oxodG was measured by high-performance liquid chromatography with electrochemical detection as previously described [11]. After RNase treatment and enzymatic hydrolysis, DNA hydrolysate were filtered through 0.22 μm cellulose acetate (Millipore) Centricon filter units and dispensed in autosampling vials. Separation of 8-oxodG was performed in a System Gold HPLC instrument (Beckman Coulter Inc., Brea, CA, USA) equipped with a diode array UV detector. The LC-18-DB (Supelco, 75 x 4.6 mm) column, equipped with a LC-18-DB guard column cartridge, was eluted with an isocratic mixture of 50 mM potassium phospahate, pH 5.5, and methanol (90%/10%) at 0.5 ml/min flow rate. Electrochemical analysis was carried out by a Coulochem II detector (ESA Inc., Chelmsford, MA, USA). Deoxyguanosine was measured in the same run of corresponding 8-oxodG.

2.8. Genotoxicity assay

About 1.5 x 10^5 Caco-2 cells were seeded in six well plates and cultured for at least 48 h before treatment in complete culture medium supplemented with 20% FCS.

Cells were treated with 1, 2, 5, 10, 20 μg/cm^2 concentrations of ZnO and TiO_2 NPs and 20 μg/cm^2 TiO_2 fine for 24 h in culture medium without FCS. Cells were washed twice with PBS and cultured in complete medium with 10% FCS and 4.5 μg/mL of Cytochalasin B for further 24 h then rinsed with PBS and harvested after 1.5 h of recovery time in complete medium with 20% FCS.

Slides were prepared according to Bazin et al. [12] and 1000 binucleated cells for each point of treatment were scored; mean values and standard deviations were calculated on the basis of three repeated experiments. Based on the frequency of mono-, bi- and multinucleated cells a Nuclear Division Index was calculated as a measure of cell proliferation.

3. Results and discussion

3.1. Particle characterization

Size distribution of ZnO NPs obtained by SEM equipped with Soft Imaging System showed an average diameter equal to 128 nm. Only 34% of particles possessed dimensions below 100 nm, the remaining particles were agglomerates ranging from 100 to 800 nm.

Size distribution showed an average diameter equal to 220 nm for TiO_2 NPs. Only 16% of particles possessed dimensions below 100 nm. Most of particles were agglomerates ranging from 100 nm to 1.4 μm.

The size distribution of TiO_2 fine showed an average diameter equal to 335 nm and a percentage of particles (10%) with dimension below 100 nm.

In table 1, the particle characteristics obtained by electron microscopy analysis are displayed.

Table 1. Characteristics of ZnO NPs, TiO_2 NPs and TiO_2 fine analyzed by electron microscopy

Particles	Primary size (nm)[a]	Average diameter (nm)[b]	Particle number (part/g) [b]	Surface area (m^2/g) [b]
ZnO	45-170	128	2.5 10^{13}	3.0
TiO_2	20-60	220	1.3 10^{13}	3.3
TiO_2 fine	80-2000	335	1.9 10^{13}	6.7

[a] determined by transmission electron microscopy
[b] determined by scanning electron microscopy

3.2. Cytotoxicity induced by ZnO and TiO_2 NP treatment on Caco-2 cells
Cell viability was measured by Neutral Red assay. No effect on Caco-2 viability was reported after treatment with TiO_2 NPs at all the concentrations tested either in presence or in absence of FCS (figure 1, top panel).

On the contrary, a dose-dependent decrease of cell viability was observed in presence of ZnO NPs already after 6 h of treatment. The presence of FCS strongly reduced zinc toxic effects, probably through its interaction with the NP surface (figure 1, bottom panel).

TiO_2 fine was always used as negative control and it never induces toxic effects until 40 μg/cm^2 (data not shown).

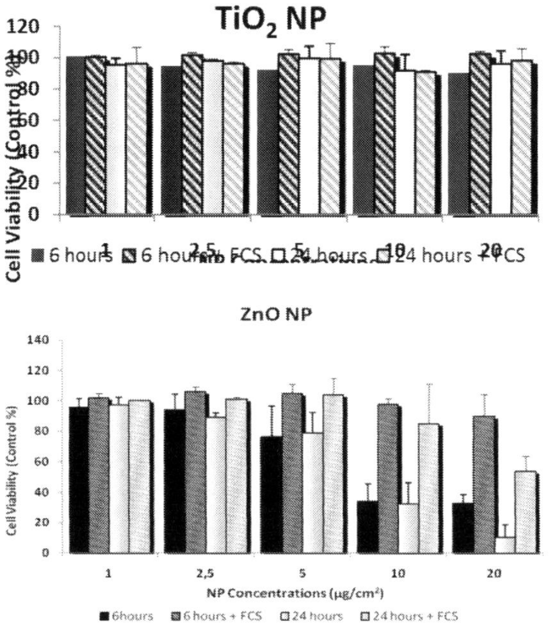

Figure 1. Effects of different concentrations of TiO_2 and ZnO NPs on cellular viability of Caco-2 cells determined by NRU assay, after 6 and 24 h of treatment with or without FCS. Data are expressed as mean ± SD of three independent experiments, performed in triplicate.

To evaluate the role of surface in the cytotoxicity induced on Caco-2 cells, Neutral Red assay was performed at several surface doses of the three types of particles examined. Our preliminary results suggest that the chemical composition played primary role in the different cytotoxic effect of ZnO and TiO_2 NPs (data not shown).

3.3. Analysis of oxidative DNA damage induced by particle exposure
8-oxodG is a highly pro-mutagenic lesion and is considered the main marker of oxidative damage to DNA. The induction of oxidative DNA damage was evaluated after ZnO and TiO_2 NP exposure (2.5 μg/cm^2 for 2-4-6 and 24 h) by measuring the levels of DNA 8-oxodG.

A significant increase in the 8-oxodG levels was observed after 6 h exposure for both NPs. Moreover, an increase in oxidation level was obtained with fine TiO_2 particles too, suggesting a probable damage induction at molecular level regardless of TiO_2 particle size. Interestingly, we noticed a decrease in the oxidation after 24 h exposure probably due to the action of the repair mechanisms involved in this type of damage (figure 2).

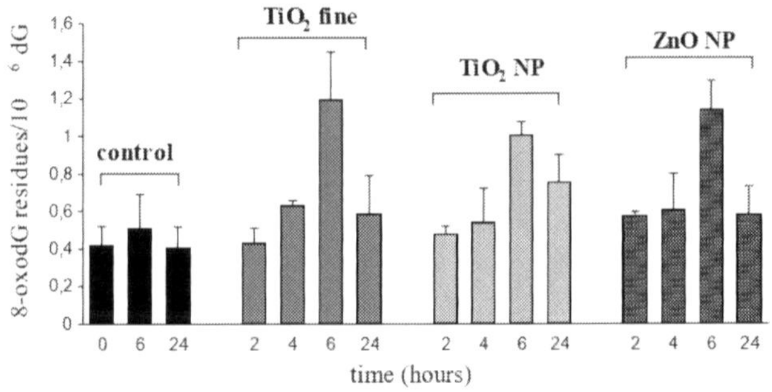

Figure 2. Levels of DNA 8-oxodG measured by HPLC/EC. Caco-2 cells were exposed to 2.5 μg/cm^2 TiO_2 fine, TiO_2 NP and ZnO NP for 2-4-6 and 24 h. Deoxyguanosine was measured in the same run of corresponding 8-oxodG and the data are mean ± SD from three experiments.

3.4. Analysis of the genotoxic effect induced by particle exposure by cytokinesis-block micronucleus (MN) assay

The percentage of Micronucleated Binucleated cells (BinMned) was determined in Caco-2 cells treated with different concentrations of ZnO and TiO_2 NPs and TiO_2 fine.

The % BinMned background was 9.03%. A slight increase of % BinMned was observed after treatment with 2 and 5 μg/cm^2 of ZnO (12.1 and 12.2% BinMned, respectively) (figure 3a). Conversely no effect was detected after treatment with TIO_2 both NP and fine (figure 3b).

The % of BinMned was not determined at the highest NP concentrations (10, 20 μg/cm^2) because of both high cytotoxicity exerts by ZnO and impossibility to perform a reliable score of cells treated with TiO_2, due to the great accumulation of NPs on cell surface.

A dose dependent decrease of Nuclear Division Index (NDI) was observed in cells treated with ZnO (figure 3c) whereas no reductions of NDI values were detected after treatment with TiO_2 NPs (figure 3d), according to Neutral Red assay results.

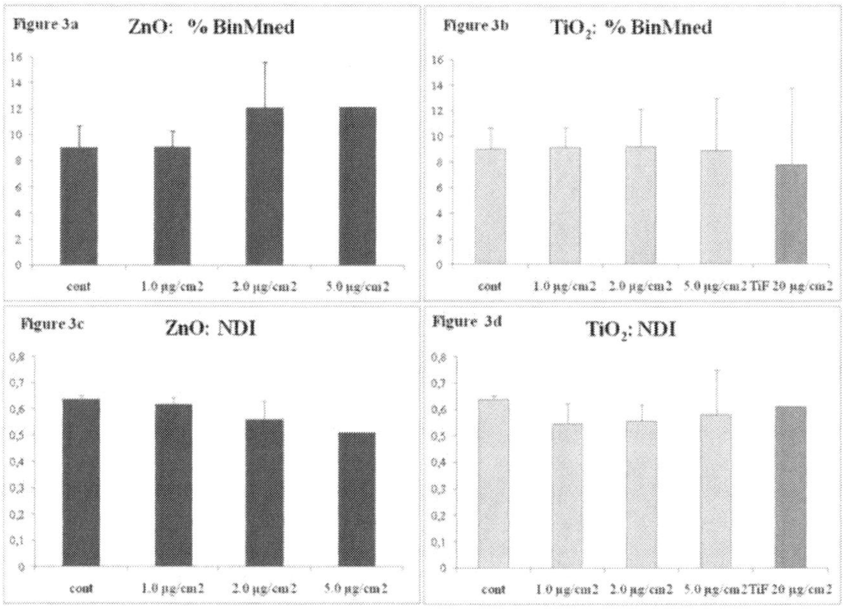

Figure 3. a and b panels show the percentage of Micronucleated Binucleated cells (BinMned) induced by 1.0, 2.0 and 5.0 µg/cm^2 of Zinc oxide (ZnO), Titanium dioxide (TiO_2) and 20 µg/cm^2 of Titanium dioxide Fine (TiF) respectively, after 24 h of treatment.
c and d panels show the Nuclear Division Index (NDI) observed after 24 h treatment with 1.0, 2.0 and 5.0 µg/cm^2 of ZnO and TiO_2 respectively, and 20µg/cm^2 of TiF. Data represent the mean of three experiments ± SD except for 5.0 µg/cm^2 of ZnO.

4. Conclusion

In this study we show that ZnO nanoparticles induced a significant cytotoxic effect on Caco-2 cells and this effect was strongly reduced in presence of FCS in culture medium. On the contrary, cell viability was not affected by the exposure to TiO_2 particles and it was independent on the particle size. Preliminary data, obtained by cell exposure to the same particle surface dose, agree with neutral red results indicating a primary role of the NP chemical composition in the different cytotoxicity induced on Caco-2 cells. Further studies will be performed in order to analyse the involvement of ion release from ZnO NPs on the reported cytotoxic effect.

Oxidative damage to genomic DNA was observed after 6 h exposure to ZnO and TiO_2 particles, suggesting a damage induction at molecular level. This damage was not dependent on particle size and chemical composition. The role of ZnO and TiO_2 NPs surface reactivity on the oxidative damage to DNA will be further investigated. Besides, a decrease in the 8-oxodG levels was observed after 24 h exposure and this is probably related to the cellular mechanisms of repair.

In order to better understand our results, the production of reactive oxygen species (ROS) will be measured in the Caco-2 cells exposed to the three different particles. Moreover, it could be interesting to study both oxidative damage to mitochondrial DNA and the expression levels of the main protein involved in the repair mechanisms.

Chromosome damage analysis, evaluated by the cytokinesis blocked micronucleus assay 24 h after NP treatment, highlighted a slight increase of micronuclei induced only by ZnO. In order to elucidate the kinetic of chromosome damage in this cell line and the possible association to oxidative damage that peaked at 6 h, a different protocol for the treatment will be considered in future experiments.

References

[1] Yang H, Liu C, Yang D, Zhang H and Xi Z 2008 *J. Appl. Toxicol.* **29** 69-78
[2] Karlsson HL, Gustafsson J, Cronholm P and Moller L 2009 *Toxicol. Lett.* **188(2)**112-8.
[3] Xia T, Kovochich M, Brant J, Hotze M, Sempf J, Oberley T, Sioutas C, Yeh JI, Wiesner MR and Nel AE 2006 *Nano Lett.* **6** 1794-807
[4] Hsin Y, Chen C, Huang S, Shih T, Lai P and Chueh PJ 2008 *Toxicol. Lett.* **179** 130-9
[5] Reeves JF, Davies SJ, Dodd NJF and Jha AJ 2008 *Mutat. Res.* **640** 113-22
[6] Sharma V, Shukla RK, Saxena N, Parmar D, Das M and Dhawan A 2009 *Toxicol. Lett.* **185** 211-8
[7] Singh S *et al* 2007 *Toxicol. Appl. Pharmacol.* **222** 141-51
[8] Koeneman BA, Zhang Y, Westerhoff P, Chen Y, Crittenden JC and Capco DG 2010 *Cell Biol Toxicol* **26** 225-38
[9] De Berardis B, Civitelli G, Condello M, Lista P, Pozzi R, Arancia G and Meschini S 2010 *Toxicol. Appl. Pharmacol.* **246 (3)** 116-27
[10] Borenfreund E and Puerner JA 1985 *Toxicol. Lett.* **24(2-3)**119-24.
[11] Goitre L, Balzac F, Degani S, Degan P, Marchi S, Pinton P and Retta SF 2010 *PLoS One* **5 (7)** e11786
[12] Bazin E, Mourot A, Humpage AR and Fessard V 2010 *Environ Mol Mutagenesis* **51(3)** 251-9.

Investigation of TiO_2 nanoparticles translocation through a Caco-2 monolayer

E. Brun[1], M.-L. Jugan[1], N. Herlin-Boime[2], D. Jaillard[3], B. Fayard[4], A.-M. Flank[5], A. Mabondzo[6] and M. Carrière[1]

[1] Laboratoire de Structure et Dynamique par Résonance Magnétique, UMR3299 CEA-CNRS, Saclay, France

[2] Laboratoire Francis Perrin, URA2453 CEA-CNRS, Saclay, France

[3] Centre Commun de Microscopie Electronique d'Orsay, UMR8195 CNRS-univ. Paris Sud, Orsay, France

[4] ID21 beamline, ESRF, Grenoble, France and Laboratoire de Physique du Solide, UMR8502 CNRS-univ. Paris Sud, Orsay, France

[5] LUCIA beamline, SOLEIL synchrotron, Saint-Aubin, France

[6] Laboratoire d'Etude du Métabolisme du Médicament, CEA, Saclay, France

E-mail : marie.carriere@cea.fr

Abstract. Nanoparticles (NPs) are introduced in a growing number of commercial products, including food and beverage but their effects on gastrointestinal tract are poorly investigated. Here we focused on the translocation of TiO_2 NPs through Caco-2 monolayers exposed to anatase and rutile NPs up to 24 h. Internalization was followed by transmission electronic microscopy and μ-XRF elemental mapping, coupled to XAS analysis of Ti atoms environment. This innovative technique is among the best techniques to get insights on NP fate after internalization. The originality of this project relies on the panel of microscopy techniques implemented to investigate digestive barrier translocation, bringing together biologists, chemists and physicists in a pluridisciplinary research program.

1. Introduction

Due to their remarkable properties, nanoparticles (NPs) are presently introduced in a fast growing number of commercial products as these have known a 379% increase since March 2006 [1]. As its counterparts, the food industry is starting to introduce various NPs as food additives, food packaging components or as dietary supplements. Among them, titanium dioxide (named E171 in Europe) is used for whitening and brightening foods [2] but can also be found in daily use hygiene products such as toothpaste or sunscreens, or orally-administered drugs. Little is known however about the effects of NPs within the gastrointestinal tract and there is still a controversial debate whether they are able to cross this physiological barrier or not. Three main routes can be thought of, for describing NPs crossing the gut. First, NPs could be toxic enough to cross the epithelium *via* holes resulting from dead cells. Second, they could damage the barrier integrity and squeeze through live cells. Last, they could be taken in charge by cells and transported from the apical to the basolateral poles in a process called transcytosis. Among the *in vitro* models of gastrointestinal barrier, Caco-2 cell monolayers are a cornerstone to probe and predict intestinal drug permeability and absorption as many of the *in vivo* brush border enzymes and transport proteins are present and functional [3].

In the present study, an original approach, combining several microscopies, has been developed to assess TiO_2 NPs transport across Caco-2 epithelium. In particular, not only microbeam X-ray fluorescence proves to be a valuable tool for elemental imaging at a single cell level but also allows *in situ* speciation analyses of NP identified inside the cells by X-rays absorption spectroscopy (XAS).

2. Materials and Methods

2.1. Nanoparticles

TiO2 NPs synthesis and characterization have already been described in [4,5]. Briefly, 95% anatase and 100% rutile NPs have mean diameters of 12 nm and 21 nm, and specific surface areas of 82 m^2/g and 73 m^2/g, respectively. To make the notations clearer, they will be referred to as A12 and R20 throughout this manuscript. NPs suspensions were prepared in ultrapure sterile water by pulsed sonication at a concentration of 10 mg/mL. Agglomeration state of NPs suspensions was determined by photon correlation spectroscopy (Malvern ZetaSizer 3000HS, Worcestershire, UK).

2.2. Cells

Caco-2 cells (ATCC HTB-37, passages from 39 to 43) were cultured in a 37°C incubator with a 5% CO_2 humidified atmosphere in Dulbecco Modified Medium supplemented with 10% (v/v) foetal bovine serum, 2 mM L-glutamine, 1% (v/v) non essential amino acids, 50 UI/mL penicillin and 50 µg/mL streptomycin. Cells were seeded at a density of 50,000 cells/cm^2 in 6-well porous transwell®-clear polyester inserts (0.4 µm pores, Corning, Costar) and were cultured to confluence over 21 days, medium being changed every other day. They were exposed on their apical pole to A12 and R20 diluted in culture medium at a concentration of 50 µg/mL for 6, 12 or 24 h.

2.3. Cytotoxicity assays

Cells were seeded in 96-well plates at a density of 30,000 cells/cm^2. After 24 h, they were exposed to 100 µL per well of NPs suspensions from 0 to 500 µg/mL up to 24 h. Cell metabolic activity, reflecting NPs cytotoxicity, was assessed by using 3-(4,5-dimethylthiazol-z-yl)-2,5-diphenyl-tetrazotium bromide (MTT) [6]. Mitochondrial dehydrogenases of viable cells reduce MTT to water-insoluble blue formazan crystals. At considered endpoints, exposure medium was replaced by 100 µL of 0.5 mg/mL MTT solution in culture medium. After 2 h at 37°C, medium was replaced by 100 µL DMSO and mixed to dissolve formazan crystals. Absorbance was measured at 550 nm and metabolic activity was determined as a percentage of the negative controls. To probe the interference of NPs with this test, absorbance of exposed cells, treated as described below but without MTT, were recorded.

2.4. TEER measurements

Millicell-ERS (Millipore) and chopstick electrodes were used to measure transepithelial electrical resistance (TEER). A blank insert without cells but immerged in medium was used to obtain the transwell resistance. TEER was calculated as follows: $TEER = (R_{cell} - R_{blank}) \times A$, where TEER is expressed in $\Omega.cm^2$, cell and blank resistances in Ω and A, the surface area of the insert, in cm^2. As electrodes height positioning had a great influence on the displayed resistance value, at least five measurements were carried out for each insert.

2.5. Electron microscopy
After exposure to NPs, cells were rinsed with phosphate saline buffer, fixed in 2% glutaraldehyde in cacodylate buffer and in 1% osmium tetroxide solution, dehydrated through a graded series of ethanol and embedded in Epon resin, taking care of preserving their polarity. Ultra-thin sections were cut, then stained with 1% uranyl acetate for observation in transmission electronic microscopy (TEM) with a Philips CM120 electron microscope operating at 80 kV.

2.6. μ-SR-XRF and XAS analysis
Embedded samples used for TEM were also analyzed by synchrotron-based micro X-ray fluorescence (μ-SR-XRF) and micro X-ray absorption spectroscopy (μ-XAS). Transversal sections (thickness: 1 μm and 3 μm) of these samples were cut using an ultramicrotome, deposited on kapton polymer and covered by a thin layer of kapton. These sections were directly analyzed.

Cartographies presented here were obtained on LUCIA beamline (SOLEIL synchrotron). Scanning parameters were fixed as follows: cartography 100 μm x 100 μm, step: 2 μm x 2 μm, counting time: 1 s on each point. Ti, K, Ca, P/Os and Cl maps were drawn. Transwell membranes could be identified because they do not contain Cl, cells were identified by mapping K or P/Os distribution. Epoxy resin did not cause any interference in P/Os and Ti fluorescence signals. XAS analyses were performed on cell regions containing high amounts of Ti. XAS spectra were registered at the Ti K-edge (4966 keV).

3. Results and discussion

3.1. TiO2 NPs cytotoxicity
Caco-2 cells were chosen as a first model of gastrointestinal barrier as they exhibit properties of the small intestine *in vitro* when fully differentiated [3]. As a prerequisite for translocation experiments, we had to determine the sub-toxic concentrations of A12 and R20 NPs. Note that NPs suspensions in culture medium led to their aggregation, as their hydrodynamic diameters, evaluated by photon correlation spectroscopy were higher than 1 μm (data not shown). Their cytotoxicities were assessed in a concentration- and time- dependent manner by evaluating the cell metabolic activity through MTT reduction. As mentioned recently by Jones and Grainger [7], spectral properties of NPs should be carefully quantified in order not to bias dye-based assays. As shown in Figure 1A, interference of R20 NPs with MTT assay was to be taken into account as the residual absorbance of TiO_2 cannot be neglected for high concentrations. This may be due to firmly adsorbed NPs on cells, remaining after washings. A12 NPs exposure led to the same observations and absorbance corrections. For both NPs, no modification of cellular metabolic activity, as measured by MTT reduction, was detected as the concentration of TiO_2 increased (from 1 to 100 μg/mL) and after the longer exposure time of 24 h. The sub-toxic concentration of 50 μg/mL was then chosen for further experiments.

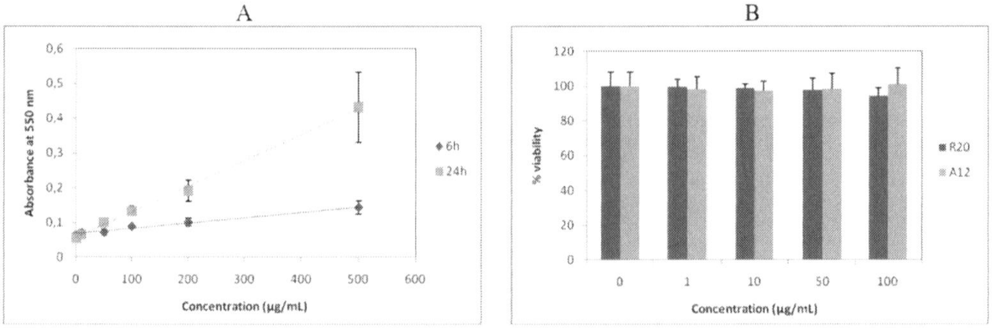

Figure 1. Caco-2 cells viability when exposed to A12 and R20 NPs in culture medium. Two time exposures were tested: 6 and 24h. Experiments were performed in n=5. Values are expressed as means and standard deviations. A: Residual absorbance of cells treated during 6 or 24 h to R20 NPs. The coefficients of determination of the linear regressions are 0.981 and 0.997 for 6 and 24 h respectively. B: Results from the MTT assay after absorbance correction for A12 and R20 after a 24h exposure.

3.2. Translocation of TiO_2 NPs across the epithelium

When cultivated on membranes over confluence, Caco-2 cells form a monolayer in which cells are maintained together by tight and adherens junctions. Its resistance testifies for its integrity. After 3 culture weeks and subsequent exposure to NPs, transepithelial electrical resistances were measured. According to Figure 2, a 50 µg/mL treatment had no significant effect on TEER for both NPs, suggesting that the epithelium remains intact, as observed previously for mixed anatase-rutile NPs [8].

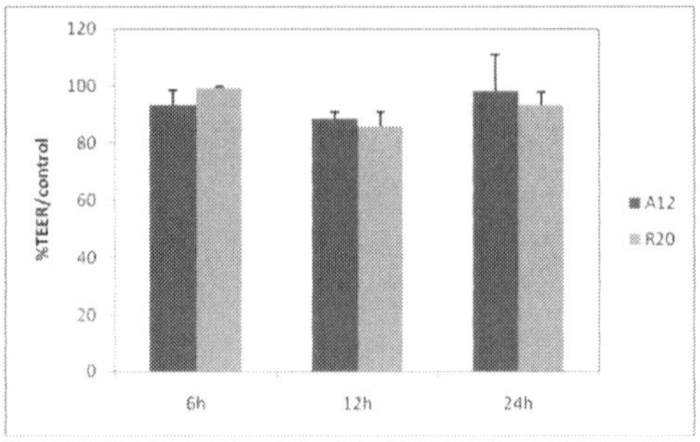

Figure 2. Transepithelial electrical resistance measurements after exposure to A12 and R20 for 6, 12 and 24 h. Three inserts were measured five times for each condition. Values are expressed as means and standard deviations.

We then analysed transversal sections of Caco-2 cells monolayers with TEM and µ-SR-XRF in parallel. The advantages of synchrotron radiation over others X-rays sources are an improvement of sensitivity and spatial resolution of fluorescence analyses, allowing a single cell level imaging [9]. Regions of Interest (ROI) were chosen in order to draw images of P/Os (Figure 3A), Ti (Figure 3B) from the X-ray fluorescence spectrum (Figure 3C). These µ-SR-XRF analyses showed that even if mapping of Ti on a 1 µm-thick section is achievable, the quality of the images is much higher when analyzing 3 µm-thick sections. Transwell membranes could be identified because the whole sample section, except the transwell region, contains high amounts of Cl, probably contained in epoxy resin whereas cells were identified by mapping Os/P distribution as fluorescence peaks of these two elements are partly superimposed. By drawing these two maps, it was thus possible to identify the apical pole exposed to NPs and the basolateral pole near the transwell region. Mapping of Ti distribution enabled to observe NPs internalized in cells. After 12h of exposure, R20 NPs were observed on cell surfaces but also deeper inside the cells (Figure 3B). However they were not detected on the basolateral pole. This distribution was also observed when cells were exposed to A12 NPs and for both NPs after a 24 h exposure.

TEM observations consolidated this statement, along with giving a sub-cellular localization of NPs. As can be seen in Figure 3C, NPs were most of the time sequestered in the upper regions of the epithelium and they rarely reached the basolateral pole of cells. This is in good agreement with µ-SR-XRF mapping. Moreover, they were mainly grouped together inside the cells into heterogeneous clusters of 1-2 µm, in intracytoplasmic vacuoles or compartments resembling late endosomes. Far few NPs were observed isolated, as already reported for A549 human type II alveolar epithelium cell line [5]. Due to their size in medium culture, agglomerates of NPs would rather be internalized through macropinocytosis, the main pathway for particles around 1 µm, even if we did not observe any characteristic membrane ruffling. As a conclusion, such TiO_2 NPs accumulated in cells in low amounts and did not cross Caco-2 monolayers during the first 24 h of exposure.

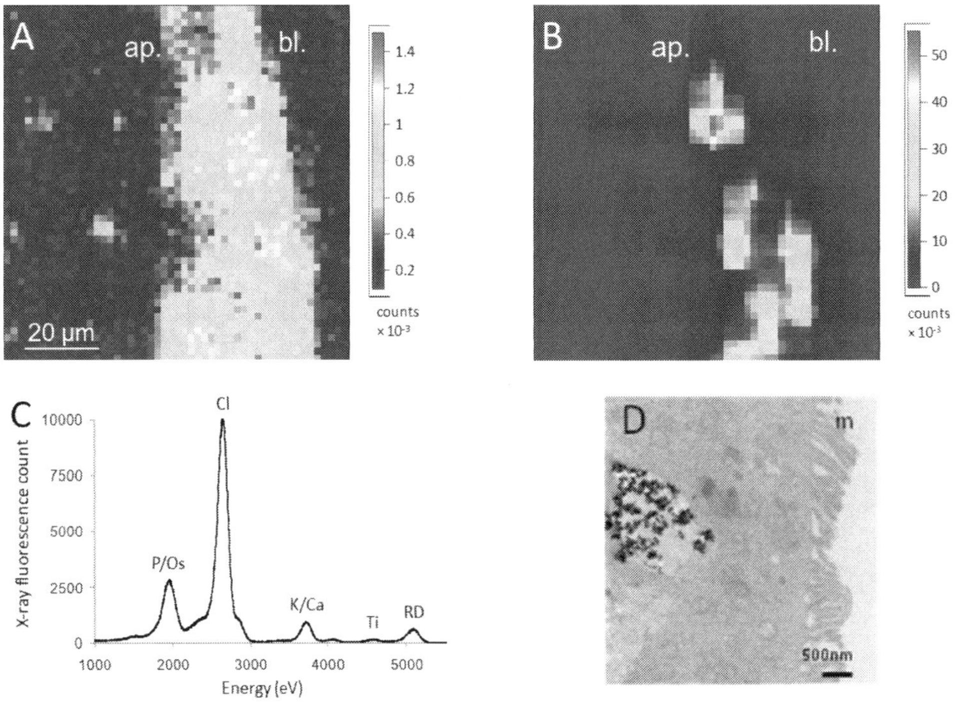

Figure 3. Ti distribution in Caco-2 cells. Cells were exposed for 12 h to R20 NPs on their apical pole (ap.). P/Os signals (A) allow cells identification. Translocation of NPs to the basolateral pole (bl.) is evaluated through mapping of Ti distribution (B) inside the cells. Elemental signals are extracted from total X-ray fluorescence spectrum (C). NPs distribution in cells is also observed by TEM (D).Note the presence of microvilli (m), one main feature of enterocytes on the apical side.

3.3. Fate of internalized NPs

Elemental mapping at the cellular level is not the only strong point of synchrotron radiation based analysis as it also provides a unique way to gain insights into internalized NPs by recording XAS spectra in Ti-rich regions of cell sections. Compared to reference compounds spectra, they inform on Ti speciation, on TiO2 crystalline phase and on the size of TiO2 NPs.

Ti pre-edge features are characteristic of the crystalline phase and size of TiO_2 NPs. Anatase displays a typical triplet feature (A1, A3 and B peaks), with a weak shoulder on the low-energy side of the central A3 peak (A2) [10]. The intensity of A2 peak is related to the distorsion of the octahedral TiO_6 unit, particularly the distorsion observed on the surface of NP. The intensity of this peak is thus related to the size of the particle: the smaller the diameter, the higher this A2 peak [11]. Meanwhile, the intensity of the A1 peak also decreases as the particle size decreases [12]. As can be seen in Figure 4, in rutile XAS pre-edge feature, no A2 peak is observed, A3 peak is more intense while B peak is shifted to +1.3 eV as compared to B peak in anatase pre-edge. Analysis of this pre-edge region may then inform on local changes of the crystalline phase, on NPs surface after cell internalization, and on their dissolution. In the present experiment, no modification of these pre-edge features was observed after cell internalization of R20 or A12, meaning that Ti is still in the nanoparticulate TiO_2 chemical form inside cells, and that their crystalline phase did not change. Moreover the size of NP remained constant after cell internalization meaning that no partial dissolution of NP occurred.

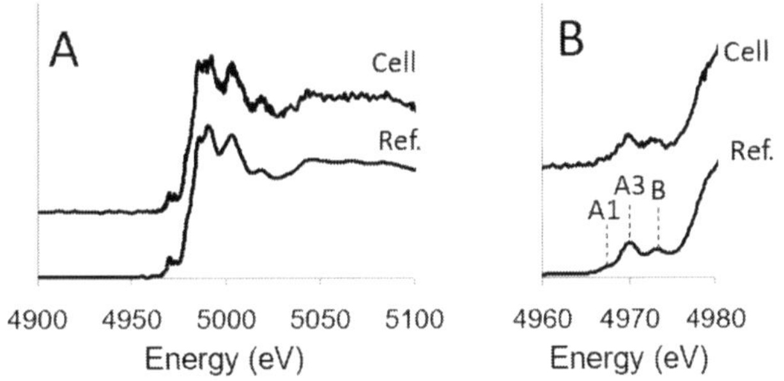

Figure 4. XAS spectra of Ti in cells exposed to TiO_2 NPs. Cells were exposed to R20 NPs. Spectra were recorded at the Ti K-edge (4.966 keV) at room temperature on locally Ti-rich regions determined on μ-SR-XRF maps (Cell), and compared to the spectrum of a reference nanopowder (Ref.). A: whole XAS spectrum, B: pre-edge feature.

4. Conclusions

In brief, the present results show that our forward-looking approach, associating TEM with μ-SR-XRF and XAS was effective and powerful to address the issue of potential translocation of TiO_2 NPs across a gastrointestinal model and to obtain *in situ* information on internalized NPs. Caco-2 epithelia were exposed to bare, pure, anatase and rutile TiO_2 NPs. NPs accumulated in cells, although in low amounts as shown in very good agreement by μ-SR-XRF mapping and TEM. Moreover Ti XAS spectra recorded on Ti-rich intracellular regions indicate that after internalization, Ti is still in the TiO_2 nanoparticulate state, that crystalline phase and speciation remain identical and that no partial dissolution occurs.

Acknowledgments

This work was financially supported by the NANOTRANS project funded by the INERIS in the post-Grenelle program frame and by the region Ile de France through the framework C'nano Ile de France.

References

[1] www.nanotechproject.org
[2] Powell J J, Faria N, Thomas-McKay E, Pele LC 2010, J. Autoimmun. 34 J226
[3] Alvarez-Hernandez X, Nichols G M, Glass J 1991 Biochim. Biophys. Acta 1070 205
[4] Pignon B, Maskrot H, Leconte Y, Coste S, Reynaud C, Herlin-Boime N, Gervais M, Guyot-

Ferreol V, Pouget T, Tranchant J F 2008 Eur. J. Inorg. Chem. 6 883
[5] Simon-Deckers A, Gouget B, Mayne-L'hermite M, Herlin-Boime N, Reynaud C, Carriere M. 2008 Toxicology 253 137
[6] Mosmann, T 1983 J. Immunol. Methods 65 55
[7] Jones C F and Grainger D W 2009 Adv. Drug Deliv. Reviews 61 438
[8] Koeneman B A, Zhang Y, Westerhoff P, Chen Y, Crittenden J C, Capco D G 2010 Cell. Biol. Toxicol. 26 225
[9] Petibois C 2010 Anal. Bioanal. Chem. 397 2051
[10] Manzini I, Antonioli G,Lottici P P, Gnappi G, Montenero A 1995 Physica B 208 607
[11] Luca V, Djajanti S, Howe R F 1998 J. Phys. Chem. 102 10650
[12] Chen L X, Rajh T, Wang Z, Thurnauer M C 1997 J. Phys. Chem. 101 10688

Functional-dependent and size-dependent uptake of nanoparticles in PC12

N Sakai[1, 2], Y Matsui[1], A Nakayama[1], A Tsuda[3] and M Yoneda[1]

[1]Department of Environment Engineering, Graduate School of Engineering, Kyoto University, 4 Kyotodaigaku Katsura, Nishikyo-ku, Kyoto 6158540, JAPAN

[2]Grant-in-Aid for Scientific Research from the Ministry of Education, Culture, Sports, Science and Technology, 20-4062, JAPAN

[3]Department of Environmental Health, Harvard School of Public Health, 665 Huntington Avenue, Boston, MA 02115, USA

E-mail: sakai@risk.env.kyoto-u.ac.jp

Abstract. It is suggested that the uptake of nanoparticles is changed by the particle size or the surface modification. In this study, we quantified the uptake of nanoparticles in PC12 cells exposed Quantum Dots with different surface modification or fluorescent polystyrene particles with different particle size. The PC12 cells were exposed three types of the Quantum Dots (carboxyl base-functionalized, amino base-functionalized or non-base-functionalized) or three types of the fluorescent particles (22 nm, 100 nm or 1000 nm) for 3 hours. The uptake of the nanoparticles was quantified with a spectrofluorophotometer. The carboxyl base-functionalized Quantum Dots were considerably taken up by the cells than the non-base-functionalized Quantum Dots. Conversely, the amino base-functionalized Quantum Dots were taken up by the cells less frequently than the non-base-functionalized Quantum Dots. The particle number of the 22 nm-nanoparticles taken up by the cells was about 53 times higher than the 100 nm-particles. However, the particle weight of the 100 nm-particles taken up by the cells was higher than that of the 22 nm-nanoparticles. The 1000 nm-particles were adhered to the cell membrane, but they were little taken up by the cells. We concluded that nanoparticles can be taken up nerve cells in functional-dependent and size-dependent manners.

1. Introduction

Previous studies suggested that nanoparticles were taken up by cells and invaded tissues [1,2]. Oberdorster *et al.* (2004) showed that Fischer-344 rats were exposed ^{13}C nanoparticles (36 nm) by inhalation and consequently the ^{13}C nanoparticles reached to the brain [3]. Although the exposure pathway was not observed, it was suggested that the nanoparticles would be translocated to the brain via the olfactory nerves. Furthermore, de Lorenzo (1970) visualized the silver-coated colloidal gold (50 nm) taken up by the olfactory nerves [4]. Therefore, the uptake of the nanoparticles in the olfactory nerves may be occurred when we inhale the nanoparticles in the air. Thus, it should be quantified to consider the biological risk to the brain.

Many studies showed that the uptake of the nanoparticles was affected by the particle size and surface modification. Rejman et al. (2004) showed that non-phagocytic mouse melanoma B16 cells took up particles up to a size of 500 nm at 37°C, but no uptake was seen of particles 1 μm in size [5]. Also, Sur et al. (2010) showed that the uptake of the nanoparticles was changed by the surface modification [6]. However, these tendencies may be not applicable to the nerve cells because the uptake of the nanoparticles can be different by the species. Although the uptake of the nanoparticles in the nerve cells was visualized [7], it is still unknown how the particle size and the surface modification affect the uptake of the nanoparticles.

In this study, we exposed three types of fluorescent particles with different sizes or three types of Quantum Dots with different surface modifications to rat pheochromocytoma (PC12) cells, as a model of nerve cells. The particle number and the particle weight of these nanoparticles which were taken up by the cells or adhered to the cell membrane were quantified with a spectrofluorophotometer. Thus, we studied how the uptake of the nanoparticles in the nerve cells was affected by the particle size and the surface modification.

2. Materials and Methods

2.1. The preparation of nanoparticles for exposure

Three types of the fluorescent polystyrene particle suspensions (MORITEX Ltd., Tokyo, Japan) and three types of the Quantum Dot suspensions (Invitrogen Ltd., Tokyo, Japan) were purchased. The particle diameter of the fluorescent particles was 22 nm, 100 nm or 1000 nm, respectively. Also, two types of the Quantum Dots were functionalized by carboxyl base or amino base and another type of the Quantum Dots were not functionalized. The characterization of these particles is shown in table 1.

Each particle suspension was mixed with RPMI 1640 medium (GIBCO Ltd., Tokyo, Japan). The concentration of the suspensions in the medium was adjusted 0.5% (v/v). The particle diameter and number distribution of these particles were analyzed with a dynamic light-scattering particle-size analyzer (ELSZ-0, Otsuka Electronics Company, Limited, Osaka, Japan) 5 minutes and 3 hours after the mixture with the medium.

2.2. Cell culture

The PC12 cell line, derived from rat pheochromocytoma (CRL-1721.1™), was obtained from American Type Tissue Culture. The PC12 cells were cultured on poly-D-lysine cellware dishes (φ 100 mm, Becton Dickinson Ltd., Tokyo, Japan) and grown in the RPMI 1640 medium supplemented with 5% fetal bovine serum, 10% horse serum (both heat inactivated) in a humidified atmosphere with 5% CO_2 at 37°C. The PC12 cells were repeatedly-subcultured when the cells reached 80% confluent and seeded in the range of 1×10^6 cells/dish to 4×10^6 cells/dish.

Table 1. Characterization of fluorescent particles and Quantum Dots.

Characteristic	Fluorescent particles			Quantum Dots		
	22 nm	100 nm	1000 nm	Carboxyl base	Amino base	Non-base
Primary diameter (nm)	22	100	1000	10—20	10—20	10—20
Composition	Polystyrene	Polystyrene	Polystyrene	CdSe	CdSe	CdSe
Shape	Spherical	Spherical	Spherical	Spherical	Spherical	Spherical
Surface modification	—	—	—	Carboxyl base	Amino base	—
Density (g/cm^3)	1.06	1.06	1.06	—	—	—
Number concentration (particles/mL)	1.7×10^{15}	1.8×10^{13}	1.8×10^{10}	4.8×10^{15} [a]	4.8×10^{15} [a]	1.2×10^{15} [a]
Maximum excitation wavelength (nm)	542	542	542	350	350	350
Maximum fluorescent wavelength (nm)	612	612	612	565	565	565

[a] The number concentration was calculated as one particle includes 1000 CdSe molecules.

2.3. Uptake test

The PC12 cells which reached 80% confluent were exposed three types of the fluorescent particles (22 nm, 100 nm or 1000 nm) or three types of the Quantum Dots (carboxyl base-functionalized particles, amino base-functionalized particles or non-base-functionalized particles). Each particle suspension was mixed with RPMI 1640 medium at 0.5% (v/v) just before the exposure. The medium in the dishes was aspirated and 2 mL of the mixed suspension was added into three dishes. As a negative control group, three dishes were added 2 mL of the RPMI 1640 medium. All the dishes were incubated in a humidified atmosphere with 5% CO_2 at 37°C for 3 hours. After the exposure, the exposure medium was aspirated and the cells collected into a centrifugal tube. The cells were washed with PBS and centrifuged at 1000 rpm for 10 minutes. The wash was repeated three times and the number of the collected cells was measured after the wash. Then, 1 mL of 1% triton-X100 containing PBS was added into the tube and homogenized to break the cell membrane. The tube was centrifuged at 1000 rpm for 10 minutes and the supernatant was separated into a micro tube. The pellet was washed gently by 1 mL of PBS three times and then suspended in 1 mL of PBS. The fluorescent intensity in the supernatant and the pellet was measured with a spectrofluorophotometer (RF-5300PC, Shimadzu Ltd., Kyoto, Japan). The standard curves of the fluorescent intensity in each particle were made by the dilution series of the particle suspensions mixed with PBS. The fluorescent intensity of the diluted particle suspensions was measured with the spectrofluorophotometer and the particle number in the supernatant and the pellet was calculated by the standard curves.

2.4. Cell viability test

The PC12 cells were seeded in a poly-D-lysine cellware 96 well Plate (Becton Dickinson Ltd., Tokyo, Japan) at 1×10^4 cells/well and incubated in a humidified atmosphere with 5% CO_2 at 37°C for 3 days. The PC12 cells were exposed 0.2 mL of the 0.01%, 0.1% or 1% (v/v) fluorescent particles containing RPMI 1640 medium or the 0.1%, 1% or 10% (v/v) Quantum Dots containing RPMI 1640 medium and incubated in a humidified atmosphere with 5% CO_2 at 37°C for 24 hours. As a negative control group, three wells were added 2 mL of the RPMI 1640 medium. After the exposure, 0.01 mL of Cell Counting Kit-8 (Dojindo Laboratories, Kumamoto, Japan) was added into each well and incubated in a humidified atmosphere with 5% CO_2 at 37°C for 3 hours. The absorbance at 450 nm in each well was measured with Microplate Reader (Bio-Rad Laboratories Inc., Tokyo, Japan).

3. Results

3.1. Particle characterization

The number distribution of the fluorescent particles or the Quantum Dots in the RPMI 1640 medium was measured with a dynamic light-scattering particle-size analyzer. Figure 1 shows the number distribution of three types of the fluorescent particles (22 nm, 100 nm or 1000 nm) 5 minutes or 3 hours after mixing with the RPMI 1640 medium. The number distribution of the 22 nm-particles 5 minutes after the mixture appeared two peaks. The mean diameter of the first peak and the second peak was 25.6 nm and 88.2 nm, respectively. The 100 nm-particles and 1000 nm-particles showed monodisperse. The mean particle diameter of the peak was 158.1 nm in the 100 nm-particles and 1089.8 nm in the 1000 nm-particles. The size distribution of these suspensions was in the range of 22.1 nm to 225.9 nm in the 22 nm-particles, 126.4 nm to 594.6 nm in the 100 nm-particles and 875.7 nm to 2795.5 nm in the 1000 nm-particles. After 3 hours, the second peak in the 22 nm-particles was disappeared and showed monodisperse. Similarly, The 100 nm-particles showed monodisperse and the spectrum was almost same as that measured 5 minutes after the mixture. However, the particle diameter became larger in the 1000 nm-particles than that measured 5 minutes after the mixture, though the spectrum was still monodisperse. The mean particle diameter of the peak was 27.3 nm in the 22 nm-particles, 155.5 nm in the 100 nm-particles and 1326.3 nm in the 1000 nm-particles. Also, the size distribution of these suspensions was in the range of 22.1 nm to 104.1 nm in the

22 nm-particles, 126.4 nm to 403.7 nm in the 100 nm-particles and 1062.7 nm to 3395.0 nm in the 1000 nm-particles.

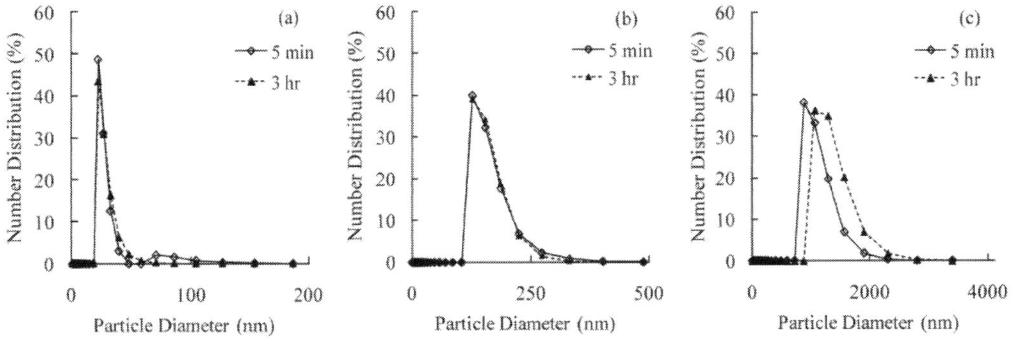

Figure 1. The number distribution of three types of the fluorescent particles 5 minutes after mixing with the RPMI 1640 medium (open rhomboid) and 3 hours after mixing with the medium (solid triangle). (a); 22 nm-particles, (b); 100 nm-particles, (c); 1000 nm-particles.

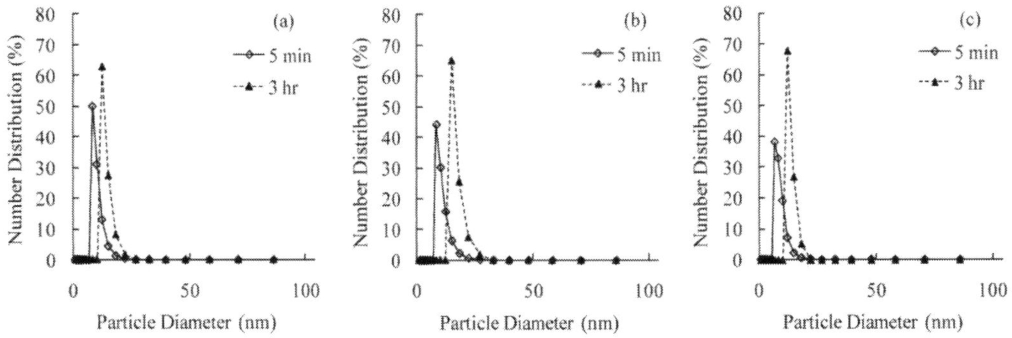

Figure 2. The number distribution of three types of the Quantum Dots 5 minutes after mixing with the RPMI 1640 medium (open rhomboid) and 3 hours after mixing with the medium (solid triangle). (a); carboxyl base-functionalized particles, (b); amino base-functionalized particles, (c); non-base-functionalized particles.

Figure 2 shows the number distribution of three types of the Quantum Dots (carboxyl base-functionalized particles, amino base-functionalized particles and non-base-functionalized particles) 5 minutes or 3 hours after mixing with the RPMI 1640 medium. The number distribution of these suspensions 5 minutes after the mixture showed monodisperse. The mean particle diameter of the peak was 10.0 nm in the carboxyl base-functionalized particles, 10.4 nm in the amino base-functionalized particles and 8.7 nm in the non-base-functionalized particles. The size distribution of these suspensions was in the range of 8.4 nm to 39.6 nm in the carboxyl base-functionalized particles, 8.4 nm to 39.6 nm in the amino base-functionalized particles and 6.9 nm to 70.7 nm in the non-base-functionalized particles. After 3 hours, the particle diameter in these suspensions became larger than that measured 5 minutes after the mixture, though the spectrum was still monodisperse. The mean particle diameter of the peak was 13.7 nm in the carboxyl base-functionalized particles, 16.6 nm in the amino base-functionalized particles and 13.4 nm in the non-base-functionalized particles. Also, the size distribution of the suspensions became broad in the range of 12.3 nm to 153.4 nm in the carboxyl base-functionalized particles, 15.0 nm to 225.9 nm in the amino base-functionalized particles and

12.4 nm to186.1 nm in the non-base-functionalized particles. The summery of the particle diameter and the size distribution of these suspensions is shown in table 2.

Table 2. The particle diameter and the size distribution after mixing with the RPMI 1640 medium.

Characteristic	Fluorescent particles			Quantum Dots		
	22 nm	100 nm	1000 nm	Carboxyl base	Amino base	Non-base
Primary diameter (nm)	22	100	1000	10—20	10—20	10—20
5 min after mixing with RPMI 1640;						
Particle diameter (nm)	25.6	158.1	1089.8	10	10.4	8.7
Size distribution (nm)	22.1—225.9	126.4—594.6	875.7—2795.5	8.4—39.6	8.4—39.6	6.9—70.7
3 h after mixing with RPMI 1640;						
Particle diameter (nm)	27.3	155.5	1326.3	13.7	16.6	13.4
Size distribution (nm)	22.1—104.1	126.4—403.7	1062.7—3395.0	12.4—153.4	15.0—225.9	12.4—186.1

3.2. Uptake test

The particle number of three types of the fluorescent particles and the Quantum Dots taken up by the PC12 cells or adhered to the cell membrane was quantified with a spectrofluorophotometer. We confirmed that there was no fluorescence in the PBS used for the third washing after the exposure. Therefore, the fluorescence in the supernatant and the pellet collected after the washing was probably derived of the particles taken up by the cells and adhered to the cell membrane, respectively.

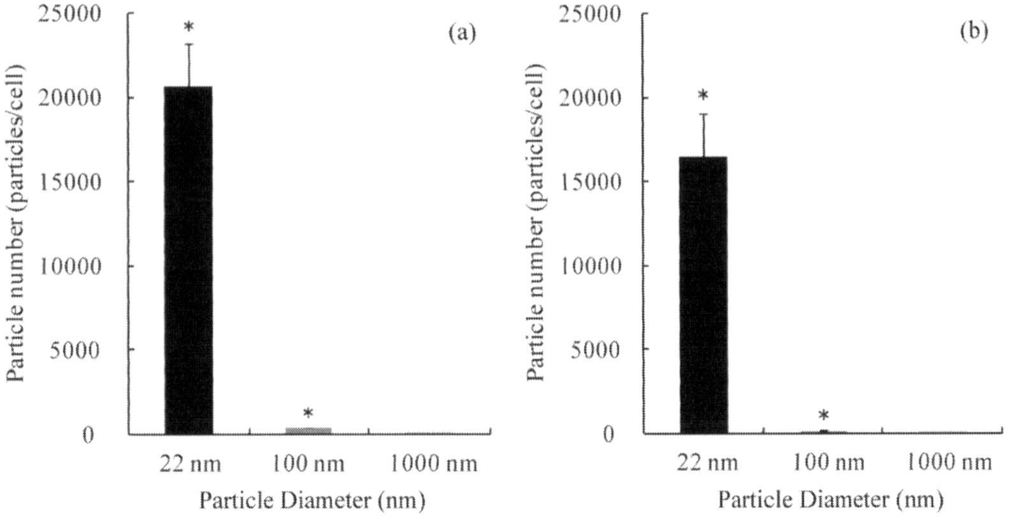

Figure 3. The particle number of three types of the fluorescent particles (22 nm, 100 nm, 1000 nm) in the supernatant including the particles taken up by the PC12 cells (a) or in the pellet including the particles adhered to the cell membrane (b). The PC12 cells were exposed 0.5% (v/v) fluorescent particles containing RPMI 1640 medium and incubated in a humidified atmosphere with 5% CO_2 at 37°C for 3 hours.

* $p < 0.05$ versus 1000 nm-particle-exposed cells.

Figure 3 shows the particle number in the supernatant and the pellet in the fluorescent particle-exposed PC12 cells. The particle number in the supernatant was considerably higher in the 22 nm-particle-exposed cells (20654 particles/cell) than that in the 100 nm-particle-exposed cells (390 particles/cell) and in the 1000 nm-particle-exposed cells (0.02 particles/cell). The total surface area of the fluorescent particles in the supernatant was also higher in the 22 nm-particle-exposed cells (31.2 μm^2/cell) than that in the 100 nm-particle-exposed cells (12.3 μm^2/cell) and in the 1000 nm-particle-exposed cells (0.1 μm^2/cell). Similarly, the particle number in the pellet was obviously higher in the 22 nm-particle-exposed cells (16468 particles/cell) than that in the 100 nm-particle-exposed cells (139 particles/cell) and in the 1000 nm-particle-exposed cells (0.23 particles/cell). The total surface area of the fluorescent particles in the pellet was calculated 25.0 μm^2/cell in the 22 nm-particle-exposed cells, 4.4 μm^2/cell in the 100 nm-particle-exposed cells and 0.7 μm^2/cell in the 1000 nm-particle-exposed cells.

However, the particle weight in the supernatant was higher in the 100 nm-particle-exposed cells (0.41 pg/cell) than that in the 22 nm-particle-exposed cells (0.23 pg/cell) and in the 1000 nm-particle-exposed cells (0.025 pg/cell), as shown in figure 4. It was calculated that the percent of the dose in the supernatant was 0.12 percent/10^6 cells in the 22 nm-particle-exposed cells, 0.22 percent/10^6 cells in the 100 nm-particle-exposed cells and 0.01 percent/10^6 cells in the 1000 nm-particle-exposed cells. On the other hand, the particle weight in the pellet was higher in the 1000 nm-particle-exposed cells (0.072 pg/cell) than that in the 22 nm-particle-exposed cells (0.056 pg/cell) and in the 100 nm-particle-exposed cells (0.044 pg/cell). The percent of the dose in the pellet was calculated 0.10 percent/10^6 cells in the 22 nm-particle-exposed cells, 0.08 percent/10^6 cells in the 100 nm-particle-exposed cells and 0.13 percent/10^6 cells in the 1000 nm-particle-exposed cells.

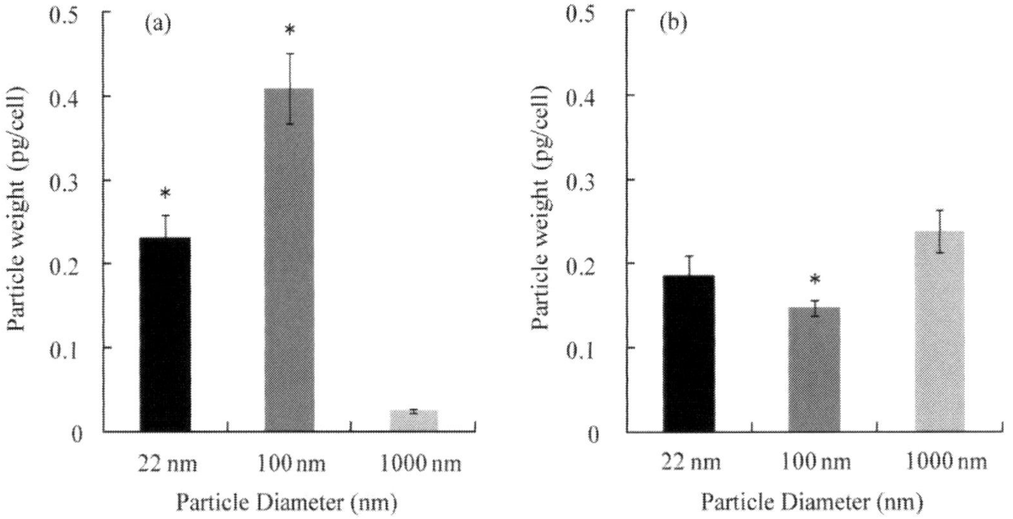

Figure 4. The particle weight of three types of the fluorescent particles (22 nm, 100 nm, 1000 nm) in the supernatant including the particles taken up by the PC12 cells (a) or in the pellet including the particles adhered to the cell membrane (b). The PC12 cells were exposed 0.5% (v/v) fluorescent particles containing RPMI 1640 medium in a humidified atmosphere with 5% CO_2 at 37°C for 3 hours.
* $p < 0.05$ versus 1000 nm-particle-exposed cells.

Figure 5 shows the particle number in the supernatant and the pellet in the Quantum Dot-exposed PC12 cells. The particle number in the supernatant was considerably higher in the carboxyl base-functionalized particle-exposed cells (26488 particles/cell) than that in the amino base-functionalized particle-exposed cells (4321 particles/cell) and in the non-base-functionalized particle-exposed cells (1769 particles/cell). Also, the percent of the dose and the total surface area of the Quantum Dots in the supernatant were higher in the carboxyl base-functionalized particle-exposed cells (0.055 percent/10^6 cells and 74.9 μm^2/cell, respectively) than those in the non-base-functionalized Quantum Dots (0.015 percent/10^6 cells and 12.2 μm^2/cell, respectively). Conversely, they were lower in the amino base-functionalized Quantum Dots than the non-base-functionalized Quantum Dots; 0.009 percent/10^6 cells and 5.0 μm^2/cell, respectively.

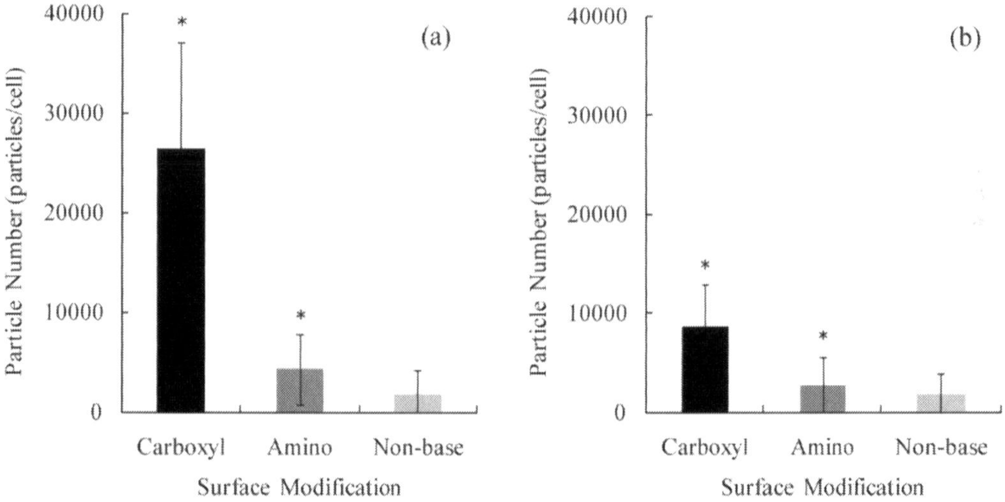

Figure 5. The particle number of three types of the Quantum Dots (carboxyl base-functionalized particles, amino base-functionalized particles, non-base-functionalized particles) in the supernatant including the particles taken up by the PC12 cells (a) or in the pellet including the particles adhered to the cell membrane (b). The PC12 cells were exposed 0.5% (v/v) Quantum Dots containing RPMI 1640 medium in a humidified atmosphere with 5% CO_2 at 37°C for 3 hours.
* $p < 0.05$ versus non-base-functionalized particle-exposed cells.

The particle number in the pellet was also higher in the carboxyl base-functionalized particle-exposed cells (8590 particles/cell) than that in the amino base-functionalized particle-exposed cells (2701 particles/cell) and in the non-base-functionalized particle-exposed cells (1859 particles/cell). The total surface area and the percent of the dose of these Quantum Dots in the pellet were calculated 24.3 μm^2/cell and 0.018 percent/10^6 cells in the carboxyl base-functionalized Quantum Dots, 7.6 μm^2/cell and 0.006 percent/10^6 cells in the amino base-functionalized Quantum Dots and 5.3 μm^2/cell and 0.015 percent/10^6 cells in the non-base-functionalized Quantum Dots.

3.3. Cell viability test

Figure 6 shows the cell viability of the PC12 cells exposed the Quantum Dots or the fluorescent particles for 24 hours. The cell viability was little changed between three types of the Quantum Dots exposed cells up to 1% particle concentration in the medium, compared to the negative control group. Similarly, the cell viability exposed three types of the fluorescent particles was almost same as the negative control until 1% particle concentration in the medium. However, the cell viability was decreased at the 10% particle concentration and it was different by the particle size. Although the 22 nm-particle-exposed cells were completely died, the 100 nm-particle-exposed cells and the 1000 nm-particle-exposed cells were still alive to some extent. The cell viability in the 100 nm-particle-exposed cells and the 1000 nm-particle exposed cells was 73% and 54%, respectively.

4. Discussion

The particle diameter of the fluorescent particles was little changed in the 22 nm-particles and the 100 nm-particles. Conversely, the 1000 nm-particles were aggregated 3 hours after the mixture with RPMI 1640 medium. It is suggested that the 1000 nm-particles were settled out in the medium and aggregated due to the high particle concentration on the bottom. According to Stokes' theorem, the sedimentation distance of these particles 3 hours after the mixture with the medium (viscosity: 0.71 cP) is calculated as 2.4×10^{-4} mm in the 22 nm-particles, 5.0×10^{-3} mm in the 100 nm-particles and 0.50 mm in the 100 nm-particles. Therefore, the 1000 nm-particles were accumulated on the bottom of the dish faster than the 22 nm-particles and the 100 nm-particles, and consequently they might be aggregated.

The particle diameter of the Quantum Dots was increased 3 hours after the mixture with RPMI 1640 medium. It is suggested that nanoparticles are likely to be aggregated in medium because of the large amount of electrolyte ions and the protein adsorption [6, 8]. Sur *et al.* (2010) showed that the particles diameter in the medium was increased in the range of 26 to 37 nm due to the protein adsorption [6]. In this study, the mean particle diameter of the Quantum Dots was also increased in the range of 3.7 nm to 6.2 nm. Therefore, the Quantum Dots were possibly adsorbed by the proteins in the medium and consequently the particle size became larger than the original size.

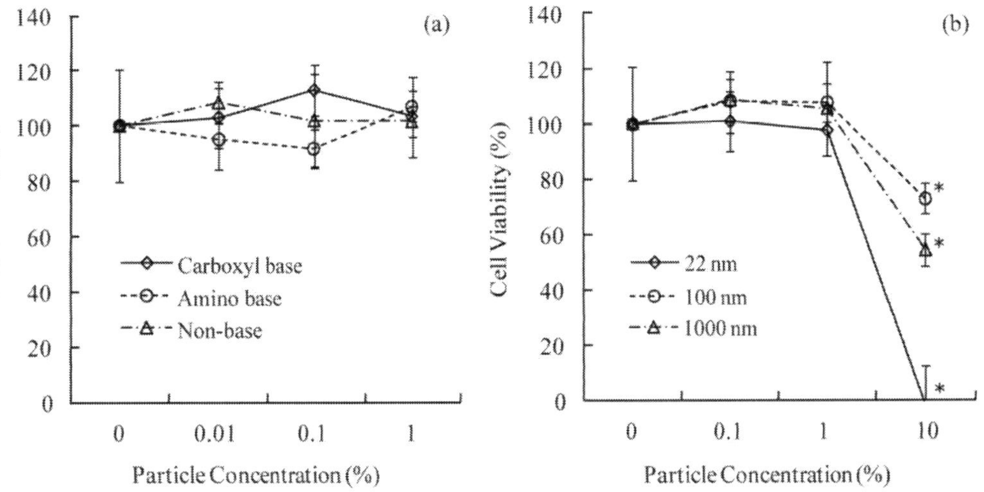

Figure 6. The viability of the PC12 cells exposed three types of the Quantum Dots or three types of the fluorescent particles. (a); The cell viability of the PC12 cells exposed the carboxyl base-functionalized particles (open rhomboid), the amino base-functionalized particles (open circle) or the non-base-functionalized particles (open triangle) mixed with the RPMI 1640 medium (particle concentration; 0.01%, 0.1% or 1%) for 24 hours. (b); The cell viability of the PC12 cells exposed the 22 nm-particles (open rhomboid), the 100 nm-particles (open circle) or the 1000 nm-particles (open triangle) mixed with the RPMI 1640 medium (particle concentration; 0.1%, 1% or 10%) for 24 hours. * $p < 0.05$ versus untreated controls.

The particle number of the fluorescent particles taken up by the PC12 cells was increased in a size-dependent manner (figure 3). The 22 nm-particles were taken up by the cells over 20000 particles in a cell and the particle number was about 53 times higher than that of the 100 nm-particles. Furthermore, the 1000 nm-particles were rarely taken up by the cells, which was a same result as Rejman's study [4]. Similarly, the particle number of the fluorescent particles adhered to the cell membrane was increased in a size-dependent manner. Therefore, we suggest that the particle number of particles taken up by nerve cells may be exponentially increased as the particle size becomes small. On the other hand, the particle weight/the percent of the dose was about 1.8 times higher in the 100 nm-particles than that in the 22 nm-particles (figure 4). It means that the 100 nm-particles were taken up by the cells more effectively than the 22 nm-particles. Rejman *et al.* (2004) showed that the effects of the uptake of the particles by caveolae-mediated endocytosis and clathrin-mediated endocytosis were differently by the particle size [4]. Therefore, the 100 nm-particles might be taken up effectively by such uptake mechanisms. In addition, the particle weight of the 1000 nm-particles in the pellet was the highest among the particle types and also higher than that in the supernatant. Therefore, it was suggested that the 1000 nm-particles were interacted with the cells and adhered to the cell membrane, but it was too large to be taken up by the cells.

The cell viability test of the fluorescent particles showed that the 22 nm-particles completely killed the cells but the 100 nm-particle-exposed cells and the 1000 nm-particle-exposed cells were still alive (figure 6). It might be due to the high surface area in the 22 nm-particles (31.2 μm^2/cell) rather than that in the 100 nm-particles (12.3 μm^2/cell) and in the 1000 nm-particles (0.1 μm^2/cell). Oberdorster *et al.* (2000) showed that the cytotoxicity of TiO_2 particles (20 nm and 250 nm) was related to the surface area [9]. However, despite the surface area of the 1000 nm-particles was lower than that of the 100 nm-particles, the cell viability in the 1000 nm-particle-exposed cells was decreased to 54% more than the 100 nm-particle-exposed cells (73%). It might be caused by the sedimentation of the 1000 nm-particles during the exposure. As shown in figure 1, the 1000 nm-particles were aggregated and these aggregates probably settled out in the dish. Consequently, unexpected higher amount of the 1000 nm-particles might be interacted with the adherent PC12 cells. Further studies are needed to determine the accurate relationship between the particle size and the cell viability.

The particle number of the Quantum Dots taken up by the PC12 cells was different by the surface modification (figure 5). The particle number and the percent of the dose of these particles in the supernatant were higher in the carboxyl base-functionalized particles but lower in the amino base-functionalized particles than those of the non-base-functionalized particles. The positively charged particles were likely to be taken up by the cells effectively because the cell membrane was charged negatively. Geys *et al.* (2008) showed that the isoelectric point of the carboxyl-functionalized Quantum Dots and the amine-functionalized Quantum Dots was pH 1.71 and 3.19, respectively, and both particles were charged negatively in the medium [10]. Therefore, the amino base-functionalized particles might be less interacted with the cells. Nevertheless, the carboxyl base-functionalized particles were taken up more frequently than the non-base-functionalized particles. Although the uptake mechanism of these particles was still unknown, it was clearly that the particles were selectively taken up by the cells according to the surface modification.

The cell viability test of the Quantum Dots showed almost no difference compared to the negative control cells (figure 6). It may be changed between the particle types because the surface area of the Quantum Dots taken up by the cells was considerably higher in the carboxyl base-functionalized Quantum Dots (74.9 μm^2/cell) than that in the amino base-functionalized Quantum Dots (5.0 μm^2/cell) and in the non-base-functionalized Quantum Dots (12.2 μm^2/cell). Similarly, Geys *et al.* (2008) showed that Quantum Dots caused pulmonary vascular thrombosis, with carboxyl-functionalized Quantum Dots being more potent in inducing this effect than amine-functionalized Quantum Dots [10]. Therefore, it should be concerned more in further studies how the surface modification of the particles affects to the cells.

5. Conclusion

In this study, the uptake of the nanoparticles in the PC12 cells was quantified and the effects of the particle size and the surface modification were tested. The large number of the 22 nm-fluorescent particles was taken up by the PC12 cells than the 100 nm-particles. However, the 100 nm-particles were taken up more frequently than the 22 nm-particles. Conversely, the 1000 nm-particles were little taken up by the cells. It is suggested that nanoparticles are likely to be taken up by the cells than micro-sized particles. Furthermore, the uptake of the Quantum Dots was changed by the surface modification. Although the carboxyl base-functionalized Quantum Dots were taken up by the cells more frequently than the non-base-functionalized particles, the amino base-functionalized Quantum Dots were less taken up by the cells. These findings suggest that the nanoparticles are taken up by cells in size-dependent and functional-dependent manners. Although the uptake mechanism of the nanoparticles is still too complicated to clarify, our present study is valuable to consider the biological risk of the nanoparticles to nerve tissues.

References

[1] Geiser M, Rothen-Rutishauser B, Kapp N, Schürch S, Kreyling W, Schulz H, Semmler M, Im Hof V, Heyder J and Gehr P 2005 *Environ. Health Perspect.* **113** 1555-60
[2] Stearns R C, Paulauskis J D and Godleski J J *Am. J. Respir. Cell Mol. Biol.* **24** 108-15
[3] Elder A, Gelein R, Silva V, Feikert T, Opanashuk L, Carter J, Potter R, Maynard A, Ito Y, Finkelstein J and Oberdörster G 2006 *Environ. Health Perspect.* **114** 1172-8
[4] de Lorenzo A J 1970 *Taste and Smell in Vertebrates* ed Wolstenholme G and Knight J (London: Churchill) chapter 9 pp 151-76
[5] Rejman J, Oberle V, Zuhorn I S and Hoekstra D 2004 *Biochem. J.* **377** 159-69
[6] Sur I, Cam D, Kahraman M, Baysal A and Culha M 2010 *Nanotechnology* **21** 175104
[7] Hussain S M, Javorina A K, Schrand A M, Duhart H M, Ali S F and Schlager J J 2006 *Toxicol. Sci.* **92** 456-63
[8] Wise J P, Goodale B C, Wise S S, Craig G A, Pongan A F, Walter R B, Thompson W D, Ng A K, Aboueissa A M, Mitani H, Spalding M J and Mason M D 2010 *Aquat. Toxicol.* **97** 34-41
[9] Oberdörster G, Finkelstein J N, Johnston C, Gelein R, Cox C, Baggs R and Elder A C 2000 *Res. Rep. Health Eff. Inst.* **96** 5-74
[10] Geys J, Nemmar A, Verbeken E, Smolders E, Ratoi M, Hoylaerts M F, Nemery B and Hoet P H 2008 *Environ. Health Perspect.* **116** 1607-13

Journal of Physics: Conference Series **304** (2011) 012050 doi:10.1088/1742-6596/304/1/012050

Assessment of phototoxicity, skin irritation, and sensitization potential of polystyrene and TiO_2 nanoparticles

Yoon-Hee Park[1], Sang Hoon Jeong[1], Sang Min Yi[1], Byeong Hyeok Choi[1], Yu-Ri Kim[2], In-Kyoung Kim[2], Meyoung-Kon Kim[2], and Sang Wook Son[1]

[1] Laboratory of Cell Signaling and Nanomedicine, Department of Dermatology and Division of Brain Korea 21 Project for Biomedical Science, Korea University College of Medicine, Seoul, Korea

[2] Department of Biochemistry & Molecular Biology, Korea University College of Medicine, Seoul, Korea

Abstract. The human skin equivalent model (HSEM) is well known as an attractive alternative model for evaluation of dermal toxicity. However, only limited data are available on the usefulness of an HSEM for nanotoxicity testing. This study was designed to investigate cutaneous toxicity of polystyrene and TiO_2 nanoparticles using cultured keratinocytes, an HSEM, and an animal model. In addition, we also evaluated the skin sensitization potential of nanoparticles using a local lymph node assay with incorporation of BrdU. Findings from the present study indicate that polystyrene and TiO_2 nanoparticles do not induce phototoxicity, acute cutaneous irritation, or skin sensitization. Results from evaluation of the HSEMs correspond well with those from animal models. Our findings suggest that the HSEM might be a useful alternative model for evaluation of dermal nanotoxicity.

1. Introduction

Despite wide use of NPs, few studies have reported on the toxicity (safety) of manufactured NPs to humans and the environment. Therefore, studies into *in vitro* and *in vivo* toxicity of NPs are critically needed.

Phototoxicity can be defined as a skin inflammatory reaction by topical application of chemicals, drugs, and subsequent exposure to light, particularly ultraviolet A (UVA) radiation (320~400 nm) [1,2].

The human skin equivalent model (HSEM) is well known as an attractive alternative model for evaluation of dermal toxicity. However, only a limited number of papers have suggested the usefulness of the HSEM as a screening method for determination of the dermal irritation potential of NPs [3,4]. To the best of our knowledge, little data are available on the usefulness of an HSEM for phototoxicity testing of NPs.

In the present study, using cultured keratinocytes, an HSEM, and an animal model, we investigated the dermal phototoxicity and irritation potential of polystyrene and TiO_2 NPs. In addition, the murine local lymph node assay has been used for evaluation of the skin sensitization potential of NPs.

2. Materials and methods

2.1. Cell culture
The human immortalized HaCaT human keratinocyte cell line was grown at 37°C and 5% CO_2 in Dulbecco's modified Eagle's medium (DMEM; Hyclone, Logan, UT, USA) supplemented with 10% fetal bovine serum (FBS; Hyclone), and 1% penicillin-streptomycin (10,000 unit /ml).

2.2. Nanoparticle preparation
Polystyrene latex beads (amine-modified 50 nm labeled with blue and fluorophores) and TiO_2 were purchased from Sigma-Aldrich (St. Lois, MO, USA). Polystyrene and TiO_2 were diluted in 1X concentration of PBS. In order to produce a stable, less-aggregated nanocrystalline suspension, suspended TiO_2 in PBS were dispersed by sonication for 30 min at 4°C immediately prior to preparation of treatment-dilutions in serum-free DMEM.

2.3. Nanoparticle characterization
To obtain photographs, diameter, size distribution, and morphology information, polystyrene and TiO_2 were morphologically characterized by transmission electron microscopy (TEM) using Tecnai 20 (FEI Co., Eindhoven, Netherlands) at an acceleration voltage of 200 kV.

2.4. MTT viability assays
The thiazolyl blue tetrazolium bromide (3-[4,5]dimethylthiazol-2,5 dephenyltetrazolium bromide (MTT)) assay was used as described to determine the cytotoxicity of polystyrene and TiO_2. HaCaT cells were plated in 96-well multititer plates (5,000 cells per well) and treated with 1~100 μg/ml polystyrene and 25~1000 μg/ml TiO_2 for 24 and 48 h. MTT (0.5 mg/ml) was then added to each well, and cells were further incubated for 4 h at 37°C. Barillet *et al.* suggested that NPs can interact with the substrate during toxicity assessment [5]. Therefore, to reduce such a potential interference of NPs in the MTT assay, NPs were allowed to sediment for 1 h and a 50 μl solution from each well was then transferred to another plate.

2.5. EpiDerm skin irritation test
3D EpiDerm^{TM} models (EPI-200), which were used as HSEMs, were obtained from the MatTek Corporation. On arrival, the gauze pad, which covered the top of the HSEMs, was removed; HSEMs were transferred into 6-well plates containing 0.9 ml of the pre-warmed assay medium. And HSEMs were incubated for 1 h at 37°C and 5% CO_2. The medium was then replaced with pre-warmed, fresh assay medium, and models were pre-incubated overnight at 37°C in a 5% CO_2 atmosphere. HSEMs were treated with 1000 μg/ml polystyrene, 100 μg/ml TiO_2, PBS as a negative control, or 5% SDS as a positive control on top of the models for 1 h.

2.6. Draize skin irritation test
Skin irritation potentials for polystyrene and TiO_2 were evaluated in rabbits using the Draize score test. Rabbits weighing 2.5 - 3.5 kg were acclimatized prior to the beginning of the study. On day 0 of the test period, backs were clipped free of hair. Several layers of skin were removed with adhesive tape from one half of the shaved back and areas of skin abrasion were created. Each rabbit was treated with 0.5 ml of the 1000 μg/ml polystyrene and 100 μg/ml TiO_2 solutions, which were applied over a ~6.25 cm^2 area of hair-free skin.

2.7. 3T3 NRU phototoxicity test
The 3T3 NRU phototoxicity test was performed as described in guidelines of the Organisation for Economic Co-operation and Development (OECD) 432 (OECD, 2004).

Absorbance was measured at 540 nm. NRU phototoxicity was assessed using photo-irritancy factor (PIF) and mean photo effect (MPE), according to a previous report [6].

2.8. EpiDerm skin phototoxicity test

HSEMs, 3D EpiDerm™ models (EPI-200), were supplied from MatTek, USA. Prior to the test, HSEMs were pre-incubated in fresh assay medium for 1 h at 37°C and 5 % CO_2. HSEMs were then transferred into 6-well plates containing 0.9 ml of the pre-warmed assay medium.

2.9. Skin phototoxicity test using an animal model

Five-week-old female, Hartley albino guinea pigs (250~300 g) were used in the study. Five female guinea pigs per group received topical application of polystyrene on the back skin, which had been clipped and shaved. Each test substance of 1000 μg/ml polystyrene and 100 μg/ml TiO_2 solutions was applied to an area measuring 1.5 cm x 1.5 cm (0.05 ml/site). Skin reactions were graded in accordance with the Draize scoring system, as follows [7,8]: Score 0: no erythema or no edema; Score 1: barely perceptible erythema or edema; Score 2: well-defined erythema or slight edema; Score 3: moderate to severe erythema or moderate edema; Score 4: severe erythema or edema. Dermal phototoxicity was tested according to the grades described in the OECD Guidelines 404 (2002).

2.10. Local lymph node assay (LLNA)

Female CBA/N mice were purchased from SLC Japan Co. Ltd (Shizuoka, Japan). Female mice were randomly allocated to four groups (four mice per group). A 25 μl volume of the concentration ranging from 10 to 1000 μg/ml polystyrene and from 10 to 1000 μg/ml TiO_2 was applied to the dorsum of both ears daily for three consecutive days. Mice were not treated with nanoparticles on day 4. A single interperitoneal injection of 0.5 ml of BrdU solution (5 mg per mouse per injection) was given to the mice on day 5. On day 6, the auricular lymph nodes from the ears of each mouse were removed, weighed, and stored at -20°C until measurement of the level of BrdU incorporation by enzyme-linked immunosorbent assay (ELISA).

2.11. Statistics

Data were analyzed using Prism 3.02 software (GraphicPad Software Inc., La Jolla, California) by one-way analysis of variance (ANOVA), followed by Dunnett's method. P values of < 0.05 were considered statistically significant.

3. Results

3.1. MTT assay

The MTT assay was used for evaluation of the cytotoxicity of polystyrene and TiO_2. HaCaT cells were treated with medium containing various concentrations of 1~100 μg/ml polystyrene and 25~1000 μg/ml TiO_2, and cell viability was determined at 24 and 48 h. Cytotoxicity of polystyrene increased in a dose-dependent manner. However, there were no differences in cell viability for TiO_2.

3.2. EpiDerm skin irritation test

To evaluate irritation of polystyrene and TiO_2 on HSEMs, HSEMs were treated with 1000 μg/ml polystyrene and 100 μg/ml TiO_2. No significant differences in viability were observed for polystyrene and TiO_2. However, 5% SDS provided as a positive control induced a significant reduction in viability.

3.3. Draize skin irritation test

Draize skin irritation tests were conducted using polystyrene and TiO_2; 1000 μg/ml polystyrene and 100 μg/ml TiO_2 were applied to rabbit skin. However, both polystyrene and TiO_2 were found to have no irritation effect on rabbits. No observable edema or erythema was observed on the abrasion site of rabbit skin after exposure for 24 or 72 h for any rabbit, and no skin irritation was observed after 24 or 72 h of treatment. No erythema or edema was observed on the non-abrasion site of rabbit skin.

3.4. 3T3 NRU phototoxicity test
The 3T3 neutral red uptake phototoxicity test (NRU PT) is designed to detect phototoxicity induced by polystyrene, TiO_2, and UVA light using an *in vitro* cytotoxicity assay in the Balb/c 3T3 mouse fibroblast cell line. Polystyrene on UV- irradiated cells was more cytotoxic than on non-UV-irradiated cells at concentrations higher than 75 μg/ml polystyrene. However, photo-irritancy factor (PIF) and mean photo effect (MPE) values of polystyrene were calculated to 1.149 and -0.005, respectively, indicating non-phototoxicity. No differences in viability were observed for TiO_2 on UV- irradiated cells and non-UV-irradiated cells. TiO_2 has an MPE of -0.046, indicating its non-phototoxicity and the PIF value of TiO_2 was not calculated. Therefore, both polystyrene and TiO_2 were found to be non-cytotoxic at the tested concentrations.

3.5. EpiDerm skin phototoxicity test
Phototoxicity testing in the HSEMs confirmed phototoxicity of the 3T3 NRU test for polystyrene and TiO_2. Polystyrene and TiO_2 did not exhibit phototoxicity in the HSEMs.

3.6. Skin phototoxicity test using an animal model
Polystyrene at 1000 μg/ml and TiO_2 at 100 μg/ml were evaluated for skin phototoxicity in guinea pigs. No visible changes with or without UV exposure were observed at any of the time points during evaluation of animals treated with polystyrene and TiO_2. No erythema or edema was detected. Results from evaluation of the HSEMs correspond well with those from animal models.

3.7. LLNA test
A non-RI LLNA test was performed for evaluation of the skin sensitization potential of polystyrene and TiO_2. The SI value of PC was in excess of 2, indicating a positive response of skin sensitization. However, SI values did not increase in polystyrene and TiO_2 and SI value of eugenol as a positive control was only increased.

4. Discussion
The skin is the largest organ of the body and an important route of entry for environmental toxicants, including NPs, into the body; as a result, assessment of skin penetration of NPs has attracted a great deal of attention. Our previous study has shown that quantum dot nanoparticles could actually penetrate through the stratum corneum of human skin [9]. Recent papers have also shown that, topically applied, NPs may permeate normal or damaged skin and induce a health risk to humans [10,11]. Despite the great potentialities of nanotechnology, few NPs have been evaluated for their dermal toxicity potential.

Using the MTT assay, we evaluated the cytotoxic effects of polystyrene and TiO_2 NPs. Cell viability decreased with increasing polystyrene NPs concentration in a dose-dependent manner. TiO_2 NPs had no effect on cell viability. Similar findings were reported by Tian Xia *et al.* (2006) who found that polystyrene NPs can induce cell death; however, TiO_2 NPs did not demonstrate cellular toxicity [12]. However, TiO_2 NPs induce allergic sensitization and lung inflammation [13]. And TiO_2 NPs have cytotoxicity and induce apoptosis [14].

Because they reflect bioavailability of phototoxic chemicals in the skin, HSEMs are regarded as a promising alternative approach designed to mimic *in vivo* models in humans as closely as possible [15,16]. HSEMs have been prevalidated in dermal studies, including phototoxicity, skin irritancy, and skin corrosivity of chemicals [16-18]. In the assessment for skin irritation and phototoxicity using HSEMs, cell viability was analyzed by MTT assay according to the protocol provided by the manufacturer.

Murine LLNA is used for measurement of proliferation in skin-draining lymph nodes following topical application of chemicals. The LLNA method should be used as an alternative to guinea pig models for the sensitization potential of chemicals [19,20]. Non-radioisotopic LLNA has recently been developed as a non-radioactive modification method based on assessment of BrdU incorporation into lymph node cells instead of radioisotopes (^{3}H-methyl thymidine) [20,21].

In conclusion, findings from the present study demonstrate that polystyrene and TiO_2 nanoparticles induce no phototoxicity, acute cutaneous irritation, or skin sensitization. LLNA results from this study indicated that polystyrene and TiO_2 NPs were not in themselves dermal sensitizers.

Acknowledgements
This research was supported by a grant (09152KFDA696) from the Korea Food & Drug Administration in 2009.

References

[1] Dijoux N, Guingand Y, Bourgeois C, Durand S, Fromageot C, Combe C, et al. Assessment of the phototoxic hazard of some essential oils using modified 3T3 neutral red uptake assay. *Toxicol In Vitro*. 2006 Jun;**20**(4):480-9.

[2] Epstein JH. Phototoxicity and photoallergy in man. J Am Acad Dermatol. 1983 Feb;8(2):141-7.

[3] Kishore AS, Surekha P, Murthy PB. Assessment of the dermal and ocular irritation potential of multi-walled carbon nanotubes by using in vitro and in vivo methods. *Toxicol Lett*. 2009 Dec 15;**191**(2-3):268-74.

[4] Park YH, Kim JN, Jeong SH, Choi JE, Lee SH, Choi BH, et al. Assessment of dermal toxicity of nanosilica using cultured keratinocytes, a human skin equivalent model and an in vivo model. *Toxicology*. 2010 Jan 12;**267**(1-3):178-81.

[5] Barillet S, Simon-Deckers A, Herlin-Boime N, Mayne-L'Hermite M, Reynaud C, Cassio D, et al. Toxicological consequences of TiO2, SiC nanoparticles and multi-walled carbon nanotubes exposure in several mammalian cell type: an in vitro study. *J Nanopart Res*. 2009.

[6] Onoue S, Kawamura K, Igarashi N, Zhou Y, Fujikawa M, Yamada H, et al. Reactive oxygen species assay-based risk assessment of drug-induced phototoxicity: classification criteria and application to drug candidates. *J Pharm Biomed Anal*. 2008 Aug 5;**47**(4-5):967-72.

[7] Draize J, Calvery GWH. Methods for the study of irritation and toxicity of substances applied topically to the skin and mucous membranes. *J Pharmacol Exp Therapeut*. 1944;**82**:377-90.

[8] Mandawgade SD, Patravale VB. Development of SLNs from natural lipids: application to topical delivery of tretinoin. *Int J Pharm*. 2008 Nov 3;**363**(1-2):132-8.

[9] Jeong SH, Kim JH, Yi SM, Lee JP, Sohn KH, Park KL, et al. Assessment of penetration of quantum dots through in vitro and in vivo human skin using the human skin equivalent model and the tape stripping method. *Biochem Biophys Res Commun*. 2010 Apr 9;**394**(3):612-5.

[10] Larese FF, D'Agostin F, Crosera M, Adami G, Renzi N, Bovenzi M, et al. Human skin penetration of silver nanoparticles through intact and damaged skin. *Toxicology*. 2009 Jan 8;**255**(1-2):33-7.

[11] Wu J, Liu W, Xue C, Zhou S, Lan F, Bi L, et al. Toxicity and penetration of TiO2 nanoparticles in hairless mice and porcine skin after subchronic dermal exposure. *Toxicol Lett*. 2009 Dec 1;**191**(1):1-8.

[12] Xia T, Kovochich M, Brant J, Hotze M, Sempf J, Oberley T, et al. Comparison of the abilities of ambient and manufactured nanoparticles to induce cellular toxicity according to an oxidative stress paradigm. *Nano Lett*. 2006 Aug;**6**(8):1794-807.

[13] Larsen ST, Roursgaard M, Jensen KA, Nielsen GD. Nano titanium dioxide particles promote allergic sensitization and lung inflammation in mice. *Basic Clin Pharmacol Toxicol*. 2009 Feb;**106**(2):114-7.

[14] Zhao J, Bowman L, Zhang X, Vallyathan V, Young SH, Castranova V, et al. Titanium dioxide

(TiO2) nanoparticles induce JB6 cell apoptosis through activation of the caspase-8/Bid and mitochondrial pathways. *J Toxicol Environ Health A*. 2009;**72**(19):1141-9.
[15] Jirova D, Kejlova K, Bendova H, Ditrichova D, Mezulanikova M. Phototoxicity of bituminous tars-correspondence between results of 3T3 NRU PT, 3D skin model and experimental human data. *Toxicol In Vitro*. 2005 Oct;**19**(7):931-4.
[16] Lelievre D, Justine P, Christiaens F, Bonaventure N, Coutet J, Marrot L, et al. The EpiSkin phototoxicity assay (EPA): development of an in vitro tiered strategy using 17 reference chemicals to predict phototoxic potency. *Toxicol In Vitro*. 2007 Sep;**21**(6):977-95.
[17] Netzlaff F, Lehr CM, Wertz PW, Schaefer UF. The human epidermis models EpiSkin, SkinEthic and EpiDerm: an evaluation of morphology and their suitability for testing phototoxicity, irritancy, corrosivity, and substance transport. *Eur J Pharm Biopharm*. 2005 Jul;**60**(2):167-78.
[18] Portes P, Pygmalion MJ, Popovic E, Cottin M, Mariani M. Use of human reconstituted epidermis Episkin for assessment of weak phototoxic potential of chemical compounds. *Photodermatol Photoimmunol Photomed*. 2002 Apr;**18**(2):96-102.
[19] Chamberlain M, Basketter DA. The local lymph node assay: status of validation. *Food Chem Toxicol*. 1996 Oct;**34**(10):999-1002.
[20] Takeyoshi M, Noda S, Yamasaki K, Kimber I. Advantage of using CBA/N strain mice in a non-radioisotopic modification of the local lymph node assay. *J Appl Toxicol*. 2006 Jan-Feb;**26**(1):5-9.
[21] Lee JK, Park JH, Park SH, Kim HS, Oh HY. A nonradioisotopic endpoint for measurement of lymph node cell proliferation in a murine allergic contact dermatitis model, using bromodeoxyuridine immunohistochemistry. *J Pharmacol Toxicol Methods*. 2002 Jul-Aug;**48**(1):53-61.

Relationship between Surface Modifications of Nanoparticle and Invasion into Suspension Cells – The sequel of Nanosafe 2008 –

Y Matsui[1], N Sakai[1], A Tsuda[2] and M Yoneda[1]

[1] Graduate School of Engineering, Kyoto University, Katsura, Kyoto 6158540, Japan

[2] Department of Environmental Health, Harvard School of Public Health, 665 Huntington Avenue, Boston, MA 02115, USA

E-mail: ymatsui@risk.env.kyoto-u.ac.jp

Abstract. Nanomaterials have a variety of properties for each material. There is little information available on which kinds of material properties have effects on toxicity and kinetics. This paper presents that a relationship between material properties and hazard data by undertaking a bibliographical survey at first. With respect to cytotoxicity, it probably depends mainly on the particle volume dose and to a certain degree on particle solubility. It can be concluded from these results that there is a relationship between material properties and hazard data. Many activities involving nano risk are occurring all over the world. Secondly, we assayed actually for cellular uptake of three kinds of Quantum dots (15 nm, 5.5×10^{12} particles/ml) to demonstrate our result of bibliographical survey. Three different surface modification quantum dots (non-modification, -COOH, -NH3) were mixed with floating Jurkat cells in each. After thirty minute, we washed these cells three times and detected fluorescence by flow cytometer. Almost all the carboxylate particles invaded a cell, about 60% aminated them also invaded and few non-modification particles were taken up. Nanomaterials are often very broadly categorized and named based upon their basic material composition or product shape. Our results confirm that we have to examine which physical-chemical properties affect some adverse effects for each nanomaterial.

1. Introduction

More than 300 products involving nanotechnology ranging from sunscreens to semiconductors are already on the global market [1]. Supported by government funding, multiple nano-related research projects are being carried out at an unprecedentedly rapid pace. Applications of such research include nano-colloidal platinum solution and nano-sized titanium powder, both wildly used in the cosmetics market. However, many companies have come to withdraw from such research because of the potential adverse effects or uncertain risks [2]. The purpose of this work is to seek out the risk factors of nanomaterials using objective methods and to develop software that can, by revealing the potential risks, support nanotechnology developers. Additionally, by using floating cells and quantum dots, we measured the dose of uptake to these cells and suggested the relationship between surface modification and uptake. Nanomaterials possess various properties of powder engineering: specific surface area, zeta potential, van der Waals forces, crystal type and so on. We attempted to determine which hazards correspond to which aspects of nanomaterials. It is well known that the diameter or surface area of nanomaterials correlates highly with inflammation response. These results appear reasonable and suit reaction kinetics. In this study, we focused on other parameters and found correlations. We then attempted to predict whether these hazardous nanomaterials could invade or transfer into the human body or not (Figure 1).

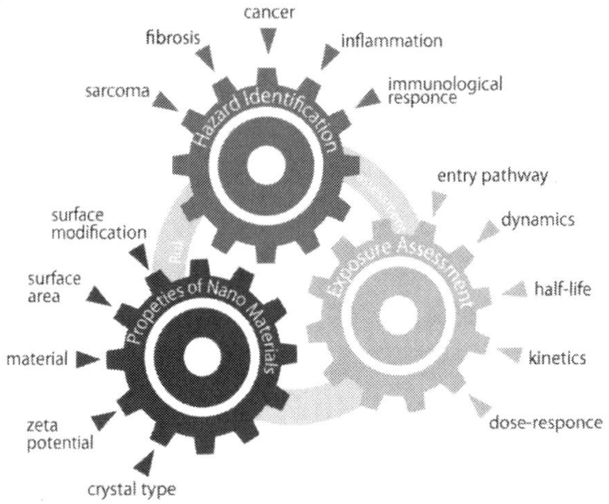

Figure 1. This shows the main concept of our survey on nano risk assessment.

Nanomaterials have many kinds of characteristics such as zeta potential, specific surface modification, and diameter (black gear). On the other hand, many kinds of hazard data were shown in past reports such as data relating to inflammation and cancer (gray gear). We predicted which nanomaterial characteristics relate to biomarker values indicating various hazards. Half life and kinetics will depend on the particular features of nanomaterials (light gray gear).

It is well known that a diameter or surface area of nanoparticles correlates highly with an inflammation around alveoli region after inhalation [3-5]. On the other hand, almost all the past research treated various nanomaterials such as titanium dioxide, silver and carbon nanotube in each report. These reports suggest that the relationship between material properties and adverse effect of human health, however, tightly correlates have not been obtained yet. We are trying to develop risk assessment software which can suggest a clear direction of controlling manufacturing processes to support nanotechnology developers from Nanosafe 2008 [6,7]. By using this software, cell mortality may have relevance to a zeta potential, surface modification and solubility. We focused on surface modification and measured a dose of uptake to floating cells actually.

2. Methods

2.1. Bibliographical survey

To better estimate the potential risk of nanomaterials, correlations were investigated between material properties and various biomarkers indicating adverse effects on humans. Nanomaterials have a variety of properties such as solubility, iso-electric point, crystal shape, BET specific surface area and so on. I and Professor Yamaguchi at the University of Tokyo predicted a relationship between material properties and hazard data by undertaking a bibliographical survey. For a viability analysis of various types of particles, many papers were selected. The selected papers consisted of experiments involving viability assays performed *in vitro* on nano and micro particles of different chemical species. Defining a research domain via a set of queries is not a rudimentary task [8]. We use the following as queries to define the relationship between material parameters and nano risks to human health:

(A and B) or (A and B) or C (1)
A: Nano* or Ultrafine
B: Toxic*
C: Exposure or Dose or Injection or Administration or Inhale
D: Nano* and Risk
*means successive words

We collected citation data on publications from the Science Citation Index (SCI) and the Social Sciences Citation Index (SSCI), compiled by the Institute for Scientific Information (ISI), because SCI and SSCI are two of the best sources for citation data. We also used the Web of Science, which is a Web-based user interface of ISI's citation databases. A total of 17,480 papers were retrieved. The data includes all ISI records, *i.e.* articles, letters, reviews, editorials, meeting abstracts, and so on. Some of these papers may be not relevant to nano risk because we collected data simply by the query described above. Therefore, we focused on the maximum connected component, which currently comprises 6,475 papers and accounts for 37.0%. In other words, we regarded papers not citing other papers in the component as sidetrack in the mainstream of nano risk research field and ignored them. The retrieved data were converted into a non-weighted, non-directed network. Subsequently, the network was divided into clusters using the topological clustering method [9,10]. Traditionally, co-citation has been used to analyze a citation network. However, because co-citation is accompanied by a time lag to create a link, and analysis of intercitation is more relevant in the similarity of pairs of documents than co-citation [11], we used intercitation as a link. The clustering algorithm is based on modularity Q, which is defined as follows [9,10]:

$$Q = \sum_{s=1}^{N_m} \left[\frac{l_s}{l} - \left(\frac{d_s}{2l} \right)^2 \right] \quad (2)$$

where N_m is the number of clusters, l_s is the number of links between nodes in cluster s, and d_s is the sum of the degrees of the nodes in cluster s. In other words, Q is the fraction of links that fall within clusters, minus the expected value of the same quantity if the links fall at random without regard for the clustered structure. Since a high value of Q represents a good division, we stopped joining when ΔQ became minus. A good partition of a network into clusters means there are many within-cluster links and minimal between-cluster links.

After clustering the network, we characterized each cluster by the titles and abstracts of papers that are frequently cited by the other papers in the same cluster. It does not mean that all papers in the cluster study the same topics as covered in these frequently cited papers. In fact, each paper studies its own topics, and each paper has its own unique focus. However, as a first approach, it is reasonable to treat these inter-cited papers as a cluster to investigate the brief structure of a research domain and to consider the frequently cited papers in the cluster as representative of the same. The clustered network is visualized by using a large graph layout (LGL) [12], which is based on a spring layout algorithm where links play the role of spring connecting nodes. Thanks to such layout, papers that cite each other and form a group can be located in closer proximity.

2.2. Dose-mortality analysis in vitro – size effect –

From the experimental results reported in the papers, dose and viability data sets were singled out and scatter diagrams were made with dose as the horizontal axis and "% death" as the vertical axis. "% death" means the mortality rate of cells exposed in vitro to nano/sub-micro/micro particles versus control cells calculated using the formulation: 100 (% of cells) - % of still viable cells. On the other hand, the kinds of unit measurements used for dosages were varied by using 4 unit measurement types. "μg/ml" is a unit measurement for particle weight over cell culture media volume. "cm^2/ml" is, in one way, a type of unit measurement for particle surface area computed by BET measurement over cell culture media volume, and, in another way, a measurement of particle surface area calculated using a particle's minor and major axes. When a particle has a spherical shape, calculated surface area (m^2/g) should be $6/\rho D$ where ρ (g/cm^3) is density and D (μm) is diameter. If a particle's form is fibrous, calculated surface area should be nearly $4/\rho D$ supposed D<<Length. "No. of particles" is a unit for particle count over cell culture media volume. In addition, in the scatter diagrams, each plot had different marks depending on size: one type of key for particles with a primary diameter less than 120 nm and another type of key for those above 400 nm. Generally "nano-sized" is said to be 1-100 nm, but in this paper the collected particle data were for particles of 1-120 nm and of 400-10 000 nm, so the separation value was determined as 130 nm.

2.3. Uptake measurement using floating cell
Three different surface modification quantum dots were mixed with floating cells in each. One quantum dot was just coated with polyethyleneglycol (non-modification; plane) and others were modified with a carboxyl (-COOH) or amino ($-NH_3$) group on polymer layer. The median diameter was 15 nm and the number concentration was 5.5×10^{12} particles/ml. Jurkat floating cells increased at confluence were washed by PBS(-) and were blended with each quantum dot solution. After thirty minutes, we washed these cells three times by PBS(-) solution and detected fluorescence of quantum dot by FACS calibur (Becton Dickinson). Lower there graphs indicate the uptake percentage of each particle in Jurkat cells. X-axis is the number of cells and Y-axis is the height of fluorescent plus; FL2-H (BP 585/42 nm). These procedures were shown in Figure 2.

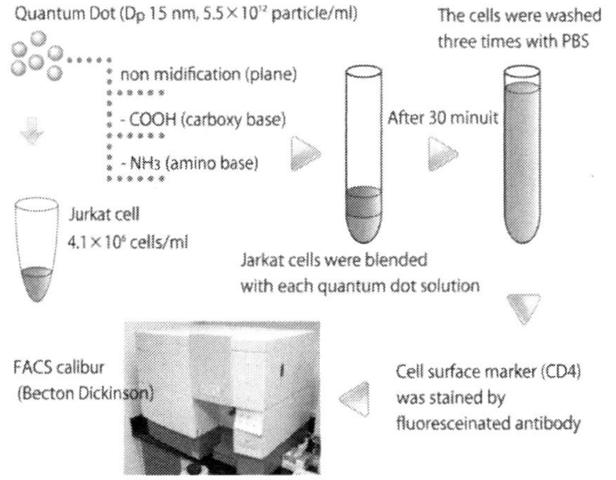

Figure 2. Three types of quantum dot were mixed with Jurkat floating cells.

3. Results
Eleven Papers [13-23] discussing in vitro experiments concerning the viability of cells except macrophages were selected from our search. These papers checked the endpoint of cell viability after 3 days of exposure. For papers in which no BET surface areas described, calculated surface areas were substituted for BET surface areas (If a particle has spherical shape, calculated surface areas (m^2/g) should be $6/\rho D$ where ρ (g/cm^3) is density and D (μm) is diameter. If it is fibrous in form, calculated surface areas should be nearly $4/\rho D$ assuming D<<Length.). The collected cell viability data were used to make scatter diagrams. Figure 3-*(a)* shows a conventional diagram whose horizontal axis is particle dose (μg/ml) and whose vertical axis is % death (100 (%) - viability (%)). Its keys were differentiated in particle diameter between 0-120 nm and 400-10 000 nm. Plots of 0-120 nm and 400-10 000 nm overlapped significantly so that the effect of size on cell viability is likely to be indicated by particle weight or particle volume (since particle weight and volume show similar values). Figure 3-*(b)* indicates a diagram whose horizontal axis is the particle dose of BET surface areas (cm^2/ml). There seems to be a split between the 0-120 nm group and the 400-10 000 nm groups. Figure 3-*(c)* shows a diagram whose horizontal axis is the particle dose of calculated surface areas (cm^2/ml). Here BET

surface areas of amorphous silica were converted to the smaller value of calculated surface areas, but the total trend of scattering plots was not altered significantly. Figure 3-*(d)* indicates a diagram whose horizontal axis is the particle dose of the number of particles (cm^2/ml). Two groups were split in two more widely and the size effect for cell viability is not likely to be the number of particles.

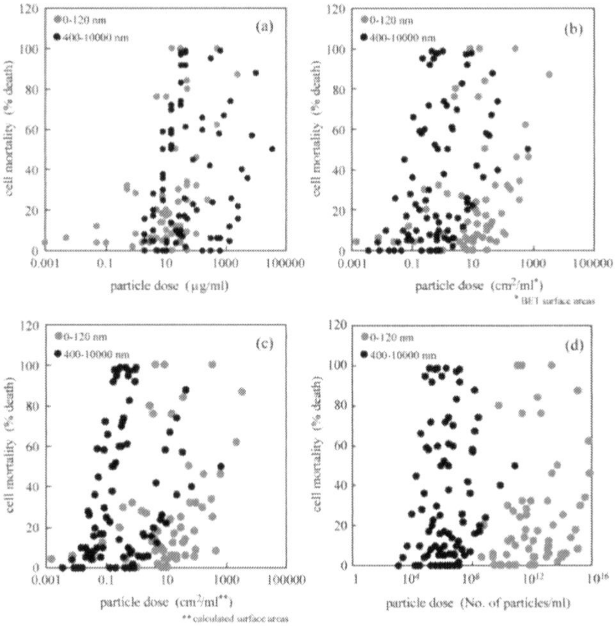

Figure 3. Relationship between the ratio of cell mortality and particle dose in a culture medium.

Particle dose was written in weight concentration *(a)*. BET specific surface area was measured in eleven selected articles. The x-axis of *(b)* is it and *(c)* is calculated surface area. Particle number was calculated by weight concentration *(d)*. There was a highly significant difference between the 0-120 nm group and the 400-1000 nm group in the ratio of cell mortality (Wilcoxon, two-tails, p <0.01).

The collected cell viability data are all shown in Figure 4 with each type of mark representing a particle species. Figure 4-*(a)*, *(b)* indicated that a different crystal type gave a different dose-response curve if these materials were the same. Of course, each material or each oxidation number is presented in a different effect level to cell mortality as shown in Figure 4-*(c)*, *(d)*. Additionally, a regression line $Y = a + b\ ln(x)$ was drawn on each particle species, and EC50 (the concentration of μg/ml that reduced cell proliferation to 50% of unexposed controls) was evaluated. Seen as Figure 5, an x-y scatter diagram was developed, with the x-axis particle solubility (g/l) and the y-axis EC50 calculated as described previously. As can be seen, a possible correlation exists: the greater the solubility, the smaller the EC50 (meaning greater cytotoxicity). At the same time, the plots for anatase TiO_2 as well as Fe, Fe_2O_3 and ZrO_2 contain results that do not clearly support such a correlation. There may be other, more important factors affecting their cytotoxicity, or the data volume may have simply been insufficient.

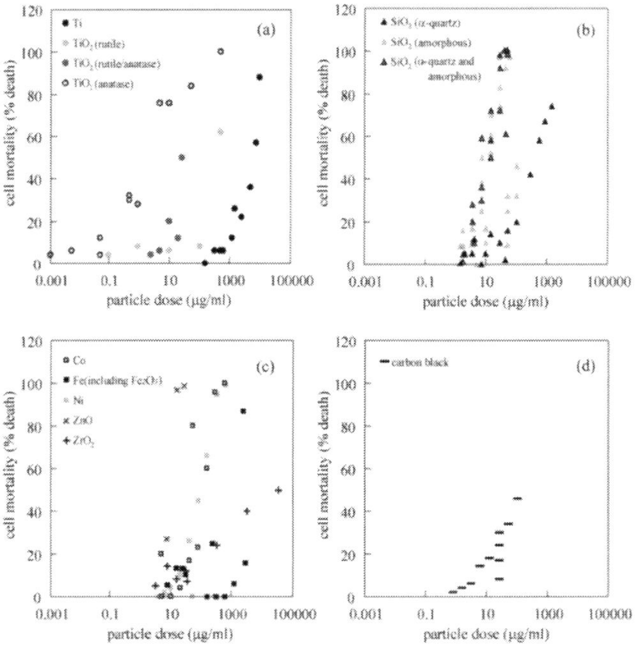

Figure 4. Relationship between the ratio of cell mortality and particle weight in a culture medium.

The ratio of cell mortality indicated a high level of linearity as the weight of the particle is high in each substance. Each substance showed a certain gradient. Toxicity will probably vary depending upon the crystal type of titanium (a).

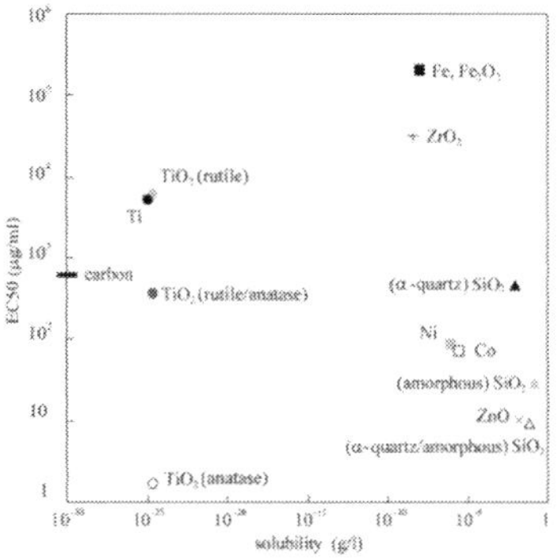

Figure 5. Relationship between solubility and EC50.

EC50 was determined by the concentration of μg/ml that reduced cell proliferation to 50% of unexposed controls.

To use a Jurkat cell which floats in the medium, we measured the dose of invasion to this cell after 30 minutes. This cell is incubated in medium and FBS generally. However, some reports showed that nanoparticles adducted to the protein floating in an alveolus surfactant. Three kind of quantum dots (plane, -COOH, -NH_3) mixed with only PBS(-) or PBS(-) plus FBS solution. Right column graphs in Figure 6 indicate the uptake percentage of each particle in Jurkat cells. X-axis is the number of cells and Y-axis is the height of fluorescent puls; FL2-H (BP 585/42 nm). Center is the result of flow cytometry after blended cell and particle with PBS(-) and right is with PBS(-) and 2% FBS. Few non-modification quantum dots were taken up to cells (0.19%). On the other hand, 99.3% carboxylated particles and 62.9% aminated them were invade to cells after thirty minutes shown in Figure 6.

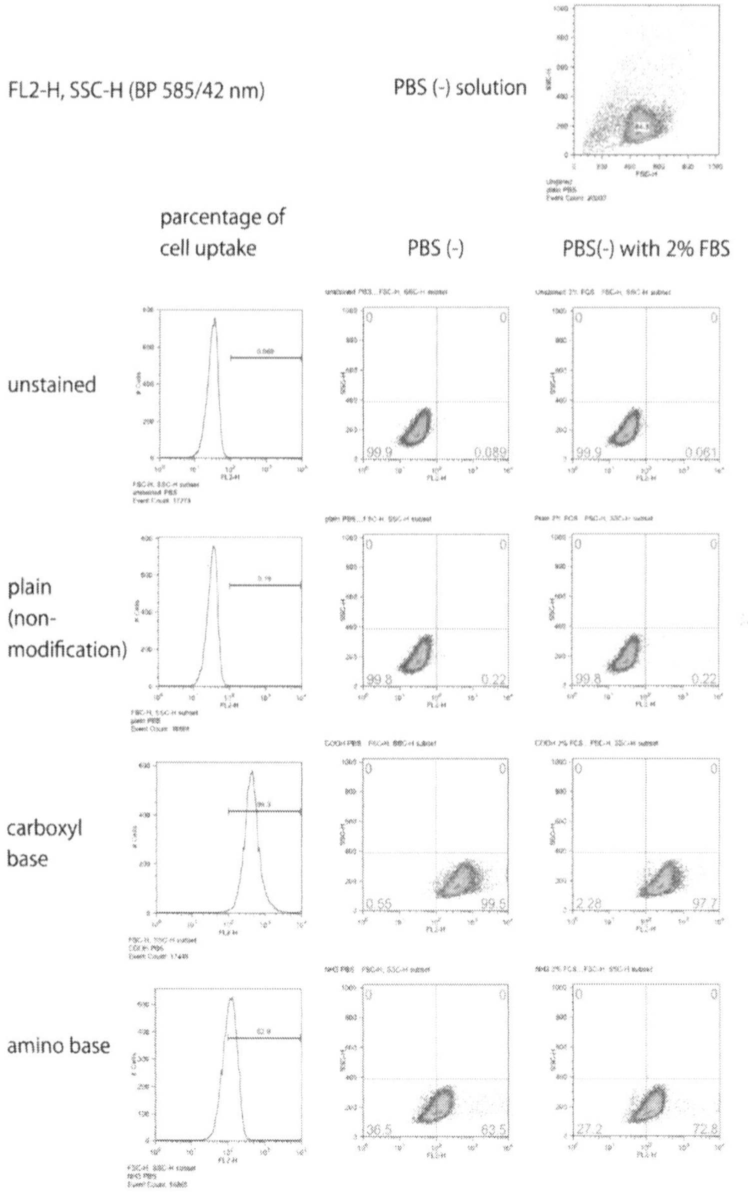

Figure 6. Uptake of 15 nm quantum dots in Jurkat cells.

Top is FSC-H/SSC-H figure of PBS solution only. Left column shows a parcentage of cell uptake and center and right columns show FL2-H/SSC-H figure. Center column is treated cell and quantum dot in a PBS(-) solution and right column is treated them in a PBS(-) with 2% FBS. Polymer coated non-modification quantum dot was mixed with Jurkat cells and measured before staining (the 1st line) and after staining (the 2nd line). Modified quantum dot of carboxyl and amino base showed in the 3rd and 4th lines.

4. Discussion

To study the effects of particle size and particle species on nanomaterial risk to health, dose-mortality analyses in vitro were performed for this paper. Our results suggest that the effect of size on cytotoxicity is due to particle weight or volume and that another cytotoxic factor is particle solubility. The effect of particle size on cytotoxicity has been argued in many papers. Nano and micro alumina and titania particles have been tested using viability assays, and viable osteoblast density has not always been smaller in nano particle cultures than micro particle cultures when particle weight dose (μg/ml) has been equivalent [14]. This trend is also seen in Figure 2 of our results showing data of additional particle species and more plots. Another paper argued that cytotoxicity does not depend on size or chemical species but, decisively, on total volume [11]. The authors introduced 5 species of ceramic particles into cell culture and measured the dose-response curve with varying doses using three criteria: particle number, particle surface area, and particle volume. Their results were quite similar to our results in terms of splitting formation of plots on scatter diagrams. On the other hand, the cytotoxic effect of aggregated nanomaterials was investigated with little correlation found between BET specific surface area and the cytotoxicity index [15]. For this paper, only BET specific surface area was chosen for determination as to whether it is a strong factor for cytotoxicity, but a number of other parameters, such as solubility, may also require study. Based upon our analysis and examination of already-published papers, the chief factor with respect to cytotoxicity appears to be the particle volume dose with a subordinate but still significant additional factor possibly being particle solubility. A possible mechanism could be that particle volume obstructs intra-cellular adherence or cell-extracellular matrix interaction so that the cell proliferation rate slows depending on particle volume. Under such conditions, a using equivalent particle dose for a given surface area does not work because smaller particles are more susceptible to being pushed away. In addition, dissolved materials in culture media may affect cellular osmotic systems, or larger surface areas of dissolving particles may increase particle volume and thereby obstruct various kinds of adherence or interactions.

From the cell culture experiment, nano particles (15 nm) can invade suspension cells in functional-dependent manners. Especially, almost all the carboxyl base modificated quantum dots were taken up to cells in a short time. The similar results were reported [24] from a few years ago. A cell surface has a negative charge and a carboxylated quantum dot also has the same charge. These cells and particles act repulsively by Coulomb's force. A drug is designed to an anionic generally because of negative surface charge. Our result and some past reports are conflicting about the surface charge. In this experiment, we are not able to distinguish an uptake and adhesion between cell and particle. We will try to reveal these two problems by Nanosafe 2012.

Acknowledgment

Research grant to University of Tokyo, National Institute of Advanced Industrial Science and Technology and National Institute for Materials Science from the Ministry of Education, Culture, Sports, Science and Technology, Japan is gratefully acknowledged. I appreciate Professor Y. Yamaguchi, K. Matsushima and my previous colleagues at Nano Material Center in the University of Tokyo.

References

[1] Andrew D. Maynard *et al.* 2006 *Nature* **444** 267-269.
[2] Robert F. 2006 *Science* **314** 45.
[3] Gosens I, Post JA, de la Fonteyne LJ, Jansen EH, Geus JW, Cassee FR, de Jong WH. 2010 Particle and fibre toxicology **7**(1) 37
[4] Gillespie PA, Kang GS, Elder A, Gelein R, Chen L, Moreira AL, Koberstein J, Tchou-Wong KM, Gordon T, Chen LC. 2010 Nanotoxicology **4**(1) 106-119

[5] Warheit DB, Sayes CM, Reed KL. 2009 Environmental Science and Technology **43**(20) 7939-7945

[6] Y. Matsui, K. Miyaoi, T. Hayashi, Y. Yamaguchi 2009 *Journal of Physics: Conference Series* **170** 012030

[7] Y. Matsui, T. Hayashi, K. Miyaoi, H. Tomobe, Y. Kajikawa, K. Matsushima, Y. Yamaguchi 2009 *Journal of Physics: Conference Series* **170** 012033

[8] Zitt M and Bassecoulard E 2007 *Information Processing and Management* **42** 1513

[9] Newman M E J 2004 *European Physical Journal B* **38**(2) 321

[10] Newman M E J and Girvan M 2004 *Physical Review E* **69**(2) 026113 Part 2

[11] Klavans R and Boyack K W 2006 *Journal of the American Society for Information Science and Technology* **57** 251

[12] Adai A T, Date S V, Wieland S and Marcotte E M *2004 Journal of Molecular Biology* **340**(1) 179

[13] Brunner TJ., Wick P., Manser P. et al.: *In vitro* cytotoxicity of oxide nanoparticles 2006 *Environmental Science & Technology* 40(**14**) 4374-4381

[14] Tian FR., Cui DX., Schwarz H. et al. 2006 *Toxicology in vitro* 20(**7**) 1202-1212

[15] Lin WS., Huang YW., Zhou XD. et al. 2006 *Toxicology and Applied Pharmacology* 217(**3**) 252-259

[16] Sayes CM., Wahi R., Kurian PA. et al. 2006 *Toxicological Sciences* 92(**1**) 174-185

[17] Pioletti DP., Takei H., Kwon SY. et al. 1999 *Journal of Biomedical Materials Research* 46(**3**) 399-407

[18] Peters K., Unger RE., Kirkpatrick CJ. et al. 2004 *Journal of Materials Science – Materials in Medicine* 15(**4**) 321-325

[19] Sakai T, Takeda S., Nakamura, M. 2002 *Dental Materials Journal* 21(**2**) 133-146

[20] Elias Z., Poirot O., Fenoglio I. et al. 2006 *Toxicological Sciences* 91(**2**) 510-520

[21] Akiko Yamamoto, Rieko Honma, Masae Sumita, Takao Hanawa 2004 *Journal of Biomedical Materials Research Part A* 68A(**2**) 244-256

[22] Pisanic TR., Blackwell JD., Shubayev VI. et al. 2007 *Biomaterials* 28(**16**) 2572-2581

[23] Hart Georgia A., Hesterberg Thomas W. 1998 *Journal of Occupational and Environmental Medicine* 40(**1**) 29-42

[24] Choi HS, Ashitate Y, Lee JH, Kim SH, Matsui A, Insin N, Bawendi MG, Semmler-Behnke M, Frangioni JV, Tsuda A 2010 Nature Biotechnology (Epub ahead of print)

Development of prognostic occupational air standards for nanoparticles

Andrey S Radilov, Anzhela V Glushkova, Sergej A Dulov and Nataliya S Khlebnikova

Research Institute of Hygiene, Occupational Pathology and Human Ecology Federal State Unitary Enterprise, Federal Medical Biological Agency, Kuz'molovsky g/p, Vsevolozhsky District, Leningrad Region, 188663 Russia

E-mail : aglushkova_rihope@hotmail.com, radilov@rihophe.ru

Abstract. The intensive progress of nanoindustry in the Russian Federation makes quite urgent the problem of development and especially express development of occupational exposure standards for nanoparticles and nanoaerosols in the workplace air. We developed an approach to comparative toxicity assessment and express calculation of occupational exposure standards for nanoaerosols, based on criteria for the development of maximum allowable concentrations (MACs) of aerosols in the workplace air. The developed approach was used to obtain prognostic MACs of certain aerosols in the workplace air, mg/m^3: nano-Ag 0.08, nano-TiO_2 0.19, and C_{60} 0.08.

1. Introduction

In view of the rapidly increasing industrial application of nanoparticles and, consequently, their possible health impact on workers and population, the problem of development of tentative occupational exposure standards for nanoparticles and, especially, nanoaerosols in the workplace air seems to be quite urgent (Nowack, B., Bucheli, T.D., 2007). Therewith, of particular importance is to assess the health impact of exposure to nanoaerosols at workplace. A key facet of this problem is express development of occupational exposure standards for nanoparticles and nanoaerosols, which may cardinally improve the situation with hazard prognosis and introduction of nanotechnologies.

2. Methods

We developed a scale of safety factors depending on the routes of entry (intragastric, intraperitoneal, intratracheal, or inhalational) (Table 1). These safety factors can be used in calculation and substantiation of hygienic standards and also in the assessment of impact of nanoaerosols on workers' health. If the TI at some route of entry is impossible to calculate, the worst scenario should be taken into account.

Depending on the exposure mode (acute, chronic) at the corresponding route of entry, the following safety factors are suggested. The suggested safety factors are based on the idea of NP bioavailability depending on the route of entry. Therewith, it should be taken into account that at the identical toxicity parameters for intragastric and intraperitoneal routes of entry the substance, independently of its hazard class, will possess a low bioavailability, and, therefore, its safety factor will equal 1.

Table 1. Determination of safety factors depending on the route of entry.

Variant no.	Relative toxicities of nano- and microparticles at the corresponding routes of entry			Safety factor
	i.g.	i.p.	i.t. or i.h.	
1.	$Tn^a \cong Tm^a \geq 5$ g/kg	$Tn \leq Tm$	$Tn \cong Tm$	1-3
2.	$Tn \leq Tm < 5$ g/kg	$Tn \leq Tm$	$Tn \leq Tm$	4-9
3.	$Tn \cong Tm$	$Tn > Tm$	$Tn \geq Tm$	10-30
4.	$Tn > Tm$	$Tn > Tm$	$Tn > Tm$	>30

Notes. [a] Here and hereinafter, *Tn* and *Tm* stand for the toxicity of nano- and microparticles of the same substance, respectively.

We suggest to calculate a toxicity index (TI) to relate the toxicity of microparticles of a substance to that of nanoparticles:

TI = toxicity parameters of microparticles / toxicity parameters of nanoparticles

The above formula can be used to estimate:
TI for acute exposure LD_{50} $(LC50)_{MPs}/LD_{50}$ $(LC50)_{NPs}$; $Lim_{ac\ MPs}/Lim_{ac\ NPs}$,
TI for chronic exposure = $Lim_{ch\ MPs}/Lim_{ch\ NPs}$.

The safety factors for different routes of entry and comparative toxicity classification are presented in Table 2.

Table 2. Safety factors variants depending on the route of entry and classification comparative toxicity

No.	TI i.g.	TI i.p.	TI i.t. or i.h.	Safety factor
1	1	1	1	3
2	1-3	1-3	1-3	3-9
3	3-10	3-10	3-10	9-30
4	>10	>10	>10	>30

The safety factors are calculated by the following formula:

$$SF = TI\ i.g. + TI\ i.p. + TI\ i.t.\ or\ i.h.$$

The MAC values for nanoparticles are calculated by the following scheme:

1. Determination of LD_{50} or LC_{50} for substances in the nano and macro forms;
2. Methods for calculation of occupational exposure standards for aerosols of metals, metal oxides, and other metal compounds in the workplace air, accepted in the Russian Federation, were used (MU 4000-85, Council for Mutual Economic Aid, 1982).

The MACs for metal aerosols were calculated by the formula:

$$MACwa = 0.0008\ x\ LD_{50} \qquad (1)$$

and for aerosols of metal oxides and other metal compounds, by the formula:

$$\log MACwa = 0.85 logLD_{50} - 3.0 + logM - logN \quad (2)$$

where LD_{50} is intraperitoneal LD_{50} for mice after a week observation period, milliatoms per kg body weight; *N*, number of metal atoms in the molecule; and *M*, molecular weight.

3. Determination of safety factors including the toxicity parameters for substances in the nano and macro forms (Table 2);
4. Estimation of MAC values for nanoparticles with the safety factors in Table 1, by the following formula:

$$MACwa\ NP = MACwa\ MP\ /\ SF$$

Below we present a calculation of a tentative hygienic standard for copper nanoparticles using the above equations for calculation of hygienic standards for metal aerosols in the workplace air, as well as toxicity of copper nanoparticles and the corresponding safety factor:

$$LD_{50}\,(\text{copper NPs}) = 413\ \text{mg/kg}$$

Safety factor for copper nanoparticles is equal 30.
(IT i.g. NPs more that MPs in 10 times, IT i.p. NPs more that MPs in 10 times and IT i.t. NPs more that MPs in 10 times),

$$MACwa\,(\text{copper NPs}) = (0.0008 \times 413) / 30 = 0.01\ \text{mg/m}^3$$

3. Results

The probability of a negative impact of NPs on the environment and population health is known to the determined of a potential hazard which is characterized, among other factors, by the toxicity of one or several components of a material contacting with live organisms (Nowack B., Mueller N.C., 2008).

Nanoparticles are not infrequently much more toxic than macro- and microparticles of the same substances. Table 3 compares the toxicities of selected nano- and microparticles (MPs).

Table 3. Comparative toxicities of nano- and microparticles.

No.	Substance	Test object / Route of entry	LD_{50}/LC_{50}	
			NPs	MPs
1	Cu	rats/i.g.	413 mg/kg [8]	5000 mg/kg [8]
2	Fe_2O_3	rats/i.g.	> 310 mg/kg [2]	98.6 g/kg [2]
3	Zn	daphnias (*Daphnia magna*)	3.2 mg/l [19]	8.8 mg/l [19]
		bacteria (*Vibrio fischeri*)	1.9 mg/l [19]	1.8 mg/l [19]

The prognostic MACs for selected nanoaerosols in the workplace air, calculated by the suggested approaches, are listed in Table 4.

Table 4. Prognostic MAC values for selected nano- and microaerosols.

Substance	Test object/ Route of entry	LD_{50}, mg/kg NPs	LD_{50}, mg/kg MPs	MAC_{wa}, mg/m^3 MPs	Prognostic MAC_{wa}, mg/m^3 NPs
Ag	rats/i.p.	5000 [3]	not attained	1 [1]	0.13
Cu	rats/ i.p.	413 [8]	5000 [10]	1 [1]	0.01
TiO_2	rats/i.p.	12000 [14]	not attained	10 [1]	0.32
Fe_2O_3	rats/i.g.	310 [2]	98.6 g/kg [2]	6 [1]	0.008
C_{60}	rats/i.p.	5000 [16]			0.13
Soots, black, industrial, benz(a)pyrene content ≤35 mg/kg			not attained	4 [1]	

4. Discussion

Developing the set of safety factors, we based on the idea that the bioavailability of a substance depends on the route of entry. We suggest that the safety factor for NPs will depend primarily of the route they enter the body. For example, whatever the hazard class, if a substance has similar (or equal) toxicity parameters for the oral and intraperitoneal routes of entry, it will possess a low bioavailability and, consequently, we assign to this substance the safety factor 1. In cases where the intraperitoneal toxicity of a substance is higher than intragastric or inhalational, we assign to this substance the safety factor 10. And, finally, if the toxicity of a substance in the nano form is higher than the toxicity of the same substance in the micro form at the intragastric, intraperitoneal, and inhalational routes of entry, we assign to this substance the safety factor 50. We would like to mention that the suggested safety factors are estimated and subject to discussion.

It is known that if experimentally substantiated LD_{50} for a substance at different routes of entry are available, the lowest (as a rule intragastric) of the available LD_{50} values are used in the calculations, relying of the worst prognosis and the corresponding safety factor.

5. Conclusion

The suggested approaches to calculation of tentative occupational exposure standards for nanoaerosols in the workplace air make it possible to estimate hazard parameters for NPs and base on these data to assess, in the look-ahead manner, health risks to workers occupied in nanoindustrial enterprises.

Thus, the algorithm for express development of occupational exposure standards for NPs and nanoaerosols in the workplace air involves the following steps:

1. Comparison of the toxicities of nano- and microparticles.
2. Choice of safety factor, based on available comparative experimental toxicity data for nano- and microparticles.
3. Calculation of a tentative MAC_{wa} value for nanoparticles by the standard formula, including the chosen safety factor.

References

[1] Predel'no dopustimye kontsentratsii (PDK) vrednykh veshchestv v vozdukhe rabochei zony: GN 2.2.5.1313-03 (Maximum Allowable Concentrations (MAC) of Hazardous Substances in the Workplace Air, Hygienic Standard 2.2.5.1313-03), Moscow, 2003.

[2] Ermakov, A.E., Poluchenie biosovmestimykh netoksichnykh nanoporoshkov dlya dostavki lekarstvennykh preparatov, diagnostiki i terapii novoobrazovanii (Preparation of Biocompatible Nontoxic Nanopowders for Drug Delivery and Diagnostics and Therapy of Neoplasms), Abstracts of papers, 1st Int. Summer School "Nanomaterials and Nanotechnologies in Live Systems," Moscow, June 29-July 4, 2009, p. 90.

[3] Kinzirskii, A.S., Bovina, E.V., Klochkov, S.G., etc., Novye dannye o toksikologii i biorapredelenii tipovykh nanochastits v organizme gryzunov (New Data on the Toxicology and Biodistribution of Typical Nanoparticles in Rodent's Body), Abstracts of papers, 1st Int. Summer School "Nanomaterials and Nanotechnologies in Live Systems," Moscow, June 29-July 4, 2009, p. 162.

[4] Metodicheskie ukazaniya po ustanovleniyu orientirovochnykh bezopasnykh urovnei vozdeistviya vrednykh veshchestv v vozdukhe rabochei zony: MU 4000-85 (Methodical Guidelines on the Establishment of Tentative Safe Exposure Levels of Hazardous Substanses in the Workplace Air: Methodical Guidelines 4000-85), Moscow, 1985.

[5] Postoyannaya komissiya po sotrudnichestvu v oblasti zdravookhraneniya (Standing Commission on Collaboration in Public Health Portection), Kaloyanova-Simeonova, F. and Sanotskii, I.V., Eds., Moscow: Council for Mutual Economic Aid, 1982.

[6] *Guidelines on Human Health Risk Assessment from Environmental Chemicals* (P 2.1.10.1920-04), Moscow, May 5, 2004.

[7] Brown C.L., Whitehouse M.W., Tiekink E.R., Bushell G.R., *Inflammopharmacology*. 2008. **16**(3), P. 133-137.

[8] Chen Z., Meng H., Xing G. et al. // The journal of physical chemistry. *Toxicology letters*. 2006 **163**,109-120.

[9] ECB. Technical Guidance Document on Risk Assessment; European Chemicals Bureau, 2003.

[10] Elder, A.C.P. The Toxicology of Nanomaterials. – Univ.of Rochester, 2007. – 37 p.

[11] Gurgueira S. A., Lawrence J., Coull B., *Environ. Health Perspect*. 2002 **110**(7) 749-755.

[12] Holsapple M.P., Farland W. H., Landry T. D. et al., *Toxicol. Sci.* 2005 **88**(1) 12-17.
[13] Nowack, B., Bucheli, T.D., *Environ. Pollut.* 2007 **150** 5–22.
[14] Nowack B., Mueller N.C., *Environ. Sci. Technol.* 2008 **42** 4447–4453.
[15] Oberdorster G., Oberdorster E., Oberdorster J., *Environ.Health Perspect.* 2005 **113** 823–839.
[16] Oberdorster G., Stone V., Donaldson K., *Nanotoxicology* 2007 **1**(1) 2-25.
[17] Ouellette J., *The Industrial Physicist*. 2003 **8** 18.
[18] Warheit D.B., Laurence B.R., Reed K.L. et al., *Toxicol. Science* 2004 **77**(1) 117-125.
[19] Wigginton N.S., Haus K.L., Hochella M.F., *J. Environ. Monit.* 2007 **9** 1306–1316.

A novel comprehensive evaluation platform to assess nanoparticle toxicity *in vitro*

C Hirsch, J-P Kaiser, F Wessling, K Fischer, M Roesslein, P Wick and H F Krug

Empa – Swiss Federal Laboratories for Materials Science and Technology, St. Gallen, Switzerland

E-mail: cordula.hirsch@empa.ch

Abstract. The amount of engineered nanomaterials (ENM) is constantly increasing. Their unique properties, compared to their bulk counterparts, render them suitable for various applications in many areas of life. Hence, nanomaterials appear in a variety of different consumer products leading to the exposure of human beings and the environment during their lifecycle. Even though results on biological effects of ENM are available, harmonized and validated test systems are still missing. One major problem concerning the reliable and robust toxicity testing arises from interactions of ENM with different assay systems. Modifications or damage to DNA can have fatal consequences, such as the formation of tumor cells and hence carcinogenesis. Therefore we focused on the re-evaluation of two genotoxicity assays concerning their nanomaterial compatibility; namely the cytokinesis-block micronucleus cytome assay (MN-assay) and the alkaline single cell gel electorphoresis assay (comet assay). We demonstrate the interference of ENM agglomerates with the read-out of both assays and discuss possibilities how to acquire relevant genotoxicity data.

1. Introduction

Engineered nanomaterials (ENM) possess new properties compared to their bulk counterparts which render them suitable for different applications in many areas of life. Hence, ENM already appear in a huge variety of consumer products and other applications. High expectations from industry as well as science accelerate their production quantity and rate. Consequently, the exposure of human beings and the environment to ENM during their life cycle is inevitable [1]. At the same time concerns arise among consumers concerning the safety aspects of this new technology. As currently toxicological evaluation standards are mostly missing, manufacturers and enterprises face the problem of how to ensure work place safety as well as the safety of their end-products.

Thus two important questions arise: "How to control for the safety of ENM?" or "How to analyze their potential toxicity?"

Even though many studies and results on biological effects of ENM are available, there is still a lack of harmonized and validated test systems. ENM, again due to their exciting special properties, often interfere with assay systems, making reliable conclusions questionable [2-7]. Furthermore, inappropriate suspending methods often lead to false results [8]. Summarizing and comparing all available data to obtain an overall evaluation of potential nano-related toxicity is thus virtually impossible. Therefore, harmonized and robust methods which create a platform for comprehensively validated tools to assess toxicological effects of nanomaterials are urgently needed.

The novel *in vitro* evaluation platform, we are currently establishing, is based on already existing ISO and OECD guidelines, addressing four key aspects of cytotoxicity, namely the viability of cells, inflammatory reactions, genotoxicity and oxidative stress. Ultimately it's supposed to offer a fast, reliable, robust, reproducible and sensitive screening tool to detect suspicious nanomaterials. To this end the suitability of at least two independent methods for each of the four parameters will be carefully reassessed and validated. Furthermore a thorough characterization of the applied nanomaterial samples will be included. The reliability and robustness of the assays is subsequently verified by interlaboratory comparisons. A successive and integrated use of this platform will allow reducing the currently prevailing ambiguity concerning the toxicity of nanoparticle containing products. This is fundamental not only for consumers but also for authorities and industry.

Special emphasis during the initial establishment period is laid on detecting interference of ENM with the assay systems themselves and the careful implementation of suitable control experiments. In particular two assays to detect DNA damage turned out to be highly susceptible for disturbance by certain nanomaterials. DNA damage is considered to be a very important parameter to be tested. This type of toxicologic effect might not be directly detectable by simple viability assays as cells are able to repair defective DNA to a certain degree and appear to be healthy for a considerable amount of time. However, accumulation of DNA lesions may later on lead to severe consequences such as cancer and/or heritable diseases.

We show the interference of three different ENM (Carbon black (CB), nano-sized titanium dioxide (TiO_2) and multi-walled carbon nanotubes (MWCNT)) with two common genotoxicity assays: the cytokinesis-block micronucleus cytome assay (MN-assay) as a biomarker of chromosome breakage and/or whole chromosome loss [reviewed in 9] and the alkaline single cell gel electorphoresis assay (comet assay) to detect DNA strand breaks [10]. Both assays rely on a fluorescent microscopy read-out and agglomerates of the ENM used clearly either quench or enhance fluorescence intensity. In conclusion critical concentrations above which ENM agglomerate massively, should not be exceeded. Furthermore, alternative methods are needed that circumvent these problems.

2. Materials and Methods

2.1. Cell culture

The human alveolar epithelial cell line A549 (ATCC, CCL-185) [11] was obtained from American Type Culture Collection and grown in RPMI-1640 cell culture medium (Sigma) supplemented with 10% fetal calf serum (FCS, Lonza), 0.2 mg/ml L-glutamine (Gibco), 50 µg/ml penicillin (Gibco), 50 µg/ml streptomycin (Gibco) and 100 µg/ml neomycin (Gibco) in humified air at 37°C and 5% CO_2.

2.2. Preparation and characterization of nanomaterials

Nano-sized TiO_2 and CB suspensions were prepared according to the standard operation procedure "Preparing suspensions of nanoscale metal oxides for biological testing" published at the NanoCare homepage www.nanopartikel.info. Their physico-chemical characterization was done as part of the NanoCare project. Main characteristics are summarized in table 1. MWCNTs (Baytubes®, Bayer Technologies Service) were suspended in 160 ppm Pluronic F127® as described before [12]. Material properties are also detailed in ref. 12.

Table 1. Nanoparticle characteristics.

material	provider	primary particle size [nm]	size distribution [nm]	BET surface [m^2/g]
TiO_2	Evonik Degussa GmbH	27	20 – 50000	52
Carbon black	Evonik Degussa GmbH	16	7 – 43	339 +/- 6

2.3. Cytokinesis-block micronucleus cytome assay

A549 cells were seeded in a 4-well chambered coverglass slides (Nunc) at a density of 100000 cells per well and cultivated over night. All samples were treated with 4 µg/ml cytochalasin B (Sigma) for 24 h to block cytokinesis. The resulting binuclearity marks cells that underwent mitosis once during the time of treatment. During the 24 h cytokinesis block period cells were treated with 11 µg/cm^2 MWCNT or 25 µg/cm^2 CB to analyze nanoparticle interference and genotoxicity. Treatment of cells with 150 µM Methylmethanesulfonate (MMS; Sigma) for 24 h served as a control substance known to induce chromosomal aberrations. Each sample was run in duplicates. After incubation for 24 h at 37°C and 5% CO_2, medium was removed and cells were fixed with 1 ml pre-cooled (-20°C) ethanol for at least 12 h. Samples can be stored in ethanol at -20°C until further processing. To stain the fixed cells, ethanol was removed and cells were air dried for several minutes. Acridine orange (Sigma) was diluted to 0.2 mM in freshly prepared Sörensen's buffer (24.6 ml 0.06 M $Na_2HPO_4 \cdot 2H_2O$ and 25.4 ml 0.06 M KH_2PO_4; pH 6.8) and applied for 10 min on a shaking platform in the dark. Cells were washed three times for 5 min with Sörensen's buffer. Pictures were taken using an Imager M1 epifluorescence microscope (Zeiss).

2.4. Alkaline single cell gel electrophoresis (comet assay)

The comet assay was performed as described before [13]. In brief, A549 cells were seeded in 6-well plates at a density of 250000 cells per well one day prior to treatment. Cells were exposed to 25 µg/cm^2 TiO_2 or CB for 24 h. Treatment with 15 mM ethylmethanesulfonate (EMS) for 30 min served as a positive control. After incubation, medium was removed, cells were washed with PBS, detached with 0.2 ml Accutase (PAA) and resuspended in 0.3 ml medium. 40 µl of the resulting 0.5 ml cell suspension were mixed with 160 µl 0.5% (w/v) low-melting point agarose in PBS (37°C). 180 µl of this mixture were applied to microsope slides precoated with 1.5% (w/v) agarose in PBS, covered with a coverslip and agarose was allowed to solidify for at least 5 min at 4°C. After removing the coverslips cells were lysed in lysis solution (basic buffer: 2.5 M NaCl, 100 mM EDTA, 10 mM TRIS, 1% Na-Sarcosinat, pH 10; freshly added before use: 10% (v/v) DMSO, 1% (v/v) Triton-X-100) for 1 h at 4°C. Slides were placed into a horizontal electrophoresis tray containing alkaline electrophoresis buffer (0.3 M NaOH, 1 mM EDTA) for 20 min without current to allow DNA unwinding. Electrophoresis was performed for additional 20 min at 24 V and 300 mA. Afterwards slides were neutralized for 5 min in TRIS buffer (0.4 M TRIS, pH 7.5), rinsed in H_2O, dehydrated for 5 min in 99% ethanol and dried over night. DNA was stained with 20 µg/ml ethidium bromide and comets were analyzed using a Nikon Eclipse TS 100 microscope (Nikon) equipped with a Intenslight C-HGFI lamp (Nikon) and a Stingray F046B IRF Jenofilt camera (Allied Vision Technologies). Quantification of head and tail regions was done using the "Comet Assay IV" software (Perceptive Instruments).

3. Results and Discussion

Genotoxicity: two assays and their limitations.

Two different genotoxicity assays were assessed for their nanomaterial compatibility: the cytokinesis-block micronucleus cytome assay (MN assay) and the alkaline single cell gel electrophoresis assay (comet assay). We chose the human alveolar lung epithelial cell line A549 and three different nanomaterials, MWCNT, CB and nano-sized TiO_2, to show typical interference problems with MN-assay as well as comet assay read-outs.

3.1. Cytokinesis-block micronucleus cytome assay.

The MN-assay is a comprehensive system to measure DNA damage, cytostasis and cytotoxicity [9]. Cytochalasin-B treatment allows DNA replication and nuclear division without cytokinesis. This results in a considerable amount of binucleated cells after 24 hours of treatment. After fixation, staining with acridine orange reveals a red fluorescent cytoplasm and green fluorescent DNA in the nuclei. By visual scoring at the fluorescence microscope DNA damage is quantified as the number of binucleated cells containing an additional micronucleus (arrowhead in figure 1A'). This separate micronucleus results from the breakage of chromosomes or the loss of whole chromosomes due to structural changes in the centromeric region. The proper distribution of these malformed chromosomes during mitosis is thus impossible. The remaining chromosome fragments are enveloped by their own nuclear membrane after completion of mitosis and appear as a small additional nucleus, the "micronucleus". Increasing numbers of cells with micronuclei could be correlated with an elevated risk to develop cancer [14]. The number of mono-, bi- and polynucleated cells and cells found in mitosis gives a measure of cytostasis. The faster a cell population proliferates the more bi- and polynucleated cells are expected. Whereas in growth inhibited cultures the number of mono-nucleated cells would be increased. To analyze cytotoxicity the number of dead cells (necrotic and apoptotic) is determined.

Figure 1 shows typical examples of MMS (A, A') MWCNT (B, B') and CB (C, C') treated cells. The black agglomerates that stick firmly to the cells impede the reliable scoring of any of the parameters mentioned above. In particular, tiny micronuclei, the essential genotoxicity characteristic, might be hidden under such agglomerates, falsifying the results drastically. Furthermore, the easily visible agglomerates render it impossible to blind the studies as the scorer recognizes the treated samples at first glance. In conclusion, the MN assay should not be used for black, or otherwise colored, ENM that agglomerate to bigger entities and therefore overlay the fluorescent signal.

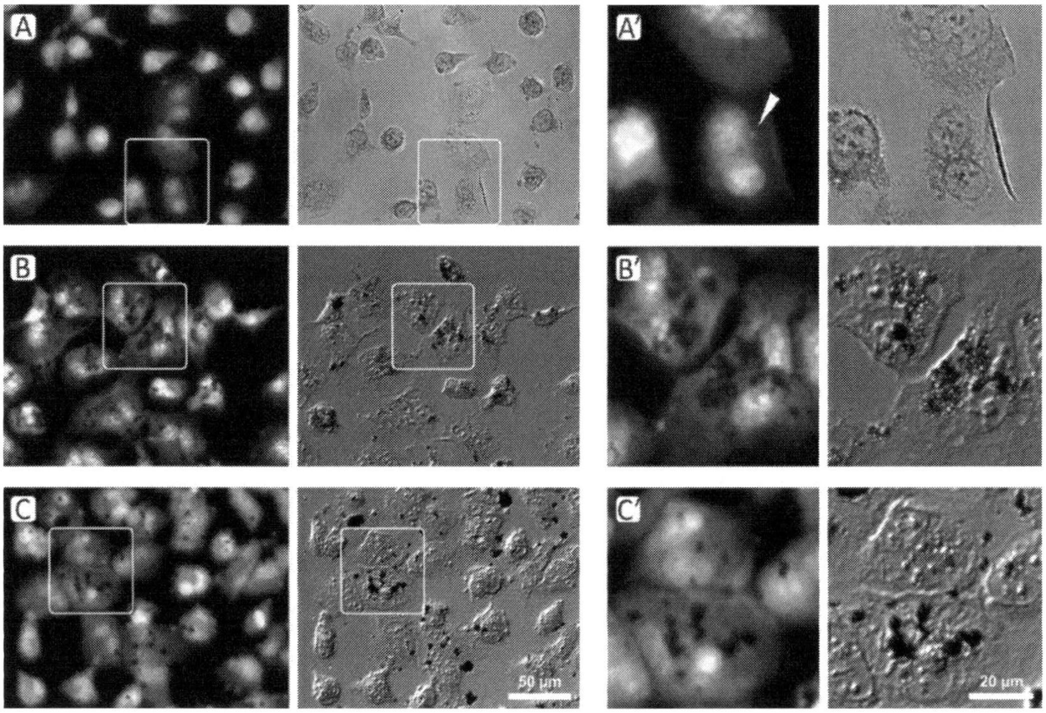

Figure 1. MWCNT and CB interfere with MN assay read out. A549 cells were treated with 4 µg/ml Cytochalasin B and 150 µM MMS (A), 11 µg/cm² MWCNT (B) or 25 µg/cm² CB (C) for 24h, respectively. After ethanol fixation cells were stained with acridine orange to visualize nuclei (green) and cytoplasm (red). Representative fluorescence as well as DIC images are shown. Boxes indicate enlarged sections shown in A' to C'. Arrowhead (A): micronucleus.

3.2. Alkaline single cell gel electrophoresis.

DNA strand breaks as well as alkali-labile sites (abasic sites, resulting from base loss or base excision) are detected in the comet assay. It is based on the electrophoretic separation of DNA fragments according to their size. Smaller fragments migrate faster and thus out of the former nucleus. Staining with an intercalating fluorescent dye (ethidium bromide) and subsequent microscopic analysis reveal the typical picture of a "comet" with intact DNA remaining in a bright spot, resembling the former nucleus and smaller DNA fragments representing the tail (figure 2B: EMS treated samples show a considerable amount of damaged DNA in the tail region). The fluorescence intensities of head and tail regions of the comets are measured using semi-automated software (Comet Assay IV, Perspective Instruments). Thus, compared to appropriate control samples (figure 2A: untreated sample; figure 2B: positive control EMS treatment), the relative amount of DNA breakage can be quantified.

Figure 2C shows A549 cells treated for 24 h with 25 μg/cm² nano-sized TiO_2. Arrowheads indicate bright spots of nanoparticle agglomerates associated with the DNA. The increase in fluorescence intensity which may result from the photocatalytic properties of these particles disturbs the quantitative image analysis. As a consequence comets that contain such particle spots cannot be included in the analysis. Treatment of A549 cells for 24 h with 25 μg/cm² CB (figure 2D) also reveals interference problems. In this case black spots of CB agglomerates reduce/quench the fluorescence signal equally falsifying the quantitative read out.

Omitting nanoparticle containing comets from the analysis poses major interpretation problems. It is unknown whether particles are able to enter the nucleus already in the intact cell, potentially leading to DNA damage there, or are later on (during the assay procedure) deposited on the naked DNA. Considering the first scenario, omitting comets that contain particle agglomerates would probably underestimate their possible genotoxic effect. This is not true if particles contact DNA after cell lysis. Furthermore, as described for the MN assay, a blinded study is not feasible due to the clearly visible deposits.

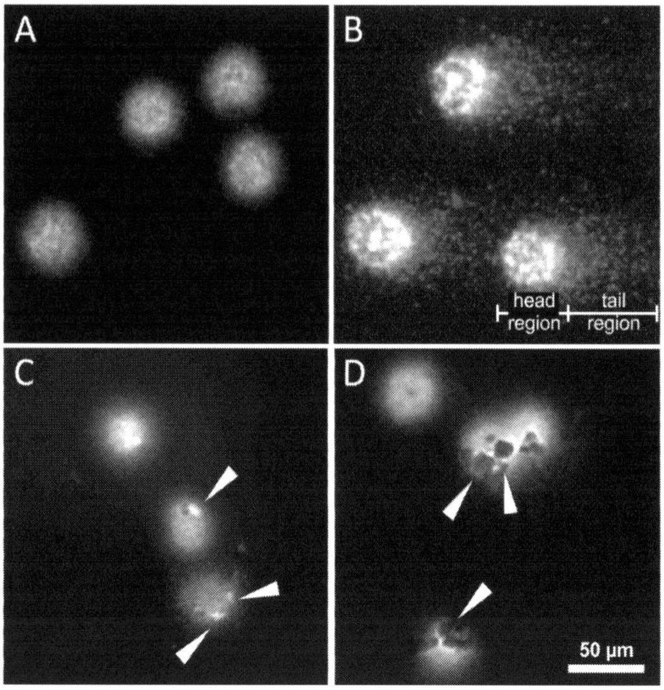

Figure 2. Carbon black and TiO_2 nanomaterial deposits interfere with quantitative Comet assay read-out. A549 cells were incubated with 25 μg/cm² TiO_2 (C) or CB (D) for 24h, treated with 15 mM EMS (B) for 30 min or left untreated (A). After lysis and gel electrophoresis DNA was stained with ethidium bromide. Representative fluorescence images are shown. Head and tail regions of a typical EMS induced comet are indicated in (B). Arrowheads in (C) and (D) point to TiO_2 and CB agglomerates, respectively.

Taken together the presence of insoluble nanomaterial agglomerates makes scoring in both assays much more difficult than with soluble chemicals. Two major problems arise from such interference: Firstly, a blinded study is almost impossible, as deposits are easily visible. Secondly, to omit cells with associated or internalized particles might result in missing the cells with the highest DNA damage and therefore lead to false negative results.

4. Conclusions

As a consequence to perform the MN and comet assay with ENM the following precautions should be made: *1)* Avoid high, cytotoxic concentrations of ENM. This not only prevents the formation of large visible agglomerates but also prevents false results due to apoptotic DNA fragmentation. The typical apoptotic nuclear fragmentation leads to DNA degradation and therefore false positive results in the comet assay. In contrast, an increase in apoptosis would lead to a false negative result in the MN assay due to the disappearance of micronucleated cells [15]. *2)* Only cells without nanomaterial deposits should be scored. However, consider the possibility that cells with highest DNA damage might be missed. *3)* If possible, results from comet and MN assay should be compared to and completed with methods that are not prone to particle interference – if those exist at all.

In any case the robust and reliable evaluation of genotoxicity caused by nanomaterials is very difficult as soon as agglomerated nanomaterials are visible. In conclusion, the development of robust alternative methods that avoid interference with nanomaterials are desirable and needed.

References

[1] Behra R, Krug HF; 2008: Nanoecotoxicology: Nanoparticles at large. *Nature Nanotechnol.* **3**: 253-254

[2] Belyanskaya L, Manser P, Spohn P, Bruinink A, Wick P; 2007: The reliability and limits of the MTT reduction assay for carbon nanotubes-cell interaction. *Carbon* **45**: 2643-2648

[3] Wörle-Knirsch JM, Pulskamp K, Krug HF; 2006: Oops they did it again! Carbon nanotubes hoax scientists in viability assays. *Nano Lett.* **6**: 1261-1268

[4] Casey A, Herzog E, Davoren M, Lyng FM, Byrne HJ, Chambers G; 2007: Spectroscopic analysis confirms the interactions between single walled carbon nanotubes and various dyes commonly used to assess cytotoxicity. *Carbon* **45**: 1425-1432

[5] Guo L, von dem Bussche A, Buechner M, Yan A, Kane AB, Hurt RH; 2008: Adsorption of essential micronutrients by carbon nanotubes and the implications for nanotoxicity testing. *Small* **4**: 721-727

[6] Monteiro-Riviere NA, Inman AO; 2006: Challenges for assessing carbon nanomaterial toxicity to the skin. *Carbon* **44**: 1070-1078

[7] Pulskamp K, Diabate S, Krug HF; 2007: Carbon nanotubes show no sign of acute toxicity but induce intracellular reative oxygen species in dependence on contaminants. *Toxicol. Lett.* **168**: 58-74

[8] Spohn P, Hirsch C, Hasler F, Bruinink A, Krug HF, Wick P; 2009: C_{60} fullerene: A powerful antioxidant or a damaging agent? The importance of an in-depth material characterization prior to toxicity assays. *Environmental Pollution* **157**: 1134-1139

[9] Fenech M; 2007: Cytokinesis-block micronucleus cytome assay. *Nature Protocols* **2**: 1084-1104

[10] Ostling O, Johanson KJ; 1984: Microelectrophoretic study of radiation-induced DNA damages in individual mammalian cells. *Biochem Biophys Res Commun* **123 (1)**: 291-298.

[11] Lieber M, Smith B, Szakal A, Nelson-Rees W, Todaro G; 1976: A continuous tumor-cell line from a human lung carcinoma with properties of type II alveolar epithelial cells. *Int J Cancer* **17 (1)**: 62-70.

[12] Thurnherr T, Sheng Su D, Diener L, Weinberg G, Manser P, Pfänder N, Arrigo R, Schuster ME, Wick P, Krug HF; 2009: Comprehensive evaluation of *in vitro* toxicity of three large-scale produced carbon nanotubes on human Jurkat T cells and a comparison to crocidolite asbestos. *Nanotoxicology* **3 (4)**: 319-338.

[13] Singh NP, McCoy MT, Tice RR, Schneider EL; 1988: A simple technique for quantitation of low levels of DNA damage in individual cells. *Exp Cell Res.* **175 (1)**:184-191.

[14] Bonassi S et al.; 2007: An increased micronucleus frequency in peripheral blood lymphocytes predicts the risk of cancer in humans. *Carcinogenesis* **28 (3)**: 625-631

[15] Decordier I, Dillen L, Cundari E, Kirsch-Volders M; 2002: Elimination of micronucleated cells by apoptosis after treatment with inhibitors of microtubules. *Mutagenesis* **17 (4)**:337-344.

Nanoparticle toxicity and cancer

T Prevenslik

QED Radiations, Discovery Bay, Hong Kong, China

E-mail: nanoqed@gmail.com

Abstract. Nanoparticles (NPs) have provided significant advancements in cancer treatment. But as in any technology, there is a darkside. Experiments have shown NPs in body fluids pose a health risk by causing DNA damage that in of itself may lead to cancer. To avoid the dilemma that NPs are toxic to both cancer cells and DNA alike, the mechanism of NP toxicity must be understood so that the safe use of NPs may go forward. Reactive oxidative species (ROS) of peroxide and hydroxyl radicals damage the DNA by chemical reaction, but require NPs provide energies of about 5 eV not possible by surface effects. Only electromagnetic (EM) radiations beyond ultraviolet (UV) levels may explain the toxicity of NPs. Indeed, experiments show DNA damage from <100 nm NPs mimic the same reaction pathways of conventional sources of ionizing radiation, Hence, it is reasonable to hypothesize that NPs produce their own source of UV radiation, albeit at low intensity. Ionizing radiation from NPs at UV levels is consistent with the theory of QED induced EM radiation. QED stands for quantum electrodynamics. By this theory, fine < 100 nm NPs absorb low frequency thermal energy in the far infrared (FIR) from collisions with the water molecules in body fluids. Since quantum mechanics (QM) precludes NPs from having specific heat, absorbed EM collision energy cannot be conserved by an increase in temperature. But total internal reflection (TIR) momentarily confines the absorbed EM energy within the NP. Conservation proceeds by the creation of QED photons by frequency up-conversion of the absorbed EM energy to the TIR confinement frequency, typically beyond the UV. Subsequently, the QED photons upon scattering from atoms within the NP avoid TIR confinement and leak UV to the surroundings, thereby explaining the remarkable toxicity of NPs. But QED radiation need not be limited to natural or man-made NPs. Extensions suggest UV radiation is produced from biological NPs within the body, e.g., enzyme induced fragmentation of epithelial tissue, exocytosis of small proteins, and ironically, the same molecular markers used to detect cancer itself.

1. Introduction

DNA damage by <100 nm NPs is now [1] considered to mimic the same reaction pathways as conventional sources of ionizing radiation. The most reasonable hypothesis is the NPs are producing their own ionizing radiation at least at UV levels, albeit at low intensity.

NPs producing low level ionizing radiation is consistent with the theory of QED induced EM radiation [2]. By this theory, NPs in body fluids produce at least UV radiation upon absorbing kT energy of colliding water molecules. Even though the UV intensity is low, DNA damage by single strand (SS) and double strand (DS) breaks may occur directly by photolysis of dry DNA or indirectly by forming ROS of peroxide and hydroxyl radicals from the water in body fluids that damage the DNA by chemical reaction.

Currently, DNA damage is quantified by the oxidative stress paradigm based on the surface area of < 100 nm natural and man-made NPs. However, experimental data [1,3-13] over the past few decades has placed this paradigm in question because greater DNA damage is found with larger particulate.

The purpose of this paper is to not only explain how QED induced radiations are related to the oxidative stress paradigm based on < 100 nm NPs, but also how larger particulate cause greater DNA damage. But DNA damage need not be limited to natural and man-made NPs. Extensions suggest DNA damage from biological NPs created within the body including enzyme induced fragmentation of submicron epithelial tissue, exocytosis of small proteins, and molecular markers long thought useful in detecting cancer.

2. Background

2.1. EM Energies

EM energies necessary to directly damage the DNA require at least photolysis at UV levels. The DNA ionization potential [3] varies from 7.5 to 10 eV. Breaking SS and DS in dry DNA requires EM radiation [4] having energies above a threshold of 7 eV. The number of DS breaks then increases monotonically to about 12 eV and then remains constant.

The indirect ionizing radiation pathway relies on photolysis to form ROS of hydroxyl and peroxide radicals in the water [5] of the body fluids that cause SS and DS breaks of DNA by chemical reaction. Since QED radiation need only exceed 5.2 eV to break the H-OH bond, the indirect pathway is the more likely DNA damage path.

2.2. Oxidative Stress Paradigm

In the 1990's, evidence that α-quartz particles (Min-U-Sil) having a mean diameter of 5 microns were capable [6] of inducing oxidation damage of biological systems. However, it is likely that some <100 nm NPs were included with the Min-U-Sil particles that actually caused the DNA damage. In silicosis, the induced hemolysis from ROS upon the interaction of silica particles with red blood cell membranes was attributed to the formation of hydrogen peroxide on the particle surface that upon reaction with metal ions by the Fenton reaction produced the hydroxyl radical. Indeed, hydrogen peroxide was detected [7] by ESR in aqueous suspensions of quartz particles. However, the source of hydrogen peroxide that produced the hydroxyl radical has never been conclusively identified.

In 2003, the NP oxidative stress paradigm as a measure of forming ROS was correlated [8] with the surface area of <100 nm NPs, although the mechanism by which the hydroxyl radicals and hydrogen peroxide form was not defined. However, the paradigm was questioned because ESR comparisons [9] of the coarse PM2.3-10 particulate produced a greater number of hydroxyl radicals than the fine PM<2.5 particulate.

Similar problems were found [10] with the oxidative stress paradigm in 2006. Natural and man-made NPs were investigated with regard to the biological consequences of ROS production. Particulate collected from the Los Angeles basin having diameters about 1500 nm and NH_2–PS spheres 1000 nm in diameter showed the clearest evidence of toxicity compared to 100 to 300 nm NPs. Similarly, pulmonary studies were conducted on rats using a wide range of α-quartz NPs that showed [11] about the same toxicity for 10-20 nm synthetic and 300-700 nm (Min-U-Sil) NPs. Nevertheless, DNA damage was attributed to surface activity.

In 2008, DNA damage [12] by silver NPs widely used as bactericidal agents was studied. Bare 25 nm silver NPs while were coated with polysaccharide to an overall diameter of 80 nm. More severe DNA damage comprising DS breaks and apoptosis/cell death was found with the larger coated NPs. Interestingly, otherwise inert gold NPs were found to generate free radicals. Of relevance to QED induced radiation, NPs naturally present [13] in *ex vivo* human skin were found to produce free radicals upon irradiation with VIS (400-700 nm) and NIR (700-1600 nm) light, although VIS and NIR light itself is not ionizing radiation.

2.3. Modified Oxidative Stress Paradigm.

Observations [9-13] suggest the oxidative stress paradigm that correlates DNA damage with the area of <100 nm NPs should be modified to account for the greater DNA damage from larger 300-1400 nm particulate.

But if so, how should the oxidative stress paradigm be modified?

By QED induced EM radiation, the larger particulate do not directly damage the DNA, but rather act to frequency up-convert the FIR inherent in body fluids to higher energy QED photons in the near infrared (NIR). Mie theory [14] shows NPs more efficiently absorb NIR than FIR radiation, and therefore large particulate enhance the UV emission of nearby < 100 nm NPs above that by the FIR alone. Hence, the NP induced oxidative stress paradigm based on <100 nm NPs itself is not modified. Instead, the larger 300-1400 nm particulate are viewed as providing enhancement of the UV radiation from the <100 nm NPs.

2.4. QED Induced Radiations

Ionizing radiation from NPs based on QED induced EM radiation was proposed [2] as an alternative to the high temperatures from heating thought to cause cancer necrosis in photodynamic therapy (PDT). Previously, gold NPs attached to cancer tumours were thought destroyed by high temperatures upon the absorption of NIR laser irradiation. However, conservation of the absorbed laser photon does not proceed by temperature increase of the NP, but rather by the emission of EM radiation at its TIR confinement frequency, typically beyond the UV. By QED theory, the UV radiation causes cancer necrosis – not high temperature.

3. Theory

The DNA may be damaged by NP induced UV radiation by absorbed EM energy from collisions with intra or extra-cellular water molecules. NPs need not enter the cell because the emitted UV radiation readily penetrates the membrane as depicted in figure 1.

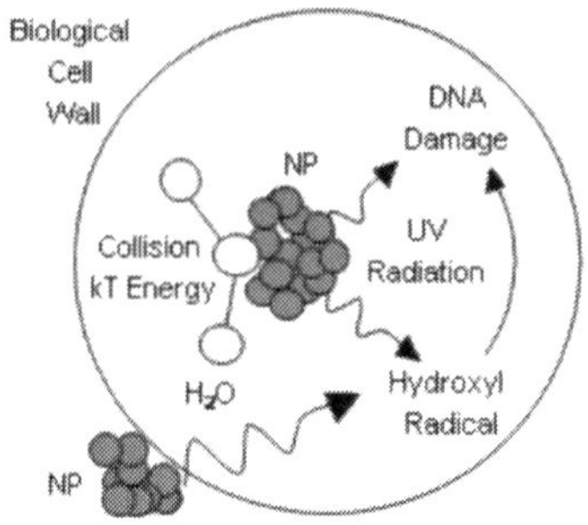

Figure 1. NP emitting UV Radiation and DNA Damage

Intra and extra-cellular water molecules continuously collide with and transfer thermal kT energy to the NPs. Since the water molecules are small compared to the NP, the collisions are inelastic and the transfer of kT energy to the NP is very efficient. The NPs lacking specific heat conserve the collision energy by the emission of EM radiation at the TIR confinement frequency of the NP, usually beyond the UV that is sufficient to produce the ROS that induce DNA damage.

3.1. EM Confinement

QED induced radiation is produced by NPs during the momentary TIR confinement of absorbed EM energy. Although NPs have diameter $D \ll \lambda$, it is instructive to consider TIR for $D \gg \lambda$. The equatorial TIR mode [15] traps absorbed EM energy at the NP surface, the number n of reflections around the QD depends on the wavelength λ of the incident radiation. As $\lambda \rightarrow D$, the ratio $\lambda/D \rightarrow 2$.

The speed of light in the NP is the speed c in the vacuum reduced by its refractive index n_r giving the frequency f,

$$f = \frac{c / n_r}{\lambda} = \frac{c}{2n_r D} \tag{1}$$

NPs having $\lambda >> D$ have the macroscopic index n_r because the speed of light c in a medium is independent of size. QED photon creation in the TIR mode is analog ous with the QM analogy of creating photons of wavelength λ by supplying EM energy to a QM box with walls separated by $\lambda/2$. For the spherical NP as a QM box of diameter D, the Planck energy E induced by TIR confinement at wavelength λ is,

$$\lambda = 2n_r D \text{ and } E = hf = hc / 2n_r D \tag{2}$$

where, h is Planck's constant.

3.2. Classical and QM Oscillators

Classical oscillators by statistical mechanics differ from their QM counterparts. The QM oscillator given by the Einstein-Hopf relation,

$$E = \frac{hc}{\lambda} \Big/ exp\left(\frac{hc}{\lambda kT}\right) - 1 \tag{3}$$

The average Planck energy E or heat capacity with wavelength λ at $T \sim 300$ K is shown in figure 2.

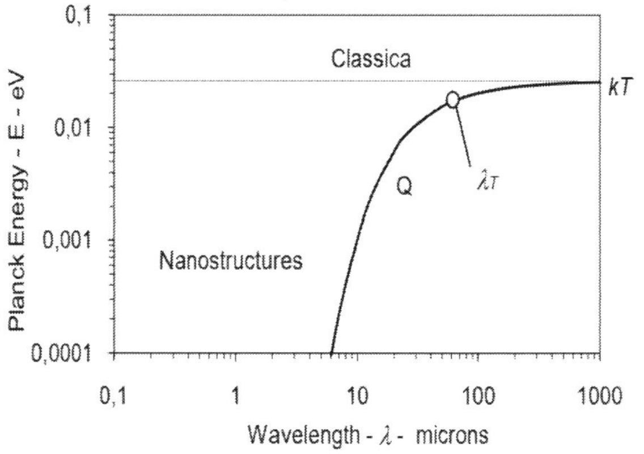

Figure 2. Harmonic Oscillator at $T \sim 300$ K

Classical oscillators have the same kT energy at all wavelengths λ as shown in figure 2. Hence, the atoms in NPs under TIR confinement have heat capacity to conserve the absorption of EM energy by an increase in temperature.

Submicron QM oscillators lack heat capacity. For $\lambda << \lambda_T = hc/kT$, figure 2 shows that an atom confined in a NP under TIR cannot conserve absorbed EM energy by an increase in temperature. Only for $\lambda > \lambda_T$, does the QM oscillator have the heat capacity of its classical counterpart.

3.3. Vanishing Specific Heat

Classical heat transfer conserves absorbed EM energy by an increase in temperature, but is not applicable to NPs because of QM restrictions on heat capacity under TIR confinement. The specific heat C of a NP may be obtained from the thermal gradient of the average Planck energy E. (equation in the inset of figure 2) depending on the TIR confinement at wavelength $\lambda = 2\ n_r D$.

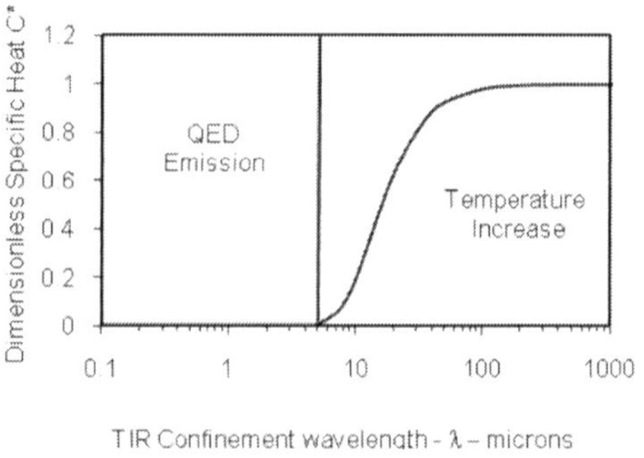

Figure 3. Dimensionless Specific Heat C^*

Differentiating the Einstein-Hopf relation for average Planck energy E with respect to temperature gives the dimensionless specific heat C^*

$$C^* = \frac{C}{3Nk} = \frac{\left(\frac{hc}{\lambda kT}\right)^2 exp\left(\frac{hc}{\lambda kT}\right)}{\left[exp\left(\frac{hc}{\lambda kT}\right) - 1\right]^2} \tag{4}$$

where, N is the number of oscillators. At 300K, C^* vanishes for $\lambda = 2n_r D < 4$ microns as shown in figure 3. For $n_r < 2$, the absorbed EM energy for $D > 1$ microns is conserved by a temperature increase while QED emission occurs for $D < 1$ micron.

3.4. Collision Power and QED Induced Photons and Rate

The collision power Q_C of water molecules of mass m transferred to NPs having diameter D is,

$$Q_C = \frac{\pi}{2\sqrt{3}} pPD^2 \sqrt{\frac{kT}{m}} \tag{5}$$

where, p is the unit probability of full kT energy transfer for inelastic collisions and P is ambient pressure. The mass m =MW/ N_{Avag} where MW = 18 and N_{Avag} is Avagadro's number. The power Q_C with NP diameter D is given in figure 4.

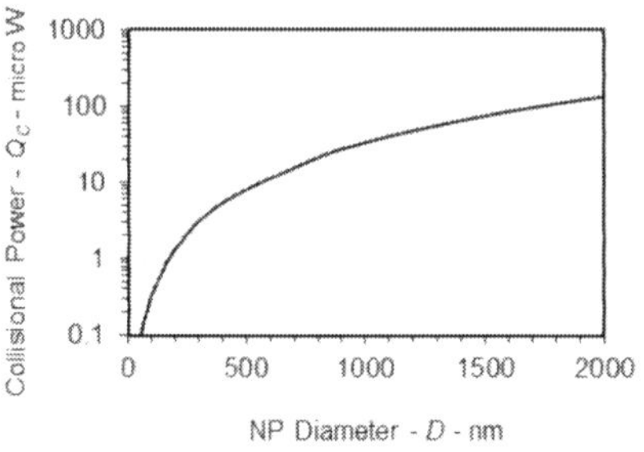

Figure 4. Collisional Power Q_C

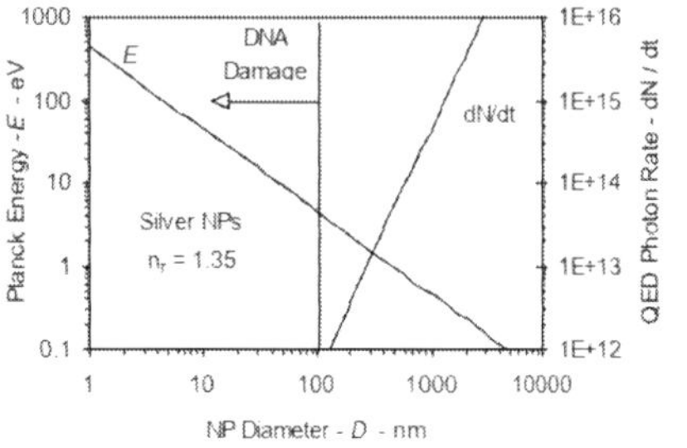

Figure 5. QED Photons Energy and Rate

Absent an increase in NP temperature, the collision power Q_C is conserved by the emission of QED induced radiation,

$$E\frac{dN}{dt} = Q_C \tag{6}$$

where, dN/dt is the rate of QED induced photons having Planck energy E created *inside* the NP. For silver having n_r = 1.35, the QED induced photon energy and rate is shown in figure 5. Silver NPs <100 nm emit ionizing radiation beyond the UV. Hydroxyl radicals form at 5.2 eV (238 and 123 nm) in d = 88 nm NPs. However, silver NPs > 100 nm emitting non-ionizing radiation in the VIS and NIR lack the Planck energy to produce hydroxyl radicals.

4. Analysis

In QED Induced EM radiation, the oxidative stress paradigm for <100 nm NPs need not be invalidated by the greater DNA damage found in coarse 300-1500 nm particulate, but rather corrected for UV enhancement. Consider an arrangement of NPs of diameter d in relation to larger particulate of diameter D shown in figure 6.

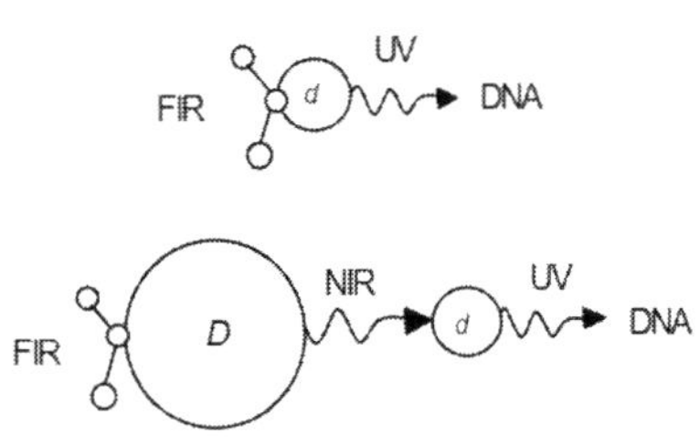

Figure 6. Large Particulate Enhancement

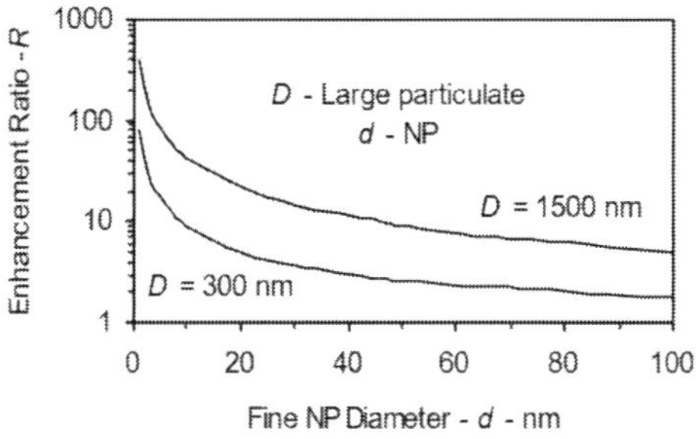

Figure 7. Enhancement Ratio R

Collisions induce the NPs to emit UV and the larger particulate to emit NIR radiation,

$$Q_{UV} = \frac{\pi}{2\sqrt{3}} pPd^2 \sqrt{\frac{kT}{m}} \text{ and } Q_{NIR} = \frac{\pi}{2\sqrt{3}} pPD^2 \sqrt{\frac{kT}{m}} \tag{7}$$

Mie theory [14] gives the absorption efficiency Q_{abs} of the NPs to the NIR radiation,

$$Q_{abs} = F\left(\frac{d}{\lambda_{NIR}}\right) \text{ and } F = \frac{24\pi ab}{\left(a^2 + b^2 + 2\right)^2 + 4a^2b^2} \tag{8}$$

The NIR wavelength $\lambda_{NIR} = 2n_r D$. Parameters a, b are the real and complex refractive indices of the NPs. The DNA absorb power Q_{UV+NIR},

$$Q_{UV+NIR} = Q_{UV} + Q_{abs} Q_{NIR} \rightarrow R = \frac{Q_{UV+NIR}}{Q_{UV}} = \left(\frac{D}{d}\right)\frac{F}{2n_r} + 1 \tag{9}$$

The UV enhancement ratio R for silver NPs having a = 1.35 and b = 4 by larger silica particulate having n_r = 1.45 is shown in figure 7. For 300 and 1500 nm particulate, DNA damage by silver NP induced hydroxyl radical at 5.2 eV occurs at d = 88 is shown enhanced by ratios R of about 2 and 8, respectively.

5. Discussion

The oxidative stress paradigm that claims the ROS correlate with the area of <100 nm NPs is not modified because of the greater DNA damage found with the larger 300-1500 nm particulate. The larger particulate themselves do not damage the DNA, but rather enhance the DNA damage caused by the <100 nm NPs.

5.1. Similarity of NP induced UV to Conventional Radiation

Air pollution [8, 9] studies give direct evidence of DNA damage by PM10 having < 50% by mass of combustion derived nanoparticles (CDNPs). The CDNPs are carbon centered NPs from automobile exhausts, but NP induced DNA damage mechanisms are currently not known.

QED induced radiation is produced in NPs at least at UV levels from which molecular mechanisms for DNA damage may be formulated. Indeed, NP induced respiratory DNA damage mimics [1] that by ionizing radiation, albeit at lower UV intensities. This means DNA damage mechanisms under ionizing radiation are applicable to NP induced DNA damage.

5.2. Oxidative Stress Paradigm

Recent pulmonary studies on rats [11] contradict the NP oxidative stress paradigm that states DNA damage is caused by toxicity that correlates with the surface area of <100 nm NPs. Indeed, the toxicity of 500 nm mined α-quartz (Min-U-Sil) particles was found equivalent to that from synthetic 12 nm quartz NPs. The haemolytic potential of α-quartz in red blood cells was attributed to the surface activity [6] caused by defects, and jagged edges in producing ROS. Silica is known to generate hydroxyl ions from hydrogen peroxide, and indeed both have been detected aqueous suspensions α-quartz. However, the specific reactions leading to the formations of hydrogen peroxide from silica have never been identified.

In contrast, QED induced EM radiation from <100 nm NPs unequivocally provides the UV to directly form the hydroxyl radicals and hydrogen peroxide. Surface activity itself has nothing to do with ROS produced by α-quartz.

5.3. Anti-Microbial Silver Nanoparticles

Silver NPs having the greatest degree of commercialization are of interest in DNA damage because of the potential treatment of inflammations in the blood. Indeed, the antimicrobial activity in controlling infections [16] and limiting bacterial growth [17] in the food industry are only a few of the many applications of silver NPs.

However, silver NPs also damage [12] the DNA. The ROS including hydroxyl radicals and hydrogen peroxide are thought produced by surface chemistry. But surface chemistry cannot be the mechanism for bactericidal action of silver NPs because polysaccharide coated silver NPs produced greater DNA damage than for bare silver NPs. EM energy is required to produce ROS that cannot be produced by surface chemistry. But QED induced EM radiation is produced in NPs beyond the UV. Indeed, the 3-fold increase in the diameter of the coated to bare NPs corresponding to EM confinement wavelengths from 68 to 200 nm suggest the NPs exposed the DNA to UV beyond 6.2 eV. To avoid DNA damage, NPs larger than 100 nm are required, but this would negate the bactericidal action of the silver NPs.

5.4. Sunscreens

The interaction of sunlight with the human skin has led to a fragile equilibrium between the EM radiation necessary for life and UV levels that damage the DNA. Prompted by the nearly epidemic increase in skin cancers over the past few decades, the European Commission has lowered the acceptable ratio UVA/UVB. Here UVA (320-400 nm) and UVB (280-320 nm). However, only about 6% of sunlight is in the UV with 52% in the VIS and 42% in the NIR suggesting the VIS and NIR0 are somehow also producing DNA damage.

Indeed, the ROS in the form of free radicals were found [13] in human skin under both UV and VIS/NIR radiation. UVB is ionizing radiation that is expected to produce free radicals, but the VIS/NIR is not. The free radicals measured were thought caused by heat from the VIS/NIR increasing the skin temperature. But there is no known mechanism by which simply raising the temperature produces free radicals.

QED induced EM radiation at UV levels produces ionizing radiation provided <100 nm NPs are present on the skin surface. Adherent sub cutis and fascia were removed, but the concentration of the remaining natural NPs was not given to assess QED induced EM radiation in producing free radicals directly from the NPs by VIS/NIR radiation *Ibid*. Nevertheless, it is highly likely NPs were in fact present to explain the free radicals observed.

Sunscreens use white color zinc oxide particles to deflect damaging UV radiation, but the zinc oxide may be made transparent and more absorbent by shrinking the particles down to <100 nm NPs. By QED induced EM radiation, the zinc oxide NPs absorb fractions of the VIS/NIR radiation only to produce higher energy UV radiation that damages the DNA. To avoid DNA damage, NPs > 100 nm would convert the UV content in sunlight to non-damaging VIS/NIR radiation. The claim that NPs are absorbed in the skin and therefore cannot cause DNA damage to the brain or liver do not consider the UV radiation to penetrate the skin and induce DNA damage in the RBC.

5.5. Cancer Therapy

In PDT, photosensitizers in the form of NPs that preferentially attach to cancer cells and activated by NIR radiation are claimed [19] to produce singlet oxygen, thereby destroying the cells by chemical reaction. But cancer cells are destroyed without photosensitizers, thereby begging the question of what actually caused cancer necrosis in PDT.

Prior to photosensitizers, high temperature was thought to induce cancer necrosis in PDT. But NPs lack the specific heat [15] to allow a temperature increase to conserve the absorbed NIR radiation, and therefore QED induces the NP to emit EM radiation beyond the UV that causes cell necrosis, thereby obviating the need for photosensitizers in PDT to activate the oxygen singlet state.

Whether DNA is not damaged by a certain frequency range of ionizing radiation that is damaging to a specific cancer is an unlikely conjecture. But if research shows otherwise, the selection of a NP size tuned solely to the frequency causing necrosis of the cancer may be possible. Only then may NPs be justified in cancer therapy.

6. Extensions to Biological NPs

Cancer research is only beginning to recognize the remarkable fact that natural and man-made NPs in body fluids emit low-level UV radiation. But UV is also emitted from biological NPs < 100 nm having a refractive index greater than the water surroundings. For cancer cells, the index varies from 1.34 to 1.38 compared to 1.33 for water. Only the process by which the biological NPs form differs. A brief description of which is as follows.

6.1. Disorganization o f Epithelial Tissue

Epithelial tissue forming the outer layers of the skin protect exterior surfaces of the body, but also provide protection for hollow organs and glands including the breast, prostate, colon, and lung from body fluids. Epithelial tissue is organized by a submicron thick < 100 nm basement membrane (BM) that provides the structural scaffold template for the extracellular matrix (ECM). Breakdown of the BM is associated with the spread of tumors.

Loss of integrity in the ECM is triggered [19] by enzymes called matrix metalloproteinases (MMPs). Indeed, MMPs induce the epithelial-mesenchymal transition (EMT) that fragment the BM and move through the body. In breast cancer, EMT allows tumor cells more mobility to penetrate barriers like the walls of lymph and blood vessels, facilitating metastasis, e.g., the MMP-3 enzyme causes normal cells to produce a protein called Rac1b that is found only in cancers.

Current thought is the Rac1b protein found in most cancers itself damages the DNA. Contrarily, the ROS are not induced by Rac1b to stimulate the development of cancer by mutation of genomic DNA. Rather, the ROS are formed in a side reaction from the QED radiation emitted from the NPs of fragmented BM. Nevertheless, the Rac1b is still a NP that emits UV and can damage DNA.

6.2. Exocytosis of Small Proteins
The exocytosis or release of fusion products into the extracellular fluid through the tumor cell membrane is known [20] to produce onco proteins. Indeed, the release of fusion products is required for the initiation and growth of malignancy. It should come as no surprise that exocytosis is linked to tumor growth. Hence, QED induced radiation at UV levels from submicron fusion products as biological NPs during exocytosis is consistent with malignancy.

6.3. Molecular Markers in Cancer Detection
Changes that occur in cancer cells compared with normal tissue can be detected in body fluids and used as molecular markers of cancer. As an epithelial tumor grows, cancer cells fragment from the organ epithelium and enter the body fluid as NPs making it possible to detect molecular markers such as DNA mutations.

Protein markers are of interest in QED induced radiation whether by BM fragmentation or exocytosis because the UV radiation produced is damaging to DNA. One such marker is the telomerase enzyme [21] expressed by almost every cancer type: head and neck, lung, breast, colon, pancreas, bladder, and prostate cancers.

7. Summary
- NPs < 100 nm produce QED induced radiations beyond the UV that damages DNA.
- NIR lasers used to activate NPs of photosensitizers are not necessary to produce UV radiation. The thermal *kT* energy of water molecules that collide with the NPs upon absorption is induced by QED to be frequency up-converted to UV levels.
- QED only induces ionizing radiation beyond the UV at NP diameters <100 nm.
- QED radiations cause the large 300-1500 NPs to produce VIS/NIR radiation that enhances the UV emission from adjacent <100 nm NPs.
- The NP induced oxidative stress paradigm that DNA damage is caused by ROS produced proportional to the area of <100 nm NPs needs to be modified to include the enhancement by the larger 300-1500 nm particulate.
- Surface activity of <100 nm NPs has nothing to do with NP toxicity.
- Sunscreens having NPs < 100 nm NPs should be banned in favor of NPs > 100 nm that would absorb UV radiation that then is frequency down-converted to DNA non-damaging VIS and NIR radiation.
- The widespread use of silver NPs in limiting bacteria in food processing increases the risk of developing cancers.
- MMP-3 disintegration of BM produces submicron biological NPs that produce low-level sources of UV radiation that readily move throughout the body.
- Submicron cancer markers are themselves a source of UV radiations.
- More study directed to DNA damage from the UV radiation produced by NPs is suggested to extend the preliminary assessment given in this paper.

8. Conclusions
QED radiation is proposed as the mechanism by which DNA is damaged by low–level UV emission from natural, man-made, and biological NPs, the DNA damage by NPs mimicking the same reaction pathways of conventional ionizing radiation. Specifically,
- The DNA damage induced by NPs is a cancer risk if not properly repaired. Given that NPs produce ionizing radiation beyond UV levels from the QED induced collision energy of water molecules, and that natural, man-made, and biological NPs are ubiquitous, the *conjecture* may be made that NPs are one of the most likely cause of cancers in man.
- The sensitivity of DNA to ionizing UV radiation should be determined to see if cancer necrosis is possible without causing DNA damage.

References

[1] Mroz R M, Schins R P F, Li H, Jimenez L A, Drost E M, Holownia A, Mac Nee W, Donaldson K 2008 Nanoparticle-driven DNA damage mimics irradiation-related carcinogenesis pathways *Euro. Respir. J* 31 241–251

[2] Prevenslik, T V 2007 Nanoparticles in Cancer Therapy presented at GEM Conference on Cancer in conjunction with *Inter Conf Mat App Tech* Singapore July 1-4

[3] Boudaiffa B, Clotier P, Hunting D, Huels MA, Sanche L 2000 Resonant Formation of DNA Strand Breaks by Low-Energy (3-20 eV) Electrons *Science* 287 1658-1660

[4] Hieda K 1994 DNA damage induced by vacuum and soft X-ray photons from Synchrotron radiation *J. Radiat Biol* 66 561

[5] Kuwarara M, Minegishi A, Takakura K, Hieda K, Ito T 1991 Photolysis of water VUV radiation and reactions with DNA and related compounds in aqueous systems In *Photobiology*, Edited by E. Riglis, New York, Plenum 355-363

[6] Razzaboni B L, Bolsaitis, P 1990 Evidence of an Oxidative Mechanism for the Hemolytic Activity of Silica Particles *Environ. Health Persp* 87 337-341

[7] Shi X, Dalal N S, Vallyathan V 1988 ESR evidence for the hydroxyl radical formation in aqueous suspension for quartz particle and its possible significance to lipid peroxidation in silicosis *J Toxicol Environ Health* 25 237-245

[8] Donaldson K, 2003 The biological effect of coarse and fine particulate matter *Occup Environ Med* 60 313-314

[9] Shi T, Knaapen A M, Begerow J, Birmili W, Borm P J A, Schins R P F 2003 Temporal variation of hydroxyl radical generation and 8-hydroxyl radical generation and 8-hydroxy-2'-deoxyguanosine formation by coarse and fine particulate matter *Occup. Environ Med* 60 315-21

[10] Xia T, Kovochich M, Brant J, Hotze M, Sempf J, Oberley T, Sioutas C, Yeh JI, Wiesner MR, Nel AE 2008 Comparison of the Abilities of Ambient and Manufactured Nanoparticles to Induce Cellular Toxicity According to the Oxidative Stress Paradigm *Nano Lett* 6 1794-1807

[11] Warheit D B, Webb T R, Colvin V L, Reed R L, Sayes C N 2007 Pulmonary Bioassay Studies with Nanoscale and Fine-Quartz Particle in Rats: Toxicity is not Dependent upon Particle Size but on Surface Characteristics *Toxicological Sciences* 95 270-280

[12] Ahmed M et al. 2008 DNA damage .response to different surface chemistry of silver nanoparticles in mammalian cells *Toxico Appl. Pharmacol* 233 404-410

[13] Zastrow L, Groth N, Klein F, Kockott D, Lademann J, Renneberg R, Ferrero L 2008 The Missing Link – Light-induced (280-1600nm) Free Radical Formation in Human Skin *Pharmacology and Physiology* 22 31-44

[14] Bohren C F, Huffman D R 1983 *Absorption and Scattering of Light by Small Particles*, J. Wiley & Sons, New York

[15] Shortt B, Carey R, Nic Chormaic 2005 Characterization of Er:ZBNA microspherical lasers *Proc. SPIE* 5827 47-57

[16] Kim J S, Kuk E, Kim J H, Park S J, and Lee H J 2007 Antimicrobial effect of silver nanoparticles *Nanomedicine* 3 95-101

[17] Chau C F, Wu S H, Yen G C 2007 The development of regulations for food nanotechnology *Trends Food Sci. Technol* 16 269-280

[18] Huang X, El-Sayed I H, Qian W, El-Sayed M A 2006 Cancer Cell Imaging and Photothermal Therapy in the Near-Infrared Region by Using Gold Nanorods *J Am Chem Soc* 2115-20

[19] Radisky D C, Levy D D, Littlepage L E, Liu H, Nelson C M, Fata J E, Leake D, Godden E L, Albertson D G, Nieto M A, Werb Z, Bissell M J 2005 Rac1b and reactive oxidative species mediate MMP-3 induced EMT and genomic instability *Nature* 436 123-7

[20] Chan A M, Weber T 2002 A putative link between exocytosis and tumor development *Cancer Cell* 2 427-8

[21] Sidransky D 2002 Emerging Molecular Markers of Cancer *Nature Reviews – Cancer* 2 210-9

Development of innovative pH sensor to evaluate phagocytosis of nanoparticles

L Leclerc[1,2,3], D Boudard[1,4,5], J Pourchez[1,2], L Marmuse[6], C Louis[6], S Palle[7], P Grosseau[3], D Bernache[2] and M Cottier[1,4,5]

[1]LINA Laboratoire Interdisciplinaire d'étude des Nanoparticules Aérosolisées, F-42023, Saint-Etienne, France

[2]Ecole Nationale Supérieure des Mines de Saint-Etienne, Centre Ingénierie et Santé, Saint-Etienne, France

[3]Ecole Nationale Supérieure des Mines de Saint-Etienne, LPMG, UMR CNRS 5148, F-42023, Saint-Etienne, France

[4]PRES Lyon - Université Jean Monnet, Faculté de Médecine, IFR Inserm 143, F-42023, Saint-Etienne, France

[5]Laboratoire d'Histologie CHU de Saint-Etienne, F-42055, Saint-Etienne, France

[6]NANO-H S.A.S., 2 place de l'Europe, Bâtiment A, Parc d'activité GVIO, F-38070, Saint Quentin Fallavier, France

[7]PRES Lyon, Université Jean Monnet, Centre de Microscopie Confocale Multiphotonique, Pôle Optique et Vision, Saint-Etienne, France

E-mail: leclerc@emse.fr

Abstract. The aim of this work was the development of pH-sensor-NP allowing the quantification of the amount of NP phagocytosed by macrophages. Two types of fluorescent NP with variable and well-characterized sizes and chemicals coatings have been synthesized:

- NP with a FITC core (FITC-NP): green fluorescence (control).
- FITC-NP functionalized with a pH sensitive probe (pH-sensor-NP): green fluorescence of the FITC and red fluorescence of the pH probe in acidic conditions.

Our pH-sensor-NP model was first validated in acellular conditions. They were then incubated with a macrophage cell line allowing distinction and quantification of internalized NP with no major effects on biological toxicity.

1. Introduction

Macrophages from the respiratory system constitute the first defense line against inhaled nanoparticles (NP) thanks to their phagocytic activity [1]. The toxicity of NP depends on their physicochemical characteristics (size, morphology, crystallinity, chemical surface composition…) [2]. The relationship between cellular activity and the number of phagocytosed NP is still unclear. However, the quantification of NP uptake by macrophages could allow a better assessment of NP toxicity. To that purpose we developed a new type of NP sensitive to pH variations as it is known that during phagocytosis, vesicles content acidifies [3].

Two types of fluorescent NP with well-characterized sizes (250 and 500 nm) and chemicals coatings (COOH) have been synthesized:

- FITC-NP: green fluorescence due to fluoresceine isothiocyanate in NP core (control NP).
- pH-sensor-NP: green fluorescence in NP core but also pHrodo™ (Invitrogen) conjugated in a porous polysiloxane shell. The red fluorescence of pHrodo probe increases as pH decreases allowing the detection of intra-lysosomal engulfed NP (yellow double fluorescence).

2. Materials and methods

2.1. Acellular validations

NP were synthesized with the collaboration of Nano-H Company. Different size characterizations were done: DLS (Zetasizer Nano S from Malvern instrument) and TEM (Jeol 2010).
Confocal spectral analyses were performed in solutions of different pH (pH4 and pH10) with a confocal microscope TCS-SP2 AOBS (Leica®).

2.2. Cellular validations

Cells from the RAW 264.7 macrophage cell line were exposed to NP (arbitrary ratio of 1000 NP for one cell) for 24h. SEM (Hitachi) and TEM observations were performed.
The fluorescence of NP was assessed in confocal microscopy and the percentage of internalized NP was calculated using image analysis.
Then toxicity assessment was performed regarding different parameters: inflammatory response (Tumor Necrosis Factor alpha (TNF-α) production), integrity of the cell membrane (Lactate Dehydrogenase (LDH) release) and the oxidative stress (production of different Reactive Oxygen Species (ROS)) [4-6].

3. Results and conclusion

Our pH-sensor-NP model was validated by fluorescence spectral analysis performed by confocal microscopy in different acellular media with decreasing pH values.
Moreover after cellular contacts it was possible to distinguish and quantify entirely engulfed pH-sensor-NP that appeared yellow (due to the colocalization of green and red fluorescences of NP in acidic vesicles) from those which were just adherent to the cell membrane (green fluorescence).
Moreover, incorporation of pHrodo had no major effects on biological toxicity of NP especially for the double fluorescent pH-sensor-NP of 250 nm.

References

[1] Aderem A, Underhill DM. 1999 Mechanisms of phagocytosis in macrophages. *Annu Rev Immunol* **17** 593-623.
[2] Oberdorster G, Oberdorster E, Oberdorster J. 2005 Nanotoxicology: an emerging discipline evolving from studies of ultrafine particles *Environ Health Perspect* **113**(7) 823-39.
[3] Vieira OV, Botelho RJ, Grinstein S. 2002 Phagosome maturation: aging gracefully *Biochem J* **366**(Pt 3) 689-704.
[4] Bruch J, Rehn S, Rehn B, Borm PJ, Fubini B. 2004 Variation of biological responses to different respirable quartz flours determined by a vector model *Int J Hyg Environ Health* **207**(3) 203-16.
[5] Sayes CM, Reed KL, Warheit DB. 2007 Assessing toxicity of fine and nanoparticles: comparing in vitro measurements to in vivo pulmonary toxicity profiles *Toxicol Sci* **97**(1) 163-80.

[6] De la Harpe J, Nathan CF. 1985 A semi-automated micro-assay for H2O2 release by human blood monocytes and mouse peritoneal macrophages *J Immunol Methods* **78**(2) 323-36.

Nanosafe2010: International Conference on Safe Production and Use of Nanomaterials IOP Publishing
Journal of Physics: Conference Series **304** (2011) 012056 doi:10.1088/1742-6596/304/1/012056

Acute toxicity of virgin and used engine oil enriched with copper nano particles in the earthworm

M Khodabandeh[1], M K Koohi[1], A Roshani[2], E Shahroziyan[1], B Badri [1], A Pourfallah[1], Gh Shams[1], R Hobbenaghi[3] and G Sadeghi-Hashjin[1]

[1]Faculty of Veterinary Medicine, University of Tehran, Tehran, Iran

[2]Industrial & Environmental Protection Division, Research Institute of Petroleum Industry (RRIPI), Tehran, Iran

[3]Faculty of Veterinary Medicine, Urmia University, Urmia, Iran

E-mail: gsadeghi@ut.ac.ir

Abstract. In spite of development of nanotechnology and creation of new opportunities for industry, new applications and products initiated by this technology may cause harmful effects on human health and environment. Unfortunately, there is no sufficient information on the harmful effects caused by application of some nano materials; the current knowledge in this field is limited solely to the nano particles but not the final products. Nano cupper particles, as one of the common materials produced in industrial scale is widely used as additives into engine oil to reduce friction and improve lubrication. However, the difference between the effects of virgin and used conventional engine oil (CEO) and the engine oil containing cupper nano particles (NEO) on the environment is not known. Earthworm, as a one of the species which could live and survive in different sorts of earth and has a certain role in protecting the soil structure and fertility, was used in this experiment. In accordance with the recommended method of OECD.1984, Filter Paper test in 24 and 48 h based on 8 concentrations in the range of 3×10^{-3} - 24×10^{-3} ml/cm^2 and Artificial Soil test in 7 and 14 days based on 7 concentrations in the range of 0.1 mg/kg – 100 g/kg were carried out to study earthworms in terms of lifetime (LC50), morphology and pathology. It was shown that the 48 h LC50 for virgin CEO, virgin NEO, used CEO(8000 km) and used NEO (8000 km) were 6×10^{-3}, 23×10^{-3}, 24×10^{-3} and 16×10^{-3} ml/cm^2, respectively. Furthermore, 14-day LC50 in artificial soil for all cases were above 100 g/kg. It is concluded that virgin CEO is more toxic than virgin NEO. Meanwhile, the CEO shows significant reduction in toxicity after consumption and the used NEO shows more toxicity in comparison to virgin product. It seems that more investigations on the effects of final products specifically after consumption is necessary because the products after consumption have the most contact with environment and subsequently human health.

1. Introduction

Conventional products are gradually replaced by nano products. Recent studies indicate the toxicity of nano materials on aquatic and rodents, however, many hidden effects of nano materials are still unknown [1]. Therefore, a greater amount of study is required on the effects of nano products on humans and the environment [2]. Currently, there is a limited understanding about side effects of nano materials in nature and wildlife [3]. Lots of data is needed for systematic studies of nano materials' toxicity in different species in water and soil environments [3]. Much of our information about toxicity of nano materials is on aquatic organisms, thus, there is a need for further studies on toxicity of nano-materials on land animals as well [1].

The earthworm, as a species which is able to live in different types of soil, is known as an important experimental model for monitoring the possible effects of environmental chemicals on the soil systems. Its higher sensitivity, limited mobility, and easy collection and rearing in the laboratory are among factors making it suitable for this kind of studies. Therefore, many studies have been accomplished using this species on the effects of chemicals, including heavy metals, pesticides, aromatic hydrocarbons, cadmium, nickel, MTBE, perchlorate, and lead compounds [4-12]. The available information indicates that engine oil is considered as a dangerous environmental pollutant and its use may cause destructive environmental effects around the world [13].

Because of substantial increase in the introduction and applications of nano materials and potential impacts of them on the environment as well as little information available in this field, the environmental effects of these substances are required to be identified. Copper nano particle is one of the nano metals which is commercially manufactured for application in various industries. For instance, due to its high speed, low cost and recycling properties, copper nano particle is used in the anode pole materials of the lithium batteries [14]. Moreover, as a bioactive cover, it is capable of inhibiting microorganisms such as *Sacromaysis*, *E. coli*, *S. aureus* and *Listeria* [15]. One of the other characteristics of copper nano particles is their plasticizing feature when mixed with used oils in the oil industry [16]. In general, it can be said that nano particles have restorative properties in abrasion surfaces, which can cause the friction to be reduced.

Nano cupper particles are widely used as additives into engine oil to reduce friction and improve lubrication. However, the difference between the effects of virgin and used conventional engine oil (CEO) and the engine oil containing cupper nano particles (NEO) on the environment is not known. The aim of the present study was to compare the effects of 4 conditions of the engine oil on the earthworm. The experiments were performed in accordance with the recommended protocols of OECD [17].

2. Materials and Methods

2.1. Test Animals

Earthworm specie used for testing was *Eisenia foetida* (Michaelsen). The worms reached to maturity level (>2 months old) and weighed 300 to 600 mg. The worms were taken from a research institute for organic vegetable wastes in the city of Tehran, Iran.

2.2. Test Substances

The engine oil used in this study was a commercial brand manufactured in Iran and enriched with copper nano particles by researchers working at Research Institute of Petroleum Industry (RRIPI), Tehran, Iran. The product was tested in 4 different conditions: virgin conventional engine oil (vCEO), used conventional engine oil (uCEO), virgin nano engine oil (vNEO) and used nano engine oil (uNEO). Copper nanoparticles had the cover in the size range of 6-20 nm and the average diameter of 12 nm. The engine oil type was SAE20W-50W with oil basis, and the density of 0.898. Used products had a performance of 8000 km.

2.3. Experimental Protocols

Acute toxicity test was done with the methods of filter paper (test) and artificial soil test according to instructions of the number 1984 (207).

2.3.1. *Filter Paper Test.* The filter paper had the following specifications: average grade, weight 80-85 g/m^2, thickness 0.2 mm, area 3×3 cm. Vials were flat-bottom glass, area 3×3 cm, height 8 cm. Two ml of the final solution (test substance in chloroform as the solvent) was inserted by pipette in each of the vials. The test substances were either 0, 0.25, 0.5, 0.75, 1.0, 1.25, 1.5, 1.75 or 2.0 ml. Thirty animals were used for each group (every concentration of every substance). Before covering, the vials were exposed to slight flow of the air to dry. To eliminate any bias due to the effect of the physico-chemical properties of the engine oil, three extra control solutions were also used, they included olive oil, olive oil + chloroform (50:50), and water. Before the test, worms were placed for 3 h in a dry dish to defecate. Thereafter, they were placed in the vial (1 worm/vial) and 1 ml of de-ionized water was added onto the paper. Then, the vials were covered by nylon. They were left in a dark place at room temperature for 48 h. The number of dead animals as well as morphological changes was evaluated after both 24 h and 48 h.

2.3.2. *Artificial Soil Test.* A mixture of 10% peat moss (mulch), 20% clay and 70% sand was used as the artificial soil. The pH of the medium was set at 6.0±0.5 by adding calcium bicarbonate. De-ionised water was added to the soil to increase the moisture (350 ml of water for 1 kg of DRY soil). Test substances were added to the final soil (750 g in 1 L glass containers) at the concentrations either 0, 1, 10, 100, 1.000, 10.000 or 100.000 ppm. The worms were kept in un-treated soil for 24 h, after which they were transferred to the test medium. Ten animals were placed in every glass container and the experiment was repeated 4 times (number of animals per group = 40). The containers were left in a light place at room temperature for 14 days. Evaluation of mortality and morphological changes were performed on days 7 and 14.

2.4. Statistical Analysis

Data of each experimental group was expressed as mean ± SEM. Differences were first evaluated by one-way analysis of variance (ANOVA) and were then compared by the post-ANOVA test, *Bonferroni's t test.* In addition, correlations between concentration of the engine oil and the death of the earthworm were evaluated and the regression factor was calculated ($-1 > r < +1$). The difference was considered to be statistically significant whenever $P < 0.05$.

3. Results

3.1. Mortality rates (LC50)

LC50 within 48 h in paper test for vCEO, vNEO, uCEO and uNEO were 6×10^{-3}, 23×10^{-3}, $>24\times10^{-3}$ and 16×10^{-3} ml/cm^2, respectively. In brief, the toxicity ration was as vCEO > uNEO > vNEO > uCEO. There were correlations between concentrations of the oil and response (death): vCEO ($r=0.924$, $P=0.025$), uNEO ($r=0.968$, $P<0.001$), vNEO ($r=0.982$, $P=0.121$), uNEO ($r=0.954$, $P<0.001$). Furthermore, LC50 within 14-day in artificial soil test 14-day LC50 in artificial soil for all cases were >100 g/kg. All together, in most concentrations, vCEO was found to be more toxic than all other products used ($P<0.05$). Figure 1 shows the details of mortality at all used concentration of 4 different types of the engine oil.

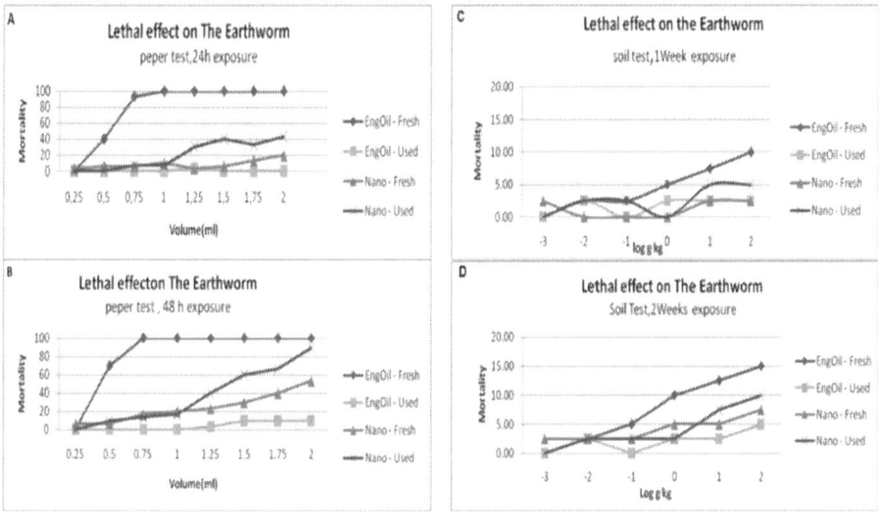

Figure 1. Charts related to mortality (deaths) in four types of oil in different concentrations:
A-Comparative effects of used & virgin NEO and CEO in paper test within 24 hours.
B-Comparative effects of used & virgin NEO and CEO in paper test within 48 hours.
C-Comparative effects of used & virgin NEO and CEO in artificial soil test within 7 days.
D-Comparative effects of used & virgin NEO and CEO in artificial soil test within 14 days.

3.2. Histopathological changes

Extensive damage to many organs were caused by engine oil in the earthworms subjected to paper test within 48 hours. Some of important observations are outlined in Table 1 and representatives of the photomicrographs are shown in Figure 2.

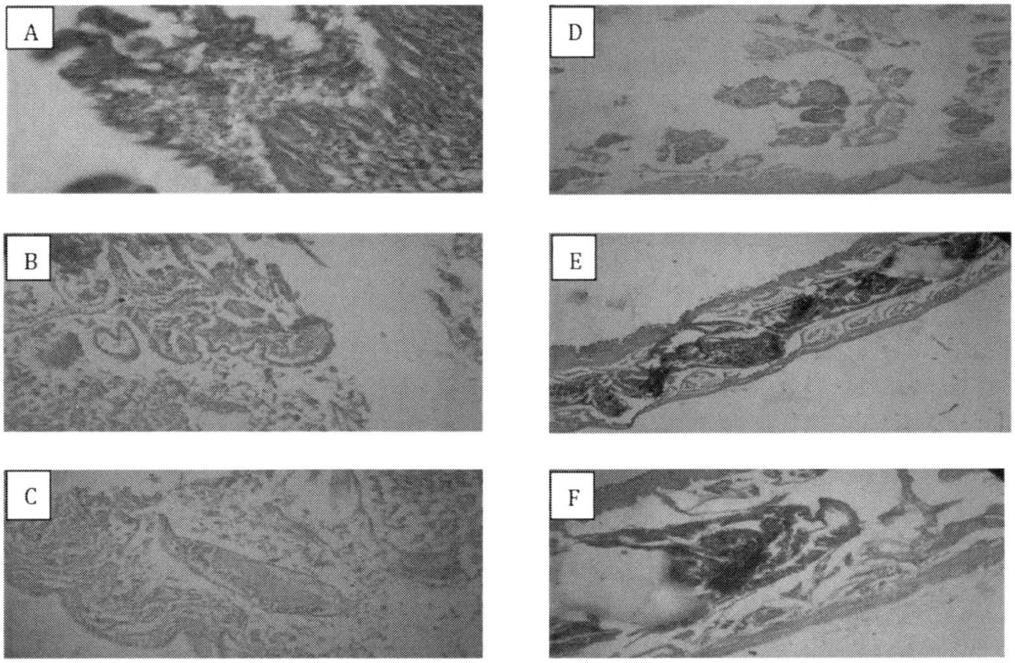

Figure 2. Histopathological Effects of vCEO: Watery degeneration, necrosis and individuation of the epidermis from its underlying tissue, along with influence the dark pigment among circular muscle cells (**A**). Extensive destruction of nephric tissue (**B**). Leucomalasia in the abdominal nerve (**C**). **Histopathological effects of vNEO:** Leucomalasia in the abdominal nerve (**D**). **Effects of uNEO:** Normal cuticle, normal circular and longitudinal muscles, and dark pigments can be seen in the digestive tract (**E**). **Histopathological effects of uCEO:** Normal cuticle, normal circular and longitudinal muscles, and dark pigments are seen in the digestive tract (**F**).

Table 1. Histopathological changes in the earthworm subjected to paper test within 48 h.

	vCEO	uCEO	vNEO	uNEO
Epidermal separation	+++	-	++	+
Loss of villi in the gut	+++	-	++	-
Nerve leukomalacia	+++	-	+	-
Tubular necrosis in the urinary tract	+++	-	+	-
Skeletal muscle necrosis	++	-	-	-

3.3. Behavioral and morphological changes

No effect was observed on the worms subjected to the artificial soil test. On the other hand, concentration-dependent decreases in response to needle and light stimulations by the survived worms were recognized during the paper test; all groups showed degrees of mucosal secretions and coiling. Except in the uCEO group, all animals showed the following signs: fragmentation, bleeding, and swelling.

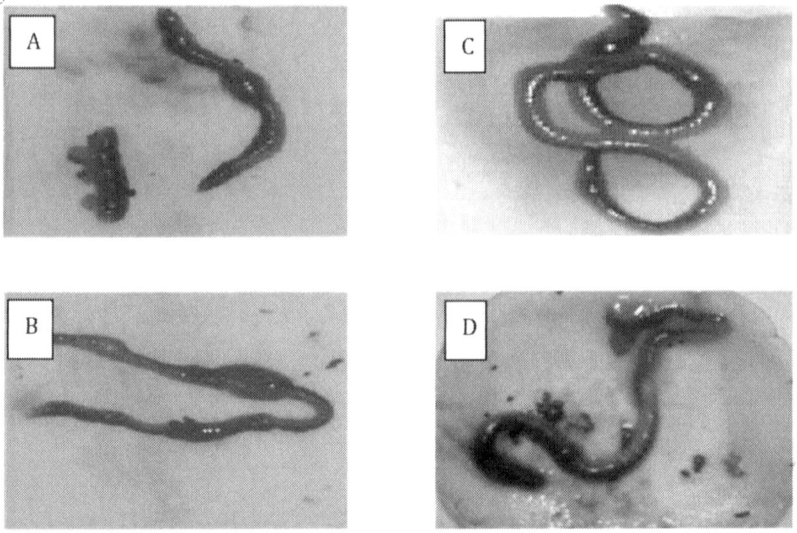

Figure 3. Morphological observations: Fragmentation (A), swelling (B), coiling (C) and bleeding (D) in the earthworms subjected to engine oil during paper test.

4. Conclusion

In the present study, a full study was performed to determine the environmental impact of engine oil. In addition to the conventional form, a commercially designed product, enriched with copper nano particles was also evaluated. As the toxicity could be either enhanced or declined with the usage of these compounds, besides the virgin product, the effects of used engine oil of both forms were also investigated. Based on the protocols imposed by OECD, the earthworm was used as the experimental model.

Although not studied extensively enough, the carcinogenic action of engine oil has been understood since a half century ago [18]. The carcinogenic effect of used engine oil in mouse skin, explained 30 years later, was shown to be related to specified adducts produced following the exposure with the reactive metabolites of the product [19]. In spite of the carcinogenic evidence, used engine oil is applied on damaged human skin by unauthorized persons in Iran. It should be taken strange that it is suggested to be used in the treatment of skin disorders of farm animals [20]. A case of 64-year-old male with longstanding occupational exposure to used engine oil was reported to present with a rare intraepithelial adenocarcinoma, *i.e.* extramammary Paget's disease of the left scrotum and groin [21]. Evidence suggest that more caution should be taken into account in the use, destruction and recycling of all kinds of engine oil throughout the world.

Finally, the outcome of the findings in this paper can be summarized as follows: (1) Virgin engine oil can be harmful for the environment as shown in the earthworm. (2) The product is detoxified after conventional use in motor vehicles. Therefore, the used products are considered to be less toxic than the virgin engine oil. (3) Engine oil becomes less toxic when enriched with copper nano particles. This is true only in case of the virgin product and, hence, it safer for professionals working with this product. (4) The used nano enriched engine oil is more toxic than the conventional used product for the environment. Therefore, it has to be disposed with particular attention.

References

[1] UNESCO. The Ethics and Politics of Nanotechnology. Fontenoy, France, 2006
[2] Roco MC (2005) *J Nanoparticle Res* **7**: 129-143
[3] Handy RD (2008) *Ecotoxicol* **17**: 315-325
[4] Sainet-Denis M, Narbonne MJF , Arnaud C, Thybaud E, Ribera D (1999) *Soil Biol Biochem* **31**: 1837-1846
[5] Luo Y, Zhang Y, Zhong Y, Kong ZM (1999) *Chemosphere* **39**: 2347-2356
[6] Chakra Reddy N, Venkateswara Rao J (2008) *Ecotoxicol Environ Safety* **71**: 574-582
[7] Lock K , Janssen CR (2002) *Chemosphere* **46**: 197-200
[8] Lock K, Janssen CR (2001) *Sci Total Environ* **280**: 79-84
[9] Youn-Joo A (2005) *Environ Pollution* **134**: 181-186
[10] Landrum M, Cañas JC, Coimbatore G, Cobb GP, Jackson WA, Zhang B, Anderson TA (2006) *Sci Total Environ* **363**: 237-244
[11] Venkateswara Rao J, Kavitha P, Padmanabha Rao A (2003) *Environ Res* **92**: 271-276
[12] Darling CTR, Thomas VG (2005) *Sci Total Environ* **346**: 70-80
[13] Duhalt V(1989) *Sci Total Environ* **79**: 1-23
[14] Guo K (2002) *Electrochem* **32**: 679–685
[15] Cioffi N (2005) *Anal Bioanal Chem* **381**: 607–616
[16] Liu G (2004) *Tribology Lett* **17**: 961–966
[17] OECD (1984) OECD Guideline for Testing of Chemicals No. 207, Paris
[18] Baldwin RW, Cunningham GJ, Pratt D (1961) *Br J Cancer* **15**: 123-126
[19] Carmichael PL, Ni Shé M, Hewer A, Jacob J, Grimmer G, Phillips DH (1992) *Cancer Lett* **64**: 137-144
[20] Dreyer K, Fourie LJ, Kok DJ (1998) *Onderstepoort J Vet Res* **65**: 275-279
[21] Virich G, Gudi V, Canal A (2008) *J Plast Reconstr Aesthet Surg* **61**: 1528-15299

Journal of Physics: Conference Series **304** (2011) 012057 doi:10.1088/1742-6596/304/1/012057

Investigation of titanium dioxide nanoparticles toxicity and uptake by plants

C Larue[1], H Khodja[2], N Herlin-Boime[3], F Brisset[4], AM Flank[5], B Fayard[6], S Chaillou[7] and M Carrière[1]

[1]Laboratoire de Structure et Dynamique par Résonance Magnétique UMR 9956 CEA-CNRS-IRAMIS, Gif-sur-Yvette, France

[2]Laboratoire d'Etude des Eléments Legers, UMR 9956 CEA-CNRS-IRAMIS, Gif-sur-Yvette, France

[3]Laboratoire Francis Perrin URA 2453 CEA-CNRS-IRAMIS, 91191 Gif-sur-Yvette, France

4Institut de Chimie Moléculaire et des Matériaux d'Orsay, UMR8182 CNRS-Univ. Paris sud, Orsay, France

[5]LUCIA beamline, SOLEIL synchrotron, Saint-Aubin, France

[6]Laboratoire de Physique du solide, Orsay, France and ID21 beamline, ESRF, Grenoble, France

[7]Unité de Nutrition Azotée des Plantes, INRA, Versailles, France

E-mail: marie.carriere@cea.fr

Abstract. Nanoparticles (NP) are introduced in a growing number of commercial products and their production may lead to their release in the environment. Plants may be a potential entry point for NP in the food chain. Up to now, results describing NP phytotoxical effects and plant accumulation are scarce and contradictory.

To increase knowledge on titanium dioxide NP (TiO_2-NPs) accumulation and impact on plants, we designed a study on three plant species, namely wheat (*Triticum aestivum*), oilseed rape (*Brassica napus*) and *Arabidopsis thaliana*. These plants were exposed in hydroponics to a panel of well-characterized TiO_2-NPs, with diameters ranging from 12 to 140 nm, either anatase or rutile. Their accumulation in plant tissues is currently being assessed by complementary imaging techniques: scanning electron microscopy (SEM), transmission electron microscopy (TEM), micro-X-ray fluorescence (SR-μ-XRF) imaging and micro-particle induced X-ray emission (μ-PIXE) imaging. Moreover, the impact of TiO_2-NP exposure on germination rate, root elongation, dry biomass and evapotranspiration is evaluated. Preliminary results are presented here, with data collected on wheat plants exposed to 12 nm and 25 nm anatase TiO_2-NPs. These results show that TiO_2-NPs are taken up by plants, and do not significantly alter their germination and root elongation. These results underline the necessity of deeper evaluation of nanoparticle ecotoxicity, and particularly on their interaction with plants.

1. Introduction

Nanoparticles (NP) are introduced in a growing number of commercial products. Among them, TiO_2 anatase nanoparticles (TiO_2-NPs) are one of the most produced NPs in the world (see Nanowerk database, www.nanowerk.com). They are used as a pigment in paint, paper, ink and plastic. They are also introduced in cosmetics such as sunscreens for their UV protective properties (see Nanowerk database, www.nanowerk.com). Finally, they are presently used in pilot water-purification reactors, because of their ultraviolet irradiation-induced bactericidal effects (photocatalysis). For these reasons, these NPs will surely be released in the environment [1]. The knowledge of their potential effects on human health is rapidly increasing, however little is known on their potential toxicological effects on environment, i.e. destabilization of the ecosystems and trophic transfer, but also on their potential transfer to the food chain via plant ingestion. The literature relating NP toxicity and accumulation in plants is scarce; authors mostly relate NP inhibitory effects on seed germination and root elongation [2-4]. A few studies relate NP uptake by plant roots [3, 5], and root to shoot transfer [5]. Concerning TiO_2-NPs, TiO_2-based nanocomposites have recently been shown to be accumulated and translocated to the shoots of Arabidopsis thaliana [6]. These translocation studies are based on elemental analysis of the ionic form of metal in plants, but neither address the presence of metal as NPs, nor the spatial distribution of NPs throughout the plant. Actually, TiO_2-NP dissolution has been demonstrated to be low in aqueous solution [7], however its occurrence has never been tested in plant exposure solution and/or during NP transfer to the roots or from the roots to the shoots. Synchrotron radiation micro X-ray fluorescence (SR-μ-XRF), coupled to micro X-ray absorption spectroscopic (μ-XAS) analysis is the ideal method to address metal distribution after plant uptake, together with speciation analysis which would permit the identification of this metal as NPs. It has already been used to assess ionic metal internalization and distribution in plants [8].

We used a combination of complementary imaging techniques: scanning electron microscopy (SEM), transmission electron microscopy (TEM), micro-X-ray fluorescence (SR-μ-XRF) imaging and micro-particle induced X-ray emission (μ-PIXE) imaging, to assess TiO_2-NP accumulation in plants, and correlated their uptake with their impact on plant development. This study is currently under process in our laboratory with several plant models and several TiO_2-NPs varying in their diameter, crystalline phase and shape; we present here some preliminary results obtained on wheat (*Triticum aestivum*), exposed in hydroponics to 12 nm and 25 nm anatase TiO_2-NPs. These results show that TiO_2-NPs are taken up by plants, and do not significantly alter their germination and root elongation. These results underline the necessity of deeper evaluation of nanoparticle ecotoxicity, and particularly on their interaction with plants.

2. Methods

2.1. Nanoparticles, characterization and suspension

For TiO_2-NP dispersion studies and impact studies on plant development, we used commercial Evonik Aeroxide® P25 TiO_2-NPs. The NPs used in plant uptake experiments were synthesized by laser pyrolysis in Laboratoire Francis Perrin (LFP) [9], then annealed under air at 400°C during 3 h. Physico-chemical characterization of these NPs has been reported in two previous studies [10, 11]. Briefly, LFP NPs are 95% anatase and 5% rutile. Their mean diameter is 12 ± 3 nm, they are spherical-shaped and their specific surface area, measured according to Brunauer, Emmett and Teller (BET) [12], is 82 ± 4 $m^2.g^{-1}$. Their point of zero charge, measured by assessment of zeta potential at increasing pH, is 6.4. Evonik P25 nanoparticles are 75% anatase and 25% rutile. Their mean diameter is 25 ± 7 nm, they are spherical-shaped and their specific surface area, measured according to Brunauer, Emmett and Teller (BET) [12], is 46 ± 1 $m^2.g^{-1}$. Their point of zero charge is 7.

These TiO_2-NPs were suspended in ultrapure water and sonicated using a high energy probe, during 30 mn, at 4°C, in pulsed mode (1 s on, 1 s off). Agglomeration status was evaluated using photon correlation spectroscopy (PCS) using a Zetasizer 3000HS (Malvern). Z-average values are given in the result section.

2.2. Plant exposure and sample preparation for imaging

For imaging, seeds were germinated on sand, and their roots were directly immersed in the NP suspension (100 $mg.L^{-1}$), and placed in a growth chamber under controlled conditions: day/night photoperiod: 16/8 h; temperature (day/night), 24/20 ± 1°C and relative humidity (day/night), 70/75%. After 7 days, plant roots were thoroughly rinsed with deionised water. For μ-XRF, SEM and μ-PIXE analysis, plant roots and leaves were cryofixed, immediately embedded in Tissue Tek OCT™ resin (Agar Scientific, Essex, UK) and cross sections were cut and freeze-dried.

The distribution of Ti, P, Cl and Ca in plant thin sections were mapped using synchrotron radiation μ-X-ray fluorescence (SR-μ-XRF), images presented in this study were recorded on LUCIA beamline (SOLEIL synchrotron, France). The spot size was 3 × 3 μm^2. XAS spectra were recorded between 4.95 and 5.05 keV, *i.e.* around Ti X-ray absorption edge (4.966 Kev). On each sample, at least 10 spectra were collected, *in situ* on map locations containing high concentrations of Ti. These spectra were then averaged and compared to XAS spectra collected on Ti standards: 12 nm, 25 nm and 140 nm TiO_2 anatase NPs, 20 and 50 nm TiO_2 rutile NPs. Spectrum was then normalized to the second-order polynomial to be equal to one, with Athena software [13]. SEM image were acquired on a Zeiss-Supra 55VP Field Emission Gun Scanning Electron Microscope (FEG-SEM) equipped with a X-ray Energy Dispersive Spectrometer (EDS), operated at 15 kV, at ICMMO, Orsay, France.

2.3. Nanoparticle impact on plant development

For NP impact on plant development assessment, 10 seeds were germinated in Petri dishes in deionised water containing 10 to 100 mg/L NPs. According to US-EPA guidelines (US-EPA, 1996), seeds with roots longer than 5 mm were considered germinated. For root elongation experiment, after 7 days, roots were measured on a picture thanks to ImageJ software.

3. Preliminary results and discussion

3.1. Characterization of TiO_2-NPs and dispersion in aqueous solution

In order to obtain properly dispersed and stable TiO_2-NP suspensions, sonication was applied to TiO_2-NP powders dispersed in water at 10 $g.L^{-1}$. For this study, TiO_2-NPs from Evonik (Aeroxide® P25) were used. Optimal TiO_2-NP concentration for proper measure was evaluated, and concentrations of 50-1000 $mg.L^{-1}$ were found to be appropriate, whereas at lower concentrations, the Zetasizer was not able to properly measure the mean hydrodynamic diameter (Table 1). At concentrations higher than 1000 $mg.L^{-1}$, the suspension is too opaque to obtain a proper measure.

Table 1. Mean hydrodynamic diameter of nanoparticle agglomerates[a]

Concentration ($mg.l^{-1}$)	10	50	100	500	1000	2000
Diameter (nm)	140 0.6	24.2	38.7	43.1	66.6	12.5
Standard deviation	333. 4	11.1	15.4	22.6	26.1	7.3

[a]Mean hydrodynamic diameter was determined by PCS after a 30 mn sonication, in water, using high energy probe. TiO_2-NP concentration in these suspensions ranged from 10 to 2000 $mg.l^{-1}$.

When diluted in Hoagland medium, these suspensions of TiO_2-NPs immediately agglomerated. To prevent agglomeration, various dispersing agents were tested. The obtained suspensions are presented in Figure 1, which shows that NP agglomeration occurs in most conditions. Mean hydrodynamic diameters of TiO_2-NP agglomerates are presented in Table 2.

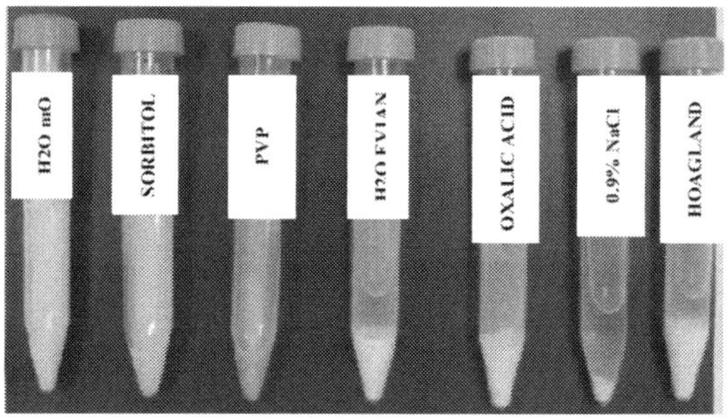

Figure 1. TiO_2-NP suspensions, use of a dispersing agent. From left to right: TiO_2-NPs sonicated in water, sorbitol, 1 $g.l^{-1}$ PVP, Evian water, oxalic acid, 0.9% NaCl, or sonicated in water and then diluted in Hoagland medium.

Table 2. Mean hydrodynamic diameter of NP agglomerates when using a dispersing agent[a]

Dispersing agent	20 mM oxalic acid	1 $g.l^{-1}$ PVP	10 $g.l^{-1}$ PVP	1 mM sorbitol	0.9% NaCl	Evian water
After sonication	1536	46	245	41	1000	1668
Hoagland	1234	896	41	853	1126	746

[a]Mean hydrodynamic diameter was determined by PCS at 50 mg/L after a 30 mn sonication, in water containing a dispersing agent (oxalic acid or polyvinylpyrrolidone (PVP) or sorbitol or in physiologic serum (0.9% NaCl) or Evian water). Diameters are determined directly after sonication, or after sonication and dilution in Hoagland medium.

As regards these dispersion results, and considering that we aimed at exposing plants in a well-dispersed TiO_2-NP suspension, we chose to expose plants in TiO_2-NPs prepared in water without addition of any dispersing agent, which would not increase suspension quality.

3.2. Impact of TiO_2-NPs on plant development

The impact of TiO_2-NP exposure on plant development was evaluated on wheat seeds and plantlets, exposed to Evonik (Aeroxide® P25) TiO_2-NPs. Exposure did not induced any statistically significant modification in wheat seed germination (Figure 2B), nor in root elongation (Figure 2C). However, during root development, TiO_2-NPs strongly stuck onto roots and agglomerated in exposure medium (water, Figure 2A).

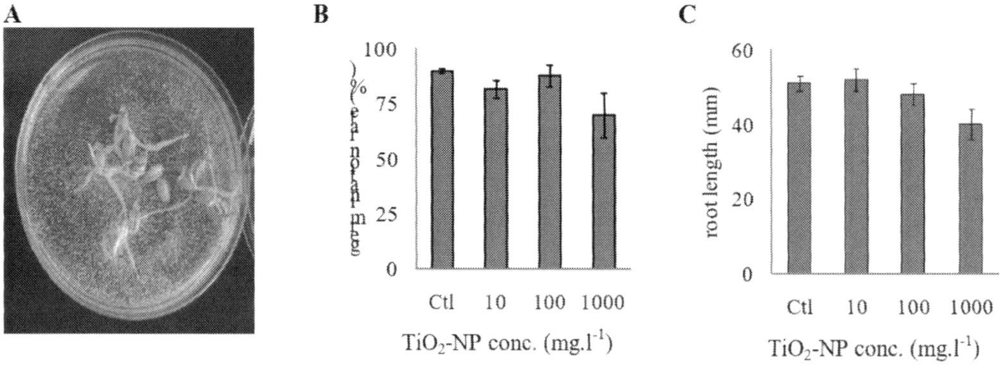

Figure 2. Seed germination test. Wheat seeds were exposed to Aeroxide® P25 TiO_2-NPs (A), and number of germinated seeds was counted (B), then the mean length of roots was measured (C).

3.3. TiO_2-NP accumulation in plants

Elemental analysis, in scanning mode, enabled to draw images of P content, representative of plant tissues, and of Ti content, representative of TiO_2-NPs, in plant roots and leaves after exposure. An example is presented in Figure 3, mapped on wheat plant roots after exposure to anatase, 12 nm TiO_2-NPs. We used these nanoparticles rather than Evonik nanoparticles because they were smaller, and thus would more probably be taken up by plants. Their agglomeration status and impact on plant development was close to that of Evonik nanoparticles (not shown). After 7 days of exposure, Ti was observed in the parenchyma of wheat roots (Figure 3, arrows), but also in the vascular cylinder, demonstrating that TiO_2-NPs were transferred from the exposure suspension to vegetal tissues.

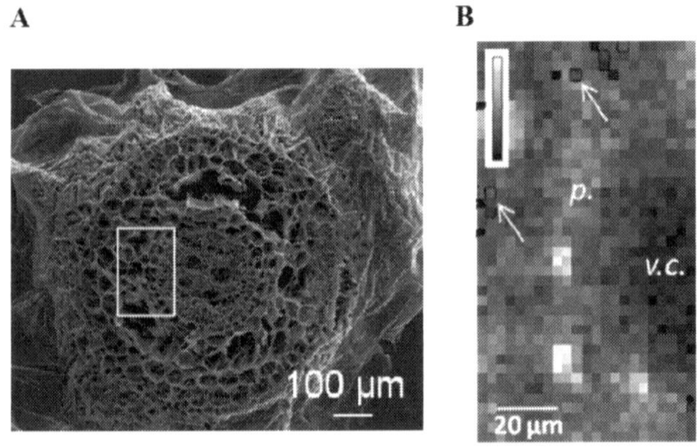

Figure 3. Ti distribution in wheat roots. Transversal sections of a root analyzed by SEM (A) and by µ-XRF image of a region related to the white square in A (B) after a 7-day exposure to 12 nm anatase TiO_2-NPs. In B: maps of P distribution (gray) and Ti distribution (red, clusters are also indicated by arrows) are superimposed. SEM image was recorded in ICMMO and µ-SR-XRF image on LUCIA beamline, SOLEIL synchrotron.

As described in several publications, Ti pre-edge features are characteristic of the crystalline phase and size of TiO_2-NPs. Anatase displays a typical triplet feature (A1, A3 and B peaks), with a weak shoulder on the low-energy side of the central A3 peak (A2) [14]. The intensity of A2 peak is related to the distortion of the octahedral TiO_6 unit, particularly the distortion observed on the surface of NP. The intensity of this peak is thus related to the size of the particle: the smaller the NP, the higher this A2 peak [15]. Meanwhile, the intensity of the A1 peak also decreases as the particle size decreases [16]. Analysis of this pre-edge region may then inform on changes in the crystalline phase of NPs after cellular internalization, and on their dissolution. In the present experiment, no modification of these pre-edge features was observed after plant internalization of anatase TiO_2-NPs (Figure 4), meaning that Ti is still in the nanoparticulate TiO_2 chemical form inside plants, and that their crystalline phase did not change. Moreover the size of NPs remained constant after plant internalization meaning that no partial dissolution of NPs occurred.

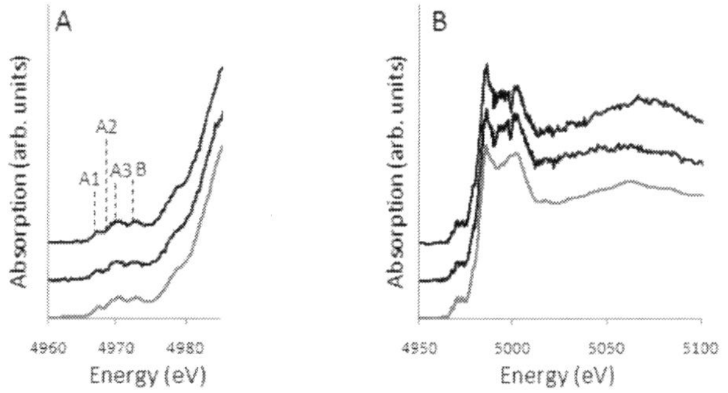

Figure 4. XANES spectra of Ti in plant roots exposed to TiO_2-NPs. Plants were exposed for 7 days to anatase TiO_2-NPs. Spectra were recorded at the Ti K-edge (4.966keV) at room temperature, locally on two Ti-rich regions determined on μ-SR-XRF maps (black spectra), and compared to the spectrum of a reference compound (red spectrum). A: pre-edge feature; B: XANES spectra. Image recorded on LUCIA beamline, SOLEIL synchrotron.

In the post-edge region, the oscillations show some differences in the samples as compared to reference compounds. However these spectra are noisy and their interpretation is difficult. Still, these changes may be due to surface modifications of TiO_2-NPs exposed to root exudates or plant fluids. A future experiment will aim at studying these changes which may occur in these conditions.

4. Conclusion

We determined sonication conditions enabling to properly disperse anatase NPs, the conclusion is that dispersion has to be performed in water, by ultra-sonication during 30 min. In these conditions, dilution in Hoagland medium leads to agglomeration, and none of the dispersing agent studied permits to maintain a stable dispersion.

Plant exposure to anatase, 25 nm NPs neither affect their seed germination rate, nor their root elongation. However 12 nm anatase NPs are efficiently taken up by plant roots and localize in the parenchymal region and vascular cylinder. Their nutrient solution-to-root transfer do not induce their dissolution, nor localized change in their crytalline structure.

μPIXE experiment are under process and will permit to have a global view of the perturbations in major elements homeostasis which may be caused by TiO_2-NP uptake. Moreover, on all these plant samples, transmission electron microscopic observations will now enable to evaluate if isolated NPs are taken up by plants, or if these NPs are grouped into clusters. The present results prove that TiO_2-NPs can be transferred to plants by their roots in case of an accidental release in the environment, and consequenltly may enter the food chain.

Acknowledgements

This work was funded by ADEME, CEA and the region Ile de France through the C'Nano framework. The authors would like to thank the European Synchrotron Radiation Facility and SOLEIL for provision of synchrotron radiation facilities at ID21 and LUCIA. This work was performed in the framework of the iCEINT research consortium.

References

[1] Gottschalk F, T Sonderer, R W Scholz, and B Nowack 2009 *Environ Sci Technol* **43**.
[2] Lin D and B Xing 2007 *Environ Pollut* **150**.
[3] Lin D and B Xing 2008 *Environ Sci Technol* **42**.
[4] Yang F, F Hong, W You, C Liu, F Gao, C Wu, and P Yang 2006 *Biol Trace Elem Res* **110**.
[5] Zhu H, J Han, J Q Xiao, and Y Jin 2008 *J Environ Monit* **10**.

[6] Kurepa J, T Paunesku, S Vogt, H Arora, B M Rabatic, J Lu, M B Wanzer, G E Woloschak, and J A Smalle 2010 *Nano Lett*
[7] Schmidt J and W Vogelsberger 2009 *J Sol Chem* **38**.
[8] Isaure M P, B Fayard, G Sarret, S Pairis, and J Bourguignon 2006 *Spectrochim Acta, Part B*
[9] Pignon B, H Maskrot, Y Leconte, S Coste, C Reynaud, N Herlin-Boime, M Gervais, V Guyot Ferreol, T Pouget, and J F Tranchant 2008 *Eur J Inog Chem* **6**.
[10] Simon-Deckers A, B Gouget, M Mayne-L'hermite, N Herlin-Boime, C Reynaud, and M Carriere 2008 *Toxicology* **253**.
[11] Simon-Deckers A, S Loo, M Mayne-L'hermite, N Herlin-Boime, N Menguy, C Reynaud, B Gouget, and M Carriere 2009 *Environ Sci Technol* **43**.
[12] Brunauer S, P H Emmett, and E Teller 1938 *J Am Chem Soc* **60**.
[13] Ravel B and M Newville 2005 *Phys Scr* **T115**.
[14] Wu Z Y, J Zhang, D Ibrahim, D C Xian, G Li, Y Tao, and T D Hu 2002 *App. Phys. Lett.* **80**.
[15] Luca V, S Djajanti, and R F Howe 1998 *J Phys Chem B* **102**.
[16] Chen L X, T Rajh, Z Y Wang, and M C Thurnauer 1997 *J Phys Chem B* **101**.

Behaviour of ceria nanoparticles in standardized test media – influence on the results of ecotoxicological tests

Nicolas Manier, Maël Garaud, Patrice Delalain, Olivier Aguerre-Chariol and Pascal Pandard

INERIS, Parc technologique ALATA, BP2, 60550 Verneuil en Halatte, FRANCE

E-mail: nicolas.manier@ineris.fr

Abstract. The main objectives of this work were to establish the behaviour of a ceria nanopowder in different ecotoxicological media commonly used in standardized aquatic ecotoxicity tests and consequently to assess the acute and chronic ecotoxicity in two micro-invertebrates: *Daphnia magna* and *Ceriodaphnia dubia* and in a freshwater green algae: *Pseudokirchneriella subcapitata*. Different approaches to disperse the ceria nanoparticles (*i.e.* stirring, use of probe sonication, addition of humic acids) were tested and the influence on the biological endpoints was investigated. Despite the agglomeration phenomena observed in all the tested media, the results obtained indicated higher stability in the lower ionic strength media with addition of humic acid (2 mg.L^{-1}, TOC). No acute toxicity were observed with *D. magna*, whatever the dispersal method performed and the nCeO$_2$ concentration tested (up to 1000 mg.L^{-1}), as no acute toxicity was recorded with *C. dubia* following exposure to the stirring suspensions. On contrary, acute toxicity was recorded in *C. dubia* with EC$_{50}$ values comprise between 11.9 and 25.3 mg.L^{-1} using the probe sonicated suspension with or without humic acids addition. Significant chronic effect on the reproduction capability was also recorded in *C. dubia*. The estimated EC$_{10}$ values were comprised between 2.1 and 2.9 mg.L^{-1}. Focusing on *P. subcapitata*, despite the different agglomerate size recorded in the tested media at the end of the exposure periods, results obtained were similar. Adverse effect on algal growth around 5 mg.L^{-1} were reported (mean EC$_{10}$ = 4 ± 1.8 mg.L^{-1}). Those results suggested the needed for standardized testing protocol concerning the aqueous media used or the sample preparation for laboratory testing.

1. Introduction

Nanomaterials have nowadays widespread applications in industries, medicine, agricultural and environmental sector [1,2]. As a consequence, new nanoparticulate forms of chemicals, of which behavior and toxicological effects are still unknown, are expected to be introduced into air, soil and water, and imply uncertainties regarding the possible environmental impacts.

Due to their specific properties, experts agree with the fact that the environmental effects of nanoparticles cannot be predicted from the known ecotoxicity of the "classical" macroscopic material and thus, specific hazard assessment of nanomaterials or nanoparticles are needed. In a regulatory point of view, if conventional ecotoxicity tests have already shown to be useful in evaluating the hazard of nanoparticles [3], questions rise concerning the relevance of the existing methods, especially when considering the properties of the tested media and sample preparation for conducting standard laboratory testing. Focusing on the conventional aquatic toxicity bioassays, some standardized test methods leave the choice of aqueous media with different physico-chemical properties. In view of nano-ecotoxicology, this can be seen as a drawback of the standard test protocols, since variability in nanomaterial properties, especially agglomeration phenomena, has been shown to depend on pH of the media, ionic strength and concentration of dissolved organic matter [4-6]. Moreover, debate about the preparation of nanoparticle suspensions has recently occurred and there is currently no broad consensus on the best approaches for preparing suspensions of nanomaterials in order to assess their toxicity towards aquatic organisms [7,8]. Suspension methods have involved use of strong solvent, bath or probe sonication, magnetic stirring for a broad range of time [7,9]. The use of different approaches prepare stock suspension for testing their toxicity towards aquatic organisms may lead to different suspension properties as different agglomeration state of the nanoparticles. It consequently may lead to different toxicity [10-12].

As many other nanomaterials, the use of ceria nanoparticles in number of applications is increasing rapidly and to date, this product can be found in different applications such as sun creams, outdoor paints or used as fuels catalyst [13]. As a result, an increase of direct and indirect release to water compartment is expected. For such reason, ceria nanoparticle is one of the NPs selected for priority testing by the Organization for Economic Cooperation and Development [14]. Nevertheless, the possible impacts of $n\mathrm{CeO_2}$ on aquatic ecosystems have not been thoroughly examined and available results are mainly related to *Daphnia magna* [11,15].

The purpose of this work was to determine whether the use of different ecotoxicological media and different protocols to suspend a ceria nanopowder can influence its physico-chemicals behavior and consequently, its ecotoxicological response. Thus, agglomerate size and surface charge (ζ-potential) of a ceria nanopowder were investigated in MilliQ-water (MQ) and in different ecotoxicological media commonly used in standardized protocols (invertebrates ISO medium and moderately hard water medium). Different protocols to disperse the ceria nanoparticles (*i.e.* magnetic stirring, use of probe sonication, addition of humic acids) were tested and their influence on the ceria nanopowder dispersion was investigated. Finally, the acute and chronic effects of ceria nanoparticles on the micro-invertebrates (*Daphnia magna* and *Ceriodaphnia dubia*) and on algae (*Pseudokirchneriella subcapitata*) were studied, and the influence of the different ways to disperse the suspension on the ecotoxicity is discussed.

2. Materials and Methods

2.1. Chemicals and ecotoxicological media

A commercially available ceria nanopowder ($n\mathrm{CeO_2}$) was purchased from Sigma Aldrich (Schnelldorf, Germany). The primary $n\mathrm{CeO_2}$ particle diameter as provided by the manufacturer was below 25 nm. A same batch was used for the all experiments.

The humic acids (HA) concentrated stock solution was prepared by dissolving a dry powder of humic acids sodium salt (Sigma Aldrich, Schnelldorf, Germany) in MilliQ-water and stirring vigorously for 30 minutes. Undissolved particles were removed by filtration through a 0.2 μm membrane. The stock solution of HA was kept at 4 ± 2°C, in the dark.

The micro-invertebrates ISO and Moderately Hard Water (MHW) media were prepared as described in the following standardized protocols: OECD 202 test guideline [16] and ISO 20665 standard [17]. The OECD algae growth medium was prepared following the OECD 201 test guideline [18]. Main components and physico-chemical parameters of these media are summarized in table 1.

Table 1. Physicochemical parameters of investigated media.

		pH	Hardness (mg/L)	Conductivity (µS/cm)	Main macro-nutrients
Micro-invertebrates media	ISO (OECD 202)	7.6	250	650	$CaCl_2$, $MgSO_4$, $NaHCO_3$
	Moderately hard water (ISO 20665)	7.7	85	300	$NaHCO_3$, $MgSO_4$, $CaSO_4$
Algae growth medium	OECD (OECD 201)	8.1	25	165	$NaHCO_3$, $CaCl_2$, NH_4Cl, $MgSO_4$

2.2. Preparation of the nanoparticle dispersions

All the suspensions of nCeO$_2$ were prepared in glass flask by weighting and directly adding the nanopowder in the different media or in MilliQ water. Addition of HA from the stock solution to the media was performed prior to the powder addition in order to obtained a final concentration of 2 mg.L^{-1}, total organic carbon (TOC). The nCeO$_2$ suspensions were dispersed using a vigorous 24h magnetic stirring or probe sonication (1 minute / 70 watts). Fresh suspensions were prepared extemporaneously.

The nCeO$_2$ suspensions were characterized for particle size (Z-average) by dynamic light scattering (DLS) particle sizer and for surface charge (ζ-potential, in mV), (Zetasizer, Malvern Instruments zetasizer). Light microscopy (Axio Imager, Zeiss®) and transmission electron microscopy (CM 12, Philips®) were used to visualize the agglomeration state of nCeO$_2$ in the different suspensions. For transmission electron microscopy (TEM), carbon coated grids were hydrophilised using an Emitech K100X glow discharger (10^{-1} mbar/40 mA/3 minutes) prior to the suspension deposition, to limit artifactual agglomeration phenomenon during the grids preparation [19].

2.3. Acute and chronic ecotoxicity tests

The ecotoxicity tests were performed on the following fresh water micro-invertebrate species: *D. magna* and *C. dubia*, and on the fresh water green algae: *P. subcapitata*. Methods, duration and followed endpoints are summarized in table 2.

Table 2. Ecotoxicity test protocols.

Organisms	*D. magna*	*C. dubia*		*P. subcapitata*
Type of toxicity	Acute	Acute	Chronic	Chronic
Test methods	OECD 202	US EPA 2002.0	ISO 20665	OECD 201
Test duration	48h	48h	8d	72h
Renewal	24h	24h	6 times *	No
Endpoint	Inhibition of mobility	Inhibition of mobility	Inhibition of reproduction	Inhibition of growth
Age of the test organisms	< 24h old	< 24h old	< 24h old	exponential growth phase
Test solution volume	10 ml	15 ml	50 ml	25 ml
Nb of organisms/flask	5	5	1	-
Condition	No stirring	No stirring	No stirring	Continuous stirring

* 6 times during exposure period (Day 0, D3, D4, D5, D6 and D7).

2.4. Statistical analysis

The Log-normal model in the REGTOX software for Microsoft Excel [20] was used for the calculation of toxicity parameters (ECx) and their confidence intervals. Values are drawn as mean ± standard deviation in the figures. ECx values are reported with 95% confidence intervals.

3. Results

3.1. Size and surface charge characterization of $nCeO_2$ in ecotoxicological media

Agglomerates state and size and surface charge (ζ-potential) of each $nCeO_2$ suspension are presented in table 3 and figure 1. The suspensions prepared using the 24h-stirring protocol without HA showed agglomerates of various sizes. As expected, the higher ionic strength media were subjected to highest agglomeration phenomena compared to the lower ionic strength media. As illustrated by the figure 1 (picture a.), the 24h-stirred suspensions in ISO medium appeared to be highly polydispersed with large agglomerate size up to hundred micrometers. Agglomerates size between 2 and 5 μm were also found. Concerning the 24h-stirred MHW and OECD media without HA, large agglomerate sizes up to several micrometers were observed (figure 1, b. and c.), submicron agglomerates around 500 nm were also noted in both media. Z-average values above 2 μm were obtained by DLS for the 24-h stirred ISO and MHW media prepared without HA. However, those values must be considered with care because of the presence of very large and settling agglomerates during the measurement and the very high polydispersity of the suspensions, which make the results less accurate. Concerning OECD media, a mean value of 1588 nm was recorded.

Not surprisingly, the use of probe sonication (1 min/ 70 watts) greatly improved the dispersion of $nCeO_2$, especially in the less ionic strength media. Z-average recorded by DLS showed mean agglomerates size of 1871 nm, 1771 nm and 864 nm in the ISO, MHW and OECD media without HA. Mean agglomerates size around 250 was recorded in the MQ water.

Relating to the surface charge of $nCeO_2$, our results showed that the |ζ-potential| of particles decrease with the increase of ionic strength of the aqueous medium used for the dispersion. A |ζ-potential| from 8 to 9 mV was recorded for the highest ionic strength ISO medium prepared without HA addition, which was two-fold lower than the value recorded in the lowest ionic strength OECD medium without HA addition.

Table 3. Mean diameter ± standard deviation (Z-average in nm) and mean value of the ζ-potential ± standard deviation (in mV) of ceria in the different ecotoxicological media and in MilliQ water using a 24h-stirring or a 1 minute-probe sonication as dispersion method. All values are reported as mean (± standard deviation) of 10 different preparations (3 measurements per preparation), except for ISO media (only one preparation).

	Aqueous media	Mean particles size (Z-average, nm)	Pdi*	|ζ-potential| mV (± S.D)	pH
24h-stirring	ISO	*2346* (a)	>1	9	7.3
	ISO + AH	*1736* (a)	>1	16	7.1
	MHW	*3180 (±817)* (a)	>1	16 (± 0.7)	7.4
	MHW + AH	931 (± 235)	0.7	19 (± 0.4)	7.4
	OECD	1588 (± 479)	0.6	18 (± 1.3)	7.7
	OECD + AH	431 (± 68)	0.5	21 (±0.5)	7.7
	MQ	863 (±482)	0.7	20 (± 6.8)	7.8
	MQ + AH	542 (± 71)	0.5	35 (± 2.6)	5.1
Probe sonication 1 minute	ISO	*1871* (a)	>1	8.6	n.a
	ISO + HA	1126	0.4	16	n.a
	MHW	1771.8 (± 425)	0.4	16 (± 0.7)	7.6
	MHW + HA	229 (± 15)	0.3	19 (± 0.4)	7.5
	OECD	864 (± 500)	0.3	18 (±0.4)	7.7
	OECD + HA	222 (± 15)	0.3	23 (±2.3)	7.7
	MQ	246 (± 26)	0.3	31 (± 3.3)	7.8
	MQ + HA	221 (± 4)	0.3	38 (± 2)	6.2

* Polydispersity index
n.a: not available
(a) to consider with care because sample may be not suitable for DLS measurement (presence of large and settling particles)

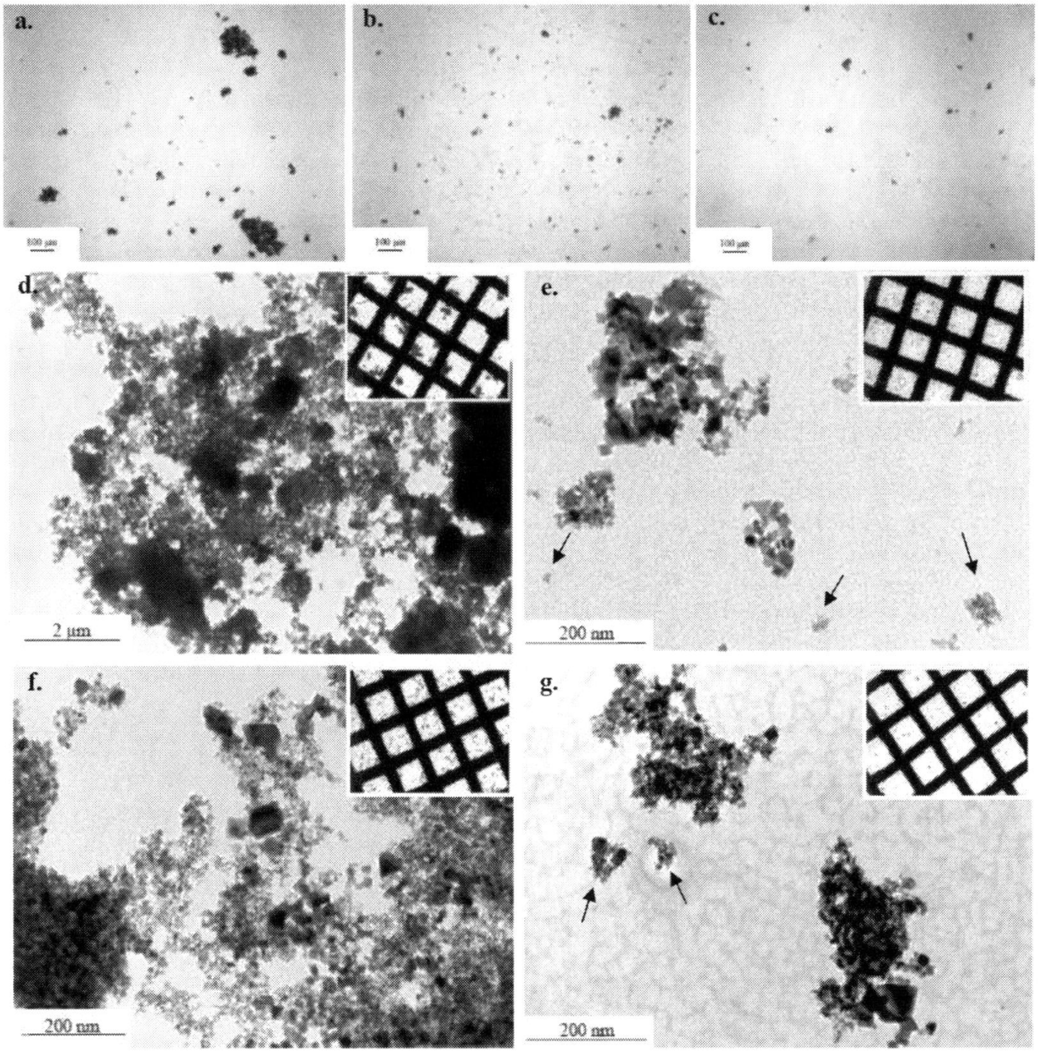

Figure 1. Light microscope images and TEM micrographs showing nCeO$_2$ agglomerates in the different ecotoxicological media, after 24h-stirring (picture a., b. and c.) or probe sonication (pictures d., e., f. and g.). After 24-h stirring, largest agglomerates are visible in the ISO medium without HA addition (picture a.). Smaller agglomerates were noted in MHW and OECD media (pictures b. and c., respectively). Pictures d. and e., show the nCeO$_2$ TEM micrographs of nCeO$_2$ suspensions dispersed using probe sonication in MHW medium without HA (picture d.) and with HA addition (picture e.). Pictures f. and g., show the nCeO$_2$ TEM micrographs of nCeO$_2$ suspensions dispersed using probe sonication in OECD medium without HA (picture f.) and with HA addition (picture g.). Black arrows highlight very small agglomerates of nCeO$_2$ below 100 nm.

Upon addition of HA (2 mg.L^{-1}, TOC) to the different media, a significant increase of the |ζ-potential| was recorded whatever the suspension investigated. The mean agglomerates size was also reduced whatever the media tested or the dispersion methods performed. That may be explaining by the adsorption of HA to the particles surface which enhances the particles charge and thus reduce the agglomeration phenomena. As illustrated by figure 2, it was shown that the increase of the |ζ-potential| above 20 mV has significantly reduced the agglomeration phenomena and the polydispersity of the different suspensions. The use of probe sonication combined with HA acid addition was particularly efficient to disperse the ceria nanopowder in MHW media, OECD media as in MQ water. Using this protocol, the Z-average recorded by DLS showed mean agglomerates size from 221 to 229 nm. Those results were also confirmed by the TEM micrographs of MHW and OECD suspensions which indicate large and loose agglomerates in the absence of HA addition (figure 1, d. and f.), when smaller agglomerates are observed in the sonicated media prepared with HA (figure 1, e. and g.). As for MQ water, the sonicated suspensions in MHW and OECD media with HA addition appeared to be less polydispersed and more stable in time, with small agglomerates which fell in the field of "nanoparticles" definition (figure 1, pictures e. and g.).

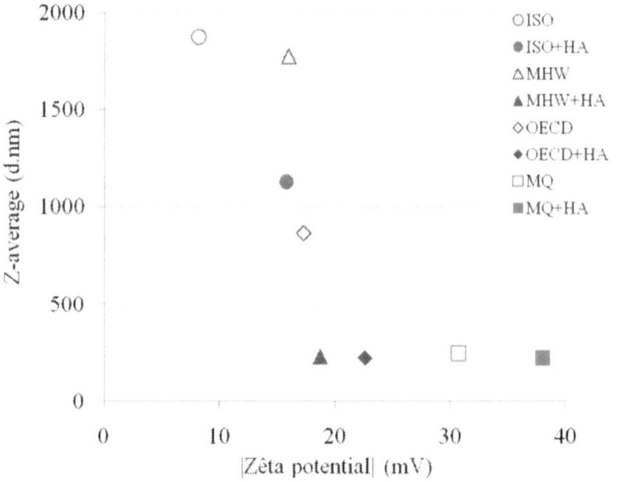

Figure 2. Physico-chemical characteristics of nCeO$_2$ in MilliQ water and in the standardized ecotoxicological media for micro-invertebrates and algae dispersed by probe sonication. All the media were prepared with or without addition of humic acids (HA ; 2 mg.L^{-1} TOC). Values are drawn as mean of 10 different preparations (3 measurements per preparation) except for ISO media (only one preparation).

3.2. Acute and chronic effects of nCeO$_2$ on D. magna and C. dubia

Data concerning acute ecotoxicity of nCeO$_2$ to micro-invertebrates are summarized in table 4. Regarding to both *D. magna* and *C. dubia*, no inhibition of the mobility was recorded after a 48h exposure whatever the medium tested and considering the 24h-stirring condition. Likewise, after a 48h exposure period to the probe sonicated suspensions with or without HA, no significant inhibition of the mobility of *D. magna* was recorded.

Opposite, we found a significant inhibition of the mobility in *C. dubia* after a 48h exposure to the probe sonicated suspensions with or without HA addition. The estimated EC$_{50}$ values ranged from 13.6 to 16.8 mg.L^{-1}, in the media prepared without HA and from 11.6 mg.L^{-1} to 25 mg.L^{-1}, in the media prepared with HA. A control containing 2 mg.L^{-1} TOC of HA without nCeO$_2$ confirmed that HA did not affect the mobility of *D. magna* and *C. dubia*.

Whatever the condition tested, binocular observation indicated that nCeO$_2$ agglomerates were clearly visible on and within the organisms (figure 3). The vast majority of nCeO$_2$ agglomerates were found to be accumulated in the gut of the organisms and also adsorbed on the shell. After a 2 hours post-exposure period in clean medium, a clearance of the nCeO$_2$ aggregates has occurred, but residual particles remain in the gut and on the shell. Same observation was made in *C. dubia* (data not shown).

Table 4. Results of the micro-invertebrates acute toxicity tests.

Method of dispersion	Organisms	Medium	Maximal concentration (mg.L^{-1})	EC_{50} (mg.L^{-1})
24h-stirring	*D. magna*	ISO	1000	> 1000
		ISO+HA	1000	> 1000
		MHW	100	> 100
		MHW+HA	100	> 100
	C. dubia	MHW	100	> 100
		MHW+HA	100	> 100
Probe sonication 1 minute	*D. magna*	MHW	100	> 100
		MHW+HA	100	> 100
	C. dubia	MHW	100	13.6 (11.2 - 14.6)
		MHW+HA	100	25.3 (18 - 33.1)
		MHW	100	16.8 (15 - 17.8)
		MHW+HA	100	11.9 (9.5 - 15)

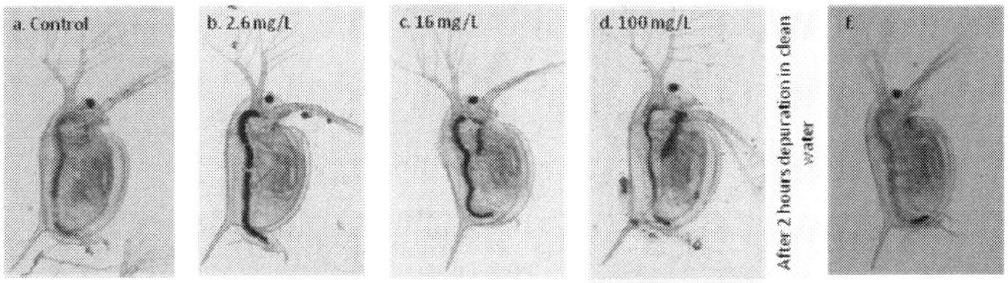

Figure 3. Binocular observation of *D. magna* after 48h exposure in nCeO$_2$ suspensions. The black color of the digestive tract shows the uptake of nCeO$_2$. Agglomerates of nCeO$_2$ were also found adsorbed on the shell. Picture f. shows *D. magna* after a 2h post-exposure period in clean medium. Red arrows highlight the accumulated nCeO$_2$.

The *C. dubia* 8d-chronic toxicity tests were performed using MHW media prepared by stirring and probe sonication without HA addition. After exposition to the 24 h-stirring suspensions, no immobility was observed during the 8 days of exposure period, when 50% immobility were recorded in the probe sonicated suspension at 100 mg.L^{-1}. Surviving adults were measured at the end of the exposure period, and indicated only a slight (but significant) decrease of the size of the adults exposed at the most concentrated suspension (100 mg.L^{-1}) prepared by probe sonication (data not shown).

However, significant adverse effects concerning *C. dubia* reproduction were observed in the tested concentration range of the magnetic-stirred suspensions. At 100 mg.L^{-1} only 3.6 neonates per exposed daphnids were observed when 15.6 neonates per exposed daphnids were recorded in the control flasks. By comparison, after exposure to the probe sonicated suspension, adverse effects were observed in the 0.3 to 100 mg.L^{-1} concentration range. Only 2.9 neonates per exposed daphnids were recorded at 3.05 mg.L^{-1}, and no neonates were observed at the highest concentrations (*i.e.* 31.25 and 100 mg.L^{-1}). At 9.77 mg.L^{-1}, only 2 neonates were recorded for all the exposed daphnids.

The ECx for reproduction in the magnetic-stirring and probe sonicated suspensions are given in table 5. An EC_{50} value of 15 mg.L^{-1} and an EC_{10} value of 2.9 mg.L^{-1} were determined after exposure to the 24 h-stirred suspensions compared to the lower estimated EC_{50} of 2.6 mg.L^{-1} and EC_{10} of 2.1 mg.L^{-1} calculated in the experiment with the probe sonicated suspensions.

Table 5. Results of the *C. dubia* chronic toxicity tests.

Methods of dispersion	Medium	EC_{50} ($mg.L^{-1}$)	EC_{10} ($mg.L^{-1}$)
24h-stirring	MHW	15.7 (6.6 - 35.2)	2.9 (0.44 - 15.2)
Probe sonication 1 minute	MHW	2.6 (1.1 - 2.7)	2.1 (0.8 - 2.2)

3.3. Effect of $nCeO_2$ on P.subcapitata growth

The results obtained in the *P. subcapitata* growth inhibition tests are presented in table 6. When possible, the evolution of $n$$CeO_2$ agglomerates during the test was followed. A re-agglomeration was observed in the suspension prepared without humic acids, leading with agglomerates around 2 μm (measured by DLS) at the end of the experiment, when no re-agglomeration was recorded in the suspension prepared with HA addition. In those suspensions with HA addition, the agglomerate sizes stay around 350 nm in the stirring suspension and around 250 nm in the probe sonicated suspension, all over the exposure period.

Furthermore, all the estimated ECx values indicated a similar level of toxicity, whatever the way the suspensions were prepared and consequently the size of the particles in the suspension. Mean EC_{10} values of 4.2 $mg.L^{-1}$ and 3 $mg.L^{-1}$ were recorded in the stirring and probe sonicated suspensions, respectively. Mean EC_{50} of 14.7 $mg.L^{-1}$ and 11.5 $mg.L^{-1}$ were recorded in the stirring and probe sonicated suspensions, respectively.

Table 6. Results of the *P. subcapitata* growth inhibition test.

Method of dispersion	Medium	Stock suspension concentration	initial agglomerate size	Agglomerate size at the end of the experiment	EC_{10}	EC_{50}
		($mg.L^{-1}$)	(d. nm)	(d. nm)	($mg.L^{-1}$)	($mg.L^{-1}$)
24h-stirring	OECD	1000	nd	nd	2.8 (2.1 - 3.6)	12 (10.6 - 13.4)
	OECD	1000	nd	nd	2.9 (2 - 4.2)	14.2 (11.9 - 16.3)
	OECD	125	nd	nd	2.6 (1.8 - 3.6)	16.2 (13.9 - 19.1)
	OECD	125	nd	nd	7.5 (5.6 - 9.5)	16.4 (13.8 - 17.9)
	OECD	125	nd	1562 ± 25	5.8 (4.4 - 7.8)	12.5 (11.1 - 14.3)
	OECD+HA	125	357 ± 5.8	374 ± 10	4.1 (3.4 - 5.0)	16.9 (15.5 - 18.5)
Probe sonication 1 minute	OECD	125	647 ± 65	2782 ± 143	2.8 (2.3 - 3.3)	11.1 (10.3 - 11.9)
	OECD	125	761 ± 12	2087 ± 632	3.9 (3.4 - 4.7)	9.9 (9.1 - 10.5)
	OECD+HA	125	238 ± 2.6	253 ± 5	2.3 (1.2 - 4.1)	13.7 (10.6 - 17.1)

4. Discussion

In this study, we looked forward to investigate whether the use of different ecotoxicological media and different protocols to suspend a ceria nanopowder can influence its behavior and consequently its ecotoxicity on freshwater micro-invertebrates and algae following the conventional bioassays guidelines and standards.

4.1. Ceria nanoparticles in the standardized ecotoxicological media

From our results, the use of different ecotoxicological media and different dispersion protocols had obviously led to different behaviors of ceria nanopowder. As expected, agglomeration phenomenon was linked to the media ionic strength. This is mainly explained by the fact that the stability of a nanoparticles suspension is governed by the balance of interaction between attractive forces (such as Van-der-Waals attraction), and the surface charges repulsion from the particles [21,22]. An increase in the suspension ionic strength shields the surface charges of particles and consequently reduces the effectiveness of the electrostatic repulsion, which enhance the agglomeration process [23,24].

The use of chemical solvents to improve the suspensions stability can be controversial in ecotoxicological experiments, as they can be inherently toxic [7,25]. Several authors have shown that humic substances form surface films on colloidal particles [23,26-28] which can improve the stability of nanoparticles as previously reported on iron oxide, carbon nanotubes and fullerenes [29,30]. Thus,

HA were used as a more relevant agent relative to chemical solvents. From our data, it was shown that HA enhance the surface charges of $n\mathrm{CeO_2}$, whatever the media investigated, which subsequently increase the electrostatic repulsion between particles. As a consequence, a decrease of agglomerates size was recorded. In the media with the lower ionic strength, a $|\zeta$-potential$|$ around 20 mV was enough to significantly improve the stability of the small particles in suspensions. At least, the use of probe sonication as a dispersal methods and the addition of HA (2 $mg.L^{-1}$, TOC), were the optimal protocol to produce small and consistent particles sizes in suspensions, with a reasonable stability over the exposure period. Moreover this protocol appears to be the most reproducible methods to disperse the ceria nanopowder in the different aqueous media.

4.2. Effect on aquatic organisms

In addition to investigate the ecotoxicity of $n\mathrm{CeO_2}$ nanoparticles, the second objective of this work was to test the influence of the suspension preparation on the ecotoxicological effect observed. The results obtained demonstrate that $n\mathrm{CeO_2}$ was capable of causing acute and chronic toxicity in micro-invertebrates and algae; however, toxicity may differ significantly with the species tested and the dispersal method performed to suspend the nanoparticles.

Concerning micro-invertebrates acute ecotoxicity tests, it is clear from the data reported above that susceptibility to ecotoxicity of $n\mathrm{CeO_2}$ varies among micro-invertebrate species. *C. dubia* tend to be more susceptible to $n\mathrm{CeO_2}$ suspensions compare with *D. magna*, especially considering exposure to the probe sonicated suspensions. Therefore, no inhibition of the mobility was recorded in *D. magna*, whatever the media used (ISO or MHW medium, with or without HA addition) or the dispersal methods performed and despite the fact that *D. magna* was clearly able to ingest a large amount of $n\mathrm{CeO_2}$. Those findings are however in line with the data obtained by Velzeboer *et al.* [15] Van Hoecke *et al.* [11] and Lee *et al.* [31], who mentioned no acute toxicity up to 100 $mg.L^{-1}$ with the same organisms and exposure period. Opposite, the findings recorded with *C. dubia* indicated significant (and dose-related) inhibition of the mobility when organisms were exposed to the probe sonicated suspension, while no inhibition was recorded after exposure to the magnetic stirring suspensions. Those findings obtained with *C. dubia* also highlight that the preparation and the dispersal methods have impacted the testing results.

In the chronic tests on *C. dubia*, data shown that the number of neonates was the most sensitive endpoint compared to the growth of adults. In our experiment on *C. dubia*, we have recorded EC_{50} values for reproduction of 15.7 $mg.L^{-1}$ (6.6 - 35.2) and 2.9 $mg.L^{-1}$ (0.44 - 15.2) and EC_{10} 2.6 $mg.L^{-1}$ (1.1 - 2.7) and 2.1 $mg.L^{-1}$ (0.8 - 2.2) using the 24h-magnetic stirring suspension or the probe sonicated suspension, respectively. One more time, difference in toxicity relating to the dispersal methods used was observed, indicating higher toxicity after exposure to the probe sonicated suspensions. Despite of its relevance, chronic effect of $n\mathrm{CeO_2}$ on micro-invertebrates has rarely been investigated, and no published data on chronic toxicity to *C. dubia* of $n\mathrm{CeO_2}$ could be found in the literature for comparison. At the best of our knowledge, only one reference was published on *D. magna* [11]. These authors also reported adverse effect of $n\mathrm{CeO_2}$ on reproduction capability with estimated EC_{50} values between 20 and 42 $mg.L^{-1}$ and EC_{10} values between 8.8 and 20 $mg.L^{-1}$, according to the primary particle size. Although those values are slightly higher, they were nevertheless in line with our results. The variation observed may be explained by the highest sensitivity of *C. dubia*, the difference in sample preparation and by difference in the primary particle size.

We hypothesize that the difference between sonicated and stirring suspension in acute or chronic ecotoxicological response recorded with *C. dubia* may be explained, at least in part, by greater fragmentation of the $n\mathrm{CeO_2}$ agglomerate within the probe sonicated suspension and consequently, a better stability of $n\mathrm{CeO_2}$, over the exposure period. Nevertheless, $n\mathrm{CeO_2}$ in aqueous suspensions is not a static system and is undergoing simultaneous agglomeration and sedimentation. Hence, during micro-invertebrates toxicity test, a large amount of the initial mass, of $n\mathrm{CeO_2}$, may be lost by sedimentation in a more or less short time, depending on the media parameters and the dispersal protocol performed. This will result in a dynamic exposure scenario, making it difficult to quantify the real exposure concentration of $n\text{CeO}_2$ in the micro-invertebrate bioassay. Furthermore, micro-invertebrates are expected not only to swim in the middle of water column, but also to crawl in the sedimented nanopowder, making exposure a more complicated question. Therefore, the real concentrations to which the organisms were exposed may be less than reported. It is obvious form the discussion above that much research is needed concerning the exposure characterization when using static or semi-static system. Further works will focused on this point in order to refine the effect concentrations.

Finally, we have demonstrated that $n\text{CeO}_2$ induced adverse effects on the growth of *P. subcapitata*. Furthermore, despite the fact that the different dispersal protocol leads with different agglomerate size of $n\text{CeO}_2$ at the end of the exposure period, the algae growth inhibition were similar (*i.e.* mean $EC_{10} = 4 \pm 1.8$ mg.L^{-1} and mean $EC_{50} = 13.9 \pm 2.6$ mg.L^{-1}). Recent results reported by Van Hoecke *et al.* [11] and Rodea-Palomares *et al.* [32] using such $n\text{CeO}_2$, but different way to prepare the testing sample (*i.e.* stabilized stock suspension in MilliQ water at low pH and drop addition in tested media, with addition of MOPS (3-(*N*-morpholino) propanesulfonic acid) buffer in the algae growth medium in the case of Van Hoecke *et al.* [11]) are also in line with our results. Van Hoecke *et al.* (2009) indicated an EC_{10} of 5.4 mg.L^{-1} and an EC_{50} of 19.1 mg.L^{-1} in *P. subcapitata* growth inhibition test, using CeO_2 nanoparticles with primary particle size around 29 nm and recorded agglomerates in the experimental media (OECD growth algae media) around 500 nm, when Rodea-Palomares *et al.* [32] reported an EC_{50} value around 10 mg.L^{-1}, using CeO_2 with primary particles size of 25 nm and recorded agglomerates around 2107 nm in the OECD growth algae media. Those findings also support the fact that the protocol to suspend $n\text{CeO}_2$ has a few influence on its toxicity when focusing on the conventional algae growth inhibition test.

5. Conclusion

In conclusion, the results above confirm that the conventional aquatic bioassays could be relevant to assess the hazard of nanoparticles, in a regulatory point of view. Nevertheless, we have demonstrated that the use of different ecotoxicological media proposed by guidelines or standards as the use of different methods to disperse nanoparticles may lead to different behaviors and consequently may influence the results obtained when performing conventional test, especially when considering the micro-invertebrates bioassays. This clearly illustrates the needed for standardized testing protocol concerning the aqueous media used or the sample preparation for laboratory testing.

Acknowledgment

This work was supported by the French Ministry in charge of Ecology and Sustainable Development, within the framework of the program IMPECNANO (Environmental impact of nanomaterials: exposure through their life cycle and ecotoxicity).

References

[1] Tiede K, Boxall ABA, Tear SP, Lewis J, David H, Hassellöv M. 2008. Detection and characterization of engineered nanoparticles in food and the environment. *Food Addit. Contam.* Part A. **25** 795-821

[2] Liu WT. 2006. Nanoparticles and Their Biological and Environmental Applications. *J. Biosci. Bioeng.* **102** 1-7

[3] SCENIHR (SCientific commitee on Emerging and Newly Indentified Health Risks). 2009. Risk assessment of product of nanotechnologies.

[4] Keller A, Wang H, Zhou D, Lenihan HS, Cherr G, Cardinal BJ, Miller R, Ji Z. 2010. Stability and Aggregation of Metal Oxide Nanoparticles in Natural Aqueous Matrices. *Environ. Sci. Technol.* **44** 1962-67.

[5] Domingos RF, Baalousha M, Ju-Nam Y, Reid MM, Tufankji N, Lead JR, Leppard GG, Wilkinson KJ. 2009. Characterizing manufactured nanoparticles in the environment: Multimethod determination of particle sizes. *Environ. Sci. Technol.* **43** 7277-84.

[6] French RA, Jacobson AR, Kim B, Isley SL, Penn RL, Baveye PC. 2009. Influence of ionic strength, pH and cation valence on aggregation kinetics of titanium dioxide nanoparticles. *Environ. Sci. Technol.* **43** 1354-59

[7] Klaine SJ, Alvarez PJJ, Batley GE, Fernandes TF, Handy RD, Lyon DY, Mahendra SY, McLaughlin MJ, Lead JR. 2009. Nanomaterials in the environment: Behavior, Fate, Bioavailability and Effects. *Environ. Pollut.* **157** 1101-09.

[8] OECD. 2010. Preliminary guidance notes on sample preparation and dosimetry for the safety testing of manufactured nanomaterials.

[9] Handy RD, Henry TB, Scown TM, Johnston BD, Tyler CR. 2008. Manufactured nanoparticles: their uptake and effects on fish - a mechanistic analysis. *Ecotoxicology*. **17** 396-409

[10] Oberdörster E, Zhu S, Blickley TM, McClellan-Green P, Haasch ML. 2006. Ecotoxicology of carbon-based engineered nanoparticles: Effects of fullerene (C60) on aquatic organisms. *Carbon*. **44** 1112-20

[11] Van Hoecke K, Quick JTK, Mankiewicz-Boczek J, De Schamphelaere A, Van der Meeren P, Barnes C, McKerr G, Vyvyan Howard C, Van De Meent D, Rydzynski K, Dawson KA, Salvati A, Lesniak A, Lynch I, Silversmit G, De Samber B, Vincze L, Janssen CR. 2009. Fate and effects of CeO_2 nanoparticles in aquatic ecotoxicity tests. *Environ. Sci. Technol.* **43** 4537-46

[12] Zhu S, Oberdörster E, Haasch ML. 2006. Toxicity of an engineered nanoparticle (fullerene, C60) in two aquatic species, *Daphnia* and fathead minnow. *Mar. Environ. Res.* **62** S5–S9

[13] Brar SK, Verma M, Tyagi RD, Surampalli RY. 2010. Engineered nanoparticles in wastewater and wastewater sludge – Evidence and impacts. *Waste Manage*. **30** 504-520

[14] OECD. 2008. Series on the safety of manufactured nanomaterials. Number 6. List of manufactured nanomaterials and list of endpoints for phase one of the OECD testing program.

[15] Velzeboer I, Hendriks AJ, Ragas AM, Van de Meent D. 2008. Aquatic ecotoxicity tests of some nanomaterials. *Environ. Toxicol. Chem.* **27** 1942-47

[16] OECD. 2004. OECD guidelines for the testing of chemicals no. 202. *Daphnia* sp., acute immobilisation test. Paris, 12p.

[17] ISO 20665. 2008. Water quality. Determination of chronic toxicity to *Ceriodaphnia dubia*. International Standard Organization.

[18] OECD. 2006. OECD guidelines for the testing of chemicals no. 201. Alga growth inhibition test. Paris, 25p.

[19] Dubochet J, Groom M, Mueller-Neuteboom S. 1982. The Mounting of Macromolecules for Electron Microscopy with particular reference to surface phenomena and the treatment of support films by Glow Discharge. *Adv. Opt. Electron Microsc*. **8** 107-135

[20] Vindimian E. MSExcel macro REGTOX_EV7.0.5.xls; 2005. http://eric.vindimian.9online.fr/.

[21] Fang J , Shan X, Wen B, Lin J, Owens G. 2009. Stability of titania nanoparticles in soil suspensions and transport in saturated homogeneous soil columns. *Environ. Pollut.* **157** 1101–09.

[22] Tadros TF. 2007. Colloid Stability: the Role of Surface Forces Part I, Wiley-VCH Verlag Gmbh & Co. kGaA, Weinheim.

[23] Christian P, Von der Kammer F, Baalousha M, Hofmann T. 2008. Nanoparticles: structure, properties, preparation and behavior in environmental media. *Ecotoxicology*. **17** 326-343

[24] Johnson BD, Scown TM, Moger J, Cumberland SA, Baalousha M, Linge K, Van Aerle R, Jarvis K, Lead JR, Tyler CR. 2010. Bioavailability of Nanoscale Metal Oxides TiO_2, CeO_2, and ZnO to Fish. *Environ. Sci. Technol*. **44** 1144-51.

[25] Chen KL and Elimelech M. 2007. Influence of humic acid on the aggregation kinetics of fullerene (C60) nanoparticles in monovalent and divalent electrolyte solutions. *J. Colloid Interface Sci*. **309** 126–134

[26] Zhang Y, Chen Y, Westerhoff P, Crittenden J. 2009. Impact of natural organic matter and

divalent cations on the stability of aqueous nanoparticles. *Water res.* **43** 4249-57.

[27] Lead JR, Muirhead D, Gibson CT. 2005. Characterization of freshwater natural aquatic colloids by atomic force microscopy (AFM). *Environ. Sci. Technol.* **39** 6930-36

[28] Baalousha M and Lead JR. 2007. Size fractionation and characterization of natural aquatic colloids and nanoparticles. Size fractionation and characterization of natural aquatic colloids and nanoparticles. *Sci. Tot. Env.* **386** 93-102

[29] Baalousha M, Manciulea A, Cumberland S, Kendall K, Lead JR. 2008. Aggregation and surface properties of iron oxide nanoparticles: Influence of pH and natural organic matter. *Environ. Tox. Chem.* **27** 1875–82

[30] Hyung H, Fortner J, Hughes H, Hongkim A. 2007. Natural Organic Matter Stabilizes Carbon Nanotubes in the Aqueous Phase. *Environ. Sci. Technol.* **41** 179-184.

[31] Lee SW, Kim SM, Choi J. 2009. Genotoxicity and ecotoxicity assays using the freshwater crustacean *Daphnia magna* and the larva of the aquatic midge *Chironomus riparius* to screen the ecotoxicological risks of nanoparticles exposure. *Environ. Tox. Pharm.* **28** 86-91

[32] Rodea-Palomares I, Boltes K, Fernandez-Pinas F, Leganés F, Garcia-Calvo E, Santiago J, Rosal R. 2010. Physicochemical characterization and ecotoxicological assessment of CeO_2 nanoparticles using two aquatic microorganisms. *Toxicol. Sci.* doi: 10.1093/toxsci/kfq311

Highly porous nanocomposites based on TiO2-noble metal particles for sensitive detection of water pollutants by SERS

M Baia[1], G Melinte[1], L Barbu-Tudoran[2], L Diamandescu[3], V Iancu[1], V Cosoveanu[4], V Danciu[4] and L Baia[1]

[1]Faculty of Physics, Babes-Bolyai University, 400084, Cluj-Napoca, Romania

[2]Faculty of Biology and Geology, Babes-Bolyai University, 400015, Cluj-Napoca, Romania

[3]National Institute of Materials Physics, P.O. Box MG-7, 77125, Bucharest-Magurele, Romania

[4]Faculty of Chemistry and Chemical Engineering, Babes-Bolyai University, 400028, Cluj-Napoca, Romania

E-mail: lucian.baia@phys.ubbcluj.ro

Abstract. Highly porous nanocomposites based on TiO2 aerogel and silver colloidal particles were prepared by different methods in order to study their capacity to detect pollutant species adsorbed on metallic nanoparticles surface from aqueous solution. The efficiency of the obtained composites to detect contaminants from water by means of SERS was evaluated using acrylamide and crystal violet as test molecules. It was found that the detection limits depend both on pollutant and composite type, and were determined to be in the range of 10^{-1}-10^{-4} M for acrylamide and around 10^{-5} M for the dye molecule. These results prove the potential of the prepared porous composites for further use in the development of new SERS-based sensors devices.

1. Introduction

Clean water is essential to human health. The world is facing formidable challenges in meeting rising requirement to clean water. Even trace pollutants can enter in human body and affect the human health. Trace detection and treatment of the pollutants become an essential problem that requires a rapid elucidation [1]. Current analysis of water pollutants include their pre-extraction using techniques such as headspace analysis, purge and trap, etc. from large quantities of water followed by chromatographic analysis. These methods are laborious, time consuming and expensive, hence unsuited for monitoring of locations where high temporal variations occur; therefore, efficient, *in situ* methods are desired [2, 3].

Due to its sensitivity and exceptional spectral selectivity, surface-enhanced Raman scattering (SERS) was thought to be a promising technique for the trace detection and analysis of environmental samples [4]. The primary attraction of SERS consists in its ability to provide large amounts of structural details. It can act as a molecular "fingerprint" and can allow simultaneous analysis of trace analytes in a complex mixture without extraction and separation steps. SERS overcomes the most significant drawback of normal Raman spectroscopy for analytical applications, i.e. the inherently weak cross section, by exciting the sample in contact with a 'plasmonic' surface with an appropriate laser line. Under such conditions, the Raman cross section and, in turn, the signal intensity are extraordinarily enhanced so that levels of detection down to the single molecule can be reached, without loosing the structural information provided by Raman spectroscopy [5].

The development in SERS detection is related to the progress in synthesis and optical characterization of new nanostructured materials able to control both the enhancement mechanism (electromagnetic and chemical) and the adsorption of the analytes of interest [6]. The preparation of optical substrates with optimized properties is a very dynamic field of research, being developed many types of substrates: metal colloids, metal island films, self-assembled monolayer-modified metal substrates, metal-coated microsphere and metal nanoparticles entrapped in sol-gel derived matrix [7,8].

Nobel metals sealed in TiO_2 sol-gel matrixes present many advantages as SERS substrates: (1) the matrix isolation against aggregation leads to the stability of entrapped metal nanoparticles, (2) the porosity encourage the diffusion of analytes in the matrix and consequently the increasing the number of metal/analytes contacts, (3) enhanced adsorption capacities and selectivity obtained by matrix functionalization [9].

In a previous work [10], we obtained by different chemical routes porous nanoarchitectures based on TiO_2 aerogel and Au colloidal particles, annealed them at 500°C for 2 hours, and tested their capability to detect by SERS contaminant species adsorbed from an aqueous solution on metallic nanoparticles surface. The composites detection efficiency was established by using thioacetamide, crystal violet and rhodamine 6G as test molecules and the lowest detectable concentrations were found to be around 10^{-2} M for thioacetamide and in the range between 10^{-5} and 10^{-7} M for dye molecules off and under resonant excitation, respectively.

Having in view the affinity of some pollutant molecules, e.g. chlorates, amino and hydroxyl, to bind to the metallic silver surface, in the present work we proposed to evaluate the potential of new types of highly porous nanocomposites based on titania aerogel and Ag colloidal particles, prepared by three different methods and annealed at 500 °C for 2 hours, to detect pollutant species adsorbed from an aqueous solution on the noble metal surfaces. Prior evidencing the detection efficiency of the composites for different pollutant molecules, the existence of their porous-like structure was proved by means of scanning electron microscopy (SEM) images and specific surface area measurements. The TiO_2 crystalline phase that built up the composites porous network was examined by X-ray diffraction (XRD) and Raman spectroscopy.

2. Experimental

2.1. Sample preparation

2.1.1. *Preparation of silver colloid.* The silver colloidal solution was prepared by the following method [11]: 100 ml of $5x10^{-3}$ M silver nitrate ($AgNO_3$) was heated to boiling under continuous stirring. 300 ml of $2x10^{-3}$ M sodium borohydride ($NaBH_4$) was added dropwise with vigorous stirring. The boiling and stirring of the reaction mixture was continued for another 60 min. Colloidal solution were stored in clean brown glass bottles until used.

2.1.2. *Preparation of titania gel.* TiO_2 gels were prepared with by sol-gel procedure using titanium isopropoxide ($Ti\{OCH(CH_3)_2\}_4$), nitric acid (HNO_3, 70%), ethanol (92%) and distilled water in molar ratio of 1/0.08/21/3.67. The formed gels were aged for 21 days.

- *Method 1- composite **TiO_2-Ag1***

The aged TiO_2 gel was dried under supercritical conditions with liquid CO_2 (T = 40°C and p = 1400 psi) in order to obtain a highly porous material, named aerogel. The TiO_2 aerogel was immersed into the silver colloidal solution where was maintained for three days. The aerogel impregnated with silver nanoparticles was three times washed with ethanol (92%) and then dried at 100°C.

- *Method 2-composite* ***TiO_2-Ag2***

The aged TiO_2 gel was immersed into the silver colloidal solution and maintained there for three days. Afterward it was washed with ethanol (92%) and dried under supercritical conditions.

- *Method 3-composite* ***TiO_2-Ag3***

The silver colloidal solution was added in the TiO_2 gel synthesis mixture instead of distilled water. The obtained gel was allowed to age and then dried under supercritical conditions.

The aged TiO_2 gel was also dried under supercritical conditions in order to obtain silver metal nanoparticles free sample and was investigated only for comparison purpose. This sample will be further denoted as ***TiO_2***.

All prepared samples were subjected to a thermal treatment at 500 °C for 2 h in order to remove the organic residues.

2.2. Sample measurements

The surface area of the as prepared samples was determined by the BET method, in a partial pressure range of $0.05<P/P_0<0.3$. The nitrogen adsorption was carried out at 77 K. Before each measurement the samples were heat cleaned at 333 K for 2 h. The total pore volume (V_{TP}) was obtained from the N_2 adsorption isotherm at a partial pressure of 0.98.

SEM measurements were performed with a Quanta 3D FEG 200/600 scanning electron microscope. A Jeol JEM 1010 TEM operating at an accelerating voltage of 100 kV equipped with a MegaViewIII CCD camera was utilized to obtain characteristic images. The samples were sonicated for 15 min and then deposited on Cu grids with carbon film evaporated on freshly cleaved mica.

The phase content and the particle dimensions were determined using a DRON X-ray powder diffractometer linked to a data acquisition and processing facility; CuK_α radiation (λ = 1.540598 Å) and a graphite monochromator were used.

SERS spectra were recorded with a WITec confocal Raman system CRM 200 equipped with a 100x/0.8 microscope objective and a 300 lines/mm grating. In the recording of the spectra, the 532 nm laser line with a power of 3 mW and a spectral resolution around 7 cm^{-1} were employed. Prior to perform the SERS measurements the composites were kept 2-3 min. into aqueous solutions of contaminants of various concentrations. In order to assure the measurements reproducibility SERS spectra were collected from different points located on the composites surface and inside the samples.

The FT-Raman spectrum of the crystal violet was recorded using a Bruker Equinox 55 spectrometer with an integrated FRA 106 Raman module, a radiation with a power of 30 mW incident on sample, and a spectral resolution of 4 cm^{-1}. A Nd-YAG laser (λ = 1064 nm) was employed for the recording of Raman spectra of the bulk pollutant molecules.

3. Results and discussion

The first argument that accounts for the existence of a porous structure is given by the values obtained from the BET surface area measurements presented in Table 1. It can be observed that the composite ***TiO_2-Ag2*** exhibits the highest BET area (157 m^2g^{-1}), while ***TiO_2-Ag3*** the smallest one (129 m^2g^{-1}). One should note that the presence of silver nanoparticles induces an increase of the pores surface area.

Table 1. Textural and structural properties of the $\mathbf{TiO_2}$ and $\mathbf{TiO_2}$**-Ag** composites from BET and XRD measurements.

Sample	Specific surface area (m^2g^{-1})	Phase composition (%)		Crystallite size (nm)	
		Anatase	Brookite	Anatase	Brookite
TiO_2	121	92.2	7.8	11	13
TiO_2-Ag1	135	76.7	23.3	10	9
TiO_2-Ag2	157	100	-	10	-
TiO_2-Ag3	129	88.9	11.1	15	9
Errors	±6	±0.5	±0.5	±1.5	±1.5

The highly porous character of the annealed aerogel like composites was tested out by SEM images. Selected images of TiO_2-Ag composites are illustrated in Figure 1. As can be observed the samples show a porous structure that allows the diffusion of pollutant molecules through the three dimensional structure and their adsorption on the silver nanoparticle surfaces.

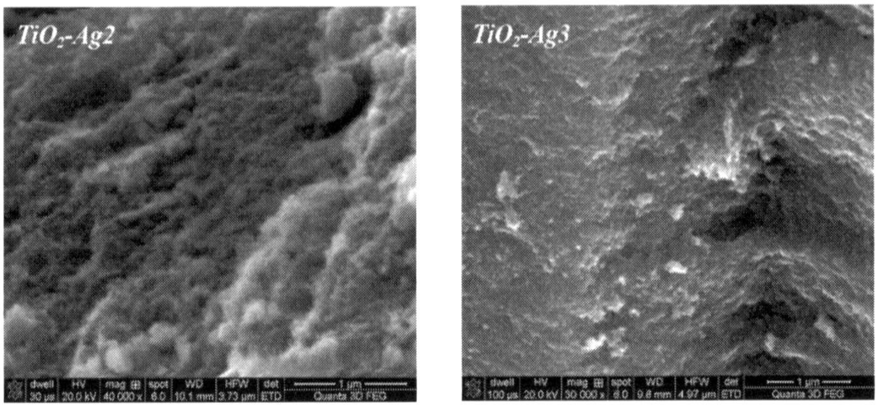

Figure 1. Selected SEM images of the $\mathbf{TiO_2}$**-Ag** composites as indicated.

Prior evaluating the composites ability to detect small concentrations of contaminant species from water by SERS, the samples were structurally examined by performing XRD measurements. XRD patterns are shown in Figure 2 and confirm the presence of anatase phase, which is accompanied for two composites by a quite important fraction of brookite phase (see Table 1).

The smallest anatase content (76.7 %), as revealed by Rietveld refinement, was obtained for the annealed composite prepared by impregnation of the TiO_2 aerogel with Ag colloidal suspension (sample ***TiO_2-Ag1***). It was found that the sample ***TiO_2-Ag2***, obtained by impregnating the aged TiO_2 gel with Ag colloidal suspension followed by thermal treatment, contains anatase as unique phase, while the sample ***TiO_2-Ag3***, prepared by replacing the water during the sol-gel process with the Ag colloidal suspension, has a structure built up from 88.9 % anatase. By comparing the percentage obtained for ***TiO_2*** sample with those derived for the composites one can infer that the preparation method of the composite ***TiO_2-Ag2*** favors the anatase phase development, while the preparation of the composites ***TiO_2-Ag1*** and ***TiO_2-Ag3*** favors the brookite phase formation.

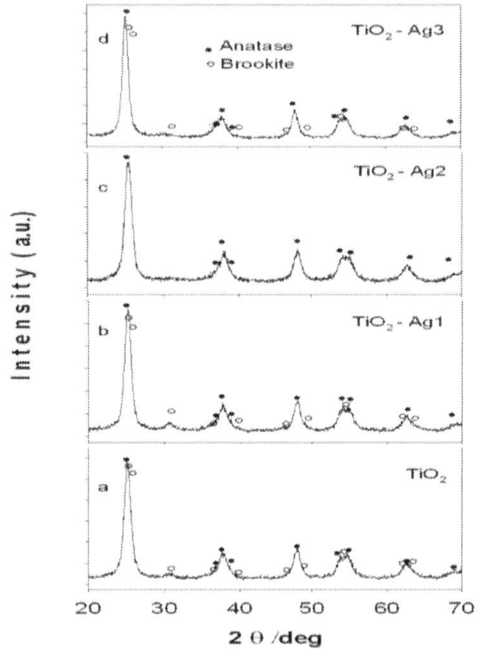

(a)

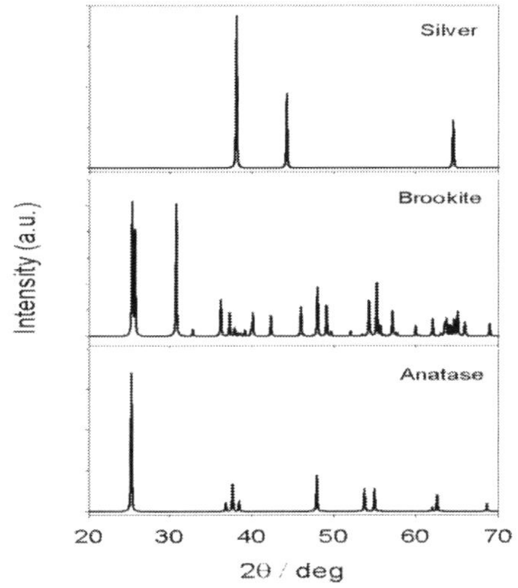

(b)

Figure 2. a) X-ray diffractograms of the $\mathbf{TiO_2}$ and $\mathbf{TiO_2}$**-Ag** composites; b) XRD patterns of anatase, brookite and silver based on ICSD.

The mean size of the TiO_2 crystallites was also derived from XRD diffractograms and is presented in Table 1. The crystallite size has been calculated by using the Scherrer equation for all reflection lines in the range of 20-70 (2θ degrees). Because of very low content (expected to be lower than 1wt %), no silver peaks were evidenced in the XRD patterns.

The obtained results were confirmed by TEM images that provided additional information related to the dimension of the Ag nanoparticles, found to be around 8.5±2.5 nm for all investigated samples (see Figure 3).

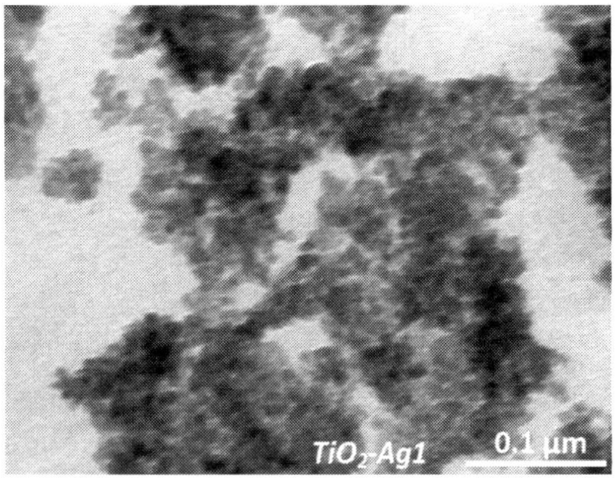

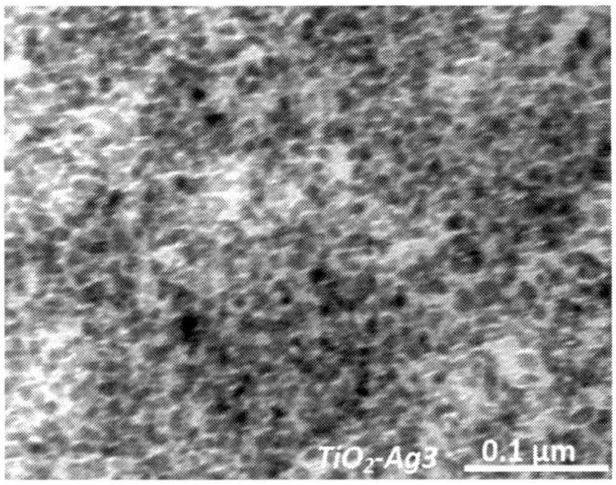

Figure 3. Selected TEM images of the $\mathbf{TiO_2}$**-Ag** composites as indicated.

Our attention was further focused on the assessment of the porous composites capacity to detect various pollutant species adsorbed on Ag nanoparticles surface. Small amounts of each composite were immersed for 2-3 min. into aqueous solution of different test molecules, i.e. acrylamide and crystal violet, and Raman spectra were recorded. Figures 4 (crystal violet) and 5 (acrylamide) show spectra recorded on porous TiO_2-Ag composites after immersion. One can recognize the most intense Raman characteristics of TiO_2 anatase phase in the low wave number region (around 144, 400, 515 and 636 cm^{-1}) that originate from vibrations of TiO_6 octahedra. Other spectral features that based on the comparison of the spectra with the Raman spectra of contaminant bulk sample can be easily associated with signals given by pollutants vibrations can be observed. Thus, in the spectra of all composites immersed in crystal violet solution (Figure 4) the bands around 1612 cm^{-1}, due to the stretching vibrations of CC from phenyl rings, and those around 1366 and 1175 cm^{-1}, attributed to the CN stretching and CH bending vibrations, respectively, occur [10,12]. The bands from the spectra of composites maintained in acrylamide solution (Figure 5) observed at 1635 and 1435 cm^{-1} are given by CC and CN stretching vibrations, and that around 1275 cm^{-1} was attributed to the CH_2 deformation vibration [13]. Taking into account the small concentration of pollutant molecules one can certainly conclude that these fingerprints are given by the vibrational modes of the adsorbed species. The changes in position and relative intensity of the SERS bands with respect to their corresponding Raman bands of the contaminant bulk samples further support this statement.

The detection sensitivity of the investigated porous composites was also established by using various solutions of predetermined concentration for each pollutant molecule. It was found that the detection limits depend on the sample preparation method and the pollutant type. Composite ***TiO_2-Ag1*** was found to exhibit the best detection sensitivity for crystal violet (10^{-5} M) and the worst one for acrylamide (10^{-1} M). Composite ***TiO_2-Ag2*** exhibited the best detection capacity for acrylamide (10^{-4} M) and a good sensitivity for crystal violet ($5x10^{-5}$ M), while the sample ***TiO_2-Ag3*** was able to detect concentrations of 10^{-3} and $5x10^{-5}$ M for acrylamide and crystal violet, respectively.

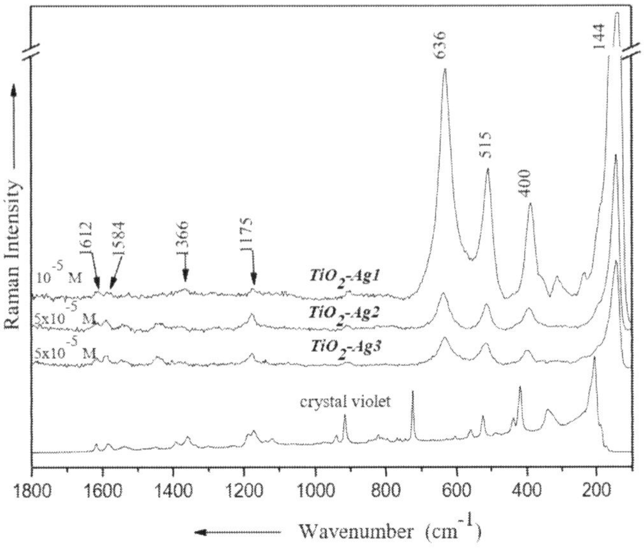

Figure 4. Raman and selected SERS spectra of crystal violet adsorbed on TiO2-Ag samples. Note that the pollutant concentrations are also indicated.

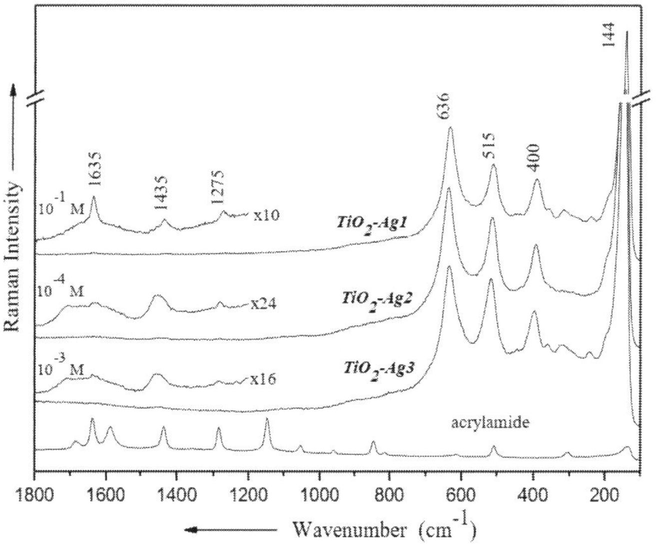

Figure 5. Raman and selected SERS spectra of acrylamide adsorbed on **TiO_2-Ag** samples. Note that the pollutant concentrations are also indicated.

Similar results were obtained by using noble metal nanoparticles sealed in porous silica matrix when the detected limit concentration was of 10^{-5} M of dye molecules [14]. In a recent study [15], we obtained porous nanoarchitectures based on TiO_2 aerogel and noble metal colloidal nanoparticles by using various methods of preparation, and tested their capability to detect by SERS chemical pollutants from water. The remarkably small detection limits (between 10^{-3} and 10^{-10} M) determined in the reported study were due to the missing of the heat treatment, on one hand, and to the recording of the SERRS (surface enhanced resonance Raman spectroscopy) spectra, on the other hand. However, there is a compromise between the porosity preserving and the organic residues removal, and therefore, in the present study a thermal treatment was applied. It should be noted that there are SERS-active substrates able to detect concentrations lower than those detected using our composites, but they are very unstable over time [7]. Therefore, it should be emphasized that the most important advantages of the TiO_2 aerogel-noble metal nanoparticles based SERS substrates over the conventional ones are their time stability and resistance to different environmental conditions.

One should emphasize that for all porous composites SERS spectra were recorded from different points inside the samples (up to 100 μm). This proves the good dispersion of the Ag nanoparticles located inside the TiO_2 pores and indicates once again that the recorded signals are indeed originating from molecules adsorbed on noble metal surface. Although the applied thermal treatment induces a clear diminution of the pores volume, the contaminants detection by SERS is still remarkable, when relatively large molecules like crystal violet adsorbed on Ag nanoparticle surfaces. Furthermore, since the stability of the obtained porous composites samples is higher in comparison with that of the silver colloids in liquid media [7], these samples can offer additional advantages besides those of detecting water pollutants by using SERS.

4. Conclusions

In the work presented herein we prepared three different types of porous composites based on TiO_2 aerogel and silver colloidal particles that are able to detect by SERS small concentrations of various pollutants adsorbed on the noble metal surfaces. Prior evidencing the detection efficiency of the composites for different pollutant molecules, the existence of their porous-like structure was proved with the help of SEM images and specific surface area measurements and, they were structurally investigated by XRD technique. The detection capacity by SERS of the porous samples obtained after the immersion of the TiO_2 gel in Ag colloidal solution followed by supercritical drying and heat treatment was superior to the other types of composites for both test molecules. The lowest detectable concentrations were found to be of 10^{-4} and $5x10^{-5}$ M for acrylamide and crystal violet, respectively. The detection limits were established and found to be in the range of 10^{-1} -10^{-4} M for acrylamide and around 10^{-5} M for crystal violet.

Acknowledgments
This work was supported by CNCSIS-UEFISCSU, project number PNII-RU-TE 81/2010.

References

[1] Zhang L., Fang M., *Nano Today* (2010) **5**:128-142.
[2] Schmidt H., Ha N. B., J. Pfannkuche, H. Amann, H. D. Kronfeldt, G. Kowalewska, *Mar. Pollut. Bull.* (2004) **49**:229–234.
[3] De Jesus M. A., Giesfeldt K. S., Sepaniak M. J., *J. Raman Spectrosc.* (2004) **35**:895–904.
[4] McCreery R. L., *Meas. Sci. Technol.* (2001) **12**:653.
[5] Kneipp J., Kneipp H., Kneipp K., *Chem. Soc. Rev.* (2008) **37**:1052–1060.
[6] Abalde-Cela S., Aldeanueva-Potel P., Mateo-Mateo C., Rodríguez-Lorenzo L., Alvarez-Puebla R. A., Liz-Marzán L. M., *J. R. Soc. Interface* (2010) **7**:435–450.
[7] Baker G. A., Moore D. S., *Anal. Bioanal. Chem.* (2005) **382**:1751–1770.
[8] Volkan M., Stokes D. L., Vo-Dinh T., *Sens. Actuators B* (2005) **106**:660-667.
[9] Bao L., Mahurin S. M., Dai S., *Anal. Chem.* (2004) **76**:4531-4536.
[10] Baia M., Danciu V., Cosoveanu V., Baia L., *Vib. Spectrosc.* (2008) **48**:206–209.
[11] Lee P. C., Meisel D., *J. Phys. Chem.* (1982) **86**:3391-3395.
[12] Kudelski A., *Chem. Phys. Lett.* (2005) **414**:271-275.
[13] Duarte A. S. R., Amorim da Costa A. M., Amado A. M., *J. Molec. Struct.: TheoChem* (2005) **723**:63-68.
[14] Gilbert R., Improved SERRS Substrate, WO/2007/110614.
[15] Baia L., Diamandescu L., Barbu-Tudoran L., Peter A., Melinte G., Danciu V., Baia M., *J. Alloy Comp*, in press.

Journal of Physics: Conference Series **304** (2011) 012060 doi:10.1088/1742-6596/304/1/012060

Fate of nanoparticles during life cycle of polymer nanocomposites

T Nguyen, B Pellegrin, C Bernard, X Gu, J M Gorham, P Stutzman, D Stanley, A Shapiro, E Byrd, R Hettenhouser and J Chin

National institute of Standards and Technology, 100 Bureau Drive, Gaithersburg, MD 20899, USA

E-mail: Tinh.nguyen@nist.gov

Abstract. Nanoparticles are increasingly used in consumer and structural polymeric products to enhance a variety of properties. Under the influence of environmental factors (e.g., ultraviolet, moisture, temperature) and mechanical actions (e.g., scratching, vibrations, abrasion), nanoparticles could potentially release from the products and thus have negative effects on the environment, health and safety. The fate of nanoparticles in polymer nanocomposites during their exposure to UV environment has been investigated. Epoxy polymer containing multi-walled carbon nanotubes (MWCNTs) and silica nanoparticles were studied. Specially-designed cells containing nanocomposite specimens were irradiated with UV radiation between 295 nm and 400 nm. Chemical degradation, mass loss and surface morphology of the epoxy nanocomposites, and release of nanoparticles were measured. Epoxy containing MWCNTs exposed to UV radiation degraded at a much slower rate than the unfilled epoxy or the epoxy/nanosilica composite. Photodegradation of the matrix resulted in substantial accumulation of nanoparticles on the composite surfaces. Silica nanoparticles were found to release into the environment, but MWCNTs formed a dense network on the composite surface, with no evidence of release even after prolonged exposure. Conceptual models for silica nanoparticle release and MWCNT retention on the surface during UV exposure of nanocomposites are presented.

1. Introduction

In the last decade, the outstanding properties of nanoparticles (defined as a particle with at least dimension <100 nm) have fueled intensive research in incorporating them in polymers (polymer nanocomposites) [1-3]. These advanced composite materials are increasingly used in many high-volume industries including consumer products, buildings, and automobiles. However, research has shown that nanoparticles pose potential risks to the environment, human health, and safety (EHS) [4-6]. Current research on the effects of nanoparticles on EHS mostly focuses on exposures during the manufacturing process. For polymer nanocomposites, whatever the application, the long-term performance of the composite product itself, the fate of the nanoparticles in the polymer matrix, and the hazard of the released nanoparticles play a key role in the acceptance and commercialization of these advanced products. This is because the polymer matrix in a nanocomposite tends to undergo degradation during service and post-service (e.g., landfills, incineration). One serious consequence of matrix degradation is that nanoparticles will be exposed on the composite surface and could be released to the environment due to mechanical vibration, rain, condensed water, and wind.

Release of nanoparticles from a polymer nanocomposite is tied closely to the degradation of the polymer matrix, which is an inherently complex process controlled by environmental conditions, processing variables, and materials properties. Nanoparticles also possess unique physical properties and create a large polymer/particle interfacial volume fraction in the nanocomposites, both of which would likely affect the rate and degradation mechanism of the matrix, and the release potential of nanoparticles during environmental exposures. Yet, little data is available on the degradation of polymer nanocomposites, the state of the embedded nanoparticles during environmental exposures, or how they will be released during their life cycles. Lack of this information hinders our ability to understand the release mechanisms, to predict the long term release, and develop strategies to mitigate this potentially serious problem. For that reason, potentially harmful effects of nanoparticles incorporated in polymers on the environment and health cannot be determined. To provide data needed to address this issue, the National Institute of Standards and Technology initiated a study to critically assess the release of nanoparticles during the life cycle of polymer nanocomposites. The study consists of four components: *1)* Development of metrologies to measure the degradation of polymer nanocomposites, release rates, and chemical composition of released particles during life cycles of polymer nanocomposites, *2)* generating microscopic and molecular experimental data to understand the mechanism of nanoparticle release, *3)* development of standardized methods to measure the nanoparticle release rate and properties of the released nanoparticles, and *4)* development of experimentally-verified model, which takes into account materials, processing and environmental factors as inputs, to predict the nanoparticle release.

The main objective of the study reported here is to investigate the fates of silica nanoparticles and carbon nanotubes (CNTs) embedded in an epoxy matrix during its exposure to ultraviolet (UV) radiation. Epoxy degradation and the fate of nanoparticles in nanocomposites were characterized by a number of microscopic and spectroscopic techniques, including field-emission scanning electron microscopy (FE-SEM), Fourier transform infrared spectroscopy (FTIR), and X-ray photoelectron spectroscopy (XPS). Analyses were also performed on particles released from the samples during exposure. Results showed that, under UV radiation having wavelengths similar to those of the sunlight, amine-cured epoxy in the nanocomposites underwent rapid photodegradation, resulting in mass loss, exposing large concentration of nanoparticles on the composite surface, and releasing silica nanoparticles into the environment. For CNT composites, matrix degradation resulted in a dense network of CNTs on the surface, which reduced the degradation of the epoxy layer underneath and essentially prevented their release. The results of this study should provide useful information to assess the potential risk of nanoparticles in an epoxy matrix during service in outdoor environment.

2. Experimental Procedures

2.1. Materials and nanocomposite preparation

The polymer matrix was an amine-cured epoxy commonly used in fiber-reinforced polymer composites, coatings and adhesives, and the nanoparticles were multi-walled carbon nanotubes (MWCNTs) and silica (SiO_2) nanoparticles (nanosilica). The epoxy was prepared as a stoichiometric mixture of a diglycidyl ether of bisphenol A (DGEBA) epoxy resin having an equivalent mass of 189 (grams of resin containing one gram equivalent of epoxide) and a polyoxypropylenetriamine curing agent. No additional additives were used in this epoxy system. MWCNTs and silica nanoparticles were obtained from commercial sources. MWCNT was a 1% mass fraction pre-dispersed product in the same epoxy resin. The silica nanoparticles were surface modified to provide good dispersion in polymers, had an individual average diameter of 7 nm, and a purity of > 99.8%. Reagent grade toluene (purity > 99.5%) was used for composite processing. Figure 1 presents the chemical structure of this amine-cured epoxy, which will help to better understand its degradation behaviour. The presence of the electron rich nitrogen atoms and residual epoxide groups in the cured, crosslinked structure plays a major role in the photodegradation of epoxy materials.

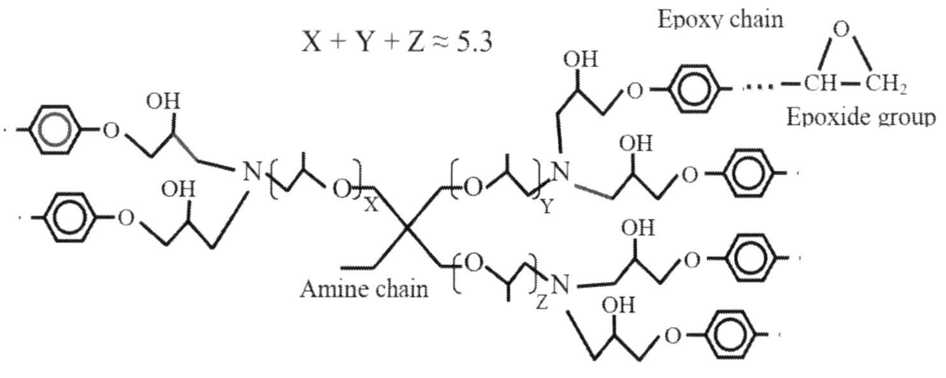

Figure 1. Chemical structure of the crosslinked amine-cured bisphenol A epoxy.

Free-standing films of *1)* unfilled epoxy, *2)* epoxy/0.72% MWCNT composite, and *3)* epoxy/5% nanosilica composite were fabricated. Nanoparticle loadings were based on mass fraction of the solid amine-cured epoxy. (Note that the 1% mass fraction MWCNT in the neat epoxy resin is reduced to 0.72% mass fraction in the amine-cured epoxy.) Unfilled epoxy and MWCNT/epoxy composite films were prepared by simply adding amine curing agent to epoxy resin or to MWCNT-containing epoxy resin, respectively, and then stirring for 1 h with a magnetic stirrer. For epoxy/nanosilica composites, SiO_2 nanoparticles were first stirred and sonicated in a large amount of toluene for 30 min. After adding epoxy resin, the nanosilica suspension was stirred and sonicated for an additional 1 h using an 80 kHz tip sonicator. The amine component was then added to the suspension and stirred for another 1 h. After the mixing step, all epoxy/amine/nanoparticles mixtures were degassed for 1 h at room temperature and then drawn down using a bar applicator on a polyethylene terephthalate sheet (PET) to produce free films having a thickness of approximately 150 µm. In addition, 6 µm thick films on CaF_2 substrates were prepared by spin casting for degradation study by transmission FTIR spectroscopy. All coated samples were cured at ambient conditions (24°C and 45% relative humidity) for three days, followed by post-curing at 110°C for 4 h in an air circulating oven.

2.2. Exposure to UV radiation

The degradation of unfilled epoxy and epoxy nanocomposites was studied using a 2 m integrating sphere-based weathering chamber, referred to as SPHERE (Simulated Photodegradation *via* High Energy Radiant Exposure) [7]. This SPHERE UV chamber utilizes a mercury arc lamp system that produces a collimated and highly uniform UV flux of approximately 480 W/m^2 in the 295 nm to 400 nm range. It can also precisely control the relative humidity (RH) and temperature. In this study, 25 mm x 25 mm specimens were exposed in the SPHERE UV chamber at 50°C and 75% RH. Specimens were removed at specified time intervals for characterization. For assessing the release of nanoparticles during UV radiation exposure, a special sample holder shown in Figure 2 was used, which collects particles released from the samples during UV exposure. It consists of a sample chamber and a container located at the bottom of the holder to collect released particles. White poly(tetrafluoroethylene) film or highly polished silicon plates was placed on the collector surface. All samples were mounted normal to the horizontal direction.

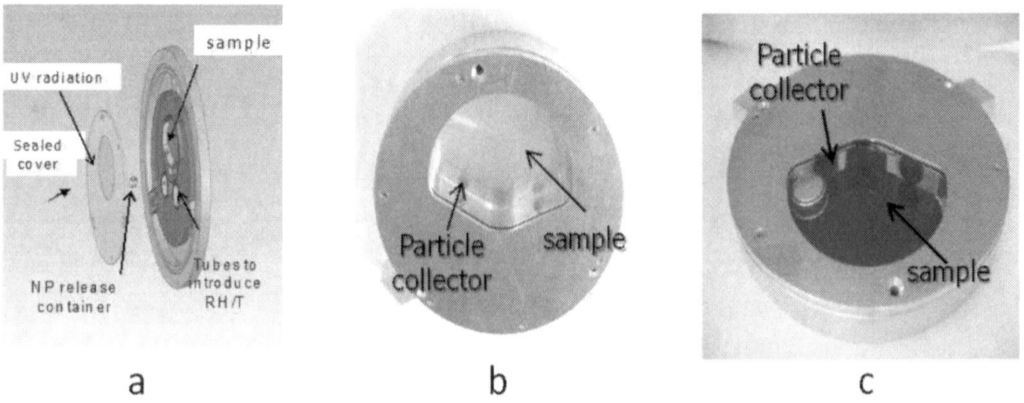

Figure 2. Sample holder to collect released particles: a) a schematic, b & c) photographs with exposed nanosilica (b) and MWCNT (c) epoxy nanocomposites.

2.3. Characterization of UV-exposed nanocomposite and fate of nanoparticles

Mass loss, surface morphology, and chemical degradation of UV-exposed specimens were characterized as a function of exposure time. Mass loss was measured using an analytical balance having a resolution of 10^{-5} g, and is expressed as: $(M_t-M_o)/M_o$ x 100 where M_t is the specimen mass at time t, M_o is the specimen mass before exposure. Surface morphology was characterized by field-emission scanning electron microscopy (FE-SEM). The released particles were verified by FE-SEM imaging and Energy Dispersive X-ray Spectroscopy (EDS). Chemical degradation was measured with transmission FTIR (t-FTIR) and XPS. FTIR spectra were recorded at a resolution of 4 cm^{-1} using dry air as a purge gas and a spectrometer equipped with a liquid nitrogen-cooled mercury cadmium telluride (MCT) detector. All spectra were the average of 128 scans. The peak height was used to represent the infrared intensity, which is expressed in absorbance, A. All FTIR results were the average of four specimens. XPS analyses were carried out using a Mg Kα X-ray source (1253.6 eV) at a 45° angle between the sample surface normal and the lens/hemispherical analyzer. Spectra were acquired at a pass energy 44.75 eV and a step size of 0.125 eV/step. All XPS spectra were fit using 100% Gaussian peaks, a Shirley baseline, and adjusted with the appropriate sensitivity factors.

3. Results and discussion

3.1. Mass loss

Figure 3 displays the mass loss of unfilled epoxy and epoxy nanocomposites a function of exposure time in the UV/50°C/75% RH environment. Except for a small increase at early exposure times, the mass loss in all materials was nearly linear with exposure time. The early increase was likely due to the moisture uptake, which was greater than the material loss induced by the degradation in the early stage. These samples were exposed to 75% RH, and thus the amount of sorbed water in the sample was substantial. The rate of mass loss was lowest for MWCNT composite and greatest for nanosilica composite. After 43 days of exposure, the maximum mass losses in unfilled epoxy, epoxy/MWCNT composite and epoxy/nanosilica composite were 2.4% ± 0.27%, 1.3% ± 0.30%, and 4.0% ± 0.95%, respectively.

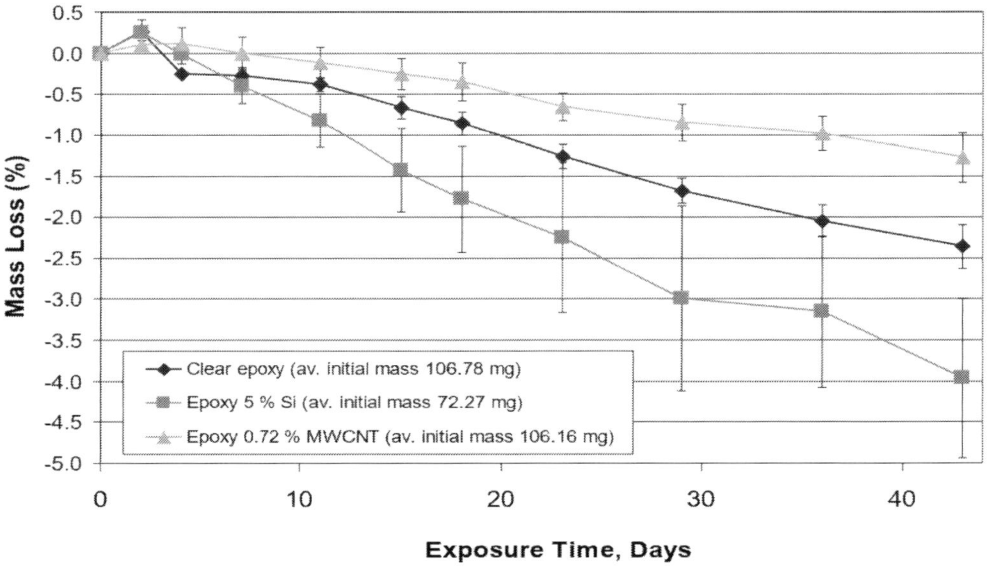

Figure 3. Mass loss *vs.* exposure time for unfilled, epoxy/0.72% MWCNT composite, and epoxy/5% nanosilica composite samples exposed to UV/50°C/75% RH. Results are the average of 6 specimens, and the error bars represent one standard deviation.

3.2. Photodegradation of the polymer matrix

Chemical degradation of the matrix was followed by FTIR analyses in the transmission mode (t-FTIR), which can provide quantitative data on changes of functional groups of the epoxy polymer. t-FTIR spectra recorded at different exposure times showed that intensities of numerous bands of the epoxy structure substantially decreased, including the bands at 1508 cm^{-1}, attributed to benzene ring, 1296 cm^{-1}, due to aliphatic ether, and new bands in the 1620-1740 cm^{-1} region due to the formation of various carbonyl groups (C=O), such as aldehydes, ketones, and amides. These changes are attributed to photo-oxidation of amine-cured epoxy by UV in the 295 nm to 400 nm wavelength, leading to extensive chain scission of the main chains of the epoxy. Ketone formation is derived from the secondary hydroxyl groups, while amide is generated from the abstraction of a methylene hydrogen adjacent to the crosslink [8-10]. Carbonyl groups formed can absorb radiation at long wavelengths and accelerate the degradation. UV-visible transmission spectra (not shown) showed a progressive increase of the absorbance intensity in the region above 300 nm, indicating that the degraded epoxy films did absorb UV light at longer wavelengths. Both the unfilled and epoxy/nanosilica composite films also exhibited visible yellowing after the UV exposure, suggesting that a substantial amount of conjugated structures had formed in this amine-cured epoxy. Our previous extensive study has demonstrated clearly that UV is the main weathering factor that causes severe degradation of amine-cured epoxies, with temperature and RH playing a minor role [11].

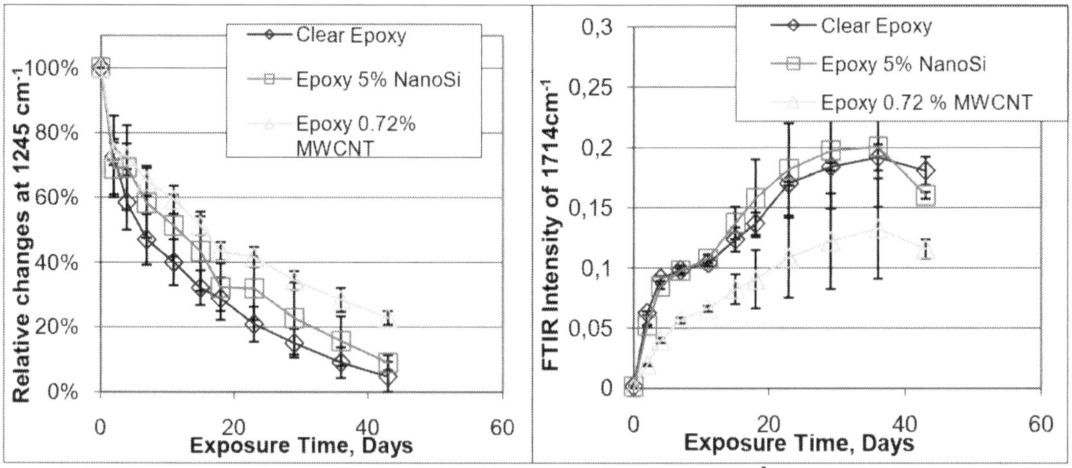

a b

Figure 4. a) Chain scission, and b) oxidation vs. time for unfilled, epoxy/0.72 % MWCNT composite, and epoxy/5 % nanosilica composite samples exposed to UV/50 °C/75 % RH. Each data point is the average of four specimens and error bars represent one standard deviation.

The bands at 1245 cm^{-1} and 1714 cm^{-1}, representing chain scission and oxidation, respectively, were used to follow various degradation processes of unfilled epoxy and epoxy nanocomposites exposed to UV radiation. The results are depicted in Figure 4. Both the rates of oxidation and chain scission of the epoxy/MWCNT composite were lower than those of the unfilled epoxy or epoxy/nanosilica material. This observation is consistent with our previous results for polyurethane/MWCNT composites [12] and is similar to that reported in the literature for PMMA/CNT composites subjected to high-energy radiation [13]. The higher UV resistance of polymer/CNT composites has been attributed to the CNT's ability to disperse and filter radiation energy as well as the strong interaction between free radicals (generated during irradiation) and CNTs [13]. Although not shown here, the intensity of the band at 1060 cm^{-1}, due to Si-O-Si, of the nanosilica composites decreased at first then started to rise after two days, suggesting that the concentration of SiO_2 nanoparticles on the composite surface has substantially increased, consistent with SEM data to be shown later.

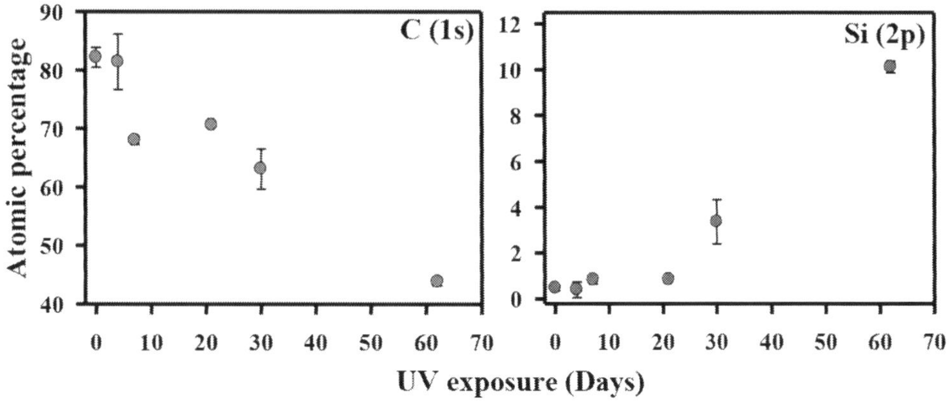

Figure 5. XPS-based C and Si concentrations on the epoxy/nanosilica composite surface Vs. time exposed to UV/50°C/75% RH. Except for the 21 d where only one specimen was used, other data consisted of two or more specimens

Degradation of the epoxy and an increase of the SiO_2 material near the nanocomposite surface is supported by XPS data displayed in Figure 5. The surface concentration of C decreased from 82.2% ± 1.7% to 43.8% ± 0.7%, while that of Si increased from 0.5% ± 0.1% to 10.1% ± 0.3% after 62 d exposure. The increase of SiO_2 on the composite surface following UV irradiation has also been confirmed and quantified by inductively-coupled plasma–optical emission spectroscopy (ICP-AES) analysis [14].

3.3. Nanoparticle fate during UV exposure of polymer nanocomposites

Surface morphological changes obtained by FE-SEM of epoxy/5% nanosilica composite exposed to UV radiation for varying time are displayed in Figure 6. The initial surface appeared smooth with evidence of some SiO_2 nanoparticles on it. After four days of exposure, numerous SiO_2 nanoparticles can be clearly seen on the composite surface. The surface concentration of nanosilica continued to increase with exposure time, and after 43 days, SiO_2 nanoparticles covered the entire composite surface.

Microscopic and spectroscopic data shown in Figures 4, 5 and 6 strongly suggested that the increase of nanosilica concentration at the composite surface with UV exposure was a result of the epoxy matrix degradation. However, these results do not answer the question: did SiO_2 nanoparticles release from the nanocomposite during its exposure to UV radiation? Analysis of particles collected at the bottom of the sample holder (i.e., Figure 2b) addresses this question. Figure 7 shows SEM images and EDS spectra of the particle collector surface before and after the sample was exposed to UV radiation. Before the exposure, the collector surface showed no evidence of particles (Figure 7a). After 43 d exposure, many particles were observed on the collector surface (Figure 7b), and numerous spherical nanoparticles can be seen with higher magnification (Figure 7c).

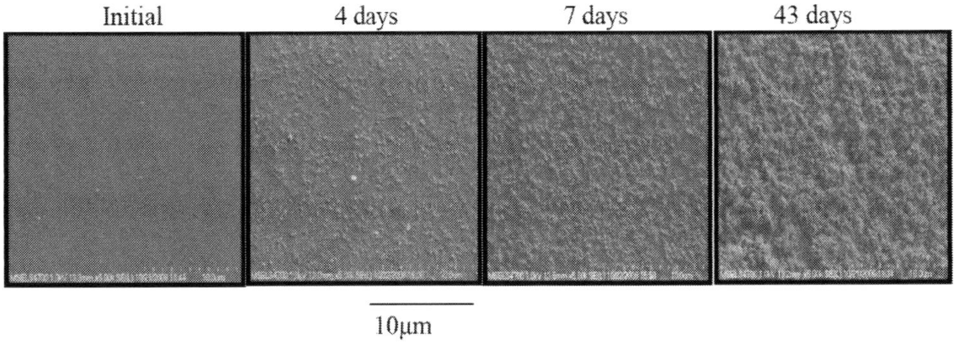

Figure 6. FE-SEM images of epoxy/5% nanosilica composite exposed to UV/75%RH/50°C for different times.

The EDS spectrum of the collector surface before the sample was exposed to UV radiation showed only fluorine (F) and carbon (C) elements (Figure 7d), as expected for a poly(tetrafluoroethylene) film. However, EDS spectrum of the collector surface after the sample was irradiated by UV for 43 days revealed the presence of silicon (Si) element (Figure 7e). The concentration of oxygen (O) also increased after the UV exposure. The increase of both Si and O elements implies that the spherical nanoparticles on the collector surface observed in Figure 7c were likely the SiO_2 compound. Results of Figure 7 have provided direct evidence that SiO_2 nanoparticles or their aggregated form were released from the epoxy nanocomposites to the surroundings during their exposure to 295 nm – 400 nm UV radiation. However, this preliminary result does not establish whether these SiO_2 nanoparticles are pristine or coated with polymer. Such information is important for assessing the potential risks of released nanoparticles, because nanomaterials that are covered with an inert polymer layer are less likely to impart an EHS concern than the pristine nanomaterials do. Work is in progress to address this question.

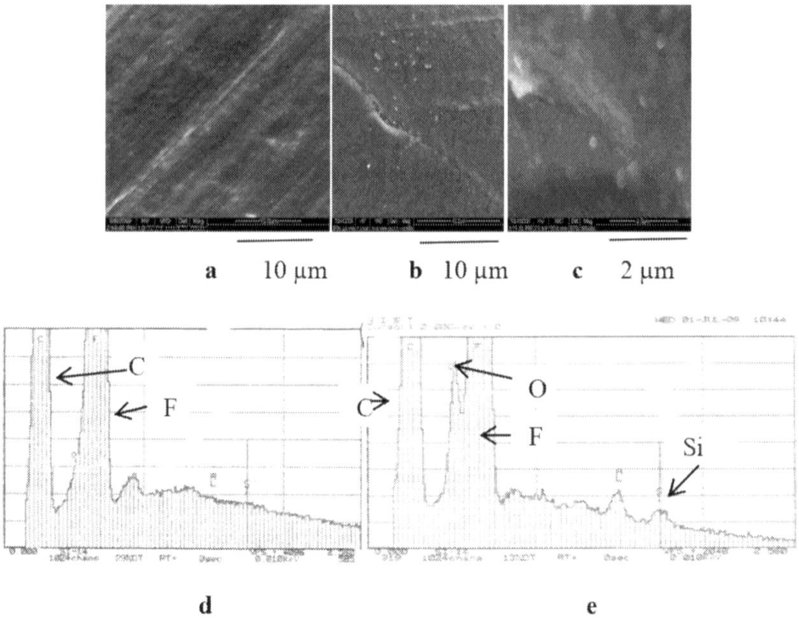

Figure 7. SEM images of particle collector surface: (a) before, and (b) after the epoxy/nanosilica composite sample was exposed to UV radiation for 43d, (c) higher magnification of b, showing numerous spherical nanoparticles; EDS spectra of the collector surface: before (d), and after the sample was exposed for 43 d (e), showing the presence of Si element.

Based on microscopic and spectroscopic evidence, SiO_2 nanoparticle release during exposure of epoxy nanocomposites to the UV condition used in this study likely followed the sequence shown in Figure 8. First, the epoxy polymer layer on the surface underwent photodegradation and was removed. This resulted in gradual increase of the number of SiO_2 nanoparticles on the composite surface with exposure time. At a critical thickness/concentration, particles containing SiO_2 fell off the vertical surface (release) likely due to gravitation force. For nanocomposites exposed to outdoor, environmental elements such as rain, condensed water, wind, mechanical vibration, and stresses resulting from materials dimensional changes likely affect the release rate of nanoparticles. In the absence of UV stabilizers, most common polymers are susceptible to degradation during exposures to the UV environment. Therefore, this conceptual model for the release of SiO_2 nanoparticles should be applicable to other polymers and to other low aspect ratio, spherical nanoparticles used in nanocomposites.

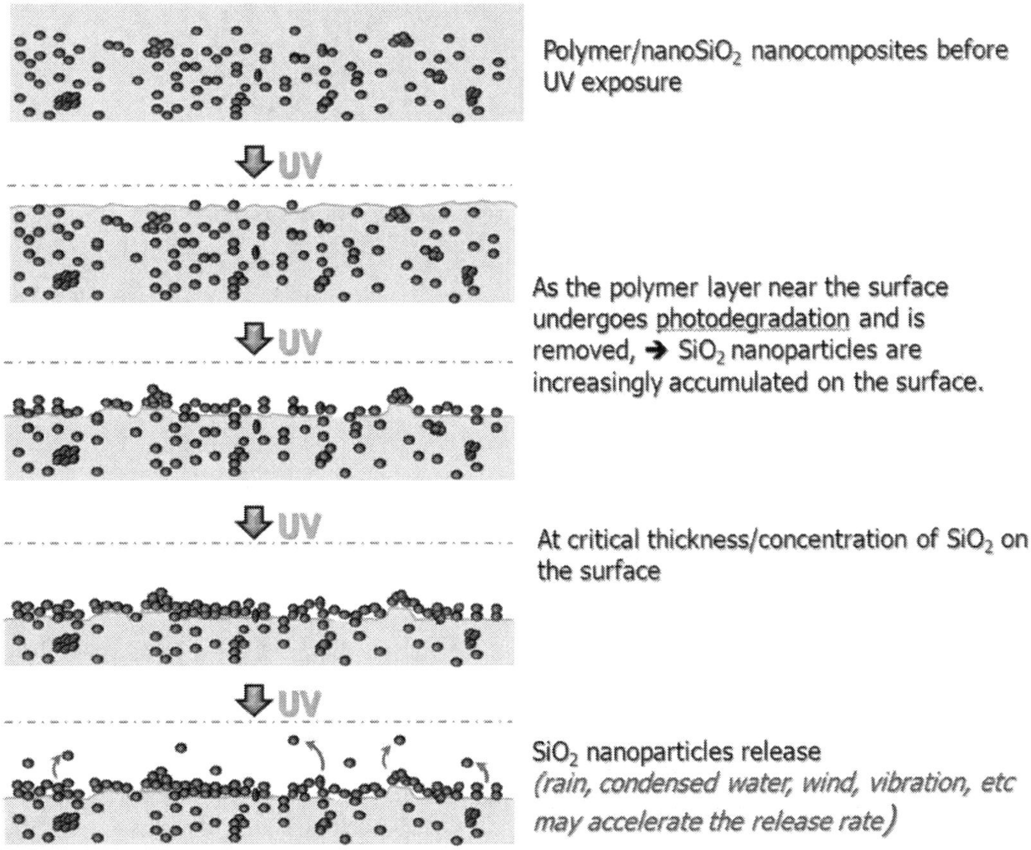

Figure 8. Conceptual model for the release of SiO_2 nanoparticles from epoxy/nanosilica composites exposed to UV radiation

The surface morphology of epoxy/MWCNT composites exposed to UV radiation at varying time is shown in Figure 9. The initial composite surface appeared smooth with no evidence of CNTs. However, MWCNTs became visible on the surface after 11 days, and their concentration on the surface continued to increase with exposure time. After 43 days, the MWCNTs formed a dense layer on the composite surface, which is clearly shown in the high magnification image. From these microscopy results and FTIR data shown in Figure 4, it is reasonable to suggest that the increased CNT concentration at the composite surface with exposure time was a result of the matrix photodegradation, similar to that for nanosilica nanocomposites. As the epoxy matrix degraded and was removed, MWCNTs were increasingly exposed on the surface.

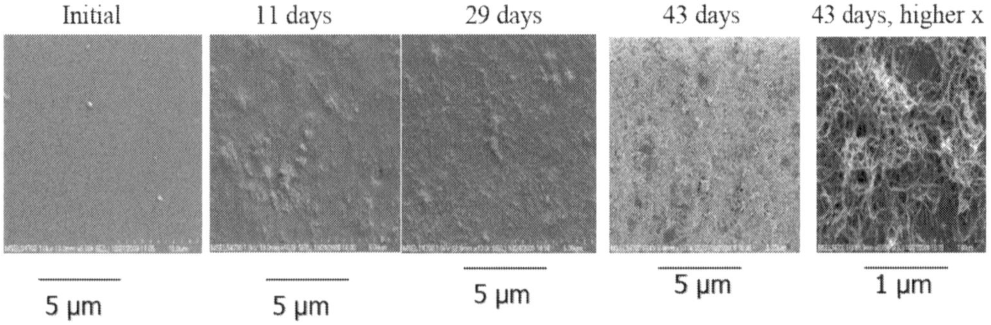

Figure 9. FE SEM images taken at different times for the epoxy/0.72 % MWCNT composite exposed to UV/50 °C/75 % RH, showing the aggregated network of MWCNTs on the composite surface after UV exposure.

The results of this study raised an important question: does the amount of MWCNTs at the surface observed at each exposure time represent its concentration at that particular depth in the epoxy matrix before exposure, or is it an amassing of CNTs from different depths as the degradation of the matrix progressed? In the first case, as soon as MWCNTs were exposed at the surface, they would be removed from it. Thus, the rate of CNT release into the environment should be similar to that of the polymer degradation. If the second case is true, then MWCNTs would form an aggregated network as they become exposed on the composite surface. These MWCNTs were highly entangled, with some of them were still strongly embedded in the matrix. Therefore, individual MWCNTs are not likely to be released to the environment to any great extent during normal exposure conditions. To better answer this question, the collector surface of the holder in which a large epoxy/MWCNT composite sample was exposed for nine months under the same conditions as for the nanosilica composites was carefully analyzed by SEM. The result showed no evidence of CNTs on the collector surface, suggesting that essentially little or no release of MWCNTs occurred during exposure to UV. Based on these data, the fate of MWCNTs in polymer/MWCNT composites is schematically illustrated in Figure 10. Except for the last step, the sequence is similar to that of the nanosilica composite. That is, photodegradation of the matrix has resulted in accumulation of large concentration of MWCNTs on the surface, which aggregated and formed a dense entangled structure. Such dense structure of CNTs decreased the photodegradation of the epoxy polymer layer underneath and minimized their release to the environment. This model should be applicable for polymer nanocomposites containing high aspect ratio, fiber-like nanoparticles.

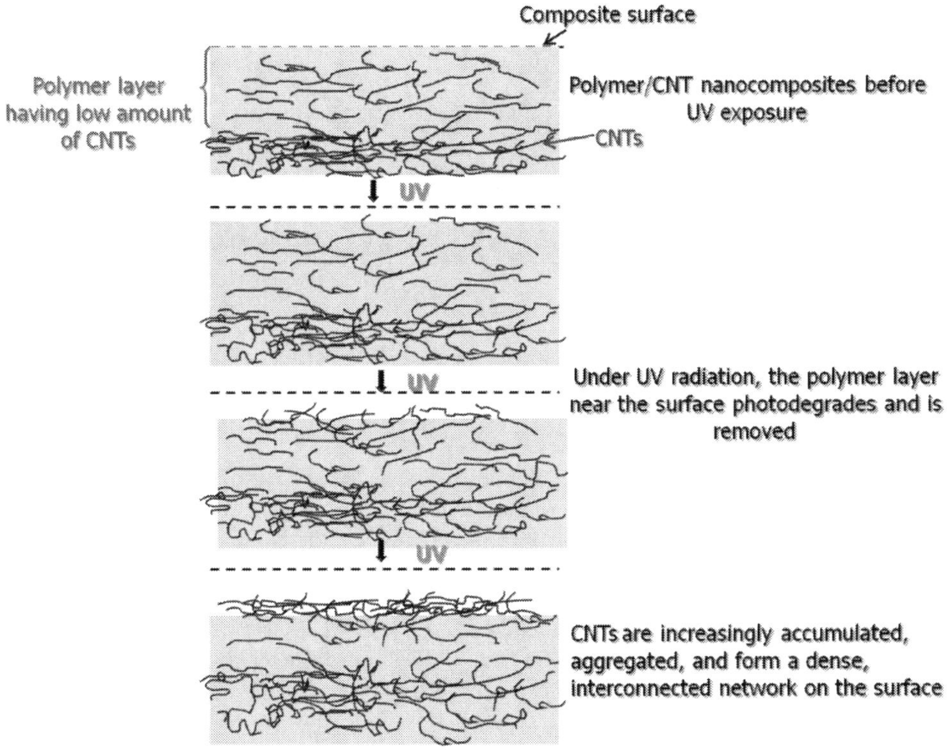

Figure 10. Conceptual model for the formation of dense MWCNT network on the epoxy nanocomposite surface during exposure to UV radiation.

4. Conclusions

The fate of MWCNTs and nanosilica in amine-cured epoxy nanocomposites exposed to 295-400 nm UV radiation at 50°C and 75% RH has been investigated. Based on experimental results, the following conclusions are provided:

- Amine-cured epoxy matrix in the nanocomposite underwent photodegradation during exposure to 295 nm to 400 nm UV radiation, resulting in substantial mass loss and marked increase in nanoparticle concentration on nanocomposite surfaces.
- The rate of mass loss and chemical degradation of MWCNT/epoxy composite was lower than that observed in unfilled epoxy or epoxy/nanosilica composite, suggesting that MWCNTs decrease the photodegradation of the amine-cured epoxy.
- MWCNTs formed a dense, entangled network on the nanocomposite surface, with no evidence of release even after prolonged exposure to UV radiation; the CNT network formed on the composite was responsible for essentially eliminating the release of CNTs to the environment during exposure to UV radiation.
- Microscopic and spectroscopic analyses showed conclusively that SiO_2 nanoparticles were released to the environment after short exposure to UV radiation.
- Conceptual models for the release of nanosilica and retention of CNTs during exposure to UV have been proposed, which should be applicable to other polymers and other types of nanoparticles.

The results of this study provide valuable information to assess the potential risk of nanoparticles release from epoxy nanocomposites during service. Work is under way to predict the nanoparticle release rates and characterize chemical composition of the released nanoparticles.

References

[1] Coleman J. N., Khan U., Blau W. J. and Gun'ko Y. K., *Carbon,* **44** (2006) 1624.
[2] Ajayan P. M. and Tour J. M., *Nature,* **447** (2007) 1066.
[3] Chen C., Justice R. S., Schaefer D. W. and Bauer J. W., *Polymer*, **49** (2008) 3805.
[4] Helland, P. Wick, A. Koehler, K. Schmid and C. Som, *Environmental Health Perspectives,* **115** (2007) 1125.
[5] Poland A., Duffin R., Kinloch I., Maynard A., Wallace W. A. H., Seaton A., Stone V., Brown S., MacNee W. and Donaldson K., *Nature Nanotechnology,* **3** (2008) 423.
[6] Maynard A.D., *Nanotoday,* **2** (2006) 22.
[7] Chin J., Byrd E., Embree N., Garver J., Dickens B., Fin T. and Martin J. W., *Review Scientific Instruments,* **75** (2004) 4951.
[8] Rabek J.F., Polymer Photodegradation: Mechanism and Experimental Methods, Chapman &Hall, NY., 1995, pp 185-216.
[9] Bellinger V., Bouchard C., Claveirolle P. and Verdu J., Photochemistry, **1** (1981) 69.
[10] Bellinger V. and Verdu J., Oxidative Skeleton Breaking in Epoxy-Amine Networks, *J. Appl. Polym. Sci.*, **30** (1985) 363.
[11] Gu X. et al., in "Service Life Prediction of Polymeric Materials, Global Perspectives", Eds. J. W. Martin, R. A. Ryntz, J. Chin, R. A. Dickie, Springer, NY., 2009, p. 1.
[12] Nguyen T., Granier A., Steffens K. L., Lee H., Shapiro A. and Martin J. W., Proceedings, International Coating Conference, Toronto, Canada, October, 2007.
[13] Najafi E. and Shin K., *Colloids and Surfaces, A: Physicochem. Eng. Aspects*, **257** (2005) 333.
[14] Rabb S., Yu L., Bernard C. and Nguyen T., Nanotech2010, June, Anaheim, CA.

Artifacts by marker enzyme adsorption on nanomaterials in cytotoxicity assays with tissue cultures

Wendel Wohlleben, Susanne N Kolle, Laura-Carolin Hasenkamp, Alexander Böser, Sandra Vogel, Bernhard von Vacano, Ben van Ravenzwaay and Robert Landsiedel

BASF SE, Experimental Toxicology and Ecology Department
and Polymer Physics Department, 67056 Ludwigshafen, Germany.

Email: wendel.wohlleben@basf.com

Abstract. We used precision cut lung slices (PCLS) to study the cytotoxicity of cobalt ferrite nanomaterials with and without bovine serum albumin (BSA) stabilization. Using mitochondrial activity as an indicator of cytotoxicity (WST-1 assay) increasing concentrations of cobalt ferrite nanomaterial caused increasing levels of cytotoxicity in PCLS irrespective of BSA stabilization. However, there was no increase in released lactate dehydrogenase (LDH) levels caused by BSA stabilized nanomaterial indicating concentration depended cytotoxictiy. Moreover, non-stabilized nanomaterial caused a decrease of background LDH levels in the PCLS culture supernatant confirmed by complementary methods. Direct characterization of the protein corona of extracted nanomaterial shows that the LDH decrease is due to adsorption of LDH onto the surface of the non-stabilized nanomaterial, correlated with strong agglomeration. Preincubation with serum protein blocks the adsorption of LDH and stabilizes the nanomaterial at low agglomeration. We have thus demonstrated the cytotoxicity of nanomaterials in PCLS does not correlate with disrupted membrane integrity followed by LDH release. Furthermore, we found that intracellular enzymes such as the marker enzyme LDH are able to bind onto surfaces of nanomaterial and thereby adulterate the detection of toxic effects. A replacement of BSA by LDH or a secondary LDH-on-BSA-corona were not observed, confirming earlier indications that the protein corona exchange rate are slow or vanishing on inorganic nanomaterial. Thus, the method(s) to assess nanomaterial-mediated effects have to be carefully chosen based on the cellular effect and possible nano-specific artifacts.

1. Introduction

Toxicological testing of nanomaterials demands test systems which take into account the unique physicochemical characteristics of nanomaterials compared to the bulk material. For human safety, respiratory exposure to dust atmospheres is considered to be the route of highest concern for nanomaterials.[1] *In vivo* inhalation studies on rats or mice have established that there is not a general hazard due to the nanoscale structure, but that the hazard potential depends on specific material properties [2,3]. Hazard assessment is hence required for each nanomaterial, at least until cross-reading and groupings become validated in future REACH revisions. Considering the large number of materials that will fall under the upcoming regulatory definitions, the reduction of animal tests is a pressing issue. Precision cut lung slices (PCLS) may offer a test system to assess the lung toxicity of nanomaterials without *in vivo* inhalation studies. This method offers the opportunity to prepare around

100 lung slices from a single *ex vivo* rat lung. Thus, 50 different conditions can be tested and this qualifies it as an alternative method to *in vivo* studies. Biophysical interactions have not been reported so far for the PCLS assay, but a sign of warning is raised by the dose-response shift of inorganic nanomaterials due to adsorption effects [4] and the generally recognized importance of characterization for in vitro assays.[5,6]. Significant artifacts have been reported for hydrophobic nanomaterials (carbon nanotubes) adsorbing MTT [7] or essential nutrients [8]. The biophysical mechanisms of particulate nanomaterials interacting with proteins was pioneered by the Dawson group with a focus on serum interactions of polymeric model particles [9-12], and first evidence emerges that similar effects occur with inorganic nanomaterials in ecological media [13; 14], in blood serum [14-18] or in alveolar environments [19,20]. Incubation of metal oxides with dysopsonins (BSA, IgA) decreased resorption [21] and prolonged circulation in the bloodstream[22], and incubation at unphysiologically high nanomaterial concentrations >10 mg/ml can even induce buffer depletion,[23] unless the adsorption was blocked by pre-incubation in serum.[24]

Here we used PCLS to study the cytotoxicity of cobalt ferrite (CoFe) nanomaterial. Bovine serum albumin (BSA) was added to prevent agglomeration of the nanomaterial during the test, but lead to the discovery that artifacts (false-negative results) occur in the determination of cytotoxicity by means of released levels of LDH depending on preincubation of the nanomaterial to be tested. The endpoint of LDH release can therefore be underestimated by adsorption of LDH onto non-stabilized CoFe nanomaterial. The in situ characterisation of the nanomaterial in PCLS revealed that the protein corona of either LDH or BSA correlates directly with the state of agglomeration and does not exchange noticeably in successive incubations. The cytotoxicity of nanomaterials in PCLS does not correlate with disrupted membrane integrity followed by LDH release unless the nanomaterials are carefully pre-incubated. These results are essential to enable the safety evaluation by alternative methods.

2. Materials and Methods

2.1. Preparation of precision cut lung slices (PCLS)

Animals (8-10 weeks old nulliparous and non-pregnant female Wistar Crl:WI (Han) rats, Charles River, Germany) were euthanized with an overdose of pentobarbital-Na. Lung tissue was prepared directly post mortem to conserve the viability of the tissue. Through the trachea the lung was carefully filled in situ with 10 mL / 200 g body weight prewarmed 1.5 % agarose – medium solution. The lung was removed and put on ice for 20 minutes, allowing the agarose to polymerize. Lung lobes were separated, placed on wax, and 8 mm tissue cylinders were subsequently prepared. The cylinders were placed into the Krumdieck tissue slicer (Alabama Research and Development, USA) filled with ice cold salt solution (EBSS) and slices with a thickness of approximately 200 μm were prepared. The PCLS were then washed three times with prewarmed DMEM/F-12 to remove the agarose.

2.2. Preparation of the cobalt ferrite nanomaterial suspension

Cobalt ferrite (CoFe) nanomaterial suspensions were prepared in serum-free DMEM/F12 medium as well as in DMEM/F12 medium with BSA (Sigma-Aldrich, Germany) to stabilize the surface of the nanomaterial and to avoid the aggregation of the particles. NFeCo132 was provided by Colorobbia Italia SPA (Vinci, Italy) in a concentration of 117 mmol/kg as suspension in water (PDI = 0,225; D (V mean) nm = 16,76; Fe/Co ratio = 1.91; ICP % w/w = 2.74 and grain size by XRD = 5 nm). For a 10 mM stock solution a total of 0.8547 g NFeCo132 and 9.1453 g were added to serum-free DMEM/F12 medium (Gibco Product No. 21041-025). For the preparation of a 10 mM stock solution with BSA, 0.8547 g of the NFeCo132 nanomaterial were suspended by shaking in 5 % BSA solution (0.5 g in ultrapure Milli-Q water) for 30 minutes. Then, 8.6453 g DMEM/F 12 medium was added to the mixture. PCLS were exposed to CoFe nanomaterial suspensions in DMEM/F12 medium (0, 0.01, 0.05, 0.1, 0.2, 0.4, 0.8, 1.0, 2.5, and 5.0 mM). In addition, a BSA solution (5 % in ultrapure Milli-Q water) was used as a negative control. For cytotoxicity assays, 1% Triton X-100 (in DMEM/F-12, Sigma-Aldrich, Germany) was used as positive control.

2.3. Cytotoxicity assays

After an 24 hours incubation of nanomaterial semi-quantitative determinations of the mitochondrial metabolic activity in PCLS using the water-soluble tetrazolium assay (WST-1, Roche, Germany) and determination of the cell membrane integrity by measuring the released levels of lactate dehydrogenase in the supernatant (LDH Cytotoxicity Detection Kit, Roche, Germany) were performed. Triton X-100 lysed PCLS were served as positive control (100% LDH release). Results were calculated as percentage of the total LDH content or maximum viability. All tests were performed in duplicate.

2.4. Incubation protocol

CoFe nanomaterial concentrations were incubated PCLS-free under different conditions for 24 hours: (1) in serum-free DMEM/F12 medium, (2) in serum-free DMEM/F12 medium containing human serum LDH (at a concentration of 0.05 U/mL, Prezinorm U, Roche, Germany) (3) non-BSA-stabilized NP suspension in serum-free DMEM/F12 containing 20% of PCLS-lysate, and (4) BSA-stabilized nanomaterial suspension in serum-free DMEM/F12 containing 20% of PCLS-lysate. As positive control (100% LDH release) PCLS were treated for 24 hours with 1% Triton X-100 and examined in the LDH assay. After a PCLS-free incubation of 24 hours the LDH content of the nanomaterial incubates was determined as described above.

2.5. Analytical ultracentrifugation with interference detection (AUC)

A mighty tool for the characterization of nano-colloids (0.5 – 10,000 nm diameter) is the Analytical Ultracentrifuge (AUC)[25-27], especially the universal interference optics Beckman XLI™ with widespread use in the proteome business[28-30], and, only to a lesser extent, also the disc centrifuges (Brookhaven Instruments XDC, CPS Instruments DC24000) with their rather limited detection optics and lower speeds. Schlieren, turbidity, interference, UV-VIS absorption and X-Ray absorption detection are published.[26,31] The optical AUC method detects the time- and radius- dependent concentration of the solutes simultaneously with the sedimentation at 600 to 60,000 rpm. Thereby, we quantify the amount and the diameter of each component independently.[25] At present, AUC is the only method that detects all components from the agglomerates, to the dispersed nanoparticles and the sub-10-nm proteins.[3]

Here we use interference optics (Beckman model *'XLI proteome lab'*) at 10,000 rpm and 44,000 rpm. The raw data is fitted by the free-ware software SedFit, resulting in a distribution of sedimentation coefficients s [32]. In order to convert the observed sedimentation coefficients s to diameters d, we rely on the simplifications of the Stokes-Einstein relation:

$$d = \sqrt{\frac{18\eta s}{\Delta\rho}} \quad (1)$$

where η is the viscosity of the solvent and $\Delta\rho$ is the density difference between solute and solvent. The concentration shares c are read directly from the interference fringe shift j with

$$j = \frac{c(dn/dc)l}{\lambda} \quad (2)$$

where dn/dc is the refractive index increment, l = 12 mm the length of the optical cell, and λ = 680 nm the wavelength of the laser. Here we evaluated the particulate signal with a density difference $\Delta\rho$=4.6 g/cm^3. The protein signal was evaluated with $\Delta\rho$=0.36 g/cm^3 and dn/dc=0.18 cm^3/g.

Samples were prepared by mixing 0.1975 g stock particle suspension with 4.7628 g DMEM. For conditioning by albumin, 0.1974 g stock particle suspension were mixed with 0.2737 BSA solution, stirred for 30 minutes, then added to 4.5659 g DMEM and stirred again. Either of these suspensions were stirred for 30 minutes, then mixed 1:1 with the PCLS lysate and stirred again. During this preparation, intermediate and final samples were measured by AUC.

2.6. Time of flight secondary ion mass spectrometry (TOF-SIMS)

Time of flight secondary ion mass spectrometry (TOF-SIMS) is a powerful surface analysis technique that allows chemical characterization of the first atomic layers of any vacuum stable solid surface. Being a fragmenting mass spectrometry technique, it provides a wealth of chemical information and allows identifying compounds with very high sensitivity. This makes TOF-SIMS ideally suited to characterize and study biomolecules on surfaces, such as proteins.[33] For SIMS measurements, the particulate fraction was separated by sedimentation at 10,000 rpm for 4 minutes. The supernatant was discarded, and the volume was refilled with DMEM, stirred shortly and pelleted again for washing of non-adsorbed proteins. The clean particle sediment was prepared for SIMS analysis by their placing on clean silicon wafers. On the thus prepared sample positions, no silicon wafer secondary ion mass signal could be detected any more, confirming that the sample layer thickness well exceeded the SIMS information depth of typically 1-3 nm.

Static TOF-SIMS spectra were recorded using a TOF-SIMS V spectrometer (Iontof GmbH, Germany). A pulsed mass-filtered primary ion beam of 25 keV singly charged bismuth (Bi^+) was used. This primary ion beam, resulting in a spot size of typically 5 μm on the sample surface, was raster scanned over an area of 250 x 250 μm to record spectra of positive and negative secondary ions. The primary ion dose density was always kept well below 10-12 ions/cm² and thus in the static SIMS regime. To prevent charging of the sample surface, a low-electron energy flood gun was used.

3. Results

3.1. CoFe nanomaterial induced cytotoxicity

Cell viability (WST-1 assay) is presented as relative mitochondrial activity normalized to the vehicle control (100% mitochondrial activity). The mean mitochondrial activity was decreased from 0.4 mM concentration to almost the value of the positive control (1% Triton X-100) when compared to the vehicle or BSA control. Therefore the WST-1 assay indicates at increasing concentrations of nanomaterial (without and with BSA stabilization) increasing levels of cytotoxicity in PCLS. However, by means of the LDH release assay, presented as percentage of the total LDH content (calculated from Triton X-100 lysed samples, 100% LDH release), was not increased when compared to the vehicle control. This was independent of incubating PCLS with or without BSA-stabilized nanomaterial. Moreover, upon incubation with non-stabilized nanomaterial decrease of background LDH levels in the PCLS culture supernatant were measured (Figure 1).

Released LDH-levels (red bars, "LDH R.") and mitochondrial metabolic activity (blue bars, "Mit. Act.") were assessed after 24 hours exposure of PCLS to CoFe nanomaterial without (solid bars, "w/o BSA") and with BSA stabilization (diagonal stripes, "w/ BSA"). Controls were PCLS incubated with BSA (horizontal stripes) and Triton X-100 with and without BSA (100% LDH release, open bars and horizontal stripes).

3.2. Biophysical interactions of CoFe nanomaterial in the PCLS assay

3.2.1. *Adsorption: Indirect assessment by quantification of non-adsorbed LDH.* We have hypothesized the discordance of the two cytotoxicity assays to be an artifact: the LDH released from the disintegrated cell membrane at particle concentrations was adsorbed to the non-stabilized nanomaterial in the supernatant. Figure 2 shows that the LDH is adsorped onto the surface of the non-stabilized nanomaterial whereas BSA stabilization blocked the absorption of LDH (Figure 2).

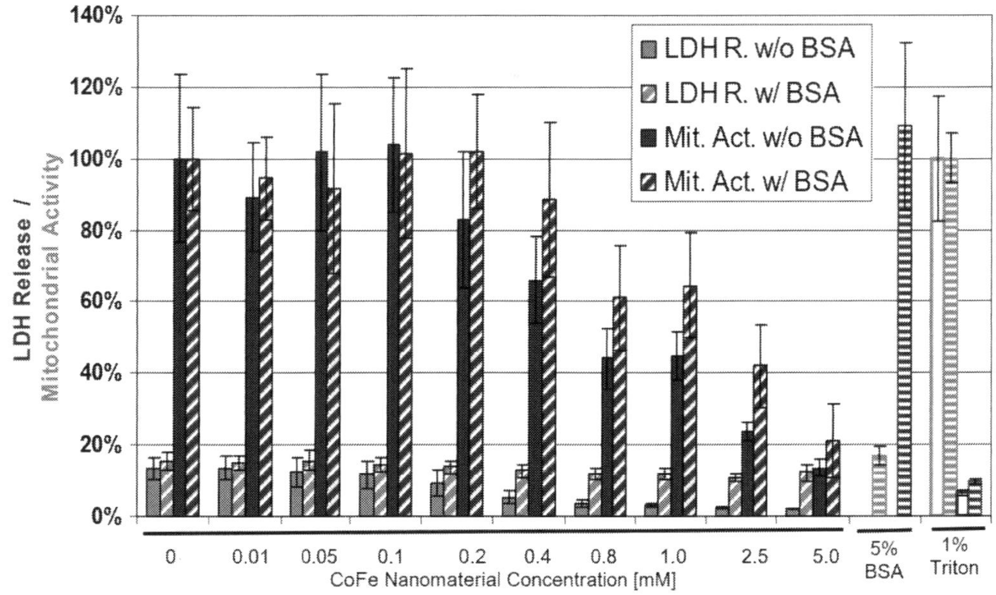

Figure 1. Assessment of LDH release and mitochondrial activity after 24 hour exposure of PCLS to different concentrations of CoFe nanomaterial.

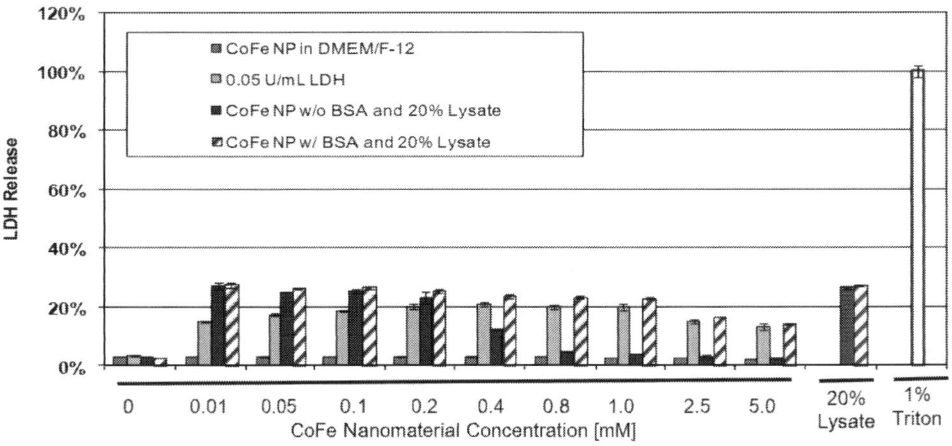

Figure 2: Assessment of LDH adsorption to CoFe nanomaterial by the LDH Detection Kit.

LDH-levels were assessed after 24 hours exposure (solid bars, "w/o BSA") and with BSA stabilization (diagonal stripes, "w/ BSA"). Controls were PCLS incubated with Triton X-100 (100% LDH release, open blue bar) and 20% PCLS lysate in medium without (purple bar, solid) and with BSA (purple bar, stripes). Assessed were CoFe nanomaterial in serum-free DMEM/F12 medium (green bars), in serum-free DMEM/F12 medium containing 0.05 U/mL human serum LDH (yellow bars), non-BSA-stabilized nanomaterial suspension in serum-free DMEM/F12 containing 20% of PCLS-lysate (blue bars, solid), and BSA-stabilized nanomaterial suspension in serum-free DMEM/F12 containing 20% of PCLS-lysate (blue bars, stripes).

As a cross-check with a complementary technique, we determined the concentrations of free proteins by analytical ultracentrifugation with interference detection. The concentration of proteins in PCLS rises from 2.7 mg/mL to 3.3 mg/mL (measured by the refractive index signal below 10 nm diameter in the AUC, data not shown) when the nanomaterial was stabilized in BSA before incubation with PCLS. The BSA signal is clearly identified by the refractive index signal at 66 kDa and 132 kDa, attributed to monomer and dimer, respectively. The concentration of non-adsorbed BSA (monomer and dimer added) did not increase when the nanomaterial was transferred from DMEM into PCLS lysate, indicating that BSA is not replaced by other ligands. These results with hydrodynamic separation of proteins by mass and unselective detection confirmed the results obtained with the enzymatic LDH measurement.

3.2.2. *Adsorption: direct assessment of the protein corona on CoFe nanomaterial.* In the context of this study, TOF-SIMS is employed to characterize the nanomaterial adsorbates and elucidate the influence of prior particle conditioning with BSA.

As reference, the untreated nanomaterial was investigated by TOF-SIMS revealing the elementary ion signals of iron and cobalt in the positive partial mass spectrum shown in **Figure** 3a (marked with blue, diagonally striped circles). In addition, bulk samples of DMEM, BSA and the PCLS lysate were analyzed for reference (not shown).

Nanomaterial treated only with PCLS lysate (non-BSA stabilized) showed a distinct change in their surface mass spectra (**Figure** 3b), which indicates the formation of an adsorbate layer. Upon comparison with the reference spectra, it can clearly be shown by distinct marker secondary ion peaks (98, 137 and 142 amu, labeled with orange circles, vertically striped) that this adsorbate stems form the PCLS lysate. On the other hand, such signals are not detected from the surface of the sample conditioned with BSA prior to exposure to PCLS lysate. In this case (Figure 3c) only clear BSA fragments secondary ions are detected, e.g. at 70, 84, 86, 110 and 120 amu (plain green circles).

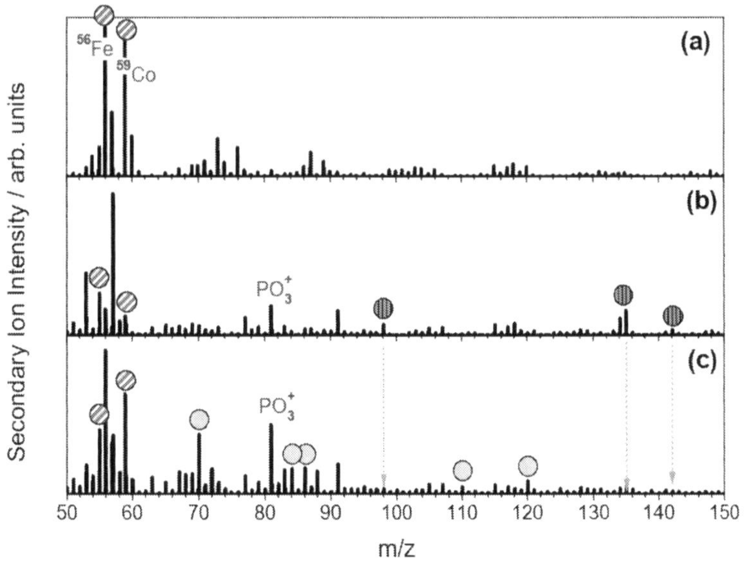

Figure 3. Secondary ion mass spectra recorded from (a) untreated reference particles, (b) non-stabilized particles exposed to PCLS lysate and (c) BSA stabilized particles exposed to PCLS lysate.

3.2.3. *State of agglomeration in situ of the PCLS assay.* The degree of nanomaterial agglomeration in physiological media highly depends on the composition of the used solvent. In DMEM the nanomaterial agglomerates with a diameter distribution between 200 and 1200 nm. When the nanomaterial is first stabilized in BSA and then added to DMEM, the degree of agglomeration decreases to a mean diameter below 200 nm (Figure 4). Transfer of BSA-stabilized nanomaterial from DMEM into PCLS lysate maintains the low degree of agglomeration. In contrast, when nanomaterial without BSA stabilization is transferred from DMEM directly into PCLS lysate, the degree of agglomeration is higher.

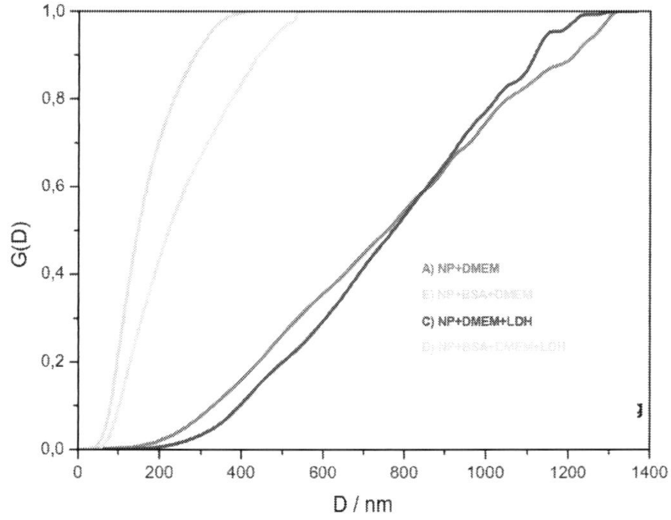

Figure 4. State of agglomeration depending on pre-incubation as determined by analytical ultracentrifugation (AUC) with interference detection. Cumulative distribution allows for direct comparison of the state of agglomeration, which is above primary particle size for all conditions tested, but only weakly agglomerated with pre-incubation in BSA solution.

4. Discussion

In comparison to the relatively constant values of the LDH release of the nanomaterial stabilized with BSA, the non-stabilized nanomaterial seemed to induce, with increasing concentrations, a reduced LDH release. This was hypothesized to be an artifact and the LDH released from the cells at particle concentrations disrupting the membrane integrity adsorbed to the non-stabilized nanomaterial in the supernatant. Therefore the detection of LDH as indicator of the membrane integrity was shown to be limited to assess the cytotoxic effects of nanomaterials.

The results of this study demonstrate that released LDH from the culture supernatant of the lysed PCLS bound to the surface of non-stabilized CoFe nanomaterial, whereas the stabilization of the nanomaterial with BSA prevented the adsorption of the marker enzyme to the particle surface. Serum-free DMEM/F12 medium had no effect on the released LDH levels, such that an artifact by the interaction of the nanomaterial with the components of the assay was excluded. The constant LDH level in the approach with defined LDH concentration in human serum can be explained by the fact that other proteins of the human serum adsorb to the particle surface and therefore serve as particle stabilization. The adsorption of proteins and organic molecules at the surface of nanomaterial has been described in the literature [4,14,15,17-20] and is used in medicine for the improvement of the efficiency of drug effects [34]. Against this background, the reduced released LDH levels determined in the cytotoxicity assay can be explained by the adsorption of released LDH at particle surfaces. We have substantiated this hypothesis by direct analysis on the nanoparticle surfaces, where the TOF-SIMS data confirms that components from PCLS lysate, while clearly adsorbing on neat nanomaterial, are not able to displace previously adsorbed BSA or form an adsorbate layer on top of the BSA conditioning film. The surface of dispersed colloids determines their mutual interactions. From this point of view, it is obvious that adsorbed proteins can act as protection colloids, if the specific protein has flexible solvated loops that stabilize the nanomaterial by steric mechanisms. Especially albumins in water or DMEM have dispersed and stabilized a wide variety of nanomaterials: CNTs,[14,35-37] metal nanoparticles,[6] metal carbide nanoparticles[38] and metal oxide nanoparticles.[37,39-43], whereas bronchoalvelolar lavage induces the opposite effect of increasing agglomeration.[19]

Since an exchange between BSA and LDH in the lysate was not observed on the 1 h-time scale of this experiment, this supports earlier indications that the protein corona exchange rate is slow or vanishing on inorganic nanomaterial.[16; 44], in contrast to nanomaterials with polymeric surface functionalization.[9; 45] It may well be that morphological changes of the proteins upon adsorption impose a strong enthalpic barrier against dissociation, leading to an effective non-equilibrium adsorption mechanism.

5. Conclusion

The results of this study demonstrate that LDH from the culture supernatant of the lysed PCLS bound to the surface of non-stabilized CoFe nanomaterial, whereas the stabilization of the nanomaterial with BSA prevented the adsorption of the marker enzyme to the particle surface. We have shown that, besides BSA, intracellular enzymes such as the marker enzyme LDH can bind at the surfaces of nanomaterial and thus falsify the detection of cytoxicic effects. We characterized the resulting hard protein corona by several complementary techniques – indirect by the loss of free LDH, direct by the nanomaterial's surface – and found a non-exchanging corona that correlates exactly also with the state of agglomeration, which is determined by the adsorbates on the colloidal surfaces. The enzymatic detection of LDH as indicator of the membrane integrity was shown to be limited to assess the cytotoxic effects of nanomaterial, but a proper preincubation of the tested nanomaterial could restore the applicability of the LDH assay. For this reason, the method for the examination of cytotoxicity of nanomaterials including the dispersion / preincubation protocol has to be carefully chosen. Even though it has been argued that preincubation 'masks' the nanomaterial surface and hence supposedly lowers physiological effects, the hypothetical disadvantage is overcompensated by the advantage of avoiding false positive results by adsorption of marker enzymes.

Acknowledgements
We thank Hans-Albrecht Hüner, Klaus Vilsmeier and Helmuth Kullmann for excellent laboratory support. This study was supported by EC FP6 funded STREP – CellNanoTox (NMP-CT-2006-032731).

References

[1] Oberdörster G, Oberdörster E and Oberdörster J . 2005. Nanotoxicology: an emerging discipline evolving from studies of ultrafine particles. *Environ Health Perspect* **113** 823-839
[2] Brain JD, Curran MA, Donaghey T and Molina RM . 2009. Biologic responses to nanomaterials depend on exposure, clearance, and material characteristics. *Nanotoxicology* **3** 174-180
[3] Landsiedel R, Ma-Hock L, Kroll A, Hahn D, Schnekenburger J, Wiench K and Wohlleben W . 2010. Testing metal-oxide nanomaterials for human safety. *Adv Mater* **22** 2601-2627
[4] Barrett EG, Johnston C, Oberdorster G and Finkelstein JN . 1999. Silica Binds Serum Proteins Resulting in a Shift of the Dose-Response for Silica-Induced Chemokine Expression in an Alveolar Type II Cell Line. *Toxicology and Applied Pharmacology* **161** 111
[5] Kroll A, Pillukat MH, Hahn D and Schnekenburger J . 2009. Current in vitro methods in nanoparticle risk assessment: Limitations and challenges. *Eur J Pharm Biopharm* **72** 370-377
[6] Hussain SM, Braydich-Stolle L, Schrand AM, Murdock RC, Yu KO, Mattie DM, Schlager JJ and Terrones M . 2009. Toxicity Evaluation for Safe Use of Nanomaterials: Recent Achievements and Technical Challenges. *Adv Mater* **21** 1-11
[7] Wörle-Knirsch J-M, Pulskamp K and Krug HF . 2006. Oops They Did It Again! Carbon Nanotubes Hoax Scientists in Viability Assays. *Nano Lett* **6** 1261-1268
[8] Guo L, von dem Bussche A, Buechner M, Yan A, Kane AB and Hurt RH . 2008. Adsorption of Essential Micronutrients by Carbon Nanotubes and the Implications for Nanotoxicity Testing. *Small* **4** 721-727
[9] Lundqvist M, Stigler J, Elia G, Lynch I, Cedervall T and Dawson KA . 2008. Nanoparticle size

and surface properties determine the protein corona with possible implications for biological impacts. *Proc Nat Acad Sci* **105** 14265-14270

[10] Cedervall T, Lynch I, Lindman S, Berggard T, Thulin E, Nilsson H, Dawson KA and Linse S . 2007. Understanding the nanoparticle–protein corona using methods to quantify exchange rates and affinities of proteins for nanoparticles. *Proc Nat Acad Sci* **104** 2050-2055

[11] Lindman S, Lynch I, Thulin E, Nilsson H, Dawson KA and Linse S . 2007. Systematic investigation of the thermodynamics of HSA adsorption to N-iso-propylacrylamide/N-tert-butylacrylamide copolymer nanoparticles. Effects of particle size and hydrophobicity. *Nano Letters* **7** 914-920

[12] Cedervall T, Lynch I, Foy M, Berggad T, Donnelly SC, Cagney G, Linse S and Dawson KA . 2007. Detailed identification of plasma proteins adsorbed on copolymer nanoparticles. *Angewandte Chemie-International Edition* **46** 5754-5756

[13] Li Z, Greden K, Alvarez PJJ, Gregory KB and Lowry GV . 2010. Adsorbed Polymer and NOM Limits Adhesion and Toxicity of Nano Scale Zerovalent Iron to E. coli. *Environmental Science & Technology* **44** 3462-3467

[14] Lin DH and Xing BS . 2008. Tannic acid adsorption and its role for stabilizing carbon nanotube suspensions. *Environmental Science & Technology* **42** 5917-5923

[15] Schulze C, Kroll A, Lehr CM, Schäfer UF, Becker K, Schnekenburger J, Schulze Isfort C, Landsiedel R and Wohlleben W . 2008. Not ready to use - overcoming pitfalls when dispersing nanoparticles in physiological media. *Nanotoxicology* **2** 51-61

[16] Schäfer J, Schulze C, Marxer EEJ, Schäfer UF, Wohlleben W, Bakowsky U and Lehr C-M . 2010. Biorelevant protein adsorption to CeO2 nanoparticles by atomic force microscopy and analytical ultracentrifugation. *Acs Nano* submitted

[17] Meissner T, Kühnel D, Busch W, Oswald S, Richter V, Michaelis A, Schirmer K and Potthoff A . 2010. Physical-chemical characterization of tungsten carbide nanoparticles as a basis for toxicological investigations. *Nanotoxicology* **4** 196-206

[18] Wasdo SC, Barber DS, Denslow ND, Powers KW, Palazuelos M, Stevens SM, Moudgil BM and Roberts SM . 2008. Differential binding of serum proteins to nanoparticles. *International Journal of Nanotechnology* **5** 92-115

[19] Schulze C, Schäfer UF, Ruge C, Wohlleben W and Lehr C-M . 2010. Interaction of metal oxide nanoparticles with lung surfactant proteins. *Eur J Pharm Biopharm* submitted

[20] Keane MJ and Wallace WE . 2005. A quantitative in vitro fluorescence imaging method for phospholipid loss from respirable mineral particles. *Inhalation Toxicology* **17** 287-292

[21] Patil S, Sanberg A, Heckert E, Self W and Seal S . 2007. Protein adsorption and cellular uptake of cerium oxide nanoparticles as a function of zeta potential. *Biomaterials* **28** 4600-4607

[22] Joshi MD and Müller RH . 2009. Lipid nanoparticles for parenteral delivery of actives. *Eur J Pharm Biopharm* **71** 161-172

[23] Horie M, Nishio K, Fujita K, Endoh S, Miyauchi A, Saito Y, Iwahashi H, Yamamoto K, Murayama H, Nakano H, Nanashima N, Niki E and Yoshida Y . 2009. Protein Adsorption of Ultrafine Metal Oxide and Its Influence on Cytotoxicity toward Cultured Cells. *Chemical Research in Toxicology* **22** 543-553

[24] Rocker C, Potzl M, Zhang F, Parak WJ and Nienhaus GU . 2009. A quantitative fluorescence study of protein monolayer formation on colloidal nanoparticles. *Nature Nanotechnology* **4** 577-580

[25] Jamison JA, Krueger KM, Yavuz CT, Mayo JT, LeCrone D, Redden JJ and Colvin VL . 2008. Size-dependent sedimentation properties of nanocrystals. *Acs Nano* **2** 311-319

[26] Mächtle W and Börger L . 2006. Analytical Ultracentrifugation of Polymers and Nanoparticles. Springer, Berlin.

[27] Cölfen H . 2005. Analytical Ultracentrifugation of Colloids. In: Scott DJ, Harding SE and Rowe AJ (eds) Analytical Ultracentrifugation. Techniques and Methods. The Royal Society of Chemistry, Cambridge, pp 501-583.

[28] Balbo A, Minor KH, Velikovsky CA, Mariuzza RA, Peterson CB and Schuck P . 2005. Studying multiprotein complexes by multisignal sedimentation velocity analytical ultracentrifugation. *Proc Nat Acad Sci* **102** 81-86

[29] Stoehr J, Weinmann N, Wille H, Kaimann T, Nagel-Steger L, Birkmann E, Panza G, Prusiner SB, Eigen M and Riesner D . 2008. Mechanisms of prion protein assembly into amyloid. *Proc Nat Acad Sci* **105** 2409-2414

[30] Demeler B . 2005. Hydrodynamic methods. In: Bühler LK and Rashidi HH (eds) Bioinformatics basics: Application in biological science and medicine. CRC Boca Raton, London, pp 226-255.

[31] Strauss H, Karabudak E, Bhattacharyya S, Kretzschmar A, Wohlleben W and Cölfen H . 2008. Performance of a fast fiber based UV/Vis multiwavelength detector for the analytical ultracentrifuge. *Colloid Polym Sci* **286** 121-128

[32] Schuck P . 1998. Sedimentation analysis of noninteracting and self-associating solutes using numerical solutions to the Lamm equation. *Biophys J* **75** 1503-1512

[33] Grainger DW and Castner DG . 2008. Nanobiomaterials and Nanoanalysis: Opportunities for Improving the Science to Benefit Biomedical Technologies. *Adv Mater* **20** 867-877

[34] Aggarwal P, Hall JB, McLeland CB, Dobrovolskaia MA and McNeil SE . 2009. Nanoparticle interaction with plasma proteins as it relates to particle biodistribution, biocompatibility and therapeutic efficacy. *Advanced Drug Delivery Reviews* **61** 428-437

[35] Nepal D and Geckeler KE . 2007. Proteins and Carbon Nanotubes: Close Encounter in Water. *Small* **3** 1259-1265

[36] Lacerda L, Herrero MA, Venner K, Bianco A, Prato M and Kostarelos K . 2008. Carbon-nanotube shape and individualization critical for renal excretion. *Small* **4** 1130-1132

[37] Vippola M, Falck GCM, Lindberg HK, Suhonen S, Vanhala E, Norppa H, Savolainen K, Tossavainen A and Tuomi T . 2009. Preparation of nanoparticle dispersions for in-vitro toxicity testing. *Human & Experimental Toxicology* **28** 377-385

[38] Bastian S, Busch W, Kühnel D, Springer A, Meißner T, Holke R, Scholz S, Iwe M, Pompe W, Gelinsky M, Potthoff A, Richter V, Ikonomidou C and Schirmer K . 2009. Toxicity of Tungsten Carbide and Cobalt-Doped Tungsten Carbide Nanoparticles in Mammalian Cells in Vitro. *Environ Health Perspect* **117** 530-536

[39] Schulze C, Kroll A, Lehr C-M, Schäfer UF, Becker K, Schnekenburger J, Schulze-Isfort C, Landsiedel R and Wohlleben W . 2008. Not ready to use – overcoming pitfalls when dispersing nanoparticles in physiological media. *Nanotoxicology* **2** 51-61

[40] Sayes CM, Reed KL and Warheit DB . 2007. Assessing toxicity of fine and nanoparticles: Comparing in vitro measurements to in vivo pulmonary toxicity profiles. *Toxicological Sciences* **97** 163-180

[41] Yokel, R. A., Florence, R., Tseng, M, Graham, U., Sultana, R., Butterfield, D. A., Wu, P., and Grulke, E. Biodistribution and toxicity of systemically-introduced nanoscale ceria. Nanotox 2008 Zürich , 87. 2008.

[42] Bihari P, Vippola M, Schultes S, Prätner M, Khandoga AG, Reichel CA, Coester C, Tuomi T, Rehberg M and Krombach F . 2008. Optimized dispersion of nanoparticles for biological in vitro and in vivo studies. *Particle and Fibre Toxicology* **5** 14

[43] Meißner T, Potthoff A and Richter V . 2009. Suspension characterization as important key to toxicological investigations. *J Physics: Conf Series* **170** 012012

[44] Lundqvist M, Sethson I and Jonsson BH . 2004. Protein Adsorption onto Silica Nanoparticles:ΓÇë Conformational Changes Depend on the Particles' Curvature and the Protein Stability. *Langmuir* **20** 10639-10647

[45] Rocker C, Potzl M, Zhang F, Parak WJ and Nienhaus GU . 2009. A quantitative fluorescence study of protein monolayer formation on colloidal nanoparticles. *Nature Nanotechnology* **4** 577-580

Nanosafe2010: International Conference on Safe Production and Use of Nanomaterials IOP Publishing
Journal of Physics: Conference Series **304** (2011) 012062 doi:10.1088/1742-6596/304/1/012062

Characterization of abrasion-induced nanoparticle release from paints into liquids and air

L Golanski[1], A Gaborieau[2], A Guiot[1], G Uzu[1], J Chatenet[2] and FTardif[1]

[1]CEA-Grenoble, Liten, 17 rue des Martyrs, F38054 Grenoble Cedex09, France

[2]Blanchon S.A., 28 rue Charles Martin, Saint Fons, 69190, France

E-mail: luana.golanski@cea.fr

Abstract. Two standard methods for the characterization of the abrasion nanoparticle release into air and liquid from coatings containing nanoparticles were developed. Details of the abrasion processes and the measurement methods are shown. Paints were formulated in an industrial facility. Standard abrasion conditions in wet environments were simulated. The size distribution of the particles abraded into liquid was analyzed by a laser granulometer: submicrometric and micrometric particles were observed, but no nanometric particles. The nanoparticles released in liquid were deposited on filters for SEM (Scanning Electron Microscopy) analysis. No free or agglomerated nanoparticles were observed by SEM: nanoparticles seem to remain embedded in the paint matrix. The same coatings were abraded in the air using another standard method. The ELPI (Electrical Low Pressure Impactor) was used to determine the number size distribution of the dust generated. Abrasion is found to produce submicrometric and micrometric particles in the air but no nanoparticles. Further characterizations by SEM confirmed that no free or agglomerated nanoparticles were emitted: nanoparticles seem to remain embedded in the paint matrix.

1. Introduction

The Nanoparticles (NP) defined as objects with three external dimensions in the nanoscale range (approximately 1 nm to 100 nm) cf. standard ISO/TS 27687, find every day new industrial applications. Nanoparticles embedded in paint coatings give them very interesting properties: eg highly resistant to bacteria, better protection against UV. But this new industry can really take off only if it takes part of a responsible development for humans and the environment. Today in the absence of reliable data on the toxicity of nanoparticles, the only way to guarantee a risk close to zero is to verify that paints containing NP do not release their NP. This implies in particular the control of the implementation of nanoparticles in consumer products throughout their life cycle: during their manufacture, use and until the end of life of these nano-products. No standard methods for the characterization of abrasion-induced nanoparticle release are currently available. Standard methods for investigation of wear resistance to abrasion in dry and wet environments are used by the paint manufacturers. However, these methods have to be adapted today for paints containing NP, in order to measure if this NP could be released by the abrasion process or not.

The Taber Abrader is used in standards methods in order to abrade a large variety of products: paint coating, metal, paper, floor coating and textiles. One method for the characterization of the abrasion induced nanoparticles release into air from surface coatings was developed using a circular

Taber abrader [1]. In this case, the total number of generated nanoparticles and submicron particles was extremely low [1]. Another method was developed using a linear Taber Abrader for the characterization of nanoparticles release from polymers coatings, fabrics and nano-products like socks [2-4]. The abrasion process was optimized to give the highest release of nanodust using standards tools [4]. To our knowledge no method for the characterization of the abrasion nanoparticle release into liquid from coatings containing nanoparticles was developed.

The objective of this work is to simulate the wear of the paints induced by domestic use and to characterize the nanoparticle abrasion-induced release into liquid and air. For this purpose two measurement methods were developed: one using an Elcometer abrader coupled to off-line colloid measurements. Second, using a circular Taber abrader coupled to appropriate aerosol measurement techniques. Results of dust release under wet and dry abrasion from paints containing TiO_2 nanoparticles are shown.

2. Materials and methods

2.1. Description of the paints

Paints containing TiO_2 nanoparticles and TiO_2 pigment of pigmentary TiO_2 were formulated and applied in an industrial facility following the conditions described by the standard NF EN ISO 11998. The biocide (Mergal 723K), dispersant (Orotan 731) and against foam Byk 024 were dispersed in water. The pulverizers were introduced by increasing order of oil intakes: nanoTiO_2, TiO_2 pigment of pigmentary TiO_2 and calcium carbonate. Four types of paint were prepared: a nanoTiO_2 paint containing calcium carbonate and nanoTiO_2, TiO_2 paint containing calcium carbonate and TiO_2 pigment of pigmentary TiO_2, paint without TiO_2 containing only calcium carbonate and nanoTiO_2 and TiO_2 paint containing nanoTiO_2 and TiO_2 pigment of pigmentary TiO_2.

A Natrosol 250 HBR gel was added in order to control the mixture viscosity. At the end, a styrene-acrylic binding agent and a coalescent were added. The paints used to perform wet abrasion tests were applied on a PVC substrate. Drying of these films was performed during 20 days. The thickness of the films was estimated around 150µm. The same paints were formulated and applied on a glass substrate in order to induce dry abrasion. Drying of these films was performed during 15 days.

2.2. Measurement method for the characterization of abrasion-induced nanoparticles from paints into liquid

2.2.1. Abrasion process. For wet abrasion an Elcometer 1720 abrader was used. The abrader takes a liquid solution from a pump and spreads it out on the coating. The abrasive buffer is linearly rubbed on the sample in order to generate a dynamic friction. The abrasion conditions applied cf. standard NF EN ISO 11 998 are summarized in Table 1.

Table 1. Standard conditions used for wet abrasion of paints.

Buffer	Abrasive action	Utilisation	Liquid solution	Speed	Cycles
Abrasive buffer	Soft	Paints	2,5 g/L of dodecylbenzenesulfonate of sodium	37 cycles/minute	200

This standard describes the resistance of a paint to abrasion and leaching through time. Usually, the abrasion resistance is determined by the mass loss obtained by weighing the sample before and after the abrasion process. In this work other aspects were investigated: the solutions emitted were collected and analyzed.

2.2.2. Measurement method. One instrument was used for measuring the number size distribution of the abraded particles in the liquid solution: a laser granulometer. It uses a laser beam passing through a sample of particles in suspension in liquids and collects light intensity at different angles. Mie light scattering theory calculations and standard mathematical methods for solving the inverse problem are applied to the intensity data to generate a distribution of particle sizes that is consistent with the observed light intensities at the observed angles. The Mastersizer allows the determination of particle number distribution in the range of 0.01 µm to 1000 µm. These granulometers give no information about the type of nanoparticle measured. To address this issue the liquid was deposited on a substrate for SEM analysis. In this way the nanoparticles released were analyzed in order to determine their type: NP in agglomerate or free form or moreover NP embedded in the paint matrix.

2.3. Measurement method for the characterization of abrasion-induced nanoparticles from paints into air

2.3.1. Abrasion process. For abrasion of surface paints in the air the Taber Abrader (type 5131, Taber Industries, USA) was used. This test reproduces the aging effect induced by a mechanical friction. As indicated on Table 2, three different tools were used simulating different mechanical actions intensities. S42 tool simulates sanding process of a surface and CS10, CS17 tools simulate abrasion process. The abrasion conditions conform standard NF EN ISO 7784-1 and NF EN ISO 7784-2 are summarized in Table 2. The composition of the wheels is shown as well (see Table 2).

Table 2. Standard conditions used for the dry abrasion of paints.

Type	Composition	Abrasive action	Weight	Number of cycles	Type
Wheel CS10	Rubber and aluminium oxide particles	Soft -	1 kg	1000	Wheel CS10
Wheel CS17	Rubber and aluminium oxide particles	Medium ++	1 kg	1000	Wheel CS17
Wheel S42	Abrasive sticker paper: 12,7mmx165mm, 180ANSI-CAMI	Hard +++	500 g	200	Wheel S42

2.3.2. Measurement method. All manipulations are performed in a sealed glove box equipped with a HEPA filter in order to limit the initial presence of particles in the atmosphere. The background noise in the sealed glove box is quantified to less than 20 part./cm^3. These special conditions are obtained by creating a vacuum at the top of the box, that will suck the air rich in particles at a rate of 150 L/min and a clean air is obtained using an absolute filter placed in the bottom of the glove box. Nanomaterials are mechanically solicited by abrasion using a standardized circular Taber equipment. Two wheels are rubbed on the sample in order to generate a dynamic friction. The friction is circular and cyclical on the surface of the sample. The particles released are sucked by a sampling line at a rate of 10.6 L/min into the ELPI. This instrument is used for measuring the number size distribution in the range of 7 nm to 10 µm. This device is capable of achieving real-time measurements and displays, for each size class, changes in concentration over time. It also allows capturing released particles on filters. These filters (hydrophilic polycarbonate membrane) are then analyzed by SEM to observe the morphology and size of the collected particles. Thus, it is possible to measure the size distribution and to perform physical-chemical characterization of the collected particles. These characterizations were performed in order to determine if nanoparticles are released by abrasion or sanding in free or agglomerated form or embedded in paint matrix.

3. Results

3.1. Size distribution of wear paint particles into liquid

The size distribution of wear particles measured by laser granulometer for nanoTiO_2 and TiO_2 paint, for nanoTiO_2 and for the paint without TiO_2 is showed on Figure 1. The *x* axis show the size, while the *y* axis shows the relative intensity of the scattered light converted, using Mie theory, to a volume distribution.

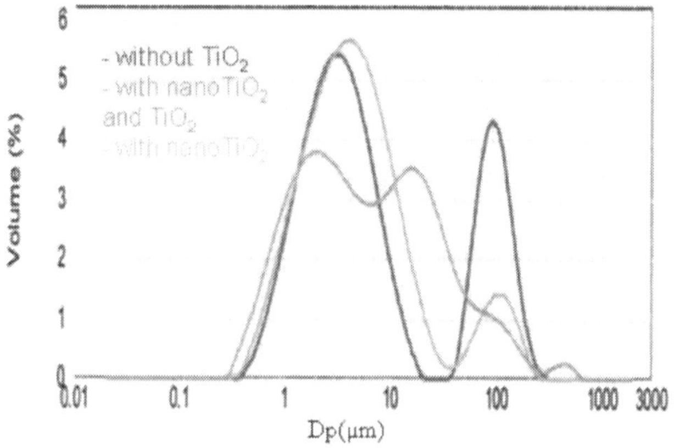

Figure 1. Size distribution of wear particles for nanoTiO_2 paint, nanoTiO_2 and TiO_2 paint and the paint without TiO_2 given by the laser Granulometer.

Particle size distribution between 0.3 µm and 700 µm are observed for the three paints. Submicrometric and micrometric particles were released. No nanoparticles released from the paints are observed whatever the composition.

3.2. Characterization of particles released into liquids

SEM analysis by retrodiffusion were performed on paint with nanoTiO_2 and TiO_2 (see Figures 2, 3a and 3b).

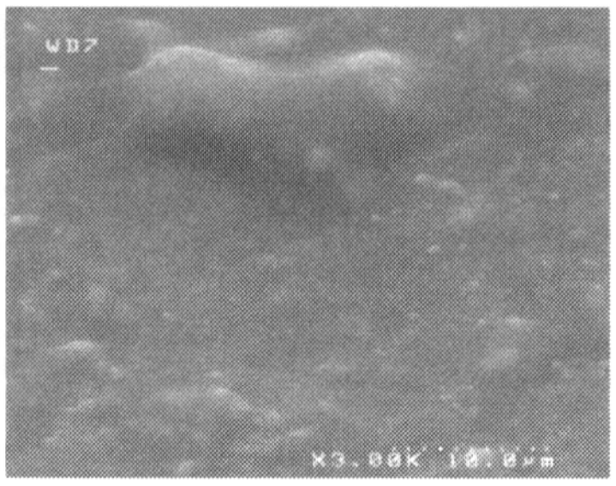

Figure 2. SEM image of the paint with TiO_2 and nanoTiO_2.

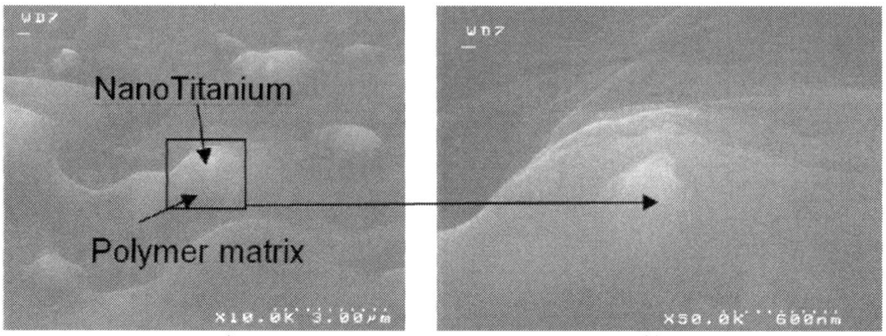

a) b)

Figure3. SEM image of a TiO_2 nanoparticle embedded in a polymer a) with a zoom X 10 and b) with a zoom X 50.

Bright TiO_2 particles with size around 200-300 nm and nanoTiO_2 particles with size around few nm were observed (Figures 2, 3a, 3b). The sizes of these particles correspond to the indications given by the supplier. Ti atoms have an atomic mass higher than carbon (contained in the polymer matrix) and thus emit more retrodiffused electrons. Therefore, TiO_2 particles and nanoparticles appear brighter on the SEM images. As represented on Figures 3a and 3b nanoparticles are embedded in the polymer matrix therefore no free or agglomerated nanotitanium particles were observed. Bigger particles with size around 20 µm could be observed as well which corresponds to $CaCO_3$ particles size given by the supplier. These particles are embedded in the polymer matrix as well (see Figure 2). SEM characterizations confirm that under wet abrasion, submicrometric and micrometric particles were released, but no free nanotitanium particles.

3.3. Paints dust emissions

The size distributions of wear particles for paint with nanoTiO_2 by comparison with the size distribution of the background of the glove box, given by the ELPI are showed on Figures 4 and 5.

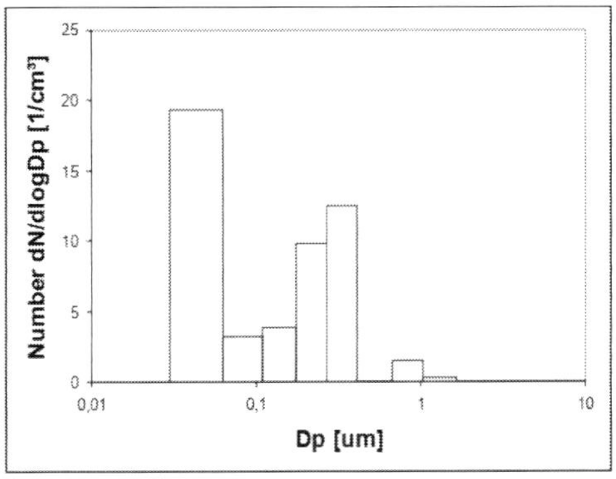

Figure 4. Size distribution of the background of the glove box.

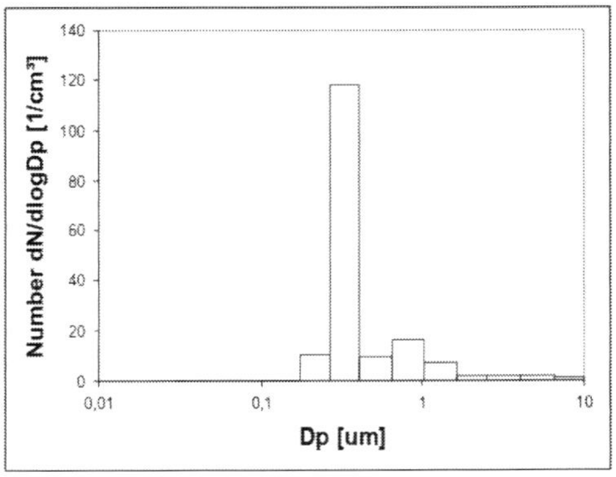

Figure 5. Size distribution from paint with nanoTiO_2 abraded using S42 wheels during 200 cycles.

The background noise is quantified less than 20 part./cm^3 (Figure 4). Low particles release was detected in the submicronic range (~ 0.3 µm) increasing from 10 particles/cm^3 to 120 particles/cm^3 during the abrasion process (Figure 5). No nanoparticles were emitted. This result cannot be explained by a low quality of the aspiration system.

Indeed when the glass substrate was abraded important release of particles coming from the substrate was measured (Figure 6).

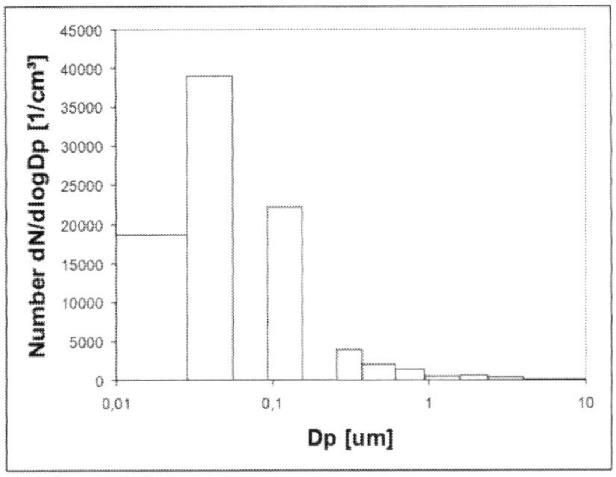

Figure 6. Size distribution from the substrate abraded using S42 wheels during 200 cycles.

Significant release: 40 000 part/cm^3 was measured on channel 2 which attests the high quality of the aspiration system.

Dust emissions from nanoTiO_2 and TiO_2 paint and the paint without TiO_2 in the air using different tools were measured using the ELPI granulometer. The total concentration released and the sizes obtained are summarized in Table 3.

Table 3. Dust emissions from nanoTiO_2 and TiO_2 paint and the paint without TiO_2 in the air using different tools

	Tools =>	**CS10**	**S42**
Paint without TiO_2	Nanometric	-	-
	Submicronic	-	-
	Micronic	-	1,021μm < D_p < 1,655μm 2,52μm < D_p < 10μm Comparable to the background
Paint with TiO_2+nanoTiO_2	Nanometric	-	-
	Submicronic	0,26μm < D_p < 0,381μm C= 55 particules/cm^3 0,611μm < Ø < 0,946μm C≤ 25 particules/cm^3	0,109μm < D_p < 0,173μm C=170 particules/cm^3
	Micronic	3,99μm < D_p < 10μm Comparable to the background	1μm < D_p < 10μm Comparable to the background

Under dry abrasion, submicrometric and micrometric particles were released. No nanometric particles were emitted. Very low release of particles was detected: maximum 170 particles/cm^3 were measured using tool S42 around 150 nm and not in the nanometric range. Negligible concentrations of particles were emitted using CS17 tool therefore results are not shown in the Table 3. In conclusion, very few particles were released whatever the paint and whatever the tool.

3.4. Characterization of particles released into the air

Paints abraded using S42 wheels were observed by SEM (see Figures 7a and 7b). These wear particles were collected on the second filter stage 3 of the ELPI.

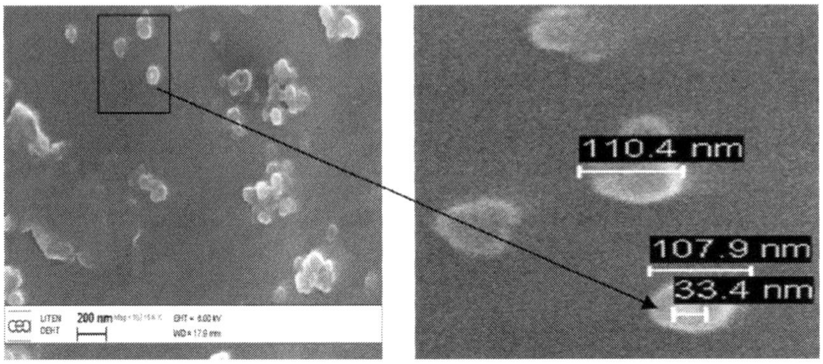

Figure7a. SEM observation of detected abraded particles on ELPI filter.

Figure7b. Zoom on abraded particles.

No free or agglomerated TiO_2 nanoparticles were observed: TiO_2 nanoparticles (~ 30 nm) seem to remain embedded in the paint matrix (see on fig.7b) which is a zoom image on abraded nanoparticles). Agglomerates of matrix particles having sizes around 110 nm are observed. The sizes vary between 100 nm and 300 nm.

4. Conclusions

Two methods for characterizing abrasion-induced nanoparticle release from paints into liquids and air were developed.

Under wet abrasion, submicrometric and micrometric particles were released, but no nanometric particles.

Tested paints present high quality in terms of degradability: under dry abrasion using different tools very low release of micronic and submicronic particles was detected while no nanometric particles were detected.

SEM images showed no free nor agglomerated nanoparticles released during abrasion in the air and liquid: nanoparticles seem to remain embedded in the paint matrix.

Acknowledgments

The authors thank the French Paint Industry Association FIPEC for the participation to the paint formulation.

References

[1] Vorbau M, Hillemann L and Stintz M 2009 *J. Aerosol Science* Vol **40** Issue 3 P 209
[2] Guiot A, Golanski L and Tardif F 2009 *Journal of Physics: Conference Series* **170** no.012014
[3] Tardif F, Guiot A and Golanski L 2009 *Proceedings of the Nanotech Conference* Boston USA
[4] Golanski L, Guiot A and Tardif F 2010 *Proceedings of the Nanotech Conference* Anaheim USA

Fate of graphene in polymer nanocomposite exposed to UV radiation

C Bernard, T Nguyen, B Pellegrin, R D Holbrook, M Zhao and J Chin

National Institute of Standards and Technology, Gaithersburg, MD 20899, USA

E-mail: tinh.nguyen@nist.gov

Abstract. Graphene is increasingly incorporated in polymers to enhance their mechanical, thermal and electrical properties. However, organic polymers are prone to degrade when exposed to UV radiation. Therefore, graphene in polymer nanocomposites could eventually be released into the environment during their life cycle, which might have a negative impact on the environment and thus presents a roadblock to their use. This study investigates the degradation of a graphene/polyurethane composite and characterizes the graphene concentration at the nanocomposite surfaces during exposure to UV radiation. The polyurethane was a one-component, water-borne polyurethane and graphene material was graphene oxide (GO) sheets. GO/WBPU composites having a thickness between 105 µm and 150 µm were exposed to 75% RH, 50°C, and UV radiation between 290 nm and 400 nm in a NIST-developed UV chamber. Chemical degradation, mass loss, and surface morphology were measured at specified exposure time using FTIR, gravimetry, SEM, AFM and LCSM techniques. Results showed that, when exposed to UV radiation having wavelengths similar to those of the sunlight, the polyurethane matrix underwent photodegradation, subsequent mass loss and accumulation of a large amount of graphene on the composite surface.

1. Introduction

The exceptional properties (e.g., mechanical, barrier, thermal) of polymer nanocomposites [1,2] make them excellent materials to be used in a wide variety of applications. Graphene – one atom thick, sheets of sp^2-bonded carbon – has emerged as a new class of nanomaterial. Interest in graphene is exploding because of its unique physical properties and its potential application to nanoelectronic systems. Our previous study showed that graphene greatly enhances the mechanical, oxygen barrier, and fire resistance properties of polyurethane polymer [3].Whatever the application, long-term performance of the nanocomposite itself and the fate of the nanoparticles in the polymer matrix have an important role in the use of these products. However, the polymer matrix is susceptible to degradation during outdoor exposure, which potentially leads to nanoparticle release into the environment. Since nanoparticles have shown potential risks to human health and environment [4,5], their release during the life cycle of polymer nanocomposites could restrain commercialization of polymer nanocomposites. Currently, little information is available about the fate of graphene nanoparticles or their potential health and environment risks during the graphene/polymer composite life cycle. Such information is sorely needed for assessing the potential risks of this new class of nanomaterial. This study investigates the fate of graphene in a graphene/water-borne polyurethane nanocomposite during its exposure to ultraviolet (UV) radiation.

2. Experimental Procedures

2.1. Materials and Nanocomposite Preparation

The polyurethane was a one-component, water-borne polyurethane (WBPU), an anionic dispersion of an aliphatic polyester urethane resin in water (50%, by mass) and in n-methyl-2-pyrrolidone (NMP) (15%, by mass) (Bayhydrol 110, Bayer MaterialScience LLC). The chemical structure and other components, such as surfactants and additives, in this material are not known. The graphene material, graphene oxide (GO), used in this study was prepared at the University of Texas, Austin, using a process described elsewhere [6]. Graphene sheets (i.e., graphene oxide sheets) produced by this process have a thickness of approximately 2 nm and a width of approximately 2 μm. They contain a variety of oxygen groups, such as hydroxyl, epoxide, and carboxylic acids, on the basal planes and at the sheet edges, which can form strong hydrogen and covalent bonds with a variety of polymer matrices. This material disperses easily in water-based polymers.

PU nanocomposites containing mass fractions of 0% (i.e., unfilled) and 2% of GO were prepared following the process illustrated in Figure 1 (because 60% of the mass of the as-received GO sample was water, the amounts added to the composites were adjusted accordingly to provide the desired loadings). The amount of GO was progressively added to a water-containing flask to provide a concentration of approximately 6 mg GO in 1 ml water. After sonication (at 40 kHz) for 8 h in an ultra-sonic bath, the water-dispersed GO was placed in a beaker containing an appropriate amount of water-borne PU. The PU/GO mixture in water and NMP was mixed in a high shear mixer at 126 rad/s (1200 rpm) for 1 h 30 min followed by degassing in vacuum for 1 h and storing at room temperature for 12 h. It is expected that the GO was mostly exfoliated into graphene oxide sheets at this stage. After undergoing further high-shear mixing for 45 min followed by degassing for 1 h 30 min, the viscous GO/WBPU mixture was applied onto a polyethylene terephthalate (Mylar) sheet, followed by film spreading using a drawdown blade. All samples were cured for 12 h at room temperature, followed by 10 min at 150°C in an air-circulating oven. The dry film thickness was between 105 μm and 150 μm. In addition, specimens of unfilled and GO-filled coated films on CaF_2 substrate having a thickness of approximately 6 μm were also prepared for chemical degradation study.

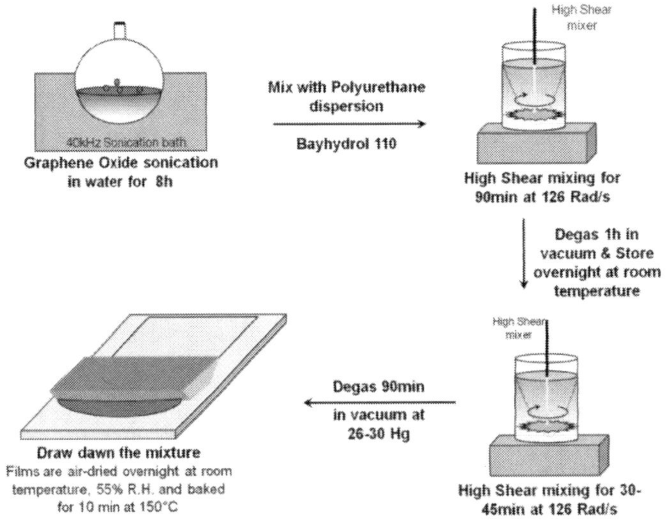

Figure 1. Steps and conditions used to prepare GO/WBPU composite.

2.2. Ultraviolet (UV) Exposure Conditions.

The UV radiation source was a 2 m integrating sphere-based environmental chamber, referred to as SPHERE (Simulated Photodegradation via High Energy Radiant Exposure) [7] This SPHERE UV chamber utilizes a mercury arc lamp system that produces a collimated and highly uniform UV flux of approximately 480 W/m^2 in the 295 nm to 400 nm range. Within the specimen chambers, the relative humidity (RH) and temperature can be precisely controlled. Free standing films were cut into 25 mm square specimens, and mounted on a 17-window exposure cell, normal to the horizontal direction. All samples were exposed to UV radiation at 50°C and 75% RH up to 137 days. Specimens were removed at specified time intervals for characterization.

2.3. Characterization of UV exposed samples.

Chemical degradation, surface morphology and mass loss of UV-exposed nanocomposites were characterized before and as a function of exposure time. Chemical degradation of the GO/WBPU films was studied by Fourier transform infrared spectroscopy in both the transmission mode (t-FTIR) and attenuated total reflection (ATR) mode (FTIR-ATR) using thin coated CaF_2 and thick free standing film samples, respectively. ZnSe prism was used for the ATR measurement. The FTIR spectrometer (Nicolet Nexus 670x) was equipped with a liquid nitrogen-cooled mercury cadmium telluride (MCT) detector. FTIR spectra were recorded at a resolution of 4 cm^{-1}. Surface morphology was characterized and imaged via field emission scanning electron microscopy (FE-SEM), contact scanning mode atomic force microscopy (AFM) and laser scanning confocal microscopy (LSCM). AFM was a Dimension 3100 (Veeco Metrology), operated in the tapping mode using commercial silicon probes. Mass loss was measured with an analytical balance.

3. Results and Discussion

3.1. Chemical Degradation

Chemical degradation of GO/WBPU films during UV exposure was characterized by t-FTIR and FTIR-ATR. FTIR spectra taken at different exposure times showed that the intensity of major bands at 1250, 1530 and 1730 cm^{-1}, assigned to amide III, N-H bending (amide II), and ester C=O, respectively, decreased with increasing exposure time. In addition, intensity in the 1765 to 1790 cm^{-1} region has increased, implying that oxidation has occurred. These results indicate that the PU has undergone photodegradation, resulting in chain scission and formation of oxidative products.

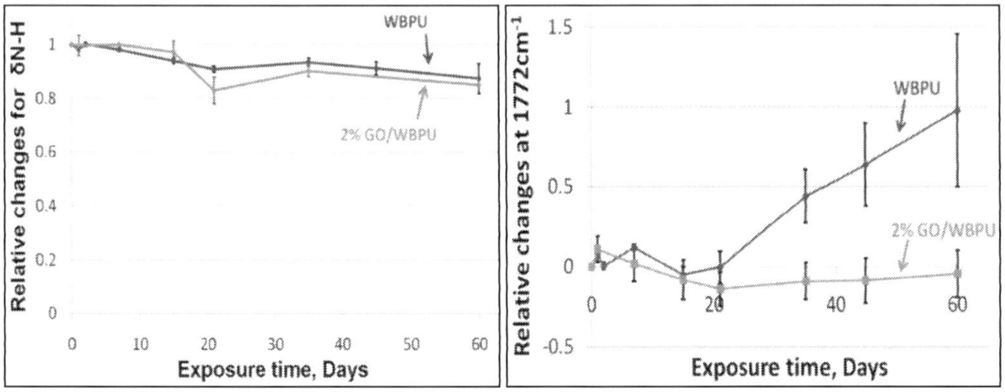

a *b*

Figure 2. Plots of t-FTIR intensity changes at (a) 1530 cm^{-1} and (b) 1772 cm^{-1} as a function of exposure time to SPHERE at 50 °C, 75 % RH for unfilled and 2% GO-filled WBPU films. Each data point was the average of four specimens, and error bars represent one standard deviation.

Figure 2a depicts t-FTIR intensity changes of the 1530 cm^{-1} band, representing chain scisison in the polymer, as a function of exposure time for 2% GO/WBPU and unfilled WBPU films. For both unfilled and GO filled film, UV exposure caused a gradual loss of this group. Figure 2b displays t-FTIR intensity changes of the 1772 cm^{-1} band, attributed to acetylurethane linkage (O=C-NH-C=O) [8]. The rate of acetylurethane formation was higher for unfilled WBPU than GO/WBPU composites, suggesting that photo-oxidation of this polymer was suppressed by the presence of GO.

The t-FTIR data did not show significant difference on the chain scission rate (i.e., loss of NH band at 1530 cm^{-1}) between unfilled and GO-filled WBPU films. However the FTIR-ATR results reveal that intensity of this band was sharply reduced after 35 days for 2% GO/WBPU samples, with the loss being much less for the unfilled WBPU after 60 days of exposure (Figure 3a). It should be noted that chemical changes detected by the ATR method originated from the composite layer near the surface, between 0.5 μm and 2.5 μm into the polymer. The 25% difference in FTIR-ATR intensity between unfilled and GO filled films is probably due to the graphene particles stacking on the surface, which reduced the penetration depth of the evanescent wave into the polymer material and consequently spectral intensity (Figure 3b).

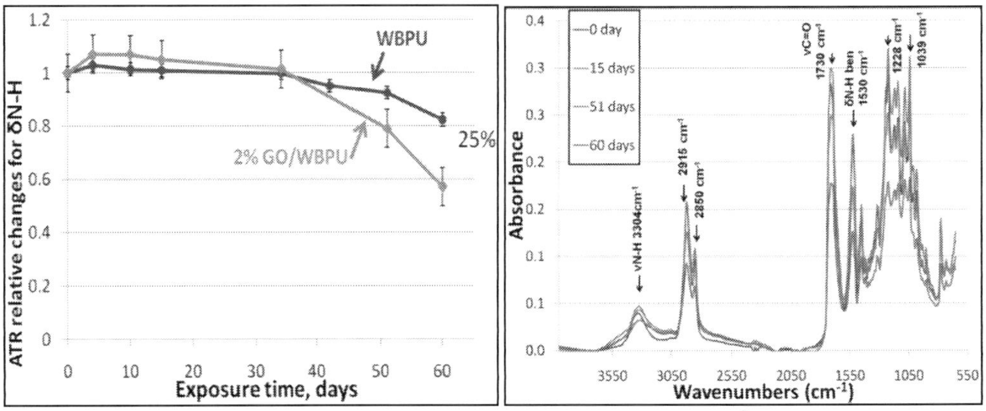

a *b*

Figure 3. (a) Plots of FTIR-ATR intensities changes at 1530cm^{-1} in function of time for unfilled and 2% GO-filled WBPU films exposed to UV/50 °C/75 % RH. (b) FTIR-ATR spectra for 2% GO/WBPU composite at different exposure time. Each data point was the average of four specimens, and error bars represent one standard deviation.

3.2. Surface Morphology and topography of UV-exposed waterborne polyurethane Nanocomposites

Figures 4 and 5 display AFM images of unfilled and 2% GO-filled polyurethane films, respectively, exposed to UV radiation at 50°C and 75% RH for different times. Both unfilled and GO-filled samples exhibited a smooth surface before UV exposure. However, different topographic features are observed between unfilled and 2% GO/WBPU films after UV exposure. Exposed unfilled WBPU film showed a surface degradation pattern with circular pits (Figure 4).

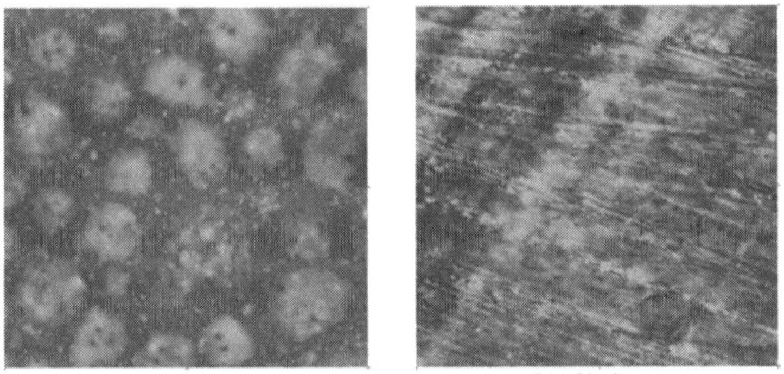

a) 0 day **b) 60 days**

Figure 4. AFM height images of unfilled WBPU films **(a)** before and **(b)** after exposure to SPHERE at 50 °C, 75 % RH for 60 days. Scan size is 50 μm × 50 μm.

For 2% GO/WBPU films exposed to the same UV condition, a visible surface morphological change is clearly observed (Figure 5). After only 15 days exposure, some GO nanoparticles started to appear and they covered most of the surface after 60 days. At this time, several graphene layers on the composite surface can be distinguished. Further, the height contrast also increased with UV exposure, from 397 nm initially to 2 μm after 137 days irradiated with UV. Particularly between 42 and 137 days, both the size and the shape of graphene particles have changed appreciably. They become larger and more protruding at longer UV exposure time. Such changes were probably due to the combined gradual removal of the polymer layer and stacking of the graphene sheets on the surface.

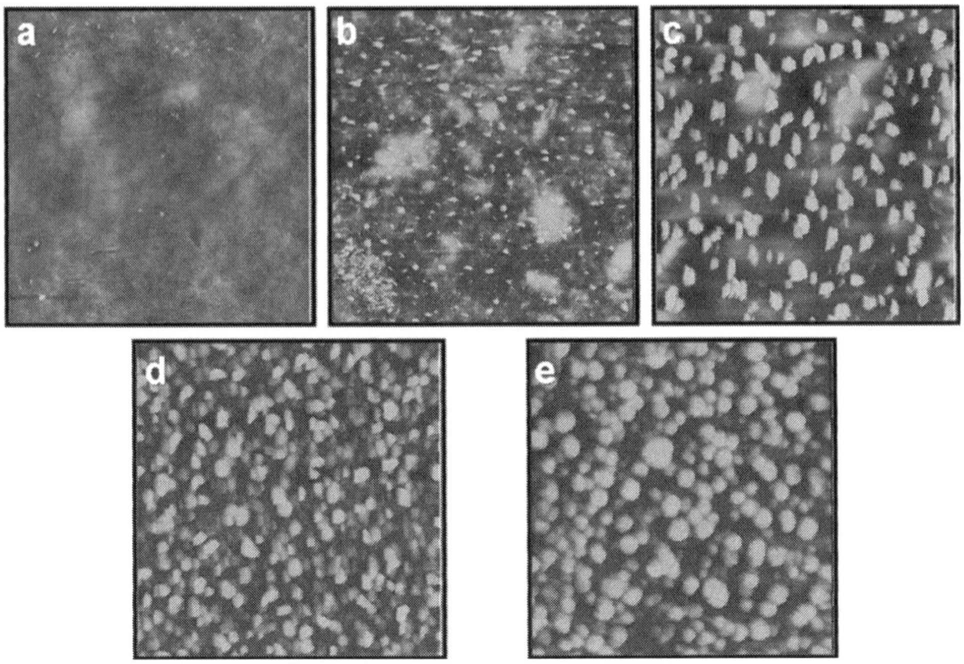

Figure 5. AFM height images of WBPU films containing 2 % mass fraction of GO nanoparticles, exposed **(a)** 0, **(b)** 15, **(c)** 42, **(d)** 60 and **(e)**137 days to SPHERE at 50°C, 75% RH. Scan size is 50 μm × 50 μm.

Roughness of the graphene composite surface also changed markedly with exposure time under UV radiation, as shown in Figure 6. Root mean square (RMS) value, which represents the average height of the scanned surface area, was taken directly from the instrument software. The RMS values were all measured from 50 µm square images. The roughness value for the surface after 137 days exposure was close to 375 nm, which is 40 times greater than before exposure, indicating that UV radiation has caused a substnatial damage to the surface. The rate of increase was different between the first 60 days and the last 77 days of exposure. The roughness increased nearly three times faster in the first 60 days exposure than thereafter. The more rapid roughness increase at the early stage may be explained due to the degradation of the polymer layer on the surface, which allowed more graphene particles to appear at the surface and thus made the roughness increase more rapidly. In the later stage, although the polymer continued to degrade but the graphene stacking process became more dominant, and therefore the roughness was increased at a slower rate.

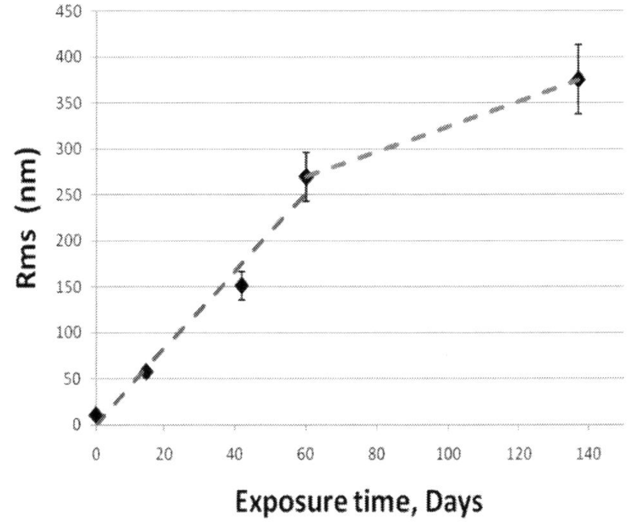

Figure 6. Roughness value obtained for 2% GO/WBPU samples during SPHERE exposure at 50°C, 75% RH. Roughness was measured from 50 µm square images.

Changes in surface morphology of the GO/WBPU composite resulting from UV irradiation can also be observed by SEM imaging as shown in Figure 7. Before exposure, the surface contained a limited amount of nanoparticles. The density of nanoparticles on the surface increased with exposure time. After 60 days exposure, almost the entire surface appeared to be covered with particles (Figure 7b). After 137 days exposure, graphene particles are clearly seen agglomerating. This result is consistent with the increase in size and height of particles observed with AFM imaging.

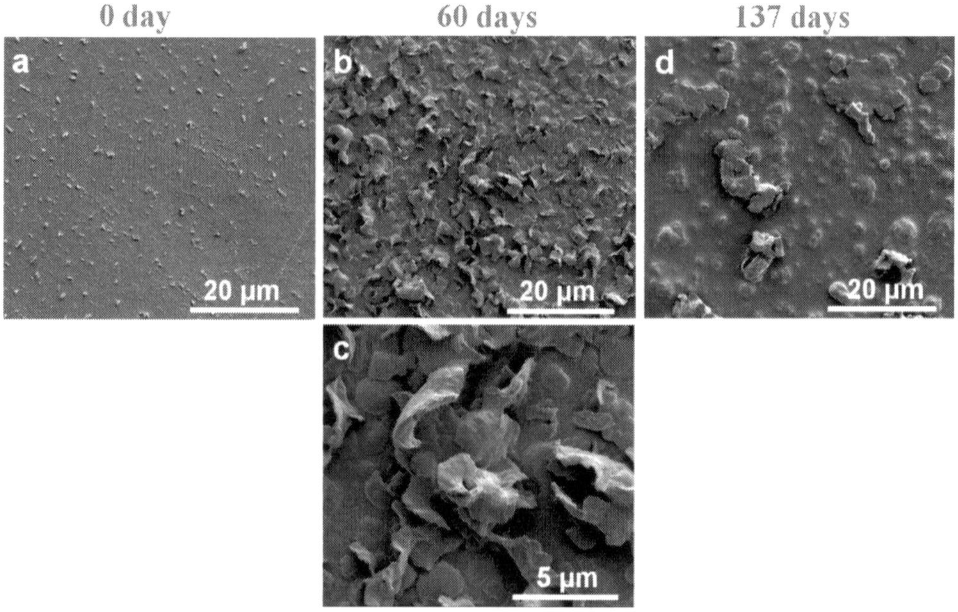

Figure 7. FE-SEM images of 2% GO/WBPU composite exposed to UV radiation for various times: **(a)** before exposure; **(b)** 60 days; **(c)** 60 days at higher magnification and **(d)** 137 days.

Figure 8 shows LSCM images of GO/WBPU composite surface at different magnifications before and after 60 days exposure. Images in Figure 8 show a smooth surface before exposure and a black surface was completely covered with graphene after 60 days. At higher magnification, we can clearly see uniform graphene dispersion on the composite surface.

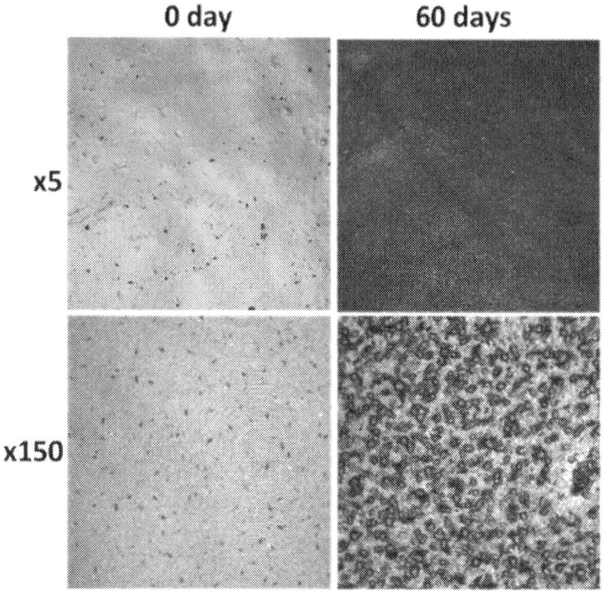

Figure 8. LSCM images of 2% GO-filled WBPU films before and after 60 days. First row are micrographs of 2 x 2 mm (x5) and second row are micrographs of 56 x 56 µm. (x150).

The results of Figures 5, 7 and 8 showed that a substantial amount of graphene nanoparticles has accumulated on the GO/WBPU composite surface as a result of exposure to UV radiation. However, further research is needed to ascertain whether any of these graphene has released to the environment during exposure.

3.3. Mass Loss

Figure 9 displays the mass loss of unfilled and 2% GO-filled WBPU versus exposure time exposed to UV/50°C/75% RH environment. The rate of mass loss of the 2% GO/WBPU composite appeared to be lower than that of the unfilled film at early exposure stage (<25 days), but both had similar mass loss thereafter. After 42 days exposure, the maximum mass losses of unfilled and 2% GO/WBPU nanocomposite films were 1.53% ± 0.22% and 1.50% ± 0.19%, respectively. The reason for the difference in mass loss between GO-filled and unfilled WBPU during the early exposure stage is unknown at this time.

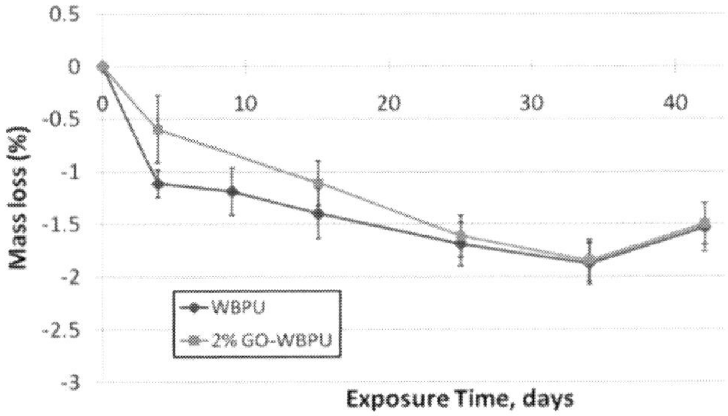

Figure 9. Mass loss vs. time for unfilled and 2% GO-filled WBPU films exposed to UV/50°C/75% RH.

4. Conclusion

The fate of GO nanoparticles in waterborne polyurethane/GO nanocomposites during exposure to UV radiation has been investigated. Analyses of chemical composition and morphology of the UV-exposed composite surfaces showed that the polyurethane matrix underwent photodegradation, resulting in accumulation of a large concentration of GO particles on the composite surface. The results of this study should provide useful information to assess the potential risk of GO in polyurethane nanocomposites during outdoor exposure.

** Certain commercial product or equipment is described in this paper in order to specify adequately the experimental procedure. In no case does such identification imply recommendation or endorsement by the National Institute of Standards and Technology, nor does it imply that it is necessarily the best available for the purpose.*

References

[1] Hussain F, Hojjati M, Okamoto M, Gorga R E 2006 *J. Composite Materials* **40** 1511
[2] Ajayan P M and Tour J M 2007 *Nature* **447** 1066
[3] Bernard C, Nguyen T, Pellegrin B, Celina M, Shapiro A J, Stanley D, Tan K T, Park S, Ruoff R, and Chin J 2010 *Adhesion Society Meeting*
[4] Helland A, Wick P, Koehler A, Schmid K, Som C 2007. Environ Health Perspect **115** 1125-1131
[5] Nel A E, et al. 2009 *Nature Materials* **8** 543 - 557
[6] Park S, et al. 2008 *Chem. Mat.* **20** 6592–6594
[7] Chin J, Byrd E, Embree N, Garver J, Dickens B, Finn T, and Martin J W 2004 *Review of Scientific Instruments* **75** 4951
[8] Nguyen T, Martin J W, Byrd E and Embree E J 2002 *Proceedings of the FSCT*

Accurate measurement of nanoparticle charge, number and size with the ELPI+™ instrument

Erkki Lamminen

Osuusmyllynkatu 13, 33700 Tampere, Finland

E-mail: erkki.lamminen@dekati.fi

Abstract. Nanoparticle characterization is mainly carried out by microscopy techniques and measurements of size and concentration. However to be able to comprehensively understand the behavior of the aerosol, information on particle charge level is critical. Particle charge has a major effect on the coagulation, deposition and transport as well as on the health effects of the particles. While aged aerosols generally are mainly neutral in terms of charge, freshly generated or resuspended particles can have high charge levels that vary depending on the size, material and generation method of the particles. Charging processes for particles are problematic as they are usually both time dependent and material dependent with sudden changes in magnitude. This makes the charge of a particle nearly impossible to predict without detailed studies and direct measurements. ELPI+™ instrument is a completely new aerosol measurement instrument based on the widely used Electrical Low Pressure Impactor (ELPI) technique. ELPI+™ measures particle concentration and size in a wide size range from 6nm to 10μm in real-time. The operation is based on first charging the particles in a corona charger and then size segregating them in a cascade impactor where all impactor stages are electrically insulated. With the ELPI+™ instrument it is possible to measure not only the particle size and concentration in real-time, but also the charge distribution of the particles. As the particle size classification is made with an impactor, the measured particles can be subsequently analyzed if needed with a suitable technique.

1. Introduction

Effective use of nanoparticles in industry benefits from thorough understanding of the processes that affect the particle transformation, transport and deposition while in aerosol form. These processes are governed mainly by particle size, concentration and shape. Freshly generated nanoparticles, especially in the case of high temperature processes or burst generation, may carry an electrical charge. This charge may have a significant effect on the behavior of the formed aerosol.

Charging processes during particle generation are sensitive to particle material, temperature, humidity. In the case of burst generation processes there may be build-up and release effects that have a significant effect on the generated particle charge level. The magnitude and/or polarity of the charge is generally very difficult to predict even if the initial conditions before are well known. Therefore in order to be able to estimate the effect of the charge it is beneficial to be able to make a direct measurement of the charge.

Electrical Low Pressure Impactor+ (ELPI+™, Dekati Ltd.) measures particle size and concentration with a dynamic range. In addition ELPI+™ is capable to measure the charge distribution in real-time and to carry out charge per number or charge per mass (Q/N, Q/M) measurement automatically.

In this work we first discuss the effect of charge on nanoparticle behavior and then present the features and specifications of the ELPI+™ instrument.

2. Effect of charge on nanoparticles

This chapter describes the effects of charge on nanoparticle behavior.

2.1. Coagulation

When two aerosol particles come into contact with each other, generally the particles become attached. This process thus reduces particle concentration, while increasing particle size. Mass concentration is not affected by this process and while the total surface area of the population decreases slightly, surface area of single particles increases significantly. Coagulation occurs in all aerosol populations and is highly dependent on aerosol concentration and on the width of the aerosol size distribution. High aerosol concentration and a large spread in particle sizes enhance coagulation.

If an aerosol population is charged it has a significant effect on the coagulation process [1]. In the case of both positive and negative particles present in an aerosol, coagulation is enhanced [2] and resulting particle size distribution is narrower than in the case of neutral aerosol. If the aerosol contains mainly particles charged by single polarity, coagulation is slowed down.

Depending on the charge state of the aerosol, coagulation may either be enhanced or slowed down. This will cause a generated aerosol to have a different number concentration, size and surface properties depending on the charge attained during the generation or dispersion process.

2.2. Space charge effect

A cloud of generated particles expands due to mixing with surrounding gas and diffusion. If the particles are charged mainly with one polarity, the cloud of particles will expand more rapidly. This effect therefore reduces the concentration of particles in a specific volume compared to uncharged particles.

Space charge may have a significant effect also on losses, if a high concentration of charged particles is transported via a transfer line. Space charge effect will cause the particles to move towards the sampling line walls due to expansion of the cloud, hence there is a high chance of deposition on the walls. If the charge state of the generated aerosol changes and the aerosol is transferred through a line, the transfer efficiency will also change.

2.3. Deposition

Charged particles move in an electric field according to their charge level and according to field strength. This electrostatic force is usually significant compared to other forces acting upon particles in an aerosol.

Electrostatic force is especially important when charged particles are transferred in lines. The transfer lines must be constructed of conductive material. Insulated lines can easily become statically charged and essentially turn into electrostatic precipitators that remove charged particles. In addition to transfer lines, any space with an electric field will have a significant effect on the movement of charged particles.

3. Measurement of particle charge

This chapter includes the operation principle, specifications and features of the ELPI+™ instrument and describes how charge and charge/number measurements are carried out.

3.1. ELPI+™ operation principle

The ELPI+™ operating principle can be divided into three major parts: particle charging, size classification in a cascade impactor and electrical detection with sensitive electrometers. The particles are first charged into a known charge level in the corona charger. After charging, the particles enter a cascade low pressure impactor with 14 electrically insulated collection stages. The particles are collected in the different impactor stages according to their aerodynamic diameter, and the electric charge carried by particles into each impactor stage is measured in real time by sensitive electrometers. This measured current signal is directly proportional to particle number concentration and size. The particle collection into each impactor stage is dependent on the aerodynamic size of the particles. Measured current signals are converted to particle size distribution using particle size dependent relations describing the properties of the charger and the impactor stages. The result is particle number concentration and size distribution in real-time. ELPI+™ operation principle is shown in Figure 1.

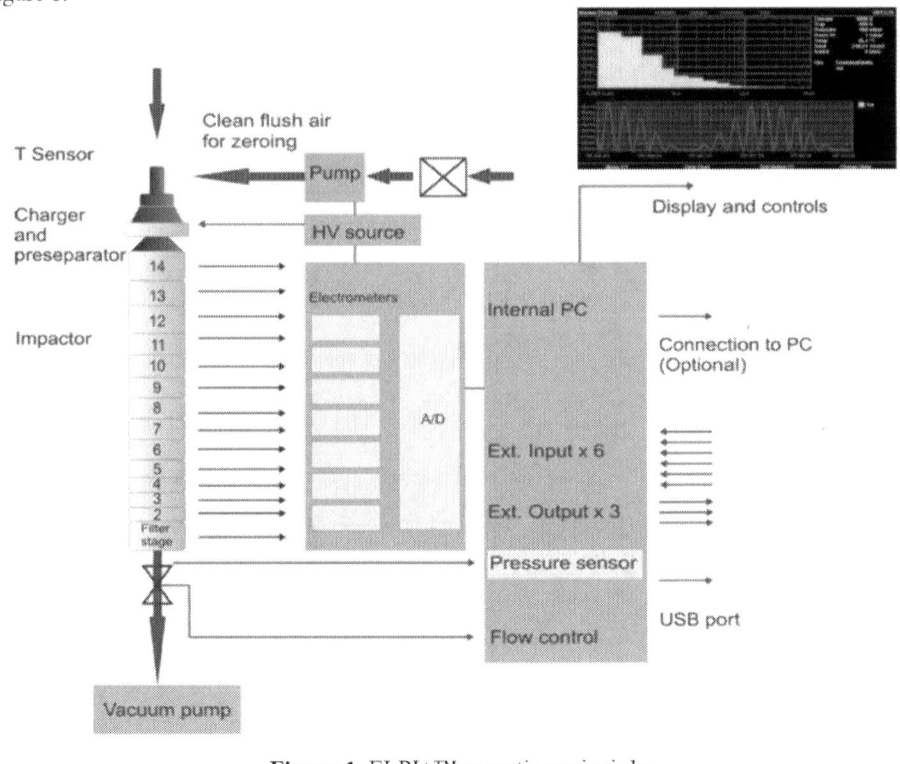

Figure 1. ELPI+™ operation principle

Journal of Physics: Conference Series **304** (2011) 012064 doi:10.1088/1742-6596/304/1/012064

3.2. ELPI+™ specifications

ELPI+™ specifications are given in Table 1.

Table 1. ELPI+™ Specifications

Measurement	
Nominal air flow	10lpm
Particle size range	0.006–10μm
Number of size channels	14
Sampling rate	10Hz
Operation conditions, instrument	
Ambient temperature	5–35°C
Ambient humidity	0–90%, non condensing
Sample conditions	
Gas temperature	< 50°C
Dimensions	
Weight	22kg
Dimensions	400 x 420 x 220 (mm)
Electrical specifications	
Electric power	100-250V, 50-60Hz, 200W
Charger voltage	3.5kV
Charger current	1μA
Computer specifications	
Optional	
Connection to ELPI+™	RS-232 serial (USB–RS-232 adapter provided with the ELPI+™ instrument), or Ethernet

Nominal impactor specifications for ELPI+™ impactor

Stage	**D50% [μm]**	**Di [μm]**	**Number min [1/cm^3]**	**Number max [1/cm^3]**	**Mass min [μg/m^3]**	**Mass max [mg/m^3]**
15	10					
14	6.800	8.4	0.10	2.4E+04	30	10000
13	4.400	5.3	0.10	2.4E+04	10	3000
12	2.500	3.2	0.15	5.4E+04	3.0	1000
11	1.600	2.0	0.3	1.1E+05	1.4	450
10	1.000	1.3	0.5	1.9E+05	0.7	210
9	0.640	0.810	1	3.5E+05	0.3	100
8	0.400	0.510	2	6.4E+05	0.1	50
7	0.260	0.330	3	1.2E+06	0.07	20
6	0.170	0.210	5	2.1E+06	0.03	10
5	0.108	0.140	10	3.7E+06	0.02	5
4	0.060	0.081	20	7.3E+07	0.005	2
3	0.030	0.042	50	1.7E+07	0.002	0.5
2	0.017	0.022	100	3.4E+07	0.001	0.25
1	0.006	0.01	250	8.3E+07	0.0004	0.13

The min and max values given in Table 1 are calculated assuming 1.5 fA and 400 000 fA current measured per stage. Please refer to Appendix A for a detailed description on the ELPI+ calculation procedure. D50% is the aerodynamic equivalent size for an impactor stage at which 50% of the particles are collected. Di is the geometric midpoint of the stage collection range.

3.3. ELPI+™ features

ELPI+™ enables measurement of real-time particle size distribution and concentration in the size range of 6nm - 10μm with 10Hz acquisition rate. ELPI+™ is based on well-known ELPI™ technology [3-5]. This allows the measurement of a wide particle size spectrum with a single device for different studies. The ELPI+™ features also include stand-alone operation, wide sample concentration range and robust structure for operation even in harsh conditions. The use of impactor technology for size classification of particles enables post-measurement chemical analysis of the collected particles.

There has been a great deal of interest in the nanotechnology community for measurement of aerosol particles via particle charging and detection of charged particles. ELPI+™ also provides a detailed measurement of this via the direct electrometer current readings with added accurate impactor size classification.

3.4. Measurement of particle charge with ELPI+™

ELPI+™ measures the particle concentration by charging particles with a unipolar charger and then size classifying the particles with an impactor into individual size bins, from which electrometers continuously measure currents. Particle charge measurements are carried out by simply turning the unipolar charger off and reading the electrometer currents. With this measurement we know the total amount of charge in each size bin in real-time. This allows detailed monitoring of the particle charge in the size range from 6 nanometers to 10 micrometers.

Charge/number and charge/mass (Q/N and Q/M) are interesting quantities which give the number of elementary charges per particle or per mass. Measurement of Q/N gives detailed information on the charging state of particles and takes into account changes both in carried charge and particle concentration. ELPI+™ measures the average number of elementary charges per number of particles or mass by switching the charger unit on and off for defined periods of time in the order of 3-5 seconds. The ELPI+™ measurement gives the charge averaged for all particles in the measurement channel and hence cannot take into account possible differences in case two particles in the measurement channel carry a different amount of elementary charges. The measured quantity will also differ from Q/N if there are both positively and negatively charged particles present in a measurement channel. In this case the ELPI+™measurement will give the resulting net charge/particle per measurement channel. During the measurement period the aerosol must remain stable in terms of concentration and size to gain reasonable results. Based on the measurements, ELPI+™ software automatically calculates Q/N and Q/M as a function of particle size. An example of measured Q/N data of particles charged with a commercial diode charger is presented in Figure 2.

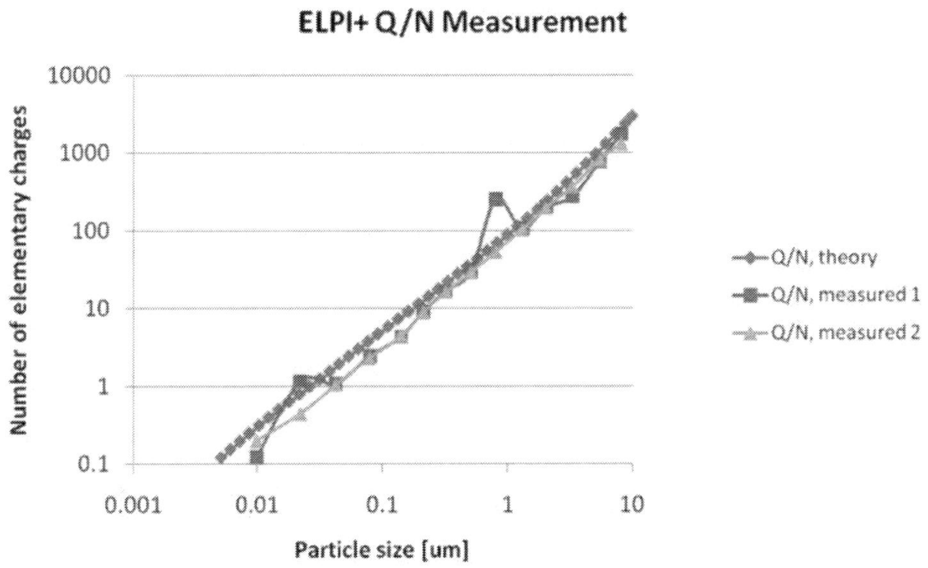

Figure 2. ELPI+™ Q/N measurement result as a function of aerodynamic diameter

4. Conclusions.

Electrical charge can have a major effect on nanoparticle behaviour and may influence particle number, size and surface area. Due to the high dependence of charge on materials and conditions, it is necessary to measure the charge level to be able to take the effect into account.

ELPI+™ instrument is capable of measuring particle number concentration, size distribution and particle charge in real time, which makes it an ideal tool for nanoparticle research including charge studies.

References

[1] Shahub, A. M. and M. M. R. Williams. Brownian Collision Efficiency. *J. Physics: D, Applied Physics*, **21**:231 (1988)

[2] Friedlander, S. K. Smoke, Dust, and Haze, *2nd Ed. Oxford University Press* (2000)

[3] Keskinen, J., Pietarinen, K. and Lehtimäki, M. (1992) Electrical Low Pressure Impactor, *J. Aerosol Sci.* **23**, 353-360

[4] Marjamäki, M., Keskinen, J., Chen, D-R. and Pui, D. Y. H. (2000) Performance Evaluation of the Electrical Low-Pressure Impactor (ELPI), *Journal of Aerosol Science* **31**:2, pp. 249-261

[5] Yli-Ojanperä, J., Kannosto, J., Marjamäki, M. & Keskinen. J. 2010. Improving the Nanoparticle Resolution of the ELPI. *Aerosol and Air Quality Research*, vol **10**, pp. 360-366

Journal of Physics: Conference Series **304** (2011) 012064 doi:10.1088/1742-6596/304/1/012064

Appendix A

This section contains an example of ELPI+™ calculation from current values to selected distribution. Note that this calculation does not consider small particle correction. Only the calculation from current values to aerosol size distribution is presented.

Few things have to be known before it is possible to calculate selected distribution from current values:

Impactor:	Aerodynamic D50% values (see impactor data sheet)
	Flowrate (see impactor data sheet)
Charger:	Efficiency curve (PneQ) (see ELPI+™ data sheet)
Particles:	Density
	Is Aerodynamic or Stokes diameter wanted
ELPI+™:	Current values
	Dilution factor

Used symbols:

Dp_a	Particle aerodynamic diameter
Dp_s	Particle stokes diameter
Di_a	Geometric mean of a channel (Aerodynamic diameter)
Di_s	Geometric mean of a channel (Stokes diameter)
Cc_a	Cunningham's slip correction vector for aerodynamic particle size
Cc_s	Cunningham's slip correction vector for stokes particle size

If we assume that the following startup values are used:

Impactor properties		1	2	3	4	5	6	7	8	9	10	11	12	13	14	15
Aerodynamic	Dp_a [μm]	0.006	0.017	0.03	0.06	0.108	0.17	0.26	0.4	0.64	1	1.6	2.5	4.4	6.8	10

Charger efficiency function					
Mult1	5.941	Mult2	1.819	Mult3	1.819
Exp1	1.637	Exp2	1.3201	Exp3	1.3201
Limit1	0.0239	Limit2	10		

Impactor type	
Impactor	10045
Flowrate	10

Density	1
Dilution	1

Then Cunningham's slip correction factor is calculated using equation

$$C_{Ca} = 1 + \left(\frac{2}{76 * Dp_a} \right) * (6.32 + 2.01 * e^{-0.1095*76*Dp_a})$$ for each D50% value (76=pressure in cmHg)

Now we have the Cunningham's slip correction factor for each D50% value:

Impactor properties		1	2	3	4	5	6	7	8	9	10	11	12	13	14	15
Aero-dynamic	Dp_a [μm]	0.006	0.017	0.03	0.06	0.108	0.17	0.26	0.4	0.64	1	1.6	2.5	4.4	6.8	10
	Cc_a	37.1057	13.4843	7.9175	4.3070	2.7393	2.0539	1.6631	1.4205	1.2603	1.1663	1.1039	1.0665	1.0378	1.0245	1.0166

We calculate also geometric mean of each channel, which is called Di (Di_n equals Di of channel n)

$$Di_n = \sqrt{Dp_n * Dp_{n+1}}$$

Impactor properties		1	2	3	4	5	6	7	8	9	10	11	12	13	14	15
Aero-dynamic	Dp_a [μm]	0.006	0.017	0.03	0.06	0.108	0.17	0.26	0.4	0.64	1	1.6	2.5	4.4	6.8	10
	Cc_a	37.1057	13.4843	7.9175	4.3070	2.7393	2.0539	1.6631	1.4205	1.2603	1.1663	1.1039	1.0665	1.0378	1.0245	1.0166
	Di_a	0.0136	0.0226	0.0424	0.0805	0.1355	0.2102	0.3225	0.5060	0.8000	1.2649	2.0000	3.3166	5.4699		

Next, we must find the same values for particle Stokes diameters. Iteration must be used to find out the stokes diameter.

$$Dp_s = \frac{Dp_a}{\sqrt{Density * \frac{Cc_s}{Cc_a}}}$$

Where Cc_s is again

$$C_{Cs} = 1 + \left(\frac{2}{76 * Dp_s} \right) * (6.32 + 2.01 * e^{-0.1095*76*Dp_s})$$

Now we have the following values for the impactor stages:

Impactor properties		1	2	3	4	5	6	7	8	9	10	11	12	13	14	15
Aero-dynamic	Dp_a [μm]	0.006	0.017	0.03	0.06	0.108	0.17	0.26	0.4	0.64	1	1.6	2.5	4.4	6.8	10
	Cc_a	37.1057	13.4843	7.9175	4.3070	2.7393	2.0539	1.6631	1.4205	1.2603	1.1663	1.1039	1.0665	1.0378	1.0245	1.0166
	Di_a	0.0136	0.0226	0.0424	0.0805	0.1355	0.2102	0.3225	0.5060	0.8000	1.2649	2.0000	3.3166	5.4699		
Stokes	Dp_s[μm]	0.006	0.017	0.030	0.0600	0.108	0.170	0.260	0.400	0.640	1.000	1.600	2.500	4.400	6.8	10
	Cc_s	37.1057	13.4843	7.9175	4.3070	2.7393	2.0539	1.6631	1.4205	1.2603	1.1663	1.1039	1.0665	1.0378	1.0245	1.0166
	Di_s	0.0100	0.0154	0.0294	0.0574	0.0997	0.1591	0.2500	0.3994	0.6391	1.0182	1.6181	2.6926	4.4508		

At this point we have to decide if aerodynamic or stokes particle size is used. Let's assume that it is aerodynamic:

Impactor properties		1	2	3	4	5	6	7	8	9	10	11	12	13	14	15
Aero-dynamic	Dp_a [μm]	0.006	0.017	0.03	0.06	0.108	0.17	0.26	0.4	0.64	1	1.6	2.5	4.4	6.8	10
	Cc_a	37.1057	13.4843	7.9175	4.3070	2.7393	2.0539	1.6631	1.4205	1.2603	1.1663	1.1039	1.0665	1.0378	1.0245	
	Di_a	0.0136	0.0226	0.0424	0.0805	0.1355	0.2102	0.3225	0.5060	0.8000	1.2649	2.0000	3.3166	5.4699		
Stokes	Dp_s[μm]	0.006	0.017	0.030	0.0600	0.108	0.170	0.260	0.400	0.640	1.000	1.600	2.500	4.400	6.8	10
	Cc_s	37.1057	19.6398	11.2747	5.8383	3.4773	2.4547	1.8844	1.5432	1.3282	1.2073	1.1287	1.0821	1.0465	1.0300	
	Di_s	0.0100	0.0154	0.0294	0.0574	0.0997	0.1591	0.2500	0.3994	0.6391	1.0182	1.6181	2.6926	4.4508		
Di [μm]		0.0136	0.0226	0.0424	0.0805	0.1355	0.2102	0.3225	0.5060	0.8000	1.2649	2.0000	3.3166	5.4699		

Then, we calculate a dlogDp multiplier vector

$$dlogDp = \log_{10} \frac{Dp_{n+1}}{Dp_n}$$

Impactor properties		1	2	3	4	5	6	7	8	9	10	11	12	13	14	15
Aero-dynamic	Dp_a [μm]	0.006	0.017	0.03	0.06	0.108	0.17	0.26	0.4	0.64	1	1.6	2.5	4.4	6.8	10
	Cc_a	37.1057	13.4843	7.9175	4.3070	2.7393	2.0539	1.6631	1.4205	1.2603	1.1663	1.1039	1.0665	1.0378	1.0245	1.0166
	Di_a	0.0136	0.0226	0.0424	0.0805	0.1355	0.2102	0.3225	0.5060	0.8000	1.2649	2.0000	3.3166	5.4699	8.2462	
Stokes	Dp_s[μm]	0.006	0.017	0.03	0.06	0.108	0.17	0.26	0.4	0.64	1	1.6	2.5	4.4	6.8	10
	Cc_s	37.1057	13.4843	7.9175	4.3070	2.7393	2.0539	1.6631	1.4205	1.2603	1.1663	1.1039	1.0665	1.0378	1.0245	1.0166
	Di_s	0.0136	0.0226	0.0424	0.0805	0.1355	0.2102	0.3225	0.5060	0.8000	1.2649	2.0000	3.3166	5.4699	8.2462	
Di [μm]		0.0136	0.0226	0.0424	0.0805	0.1355	0.2102	0.3225	0.5060	0.8000	1.2649	2.0000	3.3166	5.4699	8.2462	
dlogDp		0.4523	0.2467	0.3010	0.2553	0.1970	0.1845	0.1871	0.2041	0.1938	0.2041	0.1938	0.2455	0.1891	0.1675	

Where Di values are either stokes or aerodynamic values, depending on the selection.

Finally, we calculate the conversion vector X from current to number. Now we need the charger efficiency function.

Charger efficiency function is a three-part power fit and it is a result of calibration values P*n and constants e and Q where:
P = Charger penetration
n = average charge per particle
e = elementary charge ($1.602*10^{-19}$ C)
Q = Calibration flow rate (10lpm)

P*n values are originated from article
Marjamäki, M., Keskinen, J., Chen, D-R. and Pui, D. Y. H. (1999) Performance Evaluation of the Electrical Low-Pressure Impactor (ELPI), Journal of Aerosol Science 31 (2) (2000) pp. 249-261.

Giving us the efficiency function:

Di < 0.095 μm:

$X(Di) = 4.48Di^{1.9087}$

0.095μm < Di < 1.196

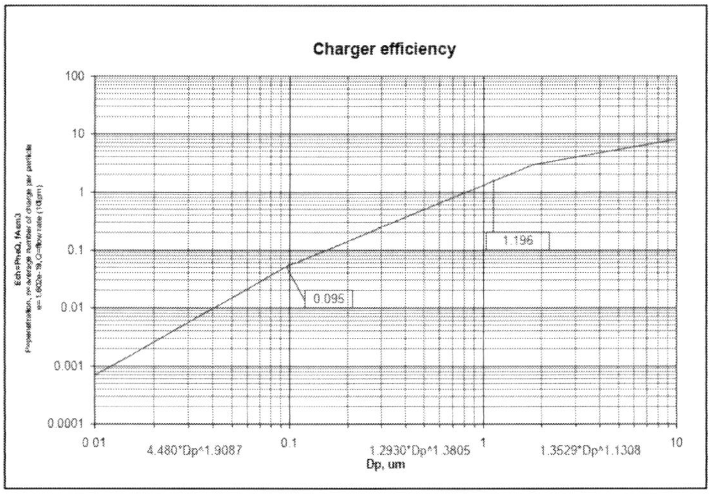

The charger efficiency curve is calibrated using flowrate of 10lpm, so we must reduce the efficiency curve to a real flow.

For example, stage 1: Di_s=0.0291 (less than 0.095), Flow rate = 9.71, Charger efficiency as above:

$$X = 4.48 * Di_s^{1.9087} * \frac{Flowrate}{Calibrationflow} = 4.48 * 0.0291^{1.9087} * \frac{9.71}{10}$$

Note that **Stokes** Di is used, because the charging process is dependent on the particles stokes diameter. That's why it is always necessary to estimate the particle density and calculate the particle stokes diameter.

Impactor properties		1	2	3	4	5	6	7	8	9	10	11	12	13	14	15
Aero-dynamic	Dp_a [μm]	0.006	0.017	0.03	0.06	0.108	0.17	0.26	0.4	0.64	1	1.6	2.5	4.4	6.8	10
	Cc_a	37.1057	13.4843	7.9175	4.3070	2.7393	2.0539	1.6631	1.4205	1.2603	1.1663	1.1039	1.0665	1.0378	1.0245	1.0166
	Di_a	0.0136	0.0226	0.0424	0.0805	0.1355	0.2102	0.3225	0.5060	0.8000	1.2649	2.0000	3.3166	5.4699	8.2462	
Stokes	Dp_s[μm]	0.006	0.017	0.03	0.06	0.108	0.17	0.26	0.4	0.64	1	1.6	2.5	4.4	6.8	10
	Cc_s	37.1057	13.4843	7.9175	4.3070	2.7393	2.0539	1.6631	1.4205	1.2603	1.1663	1.1039	1.0665	1.0378	1.0245	1.0166
	Di_s	0.0136	0.0226	0.0424	0.0805	0.1355	0.2102	0.3225	0.5060	0.8000	1.2649	2.0000	3.3166	5.4699	8.2462	
Di [μm]		0.0136	0.0226	0.0424	0.0805	0.1355	0.2102	0.3225	0.5060	0.8000	1.2649	2.0000	3.3166	5.4699	8.2462	
dlogDp		0.4523	0.2467	0.3010	0.2553	0.1970	0.1845	0.1871	0.2041	0.1938	0.2041	0.1938	0.2455	0.1891	0.1675	
X, fAcc		0.0052	0.0120	0.0281	0.0654	0.1300	0.2321	0.4083	0.7400	1.3549	2.4806	4.5417	8.8553	17.1412	29.47	

We are able to form all conversion vectors :

Number dN/dlogDp [1/cm³]	1/X*1/dlogDp*Dilution
Diameter dD/dlogDp [μm/cm³]	1/X*Di*1/dlogDp*Dilution
Area dA/dlogDp [μm²/cm³]	$1/X*Di^2 * \pi * 1/dlogDp*Dilution$
Volume dV/dlogDp [μm³/cm³]	$1/X*Di^3 * \pi *(1/6)*1/dlogDp*Dilution$
Mass dM/dlogDp [mg/m³]	$1/X*Di^3 * \pi *(1/6)*1/dlogDp*Dilution*Density*0.001$
Number N [1/cm³]	1/X*Dilution
Diameter D [μm/cm³]	1/X*Di*Dilution
Area A [μm²/cm³]	$1/X*Di^2 * \pi *Dilution$
Volume V [μm³/cm³]	$1/X*Di^3 * \pi *(1/6)*Dilution$
Mass M [mg/m³]	$1/X*Di^3 * \pi *(1/6)*Dilution*Density*0.001$

Conversion vectors	1	2	3	4	5	6	7	8	9	10	11	12	13	14
Number dN/dlogDp [1/cm³]	422.4522	337.9666	118.3627	59.9285	39.0460	23.3455	13.0898	6.6202	3.8080	1.9749	1.1360	0.4600	0.3086	0.2026
Diameter dD/dlogDp [µm/cm³]	5.7481	7.6324	5.0217	4.8241	5.2907	4.9081	4.2213	3.3496	3.0464	2.4981	2.2720	1.5255	1.6879	1.6706
Area dA/dlogDp [µm²/cm³]	0.2457	0.5415	0.6693	1.2200	2.2522	3.2417	4.2768	5.3243	7.6565	9.9271	14.2754	15.8952	29.0054	43.2800
Volume dV/dlogDp [µm³/cm³]	0.0006	0.0020	0.0047	0.0164	0.0509	0.1136	0.2299	0.4490	1.0209	2.0928	4.7585	8.7864	26.4429	59.4826
Mass dM/dlogDp [mg/m³]	0.0000	0.0000	0.0000	0.0000	0.0001	0.0001	0.0002	0.0004	0.0010	0.0021	0.0048	0.0088	0.0264	0.0595
Number N [1/cm³]	191.0741	83.3670	35.6307	15.2981	7.6930	4.3078	2.4489	1.3513	0.7381	0.4031	0.2202	0.1129	0.0583	0.0339
Diameter D [µm/cm³]	2.5998	1.8827	1.5117	1.2315	1.0424	0.9057	0.7898	0.6837	0.5905	0.5099	0.4404	0.3745	0.3191	0.2798
Area A [µm²/cm³]	0.1111	0.1336	0.2015	0.3114	0.4437	0.5982	0.8001	1.0868	1.4840	2.0263	2.7669	3.9025	5.4837	7.2490
Volume V [µm³/cm³]	0.0003	0.0005	0.0014	0.0042	0.0100	0.0210	0.0430	0.0916	0.1979	0.4272	0.9223	2.1572	4.9992	9.9628
Mass M [mg/m³]	0.0000	0.0000	0.0000	0.0000	0.0000	0.0000	0.0000	0.0001	0.0002	0.0004	0.0009	0.0022	0.0050	0.0100

Now when we multiply the current values vector with these conversion vectors we get the selected distribution. For example consider the following values:

	1	2	3	4	5	6	7	8	9	10	11	12	13	14
Di_s	0.006	0.017	0.03	0.06	0.108	0.17	0.26	0.4	0.64	1	1.6	2.5	4.4	6.8
Current values	585.69	1233.40	4571.91	12232.4	29672.2	36239.5	68540.1	80523.3	80747.2	44362.2	23538.5	8385.29	585.69	1233.40
Conversion vector from current to number distribution (dN/dlogDp) from above	422.4522	337.9666	118.3627	59.9285	39.0460	23.3455	13.0898	6.6202	3.8080	1.9749	1.1360	0.4600	0.3086	0.2026
Number distribution (dN/dlogDp)	247426	416848	541143	733069	1158580	846029	897176	533080	307485	87610	26739	3857	1404	621

And we get the number distribution.

Terms

Number distribution

Number distribution is formed by multiplying the current distribution with the conversion vector.

$$[dN] = [Ci\ ] * [(1/Xi(Di))]$$

1/dlog(Dp)

Aerosol distributions are normally given in the 1/dlog(Dp) mode. It is formed by dividing the stages measured value by the logarithmic width of the stage. In 1/dlog(Dp) mode the area of the histogram gives the value in each size range.

$$[dN/dlog(Dp)] = [dN] * [(1/dlog(Dp))]$$

Diameter distribution

Diameter distribution gives the total diameter of all particles in each size range. It is formed by multiplying the current distribution by the conversion vector and by a vector formed from the midpoint values (Di) of each stage.

$$[dD/dlog(Dp)] = [Ci\ ] * [(1/Xi(Di))] * [Di] * [(1/dlog(Dp))]$$

Area distribution

Area distribution gives the total surface area of all particles in each size range. It is formed by multiplying the current distribution by the conversion vector and by a vector formed from the surface areas of spheres having diameter equal to midpoint values (Di) of each stage.

$$[dA/dlog(Dp)] = [Ci\ ] * [(1/Xi(Di))] * [\pi * Di^2] * [(1/dlog(Dp))]$$

Volume distribution

Volume distribution gives the total volume of all particles in each size range. It is formed by multiplying the current distribution by the conversion vector and by a vector formed from the volumes of spheres having diameter equal to midpoint values (Di) of each stage.

$$[dV/dlog(Dp)] = [Ci\ ] * [(1/Xi(Di))] * [{}^1/_6 * \pi * Di^3] * [(1/dlog(Dp))]$$

Mass distribution

Mass distribution gives the total mass of all particles in each size range. It is formed by multiplying the current distribution by the conversion vector and by a vector formed from the masses of spheres having diameter equal to midpoint values (Di) of each stage. Note that mass distribution is scaled to unit mg / m^3.

$$[dV/dlog(Dp)] = [Ci\ ] * [(1/Xi(Di))] * [{}^1/_6 * \pi * Di^3 * Density] * [(1/dlog(Dp))] * 10^{-3}$$

Comprehensive Measurement of Atmospheric Aerosols with a Wide Range Aerosol Spectrometer

L. Keck, M. Pesch and H. Grimm

GRIMM Aerosol Technik GmbH & Co. KG, Dorfstrasse 9, D-83404 Ainring, Bayern, Germany

Email: lk@grimm-aerosol.com

Abstract. A wide range aerosol spectrometer (WRAS) was used for comprehensive long term measurements of aerosol size distributions. The system combines the results of an optical aerosol spectrometer with the results of a Scanning Mobility Particle Sizer (SMPS) to record essentially the full size range (5 nm – 32 µm) of atmospheric particles in 72 channels. Measurements were carried out over one year (2009) at the Global Atmospheric Watch (GAW)-Station Hohenpeißenberg, Bavaria. Total particle number concentrations obtained from the aerosol size distributions were compared to the total number concentrations measured by a Condensation Particle Counter (CPC). The comparison showed an excellent agreement of the data. The high time resolution of 5 minutes allows the combination of the measured size distributions with meteorological data and correlations to gaseous pollutants (CO, NOx and SO_2). A good correlation of particle number and CO concentrations was found for long distance transported small particles, which were probably mainly soot particles. Correlations to NOx were observed for aerosols from local sources such as traffic emissions. The formation of secondary aerosols from gaseous precursors was also observed. Episodes of relatively high concentration of particles in the range of 2-3 µm were probably caused by pollen.

1. Introduction

Comprehensive measurement of particle size distributions plays an important role in aerosol research. Examples are studies on particle source apportionment, on effects of atmospheric aerosols on human health, and on the role of particle in the climate system [1-3]. Nowadays people pay more and more attention to the airborne nanoparticles particularly because there is the suspect that these nanoparticles are, despite of their small contribution to PM Mass concentrations, relevant for the adverse health effects of atmospheric aerosol. The relatively large surface area and the higher concentration of adsorbed or condensed toxic material per unit mass [4] could contribute to the adverse effects.

Measurement of the full size range of atmospheric particles is a challenge and there is no single measurement principle, which can detect both nanoparticles and coarse particles. The most promising approach for the detection of the complete size range is the combination of *(1)* a Scanning Mobility Particle Sizer with Condensation Particle Counter (CPC) as detection system (SMPS+C) and *(2)* an optical aerosol spectrometer (OPC). The SMPS+C covers the size range of 5 – 350 nm in 44 size channels, the OPC covers the size range of 250 nm – 32 m in 31 channels. This instrument enables: *(1)* Continuous and reliable aerosol measurements over with low maintenance, *(2)* High time resolution of 5 minutes, *(3)* High dynamic range (Mass concentrations 1 µg/m³ to 5000 µg/m³; number concentrations 1 P/cm³ to 10^9 P/cm³), *(4)* Low modification of the size distribution during sampling by diffusion or impaction, agglomeration, and losses of semi volatile components (SVC), *(5)* Simultaneous measurement with meteorological sensors *(6)* Operation as a stand alone system even under rugged conditions due to stainless steel weather housing and built-in data logger.

2. Methods

The WRAS system consists of a scanning mobility particle sizer with condensation particle counter (SMPS+C) and an optical aerosol spectrometer (OPC). The GRIMM SMPS+C includes a CPC and a differential mobility analyzer (DMA). The classification occurs in the electrostatic field in the annulus between inner and outer electrodes of the DMA. A particle size distribution (electric mobility diameters) in the range of 5.5 nm – 350 nm can be obtained by changing the DMA voltage stepwise. The GRIMM OPC works on the basis of light-scattering for single particles. The scattered light is detected with a PIN-diode. The signals are amplified and classified in 31 size channels. Thus the particle size number distribution (optical diameters) in the range of 250 nm – 32 μm is measured. The two datasets from the SMPS+C and OPC are automatically synchronized and combined using the GRIMM software, there is however no fitting or adaption to smooth the transition from mobility to optical diameters. In this way the WRAS system measures particle number size distributions from 5.5 nm up to 32 μm.

The whole system is mounted in stainless steel weather protection housing with air conditioning system. The sampling probe features a Nafion dryer to reduce the humidity to ~40%, thus the particles feature a well defined diameter in the detection systems independent of the ambient humidity. Unlike heated sampling probes, this drying system causes no loss of volatile components. Meteorological sensors for wind speed, wind direction, temperature, humidity, and rain are also integrated in the weather housing.

Data were stored on a notebook in the weather casing. The newest version of the WRAS system features however also a data logger for online data transfer to the internet. All data can now be downloaded by the user from a password protected web-site.

Figure 1. Photo of the WRAS system during regular maintenance on the Hohenpeissenberg.

In this study, a GRIMM WRAS system was applied for comprehensive measurements of atmospheric aerosols at the GAW Station Hohenpeissenberg (47°48' N, 11°02' E, 985m a.s.l.), located on a hill with about 300 m elevation above a rural area in Bavaria, Germany. Measurement started on January, 13th 2009 and lasted essentially one year.

3. Results and Discussion

The typical number size distributions featured a count median diameter of 51 nm. Mass size distributions were calculated from the measured number size distributions, these mass size distribution featured a clear bimodal shape of accumulation mode and coarse mode.

To verify the results from the WRAS system, we have calculated total number concentrations from the measured size distributions and we have compared these values to the total number concentrations measured with a CPC. The two total number concentrations agreed well, the concentrations from the WRAS were however slightly lower than those from the CPC. This small difference is however partly caused by the different minimal detected particle size, being 5.5 nm for the WRAS system and 3 nm for the CPC.

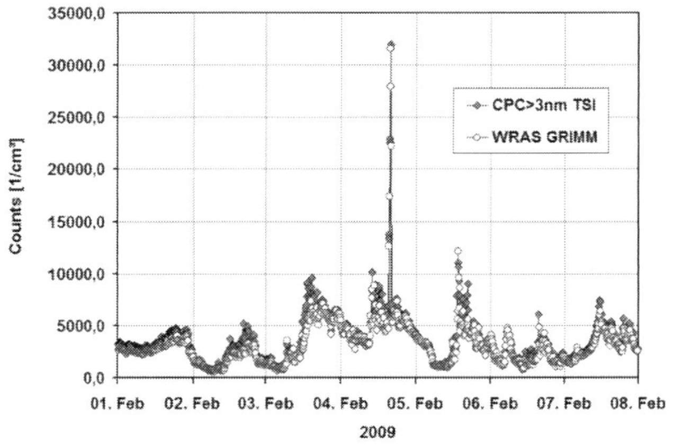

Figure 2. Comparison of total particle number concentrations (1) calculated from the aerosol size distributions measured by WRAS system and (2) measured by a TSI CPC

The size distributions showed pronounced diurnal variations. Typical features were high concentrations of very small particles right after sunrise, which show the particle formation by photochemical processes. A second feature is the increasing concentrations of rather large particles in the afternoon, probably due to the onset of convection. Figure 3 shows the aerosol size distributions on January 23-24, 2009. Although the measurement site is located on a hill with about 300 m elevation above a rural area, large numbers of ultrafine airborne particles were measured. As evident from Fig.3a and Fig.3b, large differences between the diurnal variations of size distributions occur. The high particle concentration at January 24, 21:31 (nanoparticles with CMD of 35.6 nm) result from combustion aerosol particles.

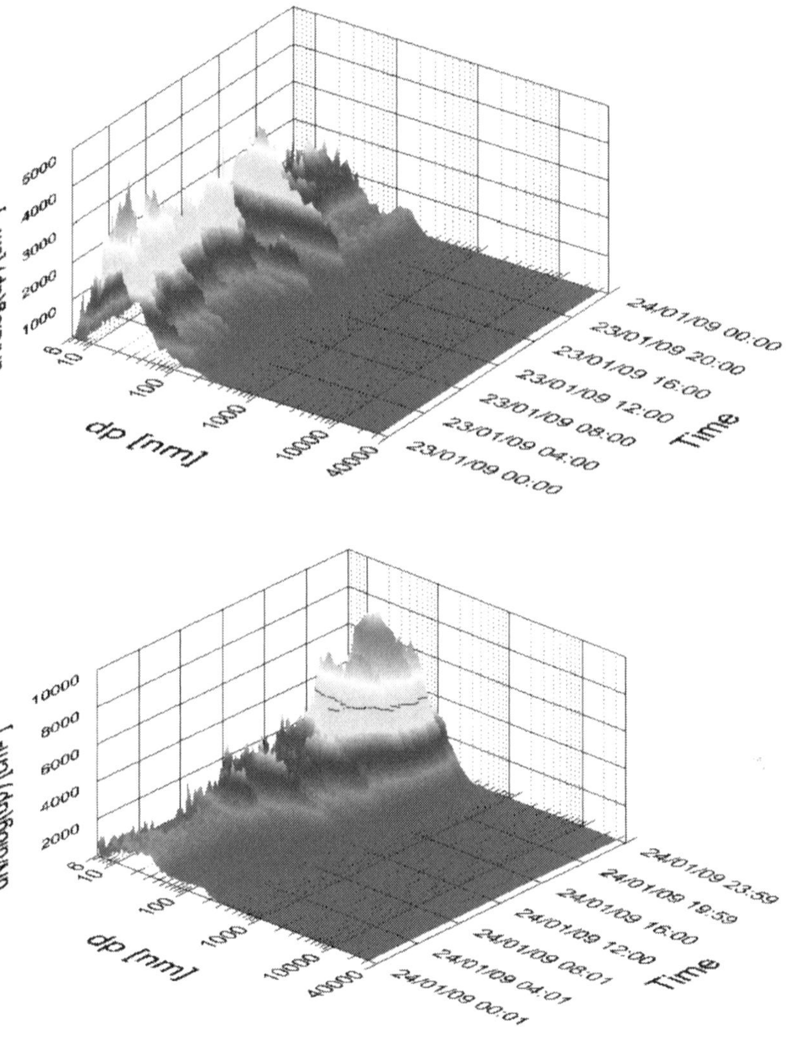

Figure 3. Particle number concentrations on a) January 23, b) January 24.

The total particle number concentrations showed a good correlation to gaseous CO, indicating that CO and particles originate both from the same combustion source. Figure 4 shows exemplarily the time series of gaseous SO_2 and total number concentrations. The strong SO_2 peak was accompanied by a nearly simultaneous small peak in the particle number concentrations and a second, much higher peak in the particle concentration occurred with approximately 1 h of delay after the SO_2 peak. We speculate that the first speak in the particle concentrations was due to the combustion particles which were released together with the SO_2, the second larger peak in the particle concentrations might consist of secondary sulfate particles, which were formed from the gaseous precursor with the consequence that the gas concentrations were lower in that air parcels. Such sulfate particles feature, unlike nitrate particles, only a accumulation mode.

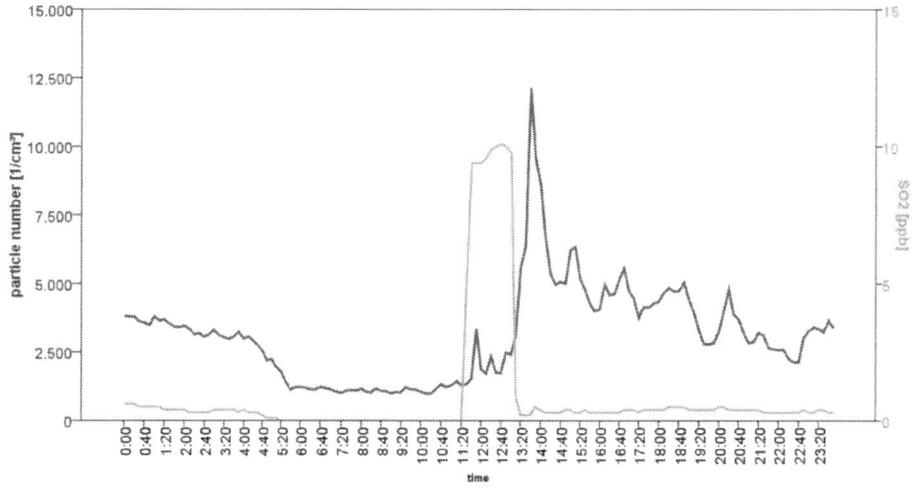

Figure 4. Comparison of total particle number concentrations and gaseous SO_2.

4. Conclusions

The Wide Range Aerosol spectrometer employed for this test made it possible to measure the complete size range of atmospheric particles over one year with a size resolution of 72 channels and a time resolution of 5 min.

Acknowledgements

We are particularly grateful for Dr. H. Flentje's assistance at the GAW Station Hohenpeissenberg, Germany.

References

[1] Anderson H.R., 2000, Differential epidemiology of ambient aerosols, Phil. Trans. R. Soc. A358: 2771-2785.

[2] Birmili W., Heinke K., Pitz M., Matschullat J., Wiedensohler A., Cyrys J., Wichmann H.-E. and Peters A., 2009, Particle number size distributions in urban air before and after volatilisation, Atmos. Chem. Phys. Discuss., 9, 9171–9220.

[3] Oberdörster G., Finkelstein J., Ferin J., Godleski J., Chang L.Y., Gelein R., Johnston C., Crapo J.D., 1996, Ultrafine particles as a potential environmental health hazard. Studies with model particles, Chest 109, 68S-69S.

[4] Sioutas C., Delfino R., and Singh M., 2005, Exposure assessment for atmospheric ultrafine particles (UFPs) and implications in epidemiologic research, Environ. Health Persp., 113, 947–955.

Development of a test method for protective gloves against nanoparticles in conditions simulating occupational use

Patricia Dolez[1], Ludwig Vinches[1], Kevin Wilkinson[2], Philippe Plamondon[3] and Toan Vu-Khanh[1]

[1]École de technologie supérieure, 1100 rue Notre-Dame Ouest, Montréal QC H3C 1K3 Canada

[2]Université de Montréal, C.P. 6128, succ. Centre-ville Montréal QC H3C 3J7 Canada

[3]École polytechnique, C.P. 6079, succ. Centre-ville, Montréal QC H3C 3A7 Canada

E-mail: patricia.dolez@etsmtl.ca

Abstract. Nanoparticle manufacture and use are in full expansion. The associated risks of occupational exposure raise large concerns due to their potential toxicity. Even if they stand as a last resort in the traditional occupational Health & Safety (H&S) risk management strategy, personal protective equipment (PPE) against nanoparticles are an absolute need in the context of precautionary principle advocated by H&S organizations worldwide. However no standard test method is currently available for evaluating the efficiency of PPE against nanoparticles, in particular in the case of gloves. A project is thus underway to develop a test method for measuring nanoparticle penetration through protective gloves in conditions simulating glove-nanoparticle occupational interaction.
The test setup includes an exposure and a sampling chamber separated by a circular glove sample. A system of cylinders is used to deform the sample while it is exposed to nanoparticles. The whole system is enclosed in a glove box to ensure the operator safety during assembly, dismounting and clean-up operations as well as during the tests. Appropriate nanoparticle detection techniques were also identified.
Results are reported here for commercial 15nm TiO_2 nanoparticles - powder and colloidal solutions in 1,2-propanediol, ethylene glycol and water - and four types of protective gloves: disposable nitrile and latex as well as unsupported neoprene and butyl rubber gloves. They show that mechanical deformations and contact with colloidal solution liquid carriers may affect glove materials. Preliminary results obtained with TiO_2 powder indicate a possible penetration of nanoparticles through gloves following mechanical deformations.

With the growing expansion of nanotechnologies worldwide, increasing risks of exposure to nanoparticles are expected. This is especially worrying in the case of occupational settings, where the level of toxicological risk has been evaluated as significant [1]. In the context of precautionary principle advocated by numerous H&S organizations around the world [2], large efforts are directed towards the development of standards [3], guides [4,5] and risk assessment and control methodologies [6] specific for nanomaterials. Indeed, it has been estimated that the number of workers involved in nano-related activities would reach 2 millions by 2015 [7]. As a result and even if they stand as a last resort in the traditional occupational H&S risk management strategy, PPE against nanoparticles are needed right away [8]. Yet, knowledge, data and test methods in this area are scarce whereas questions have been raised about the efficiency of existing PPE products against nanoparticles [9].

If some progress has been accomplished regarding respiratory PPE against nanoparticles, the question of dermal protection is still largely unexplored [9]. However, studies are increasingly showing that skin is a possible uptake route for nanoparticles, either when injured by abrasion [10], after repeated flexions [11] or even intact [12]. Pores, hair follicles and sweat may also increase the likelihood of nanoparticles percutaneous transportation [13].

A limited number of groups have reported research carried out on protective clothing and gloves against nanoparticles. Most of it involves aerosols. In the case of air-permeable fabrics, tests have been performed with oleic acid, KCl, NaCl, graphite, TiO_2 and Pt nano-aerosols as small as 10nm in diameter [14-20]. According to some authors, the variation of the nanoparticle penetration ratio through fabrics as a function of the particle diameter and air flow rate is in agreement with the filtration theory [15,16,21]. Others have reported diverging results, for example a plateau in the penetration of graphite nanoparticles larger than 50 nm through woven cotton with a 0.6 cm/s face velocity [18], and a higher penetration of 30nm graphite nanoparticles through a paper fabric than 80nm ones [9]. A much higher efficiency against nanoparticle penetration was observed for a thin high-density polyethylene non-woven membrane than with other, thicker fabrics, with and without air flow [17-20]. In the case of protective gloves, the results seem conflicting. Diffusion of 30 and 80nm graphite nanoparticles through nitrile, vinyl, latex and neoprene commercial glove samples has been reported [17] while no penetration was later measured for the same gloves with 40nm graphite and 10nm TiO_2 and Pt particles [18,19]. These data involving nano-aerosols were obtained without air flow.

Exposure to nanoparticles in occupational settings may also involve powder and colloidal solutions. This situation is especially relevant to protective gloves. Scanning electron microscopy (SEM) observations of latex and nitrile rubber gloves after static and dynamic contact with clay and alumina powders have shown that nanoparticles tend to accumulate inside micrometer-size pores on the surface of the gloves [22]. In the case of nanoparticles in colloidal solutions, some concerns have been raised about increased risks related to the liquid carrier [23]. In addition, the mechanical deformations suffered by gloves in service as well as the presence of a microclimate inside the gloves may also affect the penetration of the nanoparticles. It is thus important that test methods take into account conditions experienced by PPE while in use [21].

This paper reports on the development of a test method for measuring nanoparticle penetration through protective gloves in conditions simulating glove occupational use. It describes the setup designed for the test as well as the nanoparticle detection techniques investigated. Finally, it provides preliminary results obtained with commercial 15nm TiO_2 powder and colloidal solutions, and four types of protective gloves: disposable nitrile and latex as well as unsupported neoprene and butyl rubber.

1. Materials

1.1. Nanoparticles

Titanium dioxide (TiO_2) nanoparticles have been selected for the development and validation of the test method since it is one of the most widely used nanomaterials [24]. TiO_2 powder (15 nm, 99.7% pure anatase) was obtained from Nanostructured & Amorphous Materials, Inc. (Houston, TX). A statistical analysis (n = 174) of transmission electron microscopy (TEM) pictures (JEOL JEM-2100F) of the bulk powder sprinkled on carbon-coated copper grids revealed the presence of aggregates and agglomerates (see Figure 1). As illustrated in Figure 2, the average size of aggregates is 100 nm and some agglomerates are as large as 1200 nm. In fact, only two individual nanoparticles were counted in the analysed sample. Figure 1 also shows the presence of rod-like rutile nanoparticles in addition to the spherical anatase. A 3 to 6% ratio of rutile in the TiO_2 powder was measured by X-ray diffraction (Philips X'PERT).

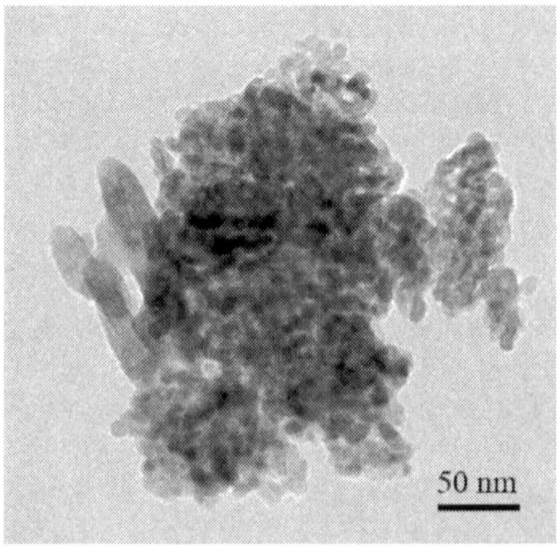

Figure 1. TEM image of the TiO_2 powder.

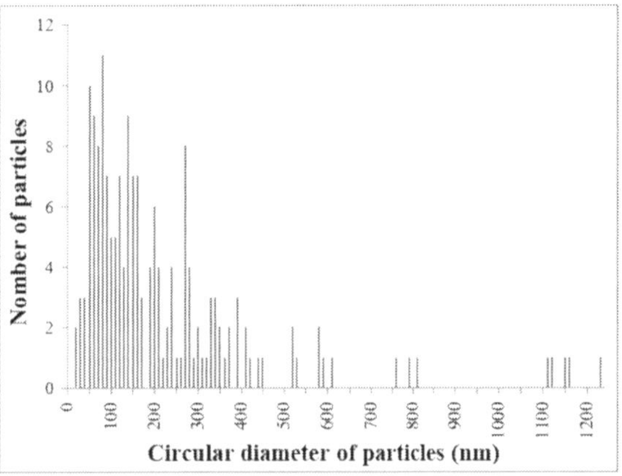

Figure 2. Size distribution of the TiO_2 powder.

TiO_2 colloidal solutions (15 nm, anatase) in water (15 wt%, Nanostructured & Amorphous Materials, Inc., Houston, TX), ethylene glycol (20 wt%, idem) and 1,2-propanediol (20 wt%, M K Impex, Mississauga, ON) were also obtained. Analysis of the colloidal solutions by fluorescence correlation spectroscopy (FCS) was performed and provided a particle size of 21 ± 2 nm for TiO_2 colloids in water and 35 ± 3 nm for TiO_2 colloids in ethylene glycol. It can be noted that no characterization of the TiO_2 colloidal solutions by microscopy was possible due to the formation of a µm-thick viscous film on the nanoparticles upon solvent evaporation. It was attributed to the presence of additives in the colloidal solutions, used for example as stabilising agents.

1.2. Gloves

Four types of protective gloves made of different elastomers were selected for the study: disposable nitrile rubber gloves (100 µm thick), disposable latex gloves (100 µm thick), non-supported neoprene gloves (450 µm thick) and non-supported butyl rubber gloves (700 µm thick). The samples were taken in the palm section of the gloves.

The external surface of the gloves, i.e. in contact with nanoparticles, was analyzed by SEM (Hitachi S3600N). Figure 3 displays examples of what was obtained for the nitrile (Figure 3.a) and neoprene (Figure 3.b) gloves. The surface of nitrile samples shows micrometer-size pores, in agreement with what has already been reported [22], while neoprene also includes the presence of cracks. In the case of latex and butyl rubber gloves, plate-like structures are observed.

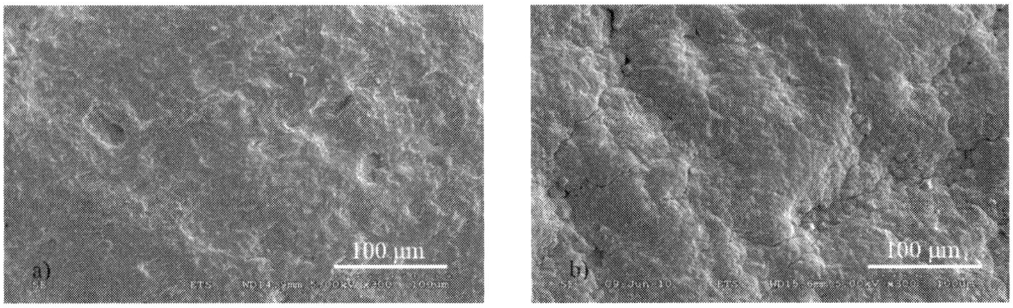

Figure 3. SEM pictures of external glove surface: a) nitrile and b) neoprene.

2. Design of the Test Setup

A setup has been designed to fulfill a series of requirements. They include the mode of application of the nanoparticles (powder and colloidal solutions), the simulation of glove use conditions (mechanical deformation, presence of a microclimate inside the glove), sampling outputs corresponding what is needed by nanoparticle detection techniques, as well as the safety of the operator.

As illustrated in Figure 4, the test setup comprises an exposure and a physiological/sampling chamber between which is secured a circular glove sample. Both chambers have a 50-mm inner diameter and a 60-mm inner height. All setup parts in contact with nanoparticles are made out of ultrahigh molecular weight polyethylene to limit the nanoparticle adsorption effect observed by FCS with other materials like anodized aluminum in the case of colloidal solutions. The nanoparticles are introduced in the exposure chamber and put in contact with the external surface of the glove samples. In case of powder nanoparticles, a second thin circular nitrile membrane is placed on top of the sample to enclose the nanoparticles and prevent them from dispersing in the exposure chamber. The physiological/sampling chamber plays the dual role of allowing the detection of nanoparticles having crossed the sample membrane as well as simulating the microclimate inside gloves. This includes the control of temperature and humidity conditions as well as a contact with physiological solutions at different pH simulating sweat. The whole system is enclosed in a glove box to ensure the operator safety during assembly, dismounting and clean-up operations as well as during the tests.

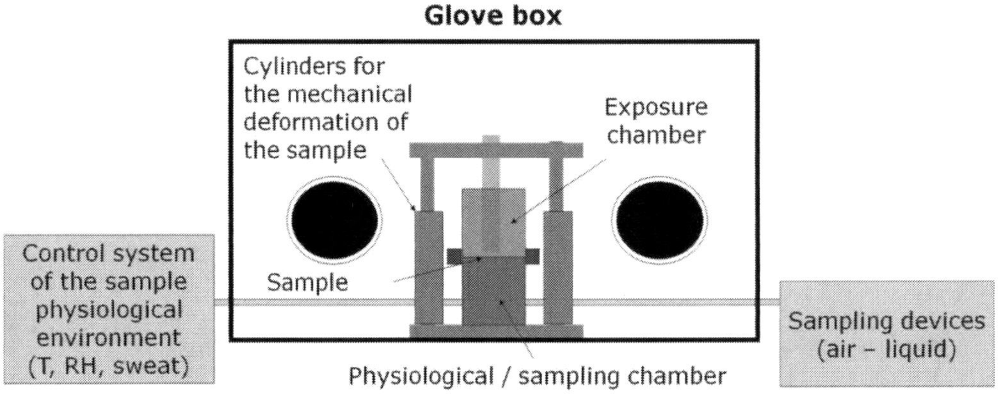

Figure 4. Schematic representation of the test setup.

Mechanical constraints subjected to gloves while in use are simulated by a system of cylinders and probe heads of various shapes and sizes. They are used to deform the sample while it is exposed to nanoparticles. A load cell and a displacement sensor allow a computer control and characterization of these mechanical stimuli. Four probe heads have been fabricated as illustrated in Figure 5. Probe A is a 35mm-diameter full cylinder applying an out-of-plane deformation to the sample while pressing on the nanoparticles. Probe B is a 35mm-diameter hollow cylinder used to deform the sample without applying any constraint on the nanoparticles. Probe C is a 35-mm cone with a spherical tip which is used to simulate biaxial deformations corresponding to hand and finger flexion [25]. Finally, probe D is a 50mm-diameter cylinder used for compression in combination with a sample supporting plate. An overflow is connected to the physiological/sampling chamber to accommodate the sampling chamber volume reduction due to sample deformation. The system is computer-controlled so that sample deformations can be applied in a static or a dynamic mode using these four probe heads.

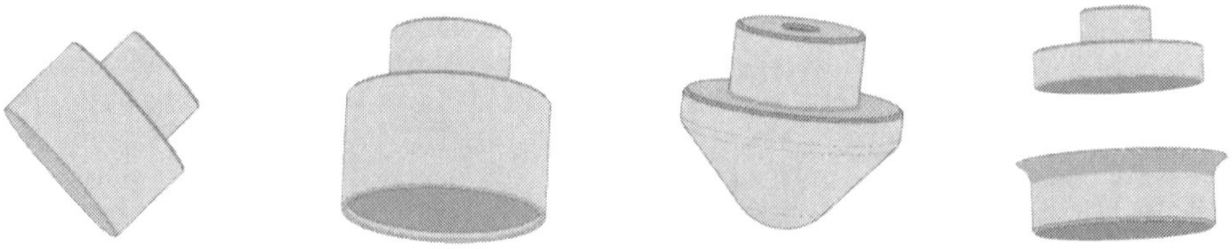

Figure 5. Setup probe heads, a) Probe A, b) Probe B, c) Probe C, and d) Probe D.

3. Investigation of Nanoparticle Detection Techniques

The suitability of a series of nanoparticle detection techniques for the test method has been investigated. This includes the development of specific sampling and sample preparation protocols.

Attempts have been made to use a scanning mobility particle sizer (SMPS, TSI) to detect the presence of nanoparticles in the sampling chamber air. Trials involved two ranges of detection: 2-150 nm and 10-1000 nm, and flow rates of up to 10 L/min. In all cases, nothing but a few isolated nanoparticle detection events were recorded, even when nanoparticles sucked up directly from a container were seen disappearing from it. The reason for the inability to detect nanoparticles using this device was attributed to electrostatic forces strongly affecting nanoparticles [26] and making TiO_2 powder stick to non-conductive surfaces, even to the SMPS graphite silicone tubing (see Figure 6).

A second set of investigated nanoparticle detection techniques involved field emission gun SEM (FEG-SEM), atomic force microscopy (AFM) and TEM for analyzing the inner (sampling chamber side) surface of glove samples. As illustrated in Figure 7 with a FEG-SEM (JEOL JSM-7600F) image of a TiO_2 aggregate on the surface of a nitrile glove, direct observation of nanoparticles on the glove sample surface happens to be highly imprecise due to the similarity between nanoparticles and glove surface features as well as to the difficulty in locating a few nanoparticles over a large sample area. Positioning TEM grids on the inner surface of glove samples (facing the physiological/sampling chamber) was also not an option because of the deformations subjected to membranes during the tests. Future attempts will be made to place charged TEM grids on a support at the bottom of the physiological/sampling chamber.

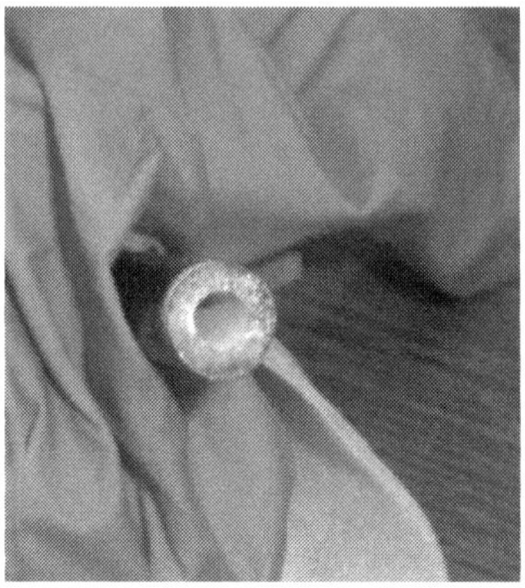

Figure 6. Picture of the SMPS sampling line inlet covered with TiO_2 powder.

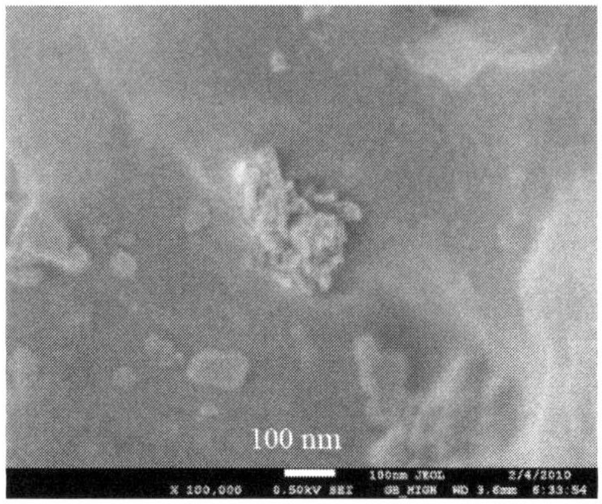

Figure 7. FEG-SEM image of TiO_2 nanoparticles on the surface of a nitrile glove.

In order to allow using microscopy techniques to detect nanoparticles in the physiological/sampling chamber despite the natural tendency of TiO2 nanoparticles to stick to the chamber walls, a special sampling protocol was developed. It involves adding a 10-mm thick layer of methanol in the physiological/sampling chamber before closing it with the sample membrane during setup assembly. Heights of membrane deformations are such that no contact between the sampling solution and the sample surface occurs during the tests. When the experiment is over and before dismounting the test setup, the chamber assembly is gently tilted and rotated so that the liquid can rinse the physiological/sampling chamber walls as well as the sample surface. The sampling solution is then collected in a vial and centrifuged on a carbon-coated copper grid or on a mica substrate for further observation by TEM, AFM or FEG-SEM [27].

The last category of investigated nanoparticle detection techniques deals directly with liquids. The first one is inductively coupled plasma mass spectrometry (ICP-MS). The sampling protocol involving a 10-mm thick layer of sampling solution rinsing the physiological/sampling chamber was also applied, but with a 1% nitric acid solution in ultra high purity (UHP) water according to the methodology developed by Kaegi et al. [28]. The second technique is nanoparticle tracking analysis (NTA). In that case, the sampling solution is made of UHP water.

4. Results and Discussion

A first set of experiments was performed to characterize the effect of mechanical deformations and liquid carriers used in TiO_2 colloidal solutions on the glove materials. The swelling behaviour of the four elastomers was assessed by weight and length change measurements on 4x50mm samples immersed in the three TiO_2 colloidal solutions as well as in solvents corresponding to the liquid carriers, i.e. water, ethylene glycol and 1,2-propanediol, over a period of two hours. Samples were cut along the glove transverse and longitudinal directions and at least 3 replicates were measured for each condition. Increases in sample weight and length have been observed in some cases, essentially with latex and nitrile rubber membranes. Figure 8 illustrates for example the variation in nitrile rubber glove transverse length ($(l-l_0)/l_0$, with l and l_0 respectively the length of the exposed and unexposed samples) as a function of immersion time in TiO_2 colloidal solutions in water and 1,2-propanediol. This increase in sample weight and length may indicate a penetration into the elastomer membranes of the solvents used as liquid carrier in the TiO_2 colloidal solutions. If these diffusing solvents act as carrier for TiO_2 nanoparticles through the polymer membranes, they could increase their penetration rate though the gloves. In some instances, a small and gradual weight loss over time was recorded after the initial increase. This may be due to a release of leachable additives from the glove material during the swelling process. In the case of latex and nitrile rubber, a difference in swelling behaviour depending on the sample orientation (transverse or longitudinal) was noted as illustrated in Figure 9 for nitrile rubber immersed in 1,2-propanediol. This phenomenon may be due to the anisotropy in the glove mechanical properties which results from the dipping manufacturing process.

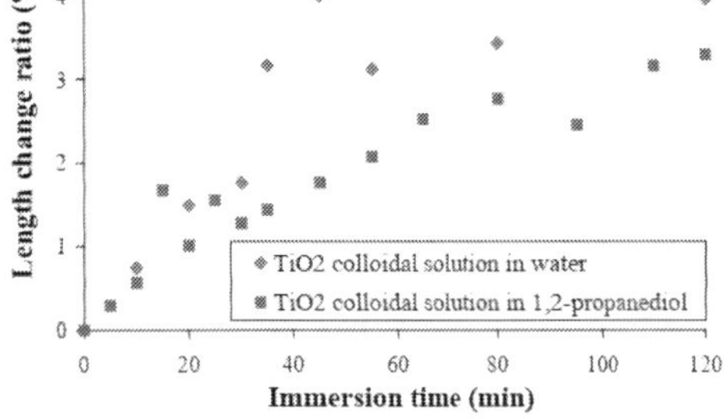

Figure 8. Variation in transverse length vs. immersion time for nitrile rubber in TiO_2 colloidal solutions in water and 1,2-propanediol.

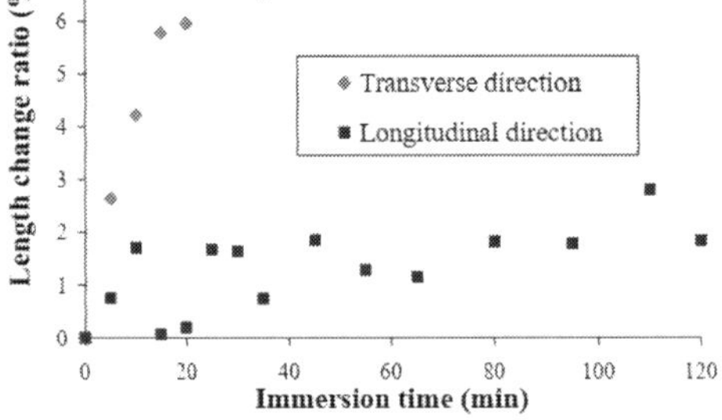

Figure 9. Variation in sample length as a function of immersion time in 1,2-propanediol for transverse and longitudinal nitrile rubber samples.

The effect of mechanical deformation and exposure to TiO2 colloidal solution liquid carriers on the glove surface morphology was investigated using the test setup described in Section 2. Dynamic biaxial deformations were applied to glove samples for 7 hours using Probe C, a maximum probe displacement of 40 mm (corresponding to a strain of about 80%) and a frequency of 0.0033 Hz (1 deformation every 5 minutes). Sample exposure to ethylene glycol and 1,2-propanediol was performed using the test setup, either by itself also for a period of 7 hours or with simultaneous application of dynamical mechanical deformations to the sample. The surface morphology of both sides of samples exposed to solvents, deformations or combination of both was analyzed by SEM (Hitachi S3600N). The quantification of the induced surface modifications compared to unexposed gloves was performed on the basis of the surface area of surface features specific to each material (described in Section 1.2): pores for nitrile rubber, cracks for neoprene and platelets for latex and butyl rubber. Figure 10 displays the result of this analysis in the case of 1,2-propanediol used as solvent and for the glove sample external surface (on the side of the exposure chamber). Both solvent exposure and deformations appear to strongly affect the external surface of latex gloves and, to a lesser extent, that of butyl rubber ones. The last two materials seem to be only affected by mechanical deformations, with a larger effect produced in the case of nitrile rubber. A similar behavior was observed with ethylene glycol. The effect of mechanical deformations observed on the glove external surface might include an abrasion contribution induced by the probe rubbing on the sample surface. This hypothesis is supported by the reduction in surface features observed when the mechanical deformations are combined with exposure to glycols playing the role of a lubricating agent. Analysis of the other side of the samples, i.e. on the glove inner surface facing the physiologic/sampling chamber, only revealed significant modifications in surface features for latex, with the same trend as for the external surface. These results show that deformations and exposure to TiO2 colloidal solution liquid carriers may induce damages in glove materials which could facilitate the penetration of nanoparticles through the membrane.

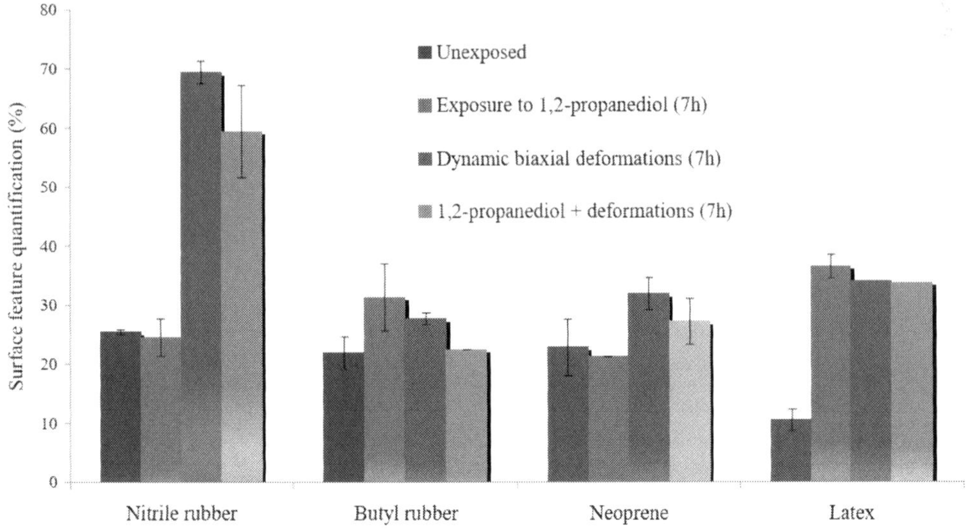

Figure 10. Effect of 7h exposure to solvent, dynamic biaxial deformations and solvent + deformations on material-specific glove surface features for nitrile rubber, butyl rubber, neoprene and latex samples.

Measurement of nanoparticle penetration through the four glove materials has been performed using the developed test setup (see Figure 4) and TiO2 powder. Glove samples were subjected to dynamical biaxial deformations (Probe C, 1 deformation every 5 minutes, 40mm maximum probe displacement) while being exposed to TiO2 powder (250 mg deposited on the sample in the exposure chamber) for periods of 1 hour 30 minutes, 3h, 5h and 7h. A 10-mm layer of sampling solution (1% nitric acid solution in UHP water) was present in the sampling chamber during the experiment and used to rinse the sampling chamber walls when the test was over. The collected sampling solutions were then analysed by ICP-MS. The results obtained with nitrile and butyl rubber samples are displayed in Figure 11. A gradual and clear increase in Titanium concentration in the sampling solutions is observed, which may indicate a penetration of TiO2 powder through the glove samples as a result of mechanical deformation. In addition, an observation by FEG-SEM of the exposed side of the sample surfaces after excess TiO2 has been removed seems to point towards a reduction in the size of nanoparticle agglomerates as test time increases. Even if these results are very preliminary and need to be replicated, they indicate a possible penetration of TiO2 nanoparticles through protective gloves when subjected to mechanical deformations over a period of time. They also demonstrate that ICP-MS in combination with the 1% nitric acid / UHP water sampling solution protocol is an appropriate detection technique for the nanoparticle penetration test method.

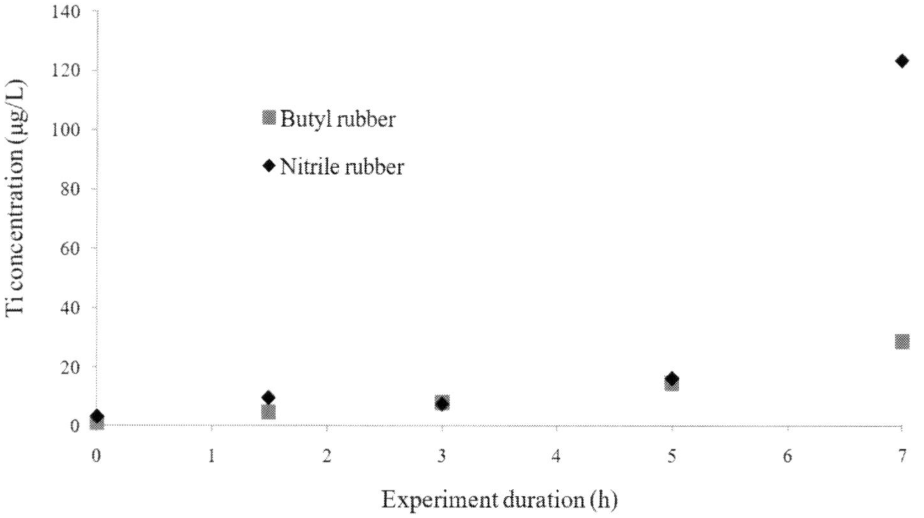

Figure 11. Variation in Ti concentration in the sampling solutions as a function of test time for nitrile and butyl rubber samples exposed simultaneously to biaxial deformations and TiO_2 powder.

5. Conclusions

A test method has been developed to measure the penetration of nanoparticles through protective gloves in conditions simulating occupational use. It involves nanoparticles applied as powder and colloidal solutions. In addition, samples can be subjected to various types of static and dynamic mechanical deformations simultaneously with nanoparticle exposure. Finally, the presence of a microclimate inside the glove is simulated by the use of a physiological solution in contact with the sample as well as a control of temperature and humidity conditions.

The development of the test method also involves the identification of appropriate nanoparticle detection techniques. A sampling protocol based on the use of a sampling solution rinsing the sampling chamber has been designed for collecting nanoparticles having crossed glove samples.

Methanol-based sampling solutions are centrifuged on grids or mica substrates for analysis by microscopy techniques. On the other hand, NTA and ICP-MS may also be used to directly detect nanoparticles in water-based sampling solutions.

Results have been obtained with commercial 15nm TiO2 nanoparticles - powder and colloidal solutions in 1,2-propanediol, ethylene glycol and water - and four protective glove materials: nitrile rubber, latex, neoprene and butyl rubber. They show that glove materials are affected by exposure to TiO2 colloidal solution liquid carriers and mechanical deformations. In addition, preliminary data obtained with the developed test setup and TiO2 powder seem to indicate a possible penetration of nanoparticles through protective gloves when subjected to mechanical deformations over a period of time. These results demonstrate the need for research in this area.

Acknowledgments

This project was made possible thanks to the support of the Institut de recherche Robert-Sauvé en santé et en sécurité du travail (IRSST) and NanoQuébec. The authors also want to acknowledge the contribution of Mr. Swann Mahé, Mrs. Alice Jambou and Mr. Félicien Deltombe during their internship at École de technologie supérieure as well as the collaboration of G. L'Espérance (École Polytechnique de Montréal), Y. Cloutier (IRSST), J. Lara (IRSST), C. Dion (IRSST), G. Truchon (IRSST) and P. Koutrakis (Harvard) on the project.

References

[1] Hänninen O, Bruske-Hohlfeld I, Loh M, Stoeger T, Kreyling W, Schmidand O, Peters A 2009 *Journal of Physics: Conference Series* **170** 012031 (NanoSafe 2008, Grenoble, France)

[2] Organisation for Economic Co-operation and Development 2010 *Current developments/activities on the safety of manufactured nanomaterials - Tour de table at the 7th meeting of the working party on manufactured nanomaterials*, Paris, France

[3] Bard D, Markand D, Mohlmann C 2009 *Journal of Physics: Conference Series* **170** 012036 (NanoSafe 2008, Grenoble, France)

[4] Amoabediny G H, Naderi A, Malakootikhah J, Koohi M K, Mortazavi S A, Naderi M, Rashedi H 2009 *Journal of Physics: Conference Series* **170** 012037 (NanoSafe 2008, Grenoble, France)

[5] Ostiguy C, Roberge B, Ménard L, Endo C-A 2009 *Best practices guide to synthetic nanoparticle management* (Montreal, QC, Canada: Institut de recherche Robert-Sauvé en santé et en sécurité du travail)

[6] Giacobbe F, Minica L, Geraci D 2009 *Journal of Physics: Conference Series* **170** 012035 (NanoSafe 2008, Grenoble, France)

[7] Freeman R, Shukla K 2008 *Science and Engineering Workforce Project Digest* **June** 1

[8] Schulte P, Geraci C, Zumwalde R, Hoover M, Kuempel E 2008 *Journal of Occupational and Environmental Hygiene* **5** 239

[9] Dolez P I, Bodila N, Lara J, Truchon G 2010 *International Journal of Nanotechnology* **7** 99

[10] Zhang L W, Monteiro-Riviere N A 2008 *Skin Pharmacology and Physiology* **21** 166

[11] Rouse J G, Yang J, Ryman-Rasmussen J P, Barron A R, Monteiro-Riviere N A 2007 *Nano Letters* **7** 155

[12] Ryman-Rasmussen J P, Riviere J E, Monteiro-Riviere N A 2006 *Toxicol. Sci.* **91** 159

[13] Hervé-Bazin B 2007 *Les nanoparticules: Un enjeu majeur pour la santé au travail ?* (EDP Sciences)

[14] Hanley J T 2006 *Proc. Elevated Wind Studies International Conference* (Arlington, VA)

[15] Hofacre K C 2006 *Proc. Elevated Wind Studies International Conference* (Arlington, VA)

[16] Huang S-H, Huang Y-H, Chen C-W, Chang C-P 2007 *Proc. 3rd International Symposium on Nanotechnology, Occupational and Environmental Health* (Taipei, Taiwan)

[17] Golanski L, Guiot A, Tardif F 2008 *Are conventional protective devices such as fibrous filter media, respirator cartridges, protective clothing and gloves also efficient for nanoaerosols ?*

(European Strategy for Nanosafety)
[18] Golanski L, Guiot A, Rouillon F, Pocachard J, Tardif F 2009 *Hum. Exp. Toxicol.* **28** 353
[19] Golanski L, Guiot A, Tardif F 2009 *Journal of Physics: Conference Series* **170** 012025 (NanoSafe 2008, Grenoble, France)
[20] Golanski L, Guiot A, Tardif F 2008 *Proc. of the Nanotechnology Conference and Trade Show Nanotech 2008* (Boston, MA)
[21] Afsset 2008 *Les nanomatériaux : Sécurité au travail* (Maisons-Alfort, France: Agence française de sécurité sanitaire de l'environnement et du travail)
[22] Ahn K, Ellenbecker M J 2006 *Proc. AIHce Conference* (Chicago, IL)
[23] EPA 2007 *Nanotechnology White Paper* (Washington, DC: United States Environmental Protection Agency)
[24] Robichaud C O, Uyar A E, Darby M R, Zucker L G, Wiesner M R 2009 *Environ. Sci. Technol.* **43** 4227
[25] Harrabi L, Dolez P I, Vu-Khanh T, Lara J, Tremblay G, Nadeau S, Lariviere C 2008 *Safety Science* **46** 1025
[26] Jankovic J T, Hall M A, Zontek T L, Hollenbeck S M, Ogle B R 2010 *Int. J. Occup. Environ. Health* **16** 429
[27] Wilkinson K J, Balnois E, Leppard G G, Buffle 1999 J *Colloids and Surfaces A: Physicochemical and Engineering Aspects* **155** 287
[28] Kaegi R, Ulrich A, Sinnet B, Vonbank R, Wichser A, Zuleeg S, Simmler H, Brunner S, Vonmont H, Burkhardt M, Boller M 2008 *Environ. Pollut.* **156** 233

Nanofiber-based filters as novel barrier systems for nanomaterial exposure scenarios

M Faccini[1], D Amantia[1], S Vázquez-Campos[1], C Vaquero[2], J M López de Ipiña[2] and L Aubouy[1]

[1]LEITAT Technological Center, Passeig 22 de Juliol, 218, 08221 Terrassa -Spain

[2]Tecnalia, P.T. Alava – C/ Leonardo Da Vinci, 11, 01510 Miñano (Álava) - Spain

E-mail: mfaccini@leitat.org

Abstract. In this article our latest advances in the development of efficient barrier systems against micro- and nano-particulate materials are presented. Nanofibrous polyamide 6 (PA6) mats were fabricated by electrospinning onto a nonwoven viscose substrate. The influence of electrospinning parameters including solution concentration, viscosity, and conductivity, were studied for the production of nonwovens with different fiber size distribution ranging from 74 to 261 nm in diameters. Moreover, the formation of nanowebs (30-40 nm) which are widely distributed among fibers was observed. By varying several process parameters, membranes with different thickness of the nanofiber layer and therefore air permeability and nanoparticle filtration efficiency were obtained.

1. Introduction

A large diffusion of nanotechnology and nanotechnology related products, including the production processes of nanomaterials, is predicted in the coming years. Free particles can emerge in all stages of nanomaterials life cycle (production, processing, use, recycling and disposal). Therefore, the development of new barrier materials and protective systems are needed to reduce the potential risks related to human and environmental exposure to nanomaterials [1]. Nanofiber webs produced by electrospinning, due to their very large specific area, very small pore size, and high porosity, have shown to improve the efficiency of conventional materials used for filtration while reducing the pressure drop across the membrane [2].

Electrospinning is a well-established and versatile process that has been used to produce ultrafine fibers including microfibers (>1 μm) or nanofibers (<1000 nm). In electrospinning, a high voltage is applied to a polymer solution or melt, which overcomes the surface tension to form a charged jet. The charged polymer solution is ejected, dried, and solidified onto a grounded substrate. The ejected polymer solutions repel each other during the travel to the grounded collector, which forms thin fibers after solvent evaporation. By controlling the spinning conditions, the resulting fibers can range from about 20 nm to a few μm. The main advantage of electrospinning process is the relative quick and simple way to fabricate a variety of materials into nanofibrous structure.

In this paper, the development of new nanofiber-based non-wovens is described, including the study of their efficiency as barrier systems against nanoparticles. Polyamide 6 (PA6) was selected as the material for electrospun filters due to its good properties such as its toughness, resilience, and easy processability. Moreover, PA6 is extensively used in membrane and textile technologies. The fiber diameter and size distribution was studied as a function of parameters such as solution concentration, viscosity, and conductivity. The nanoparticle penetration was evaluated as a function of the nanofiber layer thickness using a polydisperse NaCl aerosol.

2. Experimental procedure

2.1. Electrospinning

PA6 and formic acid were purchased from Sigma Aldrich. PA6 was dissolved in formic acid at 50°C and the solution was stirred for a sufficiently long time until it became homogeneous. Polymer solutions of several different concentrations were prepared, ranging from 10 to 30 wt. %. The solution viscosity and the electric conductivity were determined by a digital viscometer (DV-E, Brookfield Co.) and an electric conductivity meter (CRISON EC-meter BASIC) at 25°C, respectively. The PA6 solutions were electrospun onto an aluminum foil or a non-woven viscose support (28 g/m^2) by using a commercially available electrospinning setup (MECC Co. LTD., model NF-103) equipped with a rotary drum collector. Typical operating conditions were: flow rates of 0.4 mL/h, applied voltages between 20 and 30 kV, and working distance of 8-10 cm.

The nanofibrous mats were characterized using a scanning electron microscope (SEM, Hitachi H-4100FE) after coating with carbon to minimize the charging effect. Images taken by the SEM were analyzed to obtain the fiber diameter by the *ImageJ* software. At least four pictures were used to calculate the mean values of the diameter of the fibers.

2.2. Aerosol filtration

The experimental system used to determine the penetration of nanoparticles through nanofibrous PA 6 filters is shown in Figure **1**. It consists of a polydisperse aerosol generator system, a filter holder containing the fiber material and the measurement equipment. Polydisperse aerosols have been used by several authors [3] after Japuntich et al [4] demonstrated that it produces similar results to the slower method of using monodisperse aerosols. The challenge polydisperse aerosol (NaCl) is generated from a 0.3% NaCl solution using a collision atomizer (TSI 3076) and then dried by passing it through a diffusion drier (TSI 3062). The filter media is mounted and clamped in a 4 cm diameter filter holder, corresponding to a filter face velocity of 1.7 cm/s. The number concentration of particles is measured upstream and downstream the filter using first an Electrostatic Classifier (TSI series 3080) with a long DMA (TSI 3081) to classify the particles followed by a Condensation Particle Counter (TSI CPC 3775) to measure them. The penetration of particles is calculated as a ratio among the concentration downstream (c_d) and upstream (c_u) ($P\% = c_d * 100 / c_u$).

For some of the filter media the effect of particle charge state was analysed [3]. To achieve this, the penetration of charged particles (coming directly from the aerosol drier) and charge neutralized particles (equilibrium distribution) were compared. The charge neutralized aerosol was achieved by passing the dried aerosol through a Kr-85 radioactive source.

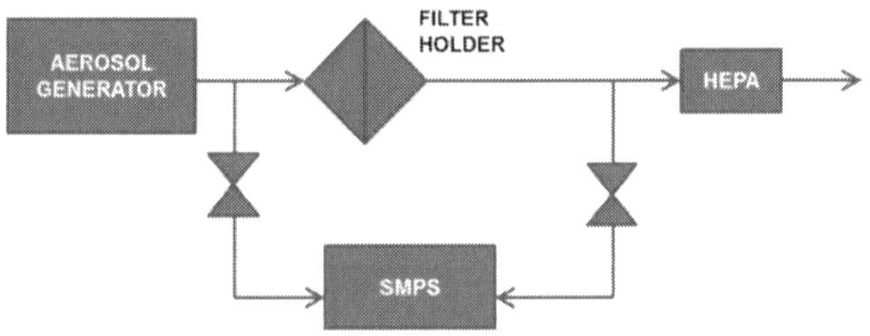

Figure 1. System used for the measurement of nanoparticle penetration through electrospun filters.

3. Results and discussion

3.1. Electrospinning

It is well known that the morphology of electrospun fibers depends on various processing parameters and environmental conditions. In order to gain control over the properties of the obtained nanofibers such as diameter, porosity, and morphology, the viscosity and the electrical conductivity were measured as function of the PA6 concentration in formic acid. Figure **2** shows that the viscosity increases steadily with polymer concentration going from 220 cP at 10 wt. % to 3840 cP at 30 wt. %. The electric conductivity of polymer solutions was between 4120 and 4880 μS/cm (Figure 2). The conductivity first increases going from 10 to 15 wt. % to then decrease with increasing polymer concentration due to the high dielectric constant (58.5 at 15°C) and dipole moment (1.41 Debye) of formic acid [5].

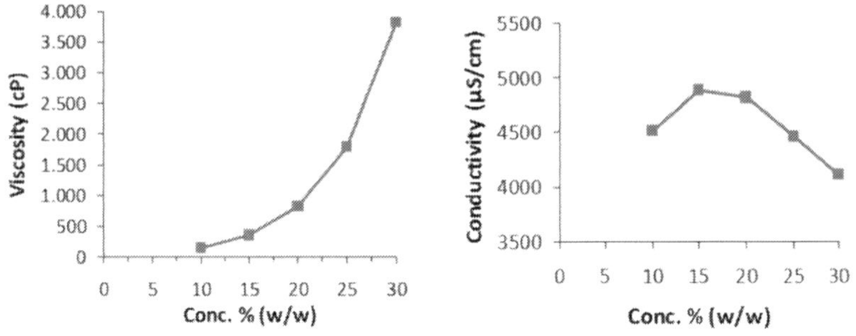

Figure 2. Solution viscosity and electrical conductivity as function of the polymer solution concentration.

The SEM images of PA6 nanofibers as function of polymer concentration are shown in Figure **3**. In general the fibers are uniform, without beads formation even at the lowest concentration of 10 wt. %. The average fiber diameter increases linearly from 74 to 261 nm going from 10 to 30 wt. % PA6 in formic acid (Figure 4). This indicates that the morphology of the nanofibers greatly depends on polymer concentration which effect viscosity. In fact, at higher polymer concentration there is more chain entanglement and less chain mobility, resulting in less extension during spinning, and therefore producing thicker fibers. Moreover, the fiber distribution is becoming gradually broader with increasing the concentration. Interestingly, the formation of a spider-net within nanofibers (nanowebs) appears at 15 wt. % concentration or above. These structures, which have a fiber diameter of 30-40 nm, although being widely present among fibers, are not homogeneously distributed over the whole nanofiber mat. The formation of such nanowebs has been investigated by Barak at et al. [6] by the additions of ionic salts to polymer solutions. The authors concluded that nanowebs formation depends upon the concentration and the ionization of the used salt. In our case the formation of spider-net structures seems to increase with polymer concentration. Nanowebs can improve the mechanical properties of the PA6 nanofibers and may have a beneficial effect on the filtration efficiency. Therefore, more research is needed to clarify the mechanism governing the generation of nanowebs and to find a reliable process for producing homogeneously distributed structures in a controlled manner.

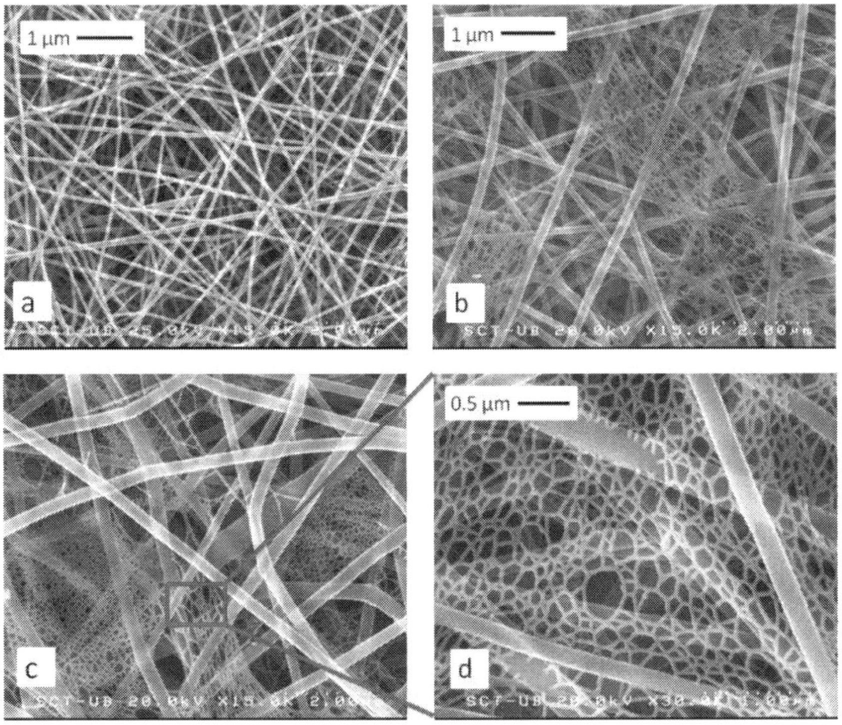

Figure 3. SEM images of electrospun PA6 nanofiber mats at different concentrations (a) 10, (b) 20, (c) 30 wt. %; and (d) PA nanowebs formation.

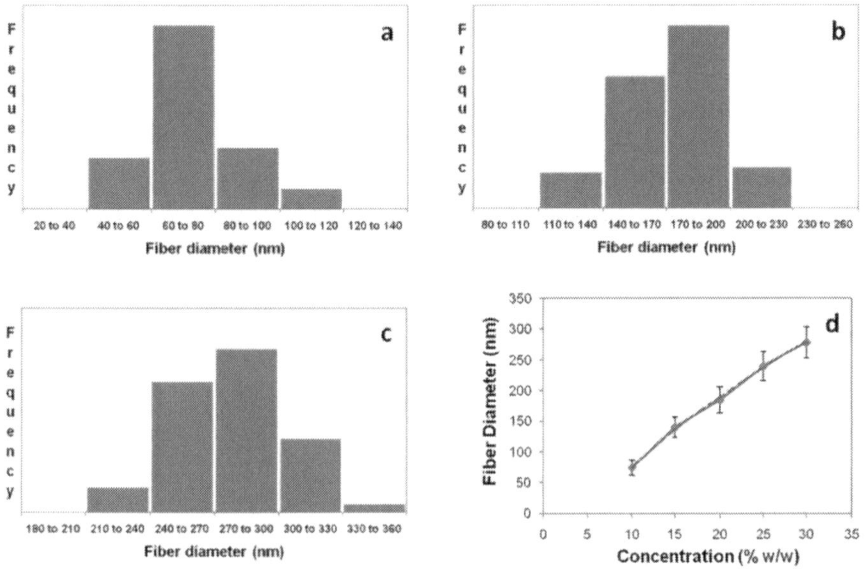

Figure 4. Fiber diameter distribution of electrospun PA6 fibers at different concentrations (a) 10, (b) 20, (c) 30 wt. %; and (d) average fiber diameter as function of polymer concentration.

3.2. Fabrication of the composite filter

For the mass production of nanofilters, the optimal conditions for steady state electrospinning have to be found. In our case a stable jet over a long period of time (several hours) was obtained using 15 wt. % of PA6 in formic acid, flow rate of 0.4 mL/h, applied voltage of 30 kV, and tip-to-collector distance of 8 cm. The viscose non-woven supports were coated with PA6 nanofibers using 5 different coating times (5, 10, 20, 30 and 60 min). By varying the nanofiber collection time allowed to generate filters with different thickness of the nanofiber layer and therefore fine tune the pore size, the air permeability, and therefore the filtration efficiency. Figure **5** shows that for shorter coating times the fibers and the structure of the viscose substrate can be clearly seen due to the small coverage and the larger pore size, while for longer coating times are almost undetectable.

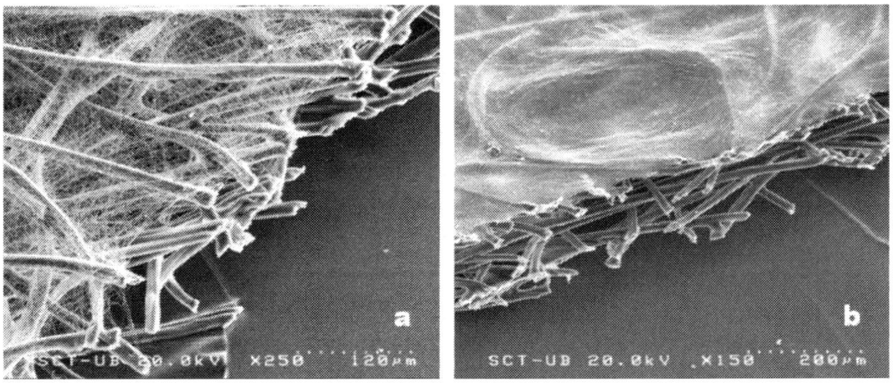

Figure 5. SEM images of the PA6 electrospun membranes with different coating times: (a) 10 min, (b) 60 min.

3.3. Nanoparticle penetration measurements

Figure **6** shows the initial results obtained for the penetration of nanoparticles (size range of 10-150 nm) through the five tested filters. The results showed that the penetration of nanoparticles through electrospun nanofibrous filters followed the classical filtration model, decreasing the penetration when decreasing the particle size, due to Brownian diffusion [7]. Figure **6** also shows that particle penetration is strongly a function of filter thickness (which is proportional to nanofiber deposition time) as predicted by other authors [8].

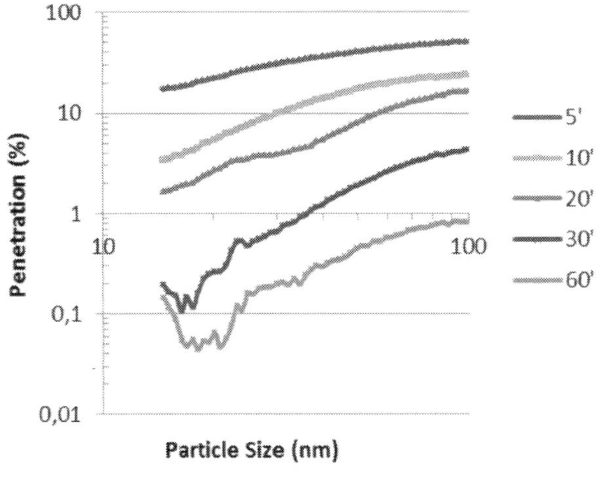

Figure 6. Penetration of nanoparticles through electrospun PA6 filters as a function of nanofiber coating time for a face velocity of 1.7 cm/s.

The study of the effect of particle charge was made comparing the penetration of charged particles versus neutralized particles (charge neutralized particles, equilibrium distribution) for some fiber samples. The results confirmed that particle penetration is independent of particle charge as predicted by other authors [4], which suggests that electrostatic interactions present between the particles and the fibers did not lead to enhancement of particle collection. Hence, indicating that, for electrospun filters, the high collection by diffusion mechanism decreases the electrostatic collection. These results, however, have to be considered preliminary and more research is on-going using electrospun filters made from different materials and using different test conditions.

4. Conclusions

PA6 nanofibrous filters were produced by electrospinning onto a nonwoven viscose substrate. The influence of electrospinning parameters including solution concentration, viscosity, and conductivity, were studied for the production of nonwovens with different fiber size distribution ranging from 74 to 261 nm in diameters. Interestingly, spider-net structures, widely distributed among nanofibers, appear when solutions with PA6 concentration above 15 wt. % in formic acid are electrospun. The formation of such nanowebs seems to depend upon polymer concentration. By varying the nanofiber collection time, filters with different thickness of the PA6 nanofiber layer were obtained and their efficiency against nanoparticles penetration was evaluated for a face velocity of 1.7 cm/s. Particle penetration through the electrospun PA6 filters was strongly dependent on filter thickness, while the electrostatic forces did not enhance particle collection which only depends on diffusion mechanism. These results showed the possibility of applying the electrospinning technique to produce filter media and personal protective equipment against nanoparticle exposure.

Acknowledgments

The financial support of this work was provided by MICINN (Spanish Ministry of Science and Innovation) and ERDF (European Regional Development Fund) (ref: PSE-420000-2008-003).

References

[1] Aussawasathien D., Teerawattananon C., Vongachariya A. 2008 *J. Membr. Sci.* **315** 11-19.
[2] Thavasi V., Singh G., Ramakrishna S. 2008 *Energy Environ. Sci.* **1** 205-221.
[3] Golanski L., Guiot A., Tardif F. 2010 *J. Nanopart. Res.* **12** 83-89.
[4] Japuntich D., Franklin L.M., Pui D.Y., Kuehn T.H., Kim S.C., Viner A.S. 2007 *J. Nanopart. Res.* **9** 93-107.
[5] Ryu Y.J., Yong Kim H., Hyung Lee K., Chon Park H., Lee D.R. 2003 *Eur. Polym J.* **39** 1883-1889.
[6] Barakat N.A.M., Kanjwal M.A., Sheikh F.A., Yong Kim H. 2009 Polymer **50** 4389-4396.
[7] Hinds W.C. 1999 *Aerosol Technology*, John Wiley & Sons, New York.
[8] Yun K.M., Hogan C.J., Matsubayashi Y., Kawabe M., Iskandar F., Okuyama K. 2007 *Chemical Engineering Science* **62** 4751-4759.

Journal of Physics: Conference Series **304** (2011) 012068 doi:10.1088/1742-6596/304/1/012068

Penetration of nanoparticles in 5 nm to 400 nm size range through two selected fibrous media

C. Brochot[1], G. Mouret[2], N. Michielsen[1], S. Chazelet[3] and D. Thomas[2]

[1]Aerosol Physics and Metrology Laboratory, Institute for Radiological Protection and Nuclear Safety, BP68 - 91192,Gif-sur-Yvette Cedex, France

[2]Laboratoire Réactions et Génie des Procédés, Nancy Université, BP 20451 - 54001 Nancy, France

[3]Laboratory of polluant and air cleaning process, National Institute for Occupational Safety and Health, Rue du Morvan CS 60027 - 54519 Vandoeuvre Les Nancy, France

E-mail: clothilde.brochot@irsn.fr

Abstract. Due to the strong development of nanotechnologies, ultrafine particles could represent a growing hazard for workers health. When it is not possible to reduce the risk at its source, filtration systems are one of the means used to limit the exposure to hazardous substances such as airborne particles. The aim of this study is to measure the penetration of nanoparticles on a very large diameter range, from the nanometer size to the most penetrating particle size (MPPS). Here we present experimental results obtained for three different types of nanoparticles. Measurements of nanoparticle penetration through two low efficiency fiberglass media are carried out using two test benches presented in this article. Penetration values for carbon, copper and NaCl nanoparticles decreases with particle size, as predicted by theory. The value of the most penetrating particle size is situated between 100 and 300 nm. No thermal rebound was observed in this particle size range. The penetration values will be used, in further studies, to determine a global penetration model.

1. Introduction

The use of nanoparticles in the industrial world has increased since the 1990s. These particles, with at least one dimension less than 100 nm, are exploited for their special properties and can cause new cases of occupational exposure. Because of their dimension, they may deposit in the lungs and are also liable to reach the nervous system via the nasal cavities. The scientific community is therefore most concerned about the impact of these particles on the environment and health. The filtration system is one of the means used to reduce aerosol exposure.

The work presented in this paper concerns filtration efficiency for two filters used in filtration systems tested with three different types of nanoparticles and using two different test benches. In the first part, the article describes the two experimental setups. In the second part, a comparison between both test benches allows us to compile the penetration values. The properties of the tested filters are described in the third part of this paper. Finally, experimental results are plotted and penetration data are presented and discussed.

2. Material and methods

2.1. Description of FANA test bench

The nanoaerosol filtration test bench FANA [1] was developed at the Institute for Radiological Protection and Nuclear Safety (IRSN) in order to measure filter penetration as accurately as possible, for particle sizes ranging from 2.5 to 50 nm. FANA is presented in figure 1.

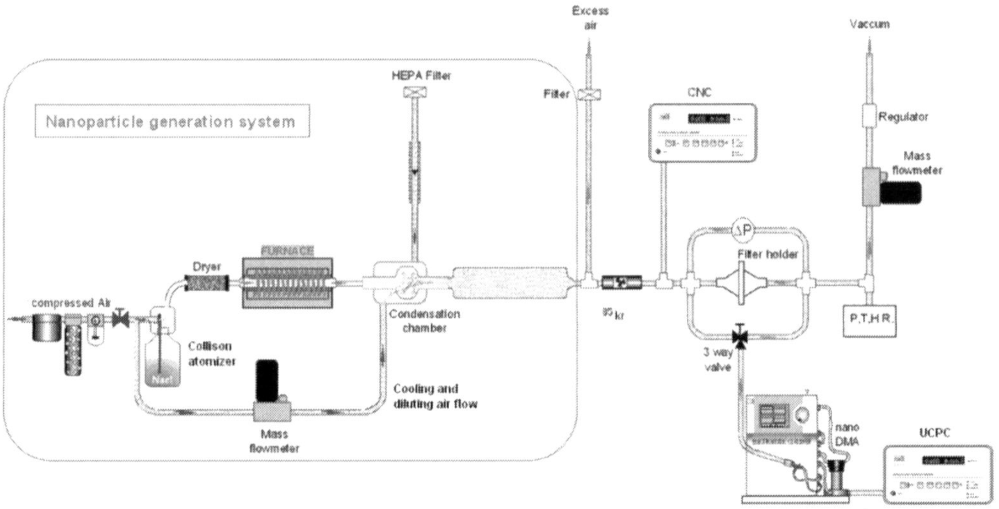

Figure 1. FANA experimental setup for nanoparticles (FANA 1)

FANA's production of nanoparticles is based on the work of Bartz et al. [2]. NaCl aerosol produced by an atomizer Collison TSI is evaporated using a tube furnace. The gas particles are then condensed in a nanometer size by mixing with cold air. The particle size distribution produced by FANA evolves according to the NaCl concentration, the furnace temperature and the air flow dilution.

For a furnace temperature of 750 °C and an air dilution of 30 L/min, the count median diameter of the generated distribution is 8.5 nm, the geometric standard deviation is 1.5 and the total concentration is 9.10^6 particles per cm^3 (figure 2). This aerosol size distribution has been characterized using a Scanning Mobility Particle Sizer (SMPS). This includes a source of ^{85}Kr and a nano-DMA (Differential Mobility Analyzer). The ^{85}Kr source neutralizes the aerosol to the Boltzmann equilibrium state.

Observations by transmission electron microscopy show that the generated NaCl particles are single particles and have a cubic shape (figure 3).

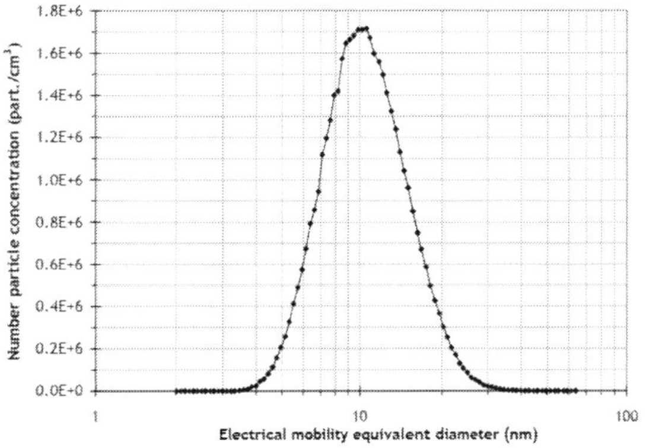

Figure 2. Particle size distribution produced by FANA test bench

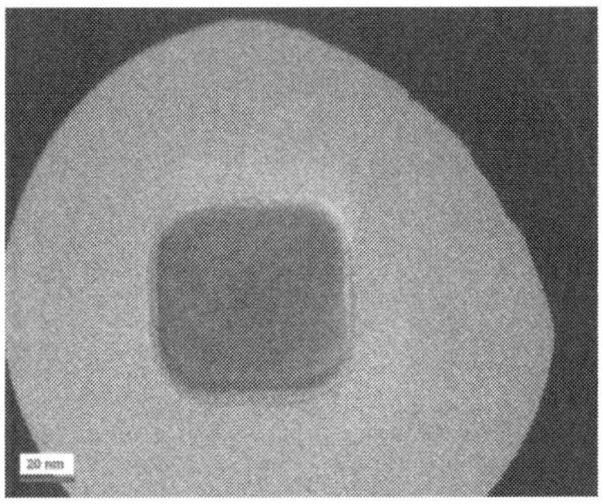

Figure 3. TEM picture of a NaCl nanoparticle

Once the aerosol has been characterized, all particles are sent to the filter holder. After the selection of particles by their electrical mobility, the experimental penetration of filters is determined according to the detection of the particles at the inlet and outlet of the filter holder.

To avoid artefact measurements, the same Ultrafine Condensation Particle Counter 3025A (UCPC) is used for measurements upstream and downstream of the filter holder. This UCPC is used in counting mode for a duration measurement of 60 minutes on average. The experimental penetration is then calculated by the ratio of the downstream number of particles to the upstream number of particles for the same volume sampled.

Temperature, pressure and relative humidity are monitored and a Condensation Nuclei Counter is installed at the exit of the nanoparticle generation system to control the stability of the generation. The filtration velocity through the filter is regulated during the test at 5.3 cm/s.

In order to measure penetration in the range of the Most Penetrating Particle Size (MPPS), the test bench was modified to generate larger nanoparticles (50-500 nm). In this case (figure 4), the generation system is composed of the atomizer Collison TSI and the dryer only.

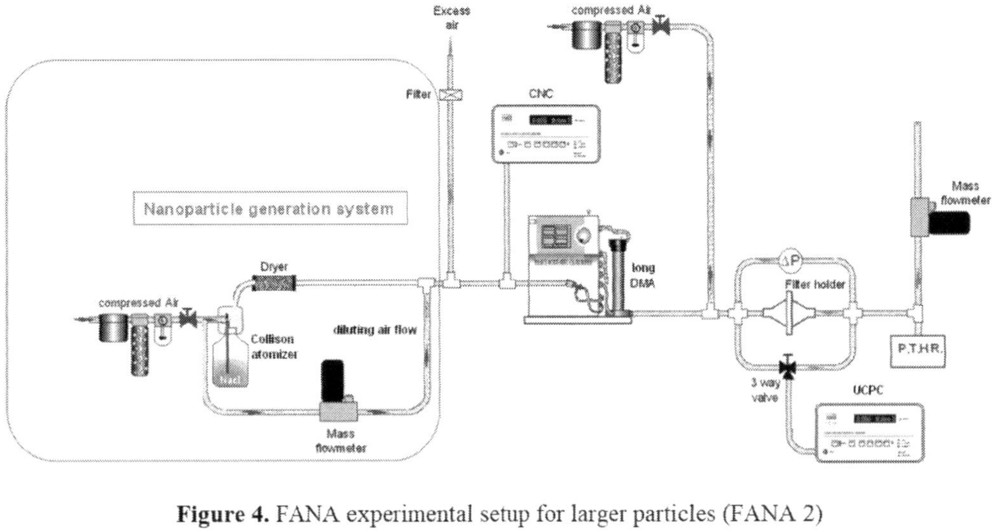

Figure 4. FANA experimental setup for larger particles (FANA 2)

This generation system produces an aerosol with a count median diameter of 70 nm, a geometric standard deviation of 1.7 and a total concentration of 1.10^6 particles per cm^3 (figure 5).

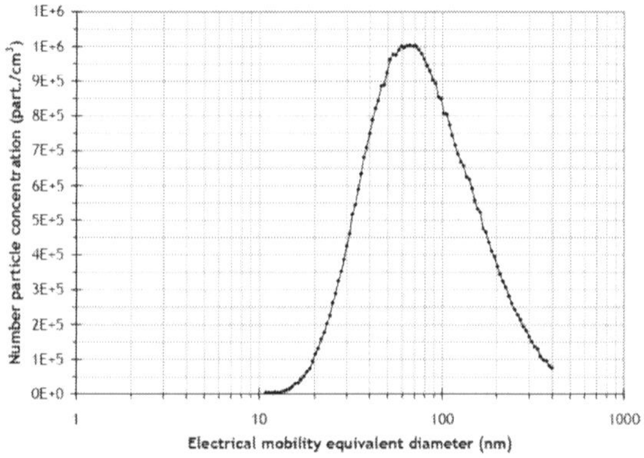

Figure 5. Particle size distribution produced by FANA test bench for larger particles

In order to avoid potential clogging problems during the measurement of penetration, the DMA is moved upstream of the filter holder. As described before, the same UCPC is used in counting mode for a duration measurement of 60 minutes on the average. The experimental penetration is calculated by the ratio of the downstream number of particles to the upstream number of particles for the same volume sampled. The filtration velocity through the filter is regulated during the test at 5.3 cm/s.

2.2. Description of MEFIANCE test bench
This experimental setup (figure 6) was developed at LRGP [3, 4].

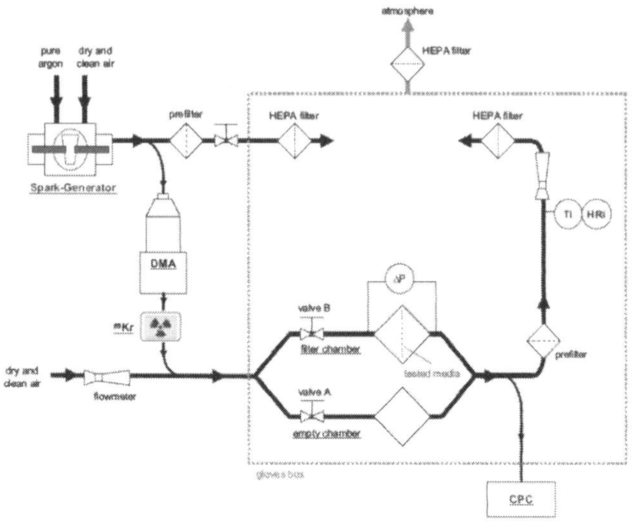

Figure 6. MEFIANCE experimental setup

The generation of aerosol on this test bench is obtained using the nanoparticle spark-generator PALAS GFG-1000. The particles are produced by electrical discharge between two electrodes. The size of the nanoparticles varies depending on the operating spark frequency of the generator. The particles are then selected according to their electrical mobility by the same type of nano-DMA as in the FANA test bench. These particles are brought to the Boltzmann equilibrium by the use of a source of ^{85}Kr.

The particles used in this experiment are generated using two types of electrodes, namely carbon and copper electrodes (figure 7 and 8). When using copper electrodes, the count median diameter is about 7 nm whereas, for carbon electrodes, it is about 70 nm.

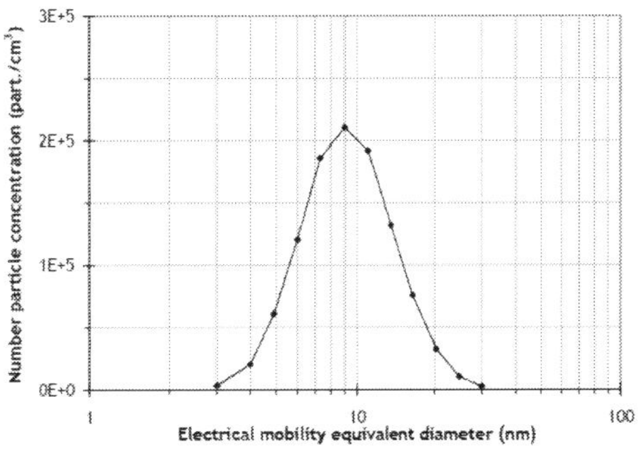

Figure 7. Particle size distribution produced by copper electrodes

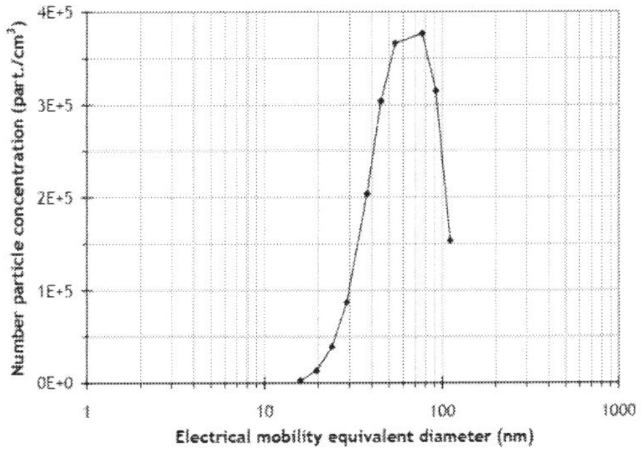

Figure 8. Particle size distribution produced by carbon electrodes

Then, particles are sent to the filtration system which is made up of two identical lines. Filtration velocity is regulated through filter at 5 cm/s. Only one Condensation Particle Counter (CPC) is used to remove artefacts. The 3022A CPC is used in the single count concentration mode. The experimental penetration is then calculated by the ratio of the outputs from the two lines.

3. Comparison of the two experimental setups

Because the two test benches are not operating in the same laboratory there are differences between the two experimental setups and protocol measurements. In this section, we will enumerate the differences in order to evaluate their potential influence on penetration measurement.

First, the configurations of the two filtration systems are not the same. Either the penetration is calculated by an alternative measure of the concentration upstream and downstream of the filter holder, with controlling of the stability of the total concentration generated, or the penetration is calculated by measurements using a double line. These two configurations were presented by Heim et al. [5], who concluded that using both of them helps to limit artefacts in measurement.

Secondly, the selection of particle size is carried out upstream of the filter holder or at the filter inlet and outlet: both configurations have been studied by Michielsen et al. [1] in the case of measuring the penetration of screens. The authors do not notice differences between the two measurement methods.

Thirdly, one can see that measurements of penetration are performed at a filtration velocity of 5.3 cm/s on the FANA test bench and a filtration velocity of 5 cm/s on the MEFIANCE test bench. This range of filtration velocity is the most commonly encountered in the literature [7, 8]. According to existing models [10], the influence of a 0.3 cm/s increase in velocity is negligible on the calculated penetration.

We can also notice that the two test benches do not use identical filter holders: the diameter of the filtration surface in FANA is 30.6 mm and the diameter of the filtration surface in MEFIANCE is 60.0 mm. Again, this should not make any differences on the results.

At last, the two test benches have been tested with calibrated screens in order to compare our penetration measurements with theoretical model [6]. The results obtained are in good agreement with the theory for both experimental setups. Figure 9 shows the penetration values through a screen as a function of particle diameter obtained with the FANA test bench.

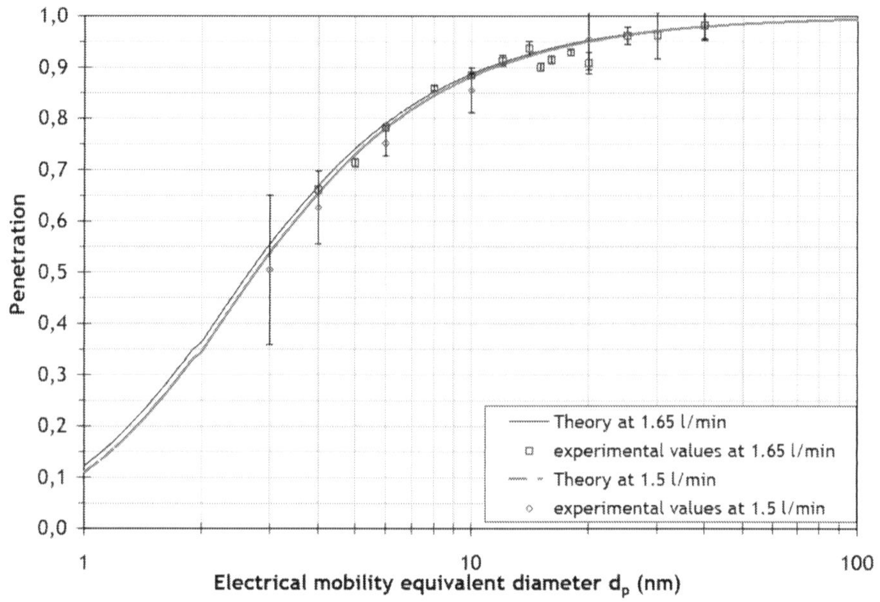

Figure 9. Penetration through a calibrated screen using FANA test bench

This validation allows us to combine penetration data on a very large diameter range for different nanoparticles.

4. Penetration results and discussion

4.1. Filter media properties

Two different commercial fibreglass filters are tested. They have a low efficiency and the table 1 shows their specifications. Physical properties have been investigated in order to determine the thickness, the pressure drop, the solidity, the Davies diameter and the geometric fiber diameter. The thickness is given by an average of microscopic observations. The solidity, α, is obtained by the following equation:

$$\alpha = \frac{G}{\rho_f Z} \quad , \tag{1}$$

where G is the filter's mass per unit area, Z is the thickness and ρ_f the fiber density.

The Davies fiber diameter, d_{Davies}, is the fiber diameter calculated from the pressure drop measurements with the following formula:

$$d_{Davies} = \sqrt{\frac{64\mu UZ\alpha^{3/2}(1+56\alpha^3)}{\Delta P}} \quad , \tag{2}$$

where μ is the air viscosity, U is the filtration velocity and ΔP the pressure drop.

The standard filter media are composed of fibers of different sizes and random orientations. The average geometric fiber diameter is obtained by observations with a TEM.

Table 1. Physical properties of media tested

	Filter 1	Filter 2
Average thickness (μm)	552 ± 50	427± 45
Solidity	0.05	0.064
Davies fiber diameter (μm)	4.23	3.25
Average geometric fiber diameter (μm)	3.19 (GSD=1.76)	5.14 (GSD=2.84)

4.2. Media penetration and discussion

Penetrations are plotted in figure 10 and 11 for the two fiberglass filters and the three different nanoparticles as a function of the electrical mobility equivalent diameter.

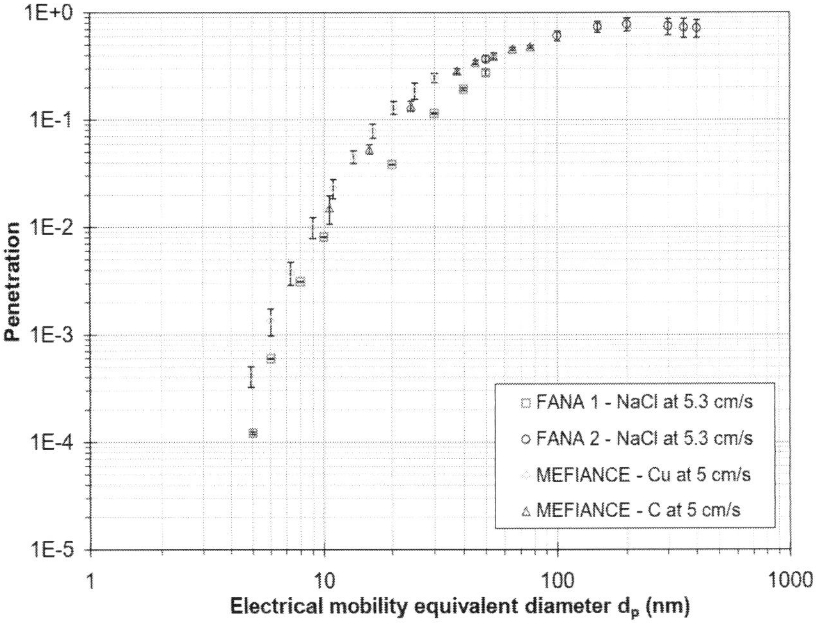

Figure 10. Spectral penetration of nanoparticles through filter 1

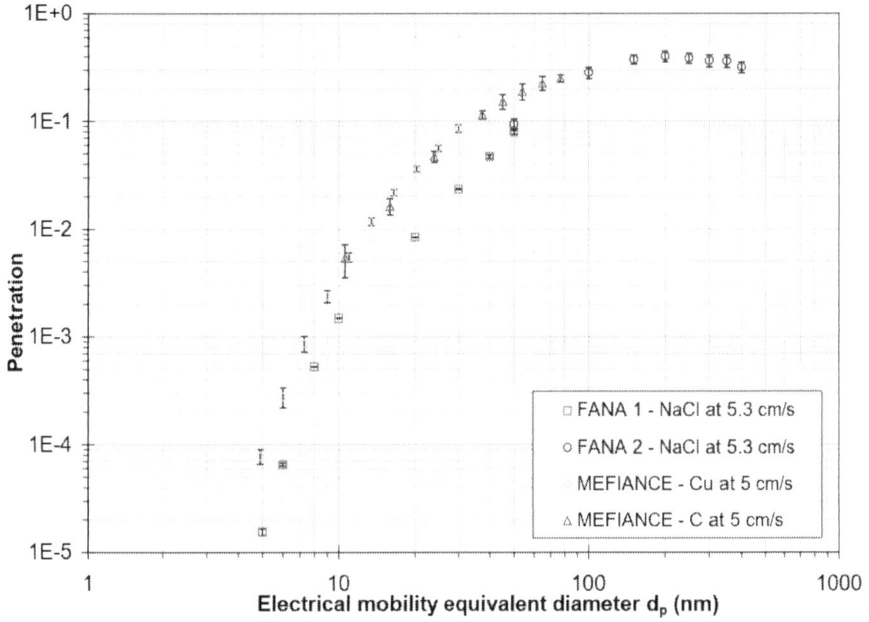

Figure 11. Spectral penetration of nanoparticles through filter 2

The data-gathering allows to obtain the penetration spectrum of the two filters studied for particle sizes ranging from 5 to 400 nm and for different kinds of particles. Each test bench generated two groups of measurements of penetration; and for each bench, the series overlap.

As expected by theory [9], the penetration decreases when diameter decreases. The value of the most penetrating particle size is situated between 100 and 300 nm. One can see that the penetration through filter 1 is slightly greater than the penetration through filter 2.

However, we can notice that there is a difference between the penetration values for NaCl nanoparticles and for C and Cu nanoparticles. One can see that, for filter 1 and diameters of 5 and 20 nm, the values of penetration for NaCl particles represent about 29 % of the values of penetration for copper particles. For filter 2, the penetration values of NaCl particles are about 23 % of the values of penetration of copper particles for diameters of 6 and 30 nm.

The morphologies of the particles are different and this may explain the gap between penetration for NaCl cubic nanoparticles and penetration for metal nanoparticles. But differences in morphology between Cu and C nanoparticles are also expected, and we do not see penetration differences. In addition, one can also notice that filters tested with FANA test bench and filters tested with MEFIANCE test bench are coming from different batches.

5. Conclusions and perspectives

After describing the two test benches used in this study, penetration measurements were performed for two different fiberglass fiber filters and three different types of nanoparticles: carbon, copper and NaCl nanoparticles.

First, the data-gathering allowed to obtain the penetration spectrum of filters studied for a very large diameter range from 5 nm to 400 nm. The penetration decreases when diameter decreases even below 10 nm. The most penetrating particle size is situated between 100 and 300 nm and the penetration through medium 1 is greater than the penetration through medium 2.

However, a comparison of the values of penetration showed a significant difference between NaCl nanoparticles and C and Cu nanoparticles penetration. Differences in the experimental protocol have been listed and did not explain the gap observed in penetrations. The difference in morphology between particles cannot be the unique cause of penetration differences. We have to keep in mind that the filters tested in this work are coming from different batches. This can also yield variations in the measured penetration.

This study gives reliable data on the penetration values on a very large diameter range. In parallel, filter characterisation have been studied. From now on, these data will be used to test existing models and to develop a penetration model.

References

[1] Michielsen N., Lelandais T., Brochot C., and Bondiguel S. (2009) Filtech conference, October 13-15 2009, Wiesbaden, Germany.

[2] Bartz H., Fissan H., and Liu B. Y. H. (1987) *Aerosol Science and Technology* **6** 163-171.

[3] Mouret G., Thomas D., Chazelet S. and Bemer D. (2008) World Filtration Congress -10, Avril 14-18 2008, Leipzig, Germany.

[4] Mouret G., Thomas D., Chazelet S., Appert-Collin J. C. and Bemer D. (2009) *Journal of Aerosol Science* **40** 762-775.

[5] Heim M., Mullins B. J., Wild M., Meyer Y. and Kasper G. (2005) *Aerosol Science and Technology* **39** 782-789.

[6] Cheng Y. S. and Yeh H. C. (1980) *Journal of Aerosol Science* **11** 313-320.

[7] Kim S. C., Harrington M. S., and Pui D. Y. H. (2007) *Journal of Nanoparticle Research* **9** 117-125.

[8] Japuntich D. A., Franklin L. M., Pui D. Y., Kuehn T. H. and Kim S. C. (2007) *Journal of Nanoparticle Research* **9** 93-107.

[9] Lee K. W. and Liu B. Y. H. (1982) *Aerosol Science and Technology* **1** 147-161.

[10] Vendel J., Laborde J. C., Michielsen N. and Gensdarmes F. (2009) Journal of Physics: Conference Series. **170** 012026.

Socio-economic analysis: a tool for assessing the potential of nanotechnologies

Jean-Marc Brignon

INERIS, Parc Technologique Alata BP2 – 60550 Verneuil-en-Halatte, France

E-mail: jean-marc.brignon@ineris.fr

Abstract. Cost-Benefit Analysis (CBA) has a long history, especially in the USA, of being used for the assessment of new regulation, new infrastructure and more recently for new technologies. Under the denomination of Socio-Economic Analysis (SEA), this concept is used in EU safety and environmental regulation, especially for the placing of chemicals on the market (REACh regulation) and the operation of industrial installations (Industrial Emissions Directive).

As far as REACh and other EU legislation apply specifically to nanomaterials in the future, SEA might become an important assessment tool for nanotechnologies. The most important asset of SEA regarding nanomaterials, is the comparison with alternatives in socio-economic scenarios, which is key for the understanding of how a nanomaterial « socially » performs in comparison with its alternatives. « Industrial economics » methods should be introduced in SEAs to make industry and the regulator share common concepts and visions about economic competitiveness implications of regulating nanotechnologies,

SEA and Life Cycle Analysis (LCA) can complement each other : Socio-Economic LCA are increasingly seen as a complete assessment tool for nanotechnologies, but the perspective between Social LCA and SEA are different and the respective merits and limitations of both approaches should be kept in mind.

SEA is a "pragmatic regulatory impact analysis", that uses a cost/benefit framework analysis but remains open to other disciplines than economy, and open to the participation of stakeholders for the construction of scenarios of the deployment of technologies and the identification of alternatives. SEA is "pragmatic" in the sense that it is driven by the purpose to assess "what happens" with the introduction of nanotechnology, and uses methodologies such as Life Cycle Analysis only as far as they really contribute to that goal. We think that, being pragmatic, SEA is also adaptative, which is a key quality to handle the novelty of economic and social effects expected from nanotechnology.

1. Introduction

The aim of this paper is to bring about some information regarding Socio-Economic Analysis and to discuss how to use it to help drive the development of nanotechnologies through a sustainable pathway.

It first describes Socio-Economic Analysis (SEA) and makes a brief overview of how it has been used in the past. More importance is devoted to discussing the current and future use of SEA in the context of the REACh regulation of chemicals in the EU, for two reasons: REACh (Regulation 1907/2006) is one of the most important piece of legislation currently requiring SEA to be used, and REACh might in the near future be a cornerstone of the regulation of nanomaterials in the EU.

We also review the state-of-the-art of the use of SEA for nanomaterials. Finally, on the basis of our experience from chemicals management and EU Research projects on nanomaterials, we make some proposals to adapt and improve SEAs for nanomaterials, and we clarify the relations between SEA and Life Cycle Analysis.

2. An overview of Socio-Economic Analysis

Socio-Economic Analysis could be defined as a variant for Regulatory Impact Analysis (terminology used in the USA) of Impact Analysis (as used by the European Commission). (Regulatory) Impact Analysis is used by public authorities "to balance the potential benefits and costs of action or lack of action", as the EU Treaty reads. (R)IA is the term used for the pre-ante assessment and preparation of legislation, and the denomination "SEA" is more often used for implementation of legislation, but the methodology is the same in the two situations.

The conceptual framework of both (R)IA and SEA, as clearly expressed in the case of the EU by the above excerpt of the Treaty, is that of cost-benefit analysis (CBA). The concept of CBA and its first tentative uses seem to stem from France in the XIX century. CBA has been systematically developed, among others, in the USA by the US Army Corps of Engineers, then in other OECD countries and in the EU, at first for public infrastructure projects [1]. Since a number of years, CBA concepts use has expanded considerably to public policy preparation and implementation and included more frequently and systematically public health and environmental issues. Several European Directives or Regulations (Water Framework Directive, Industrial Emissions Directive, REACh Regulation…), even if they leave room to Member States for choosing the means to achieve objectives of public health and ecosystems protection, do impose or strongly encourage that the choice be based on such techniques as CBA that allow to take into account as a whole the impacts on health, the environment and the economy.

The ambition of SEA in our context is to get a rational, balanced and holistic view of advantages and drawbacks of the introduction of new products and new technologies in society. For instance, to assess the impacts of introducing a new material on the market, SEA makes an inventory of all impacts of the material and its alternatives. Since past failures of regulation of chemicals and materials was sometimes explained by lack of information or attention to the existence of alternatives, in SEA the issue of the market and non-market utility of innovation, and the comparison with alternatives, is regarded as a key issue [2,3]. SEA also considers collective risks and deals with their economical dimension, and in that respect goes beyond "risk reduction strategies" that were previously carried out for chemicals in the EU for the pre-REACh regulations [4]. For instance, risk assessments and reduction strategies for chemicals were focused on PEC[1]/PNEC[2] ratios for ecosystems or exposure/NEL[3] for human health impacts, whereas SEA will concentrate on collective impacts such as overall morbidity/mortality, ecosystems damage…

2.1. Use of SEA for chemicals under the framework of the REACh Regulation

In the framework of REACH, substances "of Very High Concern" cannot be put on the market unless risks for human health and the environment associated with their use are "adequately controlled". If not (no possibility to control the risk below a significant level), industry can still request an authorization to put the substance on the market, on the grounds that the advantages of using the substance outweigh the potential unwanted impacts, and that there are no economically and technically viable alternative substances or technology. Even if not mandated by the legal text, providing an SEA with the authorization Dossier in that case seems the logical procedure that should be followed in the future, when applications procedures under REACh will actually start.

---

[1] Predicted Effect Concentration

[2] Predicted Non-Effect Concentration

[3] No-Effect Level

Alternatively, Member States (and the European Chemicals Agency upon request of the European Commission), can ask for a total or partial "restriction" of the access of an existing chemical to the market, on the mandatory condition that a SEA shows that the benefits of the restriction outweigh its social and economic cost.

Even if conducting SEAs for chemicals presents great technical and scientific challenges, as we discuss later in this paper, the European Chemicals Agency in charge of the implementation of REACh has included SEA in its work programme: a guidance for SEA has been published [5], and a Socio-Economic Analysis Committee has started elaborating opinions on whether particular chemicals should be restricted in the EU. The guidance is limited to a general process, as summed up in Figure 1, and to general information about the techniques that are at the basis of SEA: costing methods for financial impacts, description of welfare impacts indicators, and techniques for cost/efficiency, cost/benefit and multicriteria analysis.

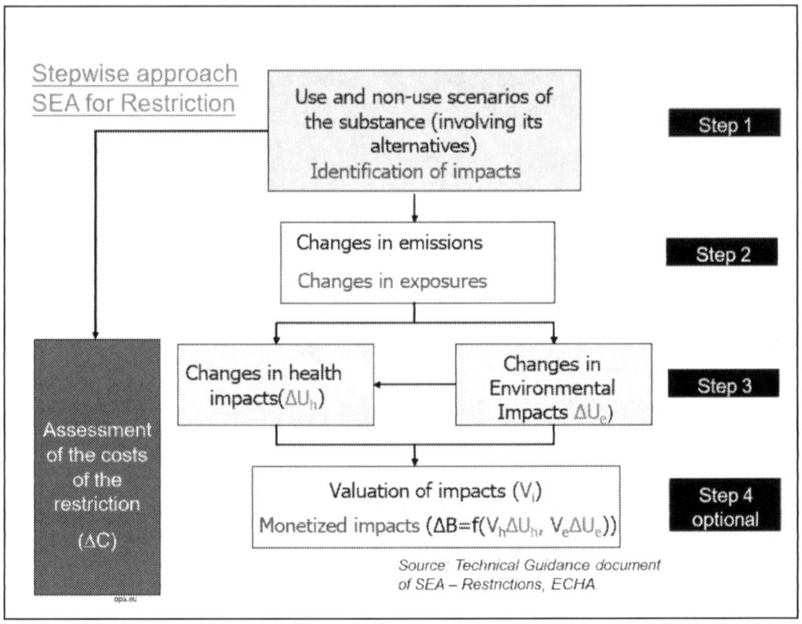

Figure 1. SEA Stepwise approach.

2.2. SEA in other environmental regulation in Europe

SEA is also used in the framework of the Directive on detergents (EC/648/2004): derogations to the prohibition of non-biodegradable surfactants in household detergents have to be based on a SEA comparing the socio-economic benefits of the surfactant to the damage caused by its release in the environment.

The future EU regulation on biocides (COM(2009) 267 final) does not request SEAs in the permitting process, but the "balance between advantages and risks" will be a criterion for decision (Article 21-3).

2.3. Nanomaterials and Safety Regulations in Europe

REACh applies to nanomaterials but currently ,without any specific provision, in particular there is no specific threshold for the registration of nanomaterials. Many nanomaterials are manufactured and imported in the EU below the thresholds applying to chemicals, and will not be registered nor assessed for their risks, unless the REACh regulation is revised and that particular aspect is changed in 2012.

Many difficulties also still exist for a definition and characterization of nanomaterials that would be effective and pragmatic under REACh, given the variety of nanomaterials and their functionalization. Other safety regulations in the EU do not either have specific provisions for nanomaterials: Directives for the protection of environmental media (air, water), Directives for Industrial Installations, Directives for biocides. In return, the new European Regulation on Cosmetics (EC 1223/2009) includes an obligation of pre-declaration and labeling of products containing nanomaterials, and the premises of a specific risk management system.

We can only at this stage conclude that SEA is just starting to play a role in risk management for nanomaterials under REACh and the cosmetic Regulations. But that this role is expected to raise significantly in the coming years, under both Regulations, and also under other environmental legislation.

3. The usefulness and the challenges of SEA for managing the risk of nanotechnology

In SEA, the central issue is the market (or consumer) and non-market (or societal) utility of nano-innovations, and its comparison with non-nano alternatives [3]. The functionality and utility of a material in relation to its alternatives will inform on the value of nanoinnovations, which can then be compared to the risks of the new nanomaterial (the comparison being also carried out for the alternatives).

The preoccupation in SEA to consider alternatives in the risk/benefit assessment and compare them with the new material under scrutiny is key to avoid some past failures of risk management. A lesson learnt for instance from the case of asbestos is that some actors might conceal available alternatives and narrow the debate on how to manage the risk of a material without questioning its social value [2]. In the current situation of high uncertainty and high stakes regarding risks of nanomaterials, utility is also a possible criteria to discriminate between promising nanomaterials and less valuable ones. Even when all risks (and a fortiori impacts) cannot be quantitatively and even qualitatively assessed, partial Socio-Economic assessments shed light on the relative merits of a nanomaterial versus its alternatives. The stake is to avoid to invest in low-value new materials (in a dense universe of competing value creation opportunities) that could consume limited financial of raw material resources or damage scarce environmental resource.

3.1. Current state-of-the art of SEAs for nanomaterials

Reviews [6] have highlighted that very few SEAs seem to have been conducted for nanomaterials or nanotechnologies until now. On the other hand, a considerably growing body of literature (scientific, grey, general public...) has noticed the high potential benefits (aquifer remediation, increased industrial safety, energy efficiency, substitution of hazardous chemicals, public health application...) and also potential risks from nanomaterials and call for socio-economic assessments.

These reviews also display the existence of a flourishing body of either optimistic or gloomy predictions regarding the future and the impacts of nanomaterials (in terms of a "nano-divide" between rich and poor countries, and in terms of employment or economic growth). These predictions are most often based on beliefs and not on evidence on the current state of our information. It is important that, as soon as feasible, as rational and transparent SEAs as possible start giving more robust basis for discussion and decision.

We give hereafter recent feedback from ongoing European Research Projects to which INERIS is taking part:

In the SAPHIR project on "Safe Manufacturing of Nanomaterials", a sustainability assessment was demanded by the European Commission at the beginning of the project. However, now that the project is towards it ends, it appears that the focus of industrial stakeholders has been put more on process design, and only on process safety (accidents) and workers health than overall sustainability issues. In the most advanced SEA case study, the cost/benefit analysis was reduced to an assessment of the financial cost of exposure reduction for workers due to lack of information and visibility on the supply chain. From this project is it also clear that data availability and data access (emissions, consumptions, description of supply and value chains...) are still a problem, probably because of confidentiality issues, and also because supply chains are not organized already at industrial scale.

In contrast, the NANOFOL project (Nanofolate biodevices for chronic inflammatory disease) started in 2009, with socio-economic impact and sustainability issues put at higher level in the project (WP). The need for a SEA of nanofolate devices, and to compare them with alternatives treatments, was identified by industry and scientists from the beginning. In this project there is therefore better hope to have useful SEAs for the nanobiodevices.

3.2. The Challenges of SEA for nanomaterials

The objective of this paper is not to discuss all the challenges and limitations of the cost/benefit, cost/efficiency or multicriteria analysis frameworks of SEAs, on which a vast body of economic literature already exists. We will just mention some features of SEA that are of importance in the case of nanomaterials:

- Many authors have highlighted the challenge of assessing the specific toxicological and ecotoxicological hazards linked to exposure to nanomaterials, and this is obviously a basic challenge for SEA.
- Going beyond "traditional" Risk Assessment (RA), SEA aims at predicting realistic estimates (and not "worst case" like in RA) of collective health and environmental impacts (and not of individual effects like in RA). Providing such impact estimates requires information on exposures along the lifecycle of nanomaterials, which is currently particularly scarce and difficult to access.
- The Cost Benefit paradigm underlying SEA supposes that benefits can provide a compensation for damages: in the case of (very) high and uncertain damages of emerging nanomaterials and technologies, such an assumption is controversial, because irreversible and long term damages could be impossible to be compensated by the benefits of the technology elsewhere.

Another less often underlined challenge is the existence of rather new impact pathways for nanomaterials, such as the ones proposed in the following examples:

- Nanosilver, given its strong and persistant biocidal activity, can impact waste water treatment works and biological activity in soils, leading to:
  - More degraded drinkingwater resource and treatment problems (because of deficiencies in waste watertreatment works upstream of surface water intakes).
  - Possible impacts on soil and food quality if nanosilver-containing sludge is spread on agricultural land.
- Some other nanomaterials (as carbon nanotubes) have a long persistence and high reactivity in the natural environment. Therefore they have been shown to adsorb chemicals like phytosanitary chemicals present in the environment, and transport them on very long distances, or to create new by-products of these chemicals [7].

3.3. Some new features of nanotechnologies and nanomaterials will make their socio-economic assessment more complex:

Rapidity: Nanotechnology is a fast-growing domain and socio-economic information on production and use is likely to become more rapidly outdated than in the case of usual chemicals.

Extensiveness: Nanotechnology is not a defined economic sector but is changing many basic technologies: the socio-economic implications of introducing a new nanotechnology or class of nanomaterials can be therefore very wide (across many economic sectors) and very deep (start at the beginning of the value/supply chain of a consumer product). Being « enabling technologies », nanotechnologies could change or create new functionalities among a wide array of industrial production sectors and consumer products. In such situations, SEA has a problem to compare the nanomaterial with alternatives that deliver the same functionality. The solution might be to adopt a wider point of view for the definition of the functionality, less focused on the technical characteristics of the nanomaterial, or to conduct ad-hoc assessments of the value attached by society to new functionalities (through economic valuation studies).

The **Novelty** of nanotechnologies and nanomaterials raises concern about the real-life applicability of risk mitigation measures. Economic actors in nanotechnology are often new players in the field, and are often small companies. They could therefore lack resources and experience to understand and correctly apply risk management measures that are themselves more complex (sophisticated cleaning and isolation procedures and techniques, clean rooms…) [8]. Theoretically adapted risk management measures could in real-life be difficult to apply or require a learning process that means a higher cost than thought. To produce realistic assessments, SEAs for nanomaterials should identify these situations based on real-life experience. This novelty also applies to some extent to the technologies to manage the nano-risks: technologies for monitoring nanoparticles, or the removal of nanoparticles from diverse environmental media are under development and their cost is difficult to assess.

The **Variety** of nanomaterials is likely to surpass the number of chemicals that regulators and industry have ever had to handle. Solutions will have to be found, and it is clear that socio-economic analysis will have to be carried out only on a selection of most important nanomaterials. Prioritization activities, which are becoming increasingly important under REACh to target resources to the most critical chemicals, will probably be even more important for nanomaterials. Screening tools using some socio-economic information will have to be developed for that purpose.

The **interpenetration between socio-economic impacts, innovation dynamics, and risk management for nanotechnologies**.

For some nanomaterials, the financial economic stakes are high for businesses and society, and competitiveness impacts of granting or not granting access to the market should be taken into account. SEAs should therefore use industrial economics methods and ratios to assess these impacts, which are seldom used in practice for regulating emerging technologies or materials and substances in the EU at present.

Industrial economics, and methods to represent innovation dynamics will also prove useful to carry out SEAs with longer term time-horizons, when the socio-economic assessment is build on foresight scenarios, and when the expected outcome of SEAs are more to highlight the far-reaching consequences of critical choices in emerging technology development. In that case, methods to integrate socio-environmental and health risks along with business risks in stakeholders decisions and strategies regarding innovation, and their uptake by the market, need to be developed, and embedded in SEAs. The use of « industrial economics » methods in SEAs would also help industry and the regulator share common concepts and visions about economic competitiveness implications of regulating nanotechnologies

4. Interactions between Life Cycle Approach and Socio-Economic Analysis

There seem to be a consensus among stakeholders on the necessity to have a Life Cycle Approach for nanomaterials (see USEPA position for instance in [13]), but the added-value for SEA and also the limitations of LCA should be underlined.

First, the evident value of Life Cycle Approach is to assess risks and impacts wherever and whenever they occur, and to address some issues important for nanomaterials : fate of nanomaterials after end-of-life of products, cross-contamination of other products, modification of recycling potential of products, overall life-cycle energy profile of products with nanomaterials designed for energy-saving applications…

The second asset of LCA is the integrated approach of environmental and human health impacts. LCA-based supplement specific impact assessment of toxic and ecotoxic effects of nanomaterials dissemination by taking into account the other impacts, such as: the toxic effect of possible use of numerous auxiliary chemicals, energy and other utility consumption… Depending on the production route used, these "annex" impacts could actually be as important as nano-specific impacts [9].

Given these assets of LCA, should SEAs systematically rely on LCA?

LCA itself has methodological weaknesses, the most important being the toxic and ecotoxic impacts, particularly important in the case of nanomaterials. Information delivered in this field by LCA is limited to global indicators of potential impact and not actual and focused enough estimates of potential impacts [8]. To inform Socio-Economic Analysis, Specific Impact Assessment on (eco)toxicological issues need to be combined with more general information from LCA for other impacts.

LCA is often regarded as attractive also because it is international and normalized, well known from industry, consultants. It is coming with well-established databases and software, and at the same time based on "hard" science. However, the ability of LCA to inform so complex, uncertain and "trans-scientific"[4] [9] questions such as the impact or sustainability of nanotechnologies is limited. One limitation is that LCA compares products or technologies of equal functionality, without questioning the utility of a nanomaterial versus its alternatives, as SEA basically does. Another limitation is a bias from LCA that it seems to give better assessments because there are numerous quantitative indicators in LCA. Another consequence is that, being methodology-driven, LCA is less adaptative in essence that the objective-driven SEA, whereas nanotechnology requires adaptative assessment frameworks [10].

Finally it is also worth mentioning the interesting developments for a "Socio-Economic LCA" (or the many analogues) [11]. Socio-Economic LCA intends to integrate new socio-economic impact categories and extends the notion of functional units so that they reflect the socio-economic world. But again, the Social LCA is a methodology-driven approach and attempts to as exhaustively as possible quantify all impacts. This carries the risk to hide or at least decrease the visibility of the most important impacts, the "what happens" that SEA essentially tries to capture.

5. Conclusion

In this paper we described the potential of Socio-Economic Analysis, for a pragmatic and open assessment of nanomaterials and nanotechnology. SEA is a "pragmatic regulatory impact analysis" [12], that uses a cost/benefit framework analysis but remains open to other disciplines than economy, and open to the participation of stakeholders for the construction of scenarios of the deployment of technologies and the identification of alternatives. SEA is "pragmatic" in the sense that it is driven by the purpose to assess "what happens" with the introduction of nanotechnology, and uses methodologies such as Life Cycle Analysis only as far as they are really contributing to that goal. We conclude that, being pragmatic, SEA is also adaptative, which is a key quality to handle the novelty of economic and social effects expected from nanotechnology.

References

[1] Pierce D., Cost Benefit Analysis and the Environment : Recent Developments, OECD, 2006
[2] Harremoës et al., Late Lessons from Early Warnings : the Precautionary Principle 1896–2000, Environmental Issue report no22, European Environment Agency, 2001
[3] Hansen S.F.et al., Late lessons from early warnings for nanotechnology, Nature Nanotechnology, Vol.3, August 2008
[4] European Commission, Technical Guidance Document on Development of Risk Reduction Strategies, 1998.
[5] ECHA, Guidance Document for Socio-Economic Analysis – Restriction, 2008
[6] Seear K. et al. (2009), The Social and Economic Impacts of Nanotechnologies: A Literature Review, Report for the Australian Department of Innovation, Industry, Science and Research, Monash University
[7] Helland A et al (2007)., Reviewing the Environmental and Human Health Knowledge Base of

[4] According to the definition given by A. Weinberg in [14], a trans-scientific" question can be correctly asked in scientific terms, by science is unable to give an answer.

Carbon Nanotubes, Environmental Health Perspectives, 115(8), 1125-1131

[8] Som et al., In press, The importance of life cycle concepts for the development of safe nanoproducts. Toxicology (2010), doi:10.1016/j.tox.2009.12.012

[9] Robichaud et al., Relative Risk Analysis of Several Manufactured Nanomaterials: An Insurance Industry Context, Environ. Sci. Technol.2005, 39, 8985-8994

[10] UNEP/SETAC, « Guidelines for the Social Life Cycle Assessment of Products », 2009.

[11] WTEC, Nanotechnology long-term impact and research directions : 2010-2010, Chapter 13 : Innovative and responsible governance of nanotechnology for societal development, 2010.

[12] Steinzor R. et al., A return to Common Sense : Protecting Health, Safety, and the Environment through the « Pragmatic Regulatory Impact Analysis », Center for Progressive Reform, White Paper N°909, 2009

[13] USEPA, 2007, Nanotechnology White Paper, EPA 100/B-07/001

[14] Weinberg, A. "Nuclear Reactions : Science and Trans-Science", American Institute of Physics, 1992

Challenges in assessing the environmental fate and exposure of nano silver.

Cherrie M Whiteley[1,2], Matteo Dalla Valle[2], Kevin C Jones[1] and Andy J Sweetman[1]

[1]Lancaster University, Lancaster Environment Centre, Lancaster, LA1 4YQ, UK.

[2]The REACH Centre, Lancaster Environment Centre, Lancaster University, LA1 4YQ, UK

E-mail: c.whiteley@lancaster.ac.uk

Abstract. There are significant challenges in assessing the fate and exposure of nano particles (NPs) owing to the lack of information on their use and potential pathways and sinks in the environment. This paper discusses these issues using nanosilver as a case study. The approach taken is to assess the production of nanosilver, the range of products that utilise its properties, potential environmental release pathways and subsequent fate. Estimates of UK nanosilver released into the environment have been made and sewage sludge identified as an important receiving compartment. This work aims to highlight the on-going challenges faced when assessing NPs in the environment. Using nanosilver as an example, difficulties in assessing production, use and release are discussed. The study also recommends a potential approach to assess the fate and behaviour assessment of nanosilver in the environment.

1. Introduction

Nanoparticles (NPs) released into the environment have the potential to cause significant impact as a result of their unique properties and potential widespread use and application [1]. However, many challenges exist in assessing their environmental fate and exposure, as well as identifying whether the release of nanoparticles into the environment could result in detrimental effects.

The approach taken in this study to assess the environmental fate and exposure of nanosilver has initially involved reviewing their production and the lifecycle of products that use nanoparticles. Reviewing the production and use of a particular NP is challenging owing to the lack of information on use of NPs. Attempts to record products that may contain NPs have been made by some organisations, such as the Woodrow Wilson Centre in the US (Project on Emerging nanotechnologies (PEN) consumer inventory)[2], and by UK Defra's Voluntary Reporting System (VRS)[3]. Clarifying whether the products actually contain the nano form or not, is an issue faced by these organisations. However, it is apparent that the use of NPs in consumer products is increasing by comparison to their bulk counterparts. The burden on the environment from widespread and extensive use is not yet known and current research is incomplete.

This study has focussed on UK use and the fate and exposure of nanosilver in the environment. The potential for large quantities of nanosilver being released into the environment raises the questions about the form of nanosilver once released. Data regarding the environmental form of silver is important when designing a best approach for environmental measurements.

Regulation covering the manufacture of nanoparticles, their use and release, lags behind rapid scientific and industrial advancements. A key example is represented by the recent EU Regulation REACH (Registration Evaluation Authorisation Restriction of Chemicals), which came into force on the 1st June 2007. The current REACH regulation applies to all chemicals manufactured or imported in quantities of more than 1 tonne per year into the EU, including chemicals manufactured in the nano form. A potential risk with regulating NPs is the low NP tonnages will fall outside the radar of regulation. There are also further limitations for REACH regulating NPs as their properties vary significantly from their bulk counterpart. Various endpoints have been suggested for testing of NPs such as specific surface area, zeta potential (surface charge) and surface chemistry.

Attaining a clearer understanding of the NPs fate and behaviour, and chemical and physical form in the environment will help support the development of effective regulation [4].

2. Nano Silver production and use

Nanosilver is increasingly applied and used in products for its antimicrobial properties. In order to better understand the impact of nanosilver upon the environment, our approach has first been to investigate global and historic uses of bulk silver and the emerging uses of nanosilver in the UK market. In this instance nanosilver refers to silver nanoparticles which are described as 'nanoscale clusters of metallic silver atoms' [5]. The total surface area maximises the effect of the silver and the presence of both metallic and ionic form will characterise many nanosilver products [1]. Commonly most commercially available silver nanoparticles have a diameter of 10 nm with specific surface area 9-11 m^2g^{-1} [6].

Silver is a familiar metal in modern society and has been in use for many hundreds of years. As a transition metal group which is soft, white, and lustrous, it has many beneficial properties. It is ductile, malleable and has high electrical and thermal conductivity which lend its use to many applications. Silver, as well as nanoparticles of silver have antimicrobial properties [7], which have been applied for uses in cosmetics, coatings, and medical applications. The primary advantage nanosilver has over its bulk counterpart is the potential to use less material in application thus providing weight and cost reductions.

Globally the use of silver has been dependent on the available supply driven by discovery of new silver mines and the application of new extraction techniques. The global demand of silver has been approximately 25,000 t/yr over the past 10 year [8]. Dips in demand have occurred, for example during 2009 a drop in fabrication has been noted due to the global financial crisis 2009, which led to fabrication figures similar to that in 2003 [8].

2.1. UK Nanosilver demand

In order to estimate the potential demand for nanosilver in the UK, we used information on existing global demand levels for silver. Global demand levels as sourced from the Silver Institute include Industrial Applications, Photography, Jewelry, Silverware, Coins & Medals. The sum of these demands is referred to as the Total Fabrication amount. The UK is assumed to use 8% of the total fabrication demand for silver, as seen in figure 1. However the maximum demand of nanosilver could be assumed to be just the silver industrial application demand, while minimum or lowest level is assumed to be the 5% of the industrial application. Figure 1 shows the output estimations for nanosilver demand in the UK.

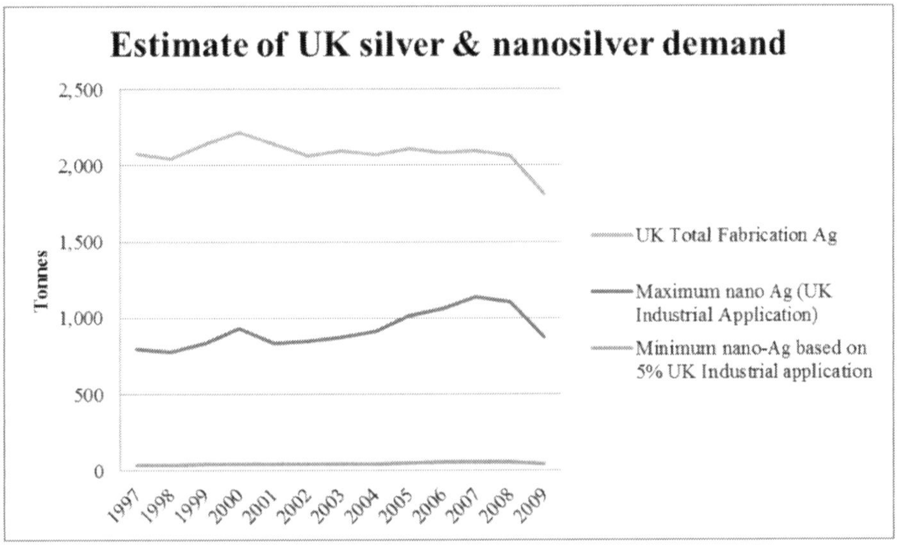

Figure 1. Estimates of UK nano silver demand.

3. Environmental release

Establishing potential or actual environmental release routes and quantities is required to better understand the nanoparticles fate and exposure. There have been a small number of studies on release. For example Geranio and Heuberger et al. (2009)[9] investigated nine textile types with nanosilver incorporated into the cotton, polyester or polyamide fibre. During washing the total Ag released varied from 1.3 to 35% of the total Ag in the fabric, although the amount lost after subsequent washes decreased. The coarse fraction >450 nm sized particles were identified as the greatest lost from textiles which were attributed to the mechanical stress during the washing process [9]. This study confirms that there would be some release into waste water from silver articles during the washing process. Kaegi et al. (2010) [10] reported the release of nanosilver from painted facades. The initial runoff from leaching from a painted façade gave a maximum concentration of 145 mg/l. The study determined that 0.5 mg/m^2 was the total amount of Ag released from the paint during one year of exposure which was about 30 % of the applied Ag [10]. Another study demonstrated that nanosilver impregnated socks lose nearly 100% of their silver content within four washings [11].

Data developed from direct studies on release during product use can be used to estimate the quantity of nanosilver that is potentially present in environmental compartments. The figure can then be compared against the natural background levels of silver in the environment. Previous work that has modeled release in the environment includes [12] and [13].

4. Environmental fate and challenges

The environmental form that nanosilver takes is one of the more complex issues in studying its behaviour in the environment. Understanding the transformation and speciation of silver in the environment is important for analysis, assessment of toxicity, and understanding its fate and behaviour [14].

Four oxidation states of silver exist: 0, 1+, 2+, and 3+; the most commonly found form in the environment being 0 and 1 [15]. Nanosilver has a propensity to change form and bind with other elements. Silver has a strong affinity to bind with sulphur and is commonly found in a sulphur form in the environment [16]. Silver can be present in the form of silver oxide, silver nitrate, silver halides (Floride, Chloride, Bromide), silver sulphate, silver thiosulphate complexes and silver sulphide. The solubility and toxicity of these forms varies and their presence in particular compartments will be dictated by the concentration of other chemicals, pH and organic matter [17].

Silver has been reported to be removed efficiently by wastewater treatment (WWT) plants (>94%) [18,16]. Settling and filtration are the main form of removal which also relates to the high particle-water partition coefficient (Kd) (Shafer et al., 1998 [18]). However, removal efficiencies vary and contamination of receiving waters from WWT plants is thought to be likely. Sewage sludge remains an important potential source of silver to the environment where it binds efficiently with sulphur. Kim (2010) [19] found nanoAgS in sludge which was hypothesised to have formed insitu via reduced sulphur reacting with the Ag NPs and soluble Ag species.

During 2004 nearly 1.4 million tons of dry solids were produced in England and Wales with 1.6 million estimated for 2010. Of this 62% of sewage sludge was applied to over 80,000 hectares of agricultural land with other disposal outlets such as incineration (19%), land reclamation (11%), other (7%) including compost, industrial crops, and landfill (1%) accounting for the rest [20].

Sludge applied to land is regulated for some metals and pathogens and maximum permitted concentrations in soils have been set by the European Community Directive 86/278/EEC (CEC, 1986), details of which have been implemented in England, Scotland and Wales by The Sludge (Use in Agriculture) Regulations. Metal concentrations within sludge must be regularly analysed with soil analysis undertaken before and after application. However, there is no concentration limit set for silver. Within the literature Beckett (1978) [21] reports a value of silver in a UK dried digested sludge at 200 mg/kg. A more recent analysis of silver in UK sludge has not been identified although a typical concentration would be approximately 25 mg/kg dry weight [22].

Soil quality is dependent on chemical, physical and biological factors [23]. The application of municipal sludge onto land for agricultural purposes has the potential to improve aspects of soil quality. The accumulation of metals on agricultural land remains a concern for soil quality for risks such as phytotoxicity. Smith (2009) [24] concluded that the 'risks to the environment, human health, crop quality and yield and soil fertility, from heavy metals in MSW-compost are minimal'. However, this review excluded silver.

5. Challenges for environmental exposure

Before considering environmental exposure it is worth briefly discussing occupational exposure to silver and nanosilver. Dermal and ingestion exposure are the two main potential human pathways during nanosilver production and use. There is also the potential for accidental ingestion and although silver is not an essential trace element in the body it is readily absorbed and commonly present in the bloodstream at low levels <2.3 μg/L [25]. There is also some evidence that inhalation exposure to airborne metals is possible in the workplace, Park et al. (2009) [26] monitored Ag NPs that were released in a commercial production facility during liquid-phase production of Ag.

Identification of nanosilver in the environment has only been confirmed in a handful of studies. The presence of nanosilver in the environment is not necessarily due to engineered nanosilver usage. However, new usage could have an impact and there are complications with existing tools and techniques for the identification of nanosilver in the environment. To better establish nanosilver in the environment we have expanded upon the potential exposure pathways from current usage.

5.1. Nanosilver exposure pathways

Products containing nanosilver that are potentially in use within the UK have been divided into a number of product categories. This is not a definitive list and the information has come from existing databases including PEN's project consumer inventory [2], Defra's VRS [3] and internet searches. These product categories were used to derive potential exposure pathways as displayed in table 1.

Table 1. The potential exposure pathways of nanosilver from products used in the UK into the environment has been summarized.

Category	Release Pathway	Number of products in category
Medical Application	Medical Incineration, Landfill	1
Health and Fitness	Wastewater, Landfill	11
Home and Garden	Wastewater, Landfill, Runoff, direct & diffuse pathway to aquatic environment	3
Food and Beverage	Landfill, Recycled, Lost, Ions to water	2
Appliances	Wastewater	1

Items with silver impregnated into their structure such as plastics are most likely to end up in landfill or be recycled. Plastic containing silver is designed to release ions for the effect of antimicrobial action to take place. The lifecycle of these plastics has been measured and assumed to be about 3 years [27]. On disposal the product may go to landfill and uncertainty exists as to whether the disposed product still contains silver. Silver ions are also used in shower filter systems which may then be released from the filter and enter the wastewater treatment system.

Nanosilver is also used in paint products for anti-mold protection and so there is potential release via runoff, direct & diffuse pathway to aquatic environment [10]. The disposal of medical equipment and clothing is controlled with hazardous waste going to incineration or other waste going to landfill.

Clothing and sheets may also release nanosilver in the indoor environment. Abrasion against skin of socks and towels during use may cause release of particles in indoor and outdoor environment. Household cleaning equipment may remove some of the released material which would normally be disposed into waste or via landfill. Socks and material washed in washing machines containing nano silver are known to be released into wastewater [9]. Once items are no longer required, clothing/material/socks maybe recycled or disposed of to landfill.

Nanosilver roll on deodorant is designed to adhere to skin but this is likely to be washed off and pass into the wastewater system. Spray deodorant is directed at the skin but may also be accidentally sprayed on other household surfaces. Deodorant may also transfer onto clothing which again may be washed and any remains may go to wastewater. Also inhalation may occur if developed into spray products.

5.2. Potential release estimates

Some assumptions have been made to calculate potential quantities of nanosilver released into sludge and water. The results are shown in table 2. Kaegi et al. 2010 [10] estimated that more than 30% of the total nanosilver is released to the environment from painted facades. This figure is applied to the nanosilver demand for the UK, as calculated from the '5% UK Industrial application'. Removal efficiencies of waste water treatment suggest that a suitable figure to apply is 90% with the remaining 10% potentially releasing into water and aquatic environments.

Table 2. Assumptions and quantities of nanosilver in UK environment 2009.

Assumption	Quantity (tonnes)
Assume nano goes in waste water system	13.1
Assume nano goes 90% to sludge	11.8
Assume nano goes 10% to water	1.3

6. Future considerations

Nanosilver is likely to continue to be developed for its antimicrobial properties and applied to many consumer products. Clarification of the environmental form of silver will enable better understanding of its fate in the environment. There are currently a lack of studies detailing environmental measurements and detection using standard equipment. Detection and measurement of silver in the environment remains challenging but will support fate studies.

The exposure scenarios for nanosilver can only be comprehensive if information on product types and use becomes available. In the meantime only estimates can be made about production of nanosilver and use. Commercially sensitive data covering production volumes remains restricted. Progressing work in field experiments, measuring release will help test exposure scenarios and support models predicting release.

Although this work has not considered the toxicity of nanosilver it is something that must acknowledged. Nanosilver has been found to be toxic to some aquatic species, detail of which can be found in the literature [28, 29]. However until further research is conducted the impact remains unclear.

It is important that the necessary understanding of the environmental fate and exposure of nanosilver is developed and can support the risk assessment process. As highlighted above, current REACH regulation applies to all chemicals manufactured or imported in more than 1 tonne per year into the EU, including chemicals manufactured in the nano form. This regulation uses a risk assessment process to protect human health and the environment. To adequately mitigate the risk, validated nano form specific data and knowledge will be required. Various endpoints have been suggested for testing NPs such as specific surface area, zeta potential (surface charge) and surface chemistry. The lack of knowledge and information available on nano particles in the environment remains an issue. Regulation and standardisation will not be achieved in the nano arena until further research has been undertaken to resolve the knowledge gaps. In the meantime industrial output will likely continue to increase.

References

[1] Panyala NR, Pena-Mendez EM, Havel J 2008 Silver or silver nanoparticles: a hazardous threat to the environment and human health? J. Appl. Biomed. **6** (3):117-29.

[2] PEN PoEN 2010 Consumer Products. [23rd April 2010]; Available from: http://www.nanotechproject.org/inventories/consumer/.

[3] DEFRA. 2008 [cited UK Voluntary Reporting Scheme for engineered nanoscale materials: Available from: http://www.defra.gov.uk/environment/quality/nanotech/documents/vrs-nanoscale.pdf.

[4] Handy RD, Owen R, Valsami-Jones E 2008 The ecotoxicology of nanoparticles and nanomaterials: current status, knowledge gaps, challenges, and future needs. Ecotoxicology. **17** (5):315-25.

[5] Marambio-Jones C, Hoek EMV 2010 A review of the antibacterial effects of silver nanomaterials and potential implications for human health and the environment. J. Nanopart. Res. **12** (5):1531-51.

[6] Navarro E, Baun A, Behra R, Hartmann NB, Filser J, Miao AJ, et al. 2008 Environmental behavior and ecotoxicity of engineered nanoparticles to algae, plants, and fungi. Ecotoxicology. **17** (5):372-86.

[7] Morones JR, Elechiguerra JL, Camacho A, Holt K, Kouri JB, Ramirez JT, et al. 2005 The bactericidal effect of silver nanoparticles. Nanotechnology. **16** (10):2346-53.

[8] The Silver Institute Demand and Supply in 2009. [3rd November 2010]; Available from: http://www.silverinstitute.org/supply_demand.php#demand.

[9] Geranio L, Heuberger M, Nowack B 2009 The Behavior of Silver Nanotextiles during Washing. Environmental Science & Technology. **43** (21):8113-8.

[10] Kaegi R, Sinnet B, Zuleeg S, Hagendorfer H, Mueller E, Vonbank R, et al. 2010 Release of silver nanoparticles from outdoor facades. Environ. Pollut. **158** (9):2900-5.
[11] Benn TM, Westerhoff P 2008 Nanoparticle silver released into water from commercially available sock fabrics (vol 42, pg 4133, 2008). Environmental Science & Technology. **42** (18):7025-6.
[12] Mueller NC, Nowack B 2008 Exposure modeling of engineered nanoparticles in the environment. Environmental Science & Technology. **42** (12):4447-53.
[13] Gottschalk F, Sonderer T, Scholz RW, Nowack B 2009 Modeled Environmental Concentrations of Engineered Nanomaterials (TiO2, ZnO, Ag, CNT, Fullerenes) for Different Regions. Environmental Science & Technology. **43** (24):9216-22.
[14] Bell RA, Kramer JR 1999 Structural chemistry and geochemistry of silver-sulfur compounds: Critical review. Environmental Toxicology and Chemistry. **18** (1):9-22.
[15] Purcell TW, Peters JJ 1998 Sources of silver in the environment. Environmental Toxicology and Chemistry. **17** (4):539-46.
[16] Kramer JR, Bell RA, Smith DS 2007 Determination of sulfide ligands and association with natural organic matter. Appl. Geochem. **22** (8):1606-11.
[17] Mukherjee B, Weaver JW 2010 Aggregation and Charge Behavior of Metallic and Nonmetallic Nanoparticles in the Presence of Competing Similarly-Charged Inorganic Ions. Environmental Science & Technology. **44** (9):3332-8.
[18] Shafer MM, Overdier JT, Armstong DE 1998 Removal, partitioning, and fate of silver and other metals in wastewater treatment plants and effluent-receiving streams. Environmental Toxicology and Chemistry. **17** (4):630-41.
[19] Kim B, Park C-S, Murayama M, Hochella MF 2010 Discovery and Characterization of Silver Sulfide Nanoparticles in Final Sewage Sludge Products. Environmental Science & Technology.null-null.
[20] Water UK 2006 Recycling of Biosolids to Land. [3rd November 2010]; Available from: http://www.water.org.uk/home/news/press-releases/biosolids-and-agriculture-/final-revised-biosolids-briefing-pack-2006-v8.pdf?s1=sludge&s2=properties.
[21] Beckett PHT 1978 An all-element analysis of digested sewage sludge. Water Pollution Control. **77** (4):539-46.
[22] Gray NF 2005 Water technology : an introduction for environmental scientists and engineers. 2nd ed. ed.(Oxford: Elsevier Butterworth-Heinemann)
[23] Wander MM, Drinkwater LE 2000 Fostering soil stewardship through soil quality assessment. Applied Soil Ecology. **15** (1):61-73.
[24] Smith SR 2009 A critical review of the bioavailability and impacts of heavy metals in municipal solid waste composts compared to sewage sludge. Environ. Int. **35** (1):142-56.
[25] Lansdown ABG 2007 Critical observations on the neurotoxicity of silver. Crit. Rev. Toxicol. **37** (3):237-50.
[26] Park J, Kwak BK, Bae E, Lee J, Kim Y, Choi K, et al. 2009 Characterization of exposure to silver nanoparticles in a manufacturing facility. J. Nanopart. Res. **11** (7):1705-12.
[27] Blaser SA, Scheringer M, MacLeod M, Hungerbuhler K 2008 Estimation of cumulative aquatic exposure and risk due to silver: Contribution of nano-functionalized plastics and textiles. Science of the Total Environment. **390** (2-3):396-409.
[28] Scown TM, Santos EM, Johnston BD, Gaiser B, Baalousha M, Mitov S, et al. 2010 Effects of Aqueous Exposure to Silver Nanoparticles of Different Sizes in Rainbow Trout. Toxicological Sciences. **115** (2):521-34.
[29] Ratte HT 1999 Bioaccumulation and toxicity of silver compounds: A review. Environmental Toxicology and Chemistry. **18** (1):89-108.

Journal of Physics: Conference Series **304** (2011) 012071 doi:10.1088/1742-6596/304/1/012071

Are classical process safety concepts relevant to nanotechnology applications?

Paul R Amyotte
Department of Process Engineering and Applied Science, Dalhousie University, 1360 Barrington Street, Halifax, NS Canada B3J 2X4

Email: paul.amyotte@dal.ca

Abstract. The answer to the question posed by the title of this paper is yes – with adaptation to the specific hazards and challenges found in the field of nanotechnology. The validity of this affirmative response is demonstrated by relating key process safety concepts to various aspects of the nanotechnology industry in which these concepts are either already practised or could be further applied. This is accomplished by drawing on the current author's experience in process safety practice and education as well as a review of the relevant literature on the safety of nanomaterials and their production. The process safety concepts selected for analysis include: (i) risk management, (ii) inherently safer design, (iii) human error and human factors, (iv) safety management systems, and (v) safety culture.

1. Introduction

Process safety deals with the prevention and mitigation of process hazards leading to fires, explosions and releases of toxic materials. Process safety is therefore focused on processing facilities and is different in scope than occupational safety, which is aimed more at protection of the individual. It should be noted, however, that process safety and occupational safety have several common features; further, the distinction between a process incident such as a fire, and an occupational incident such as a fall from height, may be quite meaningless to the affected worker. This is especially the case with nanomaterials and the accompanying concern with the uncertain nature of chronic health hazards. Nevertheless, attention to process safety issues does afford the possibility of an integrated approach to loss prevention that considers harm to people, property (asset) damage, production (business) interruption, and degradation of the natural environment.

The ensuing sections contain an analysis of the following process safety concepts in terms of their relevance to the safety of nanomaterial production and handling:

- risk management,
- inherently safer design,
- human error and human factors,
- safety management systems, and
- safety culture.

Each topic is first defined and then described from a process safety perspective. This is followed by an analysis of existing and potential applications to the field of nanotechnology. The five topics listed above were selected because, collectively, they form a body of knowledge that is at the forefront of current emphasis in process safety education, research and practice.

2. Risk management

Functional definitions of *hazard*, *risk* and *risk management* according to Wilson and McCutcheon [1] are as follows:

Hazard: The potential of a machine, equipment, process, material or physical factor in the working environment to cause harm to people, environment, assets or production.

Risk: The possibility of injury, loss or environmental incident created by a hazard. The significance of risk is a function of the *probability* of an unwanted incident and the *severity* of its consequence.

Risk Management: The complete process of understanding risk, risk assessment, and decision making to ensure effective risk controls are in place and implemented. Risk management begins with actively identifying possible hazards leading to the ongoing management of those risks deemed to be acceptable.

2.1. Process safety features

A conceptual framework of the risk management process for process safety concerns is given in figure 1; embodied in this fundamental flowchart is the cycle of risk analysis, which enables risk assessment,

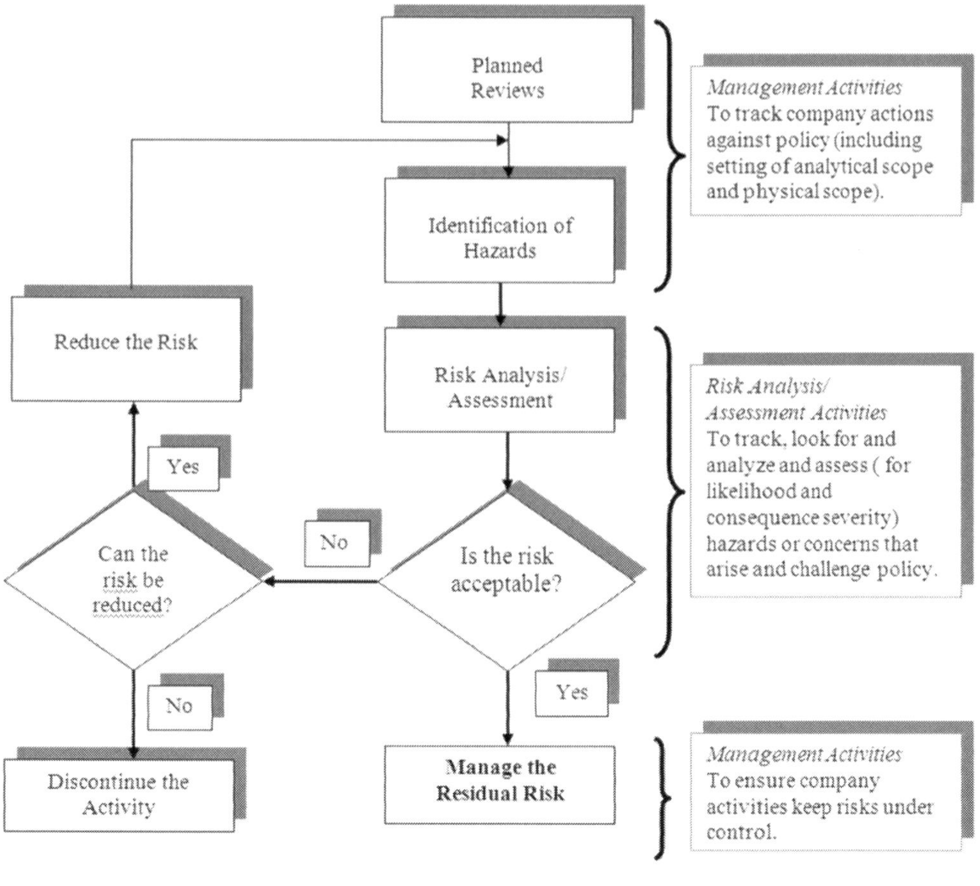

Figure 1. Risk management process (adapted from Amyotte and McCutcheon [2]).

which in turn enables risk management. In essence, process safety practitioners *analyze risk* (for probability and consequences), so they can *assess risk* (with respect to acceptability), so they can ultimately *manage risk*. As shown in figure 1, it is not possible to commence this cycle without first effectively identifying the hazards of concern.

2.2. Nanotechnology applications

The UK Health & Safety Executive's document *Reducing Risks, Protecting People* [3] comments in section 32:

...as the most obvious risks have been tackled, new and less visible hazards have emerged and gained prominence. Typical examples include those arising from technologies such as biotechnology, and processes emitting gases which contribute to global warming and ozone depletion. One frequent characteristic of these new hazards is that it can be very difficult to define precisely the risks they may give rise to, even when scientific knowledge is pushed to the limit. The processes that may give rise to risks are only partially understood with the result that regulatory decisions must frequently be based on limited data and considerable scientific and technological uncertainties. The control measures required by regulation should reflect the nature of the uncertainties and err on the side of health and safety.

Reference [3] (sections 87 and 88) distinguishes between *uncertainty* and *ignorance*, with ignorance indicating a lack of awareness of risk-influencing factors as would exist when hazard identification is incomplete. Uncertainty is defined as a state of knowledge in which, although risk-influencing factors have been identified, it is not possible to precisely describe the likelihood of any adverse effects or the magnitude of the effects themselves [3]. Figure 2 in Appendix 1 of Reference [3] identifies procedures for attempting to deal with uncertainty when assessing risks. With increasing uncertainty of likelihood, it is recommended that emphasis be placed on consequences (i.e., the event is considered by focusing solely on the consequences); with increasing uncertainty of consequences, it is recommended that the focus be on putative (supposed) consequences and scenarios. Significant uncertainty in both likelihood and consequences trends toward ignorance [3].

Considerations such as those described above play a key role in assessing the risk arsing from process hazards and explain, in part, the emphasis placed in the process industries on thorough scenario generation and hazard identification. Similar considerations apply to the nanotechnology field, as evidenced by the comments on knowledge uncertainty made in virtually every publication reviewed during preparation of the current paper (e.g., [4 – 10]). For example, Williams et al. [10] state that there is a critical lack of data concerning potential hazards, does-response relationships and exposure levels for most nanomaterials currently available. In a similar vein, Ferdous et al. [11] have demonstrated the use of fuzzy-based and evidence theory-based formulations to address data and dependency uncertainties, respectively, in fault and event tree analyses for process systems.

Williams et al. [10] also comment on the matter of risk *perception* in the case of new technologies such as nanotechnology, with perception acting essentially as a third component of risk in addition to likelihood and consequence severity. This again is not unlike the management of process risks related to activities such as waste disposal and the siting of hazardous facilities. An additional feature of process risk assessment can be found in nanotechnology applications – that of consciously defining the physical and analytical scopes for the analysis in terms of system boundaries and the nature of the hazards under consideration, respectively. This is evident in the need to describe, in the words of Williams et al. [10], the unit of analysis for which the risk is being assessed (marketed product, product components, production process, risks to the environment, or some combination of these factors). As with the process industries, nanotechnology risk assessments must consider a product's lifecycle from research and development through to manufacture and ultimately disposal [10].

While toxicity concerns can be acute or chronic, fire and explosion hazard effects are usually of short duration but of high intensity [12]. Focused efforts [13 – 20] have been undertaken recently to address the fire and explosion impacts of various nanopowders such as carbon nanotubes and nano-sized aluminium. Vignes et al. [20] have adopted a process-safety approach incorporating fault trees as well as prevention and mitigation barriers in assessing the ignitability and explosivity risk of aluminum nanopowders.

3. Inherently safer design

Inherently safer design is a proactive approach in which hazards are eliminated or lessened so as to reduce risk without over-reliance on engineered (add-on) devices and procedural measures.

3.1. Process safety features

A number of principles to facilitate inherent safety implementation in the process industries have been developed; the four basic principles identified in table 1 have gained widespread acceptance [21]. These and other principles of inherent safety work in conjunction with other means of reducing risk within a framework commonly known as the *hierarchy of controls*. Inherent safety, being the most effective approach to risk reduction, sits at the top of the hierarchy – followed in order of decreasing effectiveness by passive engineered safety, active engineered safety, and finally procedural safety.

Table 1. Key inherent safety principles.

Principle	Description
Minimization (Intensification)	Use smaller quantities of hazardous materials when the use of such materials cannot be avoided or eliminated. Perform a hazardous procedure as few times as possible when the procedure is unavoidable.
Substitution	Replace a substance with a less hazardous material or processing route with one that does not involve hazardous material. Replace a hazardous procedure with one that is less hazardous.
Moderation (Attenuation)	Use hazardous materials in their least hazardous forms or identify processing options that involve less severe processing conditions.
Simplification	Design processes, processing equipment, and procedures to eliminate opportunities for errors by eliminating excessive use of add-on safety features and protective devices.

3.2. Nanotechnology applications

There is clear evidence of inherently safer design application in the nanotechnology field. The terms *inherent safety* and *inherently safer design* are seldom used in the nanotechnology literature, but this also is similar to the process industries. Although inherent safety has been named and formalized in the process safety literature since the late 1970s, it is only in the last decade or so that the concept has been more widely quoted as such [21].

A recent publication [22] has identified five principles of *design for safer nanotechnology*. As shown in table 2, each of these design principles has at least a partial foundation in the fundamental principles of inherent safety (table 1).

Table 2. Relationship between design for safer nanotechnology [22] and inherent safety [21].

Safer Nanotechnology Design	Inherently Safer Design
Size, surface and structure	Moderation
Alternative materials	Substitution
Functionalization	Moderation
Encapsulation	Moderation
Reduce the quantity	Minimization

Other examples of inherent safety application to nanomaterial production and handling are given in table 3.

Journal of Physics: Conference Series **304** (2011) 012071 doi:10.1088/1742-6596/304/1/012071

Table 3. Inherent safety examples drawn from the nanotechnology literature.

Reference	Example	Inherent Safety Principle
[5]	Implications of form of nanomaterials (loose powder, liquid dispersion, agglomerated, aerosolized) for potential route of human and environmental exposure	Moderation
[8]	Need for process-specific engineering controls if potential hazards cannot be eliminated or substituted with less hazardous substances	Minimization Substitution
	Requirement for more stringent controls with operations having potential to aerosolize nanomaterials (e.g., handling dry powders, spray applications) than operations where nanomaterials are embedded in solid or liquid matrices	Moderation
[23]	Growth of carbon nanotubes on micron-sized Al_2O_3 particles as carriers in production of composite carbon nanotubes	Substitution Moderation
[24]	Increase in deposition range during inhalation with reduction in particle size	Moderation
	Relationship of particle toxicity to surface area	Moderation
[25]	Use of different crystal structures of TiO_2 as hazard reduction strategy (although noting reduced photocatalytic activity)	Substitution Moderation
	Efforts to limit hazard of TiO_2 particles by application of coatings and dopant ions (while again noting reduced photocatalytic activity)	Moderation
	Immobilization of TiO_2 nanoparticles on supports and substrates	Moderation
[26]	Processes that produce a carbon nanotube product that is not easily dispersed	Substitution
	Design of nanoparticles with substituent groups that minimize toxicity	Moderation
	Avoidance of manipulation of nanomaterials as free particles on lab benches	Simplification
[27]	Wet chemistry techniques for synthesis of aluminum nanopowder	Substitution
	Broad importance of understanding and describing properties of nanomaterials (underlying chemistry and physics)	Minimization Substitution Moderation Simplification
[28]	Use of alternate catalyst (e.g., iron in place of cobalt or nickel) in manufacturing processes for carbon nanotubes	Substitution
	Product form factors for carbon nanotubes and TiO_2	Moderation
[29]	Various methods for synthesis of SiO_2 aerosol particles	Substitution
[30]	Reduced likelihood of exposure to engineered nanomaterials when embedded in polymers or other matrices in given products	Substitution Moderation
	Need for hazard metrics related to surface chemistry of engineered nanomaterials	Moderation

4. Human error and human factors

Human error can be defined as any human action, or lack thereof, that exceeds or fails to achieve some limit of acceptability, where limits of human performance are defined by the system. *Human factors* are the environmental, organizational and job factors, and system design, task and human characteristics that influence behaviour and affect health and safety.

4.1. Process safety features

Human factor considerations in the process industries arise in a number of areas – e.g., the operator-process/equipment interface, use of administrative controls, and performance of human error assessments [31]. *Operator-process/equipment interface* refers to issues such as [31, 32]:

- the design of equipment increasing the potential for error (e.g., confusing equipment, positioning of dials, colour coding, different directions for on/off, etc.), and
- the need for a task analysis (a step-by-step approach to examine how a job is done) to determine what can go wrong during the task and how problem areas can be controlled.

Administrative controls require well-developed and -maintained safe work procedures, as well as an understanding of their role within the aforementioned hierarchy of controls. *Human error assessment* involves the use of appropriate techniques (both qualitative and quantitative) to estimate the likelihood and consequences of human error during operational and emergency scenarios.

4.2. Nanotechnology applications

One of the challenges for process safety in the new millennium is to overcome the fact that *organizations have no memory* [33]. Mannan et al. [33], with reference to the work of Professor Trevor Kletz, comment that organizations lose valuable incident information after about ten years because of personnel changes. One manifestation of this "brain drain" is in the lack of knowledge of facility workers with respect to the reasons for the existence of specific safety devices and protocols, as well as operating procedures, which may have been introduced following a process-related incident [33]. This can lead to human error during normal operations and emergency scenarios unless human factor issues, in particular organizational factors, are thoroughly considered.

Given the relative newness of the nanotechnology industry, it would seem opportune to heed the above lessons experienced with significant hardship by the process industries. Process-oriented techniques exist for estimation of human error probabilities (e.g., SLIM – Success Likelihood Index Methodology) and consequences (e.g., HAZOP – HAZard and OPerability analysis); these can be combined with risk matrices/graphs and safety barrier/bow-tie analyses to gain an overall picture of human error risk and the human factor considerations needed for risk reduction [34]. Application of such quantitative and semi-quantitative approaches to the processing of nanomaterials could prove valuable.

Simpler, qualitative methods widely used in the process industries are also applicable to the nanotechnology field. For example, the What-If analysis technique can be used to generate hazardous, what-if-type scenarios with an accompanying assessment of the adequacy of existing safeguards:

- What if an individual selects PPE (personal protective equipment) that is not suitable for handling nanomaterials, but which is stored in close proximity to PPE appropriate for nanomaterials (e.g., specific respirators with a flexible hood, impervious gloves, etc. [24])?
- What if material from a nanomaterial-bearing waste stream is placed in the regular trash (counter to the advice of Hallock et al. [26])?
- What if the correct information is not entered on a waste label to clearly indicate that the waste bag contains nanosized particles (again, counter to the advice of Hallock et al. [26])?

These potential human errors are rooted in inadequate consideration of the relevant human factors. The first scenario is essentially one in which the individual has been "set up for failure" by an unnecessarily complex system (recall the inherent safety principle of simplification). The second and third scenarios highlight the facts that procedural safety is the least effective measure in the hierarchy of controls, and that procedures will not work without effective training in their use.

5. Safety management systems

Safety management systems are best-practice methods recognized and accepted worldwide for managing risk (related, for example, to health, occupational safety, process safety, equipment reliability, etc.). They typically consist of 10 – 20 program elements that must be carried out to manage the risks in an acceptable manner by actively monitoring company operations for concerns and taking proactive actions to correct potential problems.

5.1. Process safety features

Having an effective management system for process-related hazards (i.e., fires, explosions and releases of toxic materials) is a critical corporate objective in the chemical process industries. An approach widely used in the Canadian chemical industries is PSM, Process Safety Management (where PSM is defined as the application of management principles and systems to the identification, understanding and control of process hazards to prevent process-related injuries and accidents). The suite of PSM elements is shown in table 4, taken from the Process Safety Management Guide of the Canadian Society for Chemical Engineering (CSCHE) [31]. The material in the CSChE PSM guide [31] is based on that developed by the Center for Chemical Process Safety (CCPS) of the American Institute of Chemical Engineers (AIChE), prior to the recent introduction by the CCPS of its 20-element risk-based process safety management system [35].

Table 4. Process Safety Management (PSM) elements [31].

Number	Element
1	Accountability: objectives and goals
2	Process knowledge and documentation
3	Capital project review and design procedures
4	Process risk management
5	Management of change
6	Process and equipment integrity
7	Human factors
8	Training and performance
9	Incident investigation
10	Company standards, codes and regulations
11	Audits and corrective actions
12	Enhancement of process safety knowledge

5.2. Nanotechnology applications

Sharatt [36] has made the somewhat provocative comments that:

There are many chemical engineers doing bio, nano, and environmental work – and in many cases doing it pretty badly. What is worrying is not that the work is being attempted, but that it is so often disconnected from the core concepts of the discipline. Concepts of mass and energy balance and a systems approach have been widely abandoned in favour of blind empiricism...

Whether or not one agrees with these statements, an appeal for a systems approach drawing on core concepts must surely include the need for a management system approach to safety. The precise nature and focus of such a management system (process safety, occupational safety, occupational hygiene, etc.) will be determined, in part, by the application areas within a given industry. With respect to nanomaterials, the following processes have been identified as posing significant exposure risk: (i) pouring or mixing liquids containing nanoparticles, (ii) generating gaseous nanoparticles in non-enclosed spaces, (iii) cleaning up spills or waste materials, (iv) machining, drilling or sanding (all of which can lead to nanoparticle aerosols), (v) spraying, weighing or blending powders containing nanoparticles, and (vi) cleaning and maintaining equipment and dust collectors [37].

Whether it is process hazards or individual exposures that are of concern, a safety management system for nanomaterial production and handling would be expected to contain elements similar to those shown in table 4 for PSM. In the current paper, reference has been made (even implicitly) to the need for consideration of *process knowledge and documentation*, *process risk management* (including hazard identification), *human factors*, *training and performance*, and *enhancement of process safety knowledge* (see table 4). Additionally, *management of change* (MOC) is a key safety management system element, particularly when the principles of inherently safer design are actively employed. Reijnders [25] has remarked on the possible release of TiO_2 nanoparticles when such particles are immobilized by linking them to large supports (e.g., due to photocatalytic degradation of carbon and synthetic organic polymer supports during application). Morose [22] has broadly commented that the use of alternative materials requires careful investigation of potential changes in hazards, product functionality and costs that may be introduced by such substitutions.

6. Safety culture

Safety culture has been defined by Hopkins [38] (with reference to an unpublished paper by Hudson) as a specific case of culture in which safety has a special place in the concerns of those who work in a given organization. The literature on safety culture places emphasis on *attitudes* and *practices* (especially *collective* practices) to varying degrees.

6.1. Process safety features

Hopkins [38] describes three concepts that address a company's cultural approach to safety and makes the argument that the three are essentially alternative ways of talking about the same phenomena:

- safety culture (with reporting, just, learning, and flexible sub-cultures),
- collective mindfulness, and
- risk awareness.

Safety culture (as defined by any of the terms in the above list) dominates much of the current discussion around process safety worldwide, driven in large measure by the BP Texas City incident in 2005 [39]. Key points emerging from this discussion include the following [21]:

- typical occupational safety indicators such as lost-time injuries (LTIs) are inappropriate as primary indicators with respect to process safety,
- leading indicators are generally viewed as more useful than lagging indicators, and
- process-safety indicators must link to, and measure the effectiveness of, the various elements of the process safety management system in use.

6.2. Nanotechnology applications

The title of the recent paper by Guldenmund [40], *(Mis)understanding Safety Culture and its Relationship to Safety Management*, is indicative of the strong and important linkage that exists between the safety management system approach and safety culture. A positive process safety culture must be the cornerstone of the first PSM element in table 4 (*accountability: objectives and goals*), as well as a unifying factor for each of the remaining elements. Quoting from a recent article [41] in *The Chemical Engineer* (a publication of the UK Institution of Chemical Engineers):

...Safety culture is how the night shift operates when it is alone without management watching... The argument goes that if managers are not sufficiently present they cannot reinforce the safety practices of their teams and this will inevitably lead to deviation from set operational routines... The result: the next Bhopal, Buncefield, or Gulf oil leak.

The nanotechnology world does not want, and neither should it need, a Bhopal, Buncefield or Gulf oil leak (all major and/or recent process/environmental incidents) to drive its safety culture. Simply put, nanotechnology industries cannot afford to ignore the hard safety lessons that have been learned and at times ignored by the chemical process industries. As noted by Stang and Sheremeta [42], whether or not one believes that the nanotechnology field raises unique ethical, economic, environmental, legal or social issues – nanotechnology will undoubtedly exacerbate and complicate these same issues. Lavoie [43], for example, has commented that the integration of some nanomaterials into products has outpaced what he has termed *diligent safety research*. A possible concern arises with TiO_2 nanoparticles in particular, which can penetrate the skin to a limited degree; contradictory research has demonstrated that some nanomaterials can be used safely with dermal contact [43]. These findings are similar to Horng's comment [23] that growing concerns about the impact of nanoparticles are likely due to a significant disconnect between environment, health and safety research efforts and business development initiatives.

As mentioned earlier in section 2.2, an issue common to the nanotechnology and process industries is that of lifecycle consideration when conducting risk assessments. Savolainen et al. [30] have noted that while most nanoparticle-containing products are not likely to cause exposure while the nanoparticles remain embedded in the product matrix, other scenarios such as product waste processing can provide a different risk profile. Production of the nanoparticles themselves at the front-end of the process must also be considered as part of the lifecycle. Considerations surrounding product stewardship and supply chain management [44] are therefore key to building a positive safety culture in facilities processing and handling nanomaterials. The Royal Society's call for a *responsible nanotechnology code* [4] is not unlike the Canadian and worldwide response to the 1984 disaster in Bhopal, India which involved the release of methyl isocyanate and the death of thousands of human beings – that of a program of Responsible Care® incorporating an ethical set of principles as well as a management system of codes and practices [44].

7. Concluding Remarks

The concepts of risk management, inherently safer design, human error and human factors, safety management systems, and safety culture are as relevant to the nanotechnology industries as they are to the process industries. Other process industry issues not discussed in the current paper (e.g., chemical security [45]) would also be expected to be pertinent to the nanotechnology field. Adaptation of specific process safety protocols to the production and handling of nanomaterials necessitates thoughtful consideration of challenges such as data uncertainty, public perception and lack of awareness of the full hierarchical suite of risk control measures.

References

[1] Wilson L and McCutcheon D 2003 *Industrial Safety and Risk Management* (Edmonton, AB, Canada: University of Alberta Press)
[2] Amyotte P R and McCutcheon D J 2006 *Risk Management – An Area of Knowlwdge for all Engineers* (Ottawa, ON, Canada: Canadian Council of Professional Engineers)
[3] Health & Safety Executive 2001 *Reducing Risks, Protecting People. HSE's Decision-Making Process* (Norwich, UK: Her Majesty's Stationery Office)
[4] The Royal Society 2006 *How Can Business Respond to the Technical, Social and Commercial Uncertainties of Nanotechnology?* (UK: The Royal Society)
[5] Environmental Defense–DuPont 2007 *Nano Risk Framework* (Wilmington, DE, USA: DuPont)
[6] Linkov I and Satterstrom F K 2008 *Nanomaterial Risk Assessment and Risk Management (Real-Time Deliberative Decision Making)* eds I Linkov et al. (New York, NY, USA: Springer Science+Business Media B.V.) pp 129–157
[7] Council of Canadian Academies 2008 *Small is Different: A Science Perspective on the Regulatory Challenges of the Nanoscale* (Ottawa, ON, Canada: Council of Canadian Academies)
[8] National Insitute for Occupational Safety and Health 2009 *Approaches to Safe Nanotechnology. Managing the Health and Safety Concerns Associated with Engineered Nanomaterials* (Washington, DC, USA: National Institute for Occupational Safety and Health)
[9] Professional Engineers Ontario 2010 *PEO Emerging Disciplines Task Force: Nanotechnology and Molecular Engineering* (Toronto, ON, Canada: Professional Engineers Ontario)
[10] Williams R A, Kulinowski K M, White R and Louis G 2010 *Risk Analysis* DOI: 10.1111/j.1539 -6924.2010.01513.x
[11] Ferdous R, Khan F, Sadiq R, Amyotte P and Veitch B 2010 *Risk Analysis* DOI: 10.1111/j.1539

-6924.2010.01475.x
[12] Lemkowitz S M, Pasman H J and Schupp B A Undated *Integration of Safety, Health, Environment and Sustainability Into Engineering Education and Industrial Training* (Delft, Netherlands: Delft University of Technology)
[13] Bouillard J, Vignes A, Dufaud O, Perrin L and Thomas D 2009 *J. Phys.: Conf. Ser.* **170** 012032
[14] Wu H-C, Chang R-C and Hsiao H-C 2009 *J. Loss Prev. Proc. Ind.* **22** 21–24
[15] Yetter R A, Risha G A, Son S F 2009 *Proc. Combust. Inst.* **32** 1819–1838
[16] Vignes A, Dufaud O, Perrin L, Thomas D, Bouillard J, Janes A and Vallieres C 2009 *Chem. Eng. Sci.* **64** 4210–4221
[17] Wu H-C, Kuo Y-C, Wang Y-H, Wu C-W and Hsiao H-C 2010 *J. Loss Prev. Proc. Indust.* **23** 308–311
[18] Wu H-C, Ou H-J, Hsiao H-C and Shih T-S 2010 *Aer. and Air Qual. Res.* **10** 38–42
[19] Bouillard J, Vignes A, Dufaud O, Perrin L and Thomas D 2010 *J. Haz. Mat.* **181** 873–880
[20] Vignes A, Munoz F, Bouillard J, Dufaud O, Perrin L, Laurent A and Thomas D 2010 *Proc. 13th Int. Symp. on Loss Prev. and Safety Prom. in the Proc. Ind. (Bruges, Belgium, 6–9 June 2010)* pp 187–194
[21] Kletz T and Amyotte P 2010 *Process Plants. A Handbook for Inherently Safer Design, Second Edition* (Boca Raton, FL, USA: CRC Press/Taylor & Francis Group)
[22] Morose G 2010 *J. Cleaner Prod.* **18** 285–289
[23] Horng J-J 2007 *Proc. Safety Env. Prot.* **85** 332–339
[24] Hoyt V W and Mason E 2008 *J. Chem. Health & Safety* **15** 10–15
[25] Reijnders L 2008 *J. Haz. Mat.* **152** 440–445
[26] Hallock M F, Greenley P, DiBerardinis L and Kallin D 2009 *J. Chem. Health & Safety* **16** 16–23
[27] Dreizin E L 2009 *Prog. Energy Combust. Sci.* **35** 141–167
[28] Dorr J 2010 *The Chem. Eng. (IChemE)* **831** 45–47
[29] Ostraat M L 2010 *Industry-Led Initiative for Occupational Health and Safety (Nanotechnology Environmental Health and Safety)* (New York, NY, USA: Elsevier Inc.) chapter 6 pp 181–246
[30] Savolainen K et al. 2010 *Safety Sci.* **48** 957–963
[31] Canadian Society for Chemical Enghineering 2002 *Process Safety Management, Third Edition* (Ottawa, ON, Canada: Canadian Society for Chemical Engineering)
[32] Amyotte P R, Goraya A U, Hendershot D C and Khan F I 2007 *Proc. Safety Prog.* **26** 333–346
[33] Mannan M S, Prem K P and Ng D 2010 *Proc. 13th Int. Symp. on Loss Prev. and Safety Prom. in the Proc. Ind. (Bruges, Belgium, 6–9 June 2010)* pp 5–13
[34] Deacon T, Amyotte P R and Khan F I 2010 *Safety Sci.* **48** 803–818
[35] Center for Chemical Process Safety 2007 *Guidelines for Risk Based Process Safety* (Hoboken, NJ, USA: John Wiley & Sons, Inc.)
[36] Sharatt P 2010 *The Chem. Eng. (IChemE)* **825** 24–25
[37] Landau E 2010 *OHS Canada* **26**(3) 48–49
[38] Hopkins A 2005 *Safety, Culture and Risk. The Organisational Causes of Disaster* (Sydney, Australia: CCH Australia Limited)
[39] Hendershot D C 2007 *J. Chem. Health & Safety* **14** 39–40
[40] Guldenmund F W 2010 *Risk Analysis* **30** 1466–1480
[41] International News 2010 *The Chem. Eng. (IChemE)* **829/830** 7
[42] Stang C and Sheremta L 2006 *Health Law Rev. (U. of Alberta)* **15**(1) 53–56
[43] Lavoie J H 2010 *Proc. Safety Prog.* **29** 182–185
[44] Topalovic P and Krantzberg G 2010 *Can. Chem. News* **62**(7) 21–25
[45] Moore D A 2010 *Proc.13th Ann. Symp., Mary Kay O'Connor Process Safety Center, Texas A&M University (College Station, TX, USA, 26–28 Oct. 2010)* pp 273–283

Can Induction Plasma Technology be Nano-safe, "Green" and Energy Efficient?

J W Jurewicz[1,2], M I Boulos[1,2], L Brochu[1], J P Crête[2], N Dignard[1], D Héraud[2], F Hudon[2] and C Ostiguy[3]

[1]Tekna Plasma Systems Inc., 2935, boul. Industriel, Sherbrooke, Qc, Canada J1L 2T9

[2]Tekna Advanced Materials Inc., 2895, boul. Industriel, Sherbrooke, Qc, Canada J1L 2T9

[3]Institut de recherche Robert-Sauvé en santé et en sécurité du travail, 505, boul. De Maisonneuve Ouest, Montréal, Qc, Canada H3A 3C2

E-mail : ostiguy.claude@irsst.qc.ca

Abstract. With the rapid development of interest in the commercial scale production of nanopowders and nano-structured materials an increasing concern is being voiced about the safety measures that are taken in the design and construction of such facilities. This paper deals with such an important issue and describes the main design and manufacturing safeguards that were implemented during the design and commissioning of a recently inaugurated Tekna Advanced Materials powder production facility in Sherbrooke, Québec, Canada.
The team of engineers from Tekna Plasma Systems was in charge of the design of multiple-units powder synthesis and processing facility to produce a wide range of nano- and micro-sized powders. The main design issues were process safety, environmental friendliness without compromising the overall processing economics. The bottom-up approach used was based on the integration of standard processing units, offered by Tekna Plasma Systems, housed in dedicated production facility equipped with process-specific hardware and operation procedures and safeguards in order to warrant the superior protection for process operators and the surrounding environment.

1. Introduction

The nanotechnology is considered as one of the four converging technologies for improving human performance [1] – the others being biotechnology, information technology and cognitive science. An intensive research effort in nanotechnology and especially in nanoparticles science is mirrored both by publication of many handbooks from general interest [2-4] to dedicated to specific powders like metallic [5] or to specific applications [6] and by increasing a number of companies launching their pilot or semi-industrial powder production.

An interest in wide spread mass-production of synthetic nanoparticles is responding to growing applications of such small size (< 100 nm) powders in different domains of industry like electronics (for Multi-Layered Ceramic Capacitors – MLCC), pigments and cosmetics (for sun screens) [7]. Some industry giants like Degussa AG, consider the nanotechnology as the most important technology of the future [8].

One of the possible technologies of nanopowders production is an induction plasma technology. The detailed description of the current development of nanopowders synthesis methods through the application of an induction plasma processing technology was published recently [9].

Tekna Plasma Systems Inc. (Sherbrooke, Qué. Canada) offers turn-key systems for both pilot scale and industrial production of wide variety of powders from micron-sized, through sub-micron- to nano-sized individual particles diameter. To answer the actual tendency in typical business model which is product's diversification, it has been decided to launch a sister company offering both powder treating services and an actual, industrial scale, powder production for world-wide market using typical induction plasma system as offered by Tekna Plasma Systems. The team of engineers from Tekna put together their effort to design, build and operate industrial scale plasma facility having in mind the three aspects of technology: process safety and environmental protection without compromising sound economic foundation. This paper describes some of the problems encountered and proposed solutions undertaken at initial stages of operation of this enterprise.

2. Designing Considerations

Any industrial activity in Canada is controlled by federal, provincial and municipal authorities through corresponding laws and regulations. These include (among others) the disposal of all post-processing spent materials like ambient air evacuated from premises, cooling water and solid wastes generated in the process. Spent heat has to be evacuated through water evaporation process. The process safety management is one of the most accentuated aspects of any industrial activity and has to be dealt with adequately.

The type of industrial activity sought – commercialization of powdered materials synthesis, implies the presence of a variety of hazards related to both the technology used as well as to treated materials'/ products' physical and chemical properties. While the risks related to the technology are known and controlled, the products considered for production have to be analysed on the on-going base together with the diversification of powders offered.

Nanoparticles of engineered materials are defined as particulates having at least one dimension between 1 and 100 nm and are characterized by very unique physical and chemical properties which significantly differ from the bulk material's ones. Their toxicities to humans and to ecosystems are currently investigated worldwide where some parameters like particle size, shape, surface area, charge, chemical properties, solubility, oxygen or nitrogen generation potential as well as degree of agglomeration may influence their toxicity.

At the initial stage of a plant design, when all of the future products to be manufactured in this facility were not clearly identified, the risks associated with the handling of ultrafine nanopowders was given major consideration independent of their chemical nature and toxicity. This required the following safety precautions:

- The control of exposure to the synthesized nanoparticles is strongly suggested [10] – such control may be practiced [11] by: lowering the dose, lowering the frequency and shortening the time of exposition as well as lowering the number of operators exposed;
- Another consequence of high specific surface area of nanoparticles is the pyrophoric nature of metallic nanopowders – this phenomenon as well as their thermal stability may be controlled by either passivation (partial superficial oxidation) process, and/or an encapsulation process [12,13];
- Collection of all residual materials from the processing units is to be handled by wet techniques during all maintenance operations.

The process safety management strategies, resourced from recently published [14,15] guidelines, suggest to apply preferably more efficient engineering techniques like isolation/containment and local ventilation than the administrative measures or a personal protection equipment. The principle of multiple layers of prevention /protection [16] was to be exercised during the design process.

3. Processing Units – Nano-powders synthesis

The nano-powders synthesis unit was designed in a way that all operations, including products' collection were to be carried out under total containment conditions (Figure 1) – the latter one either in glove box equipped with air lock to transfer canned products or (in the case of regular production) in a continuous liner packing collector.

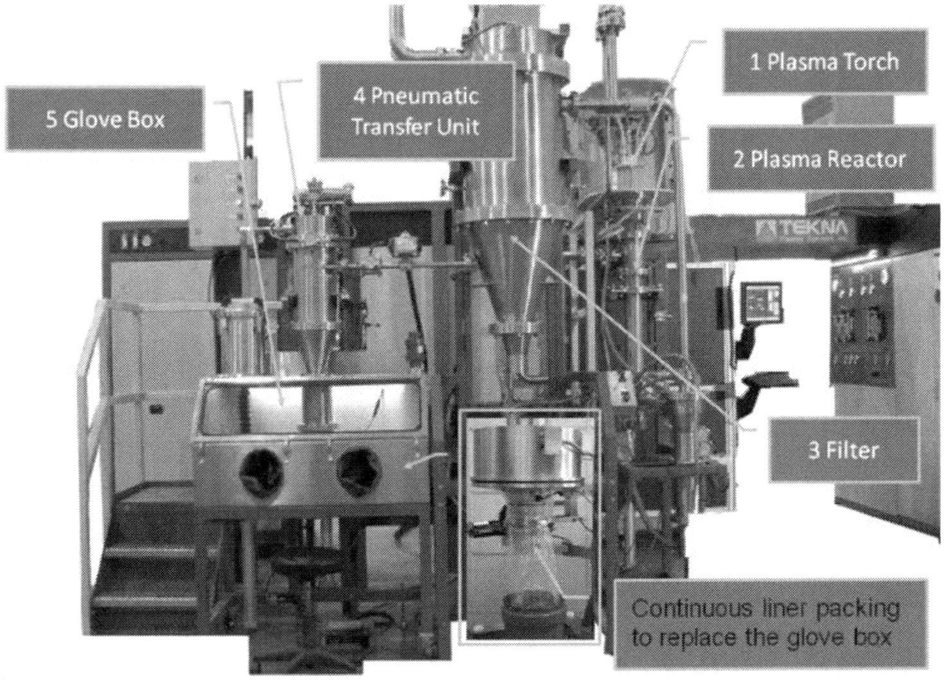

Figure 1. Image of the current 60 kW induction plasma nano-powder synthesis unit.

4. Processing Cost Reduction Strategies

The components of operational costs of thermal plasma processing plant do not significantly differ from those observed in typical chemical industries – excluding raw materials costs and waste disposal, there are three main elements: cost of processing gases, electricity and manpower. The strategies to lower these costs are discussed below.

Processing gases, pending requested products, include argon, hydrogen, oxygen, helium, etc. Some of them like oxygen or hydrogen may be partially consumed in the process while all inert gases could be easily recovered, at least partially, through their post process recompression and conditioning. Each of the plasma processing units is equipped with such gas conditioning system allowing up to 90% gas recycling. Hydrogen, usually relatively expensive among used gases, is offered on the industrial market in compressed form stocked in standard tanks. Stocking of significant quantity of hydrogen and daily handling of these tanks require specific care from the point of view of process safety. The *in-situ* hydrogen on-demand generation (by electrolysis of water) offers significant scale economy and lowers the process risk due to much smaller volume of stocked hydrogen. At the same time the gas recycling process recovers the entire hydrogen contained in gases so the *in-situ* generation replaces only the current hydrogen consumption and/or losses.

Electricity cost is strongly related to the efficiency of used processing equipment and its throughput. Very special efforts were dedicated to optimize the high frequency plasma generation as well as the raw material processing parameters.

Manpower required may be reduced by high automation of the processing units which run for hours without immediate and constant operator's surveillance as well as by bulk (high volume) handling of products – the operator takes care of products handling and final packing. The high automation has another benefit – the significant increase of process safety by computer controlled sequencing of operations such as raw materials feeding, hopper purging, products pneumatic transport and separation thus avoiding human errors.

5. Building Infrastructure

The building was designed to house two separate production halls (one for nano-powders synthesis and the other for powder spheroidization process). It had separate access to the infrastructure mechanical room for all services including cooling water, gases supply (hydrogen) and distribution equipment, incoming fresh air heating as well as all storage tanks and spent cleaning water conditioning system. Only the cooling water system is described below.

As specified above, any spent heat from the process has to be evacuated to surrounding atmosphere through evaporation of water in cooling tower. Cooling water, originating usually from natural sources like lakes or rivers and operating in closed circuit at moderate temperature, is prone to proliferation of bacterial cultures. Because their presence in water lowers thermal efficiency of the heat exchange process, the bacterial charge is usually controlled by adding some bactericide substances as well as some pH stabilisers. Unfortunately such chemically charged spent water cannot be disposed directly to the environment – it has to be treated in specific units additionally increasing the cost of whole operation. On the other hand the outside air is frequently charged with solid particulates (charged with some bacteria) which will be collected in the cooling water (cooling tower is a very efficient scrubbing device). The presence of suspended solids and bacteria may cause the fouling of heat exchangers. These two major issues were taken into account in designing the processing premises.

The chosen plant location does not allow use of any natural source of water (except rain) so it was decided to relay cooling process on city water with simultaneous collection of rain water. The characteristics of city water are the followings: low mineral charge and absence (or small, within the permitted limits) of bacterial charge - water is filtered and ozonized prior to its distribution and then slightly charged with chlorine to prevent bacterial growth in the distributing network. The temperature of such water varies from around 4-5°C during winter season to around 12°C during summertime. The rain water, collected exclusively on the building's roof with relatively abundant average but difficult to foresee actual flowrate, may be charged with dissolved gases and some solids/bacteria. The economic analysis of the cost of investment to install separate rain water tank and gain over the city water economy (maximum water consumption to dissipate total plant power of 2.5 MW is around 1 kg/sec) indicated too lengthy payback period. It was decided to collect rain water directly in an oversized (28 m^3) main cooling water tank located under the plant floor and equipped with overflow directed to outside in the case of significant rain outpour. Such solution imposed the elimination of usage of any substance to control the bacterial growth – instead the intense (> 40 000 $\mu Ws/cm^2$) ultraviolet radiation of circulated water has been applied. The solid particulates collected from cooling ambient air through scrubbing phenomenon are deposited at the tank bottom well by sedimentation and are periodically evacuated to outside by a sump pump.

The temperature of water leaving cooling tower is maintained at constant value no matter of heat load of incoming water by controlling the rotational speed of fans using frequency drive. During the heating season the incoming water heats up first the fresh air supply of production halls and then enters the cooling tower. During the high humidity periods (occurring mostly in July) where the ambient air is practically saturated with water, the city water arriving at ~ 12°C assures the adequate cooling of the plasma processing installations.

6. Safety barriers system

Process safety management principle suggests to build several sequential barrier's layers for both prevention and or protection of the operator/environment etc. Typical Tekna's processing unit is automatically controlled through the Programmable Logic Controller (PLC). All the details of such system are beyond the scope of this paper - the following additional barriers are given as example:

First layer – to prevent a leak of nanoparticles to the working atmosphere:
- each processing unit is designed to withstand a deflagration;
- each processing unit is completely sealed during operation;
- raw materials and products are pneumatically transported;
- the products are collected either in bulk containers (250 l) for spheroidized powders or in glove box or continuous liner packing system for nano-sized products;
- the size of by-product collectors are chosen to minimize frequency of their opening;
- all processing vessels are inerted/padded by argon before the start-up as well as during raw material charging to the hopper;
- all flammable/pyrophoric powders are either passivated or encapsulated (application pending) before packing;

Second level barriers – to prevent exposure of operators to nano-particles:
- Effective local ventilation system consisting of three different arrestance filters where the final filter of fibrous type is HEPA class. Such fibrous type filters are known to be very efficient for collection of particles down to 4 nm – only at this size some thermal bouncing is expected – but such small size particles powders are not produced at plant;
- Wet cleaning operations are compulsory for each processing unit – the suspension is then evacuated through sealed conduits to a collection tank, where the powder is flocculated / coagulated and separated by sedimentation while the clean water is recycled into gray water system;
- Each operator is wearing a personal protective equipment including HEPA filtered forced air supplied hood;

Third level barriers – to prevent any leak of nano-particles to an external production hall environment:
- Production hall is kept under negative pressure as compared to outside;
- Air lock with forced recirculation/filtration exit from production hall;
- HEPA class filters at each evacuation port;
- Wet off gas cleaning after HEPA filtration for process units.

7. Conclusion

The bottom-up design approach from Tekna built standard thermal plasma processing units to entire production facility allowed to address and successfully resolve all major safety and economic issues involved with the production of advanced nano-sized materials by:

- Conceiving multiple – level barriers against nano-particle spill hazard;
- Minimizing the risk of exposure to hazardous materials through (among others) high level of process automation including final products packaging and separate production halls for higher hazard level materials;
- Establishing and implementing the adequate operational procedures;
- Cooling water cost reduction by recovery of rain / snow;
- Space heating by spent heat from plasma processing units;
- Minimizing the waste disposal costs through in-house recycling procedures and elimination of water conditioning chemicals (replaced by UV radiation);
- On-site, consumption regulated, production of hydrogen;
- Recycling (after conditioning) the major part of process gas.

The new facility is currently being operated since January 2010 and specific designing considerations are being confronted with operation experience.

References

[1] Roco M C and Bainbridge W S 2003 Converging Technologies for Improving Human Performance. Nanotechnology, Biotechnology, Information Technology and Cognitive Science *Virginia : National Science Foundation* Arlington
[2] Rotello V M 2004 *Nanoparticles: building blocks for nanotechnology* Springer s.l.
[3] Schmid G 2004 *Nanoparticles: from theory to applications* Wiley-VCH s.l.
[4] Hosokawa M, Nogi K, Naito M, Yokoyama T 2007 *Nanoparticle technology handbook* Elsevier s.l.
[5] Feldheim D L, Foss C A, Feldheim L F 2001 *Metal nanoparticles: synthesis, characterization, and applications* Marcel Dekker
[6] Kumar C S S R, Hormes J, Leuschner C 2005 Nanofabrication towards biomedical applications: techniques, tools, applications, and impact Wiley_VCH s.l.
[7] Innes B, Tsuzuki T, Dawkins H, Dunlop J, Trotter G, Nearn M R and McCormick P G 2002, Nanotechnology and the cosmetic chemist, *Cosmetics, aerosols and toiletries in Australia,* **15**, 5, 10-24
[8] Gutsch A, Kraemer M, Michael G, Muehlenweg H, Pridoehl M, Zimmermann G 2002 Gas-Phase Production of Nanoparticles, *KONA Powder and Particle Journal* **20**, 24–37
[9] Jiayin G, Xiaobao F, Dolbec R, Siwen X, Jurewicz J, Boulos M 2010 Development of Nanopowder Synthesis Using Induction Plasma *Plasma Science and Technology* **12**, 2, 188-199
[10] AFSSET 2008 Les nanomatériaux. Avis de l'AFSSET Sécurité au travail. Agence française de sécurité sanitaire de l'environnement et du travail Paris 247 p
[11] Paik S Y, Zalk D M and Swuste P 2008 Application of a pilot control banding tool for risk assessment and control of nanoparticle exposures *Ann Occup Hyg.* **52**, 6, 419-428
[12] Scott A and Majetich E 1996 *Carbon Coated Nanoparticle Composites Synthesized in an rf Plasma Torch* MRS Materials Research Society. Proceedings of MRS Fall Symposium V s.l.
[13] Radloff C and Halas N J 2001 Enhanced thermal stability of silica-encapsulated metal nanoshells *Applied Physics Letters* **79**, 674-676 American Institute of Physics
[14] Ostiguy C, Roberge B, Ménard L and Endo CA 2009 Best Practices Guide to Synthetic Nanoparticle Risk Management. Technical Guide IRSST Report # **R-599**, Institut de recherche Robert-Sauvé en santé et sécurité du travail Montréal 73 p
[15] Nanotechnologies -- Health and safety practices in occupational settings relevant to nanotechnologies 2008 Technical Report ISO/TR **12885** International Organization for Standarization
[16] CCPS. *Layer of Protection Analysis: simplified process risk assessment* 2001 Center for Chemical Process Safety American Institute of Chemical Engineers 270 p

Polyfire project- an example of an industrial research project promoting safe industrial production of fire-resistant nanocomposites

C. Vaquero[1], J. López de Ipiña[1], N.Galarza[1], B. Hargreaves[2], B. Weager[2] and C. Breen[3]

[1]TECNALIA, Leonardo Da Vinci Nº 11, 01510 Miñano (Álava) Spain

[2]NetComposites Ltd, 4A Broom Business Park, Chesterfield S41 9QG, UK

[3]Materials and Engineering Research Institute, Sheffield Hallam University, Sheffield S1 1WB, UK

E-mail: celinav@leia.es

Abstract. New developments based on nanotechnology have to guarantee safe products and processes to be accepted by society. The Polyfire project will develop and scale-up techniques for processing halogen-free, fire-retardant nanocomposite materials and coatings based on unsaturated polyester resins and organoclays. The project includes a work package that will assess the Health and Environmental impacts derived from the manipulation of nanoparticles. This work package includes the following tasks: (1) Identification of Health and Environment Impacts derived from the processes, (2) Experimentation to study specific Nanoparticle Emissions, (3) Development of a Risk Management Methodology for the process, and (4) A Comparison of the Health and Environmental Impact of New and Existing Materials. To date, potential exposure scenarios to nanomaterials have been identified through the development of a Preliminary Hazard Analysis (PHA) of the new production processes. In the next step, these scenarios will be studied and simulated to evaluate potential emissions of nanomaterials. Polyfire is a collaborative European project, funded by the European Commission 7th Framework Programme (Grant Agreement Nº 229220). It features 11 partners from 5 countries (5 SMEs, 3 research institutes, 2 large companies, 1 association) and runs for three years (1st September 2009 – 31st August 2012). This project is an example of an industrial research development which aims to introduce to the market new products promoting the safe use of nanomaterials.

1. Introduction

The number of consumer goods based on nanotechnology is expected to significantly increase in future years, and with it, the number of industrial settings that incorporate nanomaterials in their production process. To date, there is a knowledge gap and also concern related to the potential health and environment impact of nanomaterials. To be accepted by society, nanotechnology based products and processes have to ensure their safety.

As toxicology science doesn't yet have conclusive results related to the potential harmful effects of nanomaterials, the current approach for occupational risk management is to characterize the exposure to nanomaterials in order to propose measures to control risks as much as possible. ENRHES project [1] showed that there are very few studies of occupational exposure assessment, with those that do exist mainly focused on carbon based nanomaterials (fullerenes and carbon nanotubes) and performed principally in lab settings. The ongoing NANEX project [2] also seems to corroborate that result, jointly with the difficulties to develop Exposure Scenarios with the actual information, and the need to develop standards for exposure risk characterization, including the measurement strategy and equipment. With regard to exposure to nano-sized clay particles specifically (focus of this work), very few studies have been found by the researches. Schnaider et al [3] addressed the study of the dustiness of powders, including bentonite and organoclays, as a potential tool to simulate personal exposure to nanopowders, while Jankowska et al [4] in the NANOSH project analyzed the emission of nanosized particles in the process of clay blending for the production of new fire-safe foams. On the other hand, related to measures to control exposure, research already done seems to show that control engineers and personal protection equipment could provide different protection levels against nanomaterials [5].

The aim of the Polyfire project is to develop and scale-up techniques for processing halogen-free, fire retardant (FR) nanocomposite materials and coatings based on unsaturated polyester resins and clays. In this project a Health and Environment (H&E) assessment of the new processes and products will be performed, focused on the study of potential exposure to nano-sized organoclay particles in order to finally propose a risk management methodology. To date a Preliminary Hazard Assessment (PHA) has been performed, where potential exposure scenarios to organoclays have been identified. In this paper we present this project, as an example of an industrial research development which aims to introduce to the market new products promoting the safe use of nanomaterials.

2. Polyfire project

The Polyfire project is structured in 8 work packages devoted to research on industrial processes for the production of new fire retardant nanocomposites and coatings based on unsaturated polyester resins and clays. A dedicated work package is focused on assessment of the main health and environmental impacts derived from these new processes. Figure 1 gives an example of the industrial process for fire retardant nanocomposite and coating production.

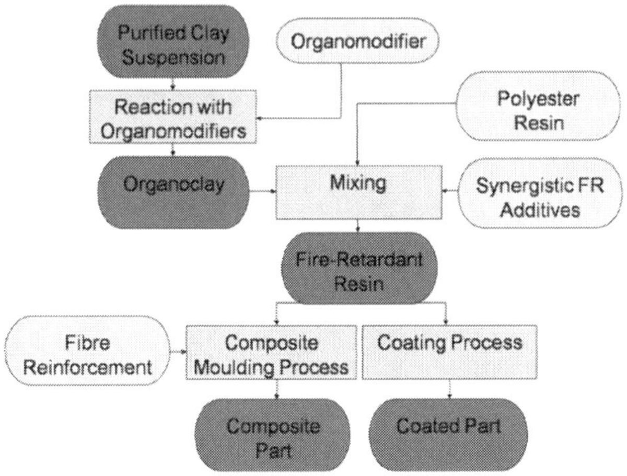

Figure 1. General processes to develop new fire retardant nanocomposites/coatings.

Polyfire is a collaborative European project, funded by the European Commission 7th Framework Programme (Grant Agreement N° 229220). It features 11 partners from 5 countries (5 SMEs, 3 research institutes, 2 large companies, 1 association) and runs for three years (1st September 2009 – 31st August 2012).

It is important to mention that to date there is no clear evidence for the presence of nanoparticles in the unsaturated polyester composites produced in this project. The x-ray diffraction evidence suggests that an intercalated nanocomposite or a microcomposite is being produced. Nevertheless, the position being taken is that there is the potential to generate nanoparticles during organoclay production, the mixing of organoclay with the resin and during handling of the molded component (for example when it is cut or sanded). Therefore, the Health and Environmental (H&E) assessment is designed to evaluate the type and nature of any nanoparticles generated, identify adequate control measures and educate relevant individuals and companies of adequate control measures.

3. The approach for the Health and Environment Assessment

Within the H&E assessment, 4 tasks will be performed: (1) Identification of Health and Environment (H&E) Impacts derived from the processes, (2) Experimentation to study specific Nanoparticle Emissions, (3) Development of a Risk Management Methodology for the process, and (4) A Comparison of the Health and Environmental Impact of New and Existing Materials. In the first year of Polyfire, task 1 has been performed.

The approach followed in task 1 (identification of H&E impacts) has been to identify potential workers exposure to organoclay particles jointly with environmental emissions. To perform it two tools have been used, a Preliminary Hazard Analysis (PHA) and an Impact Matrix (IM).

The PHA started with the analysis of the processes covering the life cycle of the new fire retardants. In the scope of this project, we have considered the production process, the professional use and the end of life of the new FR products with a main focus on the production stage. The analysis of the different industrial tasks and activities has been performed in collaboration with the partners involved in each stage of the cycle. Normal operation, emergency conditions (e.g. breakages, releases) and other tasks, such as cleaning/maintenance procedures have all been considered. The information collected in this process has been introduced in templates for each step of the processes with the key relevant information. Afterwards, for each of the steps analyzed in the previous activity, the hazards related to the potential exposure to organoclay particles have been identified. The analysis of workers hazards has focused on inhalation, although dermal exposure has been also identified. Hazard of inhalation has been ranked as High-Medium-Low based on the presentation of the organoclay particle (powder/suspension/embedded in matrix), type of system (open/closed) and the operation performed (subjected to high energy). Regarding environmental hazards, in the PHA the environmental aspects with potential significance have been identified [6].Both, the occupational and environment PHA has been collected in templates designed for it.

Finally, the results of the PHA has been summarized in an Impact Matrix (IM) which summarizes the steps of the analyzed processes involved in the life cycle of the new FR, and the identified occupational and environmental hazards (see figure 2).

					Hazards				
					Occupational		Environment		
		Step	Operation Mode	Tasks	Inhalation	Dermal	Atmospheric release	Waste Water	Waste
PRODUCTION	PROC X:	Step A Organoclay milling	N(1)	Organoclay transfer	Low				
				Grinding process	Low		X		
				Hatch opening	High	X			
			E(2)	Spilled organoclay	High	X		X	X
			O(3)	Maintenance/cleaning	High	X		X	X
		....	.....	....	....				
		Step B Organoclay+ resin mixing	N	Organoclay feeding	Low				
				Powder liquid high shear mixing	Low				
				Hatch opening- sample collection	Med	X			
			E	Spilled organoclay	High	X			X
			O	Maintenance/cleaning	Med	X			X
		....	.....						
		Step C Composite Machinig	N	Handling of composite	Low	X			
				Composite machining	Med		X		
			E	Breackage	Low	X			X
			O	Maintenance/cleaning	Low	X			X
		....	N	....	Low	X			
				....	Low				
			E	....	High	X		X	X
			O	....	High	X		X	X

(1) Normal operation; (2) Emergency; (3) Other operations (cleaning/maintenance)

Figure 2. Example of the Impact Matrix (IM) that summaries the results of the PHA

Figure 3 summarizes the targets/life cycle stages under the scope of this work and the *highest hazards* identified in the study. This figure shows that the highest occupational hazards (in red) are related to activities performed in the production stage of the FR composites/coatings, in tasks with potential dust emission (from organoclay powder). Medium hazard (in orange) is presented in the three stages (production/use/end of life) and associated to three different types of tasks: a) tasks with exposure to aerosols from organoclay suspended in volatile liquids with/without agitation, b) spraying of suspensions with organoclays and, c) machining of solid matrix with embedded organoclays.

As mentioned before, note that as yet there is no clear evidence for the presence of nanoparticles in polymeric matrixes filled with organoclays, or in the organoclays powders themselves.

The goal for the next tasks in Polyfire is to perform studies of nanoparticle emissions. From the PHA and the IM, scenarios will be grouped and selected, and experimental designs will be performed in order to get quantitative data related to the emission of nanomaterials. Greatly assist in evaluating the risks derived from the exposure to organoclay particles and finally to develop a strategy for risk management, the main goal of this workpackage in Polyfire.

TARGET / LIFE CYCLE	Worker	Environment	Consumer
Design	NC	NC	NC
Production	Tasks with potential dust emission (organoclay particles)- Potential worker inhalation Task with exposure to aerosols from nanoparticles suspended in volatile liquids: with/without agitation; spraying	Identification of environmental aspect with potential significance	NC
Use	Machining of solid matrix with organoclays embedded- Potential nanoparticles emissions/worker inhalation	Identification of environmental aspect with potential significance	NC
Maintenance	NC	NC	NC
End of life	Machining of solid matrix (or operations with dust emission) with organoclays embedded- Potential nanoparticles emissions/worker inhalation	Identification of environmental aspect with potential significance	NC

Figure 3. Summary of the highest hazards identified in the life cycle stages of the new fire retardants products.

4. Conclusions

The Polyfire project aims to introduce to the market new halogen-free fire retardant composites and coatings based on unsaturated polyester resins and organoclay particles. Although there remains a lack of clear evidence to prove nanoparticles are present in either the un-cured polyester resin or the final composite part, the Polyfire project will continue to study their potential health and environmental impact in order to promote safe industrial production. To date, processes involved in the life cycle of the new products have been analysed, identifying potential hazards for workers and the environment. The next challenges in this project are to characterize the potential emissions of nanomaterials in order to finally propose measures for risk management.

References

[1] Engineered Nanoparticles - Review of Health & Environmental Safety (ENRHES) http://nmi.jrc.ec.europa.eu/project/ENRHES.htm-
[2] Development of Exposure Scenarios for Manufactured Nanomaterials (NANEX). http://www.nanex-project.eu/
[3] Schnaider, T, (2009) Relevance of aerosol dynamics and dustiness for personal exposure to manufactured nanoparticles J Nanopart Res (2009) 11:1637–1650
[4] Jankowska, E. (2009) Emission of nanosize particles in the process of nanoclay blending. 2009 Third International Conference on Quantum, Nano and Micro Technologies.
[5] OCDE (2009) Comparison of guidance on selection of skin protective equipment and respirators for use in the workplace: manufactured nanomaterials. ENV/JM/MONO(2009)17
[6] ISO 14001:2004 Environmental management systems - Requirements with guidance for use.

Green nanoparticle production using micro reactor technology

A Kück[1], M Steinfeldt[2], K Prenzel[3], P Swiderek[3], A v Gleich[2] and J Thöming[1]

[1]UFT- Universität Bremen, Leobener Straße, 28359 Bremen, Germany

[2]Technology Design & Development-Universität Bremen, Badgasteiner Str.1, 28359 Bremen, Germany

[3]IAPC- Universität Bremen, Leobener Straße NW2, 28359 Bremen, Germany

E-mail: akueck@uni-bremen.de

Abstract. The importance and potential of nanoparticles in daily life as well as in various industrial processes is becoming more predominant. Specifically, silver nanoparticles are increasingly applied, e.g. in clothes and wipes, due to their antibacterial properties. For applications in liquid phase it is advantageous to produce the nanoparticles directly in suspension. This article describes a green production of silver nanoparticles using micro reactor technology considering principles of green chemistry. The aim is to reveal the potential and constraints of this approach and to show, how economic and environmental costs vary depending on process conditions. For this purpose our research compares the proposed process with water-based batch synthesis and demonstrates improvements in terms of product quality. Because of the lower energy consumption and lower demand of cleaning agents, micro reactor is the best ecological choice.

1. Introduction

The number of mass products that contain nanoparticles is increasing and so is the interest in greener industrial production of these materials. So far, gas phase processes are often used to synthesize aggregates of nanoparticles in ton-scale [1]. For some applications, however, well-defined nanoparticles in suspension of a (non-toxic) solvent are required, e.g. monodisperse iron oxide particles with a certain diameter is required for magnetic resonance imaging [2]. Commonly, these product characteristics can be obtained by wet chemical procedures using a solvent as a medium. When organic solvents are used, shape and size of nanoparticles can be consistently controlled. However, the cost of ensuring safety in this process can prove costly. In contrast the usage of water as green solvent results to a poor shape and size control, if the synthesis is performed in a batch reactor (table 1). The key to overcome this dilemma of defined particle size and environmental friendliness is **safer micro reactor technology**. This paper features this process by a case study, the production of silver nanoparticles, which are a mass product due to a continuously increasing application in different daily products, such as clothes and wipes. In contrast to the common batch production technology, micro reactor systems allow for continuous production improved operational safety, and, due to the highest heat and mass transfer rates, well-defined synthesis conditions in tailored reaction zones.

Table 1. Comparison of the advantages (highlighted) and disadvantages of batch reactor and a micro reactor in case of using water based synthesis or synthesis including organic solvents.

	Batch process	Micro reactor
Synthesis (water-based)	• **Environmentally friendly** • Discontinuous • Poor shape & size control • Little ton-scale (below 6 t/a)	• **Environmentally friendly** • **Continuous** • **Good shape & size control** • **Middle ton-scale (6-15 t/a)**
Synthesis (organic solvent)	• **Good shape & size control** • Discontinuous • Not environmentally friendly • Little ton-scale (below 6 t/a)	• **Continuous** • **Good shape & size control** • **Middle ton-scale (6-15 t/a)** • Not environmentally friendly

Moreover, combinations of environmentally friendly synthesis and micro reactor technology should include as many of the 12 principles of green chemistry as possible (Prevention, Atom Economy, Less Hazardous Chemical Syntheses, Designing Safer Chemicals, Safer Solvents and Auxiliaries, Design for Energy Efficiency, Use of Renewable Feedstock, Reduce Derivatives, Catalysis, Design for Degradation, Real-time analysis for Pollution Prevention, Inherently Safer Chemistry for Accident Prevention) [3]. These principles were outlined to give reference points to design new chemicals and chemical processes.
This work answers these four following leading questions.

1. Is it possible to avoid the disadvantages of aqueous synthesis by using a micro reactor?
2. From the ecological point of view: Is it beneficial to use micro reactor technology?
3. From the economic point of view: Is it efficient/ cost effective to use micro reactor technology?
4. Are there further advantages in terms of using micro reactors?

2. Product Quality

In most cases nanoparticles including organic solvents are used to get well-defined single particles because of the high shape and size control [4]. But actually these ways of producing nanoparticles are not sustainable, neither in a batch process nor in a continuous micro reactor process. Alternatively, water-based syntheses are more environmentally friendly, but the size and shape control are very poor [5]. So the aim is to avoid these disadvantages by using micro reactor technology caused by well defined reaction conditions. To survey the product quality of both processes, the analytic method of dynamic light scattering is used (DelsaNanoC, Beckmann Coulter). In the following paragraph both plants are described.

The flow sheet of the batch reactor system, figure 1a), shows a stirred tank reactor, where the first solution is placed, including the pump which transports the second solution. After the pump the tube is divided into ten smaller tubes to reduce the local gradient of concentration. The stirred tank reactor can be heated, if necessary. In figure 1b) the flow sheet of the micro reactor plant is shown. The three solid dosing feeders carry the solid reactants into the corresponding mixing tanks, where the two required starting solutions are prepared. After preparing the reactants the solutions are pumped to the entrances of the micro reactor (T-shaped micro mixer). The 2 micro reactor intakes end at the small reaction capillary, where the reaction takes place. This is called the production mode. The production mode is followed by the cleaning mode, where the cleaning agents (first HNO_3- solution; second H_2O) were pumped through the micro mixer and the capillary. The thermostat offers the possibility of heating the micro mixer.

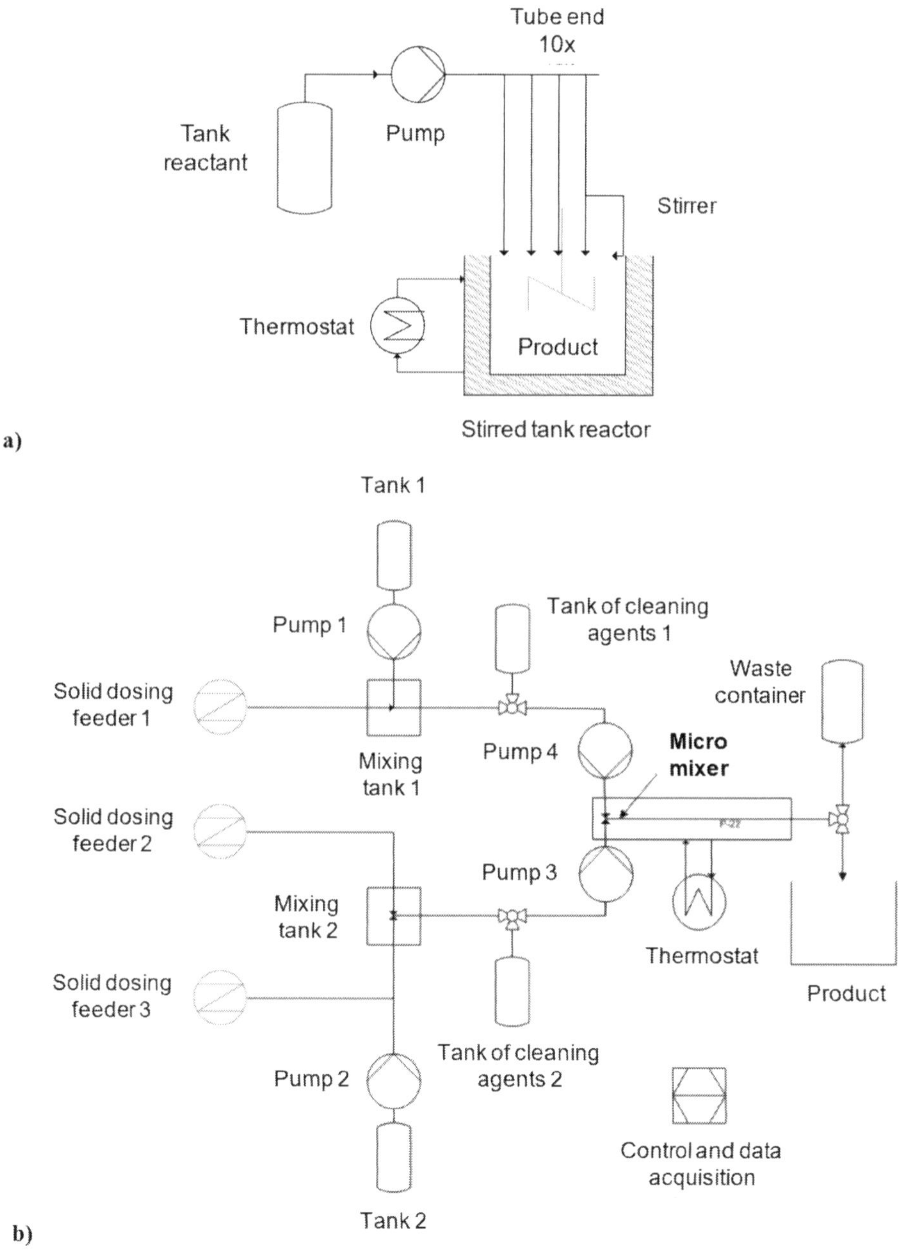

Figure 1. Flow sheet of the compared processes. (a) Batch reactor system. The first reactant solution is placed in the stirred tank reactor, while the second solution is pumped to the first solution. (b) Micro reactor plant. The reactants are automatically prepared and pumped to the micro mixer, where the silver nanoparticles are synthesized.

Our findings demonstrate that the product quality can be increased. In figure 2 a typical size distribution for batch-nanoparticles and for a micro reactor-curve are shown. Distributions were obtained by Non Negative Least Square Method. The diameters of micro reactor-nanoparticles are small and the size distribution narrower, that means a higher product quality. The small peaks, which appear in both cases, are often artefacts coming from the particle rotation [6]. Also, the shape is clearly defined as spherical particles as shown in the representative TEM (transmission electron microscope) picture, figure 2.

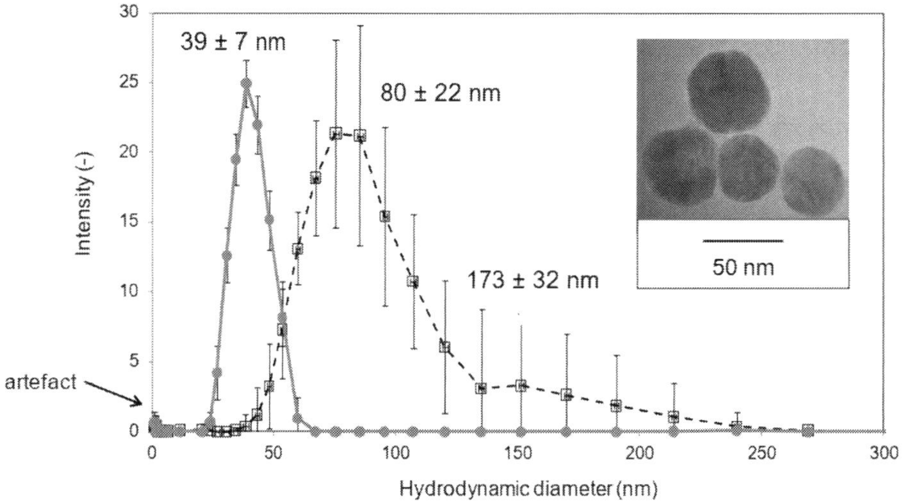

Figure 2. Particle size distributions of samples produced in a batch reactor (open squares) and a micro reactor (solid circles). Particles of the latter micro reactor-nanoparticles are smaller, and the size distribution is narrower.

3. Ecological Analysis

To determine the ecological benefit of a micro reactor a cumulative energy demand (CED) analysis is done. To realize a comparison of a batch reactor and a micro reactor the data are standardized of one year or one production cycle. So the two processes are comparable independent of the operational time.

A silver nanoparticle synthesis, which is accomplished in a standard flask, acts as a case study for obtaining a benchmark by computing the energy demand and the amount of chemicals required for a higher scale process. To accelerate the feed-in of the second reactant solution the tube is subdivided into ten tube parts. To clean the reactor the product suspension is collected and the cleaning agent (nitric acid) is filled in. After the cleaning with nitric acid the reactor is rinsed with pure water twice. The resulting energy demand is listed in table 2.

The electrical energy demand of a micro reactor is generated during the process. The measured data are shown in table 2. The different modes are mentioned in section 2.

Table 2. Process data of the two types of processes investigated.

Process	Apparatus	Power (W)	Time required	Effective consumption of electricity (Wh)
Batch reactor process				
Preparing of the solutions			20 min	manual
Production mode	Pump	15		22.5
	Stirrer	20	1h 15 min	30
	Thermostat	200		300
Further stirring	Stirrer	200	10 min	33.3
drawdown/ cleaning			20 min	manual
Sum			**2h 05min**	**385.8**
Micro reactor process				
Production mode	Solid dosing feeder 1	0.2	50 min	0.167
	Solid dosing feeder 2	1		0.833
	Solid dosing feeder 3	0.8		0.667
	Pump 1	5		4.167
	Pump 2	15		12.5
	Pump 3	5		4.167
	Pump 4	15		12.5
	Thermostat	60		50
	Control-PC	12.5		10.417
Cleaning mode (H_2O/HNO_3)	Pump 3	5	5 min	0.417
	Pump 4	5		0.417
	Control-PC	12.5		1.042
Cleaning mode (H_2O)	Pump 4	5	5 min	0.417
	Control-PC	12.5		1.042
Sum			**1 h**	**98.75**

Table 3. Cumulative energy demand of each component of both processes.

Process	Apparatus	Energy demand (Wh)	Primary energy demand (MJ)
Batch reactor process			
Production	Pump	22.5	0.23
	Stirrer	30	0.3
	Thermostat	300	3
Further stirring	Stirrer	3.3	0.33
	Sum	**385.8**	**3.86**
Micro reactor process			
Production mode	Solid dosing feeder	1.667	0.017
	Pump	33.334	0.333
	Thermostat	50	0.5
	Control-PC	10.417	0.104
Cleaning mode	Pump	1.251	0.013
	Control-PC	2.084	0.021
	Sum	**98.75**	**0.988**

The cumulative energy demand of both processes is calculated to describe the production of silver nanoparticles without the preproduction. This is considered in the analysis using **Umberto** software. Because of the same reactants and end use of both processes, further line sections of life time cycle are neglected. The degree of efficiency (energy production) is g = 36 % as average of the power plants in Germany [7]. To produce nanoparticles in a batch reactor fewer components, which need electrical energy, are required than in a micro reactor: one pump, one stirrer and one thermostat. After the actual reaction, a second stirrer mode takes place due to an improved mass flow and a higher yield. The data of each component is shown in table 3. Calculation of the CED including added end energy demand per production cycle (EEV) using equation 1 with NEV (non energetic consumption) = 0, SEI (material bounded energy content) = 0, g = 0.36) results to an energy demand of 3.87 MJ.

$$CED = \sum \frac{EEV_i}{g_i} + \sum \frac{NEV_j}{g_j} + \sum \frac{SEI_k}{g_k} \quad (1)$$

The batch reactor produces 0.096 g silver nanoparticle per production cycle (worst case: 100 μM silver, yield = 60%), that takes 2.083 h. To achieve a comparable value the energy demand is divided by the amount of produced silver nanoparticles. This quotient is called comparative quotient CQ: $\mathbf{CQ_{batch} = 37.29\ MJg^{-1}}$. For the industrial production designed micro reactor system consists of nine components, which need electrical energy: three solid dosing feeders, four pumps, one thermostat and one PC. The required power and the duration are described above. The CED is computed according to equation 1 (NEV = 0, SEI = 0, g = 0.36). Together with the production amount of 0.086 g silver nanoparticles per hour (worst case: 100μM silver, yield = 60 %) the comparative quotient for the micro reactor calculates to: $\mathbf{CQ_{micro} = 11.48\ MJg^{-1}}$. Figure 3 demonstrates the effect of "Numbering-Up". "Scaling-Up" is not computed in this article because the data is incomplete. Accomplishing a "Numbering-Up" the CQ reaches the lowest point of energy demand per gram product at a six times "Numbering-Up". Because further "Numbering-Ups" were performed a second or even more thermostat or other compounds are required, so that the CQ will increase.

Furthermore, an eco-efficiency analysis using Umberto software is done. Now the preproduction of the reactants and cleaning agents are considered. These data are taken from the **ecoinvent** (Umberto) data bank. The comparison here is not only the batch reactor versus the micro reactor. Also are compared different silver concentrations and amounts of cleaning agents (nitric acid). Figure 4 shows that the CO_2-eqivalent depends strongly on the amount of nitric acid and the electro energy production.

Using CED the CQ of the micro reactor is approximately three times lower than the CQ of the batch reactor. This result is achieved mainly due to the high heating energy. The micro reactor is also the better choice, when the CO_2- equivalents are compared because of the higher need of cleaning agents and the higher demand of electro energy.

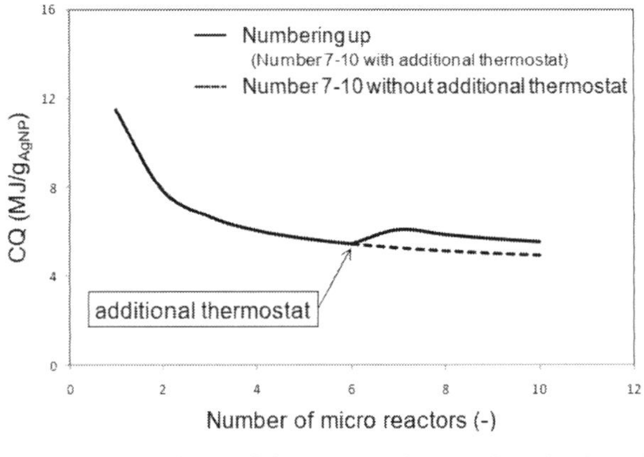

Figure 3. Dependence of the comparative quotient (CQ) on the number of parallel micro reactors determining the turnover.

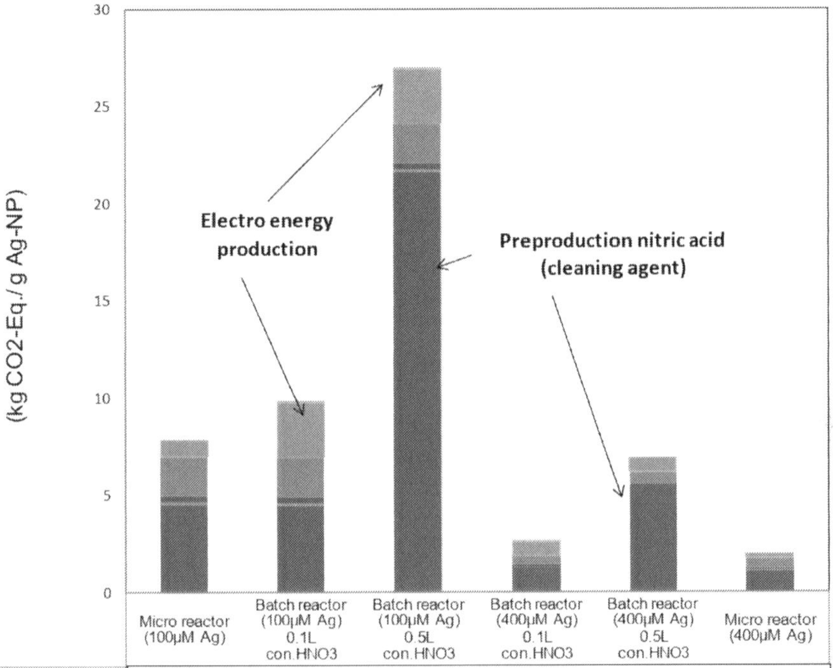

	Micro reactor (100µM Ag)	Batch reactor (100µM Ag) 0.1L con.HNO3	Batch reactor (100µM Ag) 0.5L con.HNO3	Batch reactor (400µM Ag) 0.1L con.HNO3	Batch reactor (400µM Ag) 0.5L con.HNO3	Micro reactor (400µM Ag)
Wastewater treatment	0.00	0.01	0.01	0.00	0.00	0.00
Electro energy production	0.82	2.84	2.84	0.71	0.71	0.20
Preproduction, starch	2.02	2.01	2.01	0.50	0.50	0.51
Preproduction, sodium hydroxide	0.31	0.31	0.31	0.08	0.08	0.08
Preproduction, pure water	0.10	0.14	0.14	0.03	0.03	0.03
Preproduction, nitric acid	4.37	4.30	21.47	1.07	5.37	1.09
Preproduction, silver	0.17	0.17	0.17	0.17	0.17	0.17

Figure 4. CO_2- equivalent per gram silver nanoparticle (AgNP) of a batch and a micro reactor process (column 1 and 2) including different amounts of cleaning agents for the batch reactor (column 3). CO_2- equivalent is calculated for a higher concentration of silver (column 4- 6).

4. Economic Analysis

To find out if the micro reactor is cost effective or not the comparative cost method and the period payback rule are used. Table 4 shows the data for the batch reactor and the micro reactor. The economic examination is based on the service life and therewith the depreciation and amortization, the imputed interest rate, the operational costs and the investment costs. Besides the investment costs, the operational costs present makes a significant part of the incoming costs and are calculated separately for every process. The service life for each type of reactor is appointed for five years. An imputed interest rate of 4% per year is assumed. That is the average annual rate of return of the government bond of the Federal Republic of Germany. The proceeds, which can be achieved, depend on the current price of silver. For our calculations, the prices of *PlasmaChem GmbH* are applied. 312 € are estimated or 500 mL of suspension with a silver content of 0.1 mg/mL [8]. This assumed value is recalculated for each product. For our calculations the average data for electricity cost [9], personnel costs [10] and the material costs of chemicals [11] are used.

The investment costs of the batch reactor are established according to an offer of *AP-Miniplant GmbH & Co KG* (23,354 €). The operational costs consist of the demand of chemicals, need of electrical power and working hours. The initial solutions for the batch process are manually prepared. Also the removal of the product and the cleaning of the batch reactor after every production cycle are not automatically done. For each cycle of the batch process, 0.66 h of man hours is required. It is possible to perform four cycles per 8 hour work day. Our average product exhibits a silver content of 100 μM (yield = 60%). These add up to a computed proceed of 41.60 €/L and an annual production amount of 15,000 L (250 days, one-shift operation). The investment costs of the micro reactor plant (71,910 €) are calculated by addition of every single component of our prototype plant, see figure 1. The operational costs comprise the electrical demand, material costs of the chemicals and man hours. Because of high automation, only one manual working hour per day is required. This hour is necessary to adjust the operation and inspect the production process. Per day (8 working hours) 8 production cycles may be realized. Our average product exhibits a silver content of 100μM (yield = 60%). The computed proceeds of the micro reactor add up to 41.60 €/L and an annual production amount of 26,800 L (250 days, one-shift operation).

Table 4. Data base of the batch and the micro reactor for economic analysis.

	Micro reactor	Batch reactor
Production amount per cycle	13.4 L (8 cycle/ d)	15 L (4 cycle/ d)
Operational costs/ year	128.97 €	315.58 €
Investment costs	71.91 €	23.35 €
Service life/ year	5	5
Imputed interest rate	4 % p. a.	4 % p. a.

Table 5. The costs per production unit of the micro and the batch reactor.

	Micro reactor	Batch reactor
Annual production	26 800 L	15 000 L
Annual operational costs	32 241 €	49 750 €
Annual accrual	14 382 €	4 670 €
Annual interest rate	1 438 €	467 €
Total annual costs	**48 061 €**	**54 887 €**
Costs/ production unit	**1.79 €/L**	**3.66 €/L**

Comparative cost method: Due to significantly higher investment costs of the micro reactor, the annual interest rates and the annual accrual is also very high. This mostly affects the costs per production unit, see table 5. For the batch reactor the costs per production unit depend primarily on the operational costs. The largest part of the total costs constitutes the operational costs. That is directed to both processes. In the context of the **period payback rule** the profit should be considered by the investment costs to the profit. Thus the amortization of the investment costs can be calculated for each process. The profit contribution is due to the difference of the proceeds and the operational costs per production unit, that means for 37.22 €/L the micro reactor and 37.94 €/L for the batch process. The equations E_{BR} and E_{MR} mentioned in figure 5 are for calculations beyond investment costs (BR= batch reactor, MR= micro reactor). The comparative cost method shows lower production costs per production unit using a micro reactor resulting from the higher production amount and lower man hours. The results of the period payback rules shows that the investment costs for the batch process are paid back after 10 days and after 18 days for the micro reactor, see figure 5. Higher productivity and lower operational costs affect the overall costs on micro reactors.

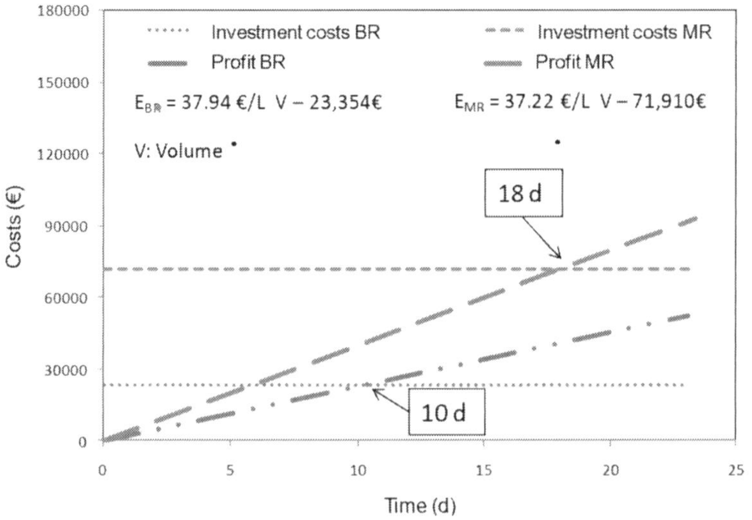

Figure 5. Payback period of micro reactor (MR) and batch reactor (BR).

For BR the payback period is 44 % shorter.

5. Advantages of micro reactors

Because of the high surface to volume ratio the thermal conditions of a process can be controlled very well. That results in a fast and intensive mass and heat transfer. Also a quick mixing can be realized by special structures in or of the micro reactor. One prime example of the advantages of the micro reactor is the prevention of a runaway, which sometimes happens in a batch reactor. Due to the high surface to volume ratio and accordingly to the quick and intensive heat transfer, the temperature profile is a steady state instead of a runaway.

6. Conclusion

The micro reactor is suitable to compensate for disadvantages of aqueous nanoparticle synthesis produced in batch reactor. At a scale of one ton per year allows for a drastically improved particle size distribution and reduced cumulative energy demand compared to a conventional production in a 15 L stirred-tank reactor. The improved environmental performance is caused by the smaller dimension of the micro reactor (*e.g.* less heating energy is required) and the low demand of cleaning agents. The eco-efficiency analysis additionally shows that also CO_2- emission is reduced in the case of a micro reactor. Furthermore, the micro reactor approach follows all relevant recommendations of the twelve principles of green chemistry.

Despite higher investment costs of a micro reactor plant compared to a batch reactor process, producing silver nanoparticles in a micro reactor system can be economically favorable. This is due to higher production amount and lower labour costs. Furthermore, efficient heat control, small reactor volume, excellent process control, and a high level of safety are possible. To sum it up micro reactor technology provides a promising avenue for sustainable production of silver nanoparticles.

Acknowledgement

We gratefully acknowledge Sebastian Tilgner for performing preliminary experiments. This project is funded by the state of Bremen (Programm Angewandte Umweltforschung) and the European Regional Development Fund ERDF 2007 - 2013. Furthermore this work is financially supported by Federal Ministry of Education and Research and Hans Böckler Stiftung (graduate school nanoToxCom).

SPONSORED BY THE

Federal Ministry of Education and Research

EUROPEAN UNION: Investing in your future – European Regional Development Fund

References

[1] Teoh W Y, Amal R, Mädler L 2010 *Nanoscale* **2** 1324- 1347
[2] Jun Y W, Huh Y M, Choi J S, Lee J H, Song H T, Kim S J, Yoon S, Kim K S, Shin J S, Cheon J 2005 *Jounal of the American Chemical Society* **127** 5732- 5733
[3] Anastas P T and Warner J C 1998 *Green Chemistry: Theory and Practice* Oxford University Press New York 30
[4] Kan C-X, Zhu J-J and Zhu X-G 2008 *Journal of Physics D: Applied* **41** 155304 (9pp)
[5] Bar H, Bhui D, Sahoo G P, Sarkar P, Pyne S, Misra A 2009 *Colloids and Surfaces A: Physicochem. Eng. Aspects* **348** 212–216
[6] Berne B J, Pecora R 1976 *Dynamic Light Scattering* (New York: Wiley) p122
[7] BMWI - Bundesministerium für Wirtschaft und Technologie 2009 *Energie in Deutschland* Berlin/ Germany p26
[8] PlasmaChem GmbH 2010 *Catalogue. Plasmachem // ADDO Galvanoadditiv DLC TRIO Nanoluminograph Sedimentation Gauge BET Carbon Analysis, Custom Synthesis.* PlasmaChem GmbH [cited 22. June 2010] http://www.plasmachem.com/onlinecatalogue/catalogue.php
[9] European Union 2010 *Eurostat - Tables, Graphs and Maps Interface (TGM) table 25. Juni 2010* Eurostat. European Union [cited 28. June 2010]
[10] European Union 2010 *Arbeitskosten Haupttab*elle *25. Juni 2010* Eurostat. European Union [cited 28. June 2010]
[11] Sigma-Aldrich 2010 *Sigma-Aldrich Product Catalog 2010* [cited 2. Juli 2010] www.sigma-aldrich.com

Are enhanced dust explosion hazards to be foreseen in production, processing and handling of powders consisting of nano-size particles?

Rolf K Eckhoff

University of Bergen, Norway

E-mail: rolf.eckhoff@ift.uib.no

Abstract. Based on experience with powders having particle sizes down to the range 1 – 0.1 μm one might expect that dust clouds from combustible nm-particle powders would exhibit extreme ignition sensitivities (very low MIEs) and extreme explosion rates (very high K_{St}-values). However, there are two basic physical reasons why this may not necessarily be the case. Firstly, the formation of clouds of well-dispersed primary particles from bulk powders consisting of nm-particles is extremely difficult to achieve, due to the comparatively very strong inter-particle cohesion forces. Secondly, should such a dispersion process nevertheless be fully successful, extremely fast coagulation processes in clouds within the explosive mass concentration range, would transform the primary nm-particles into much larger agglomerates within fractions of a second. Furthermore, for organic dusts and coal, the basic mechanism of flame propagation in the dust cloud is a further reason for not expecting extreme ignition sensitivities and explosion rates dust clouds from nm-particles. The overall conclusion is that dust clouds consisting of nm primary particles would not necessarily be expected to exhibit more severe MIE and K_{St}-values than those of dust clouds of μm primary particles. Recent experimental evidence confirms that this is in fact the case for K_{St}, whereas for MIE the values for some metals are significantly lower for nm primary particles than for μm ones.

1. Introduction

1.1. Nm-particles in a dust explosion context

As pointed out by Bouillard *et al.* [1] a range of new types of nm-particles are currently being produced and used in a number of consumer products ranging from cosmetics, sunscreens, toothpastes, pharmaceuticals and clothing, to electronics, plastics and tires. As a consequence, "nano-safety" has emerged as a new field requiring concern.

This paper addresses the question of whether production, handling and use of powders/dusts consisting of nm-particles can present an enhanced dust explosion risk compared to the risk presented by powders/dusts of larger particles. This limits the group of particle materials of concern to combustible ones only, which with few exception means

- Natural organic materials (grain, linen, sugar etc.)
- Synthetic organic materials (plastics, pigments, pesticides etc.)
- Carbon, coal and peat
- Metals (aluminium, magnesium, silicon, zinc, iron, titanium etc.)

So far most publications on production of nm-particles of combustible materials such as carbon and various metals, refer to production methods and facilities of laboratory- or fairly small pilot-scale. Kónya *et al.* [2] discussed a method for 'large scale' production of carbon nano-tubes, but their 'large scale' only implied batch charges of up to 100 g. In the common context of industrial dust explosions such small quantities of powder would be regarded as negligible. However, Wei *et al.* [3] constructed a pilot plant capable of producing high-purity multi-walled carbon nano-tubes (MWCNT) at a yield of some kg/hour, which would perhaps mean 10 – 20 tons per year on a continuous-production basis. With such quantities the dust explosion hazard would become a concern.

It may seem that the issue of a possible enhanced dust explosion risk during production and handling of nm-particle powders may sometimes be irrelevant simply because of the small scales of production and handling. On the other hand it may well be that genuine large-scale plants for production of various types of such particles may become a reality in a not too distant future.

1.2. Influence of particle size and dust concentration on ignitability and explosibility of dust clouds with primary particle sizes down to 1 μm

It has been shown experimentally decades ago that flammable dust clouds of most solid combustible materials will become more sensitive to electric spark ignition and burn with increasing violence as the size of the primary particles in the cloud decreases down to the order of 1-10 μm. Figure 1 shows how experimentally determined minimum electric spark ignition energies (MIE) of clouds in air of three different dusts decreases systematically with particle size.

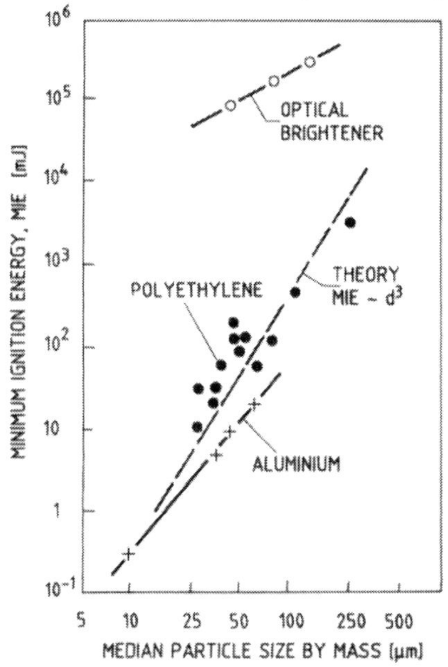

Figure 1. Minimum electric spark ignition energy of clouds in air, of 3 different types of dust, as a function of the median particle size of the dust. Experimental data from Bartknecht and theoretical line from Kalkert and Schecker. Figure 1.30 in [4]

The vertical scale in the figure is by orders of magnitude, and hence the influence of particle size on MIE is substantial. Theoretical considerations indicate that MIE for some particle materials is proportional to the cube of the particle diameter, as indicated in figure 1 by the predicted relationship for polyethylene particles.

The experimental data in figure 2 show how the combustion rate of clouds of aluminium dust in air increases systematically with the specific surface area of the dust (area of particle surface per unit mass), i.e. with decreasing particle size. A specific surface area of 6.5 m^2/g corresponds to mono-sized aluminium spheres of diameter 0.34 μm, or flakes of thickness 0.11 μm, which is in the far upper range of what would be considered as genuine nm particles. The question is then whether this trend continues into the range of genuine nm particles.

Figure 3 gives some further data for the maximum rate of pressure rise for various dusts as functions of particle size. The data for organic powders cover only the particle size range down to about 15 μm. With particles of organic materials smaller than about 10 μm the pyrolysis stage and subsequent mixing of pyrolysis gases and air, are so rapid that they will no longer be the rate controlling stages in the combustion process. Instead the controlling stage will be the homogeneous combustion of the pyrolysis gas/air mixture.

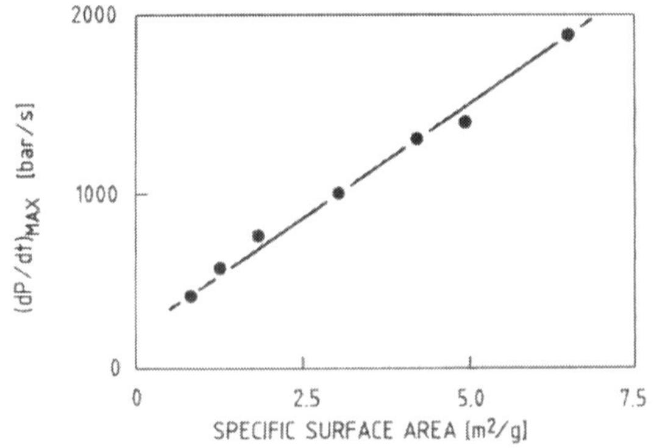

Figure 2. Influence of specific surface area of aluminium dust on maximum rate of pressure rise in dust explosions in a standard closed 1 m^3 ISO vessel. (Figure 1.23 in [4])

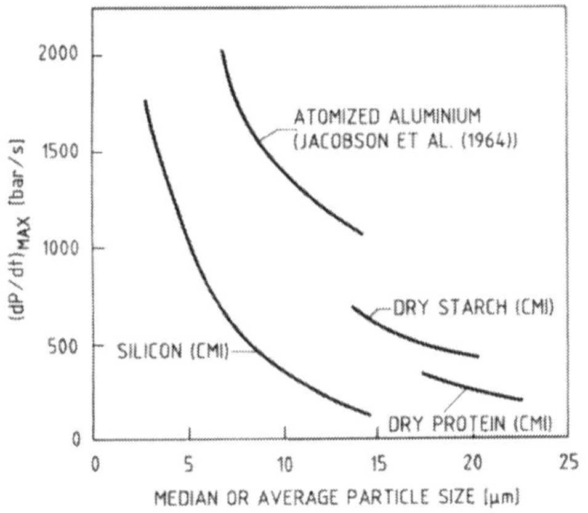

Figure 3. Influence of median/average particle size of various dusts on the maximum rate of pressure rise generated by dust explosions in a standard closed 1 litre Hartmann bomb. (Figure 1.28 in [4])

If the trends indicated in figures 1-3 for metal powders were just to be extrapolated into the nm-particle range, extreme ignition sensitivities and explosion rates would be predicted. However, two important physical processes will tend to limit the dust explosion hazard in the case of nm particles. The first is the limited ***dispersibility*** of nm-particle powders. The second is the very high ***coagulation*** rates of nano-size particles in clouds of flammable/explosive mass concentrations. In the following these two aspects will be discussed in more detail.

2. Formation of dust clouds by dispersion of powder deposits consisting of nm primary particles

2.1. Background

In some special situations such as in air jet mills, explosive dust clouds may be generated *in situ*, i.e. the dust particles become suspended in the air as they are produced. However, in most cases in industry explosive dust clouds are generated by re-entrainment and re-dispersion of powders and dusts that have been produced at an earlier stage and allowed to accumulate as deposits or layers.

There is a ratio of about 100 between the bulk density of a given powder when appearing as a deposit or a layer, and the maximum explosive concentration of a cloud of that powder in air.

Therefore, in order to obtain an explosive dust cloud from a deposit or layer, some process of re-entrainment and re-dispersion is required. Generation of powder/dust deposits may either be intentional (e.g. collection of powders and dusts in silos, hoppers and bag filters), or unintentional (e.g. deposition of dust on beams, external surfaces of process equipment or walls and floors of work and storage rooms). Also the re-entrainment and re-dispersion of such dust deposits may be either intentional (handling/transport in mixers, bucket elevators, pneumatic transport lines etc.), or unintentional (bursting of sacks and bags containing powder, leaks of powder/dust from pressurized process equipment, and dispersion by blasts of air generated by an explosion that has started elsewhere in the plant). In dust explosion hazard assessment the important role played by dispersion processes has often been overlooked. It has not always been realized that fine, cohesive powders cannot be dispersed in a gas to form a cloud of individual primary particles unless particle agglomerates are exposed to very high shear or tensile stresses. This means that the effective particle size in dust clouds in practice may be much larger than the size of the primary particles.

Professor Weber (quoted in chapter 3 in [4]), one of the pioneers of dust explosion research in Germany about 130 years ago, stressed the close relationship between dust cohesion and dust dispersibility, or dustability. In a paper on the ignitibility and explosibility of wheat flour he emphasized that cohesion of the flour, caused by inter-particle adhesion, plays an important role with respect to the ability of the flour to disperse into explosive dust clouds. Using simple but convincing laboratory experiments he demonstrated that the dispersibility of one specific flour increased systematically with decreasing moisture content in the flour. Weber suggested that two large dust explosion disasters, one in Stettin (Szczecin) and one in München, were mainly due to the low moisture content and hence the high dispersibility of the flours involved.

2.2. An overall structure of the dust dispersibility problem

A given mass of powder at equilibrium with the ambient atmosphere contains a finite number of inter-particle bonds, each of which requires a specific amount of work to be broken. The total minimum work W_{min} needed to break all these bonds in one unit mass of powder could in principle be calculated by integrating the work required for breaking all the individual inter-particle bonds. In accordance with [4], chapter 3, one could then define a theoretical upper limit value of the 'dispersibility' D_{max} for that specific powder deposit by:

$$D_{max} = 1/W_{min} \qquad (1)$$

When defined in this way, the dispersibility has the dimension mass per unit of energy or work, and is thus a measure of the quantity of powder that can be completely dispersed by spending one unit of energy from external sources in the process. However, no realistic dispersion process is one-hundred per cent efficient. This can be accounted for by incorporating an efficiency factor, k:

$$D_{real} = k/W_{min},\ 0 < k < 1 \qquad (2)$$

Clearly, the particle size distribution has a great influence on W_{min} at a given powder bulk density. Furthermore, the bulk density of a given cohesive bulk powder will vary with the degree of compaction. This means that the number of inter-particle bonds per unit mass of cohesive powder will increase with increasing compaction of the powder. Therefore W_{min} also increases with increasing compaction. Moisture increases W_{min} by increasing the strength of certain types of inter-particle bonds.

The close link between W_{min} and the nature and number density of the inter-particle bonds in a powder has given rise to detailed studies of various types of such bonds. Attempts have also been made at predicting aggregated powder-mechanical strength properties of powder deposits from microscopic inter-particle structure and forces.

However, the efficiency factor $0 < k < 1$ in Equation (2) allows D_{real} to have any value between zero and D_{max}, depending on the way in which the work W_{min} is applied to the powder to be dispersed. This in turn depends on the geometrical arrangement of the powder and the form of the energy available for the dispersion process. In the case of cohesive powders composed of very small particles, inter-particle forces play a major role and inter-particle bonds may not be broken unless the particle agglomerates are exposed to high velocity flow fields, or high-velocity impacts.

2.3. Attraction forces between particles in powder deposits

2.3.1. Van der Waals' forces

As discussed in [4], chapter 3, the theoretical van der Waals' force F_W between two spherical particles is:

$$F_W = A/a^2 \bullet (x_1 x_2)/(x_1 + x_2) \qquad (3)$$

Here A is a constant, a the smallest distance between the sphere surfaces, and x_1 and x_2 the diameters of the two spheres. Van der Waals' forces between particles are of significance as long as $x < 100$ nm. If $x_1 >> x_2$, the force is only determined by the size of the smallest particle, and equation (3) reduces to

$$F_W = A/a^2 \bullet x_2 \qquad (4)$$

Because F_W is proportional to the diameter of the smallest particle, it will play the dominating role in systems of very small particles, compared to forces due to gravity and viscous drag.

Most particles in real life are not smooth spheres, but of irregular shape and surface topography. Schubert [5] showed that F_w between a plane surface and a point on an irregular particle of diameter x, having a small elevation of radius r that touches the plane surface, is:

$$F_W = A \bullet (x/(r + a_o)^2 + 2r/a_o^2) \qquad (5)$$

The distance, a_o, is the smallest distance that can exist between two solid bodies in touch, and it is estimated at 0.4 nm.

2.3.2. Electrostatic forces

In the case of electrically conducting particles, electrostatic inter-particulate attraction between touching particles may occur even if the particles do not initially carry any net excess charge, provided their electron work functions are different. Electrons will then be transferred from one particle to the other. Different electron work functions can occur in particle systems of apparently identical materials, due to differences in impurities, oxide layers etc. Provided the smallest distance a between the two surfaces is shorter than 100 nm, i.e. the particles are in electrical contact, the electrostatic contact attraction force F_E between the two conducting particles is ([4], chapter 3):

$$F_E = \pi\varepsilon_1\varepsilon_0 \bullet U^2/a \bullet (x_1 x_2)/(x_1 + x_2) \qquad (6)$$

Here ε_o is the permittivity of vacuum and ε_1 the dielectric constant of the gas surrounding the particles. U is the contact potential between the two particle surfaces.

For electrically non-conducting particles, such as plastics, the electrostatic contact force is negligible. In this case, electrostatic attraction between particles is caused by excess charges on the particle surfaces, acquired tribo-electrically during preceding production and handling. The attraction force between two non-conducting particles having total excess opposite charges on the surfaces of q_1 and q_2, equals:

$$F_E = (q_1 q_2/4\pi\varepsilon_1\varepsilon_o) \bullet ((x_1 + x_2)/2 + a)^{-2} \qquad (7)$$

For $a >> (x_1 + x_2)$, equation (7) reduces to Coulomb's equation for attraction between two opposite point charges. If a is much smaller than the diameter of the largest particle, F_E will essentially be independent of a.

Equations (3)–(7) are all concerned with the attraction between two single particles under idealized conditions. Therefore, these equations are of limited value for predicting inter-particle attraction forces in real powders and dusts where many particles are interacting and particle shape and surface properties may be complex. In the case of electrostatic forces, realistic assessment of the particle charges q_1 and q_2 is also difficult, even for idealized particle geometries.

2.3.3. Inter-particle forces due to liquids

In agreement with the findings of professor Weber discussed above, it is common experience from practice in industry that dry dusts are usually easier to disperse than moist dusts (one exception can be heavily electrostatically charged dry plastic powders). Even small quantities of adsorbed moisture can in some cases increase the attraction forces between particles considerably. Adsorbed layers of up to 3 nm thickness can adhere firmly to the particle surface and make it smoother. This can reduce the effective distance between two touching particles appreciably. For a spherical particle of 1 μm diameter the volume of a 3 nm layer of liquid water constitutes only 2% of the particle volume. For smaller particles the percentage will increase. Adding small amounts of liquids to dusts is sometimes used intentionally in industry for reducing dust dispersibility. One application of this method is addition of soya bean oil to grain for preventing generation of grain dust clouds in grain storage plants.

At higher powder moisture contents excess water starts to form liquid bridges between particles. With further increase of the moisture content, a transition range is reached, at which the space between the particles becomes completely filled with water. Within this capillary range the capillary under pressure is the main source of the cohesion. If the water content is increased beyond this point, the system transforms from a cohesive powder to a suspension of particles in a liquid.

2.4. Dispersion of agglomerates suspended in gas, by forced flow through a nozzle

Depending on the actual dispersion process, or de-agglomeration process, the effective particle size distribution in dust clouds from a given cohesive powder can differ considerably. Because agglomerates in a dust cloud may behave as single particles of the agglomerate size when they burn, both ignition sensitivity and explosibility of such clouds will vary correspondingly.

In view of this the investigation by Yamamoto and Suganuma ([4], chapter 3) is of considerable interest. They studied the break-up of particle agglomerates in an air flow when the flow was passed through a nozzle at various velocities. A sample of the resulting dispersed dust cloud downstream of the nozzle was fed through a cascade impactor to determine the effective, aerodynamic *in situ* size distribution in the secondary dust cloud. Figure 4 gives a typical set of results.

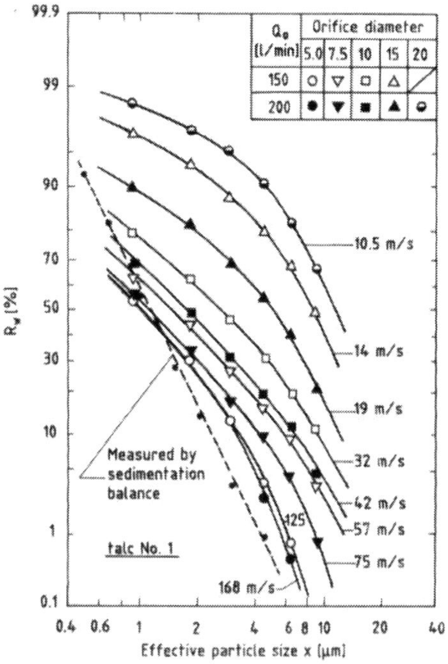

Figure 4. Effective particle size distributions of an airborne talc dust after dispersal by different orifices and air velocities. R_W is the percentage by weight of the effective 'particles' that are larger than the size x (Investigation by Yamamoto and Suganuma in 1984 quoted by [4])

The distribution of effective particle sizes is shifted systematically towards smaller particles as the dispersion process in the nozzle gets more effective, i.e. as the average air velocity through the nozzle increases.

The effect is quite dramatic. For an air velocity in the nozzle of 10.5 m/s, the median effective particle size is somewhat larger than 10 μm, whereas for velocities in the range 100–150 m/s it is only 1 μm. For the primary dust cloud, which was generated in a way that would be typical in industry, the median particle size would probably be considerably larger than 10 μm. It can be observed from figure 4 that at the highest air velocities in the nozzle, the distribution of the sizes of 'effective' particles in the secondary dust cloud approached the size distribution found by means of a sedimentation balance, after having dispersed the powder in a liquid in a way that would be expected to produce close to perfect dispersion.

2.5. Concluding remark

The understanding of how explosive dust clouds can be generated by dispersion of powder layers/deposits requires knowledge both of the nature of the powder (W_{min}) and of the actual dispersion process (value of efficiency factor k). The latter depends very much on the actual industrial situation, which is different in bucket elevators, pneumatic transport systems, fluidized beds, various kinds of mills, driers, mixers, cyclones, filters and silos. Therefore, intimate knowledge of the nature of the industrial environment in which the dust is dispersed is required to enable any reasonable estimate of k.

3. Coagulation of particles in nm-particle dust clouds in air

3.1. Introduction

The following discussion is based on the chapter on particle coagulation in the classical text by Green and Lane [6]. The point is that the presence of agglomerates in dust clouds of very small particles is not only due to incomplete particle dispersion. In addition new agglomerates are formed continually by inter-particle collision. Green and Lane named this process ***coagulation*** and pointed out that continuous and spontaneous coagulation is one of the most striking characteristics of dust clouds. The particles in the cloud, of whatever substance they are composed, will coalesce or stick together if they make contact. The process goes on continuously so that the effective particle size increases and the particles finally settle out of suspension. Because the mean inter-particle distance for a given dust mass concentration decreases proportionally with the mean particle size, the initial coagulation rate for a given mass concentration also increases with decreasing initial particle size.

According to [6] the first clear evidence of coagulation in clouds of very small particles was obtained by Tolman and his collaborators in 1919, who measured the variation with time of the intensity of a Tyndall beam in clouds of various dusts. They were able to correlate the Tyndall beam readings with both the mass concentration of dust particles in their experimental clouds, and with the average particle size. It then became clear that the observed marked decrease of the Tyndall beam intensity could only be explained by coagulation.

3.2. The coagulation constant

The nature of the coagulation process in dust clouds was, according to Green and Lane, first resolved experimentally in the pioneering research by Whytlaw-Gray and co-workers in 1932. They were able to count the particles per unit volume of dust cloud, and hence to determine the variation with time of this number. They found that coagulation of many types of small particles suspended in air followed a single simple law. When the average volume of cloud occupied by a single dust particle was plotted against the elapsed time, a straight line was obtained for a wide range of dust particles, as shown in Figure 5 for dilute smoke particles in air. If n is the number of particles per cm^3 of cloud at any time t, and n_o is the number present at the moment when the dust cloud is first formed, and before onset of coagulation, then:

$$1/n - 1/n_o = Kt \qquad (8)$$

The constant K is named the "coagulation constant". The unit on the vertical axis is 10^6 times the average volume of dust cloud occupied by one particle, the volume unit being cm^3. When converted to m^3, the number 8 on the vertical axis in figure 5 means that the average volume occupied by one particle is $8 \cdot 10^{-12}$ m^3, or the number of particles per m^3 equals $1/8 \cdot 10^{12}$. This may seem a high number, but for small particles the mass concentration will be low.

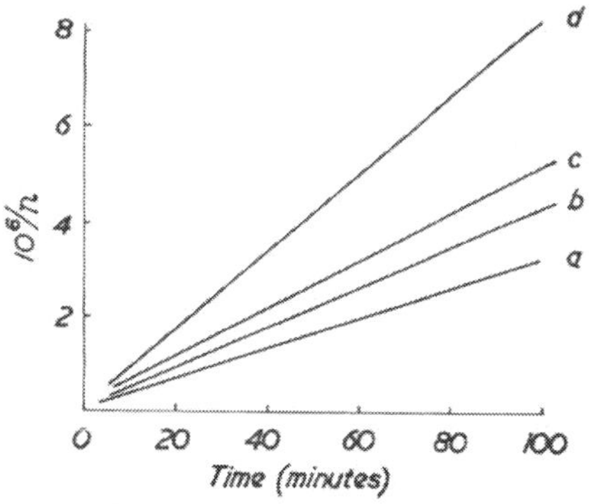

Figure 5. Average volumes of dust cloud occupied by one particle for various dust types, as functions of elapsed time: (a) stearic acid, (b) ammonium chloride, (c) cadmium oxide, (d) benzene-azo-b-naphthol. Figure 5.1 in [6].

Hence, for an organic material of density 1 g/cm^3 and particle size 1 μm, the mass concentration will only be about 0.1 g/ m^3, which is orders of magnitude below typical minimum explosive concentrations for organic powders. It is clear, therefore, that figure 5 is only relevant to explosive dust clouds of nm particles if the straight lines hold also for much smaller $10^6/n$ values than those resolvable on the vertical axis in figure 5. This assumption is the basis of the discussion in the following.

The coagulation constants for the various materials investigated by Whytlaw-Gray and co-workers are equal to the slopes K of the straight lines in figure 5. Table 1 gives experimentally determined K values for a selection of particle types, ranging from metallic oxides, via various solid organic compounds to liquid droplets of organic substances.

Table 1. Coagulation constants K for some particle materials. Table 5.1 in [6].

Substance	K (10^{-9} cm^3/s)
Ammonium chloride	0·60
Ferric oxide	0·66
Magnesium oxide	0·83
Cadmium oxide	0·80
Stearic acid	0·51
Oleic acid	0·51
Resin	0·49
Paraffin oil	0·50
p-xylene-azo-β-naphthol	0·63

In spite of the substantially different nature of the particles, the K-values in Table 1 are very similar, the ratio between the largest and the smallest value being only 1.7. Therefore Green and Lane concluded that in air at normal temperature and pressure the coagulation process does not depend significantly on the dust type.

Green and Lane [6] also pointed out that equation (8) is of the same form as that established by Rutherford for the rate of combination of small ions. The equation also applies to large atmospheric ions, and to the disappearance of nuclei. Naturally, because of the lower mobilities, K is smaller for dust particles than for ions. By diffentiation of equation (8) it follows that the rate at which the particles disappear by coagulation at any instant depends on the square of the number present at that instant and the constant K:

$$- dn/dt = Kn^2 \tag{9}$$

Coagulation is very rapid in the early stages when the number of particles n per unit volume of dust cloud is high. Using equation (8) the time interval Δt required for halving n can be assessed. Hence, for $K = 0.5 \cdot 10^{-9}$ cm^3/s and $n = 10^{11}$ particles/cm^3, Δt equals 0.02 s, whereas for $n = 10^6$ particles/cm^3 Δt equals 33 min, and for 10^5 particles/cm^3 Δt equals 5.5 hours. The time taken to generate an aerosol containing $2.5 \cdot 10^6$ particles per cm^3 by coagulation is nearly the same whether the initial number is $5 \cdot 10^{12}$ per cm^3 or $5 \cdot 10^8$ per cm^3. As pointed out by Green and Lane this explains why, when dust clouds are formed from different quantities of material, the particle number may be the same by the time the cloud is ready to be examined experimentally, whereas the average effective particle size can vary appreciably.

3.3. Effect of poly-dispersity (particle size distribution)

No coagulating dust cloud can be truly mono-disperse, although it may be nearly so at the onset of coagulation in certain cases. Most often the cloud consists of particles of a wide range of sizes, depending on its past history. According to Green and Lane [6] experiments have confirmed that the more the dust particles in a dust cloud deviate from mono-dispersity, the faster the cloud coagulates. However, considerable difficulties arise when attempts are made to calculate precisely the effect of poly-dispersity. According to theory the probability of collision between particles of unequal sizes is greater than that of collision between mono-sized particles. But because most collisions occur between particles of equal size or nearly so, the effect due to poly-dispersity is counterbalanced by the comparative rarity of collisions between particles of large size differences.

In a poly-disperse dust cloud exposed to gravitational or centrifugal forces, the larger particles will move faster than the smaller ones. The probability that large particles will collide with smaller ones is thereby increased. This type of coagulation, due to different settling rates, is negligible in clouds of primary particles of nm-size but will become more significant as the particle size increase.

3.4. Effect of particle shape

As mentioned in [6], the original coagulation equations were based on the assumption that the particles were spherical. However, in practice most particles will have irregular shapes. Slight irregularities in shape make only little difference to the mobility of a particle in a gas, and therefore the effect on the coagulation rate should be only minor. The exception is chain-like structures, for which a considerable variation in the mobility, and consequently in the coagulation rate, might be expected.

Experimental evidence confirms that the rate of coagulation of particles in dust clouds is not greatly influenced by the particle shape. The values of K for clouds of particles of stearic and oleic acids agree closely with theory, whereas that for ferric oxide of the same weight concentration is only 30 per cent greater. When these particle systems were allowed to coagulate until the particle agglomerates had grown to the order of 1 μm, the stearic acid agglomerates were of a compact, equi-dimensional form, whereas those of ferric oxide were loosely built chain-like structures. If the nm-particles possessed the same form as those of the order of 1 μm, the coagulation constants K of stearic acid and ferric oxide would have differed greatly. On the other hand electron microscopic examination showed that the chain-like ferric oxide aggregates did not appear until the particle size approached the 1 μm range.

3.5. Effect of temperature, pressure and viscosity of the air

As also pointed out by Green and Lane [6] the coagulation rate of mono-disperse particles also depends on the temperature, pressure and viscosity of the air in which the particles are suspended. Because the viscosity of air also increases with temperature, the increase of the coagulation constant K will not be proportional to the absolute temperature, but will nevertheless increase with temperature.

The sole effect of reduction of the gas pressure upon K is to increase the mean free path of the gas. This is because, except at very low pressures, the gas viscosity is practically independent of pressure. According to [6], experiments on ferric oxide dust clouds at various pressures showed that the coagulation rate increased markedly as the gas pressure was reduced.

3.6. Effect of foreign vapours mixed with the air

The experimental values for the coagulation constant for various dust clouds reported by Green and Lane [6] were in reasonable agreement with the value calculated from theory on the assumption that all collisions between the particles are irreversible, i.e. the particle were sticking together permanently after having collided. It had been suggested by some investigators that a thin layer of adsorbed foreign vapour or liquid on the particles might alter the particle surface in such a way that, on collision, the particles would not readily stick together. Systematic experiments were carried out with a number of different dust particle types in air to which vapours of phenol, oleic acid, glycerine and water had been added. In every case the measured rate of coagulation was found to be the same, within the experimental error, as the rate of coagulation in pure air, determined in control experiments.

3.7. Effect of dust cloud turbulence

One way of increasing the coagulation rate of a particulate cloud is to make it turbulent by stirring it with a fan. Eddies and swirls are then formed and the velocities of the particles relative to each other increase. Therefore the chance of collision of particles with one another must also increase, which in turn means an increased coagulation rate. As discussed in [6], this problem was investigated systematically by Langstroth and Gillespie in a study of the coagulation and surface deposition of ammonium chloride particle suspended in quiescent and turbulent air. They used a chamber of approximately 1 m^3 capacity as illustrated in figure 6. In the case of the experiments with turbulent air, a large motor-driven paddle, pierced with numerous holes, was oscillated at the desired frequency through an angle of 180° from the vertical, throughout the experiment. The readings of a small non-directional thermocouple anemometer, averaged over various positions in the dust cloud chamber, were taken as measures of the average air velocities. A set of results are reproduced in Figure 7. As can be seen, there is a systematic increase of the average volume of dust cloud occupied by one particle at any instant, with the average air velocity (turbulence).

Förster [7] suggested that accidental generation of large, explosive clouds of sprays/mists of organic liquids of high boiling points is not very likely. The reason is that the mean droplet-droplet distance in the explosive range is of the order of only 10 droplet diameters, which in a turbulent cloud makes fast coalescence of the small droplets to larger ones highly effective. The larger droplets will then "rain out" and the fuel concentration in the cloud will very soon fall below the explosive range. Förster confirmed experimentally that with a high-boiling-point liquid (octanol) it was very difficult indeed to generate an explosive spray in a 1 m^3 explosion chamber, unless a highly sophisticated spray nozzle system was adopted.

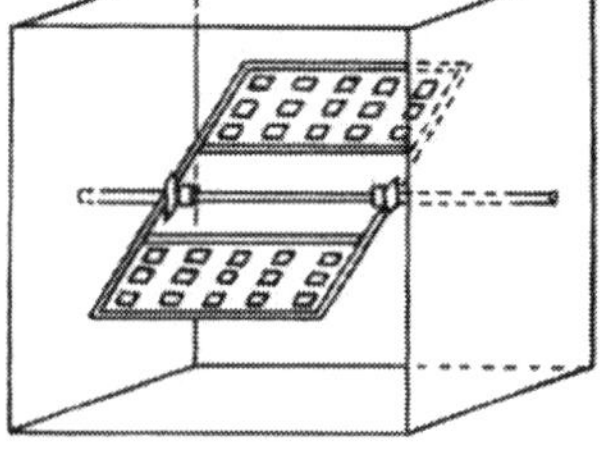

Figure 6. The 1 m^3 dust cloud chamber with oscillating paddle used by Langstroth and Gillespie. Figure 5.7 in [6].

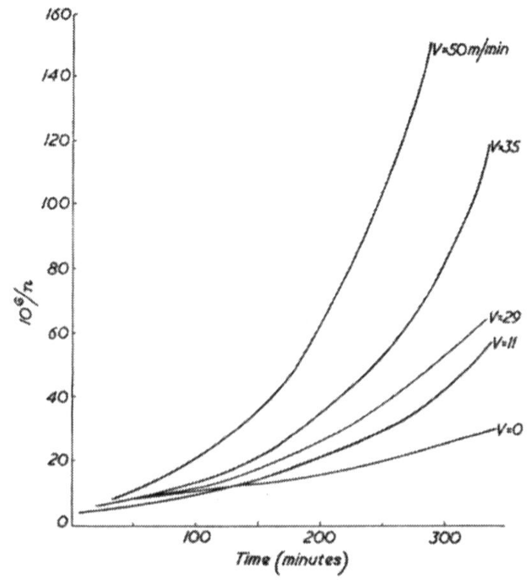

Figure 7. Average volume of dust cloud occupied by one particle as a function of time for various air turbulence levels. Ammonium chloride particles in air. Figure 5.8 in [6].

It should be pointed out that, when a dust cloud is stirred in a closed chamber, reduction of the number of particles in the cloud will not only be due to coagulation. Irreversible deposition of particles on the chamber wall will also contribute. Some of the increased volume of dust cloud per particle (reduction of the number of particles per unit volume of dust cloud) in Figure 7 is probably due to enhanced deposition of particles on the internal chamber surfaces. However, the relative importance of wall deposition decreases with increasing number of particles in the dust cloud. Therefore this affect probably plays a minor role in the case of the very high particle concentrations of interest in the context of dust explosions with nm-particles. For a more extensive discussion of the significance of the wall deposition problem reference is made to Green and Lane [6].

3.8. Coagulation of electrically charged particles

As also discussed in [6], the presence of electric charges on the particles in a dust cloud aerosol may alter the rate of coagulation appreciably. It is possible to calculate the coagulation constant K from knowledge of the amount of electric charge and the size distribution of the particles. For a dust cloud with a symmetrical electric charge distribution the influence of electric charge on the coagulation rate depends upon the fraction of negatively charged particles, the fraction of positively charged particles, the average charge of each sign and the average particle diameter. However, any change in the electric charge distribution brought about by coagulation may greatly alter the coagulation rate.

3.9. Coagulation in an acoustic field

Particles suspended in a gas coagulate readily when exposed to powerful sound waves, in particular standing waves in resonant enclosures. According to [6], various investigators have shown that coagulation occurs both at sonic and ultrasonic frequencies. Direct evidence of this effect was obtained from comparing photomicrographs of particles settled out of dust clouds which had been subjected to intense sound waves, and particles settled out of clouds that had not been exposed to such waves.

4. Expected ignition sensitivities and explosibilities of clouds of nm particles in air based on indirect evidence

4.1. Mechanism of flame propagation in clouds in air of organic and coal nm- particles

According to Hertzberg *et al.* [8-10] and Ballal [11] the following three consecutive processes take place in the propagation of a flame through a cloud of organic dust or coal dust in air:

1. Heating and pyrolysis/devolatilization of dust particles
2. Mixing of the pyrolysis gases/volatiles with air
3. Gas phase combustion of the premixed gas phase

In a subsequent stage 4. carbon residue particles burn as individual entities behind the gas flame front. The characteristic time constants for the three consecutive processes 1.– 3. may be denoted τ_d, τ_m, and τ_c. According to [8], mixing of pyrolysis gases/volatiles with air is a very rapid process and τ_m will be shorter than τ_c. Furthermore, for small particles $\tau_d << \tau_m$, and the flame propagation process will essentially be controlled by the premixed gas combustion (time constant τ_c). For smaller particles than the order of 10 μm, τ_d and τ_m are negligible compared to τ_c and particle size has no longer any influence on the burning rate, which is essentially limited by the rate of gas phase combustion.

With organic and coal powders/dusts the ultimate limits of ignition sensitivity and explosibility would probably be the values of the premixed pyrolysis/devolatilization gases/vapours only. The actual values of MIE and K_{St} would then be expected to vary with the composition of the gases/vapours. With a large fraction of the gas/vapour molecules being unsaturated compounds, quite low MIEs and high K_{St}-values would be expected, in agreement with typical values for gas groups IIB and IIC.

4.2. Mechanism of flame propagation in clouds in air of carbon and metal nano-size particles

In the case of pure carbon and various metals including aluminium and magnesium, the combustion process mainly occurs at or near the particle surface. In most cases a homogeneous combustible gas phase throughout the flame front in the cloud will not be formed in the same way with organic materials. At outset, therefore, both the ignition sensitivity and the explosibility of dust clouds of most metals and carbon would be expected to increase systematically with decreasing primary-particle size right down to the nm-range. On the other hand reduced dispersibility and fast coagulation of the very small particles would probably limit the effect of the reduced primary-particle size, as will be discussed below.

4.3. Degree of dust dispersion and particle coagulation

The minimum explosive dust concentration (MEC) for organic materials is of the order of 100 g/m^3, irrespective of particle size. For metals it is generally higher and roughly proportional with the particle density. It is then of interest to consider the number of mono-sized particles n_o that would be contained in one cm^3 of a perfectly dispersed dust cloud at MEC before any coagulation has started. For cubical particles, taking the particle edge as the particle 'size', and with a particle density of 1 g/cm^3 (organic materials), one cm^3 of dust cloud will contain 10^{17} particles of particle size 1 nm, 10^{14} particles of size 10 nm, and 10^{11} particles of size 100 nm. This means that with a coagulation constant K equal to $0.5 \cdot 10^{-9}$ cm^3/s agglomeration would have reduced the number of particles per cm^3 of dust cloud to about 10^9 per cm^3 within only 1 s irrespective of the initial particle size before onset of coagulation. The average effective particle size would then have increased to > 100 nm.

The question is then whether an ignition delay of 1 s is long or short compared with the delays between dust cloud generation and ignition in situations relevant to the practice. Two different contexts are of interest. The first is industrial scale dust clouds in process equipment, for which ignition delays of the order of 1 s would appear relatively short. In such cases coagulation will, by the time of ignition, have increased the effective particle size to the order of 0.1 – 1 μm.

However, in dust clouds in small scale laboratory test apparatuses for assessment of dust cloud ignitability and explosibility the situation will be different. In the tests for minimum ignition energy (MIE) and explosion violence (K_{St}), the experimental dust clouds are normally created by quite violent dispersion processes to ensure that even quite small particles are dispersed into individual entities in the gas. Triggering of the ignition source follows within 20 – 60 ms after onset of the dispersion process, which means that ignition occurs while the coagulation of the particles in the cloud may only be in its initial phase. On the other hand, dispersing samples of nm-particle powder into clouds in air consisting of independent primary nm-particles only would be virtually impossible by means of the dust dispersion method used in these test methods.

5. Explosibility and ignition sensitivity of nm-particle clouds in air. Experimental evidence

5.1. Explosibility

45 years ago Helwig [12], using a 43 litre closed bomb for his experiments, found that the rate of combustion, i.e. $(dp/dt)_{max}$ of clouds in air of coal dust containing 10–50% volatiles, did not increase monotonically with decreasing particle size. Instead the explosion rate for the finest fraction, of particle size 0–10 μm, was in fact lower than for the most explosive size range 20–30 μm, probably because of incomplete dispersion and/or rapid coagulation of the finest particle fraction.

More recently Vignes at al. [13] examined the explosibility of three kinds of nm-particle powders by means of the standard 20-litre explosion bomb. The data obtained were compared with those of μm-particle powders of similar materials. For aluminium it was found that down to particle sizes corresponding to a specific surface area of a few m^2/g both P_{max} and $(dp/dt)_{max}$ increased systematically with decreasing particle size. This is in agreement with the data in figure 2. However, Vignes at al. [13] also found that both P_{max} and $(dp/dt)_{max}$ in fact decreased_systematically when the specific surface area was increased further into the genuine nm-particle region. Three reasons for this were indicated. The first was uncontrolled pre-ignition of the dust cloud by friction/impact during injection/dispersion of the powder tested. The second possible reason was increased content of alumina in the aluminium with increasing specific surface area of the particles (oxidation of the particle surface). As a third possible reason was enhanced agglomeration/coagulation of the primary particles as they got very small.

Very recently Holbrow et al. [14] reported on similar experiments in a specially constructed 2-litre closed explosion bomb. The results are summarized in table 2.

Table 2. Maximum explosion pressures and maximum rates of pressure rise of clouds of nano-particle powders in air in a 2-litre explosion chamber. Table 9 in [14].

Sample number	Material	P_{max} (bar g)	dP/dt (bar/s)	Equivalent K_{St} (bar m/s)
EC/011/09	Aluminium nanopowder (210 nm)	12.5	1677	449
EC/104/07	Aluminium nanopowder (100 nm)	11.2	2000	536
EC/147/07	Iron nanopowder	2.9	68	18
EC/152/07	Zinc nanopowder	5.6	377	101
EC/148/07	Copper nanopowder	1.2	10	3
EC/042/08	Carbon nanofibre	5.2	62.5	17
EC/158/07	Carbon nanofibre	6.0	112	30
EC/116/08	Carbon nanofibre	6.9	591	158
EC/117/08	Carbon nanofibre	5.6	137	37
EC/153/07	Carbon nanofibre	6.4	339	91

The overall conclusion to be drawn from Table 2 is that the maximum explosion pressure and the maximum rate of pressure rise (K_{St}) of clouds in air of powders consisting of nm primary particles were very similar to the values of conventional μm-particle powders. However, the particle sizes of the two aluminium powders of 0.21 and 0.1 μm respectively are in the upper range of what would perhaps qualify for classification as genuine nm particle powders.

5.2. Ignition sensitivity

5.2.1. Electric/electrostatic sparks

The situation seems to somewhat different with the ignition sensitivity of the dust clouds. As illustrated by table 3 and figures 8 and 9, Wu *et al.* [15] demonstrated that experimental MIE values for nm-powders of both titanium and iron were in fact considerably lower than those of a μm-powders of the same materials. Clouds in air of nm-powders of both materials ignited readily in the Mike3 test apparatus with the lowest spark energy that this apparatus can produce, i.e. about 1 mJ. This means that dust clouds in air of some nm-powders may be more easily ignited by electric/electrostatic sparks than clouds of μm-powders of the same materials. Table 3 summarizes the results by Wu *et al.* [15].

Table 3. From [15]

Minimum ignition energy of nanoscale and micrometer scale metal powder.

Nanoscale		Micrometer scale	
Diameter	Minimum ignition energy	Diameter	Minimum ignition energy
Ti powder			
35 nm	<1 mJ	3 μm	<1 mJ
75 nm	<1 mJ	8 μm	21.91 mJ
100 nm	<1 mJ	20 μm	18.73 mJ
		45 μm	21.91 mJ
Fe powder			
15 nm	<1 mJ	150 μm	----
35 nm	<1 mJ		
65 nm	<1 mJ		

Note: '---' denotes unsuccessful ignitions by all 7 different ignition energy levels.

The question is then why the ignition sensitivity increase (MIEs decrease) when moving from μm-particle dust clouds to nm-particle clouds, whereas the explosion violence (K_{St}) does not increase? The extremely high temperatures of electric sparks compared to the temperature of the dust cloud flame could be a possible reason. Particle agglomerates than may remain agglomerates in flame propagation may disintegrate into primary nm-particles when they approach the extremely hot spark channel, and a small zone of genuine nm-particles may be generated in the vicinity of the spark and facilitate ignition of the dust cloud.

Table 4 shows that Holbrow et al. [14] obtained similar results for metals, whereas for the carbon nano-tubes the MIEs were quite high. As pointed out by Wu *et al.* [15] an alternative MIE apparatus capable of producing synchronized electric sparks of energies below 1 mJ is required to allow determination of the actual MIEs of very ignition-sensitive powders of MIEs < 1 mJ. Work in this direction was reported by Randeberg *et al.* [16], but as discussed by Eckhoff and Olsen [17], the spark generator used by Randeberg *et al.* needs to developed further to become suitable for the intended purpose.

It must be pointed out that very low MIEs have not only been measured for clouds in air of metal dusts. According to the data published by BIA [18] a number of synthetic organic substances, and sulphur, have MIEs < 1 mJ. It is not clear, however, whether the sizes of the particles in the powders tested were in the nm or in the μm range.

5.2.2. Heat from friction/impact

Further evidence of the extreme ignition sensitivity exhibited by some nano-particle metal powders was presented by Wu *et al.* [19]. They performed a detailed study of the conveyance of various types of such powders from the pre-compressed dust dispersion reservoir of a 20 litre Siwek/Kühner-type explosion bomb into the bomb itself. The powder then had to pass through the narrow tube with a bend connecting the powder reservoir to the main explosion volume of the bomb via a valve. The purpose of the experiments was to determine the maximum air velocity in the tube at which the powders could be conveyed through the tube without the dust cloud entering the bomb being spontaneously ignited by the conveyance process itself.

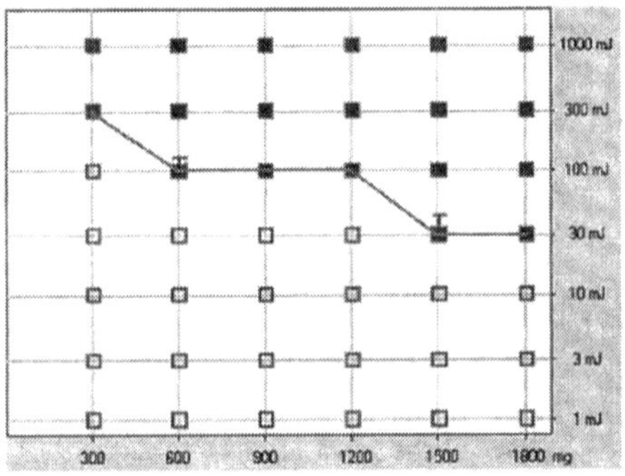

Figure 8. Minimum ignition energy (MIE) as a function of mass of dispersed powder for clouds in air of an 8 μm titanium powder determined by means of the MIKE3 apparatus.
■ = ignition, □ = no ignition in a given number of trials. From Wu et al. [19].

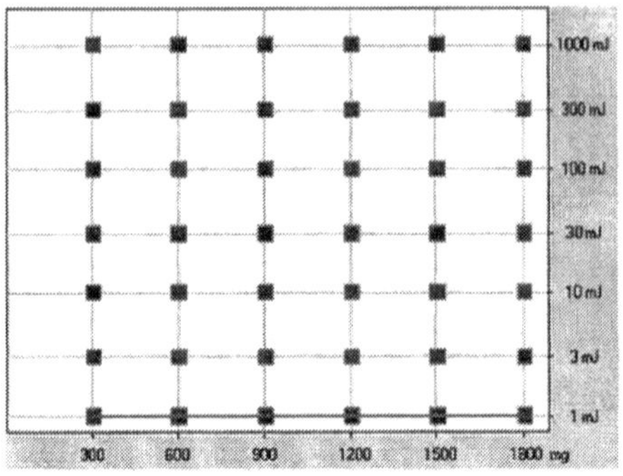

Figure 9. Minimum ignition energy (MIE) as a function of mass of dispersed powder for clouds in air of an 35 nm titanium powder determined by means of the MIKE3 apparatus.
■ = ignition, □ = no ignition in a given number of trials. From Wu et al. [19].

Wu *et al.* [19] assumed that ignition was due to the impact of the dust particles on the tube wall when passing through the bend. Four different air velocities were used, viz. 13, 8.5, 6.5, and 3.5 m/s. The powders tested were 30-nm titanium, 35-nm iron and 35-nm aluminum. The minimum ignition energy (MIE) of clouds in air of the three powders, as determined by the standard Kühner MIKE3 apparatus, which has a lower spark energy limit of 1 mJ, were all below this limit. Wu *et al.* found that the titanium powder ignited spontaneously at all the air flow velocities used. The iron powder ignited at all velocities but the lowest (3.5 m/s), whereas the aluminum powder did not ignite even at the highest velocity.

6. Consequences for the practice

6.1. Dust explosion violence

As discussed above, theoretical considerations as well as experimental evidence from laboratory-scale experiments suggest that K_{St} of powders consisting of nm primary particles will not be any higher than for powders consisting of primary particles of μm size. This means that dust explosion venting, suppression and isolation should be just as feasible with nm primary particles as with primary particles of μm size.

6.2. Dust cloud ignition sensitivity

Experimental evidence suggests that dust clouds in air of powders consisting of nm primary particles do in fact ignite more easily than μm-particle clouds of the same materials. The question is then whether this makes any difference with regard to the precautions required to prevent dust cloud ignition in industrial plants. Two different categories of ignition sources must be considered, viz. electrical/electrostatic sparks and mechanical friction/impact.

Table 4. Minimum ignition energies of clouds of nano-particle powders in air in a modified MIKE3 test apparatus. Table 10 in [14].

Sample number	Material	Minimum Ignition Energy (mJ)
EC/104/08	Aluminium nanopowder	< 1 mJ
EC/011/09	Aluminium nanopowder	< 1 mJ
EC/153/07	Carbon nanotubes	> 1000 mJ
EC/116/08	Carbon nanotubes	> 1000 mJ
EC/147/07	Iron nanopowder	< 1 mJ
EC/152/07	Zinc nanopowder	3 – 10 mJ
EC/148/07	Copper nanopowder	> 1000 mJ

As regards ***electrostatic sparks***, Glor [20] suggested that the occurrence of accidental electrostatic sparks of energies << 1 mJ may not be very likely. He pointed out that small objects of capacitances of the order of 1 pF encountered in industry normally have some sharp edges or corners, which give rise to charge drain by corona discharge as soon as the voltage of the object reaches the order of 7-10 kV. Furthermore, the time constant *RC* for the draining of any charge on such small capacitances to earth will be very short, of the order of 1 s or less even when the resistance to earth is of the order of 10^{12} Ω (tera ohms). Hence, such small objects cannot be charged to any high voltages unless the charging times are significantly shorter than the short time constants for charge drainage to earth.

If this were to be all that could be said, the conclusion would be that accidental electrostatic spark discharges of energies below the order of 1 mJ would in fact be unlikely in the process industries. Hence, the same safety procedures could be applied to any plant producing/ handling powders/dusts of MIE < 3 mJ, irrespective of the actual MIE of the powder. However, based on the description of the phenomenon of 'electrostatic induction' discussed by Lüttgens and Wilson [21] it is possible to envisage at least one situation where the charging of small objects would fast enough to generate very high voltages. Consider a small metal object (tramp metal) that enters an earthed metal silo together with the powder being fed into the silo. If the powder has acquired a significant charge, the settled powder in the silo may give rise to a strong electric field. When the small falling metal object enters this field the electrons inside it may become displaced by influence/induction. If one end of the metal object approaches the earthed silo wall whilst this electron displacement exists, a spark discharge may take place. The situation is illustrated in Figure 10. Because of the very short discharge times to be expected (< 10 μs), the spark gap length will stay approximately constant during such a spark discharge. For example, an object approaching an earthed wall at 20 m/s will move only 0.2 mm during 10 μs.

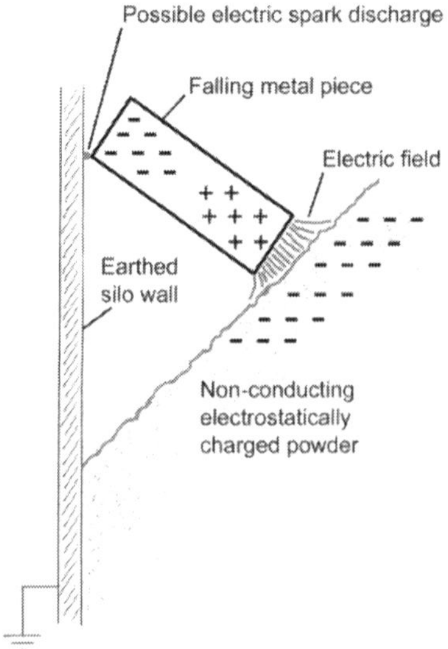

Figure 10. Illustration of how a spark discharge may be generated from an electrically neutral tramp metal piece falling into a silo, because of charge displacement in the metal piece when it enters the electric field in the silo generated by electrostatically charged powder there.

Lüttgens et al. [22] presented a brief report on a dust explosion in a silo, which was assumed to be initiated by a similar type of spark discharge. The silo was part of an assembly of a number of silos, made of aluminium, and used for intermediate storage of a plastic dust. Following several years of production without any accidents, a dust explosion unexpectedly occurred during discharge of one of the silos. At the moment of the explosion only about 20 % of the silo volume was filled with powder. After careful investigation, all ignition sources but an electrostatic spark discharge had been ruled out. However, a likely mechanism of generation of the electrostatic spark was not disclosed until part of the shaft of an aluminium spade was found close to the silo wall, just below the powder surface. It was concluded that excess charge accumulated on this object due to influence, was discharged as a spark when the object approached the silo wall.

So far indications of MIEs << 1 mJ of nm-particle dust clouds in air only exist for metals. Then the question arises whether bulk resistivities of heaps of such particles are sufficiently high to render spark discharge processes as illustrated in figure 10 possible. Two factors may be considered. Firstly, it would be expected that oxide layers on metal nm-particles, e.g. aluminium, will be comparatively thick in relation to the particle diameters. Secondly, poured bulk densities of bulk powders of nm-particles would be expected to be very low. The combined effect of these two factors might result in high bulk resistivities even for metal powders.

It should also be pointed out that MIEs of < 1 mJ have also, according to BIA [18], been found for a number of synthetic organic substances and sulphur. It is not clear, however, whether this is mainly due very small sizes of the primary particles, or to extreme reactivity of the particle substances.

At high voltages sparks of very low energies may not only be generated by very small charged capacitances. However, low-energy sparks may also result from discharge of larger capacitances that have been charged to only moderate voltages, say of the order of a one kV or even less. This would clearly also present an increased ignition risk with dusts of exceptionally low MIEs. With $C = 20$ pF, and $U = 1$ kV, the stored energy $\frac{1}{2}CU^2 = 0.01$ mJ, which might be of the order of the MIEs of some very spark-ignition sensitive dust. Clearly, with only 1 kV across the spark gap, the gap length has to be quite short, but with dusts of MIE << 1 mJ, the quenching distances are also likely to be very small.

With regard to ignition by ***mechanical friction/impact*** the findings by Wu *et al.* [19] indicate that careful precautions have to be taken to prevent ignition of clouds in air of some nm-particle metal powders by heat from mechanical impact of friction. However, further research to obtain more specific knowledge in this area is required and should be encouraged.

7. Conclusions

1. By extrapolating the trends exhibited by dust clouds having primary particles of sizes down to the order of 1 μm, dust clouds from combustible nm-particle powders would be expected to exhibit extreme ignition sensitivities (very low MIEs) and extreme explosion rates (very high K_{St}-values). However, this may not necessarily be the case for the following reasons:
   a) Formation of clouds of well-dispersed primary particles from bulk powders consisting of nm-particles is extremely difficult to achieve by dust dispersion processes normally operating in the process industries, due to the strong inter-particle cohesion forces.
   b) Should perfect dispersion nevertheless become a reality, extremely fast coagulation will transform the primary particles into much larger agglomerates within fractions of a second.
   c) For particles smaller than the order of 10 μm particle size will for most organic materials and coals particle size no longer influence the explosion violence/rate, because combustion of pyrolysis gases/volatiles becomes the rate controlling process in the flame propagation.
2. The delay between the moment of formation of a well-dispersed nm-particle dust cloud and the moment of ignition of the cloud is decisive for the extent of coagulation. In laboratory scale experiments to determine MIE and K_{St}, typical delays between dust cloud generation and ignition are of the order of 20-60 ms, and coagulation may still be only in its initial phase. Therefore, results from laboratory tests may overestimate the ignition sensitivity and explosion severity of real industrial clouds, if expected ignition delays in accidental explosion scenarios are significantly longer than in the laboratory test.
3. Dust cloud turbulence is an important parameter. Coagulation rates generally increase with increasing turbulence. This will enhance the increase of MIE with increasing turbulence. However, for the dust cloud combustion rate the net influence of turbulence is less clear. In the absence of coagulation, turbulence will increase the combustion rate. But with coagulation, the effective particle size in the cloud will increase, which will decrease combustion rate.
4. Laboratory-scale experiments have shown that nm-particle dust clouds of given materials do not have any larger K_{St}-values than clouds of dust clouds of μm-particle powders.
5. However, the MIE values of some metals are lower for dust clouds of nm primary particles than for clouds of μm primary particles. MIEs < 1 mJ have also been reported for some synthetic organic powders and sulphur, but it is not clear whether this is due to very small primary-particles sizes, or to highly reactive chemical compositions.
6. A new MIE test apparatus capable of producing synchronized electrical sparks of energies below 1 mJ is required for further clarification of this issue.
7. More research is needed before ignition sensitivities and combustion rates of all kinds of combustible nm-particle dust clouds in practical industrial accidental explosion scenarios can be reliably predicted.
8. Because K_{St} of powders consisting of nm primary particles do not appear to be any higher than of powders consisting of μm of primary particles, dust explosion venting, suppression and isolation should be just as feasible with nm primary particles as with primary particles of μm size.
9. Because of the very low MIEs (< 1 mJ) of clouds of some nm-particle dust clouds in air, the possibility of ignition by subtle electrostatic spark discharges of very low energies have to be considered.
10. In addition, careful precautions have to be taken to prevent ignition of clouds of nm-particle metal powders by heat from mechanical impact and friction.

References

[1] Bouillard, J., Vignes, A., Dufaud, O., Perrin, L., Thomas D. (2009): *Explosion risks from nano-materials*, Nanosafe 2008: International Conference on Safe Production and Use of Nano-materials, Journal of Physics: Conference Series **170** (2009) 012032, IOP Publishing

[2] Kónya, Z., Vesselenyi, I., Niesz, K., Kukovecz, A., Demortier, A., Fonseca, A., Delhalle, J.,

Mekhalif, Z., Nagy, J.B., Koós, A.A., Osváth, Z., Kocsonya, A., Biró, L.P., Kiricsi, I. (2002): Large scale production of short functionalized carbon nanotubes, *Chemical Physics Letters* Vol. 360 (2002) pp. 429-435

[3] Wei, F., Zhang, Q., Qian, W.-Z., Yu, H., Wang, Y., Luo, G.-H., Xu, G.-H., Wang, D.-Z. (2008): The mass production of carbon nano-tubes using a nano-agglomerate fluidized bed reactor: A multiscale space-time analysis. Powder Technology Vol. 183 (2008) pp. 10-20

[4] Eckhoff, R.K. (2003): Dust Explosions in the Process Industries, 3rd edition, Gulf Professional Publishing/Elsevier, Boston. ISBN 0-7506-7602-7

[5] Schubert, H. (1979): Grundlagen des Agglomerierens, Chem. –Ing. Techn. Vol. 51 (1979) pp.266-277

[6] Green, H.L. and Lane, W.R. (1964): Particulate Clouds: Dusts, Smokes and Mists, 2nd edition, E. & F.N. Spon Ltd., London. Printed by The University Press, Belfast. Catalogue No. 16/0853/7

[7] Förster, H. (1990): Private communication to R.K. Eckhoff from H. Förster, PTB, Braunschweig, Germany

[8] Hertzberg, M., Cashdollar, K. L., Daniel, L. NG., *et al.* (1982) Domains of Flammability and Thermal Ignitability for Pulverized Coals and Other Dusts: Particle Size Dependences and Microscopic Analyses. *Proc. of 19th Symp. (Internat.) on Combustion.* The Combustion Institute, pp. 1169–1180

[9] Hertzberg, M., Zlochower, I. A., and Cashdollar, K. L. (1986) Volatility Model for Coal Dust Flame Propagation and Extinguishment. *Proc. of 21st Symp. (Internat.) on Combustion.* The Combustion Institute, pp. 325–333

[10] Hertzberg, M., Zlochower, I. A., Conti, R. S., *et al.* (1987) Thermokinetic Transport Control and Structural Microscopic Realities in Coal and Polymer Pyrolysis and Devolatilization. Their Dominant Role in Dust Explosions. Prepr. Pap. *Am. Chem. Soc. Div. Fuel Chem.* **32**(3) pp. 24–41

[11] Ballal, D. R. (1983) Flame Propagation through Dust Clouds of Carbon, Coal, Aluminium and Magnesium in an Environment of Zero Gravity. *Proc. R. Soc.* London A385 pp. 21–51

[12] Helwig, N. (1965) Untersuchungen über den Einfluss der Korngroesse auf den Ablauf von Kohlenstaubexplosionen. D.82 (Diss. T. H. Aachen). Mitteilungen der Westfählischen Berggewerkschaftskasse. Heft 24

[13] Vignes, A., Traoré, M., Dufaud, O., Perrin, L., Bouillard, J., Thomas, D. (2009): Assessing explosion severety of nanopowders with the 20 l sphere, Paper presented at the 8th World Congress of Chemical Engineering, Montreal, 23-27th August (2009)

[14] Holbrow, P., Wall, M., Sanderson, E., Bennett, D., Rattigan, W, Bettis, R., Gregory, D. (2010): *Fire and explosion properties of nanopowders*, HSE Report RR782, www.hse.gov.uk

[15] Wu H.-C., Kuo, Y.-C., Wang, Y.-h., Wu, C.-W., Hsiao, H.-C. (2010): Study on safe air transporting velocity of nanograde alumimnium, iron and titanium, *Journal of Loss Prevention in the Process Industries*, Vol. 23 (2010) pp. 308-311

[16] Randeberg, E. and Eckhoff, R.K.: Measurement of minimum ignition energies of dust clouds in the < 1 mJ region, *J. Hazardous Materials* Vol. 149 (2007) pp. 237-244

[17] Eckhoff, R.K. and Olsen, W.: A new method for generation of synchronised capacitive sparks of low energies. Reconsideration of previously published findings. *J. Electrostatics,* Volume 68 (2010) pp. 73-78

[18] BIA (1997): *Combustion and explosion characteristics of dusts*, Report 13/97 from BIA (Professional Association's Occupational Safety Institute), published by HVBG (Federation of the Statutory Accident Insurance Institutions of the Industrial Sector), Germany, November 1997

[19] Wu H.-C., Kuo, Chang, R.-C., Hsiao, H.-C. (2010): Research of minimum ignition energy for nano titanium powder and nano iron powder, *Journal of Loss Prevention in the Process Industries*, Vol. 22 (2009) pp. 21-24

[20] Glor, M. (2006): *Personal communication on mechanisms of accidental electrostatic charging*, The Swiss Institute of Safety & Security, Basel, Switzerland.

[21] Lüttgens, G. And Wilson, N. (1997): *Electrostatic hazards*, Butterworth-Heinemann, Oxford, UK. ISBN 0 7506 2782 4

[22] Lüttgens, G. *et al.* (2005): *Statische Elektrizität - begreifen, beherrschen, anwenden*, 5th edition, Expert Verlag, D-71272 Renningen. ISBN: 3-8169-2506-5

Ignition and explosion of nanopowders: something new under the dust

Olivier Dufaud[1], Alexis Vignes[2], François Henry[1,2], Laurent Perrin[1] and Jacques Bouillard[2]

[1]Université de Lorraine – INPL, Laboratoire Réactions et Génie des Procédés, CNRS, ENSIC, 1 Rue Grandville, BP 20452, F-54001 Nancy Cedex, France

[2]INERIS, Parc Technologique ALATA, B.P. 2, F-60550 Verneuil-en-Halatte, France

E-mail: dufaud@ensic.inpl-nancy.fr, alexis.vignes@ineris.fr

Abstract. This work deals with the study of ignition and explosion characteristics of nanoparticles. It has been carried out on various powders: zinc, aluminum, carbon blacks... Specific behaviours have been highlighted during the first phase of this project (Nanosafe 2). For instance, it has been demonstrated that there mainly exists two combustion regimes that are either kinetically controlled, for small size particles, or diffusion controlled, for large size particles (generally with diameters greater than 1 or 2 μm). It has been found that as the particle size decreases, minimum ignition temperature and minimum ignition energy decrease (even lower than 1 mJ), indicating higher potential inflammation and explosion risks for metallic nanopowders. Moreover, the presence of agglomerates in the nanopowders could modify their reactivity. Thus, the explosion severity of Al powders tends to increase as the specific surface area decreases, before reaching a peak for 1 μm particle size.
These results are essential for industries producing or handling nanopowders in order to propose/design new and proper prevention and protection means. Nevertheless, the validity of the classical characterization tools with regard to nanopowders should be discussed. For example, the experimental laminar flame velocity of Al dusts has been compared to a theoretical one, determined by Huang's model, which assumes that the propagation of the flame is run mainly by conduction. It has shown a good agreement. However, under certain conditions, the Al flame propagation is expected to be mainly conducted by radiation. Two hypotheses can then be made. On the one hand, it can be assumed that the 20 L sphere probably disturbs the flame propagation and thermal mechanisms by absorbing radiation (wall quenching effect). On the other hand, it has been observed, thanks to the use of a high speed camera that the preheating zone is smaller for some nanopowders than for micro-particles (figure below). It could notably be explained by the fact that the flame radiation is absorbed by the cloud of unburnt Al nanopowders.
Several other factors may have an impact on the explosion severity. If these points are correctly addressed, it will be possible to get more reliable ignition and explosion characteristics.

1. Introduction

Nowadays, nanomaterials are widely used in the industrial and research fields. As the nanomaterial production and use are going to increase, there will be more and more associated risks. But the knowledge about the hazards related to these new materials is currently limited: if some studies are dedicated to the determination of their potential toxicity, only very few ones concern nanopowders fires and explosions [1]. In fact, since 2002 [2], one could quote the recent extensive studies performed by the EU NANOSAFE2 project [3] during the 2005-2009 period and more recently by Wu in Taiwan in 2009-2010 [4,5].

Obviously, dust explosion should not be neglected when the dusts are combustible [6]. However, literature studies concerning the evaluation of explosion and flammability risks of micron-sized powders do not enable to evaluate the fire and explosion risk probabilities and gravities of nanopowders.

As a consequence, the main goal of this work is to study explosion and ignition risks related to nanopowders. In fact, the tested powders are more "nanostructured particles" composed by agglomerated nanoparticles, than genuine "nanopowders" with a particle size lower than 100 nm. Existing analytical and methodological tools (such as a 20L explosion sphere, the modified Hartmann tube and specific procedures, standards, etc.) initially designed to evaluate dust explosion hazards related to micropowders have been tested in order to determine their validity to assess the ignition sensitivity and explosion severity of nanopowders. Experimental results were then compared with the explosion data obtained for micron-sized powders. We will mainly focus this paper on the cases of metallic nanopowders (aluminium and zinc). Carbon nanotubes and carbon blacks have also been studied. The main results obtained with these compounds will not be presented here in detail as they have been previously published [7]. However, they will be commented and compared with those obtained with metallic powders. Other compounds such as silicon carbide have also been tested but will not be presented here. Finally, this article will specifically focus on the evaluation of the validity of the 20L sphere. The hypotheses given in a previous paper [8] will be discussed in the light of recent experiments.

2. Materials and methods

Zinc and aluminium micron-sized powders were purchased from Goodfellow and zinc nanopowders from Inframat Advanced Materials. Carbon black, multiwalled carbon nanotubes and aluminium nanopowders were provided by industrial partners.

In order to have good reproducibility between the different tests, the carbonaceous nanomaterials and zinc powders were systematically dried at 50°C under dynamic vacuum of 1 Pa during 15 minutes and then under a static vacuum of 10 Pa during 4 hours before handling. The aluminium nanopowders were tested directly as they were initially stored under dry argon atmosphere.

Ignition sensitivity is often referred to as a combination of three parameters characterizing the ability of the particles to be ignited and to generate explosions. These parameters are the following ones: the minimum ignition energy (MIE), the minimum ignition temperature (MIT) and the minimum explosive concentration (MEC). Here, only the variations of minimum ignition energy will be considered. In our study, MIE was determined using a modified Hartmann tube (Mike 3 - Kühner AG) in accordance with IEC standard 1241-2-3.

The flame propagation has also been studied by using the same apparatus. The flame propagation process was recorded by a high speed video camera (Phantom V9 – up to 15000 pps). The relation between the measured flame speed and the burning velocity could be expressed as a function of the front flame surface and the area of the tube cross section. The flame speed dependence on the variation of the gas densities has also been taken into account in order to determine the laminar burning velocity.

The measurements of dust explosion severity, i.e. P_{max} (maximum overpressure), and $(dP/dt)_{max}$, (maximum rate of pressure rise), were performed in a 20 L spherical vessel in accordance with the EN 14034 standard. These parameters were finally obtained through a number of tests with step-wise increase in dust concentration.

3. Particles agglomeration

It was observed by TEM that carbon blacks tend to form agglomerates (weak cohesion forces) but also aggregates (strong cohesive forces) [3]. For aluminium and zinc samples, MEB observations as well as particle-size distributions measurements performed in isopropanol by a laser light diffraction analyzer (Mastersizer, Malvern Instrument), have shown that the agglomeration level is lower than for carbonaceous compounds. Nevertheless, a few agglomerates of 20 to 30 µm could be observed, for single aluminium nanoparticles of 100 nm diameter. The specific surface area was determined for the nanoparticles by BET measurements. Results are summarized in the table 1.

Table 1. Particles size characteristics of the samples

Nanopowders	BET specific surface area (m^2/g)	Particle mean diameter (µm)
Aluminium 100 nm	23	0,17
Aluminium 200 nm	10.5	0,28
Aluminium 3 - sample A	0.74	3
Aluminium 7 - sample B	0.25	7
Aluminium 27 - sample C	0.11	27
Aluminium 42 - sample D	0.08	42
Zinc 120 nm	19	120
Zn 38 µm	-	38

The powder densities have been measured with an helium pycnometer. By coupling the results obtained by BET and helium pycnometer, an average primary grain size can be calculated by considering that particles are isolated, cylindrical (or) spherical with a monodisperse size distribution. For well dispersed particles, such as metallic powders, the diameter d_{BET} could be linked to the Sauter diameter.

Other characteristics should be determined for a better understanding of the nanoparticles suspension behavior in a carrier gas. The dustiness measures could give a good estimation of the ability of the powder to be dispersed, but a force assessment (Van der Waals, electrostatic, capillary forces…) is also required to characterize the energy applied to powders to deagglomerate. It mostly depends on the surface properties of the nanoparticles (surface reactivity, particle diameter ...). A study is carried out at present in this field.

4. Nanoparticles ignition

Various authors have shown that the combustion time of large aluminium particles varies in a d^n law with n varying between 1.5 et 2 [9], which is typical of a shrinking core combustion model under diffusion control. In the case of nanometric particles and for temperatures lower than 1000°C, authors determined that the combustion time obeys a d^n law with n equals to 1.6 [10]. This model would indicate a faster reaction with regard to a d_p^2 expression which could be imparted to a specific effect of the alumina covering layer [9]. These authors explain this behaviour by a facilitated diffusion of aluminium through this layer, probably due to curvature effects. Another study of nanoparticles states that n index equals to 0.3 [11], which would indicate even faster reactions.

To sum up, the difference of combustion mechanisms between the nano and micron-sized particles can be understood as the result of two different regimes (a diffusion controlled and a kinetically controlled regimes). For aluminium nanoparticles, oxidation at low temperatures (lower than 1000°C) is ruled by diffusion mechanisms. It should be stressed that the change of combustion regime is observed above the nanometric size range [3].

Previous studies have also demonstrated that as the aluminium particle size decreases in the nanometric range, the inflammation temperature decreases (from 2300 K for 100 µm particles to 1200 K for 100 nm). It suggests that inflammation or explosion risks are inherently higher for nanopowders [12]. This behaviour has also been confirmed by minimum ignition energies measurements for aluminium and zinc particles (table 2). It could be noted that nanopowders have very high ignition sensitivities (notably low MIE, even lower than 1 mJ for aluminium) compared to those of micro-particles.

Table 2. Minimum ignition energies of Al and Zn particles

Particles	Al/42 µm	Al/3 µm	Al/200 nm	Al/100 nm	Zn/38 µm	Zn/120 nm
MIE (mJ)	440	14	7	< 1	> 1000	19

A simple model can be developed to take the variation of MIE as a function of the particle size into account. Thus, the MIE can be defined as the energy required for a given cloud of particles to sustain ignition and inflammation in a medium at an initial temperature T_u while the burning medium is at a final temperature T_{ad}, the flame adiabatic temperature [3]. Hence, one can write, for a spherical volume element of diameter D_c (critical diameter for which the "Van't Hoff and Taffanel Le Floch" conditions apply):

$$\int_{T_u}^{T_{ad}} \overline{\rho}.\frac{\pi}{6}D_c^{\ 3}.\overline{C_p}.dT = \int_0^{\tau_b}\left(Q_{generated} + Q_{ignition} - Q_{lost}\right)dt$$

where Q is an heat, T_{ad} and T_u, the adiabatic flame temperature and the unburned gases temperature, $\overline{\rho}$ and $\overline{C_p}$ are respectively the average density and heat capacity of the volume element. At ignition, $Q_{ignition}$ equals the MIE. By considering the mass conservation, the MIE could be described by the following relationship:

$$MIE \approx \frac{\pi}{6}.\overline{\rho_u}.\overline{C_p}.\frac{T_u}{T_{ad}}.(T_{ad} - T_u)D_c^3$$

By introducing τ_b, the combustion time, it can be assumed that: MIE $\propto \tau_b^{3/2}$. For the two different combustion regimes, one obtains an MIE which depends on dp^3 for kinetically controlled regime ($\tau_p \propto d_p^2$), which is consistent with Kalkert and Schecker's model [6], or on $d_p^{3/2}$ for the diffusion controlled regime ($\tau_p \propto d_p$). These two dependencies are in accordance with the variation of MIE expressed in table 2. For aluminium particles, the transition between the oxidation regimes occurs on a diameter range from 400 to 2000 nm. No experimental data are available for the carbonaceous particles, their MIE being greater than 1 J, which exceeds the limit of the modified Hartmann tube. Nevertheless, it has been demonstrated that the minimal ignition temperature of a carbonaceous particle cloud decreases and then increases as a function of the particle size [3]. The presence of this minimum value is probably due to the fact that, at lower particle sizes, there are stronger attractive interactions between particles. As a result, agglomeration/aggregation phenomena occur. This amount of energy should be added during the ignition process in order to oxidize the powders, which could explain the increase of the MIT for very small particles.

5. Explosion severity

In the literature, it was often observed that dust becomes more and more explosive up to a certain size limit at which it will plateau. For coal, [6] reports that the limiting particle size, below which there is no further increase in the maximum explosion pressure or the rate of pressure rise, is about 50 µm. [13] reports a plateau at a particle size of roughly 50 µm for flour, about 40 µm for methylcellulose, and polyethylene, but shows no sign of threshold for PVC. For metal dusts, especially the more reactive metals like aluminium and magnesium, the limiting particle size should even be smaller than for other dusts. No size limit could be found in the literature [6] for this kind of materials.

Experimental results related to the explosion severity of tested nanopowders are summarized in the table 2. The influence of the specific surface area on the explosion severity of zinc, aluminium and carbonaceous compounds is illustrated on figures 1 and 2.

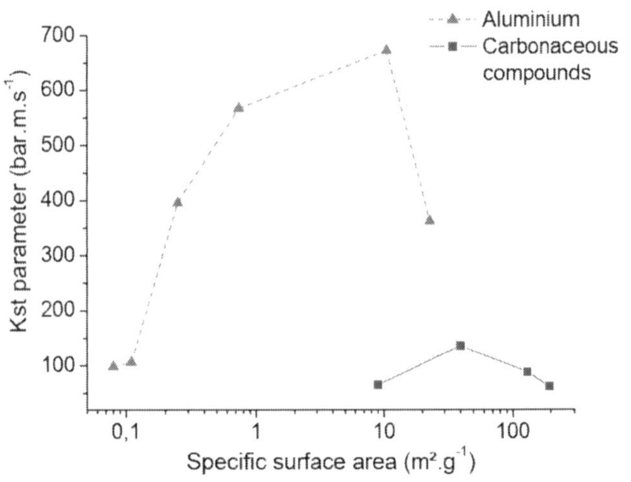

Figure 1. Evolution of the Kst parameter for aluminium and carbonaceous compounds as a function of the specific surface area

Table 3. Explosion severity parameters of metallic powders

Powders	P_{max} (barg)	$(dP/dt)_{max}$ (bar/s)	Kst ($bar.m.s^{-1}$)	Explosivity Class St
Aluminium 100 nm	8.2	1340	362	St3
Aluminium 200 nm	9.5	2420	673	St3
Aluminium 3 µm	9.8	2090	567	St3
Aluminium 7 µm	9.1	1460	396	St3
Aluminium 27 µm	7.5	400	106	St1
Aluminium 42 µm	7.2	360	98	St1
Zinc 120 nm	3.9	223	61	St1
Zn 40 µm	3.7	71	19	St1

The explosion severity tends to increase as the particle size decreases/the specific surface area decreases, before reaching a peak for 1 µm particle size in the case of aluminium powders (approximately 2 m^2/g for P_{max}) [14]. After this peak has been reached, the explosion severity finally tends to decrease as the particle size decreases/the specific surface area increases. If we focused on the aluminium nanopowders, we observed that the 200 nm particles explode more violently than the 100 nm particles. This is probably due to the degree of passivation of the aluminium that could be different for these two kinds of particles, and also due to the degree of agglomeration, which is probably more important for the 100 nm particles. A similar behavior has been observed for carbonaceous compounds: some carbon blacks with a rather high specific surface area exhibit a low explosivity, which is due to the size of their agglomerates (up to 300 µm in some cases) [3]. Once again, as for micro-particle ignition, the agglomeration phenomenon has a major influence on the particle explosivity and the explosion characteristics of nanopowders, especially their maximum rates of pressure rise are sometimes lower than those of powders with diameters lower than about 10 µm (table 3).

For aluminium, the same behavior observed for the maximum explosion pressure is obtained for the maximum rate of pressure rise, which presents a maximum at approximately 4 $m^2.g^{-1}$ of specific surface area [3]. The same evolution is observed for the carbonaceous compounds but with a peak for higher specific surface area (figure 1).

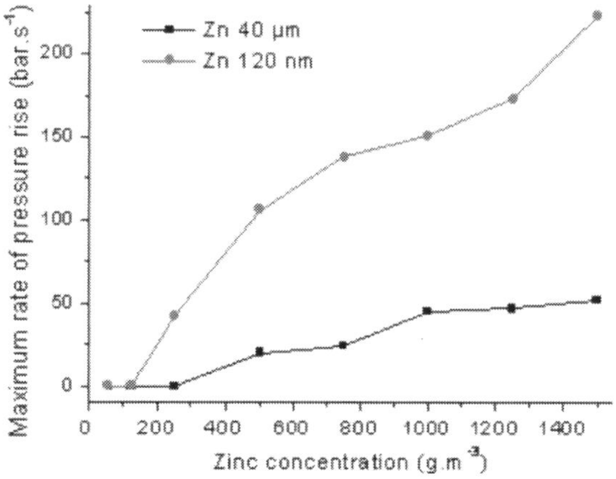

Figure 2. dP/dtm of zinc nano and micro particles versus dust concentration

Only two different particle sizes distributions of zinc have been tested. It could be seen on the figure 2 that the maximum of the dP/dt_m is not reached for concentrations ranging from 0 to 1500 $g.m^{-3}$. Nevertheless, variations of more than 300 % are observed between the maximum rates of pressure rise of zinc micro and nano powders. Tests should be carried out with 1 or 2 µm zinc powders in order to confirm the previous trends.

Flame velocities of 0.15, 0.19 and 0.69 $m.s^{-1}$ have been measured for aluminium dusts of 42, 0.12 and 3 µm diameters respectively; which confirms the previous findings (figure 3). Flame velocities of zinc particles have not been compared because the micro-particles do not ignite for ignition energies lower than 1 J, which is the maximum energy available with this apparatus.

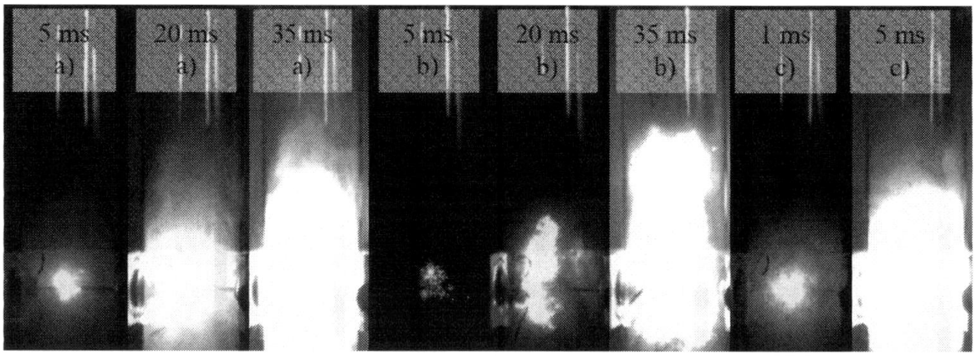

Figure 3. Flame propagation of Al dust explosion in a modified Hartmann tube. Dust concentration: 900 $g.m^{-3}$; ignition energy: 1000 mJ; a) Al 42 µm, b) Al 120 nm, c) Al 3 µm.

It should be stressed that the specific surface area of the particles should not be considered as the only factor influencing the particle explosivity, the chemical composition and the structure of the compound should also be taken into account.

6. Standards and set-up validity

So far, no literature data enables us to validate experimental data for nanopowders, it is necessary to evaluate the validity of the experimental setups and procedures, and especially the validity of the 20 L explosion sphere. Thus, some potential issues influencing the reliability of the experimental results will be developed in the following chapter [14].

6.1. Pre-ignition of the metal nanopowders.

It was observed that aluminium nanopowders were sometimes pre-ignited during the injection process into the 20 L explosion sphere [14]. Pre-ignition phenomenon may lead to an increase of the initial temperature and initial pressure as well as an increase of alumina content. However, no pressure increase was experimentally recorded. The potential heat produced by the pre-ignition phenomenon might be directly transferred through the walls of the dust storage within the vessel. Therefore, it may be considered that pre-ignition phenomenon only leads to an increase of the alumina content within the aluminium nanopowders as well as a decrease in the initial oxygen content. Consequently, it may be assumed that explosion severity may be lowered by pre-ignition.

6.2. Intrinsic oxide content.

Alumina or zinc oxides can decrease absolute explosion severity of these powders. As alumina is produced during the injection process of aluminium nanopowders in the explosion sphere, it is necessary to reduce the impact of an inadequate storage. In this way, the argon sealed bag provided by the industrial partner were opened 1 h before the explosion parameters were determined. Thus it can be considered that the experimental results are just slightly influenced by the initial alumina content.

6.3. Efficiency of the dispersion process during injection.

The degree of dispersion of nanopowders may also decrease the explosion severity. This effect was previously shown by Eckhoff [6]. The authors investigated the influence of the degree of dispersion of maize starch grains on the rate of pressure rise during explosions in the 1.2 L Hartmann bomb and showed a consistent increase of (dP/dt), as the effective particle size decreases, when the agglomerates sizes within the tested materials were lowered. Therefore, measured explosion safety parameters may only be reliable in order to assess explosion risks related to the storage or cleaning operations. The explosion risk related to the industrial production process may be underestimated as there are often well-dispersed nanoparticles within the process. However, it was also shown recently by [15,16], that agglomeration could make ignition of nanoparticles easier than for deagglomerated nanoparticles. Consequently it may be assumed that explosion severity could be increased by considering a simple flame propagation model such as Cassel's one [6].

6.4. Quenching effect of the 20L explosion sphere.

Flame propagation mechanism may also be potentially affected by nanoparticles in the dust cloud. For instance, it was shown experimentally by [17] that an aluminium flame could propagate under a mechanism mainly conducted by radiation rather than by conduction. It can then be assumed that the explosion severity is lowered due to heat losses of the flame to the walls of the explosion sphere. As it currently remains difficult to determine the accurate influence of the agglomeration, pre-ignition, and intrinsic alumina content on the explosion severity, the potential quenching effect of the 20 L sphere was studied through a theoretical investigation.

6.5. Comparison between experimental results and the modeling of the laminar flame velocity.
The explosion pressure P_m and rate of pressure rise $(dP/dt)_m$ were combined through the semi-empirical relation of Silvestrini [18] in order to calculate an experimental laminar flame velocity. It has been compared to a theoretical one, determined by the model of Huang [19], which was initially established to predict laminar flame velocity in an aluminium dust cloud. This model assumes that the propagation of the flame is run mainly by conduction.

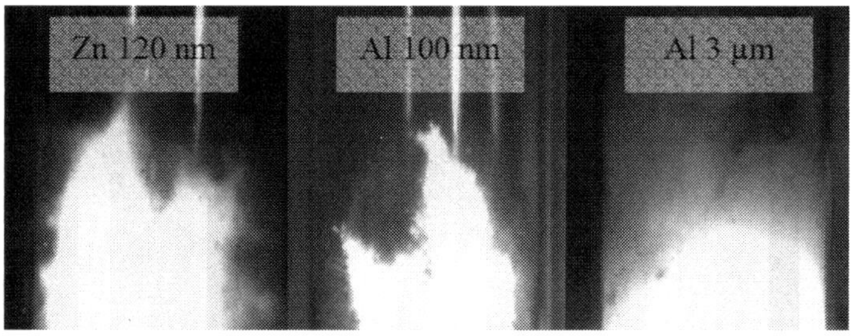

Figure 4. Flame propagation of zinc and aluminium dust explosion in a modified Hartmann tube. 900 g.m^{-3} of powder at 30 mJ for tv of 120 ms. Nano Zn - Nano Al - 15µ Al

It has been demonstrated that a Huang-like model could be used to represent our experimental data, i.e. that the propagation flame is mainly conducted by conduction within the 20L sphere. However, it should be reminded that it is experimentally [17] and theoretically [20,3] expected that the flame propagation within nanodust cloud is probably mainly conducted by radiation.

Two hypotheses can thus be made. One the one hand, it can be reasonably assumed that the 20 L sphere does not enable to get reliable data as it probably disturbs the flame propagation and thermal mechanisms by absorbing radiation (quenching due to wall effect). On the other hand, it has been observed, thanks to the use of a high speed camera and the visualization of the propagating flame edge structure (figure 4) that the preheated zone is smaller for nanopowders (100 nm) than for micro-particles. It could notably be explained by the fact that the flame radiation is absorbed by the cloud of unburnt aluminium nanopowders. A bichromatic pyrometer will also be used in order to check if this modification of the flame behavior in the visible range could be due to a change in the radiation wavelengths (flame temperature modification). Similar tests have been carried out on 200 nm particles and more radiations have been observed than in the 100 nm case.

Figure 5 sums up some key parameters governing the flame propagation phenomenon. The difference of combustion mechanisms between the nano and micron-sized particles can be understood as the result of two different regimes (a diffusion controlled regime for large powders and a kinetically controlled regime for small ones). It is clearly shown on figure 5 that the diffusion controlled regime ($1/d_p$ law) can explain the evolution of flame velocity for large particles. Concerning the heat transfers, it is interesting to consider the Mie parameter $\pi.d_p/\lambda$ (λ being the wavelength) and the balance between light scattering and absorption for nanopowders (change from Mie to Rayleigh theory). In the case of aluminium combustion, the limit is obtained for 100 to 120 nm particles. This could give information on the modification of radiation and light absorption observed between 100 and 200 nm powders. In figure 5, the flame velocities presented are those deduced from Silvestrini's law from the 20 L sphere explosivity data [18]. They differ from those previously quoted and obtained by using the modified Hartmann tube and considering both gas expansion and flame stretching. Thus, the influence of the vessel quenching could not be neglected.

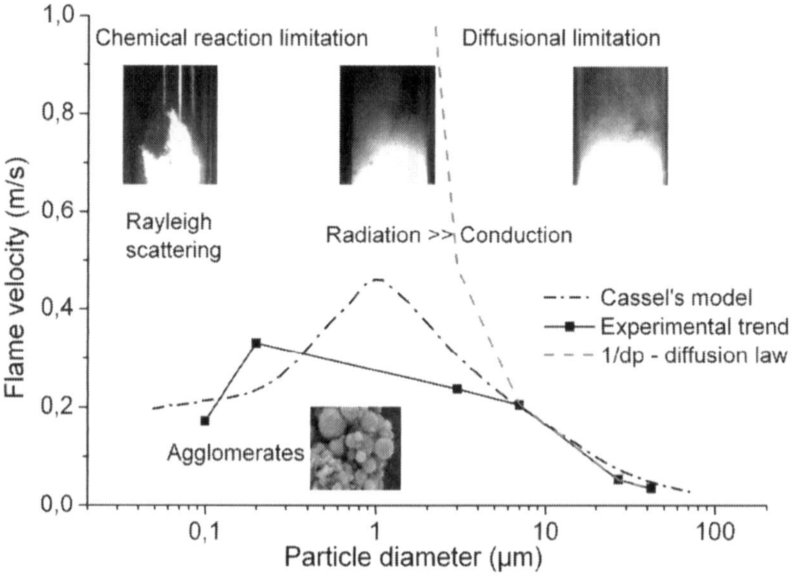

Figure 5. Influence of the Al particle size on the flame velocities of aluminium dust clouds: key parameters

7. Conclusions

During this study, it has been found that the inflammation risk of aluminium powders greatly increases with the use of nanopowders, thus delineating high potentials for ignition risks in facilities manufacturing such powders. Nevertheless, the presence of agglomerates could reduce the reactivity of the powders, and therefore, their explosion severity.

Some experimental issues should be considered in order to characterize the ignition and explosion risks of such powders. The explosion severity of nanopowders may be affected by pre-ignition phenomenon, the intrinsic alumina content of the aluminium, the agglomeration/aggregation level of the nanopowders and the quenching effect of the 20 L explosion sphere. Consequently, some improvements are needed to avoid the premature ignition of ignition sensitive nanopowders and to reproduce the agglomeration level observed in the industrial process. In all cases, it should be reminded that extrapolation of the obtained data and their use for vent sizing purpose, might be impossible by the way of the widely used "cubic law" [6]. In fact, this semi-empirical relationship assumes that the flame thickness is relatively thin; this thickness being increased by the radiation phenomenon. It induces that this law could become null and void for flame propagation mainly governed by radiation. Nevertheless, the importance of the radiation phenomenon and light absorption in the dust cloud should be discussed as a function of the particle size. Finally, the impacts of agglomeration on explosion severity and sensitivity for nanopowders are still not fully understood and should be further explored in the future. For instance, the particle size distribution of the dust dispersed in air will be determined; the results will obviously be different from those obtained by laser light diffraction in a liquid suspension or by SEM or TEM on a dust layer.

Acknowledgements

This study was carried out with the financial support of the European Commission through the Sixth Framework program for Research and Technological Development NMP2-CT-2005-515843 contract "NANOSAFE2".

References

[1] Pritchard, K. (2004). Literature Review-explosion hazards associated with nanopowders. HSL/2004/12.2004, Health and safety laboratory.

[2] Kwok, Q. S. M., R. C. Fouchard, et al. (2002). Characterization of Aluminium Nanopowder Compositions. Propell. Explos. Pyrot. 27: 229-240.

[3] Vignes, A. (2008). Evaluation of nanopowders flammability and explosivity: an essential step towards risk assessment (in French). PhD thesis, INPL - Nancy.

[4] Wu, H.C., Chang, R.C., Hsiao, H.C. (2009). Research of minimum ignition energy for nano titanium powder and nano iron powder. J. Loss. Prevent. Proc. Ind., 22: 21-24.

[5] Wu, H.C., Ou, H.U., Hsiao, H.C. , Shih, T.S. (2010). Explosion characteristics of aluminum nanopowders. Aerosol and Air Quality Research, 10 (1): 38-42.

[6] Eckhoff, R. K. (2003). Dust Explosions in the process industries (3rd edition), Gulf Professional Publishing

[7] Bouillard, J., Vignes, A., Dufaud, O., Perrin, L., Thomas, D. (2010). Ignition and explosion risks of nanopowders. Journal of Hazardous Materials, 181 (1-3), 873-880.

[8] Vignes, A., Munoz, F., Bouillard, J., Dufaud, O., Perrin, L., Thomas, D. (2010). Risk assessment of the ignitability and explosivity of aluminum nanopowders, 13th International Symposium on Loss Prevention, Brugges, 6-9 june.

[9] Beckstead, M.W., Newbold B.R., et al. (2001). A summary of aluminum combustion, in: Proceedings of 50th JANNAF Propulsion Meeting, CPIA Publication, 705 (1): 201–220.

[10] Rai, A., Park, K., Zhou, L., Zachariah, M.R. (2006). Understanding the mechanism of aluminium nanoparticle oxidation, Combustion Theory and Modelling. 10 (2006) 843.

[11] Parr, T., Johnson C. (2003). Evaluation of advanced fuels for underwater propulsion, in: Proceedings of 39th JANNAF Combustion Subcommittee Meeting.

[12] Huang, Y., Risha, G., Yang, V., Yetter, R. (2006). Flame propagation in bimodal nano/micron-sized aluminum particles/air mixtures, in: Collection of Technical Papers - 44th AIAA Aerospace Sciences Meeting, 14015-14025.

[13] Peukert, W. (1996). Trends in solids process engineering (in German). Chemie -Ingenieur-Technik, 66, 1254-1263.

[14] Vignes, A., Traoré, M., Dufaud, O., Perrin, L., Bouillard, J., Thomas, D. (2009). Assessing explosion severity of nanopowders with the 20 L sphere, 8th World Congress of Chemical Engineering, Montréal, 23-27 august.

[15] Trunov, M. A., M. Schoenitz, et al. (2005). Effect of polymorphic phase transformations in Al_2O_3 film on oxidation kinetics of aluminium powders. Combust. Flame 140(4): 310-318.

[16] Trunov, M. A., M. Schoenitz, et al. (2006). Effect of polymorphic phase transformations in Alumina Layer on ignition of Aluminium Particles. Combust. Theor. Model 10(4): 603-624.

[17] Proust, C. (2004). Formation-Inflammation-Combustion des atmosphères explosives (ATEX) et effets associés. Génie des procédés. INPL, Nancy. HDR: 278.

[18] Silvestrini, M et al. (2008). Correlations for flame speed and explosion overpressure of dust clouds inside industrial enclosures. J. Loss. Prevent. Proc. Ind. 21: 374-392.

[19] Huang, Y., Risha, G., et al. (2005). Flame propagation in Bimodal Nano/microsized aluminum particles/air mixtures. AIAA paper AIAA-2005-0739.

[20] Proust, C. (2006). A few fundamental aspects about ignition and flame propagation in dust clouds. J. Loss. Prevent. Proc. 19: 104-120.

Nano-immunosafety: issues in assay validation

Diana Boraschi[1], Gertie J. Oostingh[2], Eudald Casals[3], Paola Italiani[1], Inge Nelissen[4], Victor F. Puntes[3], Albert Duschl[2]

[1]Institute of Biomedical Technologies, National Research Council, Via G. Moruzzi 1, 56124 Pisa, Italy

[2]Department of Molecular Biology, University of Salzburg, Hellbrunnerstrasse 34, 5020 Salzburg, Austria

[3]Institut Català de Nanotecnologia, Campus de la UAB - Facultat de Ciències, Edifici CM7, 08193 Bellaterra, Spain

[4]VITO NV, Boeretang 200, BE-2400 Mol, Belgium

E-mail: diana.boraschi@itb.cnr.it

Abstract. Assessing the safety of engineered nanomaterials for human health must include a thorough evaluation of their effects on the immune system, which is responsible for defending the integrity of our body from damage and disease. An array of robust and representative assays should be set up and validated, which could be predictive of the effects of nanomaterials on immune responses. In a trans-European collaborative work, in vitro assays have been developed to this end. In vitro tests have been preferred for their suitability to standardisation and easier applicability. Adapting classical assays to testing the immunotoxicological effects of nanoparticulate materials has raised a series of issues that needed to be appropriately addressed in order to ensure reliability of results. Besides the exquisitely immunological problem of selecting representative endpoints predictive of the risk of developing disease, assay results turned out to be significantly biased by artefactual interference of the nanomaterials or contaminating agents with the assay protocol. Having addressed such problems, a series of robust and representative assays have been developed that describe the effects of engineered nanoparticles on professional and non-professional human defence cells. Two of such assays are described here, one based on primary human monocytes and the other employing human lung epithelial cells transfected with a reporter gene.

1. The importance of assessing immunosafety of engineered nanomaterials

Assessment of nanosafety, with particular regard to nanomedicines and health-related nanomaterials, should consider the possible effects on the immune system with great care. The immune system is responsible for defending the integrity of our body and maintaining its health. Possible alterations of the normal functions of the immune system, as a consequence of interaction with nanoparticles, must be avoided as they can cause severe pathological derangements. It is expected that nanotechnologies applied in medicine (nanodrugs, nano-imaging tools, nano-delivery systems, nanodevices) will greatly improve the quality of the diagnostic and therapeutic intervention in the near future, *e.g.*, allowing us to target with great precision secluded organs or tissues (brain, metastatic tumours, etc.). In order to speed up the development of such nanotools, the safety of the nanomaterials should be evaluated at every phase of development (from synthesis of the raw materials to the final formulation). To this end, it will be necessary to design and implement robust and representative methods able to predict the immune-related risk of developing diseases. Available standardised immunological assays are not suited in most cases for detecting nanoparticle effects, and should therefore be custom-adapted or re-designed to this end.

A major issue in immunosafety assay design is the need for standardisation and ease of applicability of the validated assays. In this context, *in vitro* assays are greatly preferred because of their high reproducibility, in addition to avoiding the use of large numbers of experimental animals. To increase relevance, the possibility of employing *in vitro* models of human primary cells, rather than animal tissues or transformed/tumour cell lines, is of great importance.

2. Set up and validation of nano-immunosafety in vitro assays

A series of issues will have to be considered in order to set up appropriate immunosafety assays for testing nanoparticle effects.

The first is the biological target. What we really need to know is not only if nanoparticles are toxic for immune cells or if they can activate inflammation or an allergic reaction. This is indeed important, but it is an acute effect that can be easily identified with standard assays [1]. What is much more difficult to identify is the capacity of nanoparticles to interfere with immune responses, *i.e.,* to affect the capacity of our immune system to react appropriately to dangers and protect our body's integrity. For example, will our immune cells be able to fight and destroy meningococci nesting in our throat, or will they be too weak, and thus allow the bugs to become virulent, invade our body and cause meningitis? If our immune cells are loaded with nanoparticles, even if perfectly alive, will they be able to efficiently fight the danger, *i.e.*, doing the right thing at the right moment in the right place?

Another major issue is the heterogeneity of the immunological status of the human population. It is well known that babies and elderly people have major alterations of the immune response (immature *vs.* senescent). For instance, the "normal" immune response in an 80-year old person is very different from that of a healthy adult, as the senescence of the entire body has driven a different "basal" reactivity in the immune system, to cope with the increased need of scavenging dying/anomalous cells without causing unwanted inflammation [2]. On the other hand, babies have an immature immune system, which make them almost completely unresponsive to certain types of stimuli (*e.g.*, polysaccharides). Pregnant women are a third group, besides babies and aged people, with a differently regulated "normal" immune responsiveness. Pregnant women have enhanced resistance to infections (to protect the development of the fetus) and at the same time experience a sort of immunosuppression to allow persistence and growth of a foreign tissue without rejecting it. It is evident that the impact of nanoparticles on such immune systems, which have very different regulation checkpoints, cannot be the same as in normal adults.

A final source of variability in the human immune response is disease. What is safe for healthy people may not be so for people with a disease. For instance, nanoparticles may not have effects on the immune response of a healthy adult, but they could exacerbate the hyperreactivity reaction to allergens in an asthmatic person. Conditions of frailty (*e.g.*, chronic infections, chronic diseases, malnutrition, environmental conditions), which deeply affect the immune reactivity, can also influence the possible impact of nanoparticles on the immune response.

The immunosafety assays that we aim at designing for testing the impact of nanoparticles should answer to such questions, as well as to the questions of chronic and/or cumulative exposure. These issues are summarised in the Table 1.

Table 1. Issues in setting up nano-immunosafety assays.

Issue	Description
Biological target	Direct toxicity and direct immunological effects (induction of inflammation, allergy, etc.). Interference with the normal development of the defensive immune responses (increased susceptibility to infections, tumours, autoimmune reactions, etc.)
Heterogeneity of the human population	Effects on immunologically different populations (children, elderly, pregnant women, genetically diverse populations, etc.)
Health *vs.* disease	Effects on people undergoing infections, chronic diseases (cardiovascular, pulmonary, autoimmune, chronic inflammatory, tumours, etc.), malnutrition, and other conditions of frailty, as compared to normal healthy adults
Acute *vs.* chronic or repeated exposure	Effect of continuous or cumulative exposure to low doses; effect of co-exposure to nanoparticles and other agents, both biological (bacteria, viruses, etc.) and non-biological (dusts, chemicals, etc.)

2.1. Selection of relevant endpoints/biomarkers

How to select phenotypic, molecular or functional endpoints predicting risk of developing disease is a major issue in immunosafety at large, not only for nano-immunosafety. Correlates of disease/risk must take into account the above considerations on the heterogeneity of the immune responses in different population groups, which would set different thresholds for discriminating between normal and (pre)-pathological levels of certain markers. An example is for instance the circulating levels of the inflammatory cytokine IL-18, one of the accepted early markers of renal failure [3]. IL-18 is present at measurable levels in the serum of normal individuals, and is significantly increased in renal disease, but also in several other inflammatory and autoimmune diseases, often preceding disease development [4]. Indeed, increased IL-18 may be considered as a general prognostic marker of inflammation and inflammatory disease. However, it should be noted that the circulating levels of IL-18 are different in different age groups, increasing significantly at older ages even in healthy individuals [5]. Thus, the threshold value discriminating between normality and risk of developing an inflammatory disease is not a fixed value, but depends on the age of the subjects, as well as on their general health status.

2.2. Selection of relevant cellular systems

The immune system encompasses several cell types and a multifaceted array of reactions, from early innate/inflammatory defensive reactions (*e.g.*, production of reactive oxygen species, release of inflammatory cytokines) to late, more sophisticated and specific adaptive responses (*e.g.*, antibody production, generation of memory). *In vitro* models, as we aim at developing, should be as simple as possible and able to provide surrogate markers of risk. We propose here that early defence responses, which are very reproducibly triggered by danger signals even in human subjects (not only in genetically homogeneous inbred mice), be taken as relevant endpoints for nano-immunosafety evaluation. Consequent of this, cells that are of central importance in for detecting innate/inflammatory activation can be identified in two major groups: professional defence cells such as leukocytes (in particular circulating monocytes and tissue macrophages) that first come in contact with exogenously administered nanoparticles (such as nanodrugs), and mucosal epithelial cells (such as pulmonary and gut mucosal cells) that are the mechanical and biologically active barrier lining of our inner body. It is obvious that, depending on the selected endpoint and biomarker (e.g. adaptive immunity markers), other cell types will be more relevant and should be analysed (*e.g.*, Th, Treg, Tc, B cells, etc.) [6].

2.3. Challenges in assay development

Development of reliable and representative *in vitro* assays for measuring nano-immunosafety faces a series of challenges that must be solved. In fact, they must be proven representative of the real-life situation, and devoid of artefacts that can mislead the operator and lead to false positive or false negative results.

2.3.1. *Nanoparticle characterisation.* Among issues that must be addressed for the correct interpretation of assay results, it is necessary to consider particle aggregation/agglomeration in the presence of culture medium and biological fluids (*e.g.*, serum), ageing of particles upon storage or in physiological media (particle dissolution and release of reactive ions), interaction of particles with biological molecules present in physiological media (formation of a protein or lipid corona), compatibility of particle solvents with the biological systems (chemical immunotoxicity), contamination of particles with biologically active compounds (such as bacterial endotoxin), and optical interference of particles with the assay readouts. Only particle batches devoid of secondary effects (freshly synthesized, endotoxin-free, monodispersed, suspended in biocompatible solvents at optically non-interfering concentrations) were used in the in vitro immunotoxicity assays hereafter described.

2.3.2. *Solvent contamination effects.* A significant misleading effect that leads to false interpretation of assay results is the direct effect of chemical solvents in which nanoparticles are dispersed for maintaining them in colloidal suspension. Even with solvents that turn out to be completely non-toxic for immune cells, an effect in modulating the activities of immune cells can be present. For example (Table 2), when CaCo-2 gut epithelial cells were chronically exposed to Au nanoparticles (diameter 4 nm) and CoO nanoparticles (diameter 7 nm) there is a significant release of sICAM-1, at levels comparable to those induced by the inflammatory cytokine IL-1β (positive control). However, when the effect of the solvents in which the nanoparticles were suspended was examined, it turned out that the solvent in which Au NP were suspended (sodium citrate) was as active in inducing sICAM-1 as the Au NP themselves. On the other hand, the solvent in which CoO NP were suspended (TMAOH) was completely inactive in sICAM-1 induction. Thus, the correct interpretation of these results is that only CoO NP were able to induce sICAM-1 release, while the effect observed with Au NP was a false effect caused by the solvent. It should be noted that the modulatory effect of solvents can vary depending on the biological endpoint analysed. Indeed, taking again the example of the gut mucosal epithelial cells, while the sodium citrate solvent can induce sICAM-1 release, it was unable to inducing IL-1β or CXCL4, while TMAOH did not have effect on sICAM-1 nor IL-1β induction, but it was very active in inducing CXCL4 (Table 2). The bottom line is that solvents can be active as immunomodulators also in the absence of direct toxicity, and that such activity can be observed for certain biological endpoints, but not others. Thus, in order to evaluate the true NP-induced effects, a thorough analysis of the effects of contaminants and chemicals present in the working batches of NP must be performed on every single biological endpoint under study. In the absence of such controls, results can be misinterpreted and significant effects can be wrongly attributed to inactive nanoparticles.

Table 2. Solvent contamination can be responsible for immunomodulation.

Treatment[a]	Induction (AU) of [b]		
	sICAM-1	**IL-1β**	**CXCL4**
medium	0.00	0.10	10.39
IL-1β	54.74*	20.28*	33.63*
solvent Au NP [c]	46.47*	0.86 [n.s.]	8.51 [n.s.]
Au NP (5.2 μg/ml)	58.58*	0.23 [n.s.]	25.36*
solvent CoO NP [d]	0.00 [n.s.]	0.00 [n.s.]	78.27*
CoO NP (3.5 μg /ml)	43.58*	0.00 [n.s.]	9.70 [n.s.]

[a] Differentiated human CaCo-2 gut mucosal epithelial cells were exposed for 15 days to culture medium (RPMI-1640 with 5% human AB serum) containing 9.1% of colloidal NP preparations, 9.1% of solvents without NP, or 10 ng/ml IL-1β as positive control.

[b] Soluble factors in the supernatants of 15-day cultures were assessed by Proteome Profiler Antibody Array (R&D Systems), measured densitometrically, and expressed in Arbitrary Units.

[c] Au NP were suspended in Na citrate 0.25 mM, oxidised $NaBH_4$ 0.3 mM, pH 8. Both NP and solvent were non-toxic.

[d] CoO NP were suspended in TMAOH 1 mM. Both NP and solvent were non-toxic.

* $p<0.05$ vs. medium control; n.s. not significant *vs.* medium control

2.3.3. *Endotoxin contamination.* The issue of endotoxin contamination of nanoparticles is a major source of artefacts and misinterpretation of assay results. Bacterial endotoxin is an ubiquitous contaminant of chemicals and glassware which cannot be eliminated upon sterilisation. Elimination of endotoxin can be obtained by incineration (depyrogenation), a procedure that can be applied only to few materials without risk of changing their physico-chemical characteristics. Endotoxin-free synthesis, i.e. by carrying out the full procedure with endotoxin-free glassware, materials and reagents, is the only reliable way of obtaining endotoxin-free nanoparticles [7]. Assessing the presence of trace amounts of endotoxin in every batch of engineered nanoparticles is an absolute requirement before testing them for immunotoxicity and immunomodulatory effects. Indeed, several immune cells are particularly sensitive to endotoxin, and either die in its presence or become strongly activated. This is for instance the case for human monocytes, which can react to minute amounts of bacterial endotoxin by up-regulating the expression of inflammatory genes (such as that of IL-1β). Thus, full monocyte activation may be achieved with nanoparticles containing traces of contaminating endotoxin, masking any possible effect/lack of effect due to the nanoparticles themselves [8]. The validated endotoxin detection assays used for the microbiological quality control of pharmaceuticals and medical devices are based on the Limulus Amoebocyte Lysate (LAL) and have replaced the previous in vivo test based on fever induction in the rabbit. The original LAL assay was based on endotoxin-induced gelification of the lysate; while the modern LAL assays measure the release of a dye from a substrate upon endotoxin-induced cleavage. Gelification and dye release are quantitatively correlated to the amount of endotoxin present. Assessing the presence of endotoxin in nanoparticle preparations must take into careful account the interference that optically active particles may have with the densitometric or colorimetric assay readout, as well as the possibility that nanoparticles may either directly induce or inhibit the enzymatic activity, thus mimicking or masking the presence of endotoxin [9].

3. Assay of innate/inflammatory reactivity of human primary monocytes

The innate immune system is an ancient mechanism of defence, which is quite reproducible and stable in its reactivity. In the attempt of setting up a fully *in vitro* assay for testing nano-immunosafety, we have chosen human primary blood monocytes, freshly isolated from the blood of adult healthy donors, and challenged them in culture with prototypical inflammatory stimuli (*e.g.*, bacterial endotoxin) for different times, to represent early *vs.* late inflammation. The assay was used to evaluate the ability of nanoparticles to interfere with the normal development of the innate/inflammatory reaction, by measuring inflammatory gene expression and cytokine production. It should be noted that, by intentionally stimulating monocytes with endotoxin, the possible contamination of nanoparticles with endotoxin would become biologically irrelevant. The experiment reported in Figure 1 was carried out with strictly endotoxin-free AgO nanoparticles, but in parallel trials the presence of a moderate contamination did not affect the results.

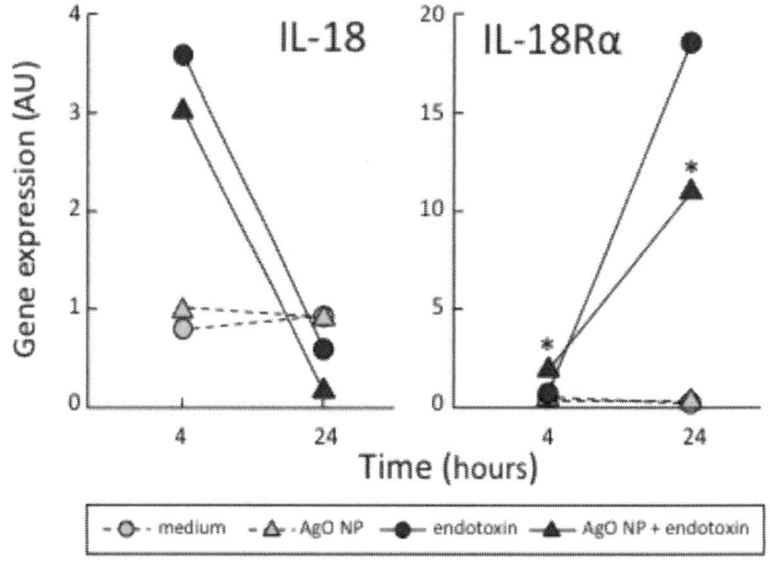

Figure 1. Differential modulation of gene expression in monocytes by nanoparticles. Human monocytes were exposed for 4 and 24 h to culture medium alone or containing 50 EU/ml endotoxin. At the beginning of the incubation, cells were exposed to endotoxin-free AgO NP (diameter 10-60 nm; 4.9 μg/ml). Controls were cells exposed to culture medium alone or to solvent (no detectable difference). Gene expression was assessed by real-time PCR, using β-actin as housekeeping gene. * $p<0.05$ vs. corresponding treatment in the absence of NP.

While nanoparticles were unable to elicit any significant reaction in quiescent monocytes (measured as expression of two inflammation-related genes, those for the cytokine IL-18 and for its receptor IL-18Rα), in endotoxin-activated cells the presence of nanoparticles could selectively alter the course of the response (no effect on endotoxin-induced IL-18, but a significant effect on receptor expression, which is upregulated at 4 h and downregulated at 24 h as compared to endotoxin alone). A kinetical *in vitro* model of monocyte activation that reproduces all phases of the inflammatory response is currently being set up and evaluated by genome-wide transcriptomic analysis. The model proves to be very reproducible from donor to donor, and describes very precisely the evolution of the defensive response, thus it can be suitable for identifying and molecularly describing the possible interfering effects caused by nanoparticles.

4. Assay of innate/inflammatory reaction of human lung epithelial cells

Epithelial cells are considered as very important non-professional defence cells, not only as barrier lining and protecting our inner body, but also as active defence effectors of innate immunity/inflammation. As a consequence, anomalous activation of epithelial cells may cause pathological derangements. For example, the interaction between bronchial epithelial cells and dendritic cells is strongly involved in the development of allergic asthma [10]. Given that bronchial epithelial cells are potentially exposed to inhaled nanomaterials, it is highly relevant, in the context of nano-immunosafety testing, including this cell type in the development of *in vitro* safety assays. The use of primary human bronchial epithelial cells is hindered by the difficulty of obtaining and isolating these cells. Therefore, the human lung epithelial cell line, A549, was used instead, after validation of the selected endpoints by comparison with primary cells. A series of reporter cell lines, based on the A549 cell line, proved useful in detecting the effects of nanoparticles in a rapid and sensitive fashion. These reporter cells react by producing a light signal (following luciferase production) upon stimuli that activate the promoters of the inflammatory cytokines TNF-α, IL-6 or IL-8, or are able to activate the inflammatory gene activator NF-κB. The effects of nanoparticles are different, depending on whether cells are quiescent or activated by inflammatory stimuli as it occurs during a defence response [11]. These reporter cell lines represent a reliable and standardised first-line screening method to determine the potential immunomodulatory capacity of novel nanomaterials. When effects are observed, further experiments have to be performed to determine the exact pathways and the effects on primary bronchial epithelial cells.

5. Conclusions and discussion

Immunosafety of engineered nanoparticles, with special reference to nanomedicines, is a relevant health issue that needs to be addressed by the Health Authorities through *ad hoc* guidelines and with appropriate investigation tools [12].

Given the high plasticity of the immune response, which can recognize minimal changes in foreign structures/molecules, the immunosafety of nanomedicines should be tested at each of the steps of pharmaceutical development, including the final formulation, since the safety of the single components of the formulation is no guarantee for the safety of their mixture.

The key issue in implementing nano-immunosafety, and in ensuring the efficient development of health-related nanotechnological products, is the availability of robust and reliable immunosafety assays. The commonly used immunotoxicity assays are ill-suited to testing nanomaterials, since the chemical-physical characteristics of the nanoparticles require in most cases to re-design the assay. Indeed, the particulate nature of the nanomaterial can cause a series of misleading artefacts resulting in false negative or false positive results. Thus, for each selected endpoint a robust assay should be selected/designed, with a readout that is not influenced by the particles. In any case, the concomitant use of two assays with different readouts for the same endpoint can usually solve the problem.

The design of *in vitro* immunosafety assays for nanoparticles requires a series of considerations. It is evident that *in vitro* assays may not fully replace the *in vivo* tests in animals. However, also by considering the global efforts in replacing animal testing with *in vitro* assays, the advantages of *in vitro* assays should be seriously considered. These include the clear focus (*e.g.*, inflammation), which should not be considered as a limitation but as the possibility of a deeper evaluation of the effect and of its mechanism of action. Then, *in vitro* assays can be developed that are fully representative of the actual human exposure, by using normal human cells in culture rather than animals or animal cells, and continuous transformed cell lines. Eventually, *in vitro* assays can be standardised and simplified, and become available for screening in a much simpler and more rapid way than *in vivo* tests will ever allow.

The use of primary cells, as described here for the innate reactivity of human blood monocytes, may suffer of the problem of donor-to-donor variability, which is one of the reasons why transformed cell lines are often preferred, together with their better accessibility and availability. It is however important that primary human cells are used beforehand, to establish relevant endpoints and markers. In the case of human monocytes, the donor-to-donor variability is minimal, and accessibility relatively easy, which allows us to detect their reactivity to challenges in an exact fashion even when using cells from different donors. However, in many cases the use of human primary cells is unfeasible or time-consuming and with very low or variable yield (*e.g.*, brain microglia, tissue macrophages, epithelial cells). In such cases, but also with easier and abundant primary cells and blood leukocytes, the use of transformed cells may be of great help for designing simple, reproducible and standardised assays, once the endpoints selected have been validated as representative of the reactivity of normal primary cells.

The use of reporter cell lines, as those described here, is a very important possibility of developing rapid, easy to use assays suitable for large screenings. The innate/inflammatory reactivity of human non-professional defence cells (alveolar epithelial cells) is assessed by using a transformed cell line (the lung carcinoma line A549), which undergoes spontaneous differentiation in culture and reacts to stimuli in a fashion overlapping with that of primary lung epithelial cells, when considering the production of a series of inflammatory factors. The cell line is transformed with a luminescent reporter gene under the control of the promoter of one of these factors, to yield a reporter cell line representing the inflammatory reactivity of the normal alveolar epithelium.

The development of robust *in vitro* assays, custom-designed for assessing nanoparticle effects, will greatly help in the sustainable knowledge-based development of nanotechnologies applied to medicine. The same assays that are being developed now for assessing immunosafety could also provide valuable information on the features of the interaction between nanoparticles and human immune cells, helping us to design a future generation of nano-immunomodulatory drugs. In this view, it is of special interest the possibility of using nanoparticles for the direct targeting of immune cells for therapeutic/preventive strategies (delivery of drugs, delivery of vaccine antigens and adjuvants). Modifying the immune response in localised settings, a very difficult task nowadays, could also be addressed with the use of nanoparticles in medical fields such as oncology, autoimmunity, chronic inflammatory diseases, and neurodegeneration.

Acknowledgements
DB was partially supported by the Italian Ministry of University and Research (MIUR) within the PRIN 2007 project prot. 2007Y84HTJ on systems biology of human macropahges.

References
[1] Dobrovolskaia MA, Germolec DR and Weaver JL 2009 *Nat. Nanotechnol.* **4** 411
[2] Boraschi D, Del Giudice G, Dutel C, Ivanoff B, Rappuoli R and Grubeck-Loebenstein B 2010 *Vaccine* **21** 3627
[3] Lock EA 2010 *Toxicol. Sci.* **116** 1
[4] Boraschi D and Dinarello CA 2006 *Eur. Cytokine Netw.* **17** 224
[5] Gangemi S, Basile G, Merendino RA, Minciullo PL, Novick D, Rubinstein M, Dinarello CA, Lo Balbo C, Franceschi C, Basili S, D'Urbano E, Daví G, Nicita-Mauro V and Romano M 2003 *Exp. Gerontol.* **38** 669
[6] Dobrovolskaia MA and McNeil SE 2008 *Nat. Nanotechnol.* **2** 469
[7] Vallhov H, Qin J, Johansson SM, Ahlborg N, Muhammed MA, Scheynius A and Gabrielsson S 2006 *Nano Lett.* **6** 1682
[8] Oostingh GJ, Casals E, Italiani P, Colognato R, Stritzinger R, Ponti J, Pfaller T, Kohl Y, Ooms D, Favilli F, Leppens H, Lucchesi D, Rossi F, Nelissen I, Thielecke H, Puntes VF, Duschl A and Boraschi D 2010 *submitted*
[9] Dobrovolskaia MA, Neun BW, Clogston JD, Ding H, Ljubimova J and McNeil SE 2010 *Nanomedicine* **5** 555
[10] Hammad H and Lambrecht BN 2008 *Nat. Rev. Immunol.* **8** 193
[11] Pfaller T, Puntes V, Casals E, Duschl A and Oostingh GJ 2009 *Nanotoxicology* **3** 46
Scientific Committee on Emerging and Newly Identified Health Risks (SCENIHR) Opinion: http://ec.europa.eu/health/ph_risk/committees/04_scenihr/scenihr_opinions_en.htm#nano

Approaches for establishing human health no effect levels for engineered nanomaterials[1]

K Aschberger and F M Christensen

Nanobiosciences Unit, Institute for Health and Consumer Protection,
European Commission-JRC, 21020 Ispra, Italy

E-mail: Karin.aschberger@ec.europa.eu

Abstract. Current Technical Guidance Documents for preparing risk assessments, like the guidance for the implementation of REACH, have limited focus on chemical substances in the particulate form and generally do not focus on substances in the nanoform.
Within the ENRHES project a comprehensive and critical scientific review of publicly available health and safety information on four types of nanoparticles was performed. Based on the identified exposure and hazard data, basic human health risk assessment appraisals were carried out for fullerenes, carbon nanotubes, nano-silver and nano-titanium dioxide. These risk assessment appraisals followed the structure of a regulatory risk assessment and if possible and relevant, it was attempted to derive indicative human no-effect levels from key studies by applying assessment factors as suggested in the technical guidance document for REACH. These assessment factors address differences and uncertainty related to exposure features between test animals and humans (time, respiratory volume), other interspecies and intraspecies differences and factors for extrapolation to chronic duration. If required, the severity of effects and the quality of the database can be addressed by additional factors.
Recently other procedures for deriving human no-effect levels have been published and these are compared to the ENRHES basic risk assessment appraisals. The main differences were observed in relation to evaluating the differences in animal and human exposure situations and interspecies differences, and in applying assessment factors for intraspecies differences for local effects. The applicability of the REACH guidance for nanomaterials is currently being investigated for possible adaptations, considering the specific behaviour and mode of action of nanomaterials.

1. Introduction

In the framework of the European Regulation REACH, registrants (importers and manufacturers) have to demonstrate that risks are controlled for manufacturing and use of their substances. This involves, where possible, derivation of DNELs (derived no-effect levels) or DMELs (derived minimal effect levels for non-threshold effects) to be compared to workplace exposures and/or exposures of the general public (consumer exposure and exposure via the environment). Risks are considered properly controlled when exposures are lower than the DNEL.

[1] The opinions expressed in this publication are those of the authors and not necessarily those of the European Commission.

The provisions of REACH apply to nanomaterials [1] and therefore the guidance for preparing a REACH chemical safety assessment [2] in principle also covers engineered nanomaterials, however it currently includes very little reference to substances in particulate form and generally do not focus on substances in the nanoform.

Within the ENRHES project (Engineered Nanoparticles: Review of Health and Environmental Safety; http://nmi.jrc.ec.europa.eu/project/ENRHES.htm) and follow up activities, basic risk assessments for four types of nanomaterials were performed. For these basic risk assessment appraisals, indicative human no-effect levels were determined based on the REACH guidance (chapter R.8) [2] for certain types of carbon nanotubes, fullerenes, nano-silver and nano-titanium dioxide (for details see Aschberger *et al.* [8] Christensen *et al.* [9]). These papers identified several challenges with deriving no-effect levels based on current methodology. Other and sometimes more sophisticated approaches for risk assessment of nanomaterials have lately appeared in the public domain. In the current manuscript, different proposed procedures for deriving human no effect levels are compared and discussed.

2. Establishing human no-effect levels.

Human exposure limit values like the DNEL or occupational exposure limits (OEL) are generally based on dose descriptors (No (lowest) observed adverse effect level/concentration N(L)OAEL/N(L)OAEC or benchmark dose) for critical effects from animal toxicity studies which then have to be modified to human exposure situations by employing assessment factors to account for differences (extrapolation) and uncertainty.

When establishing human no-effect levels, the following criteria have to be taken into account:

- Exposure route
- Key study (relevance, reliability, nanomaterial characterization)
- Toxicokinetics
- Nature and severity of effect:
    - threshold or non-threshold mechanism,
    - local – systemic effect
- Dose descriptor
- Modification to the starting point
- Assessment factors

Inhalation is expected to be the most relevant exposure route for nanoparticle exposure, especially at the workplace due to the possibilities of particles becoming airborne. So far, internationally harmonised toxicity testing procedures for nanomaterials are not available. The preferred study type for a risk assessment is an inhalation study for a sub-chronic or chronic duration. However it is technically very difficult to produce and maintain an appropriate and representative atmosphere of nanoscale particles and these studies are therefore complex and expensive and not very often performed. Further, for inhalation studies, test substances may be technically manipulated to increase the dustiness and this may not represent the real human exposure situation. Therefore hazards identified in rodents inhalation toxicity studies need to be carefully interpreted when used for establishing human no effect levels.

3. Examples for deriving human no-effect levels, following the REACH guidance

In the following we present an approach for deriving human no-effect levels for inhalation by applying the default assessment factors for local effects as described in the REACH guidance [2]. Default assessment factors should be used in the absence of substance specific or analogous data allowing the establishment of appropriate values. The presented evaluation was made based on the available information in the open literature and includes many uncertainties, e.g. related to nanomaterial characterisation. Therefore it was necessary to rely on default assessment factors and the results *should not* be used for any regulatory purposes or decision making. Consequently, to de-couple from regulatory purposes, we preferred to use the term INEL (Indicative No-effect Level) instead of DNEL. This methodology is compared to other approaches reported in the literature and the results are discussed.

In this manuscript we mainly focus on chronic occupational inhalation exposure to carbon nanotubes (CNTs) and nano-TiO_2, where most information for attempting derivation of no-effect-levels is available. The selected key studies are rat inhalation studies of high reliability and sufficient nanomaterials characterisation, where the animals were exposed for 6 hours per day and 5 days per week. For multiple walled carbon nanotubes (MWCNT) we selected 2 recent 13-week inhalation studies (OECD 413) in compliance with good laboratory practice with Nanocyl 7000 (5-15 nm x 100-1000 nm, 90% purity) [3] and Baytubes® (10 nm x 200-~1000 nm; 99.1% purity) [4]. The latter included a post-exposure period for up to 6 months. The LOAEC for Nanocyl was reported as 0.1 mg/m^3, while the same concentration was reported as a NOAEC for Baytubes ®. For nano-TiO_2 (21 nm particles) we selected a 13-week inhalation study (including assessment of pulmonary responses up to 52 weeks post-exposure) [5] with a NOAEC of 0.5 mg/m^3.

The observed effects in these toxicity studies like local inflammation and granulomas in the lung suggested a threshold mechanism and it seemed therefore relevant to derive human Indicative No-Effect Levels. However there is uncertainty with regard to possible genotoxic and/or carcinogenic effects of these nanomaterials and therefore in principle a non-threshold effect cannot be generally excluded (for further details please refer to Johnston *et al.* [6,7], Aschberger *et al.* [8] and Christensen *et al.* [9]). Under the assumption of a threshold mechanism of toxicity, modifications to the starting point and assessment factors were applied to derive INELs for the selected substances.

3.1. Modifications to the starting point

The exposure duration of the animal in the experiment (usually 6h) is corrected for the assumed daily worker exposure (usually 8h). Workers at light activity have a slightly higher breathing rate and respiratory volume (10 m^3) as compared to rest (6.7 m^3). The modifications for these conditions (6/8h*6.7/10 m^3) correspond to a factor of 2 for workers.

3.2. Assessment factors

Assessment factors are numerical values used to address differences between experimental data and the human situations taking into account the uncertainties in the extrapolation procedure and the available data set.

3.2.1. *Interspecies differences.* The interspecies assessment factors, to extrapolate effects in one species (e.g. rats) to other species (e.g. humans), address possible differences in the mechanisms influencing dose and toxicity of chemicals. It might be that these factors need to be adapted for nanosized materials considering the nanoparticle specific properties, as the behavior in the respiratory tract (deposition, persistence) and the possible different nature of toxic effects. There are different approaches for interspecies factors and some approaches split the interspecies factor into toxicokinetics and toxicodynamics, whereas in the REACH guidance they are split into allometric scaling (depending on the tested animal species) and remaining interspecies factors (ECHA 2008).

In the selected studies the observed effects were local and did not depend on metabolic rate or systemic absorption and therefore allometric scaling is not applicable. Thus, only the factor for remaining interspecies differences (default 2.5) was applied and in the case of nano-TiO_2 it was suggested to be reduced to 1.5, as it has been shown that rats were more sensitive than mice and hamsters [5].

For decisions on the extent of interspecies differences it is important to understand the nature of the observed effect. In the case of CNTs, it was postulated that the critical mechanism triggering pulmonary inflammation is dependent on the volumetric overload of alveolar macrophages [4,10]. "Lung overload", is a phenomenon more extensively seen in rats due to different deposition and clearance capacities. Symptoms caused by lung overload are not expected to be manifested to the same extend in larger mammals such as dogs, monkeys and humans [11] and therefore the rat should be considered to be more susceptible than humans for these kind of effects [12].

3.2.2. *Intraspecies differences.* For intraspecies variation default assessment factors of 5 and 10 are recommended in the REACH guidance to account for interindividual differences in workers and the general population, respectively. A factor of > 1 for workers is considered justified as some workers may have a predisposition to react more sensitive to particle exposure, e.g. asthma. However some other approaches [10,13,14] (see below) consider a factor of 1 sufficient.

3.2.3. *Exposure duration.* The assessment factors for duration can take into account the extrapolation of the duration from sub-acute (default 6) or sub-chronic (default 2) to chronic exposure, unless it is shown, that exposure duration is not relevant for the manifestation or severity of effects. In the basic risk assessment apraisal we applied the default factor of 2 for sub-chronic to chronic extrapolation. It is not known how the observed effects would develop under prolonged exposure and therefore no statement on the appropriateness of the default factor can be made. The effects could become more severe, but there could also be recovery. The extrapolation factor for the duration could change both ways if more information on the progression of the effects became available and this would thus have an impact on the results and conclusion of the risk assessment. In the case of MWCNT Baytubes® [4], where the sub-chronic study was extended by a post-exposure period of 6 months with no progression of the effects it might be justified to reduce this assessment factor.

3.2.4. *Dose-response.* If in the experiment no concentration without an effect has been determined, an assessment factor (default factor of 3) has to take account for the extrapolation from the Lowest Observed Adverse Effect Concentration (LOAEC) to a No Adverse Effect Concentration (NAEC). This factor can be changed, depending of the severity of the effects. In the inhalation study with MWCNT Nanocyl [3] the lowest tested dose of 0.1 mg/m^3 showed subclinical findings (minimal granulomatous-type inflammation in the lung and lung-associated lymph nodes) and therefore we suggested a reduced assessment factor of 2 for extrapolation from LOAEC to NAEC.

The severity of the effect can be taken into account by assessment factors, however in none of the exercises it was considered appropriate. There has also been a debate concerning the appropriateness of using nano-specific uncertainty factors to account for unknowns associated with conducting risk assessments in this field compared to the experiences with 'conventional' chemicals. However this was not applied in the basic ENRHES RA appraisals.

Further assessment factor can be applied to account for the *quality of whole database* when data are scarce or poor in quality. The overall assessment factor should be balanced in using the different factors to avoid accounting for uncertainties in multiple ways, so the whole approach would become over-precautionary.

The overall assessment factors applied to the dose descriptors (including the modification to the starting point) were 50, 100 and 30 and the calculated INELs for the investigated ENMs were 2 μg, 1 μg and 17 μg for MWCNT (Baytubes), MWCNT (Nanocyl) and nano-TiO_2 respectively.

4. Analyses of different approaches for deriving exposure limit values

The following other approaches for deriving human no effect levels have recently become available: Pauluhn [10], Lecloux and Luizi [15], NIOSH [16] and the Japanese risk assessment within the NEDO project [13,14]. They partly refer to the same substances and toxicity studies as considered in the ENRHES risk assessment appraisals.

1) Pauluhn [10] has recently published an attempt to derive an occupational exposure level (OEL) specifically for multi-walled CNT as produced by Bayer, called Baytubes®. On the basis of a repeated

(subchronic) rat inhalation studies, a mechanistic (conceptual) model was developed forming the basis for interspecies extrapolation leading to an overall extrapolation factor of 2 and an OEL value of 0.05 mg/m^3. No further factors for intraspecies differences or duration were applied.

The interspecies differences account for differences in alveolar deposition, ventilation and the time dependent particle accumulation reflecting the known species-specific differences in particle clearance half-times in rats and humans. The first step of the calculation relates to the differences in deposited dose between animals and humans. In the author's view the dosimetrics adjustment factor does not call for any deposition related dosimetrics adjustment between rats and humans (i.e. 1) because the higher respirability of particles by humans is averaged out by the increased ventilation of rats.

A comparison of the number and volumes of alveolar macrophages (AM) between rats and humans suggests that humans are six-times more resistant to attaining lung overload conditions, which are assumed to be a trigger for the observed toxic effects. Other differences addressing the lung volume, lung weight, lung surface area and total amount of surfactant [17] may be normalised by the body weight. Combining the body weight differences with number and volumes of alveolar macrophages, leads to an AF of 1.7 ~ 2 [10]. Intraspecies adjustments were not applied due to the lack of any systemic bioavailability of the nano-particles and based on the rationale that this type of portal-of-entry related toxicity (overload) is independent on metabolism. The commonly applied default value for the workers's ventilation (10 m^3/working day and adult) was considered to be implicitly overconservative [10].

2) Another risk assessment on a multi-walled CNT produced by Nanocyl for BASF was also based on a 90-day inhalation study in rats [3]. Starting from a LOAEL of 0.1 mg/m^3, an assessment factor of 40 was applied, resulting in an estimated "no effect" concentration in air of 2.5 µg/ mg/m^3 for 8 h/day exposure [15]. No further details on how this overall assessment factor was derived is presented.

3) Within the NEDO project on "Research and Development of Nanoparticle Characterisation Methods", the Japanese National Institute of Advanced Industrial Science and Technology has conducted risk assessments for 3 types of engineered nanomaterials (ENMs), including CNTs [13] and nano-TiO_2 [14]. The adjustments for human exposure situations are based on differences in lung deposition between rats and humans and no further factors were applied. No factor for toxicokinetics was considered necessary, as adverse effects related to the pulmonary inflammation, are local and depend only on the alveolar deposition fraction of particles. Differences of the deposition fraction and surface area of pulmonary alveolus had been already adjusted in the estimation of the provisional NOAELs. For toxicodynamics, they argued that rats have even greater sensitivity to particle exposure than humans.

For CNTs, Kobayashi *et al.* [13] calculated an acceptable exposure concentration of 0.21 mg/m^3 for CNTs from a NOAEC of 0.37 mg/m^3 in a 4 week-inhalation study based on CNT deposition on the pulmonary alveoli (≈ factor 0.85). The deposition into lung was calculated by multiplying the exposure concentration in air with alveolar deposition fraction (depending on particle size), exposure frequency (8h/day x 5 d/week) and breathing rate during working (36 m^3/day), divided by bodyweight (60 kg). From the air concentration of 0.37 mg/m^3, a lung dose rate was calculated at 6.0 µg/kg/day and further divided by the overall uncertainty factors (UF) of 2, resulting in 3.0 µg/kg/day, which was converted back to 0.21 mg/m^3 (time weighted average, 8 h/d, 5 d/w). The UFs cover toxicokinetics (1) and toxicodynamics (1) differences and the extrapolation of exposure period (2, sub-acute to chronic). No factor for intraspecies differences was considered.

For nano-TiO_2 [14] the amount deposited on the lungs of the experimental animals was calculated by multiplying the NOAEC of 2 mg/m^3 [5], with respiratory minute volume, exposure time and deposition fraction, and finally dividing by the bodyweight. After applying an UF of 2 (for duration: sub-chronic to chronic) the deposited dose was calculated back to an acceptable exposure value for humans of 1.2 mg/m^3 (which corresponds to an overall factor of 1.7 compared to the NOAEC).

4) The US National Institute for Occupational Safety and Health (NIOSH) [16] derived a Recommended Exposure Level (REL) for <100 nm TiO_2 of 100 µg/m^3 (for worker exposure 10 hours per day, 40 hours per week) aiming at a maximum risk for lung cancer of 1/1000 over a work life, but noted that the risk is possibly lower - perhaps even zero. This REL was derived based on data by Heinrich *et al.* [18] and using the linearised upper bound on risk from the multistage model. In this study an increased mortality and lung tumours in rats were reported following chronic inhalation of high concentrations of nano-TiO_2 (10 mg/m^3). Based on that data and a review of the literature NIOSH concluded that nano-TiO_2 is carcinogenic to rats and that it cannot be ruled out that nano-TiO_2 is also carcinogenic in humans. It is assumed that the mechanism for carcinogenicity would not be chemical specific, but caused by chronic inflammation following pulmonary overload, and thus caused by the particle nature of TiO_2.

5. Consideration on the dose metrics

In the above presented examples, the dose descriptors from the inhalation studies and the derived human no-effect levels were expressed in the mass metric. The mass metric allows a straightforward comparison of external exposure concentrations and retained cumulative doses in the lung for example for CNTs [10]. However, the same mass of nanoparticles has often been shown to be more inflammogenic than the same mass of larger particles of the same chemical composition [5]. It is thus also important to consider the influence of particle size and surface as well.

Therefore, mass metric alone does not always provide the optimal metric for the observed effects, so other metrics, like for example, the number concentrations, the surface area, or the shape (fibrous) can be better metrics for relating dose to the observed effects of a specific nanomaterial. On the other hand, some pragmatism might be needed in relation to selecting proper metrics for doing regulatory risk assessments. In conclusions, the most appropriate (unifying) metrics of dose are still unresolved [19], between others, due to the diversity of different types of engineered nanomaterials.

6. Conclusion

Different approaches for deriving human no-effect levels have been suggested by different groups.

The approach following the REACH guidance by largely applying the suggested default assessment factors usually lead to higher overall assessment factors and lower human no-effect levels as compared to other reported approaches. The higher overall assessment factors result mainly from higher factors for interspecies differences and the application of intraspecies factors for workers.

For interspecies differences other approaches have mostly calculated the differences of deposited nanoparticles doses in the lung of rodents and humans. These differences were around a factor of 2 in the OEL approach for MWCNTs [10] and are in the same range as the factor for modification to the starting point (according to the REACH guidance) taking into account the differences in exposure duration in the toxicity study and the workplace, and the respiratory volumes at rest as compared to light activity. In the Japanese approach within the NEDO project [13,14] however these differences correspond to a factor of 0.85. The additional default factor of 2.5 for remaining interspecies as in the REACH guidance [2] is the main reason for numerical differences of assessment factors in the different approaches. The possibility for reducing this factor to 1 is currently not provided in the guidance for inhalation exposure, but as chemical specific assessment factors (CSAFs) should always be given preference over default assumptions, such a reduction might be justified in certain cases. Some authors suggest that the exposure conditions for laboratory animals to particles are quite different from those for humans and may lead to an overestimation of the human risk [20]. For low-toxic particles the observed effect in rats is often inflammation secondary to lung-overload – to which rats seem to be more sensitive than humans [11,12]. However, no generalisations are possible and the definition of CSAFs requires reliable chemical specific data and good understanding of the fundamentals of the biological response to the material.

Differences were also observed in the application of intraspecies differences. While all of the approaches we have compared with the ENRHES approach considered such a factor not necessary, we do not share this view and the REACH default factor of 5 or eventually a reduced factor of 3 seems justified to us. The AFs for extrapolation to chronic duration are important, especially as some of the nanoparticle effects will only manifest after chronic exposure. If it can be shown, that effects do not depend on the duration of exposure, these factors can be reduced.

Considering the possibility that properties and mechanisms of nanomaterials may differ fundamentally from the non-nanoscale materials and chemicals, the default assessment factor values should be evaluated for their appropriateness for nanomaterials. There are currently two projects ongoing (REACH Implementation Projects RIP-oN 2 and 3) which aim at developing advice on how the REACH guidance on information requirement and chemical safety assessment could be updated to address the specific properties of nanomaterials.

References

[1] EC (European Commission) 2008. Nanomaterials in REACH. Doc. CA/59/2008 rev.1. 6th Meeting of the REACH Competent Authorities 15-15 December 2008. Brussels

[2] ECHA (European Chemicals Agency) 2008. REACH Guidance on Information Requirements and Chemicals Safety Assessment. http://guidance.echa.europa.eu/guidance_en.htm

[3] Ma-Hock L, Treumann S, Strauss V, Brill S, Luizi F, Mertler M, Wiench K, Gamer AO, van Ravenzwaay B, Landsiedel R 2009. Inhalation toxicity of multi-walled carbon nanotubes in rats exposed for 3 months Toxicol. Sci. 112 273-5

[4] Pauluhn J 2010a. Subchronic 13-week inhalation exposure of rats to multiwalled carbon nanotubes: toxic effects are determined by density of agglomerate structures, not fibrillar structures Toxicol. Sci. 113 226-42

[5] Bermudez E, Mangum JB, Wong BA, Asgharian B, Hext PM, Warheit DB, Everitt JI. 2004. Pulmonary responses of mice, rats, and hamsters to subchronic inhalation of ultrafine titanium dioxide particles Toxicol. Sci. 77 347-57

[6] Johnston HJ, Hutchison GR, Christensen FM, Peters S, Hankin S, Aschberger K, Stone V 2009b. A critical review of the biological mechanisms underlying the in vivo and in vitro toxicity of carbon nanotubes; the contribution of physicochemical characteristics Nanotoxicology 4 207-246

[7] Johnston HJ, Hutchison GR, Christensen FM, Peters S, Hankin S, Stone V 2009a. Identification of the mechanisms that drive the toxicity of TiO2 particulates; the contribution of physicochemical characteristics Part Fibre Toxicol. 6:33

[8] Aschberger K, Johnston HJ, Stone V, Aitken RJ, Tran CL, Hankin SM, Peters SAK, Christensen FM 2010. Carbon nanotubes – assessment of the feasibility and challenges for human health risk assessment based on open literature Crit. Rev. Toxicol. 40 759-90

[9] Christensen FM, Johnston HJ, Stone V, Aitken RJ, Hankin S, Peters S, Aschberger K 2010. Nano titanium–dioxide - feasibility and challenges for human health risk assessment based on open literature Nanotoxicology, Early online. DOI: 10.3109/17435390.2010.504899

[10] Pauluhn J 2010. Multi-walled carbon nanotubes (Baytubes®): Approach for derivation of occupational exposure limit. Regul. Toxicol. Pharmacol. 57 78-89

[11] Borm PJA, Robbins D, Haubold S, Kuhlbusch T, Fissan H, Donaldson K, Schins R, Stone V, Kreyling W, Lademann J, Krutmann J, Warheit D, Oberdörster E 2006. The potential risks of nanomaterials: a review carried out for ECETOC Part Fibre Toxicol 3:11

[12] Brown JS, Wilson WE, Grant LD 2005. Dosimetric comparison of particle deposition and retention in rats and humans. Inhal. Toxicol. 17 355-85

[13] Kobayashi N, Ogura I, Gamo M, Kishimoto A, Nakanishi J 2009. Risk Assessment of Manufactured Nanomaterials – Carbon Nanotubes (CNTs) - NEDO project "Research and Development of Nanoparticle Characterizations Methods (P06041)" Interim Report issued October 16, 2009; available at http://www.aist-riss.jp/main/?ml_lang=en

[14] Hanai S, Kobayashi N, Ema M, Ogura I, Gamo M, Nakanishi J 2009. Risk Assessment of Manufactured Nanomaterials – Titanium Dioxide (TiO2) - NEDO project "Research and

Development of Nanoparticle Characterizations Methods (P06041)" Interim Report issued October 16, 2009; available at http://www.aist-riss.jp/main/?ml_lang=en

[15] Lecloux AJ, Luizi F 2009. The safe use of carbon nanotubes in coating applications. presentation held at the European Commission's 3rd stakeholder dialogue on nanomaterials "Safety for success", 3-4 November 2009 in Brussels; available at: http://ec.europa.eu/health/ph_risk/documents/ev_20091103_co11_en.pdf

[16] NIOSH (National Institute for Occupational Safety and Health) 2005. NIOSH Current Intelligence bulletin: Evaluation of Health Hazard and Recommendations for Occupational Exposure to Titanium Dioxide. DRAFT. NIOSH. Available via: http://www.cdc.gov/niosh/review/public/TIo2/pdfs/TIO2Draft.pdf

[17] Oberdörster G, Oberdörster E, Oberdörster J 2007. Concepts of nanoparticle dose metric and response metric. Environ. Health Perspect. 115 A290

[18] Heinrich U, Fuhst R, Rittinghausen S, Creutzenberg O, Bellmann B, Koch W, Levsen K 1995. Chronic inhalation exposure of wistar rats and two different strains of mice to diesel engine exhaust, carbon black, and titanium dioxide. Inhal. Toxicol. 7 533-56

[19] Maynard AD 2007. Nanotechnology: the next big thing, or much ado about nothing? Ann. Occup. Hyg. 51 1-12

[20] Oller AR, Oberdörster G 2010. Incorporation of particle size differences between animal studies and human workplace aerosols for deriving exposure limit values. Regul. Toxicol. Pharmacol. 57 181-94

Re-examination of safety parameters using kinetic theory of nano-granular flows

Jacques Xavier Bouillard[1], Philippe Marchal[2], François Henry[1], Alexis Vignes[1], Olivier Dufaud[2], Laurent Perrin[2] and Edouard Plasari[2]

[1]INERIS, Parc ALATA, 60550 Verneuil en Halatte, France

[2]ENSIC, 1 rue Grandville, LRGP-UPR-CNRS-3349, 54000Nancy, France

Abstract. The origin of the kinetic theory of granular flow was originally credited to Bagnold [1]. By using a very primitive expression of the particle collision frequency, he derived an expression for the repulsive pressure of the particles in uniform shear flows. His repulsive pressure was proportional to the square of the velocity gradient and the particle diameter and directly proportional to the particle density. This theory was later extended by Savage [2] and Gidaspow [3]. Such theories provide insight on the dependence of the viscosity, and various moduli (elastic, non elastic, viscous...) in terms of the granular temperature and the associated shear-rates. Until recently, such parameters were difficult to measure because of the lack of specifically designed equipment. This challenge was successfully taken up and resolved by P. Marchal of ENSIC who designed a new rheometer for powders (figure 1). This equipment can put in evidence the importance of the granular temperature on the elastic and viscous behaviors of the granular flows. Such rheological behavior is important in risk analysis for nanopowders, because as the nanopowder may be subjected to process shear rates and stresses, its structural and topological changes, in terms of the transformation of agglomerates into primary nanoparticles, have strong impacts on emission factors of nanosized particles that can be released in the environment or into a workplace from such dense-phase nanopowder processes. Such transformation can be analyzed by studying the nano-granular rheological signature of the system. Such risk assessment approach using these new fundamental rheological safety parameters is described in this paper.

1. Introduction

The use of the concept of granular temperature for dense and dilute suspended powder has evolved since the 80s in theories and applications aiming at explaining turbulence, rheology and reactivity in such systems. The heart of the approach consists in expressing the granular temperature, θ, defined as follows [3,4]:

$$3\,\theta = (c_x^2 + c_y^2 + c_z^2)$$

where *c* is the particle fluctuating velocity, as described below.

This approach is based on the development of kinetic theory for gas–solids two-phase flow based on the theory for non-uniform dense gases described in Chapman and Cowling [5] and the pioneering paper of Lun *et al.* [4] in applying the kinetic theory of gases to granular flows. The kinetic theory approach uses one equation model to determine the turbulent kinetic energy (or granular temperature) of the particles and assumes either a Maxwellian distribution for the particles, or a non-Maxwellian distribution for both dilute and dense cases. The kinetic theory approach for granular flow allows the determination of the pressure and viscosity of the solids in place of empirical relations [6]. One of the major goals in the use of such theory was the evaluation from first principles of the viscosity of granular media.

Various kinetic theory approaches considering the effect of the interaction of particle and gas phase, and particle collisions have been applied to the study of a gas–solids flow in risers of circulating fluidized beds, chutes and dense phase mixing [7-9]. Extensions of these approached have included the modeling of agglomeration via the effects of Van der Waals forces [10]. Such approaches were successful in reproducing two-phase flow experiments performed by Miller and Gidaspow [11].

2. Gas–solids flow model description

The model adopted is based on the fundamental concept of interpenetrating continua for multiphase mixtures. According to this theory different phases can be present at the same time in the same computational volume. Such an idea is made possible by the introduction of a new dependent variable, the concentration, ε_i of each phase i. The fundamental equations of mass, momentum, and energy conservation are then solved for each considered phase. Appropriate constitutive equations have to be specified in order to describe the physical rheological properties of each phase and to close the conservation equations. In this model, solids viscosity and pressure are derived by considering the random fluctuation of particle velocity and its variations due to particle–particle collisions and the actual flow field. Such a random kinetic energy, or granular temperature, can be predicted by solving, in addition to the mass and momentum equations, a fluctuating kinetic energy equation for the particles. The solids viscosity and pressure can then be computed as a function of granular temperature at any time and position. Particles are considered smooth, spherical, inelastic, and undergoing binary collisions. The adoption of the second approximation distribution function allows us to apply the theory to both dense and dilute two-phase flows. A more complete discussion of the implemented kinetic theory model can be found in Gidaspow [3].

The basic Hydrodynamic equations governing granular flows are given below [3]:

- Continuity equation for phase k: $\frac{\partial(\rho_k \varepsilon_k)}{\partial t} + \nabla.(\rho_k \varepsilon_k v_k) = \dot{m}_k$

- Momentum equation for phase k: $\frac{\partial(\rho_k \varepsilon_k v_k)}{\partial t} + \nabla.(\rho_k \varepsilon_k v_k v_k) = \varepsilon_k \rho_k \tilde{g} + \nabla.[\tau_k] + \sum_l^N \beta(v_l - v_k) + \dot{m}_k \widetilde{v_k}$

  acceleration of phase "k" = buoyancy + stress + drag force + phase change

- Constitutive equation for stress (above min. fluidization): $[\tau_k] = [-P_k + \xi_k \nabla v_k][I] + 2\mu_k [S_k]$

$$[S_k] = \frac{1}{2}[\nabla v_k + (\nabla v_k)^T] - \frac{1}{3}\nabla.v_k[I]$$

and the granular temperature governing equations [3]:

- Fluctuating energy equation: $\frac{3}{2}\left[\frac{\partial}{\partial t}(\varepsilon_s \rho_s \theta) + \nabla.(\varepsilon_s \rho_s v_s \theta)\right] = \overline{\overline{\tau_s}}: \nabla v_s - q - \gamma_s + \beta_A \langle C_g . C_p \rangle - 3\beta_A \theta$

  Collisional energy dissipation: $\gamma_s = 3(1 - e^2)\varepsilon_s^2 \rho_s g_o \theta \left(\frac{4}{d_s}\sqrt{\frac{\theta}{\pi}} - \nabla.v_s\right)$

  Conductivity of fluctuating energy: $\kappa = \frac{2}{(1+e)g_o}\left[1 + \frac{6}{5}(1+e)g_o \varepsilon_s\right]^2 \kappa_{dil} + 2\varepsilon_s^2 \rho_s d_s g_o (1+e)\sqrt{\frac{\theta}{\pi}}$

  Dilute phase (Eddy type) granular conductivity: $\kappa_{dil} = \frac{75}{384}\sqrt{\pi \rho_s} d_s \theta^{1/2}$

- Gas-solid drag coefficients [3]:
  - Based on Ergun equation, for $\varepsilon_g < 0.8$: $\beta = 150\frac{\varepsilon_s^2 \mu_g}{\varepsilon_g (d_p \phi_s)^2} + 1.75\frac{\rho_g \varepsilon_s |v_g - v_s|}{d_p \phi_s}$
  - Based on single sphere drag, for $\varepsilon_g > 0.8$: $\beta = \frac{3}{4} C_d \frac{\rho_g \varepsilon_s \varepsilon_g |v_g - v_s|}{d_p \phi_s} \varepsilon_g^{-2.65}$

$$C_d = \frac{24}{Re_p}\left[1 + 0.15 Re_p^{0.687}\right] \quad for\ Re_p < 1000$$

$$C_d = 0.44 \quad for\ Re_p > 1000$$

where

$$Re_p = \frac{\varepsilon_g \rho_g d_p |v_g - v_s|}{\mu_g} \varepsilon_g^{-2.65}$$

From these equations the viscosity and rheological stresses (Solids pressure and viscosities) can be modeled as follows [3]:

- Solid phase stress:
  - Solid-phase pressure: $P_s = \rho_s \varepsilon_s \theta [1 + 2(1+e) g_o \varepsilon_s]$
  - Solid-phase bulk viscosity: $\lambda_s = \frac{4}{3}\varepsilon_s^2 \rho_s d_s g_o (1+e)\sqrt{\frac{\theta}{\pi}}$
  - Solid-phase shear viscosity: $\mu_s = \frac{2\mu_{s_{dil}}}{(1+e) g_o}\left[1 + \frac{4}{5}(1+e) g_o \varepsilon_s\right] + \frac{4}{5}\varepsilon_s^2 \rho_s d_s g_o (1+e)\sqrt{\frac{\theta}{\pi}}$
  - Radial distribution function [1]

$$g_o = \left[1 - \left(\frac{\varepsilon_s}{\varepsilon_{s,max}}\right)^{1/3}\right]^{-1}$$

  - Solid-phase dilute viscosity

$$\mu_{s_{dil}} = \frac{5\sqrt{\pi}}{96}\rho_p d_p \theta^{1/2}$$

These equations can be extended for mixtures of powders of different size as shown in Table 1 (see Gidaspow [3]).

To make it simple, in a vibrated shear flow, the total reorganization frequency λ^{-1} can simply be expressed as the sum of the frequencies corresponding to a process of vibration and a process of shear, so one can express the frequency as:

$$\lambda^{-1} = \frac{\dot{\gamma}}{\gamma_c} + f \qquad \textit{Equation (1)}$$

In other words, λ is the mean life time of intergranular contacts, when the sample is submitted to shear and vibrations. Hence, the granular system can be viewed as a temporary elastic network and the magnitude of impulse per grain *i*, transmitted from a fast-moving layer, in the *x* direction, to a slower adjacent one.

On can express the steady state viscosity as:

$$\eta = \frac{\sigma}{\gamma_c f + \dot{\gamma}} \qquad \textit{Equation (2)}$$

Thus, the model predicts a non-Newtonian rheological behaviour for dense-phase vibrated powders, bounded by two extreme flow regimes:

- at high shear rates , the viscosity varies as an inverse function of the shear rate,
- at low frequencies, *f*, the stress becomes independent of the shear rate (Newtonian regime).

Another way of looking at this expression is to start from the general case of the solids phase shear viscosity definition, and take the limit as ε_s becomes relatively large (that is we consider a dense phase), at a constant energy ($\varepsilon_s\theta_s$ = ct). The resulting viscosity is dependant of $\varepsilon_s\theta_s^{1/2}$ or $\theta_s^{-1/2}$ or f^{-1}, hence indicating an inverse function of the viscosity with respect to the shear rate or frequency. Note that the beauty of the modeling approach described earlier is that it provides analytical expressions of the rheological property dependency with respect to the particle diameter, its density and restitution coefficient. This is indeed these dependencies that are exploited to measure the agglomeration energy of nanoparticles via rheological routes.

Table 1. Constitutive relations for a mixture of M phases with rotation [3].

Granular temperature definition	$3m_i\theta_i = \frac{1}{2}m_i\langle c_i^2\rangle + \frac{1}{2}I_i\langle\omega_i^2\rangle$
Number of binary collisions per unit volume, *Nij*	$N_{ij} = 2\sqrt{2\pi}d_{ij}^2 n_i n_j\sqrt{\overline{\theta_{ij}}}\left[1 - 3\overline{\Delta_{ij}} + \cdots\right]$ where $\overline{\Delta_{ij}^2} = \frac{(w_i\theta_i - w_j\theta_j)^2}{(\theta_i+\theta_j)(w_i^2\theta_i + w_j^2\theta_j)}$ and the mean oscillating viscosity = $(\overline{\theta_{ij}})^{\frac{1}{2}}$ is expressed in terms of *i* and *j* granular temperatures, θ_i and θ_j and the previous groups shown below: $\overline{\theta_{ij}^{\frac{1}{2}}} = \left(\frac{1}{2\theta_i\theta_j A_{ij}}\right)^{\frac{3}{2}}\frac{1}{(2D_{ij})^2}$ $= \left(\frac{m_{ij}^2\theta_i\theta_j}{(\theta_i+\theta_j)(m_i^2\theta_i + m_j^2\theta_j)}\right)^{\frac{3}{2}}\left(\frac{m_{ij}^2\theta_i\theta_j}{m_i^2\theta_i + m_j^2\theta_j}\right)^{\frac{1}{2}}$ with the coefficients *Aij*, *Bij* and *Dij* originally obtained by Eirik Manger (1996) as follows $A_{ij} = \frac{1}{2}\left(\frac{1}{\theta_i} + \frac{1}{\theta_j}\right)$, $B_{ij} = \frac{1}{2m_{ij}}\left(\frac{m_i}{\theta_j} - \frac{m_j}{\theta_i}\right)$, and $D_{ij} = \frac{1}{2m_{ij}^2}\left(\frac{m_i^2}{\theta_j} - \frac{m_j^2}{\theta_i}\right)$
Viscosity coefficients and collisional pressure [3]	Collisional viscosity for phase *i*: $\mu_{c,i} = \sum_{k=1}^{M}\mu_{c,ij}$ $\mu_{c,ij} = \frac{3}{2\pi}(1+e)\frac{d_{ij}^4}{d_{pj}^3 d_{pj}^3}g_{ij}\varepsilon_i\varepsilon_j\left(1 + 6\Delta_{ij}^2 + \cdots\right)\overline{\theta_{ij}^{\frac{1}{2}}}$ Solid Viscosity for Dilute Flow: $\mu_{i,kinetic} = \frac{5}{16d_i^2}\left(\frac{m_i^2}{\pi}\theta_i\right)^{1/2}$
Mixture viscosity, μ_m (from sum of phase momentum balances)	$\varepsilon_{sm}\mu_m = \sum_{i=1}^{M}\varepsilon_i\mu_{i,kinetic} + \sum_{i=1}^{M}\varepsilon_i\sum_{j=1}^{M}\mu_{c,ij}$ where ε_{sm} is the sum of the volume fractions of the solids phases.

Journal of Physics: Conference Series **304** (2011) 012079 doi:10.1088/1742-6596/304/1/012079

Equation of state for mixture of M phases	$Solid\ Pressure = P_s = \sum_{i=1}^{M} \varepsilon_i \rho_i \theta_i + \sum_{i=1}^{M}\sum_{j=1}^{M} P_{c,ij}$ where the collisional pressure component is: $P_{c,ij} = \left(\frac{\overline{\theta_{ij}}^{\frac{1}{2}}}{D_{ij}^{\frac{1}{2}}}\right) \frac{\sqrt{2}\pi}{6}(1+e) d_{ij}^3 g_{ij} \frac{\varepsilon_i \varepsilon_j \rho_i \rho_j}{\frac{\pi}{6}\left(\rho_i d_{pi}^3 + \rho_j d_{pj}^3\right)} (1 - 3\overline{\Delta_{ij}} + \cdots)$
Collisional conductivity of phase i	$k_{i,coll} = \frac{1}{2}\sqrt{\frac{\pi}{2}} m_i g_{ij} d_{ij}^4 n_i n_j \left(\frac{\overline{\theta_{ij}}^{\frac{1}{2}}}{D_{ij}}\right)\left[\frac{8+3\pi}{16\theta_i} S_1 + \frac{m_j}{3m_{ij}\theta_i^2} S_3\right]$ where $S_1 = \left[1 + \frac{9\pi B^2}{2A_{ij}D_{ij}} + \cdots\right]$ *and* $S_3 = \frac{1}{A_{ij}}\left[1 + \frac{15\pi B^2}{2A_{ij}D_{ij}} + \cdots\right]$ Kinetic conductivity of phase i: $k_{i,kinetic} = \frac{15}{4}\mu_i$ $k_i = k_{i,coll} + k_{i,kinetic}$
Energy dissipation, γ_{ij}	$\gamma_{ij} = \frac{m_i m_j(\theta_i + \theta_j) N_{ij}}{2(m_i + m_j)} \times \left\{ \left[\frac{k_i k_j e_t(e_t - 1)}{k_i w_i + k_j w_j}\right] \sqrt{\frac{k_j \theta_i + k_i \theta_j}{k_i k_j(\theta_i + \theta_j)}} + \frac{(k_i k_j + k_i w_i + k_j w_j) e_t^2}{4(k_i w_i + k_j w_j)} + \frac{(k_i w_i + k_j w_j - k_i k_j) e_t}{2(k_i w_i + k_j w_j)} + \frac{k_i k_j - 3k_i w_i - 3k_j w_j}{4(k_i w_i + k_j w_j)} + e^2 - 1 + \left[\frac{k_i k_j(e_t^2 - 1)}{k_i w_i + k_j w_j}\right] \frac{3(k_j \theta_i + k_i \theta_j)}{2k_i k_j(\theta_i + \theta_j)} \right\}$

3. Description of the rheological viscometer

Considering the macroscopic nature of granular media, we have developed a powder rheometer prototype aiming at compensating for the lack of thermal Brownian motion [12]. It is constituted of a stress-imposed rheometer (AR2000, TA Instruments) (Fig. 1) equipped with a vibration exciter, which is the source of granular temperature. It is connected to an amplifier and controlled through a function generator and an accelerometer connected to a measuring amplifier and fixed to the characterization cell. Attached to the exciter, the characterization cell includes a four blades vane device (radius $= R_i = 10$ mm; length $= L = 30$ mm) installed into a cylindrical baffled cup (cup radius $= R_c = 25$ mm; baffles width $= W_b = 5$ mm). Such a configuration prevents the powder sample from slipping at the wall and at the surface of the measuring tool as it is equivalent to a tangential annular flow between two vertical coaxial cylinders constituted of the sample itself: the inner cylinder (radius $= R_i = 10$ mm) is constituted by the part of the sample dragged by the vane blades and the outer cylinder (radius $= R_e = R_c - W_b = 20$ mm) is constituted by the part of the sample entrapped between the baffles. Considering the geometrical dimensions of the cell, the space between vane and baffle extremities corresponds to a $R_e - R_i = 10$ mm effective gap, the distance between the base of the vane and the bottom of the cup has been adjusted to the same value.

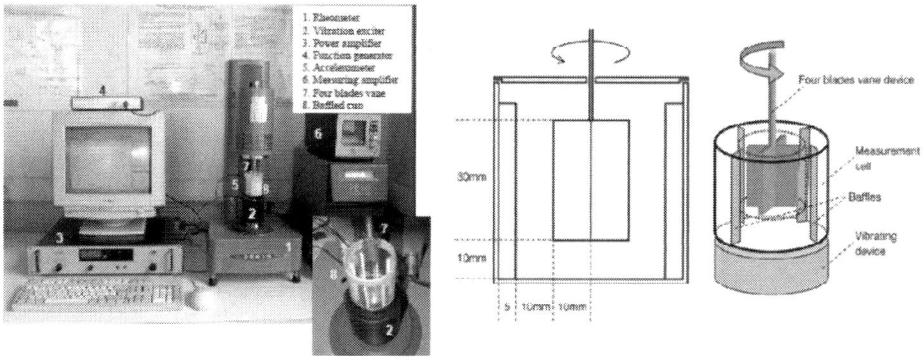

Figure 1. Powder rheometer setup (left), with its schematic representation (right)

With such apparatus, one can readily show the shear thinning rheological behaviors of the powders as shown in Figures 2 and 3, exhibiting an inverse dependency of the viscosity with respect to shear rates at high shear rates.

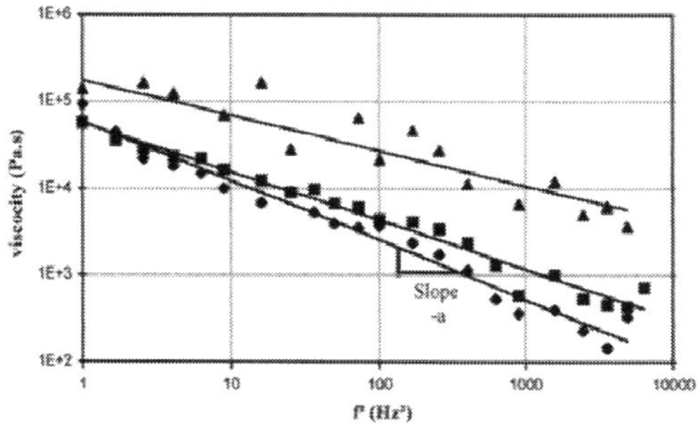

Figure 2. Reduction of viscosity with respect to the granular temperature or equivalently (f^2), for three particle sizes

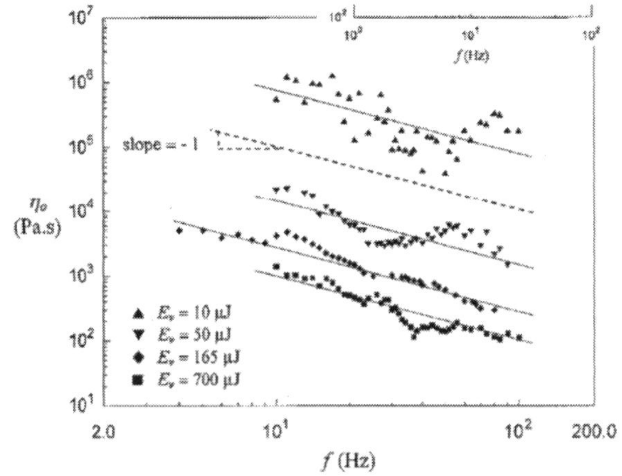

Figure 3. The inverse dependency of the viscosity with respect to the frequency

Over a whole spectrum of shear rates, typical viscosity signatures as shown below can be obtained for various vibration frequencies.

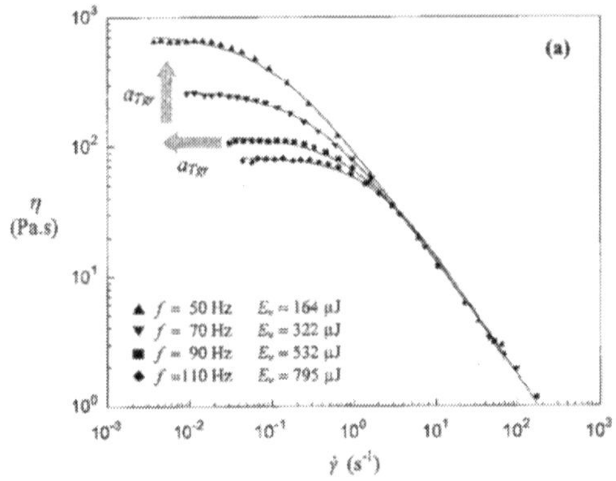

Figure 4. Typical viscosity with respect of shear rate at various vibrating frequencies (granular temperatures), showing the two regimes of Eq *(1)* and *(2)* for a given particle size.

4. Application of Nanopowder Rheological Characteristics for its Use in Risk Evaluation

As shown earlier on, the rheological behavior of a nanopowder depends mainly on four important parameters, the vibration frequency, the shear rate, the particle size and its density (see models above). Hence, if one were to use an agglomerated nanopowder and to test it with the rheometer, on would start measuring the viscosity for a powder equivalent made of large particle sizes. As the shear energy increases, the rheological signatures of the agglomerated nanopowder could significantly exhibit a change in viscous behavior trends as shown in the hypothetical continuous line below. Such change would induce a change of energy supplied by shear that can basically be approximately considered as the agglomeration energy of the powder. Such energy can directly be measured by such device.

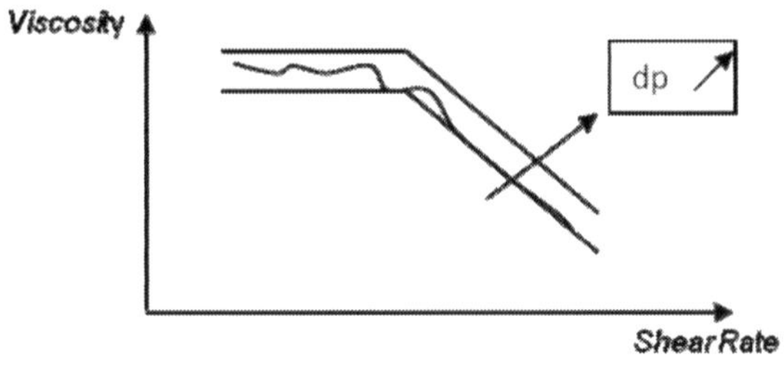

Figure 5. Typical change of the rheological signatures of two powders of different sizes of the same material. In continuous line, the possible outcome of a nanopowder that would be submitted to desagglomeration during its shearing

Two types of Al nanopowders of 120 and 200 nm were studied to evaluate the effect of their mean particle sizes. Particle size distributions of these two nanopowders are represented in Figure 7, in terms of distribution based on the numbers and on the volumes. Of course, a distribution in volume would give advantage to larger particles since they have a larger volume, while the distribution in numbers tends to privilege primary (or smaller) particles. Particle size distributions in number of Al nanopowders are presented in Table 2 which would seem to indicate that despite uncertainties, there is a low degree of agglomeration.

Table 2. Particle size distribution of Al nanopowders

	Particle sizes (μm)		
	d_{10}	d_{50}	d_{90}
Aluminum 100 nm	0,10	0,17	0,62
Aluminum 200 nm	0,15	0,28	0,77

Some clichés made by TEM of these two nanopowders are presented in Figure 6 below:

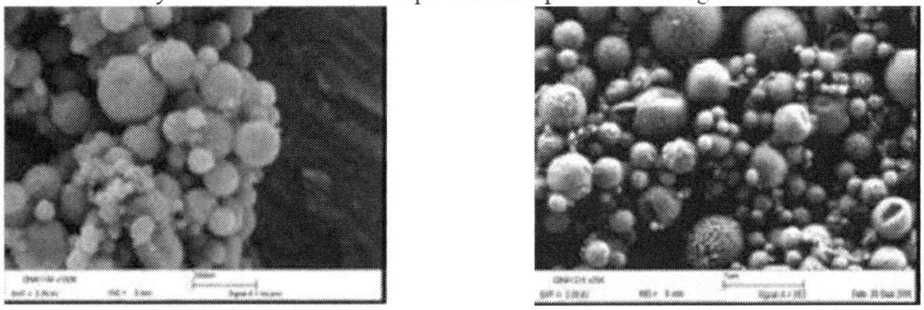

Figure 6. Typical TEM pictures of 120 nm and 200 nm Al particles.

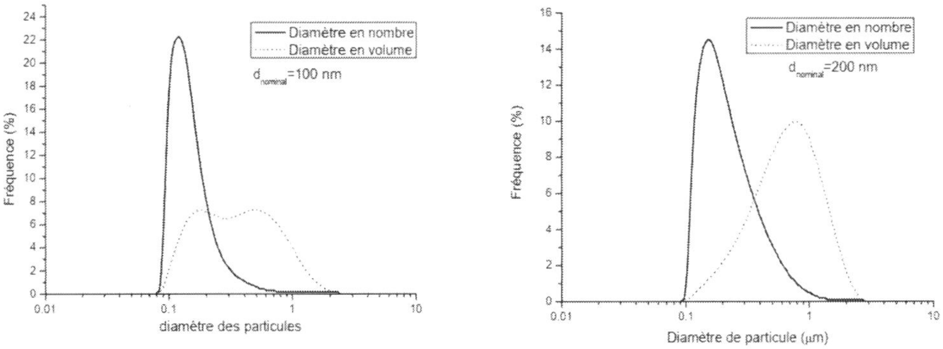

Figure 7. Particle size distribution for 120 nm and 200 nm Aluminum particle sizes.

From rheological measurements made on the two types of powders (120 and 200 nm Aluminum particles as shown in Figure 8), we can see that the two rheological signatures almost follow a same trend at high shear rates. As can also be seen, the rheological signatures differ at low shear rates. This would tend to indicate that at a moderately high shear rates, the powders behave similarly, thus indicating a low degree of agglomeration at such shear rates. This results comfort low degrees of agglomeration observed by particle size distribution measurements as shown in Figure 7. Hence the agglomeration energy is likely to be small, albeit not specifically measured in this study. This would indicate also, a high propensity for this type of powders to be released as primary particles at high shear rates and not as agglomerate particles, inducing therefore potentially higher toxicological risks.

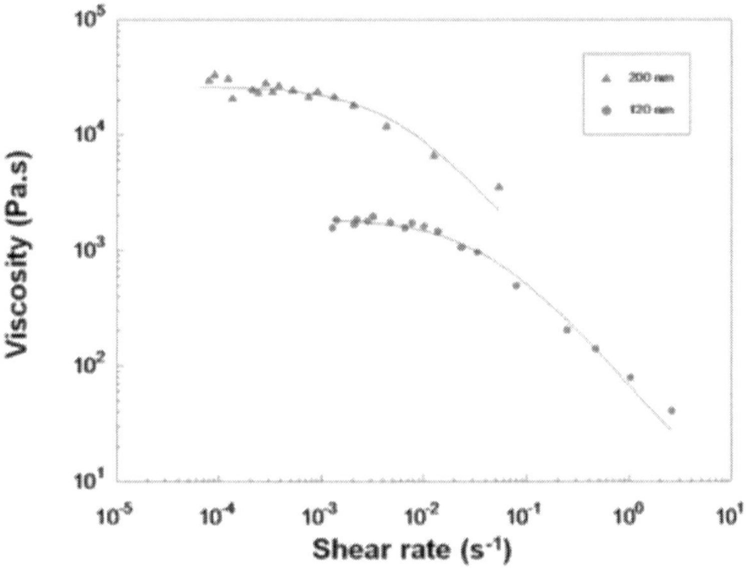

Figure 8. Typical rheological data for 200 nm and 120 nm Aluminum nanoparticles (under a 50Hz base frequency)

This agglomeration energy is therefore a critical safety parameter to consider for the assessment of nanodispersion risks and potential hazards for workers and the environment at large. To evaluate the production rate of primary nanoparticles, one could write the rate of production, r, as

$$r = \alpha \frac{Power\ supplied}{\Delta H_{agglomeration}}$$

where the power supplied is, in the case of the rheometer, the torque power, and the agglomeration energy is the one that can be measured by the technique presented above. As can be seen from this equation, a higher rate of emission of nanoparticles is obtained for powders of low agglomeration energies.

The agglomeration energy, dependent of the type of powder considered under given mechanical constraints, is a measure of the ability of an agglomerated nanopowder to liberate itself into primary nanoparticles at ambient conditions. Hence, this approach could be used for any systems in which a mechanical energy is supplied to a nanopowder, and from which a resulting emission rate is to be evaluated. Once such emission rate is established, then a classical air nanodispersion modeling, integrating reagglomeration mechanisms, can be carried out to evaluate the fate of the released nanoparticles and their potential resulting hazards /risks.

5. Conclusions

Risks induced by the emission of agglomerated particles in a process will essentially depend upon their propensity to release nanoparticles in their primary form, which will strongly depend on the agglomeration energy of the nanopowders. INERIS and LRGP (Nancy) have devised a methodology to first measure the agglomeration energy from rheological characteristics of the nanopowder, by the use of a novel dry phase nanopowder rheometer, and then use this information to evaluate potential risks due to the fate of the potential release in air into more potent primary nanoparticles. Such approach will help nanosafety engineers to evaluate the potential release of nanoparticles from a process using agglomerated nanopowders constrained to mechanical shear stresses. Such new techniques are being the object of the development of new instruments that presently remain confidential.

Nomenclature

Cd	drag coefficient
dp	particle diameter
D_{gs}	rate of energy dissipation
E	restitution coefficient of particles (= 0.999)
g	gravity
g_0	radial distribution function
k_s	effective thermal conductivity of particles
k_g	thermal conductivity of gas phase
P_g	pressure
P_s	particle pressure
Q	fluctuating energy flux
Re	Reynolds number
T	time
v_g	gas velocity
v_s	particle velocity
u_g	gas velocity
τ_g	gas stress tensor
τ_s	stress tensor of particulate phase
ξ	bulk viscosity
θ:	granular temperature
μ_g	gas viscosity
μ_s	particulate viscosity
ε_g	voidage
ε_s	solid volume fraction
ρ_s	particle density
ρ_g	gas density
β	interphase drag coefficient

References

[1] Bagnold, R. A. (1954). "Experiments on a gravity-free dispersion of large solids spheres in a Newtonian fluid under shear." *Proc. Roy. Soc.* **A225**: 49-63.

[2] Savage, S., B., (1983). Granular Flows at high shear Rates. New York.

[3] Gidaspow, D. (1994). Multiphase Flow and Fluidization: Continuum and Kinetic Theory Descriptions. New York, Academic Press.

[4] Lun, C. K. K. and D. J. J. a. N. C. S.B. Savage (1984). "Kinetic theory for granular flow: inelastic particles in Couette flow and slightly inelastic particles in a general flow field." *J. Fluid Mech.* **140**: 223-235.

[5] Chapman, S. and T. J. Cowling (1961). The Mathematical Theory of Non-Uniform Gases. London, Cambridge University Press.

[6] Campbell, C. S. (2006). "Granular material flows - An overview." *Powder Technology* **162**(3): 208-229.

[7] Huilin, L. and D. Gidaspow (2003). "Hydrodynamics of binary fluidization in a riser: CFD simulation using two granular temperatures." *Chemical Engineering Science* **58**(16): 3777-3792.

[8] Huilin, L., D. Gidaspow, et al. (2002). "Chaotic behavior of local temperature fluctuations in a laboratory-scale circulating fluidized bed." *Powder Technology* **123**(1): 59-68.

[9] Huilin, L., D. Gidaspow, et al. (2003). "Hydrodynamic simulation of gas-solid flow in a riser using kinetic theory of granular flow." Chemical Engineering Journal **95**(1-3): 1-13.
[10] Wang, S., H. Lu, et al. (2007). "Prediction of flow behavior of micro-particles in risers in the presence of van der Waals forces." *Chemical Engineering Journal* **132**(1-3): 137-149.
[11] Gidaspow, M. a. (1992). "Dense, vertical gas–solid flow in a pipe." AIChE **41**: 1801-1813.
[12] Barois-Cazenave, A., P. Marchal, et al. (1999). "Experimental study of powder rheological behaviour." *Powder Technology* **103**(1): 58-64.

Nano-silicon dioxide toxicological characterization on two human kidney cell lines

V Paget, JA Sergent and S Chevillard

Laboratory of Experimental Cancerology, Institute of Cellular and Molecular Radiobiology, CEA, Fontenay-aux-Roses, France

E-mail: sylvie.chevillard@cea.fr

Abstract. Silicon dioxide nanoparticles (n-SiO_2) have recently encountered a wide variety of applications in medicine or engineering but their toxicological effects are poorly understood. In this study, we have used SiO_2-25 nm and SiO_2-100 nm mono-dispersed nanoparticles labeled with Rhodamine B and TMPyP respectively. These two fluorophores were incorporated during synthesis in order to track nanoparticles cell incorporation. Up-to-date, no evaluation of the toxicological effects of these nanoparticles upon human kidney has been published. As kidney is one of the major traditional retention organs, the aim of our study is to evaluate the potential toxicity of these nanoparticles on two human cell lines from proximal tubule (Caki-1 and Hek293). Our results report that the two cell lines do not show similar responses after 24 hours of exposure to SiO_2-nanoparticles disregarding a similar origin in the kidney. Interestingly, our results indicate that for both tested SiO_2-nanoparticles, Caki-1 cells present a higher sensitivity in terms of cytotoxicity and genotoxicity than Hek293 cells. Furthermore, our results show that for similar concentration of exposure, SiO_2-25 nm seems to be more cytotoxic and genotoxic than SiO_2-100nm for both tested cell lines.

1. Introduction

Nanotechnology development has grown rapidly in the past decade in various fields of application such as drug delivery [1,2], environmental health [3], electronics or biomedical engineering [4]. Nevertheless, this rapid development might be also associated with unintended hazards. Despite intensive research efforts, reports of cellular responses to nanomaterials are often inconsistent and even contradictory. Furthermore, toxicological data of SiO_2 nanoparticles on human renal cells *in vitro* and their detailed molecular mechanisms still remain unknown. Indeed, to our knowledge, there are no published data reporting the study of such nanoparticles upon biological human material from kidney, which is among the more important retention organs. The aim of our study is to evaluate the potential toxicity of these nanoparticles on two human cell lines from proximal tubule (Hek293 and Caki-1).

Unfortunately, the intrinsic properties of nanomaterials have already shown some unintended interferences with some usual cytotoxic and genotoxic assays, leading to false negative or false positive results [5]. Thus it clearly appears critical to use a wide range of conventional cytotoxicity assays in order to determinate which of them are the most accurate to study a specific nanoparticle. We have decided to study the genotoxicity of two SiO_2 nanoparticles (25 and 100 nm) via five conventional cytotoxicity assays: Neutral Red, LDH release, XTT, MTT and SRB. To complete these classical approaches, we have adapted two methods based on flow cytometry and confocal microscopy to study *i)* DNA ploidy and cell mortality *ii)* DNA double strand breaks by the detection of phosphorylated γ-H2Ax-foci, together with the intracellular accumulation of nanoparticles. In order to track the nanoparticles, we have used SiO_2-25 nm and SiO_2-100 nm dispersed nanoparticles labeled with Rhodamine B and TMPyP respectively, these two fluorophores being incorporated during synthesis.

2. Experimental

2.1. Cell Culture

Human Caki-1 cells (ATCC number: HTB-46™) and human Hek293 (ATCC number: CRL-1573™) were routinely grown at 37°C in a humidified atmosphere of 5% CO_2 and 95% air, in Dulbecco's Modified Eagle Medium (DMEM) Glutamax supplemented with 10% (v/v) inactivated fetal bovine serum (FBS) and 1mM Anti-Anti (InVitrogen).

2.2. Characterization of SiO_2 nanoparticles

2.2.1. *Materials.* Triton® X100 (TX-100), Igepal CO-520, 1-hexanol anhydrous ( ), cyclohexane reagent plus® ( ), aqueous ammonia (NH_4OH) solution (25%), tetraethylorthosilicate (TEOS, 98%), ethanol, Rhodamine B, 5,10,15,20-tetrakis(1-methyl-4-pyridino)porphyrin tetra(toluene-4-sulfonate) (TMPyP) all purchased from Aldrich, were used without further purification. Water was purified with a Milli-Q system (Millipore, Bedford, MA) including a SynergyPak® unit. The exclusive Jetpore®, ultrapure grade mixed-bed ion-exchange resin, was also used in this unit. Water achieved resistivity above 18.0 MΩ.cm at 25°C. A C 3.12 centrifuge (Jouan, France), and a Sonorex Digitec sonication water-bath (Roth, France) were used. Dialysis tubes have been purchased from Roth, and possess a nominal filter rating of 3500, and a MWCO of 4000-6000 D.

2.2.2. *Synthesis of SiO_2 25 nm.* Silica nanoparticles were synthesized using a reverse micro-emulsion method, as described by Boudreau et al. in the literature [6]. To a clean 50 mL erlenmeyer, Igepal CO-520 (8 mL), cyclohexane (20 mL) and an aqueous solution or rhodamine B (650 µL at 0.01M) were mixed and stirred at room temperature to form a homogeneous micro-emulsion. After 10 minutes of equilibration, aqueous ammonia NH_4OH (35 µL) was introduced in the micro-emulsion as the catalyst in the synthesis of the silica shell. After another 10 minutes of equilibration, TEOS (60 µL) was added. After a 24 hours aging period, the nanoparticles were centrifuged (8000 g for 15 minutes) and washed several times with ethanol to remove unreacted and untrapped chemical species. Ultra-sonification (35 Hz) was used in order to disperse nanoparticles aggregated into the washing solvent and to increase the desorption rate of surfactant from the surface of the synthesized nanoparticles. Then the sample is subjected to dialysis against deionised water for a week.

2.2.3. *Synthesis of SiO_2 100 nm.* Silica nanoparticles were synthesized using a reverse micro-emulsion method, as described by Tan et al. in the literature [7]. Consequently a quaternary micro-emulsion consisted in mixing Triton X-100 (12.6 mL, 13.5 g, d = 1.07), 1-hexanol (12.3 mL), and cyclohexane (57 mL) under a vigorous stirring at room temperature, followed by additions of a concentrated aqueous solution of TMPyP dye in water (600 µL at 0.1M), water (1.2 mL), aqueous ammonia NH_4OH (725 µL at 25%) and, TEOS (725 µL) in that order. The mixture was allowed to stir for 24 h at room temperature and a subsequent addition of ethanol (100 mL) disrupted the inverse micelles. Particles were recovered by centrifugation (8000 g for 15 minutes) and washed thoroughly three times with ethanol and one time with water. Ultrasonification (35 Hz) was used in order to disperse nanoparticles aggregated into the washing solvent and to increase the desorption rate of surfactant from the surface of the synthesized nanoparticles. Then the sample is subjected to dialysis in deionised water for a week

2.2.4. *Transmission Electron Microscopy.* The morphologies and sizes of dye-doped silica nanoparticles were characterized using a transmission electron microscope (JEOL 2000 FX). The sample for transmission electron microscopy (TEM) were prepared by plunging a 200 mesh carbon-coated copper grid, 30-50 nm thickness, (Euromedex, France) in the desired nanoparticle-containing aqueous solution just after dispersion by ultra-sonification. Further to the evaporation of the water, the particles were observed at an operating voltage of 200 kV. Once the samples were imaged, TEM micrographs of dye-doped silica nanoparticles were converted to digitize images using imaging software (IMIX, PGT). Furthermore elemental analysis of the samples could be performed by Energy Dispersion RX Spectroscopy (EDS).

2.2.5. *Particle Sizing.* The hydrodynamic diameter and dispersivity of the silica nanoparticles were determined by Dynamic Light Scattering (DLS) technique using a Zetasizer Nano ZS from Malvern Instruments. The light scattering measurements were performed using a 633 nm red laser in a back-scattering geometry ($\theta = 180°$). The particle size was analyzed using a dilute suspension of particles in deionized (or ultrapure) water and in culture media.

2.2.6. *Fluorescence measurements.* All fluorescence measurements were performed at room temperature on a steady-state FS920 spectrofluorimeter (Edinburgh Instruments, UK) with a high spectral resolution (signal to noise ratio > 6000:1), using water as the solvent, and a 1 cm cell, the latter oriented at to the direction of the excitation light beam. The spectrofluorimeter covers the wavelength range from 200 to 1670 nm using two detectors: a photomultiplier R928 for UV-Vis scans (up to 870 nm) and a solid InGas TE G8605-23 detector for IR scans. The excitation source is a continuous Xenon Arc lamp (450 W) coupled to two Czerny-Turner DMX300X 1800 tr/mn monochromators, one for UV excitation (focal length 300 mm) and one for visible wavelength (focal length 500 mm). Fluorescence intensity values were integrated over the wavelength region specified. Data were recorded in a comparative manner, *ca.* using the same aperture of slits.

2.3. Cytotoxicity
Caki-1 and Hek293 human cells were exposed to SiO_2-25 and 100 nm at concentrations of 0; 0.1; 1; 10; 50 and 150 µg/ mL for 24 h at 37°C in 96-well micro-plates. Cytotoxicity was evaluated according to the Borenfreund protocol [8] for Neutral Red protocol, with a LDH assay (Roche) according to the Decker protocol [9], for LDH release with a XTT assay (Sigma) according to the Scuderio protocol [10], with a MTT assay (Sigma) according to the Alley protocol [11], and with a SRB assay (Sigma) according to the Skehan protocol [12].

2.4. Flow Cytometry
After 24 h treatments, cells were washed and trypsinised for 5 minutes. Trypsine was inactivated by complete DMEM, cells were centrifuged for 5 minutes at 300 g and then resuspended in 500 µL of DMEM with serum in flow cytometry compatible tubes (BD 352058). Multi-parametric analyses were performed on BD Facscalibur using FlowJo 7.5.5 software. A first analysis was realized on size/granulometry to collect living and dead cells and to remove fragmented cells. Rhodamine-doped SiO_2-25 nm nanoparticles events are then collected on FL2 and TMPyP-doped SiO_2-100 nm on FL3, respectively corresponding to both optimal fluorescence emission. FL2 and FL3 signals are collected after excitation by Argon laser at 488 nm, between 564-606 nm and superior to 670 nm, respectively. To-Pro3 (Molecular Probes, Invitrogen) signal was collected on FL4 and was used for the analysis of cell viability since this dye requires a 635 nm red laser, which excitation is compatible without interference with equipment and nanoparticles detection. FL4 is collected between 656 and 667 nm after He-Ne laser excitation at 636 nm. The results are reported as the mean distribution of cell fluorescence, obtained on 3 replicas with at least 20 000 events per replica. Measurements are given on a bi-parametric representation divided in quarters and are associated to four different cell populations: living cells without nanoparticles (To-Pro3/Nanoparticles negative cells), living cells with nanoparticles (To-Pro3 negative/Nanoparticles positive cells), dead cells with nanoparticles

(ToPro3 positive/Nanoparticles positive cells) and dead cells without nanoparticles (ToPro3/Nanoparticles positive cells).

2.5. Genotoxicity

2.5.1. *Phosphorylated-γ-H2Ax-foci.* Thirty thousand cells were seeded on Lab-Tek™ II Chamber Slide™ 8 wells (Nunc) 24 hours before exposure and cells were treated for 24 hours with nanoparticles at doses of 10 and 50 μg/mL. After treatment, cells were washed and fixed for 15 minutes with PFA 4%, washed twice with 200 μL of 1X PBS and then permeabilized for 10 minutes at room temperature in [1X PBS and Triton 0.1%]. Due to the number of different dyes used in this method, we paid attention in the choice of dyes for DNA, actin and γ-H2Ax foci labelling to avoid any interference. As SiO_2 nanoparticles 25 and 100 nm are respectively labelled with Rhodamine B (λ_{ex}: 557 nm, λ_{em}: 577 nm) and TMPyP (λ_{ex}: 420 nm, λ_{em}: 660 nm), we labelled nuclei with Hoechst 33342 (λ_{ex}: 345 nm, λ_{em}: 480 nm), γ-H2Ax with Alexa Fluor® 488 (λ_{ex}: 495 nm, λ_{em}: 519 nm. Cells were blocked in [1X PBS, 0.025%, 10% of goat serum (Jackson ImmunoResearch)] for one hour at room temperature, then incubated for 75 minutes at room temperature with 1:500 of monoclonal γ-H2Ax antibody (Anti-phospho-Histone H2A.X (Ser139), clone JBW301, 05-636, Upstate Millipore), washed with 3×300 μL of [1X PBS, Triton 0.025%] and then incubated at room temperature for 45 min with 1:500 dilution of Alexa Fluor® 488 goat anti-mouse IgG (H+L) (Molecular Probes, InVitrogen) as a secondary antibody. Samples were washed with 3×300 μL of [1X PBS, Triton 0.025%] and cells were incubated at 37°C for 20 min with 200 μL of [1X PBS and Hoechst 33342 at 0.1 μg/mL (Molecular Probes, InVitrogen)]. Finally, Lab-Tek™ Chamber Slide™ 8 wells were washed with 3×300 μL of [1X PBS, Triton 0.025%] before mounting in ProLong® Gold antifade reagent (Molecular Probes, InVitrogen) in order to proceed for confocal microscopy visualization.

2.5.2. *Confocal microscopy.* Fixed and labeled cells were photographed with a 40x PlanApo under a fluorescence confocal microscope (Leica TCS SP2, Wetzlar, Germany) equipped with lasers at 364 nm (Hoechst and TMPyP labelled SiO_2-100 nm nanoparticles), 488 nm (Alexa fluor 488) and 543 nm (Rhodamine B labelled SiO_2-25 nm). Hoechst signal was collected between 410-450 nm, γ-H2Ax foci signal between 500-540 nm while SiO_2-25 nm and 100 nm signals were collected between 540-570 nm and 670-725 nm, respectively. Each analysis was made on at least 100 cells and at least three images of each condition were analyzed. Confocal microscopy optical slice sections of 8 to 20 μm were made from the luminal to the basal pole of the cells, each acquisition containing nine stacks. Cell Profiler software [13] was used for the detection of foci and scoring in Alexa Fluor images.

2.5.3. *Statistical analysis.* To test whether the basal number of γ-H2Ax-foci observed in control cells was significantly different from that observed in nanoparticle exposed-cells, a Wilcoxon rank test based on at least 100 observations for each condition was performed.

3. Results

Among the five cytotoxicity assays which were used in this study, no interferences between the reading wavelengths and the intrinsic properties of both nanoparticles were observed. Even though Caki-1 and Hek293 human cell lines derivate from the same localization (proximal tubule), their responses after of SiO_2-nanoparticles exposure are different. The cytoxicity study assessed on both cell lines has shown that Caki-1 cells are more sensitive than Hek293 cells after a 24 hours of exposure to SiO_2-25 nm nanoparticles (figure 1).

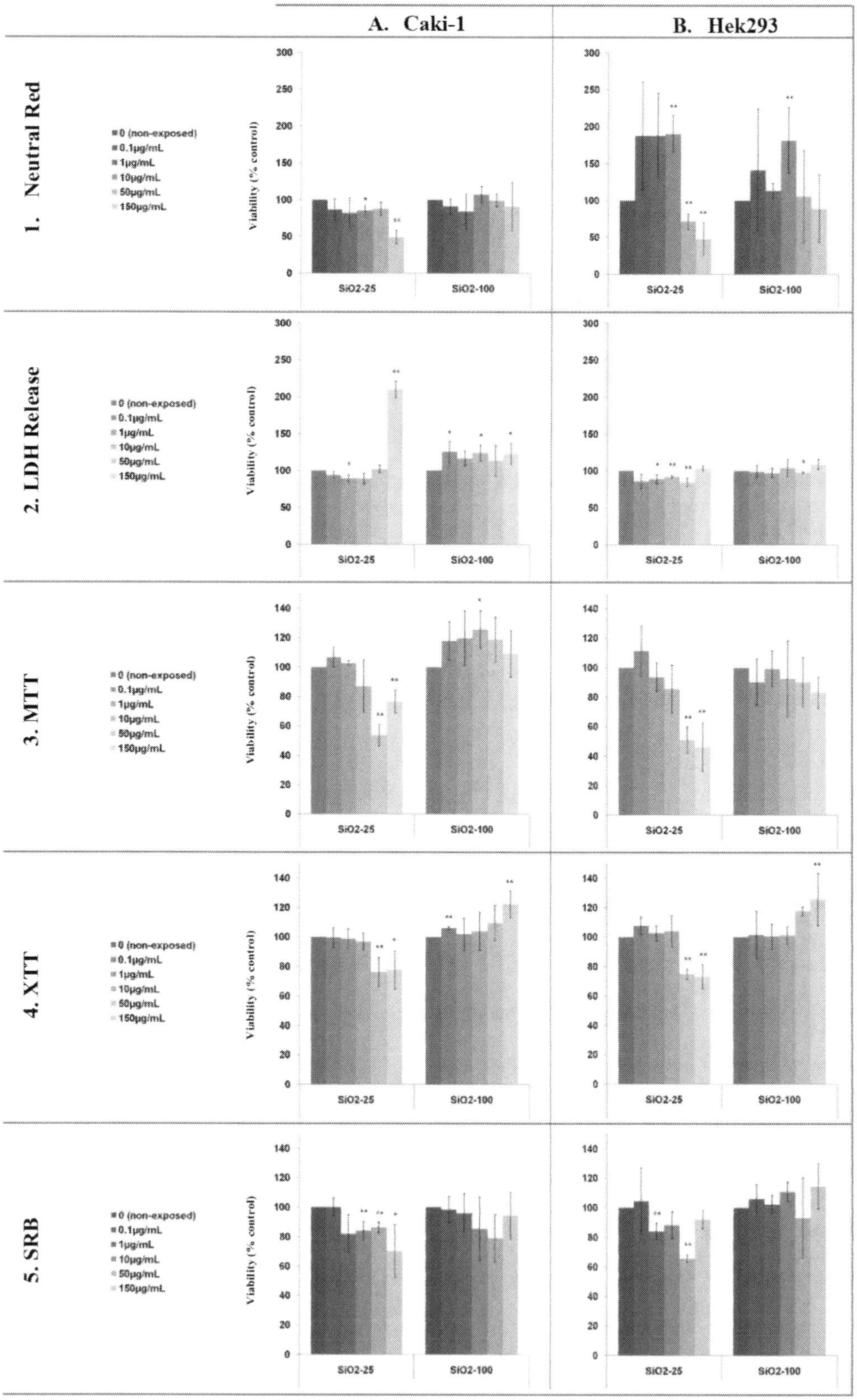

Figure 1. Cytotoxicity of SiO_2-25 nm and SiO_2-100 nm nanoparticles on Caki-1 and Hek293 human cell lines exposed for 24 hours: Neutral Red (1), LDH release (2), MTT (3), XTT (4) and SRB (5).

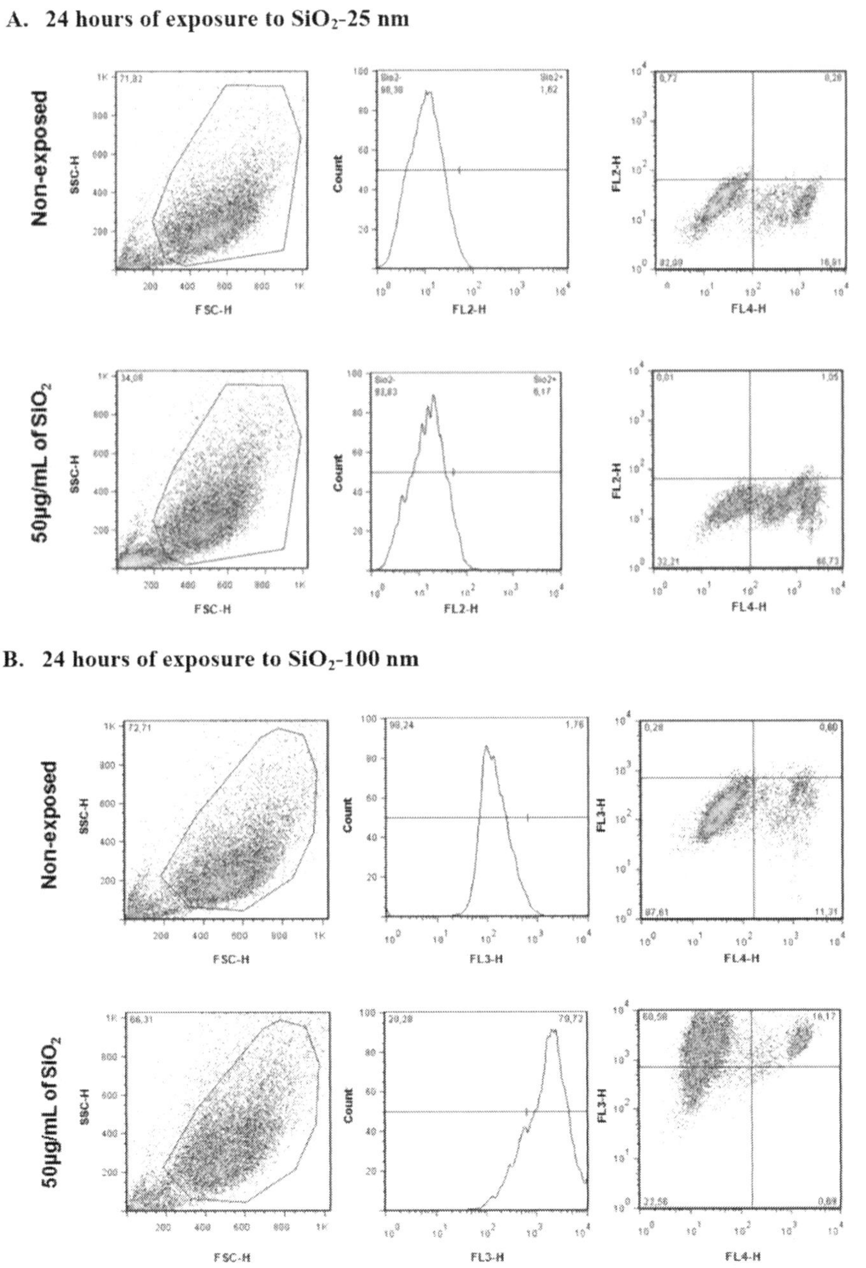

Figure 2. Flow cytometry analysis of Caki-1 control cells and cells exposed for 24 hours to SiO_2-25 nm (A) and SiO_2-100 nm (B) nanoparticles. On each series of three images, the first one correspond to the FSC vs SSC signal, the second to FL2 vs number of events for SiO_2-25 nm or FL3 vs number of events for SiO_2-100 nm, and the last one correspond to FL4 vs FL2 for SiO_2-25 nm or FL4 vs FL3 for SiO_2-100 nm.

A. 24 hours of exposure to SiO_2-25 nm

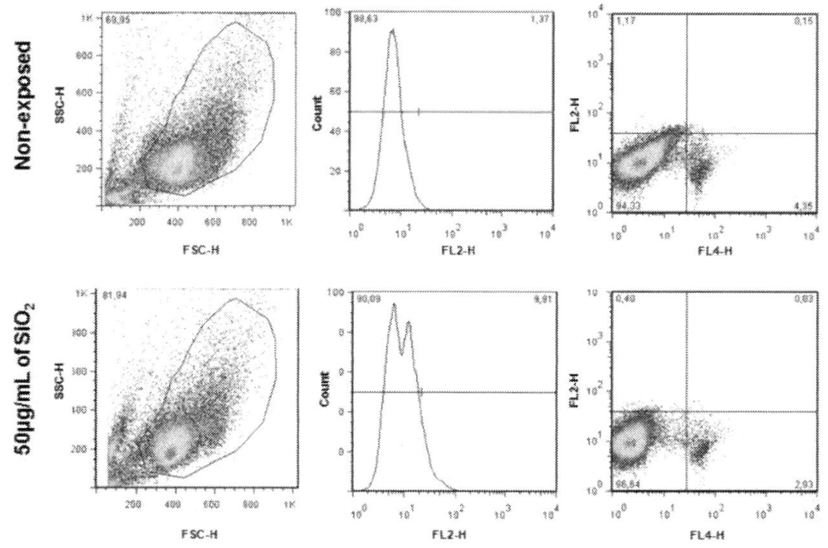

B. 24 hours of exposure to SiO_2-100 nm

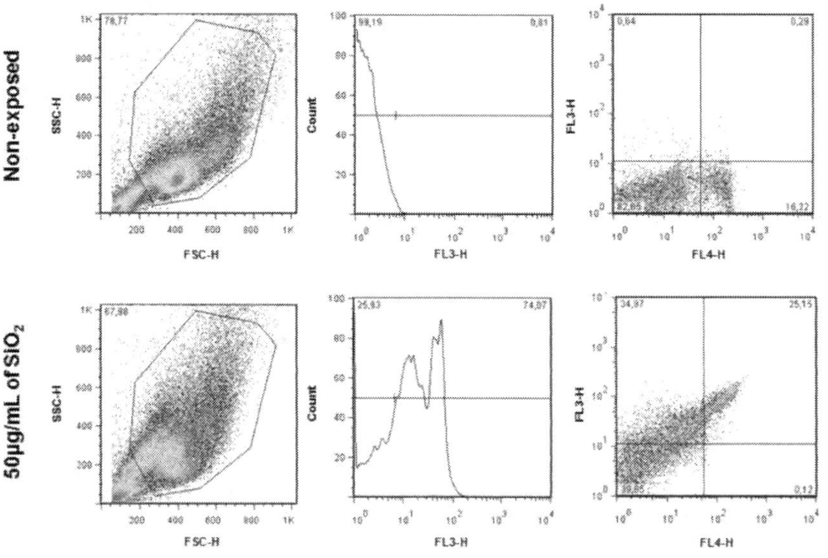

Figure 3. Flow cytometry analysis of Hek293 control cells and cells exposed for 24 hours to SiO_2-25 nm (A) and SiO_2-100 nm (B) nanoparticles. On each series of three images, the first one correspond to the FSC vs SSC signal, the second to FL2 vs number of events for SiO_2-25 nm or FL3 vs number of events for SiO_2-100 nm, and the last one correspond to FL4 vs FL2 for SiO_2-25 nm or FL4 vs FL3 for SiO_2-100nm.

Moreover this result is confirmed with the five other cytotoxicity assays : LDH activity (reported by LDH release) ; the mitochondrial activity (reported by MTT and XTT) and the metabolic activity (reported by SRB) (figure 1). On the other hand, similar experiments done after a 24 hours of exposure to SiO_2-100 nm have not shown a significant increase of cytotoxicity (figure 1) when comparing the two cell lines. This cytotoxicity study was extended by flow cytometry analysis, which allows to distinguish dead cells and living cells. Moreover, flow cytometry is a powerful and very sensitive approach to visualize incorporation of traceable nano-objects. Recent studies [14-16] have suggested that flow cytometry is of interest with regard to incorporation of nanoparticles of environmental particles and with regards to light scattering analysis. We propose an analysis of the nanoparticle incorporation, not based on light scattering but on the direct fluorescence emitted by nanoparticles. Indeed flow cytometry allows bi-parametric analysis using multi-channel detection which permits coupling the viability data to the presence of nanoparticles. Based on that principle, flow cytometry was performed upon Caki-1 and Hek293 human cells exposed to SiO_2-25 and 100 nm nanoparticles using a bi-parmetric analysis: specific fluorochrome nanoparticles detection (FL2 channel for SiO_2-25 nm and FL3 channel for SiO_2-100 nm) and mortality evaluation (FL4 channel with ToPro-3). Figure 2 shows that Caki-1 cells exposed for 24 hours to SiO_2-25 or 100 nm do not present the same profile. Caki-1 cells mostly die without detectable intracellular nanoparticles at 24 hours after exposure to 50 µg/mL of SiO_2-25 nm (figure 2.A), while at the same time, Hek293 cells are mostly living with detectable intracellular nanoparticles after an exposure to 50 µg/mL of SiO_2-100 nm (figure 3.B). These kinetics differencies of nanoparticles penetration are depicted on figure 2, which reports an absence of gap in the FL2 channel for Caki-1 cells exposed to 50 µg/mL of SiO_2-25 nm nanoparticles compared to the signal obtained in non-exposed cells (figure 2.A. and figure 3.A.). On the other hand, a high gap in the FL3 channel was observed for the two cell lines exposed to 50 µg/mL of SiO_2-100 nm compared to the signals obtained in non-exposed cells (figure 2.B. and figure 3.B.). A second bi-parametric analysis, which split up the signal into FL2 vs FL4 (for SiO_2-25 nm) and into FL3 vs FL4 (for SiO_2-100 nm), allowed us *i)* to directly correlate the absence of gap on the FL2 channel compared to control cells to the poor level of SiO_2-25 nm incorporation and to the high level of cell mortality, especially for Caki-1 cells *ii)* to directly correlate the high gap on the FL3 channel compared to control cells to the high level of SiO_2-100 nm incorporation in exposed cells, especially for Hek293 (figure 2.B and Figure 3.B.).

In order to investigate the genotoxicity of these two nanoparticles, we have adapted a second method based on phosphorylated γ-H2Ax-foci detection by confocal microscopy. The two tested concentrations for both nanoparticles and cell lines show a highly significant difference between the numbers of γ-H2Ax-foci observed in controls cells compared to those observed in exposed-cells (figure 4; $p < 0.001$ for all conditions excepted for Hek293 exposed to 10 µg/mL of SiO_2-100 nm for which $p < 0.05$). Furthermore, the two human cell lines used in this study, do not present the same sensitivity in terms of DNA breaks : Caki-1 cell line seems to be once again more sensitive for an exposure to SiO_2-nanoparticles (higher Med values observed in Caki-1 cells) than Hek293 cells for the two concentrations and for both sizes of nanoparticles. Contrary to the cytotoxicity data (where no significant differences were observed between the two cells lines for SiO_2-100 nm nanoparticles (Figure 4), the number of γ-H2Ax-foci indicates that the sizes of nanoparticles are more genotoxic for Caki-1 cells than for Hek293 cells (figure 4). Moreover, the data obtained for Caki-1 exposed to SiO_2-25 nm nanoparticles seem to suggest a dose response (Med = 9 foci per nuclei and Med = 18, for 10 µg/mL and 50 µg/mL, respectively). To confirm this point, it could be interesting to do same experiments using larger range of concentrations.

Journal of Physics: Conference Series **304** (2011) 012080 doi:10.1088/1742-6596/304/1/012080

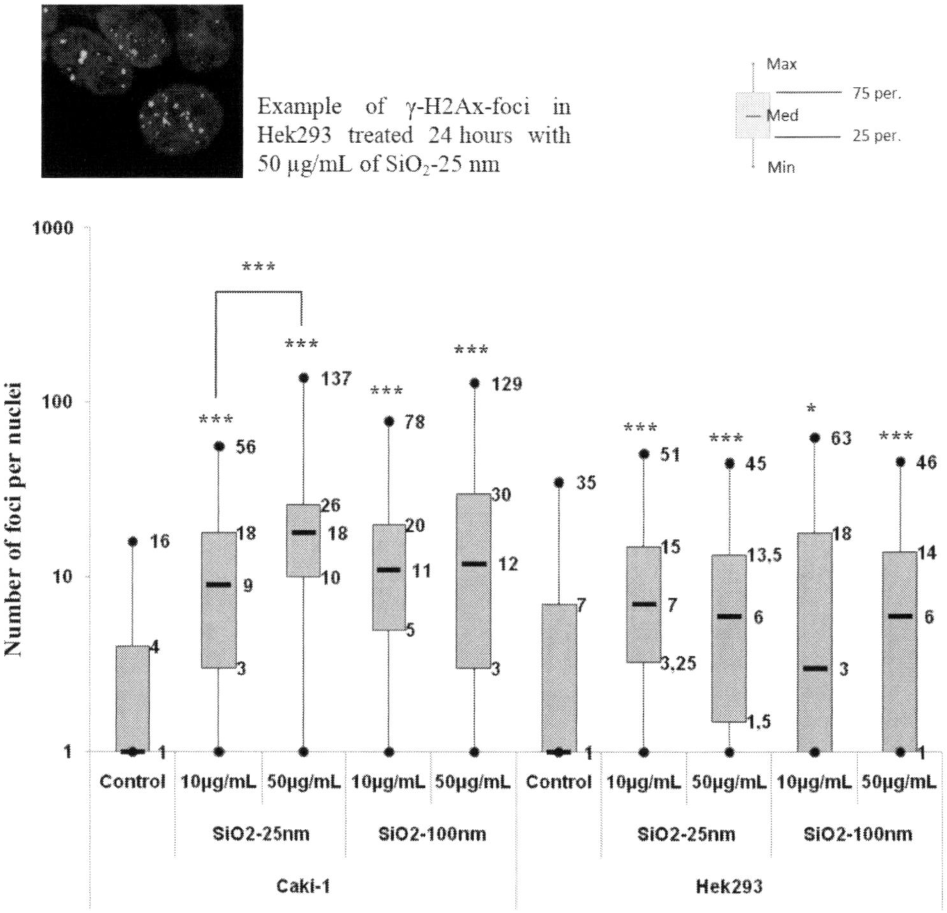

Figure 4. Number of H2Ax-foci per nuclei in Caki-1 and Hek293 cells exposed for 24 hours to SiO_2-25 nm and 100 nm nanoparticles at 10 µg/mL and 50 µg/mL compared to the control. (Statistical analysis: Willcoxon rank test: *: $p< 0.05$; ***: $p < 0.001$; Max: maximum ; Min: minimum ; Med: median ; 75 per: 75th percentile ; 25 per: 25th percentile).

4. Conclusions

Overall, no interference was detected between labeled SiO_2 nanoparticles and conventional toxicological assays used in our study. Furthermore the two adapted confocal miscroscopy and flow cytometry methods, used to track intracellular nanoparticles together with the analysis of cell survival and genotoxicity, are very useful techniques.

Our results indicated that SiO_2-25 nm nanoparticles are more cytotoxic and genotoxic than SiO_2-100 nm on both tested cell lines. On the other hand, Caki-1 cells are more sensitive to both nanoparticles than Hek293 cells, which are able to survive with intracellular nanoparticles.

More generally, these approaches could be routinely applied to others cell lines and to all others nanoparticles in mass dye labeled during synthesis.

Acknowledgements
This work was supported by the NanoSciences CEA-Transverse program and C'Nano Ile-de-France.

References

[1] Vega-Villa KR, Takemoto JK, Yáñez JA, Remsberg CM, Forrest ML, Davies NM. (2008) Clinical toxicities of nanocarrier systems. *Adv Drug Deliv Rev.* 60(8): 929-38
[2] Sung, J.C., B.L. Pulliam, D.A. Edwards. (2007) Nanoparticles for drug delivery to the lungs. *Trends Biotechnol.* 25(12): 563-70
[3] Kahru A, Dubourguier HC. (2010) From ecotoxicology to nanoecotoxicology. *Toxicology* 269(2-3): 105-19
[4] Fadeel B, Garcia-Bennett AE. (2010) Better safe than sorry: Understanding the toxicological properties of inorganic nanoparticles manufactured for biomedical applications. *Adv Drug Deliv Rev.* 62(3): 362-74
[5] Laaksonen T, Santos H, Vihola H, Salonen J, Riikonen J, Heikkilä T, Peltonen L, Kumar N, Murzin DY, Lehto VP, Hirvonen J. (2007) Failure of MTT as a toxicity testing agent for mesoporous silicon microparticles. *Chem Res Toxicol.* 20(12): 1913-18
[6] Viger ML, Live LS, Therrien OD, Boudreau D. (2008) Reduction of self-quenching in fluorescent silica-coated silver nanoparticles. *Plasmonics* 3(1): 33-40
[7] Bagwe RP, Yang C, Hilliard LR, Tan W. (2004) Optimization of Dye-Doped Silica Nanoparticles Prepared Using a Reverse Microemulsion Method. *Langmuir 20*: 8336-42
[8] Borenfreund E, Puerner JA (1985) Toxicity determined in vitro by morphological alterations and neutral red absorption. *Toxicol Lett.* 24(2-3): 119-24
[9] Decker T, Lohmann-Matthes ML. (1988) A quick and simple method for the quantitation of lactate dehydrogenase release in measurements of cellular cytotoxicity and tumor necrosis factor (TNF) activity. *J Immunol Methods* 115(1): 61-69
[10] Scudiero DA, Shoemaker RH, Paull KD, Monks A, Tierney S, Nofziger TH, Currens, Seniff D, Boyd MR. (1988) Evaluation of a soluble tetrazolium/formazan assay for cell growth and drug sensitivity in culture using human and other tumor cell lines, *Cancer Res.* 48 (1988) 4827-33
[11] Alley MC, Scudiere DA, Monks A, Czerwinski M, Shoemaker RH, Boyd MR. (1986) Validation of an automated microculture tetrazolium assay (MTA) to assess growth and drug sensitivity of human tumor cell lines, *Proc. Am. Assoc. Cancer Res.* 27: 389
[12] Skehan P, Storeng R, Scudiero D, Monks A, McMahon J, Vistica D, Warren JT, Bokesch H, Kenney S, Boyd MR. (1990) New colorimetric cytotoxicity assay for anticancer-drug screening. *J. Natl. Cancer Inst.* 82: 1107-12
[13] Carpenter AE, Jones TR, Lamprecht MR, Clarke C, Kang IH, Friman O, Guertin DA, Chang JH, Lindquist RA, Moffat J, Golland P, Sabatini DM. (2006) CellProfiler: image analysis software for identifying and quantifying cell phenotypes. *Genome Biology* 7: R100
[14] Zucker RM, Massaro EJ, Sanders KM, Degn LL, Boyes WK. (2010) Detection of TiO2 nanoparticles in cells by flow cytometry. *Cytometry* 77(7): 677-85
[15] Stringer B, Imrich A, Kobzik L. (1995) Flow cytometric assay of lung macrophage uptake of environmental particulates. *Cytometry* 20(1): 23-32
[16] Suzuki H, Toyooka T, Ibuki Y. (2007) Simple and easy method to evaluate uptake potential of nanoparticles in mammalian cells using a flow cytometric light scatter analysis. *Environ Sci Technol.* 41(8): 3018-24.

NANODEVICE: Novel Concepts, Methods, and Technologies for the Production of Portable, Easy-to-use Devices for the Measurement and Analysis of Airborne Engineered Nanoparticles in Workplace Air

Sari Sirviö and Kai Savolainen

Finnish Institute of Occupational Health, Topeliuksenkatu 41 a A, 00250, Helsinki, Finland

E-mail: sari.sirvio@ttl.fi

Abstract. NANODEVICE is a research project funded by the European Commission in the context of the 7th Framework Programme. The duration is 48 months starting 1st of April 2009. Due to their unique properties, engineered nanoparticles (ENP) are now used for a myriad of novel applications, and have a great economic and technological importance. However, some of these properties, especially their surface reactivity, have raised health concerns due to their potential health effects. There is currently a shortage of field-worthy, cost-effective ways - especially in real time - for reliable assessment of exposure levels to ENP in workplace air. NANODEVICE will provide new information on the physico-chemical properties of engineered nanoparticles (ENP) and information about their toxicology. The main emphasis of the project is in the development of novel measuring devices to assess the exposure to ENP´s from workplace air. The purpose of the project is also to promote the safe use of ENP through guidance, standards and education, implementing of safety objectives in ENP production and handling, and promotion of safety related collaborations through an international nanosafety forum. The main project goal is to develop innovative concepts and reliable methods for characterizing ENP in workplace air with novel, portable and easy-to-use devices suitable for workplaces.

1. Background

One of the important challenges for assuring a safe production of engineered nanomaterials (ENM) and the promotion of the use of nanotechnologies is reliable assessment of exposure of workers to these materials in the occupational setting, i.e. during the production, storage, handling and preparation of nanotechnology-based products, see Schulte et al. [1]; Woskie et al. [2]. This information supports the assessment of potential hazards and risks of ENM, and subsequent management of possible risks associated with ENM [3,4].

2. Overall project goals

This four year project aims at developing several portable easy-to-use devices for the monitoring of the levels of ENM workplace aerosols with an emphasis on on-line monitoring of the exposure, and to effectively reach all the project goals through efficient management and coordination of the project activities.

The key goals of the project include 1) characterization of chosen ENM such as carbon nanotubes (CNT) and metal oxides; 2) understanding of the association of ENM characteristics and their potential toxicity; 3) generation of realistic exposure scenarios for workplace exposure to ENM; 4) development of monitoring devices and measurement principles to capture the features and concentrations of ENM in aerosols in workplaces; 5) to develop testing and calibration processes to optimize the functioning of the devices in the measurement of exposure; and 6) to assure a major impact of the project and its results on the safe handling of ENM and the safety of nanotechnologies. Some of these goals were identified by [5] but the goals suggested in that paper have not been achieved so far.

The impact in the promoting of safe use of ENM and safe nanotechnologies will take place through training and dissemination information on the benefits of safe use of ENM handling and production by providing education and training, promoting international standardization of processes related to ENM, publishing a nanosafety handbook, organizing an international congress and other international events, and promoting international networking in the field of safe nanotechnologies. The impact will be focused on different user groups of ENM including the industry, social partners, workplaces, regulators, the scientific community, and the public at large. These groups will be reached by establishing a stakeholder group that consists of the representatives of these groups. The essence of the project and its goals are presented in Figure 1.

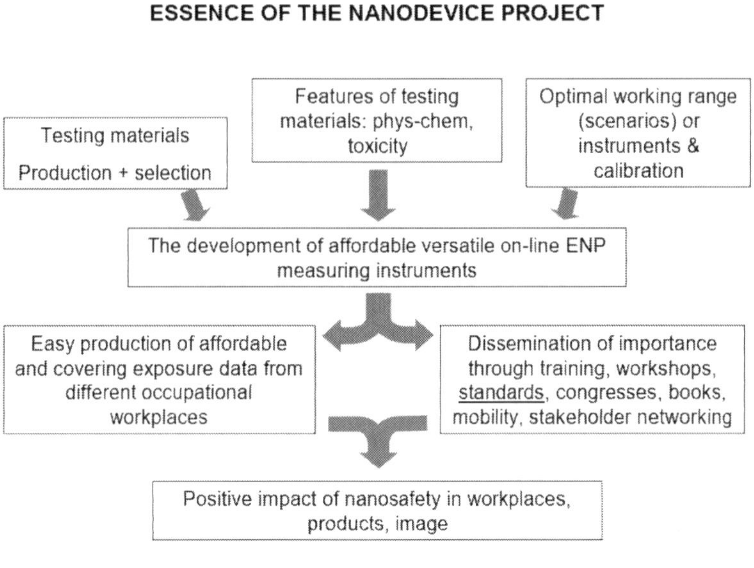

Figure 1. Essence of the NANODEVICE project.

The main project elements include 1) identification of reference and other testing materials for devices and toxicity; 2) identification of association between physical chemical metrics of the materials and their toxicity and mechanisms of toxicity; 3) identification of the range of concentrations in which the functioning of the devices to be developed should be optimal in workplaces to capture the critical range of concentrations of aerosols of engineered nanomaterials in workplaces; 4) development of affordable easy-to-use instruments to carry out exposure assessment to engineered nanomaterial aerosols in workplaces; 5) laboratory and field calibration of the testing of the instruments to assure their reliability; and 6) promotion of nanosafety in workplaces.

3. Work carried out during the course of the project to reach its goals

The work has been divided into several work packages (21) collected into six larger subprojects (SP) on 1) characterization; 2) association between ENM features and toxicity; 3) exposure scenarios in industrial processes associated with nanotechnologies; 4) development of instruments; 5) calibration and testing of the novel instruments; and 6) assuring impact of the results of the project.

Management of the project aims at assuring effective and fluent flow of the work within the project and appropriate interactions between work packages and subprojects, and leadership, of the undertaking.

In the characterization subproject, work has been carried out to identify and characterize reference and model ENM aerosols and to build a nano-metal oxide reactor. Emphasis has also been on the production of well characterized carbon nanotubes (CNT). A part of the project has focused on exploring on the characteristics of different ENM by using visual technologies such as electron microscopy, and characterizing of ENM through using non-imaging techniques to delineate physical-chemical properties of ENM, see Borm et al. [6]. Association between ENM characteristics and their association with toxicity has been explored with a large variety of methods. This subproject has also provided a list of test aerosols to be used for the performance testing [7] of the instruments to be developed within the project.

One part of the project has been generating performance requirements to set specifications of the easy-to-use devices and portable devices to assess ENM-based evaluation of the limitations of presently available static devices. A part of this goal is to evaluate existing and future work place exposure scenarios. Furthermore the subproject has created sampling strategies for appropriate assessment of exposure to ENM. A key-part of the project has focused on the development of novel on-line ENM measurement instruments with the capability to capture essential ENM features, as well as to produce innovative and new measurement principles aiming at discovering novel technologies capable of revealing essential ENM features with an association of toxic properties of given ENM, see Maynard et al. [5]; Murashov [3]. Several approaches have been chosen and work has been carried out to develop such instruments including the following principles: a) on-line, inexpensive, portable, on-line surface monitor; b) size discriminating number and surface area aerosol monitor and sampler; c) wide-range size-resolving aerosol monitoring and sampling system based on a personal size-resolved ENM selective sampler and an on-line monitor; d) high-sensitivity optical sensor for single particle number and mass allowing particle measurements at a 50-80 nm size range; e) high-sensitivity MEMS-based sensor for single-particle number and mass down to 10 nm; f) catalytic and surface-chemical aerosol monitor based on novel device concepts for new metrics proving information in a highly sensitive fashion and near real-time monitoring of catalytic activity and surface chemical groups of ENM; and g) nanofiber monitor that is a robust, portable, easy-to-use quasi real-time device to monitor airborne fibers in workplace environments [2,4,8].

A part of the project focusing on testing and calibration of the instruments has carried out work to establish the facilities and processes for testing and calibration of the instruments to be developed within the part of the project aiming at development of monitoring instruments both in laboratory conditions and in the field in work places A part of the project has aimed at maximizing the dissemination of the results of the project to a wide variety of target audiences, and finding ways to assure an impact of the project to promote safety of ENM and nanotechnologies, see Kahru and Savolainen [9] . Hence, tailored training courses and good practice guidelines have been prepared. A data-base supporting risk assessment of ENM based on the project results is under development by several partners and in collaboration with other EU-funded projects. Also information leaflets, posters and a project newsletter have been developed. Establishment of a strategy for the European and international standardization of all project results is under intensive preparation with different EU-funded projects and European Union stakeholders in this area. An advisory forum - 'Annual Forum for Nanosafety' - with the representation of leading experts and most qualified organizations from five continents has been supporting the project. Organizing an international congress on safety of engineered ENM and nanotechnologies is under preparation, and decision has been made to publish a state-of-the-art nanosafety handbook based on the project. An ongoing internal mobility program has been established, and increased the internal cohesion of the project.

4. Main results achieved so far

4.1. Generation of well-characterized test aerosols of nanoparticles

4.1.1. Characterization: Reference materials have been characterized and the production of tailored ENM materials initiated. Commercially available ENM have been purchased and characterized. Iron catalysts particles have been characterized and preliminary tests have been performed of single walled carbon nanotube (SWCNT) synthesis with the concept reactor. A portable metal oxide particle generator, carrying out test measurements with TiO_2 and SiO_2 has been built; see Rossi et al. [10].

4.2. Association between physical, chemical and toxicological properties of ENP

Association between ENM properties and health effects: Methods for generating, analyzing, characterizing and reporting the electron microscopy analysis of nano-objects and agglomerates is in progress. Methods for dispersing powders for in-vitro testing have been circulated and used. Characterization included morphological electron microscopic analysis and EDS analysis on composition. Nanomaterials induced pro-inflammatory effects in the target cells with a marked potential to differentiate between toxic and less toxic materials. All ENM showed cytotoxicity which was relatively minor. TiO_2 induced a dose-dependent increase in DNA damage [10].

4.3. Processes in ENP production, use and exposure to ENP

For industrial processes, a survey was made and sent to achieve "Requirements for the newly developed devices". The questionnaire has been summarized and documented. For the description/evaluation of existing and future workplace exposure scenarios with an emphasis on exposure conditions, concurrent exposures, a preliminary analysis has been made using data from the EU FP6-funded project "NANOSH". Knowledge gaps were identified, with focus on the conceptual model. It was concluded that the input value of parameters for (both mathematical and CFD) modeling should be as realistic as possible. Furthermore, more data were needed with respect to 'background' aerosols, e.g. ranges of size distributions, particle number concentrations [2,11].

4.4. Device development

4.4.1. For Portable active surface area aerosol monitor, it was decided that the corona discharge is a way to charge the particles in the developed portable aerosol monitor. A prototype of a small corona charger has been built and evaluated. The prototype was also modeled and the model was verified against the experimental results. During this period, new concepts for a high voltage power supply for ECT charger, an isolated power supply design and electrometer for current measurement have been developed. These concepts are currently being tested.

4.4.2. For Size-discriminating number and surface area aerosol monitor and sampler, a pre-separator has been developed that includes a virtual impactor and a cyclone. Optimal settings for the pre-separator were developed. The charging efficiency has been tested and found to be in very good agreement with literature data.

4.4.3. For wide-range size-resolving aerosol monitoring & sampling system, 1) several numerical models for particle deposition in tubes, on plates and on nets for different geometrical and material setups have been developed; 2) several numerical models for particle deposition in a cyclone as a function of cyclone geometry and operating conditions have been developed; 3) the first prototype of the size fractionating nanoparticle personal ENM selective sampler has been manufactured; 4) preliminary tests show that the performance is in the expected range of parameters; and 5) the first draft of operation instruction has been prepared.

4.4.4. For the High-sensitivity optical sensor for single-particle number and mass, an optimized laser source and a highly sensitive receiver has turned out to be the most promising approach to detect the scattered light originating from small particles in the 100 nm-regime.

4.4.5. For High-sensitivity MEMS-based sensor for single-particle number and mass, 1) algorithm for detection and weighing of several particles with resonant beams has been developed; 2) proof of sensor principle by measuring micro particles with cantilevers and strings has been obtained; 3) Fabrication process for micro beams with perfect clamping has been established; and 4) an analytical model for actuation of cantilevers made from arbitrary materials has been established.

4.4.6. For Catalytic and surface-chemical aerosol monitors to detect catalytically active ENM, 1) several model ENP materials and suitable test reactions have been selected. The proof of principle for a simultaneous determination of particle size and a fluorescence signal on single particles was accomplished on a laboratory setup. A functional relationship between the amounts of a fluorescent dye on the surface of micron sized test particles could be established. Further work is planned using a fluorescent marker molecule specific to redox-active sites.

4.4.7. For Nanofibre monitor, 1) a working setup of a AFM measurement station to measure deposited nanoparticles on surfaces situated in a Clean room ISO class 1 has been established; 2) a selection of suitable surfaces for the later usage in the fiber monitor has been made; 3) first successful measurements with nanoparticles on the surfaces have been carried out; and 4) a CNT generator to produce CNTs for the later usage in the development of the fiber monitor has been developed.

4.5. Calibration and testing of the functioning of new concepts, methods and pre-prototypes for measurement and analysis of airborne ENP

4.5.1. For calibration, testing, and background distinction, 1) a report on the adjustment of the nanoparticle calibration system for NANODEVICE which is in itself ready to use by developers; 2) a ready to use version of a primary discontinuous calibration tool for coagulation aerosols and an additional prototype for a continuously working similar system; and 3 the ready to use (for urban aerosol) Nano Test Facility, see Brouwer [11]; Woskie et al. [2].

4.5.2. For testing of the devices in the field 1) current existing instruments for measuring ultrafine and nanoparticles at workplaces were reviewed and checked for parallel application during the planned workplace comparisons; 2) Contact with the group of external technicians in Germany has been established and procedures how to distribute devices have been discussed. The draft of the questionnaire will be distributed among them to clarify the meaning of the questions; 3) a document on the sampling strategy has been drafted in a core group and discussed with external stakeholders in Germany, which also do measurements. The draft will be discussed within the consortium; and 4) it has been agreed that the partners concerned will deliver documents on the justification of their work on the pre-prototypes till summer 2010 [2,7].

4.6. SP6 Dissemination of the results and assuring impact of the NANODEVICE project
The overall goal of these activities is to fill the knowledge gaps that exits in our knowledge related to the safety of engineered nanomaterials, see e.g. Kahru and Savolainen [9]. Especially the role of dissemination of information and standardization will be emphasized in the exploitation of the results of the project [3,4,12].

Dissemination of the results and assuring impact Educational materials for nanosafety promotion and courses for different targeted audiences have been prepared. Brochures and posters on the project have been produced and newsletter is in a planning stage. Data-base to support risk assessment of ENM with other partners and other EU funded projects is underway. As to standardization, a scientifically categorized list (4 categories) of standards related to the NANODEVICE project outputs has been established, and a prioritized list of standards related to the NANODEVICE project outputs has been identified. Contacts to various relevant, national and international standardization committees have been established. Presentation on the standardization issues in general and on the status of work will take place during NANOSAFE Congress in Grenoble and during the Annual Safety Forum meeting in Berlin in March 2011. The advisory group 'Annual Forum for Nanosafety' with experts and organizations from five continents has been established. Preparations to organize an international congress on safety of engineered nanomaterials and nanotechnologies (SENN2012) are progressing well. The program committee is working actively, and the initial contents of the congress have been identified. Planning of a nanosafety handbook is ready, authors chosen and committed themselves. Currently, negotiations on the publications are ongoing with Springer. Human mobility program within NANODEVICE project has been initiated and actively ongoing.

5. The expected final results and their potential impact on safe nanotechnologies

The project results will have a marked societal impact especially on workplace safety in small and medium size enterprises and microenterprises through bringing into market affordable on-line measuring devices providing reliable exposure information on engineered nanomaterials. This will not only greatly increase our knowledge and understanding on the true workplace exposure to these materials. It will also support regulators and decision makers in making evidence based conclusions and actions on necessary measures to protect the workers from excessive exposure, especially the setting of occupational exposure limits for different types of engineered nanomaterials. One can expect remarkable savings to the industry through reliable exposure information at low or affordable expenses.

These outcomes of the project will also enable safe use of these materials, and promote safe production of engineered nanomaterials, and thereby increase the competitive edge of nanotechnologies through emphasizing safety aspects in all areas of nanotechnology applications.

The project has successfully moved toward its impact objectives as demonstrated by the successful progress of the technical and impact work. Stakeholder contacts will assure a marked further impact through making the project results more visible among ETUC and Business Europe partners, occupational safety and health and research community, among governments, and the Commission as well global organizations such as OECD, World Health Organization, ISO and CEN.

Acknowledgments

The authors wish to thank Ms. Leila Ahlström for the finalization of this manuscript. The NANODEVICE project has been funded by the European Union 7th Framework programme (grant agreement NMP4-LA-2009-211464). The views and opinions expressed in the paper do not necessarily reflect those of the European Commission.

References

[1] Schulte PA, Schubauer-Berigan MK, Mayweather C, Geraci CL, Zumwalde R and McKernan JL 2009 Issues in the development of epidemiologic studies of workers exposed to engineered nanoparticles J. Occup. Environ. Med. 51 323-35.

[2] Woskie SR, Bello D, Abbas Virji M and Stefaniak AB 2010 Understanding Workplace Processes and Factors that Determine Exposures to Engineered Nanomaterials Int. J. Occup. Environ. Health 16 365-77.

[3] Murashov V 2010 Human and Environmental Exposure Assessment for Nanomaterials: An Introduction to this Issue Int. J. Occup. Environ. Health 16 363-4.

[4] Savolainen K, Alenius H, Norppa H, Pylkkänen L, Tuomi T and Kasper G 2010 Risk Assessment of Engineered Nanomaterials and Nanotechnologies - A Review Toxicology 269 92-104.

[5] Maynard AD, Aitken RJ, Butz T, Colvin V, Donaldson K, Oberdörster G, Philbert MA, Ryan J, Seaton A, Stone V, Tinkle SS, Tran L, Walker NJ and Warheit DB 2006 Safe handling of nanotechnology Nature 444 267-9.

[6] Borm PJA, Robbins D, Haubold S, Kuhlbusch T, Fissan H, Donaldson K, Schins R, Stone V, Kreyling W, Lademann J, Krutmann J, Warheit D and Oberdörster E 2006 The potential risks of nanomaterials: a review carried out for ECETOC Part. Fibre Toxicol. 3 11.

[7] Seipenbusch M, Binder A and Kasper G 2008 Temporal evolution of nanoparticle aerosols in workplace exposure Ann. Occup. Hyg. 52 707-16.

[8] Kuhlbusch TAJ, Fissan H and Asbach C 2009 Nanotechnologies and Environmental Risks: Measurement Technologies and Strategies Nanomaterials: Risks and Benefits Ed I Linkov, J Steevens (Springer, Dordrecht) pp 233-43.

[9] Kahru A and Savolainen K 2010 Potential hazard of nanoparticles: From properties to biological and environmental effects Toxicology 269 89-91.

[10] Rossi EM, Pylkkänen L, Koivisto AJ, Vippola M, Jensen KA, Miettinen M, Sirola K, Nykäsenoja H, Karisola P, Stjernvall T, Vanhala E, Kiilunen M, Pasanen P, Mäkinen M, Hämeri K, Joutsensaari J, Tuomi T, Jokiniemi J, Wolff H, Savolainen K, Matikainen S and Alenius H 2009 Airway exposure to silica coated TiO2 nanoparticles induces pulmonary neutrophilia in mice Toxicol. Sci. 113 422-33.

[11] Brouwer D 2010 Exposure to manufactured nanoparticles in different workplaces Toxicology 269 120-7.

[12] Raivio T, Pessala P, Hjelt M, Mikkanen P and Kahelin H 2008 Nanosafety in Finland - a summary report Tekes Review 224/2008, Helsinki, Finland.

Capacity building in safe nanotechnologies

Markus Keller and Udo Gommel

Fraunhofer Institute for Manufacturing Engineering and Automation IPA; Ultraclean Technology and Micromanufacturing, Nobelstr 12, D-70569 Germany

E-mail: markus.keller@ipa.fraunhofer.de

Abstract. In all places where engineered Nanoparticles (ENPs) are produced, used or handled, adequate workplace safety precautions should be implemented due to the protection of workers and the surrounding environment. Any possible accidental release of ENPs should be evaluated. Thereby detected potential risks have to be eliminated as far as possible. An implemented reasonable safety culture in each ENP-related company will help to meet this challenge. Different infrastructures and workplace design can help to reduce the risk of an accidentally contact of the workers with ENPs: Transferable examples will be shown from the semiconductor and life-science Industry. These systems like clean rooms, glove boxes, fume cupboards, filter and suction systems and other restricted area barrier access systems (RABS) are mainly being developed to protect sensitive products, but they can also be used to protect working personnel. Clean environments regarding airborne particulate contaminations can be classified according to ISO 14644-1. A short insight into this ISO-classification will be given. But overall, a simple and reasonable workplace and workflow organization will reduce the risk of an accidental release of ENPs largely. This may lead to a therefore necessary adaption of existing workflow patterns. The workers have to get aware about the potential risks! This can be done with appropriate education materials, leaflets, posters and brochures. These are some of the later outcomes from the NanoDevice dissemination and handbook work package.

1. Workplace safety

In this article, the focus will be on the production and handling of airborne engineered nanoparticles (ENP). Looking at different exposure scenarios, several infrastructures and equipment can help to fulfill a necessary workplace safety to protect workers from accidentally contact with these potentially harmful substances. Starting from using gloves and safety goggles in nearly all laboratory settings from using different kind of breathing protections working with harmful aerosols up to the full set of clothing and breathing apparatus used by firefighters, a whole set of commercially available devices can be used for worker protection.

Another issue is the protection of the environment from any accidental release of harmful airborne substances. This can be achieved using air filters or air washers based in the air conditioning system of a laboratory setup. Also an useful laboratory layout design itself can help to reduce any risk of an accidental release of ENPs. Differential pressure from the environment with atmospheric pressure down to the laboratory settings or production plants with reduced pressures minimizes the accidental release of ENPs even if there is only a minor leak in the system. By constant monitoring the actual differential pressure, any fault in the system can for example lead to an instant emergency stop and/or the activation of other safety systems like for example additional fume extraction.

2. Existing examples of safety systems focused on workplace and product safety

Different already long established disciplines developed specific safety precautions to protect either the worker or the handled product. From each branch, some systems can be well used for the set-up of a laboratory setting or production plant with an appropriate workplace safety level regarding ENPs. Therefore, a short introduction will be given to each controlled environment from the different disciplines/industries.

2.1. Biological safety labs

Biological safety labs have to meet special criteria for each necessary safety level. These safety levels are set up to protect the worker and as well to protect environment and therefore human population from accidental release.

2.1.1. *Classification.* Biological labs are classified according to their biosafety level from S1 (lowest) to S4 (highest). The necessary level depends on the used organisms worked with. A biosafety level is the level of the biocontainment precautions required to isolate dangerous biological agents in an enclosed facility. In the United States, the Centers for Disease Control and Prevention (CDC) have specified these levels [1]. In the European Union, the same biosafety levels are defined in a directive [2].

Organisms generally regarded as safe (GRAS) like *Saccharomyces cerevisiae*, *Lactobacillus plantarum* or *Escherichia coli* can be handled in a S1 setting. S2-settings are suitable for work involving agents of moderate potential hazard to personnel and the environment, like Salmonella sp., Pseudomonas aeruginosa or HIV virus. S3-level is necessary for agents which may cause serious or potentially lethal disease after inhalation. It includes various bacteria, parasites and viruses that can cause severe to fatal disease in humans, but for which vaccines or other treatment exist, such as *Leishmania donovani*, *Mycobacterium tuberculosis*, *Bacillus anthracis* and *yellow fever virus*.

The safety level S4 is required for work with dangerous agents that pose a high individual risk of aerosol-transmitted laboratory infections, agents which cause severe to fatal disease in humans for which vaccines or other treatments are not available, such as Bolivian and Argentine hemorrhagic fevers, Marburg virus and Ebola virus.

2.1.1. *Necessary facilities and precautions for each safety level.* This incomplete list should give the reader only a very short overview about the main precautions. Some precautions can be easily adapted for a setting working with ENPs.

S2:

1. Laboratory personnel have specific training and are directed by scientists with advanced training
2. Access to the laboratory is limited when work is being conducted
3. Extreme precautions are taken with contaminated sharp items
4. Certain procedures in which infectious aerosols or splashes may be created are conducted in biological safety cabinets or other physical containment equipment.

S3: All S2 prerequisites plus

1. Double-door access zone and sealed penetrations
2. The filtered exhaust air from the laboratory room is discharged to the outdoors
3. The ventilation to the laboratory is balanced to provide directional airflow into the room
4. Access to the laboratory is restricted when work is in progress

S4: All S3 prerequisites plus

1. Self-contained oxygen supply is mandatory
2. The facility is either in a separate building or in a controlled area within a building, which is completely isolated from all other areas of the building

Journal of Physics: Conference Series **304** (2011) 012082 doi:10.1088/1742-6596/304/1/012082

3. Multiple airlocks are employed and are electronically secured to prevent both doors opening at the same time
4. All air and water going to and coming from a biosafety level 4 lab will undergo special decontamination procedures to eliminate the possibility of an accidental release

2.1.2. *Outcome.* For handling potentially hazardous ENPs, it will be useful to have restricted access to the production plant or lab, handle open processes in physical containment equipment or restricted area barrier access systems (RABS) and use pressure drops from the environment to the area with possible ENP exposure.

2.2. GMP controlled environments

Environments controlled under the Good Manufacturing Guidelines (GMP) for the production of sterile or aseptic products are regulated in the EU-GMP Annex 1 [3]. There, the different GMP cleanliness classes are defined according to their particulate and microbiological level airborne and on surfaces. Additionally, the guideline claims as follows: *«...in clean areas, all surfaces should be smooth, imperious and unbroken in order to minimize the shredding or accumulation of particles or microorganisms and to permit repeated application of cleaning agents and desinfectants where used...The manufacture of sterile products is subject to special requirements in order to minimize risks of microbiological contamination, and of particulate or pyrogen contamination...».* This means, that for all tribological elements, like joints, bearings and other material pairings, the material selection has to be carefully considered regarding their particulate emission during operation. Some material pairings tend to produce an immense load of particulate contamination released into the surrounding air, they have to be avoided. Some plastics are so called "self-lubricant" and show a reduced particulate emission during operation, like Polytetrafluorethylene (PTFE) plastics. Used materials in GMP controlled environments have to be resistant to all applied chemicals present in cleaners and disinfectants. The biological inertness is another important issue. Materials which tend to be metabolized from fungi and bacteria can have a major negative impact if microbial colonies will establish due to poor hygienic precautions. All materials should be resistant to the action of microorganisms [4]. The chemical microbiological resistance has to be evaluated prior using the material in GMP controlled environments. Relevant test procedures are described in ISO 846 [5] and ISO 2812-1 [6]. A possible classification is outlined in the guideline VDI 2083-17 [7] and -18 [8].

2.2.1. *Classification.* The EU-GMP Annex 1 defines the cleanliness classification as follows:

Table 1. Limits of particulate contamination defined in EU-GMP Annex 1.

Cleanliness Class	Maximum permissible particle count per m³ in a resting state		Maximum permissible particle count per m³ in an operating state	
	> 0.5 µm	> 5 µm	> 0.5 µm	> 5 µm
A	3,520	20	3,520	20
B	3,520	29	352,000	2,900
C	352,000	2,900	3,520,000	29,000
D	3,520,000	29,000	Not fixed	Not fixed

2.2.2. *Outcome.* For handling potentially hazardous ENPs, it will be useful to have well cleanable surfaces being resistant to all applied chemicals. No corrosion should minimize the cleaning efficiency. The microbiological resistance is of minor importance for the safe handling of ENPs; nevertheless it should be regarded due to personnel worker safety in the corresponding workplace settings.

2.3. Cleanrooms for semiconductor industry

Cleanrooms for semiconductor industry are used to gain a controlled surrounding mainly regarding airborne particulate contamination. They have a very narrow operating window regarding climatic conditions of about 22 +/- 0.5 °C and 45 +/- 5 % relative humidity. ISO 14644-1 defines the airborne particulate cleanroom classification from ISO 9 to ISO 1, the best defined ISO-class [9]. ISO 5 equals the EU-GMP Annex 1 cleanliness class A regarding particles with 0.5 µm diameter or larger. Nanoscale relevance has the ISO 14644-1 classes 6 to 1 as there, the limits for nanoscale particles down to 100 nm are defined as well.

In an average urban surrounding, between 5 and 35 trillion particles with a diameter of > 0.5 µm are present in 1 m^3 air. With an immense particulate filtration effort using pre-filtration and laminar air flow from filter-fan units installed in the ceiling to the bottom with an airflow of 0.45 m/s, the particulate contamination can be reduced below 1 particle (size >0.5 µm) in 1 m^3 cleanroom air. This equals an ISO class 1 cleanroom according to ISO 14644-1.

Airborne molecular contaminations (AMC) as defined in ISO 14644-8 [10] play an increasing role in semiconductor industries. Some critical compounds are even completely forbidden in controlled environments during open process production steps, like organophosphates, phthalates, amines and siloxanes. Organophosphates condensed on a wafer can result in an incorrect random doping process; siloxanes condensed on electric conductive surfaces can reduce connectivity of these electrical contacts during wafer testing. Volatile organic compounds (VOC) are another major concern as they may condense on critical surfaces as the lenses from the lithography tool etc. [11]. Therefore, in a cleanroom setting for the semiconductor industries, only low VOC-emitting materials without the presence of critical compounds are allowed to be used. Prior application, materials should be tested according to the guideline VDI 2083-17 [7].

2.3.1. *Classification.* The air cleanliness classes according to ISO 14644-1, is defined as follows. For comparison reasons, the values defined in the EU-GMP Annex 1 and the withdrawn US Fed. Standard 209E are listed accordingly:

Table 2. Limits of particulate contamination defined in ISO 14644-1, EU-GMP Annex 1 and the withdrawn US Fed. Standard 209E.

Regulatory				Limiting values of each air cleanliness class for differing particle sizes and reference volumes (acc. to ISO 14644-1)											
DIN EN ISO 14644-1	EU-GMP "at rest"	EU-GMP "in operation"	US Fed. Standard 209E*	0.1 µm		0.2 µm		0.3 µm		0.5 µm		1.0 µm		5.0 µm	
				per m³	per cbf	per m³	per cbf	per m³	per cbf	per m³	per cbf	per m³	per cbf	per m³	per cbf
1				10	0,3	2	0,1								
2				100	3	24	1	10	0,3	4	0,1				
3				1.000	30	237	7	102	3	35	1	8	0,2		
			1	1.240	35	265	8	106	3	35	1				
4				10.000	300	2.370	67	1.020	29	352	9,9	83	2		
			10	12.000	340	2.650	75	1.060	29	353	10				
5				100.000	2.833	23.700	671	10.200	289	3.520	100	832	24	29	0,8
	A	A								3.520	100			20	0,6
	B									3.520	100			29	0.8
			100			26.500	750	10.600	300	3.530	100				
6				1.000.000	28.329	237.000	671	102.000	2.890	35.200	997	8.320	235	293	8
			1.000							35.300	1.000			247	7
7										352.000	9.972	83.200	2.357	2.930	83
	C									352.000	9.972			2.900	82
		B								352.000	9.972			2.900	82
			10.000							353.000	10.000			2.470	70
8										3.520.000	99.716	832.000	23.569	29.300	830
	D									3.520.000	99.716			29.000	821
		C								3.520.000	99.716			29.000	821
			100.000							3.530.000	100.000			24.700	700
9										35.200.000	997.167	8.320.000	235.694	293.000	8.300

2.3.2. *Outcome.* For open process handling of potentially hazardous ENPs, it might be useful to have a cleanroom setting to protect the worker from potentially inhalable ENP dust. There is already an existing standard ISO 14644-1 defining nanoscale particulate reduction for some cleanroom classes down to 100 nm particles which eventually can be useful in the future for safe ENP handling. A laminar flow cleanroom can guide the contaminated air directly to the bottom. The necessary filtration system will retain the ENPs in the filter media. But mostly, the usage of a restricted area barrier access system (RABS), a mini-environment or a glovebox is adequate and there is no need to install a complete cleanroom system. Possibly useful additional safety equipment can be an emergency fume extractor for the whole lab activated through a specially labelled emergency switch in the laboratory setting itself. An important issue is the filtration of the outlet air prior release into the environment!

3. Safety equipment for workers

A lot of different safety equipment for workers - single or multiple usages - is already on the market and well established in different laboratory and industrial settings. Most of them can be useful to be applied in all handling operations regarding ENPs. The relevant safety equipment with focus to ENP handling is outlined in brief below.

3.1. Gloves

Single use gloves are widespread used in different laboratory settings. Standard latex gloves, powder-free latex gloves, nitrile gloves and gloves with reduced particulate contamination for cleanroom applications as well as gloves for heat protection and other purposes are used to protect the worker from the individual contamination sources. It is reported that nanoparticles can penetrate human skin based on minor scratches and injuries. Thus, it is advisable to use at minimum standard latex gloves when handling ENPs all the time. After usage, they have to be discharged accordingly.

3.2. Breathing protection

Single use breathing protections can be used to protect the worker as well as to protect the equipment or hospital patients. They are classified according to their particulate retention ability from filtering face piece (FFP) FFP 1 to FFP 3 according to EN 149 [12]. Class FFP 3 masks filter at least 99 % of airborne particles and the inward leakage between mask and face is less than 2 % (ten human subjects perform five exercises each and for eight of these individuals the average measured inward leakage listed above must not be exceeded). Other standards describing the retention ability of human face masks are EN 143 defining classes P1 to P3 for human face masks particle filters [13]. P3 masks filter at least 99.95 % of airborne particles. The US-American National Institute for Occupational Safety and Health (NIOSH) announced its own classification system. Face masks class 100 according to NIOSH filter at least 99.97 % of airborne particles. In general, the usage of the highest particle retention ability is advisable all the time. As basic rule for handling ENPs, the usage of a FFP3 mask is advisable.

3.3. Lab coat

Entering a laboratory setting, wearing a lab coat is mostly mandatory. Depending on the safety level of the different labs, it can start from standard multiple use cotton lab coats up to a full protection with internal breathing device used in a biological safety lab S4. For an ENP laboratory setting, the necessary safety level and its definitions still needs to be defined. From the actual point of view, if only an accidental release of ENPs is expected, a standard multiple use lab coat might be enough. It has to be made sure that the cleaning of these lab coats is done in a controlled manner. The used coats have to be collected in specially labelled containers to make following personnel aware of the potential risk dealing with used clothing.

4. Workplace and workflow organization

A sensible workflow organization can play a vital role during controlled handling of potentially dangerous goods such as ENPs. Some basic rules, adapted from the workflow advises for semiconductor fabs, are outlined below:

- Reduce the necessary handling steps and avoid unnecessary double handling steps
- Keep the potentially risky substances as far away as possible from human workers
- If handling ENPs in an open process, use fume extractors, minienvironments or RABS systems as far as possible
- A logical workflow setup from clean to dirty areas minimizes possible cross contamination
- Label all containers filled with ENP-substances clearly with substance, filling date and name of the worker
- Storage areas should be labeled accordingly
- The whole lab area should be restricted access only with an appropriate warning sign at the entrance doors for "nanorisk" (see warning signs for biological risk or radiation as possible basis of a ENP warning sign).
- All basic and mandatory lab requirements need to be addressed as well: Emergency shower? Eye cleaning devices? First aid kit? Evacuation plan? Alert plan? Regularly training of personnel? Lab area and workers visible from outside? Emergency phone numbers? Remind that an ENP lab is mainly an extension to a standard chemical lab with the necessarily of fume extractors and all mentioned bullet points above.
- Single usage disposable cleaning cloths and mops should be preferred.

5. Waste treatment – an open topic about the life cycle of ENPs?

As the potential risk from ENP is still in discussion and has to be either neglected or confirmed. This is an actually widely discussed topic with open outcome depending on the defined risk potential of

ENPs. Following questions will arise if the existence of a potential health risk from ENPs is proved from ongoing projects and projects in the future:

- How should lab or industrial wastewater with possible ENP contamination be treated?
- How should the exhaust air from air conditioning systems and fume extractors be filtered?
- Which filter system has to be used to retrain the focused ENPs?
- What about the whole life cycle of a product containing ENPs?
- What happens during thermal treatment in a waste disposal plant?
- How will ENPs being introduced in the environment behave after release? Will accumulation or even bioaccumulation occur inside microorganisms or higher living organisms?
- Which other organisms can be negatively influenced from ENP introduction into the environment (plants, vertebrates, insects,...)?
- Will biodegradation or chemical degradation occur to some ENP groups, like carbon nano tubes (CNT)?

If the later result from the ongoing toxicological experiments also done during the NanoDevice project states that ENPs are generally regarded as safe (GRAS), the long-term question about the life cycle of ENPs is still not solved. Therefore, it is mostly advisable to deal with all ENP-relating topics in a responsible and sustainable way.

6. Get aware of the potential risks!

One major issue in risk management is the constant education and training of the employees. This may already start during the education period at schools and universities and other education bodies. Knowing about a potential risk is the first step in controlling and reducing it. It does not need to be taught in depth, but a first insight into the subject of potential risks from ENPs and its risk management helps to remind the worker even after years of what to think about if handling ENPs.

6.1. NanoDevice dissemination

In the project NanoDevice, different dissemination activities will support the nano-related community regarding safe handling of ENPs. First, the scientific community will keep informed about the outcome of the project using the public available internet source www.nano-device.eu. This page gives an overview about all project partners, their main tasks during the project and involved personnel. All project-relevant conferences, seminars and workshops are addressed accordingly. The link to the restricted access project domain disseminates all relevant information to the project partners. The presence at different conferences together with a project leaflet will help to get the project known in the scientific community. It is planned to establish a joint workshop between different existing nanoscale-relevant other EU FP7 projects and platforms in spring 2011, as EUMINAfab, Minam (www.micronanomanufacturing.eu), NanoCom (www.nanocom-eu.org) and NanoFutures (www.nanofutures2010.eu).

A first draft of an internationally recognized warning sign similar to the biorisk sign was designed and was open for discussion in the scientific community [14].

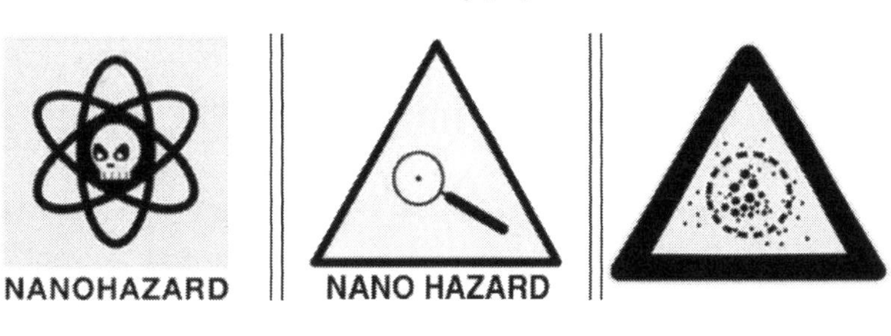

Figure 1. First draft of a possible sign " nanorisk"

7. Summary

Dealing with ENPs, the still unsure potential danger if inhaled or get into contact with these particles has to be addressed in a responsible manner. Standard precautions already used in other industrial areas can reduce the risk of getting accidentally into contact with ENPs intensively. RABS systems or other mini environments for open process handling seem to be the major safety precaution dealing with nanorisk. Usually, the worker should wear gloves, a lab coat kept in the laboratory setting and a Class FFP3 breathing protection. Fume extractors may help to raise the nano-related safety to an even higher level. All areas with possible ENP release should be labeled with a well recognized warning sign as well as all containments with ENPs inside. Restricted access is mandatory. Well organized workflows design replenishes the actual necessary precautions dealing with ENPs. All exhaust air from the nanorisk-controlled areas has to be filtered with appropriate filtration systems to keep the environment free from ENPs as far as possible. Last but not least a personnel training plays a crucial role in controlling and reducing nano-related exposure risks. Please visit www.nano-device.eu for further information.

Acknowledgements

The research leading to these results has received funding from the European Commission under grant FP7/(2007-2013)-211464-2 (NANODEVICE).

References

[1] Richmond J.Y. & McKinney R.W. (ed): Biosafety in Microbiological and Biomedical Laboratories, Government Printing Office, Washington, 1999.
[2] Council Directive 90/679/EEC: Protection of workers from risks related to exposure to biological agents at work, OJ No. L 374, p. 1, 1990.
[3] EU-GMP Guide to Good Manufacturing Practice, Annex 1. *Manufacture of sterile medicinal products*. Brussels : European Commission, 2003.
[4] Keller, M.: Reinraumtechnik in der Lebensmittelindustrie – Materialauswahl für hygienische Fertigungsumgebungen am Beispiel Wände, Böden und Fugen. *Der Lebensmittelbrief*. Lampertheim: Lebensmittelinformationsdienst GmbH, 2010. Vol. 21, 11/12, pp. 12-17.
[5] ISO 846 (1999): Plastics action of microorganisms, Beuth Verlag, Berlin.
[6] ISO 2812-1 (2007): Paints and varnishes - Determination of resistance to liquids - Part 1: Immersion in liquids other than water, Beuth Verlag, Berlin.
[7] VDI 2083-17 (draft 2010): Reinraumtechnik – Reinraum- und Reinheitstauglichkeit von Werkstoffen, Beuth Verlag, Berlin.
[8] VDI 2083-18 (draft 2010): Biocontamination, Beuth Verlag, Berlin.
[9] ISO 14644-1 (1999): Cleanroom and associated controlled environments - Part 1: Classification of air cleanliness, Beuth Verlag, Berlin.
[10] ISO 14644-8 (2006): Cleanrooms and associated controlled environments - Part 8: Classification of airborne molecular contamination, Beuth Verlag, Berlin
[11] Keller, M.: Emissionen reinraumtauglicher Materialien. *ReinRaumTechnik*. Darmstadt : GIT Verlag GmbH & Co. KG, 2010. Vol. 12, 3, pp. 14-17.
[12] EN 149 (2009): Respiratory protective devices - Filtering half masks to protect against particles - Requirements, testing, marking, Beuth Verlag, Berlin.
[13] EN 143 (2000): Respiratory protective devices - Particle filters - Requirements, testing, marking, Beuth Verlag, Berlin.
[14] ETC Group (2007): Winners of Nano-Hazard Symbol Contest Announced at World Social Forum, Nairobi,Kenya, see http://www.etcgroup.org/upload/publication/604/02/wsffinalcontestjan24_07.pdf, accessed november, 2010.

Journal of Physics: Conference Series **304** (2011) 012083 doi:10.1088/1742-6596/304/1/012083

Consolidating nanoparticles in micron-sized granules using spray drying

J S Lindeløv and M Wahlberg

GEA Niro Research and Development, GEA Process Engineering A/S, Gladsaxevej 305, DK-3520 Søborg, Denmark

E-mail: jesper.lindeloev@geagroup.com

Abstract. Suspensions of nanoparticles (SiO_2, SiC, TiO2, CNT, Nanoclay and Hydroxyapatite) were spray dried to produce dry granulated products. The nanoparticles were consolidated in granules making them more convenient and safer to use in further processing compared to handling of nanopowders.

1. Introduction

Vast amount of products are expected to emerge utilising the special properties offered by nanoscale building blocks. New production facilities are needed, designed to handle the nanoscale ingredients in the processing of the new products. The nanoscaled ingredients need to be controlled to ensure safety of the operators by avoiding nanoscale ingredients being released to the environment. This sets new challenges for production equipment and also sets requirements to the engineering of the nanoscaled ingredients such that they become convenient to control and safer to use in the production process. The nanoscale ingredients can be suspended in a liquid and consolidated to micron-size granules using a granulation process like spray drying. Being an intermediate product micron-sized granules of the nanoscale ingredients offer a way to ease the use of nanoscale ingredients in a subsequent production process. Further, micron-sized nanostructured granules can be engineered in such a way that after re-suspending the granules they decompose into individual nanosize particles [1]. Hence, micron-size granules of the nanoscale particles may produce an important link between producers of nanoparticles and their customers using nanoscaled ingredients in their products.

One of the main objectives of the EU co-founded research project SAPHIR (www.saphir-project.eu) has been on developing safe processes covering the entire production line from the nanoparticle synthesis through to the production of final products. Together with the 22 partners in the project GEA Niro has been working on different steps of the production line to develop new technologies to ensure safety and control over the entire process. GEA Niro has been using its spray drying capabilities to produce micron-size granules of a large range of different nanoparticles. Spray drying is a quick and relatively low temperature process and the promising features of the nanocomponents are preserved. As part of the SAPHIR project GEA Niro has developed ways to ensure that the granulation process can occur without release of nanoingredients to the environment preventing that operators become exposed to the nanoscale ingredients [2]. This makes the use of spray drying in a nanoparticle production line a viable solution for producing an intermediate product. This intermediate product can be traded for further use in the production of final products utilizing the more convenient properties of micron-sized granules ensuring easier handling and safer production.

It is the aim of this article to present selected results obtained during the SAPHIR project on producing micron-sized granules of different nanoparticles by use of the spray drying process.

2. Materials and Methods

2.1. Materials
Nanoparticles were supplied by SAPHIR partners (SiC, SiC + CNT) or commercial suppliers (SiO_2, TiO_2, CNT, Nanoclay and Hydroxyapatite) as suspensions or as dry nanopowders. Pumpable and sprayable suspensions were realized by addition of solvent and dispersants. Sieving of the suspensions was conducted to ensure that the suspensions did not contain lumps. The suspensions were continuously stirred until spray drying.

2.2. Spray drying
Spray drying was performed in a size Mobile Minor™ spray dryer having a capacity up to 5 kg/h. Different configurations of the spray dryer was used as shown in Figure 1. The Mobile Minor™ plant has a diameter of 0.8 meters with a cylindrical height of 0.65 meters and a 40° cone angle. Co-current drying was conducted with the two fluid atomization nozzle positioned in the centre of the hot gas disperser. A pneumatic hammer was mounted on the conical part of the chamber and was in operation during the experiments. Different means of separating the granules from the air stream has been performed as indicated in Figure 1. In all cases the exhaust gas was filtered by a HEPA filter before emission.

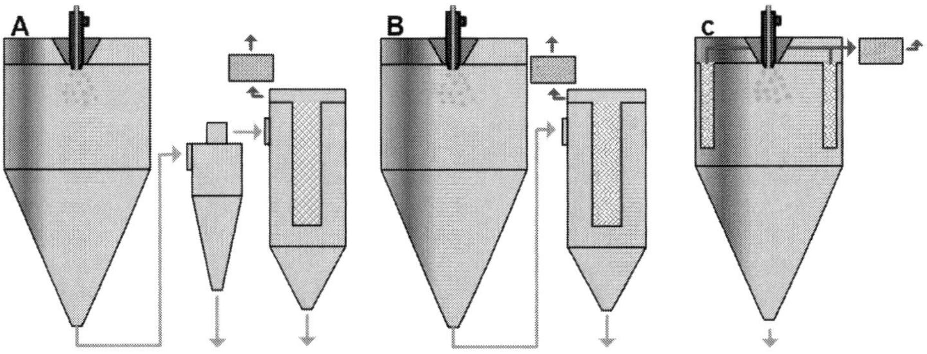

Figure 1. Different spray drying configurations used in the experiments. HEPA type filters are used as police filters. A: Co-current spray drying with powder air separation in cyclone and bag filter. B: Co-current spray drying with powder air separation done in a bag filter. C: Co-current spray drying with powder air separation by integrated bag filters.

2.3. Granule characterisation
Residual moisture content was measured using a Halogen Moisture Analyser (HR73, Mettler Toledo) at 105°C. The granule size was measured using a Mastersizer 2000 (Malvern Instruments) equipped with a dry feeding unit (Scirocco) dispersing the sample at 0,5 bar. Three size values are reported: d10 is the size where 90% (volume) of the population are larger, d50 is the median size and d90 is the size where 10% of the population is larger. Granule morphology was investigated at Danish Technological Institute, Center for Micro Technology and Surface Analysis using a scanning electron microscope (Ultra55, Zeiss).

3. Results and discussions

Experiments were conducted using the methodology and facilities available at GEA Niro's Test Center. In the planning of experiments safety aspects of handling the materials were investigated based on available information sources. In each case a safety evaluation was conducted detailing which precautions should be taken in all stages of the experiments. With nanomaterials having health aspects which are either not available or not fully determined safety precautions were implemented to ensure that experiments were conducted without exposing operators and environment to unknown risks.

Suspensions were prepared using high shear mixing (combined with ultrasonic treatment if needed) to ensure that atomization would be efficient during the spray drying operation. If needed dilution of the suspensions was done to reduce viscosity such that it became sprayable. Finally sieving was done to ensure removal of large aggregates that would otherwise block the atomization nozzle. In preparation of the spray drying experiments the equipment was assembled and safety devices checked. The spray dryer was heated and stable process parameters were obtained while running on pure solvent. When the plant was thoroughly heated and running stable spray drying of the suspensions was initiated. Process parameters were continuously recorded during the tests and if needed process parameters were adjusted to ensure stable operation. Dry granules were obtained in special containers connected to the product outlet of the spray dryer. Yield, residual moisture content and granule size distribution was measured and morphology determined. Finally, granules were sent for further processing by other SAPHIR partners. Table 1 shows results from selected experiments conducted during the SAPHIR project using the spray drying process to produce micron-size granules from nanoparticle suspensions. In the duration of the project experiments were done on a large range of nanomaterials and Table 1 is merely a selection of obtained results.

Table 1. Results from spray drying experiments with suspensions of different nanomaterials.

	Nano-material			SiO_2	SiC	TiO_2	CNT	Nanoclay	Hydroxyapatite
Suspension	Surface area, S_{bet}		m^2/g	220	59	50	-	-	-
	Particle size		nm	-	33	30	-	<200	10
	Solvent			Water	Ethanol	Water	Water	Water	Water
	Dispersant/Binder			PVA	PEI	Dispex N40	PP	Dispex N40	-
	Concentration		% TS	40	5	36	15	10	6
Spray drying	Plant setup (Figure 1)			B	A	C	A	B	A
	Nozzle			NTF-EP	NTF-E	NTF-EP	NTF-E	NTF-EP	NTF-E
	Temperatures	in/out	°C	180/90	115/60	171/82	200/81	180/90	200/100
	Drying gas		kg/h	80	79	79	80	80	64
	Feed		kg/h	2.9	2.8	3.6	3.2	3.7	1.7
	Atomization gas		kg/h	8	6	3	7		9
Granulates	Amount		g	330	42	650	68	190	110
	Residual moisture		%	4.6	1.5	1.8	3.6	4.7	2.6
	Particles size	d10	µm	14	3	10	9	11	ND
		d50	µm	43	12	29	16	20	ND
		d90	µm	103	39	77	59	50	ND
	Morphology			Ceno-spherical	Ceno-spherical	Sherical	Ceno-spherical	Sherical	Sherical

ND: Not determined

3.1. SiO_2

Nanosilica is commercial available at a relatively low cost and was obtained as a concentrated suspension (40% total solids). Nanosilica was selected as a model material in experiments designed to compare two granulation methods used in the SAPHIR project. Those were Spray Freeze Drying and

Spray Drying. The spray drying produced an interesting result as the granules had a relatively high residual moisture content. That the residual moisture content consistently is higher for granules produced from nanoparticle suspensions than from granules produced from submicron-sized particle suspensions are believed to be caused by the large surface area of the nanoparticles giving a larger binding of solvents. Figure 2 shows a microscope picture of the granules.

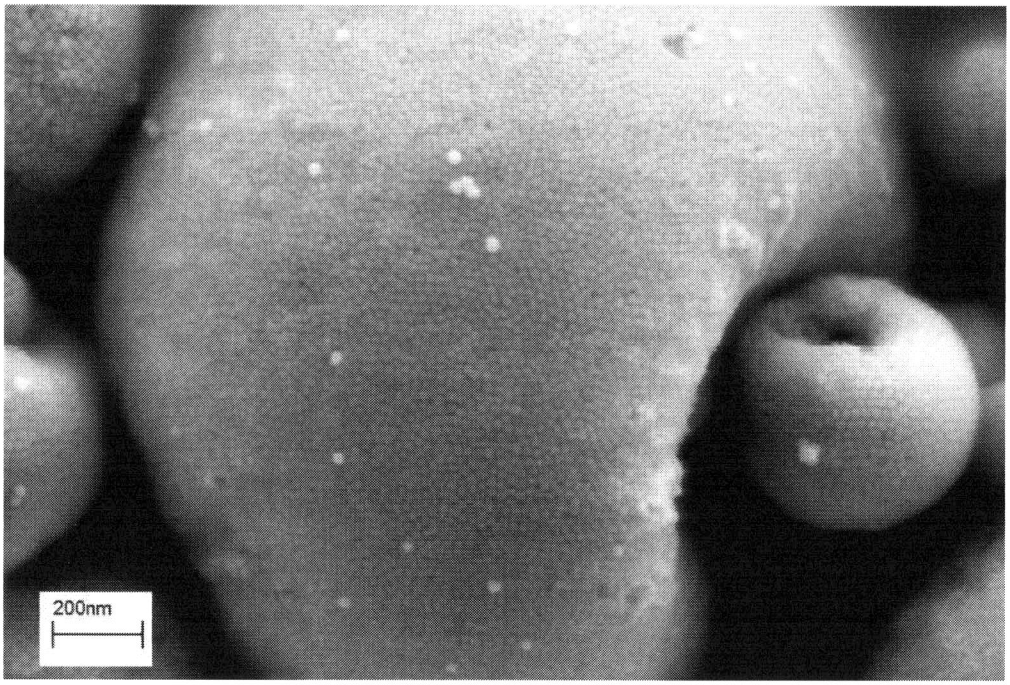

Figure 2. Microscope picture of SiO_2 granules (SEM by Danish Technological Institute)

3.2. SiC

SiC was supplied by a SAPHIR partner as a suspension using ethanol as solvent to protect the nanomaterial from oxidation. Spray drying was conducted using inert gas and product containers were sealed to avoid exposing product to atmospheric air. The aim was to produce a superior pressing powder which could be used in the production of advanced ceramic components. At a solids content of 5% the suspensions had a high viscosity. Vigorous agitation and ultrasonic treatment was implemented in order to ensure that viscosity would not limit atomization in the spray drying process. Granule morphology was ceno-spherical.

3.3. TiO_2

During the SAPHIR project vast amount of TiO_2 granules was produced using spray drying as this was used by another SAPHIR partner in production of self cleaning coated surfaces utilising the photocatalytic behaviour of titanium oxide [3]. Commercial available TiO_2 ((Degussa P25) was applied, suspensions prepared and spray dried. TiO_2 was also used as a model product in the experiments conducted with the aim of finding the optimal spray drying configuration for safe granulation of nanomaterials [2]. Furthermore, extensive work was done on the suspensions preparation with the aim of finding a dispersant that would ensure that the spray dried granules would re-disperse completely in water [1].

3.4. CNT
Carbon nanotubes (CNT) have attracted much interest due to its interesting characteristics. However, CNT health effects in humans have still not been fully determined. A spray drying experiment was conducted with the aim of producing CNT consolidated into micron-sized granules as a mean of increasing the safety during handling of this product. CNT was obtained from a commercial supplier. Atomizable suspensions could only be realized at relatively low concentrations. Spray drying was conducted with ease and a very black granulated product was obtained having a relatively good flowability.

3.5. Nanoclay
Nanoclay is another nanomaterial which is receiving much interest from science. Being available commercially it was used to test whether spray dried granules could be produced from nanoclay suspensions such that the granules would be redispersible. The shear thinning behaviour of nanoclay suspensions was utilised to ensure a low apparent viscosity of the suspensions for atomization. Spray drying was conducted and nice spherical granules were produced. Redispersability was good but further research into different dispersant systems could improve redispersability even further.

3.6. Hydroxyapatite
Hydroxyapatite nanoparticles have important dental and medical applications. Spray drying has been conducted and spherical granules were obtained. Detailed studies into its drying behaviour showed that size reduction continued during the entire drying process producing solid granules of high density.

4. Conclusions
GEA Niro has during the SAPHIR project consolidated nanoparticles into micron-size granules using the spray drying process. This has been performed on a large range of nanomaterials and in all cases micron-size granules were produced which were easy to handle compared to a nanopowder.

Acknowledgement
We gratefully acknowledge funding obtained from the European Commission through the 6th framework programme as part of the SAPHIR project (http://www.saphir-project.eu/). SAPHIR partners are acknowledged for their contribution to this work.

References
[1] Faure B., Lindeløv J.S., Wahlberg M., Adkins N., Jackson P., Bergström L., 2010 Spray drying of TiO2 nanoparticles into redispersible granules, *Powder Technology*, Vol: **203**, Issue: 2, Start page: 384-389
[2] Lindeløv, J.S., Wahlberg M., 2009 Spray drying for processing of nanomaterials. *J. Phys.: Conf. Ser.* **170** 012027
[3] Toma F.L., Berger L.M, Jacquet D., Wicky D., Villaluenga I., de Miguel Y.R., Lindeløv, J.S., 2009 Comparative study on the photocatalytic behaviour of titanium oxide thermal sprayed coatings from powders and suspensions, *Surface & Coatings Technology* **203**, 2150–2156

Nanoparticle risk management and cost evaluation: a general framework

Dominique Fleury[1], João A S Bomfim[2,3], Sébastien Metz[1], Jacques X Bouillard[1] and Jean-Marc Brignon[1]

[1] INERIS, Parc Technologique Alata BP2 – 60550 Verneuil-en-Halatte, France

[2] Centro Ricerche Plast-optica (CRP), Via Jacopo Linussio 1 – 33020 Amaro (UD), Italy

E-mail: dominique.fleury@ineris.fr

Abstract. Industrial production of nano-objects has been growing fast during the last decade and a wide range of products containing nanoparticles (NPs) is proposed to the public in various markets (automotive, electronics, textiles…). The issues encountered in monitoring the presence of nano-objects in any media cause a major difficulty for controlling the risk associated to the production stage. It is therefore very difficult to assess the efficiency of prevention and mitigation solutions, which potentially leads to overestimate the level of the protection barriers that are recommended. The extra costs in adding nano-objects to the process, especially that of nanosafety, must be estimated and optimized to ensure the competitiveness of the future production lines and associated products. The risk management and cost evaluation methods presented herein have been designed for application in a pilot production line of injection-moulded nanocomposites.

1. Introduction

Nanotechnology is the name given to a wide range of new technologies which were mostly developed in the last decade and which, according to ISO nomenclature standard [1], explore nanoscale phenomena and structures. Most of today's technology explores the use of nano-objects, especially manufactured engineered nanomaterials. This new class of anthropogenic products can lead to better performances of existing materials and composites, improving mechanical resistance, thermal or electrical conductivity or optical properties, to name just a few. However, as nano-objects appear to show properties which differ from their bulk parent material, there is an uncertainty on how to evaluate the risk associated with handling them. Indeed, there is an ongoing debate about the potentially increased toxicity of nanoscale materials in comparison with conventional ones, which even led to the creation of a new field, called nanotoxicology [2,3].

One of the proposed answers to deal with the risk evaluation and management of nanomanufacturing is to use the precautionary principle, and instead of using threshold limit values

[3] João A S Bomfim is now with Centre de Recherche Public Henri Tudor, AMS/MAT, 66 rue du Luxembourg, Esch-sur-Alzette, L-4002 Luxembourg

(which are undefined for nanomaterials), through an ALARA (as low as reasonably achieved) approach, prevent workers exposure to nano-objects [4].

The plastics industry is considered to be one of the main field in which the new nanotechnologies can play a key role. For many decades now, common polymeric materials have been reinforced with suitable additives to prepare reinforced plastics and polymer composites. The main examples are fibre-reinforced materials and laminate composites, in which embedded components (glass or carbon fibres, metallic panels) impart mechanical strength and polymeric components (epoxies, thermoplastics) provide flexibility, lightweight and/or isolation properties. The possibility to incorporate engineered nanomaterials, tailored to a specific end, can represent a major breakthrough.

The present work was developed inside the European Project "SAPHIR – Safe, integrated and controlled production of high-tech multifunctional materials and their recycling". It consists in developing a framework to analyse the risk associated with nanomanufacturing of polymer nanocomposites and evaluating the potential cost increase due to prevention and mitigation barriers to be put in place. CRP's pilot production line [5], which is used for producing innovative prototype series of automotive lighting devices, served as a model for the study.

2. Risk assessment according to European directives

For the health and safety domains, the term risk refers to the possibility of negative consequences [6]. In risk assessment, risk is a combination of the probability that a harm occurs by the severity of this harm. The source of risk is a hazard or a hazardous situation, while its consequence (harm) can be a physical injury, damage to the health of people or damage to property or to the environment. Through risk analysis, a chain of interdependence (event sequence) can be established linking hazards to harms.

In industrial safety, the risk can be classified either as chronic or accidental. Chronic risk is linked to the expected impacts on human health and on the environment of pollution and releases resulting from the facility operations in normal conditions. The Chronic risks can cause long-term impacts, then, the cumulative and time factors are of importance. On the other hand, accidental risk is the possibility of an undesirable (and sometimes unexpected) event on a site using substances or processes which can cause a hazard, resulting in an immediate harm. Principal accidental risks cause explosions, fire or emissions of toxic substances. Concerning nanomaterials, and more specifically nano-objects, both chronic and accidental risks are relevant and should be estimated

A literature review of risk assessment methodologies reveals a number of methods and procedures, such as ATEX[4] [7] for explosion risks, SEVESO II [8] for general chemical hazard, or IPPC [9] for chronic health and environmental risks. The generally adopted framework of risk estimation in these methods ("classical risk assessment approach") comprises several successive steps, starting with (i) define the scope of risk assessment, (ii) identify the hazards, (iii) set the links between hazards and harm severity, (iv) through exposure scenarios estimate exposure doses, and (v) perform an quantitative estimation of the considered risk. This quantitative "substance-based" approach is hardly, if at all, applicable to engineered nanomaterials. Anthropogenic nano-objects exist with a wide range of properties, ranging from simple chemical composition, to different particle size, surface treatment, particle shape and state of aggregation. In other words, the existing methodologies are not compatible with the present uncertainty of the harm associated to nanomaterials.

3. Basis of proposed risk assessment

The methodology proposed (Figure 1) is based on a broad qualitative analysis and tries to cover effects on human beings (on site and off site), environment (pollution of air, water, soil) and to converge towards a unique method to assess both chronic and accidental effects. In line with the objectives of the SAPHIR project, this method uses concepts already developed and used in standard industries or developed in the scope of regulations and European directives. In this way, the "SAPHIR

[4] ATEX is the name commonly given to the framework for controlling explosive atmospheres and the standards of equipment and protective systems used in them.

methodology" is based on well-known techniques already used by industries to identify and reduce risks. Unlike existing procedures, however, each step has to assume a new focus to deal with the uncertainty associated to nanomaterial-related risk, and the lack of usual toxicity (MSDS, dose-response) or emission data.

3.1. Step 1: creation of a working group

With the help of a group of stakeholder [6], experts will carry out the risk assessment of the pilot lines. The working group must be composed at least of (i) a person in charge of the process: his knowledge, especially on the properties and characteristics of nanomaterials and the nanomanufacturing process is vital for the completion of the risk assessment, (ii) a person responsible for the maintenance of the plant/unit, (iii) a person in charge of HSE prevention: his knowledge of the existing situation will allow to evaluate the need for improvements of the safety measures, (iv) a person that will lead the risk assessment, and (v) experts having an extensive knowledge in risk assessment.

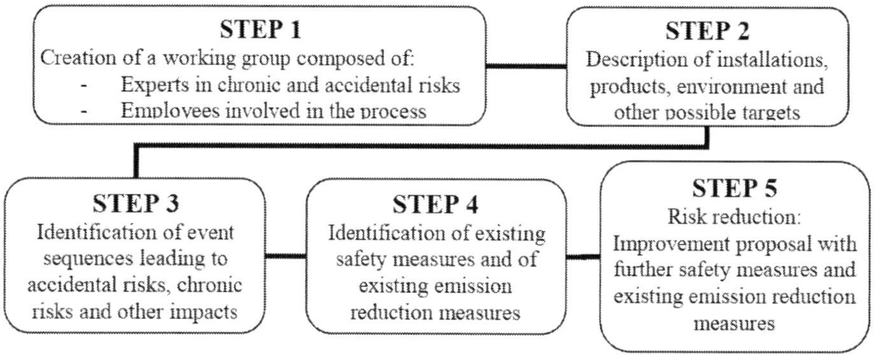

Figure 1. Main steps of SAPHIR risks assessment methodology.

3.2. Step 2: Description of installations, products, environment and other possible targets

The working group, to identify each of its elements, should process an inventory of the studied installation. The analysis will be carried out for each part of it. The list of the workplaces, threatened persons and objects, should be made during this step, in order to be able to estimate consequences for workers, people and environment.

The perimeter of the process investigation is established at this stage, especially regarding the possible on-site further processing steps of the nanomaterial produced (for instance, incorporation of manufactured nanoparticles (NPs) or nano-objects into another material).

3.3. Step 3: Identification of event sequences leading to accidental and chronic risks

The objective of this step is to provide a single framework to identify the accident event sequence from a cause to a consequence expressed in terms of a damage and to inventory all consumption (energy, water,...) and chronic pollutants (including but not limited to nanomaterials) released in the environment.

This is a critical part of the risk evaluation, where the knowledge of both the process and the risk analysis is required. To limit the risk of communication problems between process experts and risk analysts, some tools are presented in the next sessions, which can be of help in structuring the analysis.

3.3.1. The use of a table. The sequence of event will be studied through a risk assessment table used both for accidental risk and for chronic risk (Figure 2a). This table is frequently compiled in an exchange session where hazards and harms are listed and interconnected. Another important step of the risk analysis is the estimation of a probability (or occurrence frequency) for a given event [6], allowing a suitable risk management by the definition of priorities [10,11].

3.3.2. The cause tree and the "bow-tie" diagram. Through the analysis of hazards and hazardous situations listed on the table, and the critical events they lead to, it is possible to identify event chains. Through a cascade of other events (indirect and direct causes), these chains go from an initial cause, to the critical events that result in consequences, and possibly dangerous phenomena. The multiple chains of events that result in the top one (critical event) are called the fault tree, while the different consequences of a given top event are grouped in an "event tree" (Figure 2b).

Causes can be either internal, for example the cause explaining a high / low pressure in a capacity, or external: work nearby the installation, flooding, etc. For accidental risks, the method will also include assessing any deviation of parameters like pressure, temperature, level, flow, contamination. The search for internal causes will rely on a HAZOP-based approach [13] (only the safety aspects of the HAZOP are considered). For chronic risks, the initial causes are identified by screening throughout the process all the steps that cause externally provided utilities, including water and energy sources and carriers, or pollutant emission. Then, the cause is the consumption or the emission itself.

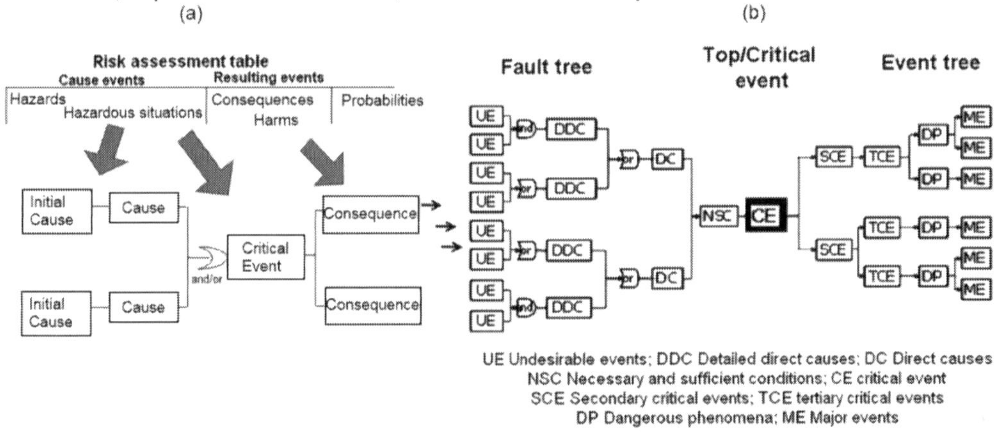

Figure 2. Risk analysis through table (a), and cause/event trees in the form of a bow-tie diagram (b)

3.3.3. The critical events and their foreseen consequences. Generally, two types of events can be identified during the risk analysis: events that may lead to accidental consequences (as defined in the SEVESO II directive [8]), or events that lead to a chronic disease or pollution (in the meaning of the IPPC directive [9]).

A loss of containment can cause a release of NPs in the workplace or outside the plant and lead to chronicle or accidental consequences with short or long term effects, for instance: disease for workers or for population, and pollution of the environment. Accumulation of NPs during the process can lead to fire or explosion events. In such cases, foreseen consequences can be thermal and / or overpressure effects. In case of the presence of an ATEX, what is feared is the production of an explosion with thermal and overpressure effects.

3.4. Step 4: Identification of existing or recommended safety and emission reduction measures

Safety and emission reduction measures are identified during risk assessment of NPs processes. For existing processes, only existing measures are listed; for processes still on project phase, only recommended measures are identified. Measures related to nanotechnologies that can be identified are defined in the following sections of this article.

3.5. Step 5: Risk estimation, risk reduction and risk control

A hazardous critical event or a chronic exposure, as explained in previous steps of this method, results from the get-together of a hazard and a target (human or environment). After the identification of the hazardous, risk estimation assigns values to the probability and the consequences of each risk. In classical methods of accidental risks assessment, this step of risk severity estimation and risk probability estimation, aims at prioritising risks and setting up reduction measures.

The probability for an accidental event is linked to the frequency to which an event is likely to occur while for a chronic risk, it is embedded in the dose-response function [6]. A chronic event is such if it is certain and is occurring continuously. Indeed, exposure takes into account the lasting time of the event which occurrence probability is 1. The accidental risk probability is the dangerous event probability (NP release) modulated by the probability to have an exposure. Estimating the probability for an accidental event to occur requires considering safety barriers or measures.

Concerning severity estimation, two parameters are highlighted by the classical methods: severity of the event or of the exposure, and vulnerability of the target. According to the probability and the severity of the event sequence, an acceptability matrix is designed to identify and set priorities to the risks that need to be reduced through further safety measures or environmental control measures.

In the scope of developing a unique method for both accidental and chronic risks, adapting this system with probability and severity classes is tricky for nanotechnologies as dose-response and effects are mostly unknown.

4. Existing measures and control options for preventing and mitigating consequences

NP manufacturing is in its infancy and mainly focuses on production problems. Therefore it is not surprising that current Environmental and Health Safety (EHS) practices – including selection of engineering controls, Personal Protective Equipment (PPE), cleanup methods, and waste management – do not significantly depart from practices for handling conventional chemicals.

Additionally, the intrinsic problems in monitoring NPs/nano-objects in any media cause a major difficulty for controlling the associated risks. That makes very difficult to assess experimentally the efficiency of prevention and mitigation systems.

4.1. Combating risks at source – Process design and containment

Process design is generally the most cost-effective way to prevent risks and should foresee the inclusion safety and chronic risk evaluation still at the design phase whenever possible. Process design options such as enclosure, isolation, and closed circuits can reduce workplace exposure and risks in general.

Recently, Morose synthesised the basic principles of safe process design applied to nanotechnologies [12] through the "SAFER" acronym: Size/Surface/Structure, Alternative materials, Functionalisation, Encapsulation, Reduce quantity. This approach can be considered an application of the "Eliminate, Substitute, Isolate" steps in the common ALARA methodology [4].

Most of NPs and nanomaterials processes involve already some degree of containment, be it for the need of controlled atmosphere or high temperatures involved. This has of course also a beneficial effect of minimising the release of NPs into the workplace during production, but if there are any leaks in the system then NPs may pass through and disperse into the workplace. With time, the NPs may form agglomerates that are not easily dispersed into the air. However, inhalation exposure to these agglomerates is still a hazard.

To a certain extent, laminar flow booths or clean rooms are an alternative containment solution for powders, already used (but not for NPs) in the pharmaceutical industry. The common system configuration is designed to extract powders downwards away from the operators and to air intakes, and then direct the flow to a filtering system. The main filters can be bag-sealed on removal from the system to prevent the release of powders. Laminar flow booths can also be used in addition to conventional containment as an additional safety measure to prevent emissions.

Finally, using dispersions, pastes and compounds (commercially sold as *premixes*, *concentrates* or *masterbatches* – the "masterbatch approach") instead of neat powders as far as possible can reduce workplace exposure, but this option may however bring the need to reconsider process design, and can lead to a higher material cost.

The storage of NPs may also involve specific conditions to conserve the products. On the one hand, NPs tend to agglomerate and aggregate, and therefore lose their nanoscale structure and unique properties. On the other hand, the small size of the particles provides a very large surface area in contact with the surrounding air, thereby facilitating chemical reactivity. Possible solutions include storage in inert gas or in anhydrous conditions. It may also be possible to embed the NPs in a protective layer of salts or polymers (as in the *masterbatch* approach). These layers can be either discarded before using the product or, if possible, incorporated into it.

4.2. Process control and "good housekeeping"

The preferred control option is "no control option", in the sense that preventing or avoiding exposure is better than to have to control exposure. A possible solution in this respect is to avoid human intervention with automation, which can reduce or suppress exposure of workers. If such a solution is not possible, working period in contaminated areas should be reduced to a minimum.

The regular cleaning of working areas should take place at the end of each work shift or more often. It avoids returning the deposited particles into suspension, and so further exposure. Dry nanomaterials should be wetted with soapy solutions or cleaning oils to prevent release, as wet instead of dry sweeping favours agglomeration). Absorbent traps should be used to isolate spills. The use of commercially available wet or electrostatic microfiber cleaning cloths may also be effective in removing particles through adsorption without release to the air.

Uncontrolled cleaning will not always reduce emissions of nanomaterials to the environment but rather shift the contamination from air to water. Adapted procedures must be put in place to ensure that waste material is removed from the system as soon as the cleaning operation is complete, and placed in suitable storage containers waiting for controlled disposal. A list of other good practices includes (i) preventing the storage or consumption of food or beverages in places where nanomaterials are handled, (ii) providing hand-washing facilities and encouraging workers to use them before eating, smoking, or leaving the worksite, (iii) providing facilities for showering and changing clothes to prevent the inadvertent contamination of other areas (including take-home) caused by the transfer of NPs on clothes and skin, and (iv) training personnel in the safe handling and understanding of powders. It is also important to ensure that Material Safety Data Sheets, if existing, are readily available for the emergency services and should be located with fire evacuation documents.

4.3. Prevention of nanoparticles emission

In addition to process containment, other measures can be taken to prevent NPs emissions in the workplace. One of the main prevention barriers is the use of local exhaust ventilation (LEV) systems. LEV enclosures and fume hoods are commonly used to control emissions from materials handling processes such as bagging, mixing and weighing. For NPs in well-dispersed, suspended aerosol form, the specification and quality of these systems should be similar to that used for vapours. Heaviest NPs or agglomerated ones can depose and stick to surfaces, not being captured in an air flow, contaminating the workspace or even the workers. For this, it is essential that the LEV entry hood is always positioned correctly and adequate capture velocity is maintained.

Maintenance and cleaning of the systems may pose the highest risk of exposure. There is frequent documentation of significant exposure levels in carbon black production, because this equipment is not designed or maintained adequately [14]. The evacuated air could have to go through several filtration stages and possibly capture in wet scrubbers or electrostatic precipitators in the final stage, to avoid significant release of NPs in ambient air and possible recirculation.

LEV systems equipped with flexible ducts or telescopic arms should be positioned closely above or adjacent to the NPs source. The emission source should also be as enclosed as possible.

4.3.1. LEV-like systems. Fume hoods, glove boxes and clean rooms can be considered LEV-like systems, but they exhibit distinctly different efficiencies and applications. Fume hoods are a classical equipment of chemical laboratories but not much information is available on their effectiveness against NPs aerosols.

At times some chemical or materials companies use glove boxes, primarily to protect the integrity of the material sample rather than out of concern for worker exposure. The main advantage of using glove boxes for nanoparticle production is that all the work can be carried in an inert gas. Glove boxes are also assimilated to containment protection, and thus the pertinent doubt about the outcome of containment loss in the glove box. Potentially explosive material, or substances sensitive to normal atmosphere should not be stored inside a glove box if loss of containment risk exists. For long shifts, glove boxes can be difficult to work with, resulting in supplementary operator fatigue. Glove bags are a smaller cheaper version of glove boxes, but provide a much less reliable protection, only assimilated to containment.

Clean rooms are isolated compartments in which a laminar flow is generated, like in laminar flow booths, the main difference being that the clean room provides better containment through physical isolation from the outside. Accordingly, clean room installation and running leads to high costs, both fixed and variable, noteworthy from the electrical power to run air pumps and the HVAC systems.

4.3.2. Is filtration efficient for nanoparticles? A frequent question related to nanosafety and safe nanomanufacturing is the filtration efficiency for nanoparticles. For large particles, impaction and interception forces dominate filtration mechanisms, but as particle size decreases, efficiency drops off to a minimum at roughly 200 nm, size limit at which the diffusion forces start to become significant. For a given filtration medium, there is a minimum efficiency at a particle size known as the most penetrating particle size (MPPS), and most filter efficiency tests are carried out using aerosols with mass median diameters equal to the MPPS [15]. So, it is expected that correctly specified fibrous filters will be good collectors of NPs. Problems may occur however if the filter material has pinhole leaks or if the filter housing has poor seals, because NPs with behaviour close to gases will penetrate these with ease. It has been recently suggested that filter efficiency drops again for very small particle size (under 2-5 nm) because of the "thermal rebound effect" [16]. Work needs to be carried out to investigate this suggestion. This is particularly relevant for cleaning systems in materials handling areas and filters installed into collection systems that circulate the air back into the workplace. Alternative systems may include modified axial-flow cyclones, which can attain a particle diameter cut-off of 20 nm [17], or magnetic filter systems, especially useful for magnetic nanoparticles.

When designing a filtration system at the workplace, there is a need to take into account the impact of change of ventilation methods on the aerosol size distributions at different locations: close to the source, in the vicinity of the worker, and in the exhaust duct.

4.4. Mitigation of nanoparticles emission

4.4.1. Personal protective equipment for inhalation exposure. Respirators for control of exposure to airborne dust use low-pressure drop filters with a range of efficiencies dependent upon the particle size distribution of the airborne dust and the specific hazard that it poses. For nanoparticles, P3 or FFP3 filters are the most relevant [15]. At this level, half-face mask will provide a protection factor of 20; better protection will be gained from full-face masks (PF = 40) and more comfort using powered air-fed masks. The same problems previously discussed for filter efficiency apply to respirators. For respirators, however, the problem of seal leakage is more severe around the face [18].

Intermediate protection is assured by using a powered air-purifying respirator (PAPR), which includes high-efficiency filtration and a pump supplying a full-face mask. The air current generated on the wearer's face increases the level of protection while maintaining positive pressure inside the mask. This results in greater comfort for the worker and minimizes exposure when the mask seal is imperfect. In cases where high-efficiency dust filters are insufficient, airline respirators or self-contained breathing apparatus are necessary. Wearing a full-face mask with high-performance filters or with a powered air purifying respirator system provides the highest protection level.

4.4.2. Personal protective equipment for dermal exposure. From now, less emphasis has been placed on dermal exposure as a route of exposure. One consequence is that the effectiveness of control measures to prevent dermal exposures is poorly understood. The facility with which NPs get airborne is such that there is a strong probability of cutaneous exposure. The product recovery and packaging stages, and general maintenance, all provide opportunities for skin contact. Evidence has been emerging that nano-objects can penetrate through the skin [19]. As an example, it has been shown that fluorescent CdSe nanoparticles (5-8 nm) can penetrate not only into the *stratum corneum* of pigskin, but can also migrate into the hair follicles, from where it is possible to diffuse into the body [20]. Moreover, chemical constituents of NPs could penetrate the skin by dissolution, and the effect of nanoparticles on damaged skin or during physical activity is still unknown.

Given the context, it would be desirable to use disposable clothing made of nonwoven impermeable materials. For example, the use of hooded coveralls, aprons and shoe covers should provide good skin protection; the same principle applies to gloves. However, there is currently no guideline for selecting protective suit materials based on nanoparticle penetration. Schneider *et al* proved that there are many factors that contribute to the inefficiency of skin protection equipment [21]. The main factors include, notably: (i) direct penetration of solid material or permeation of a liquid through the clothing, and (ii) the transfer of substances through direct contact between the equipment surfaces and the skin. Work should be carried to determine the real effectiveness of skin protection equipment with particular attention to weak points such as the hands, cuffs and opening areas.

Dermal exposure is also correlated to the risk of nanoparticles ingestion. Exposure by ingestion is likely to arise where dermal exposure to the hands is present, linked to hand to mouth contact. Consideration of exposures in relation to ingestion is at a very early stage. As with dermal exposure, it is unlikely that current preventive methods for exposure by ingestion will be more effective for NPs than for larger particles.

5. Framework for cost analysis of new nanomanufacturing processes

Analysing only the materials cost, the added-value character of nanometric fillers (such as carbon nanotubes) makes them not competitive with conventional materials from the economic-only point-of-view. Nanotechnology devices are economically competitive if they present a remarkable technical advantage, such as a performance unmatched by conventional materials (also as a much lower filler loading) or a multifunctional performance (replacing multiple conventional materials).

In the present example, polymer nanocomposites can have a whole new set of properties if compared to conventional polymers, such as thermal and electrical conductivity. On the other hand, if compared to more traditional materials, such as ceramics or metals, they excel in lightweight and ease of moulding and assembling. From the industrial point of view, thermoplastic nanocomposites parts can be of interest both due to their performance and their ability to replace both plastics and metals.

5.1. Additional costs due to nanomanufacturing

A first order approach to evaluate the economic implications of the inclusion of nano-objects in an industrial process consists in analysing how the manufacturing process is altered to include NPs as substitutes for a certain number of inputs.

First of all, the pre-process might be altered to include some adjuvants, activators or surfactants. Secondly, the process itself might be modified, and energy and fluids consumption might be impacted. Accordingly, existing protection measures might have to be reinforced and extended, and new measures may also need to be implemented. Finally, the waste generated by the process might evolve: NP dust generation or presence in effluents and waste (figure 3).

The cost of the control of additional risk generated by including NPs in the process can then be estimated by combining different costs, from that due to additional protection or mitigation measures required, including all the consumables (energy, fluids, filters...) and the manpower needed to implement them (management, training, monitoring, maintenance...), to the costs generated by the waste treatment due to the presence of NPs (or the adjuvant used) in the different waste flows, including the monitoring for detection of potential leaks of NPs and/or adjuvant.

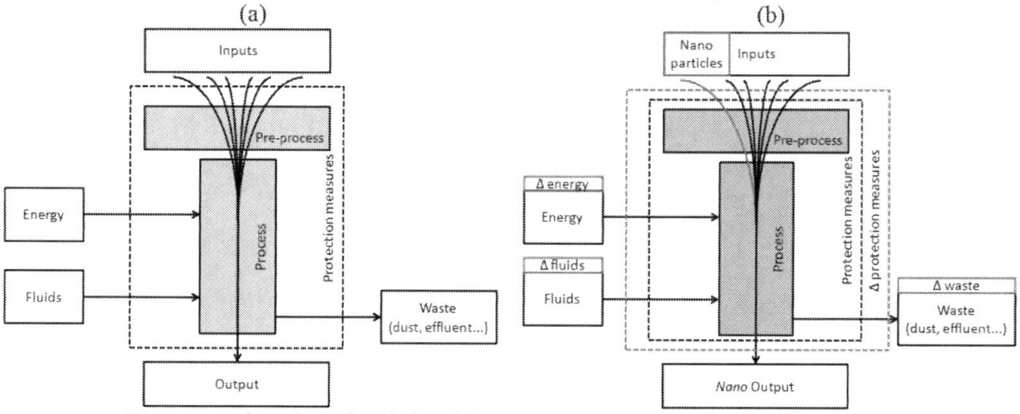

Figure 3. Alteration of an industrial process by the inclusion of nanoparticles

5.2. Effluent treatment and end-of-life issues

Brar and colleagues analysed the scientific publications on nanoparticles [22]. The majority of papers published on the subject deals with the synthesis of NPs (79%). Other two major subjects are the behaviour of these particles in the air (15%) and their toxicity (5%). Very few publications are focusing on the issue of NPs in wastewaters or environmental remediation. Some publications already indicate that it is possible to filtrate and flocculate NPs and nanotubes during their production [23] or in wastewater [24]. There is however no indication of industrial applications and very few researches focus on the behaviour of NPs in water or on the techniques of filtration to concentrate these particles by filtration. The added cost of effluent treatment depends largely if the current systems can be used for nano-effluents and should still be considered in the cost calculation.

The end-of-life issues related to nanostructured materials, or materials which contain engineered nano-objects, are still recent and yet to be solved [25-27]. Most of the difficulties related to nanomaterial end-of-life and waste management relates to the multiplicity of matrices and NPs to deal with, each combination displaying different physico-chemical properties. It is mandatory for European car producers to take into account end-of-life aspects in their production costs [28], and they should be considered within the development of every nanoproducts since a release may occur during machining, maintenance or recycling. The task is not easy because only limited information is available about this aspect. Higher costs due to additional treatments should be expected, but can only be evaluated on a very specific level. The authors are involved in an ongoing study on the end-of-life possibilities (recycle, thermal valorisation) of automotive nanocomposites.

5.3. Beyond the industrial aspects

A thorough economic analysis should include more aspects than the industrial part only. It would not be limited to comparing the production costs of nano and non-nano products, but should follow a life cycle approach, evaluating risks, costs and benefits from "cradle to grave". Societal considerations should thus be taken into account, such as final users' potential risks and real benefits, potential damages to the environment (accidental releases of nanoparticles, explosion...) as well as potential irreversibility. These aspects were not part of SAPHIR scope but are to be developed further by the authors.

6. Conclusion

The framework presented here is not based on specific nanomaterial properties or toxicity, and thus can be applied to different systems. This work gives a general framework methodology for assessing the nanomaterials-associated risks in the context of the European directives and describes a new methodology adapted to the nanoparticles. The difficulties that arise in trying to identify the link between physicochemical properties of nanoparticles and the associated risks are also discussed, as well as a synthetic view of cost analysis strategy, which is also presented. Because of the lack of data, it is difficult to set a method for risk estimation and also risk criteria for risk evaluation. Interviews of existing nanotechnology industries will allow to identify keys parameters for risk estimation and key risk criteria for risk evaluation. This whole methodology is being exploited for the particular case of the pilot production line of *Centro Ricerche Plast-optica* (CRP), which will be the target of a forthcoming communication.

Acknowledgements
This work was performed within the framework of European Project "SAPHIR – Safe, integrated and controlled production of high-tech multifunctional materials and their recycling", supported by the European Commission through the Sixth Framework Program for R&D (Contract No. 026666-2).

References

[1] ISO/TS 80004-1:2010, Nanotechnologies - vocabulary
[2] Nel A, Xia T, Mädler L and Li N 2006 *Science* **3** 622
[3] Buzea I, Pacheco I I and Robbie K 2007 *Biointerphases* **2** MR17
[4] ISO/TR 12885:2008, Nanotechnologies – Health and safety practices in occupational settings relevant to nanotechnologies
[5] Centro Ricerche Plast-optica, info available at http://www.crpo.it
[6] ISO/IEC Guide 73:2002. Risk management – Vocabulary
[7] Guide of Good Practice of the European Commission
[8] Directive 2003/105/EC of the European Parliament and of the Council on the control of major-accident hazards involving dangerous substances.
[9] European Commission, Integrated Pollution Prevention and Control Reference Document on Economics and Cross-Media Effects, July 2006.
[10] Delvosalle C, Fiévez C, Pipart A, Debray B 2006 *J Hazard Mater* **130** 200
[11] Dianous V, Fiévez C 2006 *J Hazard Mat* **130** 220
[12] Morose G 2010 *J Cleaner Prod* **18** 285
[13] Dunjó J, Fthenakis V, Vílchez J A, Arnaldos J 2010 *J Hazard Mater* **173** 19
[14] Van Tongeren M J A and Gardiner K 2001 *Appl Occup Environ Hyg* **16** 237
[15] Rengasamy S, Eimer B C and Shaffer R E 2009 *Ann Occup Hyg* **53** 117
[16] Dolez P I, Bodila N, Lara J and Truchon G 2010 *Int J Nanotechnol* **7** 99
[17] Chen S-C and Tsai C-J 2007 *J Nanopart Res* **9** 71
[18] Shaffer R E and Rengasamy S 2009 *J Nanopart Res* **11** 1661
[19] Borm P J A, Robbins D, Haubold S, Kuhlbusch T, Fissan H, Donaldson K, Schins R, Stone V, Kreyling W, Lademann J, Krutmann J, Warheit D and Oberdorster E 2006 *Part Fibre Toxicol* **3** 11
[20] Monteiro-Riviere N A and Riviere, J E 2009 *Nanotoxicology* **3** 188
[21] Schneider T, Vermeulen R, Brower D H, Cherrie J W, Kromhout H and Fogh C L 1999 *Occup Environ Med* **56** 765
[22] Brar S K, Verma M, Tyagi R D and Surampalli R Y 2010 *Waste Manage* **30** 504
[23] Bonard J-M, Stora T, Salvetat J-P, Maier F, Stöckli T, Duschl C, Forró L, Heer W A, and Châtelain A 1997 *Adv Mater* **9** 827
[24] Chang M R, Lee D J and Lai J Y 2007 *J Environ Manage* **85** 1009
[25] Ostertag K and Hüsing B 2008 *J Cleaner Prod* **16** 938
[26] Reinhart D R, Berge N D, Santra S and Bolyard S C 2010 *Waste Manage* **30** 2020
[27] Musee N 2010 *Waste Int* in press (doi:10.1016/j.envint.2010.08.005)
[28] Directive 2000/53/EC of the European Parliament and of the Council on end-of life vehicles.

Journal of Physics: Conference Series **304** (2011) 012085 doi:10.1088/1742-6596/304/1/012085

Mechanical Spectroscopy of Nanostructured Composite Materials.

Daniele Mari, Robert Schaller and Mehdi Mazaheri

Ecole Polytechnique Fédérale de Lausanne, Laboratoire de Physique de la Matière Complexe, Groupe de Spectroscopie Mécanique, CH-1015 Lausanne, Switzerland.

E-mail: daniele.mari@epfl.ch

Abstract. The thermo-mechanical behavior of different nano-structured composite materials, which were processed within the SAPHIR European Integrated Project, has been characterized by mechanical spectroscopy. The obtained results show clearly that creep resistance of fine grain ceramics such as zirconia can be improved by carbon nano-tube (CNT) reinforcements. On the other hand the elastic modulus and the damping capacity of aluminum matrix composites were increased by SiC nano-particle additions. It has also been observed that CNT additions are responsible for a better thermal stability of polymer such as ABS (Acrylonitrile-Butadiene-Styrene) used in automotive industry.

1. Introduction

Industrial equipments, particularly advanced technologies, operate more and more in extreme conditions and must be resistant to extreme temperatures, high pressure, shocks or mechanical vibrations. Examples of applications requiring highly resistant materials are turbines, engines, heat exchangers, automotive and aerospace components.

Research is based on the statement that all the existing materials are far from being tailored to the new stringent environments and consequently is focused on developing new better adapted materials. The materials, which have been investigated in the SAPHIR project, are nano-structured composites with different type of matrices: ceramic, metal or polymer. This kind of materials can bring a solution for optimizing different properties in the same material such as toughness and creep resistance, high temperature stability and high thermal conductivity, high hardness and high wear resistance. New nano-structured materials have been processed, which exhibit improved physical and mechanical properties at room temperature. However it was very important to characterize their thermo-mechanical behavior.

Mechanical spectroscopy [1, 2], which consists in measuring the complex elastic modulus, is a non-destructive technique well suited to the study of the temperature behavior of solids. For instance, the technique has produced very interesting results concerning the mechanisms that control high temperature plasticity of ceramics [3]. Moreover it measures the damping capacity, which may play an important role in the toughness of hard materials [4]. A high damping capacity is good for decreasing the noise emitted by an engine, but also for improving the fatigue resistance of the vibrating components. It has been observed that new nano-structured ceramic matrix composites exhibit a high temperature mechanical loss spectrum composed of a low exponential background as an indicator of a good creep resistance, and possibly a superimposed damping peak accounting for toughness improvement or noise reduction. In metal matrix composites, the relaxation of the interface thermal stresses may be studied by the transient components of the mechanical loss spectrum [5] and the interface damage accumulation by the evolutions of the elastic modulus.

2. Mechanical spectroscopy.

2.1. Anelasticity

Mechanical spectroscopy is primarily associated with the anelastic behavior of solids. Anelasticity [1,2] is evidenced in the following simple experiment (Figure 1a). A stress σ, of low intensity, is applied abruptly to a specimen at time, t = 0, and held constant while the strain, ε, is recorded as a function of time. The instantaneous elastic strain is given by: $\varepsilon_e = J_u\sigma$, where J_u is the unrelaxed compliance. The anelastic strain ε_a increases with time from zero to an equilibrium value ε_a^∞. When equilibrium is reached, $\varepsilon = \varepsilon_e + \varepsilon_a^\infty = J_r\sigma$, where J_r is the relaxed compliance. This evolution from one equilibrium state to a new one, under an applied stress, is called anelastic relaxation, and may be defined by two parameters, the relaxation time τ, and the relaxation strength Δ:

$$\Delta = \frac{\varepsilon_a^\infty}{\varepsilon_e} = \frac{J_r - J_u}{J_u} \tag{1}$$

If, the stress is removed after a certain delay, one can observe the instantaneous recovery of the elastic strain and with the relaxation time τ, the recovery of the anelastic strain. In such an experiment, strain is completely recoverable.

From a microscopic point-of-view, the anelastic strain can be interpreted as being due to the movements of structural defects (elastic dipoles, dislocations, interfaces) from one equilibrium position defined at $\sigma = 0$ to another one defined at σ. The relaxation strength, Δ, is then proportional to the concentration of defects, which are relaxing, the relaxation time τ accounting for their mobility.

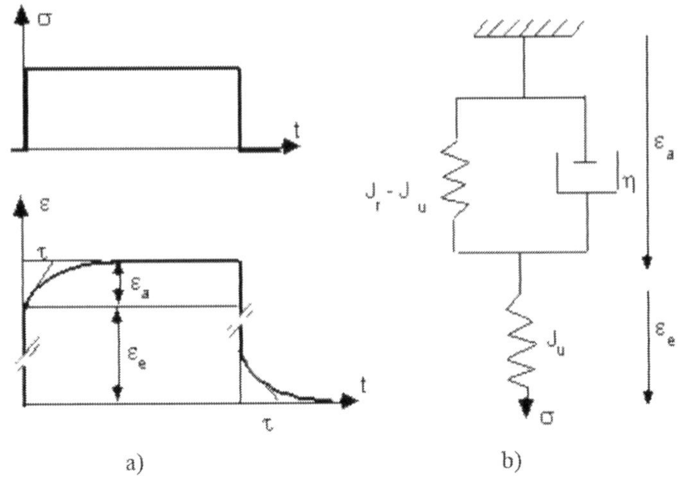

Figure 1. a) Elastic ε_e and anelastic ε_a strain of a solid submitted to stress σ.
b) Rheological model corresponding to the anelastic behavior of Figure 1a.

From a rheological point-of-view, the solid can be represented by the model shown in Figure 1b. The spring "J_r - J_u" in parallel with the dashpot of viscosity η is essential to the recoverable nature of the anelastic strain. If this spring does not operate, the anelastic strain does not reach an equilibrium value and is then no longer completely recoverable: onset of plasticity.

The anelastic behavior in figure 1 is representative of the standard anelastic solid [1,2], the equation of which is:

$$\varepsilon + \tau \cdot \dot{\varepsilon} = J_r \cdot \sigma + \tau \cdot J_u \cdot \dot{\sigma} \tag{2}$$

Measurements performed in the way described in figure 1 are very delicate, and it is more convenient to use in practice dynamic methods for measuring the relaxation parameters. If an alternative stress, of circular frequency ω, $\sigma = \sigma_0 \exp(i\omega t)$ is applied to the system, the linearity of the stress-strain relationships assures us that strain ε is periodic with the same frequency: $\varepsilon = \varepsilon_0 \exp[i(\omega t-\delta)]$, where δ is the phase lag of strain behind stress (due to anelasticity). Introducing these expressions of σ and ε in equation (2) leads to the following relationship:

$$\varepsilon = \left[\left(J_u + \frac{J_r - J_u}{1+\omega^2\tau^2}\right) - i \cdot \left(\frac{(J_r - J_u)\cdot \omega\tau}{1+\omega^2\tau^2}\right)\right] \cdot \sigma = [J_1(\omega) - iJ_2(\omega)] \cdot \sigma = J^*(\omega) \cdot \sigma \tag{3}$$

where J_1 and J_2 are, respectively, the real and the imaginary part of the complex compliance J*.
In metals and ceramics $J_r - J_u << J_u$, and the mechanical loss angle δ is given by:

$$\tan\delta = \frac{J_2}{J_1} = \Delta \cdot \frac{\omega\tau}{1+\omega^2\tau^2} \tag{4}$$

and the variation of the dynamical compliance $\delta J/J$ due to anelasticity by:

$$\frac{\delta J}{J} = \frac{J_1(\omega) - J_u}{J_u} = \Delta \frac{1}{1+\omega^2\tau^2} \tag{5}$$

The mechanical loss, tanδ, presents a maximum as a function of $\omega\tau$, centered at $\omega\tau = 1$. This maximum or peak provides the relaxation strength, Δ, (height of the peak) and the relaxation time, τ, (position of the peak on the $\omega\tau$ axis). If several relaxation mechanisms are activated, the material exhibits a mechanical loss spectrum composed of damping peaks, which inform about microstructure dynamics. As most of the relaxation mechanisms are thermally activated,

$$\tau = \tau_0 \cdot \exp\left(H/kT\right) \tag{6}$$

H is the activation enthalpy. As a consequence, tanδ can also be measured as a function of temperature by holding ω constant and the curve tanδ = tanδ (T) is the mechanical loss spectrum described as a function of temperature.

2.2. Mechanical loss spectrometer: the torsion pendulum

The most commonly used spectrometer for measuring the mechanical loss at low frequency is the inverted torsion pendulum (Figure 2). The torsion excitation is applied to the pendulum axis by means of Helmholtz coils acting on a permanent magnet fixed on the upper rod. Strain of the sample is measured by the deflection of a laser beam reflected on a linear photo detector by a mirror fixed above the sample. The photo detector is placed on a translation stage, which enables the correction of the static drift of the laser spot due to plastic deformation of the specimen. The sample is placed in a non-inductive tubular resistance furnace cooled by refractory screening and water circulation. The whole apparatus (except the lasers and photo detectors) is maintained under secondary vacuum (10^{-3} Pa). In order to avoid external mechanical vibrations, the pendulum is mounted on a damping frame. In a forced pendulum, the mechanical loss, tan (δ), and the shear modulus, G, are measured from the phase lag and the amplitude ratio between the stress and strain signals, respectively. In a free pendulum, the mechanical loss is deduced from the logarithmic decrement of the free decay of the oscillations and the shear modulus from the resonant frequency of the pendulum.

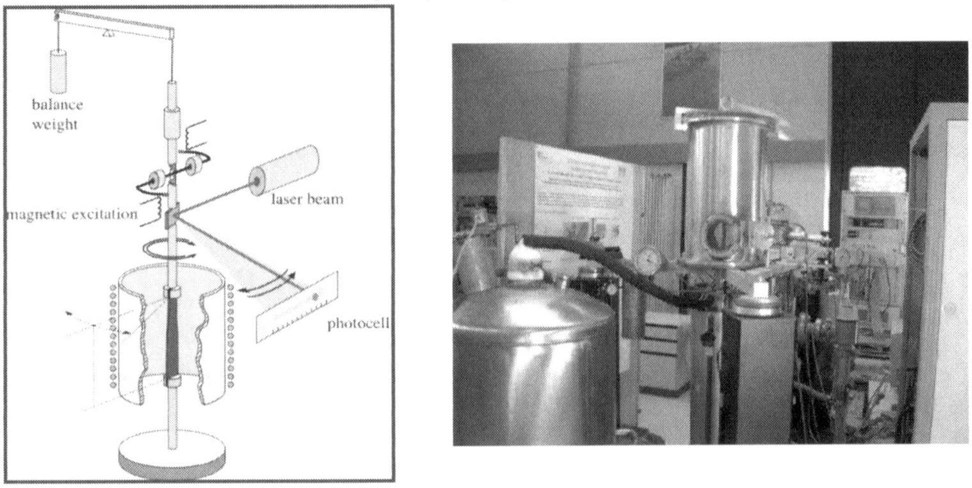

a) b)

Figure 2. Schematic drawing a) and picture b) of an inverted forced torsion pendulum.

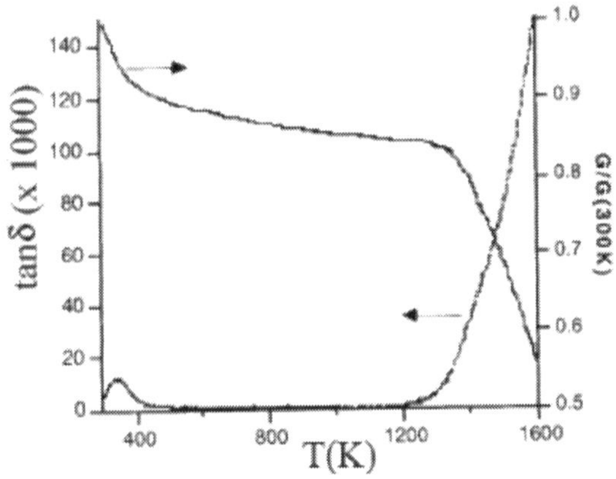

Figure 3. Mechanical loss (tanδ) and elastic shear modulus (G) as functions of temperature in 2Y-TZP zirconia as they can be measured in an inverted torsion pendulum.

As an example, Figure 3 shows a typical mechanical loss spectrum (tanδ and elastic shear modulus G=1/J) measured as a function of temperature in fine-grained zirconia (2Y-TZP: fully tetragonal zirconia stabilized by 2 mol % of yttria) [9,10]. The mechanical loss spectrum of TZP zirconia is composed of a damping peak at about 350 K associated with a modulus defect, and of an exponential increase in damping associated with a steep decrease in the shear modulus at high temperature.

The loss peak at 350 K (Figure 4a) and the modulus variation occur as well upon heating as upon cooling without any hysteresis between the curves, which means that the peak is due to an anelastic relaxation phenomenon. Moreover, it is thermally activated and it was found that the activation enthalpy and the limit relaxation time are characteristic for point defect relaxation [6]. The current interpretation invokes the reorientation of elastic dipoles of the type "oxygen vacancy- yttrium cation" under the influence of the applied stress [6].

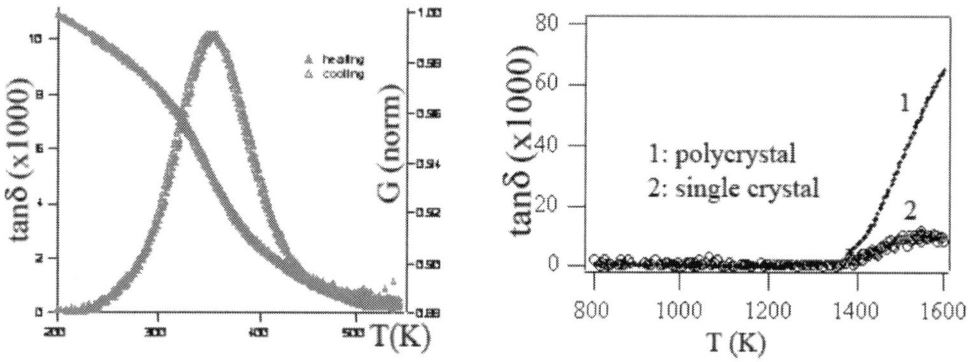

a) b)

Figure 4. a) Low temperature mechanical loss peak and modulus variation in zirconia as a function of temperature (measurements during heating and cooling).
b) Mechanical loss as a function of the temperature in poly- and in single- crystalline zirconia, curve 1 and curve 2, respectively.

In order to determine the origin of the high temperature damping, it is possible to compare the mechanical loss spectrum of the ceramic with the spectra of the constituents: grains and grain boundaries (GB). The comparison in Figure 4b of the high temperature spectra of zirconia poly- and single-crystals shows clearly that the high temperature background is due to GB.

The exponential increase in damping accounts for the appearance of ductility in a hard material. In fact, an exponential increase in tanδ can be justified if the anelastic strain does not reach an equilibrium value. In the rheological model of Figure 1b, the spring in parallel with the dashpot would have a stiffness going to zero at high temperature. As a consequence, the anelastic strain is no more limited and creep takes place. In fine grained ceramics, creep has been interpreted as primarily due to GB sliding. As a consequence, the high temperature exponential increase in the mechanical loss would be due to extensive GB sliding.

3. Results obtained in the SAPHIR project.

3.1. Nano-structured Ceramic Matrix Composites (CMCs)

If high temperature plasticity in fine grained ceramics is due to GB sliding, one way to limit GB sliding would be to add extra pinning centers to reinforce the GB triple junctions. This was the goal of reinforcing zirconia with carbon nanotubes (CNTs). 3Y-TZP zirconia was then mixed with multiwalled CNTs and sintered by SPS (Spark Plasma Sintering). Figure 5 shows a fracture surface of one specimen containing 5 wt% CNTs after processing by SPS. One can see that the CNTs are well distributed among the zirconia grains.

Figure 6a shows the variation of mechanical loss (tanδ) as a function of temperature for CNT/3Y-TZP composites with different amounts of CNTs in comparison with monolithic zirconia. The spectra are characterized by an exponential increase in mechanical loss at high temperature. Clearly, the presence of CNTs shifts the exponential increase at higher temperature, which for 3YSZ/5wt.%CNT seem to start at 1400 K.

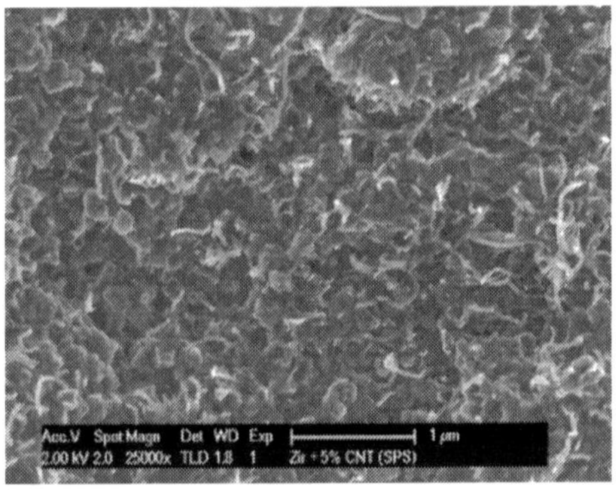

Figure 5. High Resolution Scanning Electron Microscopy picture of the composite CNT/3Y-TZP after SPS (fracture surface).

Another interesting point to notice in Figure 6b is the higher shear modulus of 3YSZ/CNTs nanocomposite in comparison with monolithic zirconia in the whole range of test temperature.

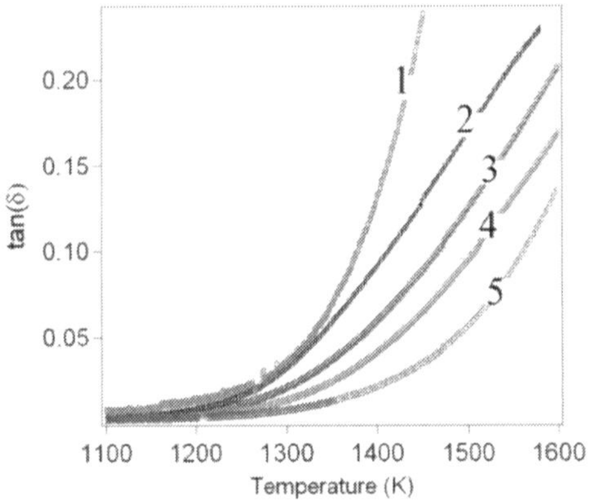

a)

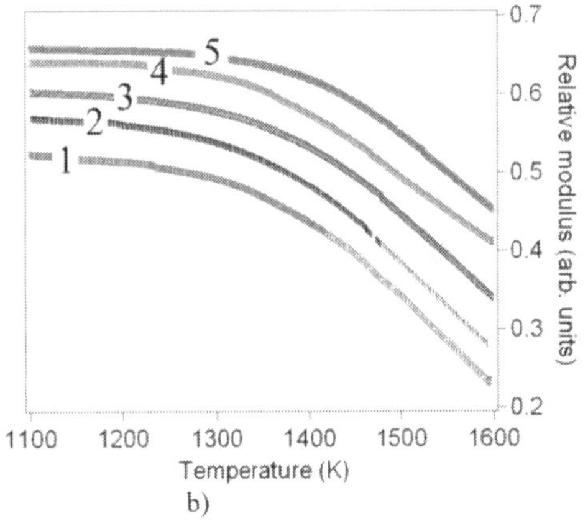

b)

Figure 6. a) High temperature mechanical loss and b) shear modulus spectra of CNT/3Y-TZP composites with different CNT content: 1 ⇔ 0 wt%, 2 ⇔ 0.5 wt%, 3 ⇔ 1.5 wt%, 4 ⇔ 3 wt%, 5 ⇔ 5 wt%.

A theoretical model was developed by Lakki [8] to account for the high temperature mechanical loss in fine grained zirconia as due to the relative sliding of grains of hexagonal shape. Taking into account the relative displacement of two grains, which are sliding over an inter-granular layer of thickness δ and viscosity η and the motion of which is limited by a restoring force due to the limiting grains at triple junctions (coefficient K), the mechanical loss tanϕ was found to be:

$$\tan\delta = \frac{G}{d}\cdot\frac{\omega\cdot\eta/\delta}{\left(\frac{KG}{d}+K^2\right)+\left(\omega\cdot\eta/\delta\right)^2} \tag{7}$$

$tan\delta$ has the form of a peak with a maximum at

$$\omega_p = \frac{\delta}{\eta}\cdot\sqrt{\frac{KG}{d}+K^2} \tag{8}$$

However, the high temperature mechanical loss spectrum in zirconia does not exhibit a peak, but an exponential increase with temperature (Figures 3 and 6). It is possible to account for this spectrum by considering that the restoring force is reduced at high temperature, where creep takes place. The transition from the peak to the exponential background would be due to the decrease in the restoring force coefficient "K" due to the weakening of the GB triple junctions. As a consequence, reinforcing the triple junctions with nanosized objects such as the CNTs increase the restoring force coefficient "K", which gives rise to a decrease of the high temperature mechanical loss (Figure 6) and of the creep strain in similar stress regime.

Indeed Figure 7 shows the effect of CNTs on the creep strain of zirconia matrix nanocomposites measured in the low stress regime (6 MPa). In general, the curves indicate a typical strain-hardening creep curve, i.e. strain increases and the strain rate decreases with time. Obviously, creep is strongly reduced in the presence of CNTs even with the addition of small amounts of them (0.5 wt.%). With higher amounts of CNT, creep is even suppressed after a certain incubation time. More interestingly, with further addition of CNTs (e.g. 3 or 5 wt% CNTs), after a transient creep during a few minutes, true strain reaches a constant value.

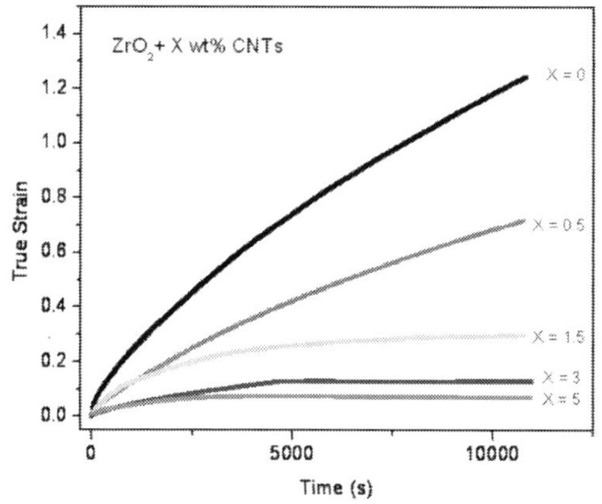

Figure 7. Creep strain in CNT/3Y-TZP composites (X=0 ⇔ 0 wt% CNT, X=0.5 ⇔ 0.5 wt% CNT, X=1.5 ⇔ 1.5 wt% CNT, X=3 ⇔ 3 wt% CNT, X=5 ⇔ 5 wt% CNT) submitted to a compression stress of 6 MPa at 1600 K.

3.2. Nano-structured Metal Matrix Composites (MMCs)

Figure 8 shows the mechanical loss tanδ spectrum and the vibration frequency f of three aluminum matrix composites as measured during thermal cycling (1 K/min) in a torsion pendulum. A2009-SiC is a commercial reference that contains 15%SiC particles with a size of 6 μm. AMC625 is a commercial reference of alloy 6061 with 25%SiC with size of 2 μm. A2009-SiC S was produced in the frame of SAPHIR with alloy 2009 and 15%SiC nanoparticles. In all cases, the mechanical loss increases exponentially with temperature, but new damping peaks appear both in AM625 and A2009-SiC S.

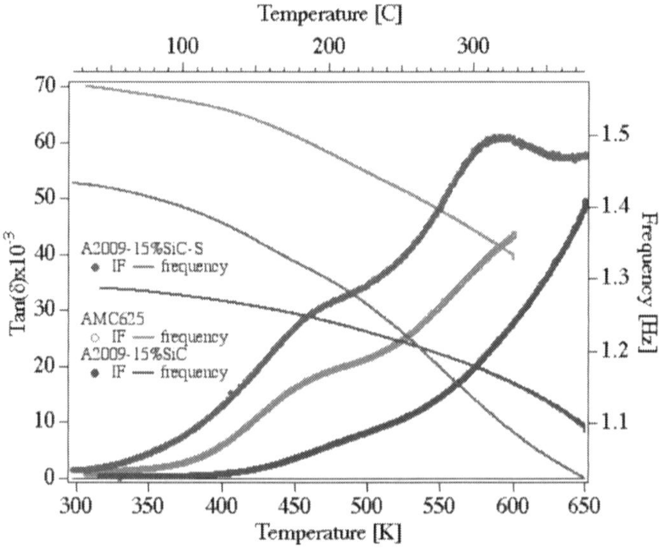

Figure 8. Mechanical loss spectra measured in a free oscillating pendulum and eigenfrequency of three aluminium based MMCs: AMC625 and A2009-15%SiC and A2009-15%SiC S.

The vibration frequency shows a decreasing trend when temperature increases. As the oscillation frequency is proportional to the square root of shear modulus, the decrease of the frequency is a common behavior. However, the comparison of the reinforced alloys 2009 is particularly interesting since with the same matrix a higher damping level is obtained in the nanostructured materials. Moreover, an increase of the oscillation frequency is clearly observed up to 500 K. The same trends are observed in AMC625 but the modulus increase should be attributed to the higher particle content. The exponential increase in mechanical loss at high temperature is related to a bad creep resistance of the material, because the anelastic or microplastic strain has no limit. On the other hand, damping mechanisms are positive factors for improving toughness. As a mechanical loss peak is associated with a limited motion of structural defects, it cannot be associated with the onset of creep. It accounts for the ability of the material to dissipate locally a part of the vibration energy and hence to improve toughness by crack propagation blunting. In the present case, the higher level of the mechanical loss in AMC625 is due to a higher peak at about 450K and not to a higher exponential background at 650 K (Figure 8). In the nanostructured composite A2009-15%SiC S, two damping peaks are found over an exponential background quite similar to that of A2009-15%SiC. As a consequence, one can suppose that A2009-15%SiC S new composite, which has a higher modulus and higher damping capacity than A 2009–15 % SiC, exhibits a higher stiffness and higher toughness with a similar refractory character at least up to 500 K. Instead, AMC625 should exhibit good properties over a wider range of temperature but the higher content in SiC particles may lead to a higher brittleness.

ODS steels to be used for pelleting nuclear fuel for next generation of fission power plants has been analysed by mechanical spectroscopy. A traditional steel F-70 and a new one processed by CEA (10078) both with 18%Cr were studied. The mechanical loss spectra show a very different behavior (Figure 9). The new one has a high damping peak at 860 K (590 °C) just visible in F-70. Moreover, the damping background at high temperature is much lower in the CEA steel. This difference let us envisage that the new process could lead to a tough and refractory steel in the domain of nuclear application from 600 °C to 900 °C (900 -1200 K).

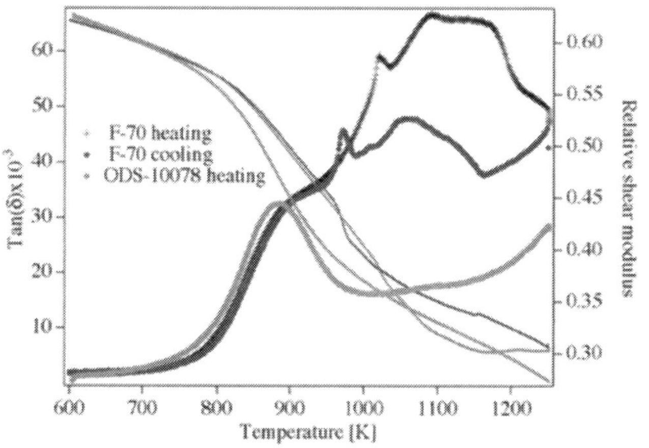

Figure 9. Mechanical loss spectra and relative shear modulus of two ODS: F 70 and 10078

3.3. Nano-structured Poymer Matrix Composites (PMCs)

Mechanical spectroscopy has been used to quantify the improvement of the mechanical properties of polymer based materials by the introduction of nanoparticles. For example, the mechanical loss and shear modulus spectra of Acrylonitrile-Butadiene-Styrene (ABS) are shown in Figure 10a. We notice the typical peaks of polymer spectra each one characterizing the onset of a molecular motion [1]. At low temperature, we find the β relaxation that is attributed to the rotation of side radicals or groups along the polymer chain and, at high temperature, the α relaxation that marks the glass transition of the polymer. At the glass transition, the elastic modulus drops to almost zero.

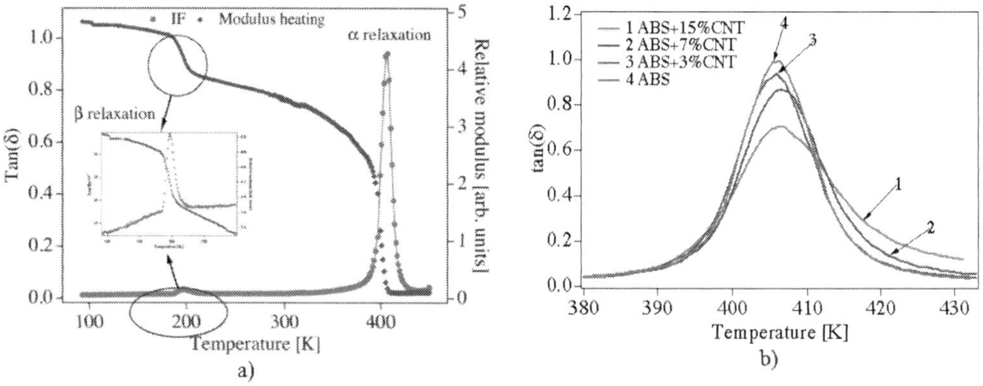

Figure 10. a) Mechanical loss spectrum measured as a function of temperature in ABS. The two main relaxations α and β are marked by tan(δ) peaks. b) Mechanical loss spectra around the α transition of ABS with different CNT content.

In the second step of this study, we have analysed the spectra of polymers reinforced with nanoparticles or nanotubes. Composite samples with ABS matrix and CNT (Arkema) reinforcement were processed by CRP (Centro Ricerche Plast-Optica, Amaro, Italy). The CNT additions were 3%, 7% and 15%. In Figure 10b, the spectra around the α transition are shown. We notice that the glassy transition is broadened by the presence of CNT. Moreover, it is slightly shifted to higher temperatures. The analysis of the microstructure shows that the α-transition occurs in the StyreneAcrylonitrile (SAN) phase. Therefore, the peak broadening is related to the interaction of CNT with such phase. The

CNT also produce a significant stiffening of the material reducing flow in the viscous phase. Such behavior can be seen by the observation of the samples heated up to 430 K in the pendulum. The pure ABS sample flows while the CNT containing sample remains stable (Figure 11).

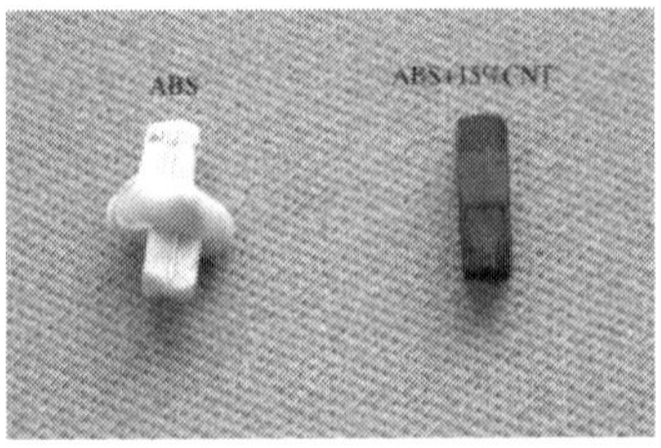

Figure 11. Deformation of samples after a thermal cycling above the glass transition in the mechanical spectrometer. A clear difference in material flow is observed.

4. Conclusions.

Mechanical spectroscopy has shown that most of the new nano-structured composites, which have been developed within the SAPHIR project, exhibit better thermo-mechanical behavior than the classical materials. For instance, it has been observed that reinforcing fine grained zirconia with multiwall carbon nanotubes (MWCNTs) improves simultaneously the creep resistance and the toughness of these ceramics [9]. In aluminum matrix composites, nanosized SiC reinforcements give rise to a simultaneous increase of the elastic modulus and of the damping level from room temperature up to 500 K that indicates a higher stiffness and a higher toughness of these MMCs. The introduction of MWCNTs in a polymer matrix such as ABS significantly improves the thermal stability of the polymer, as it can be seen by the temperature broadening of the alpha relaxation peak (widening of temperature range the glass transition).

Acknowledgements

This work has received funding from the European Community's Sixth Framework Program under grant agreement n° NMP-2-CT-2006-026666 and also from the Swiss National Science Foundation.

References

[1] Mechanical Spectroscopy Q^{-1} 2001, Schaller R, Fantozzi G and Gremaud G, Eds, 2001 *Mater. Sci. Forum* **366-368**
[2] Nowick A S and Berry B S 1972 *Anelastic Relaxation in Crystalline Solids,* (New York: Academic Press)
[3] Schaller R, Daraktchiev M and Testu S 2004 *Mater. Sci. Eng. A* **387-389** 687.
[4] Schaller R 2000 *J. Alloys and Compounds* **310** 7.
[5] Mayencourt C and Schaller R 1998 *Acta Mater.* **46** 6103.
[6] Weller M 1994 *Journal of Alloys and Compounds* **211/212** 66.
[7] Schaller R 2006 *Mater. Sci. Eng. A* **442** 423.
[8] Lakki A, Schaller R, Nauer M and Carry C 1993 *Acta Metall. Mater.* **41** 2845.
[9] Mazaheri M, Mari D, Schaller R, Bonnefont G. and Fantozzi. G to be published in the *J. Eur. Ceram. Soc.*

The International Team in NanosafeTy (TITNT): A Multidisciplinary group for an improvement of Nanorisk Assessment and Management

C Emond[1,2], C Rolando[3], S Hirano[4], F Schuster[5], O Jolliet[6], K Maghni[2], A Meyer-Plath[7], S Hallé[8], L Vandelac[9], C Sentein[5] and C Torkaski[3]

[1]BioSimulation Consulting Inc 220E. Delaware Ave. # 1182, Newark DE, 19711, USA,

[2]University of Montreal, Department of Environmental and Occupational Health Department, PO Box 6128 Main Station, Montreal, Qc, Canada.

[3]MSAP, USR CNRS 3290, and Protéomique, Modifications Post-traductionnelles et Glycobiologie, IFR 147, Université de Lille 1 Sciences et Technologies, 59655 Villeneuve d'Ascq Cedex, France

[4]National Institute for Environmental Studies16-2 Onogawa, Tsukuba, Ibaraki 305-8506, Japan

[5]Commissariat à l'Énergie Atomique (CEA) Saclay, France

[6]University of Michigan, School of Public Health, Michigan, USA

[7]BAM – Federal Institute for Materials Research and Testing, Berlin, Germany

[8]École de Technologie Supérieure, Montreal, Qc, Canada

[9]Département de sociologie, Institut des sciences de l'environnement, Université du Québec à Montréal, Montréal, Qc, Canada

Email: claude.emond@biosmc.com

Abstract. Nanotechnology allows the ability to design many new materials and devices with multiple applications, such as in medicine, electronics, and energy production. However, nanotechnology also raises several concerns about the toxicity and environmental impact of nanomaterials. A report published by the Council of Canadian Academies points out the necessity to respond about many uncertainties associated with risk assessment for ensuring the safety of health and environment. Nanotoxicology (or Nanosafety) is a part of the toxicology science that aims to study adverse effects of nanomaterials or nanoparticles on living organisms. This field includes different aspects from workers prevention to the environment protection. Group of researchers have initiated an international powerful interactive milieu for researchers to work in concert for a global and integrated study of many aspects of nanotoxicology. The International Team in NanosafeTy (TITNT) is composed of research scientists from 5 different countries (Canada, USA, Japan, France and Germany) working together on 6 different specific thematics, and organized as 9 different technology platforms (www.titnt.com). TITNT aims to study different features of nanomaterials related to nanosafety, such as *in vivo* and *in vitro* studies, life cycle, occupational protections and monitoring, early biomarkers detection, characterization and nanotoxicokinetic/dynamic assessment during and after nanoparticles synthesis and the societal, public policy and environmental aspects. While the rapid growth of nanotechnology is opening up a floodgate of opportunities, the legislation related is lagging behind mainly because of a lack of knowledge in the biosafety of most nanomaterials. The main goal of TITNT is to improve knowledge in nanosafety science for the benefit of the discipline, for better public policies and for the public itself.

The International Team in NanosafeTy (TITNT) is an initiative of international research scientists interested in different aspects of the risk of toxicity associated to nanoparticles exposure. TITNT is composed of researchers from 5 different countries (Canada, USA, Japan, France and Germany) working together on 6 different specific thematics, and organized as 9 different platforms. See the figure 1.

The main objectives of TITNT are to initiate an international powerful interactive milieu for researchers to work in concert to:

1. Evaluate potential risks associated to dermal or inhalation exposure to nanoparticles
2. Determine the safety aspects of chemically functionalized nanomaterials and the life cycle risks of nanomaterials
3. Describe the kinetic and dynamic mode of action of nanoparticles with specific organs (*e.g.* the lungs) or biological fluid
4. Improve the understanding of interactions between nanoparticles and living organisms (mainly mammalians)
5. Identify specific biomarkers of exposure to nanoparticles

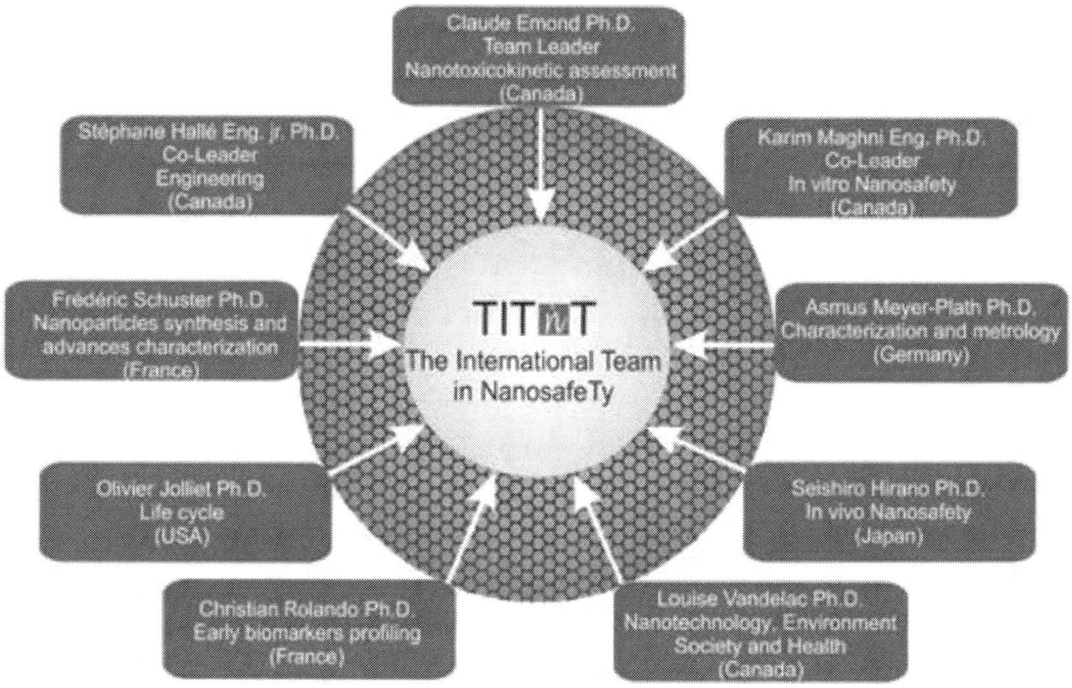

Figure 1. TITNT Platform Structure Interdisciplinary

Nanosafe2010: International Conference on Safe Production and Use of Nanomaterials IOP Publishing
Journal of Physics: Conference Series **304** (2011) 012087 doi:10.1088/1742-6596/304/1/012087

Status of characterization techniques for carbon nanotubes and suggestions towards standards suitable for toxicological assessment

Florian F Schweinberger[1,2] **and Asmus Meyer-Plath**[1,3]

[1]Member of The International Team in Nanosafety (TITNT)

[2]Technische Universität München, Catalysis Research Center, Chair of Physical Chemistry, Lichtenbergstr. 4, 85748 Garching, Germany

[3]BAM – Federal Institute for Materials Research and Testing, Division VI.5 – Polymer Surfaces, Unter den Eichen 87, 12205 Berlin, Germany

E-mail: florian.schweinberger@tum.de

Abstract. Nanotechnologies promise to contribute significantly to major technological challenges of the upcoming century. Despite profound scientific progress in the last decades, only minor advances have been made in the field of nanomaterial toxicology. The International Team in Nanosafety (TITNT) is an international and multidisciplinary group of scientists, which aims at better understanding the risks of nanomaterials. Carbon nanotubes (CNT) account for one of the most promising nanomaterials and have therefore been chosen as representative material for nanoscaled particles. They are currently investigated by the different platforms of TITNT. As a starting point, the present report summarizes a literature-based study on the physico-chemical properties of CNT, as they are closely linked with toxicological properties. A brief introduction to synthesis, purification and material properties is given. Characterization methods for CNT are discussed with respect to their reliability and the information content on chemical properties. Recommendations for a set of standard characterizations mandatory for toxicological assessment are derived.

1. Introduction

Nanotechnologies are denoted to become key technologies of this century. Their cross-disciplinary character has the potential to change everyday life to a dramatic extend. Despite profound scientific and technological advances in the recent decades, only minor progress in understanding fundamental aspects of the toxicity of nanomaterials has been made [1-3]. Considering the large number of nanomaterial applications reaching the markets [4-6] significant progress in the field of nanomaterial release, environmental exposure and the toxicology [7] becomes more and more urgent.

TITNT (The International Team in Nanosafety) is an international and multidisciplinary team of scientists focusing on the investigation of potential risks of nanomaterials by assessing toxicity and exposure. Carbon nanotubes (CNT) were selected as representative class of nanoparticles (NP), as numerous applications in consumer products are under development and already surpassed the multi-ton production scale [8]. CNT are of particular interest as their toxicity is discussed very controversially. On the one hand, there is the believe that CNT could behave toxicologically similar to asbestos [7,9,10] due to a comparable aspect ratio and physical resemblance, on the other hand, some studies indicate low toxicity [11-15].

The determination of characteristic physico-chemical properties with appropriate characterization techniques is generally accepted as a mandatory basis for any investigation on toxicological properties of nanomaterials. Due to an excess in surface-to-volume ratio for nano-scaled materials, biological interactions can also depend on physico-chemical properties, which are conventionally not considered in toxicological screenings of materials [16].

The present report summarizes results of a literature study on the physico-chemical properties of CNT and suitable characterization techniques. Both multi-walled and single-walled CNT were considered, as recommended in the literature [3,17]; however the study was limited to pristine CNT. Briefly an introduction to the synthesis and purification of CNT, as well as physico-chemical features limited to CNT are given. Applicable characterization techniques are ranked according to their information content. The report concludes with a recommendation for a set of standards and suitable characterization techniques for CNT with prospect towards later testing of toxicological properties.

2. Physico-chemical properties of CNT

Graphite nanofibres and nano-scaled filaments have long been known [18-19] before, in 1991, the observations of CNT by Iijima *et al.* raised interest in structural aspects of such materials [20]. Analysis by means of transmission electron microscopy (TEM) and electron diffraction allowed for structural characterization. Progress in large-scale synthesis of CNT and synthesis of SWCNT with well-defined morphology in 1993 [21] enabled further, more detailed investigations [22].

2.1. Synthesis methods

2.1.1. *Carbon arc discharge (CA).* Nanotubes were first observed in a DC electric arc discharge during the synthesis of fullerenes. The MWCNT grow at the cathode surface; the highest possible yield of CNT and other NPs is in the range of 25 % of the evaporated anode material. Obtaining samples with high quality is dependent on various parameters such as the current density, cooling or the stability of the formed plasma. The CNT are distributed bi-modal in bundles of 10 to 100 tubes perfectly aligned with nearly the same length [20,22-24]. Using additionally 1-10 % by weight of catalyst (i.e. Co, Ni), also SWCNT can be produced via arc discharge, providing yields of 75 % and more [21]. Occurring impurities of this process are carbon NPs, glassy carbon, amorphous carbon, and catalyst particles, making subsequent purification challenging and resulting in low yields (1%) [5,23].

2.1.2. *Chemical vapour deposition (CVD).* The classic CVD method is based on heating catalytic particles of transition metals (Co, Fe, Ni, etc.) in a furnace and flowing hydrocarbon gas through the reactor; thereby the catalytic decomposition of hydrocarbons as carbon sources towards production of MWCNT is achieved [3,25-26]. The uniformity in length and diameter is controlled by the catalyst particle size and its thermal stability. Presently, in most cases, CNT from CVD need to be size-selected and purified further, as the process may deposit substantial amounts of amorphous carbon on the surface of the tubes [4,27] and catalyst metals may be incorporated. The diversity of different CVD methods is well described in [28]. A modified process, the so called *High-pressure CO (HiPCO) decomposition* is based on the gas-phase growth from CO as carbon source (hereby the iron carbonyl complex $Fe(CO)5$ serves as catalytic precursor). It is reported to give CNT samples with highest yields and remarkable purity (97% for SWCNT) and even has been refined to a commercialized process [29-30].

2.1.3. *Pulsed-laser vaporization (PLV).* Another way to produce exclusively SWCNT in high yields (over 80%) is using a laser evaporation technique, where a laser evaporates graphite electrodes with small amounts of transition metals (Co, Ni etc.) at temperatures of 1200 °C and under helium atmosphere [24]. The condensed material is collected at a water-cooled target (i.e. a copper rod) [27].

2.1.4. Comparison, limitations and potential of methods. The extreme sensitivity of the nanotube's structure-property relations are a major challenge for the synthesis of CNT with controlled diameter, length, and chirality [26,31]. Gas phase techniques such as CVD possess the largest scale up potential for CNT production [27-28]. It can be concluded that depending on the requirements on the CNT properties, different methods must be balanced with respect to advantages and disadvantages since no optimal synthesis strategy for all types of CNT exists. With respect to toxicological studies, the most important aspects beside the CNT morphology and chemical functionality are the purity of the CNT sample and possible structural changes induced during purification.

2.2. Purification Methods

In general, all CNT materials prepared by the methods mentioned above contain impurities. Since both, amorphous carbon and metal impurities are of great toxicological concern [32], the purification of CNT materials is crucial for the acceptance of this new class of materials. Typical impurities consist of catalyst particles and of carbon in different modifications such as amorphous carbon and non tubular fullerenes [27,33-34]. The methods of purification can be divided into three categories, namely physical, chemical, and a combination of both as briefly summarized in the following as well as in reference [33].

2.2.1. *Chemical purification.* Chemical purification processes either focus on the removal of carbon or metal and metal oxide impurities. *The carbon-removing techniques* base on chemical oxidation by thermal or plasma gas phase oxidation [35], liquid phase oxidation [36], or electrochemical oxidation and rely on the fact that impurities like soot, fullerenes, of amorphous carbons [23] usually exhibit lower oxidation resistance than CNT. For effective removal of impurities, these oxidation techniques may be accompanied by considerable loss of CNT and oxidative opening of CNT endcaps [4,35]. Especially during liquid oxidation, functional groups may be created or oxidized debris may attach to CNT [37], which may affect the chemical reactivity and toxicological relevance of purified CNT [24]. *Wet-chemical conversion* of metallic impurities to water-soluble salts by mineral acids, which is the most important technique to remove metals and metal oxides from catalysts and their support, may profit from a preceding oxidative treatment to open the shell of carbon-encapsulated metallic particles that may prevent their dissolution. Since metal impurities, such as Co or Y are of great toxicological concern, effective purification processes are crucial to exclude possible effects of those metals.

2.2.2. *Physical purification.* Physical purification methods are based on a separation of CNT from impurities by means of physical property differences, both, among CNT, and between CNT and impurities, i.e. aspect ratio, size, solubility, gravity, magnetism, evaporation temperature etc. The methods include filtration, chromatography, centrifugation [24], field flow fractionation, electrophoresis, magnetophoresis [38], as well as high temperature annealing [31,39]. Physical purification methods may allow preserving the intrinsic structure of CNT. Unfortunately, most of the techniques, except high-temperature annealing, are not very effective and require the CNT to be in a well-dispersed form, which changes their degree of entanglement and may shorten them during mechanical disentangling.

2.2.3. *Multi-step purification.* Only a combination of chemical and physical methods allows optimizing the purification in a way to obtain a high-purity, predominantly undamaged CNT sample under mild conditions. Since the morphology, content and structure of impurities is strongly related to the synthesis technique, no standardized method exists. Depending on material structure and purity, sample-specific optimization is necessary of efficient purification [33,40].

2.2.4. *Conclusion and challenges.* First of all, a precise definition of the term purity has to be made, since it may depend on the point of view. It can either relate to the impurity content or to the structure homogeneity of CNT samples [33]. Another common problem is that most purification methods known are tedious, low in yield, and result in damaged CNT [31]. Therefore the additional use of size exclusion chromatography and cascade filtration could be useful. However, such diameter-, length-, or chirality-related purifications steps are not easily scaled up yet [4]. Nevertheless, for some SWCNT, efficient multi-step purification methods leading to highly pure samples have been developed and are investigated for scale-up [41]. Summarizing, effective purification requires synthesis type or even batch dependent optimization of several purifications steps involving different techniques. There is a need for a CNT purity assessment standard for evaluating and systematically improving methods of purification [33].

2.3. Properties and particularities

2.3.1. *General aspects.* In a CNT each carbon atom is bonded to three neighbouring atoms through sp^2 hybridization to form a closed tubular shell of helical structure [23]. In SWCNT, every atom position is precisely determined and the periodicity in the lattice can be reduced to a unit cell type, allowing for predictions of its properties. CNT possess superior thermal properties with high thermal conductivity and the high stability of graphitic materials of up to 2800 °C in vacuum. From an electronic point of view, SWCNT can be considered as 1-dimensional conductive structure [23]. Due to the one-dimensional nature of the conduction electron states in CNT, well-spaced and symmetric structures, so called van Hove singularities appear in the density of states (DOS) of nanotubes [42]. They can be detected with scanning tunnelling microscopy (STM) and UV/optical spectroscopy. Doping of CNT allows controlling their electronic conductivity. Boron-doped CNT exhibit exclusively metallic conductivity, whereas the properties of un-doped CNT vary between metallic and semiconducting depending on chirality [43]. The length and diameter of CNT vary significantly depending on the synthesis method [34].

2.3.2. *Chirality.* CNT can be thought as being made of a sheet of graphite (graphene) wrapped into a cylinder, so that the long edges match and form a seamless structure. However, there are several directions of rolling of a sheet, which all lead to matching of edges. Translational shifts along the edges will lead to different helical orientation of carbon atoms along the tube and thus introduce the concept of chirality. The chirality parameters, called chirality indices, describe the deviation of from parallel to normal orientation of the ring arrays of the honeycomb structure with respect to the tube axis. Two extremes confirmations of CNT can be easily distinguished: The so-called «armchair» and «zigzag» confirmations with chiral angles of 30° and 0°, respectively [27]. Between these extremes, CNT with various degrees of helicity can be found [23]. A more detailed description of the concept of helicity including the definition by the Bravais lattice vectors is given in references [4,24,44]. A nanotube's chirality characteristically determines the electronic, optical, and phononic properties of CNT [27]. For a single perfect SWCNT, the chirality indices (n, m) and the length completely determine the nanotube as a macromolecule. For MWCNT the situation is more involved since different layers may show different chirality. The determination of the chiralities of MWCNT is a complex problem [45-46].

2.3.3. *Tube ends.* Depending on synthesis and purification method, CNT can exhibit closed tube ends [20,22]. For a fullerene-like hemisphere, 6 pentagons are necessary to close the tube [24], however defects and faceted CNT walls allow for alternative configurations of the tube caps. The curvature of the caps reduces their thermal stability and enhances the reactivity of CNT. Closed SWCNT can, for instance, be opened easily by heating in air or oxygen at 750 °C.

2.3.4. *Material Variability*. Variation of the chirality indices *n, m* together with CNT length lead to an infinite manifold of different SWCNT [26]. Coaxial stacking of SWCNT to MWCNT further enhances the spectrum of ideal CNT structures, whereas real-world materials exhibit defects, structure irregularities, and impurities, which make the material spectrum even broader. As a consequence, defective, i.e. real-world CNT require many parameters for a comprehensive characterization. For ensembles of defective CNT, such characterization requires determination of particle property distributions. For toxicological studies of CNT materials, it has to be considered that structural complexity of ensembles of defective CNT cannot be comprehensively investigated by contemporary analytical techniques.

2.3.5. *Solubility and dispersibility*. Both pristine SWCNT and MWCNT show poor dispersibility in polar media like water and many organic solvents, imposing limitations towards their applications [47]. CNT exhibit a strong tendency to aggregate in bundles or ropes, due to easy polarizability and high molecular weights which promote dispersive interaction forces [48]. The different strategies to overcome the problem of poor dispersibility can be divided into covalent and *non-covalent approaches*. The non-covalent approach uses surfactants which only weakly interfere with the electronic CNT structure [49]. Also wrapping of CNT by polymers, peptides, or DNA can help to suspend CNT [49-50]. The *covalent approaches* involve the functionalization of CNT, for instance by controlling the oxidation degree of the CNT, and affect their electronic properties. Many different functionalities like peptides, acid or amino groups, or polymers have been successfully bound covalently to the sidewalls of CNT [48]. Uniform dispersion of entangled or agglomerated CNT in solvents generally requires techniques that allow exerting high shear forces to the mixture. If disentangling is unavoidable, the method will have to break and shorten tubes. The energy required for CNT dispersion can for instance be provided by ultra sonic baths or probes, high pressure-spraying, and roller mills. Solvents adapted to the CNT functionalization or surfactant chemistry are required to obtain stable suspensions [7,48,51].

3. Conclusion and recommendation

Based on the previous sections, materials selection and characterization standards are suggested that should be applied to CNT materials that are subjected to toxicological testing. For this, the different techniques of production, purification, and characterization will be reviewed to derive guidelines for method selection.

3.1. General aspects on sample selection and testing requirements

For toxicological studies aiming at investigating effects characteristic for CNT, specially tailored samples of high purity, structural uniformity, and ensemble homogeneity should be chosen and be characterized physico-chemically as comprehensively as possible.

The alternative aim to generally assess the toxicology of "real-world" CNT materials should be based on materials from industrial mass production that should likewise be characterized as comprehensively as possible. It is essential that all available information on producer, production process, batch number, storing conditions, and sample age should be compiled.

A part of the sample should be stored under dark, dry and inert conditions for possible future studies.

3.2. Production

For the sake of relevance and comparability of toxicological testing, CNT materials from well-established or highly reproducible production processes should be selected. In order to facilitate the identification of CNT-related characteristics, emphasis should be put on sample uniformity and material purity. For SWCNT, the HiPCO process evolved to a well-established, commercialized process of high structural uniformity and mature purification strategies for remaining catalyst metals [28-30]. Commercial HiPCO materials may be of environmental relevance. However, CVD synthesis techniques for SWCNT are rapidly evolving in structure and chirality control. They show a general advantage in growth efficiency, leading to lower metal impurity content, especially for fluidized-bed and for vertically-aligned CVD synthesis of SW- and MWCNT with subtracted catalysts [33]. CNT from fluidized bed synthesis contemporarily suffer from curling and entangling due to CNT by agglomerate collisions, whereas CNT vertically-aligned CVD synthesis are straight and more facile dispersed. MWCNT from CVD meanwhile can reach good uniformity in size and very high length but may suffer from amorphous carbon adsorbents [4,16,21]. MWCNT from CA may show a high degree of crystallinity and straightness that can generally be obtained by CVD synthesis only after a graphitization step [26-27]. Since the CVD method is highly suitable for mass production, the relevance of CNT from CVD for emission scenarios is higher and may justify the costs of toxicological investigations [31]. For an extended toxicological study it could be interesting to study effects of different synthesis methods, which highly affects type and amount of impurities as well as length and diameter distributions etc. [52].

3.3. Purification

For both SWCNT and MWCNT a purification procedure should be used which is tailored to the later requirements of the sample. It should be based on a multi-step purification protocol, taking advantage of both chemical and physical methods [40,33]. A few aspects have to be taken into account while choosing a purification protocol. It is important to obtain samples with lowest possible metal content, since in many cases due to inadequate assessment, the observed toxicity indices may reflect these byproducts or residues rather than the primary material structure [53,15-16]. Furthermore, the following characteristics of CNT have been reported to be altered due to modifications induced by post synthesis treatments, including: absorptive nature, aspect ratio, surface reactivity, hydrophilic properties and surface functionalization [53]. Thus, since purification also destroys CNT, the removal of impurities must be balanced against the introduction of defects into tubes [34].

3.4. Metal and amorphous carbon impurities

As mentioned previously, metal and amorphous carbon impurities can be of higher toxicological relevance than CNT-related properties. Frustrated phagocytosis in macrophages resulting from the physical CNT properties may be of higher relevance than chemical composition of fibres [9]. The relevance of radicals, oxidative potential, as well as the chemical functionality of the CNT on the reactivity as well as on the dispersibility must not be underestimated since both affect the bioavailability of CNT. Especially, the role of CNT surface area, surface functionality and polarity in immobilization of nutrients and biomolecules deserves special attention in cell tests, since serum depletion by contact to nanoparticles of large surface area has been reported. Cross checks with supernatants of centrifuged serum-CNT dispersions are considered indispensible.

3.5. Dispersion

Dispersion of CNT materials for toxicological testing is an almost philosophic problem: Are fundamental interactions of high-purity CNT and cells or organs in focus, which may require fully-dispersed CNT; or is a realistic exposure scenario planned, which may require un-purified industrially-produced CNT with a dispersion degree adapted to the emission source?

3.6. Commonly applied characterization techniques

Description of analytical capabilities of characterization techniques for CNT would exceed the scope of this work, therefore the reader is referred to the literature.

Figure 1 shows a statistic analysis of methods for CNT mentioned in 58 articles consulted for this work, excluding books and reviews. The presented characterization techniques can also be sorted according to their information content on the physico-chemical properties of CNT. The following metrics, covering the information from all techniques, can be extracted: size (aspect ratio, distribution), shape, surface area (area/ mass ratio), composition (element analysis), surface chemistry (functionalization), crystallinity, agglomeration (dispersion), porosity, heterogeneity, stability (thermal), impurities, defects, solubility, chirality and conductivity.

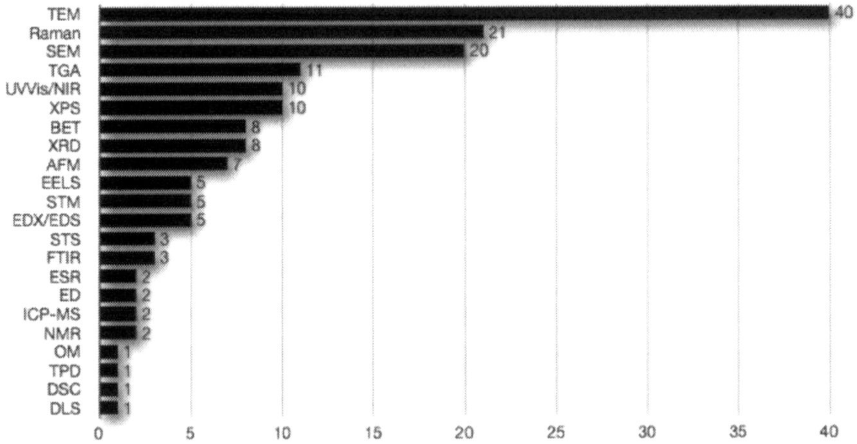

Figure 1. Number of mentions of technique per reviewed literature, including 58 consulted papers with mentions of at least one of the stated techniques

3.7. Literature recommendations for characterization methods

Literature suggestions on minimum requirements for the characterization of CNT samples are compiled in Table 1.

Table 1. Requirements for minimal characterization of CNTs from the literature.

Source	Metrics
Safadi *et al.* [54]	density, diameter, length, aspect/ratio, modulus.
Müller *et al.* [53]	surface area, size, composition, shape, porosity, surface, dispersion.
Yang *et al.* [55]	surface area, pore volume, surface chemistry, structural defects, fraction of open ends.
Hurt *et al.* [15]	element analysis, surface area, morphology (aspect ratio, secondary carbon forms, metals location), functionalization, crystallinity, defects.
Arepalli *et al.* [52]	homogeneity, purity, thermal stability.
Oberdörster *et al.* [16]	size (distribution), agglomeration, shape, crystal structure, chemical composition, surface area, surface chemistry, surface charge, porosity, particle number

3.8. Suggestions for combinations of characterization methods

Considering the fact that previously acquired data on CNT material properties must be comparable at least with reference to the measurement method, the following recommendations can be made. Raman spectroscopy, SEM, and particularly TEM can be considered as base techniques for CNT characterization. Additionally, TGA, UV-Vis/NIR, and XPS are necessary assets for a complete characterization. ED and EDX/EDS are usually included in modern TEM devices and may thus be available without additional instrumentation. Local probe techniques that characterize CNT on an individual basis such as AFM, STM, and STS have not been included here, as they are not applicable for a routine characterization of CNT.

Additionally, the capabilities of the different analytical techniques and the requirements from the literature regarding the minimum necessary metrics from table 1 have to be matched. A complete characterization will be impractical, therefore a consensus on the minimum set of characteristics is necessary. The set should also comply with the condition imposed by Oberdörster *et al.* that the set has to comprise information on all *potentially* significant material characteristics of CNT. This should enable even a retrospective interpretation of toxicological datasets in the light of new findings [16]. In the following Tab. 2, a recommendation for a set of characterization techniques which balances the different requirements and the feasibility of the method [16,33] is given, sorted by relevance of the method.

Table 2. Recommendation for a set of CNT characterization techniques for toxicological studies. Quantitative information is marked by *, important characteristics are typeset in bold.

Technique	Metrics
TEM (EDS/EDX; ED)	**diameter***, **length***, **agglomeration**, composition, **impurities**, **surface contaminations**, **uniformity**, **crystallinity**, chirality
SEM (EDX)	**diameter***, **length***, **agglomeration**, composition, **impurities**, **surface contaminations**, **uniformity**
Raman	chirality, diameter*, **crystallinity***: graphitization degree vs. structure defects
TGA	**stability***, composition*, uniformity*, **purity***, **amorphous carbon surface contaminations***, surface chemistry
BET	porosity*, **surface area***
XPS	composition*, **surface chemistry***, **impurities***, stability
UV/VIS/NIR	agglomeration, size, uniformity, conductivity

3.9. Aging of CNT

The characterization of CNT is additionally complicated by aging phenomena. It has been reported that aging of CNT occurs with respect to surface area and pore volume, surface oxygen, and structural defects. It has therefore been suggested that physico-chemical properties can be reliably characterized only after a quasi-stable conditions were obtained, which typically occurs after storing for 9-15 month [55]. Consequently, the post-synthesis stability, details of storage conditions like humidity, temperature, exposure to light, and atmosphere conditions, and aging of materials under non equilibrium and under ambient conditions [15] is an important issue, which has to be addressed and investigated prior to subjecting CNT samples extended toxicological studies [16].

4. Sample proposition and summary

Concluding the above mentioned, the following set of CNT samples should be promoted to a standard set of materials for toxicological studies. The aim is to allow correlation of toxicological data generated from future studies [48]. The ideal CNT samples should exhibit a mono-modal distribution in length and diameter, a uniform chemical functionalization and crystal structure [56].

The following set of standard CNT should be prepared, fulfilling the aforementioned requirements:
1. MWCNT from CVD process; as prepared
2. MWCNT from CVD process; optimized purification and uniform aspect-ratio (length, diameter)
3. SWCNT from HiPCO process; as prepared
4. SWCNT from HiPCO process; optimized purification and uniform aspect-ratio

Prior to toxicological testing, the samples should additionally be characterized with respect to
1. Aging behaviour of samples, until the samples remain stable (9-15 month)
2. Dispersion quality and stability of samples in aqueous solutions [51].

Minimum characterization requirements for of material properties have been presented in table 2 and include TEM, EDX, ED, Raman, SEM, TGA, BET, UV/Vis-NIR and XPS.

Additionally, for a complete assessment of the toxicological effects of CNT, tubes from different processes and different manufacturers could be investigated [48], however, this will most certainly go beyond the scope of capabilities and possibilities.

Acknowledgements
This work was supported by the Nanotechnology, Environment, Society and Health platform of the TITNT group and funded by the Canadian "Social Sciences and Humanities Research Council", under the supervision of Prof. Louise Vandelac at UQÀM, Montreal, Canada.

References

[1] Saleh N, Pfefferle L and Elimelech M 2008 *Environmental Science & Technology*, **42**, 7963
[2] Donaldson K, Tran L, Jimenez L, Duffin R, Newby D, Mills N, MacNee W and Stone V 2005 *Particle and Fibre Toxicology* **2** 10
[3] Warheit D 2006 *Carbon* **44** 1064
[4] Ajayan P 1999 *Chemical Reviews* **99** 1787
[5] Jorio A, Dresselhaus G and Dresselhaus M, 2007 *Carbon Nanotubes*, 1st ed.; Springer, Berlin
[6] Poole C J and Owens F 2003 *Introduction to Nanotechnology: Selected Topics*, 1st ed.; Wiley
[7] Magrez A, Kasas S, Salicio V, Pasquier N, Seo J, Celio M, Catsicas S, Schwaller B and Forró L 2006 *Nano Letters* **6** 1121
[8] Cheng C, Porter A E, Muller K, Koziol K, Skepper J N, Midgley P and Welland M 2009 *J. Phys.: Conference Series* **151** 012030
[9] Poland CA, Duffin R, Kinloch I, Maynard A, Wallace WAH, Seaton A, Stone V, Brown S, MacNee W and Donaldson K 2008 *Nat Nano* **3** 423
[10] Aillon K, Xie Y, El-Gendy N, Berkland C and Forrest M 2009 *Adv. Drug Del. Rev.* **61** 457
[11] Pauluhn J 2009 *Toxicological Sciences* **113** 226
[12] Pauluhn J 2010 *Regulatory Toxicology and Pharmacology* **57** 78
[13] Kagan VE, Kondurun NV, Feng W, Allen BL, Conroy J, Volkov Y, Vlasova II, Belinkova NA, Yanamala N, Kapralov A, Tyurina YY, Shi J, Kisin ER, Murray AR, Franks J, Stolz D, Gou P, Klein-Seetharaman J, Fadeel B, Star A and Shvedova AA 2010 *Nat Nano* **5**, 354
[14] Huczko A, Lange H, Calko E, Grubek-Jaworska H and Droszcz P 2001 *Full. Sci. Tech.* **9** 251
[15] Hurt R H, Monthioux M and Kane A 2006 *Carbon* **44** 1028
[16] Oberdörster G, Maynard A, Donaldson K, Castranova V, Fitzpatrick J, Ausman K, Carter J, Karn B, Kreyling W, Lai D, Olin S, Monteiro-Riviere N, Warheit D and Yang H 2005 *Particle and Fibre Toxicology* **2** 8
[17] Jia G, Wang H, Yan L, Wang X, Pei R, Yan T, Zhao Y and Guo X 2005 *Envir. Sci. Tech.* **39** 1378
[18] Oberlin A, Endo M and Koyama T 1976 *Journal of Crystal Growth* **32** 335
[19] Endo M 1988 *CHEMTECH* **18**, 568
[20] Ijima S 1991 *Nature* **354** 56
[21] Iijima S and Ichihashi T 1993 *Nature* **363** 603
[22] Ebbesen T and Ajayan P 1992 *Nature* **358** 220
[23] Ebbesen T 1994 *Annual Review of Materials Science* **24** 235
[24] Ajayan P and Ebbesen T 1997 *Rep. Prog. Phys.* **60** 1025
[25] Magrez A, Seo J, Mikó C, Hernádi K and Forró L 2005 *J. Phys. Chem. B* **109** 10087
[26] Dai H 2002 *Accounts of Chemical Research* **35** 1035
[27] Thostenson E, Ren Z and Chou TW 2001 *Composites Sci. Tech.* **21** 1899
[28] Terranova M, Sessa V and Rossi M 2006 *Chemical Vapor Deposition* **12** 315
[29] Nikolaev P, Bronikowski M, Bradley R, Rohmund F, Colbert D, Smith K and Smalley R 1999 *Chem. Phys. Lett.* **313** 91
[30] Bronikowski M, Willis P, Colbert D, Smith K and Smalley R 2001 *J. Vac. Sci. Technol. A* **19** 1800

[31] Andrews R, Jacques D, Qian D and Rantell T 2002 *Acc. of Chem. Research* **35** 1008
[32] Jakubek LM, Marangoudakis S, Raingo J, Liu X, Lipscomb D and Hurt RH 2009 *Biomaterials* **30** 6351
[33] Hou PX, Liu C, Liu HM and Cheng HM 2008 *Carbon* **46** 2003
[34] Donaldson K, Aitken R, Tran L, Stone V, Duffin R, Forrest G and Alexander A 2006 *Toxicological Sciences* **92** 5
[35] Tsang S, Harris P and Green M 1993 *Nature* **362** 520
[36] Hiura H, Ebbesen T and Tanigaki K 1995 *Advanced Materials* **7** 275
[37] Wang Z, Korobeinyk A, Whitby RLD, Meikle ST, Mikhalovsky SV, Acquah SFA and Kroto HW 2010 *Carbon* **48** 916
[38] Kang J and Park J 2007 *Small* **3** 1784
[39] Bougrine A, Dupont-Pavlovsky N, Naji A, Ghanbaja J, Marêché J and Billaud D 2001 *Carbon* **39** 685
[40] Hou P, Bai S, Yang H, Liu C and Cheng HM 2002 *Carbon* **40** 81
[41] Rinzler A, Liu J, Dai H, Nikolaev P, Huffman C, Rodriguez-Mac F, Boul P, Lu A, Heymann D, Colbert D, Lee R, Fischer J, Rao A, Eklund P and Smalley R 1998 *Applied Physics A: Materials Science & Processing* **67** 29
[42] Chiang IW, Brinson B, Smalley R, Margrave J and Hauge R 2001 *J. Phys. Chem. B* **105** 1157
[43] Golberg D, Bando Y, Bourgeois L, Kurashima K and Sato T 2000 *Carbon* **38** 2017
[44] Reich S, Thomsen C and Maultzsch J 2004 *Carbon Nanotubes: Basic Concepts and Physical Properties*, 1st ed.; John Wiley & Sons
[45] Lambin P, Meunier V, Henrard L and Lucas A 2000 *Carbon* **38** 1713
[46] Qin LC 2007 *Phys. Chem. Chem. Phys.* **9** 31
[47] Kim HS, Yoon S, Kwon SM and Jin HJ 2009 *Biomacromolecules* **10** 82
[48] Smart S, Cassady A, Lu G and Martin D 2006 *Carbon* **44** 1034
[49] Nepal D and Geckeler K 2006 *Small* **2** 406
[50] Lian Y, Maeda Y, Wakahara T, Nakahodo T, Akasaka T, Kazaoui S, Minami N, Shimizu T and Tokumoto H 2005 *Carbon* **43** 2750
[51] Sayes C, Liang F, Hudson J, Mendez J, Guo W, Beach J, Moore V, Doyle C, West J, Billups W, Ausman K and Colvin V 2006 *Toxicology Letters* **161** 135
[52] Arepalli S, Nikolaev P, Gorelik O, Hadjiev V, Holmes W, Files B and Yowell L 2004 *Carbon* **42** 1783
[53] Muller J, Huaux F, Fonseca A, Nagy J, Moreau N, Delos M, Raymundo-Piñero E, Béguin F, Kirsch-Volders M, Fenoglio I, Fubini B and Lison D 2008 *Chemical Research in Toxicology* **21** 1698
[54] Safadi B, Andrews R and Grulke E 2002 *Journal of Applied Polymer Science* **84** 2660
[55] Yang L, Kim P, Meyer H and Agnihotri S 2009 *J. of Coll. and Interface Science* **338** 128
[56] Yon JN and Lead J 2008 *Science of The Total Environment* **400** 396

Journal of Physics: Conference Series **304** (2011) 012088 doi:10.1088/1742-6596/304/1/012088

Proteomic study of human bronchial epithelial cells exposed to SiC nanoparticles

Caroline Tokarski[1], Seishiro Hirano[2], Christian Rolando[1]

[1]Miniaturisation pour l'Analyse, la Synthèse & la Protéomique (MSAP), USR CNRS 3290, and Protéomique, Modifications Post-traductionnelles et Glycobiologie, IFR 147, Université de Lille 1 Sciences et Technologies, 59655 Villeneuve d'Ascq Cedex, France

[2]National Institute for Environmental Studies, 16-2 Onogawa, Tsukuba, Ibaraki 305-8506, Japan

E-mail: Christian.Rolando@univ-lille1.fr

Abstract. The presented work proposes an optimized methodology for the study of cell exposure to nanomaterials at protein level. The study was investigated on proteins extracted from human bronchial epithelial cells exposed and non-exposed to silicon carbide nanoparticles (SiC). The analytical strategy was based on high resolution measurement using Fourier transform mass spectrometer 9.4 T. The methodology proposed succeeds in identifying over 300 proteins; most of the identified proteins are present in both exposed and non exposed cells to SiC nanoparticles. More interestingly, cytokines as Macrophage migration inhibitory factor protein could be identified only in the cells exposed to SiC nanoparticles indicating cell inflammatory response.

1. Introduction

Nanotechnology allows the ability to design many new materials [1-4] and devices with multiple applications, such as in medicine e.g. nanoparticles can represent drug delivery vectors [5]. However, nanotechnology also raises several concerns about the environmental impact [6] and the toxicity [7] of nanomaterials. In particular, the literature showed that toxicity of nanoparticles is highly related to the particle size and its chemical composition. For example, cytotoxicity studies related to the *in vitro* response of eukaryotic cells to TiO_2 nanoparticles exposure, showed that intracellular reactive oxygen species are generated [8,9]. Other study exposing human aortic endothelial cells to different metal oxide nanoparticles as iron oxide, yttrium oxide or zinc oxide showed induced cell inflammation depending on particle composition [10]. In this way, transcriptomic studies using cobalt nanoparticles pointed out increased transcription and activities of some matrix metalloproteinases [11]. Additional study based on electron microscopy shown that multi-walled carbon nanotubes associate with the plasma membrane of macrophages and disrupt the integrity of the membrane [12].

The presented work aims to highlight consequences of cell exposure to nanomaterials at the protein level using adapted proteomics methodology. The study was investigated on proteins extracted from human bronchial epithelial cells (BEAS-2B) exposed and non-exposed to silicon carbide nanoparticles

(SiC). As low abundant proteins like pro-inflammatory cytokine are mainly researched, very sensitive methodology was developed based on high resolution Fourier transform mass spectrometer 9.4 T.

2. Experimental section

2.1. Silicon carbide nanoparticles

The SiC nanoparticles were obtained from the Commissariat à l'Energie Atomique, CEA Saclay, Gif-sur-Yvette, France. The nanoparticle composition was 69.2% Si, 30.1% C and 0.88% O. The particle size and aggregation measurements were done by dynamic light scattering. In distilled water, depending on the ultrasonic duration from 2' to 6', mono disperse populations were measured with an average size from 222 nm to 182 nm. The elementary size of the particle is 36.8 nm. The density was measured at 3.09 g/cm^3. The BET surface was 52.75 m^2/g.

2.2. Cells

A human bronchial epithelial cell line BEAS-2B was cultured to early confluence in Dulbecco's modified minimum essential medium (DMEM) according to the detailed procedure published by Dr Seishiro Hirano et al. [13]. BEAS-2B cells were exposed to SiC at a concentration of 10 µg/ml for 24 hours according to the detailed procedure [13].

2.3. Protein extraction

Cells were washed 3 times using PBS buffer. Proteins were extracted using lysis solution containing 0.05 M tris(hydroxymethyl)aminomethane pH 8, sodium dodecyl sulfate 2%, 0.2 M dithiothreitol, 50 mM glycerol-2-phosphate, 5 mM sodium fluoride, 1 mM sodium orthovanadate, 1 tablet of protease inhibitor for 50 mL of buffer. A solution containing 0.05 M $MgCl_2$, DNase 2000U/mL, RNase 750 U/mL, 0.05 M tris(hydroxymethyl)aminomethane pH 8 was added in 1/10 proportion (v/v). The supernatant was separated from the cell membranes using centrifugation at 15 000 rpm during 30 min. Proteins were quantified using the commercial PlusOne 2-D Quant kit (GE Healthcare, Uppsala, Sweden); bovine serum albumin was used as calibrating agent.

2.4. Gel electrophoresis

The extracted proteins from the BEAS-2B cells exposed and non exposed to SiC nanoparticles were dissolved in Laemmli buffer [14]. The stacking gel was 4% and the separating gel was 12%. A quantity of 40 µg of extracted proteins per condition was run on 1D gel. Electrophoresis was carried out with running buffer containing 25 mM tris(hydroxymethyl)aminomethane, 0.1% sodium dodecyl sulfate (w/v), and 192 mM glycine, at 10 mA per gel until the bromophenol blue front had reached the bottom of the gel. Proteins were stained with SYPRO Ruby. Stained 1D gels were digitized using a Typhoon 9000 scanner (GE Healthcare). Image acquisition was performed at 550 V with a blue laser (λ_{exc}= 488 nm/λ_{em}= 610 nm).

2.5. Protein digestion

The protein digestion was investigated with preliminary strong protein reduction and alkylation [15]. Briefly, the gels were sliced all over in 40 bands. Each band was washed with 100 µl of 50 mM ammonium bicarbonate, pH 8.8, by agitation for 15 min with a vortex mixer. The proteins were reduced with 10 mM dithiothreitol for 2 h and alkylated with 55 mM iodoacetamide in the dark for 1 h. The supernatant was removed and the gel pieces were dehydrated with 100 µl of acetonitrile for 15 min and washed with 100 µl of 50 mM ammonium bicarbonate, pH 8.8, for 15 min. This operation was repeated three times. Gel pieces were completely dried with a Speed Vacuum (Concentrator 5301; Eppendorf, Hamburg, Germany) for 20 min. The dried gel pieces were rehydrated with 50 µl of 50 mM ammonium bicarbonate, pH 8.8, 5% acetonitrile containing 20 µg/µl of trypsin. The supernatant was removed and replaced by 50 mM ammonium bicarbonate, pH 8.8, 5% acetonitrile.

The hydrolysis was done overnight. Samples were completely dried with Speed Vacuum and 5 μl of deionized water, 0.1% formic acid, and 5% acetonitrile were added.

2.6. Peptide analysis

The peptide sequence spectra were obtained using nanochromatography (Ultimate LC system, Dionex, LC-Packings, Amsterdam, Netherlands) on-line with an Apex Qe 9.4 T Fourier Transform Ion Cyclotron Resonance Mass Spectrometer (Bruker Daltonics, Bremen, Germany). Starting from a volume of 1 μl of the protein digest solution, peptides were desalted and concentrated on a C18 preconcentration column (0.5 cm x 300 μm) and separated on a Pepmap C18 column (15 cm x 75 μm) at 200 nl/min solvent flow. The elution was performed using gradients of solvent A (95% H_2O, 5% ACN, 0.1% formic acid) and solvent B (20% H_2O, 80% ACN, 0.1% formic acid): 15 min in 100% solvent A, solvent B was increased to 100% in 130 min, then kept at 100% for 15 min, and finally decreased to 0% in 5 min. The column was allowed to equilibrate for 15 min between runs. The FT-ICR mass spectrometer is equipped with a nano-electrospray source. Detection was carried out in positive mode. A potential of 1.6 kV was applied on the needle. The detection parameters are broadband detection, 512 K acquisition size, and start mass at m/z 300. Fragmentation was triggered for any multiply charged species with a relative collision energy calculated in function of the selected species charge state by the Bruker AutoMS method in ApexControl software. Once selected and fragmented, eluted species were excluded from the MS/MS selection for 10 min.

2.7. Protein identification

The peptide monoisotopic mass was extracted from each MS and MS/MS spectrum using DataAnalysis software (Bruker Daltonics, Bremen, Germany). The generated mass data in Mascot format were post calibrated using an in-house program before extensive database search. Calibrated files were then submitted to local MASCOT software containing SwissProt database (release 56.8 of 10 February 2009: 410,518 sequences) for protein identification. The interrogations were performed with a restriction to the human species. Mass tolerance was set at 5 ppm on precursor ions and 5 mmu on fragment ions. A tolerance of three missed cleavage sites was set. Methionine oxidation and carbamidomethylation were entered as variable modifications.

The peptides related to the Macrophage migration inhibitory factor protein (Human species, MIF_HUMAN, P14174) from *in silico* trypsin digestion were search in the MS and MS/MS spectra from experiments related to cells non exposed to SiC using Extracted Ion Chromatograms (DataAnalysis software).

3. Results and discussion

The proposed study is based on multi-steps proteomic analysis including a first separation of extracted protein according to their molecular weights using 1D gel electrophoresis. To identify low abundant proteins, each step of the methodology had to be optimized. In details, proteins were extracted from human bronchial epithelial cells (BEAS-2B) exposed and non exposed to SiC nanoparticles using adapted protocol detailed in the experimental section part 2.3; it can be noted that sodium dodecyl sulfate, a strong detergent, was used due to its very high solubilizing power. The extracted proteins from exposed and non exposed cells to SiC were separated independently on 1D SDS-PAGE gel. The Figure 1 presents the separated proteins extracted from BEAS-2B cells exposed to SiC at a concentration of 10 μg/ml for 24 h. The gel was totally sliced in 40 band spots for each condition (exposed and non exposed cells) and each band was enzymatically hydrolyzed using trypsin; the digestion protocol was optimized, in particular, strong protein reduction and alkylation before digestion were used as detailed in the part 2.5. The generated peptides were analyzed using on-line nanoLC nanoESI-Qh-FT-ICR MS/MS. The nanoLC column used in this study was a C18 column of 75 μm inner diameter; C18 precolumn was also used to desalt and concentrate the sample. The chromatographic run was particularly optimized to obtain the highest number on identified peptides for each condition. The best results were obtained using 3 h experiment for each gel band as detailed in the experimental section part 2.6. The mass spectrometer settings were adapted to the sample composition, i.e., time cycle of an experiment (MS or MS/MS) including accumulation, transfer, excitation, detection and quench, run on ~3 s. In details, ions were accumulated 1 s in the hexapole, 0.01 and 2 s in the quadrupole collision cell respectively for MS and MS/MS; 0.0016 s was set for optics transfer and 0.01 s for electronic dwell time. The detection parameters were broadband detection, 512 K acquisition size, start mass at m/z 200 leading to 0.5243-s transient duration, allowing theoretical resolution of 190 000 at m/z 400. The obtained spectra were calibrated and treated using the bioinformatic tools (see part 2.7) for protein identification.

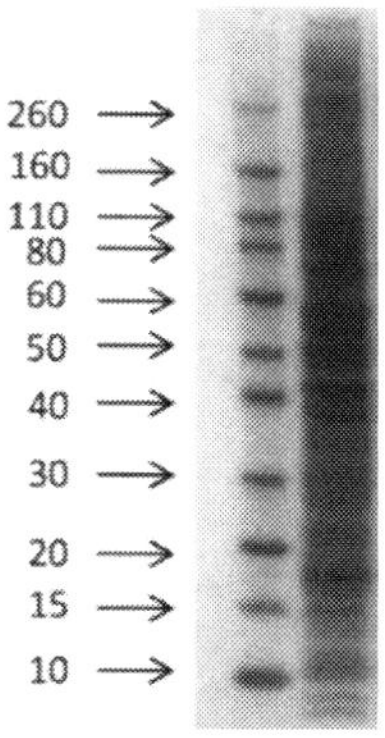

Figure 1. SDS-PAGE gel presenting the BEAS-2B cells exposed to SiC at a concentration of 10 µg/ml for 24 h (on the right) and molecular weight markers (on the left).

The molecular weights of marker proteins are indicated in kDa from 10 to 260 kDa.

The optimized proteomic methodology detailed here succeeds in identifying more than three hundred proteins. Most of the identified proteins were present in exposed and non exposed cells to SiC nanoparticles. However, some proteins were identified only in the cells exposed to SiC nanoparticles as for example pro-inflammatory cytokines as the Macrophage migration inhibitory factor protein (MIF). The Figure 2 presents the MS/MS spectrum allowing the identification of the sequence PMFIVNTNVPR relating to the position 2-12 of the MIF sequence using y_4^+ to y_{10}^+ ions. It can be noted that previous antibody-based study on BEAS-2B cells exposed to multi-walled carbon nanotubes (MWCNT) showed increased MIF concentrations in a MWCNT-dose dependent manner [13]. Based on high accuracy measurement MS and MS/MS data ($\Delta m < 1$ ppm on precursor ions and fragment ions), the lack of MIF in the cells non exposed to SiC was confirmed using extracted ion chromatograms of MIF constitutive peptides.

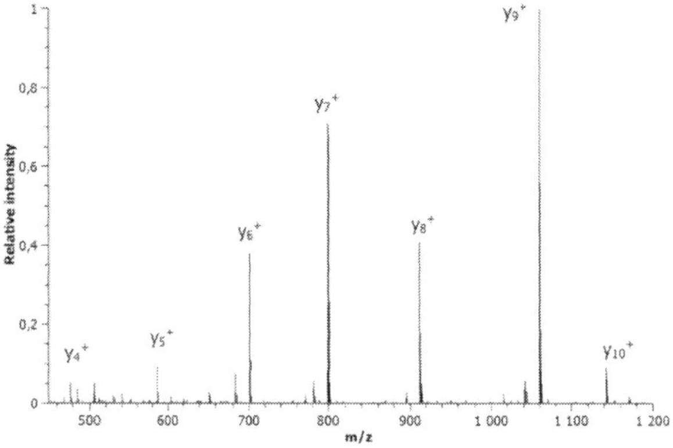

Figure 2. MS/MS spectrum of the peptide eluted at 14.7 min allowing the identification sequence PMFIVNTNVPR (precursor m/z 652.345, $\Delta m < 1$ ppm). The peptide represents the position 2 - 12 of the Macrophage migration inhibitory factor from Human species (MIF_HUMAN). The detected peptide is oxidized on the methionine amino acid.

4. Conclusion

The presented optimized proteomics methodology based on high resolution mass spectrometry allowed identifying cytokines in cells exposed to SiC nanoparticles indicating the pro-inflammatory reaction.

Acknowledgements

The Mass Spectrometry facilities used for this study are funded by the European community (FEDER), the Région Nord-Pas de Calais (France), the IBISA (Infrastructures en Biologie Santé et Agronomie) network, the CNRS, and the Université Lille 1 Sciences et Technologies. We acknowledge Carole Sentein and Frédéric Schuster for providing nanoparticles and The International Team in NanosafeTy members (TITNT).

References

[1] Sentein C, Guizard B, Giraud S, Yé C and Ténégal F 2009 *J. Phys.: Conf. Ser.* **170** 012013
[2] Schuster F 2010 *Clefs C.E.A. Eds Commissariat à l'énergie atomique, Paris* **59** 4
[3] Maximilien R, Schuster F, Sicard Y and Tardif F 2010 *Clefs C.E.A. Eds Commissariat à l'énergie atomique, Paris* **59** 84
[4] Sentein C, Schuster F and Tardif F 2009 *J. Phys.: Conf. Ser.* **170** 011001
[5] Ibeth G L, Czosnek C, Smycz A, Janik J F and Kozik A 2009 *J. Phys.: Conf. Ser.* **146** 012022.
[6] Nowack B and Bucheli T D 2007 *Environ. Pollution* **150** 5
[7] Hirano S 2009 *Environ. Health Prev. Med.* **14** 223
[8] Barillet S, Jugan M L, Simon-Deckers A, Leconte Y, Herlin-Boime N, Mayne-l'Hermite M, Reynaud C and Carrière M 2009 *J. Phys.: Conf. Ser.* **170** 012016
[9] Barillet S, Simon-Deckers A, Herlin-Boime N, Mayne-L'Hermite M, Reynaud C, Cassio D, Gouget B and Carrière M 2010 *J. Nanopart. Res.* **12** 61
[10] Gojova A, Guo B, Kota R S, Rutledge J C, Kennedy I M and Barakat A I 2007 *Environ. Health Perspect.* **115** 403
[11] Wan R, Mo Y, Zhang X, Chien S, Tollerud D J and Zhang Q 2008 *Toxicol. Appl. Pharmacol.* **233** 276
[12] Hirano S, Kanno S and Furuyama A 2008 *Toxicol. Appl. Pharmacol.* **232** 244
[13] Hirano S, Fujitani Y, Furuyama A and Kanno S 2010 *Toxicol. Appl. Pharmacol.* **249** 8
[14] Tokarski C, Cren-Olivé C, Fillet M and Rolando C 2006 *Electrophoresis* **27** 1407
[15] Tokarski C, Fillet M and Rolando C 2011 *Anal. Biochem.* **410** 98

Kinetic Behaviour of Nanoparticles Across the Biological Physiology

Claude Emond[1,2]

[1]BioSimulation Consulting Inc., 220E. Delaware Avenue #1182, Newark, DE, USA, 19711

[2]University of Montreal, Department of Environmental and Occupational Health, P.O. Box 6128, Main Station, Montreal, Quebec, Canada, H3C 3J7

E-mail: claude.emond@biosmc.com

Abstract. Nanotoxicokinetics is a subsection of the toxicology field that involves the study of kinetic displacement of nanoparticles (NPs) in an organism. Four different steps, namely absorption, distribution, metabolism and elimination (ADME), are involved in nanotoxicokinetics. However, only ADE will be covert in this mini review. Because of their size, NPs react differently than particulate matter larger than the nanometre unit in diameter. In the organism, a closer interaction between NPs and biological matrices, called nanotoxicodynamics, might increase the health effects. (Animals are usually in studies to evaluate the global interaction of NPs and biological matrices and to control and reduce the bias.) Understanding the different steps of kinetics is very important to increase the confidence of the amount of NP delivery in the target organ and to assess the level of risk. The objective of this work was to review the behaviour of the NPs interacting with the biological kinetic steps of the ADME and their limitations and constraints. Specifically, it was reviewed the impact of each of the four steps of nanotoxicokinetics, from exposure to elimination in the organism. Recent publications have provided some information on this issue, allowing for a better understanding on how the NPs behave across physiology; however, information is still lacking. We also systematically reviewed the ADME process, and supported our review with examples from the literature. We reviewed the two major factors that influence the absorption of NPs: enumerated biotransformation and elimination limitations. One of the focuses of this study was the interaction between NPs and biological matrices because the morphology and chemical properties may drive the potential for exposure. This paper present different examples of interactions find from literature. To study these interactions, we used a classical pharmacokinetic approach employed in the pharmaceutical industry and compared it to a dynamic predictive tool called the physiologically based pharmacokinetic model. This review would allow us to better interpret the behaviour of NPs. This review would also provide a better insight about the intake, site, and the disposition of NPs and would help identify the major consequences of the interaction of NPs with biological matrices. These interactions might have reversible or irreversible consequences for the integrity of the organism.

1. Introduction

The toxicology field is currently well positioned to understand the biological relevance between nanoparticles (NPs) and biological issues. Toxicological studies provide information on the defined disease mechanisms, based on dose–response relationships and related to the NPs characteristics that influence toxicity, including the size, surface area, chemistry or reactivity, solubility, and shape. Nanotoxicokinetics is a branch of toxicology and a new area in which this field evaluates NPs at the local and systemic levels. Nanotoxicokinetics is focused on understanding and predicting what is going to happen to the xenobiotic in the biota. Classically, we segment nanotoxicokinetics in four different steps, namely absorption, distribution, metabolism, and excretion (ADME). In the past, we did not imagine that particles might affect tissues at the systemic level. Toxicology studies aim to determine the biological plausibility of health effects from NPs and identify cascades of mechanisms that are causal for the gradual transition from the physiological status toward pathophysiological alterations, and eventually chronic diseases. In addition this field is considering the interaction between insoluble NPs and biological systems (e.g., body fluids, proteins, cells).

Many years later, we recognize that NPs range from 1 to 100 nanometres (nm), they can cross the biological membrane, and they can be translocated inside of cell organelles or entities such as the mitochondria, lysosome, nucleus, and others. Therefore, potential health risk will depend on the magnitude and nature of the exposures with NPs and on the physical behaviour of this particle related to dispersion, translocation and deposition. NPs have posed a new dilemma because they can migrate to body compartments away from the application site or deposition sites. In particular, because of their low uptake by macrophages, NPs are absorbed by endothelial cells, and they have access to cells in the epithelium, the interstitium, and the vascular walls [1]. Considering these challenges, it is important to understand the route taken by this new xenobiotic particulate structure

The objectives of our study were to (1) highlight some limiting steps related to nanotoxicokinetics and (2) emphasise the difficulty of finding pertinent information because there is a lack of data in this field. To respond to these objectives our work focused on the ADME steps. The absorption step explored the different route by which NPs can reach the systemic circulation and then target tissue in the organism. The major exposure routes were oral, inhalation, and the cutaneous routes. Following absorption, NPs are distributed in the organism, are free bind to proteins, and are incorporated in specialised cells. Biotransformation will only barely be introduced because commercial NPs are not easily degraded unless they are functionalised with a long chain peptide or unstable group NPs. Section 4 will discuss excretion or externalisation of the NPs and the impact of this action.

2. Absorption

Absorption is the procedure by which a xenobiotic crosses the biological barrier from the environment into the organism. For NPs, absorption can be occurring via oral route from gastrointestinal tract (GIT), in the lung (inhalation), and across the skin (cutaneous). These routes have different properties and degrees of penetration depending on the characteristics of the NPs.

2.1. Oral absorption

According to the literature, mice, rats, sheep, pigs, and cows can absorb NPs from the GIT. Florence observed that the ability of intact microparticles and NPs to be absorbed through the gut walls was different [2]. A literature consensus mentioned that the absorption increases with decreasing particle diameter. NPs such as polystyrene latex have been recognised as a useful model because they do not easily degrade. This study confirmed that NPs ranging from 50 to 100 nm in diameter reached the maximal absorption rate. However, NPs greater than 1 micrometre (μm) are being trapped in the Peyer's patches (PP) [3], [4]. It seems that at this size, NPs are not translocated to the systemic circulation. Several reviews on the oral uptake of NPs have been published.

Oral absorption is influenced by different characteristics related to the NPs (e.g., diameter, surface chemistry, surface ligands, shape and elasticity, physical and chemical stability) [5]. In general, smaller particles lead to a higher absorption rates below 1 μm. Particles greater than 3 μm are phagocytes and stay sequestered in the GIT cells. This is what we should observe in theory, but not necessarily the experimental observation. For example, even for NPs, the surface charge might limit absorption compared to non-ionic NPs. Biological surfaces such as an epithelia-containing receptor at the surface can express higher absorption rates. Shape and elasticity facilitate the passage across the barrier [3]

In the GIT, it seems well established that the PP would be mostly implicated in the process of particle uptake via the specialised epithelial cells. However, these specialised cells represent only approximately 10% of the cells cover the dome of these patches. Florence mentioned that these immunologic cells have their analogues in the bronchus-, larynx-, and nose-associated lymphoid tissue regions (referred to as BALT, LALT, and NALT, respectively) [3]. It is not surprising that PP part of the gut-associated lymphoid tissue (GALT) structure of the GIT wall, is considering that the M cells, as part of the lymphoid tissue, adapted a large range of materials to phagocytes. The uptake of particles, micro-organisms, and macromolecules by M cells occurs through adsorptive endocytosis by an approach of clathrin-coated pits and vesicles formation through an endocytosis and phagocytosis process [6]. This route has a limitation by itself because M cells (lymphoid tissues) occupy a relatively small region of the total GIT surface area, which is not compensated for by enhanced affinities. It has been shown that particulate systems can gain entry through normal gut epithelial cells (enterocytes); however, this absorption rate is not significant compared to the M cells. In addition, gastric mucus covers the entire wall of the GIT. Some authors have reported that mucus might inhibit the uptake of NPs through a dense structure in the viscous glycoprotein gel. However, there is a thought that entrapment of NPs in mucus actually delays the transit of these NPs down the gut and brings them closer to the absorption sites [3].

In attempts to increase the GIT absorption rate, many researchers have tried to improve this rate by increasing the complexities of the NPs using different approaches such as functionalisation, with a group change regarding the surface properties, and increasing the propensity for absorption. The functionalisation approach may help increase the absorption rate, but it may also increase the size of the NPs, which reduced absorption based on our observations. An equilibrium needs to be reached because when the chemical surface of the NP is altered, the possible interaction of the NPs with the target organ and the toxicity of the NPs also change. In this scenario, even if the absorption rates increase; the new NP must be characterised. Other techniques have been proposed such as coating NPs with molecules or peptides such as lectins, invasin, or internalin fragments. Even the results from such manipulation of the surface properties are often clear in vitro, the adhesion to the cell increases, and there is some evidence that absorption is enhanced [3]. The attempt is to reach the target organ, cells (e.g., M cells or normal epithelial cells), or tissues. Thus, a lack of clarity of the outcome might result from the chemical and physical instabilities of the coated particles. The increased adhesion from the interaction of different receptors may lead to enhanced uptake into cells, but this does not necessarily translate into increased transport through and out of the cell.

In the situation in which the epithelium itself is not the target, movement of particles through the M cells or epithelial cells is only the first part of this voyage, which involves passage through the mesenteric lymph, filtration in the lymph nodes, and transfer to the blood and perhaps extravasation [3]. In a study by Carr and colleagues, the researchers said that estimation between the uptake and the volume of NP is not straightforward [7]. Other factors such as maceration may overestimate the uptake into the tissue. This observation was clearly demonstrated by microscopy showing that in the small intestine, most NPs are luminal or on the surface [7]. In this study, the researchers did not find a significant different in the uptake between particles; however, the sizes of the particles were greater than 1 μm.

Transcytosis of NPs across normal absorptive epithelium may be possible, even though the capacity is considered to be limited. The transcellular transport of NPs generally begins with the uptake by one of the following endocytotic mechanisms: pinocytosis, macropinocytosis, or endocytosis [4]. After the absorption, the amount of NPs can reach the blood circulation, and if enough of an amount is present, then this can induce a therapeutic or a toxic effect [4]. The amount of

NPs absorbed is variable and depends of the size and the surface chemistry of NPs. Several studies show that between 2% and 3% of NPs orally exposed are absorbed by the GIT tissue [2]. For researchers who want to increase the intestinal absorption, they need to understand and optimise the linkage between surface chemistry of the NPs and the M cells [4]. In general, the efficiency of the uptake for NPs is less than or equal to 100-nm in size and is approximately 15- to 250-fold higher compared to a larger size [8]. In addition, from the same region, PP tissue has 200-fold higher uptakes than a non-PP tissue from the same region.

In a histological study by Desai and colleagues, they demonstrated that NPs of 100 nm were diffused throughout the submucosal layers as opposed to the larger sizes, which were predominantly found in the tissue surface [8]. Jani and colleagues exposed rat GIT mucosa to different sizes of NP (i.e., 50 nm, 100 nm, 300 nm, 500 nm, 1 μm, 3 μm) every day for 10 days [9]. After 48 hours post-exposure, the researchers measured the amount of NPs in different regions. Several authors discovered that NPs less than 100 nm accumulated in GIT tissue, but if the sizes were greater than this cut off, then the accumulation rate was significantly lower [8, 9]. In another study, colloidal gold was administered in water for 7 days to mice; the results showed a large distribution across several organs, including kidney, spleen, liver, blood, and brain [10]. Even after 12 hours of post-exposure, NPs were still measured in the GIT which suggests sequestration in cells or at the surfaces.

2.2. Inhalation

As describe by the International Commission on Radiological Protection (ICRP) in its document, the Commission measured the distribution probability of particles related to their size [11]. The ICRP study showed that NPs with a diameter less than 10 μm have a greater probability of penetrating beyond the head airways. Particles less than 100 nm in aerodynamic diameter have a significant probability of reaching the alveolar region of the lungs. In fact, there is at least a 50% probability that particles less than 4 μm in aerodynamic diameter will reach this region [12]. Regarding the smaller diameter sizes of particles, another factor called inertia is secondary to Brownian diffusion in determining deposition, leading to particles penetrating deep into the lungs and diffusing to a large lung surface area presented in the alveolar region [12]. According to the ICRP, at the 1-nm-in-diameter size, almost everything will be deposited on the head and tracheobronchial region of the lung, which means that almost nothing will reach the alveolar region because the negligible mass significantly reduces the velocity of the NPs.

For NPs to deposit in the inner wall of the lungs, the particles may induce two types of toxicities (i.e., a local toxicity or a systemic toxicity), depending on their intrinsic properties. In addition to the deposition site, different factors will influence clinical observations. For instance, the deposition site will depend on the size and concentration of NPs, the durability of NPs (i.e., insoluble aqueous), and the stability of the NPs (stable NPs will have higher durability and defence immune system in the lung area). The last important factor focuses on mucociliary clearance (acting in the upper airways) and the macrophages' immune system (acting in the lower airways and alveolar region), which tries to actively remove the deposition of NPs.

The local toxicities may cause inflammation, oxidative stress, tissue damage, and disease. Many of the biological mechanisms observed in the literature involved particle-related lung diseases (e.g., oxidative stress, inflammation, production of cytokines, chemokines, cell growth factors) [13]. Much of our understanding about the key factors that influence the biological reactivity and toxicity of airborne particulate matter has come from animal toxicological studies. These factors, including size, surface area, surface chemistry, solubility, and shape, will influence both the deposition of NPs in the lungs and the biological responses observed. Maynard and Kuempel reported that NPs, including carbon black (12 nm), elemental carbon (90 nm), and diesel exhaust particulate (120 nm), caused various cytoskeletal dysfunction [12]. These dysfunctions included impaired phagocytosis, inhibition of cell proliferation, and decreased cell viability in primary alveolar macrophages from dogs and mice from alveolar macrophage cell lines in 24 hours depending on the dosages [12]. Monteiro-Riviere and colleagues reported an important observation, saying that if the carbon nanotubes reach the lung tissue, then they were more toxic than similarly chemically composed NPs such as carbon black or quartz dust (actually these two NPs are known for their lung toxicities) [14]. Monteiro-Riviere and colleagues argued that the observation was due to the tendency of carbon nanotubes to self-aggregate, and then remove themselves from the controlled condition [14]. These agglomerated NPs have a much higher residence time than dispersed NPs.

NPs that leave the lungs might enter the blood circulation, and then may cause endothelial cell injury (of the blood vessels) and prothrombotic effects [15]. Recent studies have indicated that NPs depositing in the nasal region may be transported to the olfactory bulb via the olfactory nerves [16]. The size of NPs may also allow cells and cellular organelles to enter. In a study of concentrated NPs from air pollution in human bronchial epithelial cells and mouse alveolar macrophages, the ultra-fine fraction (less than 100 nm) was found to penetrate into cells and localise in mitochondria, causing oxidative damage to mitochondrial membranes [17]. Unlike low molecular–weight drugs, which penetrate cells easily, the cellular uptake of polymeric pro-drugs is restricted to the endocytic route, which basically means that pro-drugs are delivered to the cellular lysosomes [1]. However, other studies of drug delivery across the blood–brain barrier further confirmed the importance of surface properties, showing that particle surface components may bind to the apolipoprotein E (ApoE) receptor, which mediates crossing of this highly complicated tight barrier [1]. The olfactory neuronal pathway represents a significant exposure route of Central Nervous System (CNS) tissue to inhaled solid manganese oxide NPs. In rats, which are obligatory nose breathers, translocation of inhaled NPs along neurons seems to be a more efficient pathway to the CNS, than via the blood circulation across the blood–brain barrier for humans. Given that this neuronal translocation pathway was also demonstrated in nonhuman primates, it is likely to be operative in humans as well [18].

2.3. Cutaneous route

The cutaneous route of exposure represents an important potential route of exposure for NPs because of the skin surfaces with workplace environment [14], [19]. The risk caused by NPs or every other chemical especially after an in vivo exposure would depend upon the skin absorption if no irritant NPs are present. The process of passage across the skin layer will depend on passive diffusion, and structural organisation will influence this diffusion. The stratum corneum provides a slow diffusion capability; however, dermis and epidermis have easier diffusion capabilities. In general, literature reported that normal skin diffusion is limited to NPs that are 100 nm in diameter. Intrinsic factors such as the chemistry surface shape of NPs can also influence the diffusion. However, more research should be conducted on this issue because there is no real consensus in existing literature. Kohli and Alpar conducted a study on percutaneous NPs and found that particles were able to permeate skin if they had a negative charge and were 50 and 500 nm in diameter [20]. This result suggests that negative particles with sufficient charges may be the ideal carriers for drugs. This is different from what we observe; usually the best particles for percutaneous permeation would be when they are lipophilic neutral and less than 100 nm in diameter [20]. The vehicle containing NPs is also a parameter that can influence the toxicity. However, at least one group of researchers (i.e., Schulz and colleagues) found that cutaneous absorption interaction was limited to the stratum corneum [21]. Multiwalled carbon nanotubes are capable of entering the keratinocytes and driving interleukin-releasing effects.

Diffusion across the stratum corneum is normally the rate-limiting step for percutaneous absorption. A pig is a good model because its structure is similar to a human, and the technique of porcine preparation has been validated by numerous investigators. The permeation of NPs throughout the models seems to be an accurate predictive tool for human extrapolation. Tan and colleagues conducted a human study on the percutaneous absorption of titanium dioxide (TiO_2) from sunscreens [22]. This study was the first one to present a direct linkage between the utilisation of sunscreen and the absorption of TiO_2 in humans.

3. Distribution

After the absorption from oral, inhalation, or cutaneous routes, these NPs are then translocated in the systemic circulation and delivery to different organs, including the liver, the spleen, the kidneys, the heart, and the brain, where they may be deposited [18]. Many target organs are being studied to measure the impact of these exposures; however, only a few groups are working on the kinetic distribution of NPs in these organs.

A study conducted by Kreyling and colleagues indicated that 1 week after inhalation, only a very small fraction of ultra-fine iridium particles (18 and 80 nm in diameter) had access to systemic circulation and extrapulmonary organs [23]. Chemical composition and physical structure of the NPs surface may be an important determinant in influencing systemic translocation of ultra-fine particles [23]. De Jong and colleagues exposed rats aged 6 to 8 weeks to a single oral exposure of polystyrene NPs from 10 to 250 nm in diameter at concentrations varying from 77 to 108 microgrammes per litre (μg/mL) [19]. NPs less than 100 nm were found mainly in the blood and liver 24 hours post-exposure. In addition, De Jong group's found a lower amount of NPs expressed in percentage of doses in the spleen and kidney and only trace levels in the lungs, the testis, the heart, the brain, and the thymus [19].

These studies demonstrated that NPs, after deposition, are rapidly distributed across tissue compartments of the lung, and that these NPs may move between the tissue compartments. The NPs eventually reach the capillary lumen and may penetrate into circulating cells and constituents (e.g., erythrocytes, blood macrophages). Thereby, they may be distributed into other organs of the body, such as the liver, the heart, the kidneys, and even the brain [9, 19]. Nemmar and colleagues conducted an interesting human study in which they used the inhalation of radiolabelled technetium 99 carbon NPs less than 100 nm in diameter. The findings showed that radiolabelled technetium 99 carbon was detected in the blood 1 minute after inhalation, reached a maximum after 10 to 20 minutes, and remained for more than 60 minutes [24]. A gamma camera showed substantial radioactivity in the liver and other organs.

4. Excretion

In a study of rats by Peters and colleagues, a single intravenous dose of a fullerene labelled with technetium 99m showed a rapid distribution in different organs [18]. The researchers measured the biodistribution in different organs in mice and rabbits. They observed a decrease of the NPs in the organ tissues only between 1 and 3 hours post-exposures, except for liver at 3 hours, where it was higher than at 1 hour post-exposure. The elimination of the fullerene was mainly via urinary route [25]. In another comparison, Ogawara and colleagues administered a single intravenous dose of polystyrene microsphere (50 and 500 nm in diameter) into rats [26]. The distributions were independent of the size. The researchers measured the half life (T1/2) for the alpha phase and terminal phase at approximately 1 and 53 hours respectively [26]. They believed that urinary elimination seemed to be the major elimination route. Singh and colleagues exposed mice to a single intravenous dose of single walled carbon nanotubes (1 nm in diameter and a length of 300 to 1,000 nm) [27]. Distribution was very quick across the organs. Excretion occurred via the urine pathway. In the same experiment, the researchers also studied the impact of the exposure route on the elimination half-life. Depending on the absorption route, the other local mechanism previously mentioned may also contribute to the elimination. In a study by Furumoto and colleagues, the researchers demonstrated that NPs can also be excreted via the biliary tract [28].

5. Discussion and conclusion

The interaction between NPs and the biological system is a huge challenge for toxicokinetic. The toxicokinetics discipline describes the absorption, the distribution the metabolism process, and the excretion of the xenobiotic (the NPs). At this point, our focus is to apply the approach that we have in place in case we need to use the same kinetic approaches or develop new tools. However, it is important to point out that complementary approaches will be required to be able to adequately describe the experimental observation.

At the beginning of these experimental observations, we believed that inhalation was the major exposure route, with negligible contribution from the skin. However, observations such as those by Ryman-Rasmussen and colleagues [29] or results from Monteiro-Riviere and colleagues [14] clearly suggest that skin is surprisingly permeable to nanomaterials with diverse physicochemical properties and may serve as a portal of entry for localised, and possibly systemic, exposure of humans to quantum dots and other engineered nanoscale materials [29]. For skin absorption, these findings suggest that the toxicology of these structures must be assessed before widespread public exposure so that appropriate protective measures can be developed [14]. The status of this absorption will change with the arrival of new transformer NP (NP with exotic functionalisation) on the market.

This short review also points out that NPs can rapidly translocate from the lung into the cells, and then reach the systemic circulation. There is increasing evidence that the accumulation of NPs in different organs (e.g., heart, brain) shows sign of oxidative stress or other stress toxic effects. Even oxidative stress can occur for many other reasons, this significant addition may cause a long-term health effect. A long-term study is required to quantify the long-term risk. Given that this neuronal translocation pathway was also demonstrated in nonhuman primates, it is also likely to be operative in humans [30]. New technology approaches, such as a proteomic analysis profiling approach in which we detected much below the clinical sign effect of protein modification, will be useful to observe before the health effect is irreversible. However, these sophisticated approaches will only work if they are supported by a rigorous characterisation of the NPs in media.

References

[1] Borm, P.J. and W. Kreyling, Toxicological hazards of inhaled nanoparticles--potential implications for drug delivery. J Nanosci Nanotechnol, 2004. 4(5): p. 521-31.

[2] Florence, A.T., The oral absorption of micro- and nanoparticulates: neither exceptional nor unusual. Pharm Res, 1997. 14(3): p. 259-66.

[3] Florence, A.T., Nanoparticle uptake by the oral route: Fulfilling its potential? Drug Discovery Today: Technologies, 2005. 2(1): p. 75-81.

[4] des Rieux, A., et al., Nanoparticles as potential oral delivery systems of proteins and vaccines: a mechanistic approach. J Control Release, 2006. 116(1): p. 1-27.

[5] Hussain, N., V. Jaitley, and A.T. Florence, Recent advances in the understanding of uptake of microparticulates across the gastrointestinal lymphatics. Adv Drug Deliv Rev, 2001. 50(1-2): p. 107-42.

[6] Qaddoumi, M. G., Gukasyan, H. J., Davda, J., Labhasetwar, V., Kim, K. J., and Lee, V. H. (2003). Clathrin and caveolin-1 expression in primary pigmented rabbit conjunctival epithelial cells: role in PLGA nanoparticle endocytosis. Mol Vis. 9, 559-568.

[7] Carr, K.E., et al., The effect of size on uptake of orally administered latex microparticles in the small intestine and transport to mesenteric lymph nodes. Pharm Res, 1996. 13(8): p. 1205-9.

[8] Desai, M.P., et al., Gastrointestinal uptake of biodegradable microparticles: effect of particle size. Pharm Res, 1996. 13(12): p. 1838-45.

[9] Jani, P., et al., Nanoparticle uptake by the rat gastrointestinal mucosa: quantitation and particle size dependency. Journal of Pharmacy and Pharmacology, 1990. 42(12): p. 821-6.

[10] Hillyer, J.F. and R.M. Albrecht, Gastrointestinal persorption and tissue distribution of differently sized colloidal gold nanoparticles. J Pharm Sci, 2001. 90(12): p. 1927-36.

[11] ICRP. Human respiratory tract model for radiological protection. publication 66, 1-459. 1994. Pergamon. Annals of ICRP.

[12] Maynard, A.D. and E.D. Kuempel, Airborne nanostructured particles and occupational health. Journal of Nanoparticle Research, 2005. 7(6): p. 587-614.

[13] Castranova, V., and Vallyathan, V. (2000). Silicosis and coal workers' pneumoconiosis. Environ Health Perspect 108 Suppl 4, 675-684.

[14] Monteiro-Riviere, N. A., Nemanich, R. J., Inman, A. O., Wang, Y. Y., and Riviere, J. E. (2005). Multi-walled carbon nanotube interactions with human epidermal keratinocytes. Toxicology letters 155(3), 377-384.

[15] Nemmar, A., Hoylaerts, M. F., Hoet, P. H. M., and Nemery, B. (2004). Possible mechanisms of the cardiovascular effects of inhaled particles: systemic translocation and prothrombotic effects. Toxicology letters 149(1-3), 243-253.

[16] Oberdorster, G., Sharp, Z., Atudorei, V., Elder, A., Gelein, R., Kreyling, W., and Cox, C. (2004). Translocation of inhaled ultrafine particles to the brain. Inhal. Toxicol 16(6-7), 437-445.

[17] Li, N., Sioutas, C., Cho, A., Schmitz, D., Misra, C., Sempf, J., Wang, M., Oberley, T., Froines, J., and Nel, A. (2003). Ultrafine particulate pollutants induce oxidative stress and mitochondrial damage. Environ Health Perspect 111(4), 455-460.

[18] Peters, A., et al., Translocation and potential neurological effects of fine and ultrafine particles a critical update. Part Fibre Toxicol, 2006. 3: p. 13.

[19] De Jong, W.H., et al., Particle size-dependent organ distribution of gold nanoparticles after intravenous administration. Biomaterials, 2008. 29(12): p. 1912-9.

[20] Kohli, A.K. and H.O. Alpar, Potential use of nanoparticles for transcutaneous vaccine delivery: effect of particle size and charge. International Journal of Pharmaceutics, 2004. 275(1-2): p. 13-7.

[21] Schulz, J., et al., Distribution of sunscreens on skin. Adv Drug Deliv Rev, 2002. 54 Suppl 1: p. S157-63.

[22] Tan, M.H., et al., A pilot study on the percutaneous absorption of microfine titanium dioxide from sunscreens. Australas J Dermatol, 1996. 37(4): p. 185-7.

[23] Kreyling, W. G., Semmler, M., Erbe, F., Mayer, P., Takenaka, S., Schulz, H., Oberdorster, G., and Ziesenis, A. (2002). Translocation of ultrafine insoluble iridium particles from lung epithelium to extrapulmonary organs is size dependent but very low. J Toxicol Environ Health A 65(20), 1513-1530

[24] Nemmar, A., Hoet, P. H. M., Vanquickenborne, B., Dinsdale, D., Thomeer, M., Hoylaerts, M. F., Vanbilloen, H., Mortelmans, L., and Nemery, B. (2002). Passage of Inhaled Particles Into the Blood Circulation in Humans. Circulation 105(4), 411-414.

[25] Qingnuan, L., et al., Preparation of (99m)Tc-C(60)(OH)(x) and its biodistribution studies. Nucl Med Biol, 2002. 29(6): p. 707-10.

[26] Ogawara, K., et al., Hepatic uptake of polystyrene microspheres in rats: effect of particle size on intrahepatic distribution. J Control Release, 1999. 59(1): p. 15-22.

[27] Singh, R., et al., Tissue biodistribution and blood clearance rates of intravenously administered carbon nanotube radiotracers. Proc Natl Acad Sci U S A, 2006. 103(9): p. 3357-62.

[28] Furumoto, K., et al., Biliary excretion of polystyrene microspheres depends on the type of receptor-mediated uptake in rat liver. Biochim Biophys Acta, 2001. 1526(2): p. 221-6.

[29] Ryman-Rasmussen, J.P., J.E. Riviere, and N.A. Monteiro-Riviere, Penetration of intact skin by quantum dots with diverse physicochemical properties. Toxicol Sci, 2006. 91(1): p. 159-65.

[30] Elder, A., et al., Translocation of inhaled ultrafine manganese oxide particles to the central nervous system. Environ Health Perspect, 2006. 114(8): p. 1172-8.

The Ne^{3}LS Network, Québec's initiative to evaluate the impact and promote a responsible and sustainable development of nanotechnology

Charles-Anica Endo[1], Claude Emond[2], Renaldo Battista[3], Marie-Hélène Parizeau[4] and Catherine Beaudry[5]

[1]CHU Sainte-Justine Research Center, 3175, chemin de la Côte-Sainte-Catherine, bureau 8228, Montréal, Québec, Canada H3T 1C5

[2]Université de Montréal, Département de santé environnementale et santé au travail. Pavillon Marguerite-d'Youville, 2375, chemin de la Côte Ste-Catherine, local 4095 Montréal, Québec, Canada H3T 1A8

[3]Université de Montréal, Département d'administration de la santé, Pavillon Mont-Royal 1420, boul. Mont-Royal, Montréal, Québec, Canada H2V 4P3

[4]Université Laval, Bureau 536, Pavillon Félix-Antoine-Savard, 2325, rue des Bibliothèques, Québec, Canada, G1V 0A6

[5]École Polytechnique de Montréal, Département de mathématiques et de génie industriel, C.P. 6079, succ. Centre-ville, Montréal, Québec, Canada, H3C 3A7

E-mail: claude.emond@umontreal.ca

Abstract. The spectacular progress made by nanosciences and nanotechnologies elicits as much hope and fear. Consequently, a great number of research and training initiatives on the **e**thical, **e**nvironmental, **e**conomic, **l**egal and **s**ocial issues regarding **n**anotechnology development (Ne^{3}LS) are emerging worldwide. In Québec, Canada, a Task Force was mandated by NanoQuébec to conceive a Ne^{3}LS research and training strategy to assess those issues. This Task Force brought together experts from universities, governments or industry working in nanosciences and nanotechnologies or in Ne^{3}LS. Their resulting action plan, made public in November 2006, contained several recommendations, including the creation of a knowledge network (Ne^{3}LS Network). In the following years, after consulting with numerous key players concerned with the possible impacts of nanosciences and nanotechnologies in Québec, the Ne^{3}LS Network was launched in January 2010 in partnership with the Fonds québécois de la recherche sur la nature et les technologies, the Fonds québécois de la recherche sur la société et la culture and the Fonds de la recherche en santé du Québec, NanoQuébec, the Institut de recherche Robert-Sauvé en santé et en sécurité du travail as well as the University of

Montreal. Its objectives are to 1) Foster the development of Ne^3LS research activities (grants and fellowships); 2) Spearhead the Canadian and international Ne^3LS network; 3) Take part in the training of researchers and experts; 4) Encourage the creation of interactive tools for the general public; 5) Facilitate collaboration between decision-makers and experts; 6) Involve the scientific community through a host of activities (symposium, conferences, thematic events); 7) Build multidisciplinary research teams to evaluate the impact of nanotechnology.

Introduction

Nanoscience and nanotechnology raise many challenges for stakeholders involved in the innovation process. Challenges are even greater if we consider the tremendous potential of this enabling technology. As a matter of fact, "many experts, researchers, engineers, social scientists and policy makers believe that nanotechnology will lead to important changes in society and that these changes may be of the same magnitude as was the introduction of computers." (Science Metrix, 2005) However, technological breakthroughs need more than a strong base of science and favorable economic context to create wealth. In fact, the 20th century has shown with the turmoil surrounding biotechnology, especially in the case of genetically modified organisms (GMO), that we also need new approaches for integrating technology into society. When it comes to nanotechnology, the specter of GMO is now omnipresent in the minds of policymakers, scientists and the general population. How can we avoid a crisis that could ruin the emergence of a technology and its collective efforts of achieving good for society? One of the solutions to avoiding the pitfalls of nanotech development is to target our interventions toward a more responsible framework. As stated by the Commission de l'éthique de la science et de la technologie (CEST) in Québec, "Nanotechnology is an emerging sector of activity, and one that draws on a host of disciplines. For these reasons, steps should be taken to build the foundation of systematic and responsible development." (CEST 2006) Because nanotechnology has not reached its full potential and has not yet been fully integrated into our markets, stakeholders can still make choices and try to address concerns. Though, far from being ideal, the present nanotechnology situation can still be seen as advantageous in comparison with the one with GMO. It remains possible to better position ourselves toward value creation and responsible development.

In regards to this situation, many countries around the globe are investing in social and safety aspects of nanotechnology development. For example, the United States, through the NNI, will allocate US$ 116.9 M in 2011 on environmental, health and safety issues, as well as US$ 34.8 M for education and societal dimensions. (NNI 2010) This represents around 8.6% of the overall NNI budget of US$ 1761.6 M.

Despite interesting initiatives[1], the situation in Canada still lags behind many European countries, the U.S. and Japan. The Canadian research community is still waiting for a national strategy on nanotechnology and a clear commitment from the government to assess crucial questions about nanotechnology development. In Canada, the acronym NE^3LS is used for these questions. As defined by Lorraine Sheremeta "NE^3LS is an acronym that stands for the Ethical, Environmental, Economic, Legal and Social issues associated with Nanotechnology. This acronym is a simple adaptation of the GE^3LS acronym that relates the same issues to Genetics and Genomics." (Sheremeta 2005) However, in Québec our scope also includes health and safety issues regarding nanotechnology.

[1] For example, University of Alberta and University of Toronto joint initiative, Position statement by the Commission de l'éthique de la science, NanoQuébec's task force on ethics environment and society.

Many years have passed since Mnyusiwalla, Daar and Singer (Mnyusiwalla & al. 2003) urged us to mind the gap between ethics and nanotechnologies, and since Sheremeta stated: "Though there is evidence of *ad-hoc* progress in the area of NE³LS research, there remains a need for a well-defined, highly coordinated and appropriately funded NE³LS research agenda in Canada. The Federal Government, in cooperation with the governments of all Canadian provinces and territories must joint with industry to support the development of a research agenda that will address the unique issues, both foreseen and unexpected, that nanoscience and nanotechnology will raise."(Sheremeta 2005) Presently, the Canadian government has not yet established a national initiative to coordinate research in this field and assess Ne³LS questions. The momentum lies mainly on a provincial level.

The province of Québec is the first province in Canada to launch various initatives to assess NE³LS- related issues, supported by several agencies. One of the cornerstones of the work done in Québec was the position statement *Ethics and nanotechnology: a basis for action*, released by the CEST. Among the CEST recommendations is a proposal to create a multidisciplinary research program to assess the impact and risks associated with nanotechnology. In the meantime, NanoQuébec, the provincial organization responsible for structuring nanotechnology, in partnership with the granting agencies of Québec (Fonds québécois de la recherche sur la nature et les technologies (FQRNT), Fonds québécois de la recherche sur la société et la culture (FQRSC), Fonds de recherche en santé du Québec, (FRSQ) and the Institut de Recherche Robert-Sauvé en santé et en sécurité du travail (IRSST) developed its own framework. (Groupe de réflexion 2006). The framework was conceived with a panel of experts and included a strategy for NE³LS research and training as well as recommended actions to build capacity of NE³LS expertise in Québec. This process resulted in the creation of a Task Force on society, environment and ethics. (Groupe de réflexion 2006). This Task Force, in accordance with the CEST's recommendations, then proposed the creation of a NE³LS Knowledge Network. As a result of this initiative, the Québec Government, via its funding agencies (FQRNT, FQRSC, FRSQ), has dedicated funds for NE³LS research in Québec through special scholarships and research grants. The Ministère du Développement économique, de l'Innovation et de l'Exportation (MDEIE) and the IRSST are also providing financial support to the network. (Groupe de démarrage NE³LS, 2007).

The Ne³LS Network

In January 2010, after consulting with numerous key players concerned with the possible impact of nanosciences and nanotechnologies in Québec, the Ne³LS Network was launched in partnership with the FQRSC, the FQRNT, the FRSQ, NanoQuébec, the IRSST as well as the University of Montreal.

Its objectives are to:

1. Foster the development of Ne³LS research activities (grants and fellowships);
2. Spearhead the Canadian and international Ne³LS network;
3. Take part in the training of researchers and experts;
4. Encourage the creation of interactive tools for the general public;
5. Facilitate collaboration between decision-makers and experts;
6. Involve the scientific community through a host of activities (symposium, conferences, thematic events);
7. Build multidisciplinary research teams to evaluate the impact of nanotechnology.

Based at the CHU Sainte-Justine Hospital, which is affiliated with the University of Montreal, the Ne³LS Network aims at establishing a knowledge network. This knowledge network will connect with different actors of nanotech development (academia, industry, governments and public) and become a sharing platform. Consequently, all parties in this field will exchange knowledge and will eventually help to better shape public policies in this field. However, as a foundation of this initiative, a research network will be created to stimulate the production of knowledge in fields covered by NE^{3}LS. The main scope of this network is to monitor the nanotech development process and supply it with an appropriate knowledge in social sciences as well as risk assessment, thereby maximizing its benefits while answering its key issues. Consequently, we aim at developing a balanced approach geared toward responsible and sustainable development.

This network relies on three mains research axes, which are defined as follows:

Axis 1: Nanoscience and Nanotechnologies: How do we Define Them?
This axis examines the importance of clearly defining nanosciences and nanotechnologies, taking into account their scientific, social, political, economic and philosophical dimensions. Labelling a practice or a product as nanotechnological is not without consequence depending on the socioeconomic, cultural or ideological context. Therefore, it becomes all the more important to identify the impact that terminology has on society and perception, as these choices provide reference points for a technological development that will either align or disconnect with societal values worldwide. Moreover, close examination of the dialogues between technology actors such as government, industry, science and society, can prove beneficial.

Axis 2: Interdisciplinary Risk Analysis and Acceptability
With every technological breakthrough comes doubts posing as potential risks. Whether financial, legal, technological, environmental or health-related, these risks must be assessed. Each discipline uses its own definitions and assessment strategies to measure risk limits. Nanotechnology is particularly invested with great possibilities, be it as a means of producing new medication, improving diagnostic tests or creating greener or better performing materials. Scientific and industrial literature is teeming with discoveries in the field of nanotechnology. However, this progress must always be made with due consideration to humans and their environment. There are over 1,000 "nanoproducts" on the market today, even though studies measuring their effects on humans and the environment are yet to be completed. Furthermore, national and international regulations need to be put in place. Risk assessment thus represents a massive undertaking, requiring a multidisciplinary approach aimed at analyzing our ideas and perceptions on risk and its acceptability. In this sense, risk and risk assessment criteria will need to be clearly defined.

Axis 3: Governance of Nanotechnology
This axis investigates the regulatory modes governing nanotechnology practices as defined by the various players involved in its economic and technological development, such as governments, markets, the general public and the scientific community. Advances in nanotechnology have not only prompted questions concerning legislation and personal accountability, but have also given rise to issues related to democracy and ethics. Hence, this axis proposes not only to analyze the rules of economic and technological development of nanotechnologies, but also to understand their implications. Knowledge translation and dialogues between science and society are also explored. Research capacity building in this field was the clear mandate given by the Government of Québec to the Ne^{3}LS network. In order to meet this expectation and promote activities and expertises, the network relies on research grants and scholarships. As previously mentioned, the funding agencies of Québec have created 6 dedicated scholarships per year (3 at the Doctorate/Ph.D. level and 3 at the Master's level). In addition, the network is financing pilot projects with seed money, hoping to leverage more important provincial and federal funding.

At present, the network's start-up phase is focused on maximizing interdisciplinary research and supporting educational activities. The network then seeks to enhance knowledge transfer activities, thus enabling it to operate at full potential in the years to come. The network is seeking collaborators in all fields related to NE^3LS and would like to establish links with Asia, USA and Europe.

References

[1] Commission de l'éthique de la science et de la technologie. Position statement, Ethics and nanotechnology: a basis for action 2006 [online] http://www.ethique.gouv.qc.ca/index.php?option=com_docman&task=cat_view&gid=18&Itemid=14&phpMyAdmin=bb1c4a2d0bactf758937
[2] Groupe de démarrage NE^3LS. 2007. Réseau de connaissance sur les NE^3LS, Rapport final et recommandations du groupe de démarrage NE^3LS. Québec, 8p.
[3] Groupe de réflexion. 2006. Plan d'action 2006-2009 : Proposition du groupe de réflexion pour une stratégie de recherche et de formation concernant les aspects sociaux, économiques, éthiques et environnementaux de l'innovation fondée sur les nanotechnologies. Montreal, 26 p.
[4] Mnyusiwalla, Anisa, Abdallah S. Daar and Peter A. Singer, "'Mind the Gap': Science and Ethics in Nanotechnology", Nanotechnology, 14: R19-R13, 2003.
National Nanotechnology Initiative. Supplement to the president's FY 2011 budget. February 2010 [online] http://www.nano.gov/NNI_2011_budget_supplement.pdf
[5] Science Metrix. Canadian Stewardship Practices for Environmental Nanotechnology, Stéphane Bergeron and Éric Archambault, on behalf of Environment Canada, March 2005 [online] http://www.ethique.gouv.qc.ca/index.php?option=com_docman&task=cat_view&gid=18&Itemid=14&phpMyAdmin=bb1c4a2d0bactf758937
[6] Sheremeta, Lorraine. Nanoscience and Nanotechnology: The Ethical, Environmental, Economic, Legal and Social Issues (NE3LS) – An Overview of NE3LS Research in Canada, paper prepared for the Office of the National Science Advisor, September 13, 2005.

AUTHOR INDEX

AUTHOR INDEX

AUTHOR INDEX

AUTHOR INDEX

Institute of Physics
Dirac House, Temple Back
Bristol BS1 6BE UK

CURRAN ASSOCIATES INC
proceedings.com